Contents
in Brief

P9-DED-652

The Curriculum Focal Points identify key mathematical ideas for this grade. They are not discrete topics or a checklist to be mastered; rather, they provide a framework for the majority of instruction at a particular grade level and the foundation for future mathematics study. The complete document may be viewed at www.nctm.org/focalpoints.

KEY

G6-FP1 Grade 6 Focal Point 1

G6-FP2 Grade 6 Focal Point 2

G6-FP3 Grade 6 Focal Point 3

G6-FP4C Grade 6 Focal Point 4 Connection

G6-FP5C Grade 6 Focal Point 5 Connection

G6-FP6C Grade 6 Focal Point 6 Connection

G6-FP1 Number and Operations: **Developing an understanding of and fluency with multiplication and division of fractions and decimals**

Students use the meanings of fractions, multiplication and division, and the inverse relationship between multiplication and division to make sense of procedures for multiplying and dividing fractions and explain why they work. They use the relationship between decimals and fractions, as well as the relationship between finite decimals and whole numbers (i.e., a finite decimal multiplied by an appropriate power of 10 is a whole number), to understand and explain the procedures for multiplying and dividing decimals. Students use common procedures to multiply and divide fractions and decimals efficiently and accurately. They multiply and divide fractions and decimals to solve problems, including multistep problems and problems involving measurement.

G6-FP2 Number and Operations: **Connecting ratio and rate to multiplication and division**

Students use simple reasoning about multiplication and division to solve ratio and rate problems (e.g., "If 5 items cost \$3.75 and all items are the same price, then I can find the cost of 12 items by first dividing \$3.75 by 5 to find out how much one item costs and then multiplying the cost of a single item by 12."). By viewing equivalent ratios and rates as deriving from, and extending, pairs of rows (or columns) in the multiplication table, and by analyzing simple drawings that indicate the relative sizes of quantities, students extend whole number multiplication and division to ratios and rates. Thus, they expand the repertoire of problems that they can solve by using multiplication and division, and they build on their understanding of fractions to understand ratios. Students solve a wide variety of problems involving ratios and rates.

G6-FP3 Algebra: **Writing, interpreting, and using mathematical expressions and equations**

Students write mathematical expressions and equations that correspond to given situations, they evaluate expressions, and they use expressions and formulas to solve problems. They understand that variables represent numbers whose exact values are not yet specified, and they use variables appropriately. Students understand that expressions in different forms can be equivalent, and they can rewrite an expression to represent a quantity in a different way (e.g., to make it more compact or to feature different information). Students know that the solutions of an equation are the values of the variables that make the equation true. They solve simple one-step equations by using number sense, properties of operations, and the idea of maintaining equality on both sides of an equation. They construct and analyze tables (e.g., to show quantities that are in equivalent ratios), and they use equations to describe simple relationships (such as $3x = y$) shown in a table.

Connections to the Focal Points

G6-FP4C **Number and Operations:** Students' work in dividing fractions shows them that they can express the result of dividing two whole numbers as a fraction (viewed as parts of a whole). Students then extend their work in grade 5 with division of whole numbers to give mixed number and decimal solutions to division problems with whole numbers. They recognize that ratio tables not only derive from rows in the multiplication table but also connect with equivalent fractions. Students distinguish multiplicative comparisons from additive comparisons.

G6-FP5C **Algebra:** Students use the commutative, associative, and distributive properties to show that two expressions are equivalent. They also illustrate properties of operations by showing that two expressions are equivalent in a given context (e.g., determining the area in two different ways for a rectangle whose dimensions are $x + 3$ by 5). Sequences, including those that arise in the context of finding possible rules for patterns of figures or stacks of objects, provide opportunities for students to develop formulas.

G6-FP6C **Measurement and Geometry:** Problems that involve areas and volumes, calling on students to find areas or volumes from lengths or to find lengths from volumes or areas and lengths, are especially appropriate. These problems extend the students' work in grade 5 on area and volume and provide a context for applying new work with equations.

Authors

Roger Day, Ph.D.
Mathematics Department Chair
Pontiac Township High School
Pontiac, Illinois

Patricia Frey, Ed.D.
Math Coordinator at Westminster Community Charter School
Buffalo, New York

Arthur C. Howard
Mathematics Teacher
Houston Christian High School
Houston, Texas

Deborah A. Hutchens, Ed.D.
Principal
Chesapeake, Virginia

Beatrice Luchin
Mathematics Consultant
League City, Texas

Kay McClain, Ed.D.
Assistant Professor
Vanderbilt University
Nashville, Tennessee

Math Online Meet the Authors at glencoe.com

Rhonda J. Molix-Bailey
Mathematics Consultant
Mathematics by Design
DeSoto, Texas

Jack M. Ott, Ph.D.
Distinguished Professor of Secondary Education Emeritus
University of South Carolina
Columbia, South Carolina

Ronald Pelfrey, Ed.D.
Mathematics Specialist
Appalachian Rural Systemic Initiative and Mathematics Consultant
Lexington, Kentucky

Jack Price, Ed.D.
Professor Emeritus
California State Polytechnic University
Pomona, California

Kathleen Vielhaber
Mathematics Consultant
St. Louis, Missouri

Teri Willard, Ed.D.
Assistant Professor
Department of Mathematics
Central Washington University
Ellensburg, Washington

Contributing Author

FOLDABLES™ **Dinah Zike**
Educational Consultant
Dinah-Might Activities, Inc.
San Antonio, Texas

Consultants

Glencoe/McGraw-Hill wishes to thank the following professionals for their feedback. They were instrumental in providing valuable input toward the development of this program in these specific areas.

Mathematical Content

Viken Hovsepian
Professor of Mathematics
Rio Hondo College
Whittier, California

Grant A. Fraser, Ph.D.
Professor of Mathematics
California State University, Los Angeles
Los Angeles, California

Arthur K. Wayman, Ph.D.
Professor of Mathematics Emeritus
California State University, Long Beach
Long Beach, California

English Language Learners

Josefina V. Tinajero, Ph.D.
Dean, College of Education
The University of Texas at El Paso
El Paso, Texas

Gifted and Talented

Ed Zaccaro
Author and Consultant
Bellevue, Iowa

Graphing Calculator

Ruth M. Casey
National Mathematics Consultant
National Instructor, Teachers Teaching with Technology
Frankfort, Kentucky

Learning Disabilities

Kate Garnett, Ph.D.
Chairperson, Coordinator
Learning Disabilities
School of Education
Department of Special Education
Hunter College, CUNY
New York, New York

Mathematical Fluency

Jason Mutford
Mathematics Instructor
Coxsackie-Athens Central School District
Coxsackie, New York

Pre-AP

Dixie Ross
Mathematics Teacher
Pflugerville High School
Pflugerville, Texas

Reading and Vocabulary

Douglas Fisher, Ph.D.
Professor of Language and Literacy Education
San Diego State University
San Diego, California

Lynn T. Havens
Director of Project CRISS
Kalispell, Montana

Reviewers

Each Reviewer reviewed at least two chapters of the Student Edition, giving feedback and suggestions for improving the effectiveness of the mathematics instruction.

Sheila J. Allen
Mathematics Teacher
A.I. Root Middle School
Medina, Ohio

Paula Barnes
Mathematics Teacher
Minisink Valley CSD
Slate Hill, New York

Deborah Barnett
Mathematics Consultant
Lake Shore Public Schools
St. Clair Shores, Michigan

Laurel W. Blackburn
Teacher/Mathematics Department Chair
Hillcrest Middle School
Simpsonville, South Carolina

Drista Bowser
Mathematics Teacher
New Windsor Middle School
New Windsor, Maryland

Matthew Bowser
Teacher
Oil City Middle School
Oil City, Pennsylvania

Susan M. Brewer
Mathematics Teacher
Brunswick Middle School
Brunswick, Maryland

Patricia A. Bruzek
Mathematics Teacher
Glenn Westlake Middle School
Lombard, Illinois

Luanne Budd
Supervisor of Mathematics
Randolph Township
Randolph, New Jersey

Ella Violet Burch
Mathematics Teacher
Penns Grove High School
Carneys Point, New Jersey

Hailey Caldwell
7th Grade Mathematics Teacher
Greenville Middle Academy of Traditional Studies
Greenville, South Carolina

Linda K. Chandler
7th Grade Mathematics Teacher
Willard Middle School
Willard, Ohio

Debra M. Cline
7th Grade Mathematics Teacher
Thomas Jefferson Middle School
Winston-Salem, North Carolina

Randall G. Crites
Principal
Bunker R-3
Bunker, Missouri

Rose Dickinson
Science and Mathematics Teacher
Seneca Middle School
Clinton Township, Michigan

Joyce Wolfe Dodd
6th Grade Mathematics Teacher
Bryson Middle School
Simpsonville, South Carolina

John G. Doyle
Middle School Chairperson/ Mathematics Teacher
Wyoming Valley West School District
Kingston, Pennsylvania

Katie England
Secondary Mathematics Resource Teacher
Carroll County Public Schools
Westminster, Maryland

Carol A. Fincannon
6th Grade Mathematics Teacher
Southwood Middle School
Anderson, South Carolina

Sally J. Fulmer
7th Grade Mathematics Teacher/ Department Chair
C.E. Williams Middle School
Charleston, South Carolina

Marian K. Geist
Mathematics Teacher/Leadership Team
Baker Prairie Middle School
Canby, Oregon

Becky Gorniack
Middle School Mathematics Teacher
Fremont Middle School
Mundelein, Illinois

Donna Tutterow Hamilton
Curriculum Facilitator
Corriher Lipe Middle School
Landis, North Carolina

Danny Liebertz
8th Grade Mathematics Instructor
Fowler Middle School
Tigard, Oregon

Marie Merkel
Learning Support
North Pocono School District
Scranton, Pennsylvania

Tonda North
Algebra 1/8th Grade Mathematics Teacher
Indian Valley Middle School
Enon, Ohio

Natasha L.M. Nuttbrock
7th Grade Mathematics Teacher
Ferguson Middle School
Beavercreek, Ohio

Paul Penn
Curriculum Team Leader, Mathematics
Lima City Schools
Lima, Ohio

Casey Condran Plackett
7th Grade Mathematics Teacher
Kennedy Junior High School
Lisle, Illinois

E. Elaine Rafferty
Mathematics Consultant
Summerville, South Carolina

Edward M. Repko
Mathematics Teacher
Kilbourne Middle School
Worthington, Ohio

Alfreda Reynolds
Teacher
Charlotte-Mecklenburg School System
Charlotte, North Carolina

Alice Roberts
Mathematics Teacher
Oakdale Middle School
Ijamsville, Maryland

Jennifer L. Rodriguez
Mathematics Teacher
Glen Crest Middle School
Glen Ellyn, Illinois

Natalie Rohaley
6th Grade Mathematics
Riverside Middle School
Greer, South Carolina

Annika Lee Schilling
Mathematics and Science Teacher
Duniway Middle School
McMinnville, Oregon

Sherry Scott
Mathematics Teacher
E.A. Tighe School
Margate, New Jersey

Eli Shaheen
Mathematics Teacher/ Department Chair
Plum Senior High School
Pittsburgh, Pennsylvania

Kelly Eady Shaw
7th Grade Mathematics Teacher
Rawlinson Road Middle School
Rock Hill, South Carolina

Evan J. Silver
Mathematics Teacher
Walkersville Middle School
Frederick, Maryland

Charlotte A. Thore
6th/7th Grade Mathematics Teacher
Northwest School of the Arts
Charlotte, North Carolina

Gene A. Tournoux
Mathematics Department Head
Shaker Heights High School
Shaker Heights, Ohio

Pamela J. Trainer
Mathematics Teacher
Roland-Grise Middle School
Wilmington, North Carolina

David A. Trez
Mathematics Teacher
Bloomfield Middle School
Bloomfield, New Jersey

Pauline D. Von Hoffer
Mathematics Teacher
Wentzville School District
Wentzville, Missouri

Kentucky Consultants

Jenn Crase
8th Grade Mathematics Teacher/ Department Chair
South Oldham Middle School
Crestwood, Kentucky

Max DeBoer Lux
8th Grade Mathematics
Summit View Middle School
Independence, Kentucky

Jennifer Wells Phipps
Middle School Mathematics Teacher
Corbin Middle School
Corbin, Kentucky

Bea Torrence
Teacher/Mathematics Content Leader
Camp Ernst Middle School
Burlington, Kentucky

J. Ron Vanover
Advanced Placement Calculus
Boone County High School
Florence, Kentucky

Contents

Start Smart

Unit 1

Number, Operations, and Statistics

Algebra: Number Patterns and Functions

Focal Points and Connections
See page iv for key.

G6-FP3 Algebra
G6-FP5C Algebra

TEST PRACTICE

H.O.T. Problems
Higher Order Thinking

Statistics and Graphs

G6-FP4C Number and Operations

TEST PRACTICE

H.O.T. Problems
Higher Order Thinking

Unit 2

Number and Operations: Decimals and Fractions

Operations with Decimals

G6-FP1 Number and Operations

TEST PRACTICE

H.O.T. Problems
Higher Order Thinking

Fractions and Decimals

G6-FP1 Number and Operations
G6-FP4C Number and Operations

TEST PRACTICE

H.O.T. Problems
Higher Order Thinking

Chapter 5 Operations with Fractions

Focal Points and Connections
See page iv for key.

G6-FP1 Number and Operations
G6-FP4C Number and Operations

TEST PRACTICE

H.O.T. Problems
Higher Order Thinking

Unit 3

Patterns, Relationships, and Algebraic Thinking

Ratio, Proportion, and Functions

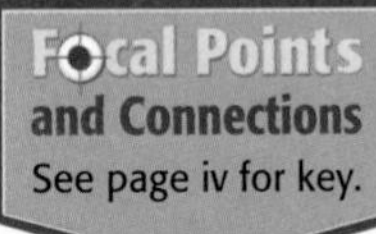

G6-FP2 Number and Operations
G6-FP4C Number and Operations

TEST PRACTICE

- Extended Response 361
- Multiple Choice 319, 327, 333, 339, 345, 346, 348, 353
- Short Response/Grid In 327, 339, 361
- Worked Out Example 345

H.O.T. Problems
Higher Order Thinking

- Challenge 318, 327, 333, 338, 353
- Find the Error 319, 338
- Number Sense 327
- Open Ended 318, 333, 338, 347, 353
- Reasoning 338
- Select a Technique 347

CHAPTER 7 Percent and Probability

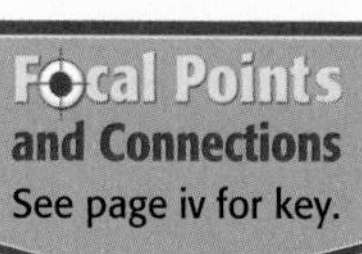

Focal Points and Connections
See page iv for key.

G6-FP4C Number and Operations

TEST PRACTICE

H.O.T. Problems
Higher Order Thinking

Unit 4

Measurement and Geometry

Systems of Measurement

Focal Points and Connections
See page iv for key.

G6-FP1 Number and Operations
G6-FP2 Number and Operations

TEST PRACTICE

- Extended Response 467
- Multiple Choice 420, 421, 423, 429, 436, 441, 449, 454, 458
- Short Response/Grid In 429, 449, 454, 467
- Worked Out Example 420

H.O.T. Problems
Higher Order Thinking

- Challenge 422, 428, 435, 441, 448, 453, 458
- Find the Error 423, 448
- Number Sense 441
- Open Ended 422, 428, 435, 441, 448, 454
- Reasoning 422, 454
- Select a Technique 428
- Select a Tool 448
- Which One Doesn't Belong? 454

Geometry: Angles and Polygons

G6-FP2 Number and Operations

TEST PRACTICE

H.O.T. Problems
Higher Order Thinking

Table of Contents

Measurement: Perimeter, Area, and Volume

G6-FP6C Measurement and Geometry

TEST PRACTICE

H.O.T. Problems
Higher Order Thinking

Unit 5

Number, Operations, and Algebraic Thinking

Integers and Transformations

G6-FP3 Algebra

TEST PRACTICE

H.O.T. Problems

Higher Order Thinking

Algebra: Properties and Equations

Focal Points and Connections
See page iv for key.

G6-FP3 Algebra
G6-FP5C Algebra

TEST PRACTICE

H.O.T. Problems
Higher Order Thinking

To the Student

As you gear up to study mathematics, you are probably wondering, "What will I learn this year?" You will focus on these three areas:

- **Number and Operations:** Multiply and divide fractions and decimals.
- **Number and Operations:** Connect ratio and rate to multiplication and division.
- **Algebra:** Write, interpret, and use mathematical expressions and equations.

Along the way, you'll learn more about problem solving, how to use the tools and language of mathematics, and how to THINK mathematically.

How to Use Your Math Book

Have you ever been in class and not understood all of what was being presented? Or, you understood everything in class, but got stuck on how to solve some of the homework problems? Don't worry. You can find answers in your math book!

- **Read** the **MAIN IDEA** at the beginning of the lesson.
- **Find** the **New Vocabulary** words, **highlighted in yellow**, and read their definitions.
- **Review** the **EXAMPLE** problems, solved step-by-step, to remind you of the day's material.
- **Refer** to the **HOMEWORK HELP** boxes that show you which examples may help with your homework problems.
- **Go** to **Math Online** where you can find extra examples to coach you through difficult problems.
- **Review** the notes you've taken on your **FOLDABLES™**.
- **Find** selected solutions and the answers to odd-numbered problems in the back of the book. Use them to see if you are solving the problems correctly.

Scavenger Hunt

Let's Get Started

Use the Scavenger Hunt below to learn where things are located in each chapter.

1. What is the title of Chapter 1?
2. How can you tell what you'll learn in Lesson 1-1?
3. What is the title of the feature in Lesson 1-2 that tells you that you can choose any pair of whole number factors to find the prime factorization of a number?
4. There is a Real-World Career mentioned in Lesson 1-3. What is it?
5. Suppose you're doing your homework on page 35 and you get stuck on Exercise 24. Where could you find help?
6. Sometimes you may ask "When am I ever going to use this?" Name a situation that uses the concepts from Lesson 1-4.
7. What is the key concept presented in Lesson 1-4?
8. How many examples are presented in Lesson 1-4?
9. In the margin of Lesson 1-5, there is a Vocabulary Link. What can you learn from that feature?
10. What is the Web address where you could find extra examples?
11. What problem-solving strategy is presented in the Problem-Solving Investigation in Lesson 1-7?
12. List the new vocabulary words that are presented in Lesson 1-8.
13. What is the Web address that would allow you to take a self-check quiz to be sure you understand the lesson?
14. On what pages will you find the Study Guide and Review for Chapter 1?
15. Suppose you can't figure out how to do Exercise 19 in the Study Guide on page 70. Where could you find help?

MATH? SYMBOLS

Start Smart

Let's Review!

Bald eagle

A Plan for Problem Solving

American Rivers

The great history of the United States can be told along its many rivers. The seven longest rivers in the United States are shown in the table.

How much longer is the Missouri River than the Arkansas River?

River	Length (mi)
Missouri	2,540
Mississippi	2,340
Yukon	1,980
Rio Grande	1,900
St. Lawrence	1,900
Arkansas	1,460
Colorado	1,450

You can use the four-step problem-solving plan to solve many kinds of problems. The four steps are Understand, Plan, Solve, and Check.

Understand

- **Read the problem carefully.**
- **What facts do you know?**
- **What do you need to find?**

You know the length of the seven longest rivers. You need to find how much longer the Missouri River is than the Arkansas River.

Plan

- **How do the facts relate to each other?**
- **Plan a strategy to solve the problem.**

The Missouri River is 2,540 miles long, and the Arkansas River is 1,460 miles long. To find the difference between the two lengths, subtract the length of the Arkansas River from the length of the Missouri River.

Solve

- **Use your plan to solve the problem.**

$$\begin{array}{r l} \overset{4\,14}{2{,}\not5\not40} & \text{length of Missouri River} \\ -\ 1{,}460 & \text{length of Arkansas River} \\ \hline 1{,}080 & \end{array}$$

So, the Missouri River is 1,080 miles longer than the Arkansas River.

Check

- **Look back at the problem.**
- **Does your answer make sense?**
- **If not, solve the problem another way.**

Estimate. Round 2,540 to 2,500 and 1,460 to 1,500.
2,500 – 1,500 = 1,000

Since 1,000 is close to 1,080, the answer is reasonable.

CHECK Your Understanding

1. The Nile River in Africa is the longest river in the world. It is 4,132 miles long. About how many times longer is the Nile River than the Yukon River?

2. **WRITING IN MATH** Explain how you would use the four-step plan to determine how long the Colorado River is in yards. (*Hint:* There are 1,760 yards in one mile.)

Number and Operations

A Great American Pastime

Baseball is one of America's truly unique sports. Major League Baseball began in 1901 with 16 teams. Today, there are 30 teams throughout the United States, each playing in a different ballpark. The average costs of items at different ballparks are shown in the table below.

Cost of Items at Major League Baseball Parks

Baseball Field, Location	Average Ticket ($)	Hot Dog ($)	Drink ($)
Busch Stadium, St. Louis	29.78	3.50	4.50
Fenway Park, Chicago	46.46	4.00	2.75
Jacob's Field, Cleveland	21.54	2.50	2.25
Minute Maid Park, Houston	26.66	4.00	4.00
Safeco Field, Seattle	24.01	3.25	2.50
Wrigley Field, Chicago	34.30	2.75	2.50

Source: Team Marketing Report

Comparing and Ordering Money

When comparing and ordering money, first line up the decimal points. Then, compare the digits in each place to order the amounts.

Use the information in the table on page 6 to answer each question.

1. In which ballpark(s) does a drink cost the most? the least?

2. In which ballpark(s) does a hot dog cost the most? the least?

3. List the parks in order from greatest average ticket price to least average ticket price.

Estimation

You can use rounding to estimate the sum or difference of a problem.

Use the table on page 6 to answer each question.

4. About how much would a hot dog and a drink cost at Wrigley Field?

5. About how much more is the average cost of a ticket at Fenway Park than Jacob's Field?

6. If a family of four goes to see a baseball game at Safeco Field, about how much will it cost to buy 4 tickets, 4 hot dogs, and 4 drinks?

7. **WRITING IN MATH** The costs of different souvenirs at the team shop at a ballpark are shown in the table. Use the information to write and solve a real-world problem about the costs of souvenirs.

Souvenir	Price ($)
Cap	22.99
Jersey	55.49
Pennant	15.95
T-Shirt	24.75
Baseball	19.95

Algebra

State Emblems

Many states have chosen an animal or animals to represent their state. The state animals of several states and their average weights are given in the table.

State	State Animal	Average Weight (lb)
Alaska	moose	1,150
Kansas	American buffalo	1,450
Nevada	desert bighorn sheep	200
Oregon	beaver	50
Utah	elk	850
Wisconsin	badger	20

Expressions

Use the table on page 8 to answer the following questions.

1. Write an expression to show that the average weight of the American buffalo is p pounds more than the average weight of the moose.

2. Write an expression to show that the average weight of the desert bighorn sheep is 10 times as great as the weight of the badger.

3. Suppose the average weight of the state animal of South Carolina, the white-tailed deer, is w pounds less than the desert bighorn sheep. Write an expression to show the average weight of the white-tailed deer.

4. Using the expression from Exercise 3, if w is equal to 30, what is the average weight of the white-tailed deer?

For Exercises 5 and 6, use the table and the information below.

In addition to animals, birds are also chosen to represent some states. The table shows some state birds and their average wingspans.

State	Bird	Average Wingspan (in.)
Louisiana	brown pelican	79
Minnesota	common loon	46
New Hampshire	purple finch	9
North Carolina	cardinal	11
Pennsylvania	ruffed grouse	22

5. Write an expression to show that the average wingspan of the ruffed grouse is twice the average wingspan of the cardinal.

6. Write an expression to show that the brown pelican has an average wingspan that is p inches greater than the average wingspan of the common loon.

7. **WRITING IN MATH** Write a real-world problem to represent the expression $10 + x$.

Geometry

Architectural Wonder

Geometry is used in all aspects of architecture. Different shapes, lines, and points are used to draw attention to features of buildings. Many excellent examples of geometry in architecture are found in Washington, D.C. In the photo of the White House above, you can see how the design uses symmetry, parallel and perpendicular lines, and different shapes.

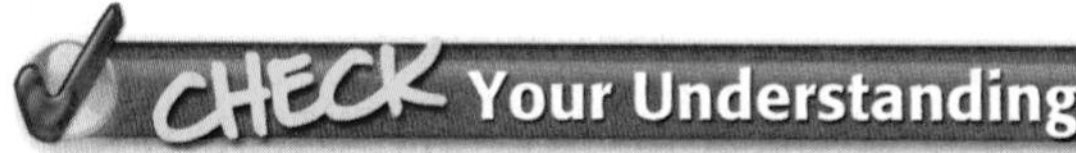

CHECK Your Understanding — Geometry

Use the photo of the White House to answer the following questions.

1. Give an example of symmetry that is found in the photo of the White House.

2. Congruent sides or angles are sides or angles that are the same size and shape. Give an example of congruent sides or angles found in the photo of the White House.

Use the photo of the Washington Monument at the right to answer the following questions.

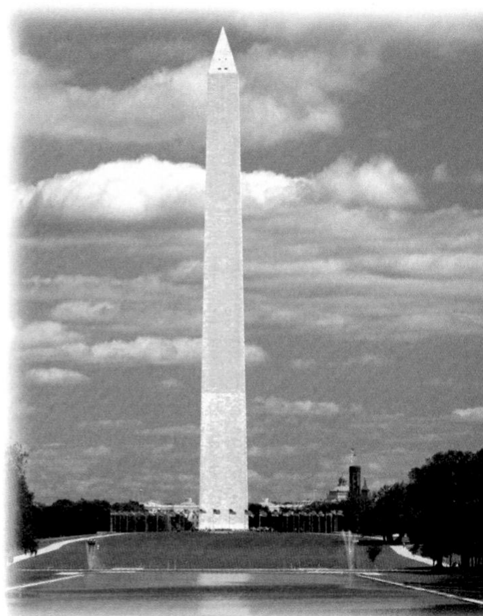

3. Describe the shapes that make up one side of the Washington Monument.

4. If you were to look at the Washington Monument from above, what shape would you see?

5. The base of the monument is a square that measures about 55 feet on each side. What is the approximate distance around the base of the Washington Monument?

6. **WRITING IN MATH** Create and solve a problem involving geometry about the National Museum of Natural History shown below.

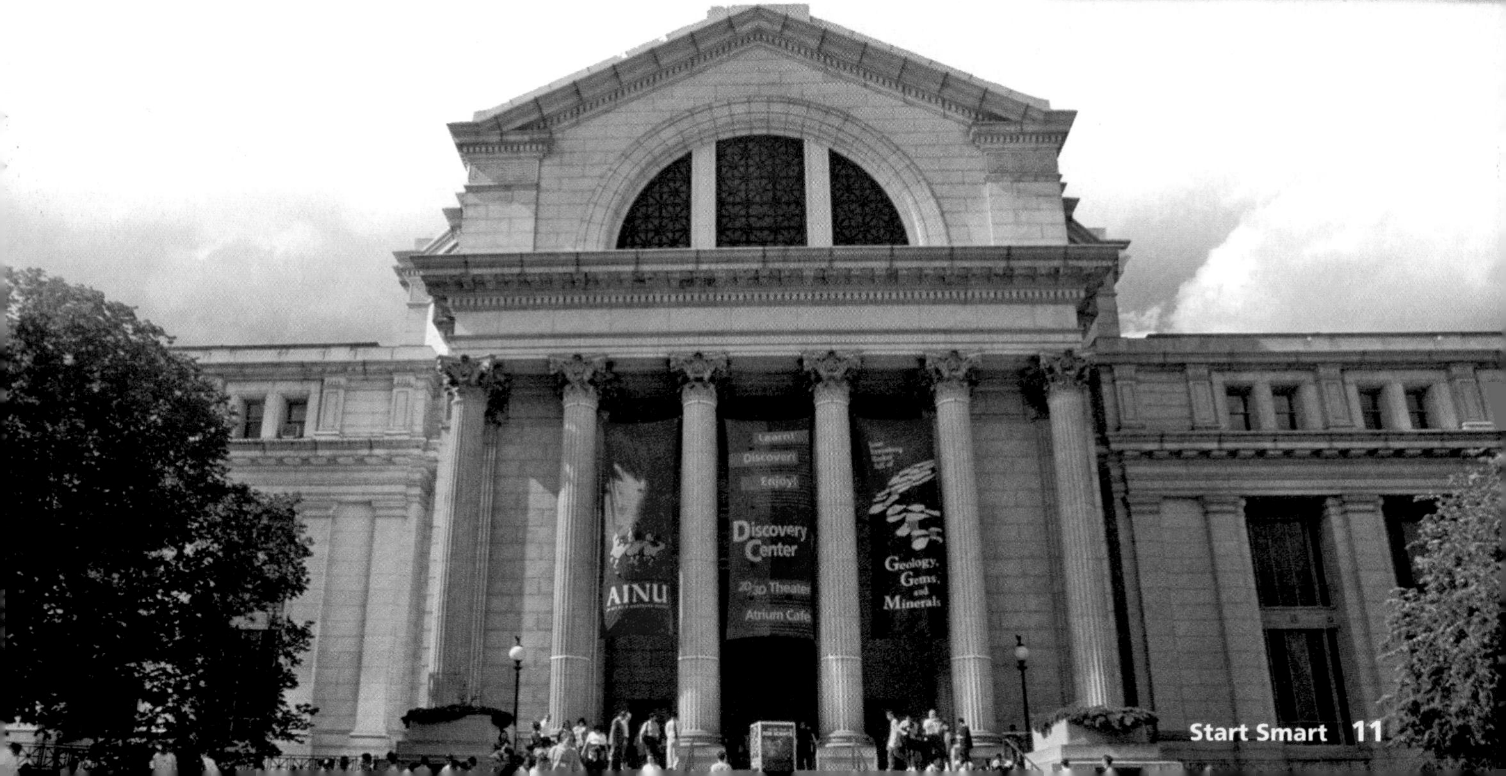

Start Smart 5

Measurement

A Record-Breaking Experience

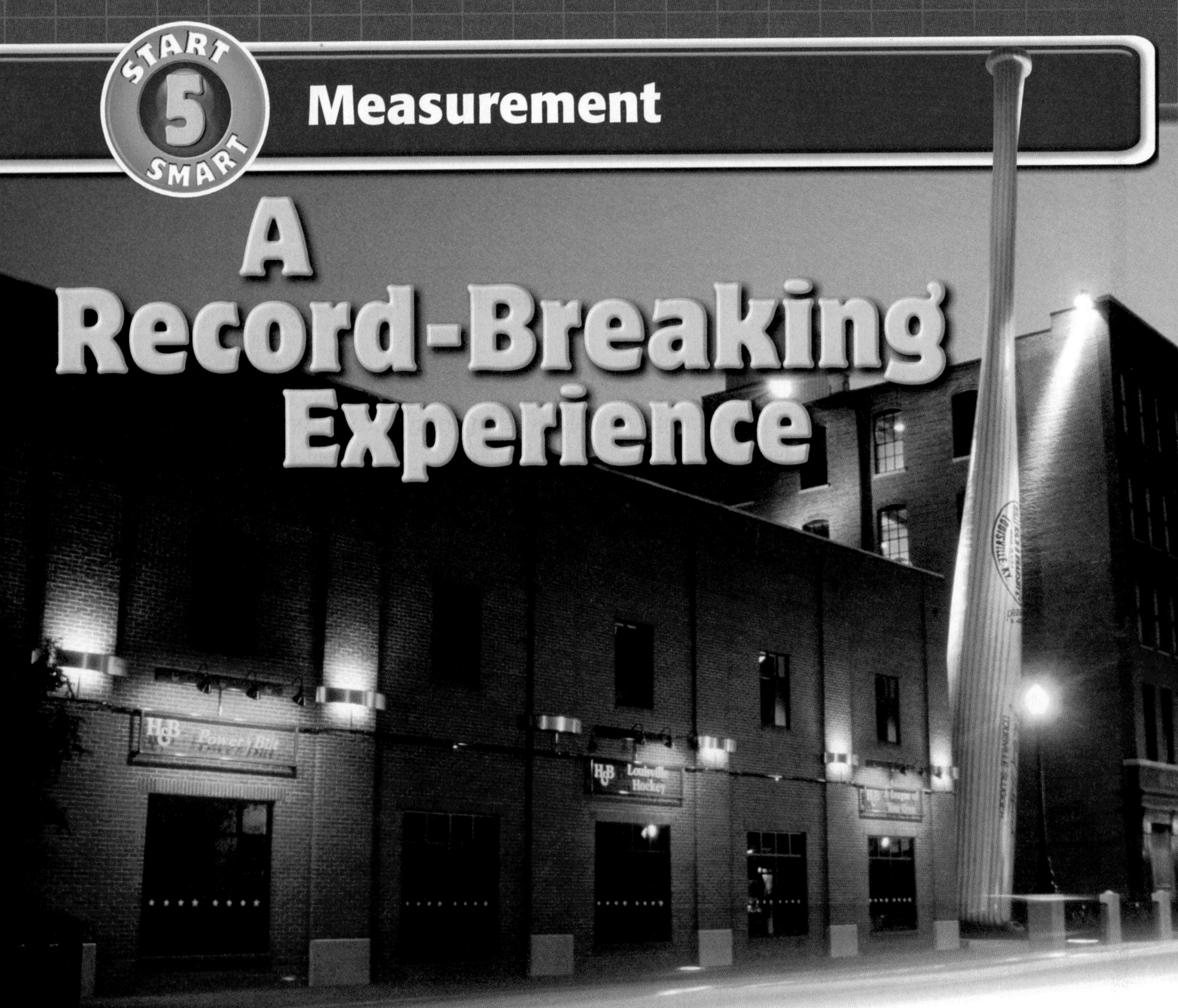

Touring America, you can see many large, unusual items. For example, the world's largest baseball bat, in Louisville, Kentucky, is 120 feet long. An actual Louisville Slugger baseball bat is only 34 inches long. Other record-breaking items are listed in the table.

Item	Length (ft)	Width (ft)	Dimensions of Actual Object (in.)	Location
wagon	27	13	36 × 17	Spokane, WA
skateboard	10	4	20 × 8	San Diego, CA
basket	208 (at the top)	142 (at the top)	15 × 10	Newark, OH

Understanding Units

Standard units of measurement are most commonly used in the United States. They include the *inch, foot, yard,* and *mile.*

1. There are 12 inches in 1 foot. How many inches long and wide is the largest skateboard?

2. There are 3 feet in 1 yard. About how many yards long and wide is the largest basket?

3. Is 15 inches or 15 feet a more reasonable measurement for the height of the tire?

4. Is 15 inches or 15 feet a more reasonable measurement for the height of the flagpole?

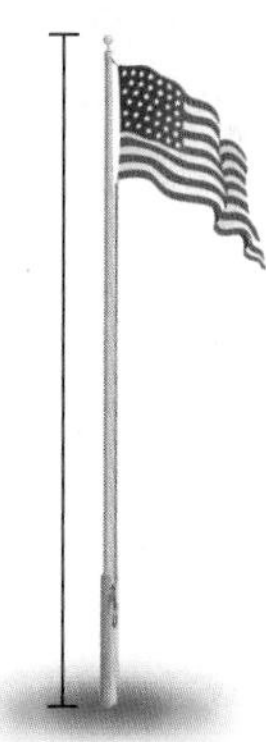

5. How does the length of the largest baseball bat compare with the length of a regular bat?

6. **WRITING IN MATH** Use the Internet or another source to find the dimensions of another record-breaking object in the United States. Write a problem involving standard units using the data that you found.

START SMART 6

Data Analysis

The Stories I Could Tell You

The first skyscraper was built in Chicago in 1884 and was 10 stories tall. Skyscrapers have come a long way since then and are now in every major city throughout the world. The table at the right shows the number of stories in the tallest buildings in the United States.

Number of Stories in the Tallest Buildings in the U.S.							
55	60	72	76	64	75	60	102
49	56	100	60	71	77	57	64
60	48	60	61	60	42	50	57
65	75	73	66	83	64	54	56
72	58	66	64	57	60	74	110
69	60	57	48	71	59	52	62

These data can be arranged in a frequency table. A *frequency table* shows the number of pieces of data that fall within the given intervals.

1. Complete the frequency table using the data about buildings in the United States.

2. Which interval has the greatest number of buildings?

3. Which interval has the least number of buildings?

4. How many buildings have 70 or more stories?

Frequency Tables

Number of Stories in the Tallest Buildings in the U.S.

Number of Stories	Tally	Frequency
40–49		
50–59		
60–69		
70–79		
80–89		
90–99		
100–109		
110–119		

5. **WRITING IN MATH** Choose an appropriate interval size for a frequency table that will represent the data in the table below. Explain your reasoning.

Heights of Tallest Buildings in the U.S. (m)

319	260	303	265	312
242	283	262	288	253
290	248	381	293	289
241	275	247	265	241
305	240	279	241	344
250	319	241	307	237
265	237	297	238	262
310	258	238	270	296
246	346	253	248	259

Data File

The following pages contain data that you'll use throughout the book.

The Grand Canyon

- 277 miles long
- At its deepest point, it is 6,000 feet deep.
- At its widest point, it is 15 miles wide.
- Some of the rocks are 2 billion years old.
- It covers 1,218,375 acres.

Source: National Parks Service

Kentucky Rumbler

Source: Beech Bend Park

Height: 96 feet
Length: 2,827 feet
Top Speed: 47.7 mph
Trains: one 24-passenger train
Crossovers: 30
Opened: 2006

Shedd Aquarium

SHEDD AQUARIUM

Type of membership	Donation ($)	Number of Adult Admissions	Number of Children's Admissions
Sponsor	500–1,499	4	4
Associate	250–499	4	4
Partner	175–249	4	4
Family	125	2	4
Family Plus	140	2	5
Individual	80	1	1(guest)

Source: Shedd Aquarium

Boston, Massachusetts

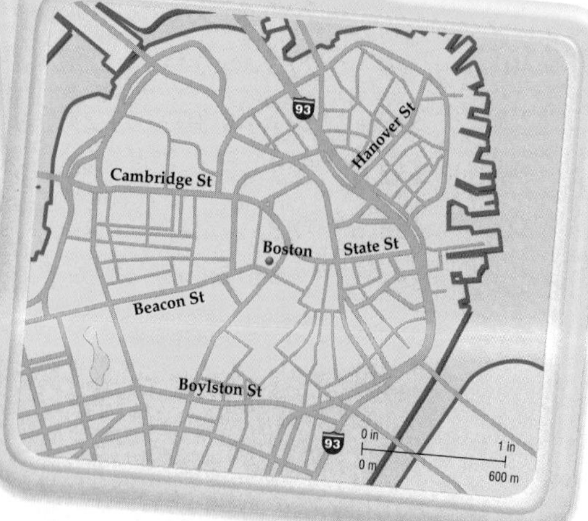

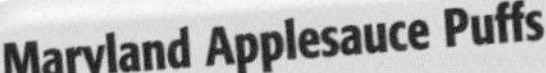

Maryland Applesauce Puffs

Maryland Applesauce Puffs

- **2** cups biscuit mix
- $\frac{1}{4}$ cup sugar
- $\frac{1}{2}$ tsp cinnamon
- $\frac{3}{4}$ cup applesauce
- **3** Tbsp milk
- **1** egg
- **2** Tbsp salad oil
- **2** Tbsp melted butter

Source: Maryland Apple Promotion Board

Gerald R. Ford International Airport

- 2nd busiest airport in Michigan
- About 239,000 pounds of air cargo pass through the airport each day.
- 3 runways:
  East/West – 150 ft wide, 10,000 ft long
  East/West – 100 ft wide, 5,000 ft long
  North/South – 150 ft wide, 8,501 ft long
- There are 2,000 acres of grass to mow.

Source: Gerald R. Ford International Airport

Seattle Space Needle

- It is 605 feet tall.
- It weighs 3,700 tons.
- The diameter of the halo is 138 feet.
- There is a revolving restaurant at 500 feet.
- The Space Needle sways about 1 inch for every 10 miles per hour the wind blows.
- It takes 43 seconds to travel from ground level to the top-house in an elevator that travels 800 feet per minute.

Source: Space Needle Corporation

University of Missouri-Columbia Facts

- Tuition, Missouri resident: $7,308
- Tuition, nonresident: $16,890
- Enrollment: 28,253
- Number of Degree Programs: 274
- Mascot: Bengal Tiger
- Colors: black and gold
- Nickname: Mizzou

Source: University of Missouri

Broadway

Longest Running Broadway Plays

Play	Number of Shows*	Opening Date	Closing Date
The Phantom of the Opera	8,061	1/26/1988	still running
Cats	7,485	10/7/1982	9/10/2000
Les Misérables	6,680	3/12/1987	5/18/2003
A Chorus Line	6,137	7/25/1975	4/28/1990
Oh! Calcutta! (Revival)	5,959	9/24/1976	8/6/1989

*Through May 2007

Source: Playbill

Extreme Temperatures

State	Record High (°F)	Record Low (°F)
Alaska	100	−80
California	134	−45
Colorado	114	−61
Idaho	118	−60
Montana	117	−70
Nevada	125	−50
North Dakota	121	−60
South Dakota	120	−58
Utah	117	−69
Wyoming	115	−66

Source: National Climatic Data Center

Most Popular Dog Breeds in United States

Breed	Number Registered in U.S.
Labrador Retriever	123,760
Yorkshire Terrier	48,346
German Shepherd	43,575
Golden Retriever	42,962
Beagle	39,484
Dachshund	36,033
Boxer	35,388
Poodle	29,939
Shih Tzu	27,282
Miniature Schnauzer	22,920

Source: American Kennel Club

Pennsylvania 500

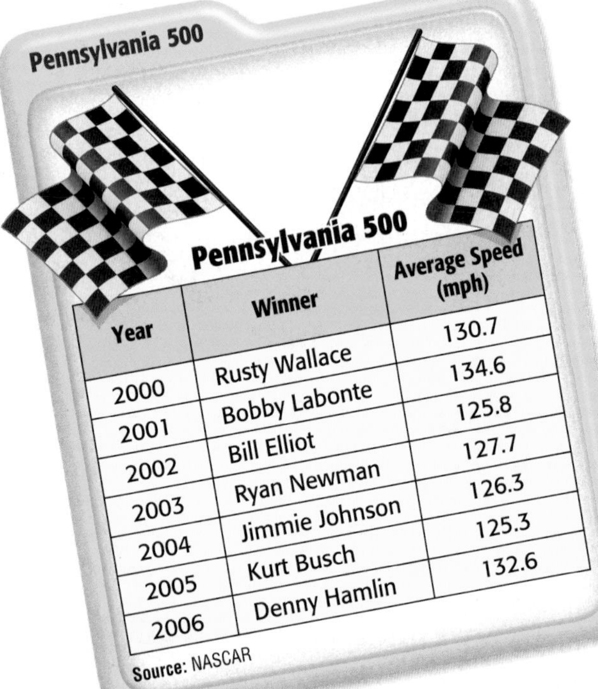

Pennsylvania 500

Year	Winner	Average Speed (mph)
2000	Rusty Wallace	130.7
2001	Bobby Labonte	134.6
2002	Bill Elliot	125.8
2003	Ryan Newman	127.7
2004	Jimmie Johnson	126.3
2005	Kurt Busch	125.3
2006	Denny Hamlin	132.6

Source: NASCAR

Cape Hatteras Lighthouse

- It was completed in 1870.
- It was built with 1,250,000 bricks.
- Its light can be seen from 20 miles off shore.
- It is 210 ft (64 m) tall.

Humpback Whale

Length: 40–50 feet (12.2–15.2 meters)

Weight: 25–40 tons (22,680–36,287 kilograms)

Fluke (Tail): 18 feet (5.5 meters)

Source: American Cetacean Society

South Carolina Cities

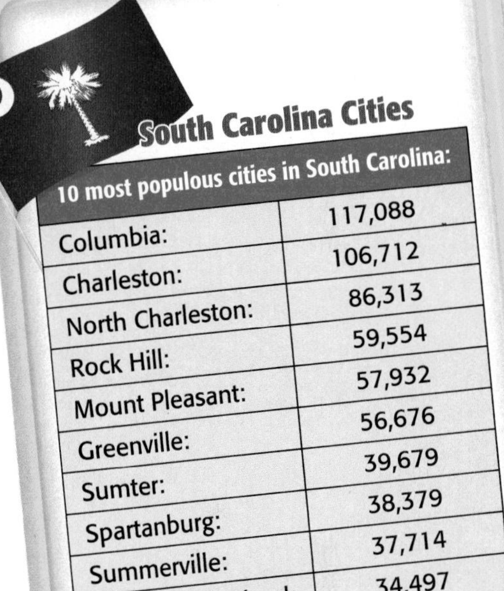

South Carolina Cities

10 most populous cities in South Carolina:	
Columbia:	117,088
Charleston:	106,712
North Charleston:	86,313
Rock Hill:	59,554
Mount Pleasant:	57,932
Greenville:	56,676
Sumter:	39,679
Spartanburg:	38,379
Summerville:	37,714
Hilton Head Island:	34,497

Minnesota

A recent survey shows that, of the people in Minnesota over the age of 15:

- 29% go fishing.
- 14% go hunting.
- 30% visit a Minnesota state park.
- 41% use boats for recreation and fishing.
- 33% use boats only for recreation.
- 52% watch wildlife within a mile of home.
- 3% travel over a mile from home to watch wildlife.

Source: Minnesota Department of Natural Resources

Tennessee Lady Vols Basketball

Player	Points per Game	Offensive Rebounds	Defensive Rebounds	Free Throw Percentage
Candace Parker	30.3	2.57	7.29	0.713
Sidney Spencer	28.4	2.14	2.14	0.897
Alexis Hornbuckle	30.6	1.89	3.19	0.731
Shannon Bobbitt	25.7	0.46	1.06	0.800
Nicky Anosike	26.9	2.86	3.06	0.629
Alex Fuller	20.2	1.72	2.53	0.776

Source: University of Tennessee Lady Vols

Unit 1

Number, Operations, and Statistics

Focus

Write, interpret, and use mathematical expressions and equations.

CHAPTER 1
Algebra: Number Patterns and Functions

BIG Idea Write mathematical expressions and equations.

BIG Idea Use variables to represent numbers.

CHAPTER 2
Statistics and Graphs

BIG Idea Construct and analyze statistical representations of data.

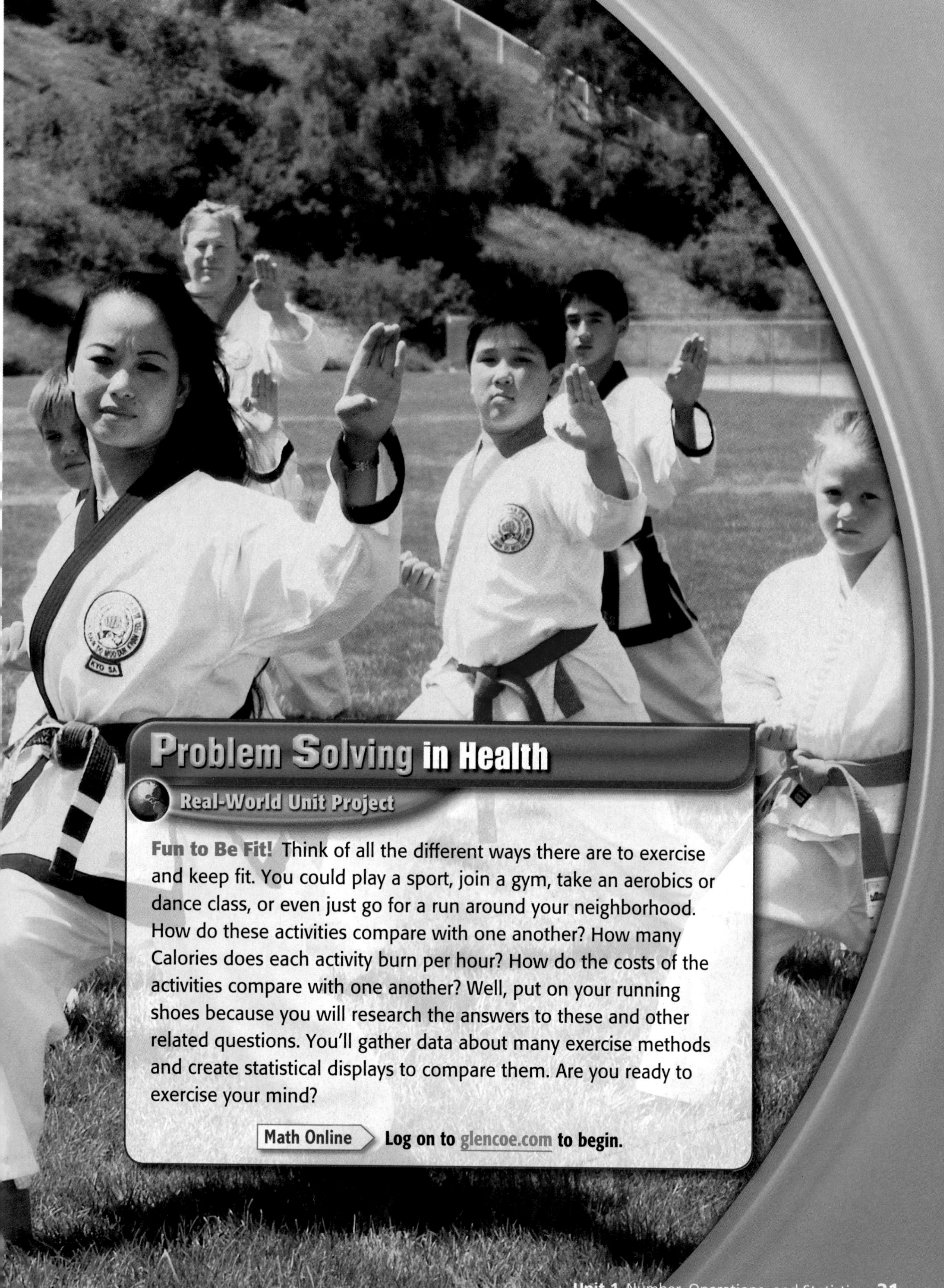

Problem Solving in Health

Real-World Unit Project

Fun to Be Fit! Think of all the different ways there are to exercise and keep fit. You could play a sport, join a gym, take an aerobics or dance class, or even just go for a run around your neighborhood. How do these activities compare with one another? How many Calories does each activity burn per hour? How do the costs of the activities compare with one another? Well, put on your running shoes because you will research the answers to these and other related questions. You'll gather data about many exercise methods and create statistical displays to compare them. Are you ready to exercise your mind?

Math Online **Log on to glencoe.com to begin.**

Algebra: Number Patterns and Functions

- Write mathematical expressions and equations.
- Use variables to represent numbers.

Key Vocabulary

area (p. 63)
evaluate (p. 42)
function (p. 49)
variable (p. 42)

Real-World Link

Stadiums Ohio Stadium, home of The Ohio State University Buckeyes, has a seating capacity of 101,568. You can use the equation $x + 35{,}358 = 101{,}568$ to find the value of x, the seating capacity of Ohio Stadium on opening day in 1922.

Algebra: Number Patterns and Functions Make this Foldable to help you organize your notes. Begin with five sheets of notebook paper.

1. **Stack** the pages, placing the sheets of paper $\frac{3}{4}$ inch apart.

2. **Roll** up bottom edges. All tabs should be the same size.

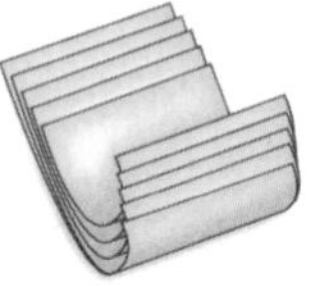

3. **Crease** and staple along the fold.

4. **Label** the tabs with the topics from the chapter.

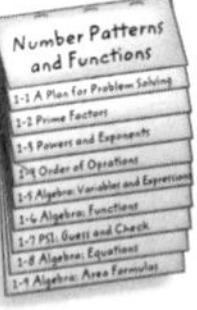

GET READY for Chapter 1

Diagnose Readiness You have two options for checking Prerequisite Skills.

Option 2

Math Online Take the Online Readiness Quiz at glencoe.com.

Option 1

Take the Quick Quiz below. Refer to the Quick Review for help.

QUICK Quiz

Add. (Prior Grade)

1. 83 + 129
2. 99 + 56
3. 67 + 42
4. 79 + 88
5. 78 + 97
6. 86 + 66

QUICK Review

Example 1 **Find 359 + 88.**

Line up the digits at the ones place.

$$\begin{array}{r} {}^{1\,1} \\ 359 \\ +\ 88 \\ \hline 447 \end{array}$$

Add the ones. Put the 7 in the ones place and place the 1 above the tens place.

Add the tens. Put the 4 in the tens place and place the 1 above the hundreds place. Then add the hundreds.

Subtract. (Prior Grade)

7. 43 − 7
8. 75 − 27
9. 128 − 34
10. 150 − 68
11. 102 − 76
12. 235 − 126
13. **MONEY** Ariana bought three shirts for a total of $89. If one shirt costs $24 and another costs $31, how much did the third shirt cost?

Example 2 **Find 853 − 79.**

Line up the digits at the ones place.

$$\begin{array}{r} 7\,14\,13 \\ \not{8}\,\not{5}\,\not{3} \\ -\ 79 \\ \hline 774 \end{array}$$

Since 9 is larger than 3, rename 3 as 13. Rename the 5 in the tens place as 14 and the 8 in the hundreds place as 7. Then subtract.

Multiply. (Prior Grade)

14. 25 × 12
15. 18 × 30
16. 42 × 15
17. 27 × 34
18. 50 × 16
19. 47 × 22

Example 3 **Find 15 × 23.**

$$\begin{array}{r} 15 \\ \times\ 23 \\ \hline 45 \\ +\ 300 \\ \hline 345 \end{array}$$

45 — Multiply. 15 × 3 = 45

300 — Multiply. 15 × 20 = 300

345 — Add. 45 + 300 = 345

Divide. (Prior Grade)

20. 72 ÷ 9
21. 84 ÷ 6
22. 126 ÷ 3
23. 146 ÷ 2
24. 208 ÷ 4
25. 504 ÷ 8

Example 4 **Find 318 ÷ 6.**

$$\begin{array}{r} 53 \\ 6\overline{)318} \\ -\ 30 \\ \hline 18 \\ -\ 18 \\ \hline 0 \end{array}$$

Divide in each place-value position from left to right.

Since 18 − 18 = 0, there is no remainder.

A Plan for Problem Solving

MAIN IDEA

Solve problems using the four-step plan.

Math Online

glencoe.com

- Extra Examples
- Personal Tutor
- Self-Check Quiz

GET READY for the Lesson

CRAFTS Michelle is making 8 necklaces by stringing beads together. To make one necklace, she will repeat the pattern of beads shown four times.

1. How many purple and yellow beads are used to make one necklace?
2. How many purple and yellow beads will be needed to make all eight necklaces?
3. Explain how you found the number of each color of beads needed to make all eight necklaces.

When solving math problems, it is often helpful to have an organized problem-solving plan. The four steps below can be used to solve any problem.

Understand
- Read the problem carefully.
- What facts do you know?
- What do you need to find out?
- Is enough information given?
- Is there extra information?

Plan
- How do the facts relate to each other?
- Plan a strategy for solving the problem.
- Estimate the answer.

Solve
- Use your plan to solve the problem.
- If your plan does not work, revise it or make a new plan.
- What is the solution?

Check
- Reread the problem.
- Does the answer fit the facts given in the problem?
- Is the answer close to your estimate?
- Does the answer make sense?
- If not, solve the problem another way.

Study Tip

Reasonableness In the last step of this plan, you check the reasonableness of the answer by comparing it to the estimate.

Some problems can be easily solved by adding, subtracting, multiplying, or dividing. Key words and phrases play an important role in deciding which operation to use.

Addition	Subtraction	Multiplication	Division
plus	minus	times	divided by
sum	difference	product	quotient
total	less	multiplied by	
in all	subtract	of	

EXAMPLE 1 Use the Problem-Solving Plan

1 GOLF Refer to the graph below. How many more official career wins did Kathy Whitworth have than Nancy Lopez?

Official Career Wins on the LPGA Tour*

Golfer	Number of Wins
Kathy Whitworth	88
Annika Sorenstam	69
Louise Suggs	58
Nancy Lopez	48
Sandra Haynie	42
Karrie Webb	35

Source: LPGA

* As of 2006 season

Understand Extra information is given in the graph. You know the number of career wins made by many golfers. You need to find how many more wins Kathy Whitworth had than Nancy Lopez.

Plan To find the difference, subtract 48 from 88. Since the question asks for an exact answer, use mental math or paper and pencil. Before you calculate, estimate.

Estimate $90 - 50 = 40$

Solve $88 - 48 = 40$

Kathy Whitworth had 40 more career wins than Nancy Lopez.

Check Compared to the estimate, the answer is reasonable. Since $40 + 48$ is 88, the answer is correct.

Real-World Link
Annika Sorenstam was the first female golfer to surpass the $20 million mark for career earnings.
Source: LPGA

CHECK Your Progress

a. **GOLF** Refer to the graph above. The number of tournaments Annika Sorenstam participated in is about 4 times the number of tournaments she actually won. About how many tournaments did Annika participate in?

2 **ALLOWANCE The table shows how Kaylee's weekly allowance increases based on her age. If the pattern continues, how much allowance will Kaylee receive when she is 13 years old?**

Age	9	10	11	12	13
Weekly Allowance	\$3.25	\$4.00	\$4.75	\$5.50	■

Understand You know Kaylee's weekly allowance by age. You need to find her weekly allowance at age 13.

Plan Since an exact answer is needed and the question contains a pattern, use mental math.

Solve \$3.25 \$4.00 \$4.75 \$5.50 ?

+\$0.75 +\$0.75 +\$0.75

The values increase by \$0.75 each time. The next value should increase by \$0.75. So, when Kaylee is 13 years old, her allowance will be \$5.50 + \$0.75, or \$6.25.

Check Start with \$6.25 and subtract \$0.75. Continue subtracting to see if she would earn \$3.25 when she is 9 years old. Since she does earn \$3.25 at 9 years old, the answer is correct.

Study Tip

Method of Computation To solve a problem, some methods you can choose are paper and pencil, mental math, a calculator, or estimation.

CHECK Your Progress

b. **TRACK** Julian is on the track team. The table shows the number of sprints he runs in the first four days of practice. If the pattern continues, how many sprints will he run on Friday?

Day	Monday	Tuesday	Wednesday	Thursday	Friday
Sprints	2	4	7	11	■

CHECK Your Understanding

For Exercises 1 and 2, use the four-step plan to solve each problem.

Example 1 (p. 25)

1. **BEARS** An adult male brown bear weighs about 1,380 pounds. An adult female brown bear weighs about 630 pounds. How much less does an adult female brown bear weigh than an adult male brown bear?

Example 2 (p. 26)

2. **POOLS** The table shows the total amount of water in a swimming pool that is being filled. At this rate, how much water will be in the swimming pool after 30 minutes?

Time (min)	5	10	15	20	25	30
Water (gal)	75	150	225	300	■	■

Practice and Problem Solving

HOMEWORK HELP	
For Exercises	See Examples
3, 4	1
5, 6	2

For Exercises 3–8, use the four-step plan to solve each problem.

3. **RIVERS** The longest river in the world is the Nile River. It is 4,132 miles long. The longest river in the United States is the Missouri River. It is 2,540 miles long. How much longer is the Nile than the Missouri?

4. **ANALYZE GRAPHS** Refer to the graph. How many more people use the Internet in Europe than in Africa?

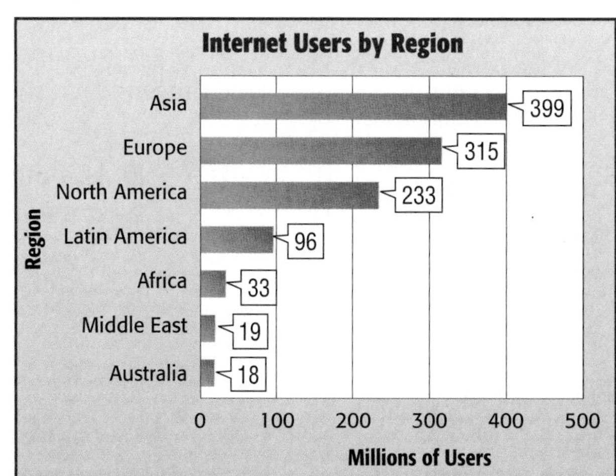

Source: Internet World Stats

5. **PATTERNS** Complete the pattern: 5, 11, 17, 23, ■, ■, ■.

6. **SCHOOL** The first five bells at Ed's middle school ring at 8:50 A.M., 8:54 A.M., 9:34 A.M., 9:38 A.M., and 10:18 A.M. If this pattern continues, when should the next three bells ring?

7. **MONEY** The Hamres are buying a new car. They will pay $350 per month for 4 years. How much will they spend in all for the car?

EXTRA PRACTICE See pages 672, 706.

8. **WALKING** Megan uses a pedometer to find how many steps she takes each school day. If she takes 6,482 steps on Monday, about how many steps will she take the entire school week?

H.O.T. Problems

9. **CHALLENGE** Complete the pattern: 3, 3, 6, 18, 72, ■.

10. **WRITING IN MATH** When using the four-step plan, explain why you should compare your answer to your estimate.

TEST PRACTICE

11. Michael can swim 8 laps in 4 minutes. At this rate, how long will it take him to swim 40 laps?

    A 24 min
    B 20 min
    C 15 min
    D 10 min

12. Find the next three numbers in the pattern below.

    57, 49, 41, 33, . . .

    F 25, 17, 9
    G 26, 18, 10
    H 25, 18, 11
    J 26, 17, 8

GET READY for the Next Lesson

PREREQUISITE SKILL Divide. (Page 658)

13. $42 \div 3$
14. $126 \div 6$
15. $49 \div 7$
16. $118 \div 2$

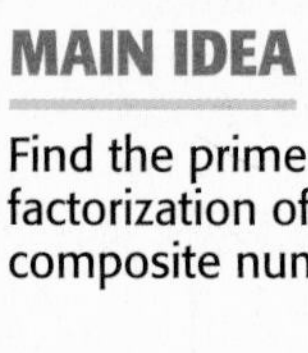

1-2 Prime Factors

MAIN IDEA

Find the prime factorization of a composite number.

New Vocabulary

factor
prime number
composite number
prime factorization

Math Online

glencoe.com

- Concepts in Motion
- Extra Examples
- Personal Tutor
- Self-Check Quiz
- Reading in the Content Area

MINI Lab

Any given number of squares can be arranged into one or more different rectangles.

STEP 1 Use a geoboard to make as many different rectangles as possible using two squares. Then repeat with four squares.

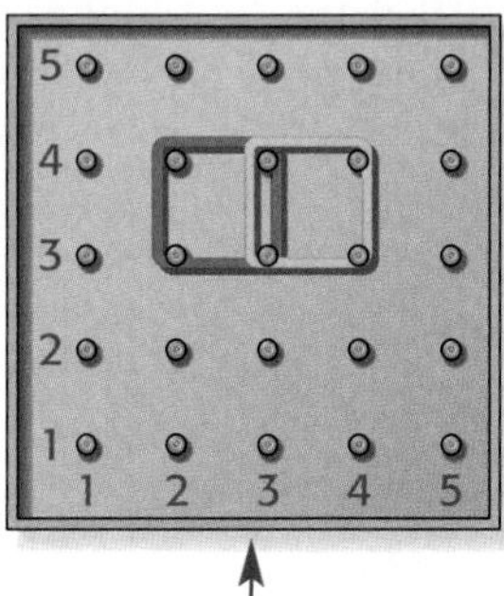

Only one rectangle can be made using two squares. The dimensions of this rectangle are 1 × 2.

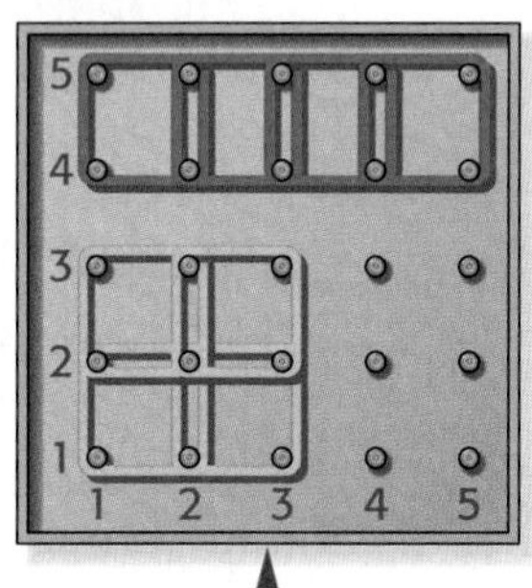

Two different rectangles can be made using four squares. The dimensions of these rectangles are 1 × 4 and 2 × 2.

STEP 2 Copy and complete the table using 2, 3, 4, . . ., 20 squares. Use a geoboard to help you.

Number of Squares	Dimensions of Each Rectangle
2	1 × 2
3	
4	1 × 4, 2 × 2

1. For what numbers can more than one rectangle be formed?
2. For what numbers can only one rectangle be formed?
3. For the numbers in which only one rectangle is formed, what do you notice about the dimensions of the rectangle?

When two or more numbers are multiplied, each number is called a **factor** of the product.

$1 \times 7 = 7$ (factors of 7: 1, 7)

$1 \times 6 = 6, 2 \times 3 = 6$ (factors of 6: 1, 6, 2, 3)

A whole number that has exactly two unique factors, 1 and the number itself, is a **prime number**. A number greater than 1 with more than two factors is a **composite number**.

Reading Math

Infinite *Infinite* means *endless*.

Prime and Composite — Key Concept

Number	Definition	Examples
prime	A whole number that has exactly two factors, 1 and the number itself.	11, 13, 23
composite	A number greater than 1 with more than two factors.	6, 10, 18
neither prime nor composite	1 has only one factor. 0 has an infinite number of factors.	0, 1

EXAMPLES Identify Prime and Composite Numbers

Tell whether each number is *prime, composite,* or *neither*.

1 **12**

Factors of 12: 1, 2, 3, 4, 6, 12

Since 12 has more than two factors, it is a composite number.

2 **19**

Factors of 19: 1, 19

Since there are exactly two factors, 19 is a prime number.

CHECK Your Progress

Tell whether each number is *prime, composite,* or *neither*.

a. 28 b. 11 c. 81

Every composite number can be expressed as a product of prime numbers. This is called a **prime factorization** of the number. A *factor tree* can be used to find the prime factorization of a number.

EXAMPLE Find Prime Factorization

3 **Find the prime factorization of 36.**

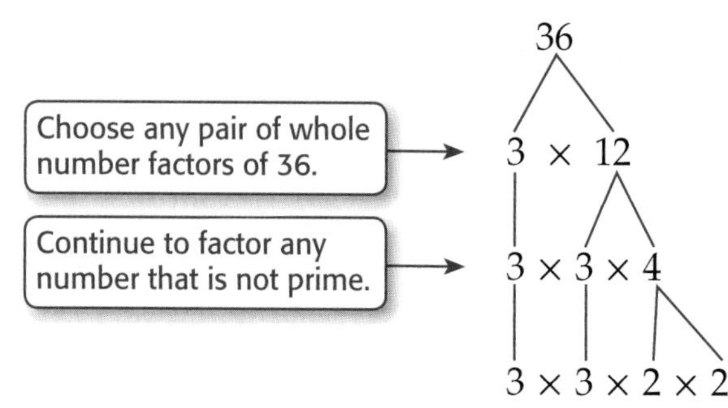

36
Choose any pair of whole number factors of 36.
3 × 12
Continue to factor any number that is not prime.
3 × 3 × 4
3 × 3 × 2 × 2

Study Tip

Prime Factors You can choose any pair of whole number factors, such as 4 × 9 or 2 × 18. Except for the order, the prime factors are the same.

CHECK Your Progress

Find the prime factorization of each number.

d. 54 e. 72

CHECK Your Understanding

Examples 1, 2 (p. 29)

Tell whether each number is *prime, composite,* or *neither.*

1. 10
2. 3
3. 1
4. 61

Example 3 (p. 29)

Find the prime factorization of each number.

5. 36
6. 81
7. 65
8. 19

9. **GEOGRAPHY** The state of South Carolina has 46 counties. Write 46 as a product of primes.

Practice and Problem Solving

HOMEWORK HELP	
For Exercises	See Examples
10–21, 36, 37	1, 2
22–35	3

Tell whether each number is *prime, composite,* or *neither.*

10. 17
11. 0
12. 15
13. 44
14. 23
15. 57
16. 45
17. 29
18. 56
19. 93
20. 53
21. 31

Find the prime factorization of each number.

22. 24
23. 18
24. 40
25. 75
26. 27
27. 32
28. 49
29. 25
30. 42
31. 104
32. 55
33. 77

ANALYZE TABLES **For Exercises 34–38, use the table that shows the average weights of popular dog breeds.**

Breed	Weight (lb)	Breed	Weight (lb)
Cocker Spaniel	20	Siberian Husky	50
German Shepherd	81	Boxer	60
Labrador Retriever	67	Rottweiler	112
Beagle	25	Dalmatian	55
Golden Retriever	70	Poodle	57

Source: Dog Breed Info Center

34. Which weight(s) have a prime factorization of exactly three factors?
35. Which weight(s) have a prime factorization with factors that are all the same number?
36. Which dog breeds have weights that are prime numbers?
37. Of the Beagle, Golden Retriever, Siberian Husky, Rottweiler, and Dalmatian breeds, which have weights that are composite numbers?
38. Name three weights that have exactly two prime factors in common.

See pages 672, 706.

Tell whether each number is *prime, composite,* or *neither.*

39. 125
40. 114
41. 179
42. 291

43. **POSTCARDS** Juliana bought packs of postcards that each had the same number of postcards. If she bought 20 postcards, find three possibilities for the number of packs and the number of postcards in each pack.

H.O.T. Problems

44. **OPEN ENDED** Select two prime numbers that are greater than 50 but less than 100.

45. **REASONING** All odd numbers greater than or equal to 7 can be expressed as the sum of three prime numbers. Which three prime numbers have a sum of 59? Justify your answer.

46. **NUMBER SENSE** *Twin primes* are two prime numbers that are consecutive odd integers such as 3 and 5, 5 and 7, and 11 and 13. Find all of the twin primes that are less than 100.

47. **CHALLENGE** A *counterexample* is an example that shows a statement is not true. Find a counterexample for the statement below. Explain your reasoning.

All even numbers are composite numbers.

48. **WRITING IN MATH** Explain how you know a number is prime.

TEST PRACTICE

49. Find the prime factorization of 225.

A $2 \times 3 \times 5 \times 5$

B $3 \times 3 \times 3 \times 5 \times 5$

C $3 \times 3 \times 5 \times 5$

D $3 \times 5 \times 5 \times 7$

50. Which number is *not* composite?

F 15

G 29

H 35

J 64

51. The volume of a rectangular prism can be found by multiplying the length, width, and height of the prism. Which of the following could be the possible dimensions of the rectangular prism below?

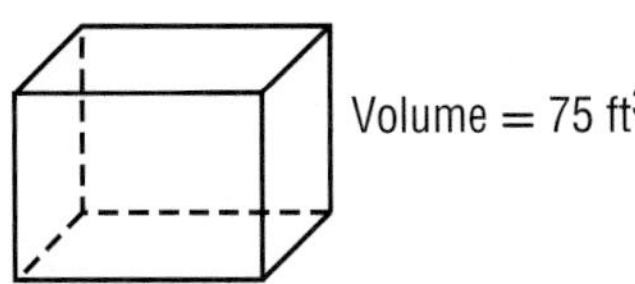

A 2 ft × 6 ft × 6 ft

B 3 ft × 5 ft × 7 ft

C 5 ft × 5 ft × 7 ft

D 3 ft × 5 ft × 5 ft

Spiral Review

52. **PATTERNS** Complete the pattern: 5, 7, 10, 14, 19, ■. (Lesson 1-1)

53. **TIME** The Pintos family left their home at 11:45 A.M. They traveled 325 miles at 65 miles per hour. If they stopped for an hour to eat lunch, how many hours did it take them to reach their destination? (Lesson 1-1)

GET READY for the Next Lesson

PREREQUISITE SKILL **Multiply.** (Page 658)

54. $2 \times 2 \times 2$

55. 5×5

56. $4 \times 4 \times 4$

57. $10 \times 10 \times 10$

1-3 Powers and Exponents

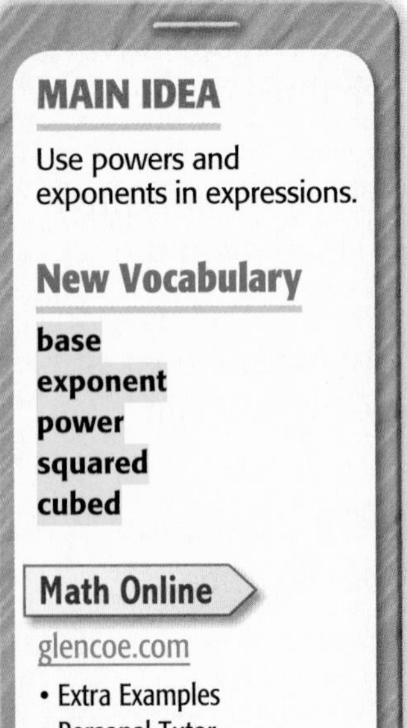

MAIN IDEA

Use powers and exponents in expressions.

New Vocabulary

base
exponent
power
squared
cubed

Math Online

glencoe.com

- Extra Examples
- Personal Tutor
- Self-Check Quiz

MINI Lab

Any number can be written as a product of prime factors.

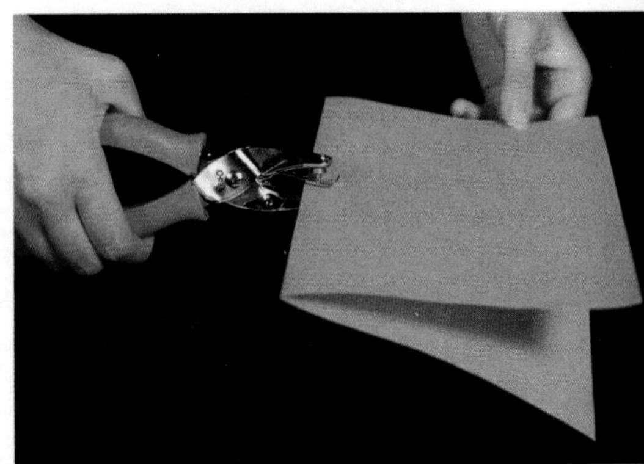

STEP 1 Fold a piece of paper in half and make one hole punch. Open the paper and count the number of holes. Copy the table below and record the results.

Number of Folds	Number of Holes	Prime Factorization
1		
⋮		
5		

STEP 2 Find the prime factorization of the number of holes and record the results in the table.

STEP 3 Fold another piece of paper in half twice. Then make one hole punch. Complete the table for two folds.

STEP 4 Complete the table for three, four, and five folds.

1. What prime factors did you record?
2. How does the number of folds relate to the number of factors in the prime factorization of the number of holes?
3. Write the prime factorization of the number of holes made if you folded it eight times.

A product of identical factors can be written using an exponent and a base. The **base** is the number used as a factor. The **exponent** indicates how many times the base is used as a factor.

$$\underbrace{2 \times 2 \times 2 \times 2 \times 2}_{\text{5 factors}} = 2^5$$

← exponent; ↑ base (the 2 in 2^5)

When no exponent is given, it is understood to be 1. For example, $5 = 5^1$.

Numbers expressed using exponents are called **powers**. Numbers raised to the second or third power have special names.

Powers	Words
2^5	2 to the fifth power
3^2	3 to the second power or 3 **squared**
10^3	10 to the third power or 10 **cubed**

Study Tip

Calculator You can use a calculator to evaluate powers. To find 3^4, enter 3  4 ENTER 81. The value of 3^4 is 81.

EXAMPLES Write Powers and Products

1 Write 3 × 3 × 3 × 3 using an exponent.

The base is 3. Since 3 is used as a factor four times, the exponent is 4.

$3 \times 3 \times 3 \times 3 = 3^4$ Write as a power.

2 Write 4^5 as a product of the same factor. Then find the value.

The base is 4. The exponent is 5. So, 4 is used as a factor five times.

$4^5 = 4 \times 4 \times 4 \times 4 \times 4$ Write 4^5 as a product.

$= 1{,}024$ Multiply.

CHECK Your Progress

Write each product using an exponent.

a. $7 \times 7 \times 7$

b. $10 \times 10 \times 10 \times 10 \times 10$

Write each power as a product of the same factor. Then find the value.

c. 2^3

d. 8^2

Real-World Career...
How Does an Environmentalist Use Math?
An environmentalist uses math to collect and analyze data from the environment they are studying.

Math Online
For more information, go to glencoe.com.

Real-World EXAMPLE

3 ENVIRONMENT In a recent year, about 10^4 youth across the United States participated in activities and events to care for Earth's environment. What is this number?

$10^4 = 10 \times 10 \times 10 \times 10$ Write 10^4 as a product.

$= 10{,}000$ Multiply.

So, about 10,000 youth participated in these events.

CHECK Your Progress

e. **COASTLINES** Georgia has 10^2 miles of coastline. What is the value of 10^2?

f. **TESTS** A multiple-choice test has 7 questions. If each question has 4 choices, there are 4^7 ways the test can be answered. What is the value of 4^7?

Exponents can be used to write the prime factorization of a number. Remember to write the prime factors in ascending order, that is, from least to greatest.

EXAMPLES Prime Factorization Using Exponents

Write the prime factorization of each number using exponents.

4 **72**

$72 = \underbrace{2 \times 2 \times 2} \times \underbrace{3 \times 3}$ Write the prime factorization.

$= 2^3 \times 3^2$ Write products of identical factors using exponents.

5 **135**

$135 = \underbrace{3 \times 3 \times 3} \times 5$ Write the prime factorization.

$= 3^3 \times 5$ Write products of identical factors using exponents.

6 **300**

$300 = \underbrace{2 \times 2} \times 3 \times \underbrace{5 \times 5}$ Write the prime factorization.

$= 2^2 \times 3 \times 5^2$ Write products of identical factors using exponents.

CHECK Your Progress

Write the prime factorization of each number using exponents.

g. 24 h. 45 i. 120

CHECK Your Understanding

Example 1 (p. 33) **Write each product using an exponent.**

1. $2 \times 2 \times 2 \times 2$
2. $6 \times 6 \times 6$

Example 2 (p. 33) **Write each power as a product of the same factor. Then find the value.**

3. 2^6
4. 3^7

Example 3 (p. 33)

5. **ANIMALS** There are nearly 3^5 species of monkeys on Earth. What is the value of 3^5?

6. **POPULATION** An estimated 10^5 people live in Charleston, South Carolina. About how many people live in Charleston?

Examples 4–6 (p. 34) **Write the prime factorization of each number using exponents.**

7. 20
8. 48
9. 90

Practice and Problem Solving

HOMEWORK HELP	
For Exercises	See Examples
10–15	1
16–23	2
24, 25	3
26–33	4–6

Write each product using an exponent.

10. 9×9
11. $8 \times 8 \times 8 \times 8$
12. $3 \times 3 \times 3 \times 3 \times 3 \times 3 \times 3$
13. $5 \times 5 \times 5 \times 5 \times 5$
14. $11 \times 11 \times 11$
15. $7 \times 7 \times 7 \times 7 \times 7 \times 7$

Write each power as a product of the same factor. Then find the value.

16. 10^3
17. 3^2
18. 5^4
19. 10^5
20. 9^3
21. 6^5
22. 10^1
23. 1^7

24. **FOOD** The number of Calories in two pancakes can be written as 7^3. What whole number does 7^3 represent?

25. **TEETH** A single tusk that weighed just over 2^8 pounds from an African elephant is the largest tooth ever recorded from any modern animal. About how many pounds did the tusk weigh?

Write the prime factorization of each number using exponents.

26. 25
27. 56
28. 50
29. 68
30. 88
31. 98
32. 560
33. 378

34. **BIRDS** To find the amount of space a cube-shaped bird cage holds, find the *cube* of the measure of one side of the bird cage. Express the amount of space occupied by the bird cage shown as a power. Then find the amount in cubic units.

Write each power as a product of the same factor. Then find the value.

35. seven squared
36. eight cubed
37. four to the fifth power

38. **GARDENING** Mrs. Locaputo's garden is organized into 6 rows. Each row contains 6 vegetable plants. How many total vegetable plants does Mrs. Locaputo have in her garden? Write using exponents, and then find the value.

39. **HOBBIES** A knitted scarf is made by joining 20 square blocks that are each made up of 20 rows of 20 stitches. How many total stitches does the scarf contain? Write using exponents, and then find the value.

H.O.T. Problems

40. **OPEN ENDED** Write a power whose value is greater than 100.

41. **NUMBER SENSE** Which is greater: 3^5 or 5^3? Explain your reasoning.

42. **FIND THE ERROR** Marissa and Rashaun are finding the value of 7^3. Who is correct? Explain your reasoning.

Marissa

Rashaun

CHALLENGE For Exercises 43–45, refer to the table at the right.

Powers of 3	Powers of 5	Powers of 10
$3^4 = 81$	$5^4 = 625$	$10^4 = 10{,}000$
$3^3 = 27$	$5^3 = 125$	$10^3 = 1{,}000$
$3^2 = 9$	$5^2 = 25$	$10^2 = 100$
$3^1 = 3$	$5^1 = 5$	$10^1 = ■$
$3^0 = ■$	$5^0 = ■$	$10^0 = ■$

43. Describe the pattern for the powers of 3. Find 3^0.

44. Describe the pattern for the powers of 5. Find 5^0.

45. Describe the pattern for the powers of 10. Find 10^1 and 10^0.

46. **WRITING IN MATH** Explain how to find 10^6 mentally.

TEST PRACTICE

47. If the pattern of figures continues, which value represents the seventh figure in the pattern?

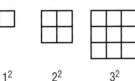

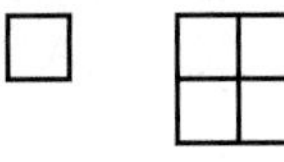

A 7^2
B 1^7
C 7^7
D 3^7

48. Which is the prime factorization of 360?

F $2^2 \times 3 \times 5^2$
G $2^3 \times 3^2 \times 5$
H $2^2 \times 3^3 \times 5$
J $2 \times 3^2 \times 5$

Spiral Review

Tell whether each number is *prime, composite,* or *neither.* (Lesson 1-2)

49. 63
50. 0
51. 29
52. 71

53. **TIME** Find the number of seconds in a day if there are 60 seconds in a minute. (Lesson 1-1)

GET READY for the Next Lesson

PREREQUISITE SKILL Divide. (Page 744)

54. $36 \div 3$
55. $45 \div 5$
56. $104 \div 8$
57. $120 \div 6$

1-4 Order of Operations

MAIN IDEA

Find the value of expressions using the order of operations.

New Vocabulary

numerical expression
order of operations

Math Online

glencoe.com

- Extra Examples
- Personal Tutor
- Self-Check Quiz

GET READY for the Lesson

SNACKS The table shows the cost of different snacks at a concession stand.

Item	Price ($)
Box of Popcorn	2
Juice Pop	1
Sandwich	4

1. How much would 3 boxes of popcorn cost? 4 sandwiches?
2. Find the total cost of buying 3 boxes of popcorn and 4 sandwiches.
3. What two operations did you use in Questions 1 and 2? Explain how to find the answer to Question 2 using these operations.

A **numerical expression** like $3 \times 2 + 4 \times 4$ is a combination of numbers and operations. The **order of operations** tells you which operation to perform first so that everyone finds the same value for an expression.

Order of Operations

Key Concept

1. Simplify the expressions inside grouping symbols, like parentheses.
2. Find the value of all powers.
3. Multiply and divide in order from left to right.
4. Add and subtract in order from left to right.

EXAMPLES Use Order of Operations

Find the value of each expression.

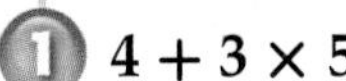

1 $4 + 3 \times 5$

$4 + 3 \times 5$
$= 4 + 15$ Multiply 3 and 5.
$= 19$ Add 4 and 15.

2 $10 - 2 + 8$

$10 - 2 + 8$
$= 8 + 8$ Subtract 2 from 10 first.
$= 16$ Add 8 and 8.

CHECK Your Progress

Find the value of each expression.

a. $10 + 2 \times 15$

b. $16 \div 2 \times 4$

EXAMPLES Parentheses and Exponents

Find the value of each expression.

3 $20 \div 4 + 17 \times (9 - 6)$

$20 \div 4 + 17 \times (9 - 6) = 20 \div 4 + 17 \times 3$	Subtract 6 from 9.
$= 5 + 17 \times 3$	Divide 20 by 4.
$= 5 + 51$	Multiply 17 by 3.
$= 56$	Add 5 and 51.

4 $3 \times 6^2 + 4$

$3 \times 6^2 + 4 = 3 \times 36 + 4$	Find 6^2.
$= 108 + 4$	Multiply 3 and 36.
$= 112$	Add 108 and 4.

CHECK Your Progress

Find the value of each expression.

c. $25 \times (5 - 2) \div 5 - 12$

d. $24 \div 2^3 + 6$

Real-World EXAMPLE

5 **SHOPPING** **A bath and body store sells lotions for $5, candles for $7, and lip balms for $2. Write an expression for the total cost of 3 lotions, 2 candles, and 4 lip balms. Then find the total cost.**

Cost of Items			
Item	lotion	candle	lip balm
Cost ($)	5	7	2

To find the total cost, write an expression and then find its value.

Words	cost of 3 lotions	plus	cost of 2 candles	plus	cost of 4 lip balms
Expression	3 × $5	+	2 × $7	+	4 × $2

3 × $5 + 2 × $7 + 4 × $2

= $15 + 2 × $7 + 4 × $2	Multiply 3 and 5.
= $15 + $14 + 4 × $2	Multiply 2 and 7.
= $15 + $14 + $8	Multiply 4 and 2.
= $37	

The total cost of the items is $37.

Real-World Link The largest hot pretzel ever baked weighed 40 pounds and was 5 feet across.

Source: Der Pretzel Haus

CHECK Your Progress

e. **SNACKS** Alexis and 3 friends are shopping at the mall. They decide to stop for a snack. Each person buys a hot pretzel for $3, a dipping sauce for $1, and a drink for $2. Write an expression for the total cost of the snacks. Then find the total cost.

CHECK Your Understanding

Examples 1–4 (pp. 37–38)

Find the value of each expression.

1. $9 + 3 - 5$
2. $10 - 3 + 9$
3. $(26 + 5) \times 2 - 15$
4. $18 \div (2 + 7) \times 2 + 1$
5. $5^2 + 8 \div 2$
6. $19 - (3^2 + 4) + 6$

Example 5 (p. 38)

7. **THEATER** Tickets to a play cost \$10 for members and \$24 for nonmembers. Write an expression to find the total cost of 4 nonmember tickets and 2 member tickets. Then find the total cost.

Practice and Problem Solving

HOMEWORK HELP

For Exercises	See Examples
8–11	1, 2
12–17	3
18–21	4
22, 23	5

Find the value of each expression.

8. $8 + 4 - 3$
9. $9 + 12 - 15$
10. $38 - 19 + 12$
11. $22 - 17 + 8$
12. $7 + 9 \times (3 + 8)$
13. $(9 + 2) \times 6 - 5$
14. $63 \div (10 - 3) \times 3$
15. $66 \times (6 \div 2) + 1$
16. $27 \div (3 + 6) \times 5 - 12$
17. $55 \div 11 + 7 \times (2 + 14)$
18. $5^3 - 12 \div 3$
19. $26 + 6^2 \div 4$
20. $15 - 2^3 \div 4$
21. $22 \div 2 \times 3^2$

22. **TICKETS** Admission to a circus is \$16 for adults and \$10 for children. Write an expression to find the total cost of 3 adult tickets and 4 children's tickets. Then find the total cost.

23. **MOVIES** Tyree and four friends go to the movies. Each person buys a movie ticket for \$7, a snack for \$3, and a drink for \$2. Write an expression for the total cost of the trip to the movies. Then find the total cost.

Find the value of each expression.

24. $8 \times (2^4 - 3) + 8$
25. $12 \div 4 + (5^2 - 6)$
26. $9 + 4^3 \times (20 - 8) \div 2 + 6$
27. $96 \div 4^2 + (25 \times 2) - 15 - 3$

28. **APPLES** Addison is making caramel covered apples for 15 friends. She has covered 3 dozen apples. If she wants each friend to receive exactly 3 apples and have no apples left over, write an expression to find how many more apples she should cover. Then find this number.

Write a numerical expression for each verbal expression. Then find its value.

See pages 673, 706.

29. the product of 7 and 6, minus 2
30. the cube of the quotient of 24 and 6

H.O.T. Problems

31. **CHALLENGE** Create an expression with a value of 10. It should contain four numbers and two different operations.

32. **FIND THE ERROR** Miranda and Dalton are finding $9 - 6 + 2$. Who is correct? Explain your reasoning.

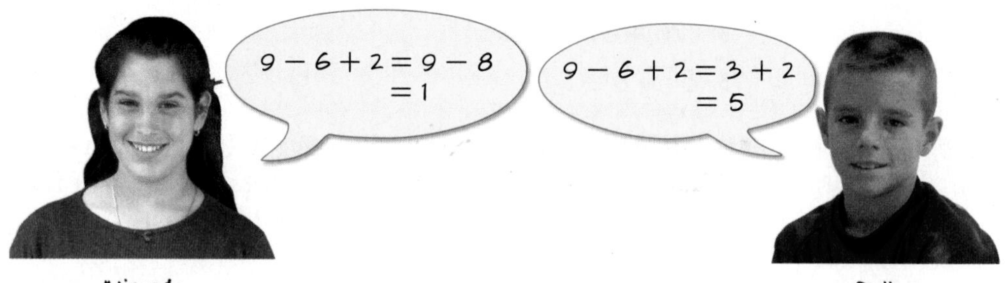

33. **WRITING IN MATH** Write a real-world problem that can be solved using order of operations. Then solve the problem.

TEST PRACTICE

34. Arleta is 2 years younger than Josh, and Josh is 5 years older than Monica, who is 9 years old. Which table could be used to find Arleta's age?

A

Name	Age (years)
Arleta	9 + 5
Josh	9 + 5 − 2
Monica	9

B

Name	Age (years)
Arleta	2
Josh	5
Monica	9

C

Name	Age (years)
Arleta	5
Josh	4
Monica	9

D

Name	Age (years)
Arleta	9 + 5 − 2
Josh	9 + 5
Monica	9

Spiral Review

35. **PHONE TREE** Four members of a certain phone tree are each given 4 people to contact. If the phone tree is activated, the total number of calls made is 4^4. How many calls is this? (Lesson 1-3)

Find the prime factorization of each number. (Lesson 1-2)

36. 42
37. 75
38. 110
39. 130

GET READY for the Next Lesson

PREREQUISITE SKILL **Add.** (Page 743)

40. 26 + 98
41. 23 + 16
42. 61 + 19
43. 54 + 6

CHAPTER 1

Mid-Chapter Quiz

Lessons 1-1 through 1-4

1. **BOOKS** Hugo needs to finish reading a 465-page book by Sunday. The number of pages he read each day are shown in the table. How many pages will he need to read on Saturday and Sunday in order to finish the book in time? (Lesson 1-1)

Day	Number of Pages
Monday	60
Tuesday	72
Wednesday	59
Thursday	85
Friday	67

2. **MULTIPLE CHOICE** A school has 144 computers in 24 classrooms. How many computers are in each classroom if each classroom has the same number of computers? (Lesson 1-1)

   A 6

   B 24

   C 120

   D 3,456

Tell whether each number is *prime, composite,* or *neither.* (Lesson 1-2)

3. 57

4. 97

5. 0

6. **BOOKS** Can a group of 41 books be placed onto more than one shelf so that each shelf has the same number of books and has more than one book per shelf? Explain your reasoning. (Lesson 1-2)

Write each power as a product of the same factor. Then find the value. (Lesson 1-3)

7. 3^4 8. 6^3

Write the prime factorization of each number using exponents. (Lesson 1-3)

9. 22 10. 40 11. 75

12. **DOGS** The average annual cost of food for a dog is about 3^5 dollars. What is this cost? (Lesson 1-3)

Find the value of each expression. (Lesson 1-4)

13. $10 - 6 + 20$

14. $25 \div (15 - 10) \times 2$

15. $3^2 + 32 \div 2$

16. $12 - (4^3 \div 8) + 1$

17. **MULTIPLE CHOICE** Mr. and Mrs. Murphy and their 4 children went to the county fair. Admission to the fair was $7.75 for an adult and $5.50 for a child. Arrange the problem-solving steps below in the correct order to find the total cost of the tickets.

    Step K: Multiply the cost of a child's ticket by the number of children.

    Step L: Add the two products together.

    Step M: Multiply the cost of an adult ticket by the number of adults.

    Step N: Write down the number of adults and the number of children that are going to the county fair.

    Which list shows the steps in the correct order? (Lesson 1-4)

    F N, L, M, K

    G N, M, K, L

    H K, M, N, L

    J M, K, N, L

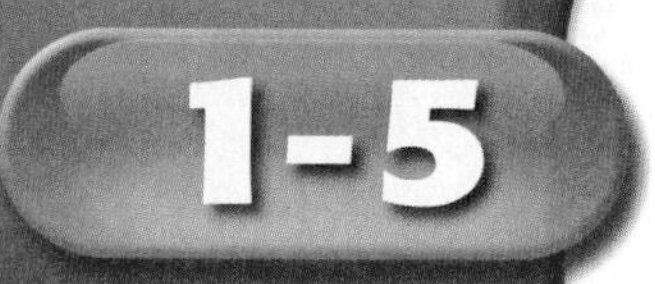

Algebra: Variables and Expressions

MAIN IDEA

Evaluate algebraic expressions.

New Vocabulary

algebra
variable
algebraic expression
evaluate

Math Online

glencoe.com

- Extra Examples
- Personal Tutor
- Self-Check Quiz

GET READY for the Lesson

ART SUPPLIES A box contains some crayons. There are also two crayons outside of the box. The total number of crayons is *the sum of two and some number.* The two crayons represent the value 2, and the box represents the unknown value.

1. What does *some number* represent?
2. Find the value of the expression *the sum of two and some number* if *some number* is 14.
3. Assume you have two boxes of crayons each with the same number of crayons inside. Write an expression that represents the total number of crayons in both boxes.

Algebra is a language of symbols, including variables. A **variable** is a symbol, usually a letter, used to represent a number. The expression $2 + n$ represents *the sum of two and some number.*

Algebraic expressions are combinations of variables, numbers, and at least one operation.

$2 + n$ ← Any letter can be used as a variable.

The letter x is often used as a variable. It is also common to use the first letter of the value you are representing.

The variables in an expression can be replaced with any number. Once the variables have been replaced, you can **evaluate**, or find the value of, the algebraic expression.

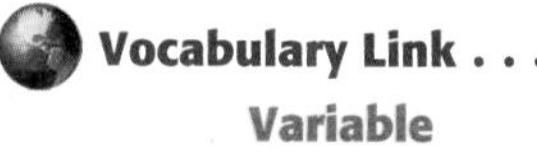

Vocabulary Link
Variable
Everyday Use able to change or vary, as in variable winds
Math Use a symbol used to represent a number

In addition to the symbol ×, there are other ways to show multiplication.

$2 \cdot 3$	$5t$	st
2 times 3	5 times t	s times t

EXAMPLES Evaluate Algebraic Expressions

1 Evaluate $16 + b$ if $b = 25$.

$16 + b = 16 + 25$ Replace b with 25.

$= 41$ Add 16 and 25.

2 Evaluate $x - y$ if $x = 64$ and $y = 27$.

$x - y = 64 - 27$ Replace x with 64 and y with 27.

$= 37$ Subtract 27 from 64.

Study Tip

Multiplication In algebra, the symbol • is often used to represent multiplication, as the symbol × may be confused with the variable x.

3 Evaluate $5t + 4$ if $t = 3$.

$5t + 4 = 5 \cdot 3 + 4$ Replace t with 3.

$= 15 + 4$ Multiply 5 and 3.

$= 19$ Add 15 and 4.

CHECK Your Progress

Evaluate each expression if $a = 6$ and $b = 4$.

a. $a + 8$ b. $a - b$ c. $a \cdot b$ d. $2a - 5$

TEST EXAMPLE

4 An expression for finding the area of a triangle that has a height 3 units longer than its base is $(b + 3) \cdot b \div 2$, where b is the measure of the base. Find the area of such a triangle with a base 8 units long.

A 20 units^2 **B** 25 units^2 **C** 44 units^2 **D** 88 units^2

Test-Taking Tip

Preparing for the Test In preparation for the test, it is often helpful to familiarize yourself with important formulas or rules such as the rules for order of operations.

Read the Item

You need to find the value of the expression given $b = 8$.

Solve the Item

$(b + 3) \cdot b \div 2 = (8 + 3) \cdot 8 \div 2$ Replace b with 8.

$= 11 \cdot 8 \div 2$ Add 8 and 3.

$= 88 \div 2$ Multiply 11 and 8.

$= 44$ Divide 88 by 2.

The area of the triangle is 44 units^2. The answer is C.

CHECK Your Progress

e. If admission to a fair is \$7 per person and each ride ticket costs \$2, the total cost for admission and t ride tickets is $7 + 2t$. Find the total cost for admission and 5 ride tickets.

F \$9 **G** \$17 **H** \$35 **J** \$45

CHECK Your Understanding

Examples 1–3 (p. 43)

Evaluate each expression if $m = 4$ and $z = 9$.

1. $3 + m$
2. $z + 5$
3. $z - m$
4. $m - 2$
5. $4m - 2$
6. $2z + 3$

Example 4 (p. 43)

7. **MULTIPLE CHOICE** The amount of money that remains from a \$20 dollar bill after Malina buys 4 party favors for p dollars each is $20 - 4p$. Find the amount remaining if each favor costs \$3.

   A \$4 **C** \$17

   B \$8 **D** \$48

Practice and Problem Solving

HOMEWORK HELP	
For Exercises	See Examples
8–19	1, 2
20–27	3
51–53	4

Evaluate each expression if $m = 2$ and $n = 16$.

8. $m + 10$
9. $n + 8$
10. $9 - m$
11. $22 - n$
12. $n \div 4$
13. $12 \div m$
14. $n \cdot 3$
15. $6 \cdot m$
16. $m + n$
17. $n + m$
18. $n - 6$
19. $m - 1$

Evaluate each expression if $a = 4$, $b = 7$, and $c = 11$.

20. $b - a$
21. $c - b$
22. $5c + 6$
23. $2b + 7$
24. $3a - 4$
25. $4b - 10$

26. **BAMBOO** To find the amount a bamboo plant can grow in a certain amount of time, use the expression rt, where r represents rate and t represents time. How many feet can a bamboo plant grow in seven days at a rate of 3 feet per day?

27. **RACING** To find the average speed of a racecar, use the expression $d \div t$, where d represents distance and t represents time. Find the speed s of a racecar that travels 508 miles in 4 hours.

Real-World Link. . . The average speed at the 2007 Daytona 500 was 149.3 mph.

Source: NASCAR

Evaluate each expression if $a = 4$, $b = 15$, and $c = 9$.

28. $c^2 + a$
29. $b^2 - 5c$
30. $3a \div 4$
31. $4b \div 5$
32. $5b \cdot 2$
33. $2ac$

34. If $y = 4$, what is the value of $5y - 3$?

35. What is the value of $st \div 6r$ if $r = 5$, $s = 32$, and $t = 45$?

36. **PLANES** The expression $500t$ can be used to find the distance traveled by a DC-10 aircraft. The variable t represents time in hours. How far can a DC-10 travel in 4 hours?

Evaluate each expression if $x = 3$, $y = 12$, and $z = 8$.

37. $4z + 8 - 6$

38. $6x - 9 \div 3$

39. $15 + 9x \div 3$

40. $7z \div 4 + 5x$

41. $y^2 \div (3z)$

42. $z^2 - (5x)$

43. **GEOMETRY** To find the area of a rectangle, use the expression ℓw, where ℓ represents the length, and w represents the width of the rectangle. What is the area of the rectangle shown?

44. **MUSIC** As a member of a music club, you can order CDs for $15 each. The music club also charges $5 for each shipment. The expression $15n + 5$ represents the cost of n CDs. Find the total cost for ordering 3 CDs.

45. **ANALYZE TABLES** To change a temperature given in degrees Celsius to degrees Fahrenheit, first multiply the Celsius temperature by 9. Next, divide the answer by 5. Finally, add 32 to the result. Write an expression that can be used to change a temperature from degrees Celsius to degrees Fahrenheit. Then use the information in the table below to find the difference in average temperatures in degrees Fahrenheit for San Antonio from January to April. (*Hint:* Convert to degrees Fahrenheit first.)

Average Monthly Temperature for San Antonio, Texas

Month	Temp. (°C)
January	10
April	20
July	29

H.O.T. Problems

46. **OPEN ENDED** Create two algebraic expressions involving multiplication that have the same meaning.

47. **CHALLENGE** Marcus and Yvette each have a calculator. Yvette starts at 100 and subtracts 7 each time. Marcus starts at zero and adds 3 each time. Suppose Marcus and Yvette press the keys at the same time. Will their displays ever show the same number? If so, what is the number? Explain your reasoning.

48. **SELECT A TECHNIQUE** Ichiro is evaluating $x^2 - z$, where $x = 3$ and $z = 8$. Which of the following techniques might Ichiro use to evaluate the expression? Justify your selection(s). Then use the technique(s) to solve the problem.

mental math | number sense | estimation

49. **Which One Doesn't Belong?** Identify the expression that does not belong with the other three. Explain your reasoning.

$7y$	$6 + 8$	xy	$3a + 2$

50. **WRITING IN MATH** Compare and contrast numerical expressions and algebraic expressions. Use examples in your explanation.

TEST PRACTICE

51. The height of the triangle below can be found using the expression $48 \div b$ where b is the base of the triangle. Find the height of the triangle.

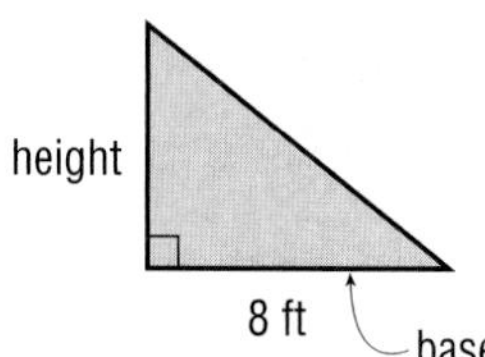

A 4 ft
B 6 ft
C 8 ft
D 10 ft

52. **SHORT RESPONSE** The expression $4s$ can be used to find the perimeter of a square where s represents the length of a side. What is the perimeter in inches of a square with a side length of 26 inches?

53. The table shows the total medal counts for different countries from the 2006 Winter Olympic games.

Total Medal Count	
Country	**Number of Medals**
Germany	29
United States	25
Canada	x
Austria	23
Russia	22
Norway	19

Source: International Olympic Committee

Which expression represents the total number of medals earned by all the countries listed in the table?

F $118 - x$
G $2x + 118$
H $x - 118$
J $118 + x$

Spiral Review

Find the value of each expression. (Lesson 1-4)

54. $12 - 8 \div 2 + 1$

55. $5^2 + (20 \div 2) - 7$

56. $21 \div (3 + 4) \times 3 - 8$

57. **LANGUAGE** An estimated 10^9 people in the world speak Mandarin Chinese. About how many people speak this language? (Lesson 1-3)

58. **TESTS** On a test with 62 questions, Trey missed 4 questions. How many did he get correct? (Lesson 1-1)

GET READY for the Next Lesson

PREREQUISITE SKILL Add or subtract. (Page 743)

59. $18 - 9$

60. $5 + 18$

61. $14 + 7$

62. $21 - 15$

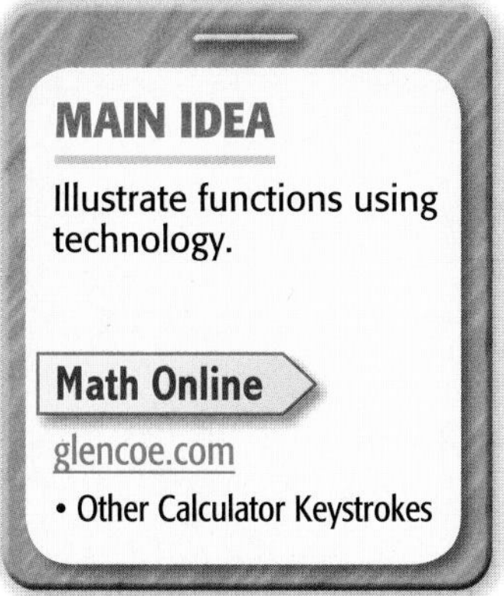

Graphing Calculator Lab
Function Machines

A *function machine* takes a value called the *input* and performs one or more operations on it according to a rule to produce a new value called the *output*.

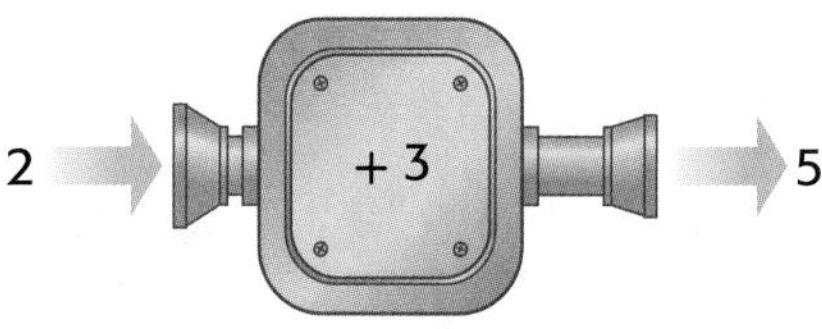

Another way to write the rule of a function machine is as an algebraic expression. For the function machine above, an input value of x produces an output value of $x + 3$. You can use the TI-83/84 Plus graphing calculator to model this function machine.

ACTIVITY

Use a graphing calculator to model a function machine for the rule $x + 3$. Then use this machine to find the output values for the input values 2, 3, 4, 9, and 12.

The graphing calculator uses X for input and Y for output values.

STEP 1 Enter the rule for the function into the function list. Press [Y=] to access the function list. Then press [X,T,θ,n] [+] 3 to enter the rule.

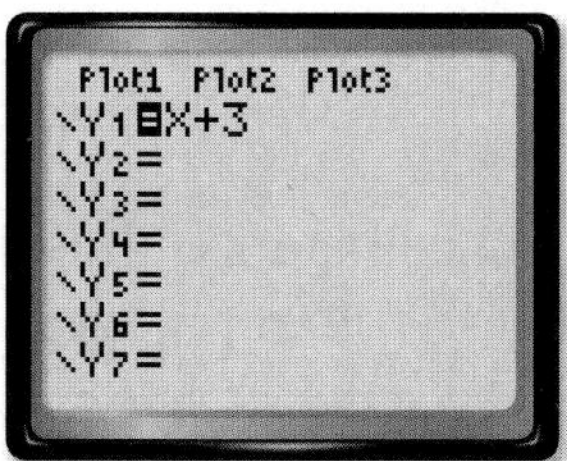

STEP 2 Next, set up a table of input and output values. Press [2nd] [TBLSET] to display the table setup screen. Press [↓] [↓] [→] [ENTER] to highlight **Indpnt: Ask.** Then press [↓] [ENTER] to highlight **Depend: Auto.**

(continued on the next page)

STEP 3 Access the table by pressing [2nd] [TABLE]. The calculator will display an empty function table.

STEP 4 Now key in your input values, pressing [ENTER] after each one.

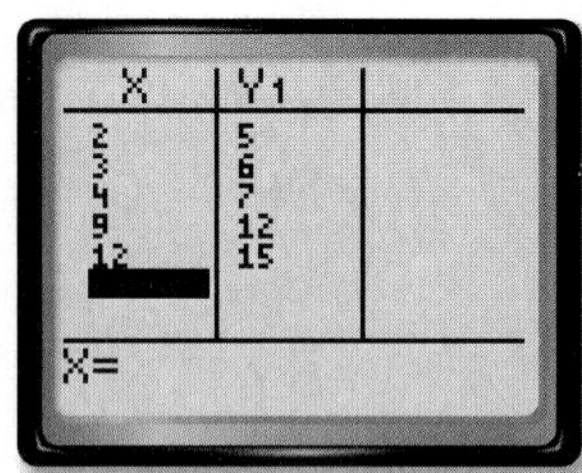

CHECK Your Progress

Use a graphing calculator to model a function machine for each of the following rules. Use the input values 5, 6, 7, and 8 for x. Record the inputs and their corresponding outputs in a table.

a. $x - 4$
b. $x + 5$
c. $x - 2$
d. $x - 3$
e. $x \cdot 2$
f. $x \cdot 3$

ANALYZE THE RESULTS

1. Examine the columns of inputs and outputs for Exercises a–d. What pattern do you observe in the column of inputs? What pattern do you observe in each column of outputs?

2. How would each column of outputs change if the order of the inputs was reversed to be 8, 7, 6, and 5?

3. Examine the columns of inputs and outputs for Exercises e and f. What patterns do you observe in the column of outputs?

4. Compare the patterns you observed in Exercise 3 to the rules given for Exercises e and f. What do you notice?

MAKE A CONJECTURE **Based on your observations from Exercises 1–4, make a conjecture about the rule for each set of input and output values. Explain your reasoning.**

5.

Input (x)	Output (y)
10	2
11	3
12	4
13	5
14	6

6.

Input (x)	Output (y)
2	12
3	18
4	24
5	30
6	36

1-6 Algebra: Functions

MAIN IDEA

Complete function tables and find function rules.

New Vocabulary

function
function table
function rule
defining the variable

Math Online

glencoe.com

- Extra Examples
- Personal Tutor
- Self-Check Quiz

GET READY for the Lesson

SCIENCE A ruby-throated hummingbird beats its wings about 52 beats per second.

1. Write an expression to represent the number of times this bird beats its wings in 2 seconds, in 6 seconds, and in *s* seconds.

A **function** is a relationship that assigns exactly one output value to one input value. The number of wing beats (output) depends on the number of seconds (input). You can organize the input-output values in a **function table**.

Input	Function Rule	Output
Number of Seconds (s)	$52s$	Wing Beats
1	52(1)	52
2	52(2)	104
3	52(3)	156

The **function rule** describes the relationship between each input and output.

EXAMPLE Complete a Function Table

1 The output is 7 more than the input. Complete a function table for this relationship.

The function rule is $x + 7$. Add 7 to each input.

Input (x)	Output ($x + 7$)
10	■
12	■
14	■

→

Input (x)	Output ($x + 7$)
10	17
12	19
14	21

CHECK Your Progress

Copy and complete each function table.

a.

Input (x)	Output ($x - 4$)
4	■
7	■
10	■

b.

Input (x)	Output ($3x$)
0	■
2	■
5	■

Study Tip

Check for Reasonableness To see if your function rule is correct, test more than one input value.

EXAMPLE Find the Rule for a Function Table

2 Find the rule for the function table.

Input (x)	Output (■)
2	6
5	15
7	21

Study the relationship between each input and output. Each output is three times the input.

So, the function rule is $3 \cdot x$, or $3x$.

CHECK Your Progress

Find the rule for each function table.

c.

Input (x)	Output (■)
0	0
4	1
16	4

d.

Input (x)	Output (■)
4	1
8	5
10	7

When you write a function rule that represents a real-world situation, you first choose a variable to represent the input. This is called **defining the variable**.

Real-World EXAMPLE

3 MUSIC A local band charges \$70 for each hour it performs. Define a variable. Then write a function rule that relates the total charge to the number of hours it performs.

Real-World Link Percussion instruments are musical instruments that are sounded by striking, shaking, plucking, or scraping. There are over 52 different kinds of percussion instruments.

Source: Virginia Tech Multimedia Music Dictionary

Determine the function rule. The cost of the performance depends on the number of hours. Let h represent the number of hours.

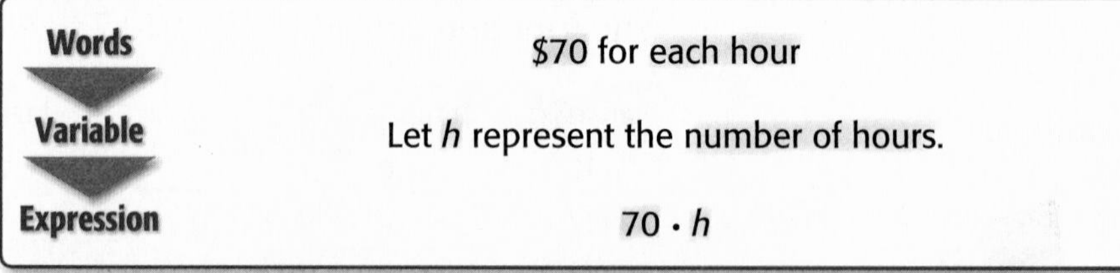

Words	\$70 for each hour
Variable	Let h represent the number of hours.
Expression	$70 \cdot h$

The function rule is $70h$.

CHECK Your Progress

e. **SHOPPING** A department store is deducting \$10 off the total purchase for shoppers from 6 A.M. to 7 A.M. Define a variable. Write a function rule that relates the final cost to the total purchase amount.

CHECK Your Understanding

Example 1 (p. 49) **Copy and complete each function table.**

1.

Input (x)	Output (x + 3)
0	■
2	■
4	■

2.

Input (x)	Output (4x)
1	■
3	■
6	■

Example 2 (p. 50) **Find the rule for each function table.**

3.

x	■
1	0
3	2
5	4

4.

x	■
0	0
3	6
6	12

Example 3 (p. 50)

5. **JELLY BEANS** Lamar is buying jelly beans for a party. He can buy them in bulk for $3 a pound. Define a variable. Write a function rule that relates the total cost of the jelly beans to the number of pounds he buys.

Practice and Problem Solving

HOMEWORK HELP

For Exercises	See Examples
6–7	1
8–13	2
14, 15	3

Copy and complete each function table.

6.

Input (x)	Output (x − 4)
4	■
8	■
11	■

7.

Input (x)	Output (x ÷ 3)
0	■
3	■
9	■

Find the rule for each function table.

8.

x	■
0	2
1	3
6	8

9.

x	■
7	2
9	4
15	10

10.

x	■
2	4
5	10
8	16

11.

x	■
0	0
4	20
7	35

12.

x	■
5	1
15	3
25	5

13.

x	■
6	3
22	11
34	17

14. **AGES** Ricardo is 8 years older than his sister. Define a variable. Write a function rule that relates Ricardo's age to his sister's age.

15. **FOOD** Whitney has a total of 30 cupcakes for her guests. Define a variable. Write a function rule that relates the number of cupcakes per guest to the number of guests.

Find the rule for each function table.

16.

x	■
2	2
3	5
4	8
5	11

17.

x	■
0	1
1	7
2	13
3	19

18.

x	■
3	13
6	28
9	43
12	58

For Exercises 19–21, define a variable and write a function rule. Then solve each problem.

19. **ANIMALS** Moose can swim up to 6 miles per hour. At this rate, find the total number of miles a moose could swim in two hours.

20. **MONEY** Kyle is buying 7 greeting cards that cost $2 each. If he has a coupon for $3 off his total purchase, how much will he spend for the greeting cards?

21. **MUSIC** An Internet company charges $10 a year to be a member of its music program. It also charges $1 for each song you download. How much will it cost if you download 46 songs in a year?

22. **FIND THE DATA** Refer to the Data File on pages 16–19. Choose some data and write about a real-world situation that can be described by a function rule.

23. **TICKETS** The science club is going on a field trip to the zoo. Student tickets are $6.00 each and adult tickets are $9.00 each. Write a function rule to represent the total cost of s student tickets and a adult tickets. Then use the function rule to find the cost for 8 students and 3 adults.

Zoo Admission Rates

Ticket	Price
Adult	$9.00
Student	$6.00

EXTRA PRACTICE
See pages 673, 706.

H.O.T. Problems

24. **OPEN ENDED** Create a function table. Then write a function rule. Choose three input values and find the output values.

25. **FIND THE ERROR** Nadia and Caitlyn are finding the function rule when each output is 3 less than the input. Who is correct? Explain.

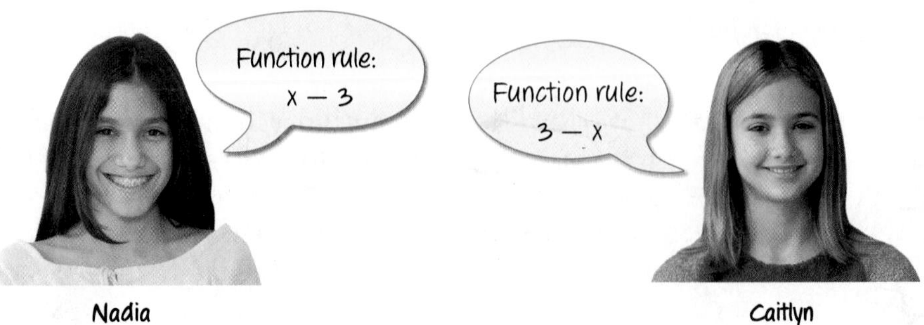

26. **CHALLENGE** Around 223 million Americans keep containers filled with coins in their home. Suppose each of the 223 million people started putting their coins back into circulation at a rate of \$10 per year. Create a function table that shows the amount of money that would be recirculated in 1, 2, and 3 years.

27. **SELECT A TOOL** Courtney is evaluating the function rule $43x - 6$ for an input of 4. Which of the following tools might Courtney use to determine the output? Justify your selection(s). Then use the tool(s) to solve the problem.

real objects | graphing calculator | paper/pencil

28. **WRITING IN MATH** Explain how to find a function rule given a function table.

TEST PRACTICE

29. Which expression best represents the y values in terms of the x values?

x	1	2	3	4	5	6
y	5	7	9	11	13	15

A $2x + 3$ C $3x - 2$

B $x + 3$ D $6 - x$

30. A store makes a profit of \$5 for each shirt sold. Which expression best represents the profit on 25 shirts?

F 5×25 H $25 \div 5$

G $5 + 25$ J $25 - 5$

Spiral Review

Evaluate each expression if $a = 3$, $b = 6$, and $c = 10$. (Lesson 1-5)

31. $b - a$ 32. $3c + a$ 33. $bc + 12$

34. **FOOD** A deli sells wraps for \$5 and soup for \$3 a bowl. Write and solve an expression for the cost of 3 wraps and 2 bowls of soup. (Lesson 1-4)

35. **AREA CODES** California has 5^2 area codes. What is the value of 5^2? (Lesson 1-3)

GET READY for the Next Lesson

36. **PREREQUISITE SKILL** The table represents the average amounts consumers spent on back-to-school merchandise in a recent year. How much more did consumers spend on clothing, accessories, and shoes than on school supplies? Use the four-step plan. (Lesson 1-1)

Back-to-School Spending

Merchandise	Amount (\$)
Clothing/Accessories	219
Electronic Equipment	101
Shoes	90
School Supplies	73

Source: *USA Today*

1-7 Problem-Solving Investigation

MAIN IDEA: Solve problems by using the guess and check strategy.

P.S.I. TEAM +

e-Mail: GUESS AND CHECK

DARIUS: For my birthday, I received $100 from my relatives. All of the money was in $10 bills and $20 bills. When I put the money in my savings account, I deposited 8 bills.

YOUR MISSION: Use guess and check to find how many of each bill Darius received for his birthday.

Understand	You know that Darius has $100 in $10 bills and $20 bills. You need to find the number of each bill he has.
Plan	Make a guess until you find an answer that makes sense for the problem.
Solve	(see table below) So, Darius received six $10 bills and two $20 bills.
Check	Six $10 bills equals $60 and two $20 bills equals $40. Since $60 + $40 = $100, the answer is correct.

Number of $10 bills	Number of $20 bills	Total Amount	
7	1	7($10) + 1($20) = $90	too low
4	4	4($10) + 4($20) = $120	too high
5	3	5($10) + 3($20) = $110	still too high
6	2	6($10) + 2($20) = $100	✓

Analyze The Strategy

1. Explain when to use the *guess and check* strategy to solve a problem.
2. **WRITING IN MATH** Write a problem that can be solved using guess and check. Then tell the steps you would take to find the solution of the problem.

Mixed Problem Solving

See pages 674, 706.

Use the *guess and check* strategy to solve Exercises 3–6.

3. **COMICS** A comic book store sells used comic books in packages of 5 and new comic books in packages of 3. If Monica buys a total of 16 comic books, how many packages of new and used comic books did she buy?

4. **QUIZZES** On a science quiz, Ivan earned 18 points. If there are 6 problems worth 2 points each and 2 problems worth 4 points each, find the number of problems of each type Ivan answered correctly.

5. **NUMBERS** Antonio is thinking of four numbers from 1 through 9 with a sum of 18. Find the numbers.

6. **MONEY** Liviana has $2.20 in coins in her change purse. If there is a total of 15 coins, how many quarters, dimes, nickels, and pennies does she have?

Use any strategy to solve Exercises 7–14. Some strategies are shown below.

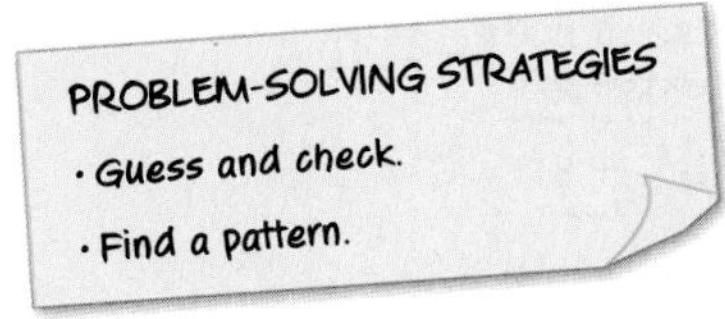

7. **ANALYZE TABLES** How much deeper is Crater Lake than Lake Superior?

Lake	Depth (ft)
Crater Lake	1,943
Lake Tahoe	1,685
Lake Chelan	1,419
Lake Superior	1,333

8. **SCIENCE** Mars orbits the Sun at a rate of 15 miles per second. How far does Mars travel in one day?

9. **NUMBERS** The sum of two prime numbers is 20. Find the numbers.

10. **PATTERNS** Draw the next figure in the pattern.

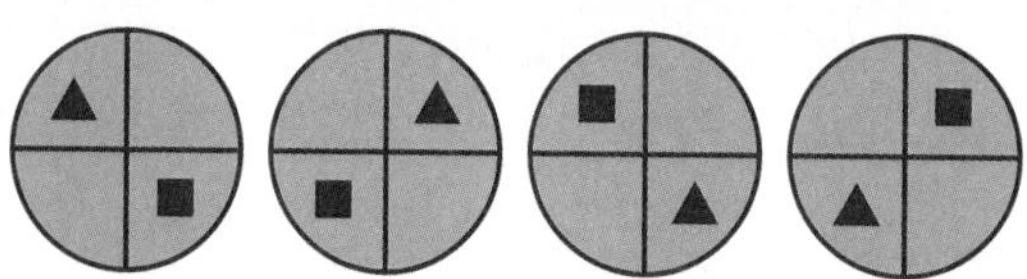

11. **ORDER OF OPERATIONS** Use the symbols $+$, $-$, $\times$, or $\div$ to make the following math sentence true. Use each symbol only once.

$$3 \blacksquare 4 \blacksquare 6 \blacksquare 1 = 18$$

12. **SCHEDULES** The schedule for a shuttle bus is shown in the table. If the pattern continues, what time will the sixth bus arrive and depart?

Bus	Arrival Time	Departure Time
1	8:42	8:52
2	9:12	9:22
3	9:42	9:52
4	10:12	10:22

13. **ANALYZE TABLES** How many fewer students were on the 6th grade honor roll in the 3rd grading period than in the 1st grading period?

6th Grade Honor Roll Students	
1st grading period	40
2nd grading period	37
3rd grading period	31

14. **MONEY** Nathaniel is saving money to buy a new graphics card for his computer that costs $250. If he is saving $20 a month and he already has $160, in how many more months will he have enough money for the graphics card?

READING to SOLVE PROBLEMS

Topic Sentences

A topic sentence is a sentence that expresses the main idea in a paragraph. It is usually found near the beginning of the paragraph and is followed by supporting details. Here's the beginning of a paragraph about Mrs. Garcia's math class.

Topic sentence

Mrs. Garcia's math class was doing research about wild horses living on public lands. They found that there are about 30,000 wild horses living in Nevada, 4,000 living in Wyoming, and 2,000 living in California.

In a word problem, the "topic sentence" is usually found near the end. It is the sentence or question that tells you what you need to find. Here's the same information, written as a word problem.

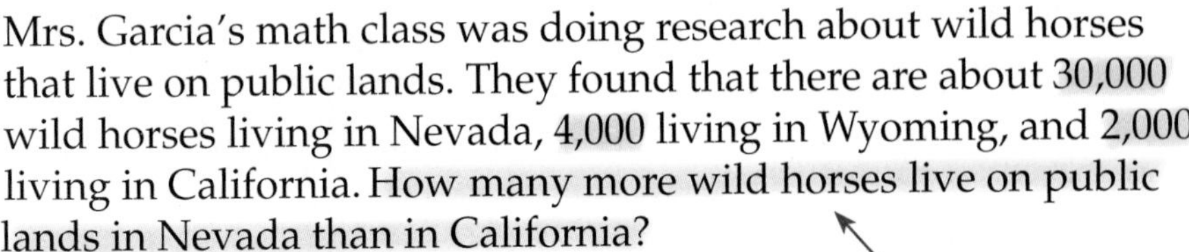

Mrs. Garcia's math class was doing research about wild horses that live on public lands. They found that there are about 30,000 wild horses living in Nevada, 4,000 living in Wyoming, and 2,000 living in California. How many more wild horses live on public lands in Nevada than in California?

Topic sentence

When you start to solve a word problem, follow these steps.

Step 1 Skim through the problem, looking for the "topic sentence."

Step 2 Go back and read the problem more carefully, looking for the supporting details you need to solve the problem.

PRACTICE

Refer to pages 59 and 60. For each exercise below, write the "topic sentence." Do not solve the problem.

1. Exercise 29
2. Exercise 30
3. Exercise 32
4. Exercise 39
5. Exercise 40
6. Exercise 41

1-8 Algebra: Equations

MAIN IDEA

Solve equations by using mental math and the guess and check strategy.

New Vocabulary

equation
equals sign
solve
solution

Math Online

glencoe.com

- Extra Examples
- Personal Tutor
- Self-Check Quiz

MINI Lab

When the amounts on each side of a scale are equal, the scale is balanced.

STEP 1 Place four centimeter cubes and a paper bag on one side of a scale.

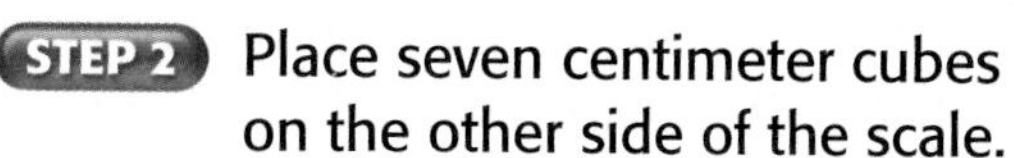

STEP 2 Place seven centimeter cubes on the other side of the scale.

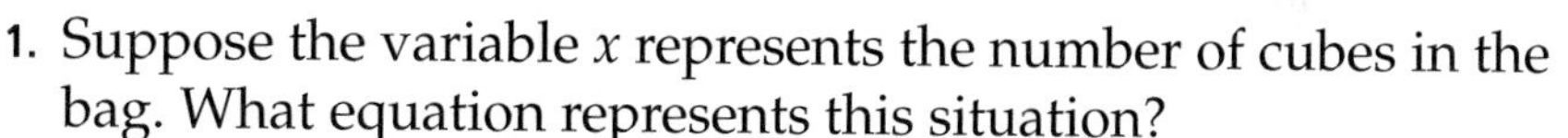

1. Suppose the variable x represents the number of cubes in the bag. What equation represents this situation?
2. Replace the bag with centimeter cubes until the scale balances. How many centimeter cubes did you need to balance the scale?

Let x represent the bag. Model each sentence on a scale. Find the number of centimeter cubes needed to balance the scale.

3. $x + 2 = 5$
4. $x + 5 = 7$
5. $x + 3 = 4$
6. $x + 6 = 6$

An **equation** is a sentence that contains an **equals sign**, =. A few examples are shown below.

$2 + 7 = 9$ $\qquad$ $10 - 6 = 4$ $\qquad$ $14 = 2 \cdot 7$

Some equations contain variables.

$2 + x = 9$ $\qquad$ $4 = k - 6$ $\qquad$ $15 \div m = 3$

When you replace a variable with a value that results in a true sentence, you **solve** the equation. That value for the variable is the **solution** of the equation.

$$2 + x = 9$$
$$2 + 7 = 9$$
$$9 = 9$$

This sentence is true.

The value for the variable that results in a true sentence is 7. So, 7 is the solution.

EXAMPLES Solve an Equation Mentally

1 **Is 3, 4, or 5 the solution of the equation $a + 7 = 11$?**

Value of a	$a + 7 \stackrel{?}{=} 11$	Are Both Sides Equal?
3	$3 + 7 = 11$ $10 \neq 11$	no
4	$4 + 7 = 11$ $11 = 11$	yes ✔
5	$5 + 7 = 11$ $12 \neq 11$	no

The solution is 4 since replacing a with 4 results in a true sentence.

2 **Solve $12 = 3h$ mentally.**

$12 = 3h$ **THINK** 12 equals 3 times what number?
$12 = 3 \cdot 4$ You know that $12 = 3 \cdot 4$.
$12 = 12$

The solution is 4.

CHECK Your Progress

a. Is 2, 3, or 4 the solution of the equation $4n = 16$?

b. Solve $24 \div w = 8$ mentally.

Real-World EXAMPLE

3 **SKATING The total cost of a pair of in-line skates and a set of kneepads is \$63. The skates cost \$45. Solve the equation $45 + k = 63$ to find k, the cost of the kneepads.**

Use the *guess and check* strategy.

Try 14.
$45 + k = 63$
$45 + 14 \stackrel{?}{=} 63$
$59 \neq 63$

Try 16.
$45 + k = 63$
$45 + 16 \stackrel{?}{=} 63$
$61 \neq 63$

Try 18.
$45 + k = 63$
$45 + 18 \stackrel{?}{=} 63$
$63 = 63$ ✔

So, the kneepads cost \$18.

Real-World Link An ostrich has the largest eye of any land animal. It is about 2 inches (5 centimeters) across.

Source: San Diego Zoo

CHECK Your Progress

c. **ANIMALS** The difference between an ostrich's speed and a chicken's speed is 31 miles per hour. An ostrich can run at a speed of 40 miles per hour. Solve the equation $40 - c = 31$ to find c, the speed a chicken can run.

CHECK Your Understanding

Example 1 (p. 58) **Identify the solution of each equation from the list given.**

1. $9 + w = 17$; 7, 8, 9
2. $d - 11 = 5$; 14, 15, 16
3. $4 = 2y$; 2, 3, 4
4. $8 \div c = 8$; 0, 1, 2

Example 2 (p. 58) **Solve each equation mentally.**

5. $x + 6 = 18$
6. $n - 10 = 30$
7. $15k = 30$

Example 3 (p. 58)

8. **CIVICS** Mississippi and Georgia have a total of 21 electoral votes. Mississippi has 6 electoral votes. Solve the equation $6 + g = 21$ to find g, the number of electoral votes Georgia has.

Practice and Problem Solving

HOMEWORK HELP

For Exercises	See Examples
9–16	1
17–28	2
29, 30	3

Identify the solution of each equation from the list given.

9. $a + 15 = 23$; 6, 7, 8
10. $29 + d = 54$; 24, 25, 26
11. $35 = 45 - n$; 10, 11, 12
12. $19 = p - 12$; 29, 30, 31
13. $6w = 30$; 5, 6, 7
14. $63 = 9k$; 6, 7, 8
15. $36 \div s = 4$; 9, 10, 11
16. $x \div 7 = 3$; 20, 21, 22

Solve each equation mentally.

17. $j + 7 = 13$
18. $m + 4 = 17$
19. $22 = 30 - m$
20. $12 = 24 - y$
21. $15 - b = 12$
22. $25 - k = 20$
23. $5m = 25$
24. $10t = 90$
25. $22 \div y = 2$
26. $d \div 3 = 6$
27. $54 = 6b$
28. $24 = 12k$

29. **BASKETBALL** One season, the Cougars won 20 games. They played a total of 25 games. Solve the equation $20 + g = 25$ to find g, the number of games the team lost.

30. **MONEY** Five friends earn a total of \$50 doing yard work in their neighborhood. Each friend earns the same amount. Solve the equation $5f = 50$ to find f, the amount that each friend earns.

31. **ANIMALS** A bottlenose dolphin is 96 inches long. There are 12 inches in 1 foot. Solve the equation $12d = 96$ to find d, the length of the bottlenose dolphin in feet.

32. **SCHOOL** Last year, 700 students attended Walnut Springs Middle School. This year, there are 665 students. Solve the equation $700 - d = 665$ to find d, the decrease in the number of students from last year to this year.

33. **REASONING** If x is a number that satisfies $4x + 3 = 18$, can x be equal to 3? Explain.

H.O.T. Problems

34. **OPEN ENDED** Give an example of an equation that has a solution of 5.

35. **REASONING** Tell whether the statement below is *sometimes*, *always*, or *never* true.

Equations like $a + 4 = 8$ and $4 - m = 2$ have exactly one solution.

CHALLENGE For Exercises 36 and 37, tell whether each statement is *true* or *false*. Then explain your reasoning.

36. In $m + 8$, the variable m can have any value.

37. In $m + 8 = 12$, the variable m can have any value and be a solution.

38. **WRITING IN MATH** Create a real-world problem in which you would solve the equation $a + 12 = 30$.

TEST PRACTICE

39. The graph shows the life expectancy of certain mammals. Which equation can be used to find the difference d between the number of years a blue whale lives and the number of years a gorilla lives?

A $d + 35 = 80$

B $d - 35 = 80$

C $80 + 35 = d$

D $d - 80 = 35$

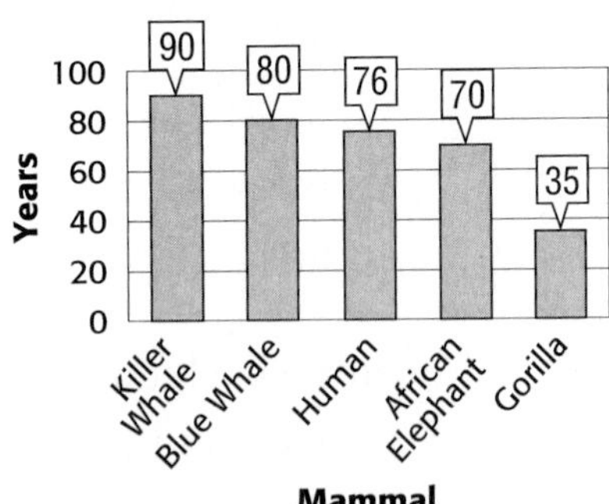

Source: *Scholastic Book of World Records*

Spiral Review

40. **SCIENCE** You have 27 bones in your hand. There are 6 more bones in your fingers than in your wrist. There are 3 fewer bones in your palm than in your wrist. How many bones are in each part of your hand? (Lesson 1-7)

41. **CHORES** Sophia earns a weekly allowance of $4. Define a variable. Write a function rule that relates the total allowance to the number of weeks. Find the total allowance she earns in 8 weeks. (Lesson 1-6)

Evaluate each expression if $r = 2$, $s = 4$, and $t = 6$. (Lesson 1-5)

42. $3rst + 14$

43. $9 \div 3 \cdot s + t$

44. $4 + t \div r \cdot 4s$

GET READY for the Next Lesson

PREREQUISITE SKILL Multiply. (Page 744)

45. 8×12

46. 6×15

47. 4×18

48. 5×17

Explore 1-9

Algebra Lab
Writing Formulas

MAIN IDEA

Use tables of data to generate formulas.

The number of square units needed to cover the surface of a figure is called its *area*. In this activity, you will explore how the area and side lengths of rectangles and squares are related. You will then express this relationship as an equation called a *formula*.

ACTIVITY

1 **STEP 1** On centimeter grid paper, draw, label, and shade a rectangle with a length of 2 centimeters and a width of 3 centimeters.

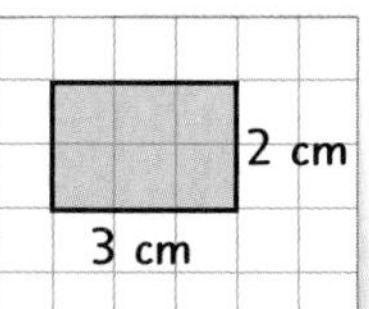

STEP 2 Count the number of squares shaded to find the area of the rectangle. Then record this information in a table like the one shown.

Rectangle	Length (cm)	Width (cm)	Area (cm^2)
A	2	3	
B	2	4	
C	2	5	
D	3	4	
E	4	4	
F	5	4	

STEP 3 Repeat Steps 1 and 2 for rectangles B, C, D, E, and F, whose dimensions are shown in the table.

Analyze the Results

1. Describe one or more patterns in the table.

2. Describe the relationship between the area of a rectangle and its length and width in words.

3. **MAKE A CONJECTURE** What would be the area of a rectangle with each of the following dimensions? Test your conjecture by modeling each rectangle and counting the number of shaded squares.

   a. length, 2 cm; width, 8 cm b. length, 9 cm; width, 4 cm

4. **WRITE A FORMULA** If A represents the area of a rectangle, write an equation that describes the relationship between the rectangle's area A, length ℓ, and width w.

Reading Math

Units Area is measured in square units. Read cm^2 as *square centimeters*.

ACTIVITY

2 For each step below, draw new rectangles on grid paper and find the areas. Organize the information in a table.

STEP 1 Using the original rectangles in Activity 1, double each length, but keep the same width.

STEP 2 Using the original rectangles in Activity 1, double each width, but keep the same length.

STEP 3 Using the original rectangles in Activity 1, double both the length and width.

ANALYZE THE RESULTS

Compare the areas you found in each step to the original areas. Write a sentence describing how the area changed. Explain.

5. Step 1
6. Step 2
7. Step 3

ACTIVITY

Study Tip

Squares A square has all four sides of equal length. So, to create the table, you only need to list the side length in one column.

3 **STEP 1** On centimeter grid paper, draw, label, and shade a square with a length of 2 centimeters.

2 cm

STEP 2 Count the number of squares shaded to find the area of the square. Record this information in a table like the one shown.

STEP 3 Repeat Steps 1 and 2 for squares B and C, whose dimensions are shown in the table.

Square	Side Length (cm)	Area (sq cm)
A	2	
B	3	
C	4	

ANALYZE THE RESULTS

8. Describe a pattern in the rows of the table.
9. **MAKE A CONJECTURE** What would be the area of a square with side lengths of 8 centimeters? Test your conjecture.
10. **WRITE A FORMULA** If A represents the area of a square, write an equation that describes the relationship between the square's area A and side length s.

1-9 Algebra: Area Formulas

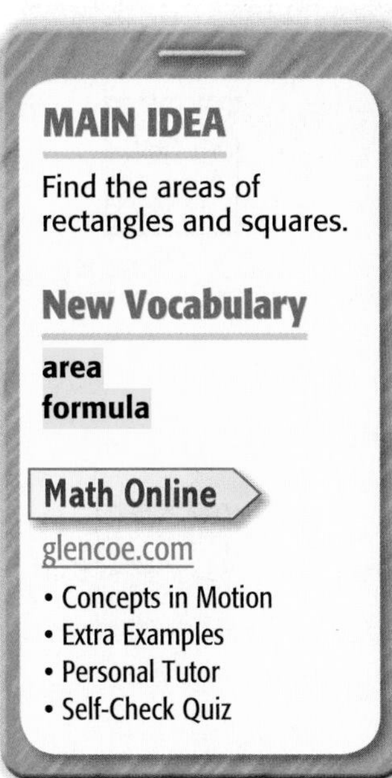

MAIN IDEA

Find the areas of rectangles and squares.

New Vocabulary

area
formula

Math Online

glencoe.com

- Concepts in Motion
- Extra Examples
- Personal Tutor
- Self-Check Quiz

MINI Lab

The rectangle at the right has an area of 20 square units. The distance around the rectangle is $5 + 4 + 5 + 4$, or 18 units.

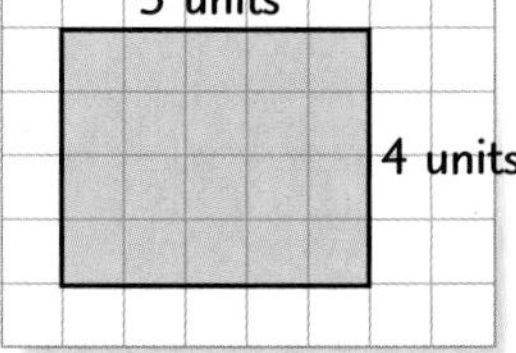

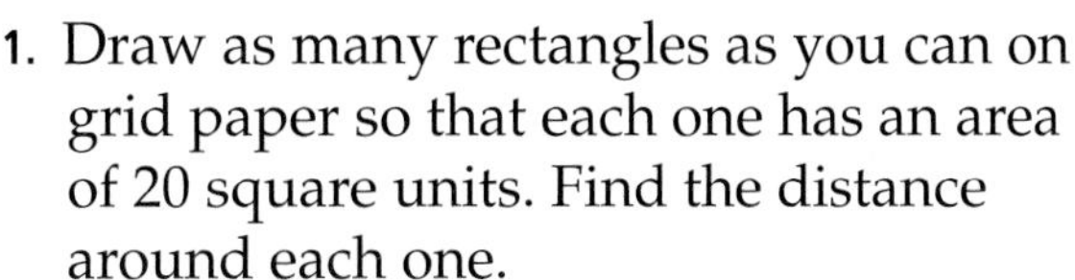

1. Draw as many rectangles as you can on grid paper so that each one has an area of 20 square units. Find the distance around each one.
2. Which rectangle from Question 1 has the greatest distance around it? the least?

The **area** of a figure is the number of square units needed to cover a surface. You can use a formula to find the area of a rectangle. A **formula** is an equation that shows a relationship among certain quantities.

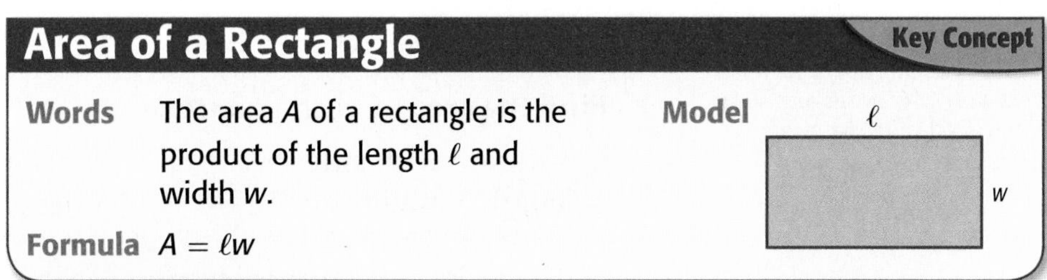

Area of a Rectangle — Key Concept

Words The area A of a rectangle is the product of the length ℓ and width w.

Model

Formula $A = \ell w$

EXAMPLE Find the Area of a Rectangle

1 Find the area of a rectangle with a length of 8 inches and a width of 6 inches.

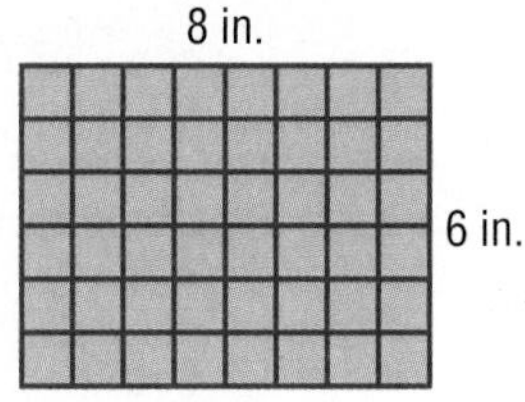

$A = \ell w$ Area of a rectangle

$A = 8 \cdot 6$ Replace ℓ with 8 and w with 6.

$A = 48$ Multiply.

The area is 48 square inches.

CHECK Your Progress

Find the area of each rectangle.

a. 12 in. by 6 in. rectangle

b. a rectangle with a length of 10 meters and a width of 2 meters

In Lesson 1-3, you wrote products as powers by using exponents. The formula for the area of a square is also written with an exponent.

Area of a Square

Key Concept

Words The area A of a square is the length of a side s squared.

Formula $A = s^2$

Model

EXAMPLE Find the Area of a Square

2 **Find the area of a square with side length 9 inches.**

$A = s^2$ Area of a square

$A = 9^2$ Replace s with 9.

$A = 81$ Multiply.

The area is 81 square inches.

CHECK Your Progress

Find the area of each square.

c. a square with side length 5 meters

d. a square with side length 7 feet

Real-World EXAMPLE

3 **GYMNASTICS Use the information at the left. What is the area of a gymnastics floor routine mat?**

The length of one side is 40 feet.

$A = s^2$ Area of a square

$A = 40^2$ Replace s with 40.

$A = 1{,}600$ Multiply.

The area of a floor routine mat is 1,600 square feet.

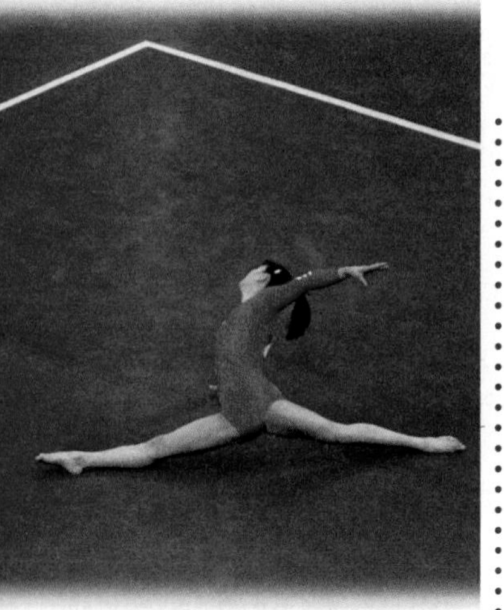

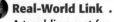

Real-World Link
A tumbling mat for a gymnastics floor routine is a square that measures 40 feet on each side.

CHECK Your Progress

e. **SPORTS** A high school basketball court measures 84 feet long and 50 feet wide. What is the area of this court?

CHECK Your Understanding

Example 1 (p. 63) **Find the area of each rectangle.**

1.
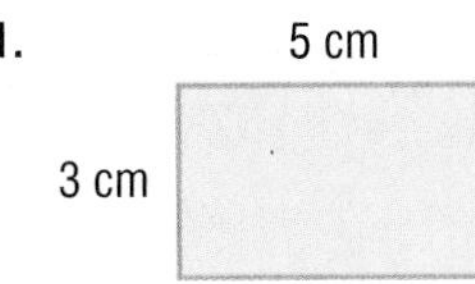

2.
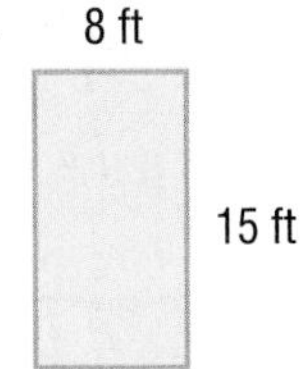

Example 2 (p. 64) **Find the area of each square.**

3.
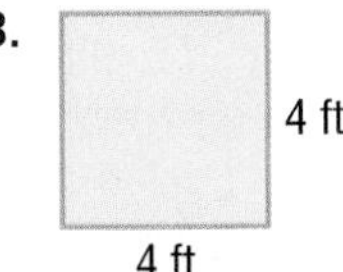

4.
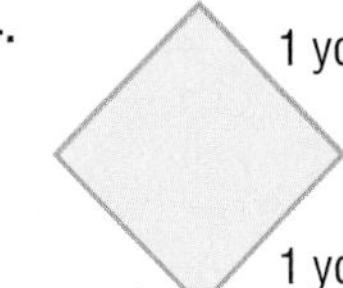

Example 3 (p. 64)

5. **DISHES** A glass baking dish measures 9 inches by 13 inches. What is the area of the baking dish?

Practice and Problem Solving

HOMEWORK HELP	
For Exercises	See Examples
6–13	1
14–17	2
18, 19	3

Find the area of each rectangle.

6.
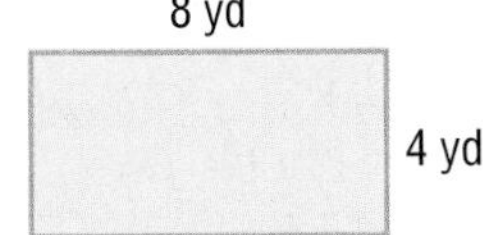

7.
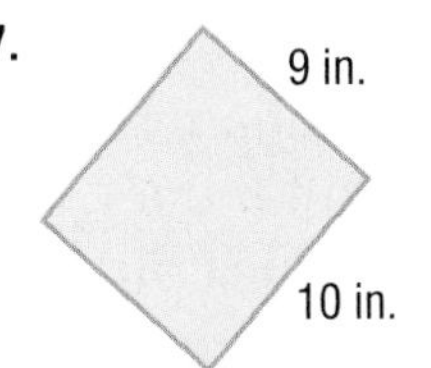

8.

9.
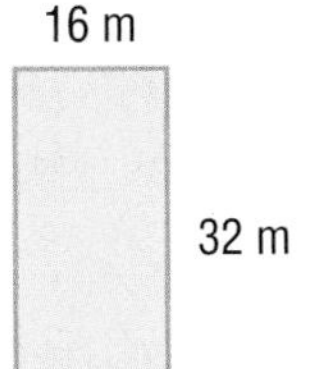

10.
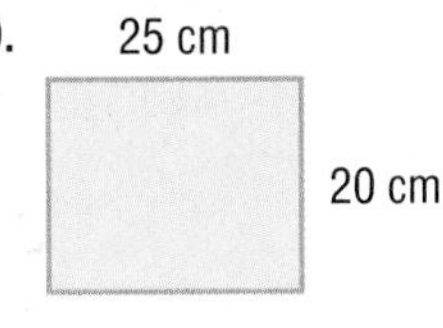

11.
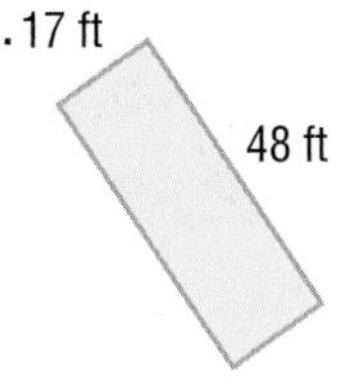

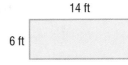

12. Find the area of a rectangle with a length of 26 inches and a width of 12 inches.

13. What is the area of a rectangle with a length of 40 centimeters and a width of 30 centimeters?

Find the area of each square.

14.
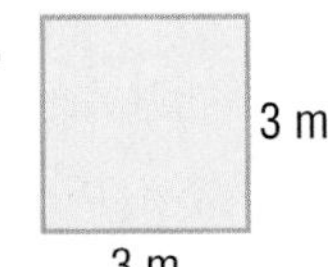

15.
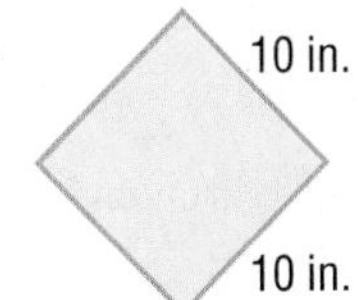

16.
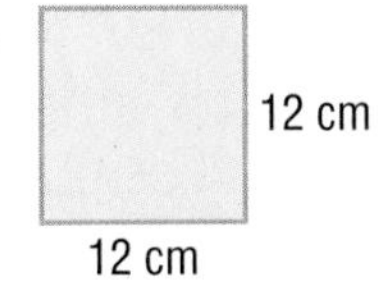

17. What is the area of a square with a side length of 22 feet?

18. **TENTS** The floor of a domed camping tent measures 7 feet by 9 feet. What is the area of the floor of the tent?

19. **HOBBIES** Meagan and her friends are knitting small squares to join together to form a blanket. The side length of each square must be 7 inches. What is the area of each square?

Find the area of each shaded region.

20. 12 ft, 8 ft, 9 ft, 4 ft

21. 5 m, 5 m, 11 m, 7 m

22. 15 cm, 15 cm, 8 cm, 7 cm

23. **FIND THE DATA** Refer to the Data File on pages 16–19. Choose some data and write a real-world problem in which you would find the area of a square or a rectangle.

24. **REMODELING** The Junkins are replacing the flooring in their kitchen with ceramic tiles. They are deciding between 12-inch square tiles and 6-inch square tiles. What is the difference in the area of the two tiles they are considering?

25. **ANIMALS** The floor spaces of two cages are shown. The square footage of Cage 1 is large enough for one guinea pig. For each additional guinea pig, the cage should be 1 square foot larger. How many guinea pigs should be kept in Cage 2?

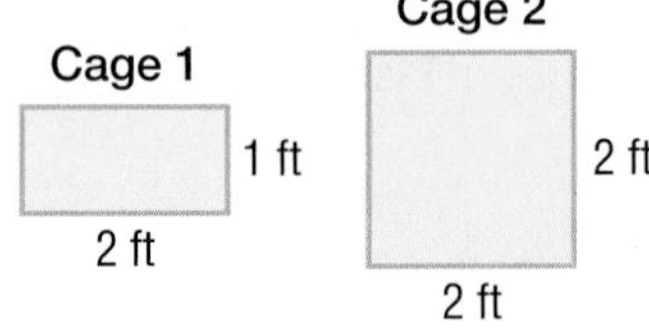

EXTRA PRACTICE
See pages 674, 706.

H.O.T. Problems

26. **OPEN ENDED** Draw and label a rectangle that has an area of 48 square units.

27. **NUMBER SENSE** Give the dimensions of two different rectangles that have the same area.

28. **FIND THE ERROR** James and John are finding the area of the square with a side of 8 feet. Who is correct? Explain your reasoning.

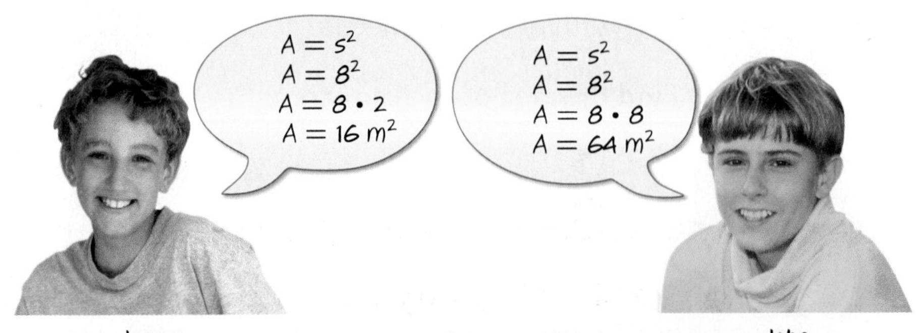

James

John

29. **CHALLENGE** Suppose opposite sides of a rectangle are increased by 5 units. Would the area of the rectangle increase by 10 square units? Use a model in your explanation.

30. **WRITING IN MATH** Explain how to use the formula for the area of a square. Include the formula for the area of a square in your explanation.

TEST PRACTICE

31. Which rectangle has an area of 54 square units?

A 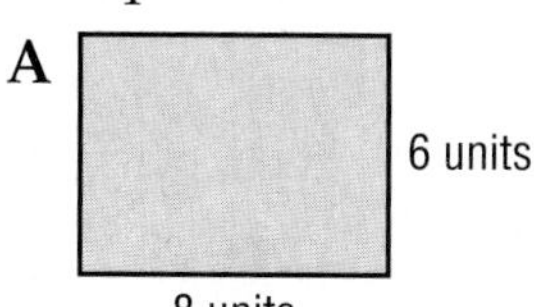

B 

C 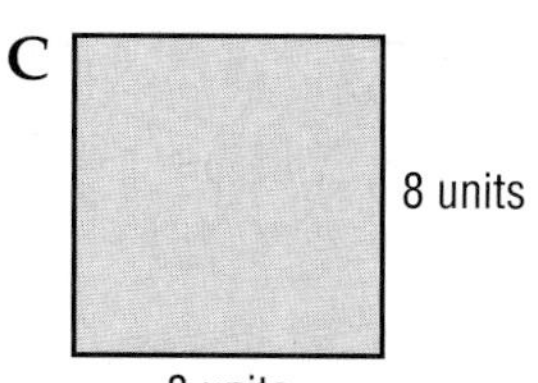

D 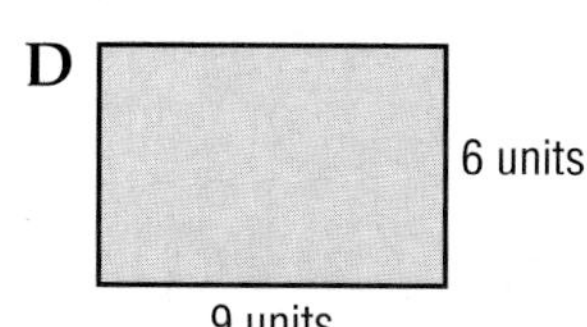

32. A family has a rectangular vegetable garden in their backyard and planted grass in the rest of the yard. The rectangular backyard is 110 feet by 70 feet, and the garden is 14 feet by 6 feet. What is the area of the backyard that is planted with grass?

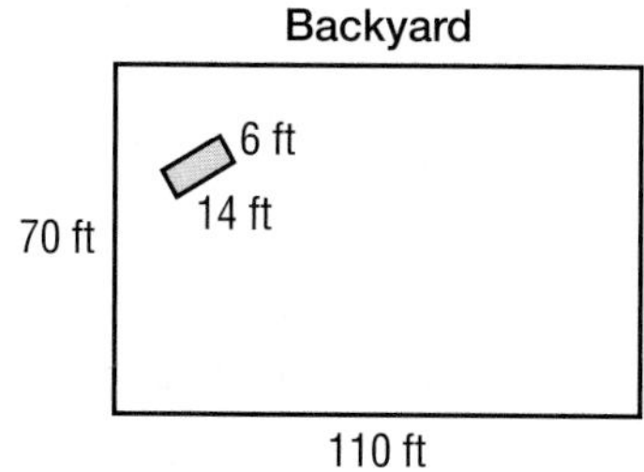

F 360 sq ft

G 7,616 sq ft

H 7,700 sq ft

J 7,804 sq ft

Spiral Review

Solve each equation mentally. (Lesson 1-8)

33. $x + 4 = 12$

34. $9 - m = 5$

35. $k - 8 = 20$

Copy and complete each function table. (Lesson 1-6)

36.

Input (x)	Output ($5 + x$)
0	■
3	■
5	■

37.

Input (x)	Output ($x \div 2$)
2	■
4	■
8	■

38. What is the value of $n^3 + 5n$ if $n = 2$? (Lesson 1-5)

39. **SCIENCE** The Milky Way galaxy is about 10^5 light years wide. What is the value of 10^5? (Lesson 1-3)

CHAPTER 1 Study Guide and Review

Math Online glencoe.com
• STUDY TO GO
• Vocabulary Review

GET READY to Study

Be sure the following Big Ideas are noted in your Foldable.

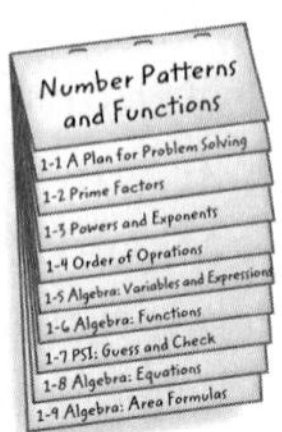

Number Patterns and Functions
1-1 A Plan for Problem Solving
1-2 Prime Factors
1-3 Powers and Exponents
1-4 Order of Operations
1-5 Algebra: Variables and Expressions
1-6 Algebra: Functions
1-7 PSI: Guess and Check
1-8 Algebra: Equations
1-9 Algebra: Area Formulas

BIG Ideas

Prime and Composite Numbers (Lesson 1-2)

- A prime number has exactly two factors, 1 and the number itself.
- A composite number is a number greater than 1 with more than two factors.
- 1 has only one factor and is neither prime nor composite. 0 has an infinite number of factors and is neither prime nor composite.

Order of Operations (Lesson 1-4)

Step 1 Simplify the expression inside grouping symbols, like parentheses.
Step 2 Find the value of all powers.
Step 3 Multiply and divide in order from left to right.
Step 4 Add and subtract in order from left to right.

Area Formulas (Lesson 1-9)

- The area A of a rectangle is the product of the length ℓ and width w.

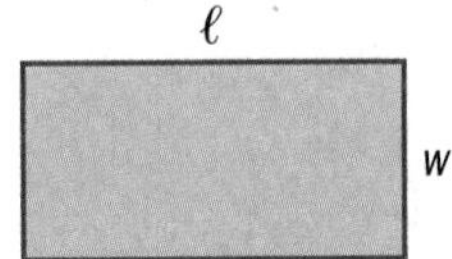

- The area A of a square is the length of a side s squared.

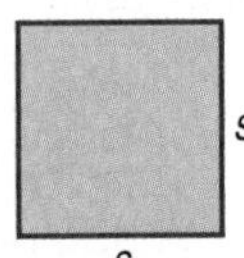

Key Vocabulary

algebra (p. 42)	function (p. 49)
algebraic expression (p. 42)	function rule (p. 49)
area (p. 63)	function table (p. 49)
base (p. 32)	numerical expression (p. 37)
composite number (p. 28)	order of operations (p. 37)
cubed (p. 33)	power (p. 33)
defining the variable (p. 50)	prime factorization (p. 29)
equals sign (p. 57)	prime number (p. 28)
equation (p. 57)	solution (p. 57)
evaluate (p. 42)	solve (p. 57)
exponent (p. 32)	squared (p. 33)
factor (p. 28)	variable (p. 42)
formula (p. 63)	

Vocabulary Check

State whether each sentence is *true* or *false*. If *false*, replace the underlined word or number to make a true sentence.

1. A <u>formula</u> is used to find the area of a rectangle.
2. When two or more numbers are multiplied, each number is called a <u>factor</u>.
3. The <u>base</u> of a figure is the number of square units needed to cover a surface.
4. A <u>function</u> represents an unknown value.
5. A <u>variable</u> is a relationship that assigns exactly one output value to one input value.

Lesson-by-Lesson Review

1-1 A Plan for Problem Solving (pp. 24–27)

Use the four-step plan to solve each problem.

6. **MOVIES** The times that a new movie is showing are 9:30 A.M., 11:12 A.M., 12:54 P.M., and 2:36 P.M. If the pattern continues, what are the next three show times?
7. **FUND-RAISING** Keith's goal is to collect \$145 for a school trip. So far, he has collected \$20 each from two people and \$10 each from six people. How far away is he from his goal?

Example 1 **Juanita works after school at an ice cream shop. Her hourly wage is \$7. If Juanita works 25 hours every week, how much does she earn?**

Understand	You need to find the total amount she earned.
Plan	Multiply \$7 by 25.
Solve	$\$7 \times 25 = \$175$ So, Juanita earned \$175.
Check	Since $175 \div 7 = 25$, the answer makes sense.

1-2 Prime Factors (pp. 28–31)

Tell whether each number is *prime*, *composite*, or *neither*.

8. 44
9. 67

Find the prime factorization of each number.

10. 42
11. 75
12. 96
13. **CODES** Cryptography uses prime numbers to encode secure bank account information. Suppose Suki's bank account was encoded with the number 273. What are the prime number factors of this code?

Example 2 **Find the prime factorization of 18.**

Make a factor tree.

18	Write the number to be factored.
2×9	Choose any two factors of 18.
$2 \times 3 \times 3$	Continue to factor any number that is not prime until you have a row of prime numbers.

The prime factorization of 18 is $2 \times 3 \times 3$.

1-3 Powers and Exponents (pp. 32–36)

Write each product using an exponent. Then find the value of the power.

14. $5 \times 5 \times 5 \times 5$
15. $12 \times 12 \times 12$
16. **ANIMALS** The average brain weight in grams for a walrus is 2^{10}. Find this value.

Example 3 **Write $4 \times 4 \times 4 \times 4 \times 4 \times 4$ using an exponent. Then find the value of the power.**

The base is 4. Since 4 is a factor 6 times, the exponent is 6.

$4 \times 4 \times 4 \times 4 \times 4 \times 4 = 4^6$ or 4,096

CHAPTER 1 Study Guide and Review

1-4 Order of Operations (pp. 37–40)

Find the value of each expression.

17. $4 \times 6 + 2 \times 3$
18. $8 + 3^3 \times 4$
19. $10 + 15 \div 5 - 6$
20. $11^2 - 6 + 3 \times 15$
21. **TRAVEL** On a family trip to The Motorcycle Museum, Maria counted 3 groups of motorcycles, each with 5 motorcycles, and an additional 7 lone motorcycles. Write an expression for the number of motorcycles Maria saw. Then find the number of motorcycles she saw.

Example 4 **Find the value of $28 \div 2 - 1 \times 5$.**

$28 \div 2 - 1 \times 5$	
$= 14 - 1 \times 5$	Divide 28 by 2.
$= 14 - 5$	Multiply 1 and 5.
$= 9$	Subtract 5 from 14.

The value of $28 \div 2 - 1 \times 5$ is 9.

1-5 Algebra: Variables and Expressions (pp. 42–46)

Evaluate each expression if $a = 18$ and $b = 6$.

22. $a \times b$
23. $a^2 \div b$
24. $3b^2 + a$
25. $2a - 10$

Evaluate each expression if $x = 6$, $y = 8$, and $z = 12$.

26. $2x + 4y$
27. $3z^2 + 4x$
28. $z \div 3 + xy$
29. **HOME REPAIR** Joe will tile a square kitchen floor with square ceramic tile. He knows the number of tiles needed is equal to $a^2 \div b^2$, where a is the floor length in inches and b is the length of the tile in inches. If $a = 96$ and $b = 8$, how many tiles are needed?

Example 5 **Evaluate $9 - k^3$ if $k = 2$.**

$9 - k^3 = 9 - 2^3$	Replace k with 2.
$= 9 - 8$	$2^3 = 8$
$= 1$	Subtract 8 from 9.

Example 6 **Evaluate $10 + mn$ if $m = 3$ and $n = 5$.**

$10 + mn = 10 + (3)(5)$	$m = 3$ and $n = 5$
$= 10 + 15$	Multiply 3 and 5.
$= 25$	Add 10 and 15.

Mixed Problem Solving
For mixed problem-solving practice, see page 706.

1-6 Algebra: Functions (pp. 49–53)

Copy and complete each function table.

30.

Input (x)	Output ($x - 1$)
1	■
6	■
8	■

31.

Input (x)	Output ($3x$)
0	■
5	■
7	■

32.

Input (x)	Output ($x \div 2$)
0	■
4	■
10	■

Find the rule for each function table.

33.

x	■
2	8
7	13
12	18

34.

x	■
0	0
3	6
5	10

35. **TRAVEL** Tina drove 60 miles per hour to Tucson. Define a variable. Write a function rule that relates the number of miles traveled to the hours driven.

36. **AGES** A boy is 5 years older than his sister. Define a variable. Write a function rule that relates the age of the boy to the age of his sister.

Example 7 Copy and complete the function table.

Input (x)	Output ($x + 5$)
0	■
4	■
9	■

The function rule is $x + 5$. Add 5 to each input.

Input (x)	Output ($x + 5$)
0	5
4	9
9	14

Example 8 Find a rule for the function table.

x	■
6	2
12	4
15	5

Study the relationship between each input and output. Divide each input by 3 to find the output.
So, the function rule is $x \div 3$.

1-7 PSI: Guess and Check (pp. 54–55)

Solve. Use the *guess and check* strategy.

37. **PRODUCTION** A company makes toy cars. It sells red cars for \$2 each and black cars for \$3 each. If the company sold 44 cars total and made \$105, how many red cars were sold?

38. **NUMBERS** The sum of two numbers is 22 and their product is 117. Find the numbers.

39. **FISHING** On a fishing trip with his friends, Alex caught 3 more catfish than he did trout. If the total number of catfish and trout was 19, how many catfish did he catch?

Example 9 **Owen is 8 inches taller than his sister, Lisa. If the sum of their heights is 124 inches, how tall is Owen?**

Make a guess. Check to see if it is correct. Adjust the guess until it is correct.

Owen's Height	Lisa's Height	Sum of Heights	
60 inches	52 inches	112 inches	*too low*
68 inches	60 inches	128 inches	*too high*
66 inches	58 inches	124 inches	

So, Owen is 66 inches tall.

1-8 Algebra: Equations (pp. 57–60)

Solve each equation mentally.

40. $p + 2 = 9$
41. $20 + y = 25$
42. $40 = 15 + m$
43. $16 - n = 10$
44. $27 = x - 3$
45. $17 = 25 - h$

46. **AGE** The equation $18 + p = 34$ represents the sum of Pedro's and Eva's ages, where p represents Pedro's age. How old is Pedro?

Example 10 **Solve $x + 9 = 13$ mentally.**

$x + 9 = 13$ What number plus 9 is 13?

$4 + 9 = 13$ You know that 4 + 9 is 13.

$x = 4$ The solution is 4.

1-9 Algebra: Area Formulas (pp. 63–67)

47. Find the area of the rectangle below.

13 cm

2 cm

48. **PAINTINGS** Find the area of a painting that measures 4 feet by 4 feet.

Example 11 **Find the area of the rectangle.**

13 m

5 m

$A = \ell w$ Area of a rectangle

$= 13 \cdot 5$ Replace ℓ with 13 and w with 5.

$= 65$

The area is 65 square meters.

CHAPTER 1 Practice Test

Math Online glencoe.com
• Chapter Test

1. **MULTIPLE CHOICE** Justin earned $308 by mowing lawns and raking leaves for a total of 43 hours. He raked leaves for 18 hours and earned $108. Arrange the steps below in a correct order to find how much he earned per hour mowing lawns.

   Step P: Find the difference between $308 and the amount Justin earned raking leaves.

   Step Q: Find the quotient of $200 and the number of hours Justin spent mowing lawns.

   Step R: Find the number of hours Justin spent mowing lawns.

   Which list shows the steps in the correct order?

   A P, Q, R
   B R, Q, P
   C Q, R, P
   D R, P, Q

Tell whether each number is *prime, composite,* or *neither.*

2. 57
3. 1
4. 31

5. Find the prime factorization of 68.

6. **BIRTHDAYS** Miranda told 3 friends that it was her birthday. Each of those 3 friends told 3 other students. By noon, 3^5 students knew it was Miranda's birthday. Write this number as a product of the same factor. Then find the value.

Find the value of each expression.

7. $12 - 3 \times 2 + 15$
8. $72 \div 2^3 - 4 \times 2$

Evaluate each expression if $a = 4$ and $b = 3$.

9. $a + 12$
10. $27 \div b$
11. $a^3 - 2b$

12. **MULTIPLE CHOICE** Latisha and Raquel ordered two beverages for $1.50 each, two dinners for $12.99 each, and a dessert for $3.50. Which of the following expressions can be used to find the amount each should pay, not including tax?

    F $1.50 + 2 \times 12.99 + 3.50 \div 2$
    G $(2 \times 1.50 + 2 \times 12.99 + 3.50) \div 2$
    H $2 \times (1.50 + 12.99 + 3.50)$
    J $(2 \times 1.50 + 12.99) + 3.50 \div 2$

Find the rule for each function table.

13.

x	■
3	8
7	12
11	16

14.

x	■
0	0
8	1
16	2

15. **NUTRITION** A medium potato has 26 grams of carbohydrates. Define a variable. Write a function rule that relates the amount of carbohydrates to the number of potatoes.

16. **MONEY** Diego has $1.30 in quarters, dimes, and nickels. He has the same amount of nickels as quarters, and one more dime than nickels. How many of each coin does he have?

Solve each equation mentally.

17. $d + 9 = 14$
18. $56 = 7k$

19. Find the area of the rectangle.

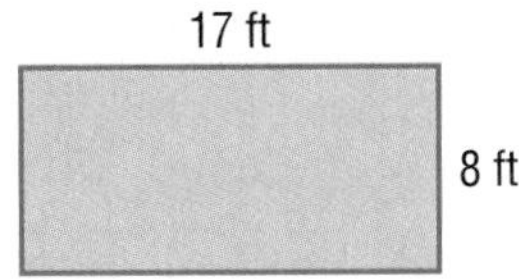

20. **RUGS** Benito has a square rug in his dining room. The length of each side of the rug is 42 inches. Find the area of the rug.

Test Practice

Cumulative, Chapter 1

Math Online glencoe.com
• Test Practice

PART 1 Multiple Choice

Read each question. Then fill in the correct answer on the answer sheet provided by your teacher or on a sheet of paper.

1. At a large middle school, there are 18 sixth-grade homerooms and approximately 22 students in each homeroom. About how many sixth-grade students attend the middle school?

   A 250

   B 325

   C 400

   D 650

2. Samantha drives 585 miles to reach her vacation spot. Her total drive time is nine hours. How would you find Samantha's average speed for the trip?

   F Add Samantha's total miles driven to her total drive time.

   G Subtract Samantha's total drive time from the total miles driven.

   H Multiply Samantha's total miles driven by the total drive time.

   J Divide Samantha's total miles driven by her total drive time.

3. The cost of renting roller blades is $4 plus $3.50 for each additional hour that the roller blades are rented. Which equation can be used to find c, the cost in dollars of the rental for h hours?

   A $c = 4h + 3.5$

   B $c = 3.5 - 4h$

   C $c = 3.5(h + 4)$

   D $c = 3.5h + 4$

4. Mr. Weiss started painting the kitchen at 8:45 A.M. and finished painting at 12:00 P.M. About how many hours elapsed between the time he started painting and the time he finished painting the kitchen?

   F 2 h

   G 3 h

   H 4 h

   J 5 h

5. Which is the prime factorization of 360?

   A $3^3 \cdot 5^2$

   B 2^7

   C $2^3 \cdot 3^2 \cdot 5$

   D $3^2 \cdot 5 \cdot 7^2$

6. Jeremy was asked to find two integers that have a difference of 3 and a sum of 71. He said that the integers were 39 and 36. Why was Jeremy's answer incorrect?

   F The difference between 39 and 36 is not 3.

   G The difference between 39 and 36 is 3.

   H The sum of 39 and 36 is not 71.

   J The sum of 39 and 36 is 71.

7. Amanda planted a square garden. The length of each side of the garden was 8 feet. Find the area of the garden.

   A 16 ft^2

   B 32 ft^2

   C 64 ft^2

   D 80 ft^2

TEST-TAKING TIP

Question 7 Review any terms and formulas that you have learned before you take the test.

Preparing for Standardized Tests

For test-taking strategies and practice, see pages 718–735.

8. The table shows Molly's age and Max's age over 4 consecutive years.

Molly's Age, x (years)	Max's Age, y (years)
2	5
3	6
4	7
5	8

Which expression best represents Max's age in terms of Molly's age?

F $y + 3$ **H** $x + 3$

G $3x$ **J** $3y$

9. At his job, Jack uses about 3 boxes of computer paper every 5 working days. About how many boxes of computer paper does Jack use in 36 working days?

A 8 **C** 21

B 15 **D** 108

10. A sub shop is keeping track of the number of meatball subs sold each day.

Day	Number of Meatball Subs Sold
Monday	40
Tuesday	25
Wednesday	30
Thursday	45
Friday	65
Saturday	70
Sunday	50

About how many subs were sold during that week?

F 150 subs **H** 250 subs

G 200 subs **J** 350 subs

PART 2 Short Response/Grid In

Record your answers on the answer sheet provided by your teacher or on a sheet of paper.

11. What is the value of $45 \div (7 + 2) - 1$?

12. Lynette is painting a 15-foot by 10-foot rectangular wall that has a 9-foot by 5-foot rectangular window at its center. How many square feet of wall will she paint?

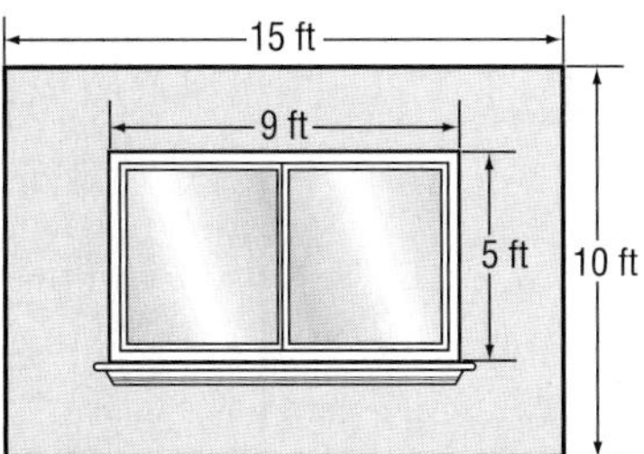

PART 3 Extended Response

Record your answers on the answer sheet provided by your teacher or on a sheet of paper. Show your work.

13. The following figures are made of toothpicks.

a. Make a table that shows the number of toothpicks needed for the first 5 figures.

b. Write an expression to find the number of toothpicks for any figure. Explain your reasoning.

NEED EXTRA HELP?													
If You Missed Question...	1	2	3	4	5	6	7	8	9	10	11	12	13
Go to Lesson...	1-1	1-1	1-5	1-1	1-3	1-1	1-9	1-6	1-1	1-1	1-4	1-9	1-1

CHAPTER 2 Statistics and Graphs

BIG Idea

- Construct and analyze statistical representations of data.

Key Vocabulary

frequency (p. 81)
graph (p. 81)
integers (p. 121)
mean (p. 102)

Real-World Link

Flowers At least 25 types of the Illinois state flower, the native violet, are found in the Chicago area. Most species have small flowers that range from 1 inch to $1\frac{1}{2}$ inches wide.

Statistics and Graphs Make this Foldable to help you organize information about statistics and graphs. Begin with five sheets of graph paper.

1. **Fold** each sheet of graph paper in half along the width.

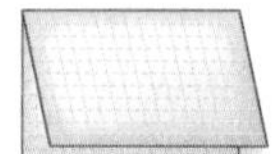

2. **Unfold** each sheet and tape to form one long piece.

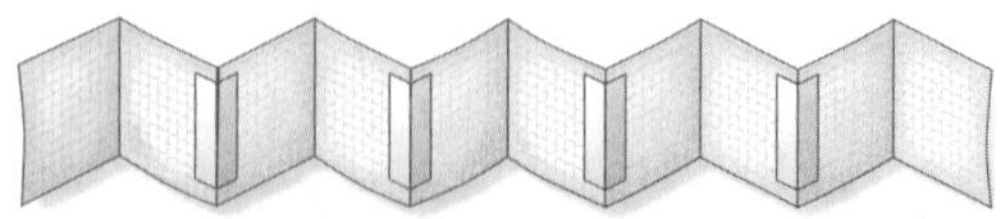

3. **Label** the pages with the lesson numbers as shown.

4. **Refold** the pages to form a journal.

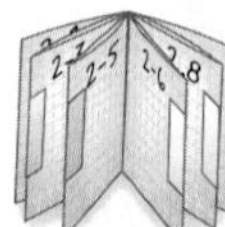

GET READY for Chapter 2

Diagnose Readiness You have two options for checking Prerequisite Skills.

Option 2

Math Online Take the Online Readiness Quiz at glencoe.com.

Option 1

Take the Quick Quiz below. Refer to the Quick Review for help.

QUICK Quiz

Add. (Prior Grade)

1. $16 + 28$
2. $39 + 25 + 11$
3. $63 + 9 + 37$
4. $74 + 14$
5. $8 + 56 + 10 + 7$
6. $44 + 18 + 5$
7. **MONEY** Jeffrey mows lawns after school. He earned \$46 on Monday, \$24 on Tuesday, \$32 on Thursday, and \$18 on Friday. How much money did he earn altogether?

QUICK Review

Example 1

Find 112 + 44 + 7.

```
   1
 112
  44
+  7
 163
```

Line up the digits at the ones place.

Add the ones. Put the 3 in the ones place and place the 1 above the tens place.

Add the tens. Then add the hundreds.

Divide. (Prior Grade)

8. $132 \div 11$
9. $96 \div 8$
10. $84 \div 2$
11. $102 \div 6$
12. $125 \div 5$
13. $212 \div 4$
14. **FOOD** A can of green beans contains 141 Calories. If there are 3 servings in the can, how many Calories does a single serving of green beans have?

Example 2

Find 183 ÷ 4.

```
   45.75
4)183.00
 - 16
    23
  - 20
     30
   - 28
      20
    - 20
       0
```

Divide in each place-value position from left to right.

Add zeros to the dividend as needed.

Find the value of each expression. (Lesson 1-4)

15. $15 - 4 + 2$
16. $6 + 35 \div 7$
17. $30 \div (8 - 3)$
18. $(2^5 \div 4) - 5$
19. $5^2 \times 2 - (5 \times 4)$
20. $7 \times (4 \div 2) + 3^3$

Example 3

Find the value of $6 + (4^3 \div 8)$.

$6 + (4^3 \div 8) = 6 + (64 \div 8)$ Find 4^3.

$= 6 + 8$ Divide 64 by 8.

$= 14$ Add 6 and 8.

Problem-Solving Investigation

MAIN IDEA: Solve problems by making a table.

P.S.I. TEAM +

e-Mail: MAKE A TABLE

LISETA: I took a survey of all my classmates to find out which of four food choices is their favorite. Using P for pizza, T for taco, H for hamburger, and C for chicken, the results are shown below.

C T H P P C H P T P H P P P P P H H C T

YOUR MISSION: Make a table to find how many more people in Liseta's class chose hamburger as their favorite food than chicken.

Understand	You need to find the number of students who chose hamburger and the number of students who chose chicken. Then find the difference.
Plan	Make a frequency table of the data.
Solve	Draw a table with three columns as shown. In the first column, list each food. Then complete the table by indicating the *frequency* or number of times each food occurs. 5 people chose hamburger and 3 chose chicken. So, 5 − 3 or 2 more students chose hamburger than chicken.
Check	If you go back to the list, there should be 5 students who wrote an H for hamburger and 3 students who wrote a C for chicken. So, an answer of 2 students is correct.

Favorite Food

Food	Tally	Frequency
pizza	卌 IIII	9
taco	III	3
hamburger	卌	5
chicken	III	3

Analyze The Strategy

1. Explain when to use the *make a table* strategy to solve a problem.
2. Tell the advantages of organizing information in a table.
3. **WRITING IN MATH** Write a real-world problem that can be solved using the *make a table* strategy. Then show how to solve the problem.

Mixed Problem Solving

See pages 675, 707.

Use the *make a table* strategy to solve Exercises 4 and 5.

4. **EYE COLOR** The table shows the eye color of each student in a sixth-grade class. Make a frequency table of the data. How many more students have brown eyes than green eyes?

Eye Color						
BR	G	BR	H	BL	BR	BR
BL	BR	H	G	BR	BR	BL
G	H	BR	BL	BR	G	H

BR = brown eyes BL = blue eyes
G = green eyes H = hazel eyes

5. **MUSIC** The table shows the number of songs downloaded in one month by each student in Mr. Jordan's class. Make a frequency table of the data. How many students downloaded at least 10 songs?

Number of Songs Downloaded							
0	5	12	8	11	15	8	9
2	23	9	3	0	6	12	7
2	4	3	0	19	1	6	13

Use any strategy to solve Exercises 6–14. Some strategies are shown below.

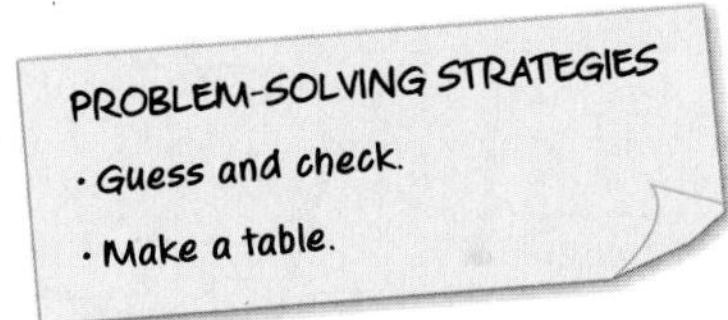

6. **NUMBERS** Emma is thinking of three numbers from 1 to 9 that have a sum of 20. Find all of the possible numbers.

7. **CARS** The table shows the color of the cars parked in a parking lot. How many more silver cars were in the parking lot than red?

Cars Parked in the Parking Lot							
W	R	S	G	S	S	G	B
B	S	S	R	R	G	W	S
R	W	G	B	S	W	S	B
B	S	W	S	W	S	B	B

S = silver R = red B = black G = green W = white

8. **GYMNASTICS** The table shows the number of years that each team member on the gymnastics team has practiced gymnastics. How many team members have practiced for less than 3 years?

Years of Experience on the Gymnastics Team										
0	6	3	4	2	4	1	3	5	1	5
1	0	1	5	1	2	3	5	2	1	2
2	1	2	7	3	4	8	6	2	3	4

9. **MAIL** Each day, Monday through Saturday, about 2,300 pieces of mail are delivered by each residential mail carrier in a certain town. About how many pieces of mail are delivered by each carrier in five years?

10. **SALES** Lavina sold some wrapping paper for $7 a roll and some gift bags for $8 a set. If she sold a total of 17 items for a total of $124, how many rolls of wrapping paper and sets of gift bags did she sell?

11. **PATTERNS** Find the next figure in the pattern.

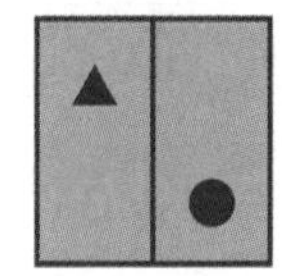 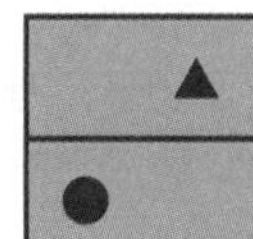 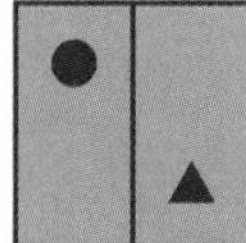 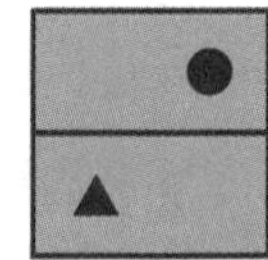

12. **MONEY** How much money will Rob save if he saves $2 a day for 25 weeks?

13. **SCHOOL** Of the 150 students at Lincoln Middle School, 55 are in the orchestra, and 75 are in marching band. Of these students, 25 are in both orchestra and marching band. How many students are neither in orchestra nor in marching band?

14. **MONEY** Jorge has $125 in his savings account. He deposits $20 every week and withdraws $25 every four weeks. What will his balance be in 8 weeks?

READING to SOLVE PROBLEMS

Large Numbers

There are more than 250,000,000 people in the United States. So, numerical data about the population may be too large to show in a table. If so, *unit indicators* are used to save space.

The *unit indicator* is millions.

Appliances Used by Households by Region (millions)

Type of Appliance	Households	Region			
		Northeast	Midwest	South	West
Color Television	106	20	24	38	23
Large Screen TV	37	7	8	13	8
Cable/Satellite Dish	82	16	19	30	17
VCR and DVD Players	96	18	23	35	21
Stereo Equipment	80	15	18	29	19

Source: U.S. Census Bureau

Suppose you need to find the total number of households in the South and the West that use a color television. Use the following steps.

Step 1 Locate the numbers in the table. The number for the South is 38, and the number for the West is 23.

Step 2 Determine the unit indicator. The unit indicator is millions.

Step 3 Multiply to find the data value. Then, add.

38 × 1,000,000	→	38,000,000	South
23 × 1,000,000	→	+ 23,000,000	West
		61,000,000	Total

PRACTICE

1. How many more households in the South use a cable/satellite dish than in the Midwest? Use the table above.

For Exercises 2–5, use the table at the right.

2. How many licensed drivers are in Indiana?
3. How many more licensed drivers are in Indiana than in Kentucky?
4. How many total licensed drivers are in North and South Carolina?
5. Wyoming has 371 thousand licensed drivers. How many more licensed drivers are in Virginia?

Licensed Drivers (thousands)

State	Number
Indiana	4,246
Kentucky	2,861
North Carolina	6,228
South Carolina	2,988
Virginia	5,178

Source: U.S. Census Bureau

2-2 Bar Graphs and Line Graphs

MAIN IDEA

Display and analyze data using bar graphs and line graphs.

New Vocabulary

data
graph
bar graph
scale
vertical axis
interval
horizontal axis
frequency
line graph

Math Online

glencoe.com

- Extra Examples
- Personal Tutor
- Self-Check Quiz
- Reading in the Content Area

GET READY for the Lesson

ANIMALS The table lists different animal species and how many of each type are endangered.

U.S. Endangered Species	
Species	**Frequency**
amphibians	13
birds	76
fish	74
mammals	69
reptiles	14

Source: Fish and Wildlife Service

1. Which species has the most endangered animals?
2. Which species has the least endangered animals?
3. What might be an advantage to organizing data in a table?
4. Are there any disadvantages to organizing data in a table?

Data are pieces of information that are often numerical. Data are often shown in a table. A **graph** is a more visual way to display data. A **bar graph** is used to compare categories of data.

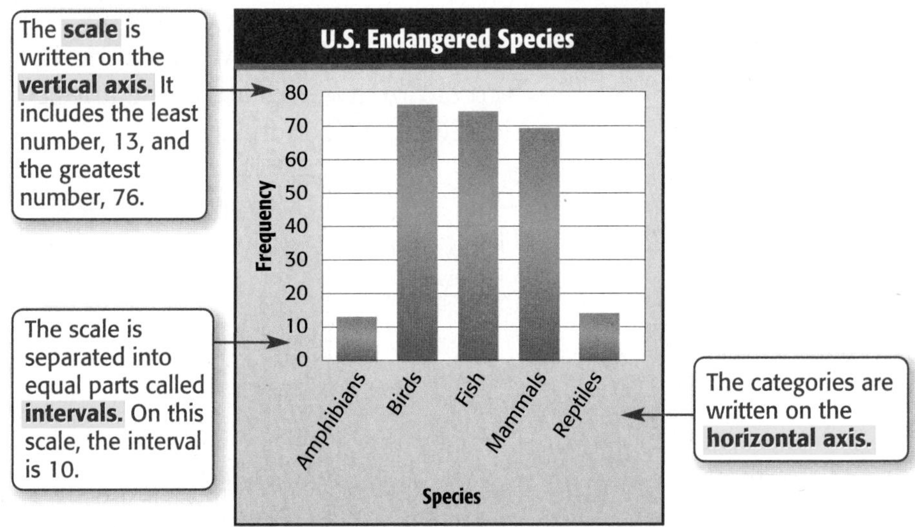
The **scale** is written on the **vertical axis.** It includes the least number, 13, and the greatest number, 76.
The scale is separated into equal parts called **intervals.** On this scale, the interval is 10.
U.S. Endangered Species
Frequency
80
70
60
50
40
30
20
10
0
Amphibians
Birds
Fish
Mammals
Reptiles
Species
The categories are written on the **horizontal axis.**

The height of each bar represents the frequency of each category of data. The **frequency** is the number of times an item occurs. For example, the frequency of endangered birds in the United States is 76.

EXAMPLE Analyze a Bar Graph

Study Tip

Bar Graphs The graph shown in Example 1 is called a vertical bar graph. Bar graphs can also be horizontal, with the categories written on the horizontal axis and the scale on the vertical axis. For horizontal bar graphs, the length of each bar represents the category's frequency.

1 **COMMUNICATION Make a bar graph of the data. Then compare the number of students who prefer e-mail to the number of students who prefer text message.**

Preferred Form of Communication	
E-mail	10
Phone	12
Text Message	4
Instant Message	2

Step 1 Decide on the scale and the interval. The data include numbers from 2 to 12. So, a scale from 0 to 14 and an interval of 2 are reasonable.

Step 2 Label the horizontal and and vertical axes.

Step 3 Draw bars for each form of communication. The height of each bar shows the number of students.

Step 4 Label the graph with a title.

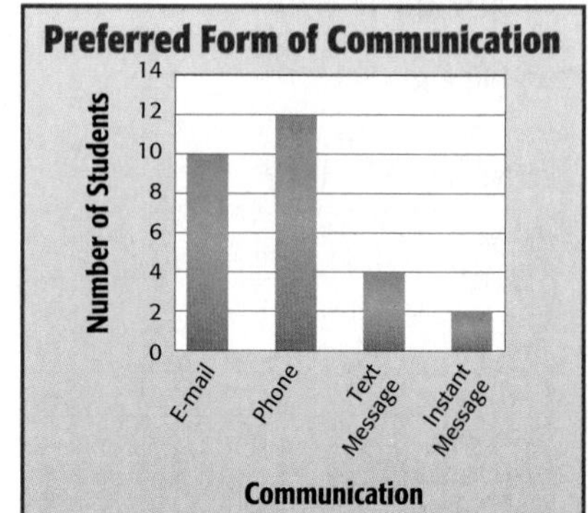

The height of the bar for e-mail is about twice as tall as the one for text message. So, about twice as many students prefer e-mail than text message.

CHECK Your Progress

a. **ICE CREAM** Make a bar graph of the data. Compare the number of students who chose chocolate to the number who chose strawberry.

Favorite Ice Cream Flavor	
Flavor	**Frequency**
chocolate	12
chocolate chip	7
strawberry	4
vanilla	9

Another type of graph is a line graph. A **line graph** is used to show how a set of data changes over a period of time. By observing the upward or downward slant of the lines connecting the points, you can describe trends in the data.

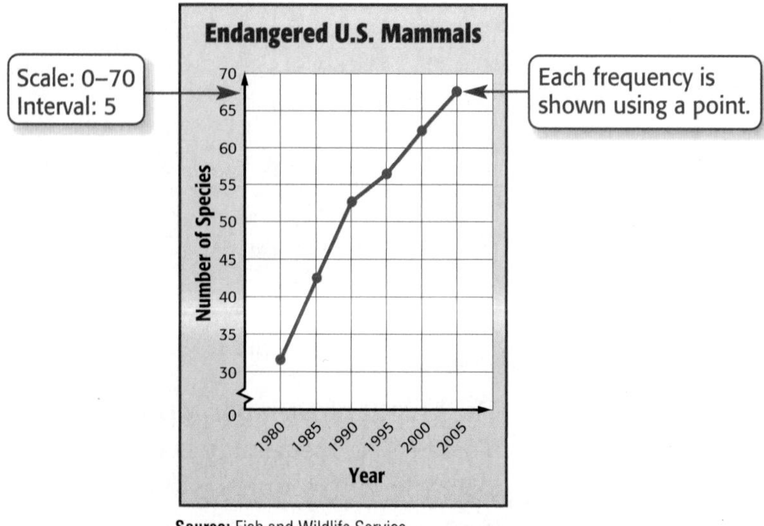

Source: Fish and Wildlife Service

Real-World Link

Earth's Population	
Year	**Population (millions)**
1750	790
1800	980
1850	1,260
1900	1,650
1950	2,555
2000	6,080

Source: United Nations and U.S. Census Bureau

EXAMPLE Analyze a Line Graph

2 POPULATION Make a line graph of the data at the left. Describe the change in Earth's population from 1750 to 2000.

Step 1 The data include numbers from 790 million to 6,080 million. So, a scale from 0 to 10,000 million and an interval of 1,000 million are reasonable.

Step 2 Label the horizontal and vertical axes.

Step 3 Draw and connect the points for each year.

Step 4 Label the graph with a title.

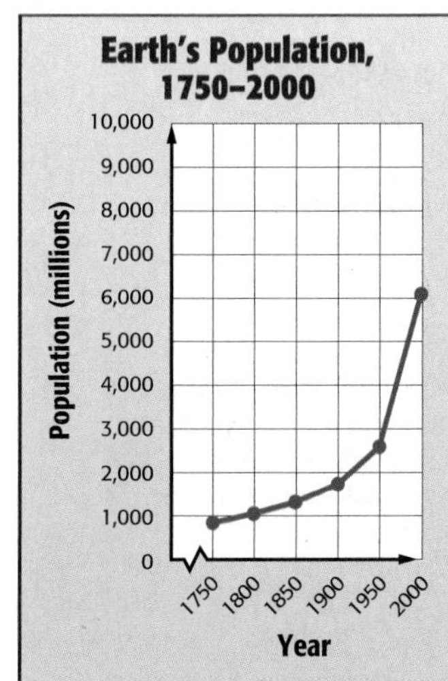

Earth's population has increased drastically from 1750 to 2000.

CHECK Your Progress

b. **FARMING** Make a line graph of the data. Describe the change in the number of North Carolina farms from 2001 to 2005.

North Carolina Farms					
Year	2001	2002	2003	2004	2005
Number of Farms	55,600	54,200	53,800	52,000	50,000

CHECK Your Understanding

Examples 1, 2 (pp. 82, 83)

1. **ROLLER COASTERS** Make a bar graph of the data. How do the number of steel coasters and wood coasters compare?

Types of Roller Coasters in the United States	
Type	**Frequency**
steel	516
wood	112
inverted	43
stand up	8
suspended	7
bobsled	4

Source: Roller Coaster Database

2. **MONEY** Make a line graph of the data. Then describe the change in the total amount Felisa saved from Week 1 to Week 5.

Felisa's Savings	
Week	**Total Amount ($)**
1	50
2	54
3	75
4	98
5	100

Practice and Problem Solving

HOMEWORK HELP	
For Exercises	See Examples
3, 4	1
5, 6	2

3. **GEOGRAPHY** Make a bar graph of the data below. Then compare the amount of shoreline for Florida and Texas.

U.S. Gulf Coast Shoreline	
State	Amount (mi)
Alabama	607
Florida	5,095
Louisiana	7,721
Mississippi	359
Texas	3,359

Source: NOAA

4. **PLANETS** Make a bar graph of the data below. How do the number of moons of Saturn compare to the number of moons of Neptune?

Number of Moons	
Planets*	Moons
Earth	1
Mars	2
Neptune	13
Uranus	27
Saturn	47
Jupiter	63

Source: *The World Almanac for Kids*
*Mercury and Venus each have 0 moons.

5. **PETS** Make a line graph of the data below. Describe the change in the percent of U.S. households owning cats from 1990–2006.

Percent of U.S. Households Owning Cats	
Year	Percent
1990	33
1994	30
1998	32
2002	34
2006	34

6. **MOVIES** Make a line graph of the data below. Describe the change in the online sales of movie tickets for Weeks 1 to 5.

Online Sales of Movie Tickets	
Week	Number of Tickets
1	1,200
2	1,450
3	1,150
4	1,575
5	1,750

WEATHER **For Exercises 7–10, refer to the table.**

Average Temperatures (°F), Minneapolis, Minnesota			
Month	Temp.	Month	Temp.
Jan.	22	July	83
Feb.	29	Aug.	80
Mar.	41	Sept.	71
Apr.	57	Oct.	58
May	70	Nov.	40
June	79	Dec.	26

Source: The Weather Channel

7. Choose an appropriate scale and interval for the data set.
8. Would these data be best represented by a bar graph or line graph? Explain your reasoning.
9. Make a graph of these data.
10. Write one question that can be answered using your graph. Then answer your question and justify your reasoning.

EXTRA PRACTICE
See pages 675, 707.

H.O.T. Problems

11. **COLLECT THE DATA** Make an appropriate display showing the amount of time you spend doing a particular activity each day for one week. Then write at least two statements that analyze your data.

12. **CHALLENGE** Can changes to the vertical scale or interval affect the look of a bar or line graph? Justify your reasoning with examples.

13. **WRITING IN MATH** Compare and contrast bar and line graphs.

TEST PRACTICE

14. The table at the right shows the gross income of five of the highest-grossing country music tours in 2000. Which graph most accurately displays the information in the table?

Artist	Gross Income (millions of dollars)
A	49.6
B	46.1
C	21.0
D	10.7
E	9.0

A

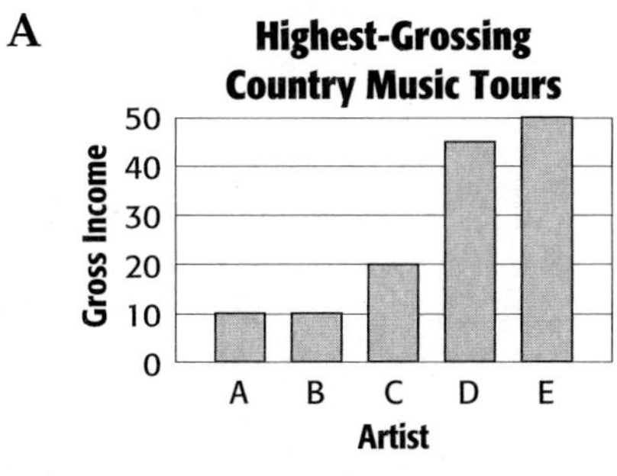

C

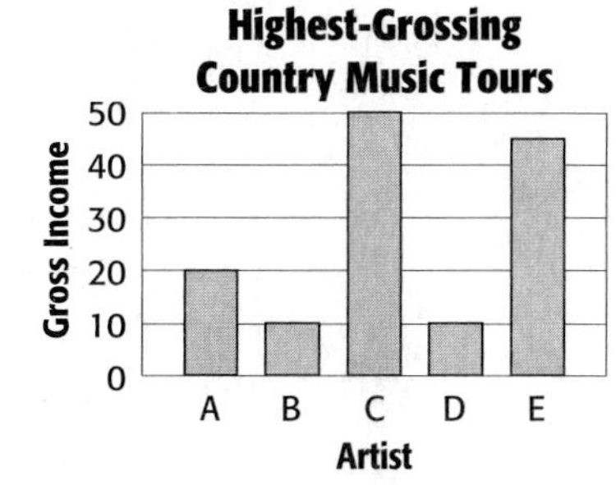

B

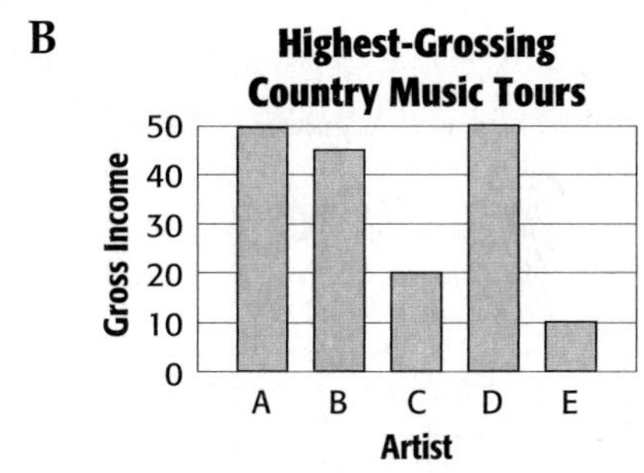

D

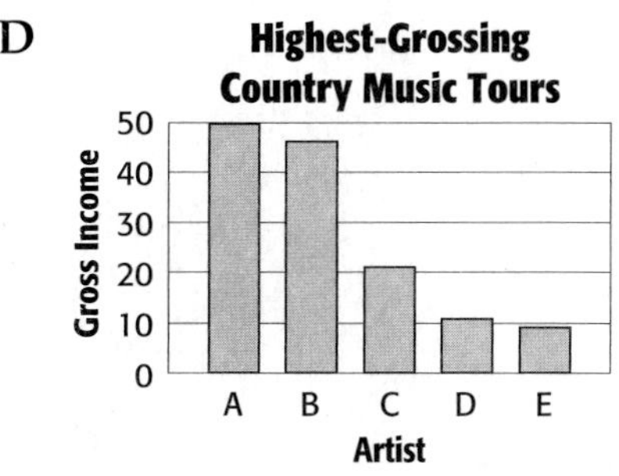

Spiral Review

15. **SURVEY** The table shows the favorite colors of a group of students. Make a frequency table of the data. How many more students chose blue than yellow? (Lesson 2-1)

Favorite Color					
R	G	B	R	R	K
B	P	P	Y	R	B
P	R	Y	K	B	Y
B	B	P	B	P	R

R = red B = blue Y = yellow
G = green P = purple K = pink

Find the area of each rectangle described. (Lesson 1-9)

16. length: 4 feet, width: 6 feet

17. length: 12 yards, width: 7 yards

GET READY for the Next Lesson

PREREQUISITE SKILL **Add.** (Page 743)

18. $13 + 41$
19. $57 + 31$
20. $5 + 18 + 32$
21. $14 + 45 + 27$

Extend 2-2

Spreadsheet Lab
Double-Line and -Bar Graphs

MAIN IDEA

Use a spreadsheet to make a line graph and a bar graph.

You can use a Microsoft Excel® spreadsheet to make double-line and -bar graphs. These graphs are used to compare two sets of data.

ACTIVITY

1 Jared and Adelina each recorded their heart rate every five minutes during a long-distance run. The table shows the results.

Heart Rate (beats per min)		
Time (min)	**Jared**	**Adelina**
0	72	76
5	120	126
10	128	132
15	134	138
20	144	146
25	148	152
30	150	158

To graph the data, first set up a spreadsheet like the one shown below.

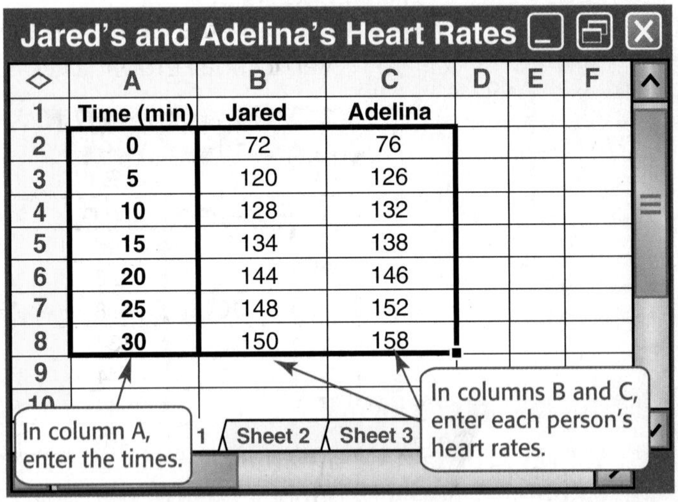

Jared's and Adelina's Heart Rates

	A	B	C	D	E	F
1	Time (min)	Jared	Adelina			
2	0	72	76			
3	5	120	126			
4	10	128	132			
5	15	134	138			
6	20	144	146			
7	25	148	152			
8	30	150	158			
9						

Sheet 1 | Sheet 2 | Sheet 3

In column A, enter the times.

In columns B and C, enter each person's heart rates.

The next step is to "tell" the spreadsheet to make a line graph of the data in columns B and C.

1. Highlight the data in columns B and C, from B2 through C8.
2. Click on the Chart Wizard icon, choose the line graph, and click Next.
3. To set the x-axis, choose the Series tab and press the icon next to the Category (X) axis labels.
4. On the spreadsheet, highlight the data in column A, from A2 through A8.

5. Press the icon on the bottom of the Chart Wizard box to automatically paste the information.
6. Click Next and enter the chart title and labels for the x- and y-axes. Click Next again and then Finish.

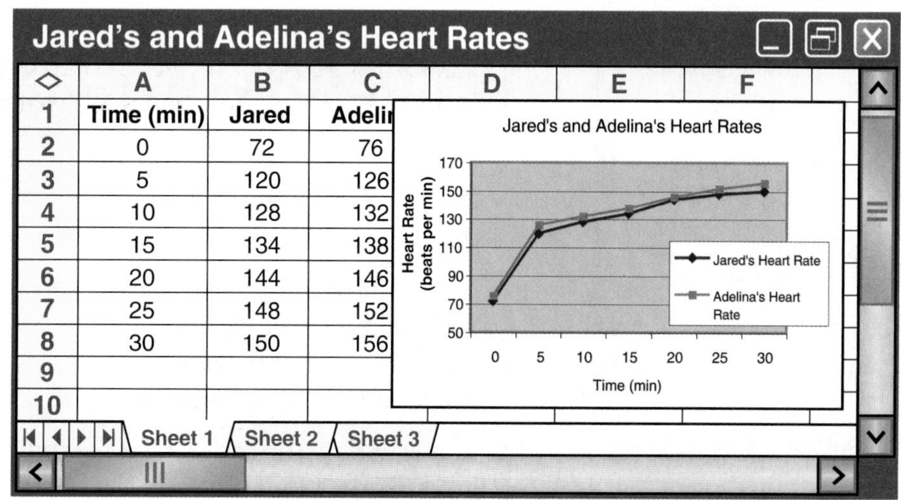

ACTIVITY

2 Set up the same spreadsheet in Activity 1.

- Highlight the data in columns B and C.
- Click on the Chart Wizard icon, Column, and Next to choose the vertical bar graph. Then complete steps 3–6 from Activity 1.

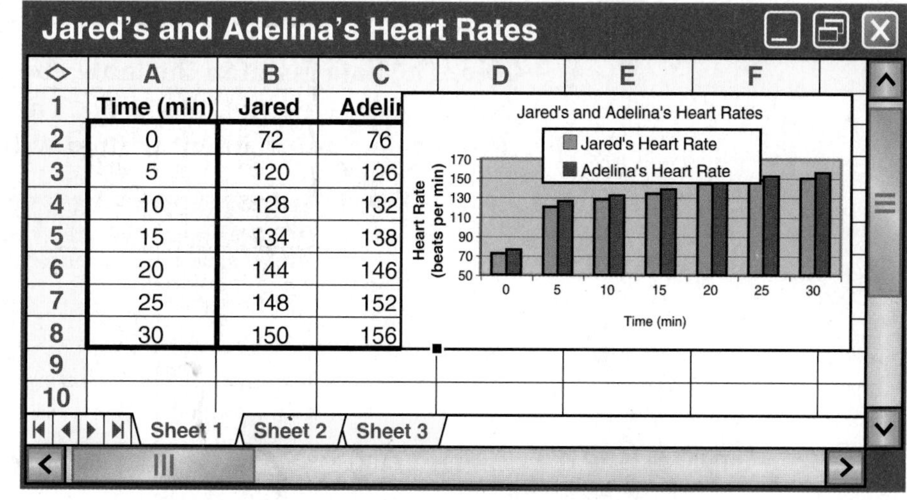

ANALYZE THE RESULTS

1. Between what two times did Jared's and Adelina's heart rates increase the most? Explain how you know.
2. **COLLECT THE DATA** Collect two sets of data that could be represented by a double-line or -bar graph. Then use a spreadsheet to make a double-line or -bar graph of the data.

2-3 Interpret Line Graphs

MAIN IDEA

Interpret line graphs.

Math Online

glencoe.com

- Extra Examples
- Personal Tutor
- Self-Check Quiz

GET READY for the Lesson

GOLF Analyze the table.

1. Describe the trends in the winning amounts.
2. Predict how much the 2008 winner received. Research and compare to the actual 2008 amount.
3. The Masters Tournament is held once a year. If a line graph is made of these data, will there be any realistic data values between years? Explain.

Money Won by Masters Tournament Winners, 2002–2007

Year	Amount ($)
2002	1,000,000
2003	1,080,000
2004	1,125,000
2005	1,170,000
2006	1,225,000
2007	1,305,000

Source: Professional Golf Association

Line graphs are often used to predict future events because they show trends over time.

EXAMPLES Interpret Line Graphs

1 GOLF The data given in the table above are shown in the line graph below. Describe the trend. Then predict how much the 2010 Masters Tournament winner will receive.

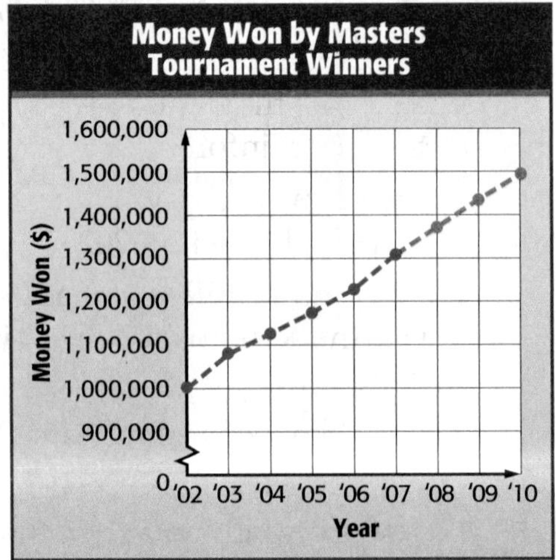

Notice that the increase is fairly steady. By extending the graph, you can predict that the winner of the 2010 Masters Tournament will receive about $1,500,000.

Study Tip

Broken Line Graphs
In Example 1, the points are connected by a dotted line because there are no realistic data values between years.

2 SKATEBOARDING What does the graph tell you about the popularity of skateboarding?

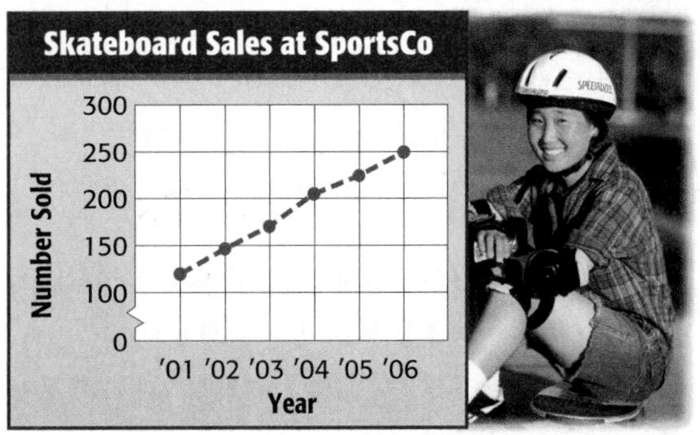

The graph shows that skateboard sales have been increasing each year. You can assume that the popularity of the sport is increasing.

ELECTRONICS **The graph shows the factory sales of VCRs in the U.S. from 2001–2007.**

a. Predict the amount of factory sales for VCRs in 2008.

b. Describe how the popularity of VCRs has changed since 2000.

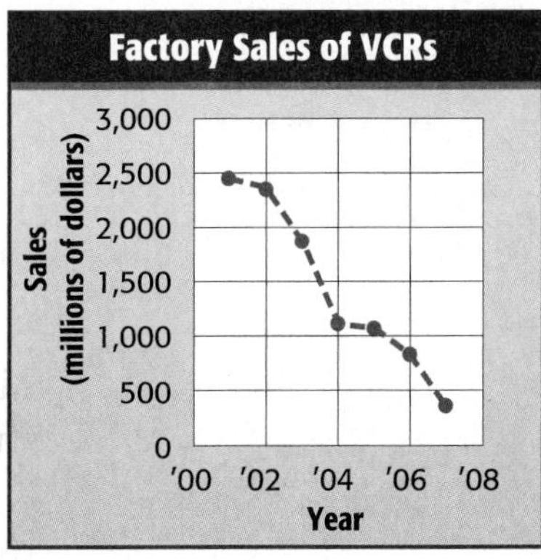

CHECK Your Understanding

RAINFORESTS **For Exercises 1–6, use the graph at the right.**

Example 1 (p. 88)

1. Describe the trend in the remaining tropical rainforests.
2. Predict how many millions of acres there will be left in 2020.
3. How many millions of acres do you think there will be left in 2030?
4. Make a prediction for the number of millions of acres left in 2040.

Example 2 (p. 89)

5. What does the graph tell you about future changes in the remaining rainforests?
6. Describe the change in the world's remaining rainforests from 1940 to 2010.

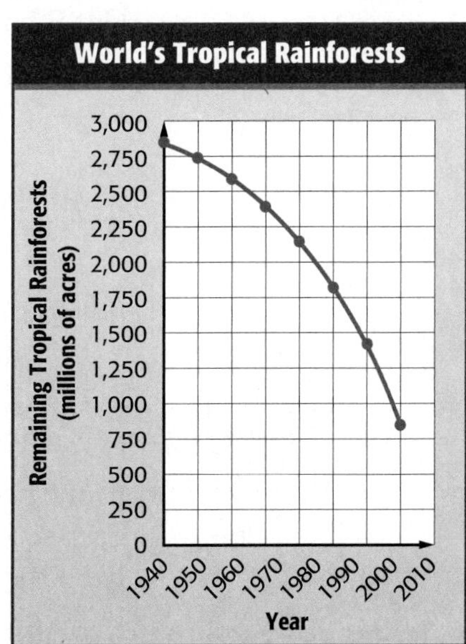

Source: Tropical American Tree Farms

Practice and Problem Solving

HOMEWORK HELP

For Exercises	See Examples
7–16	1, 2

SPORTS For Exercises 7 and 8, refer to Example 2 on page 89.

7. Predict the number of skateboards sold in 2009. Explain.
8. About how many skateboards were sold in 2000? How did you reach this conclusion?

IRONMAN CHAMPIONSHIP For Exercises 9–11, use the graph at the right.

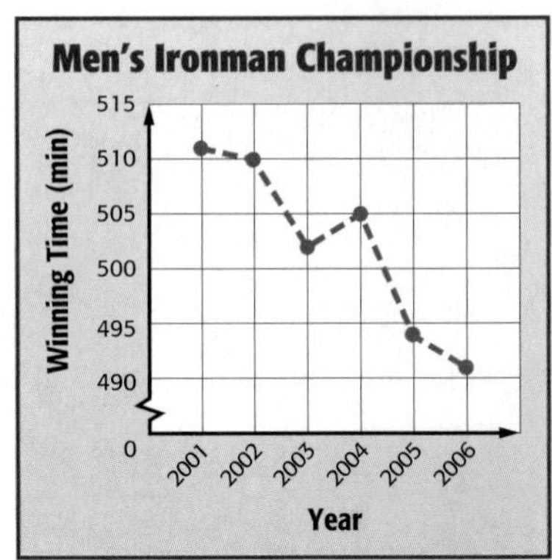

9. Describe the change in the winning times from 2001 to 2006.
10. Predict the winning time in 2012.
11. Predict when the winning time will be less than 450 minutes.

Real-World Link The Ironman Championship is held every year in Hawaii and consists of a 2.4-mile swim, a 112-mile bicycle ride, and a 26.2-mile run.
Source: *ESPN Almanac*

TRAVEL For Exercises 12–16, use the graph that shows the distance traveled by two cars on the same freeway headed in the same direction.

Distance Traveled by Two Cars

Distance (mi): 0, 50, 100, 150, 200, 250, 300, 350

Time (h): 0, 1, 2, 3, 4, 5

Car A

Car B

12. Predict the distance traveled by Car A after 5 hours.
13. Predict the distance traveled by Car B after 5 hours.
14. How many miles do you think Car A will have traveled after 8 hours?
15. Based on the graph, after how many hours will Car B have traveled about 360 miles?
16. Based on the graph, which car will reach a distance of 500 miles first? Explain your reasoning.

BASEBALL For Exercises 17–19, use the table below.

Pine Ridge Panthers Baseball Wins

Year	2000	2001	2002	2003	2004	2005	2006	2007
Games Won	46	52	25	33	35	45	42	30

17. Make a line graph of the data.
18. In what year did the team have the greatest increase in the number of games won? the greatest decrease? How can you find this information from the line graph?
19. Explain the disadvantages of using this line graph to make a prediction about the number of games that the team will win in 2010, 2011, 2012, and 2013.

EXTRA PRACTICE
See pages 675, 707.

H.O.T. Problems

20. **COLLECT THE DATA** Use the Internet or another source to collect a set of sports data that can be displayed using a line graph. Then use your graph to make a prediction.

21. **CHALLENGE** Refer to the graph for Exercises 12–16. What can you conclude about the point at which the red and blue lines cross?

22. **WRITING IN MATH** Explain why line graphs are often used to make predictions.

TEST PRACTICE

23. Every Sunday, Kailey saves a portion of her weekly earnings. The table shows the total amount of money she has saved each week. What is the best prediction for the total amount she will have saved after Week 8?

Week	Total Amount Saved ($)
1	15
2	34
3	42
4	60
5	78

A $100
B $130
C $150
D $170

24. For which year would the winning time in the Olympic Women's 3,000-Meter Speed Skating Relay have been the most difficult to predict based on previous results?

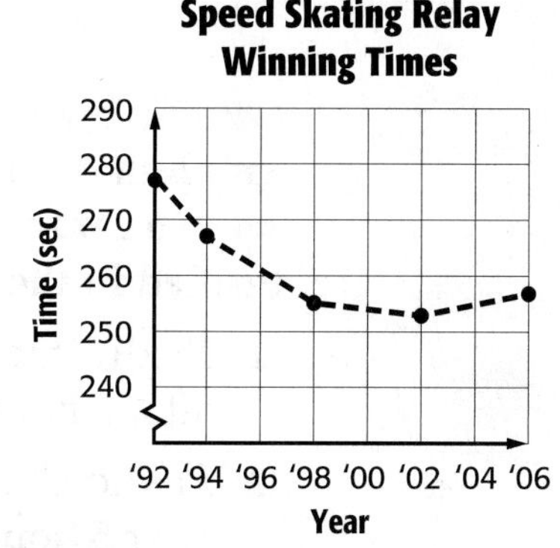

Source: *The World Almanac*

F 1994
G 1998
H 2002
J 2006

Spiral Review

25. **SCUBA DIVING** Make a line graph of the data in the table. (Lesson 2-2)

26. **SOFTBALLS** If g softballs cost $48, solve the equation $16g = 48$ to find the number of softballs purchased. (Lesson 1-8)

Depth of Scuba Diver below Sea Level	
Minutes	**Depth (ft)**
10	22
15	26
20	32
25	38
30	34
35	42

GET READY for the Next Lesson

PREREQUISITE SKILL Order each set of data from least to greatest. (Page 740)

27. Change in pocket (¢): 50, 45, 75, 55, 25, 60, 40, 80

28. Roller coaster speeds (mph): 40, 42, 48, 60, 72, 93, 57

2-4 Stem-and-Leaf Plots

MAIN IDEA

Display and analyze data using a stem-and-leaf plot.

New Vocabulary

stem-and-leaf plot
stems
leaves
key

Math Online

glencoe.com

- Concepts in Motion
- Extra Examples
- Personal Tutor
- Self-Check Quiz

GET READY for the Lesson

INSTANT MESSAGING Analyze the table.

1. What were the least and greatest number of instant messages sent?
2. Which number of instant messages occurred most often?

Number of Instant Messages Sent Each Day for 21 Days						
35	21	14	32	25	10	5
27	12	33	20	45	21	31
17	24	21	27	2	3	7

You can use a stem-and-leaf plot to organize large data sets so that they are easier to analyze and interpret. In a **stem-and-leaf plot**, the data are ordered from least to greatest and organized by place value.

EXAMPLE Construct a Stem-and-Leaf Plot

1 **INSTANT MESSAGING Make a stem-and-leaf plot of the data in the table above.**

Step 1 Order the data from least to greatest.

Step 2 Draw a vertical line and write the tens digits from least to greatest to the left of the line. These digits form the **stems**. Since the least value is 2 and the greatest value is 45, the stems are 0, 1, 2, 3, and 4.

Step 3 Write the ones digits in order to the right of the line with the corresponding stem. The ones digits form the **leaves**.

Number of Instant Messages Sent Each Day for 21 Days

Stem	Leaf
0	2 3 5 7
1	0 2 4 7
2	0 1 1 1 4 5 7 7
3	1 2 3 5
4	5

2 | 7 = 27 text messages

In these data, the tens digits form the stems.

The ones digits of the data form the leaves.

Step 4 Include a **key** that explains the stems and leaves.

CHECK Your Progress **Make a stem-and-leaf plot of the data.**

a. Calcium per serving in selected vegetables (mg): 14, 19, 10, 38, 32, 33, 40, 61, 34, 38, 55, 27, 14, 48

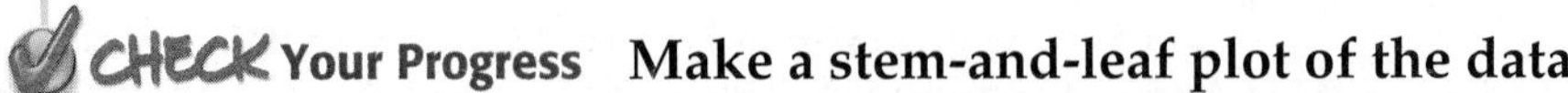

Real-World Link The tallest waterfall in the world is Angel Falls in Venezuela. It is 15 times higher than Niagara Falls.

Source: Salto-Angel

Stem-and-leaf plots are useful in analyzing data because you can see all the data values, including the greatest and least.

Real-World EXAMPLE Analyze Plots

2 **WATERFALLS The stem-and-leaf plot shows the approximate height of the twenty tallest waterfalls in the world. Write a few sentences that analyze the data.**

Approximate Height of the 20 Tallest World Waterfalls

Stem	Leaf
4	6 7 9 9
5	0 3 6 8
6	0 1 1 5 6
7	0 4 6 7
8	0
9	5 8 *4 | 6 = 460 meters*

The tallest waterfall in the world is about 980 meters. Two of the waterfalls listed are about 490 meters tall. Half of the waterfalls are at least 610 meters tall.

CHECK Your Progress

b. **GRADES** Mrs. Hudson made the stem-and-leaf plot to display the results of the math test scores for her students. Write a few sentences that analyze the data she displayed.

Math Test Scores

Stem	Leaf
6	2 7 8
7	3 5 5 9
8	0 3 5 5 5 5 5 8 9
9	1 1 3 3 5 5 8 8
10	0 0 *8 | 5 = 85%*

CHECK Your Understanding

Example 1 (p. 92)

Make a stem-and-leaf plot of each set of data.

1. Minutes spent on homework: 37, 28, 25, 29, 31, 45, 32, 31, 46, 39
2. Miles traveled to reach weekend vacation destination: 81, 76, 55, 90, 71, 80, 83, 85, 79, 99, 70, 75, 70, 92

Example 2 (p. 93)

SNACKS For Exercises 3–5, use the stem-and-leaf plot below.

Number of Calories in Selected Snack Foods

Stem	Leaf
24	0 4 4 8
25	0 0 5 7 8
26	4 5
27	5
28	4
	24 | 4 = 244 Calories

3. How many snack foods listed have more than 250 Calories?
4. How many Calories are in the snack food listed with the most number of Calories?
5. Write a few sentences that analyze the data.

Practice and Problem Solving

HOMEWORK HELP	
For Exercises	See Examples
6–9	1
10–13	2

Make a stem-and-leaf plot of each set of data.

6. Bus ride in minutes: 24, 14, 25, 28, 47, 13, 9, 17, 30, 35, 16, 39
7. Video game score: 53, 64, 15, 22, 16, 42, 12, 38, 68, 63, 23, 35, 30, 33, 34, 35
8. Calories per serving: 62, 65, 67, 67, 62, 67, 51, 73, 72, 70, 63, 72, 78, 61, 54
9. Test scores: 76, 82, 70, 93, 71, 80, 63, 73, 90, 92, 74, 79, 82, 91, 95, 93, 75

TUNNELS For Exercises 10 and 11, use the plot below.

World's Longest Tunnels

Stem	Leaf
6	3 8 9 9
7	0 1
8	0 7
9	
10	2
11	2
12	
13	
14	
15	2 *6 | 3 = 6.3 mi*

10. How long is the world's longest tunnel?
11. Write a few sentences that analyze the data.

CONCERTS For Exercises 12 and 13, use the plot below.

Shows Performed by the Top 25 Musical Tours

Stem	Leaf
2	3 4 5
3	8 8
4	3 3 3
5	0 1 2 5 6 6 9
6	7
7	0 3 3 5 5 7
8	0 5
9	9 *3 | 8 = 38 shows*

12. How many tours performed more than 50 shows?
13. Write a few sentences that analyze the data.

For Exercises 14 and 15, use the back-to-back stem-and-leaf plot at the right and the information below.

A *back-to-back stem-and-leaf plot* can be used to compare two sets of data. The leaves for one set of data are on one side of the stem and the leaves for the other set of data are on the other side of the stem.

Points Scored

Tigers	Stem	Eagles
8 4 2	5	0 3 4 8 8
6 5	6	2 6 7
8 7 4 0	7	0 2 5
9 8 6 0	8	2
5 | 4 = 54 points		*5 | 0 = 50 points*

14. How many games were there in which the Tigers scored more than 75 points? the Eagles?
15. Write a few sentences comparing the points scored by each team.

See pages 676, 707.

H.O.T. Problems

16. **FIND THE ERROR** Miguel and Anita are writing the stems for the data set 5, 45, 76, 34, 56, 2, 11, and 20. Who is correct? Explain.

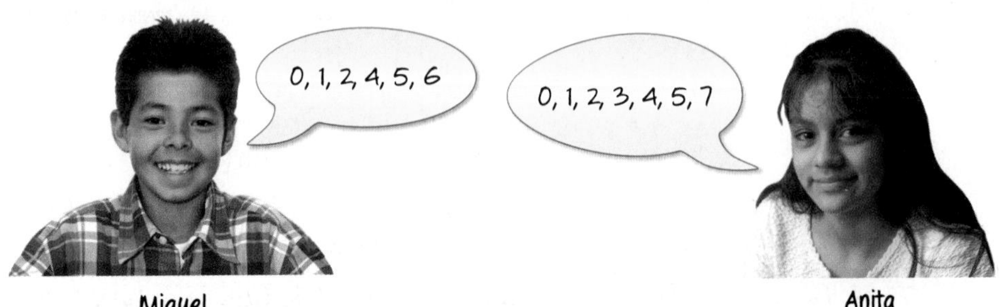

17. **COLLECT THE DATA** Display the height, in inches, of your classmates in a stem-and-leaf plot. Then write a few sentences that analyze the data.

18. **CHALLENGE** Refer to Exercises 3–5. Describe how you could create a bar graph from the stem-and-leaf plot.

19. **WRITING IN MATH** Describe an advantage of displaying a set of data in a stem-and-leaf plot instead of a bar or a line graph.

TEST PRACTICE

20. Chloe's math test scores are 74, 96, 85, 100, 96, and 87. Which stem-and-leaf plot most accurately displays her test scores?

A **Math Test Scores**

Stem	Leaf
9	6 6
7	4
8	5 7
1	0 0 $8 \mid 5 = 85$

B **Math Test Scores**

Stem	Leaf
1	0 0
7	4
8	5 7
9	6 6 $8 \mid 5 = 85$

C **Math Test Scores**

Stem	Leaf
7	4
8	5 7
9	6 6
10	0 $8 \mid 5 = 85$

D **Math Test Scores**

Stem	Leaf
7	4
8	5
8	7
9	6
9	6
10	0 $8 \mid 5 = 85$

Spiral Review

SPORTS For Exercises 21–24, use the graph.
(Lessons 2-2 and 2-3)

21. What was the approximate winning time in 1970?

22. What does the graph tell you about the winning times?

23. In which decade was the greatest decrease in the winning time?

24. Predict the winning time for the Boston Marathon in 2020.

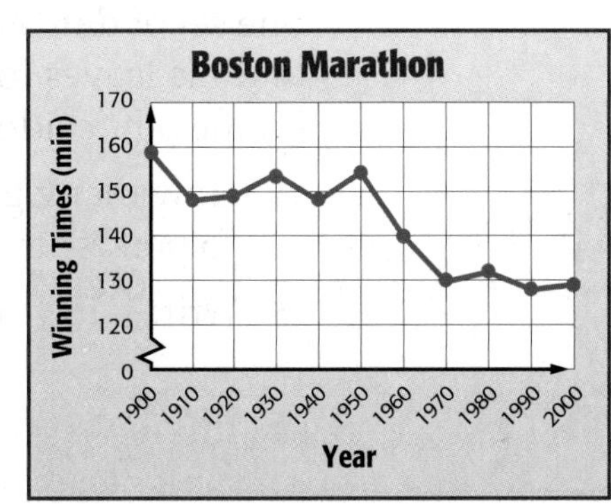

Source: Boston Athletic Association

GET READY for the Next Lesson

PREREQUISITE SKILL Order each set of data from least to greatest. (Page 740)

25. 64, 65, 63, 55, 77, 51, 54, 52, 78

26. 101, 123, 117, 120, 105, 98, 114, 113

2-5 Line Plots

MAIN IDEA

Display, analyze, and interpret data using line plots.

New Vocabulary

line plot

Math Online

glencoe.com

- Extra Examples
- Personal Tutor
- Self-Check Quiz

GET READY for the Lesson

ANIMALS The table shows the life expectancy of several animals.

1. How many of the animals have a life expectancy of 15 years?
2. How many animals have a life expectancy from 5 to 10 years, including 10?
3. What is the longest life expectancy represented?
4. What is the shortest life expectancy represented?

Animal	Years
Black bear	18
Cat (pet)	12
Chimpanzee	20
Cow	15
Dog (pet)	12
Giraffe	10
Horse	20
Leopard	12
Lion	15
Mouse	3
Pig	10
Rabbit (pet)	5

Source: *The World Almanac*

A **line plot** is a diagram that shows the frequency of data on a number line. An '×' is placed above a number on a number line each time that data value occurs.

EXAMPLE Display Data in a Line Plot

1 **ANIMALS** **Make a line plot of the data above.**

Step 1 Draw a number line. The smallest value is 3 years and the largest value is 20 years. So, you can use a scale of 0 to 20. Other scales could also be used.

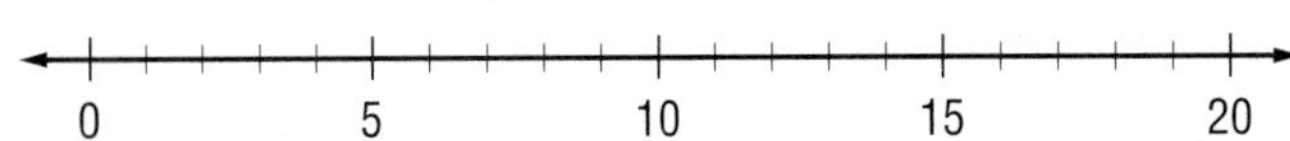

Step 2 Put an × above the number that represents the life expectancy of each animal. Add a title.

Animal Life Expectancies in Years

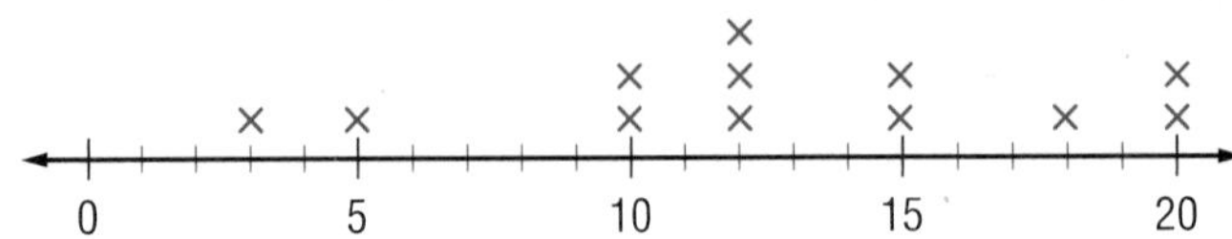

Study Tip

Line Plots It is often easier to identify the number of data that fall in a certain category with a line plot, as opposed to counting all those values in a table.

CHECK Your Progress **Make a line plot of the data.**

a. Monthly cell phone usage (min): 35, 40, 45, 27, 30, 32, 25, 30, 35, 50, 27, 32, 31, 40, 32, 28, 45, 32, 40, 32

A line plot allows you to easily analyze the *distribution* of data, or how data are grouped together or spread out.

EXAMPLES Analyze a Line Plot

EXERCISE The line plot displays the number of Calories Justin burns per minute doing activities that he enjoys.

Calories Burned per Minute for Various Activities

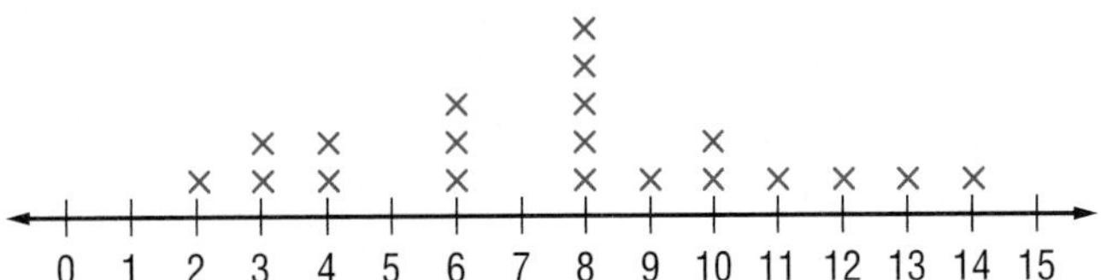

2 How many of the activities represented on the line plot burn 8 Calories per minute?

Locate 8 on the number line and count the number of ×'s above it. There are 5 activities that burn 8 Calories per minute.

3 What is the difference between the least and the greatest number of Calories burned per minute by an activity represented in the line plot?

The least number of Calories burned per minute is 2.
The greatest number of Calories burned per minute is 14.

$14 - 2 = 12$ Subtract to find the difference.

The difference is 12 Calories per minute.

4 If Justin exercises by doing each of these activities once a month, write one or two sentences that analyze the data.

Each month Justin typically burns about 6 or 8 Calories per minute exercising.

Real-World Career
How Does a Physical Therapist Use Math? Physical therapists examine the number of Calories their patients burn while doing physical activities to restore their mobility.

Math Online
For more information, go to glencoe.com.

CHECK Your Progress

VIDEO GAMES The line plot displays the number of video games owned by students in a classroom.

Number of Video Games Owned

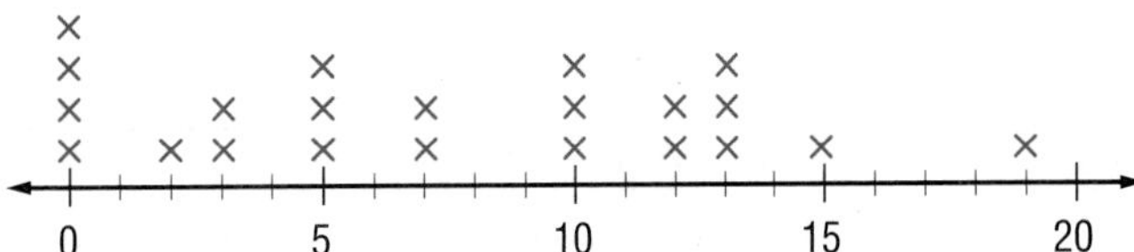

b. How many students own 3 video games?

c. How many students own 10 or more video games?

d. Write one or two sentences that analyze the data.

CHECK Your Understanding

Example 1 (p. 96)

1. **ROLLER COASTERS** The table at the right shows the speeds of the world's fastest wooden roller coasters. Make a line plot of the data.

World's Fastest Wooden Roller Coasters (miles per hour)				
66	65	66	75	65
65	64	65	78	63

Source: Coaster Grotto

Examples 2–4 (p. 97)

BASEBALL For Exercises 2–4, use the line plot at the right.

Record Number of Wins by the Leading National League Pitcher, 1993–2006

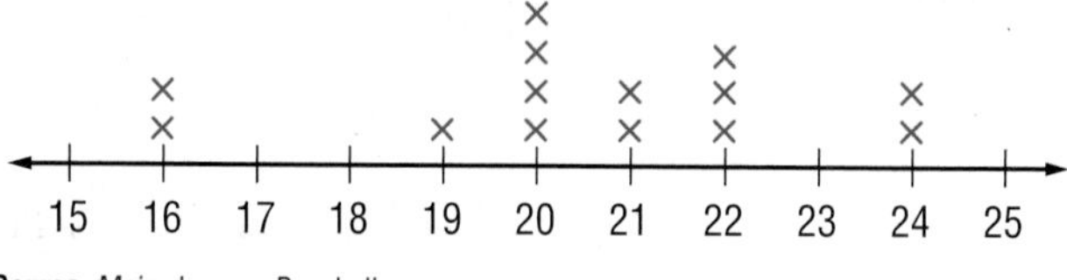

Source: Major League Baseball

2. What number of games was won by the most leading pitchers from 1993 to 2006?
3. How many of the leading pitchers won 22 or more games?
4. Write one or two sentences that analyze the data.

Practice and Problem Solving

HOMEWORK HELP	
For Exercises	**See Examples**
5, 6	1
7–10	2–4

Make a line plot of each set of data.

5.

Test Scores			
78	95	80	85
70	88	95	90
95	85	88	78
75	90	85	82
76	75	82	80

6.

Number of Stories in World's Tallest Buildings			
101	88	88	110
88	88	80	69
102	78	72	54
85	80	73	100

SPACE SCIENCE For Exercises 7–10, use the line plot below.

Ages in Years of the 20 Oldest Astronauts on Launch Day

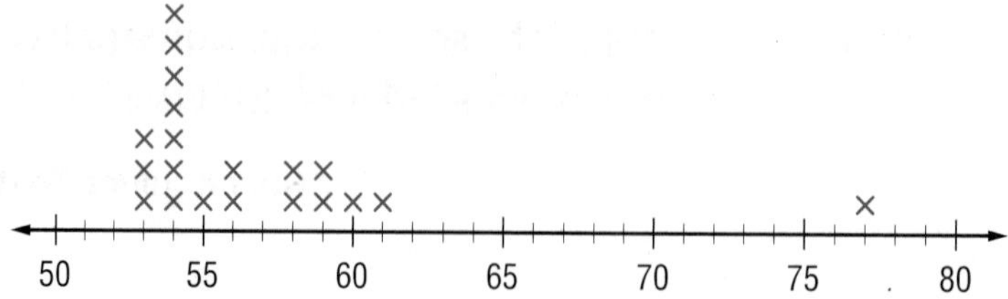

7. How many astronauts were 56 years old on launch day?
8. Which age was most common among the oldest astronauts on launch day?
9. What is the difference between the age of the oldest and youngest astronaut represented in the line plot?
10. Write one or two sentences that analyze the data.

FOOD **For Exercises 11–14, use the line plot below.**

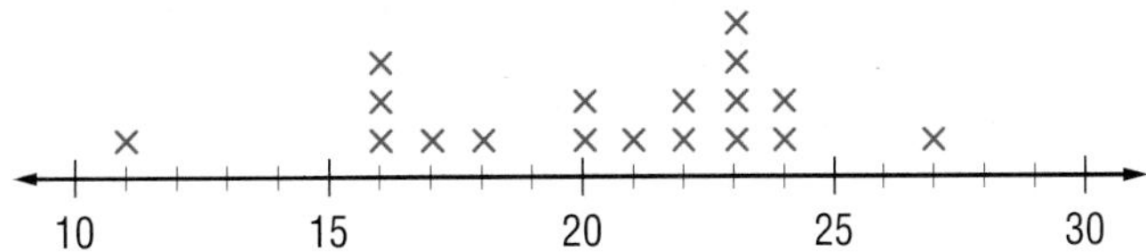

11. How many more types of fish, meat, and poultry have 23 grams of protein than 17 grams of protein?
12. A *peak* of a line plot represents the most frequently occurring value. Identify any peaks in the line plot.
13. Write one or two sentences that analyze the data.
14. **ANALYZE GRAPHS** A graph is *symmetric* if the left side looks like the right side. Is the line plot symmetric? Explain.
15. **ANALYZE GRAPHS** Refer to the line plot you made in Exercise 6. Write one or two sentences that analyze the data.

TRAVEL **For Exercises 16–18, use the table at the right that shows the average travel time to work for select cities.**

16. Make a line plot, a stem-and-leaf plot, and a bar graph of the data.
17. Which display allows you to easily determine the number of cities that have an average commute time of 29 minutes? Explain.
18. Which graph allows you to easily compare the average commute time in Atlanta with the average commute time in Omaha? Explain.

Average Travel Time to Work

City	Minutes
Anaheim, CA	24
Atlanta, GA	27
Fort Worth, TX	24
New Orleans, LA	24
New York, NY	38
Oakland, CA	29
Omaha, NE	17
Washington, D.C.	29

Source: U.S. Census Bureau

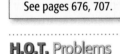

See pages 676, 707.

H.O.T. Problems

19. **COLLECT THE DATA** Create a line plot that displays the shoe size of students in your class. Then write one or two statements that analyze the data. Identify any peaks or symmetry.
20. **DATA SENSE** The science test scores of two students are shown below. Describe the shape of each graph.

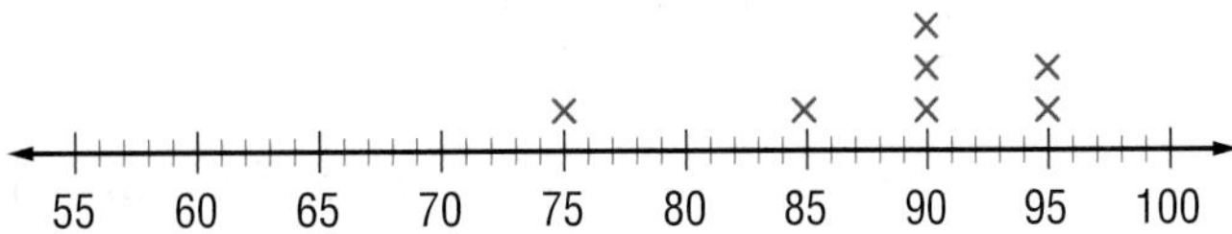

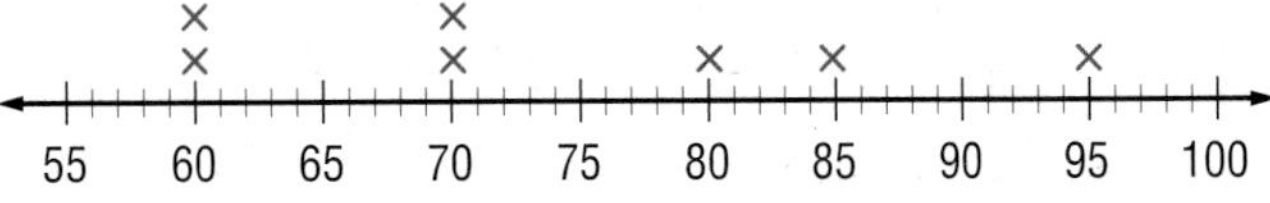

21. **CHALLENGE** *Clusters* are data that are grouped closely together. Identify the clusters in the following set of data that describe the ages of people in a movie theater.

22, 23, 11, 12, 12, 13, 12, 14, 40, 12, 30, 26

22. **WRITING IN MATH** Compare and contrast line plots and stem-and-leaf plots.

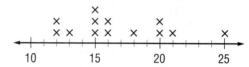

TEST PRACTICE

23. The table shows the prices, in dollars, of select DVDs. Which line plot correctly displays the data in the table?

DVD Prices ($)						
24	16	18	14	16	21	15
15	12	15	15	20	20	12

A

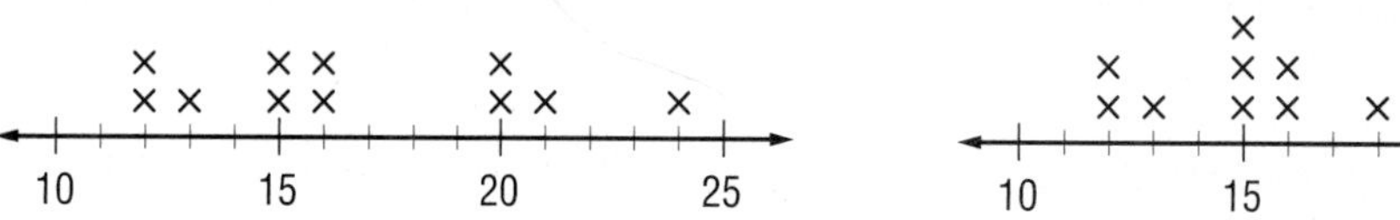

C

B

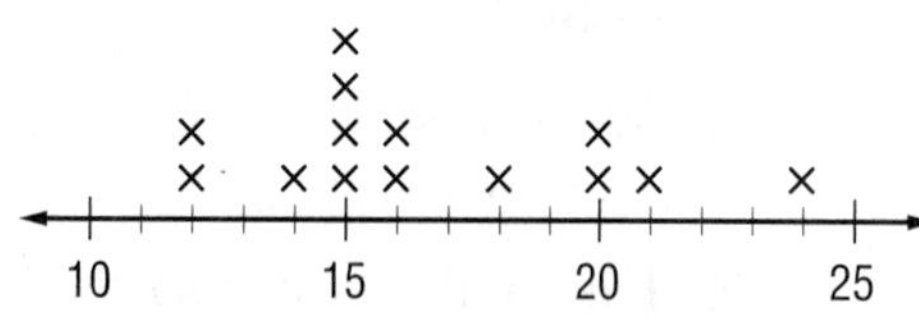

D

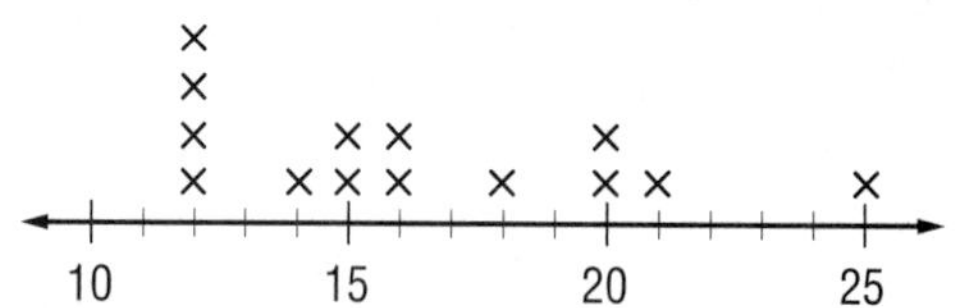

Spiral Review

Make a stem-and-leaf plot of each set of data. (Lesson 2-4)

24. Minutes spent exercising: 18, 40, 25, 30, 27, 35, 22, 25, 15, 20, 30, 27

25. Points scored in basketball games:
55, 47, 62, 38, 45, 50, 58, 60, 64, 49, 55, 45, 52

OLYMPICS For Exercises 26 and 27, use the line graph at the right. (Lesson 2-3)

26. Between which years did the winning time change the most? Describe the change.

27. Make a prediction of the winning time in the 2012 Olympics. Explain your reasoning.

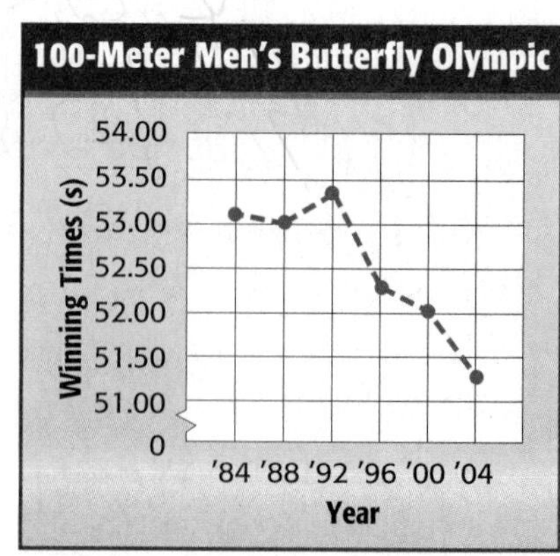

Source: *The World Almanac*

GET READY for the Next Lesson

PREREQUISITE SKILL Find the value of each expression. (Lesson 1-4)

28. $(15 + 17) \div 2$

29. $(4 + 8 + 3) \div 3$

30. $(10 + 23 + 5 + 18) \div 4$

CHAPTER 2 Mid-Chapter Quiz

Lessons 2-1 through 2-5

1. **SKATEBOARDS** Make a frequency table of the data below. How many skateboards cost between \$50 and \$69? (Lesson 2-1)

Cost (\$) of Various Skateboards				
99	67	139	63	75
59	89	59	70	78
99	55	125	64	110

2. **PLANTS** Make a bar graph of the data at the right. Compare the number of vegetables to the number of herbs. (Lesson 2-2)

Greenhouse Plants	
Type	**Frequency**
Vegetables	38
Flowers	27
Herbs	13
Other	9

3. **MULTIPLE CHOICE** Solana collected the following data from five local restaurants.

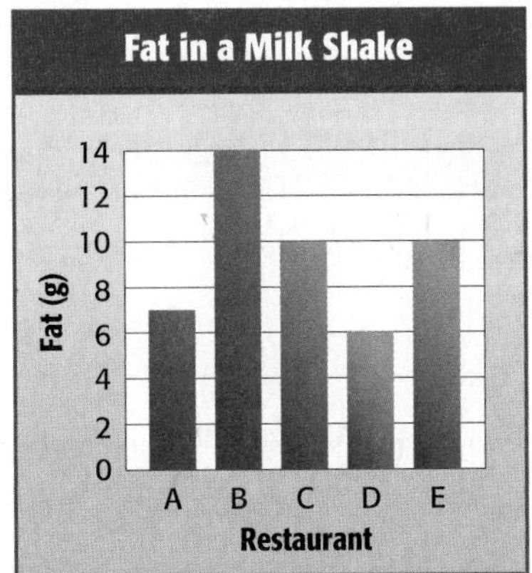

Which statement is supported by the graph? (Lesson 2-2)

A Restaurant E's milkshake contains the most fat.

B Restaurant D's milkshake contains half as many grams of fat as Restaurant C's.

C Restaurant A's milkshake contains 8 grams of fat.

D Restaurant B's milkshake contains twice as many fat grams as Restaurant A's.

4. **MULTIPLE CHOICE** Kenya's total savings for seven weeks is shown in the graph.

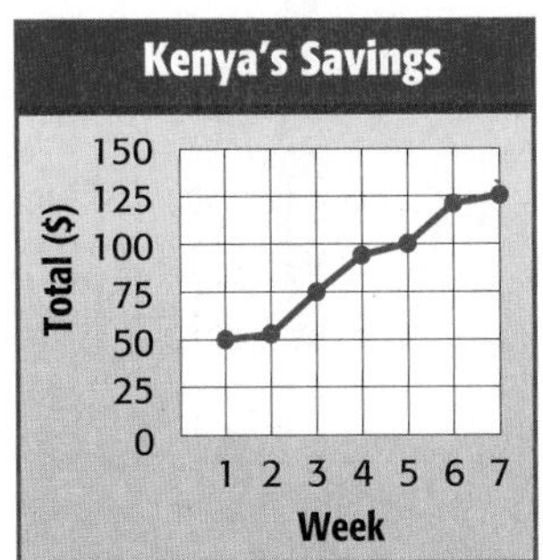

Which statement is supported by the graph? (Lesson 2-3)

F Kenya's total savings decreased from Week 2 to Week 5.

G By the end of Week 7, Kenya's total savings was \$125.

H The total amount saved by the end of Week 8 should be about \$200.

J Kenya saved more money from Week 6 to Week 7 than any other week.

5. **SPORTS** Use the stem-and-leaf plot to determine how many times the National League leader hit 50 or more home runs. (Lesson 2-4)

Home Runs by the National League Leaders, 1990–2006

Stem	Leaf
3	5 8
4	0 0 3 6 7 7 8 9 9
5	0 1 8
6	5
7	0 3 *4|3 = 43 home runs*

Source: Major League Baseball

6. **MUSICIANS** The ages in years of the youngest solo singers with a #1 U.S. single are listed below.

15, 17, 18, 15, 17, 16, 18, 17, 17, 18, 14, 16, 16, 18, 13, 13, 14, 17, 19, 19

Make a line plot of these data. (Lesson 2-5)

2-6 Mean

MAIN IDEA

Find the mean of a data set.

New Vocabulary

average
mean
outlier

Math Online

glencoe.com

- Extra Examples
- Personal Tutor
- Self-Check Quiz

MINI Lab

In five days, it snowed 4 inches, 3 inches, 5 inches, 1 inch, and 2 inches.

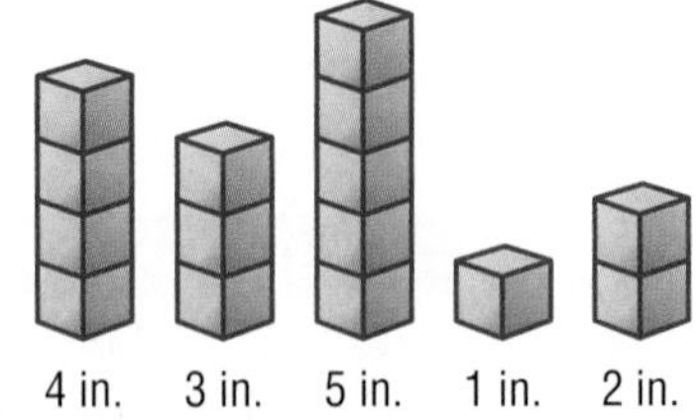

- Make a stack of centimeter cubes to represent the snowfall for each day, as shown at the right.
- Move the cubes until each stack has the same number of cubes.

1. On average, how many inches did it snow per day in five days? Explain your reasoning.
2. Suppose on the sixth day it snowed 9 inches. If you moved the cubes again, how many cubes would be in each stack?

When analyzing data, it is helpful to use a single number to describe the whole set. In the Mini Lab above, a good choice would be the number 3, the mean or **average** number of cubes in each stack that results from equally distributing all the cubes. The mean can be interpreted as a balancing point for a set of data. The mean of a set of data can also be calculated.

Mean — Key Concept

Words The **mean** of a set of data is the sum of the data divided by the number of pieces of data.

Example Data set: 4, 3, 5, 1, 2 → mean: $\frac{4 + 3 + 5 + 1 + 2}{5} = \frac{15}{5}$ or 3

EXAMPLE Find the Mean

1 **CIVICS** **Find the mean number of Representatives for the four states shown in the pictograph.**

2007 Representatives to U.S. Congress

State	Representatives
Tennessee	🯅🯅🯅🯅🯅🯅🯅🯅🯅
Kentucky	🯅🯅🯅🯅🯅🯅
Virginia	🯅🯅🯅🯅🯅🯅🯅🯅🯅🯅🯅
South Carolina	🯅🯅🯅🯅🯅🯅

METHOD 1 **Move the figures.**

2007 Representatives to U.S. Congress

Tennessee

Kentucky

Virginia

South Carolina

Move the figures to equally distribute the total number of Representatives among the four states.

METHOD 2 **Write and simplify an expression.**

$$\text{mean} = \frac{9 + 6 + 11 + 6}{4} \quad \begin{matrix} \leftarrow \text{sum of the data} \\ \leftarrow \text{number of data items} \end{matrix}$$

$$= \frac{32}{4} \text{ or } 8 \qquad \text{Simplify.}$$

Each state has a mean or average of 8 representatives.

Study Tip

Including Data Even if a data value is 0, it still should be counted in the total number of pieces of data.

CHOOSE Your Method

a. **MUSIC** The bar graph shows the number of CDs bought by a group of friends. Find the mean number of CDs bought by the group.

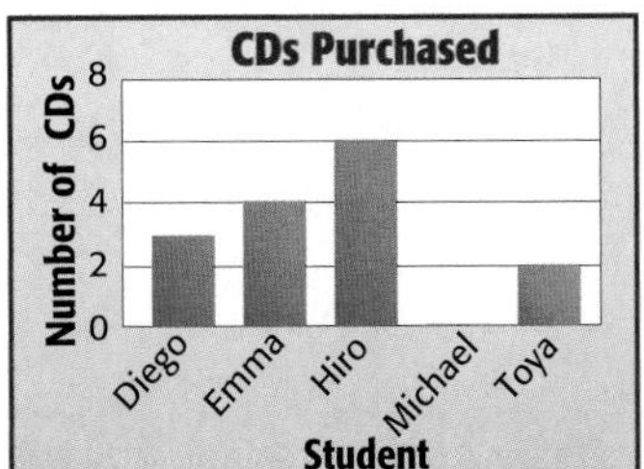

Values that are much higher or lower than others in a data set are **outliers**.

EXAMPLE Determine How Outliers Affect Mean

2 **CELL PHONES The number of minutes Mary Anne spent talking on her cell phone each month for the past five months were 494, 502, 486, 690, and 478. Identify the outlier(s) in the data. Find the mean with and without the outlier. Then describe how the outlier affects the mean.**

Compared to the other values, 690 is extremely high. So, it is an outlier. Find the mean with and without the outlier.

with outlier

$$\frac{494 + 502 + 486 + 690 + 478}{5}$$

$$= \frac{2{,}650}{5} \text{ or } 530$$

without outlier

$$\frac{494 + 502 + 486 + 478}{4}$$

$$= \frac{1{,}960}{4} \text{ or } 490$$

With the outlier, the mean is greater than all but one of the data values. Without it, the mean better represents the data.

CHECK Your Progress

b. Identify the outlier in these costs: $110, $120, $110, $135, $140, $120, $105, and $440. Describe how it affects the mean.

CHECK Your Understanding

Example 1 (pp. 102, 103)

Find the mean of the data represented in each model.

1.

2.

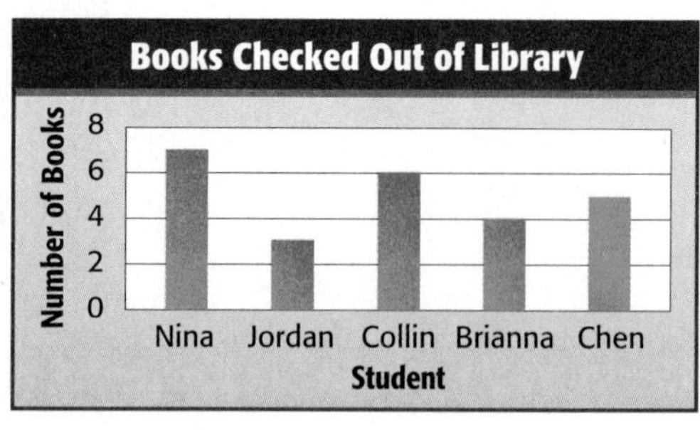

Example 2 (p. 103)

GEOGRAPHY For Exercises 3–5, use the table at the right. It lists the greatest depths of the oceans.

3. What is the mean of the data?
4. Which depth is an outlier? Explain.
5. How does this outlier affect the mean?

Ocean	Depth (ft)
Pacific	15,215
Atlantic	12,881
Indian	13,002
Arctic	3,953
Southern	14,749

Source: Enchanted Learning

Practice and Problem Solving

HOMEWORK HELP

For Exercises	See Examples
6–9	1
10–16	2

Find the mean of the data represented in each model.

6.

Number of Popcorn Bags Sold

Pilar

Marisa

Maurice

Irene

= 2 Popcorn Bags

7.

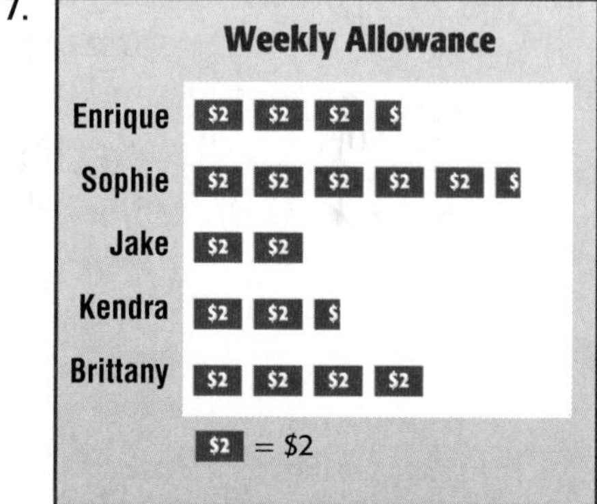

8.

Height of Students

Height (in.): 0, 50, 52, 54, 56, 58, 60

Andrea 54, Kareem 59, Jeff 57, Sonia 54

Student

9.

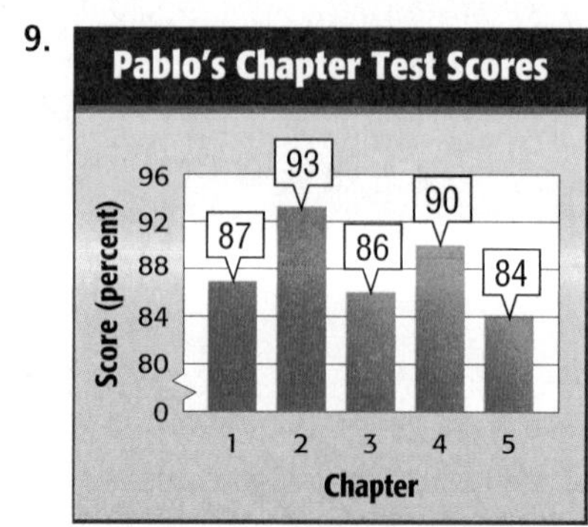

NATURE For Exercises 10–13, use the table that shows the approximate heights of some of the tallest U.S. trees.

Tallest Trees in U.S.	
Tree	**Height (ft)**
Western Red Cedar	160
Coast Redwood	320
Monterey Cypress	100
California Laurel	110
Sitka Spruce	200
Port-Orford-Cedar	220

Source: *The World Almanac*

10. Find the mean of the data.
11. Identify the outlier(s).
12. Find the mean if the outlier(s) is not included in the data set.
13. How does the outlier affect the mean of the data?

MONEY For Exercises 14–16, use the following information.

Jamila earned \$15, \$20, \$10, \$12, \$20, \$16, \$80, \$18, and \$25 baby-sitting.

14. What is the mean of the amounts she earned?
15. Identify the outlier(s).
16. How does the outlier(s) affect the mean of the data?

Find the mean for each set of data. Explain the method you used.

17. Number of songs on an MP3 player: 145, 87, 150, 122, 96
18. Money saved each month: \$28, \$30, \$32, \$21, \$29, \$28, \$28
19. Age of camp counselors (in years): 13, 17, 14, 16, 16, 14, 16, 14
20. Number of votes received: 70, 35, 64, 98, 42

See pages 676, 707.

21. **WEATHER** The graphic at the right shows the 5-day forecast as shown on the local news. What is the difference between the mean high and mean low temperature for this 5-day period? Justify your answer.

H.O.T. Problems

22. **REASONING** Tell whether the following statement is *sometimes, always,* or *never* true. Justify your answer.

    The mean of a set of data is one of the values in the data set.

23. **SELECT A TOOL** The number of people dining at a certain restaurant for several days was 319, 127, 244, 398, 427, and 261. Which of the following tools might you use to find the mean of the data? Justify your selection. Then use the tool to solve the problem.

    draw a model | calculator | real objects

24. **CHALLENGE** Find a value for n such that the mean of the ages 40, 45, 48, n, 42, and 41 is 45. Explain the method or strategy you used.

25. **OPEN ENDED** Create a set of data that has five values and a mean of 34.

26. **WRITING IN MATH** The mean amount of precipitation from January to June for a certain city was about 4 inches per month. Without doing any calculations, determine how the mean would be affected if the total precipitation for the month of July for this city is 3 inches, 5 inches, or 4 inches. Explain your reasoning.

TEST PRACTICE

27. Student Council sells school calendars each year as a fundraiser. David was on Student Council from 2007 to 2010. The bar graph shows the number of calendars he sold each year.

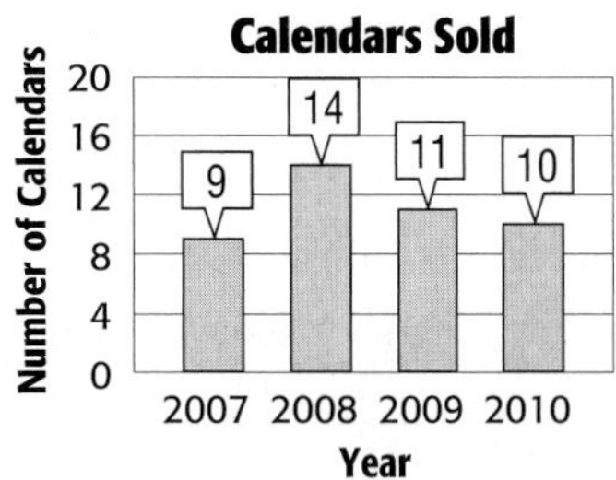

What is the mean number of calendars David sold each year?

A 9
B 10
C 11
D 14

28. The table shows the money raised by each booth at a craft sale.

Northside Craft Sale	
Booth	**Money Raised ($)**
Artwork	58
Candles	47
Holiday decorations	54
Jewelry	70
Picture frames	45
T-shirts	80

What was the mean amount raised at each booth?

F $59
G $60
H $61
J $62

Spiral Review

29. **BASEBALL** The table shows the number of players on each team in a baseball league. Make a line plot of the data. (Lesson 2-5)

Players Per Team					
16	15	16	15	18	19
12	15	16	14	18	14

30. **CARS** The average number of miles per gallon of gasoline for selected cars is shown in the stem-and-leaf plot. Into what interval(s) do most of the data lie? (Lesson 2-4)

Average Gas Mileage

Stem	Leaf
1	9
2	2 4 4 6 7 8
3	0 1 2 4 5 6 7 7
4	1

1|9 = 19 miles per gallon

31. **CARPETING** How many square feet of carpeting are needed to cover a room that is 11 feet by 16 feet? (Lesson 1-9)

GET READY for the Next Lesson

PREREQUISITE SKILL **Subtract.** (Page 743)

32. 75 − 64
33. 102 − 39
34. 571 − 218
35. 1,206 − 809

Extend 2-6

Spreadsheet Lab
Spreadsheets and Mean

MAIN IDEA

Use a spreadsheet to find the mean.

Spreadsheets can be used to find the mean of a data set.

ACTIVITY

The table shows the lengths of four insects of each type kept in a science classroom.

To find the mean length of each type, set up a spreadsheet as shown.

Insect Length (cm)				
Type	Insect 1	Insect 2	Insect 3	Insect 4
Dragonfly	2.3	2.1	2.4	2.6
Grasshopper	3.7	4.4	7.6	5.2
Walking Stick	11.0	16.2	14.8	19.6

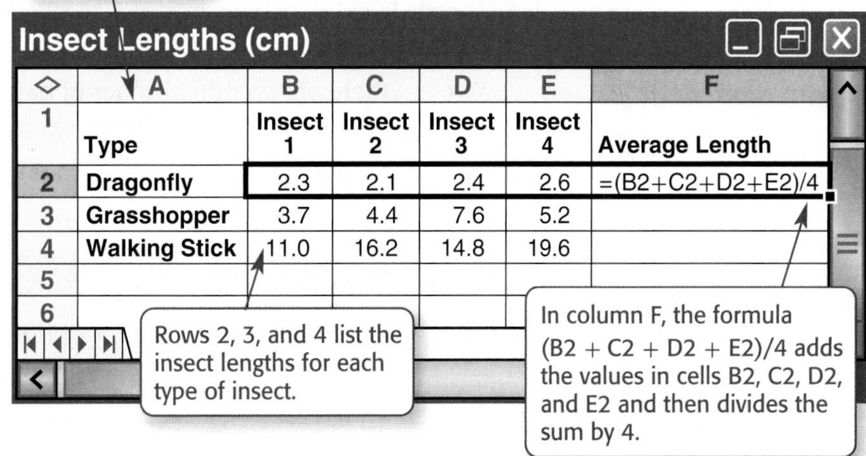

Column A lists the insect types.
Insect Lengths (cm)
A
B
C
D
E
F
1
Type
Insect 1
Insect 2
Insect 3
Insect 4
Average Length
2
Dragonfly
2.3
2.1
2.4
2.6
=(B2+C2+D2+E2)/4
3
Grasshopper
3.7
4.4
7.6
5.2
4
Walking Stick
11.0
16.2
14.8
19.6
5
6
Rows 2, 3, and 4 list the insect lengths for each type of insect.
In column F, the formula (B2 + C2 + D2 + E2)/4 adds the values in cells B2, C2, D2, and E2 and then divides the sum by 4.

Now, enter the formula in cell F2 and press the Enter key.

Analyze the Results

1. What formulas should you enter to find the mean grasshopper length? mean walking stick length?
2. Use the spreadsheet to find the mean grasshopper length and mean walking stick length. Round to the nearest tenth.
3. **MAKE A CONJECTURE** About how long would a fifth walking stick need to be in order to have a new mean length of 15.8 centimeters? Explain your reasoning.
4. Explain how you could adapt and use the spreadsheet above to check your answer to Exercise 3. Then check your answer and adjust your conjecture if necessary to determine the needed length.

2-7 Median, Mode, and Range

MAIN IDEA

Find and interpret the median, mode, and range of a set of data.

New Vocabulary

measures of central tendency
median
mode
range

Math Online

glencoe.com

- Extra Examples
- Personal Tutor
- Self-Check Quiz

GET READY for the Lesson

HURRICANES The table shows the number of Atlantic hurricanes in different years.

Atlantic Hurricanes						
5	15	9	7	4	9	8

Source: National Hurricane Center

1. Order the data from least to greatest. Which piece of data is in the middle of this list?
2. Compare this number to the mean of the data.

A data set can also be described by its median or its mode. The mean, median, and mode are called **measures of central tendency** because they describe the *center* of a set of data.

Median — Key Concept

Words The **median** is the middle number of the ordered data when there are an odd number of data, or the mean of the middle two numbers when there are an even number of data.

Examples data set: 3, 4, (8), 10, 12 → median: 8

data set: 2, 4, (6, 8), 11, 12 → median: $\frac{6+8}{2}$ or 7

Mode

Words The **mode** is the number or numbers that occur most often.

Example data set: 12, 23, (28, 28), 32, (46, 46) → modes: 28 and 46

EXAMPLE Find the Median and the Mode

1 **MONKEYS The table shows the number of monkeys at eleven different zoos. Find the median and mode of the data.**

Number of Monkeys					
28	36	18	25	12	44
18	42	34	16	30	

Order the data from least to greatest.

median: 12, 16, 18, 18, 25, (28), 30, 34, 36, 42, 44 — 28 is in the middle.

mode: 12, 16, (18, 18), 25, 28, 30, 34, 36, 42, 44 — 18 occurs most often.

The median is 28 monkeys. The mode is 18 monkeys.

CHECK Your Progress

a. **BUILDINGS** The list shows the number of stories in the 11 tallest buildings in Springfield. Find the median and mode of the data.

40, 38, 40, 37, 33, 30, 20, 24, 21, 17, 19

Study Tip

Mean, Median, Mode, and Range The mean, median, and mode of a data set describe the center of a set of data. The range of a set of data describes how much the data vary.

The **range** of a set of data is the difference between the greatest and the least values of the set. When compared to the values in the data set, a large range indicates that the data are spread out. A small range indicates that the data are close in value.

EXAMPLE Find the Range

2 COINS Ella collected 125, 45, 67, 150, 32, 45, and 12 pennies each day this week for a school fundraiser. Find the range of the data. Then write a sentence that describes how the data vary.

The greatest number of pennies is 150. The least number of pennies is 12. So, the range is 150 – 12 or 138. The range is relatively large, so the data are spread out.

CHECK Your Progress

b. **TESTS** Santos' science test scores this school year were 98, 83, 75, 74, 70, 82, 95, and 88. Find the range of the data. Then write a sentence describing how the data vary.

Real-World EXAMPLE

3 WEATHER Find the mean, median, mode, and range of the temperatures displayed in the graph.

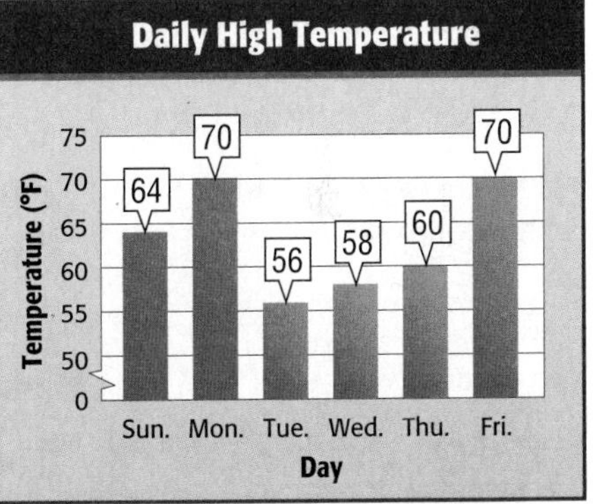

mean: $\dfrac{64 + 70 + 56 + 58 + 60 + 70}{6}$

$= \dfrac{378}{6}$ or 63°

median: 56, 58, 60, 64, 70, 70

$\dfrac{60 + 64}{2} = \dfrac{124}{2}$

$= 62°$

There are an even number of data values. So, to find the median, find the mean of the two middle numbers.

mode: 70°

range: 70 – 56 = 14°

CHECK Your Progress

c. **BACKPACKS** Find the mean, median, mode, and range of the costs in the stem-and-leaf plot.

Cost of Backpacks

Stem	Leaf
1	5 9
2	0 4 5 9
3	2 5 8 9
4	9
5	8 9

$5|8 = \$58$

TEST EXAMPLE

Real-World Link
The tallest recorded sand dune in the world is found in the Sahara Desert at 1,410 feet tall. This is just short of the height of the Sears Tower at 1,450 feet.
Source: California Association of 4 Wheel Drive Clubs

4 DESERTS The table shows the approximate sizes of the world's largest subtropical deserts. Which statement is supported by the data?

World's Largest Deserts	
Desert	Size (square miles)
Sahara	3,500,000
Arabian	1,000,000
Great Victoria	250,000
Kalahari	220,000
Chihuahuan	175,000

A Half of the deserts listed are larger than 220,000 square miles.

B The most common desert size listed is 220,000 square miles.

C The desert sizes are very spread out.

D If the total area was divided equally among these deserts, the size of each desert would be about 850,000 square miles.

Read the Item

The answer choices refer to the median, mode, range, and mean.

Solve the Item

median: The middle number is 250,000.
mode: no mode
range: 3,500,000 – 175,000 = 3,325,000
mean: The sum of the data is 5,145,000. Dividing 5,145,000 by five data values gives 1,029,000.

Determine which measure is referred to in each answer choice.

Choice A refers to the median, but the correct median is 250,000.

Choice B refers to the mode, but there is no mode.

Choice C refers to the range. The data are very spread out.

Choice D refers to the mean, but the correct mean is 1,029,000.

The correct answer is **C.**

Test-Taking Tip

Examine Answer Choices It is best to examine all the answer choices when determining which one best supports a given set of data or criteria.

CHECK Your Progress

d. **HOCKEY** The number of goals scored by each player on a high school hockey team is shown. Which statement is supported by the data in the table?

Goals Scored in Regular Season, Per Player				
3	1	2	0	4
1	5	0	3	5
4	0	2	15	0

F If the number of goals were equally distributed among all the players, each player would have scored 3 goals.

G Half the players scored more than 3 goals, and half scored fewer than 3 goals.

H Most of the players scored 2 goals.

J The range is 13 goals.

CHECK Your Understanding

Examples 1, 2 (pp. 108, 109)

Find the median, mode, and range for each set of data.

1. Points scored by football team: 15, 20, 23, 13, 17, 21, 17
2. Monthly spending: \$46, \$62, \$63, \$57, \$50, \$42, \$56, \$40

Example 3 (p. 109)

Find the mean, median, mode, and range of the data represented.

3. Cost of CDs (dollars)

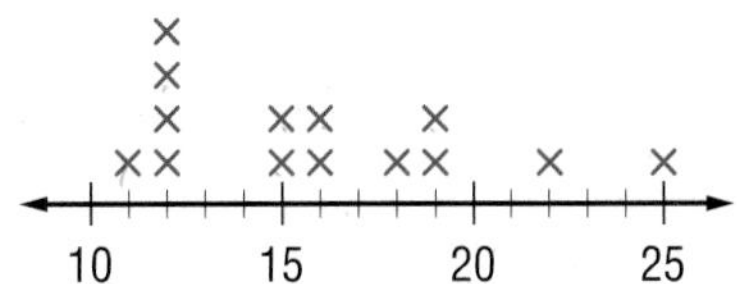

4. 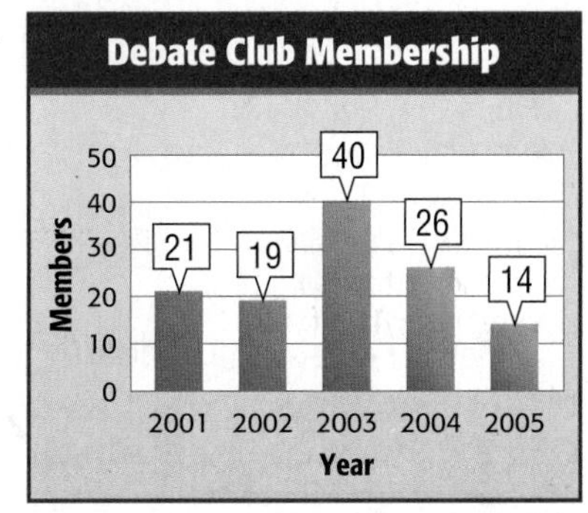

Example 4 (p. 110)

5. **MULTIPLE CHOICE** The lengths of the longest underwater car tunnels are shown. Which statement is supported by the data?

Five Longest Underwater Car Tunnels in U.S.					
State	NY	NY	MA	NY	VA
Length (ft)	8,220	8,560	8,450	9,120	8,190

Source: *The World Almanac*

A If the lengths of the tunnels were distributed equally among all 5 tunnels, each would measure 8,450 feet in length.

B There is no tunnel length that occurs more often than another.

C The lengths of the tunnels have a range of 900 feet.

D The majority of the tunnels are greater than 8,500 feet in length.

Practice and Problem Solving

HOMEWORK HELP	
For Exercises	**See Examples**
6–9	1, 2
10–13	3
22, 23	4

Find the median, mode, and range for each set of data.

6. Age of employees: 23, 22, 15, 36, 44
7. Minutes spent on homework: 18, 20, 22, 11, 19, 18, 18
8. Math test scores: 97, 85, 92, 86
9. Height of trees in feet: 23, 27, 24, 26, 26, 24, 26, 24

ANALYZE DISPLAYS For Exercises 10 and 11, find the mean, median, mode, and range of the data represented.

10. Average Speeds (mph)

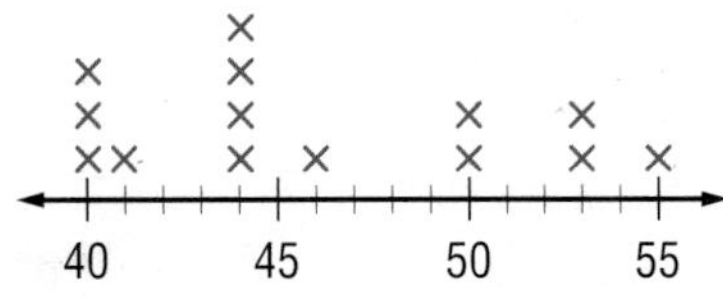

11. **Test Grades**

Stem	Leaf
6	5 7
7	0 2 2 5 7
8	0 2 2 5 5 5 5 7 8
9	0 0 2 5 5 7
10	0 0 *8|2 = 82%*

ANALYZE DISPLAYS For Exercises 12 and 13, find the mean, median, mode, and range of the data represented.

12.

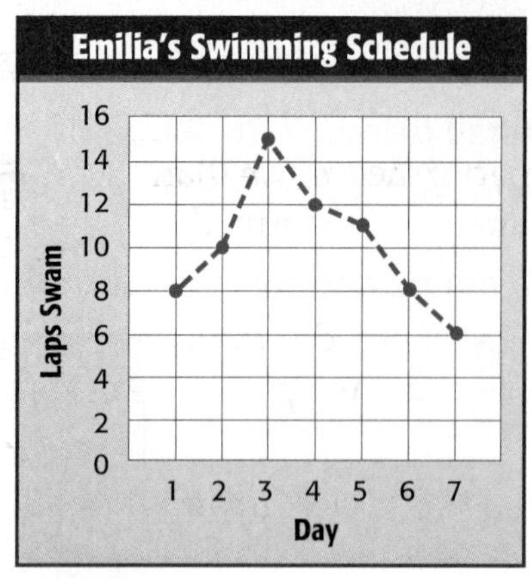

13.

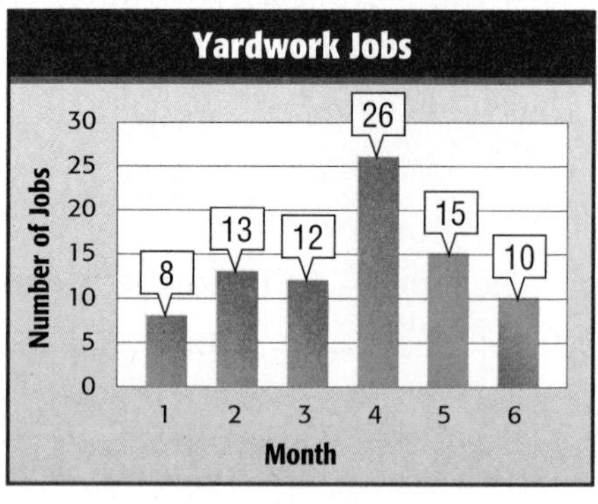

14. **MUSIC** Marjorie's friends bought CDs for \$12, \$14, \$18, \$10, \$14, \$12, \$12, and \$12. Which measure, mean, median, or mode, best describes the cost of the CDs? Explain your reasoning.

15. **ANALYZE TABLES** A Louisville newspaper claims that during seven days, the high temperature in Lexington was typically 6° warmer than the high temperature in Louisville. What measure was used to make this claim? Justify your answer.

Daily High Temperatures (°F)	
Louisville	**Lexington**
75 50 80 72 70 84 70	80 73 75 74 71 76 76

EXTRA PRACTICE
See pages 677, 707.

16. **FIND THE DATA** Refer to the Data File on pages 16–19. Choose some data that is best described by its median value. Explain your reasoning.

H.O.T. Problems

17. **COLLECT THE DATA** Record the number of students in your math class each day for one week. Then describe the data using the mean, median, and mode.

18. **CHALLENGE** The ticket prices in a concert series were \$12, \$37, \$45, \$18, \$8, \$25, and \$18. What was the ticket price of the eighth and final concert in this series if the set of 8 prices had a mean of \$23, a mode of \$18, a median of \$19.50, and a range of \$37?

REASONING One evening at a local pizzeria, the following number of toppings were ordered on each large pizza.

3, 0, 1, 1, 2, 5, 4, 3, 1, 0, 0, 1, 1, 2, 2, 3, 6, 4, 3, 2, 0, 2, 1, 3

Determine whether each statement is *true* or *false*. Explain your reasoning.

19. The most number of people ordered a pizza with 1 topping.

20. Half the customers ordered pizzas with more than 3 toppings, and half the customers ordered pizzas with less than 3 toppings.

21. **WRITING IN MATH** In the data set {3, 7, 4, 2, 31, 5, 4}, which measure: mean, median, or mode, best describes the set of data? Explain your reasoning.

TEST PRACTICE

22. The table shows the number of concerts performed by The Quest. Which statement is supported by the data in the table?

The Quest	
Year	**Number of Concerts**
2003	142
2004	142
2005	136
2006	136
2007	124
2008	138
2009	136
2010	150

   A Half of the years The Quest performed more than 142 concerts, and half the years they performed fewer than 142 concerts.

   B If the number of concerts were equally distributed among each year, The Quest would have performed 136 concerts each year.

   C The number of concerts performed varies greatly from year to year.

   D The most common number of concerts The Quest performed in one year is 136.

23. **SHORT RESPONSE** At the Town Diner, Aiden was deciding on the turkey dinner for \$9, the cheeseburger meal for \$6, the chicken salad for \$5, or the spaghetti with meatballs for \$8. What was the range of prices in dollars for the meals he was considering?

Spiral Review

24. **CELL PHONE** Find the mean number of cell phone minutes Samuel used each month this year. (Lesson 2-6)

Month	Jan.	Feb.	Mar.	Apr.	May	June	July	Aug.	Sep.	Oct.	Nov.	Dec.
Minutes Used	49	65	20	37	55	68	75	50	24	37	42	30

25. Display the following set of data in a line plot. (Lesson 2-5)
    Number of miles biked: 27, 31, 25, 19, 31, 32, 25, 26, 33, 31

Evaluate each expression if $x = 3$, $y = 12$, and $z = 8$. (Lesson 1-5)

26. xyz

27. $2x + z^2$

28. $(2z)^2 + 3x^2 - y$

GET READY for the Next Lesson

PREREQUISITE SKILL For Exercises 29–31, use the graph. (Lesson 2-2)

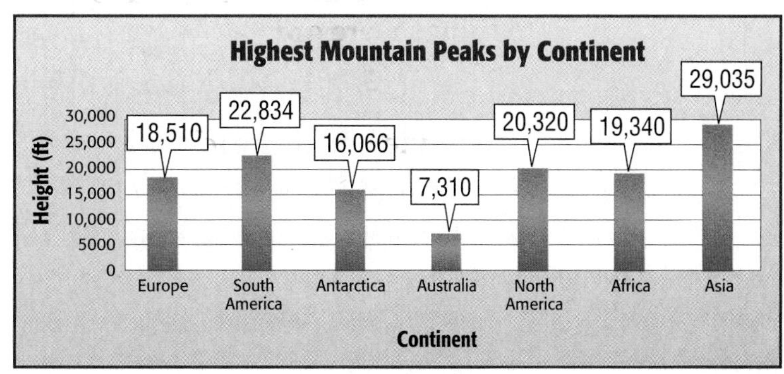

29. Which continent has the highest mountain peak?

30. Compare the highest peak in Antarctica to the highest peak in Australia.

31. About how much taller is the highest peak in Asia than the highest peak in Africa?

2-8 Selecting an Appropriate Display

MAIN IDEA

Select an appropriate display for a set of data.

Math Online

glencoe.com

- Extra Examples
- Personal Tutor
- Self-Check Quiz

GET READY for the Lesson

ANIMALS The displays show the maximum speed of eight animals.

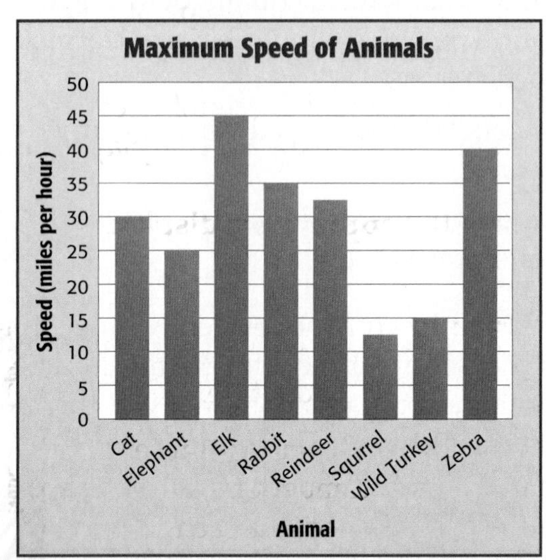

Source: *The World Almanac*

Maximum Speed of Animals

Stem	Leaf
1	2 5
2	5
3	0 2 5
4	0 5 *3|0 = 30 mi/h*

1. Which display allows you to find a rabbit's maximum speed?
2. In which display is it easier to find the range of the data?

Data can often be displayed in several different ways. The display you choose depends on your data and what you want to show.

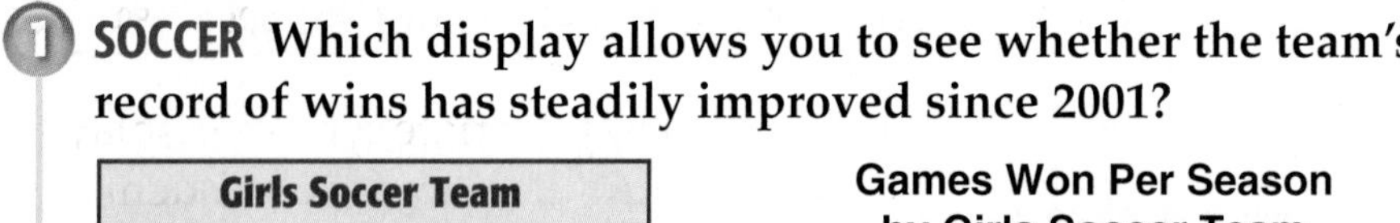

EXAMPLE Choose Between Displays

1 **SOCCER Which display allows you to see whether the team's record of wins has steadily improved since 2001?**

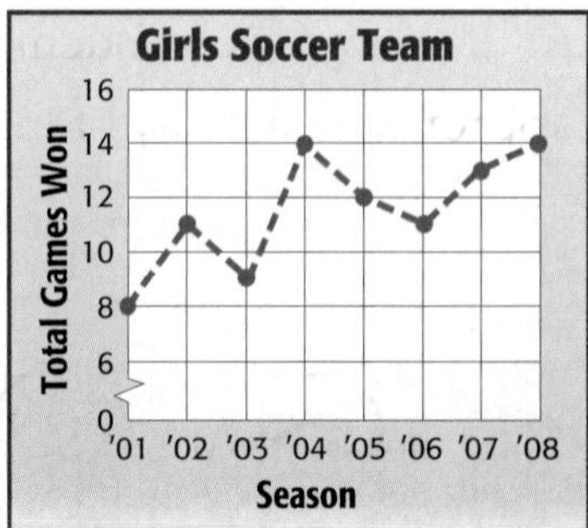

Games Won Per Season by Girls Soccer Team

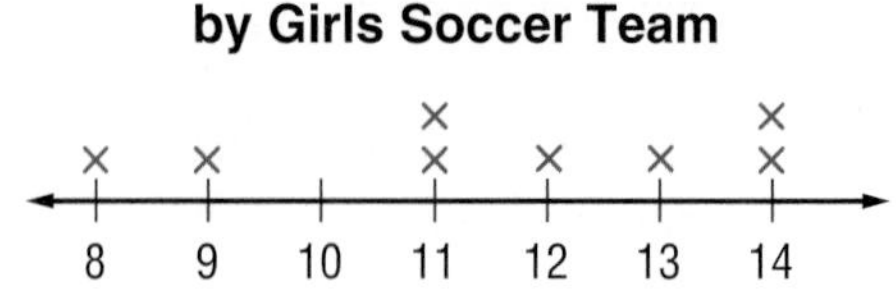

The line graph shows the change in games won from season to season, revealing some declines in the number of wins.

CHECK Your Progress

a. **SOCCER** Which of the above displays allows you to easily find the number of seasons in which the team won 11 or more games?

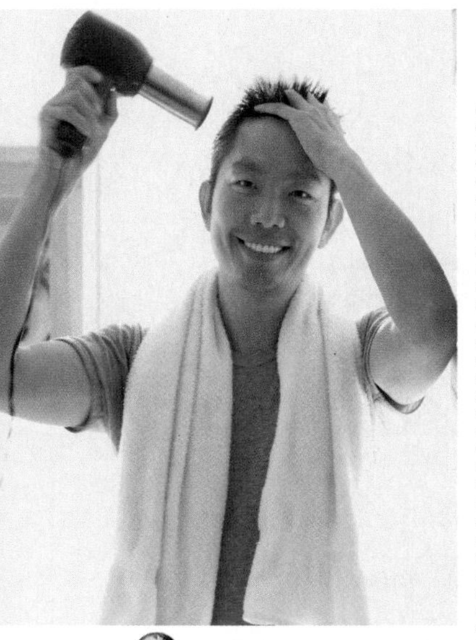

Real-World Link Men in the United States spend more than $4 billion per year on grooming products.

Source: Skin Care Daily

Real-World EXAMPLES

2 **MARKETING** **A market researcher conducted a survey to compare different brands of hair shampoo. The table shows the number of first-choice responses for each brand.**

Favorite Shampoo Survey			
Brand	Responses	Brand	Responses
A	35	D	24
B	12	E	8
C	42	F	11

Select an appropriate type of display to compare the number of responses for each brand of shampoo.

These data show the number of responses for each brand. A bar graph would be the best display to compare the responses.

3 **Make the appropriate display of the data.**

Step 1 Draw and label horizontal and vertical axes. Add a title.

Step 2 Draw a bar to represent the number of responses for each brand.

CHECK Your Progress

GRADES **The table shows the quiz scores of Mr. Vincent's sixth-grade math class.**

Sixth-Grade Math Quiz Scores											
70	70	75	80	100	85	85	65	75	85	95	80
90	100	85	90	90	95	80	85	90	85	90	75

b. Select an appropriate type of display to allow Mr. Vincent to easily count the number of students with a score of 85.

c. Make the appropriate display of the data.

Statistical Displays

Concept Summary

Type of Display	Best Used to
Bar Graph	show the number of items in specific categories
Line Graph	show change over a period of time
Line Plot	show how many times each number occurs in the data
Stem-and-Leaf Plot	list all individual numerical data in a condensed form

CHECK Your Understanding

Example 1 (p. 114)

1. **FARMING** Which display makes it easier to determine the highest average price received for a hog from 1940 to 2000?

Average Hog Price, 1940–2000

Stem	Leaf
0	5
1	5 8
2	3
3	8
4	2
5	4 *4|2 = $42 per 100 lb*

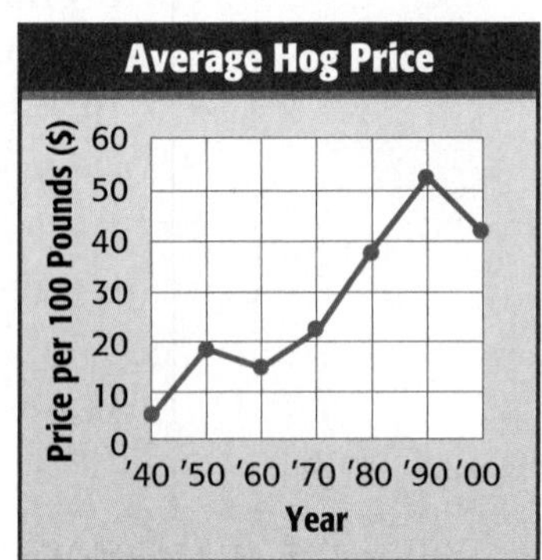

Example 2 (p. 115)

Select an appropriate type of display for data gathered about each situation.

2. the favorite cafeteria lunch item of the sixth-grade students
3. the daily high temperature over the past seven days

Example 3 (p. 115)

4. **FITNESS** Select and make an appropriate display for the following data.

Number of Push-Ups Done by Each Student											
15	20	8	11	6	25	32	12	14	16	21	25
18	35	40	20	25	15	10	5	18	20	31	28

Practice and Problem Solving

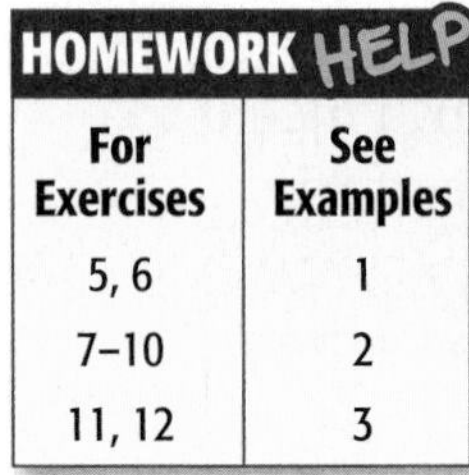

HOMEWORK HELP

For Exercises	See Examples
5, 6	1
7–10	2
11, 12	3

5. **ROLLER COASTERS** Which display makes it easier to compare the maximum speeds of Top Thrill Dragster and Millennium Force?

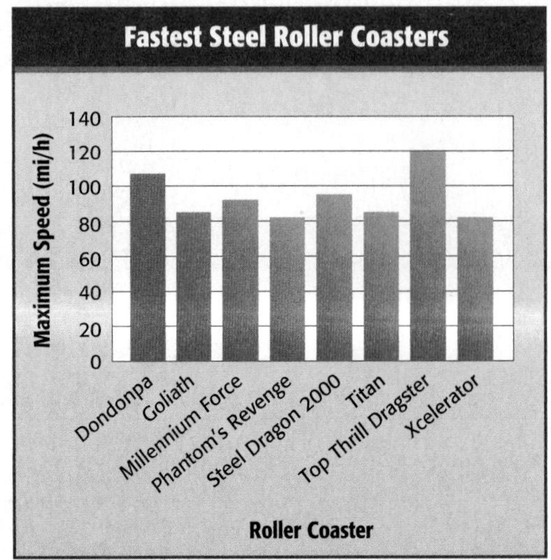

Maximum Speed of Fastest Steel Roller Coasters

Stem	Leaf
8	2 2 5 5
9	2 5
10	7
11	
12	0 *10|7 = 107 mi/h*

6. **OLYMPICS** Which display makes it easier to see how many times the winning distance of the javelin throw was 90 meters?

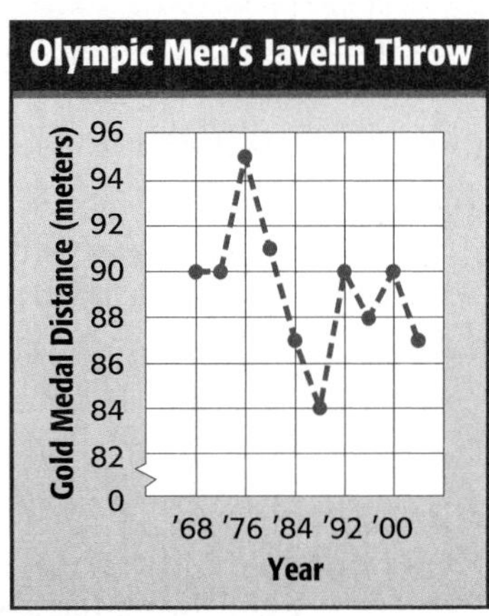

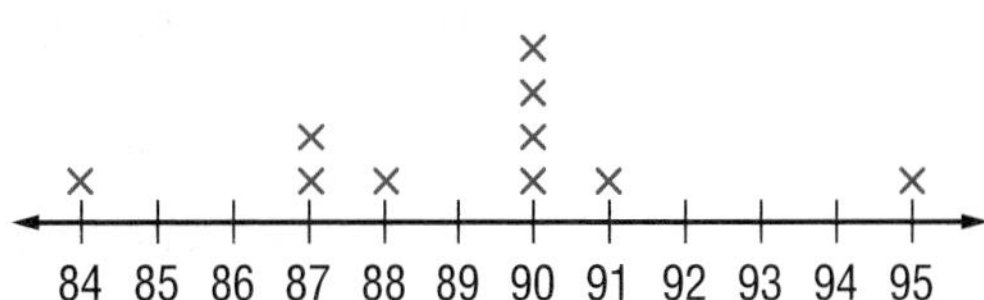

Select an appropriate type of display for data gathered about each situation.

7. the amount of sales a company has for each of the past 6 months
8. the test scores each student had on a language arts test
9. the prices of five different brands of tennis shoes at an athletic store
10. Abigail's height on her birthday over the past 10 years

Select and make an appropriate type of display for each situation.

Real-World Link ... The 14 neighbors of China are Afghanistan, Bhutan, India, Kazakhstan, Kyrgyzstan, Laos, Mongolia, Myanmar, Nepal, North Korea, Pakistan, Russia, Tajikistan, and Vietnam.

11.

Latin American Country	Year of Independence
Argentina	1816
Bolivia	1825
Chile	1818
Colombia	1819
Ecuador	1822
México	1821
Peru	1824
Venezuela	1821

Source: *The World Almanac*

12.

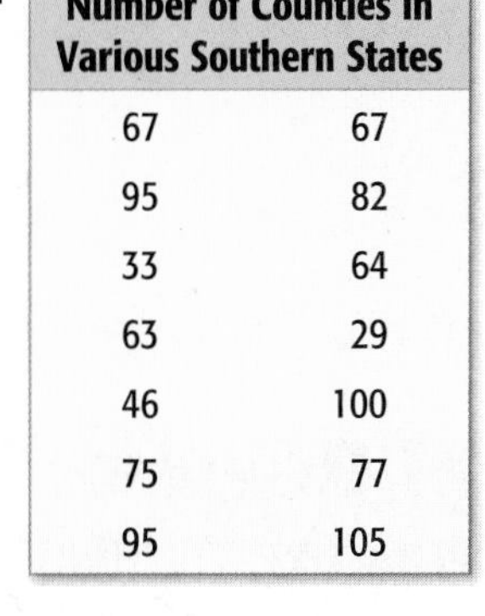

Number of Counties in Various Southern States	
67	67
95	82
33	64
63	29
46	100
75	77
95	105

13. **GEOGRAPHY** Display the data in the bar graph using another type of display. Compare the advantages of each display.

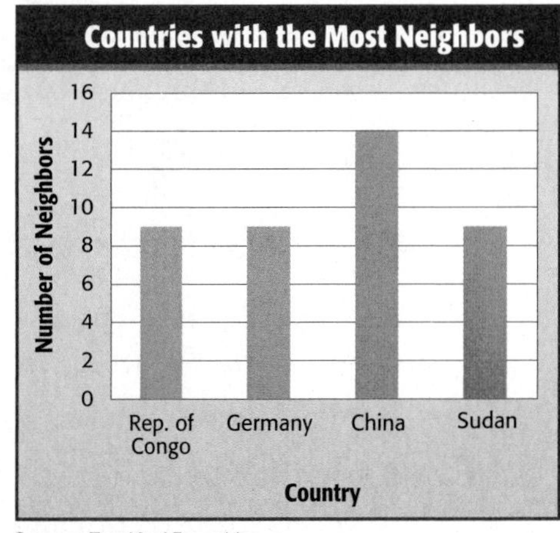

Source: *Top 10 of Everything*

14. **RESEARCH** Use the Internet or another source to find a set of data that is displayed in a bar graph, line graph, stem-and-leaf plot, or line plot. Was the most appropriate type of display used? What other ways might these same data be displayed?

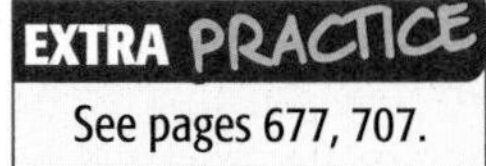

H.O.T. Problems

15. **REASONING** Determine whether the following statement is *true* or *false*. If true, explain your reasoning. If false, give a counterexample.

    Any set of data can be displayed using a line graph.

16. **CHALLENGE** Which type of display allows you to easily find the mode of the data? Explain your reasoning.

17. **WRITING IN MATH** Write about a real-world situation in which you would have to choose an appropriate display for a set of data.

TEST PRACTICE

18. Which of the following situations would involve data that are best displayed in a line graph?

    A the favorite subject of the students in Mrs. Ling's homeroom

    B a company's yearly profit over the past 10 years

    C the number of hits Dylan got in each game this baseball season

    D the number of miles each student travels to school

19. The table shows the prices of the skateboards Jacy might buy.

Skateboards	
Brand	**Price**
Blackbird	$55
Earth Bound	$68
Element Skateboards	$44
Venus Boards	$61
ZoomFast	$75

Which type of display would help Jacy best compare the prices of these skateboards?

F bar graph

G line graph

H line plot

J stem-and-leaf plot

Spiral Review

20. **TEMPERATURE** The daily high temperatures last week in Charleston, South Carolina, were 68°, 70°, 73°, 75°, 76°, 76°, and 82°. Find the median, mode, and range of these temperatures. (Lesson 2-7)

Find the mean for each set of data. (Lesson 2-6)

21. Length of television program (in minutes): 25, 20, 19, 22, 24, 28

22. Ages of cousins (in years): 25, 14, 21, 16, 19

GET READY for the Next Lesson

PREREQUISITE SKILL For Exercises 23 and 24, refer to the table showing the time Mia spent studying. (Lesson 2-2)

23. Make a line graph of the data.

24. Describe the change in the number of hours Mia studied from Wednesday to Thursday.

Day	Hours
Monday	1
Tuesday	2
Wednesday	2
Thursday	1
Friday	4

Extend 2-8

Statistics Lab
Collecting Data to Solve a Problem

MAIN IDEA

Solve a problem by collecting, organizing, displaying, and interpreting data.

In this lab, you will collect, organize, display, and interpret data in order to solve a problem.

ACTIVITY

1 *Rating scales,* like the one below, are often used on surveys to find out about people's opinions. Participants indicate how strongly they agree or disagree with a specific statement.

5	4	3	2	1
Strongly Agree	Agree	No Opinion	Disagree	Strongly Disagree

Consider each of the following topics.

- yearbooks in interactive DVD format
- voting age for presidential election
- school uniforms
- fast food in school

STEP 1 Make a data collection plan.

- Choose one of the topics above and a group to survey.
- Write one or more survey questions that include the rating scale shown above to determine student opinion on this topic.
- Identify an audience for your results.

STEP 2 Collect the data.

- Conduct your survey and record the results.
- Collect responses from at least 10 people in the population you chose.
- Record your results in a frequency table.

STEP 3 Create a display of the data.

Choose an appropriate type of display and scale for your data. Then create an accurate display.

Analyze the Results

1. What are the mean, median, mode, and range of your data?
2. Use your display to describe the distribution of your data.
3. How would you summarize the opinions of those you surveyed? Include only those statements that are clearly supported by the data.
4. Based on your analysis, what course of action would you recommend to the group interested in your data?
5. Present your findings and recommendation to the whole class. Include poster-size versions of both your displays and a written report of your data analysis.
6. **MAKE A CONJECTURE** What other factors might influence the results of your survey?

Study Tip

Bias Survey questions that favor a particular answer are biased. Be sure your survey question is worded as to avoid bias.

ACTIVITY

2 A *log* is an organized list that contains a record of events over a specified amount of time. Consider each of the following topics.

- the amount of time you spend watching television each day
- the outside air temperature over a period of 2, 4, 6, or 8 hours

STEP 1 **Make a data collection plan.**

- Choose one of the topics above and a reasonable period of time over which to collect the data.
- Create a log that you can use to collect the data.

STEP 2 **Collect the data and create an appropriate display.**

Record the necessary data in your log. Then choose and create an appropriate display for the data.

Analyze the Results

7. What type of display did you choose?
8. Describe the change in your data over the time period you chose.
9. If possible, use your display to make a prediction about future data. Explain your reasoning. If not possible, explain why not.

2-9 Integers and Graphing

MAIN IDEA

Use integers to represent real-world data.

New Vocabulary

negative numbers
positive numbers
opposites
integers
graph

Math Online

glencoe.com

- Extra Examples
- Personal Tutor
- Self-Check Quiz

GET READY for the Lesson

MONEY The bar graph shows the amount of money remaining in the clothing budgets of four students at the end of one month. A value of −$4 means that someone overspent the budget and owes his or her parents 4 dollars.

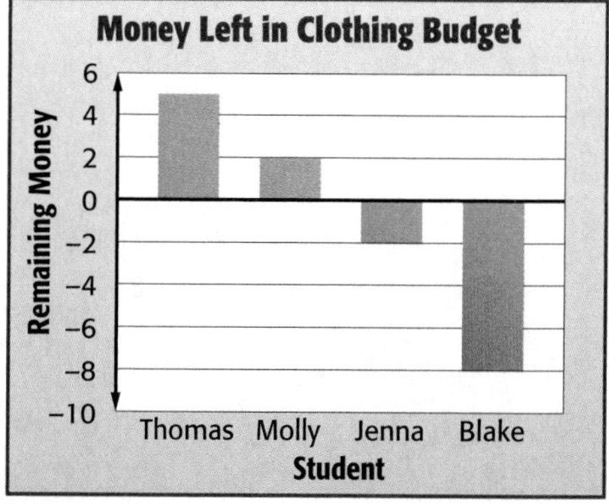

1. What number represents owing 5 dollars? What number represents having 8 dollars left?
2. Who has the most money left? Who owes the most?

To represent data that are less than a 0, you can use **negative numbers**. A negative number is written with a − sign. Data that are greater than zero are represented by **positive numbers**.

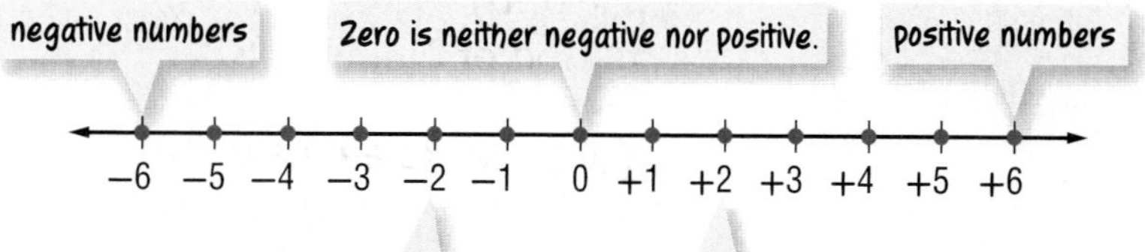

Positive whole numbers, their opposites, and zero are called **integers**.

EXAMPLES Use Integers to Represent Data

Write an integer to represent each piece of data.

1 Marjorie deposited $48 into her savings account.

The word *deposited* represents an increase. The integer is 48 or +48.

2 A scuba diver swam to a depth of 8 meters below sea level.

The word *below* here means *less than*. The integer is −8.

CHECK Your Progress

a. The team gained 12 points.
b. The temperature dropped 32°.

To **graph** a number on a number line, draw a dot at the location on the number line that corresponds to the number.

EXAMPLE Graph an Integer on a Number Line

3 Graph −7 on a number line.

Draw a number line. Then draw a dot at the location that represents −7.

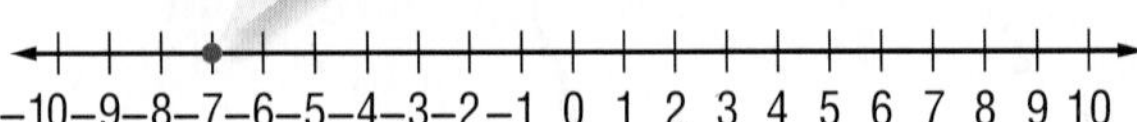

 **CHECK Your Progress**

Graph each integer on a number line.

c. −1 d. 4 e. 0

You can create line plots of data that involve positive and negative integers.

 **Real-World EXAMPLE**

4 GOLF The table shows the scores, in relation to par, of the top 25 leaders of a recent Masters Tournament in Augusta, Georgia. Make a line plot of the data.

Masters Tournament, Top 25 Scores (in relation to par)				
0	+1	0	−3	+2
−1	−4	−2	−2	−3
−1	0	−1	−1	−1
+5	−3	−1	0	−6
−1	0	−1	+3	−2

Source: Professional Golf Association

Draw a number line. You can use a scale of −10 to 10 because it contains all the data. Put an × above the number that represents each score in the table.

Real-World Link
Between 1991 and 2006, Phil Mickelson has played 54 rounds. Thirty-one of those rounds have been under par.
Source: Augusta National, Inc.

Masters Tournament, Top 25 Scores

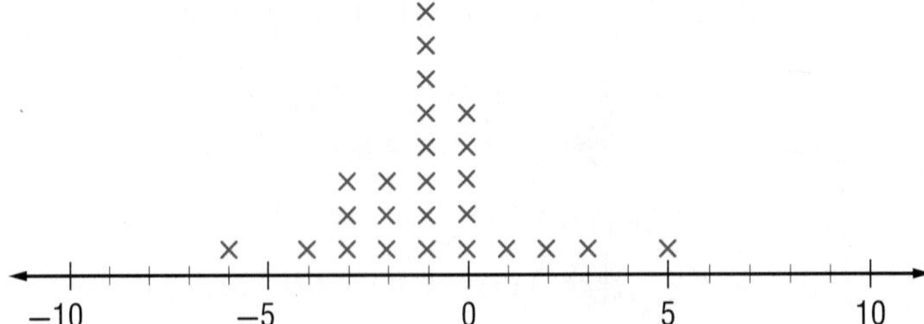

CHECK Your Progress

f. **MOVIES** The table shows the change in ranking from the previous week of the top ten movies. Make a line plot of the data.

Movie	A	B	C	D	E	F	G	H	I	J
Change in Ranking	−2	−3	+1	−1	−1	+2	−1	−1	0	−3

CHECK Your Understanding

Examples 1, 2 (p. 121)

Write an integer to represent each piece of data.

1. Mei's dog gained 3 pounds.
2. Kenji withdrew $15 from his savings account

Example 3 (p. 122)

Graph each integer on a number line.

3. −7
4. 10
5. −4
6. 3

Example 4 (p. 122)

7. **GAMES** The table at the right shows the number of points Delaney scored on each hand of a card game. Make a line plot of the data.

Points Scored			
+25	0	−15	+30
−5	+25	0	−15
+25	+20	−5	+20

Practice and Problem Solving

HOMEWORK HELP	
For Exercises	See Examples
8–11	1, 2
12–19	3
20, 21	4

Write an integer to represent each piece of data.

8. Wesley swam 5 feet below sea level.
9. Toya deposited $30 into her savings account.
10. Edmund moved ahead 4 spaces on the game board.
11. The stock market lost 6 points on Thursday.

Graph each integer on a number line.

12. 1
13. 5
14. 3
15. 7
16. −5
17. −10
18. −6
19. −8

Make a line plot of the data represented in each table.

20.

Elevations of Valleys (in reference to sea level)			
−50	25	20	−45
−15	20	0	−5
10	5	−15	25
20	−50	−15	0
20	25	−45	0

21.

Golf Scores (in reference to par)			
−1	0	5	3
−5	0	4	−6
−5	1	2	5
7	−3	−2	0
1	2	5	−2

Write the opposite of each integer.

22. −8
23. −3
24. 19
25. 42

26. **GEOGRAPHY** Death Valley, California, is 282 feet below sea level. Represent this altitude as an integer.

27. **SCIENCE** Water boils at 212°F. Represent this temperature as an integer.

Real-World Link Blue marlins can live up to 15 years.

Source: ESPN

28. **ANIMALS** Some sea creatures live near the surface while others live in the depths of the ocean. Make a drawing showing the relative habitats of the following creatures.
    - blue marlin: 0 to 600 feet below the surface
    - lantern fish: 3,300 to 13,200 feet below the surface
    - ribbon fish: 600 to 3,300 feet below the surface

29. **WEATHER** The table below shows the record low temperatures for select states. Make a line plot of the data. Then explain how the line plot allows you to determine the most common record low temperature among the given states.

Record Low Temperature by State											
State	AZ	DE	GA	KS	KY	MD	MS	MO	NC	RI	SC
Temp (°F)	−40	−17	−17	−40	−37	−40	−19	−40	−34	−23	−19

30. **VIDEO GAMES** The table shows the number of points earned for each action in a video game. While playing the video game, Kevin fell in the water 3 times, found 4 gemstones, jumped over 2 rocks, climbed over a mountain, touched a cactus 4 times, and saved the princess. Make a line plot of the points Kevin scored. Then explain how the line plot can be used to determine if Kevin gained points more times than he lost points.

Action	Points	Action	Points
fall in water	−10	find gemstone	+25
walk over bridge	+5	jump over rock	+5
step on snake	−10	walk through quicksand	−15
climb mountain	+10	touch cactus	−15
fall off cliff	−10	save princess	+50

EXTRA PRACTICE See pages 677, 707.

H.O.T. Problems

31. **OPEN ENDED** Write about a time that you used integers to describe a real-world situation.

32. **REASONING** Determine whether the following statement is *true* or *false*. Explain your reasoning.

    The number $-2\frac{1}{4}$ *is an integer.*

33. **CHALLENGE** Two numbers, *a* and *b*, are opposites. If *a* is 4 units to the left of −6 on a number line, what is the value of *b*?

34. **REASONING** The temperature outside is 15°F. If the temperature drops 20°, will the outside temperature be represented by a positive or negative integer? Explain your reasoning.

35. **WRITING IN MATH** Describe the characteristics of each set of numbers that make up the set of integers.

TEST PRACTICE

36. The record low temperature for New Mexico is 50 degrees below zero Fahrenheit. The record low temperature for Hawaii is 12 degrees above zero Fahrenheit. What integer represents the record low temperature for New Mexico?

    A 50

    B 38

    C −38

    D −50

37. On Friday, a school spirit shop gave away a free T-shirt with each purchase over \$50. There were 47 purchases over \$50. Which integer represents the change in the number of free T-shirts the spirit shop had in stock at the end of the day on Friday?

    F −50

    G −47

    H 47

    J 50

Spiral Review

38. **GRADES** The table shows the test grades of Mrs. Owens' science class. Select an appropriate type of statistical display for the data. Then make a display. (Lesson 2-8)

Science Test Scores				
83	90	74	95	80
75	92	88	85	81
98	100	70	78	85
85	90	92	76	88

Find the median, mode, and range for each set of data . (Lesson 2-7)

39. Height of students (in inches): 55, 60, 57, 55, 58, 55, 56, 60, 59

40. Points scored in football games: 14, 24, 7, 21, 21, 14, 21, 35

Solve each equation mentally. (Lesson 1-8)

41. $16 + h = 24$

42. $15 = 50 - m$

43. $14t = 42$

44. **PLANES** The expression $d \div t$, where d represents distance and t represents time, can be used to find the speed of a plane. Find the speed of a plane that travels 3,636 miles in 9 hours. (Lesson 1-5)

45. **GEOGRAPHY** New Hampshire has a total land area of about 10^4 square miles. What is the value of 10^4? (Lesson 1-3)

Problem Solving in Health

Fun to be Fit! It's time to complete your project. Use the information you have gathered about different forms of exercise to prepare a booklet, poster, or slide presentation using the computer. Be sure to include all the required statistical displays.

Math Online **Unit Project at** glencoe.com

CHAPTER 2 Study Guide and Review

Math Online glencoe.com
• STUDY TO GO
• Vocabulary Review

GET READY to Study

Be sure the following Big Ideas are noted in your Foldable.

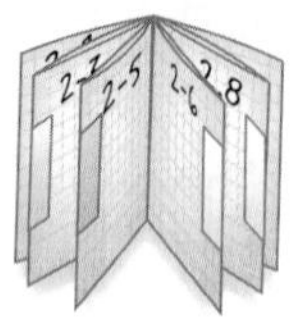

BIG Ideas

Mean, Median, Mode, and Range
(Lessons 2-6, 2-7)

- The mean of a set of data is the sum of the data divided by the number of items in the data set.
- The median of a set of data is the middle number of the ordered data, or the mean of the middle two numbers.
- The mode of a set of data is the number or numbers that occur most often.
- The range of a set of data is the difference between the greatest and the least values of the set.

Appropriate Graphs (Lessons 2-1 and 2-5)

- A bar graph is best used to show the number of items in specific categories.
- A line graph is best used to show change over a period of time.
- A line plot is best used to show how many times each number occurs in the data.
- A stem-and-leaf plot is best used when you want to list all individual numerical data in a condensed form.

Integers (Lesson 2-9)

- Positive numbers are greater than zero, while negative numbers are less than zero.
- The set of positive whole numbers, their opposites, and zero are called integers.

Key Vocabulary

average (p. 102)
bar graph (p. 81)
frequency (p. 81)
graph (p. 81)
horizontal axis (p. 81)
integers (p. 121)
key (p. 92)
line graph (p. 82)
line plot (p. 96)
mean (p. 102)
measures of central tendency (p. 108)
median (p. 108)
mode (p. 108)
negative numbers (p. 121)
opposites (p. 121)
outlier (p. 103)
positive numbers (p. 121)
range (p. 109)
stem-and-leaf plot (p. 92)
vertical axis (p. 81)

Vocabulary Check

State whether each sentence is *true* or *false*. If *false*, replace the underlined word or number to make a true sentence.

1. The sum of a set of data divided by the number of pieces of data is called the median.
2. The mean of a data set is also called the average.
3. The opposite of 4 is −4.
4. The middle number of a set of data is called the outlier.
5. Negative numbers can be written with or without a + sign.
6. The range of a stem-and-leaf plot explains the meaning of the stems and leaves.
7. A stem-and-leaf plot shows how data changes over time.
8. The number of times an event or item occurs is called the mode.

Lesson-by-Lesson Review

2-1 PSI: Make a Table (pp. 78–79)

9. **INTERNET** Make a frequency table of the data below. How many more students spent 2 hours on the Internet than 3 hours?

Number of Hours on the Internet								
0	2	3	1	0	2	2	2	1
2	1	3	4	2	3	0	1	2

10. **ADVERTISING** The table shows the number of ads displayed on the clothing of the students surveyed. Make a frequency table of the data. How many wore 3 or more ads?

Number of Ads Worn								
1	0	5	0	4	3	0	2	0
0	1	0	3	2	0	1	5	1

Example 1 Make a frequency table of the data below. How many more students received a B than an A?

Class Test Grades			
A	C	A	B
C	A	C	B
B	B	B	B

Draw a table with three columns.

Class Test Grades		
Grade	Tally	Number
A	III	3
B	𝍸 I	6
C	III	3

So, 6 – 3 or 3 more students received a B than did an A on the test.

2-2 Bar Graphs and Line Graphs (pp. 81–85)

11. **PRODUCE** In a recent year, the U.S. produced 44 million tons of bananas, 60 million tons of tomatoes, 36 million tons of apples, 34 million tons of oranges, and 22 million tons of watermelon. Display this data in a bar graph. Which item was produced about half as many times as bananas?

12. **DRIVING** Make a line graph of the data. Describe the change in the total distance traveled from Day 1 to Day 5.

Total Distance Driven on Vacation	
Day	Distance (mi)
1	465
2	618
3	657
4	783
5	1,185

Example 2 The U.S. states with the five highest birth rates per 1,000 people are shown on the bar graph below. Compare the birth rates in California and Arizona.

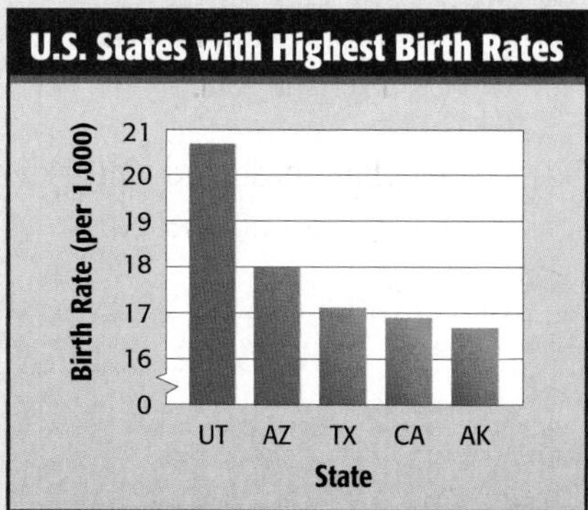

Arizona has about 1 more birth per 1,000 people than California.

2-3 Interpret Line Graphs (pp. 88–91)

CLIMBING For Exercises 13 and 14, refer to the graph.

Mountain Climber's Descent

Height (ft): 0, 200, 400, 600, 800, 1000, 1200, 1400

Time (min): 10, 20, 30, 40, 50, 60

13. Describe the trend in the mountain climber's height.

14. Predict the climber's height at 90 minutes.

SCHOOL DANCE For Exercises 15 and 16, refer to the graph.

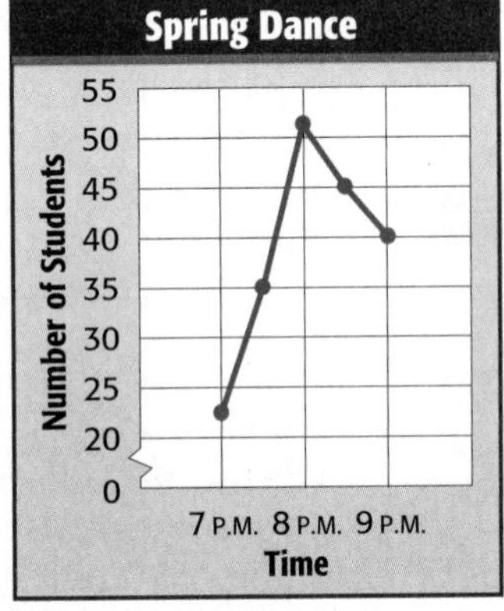

15. Describe the change in the number of students at the spring dance from 7:30 P.M. to 9 P.M.

16. Predict the number of students still at the dance when it ended at 9:30 P.M.

Example 3 Refer to the graph. Predict the number of community theater tickets that will be sold in 2010.

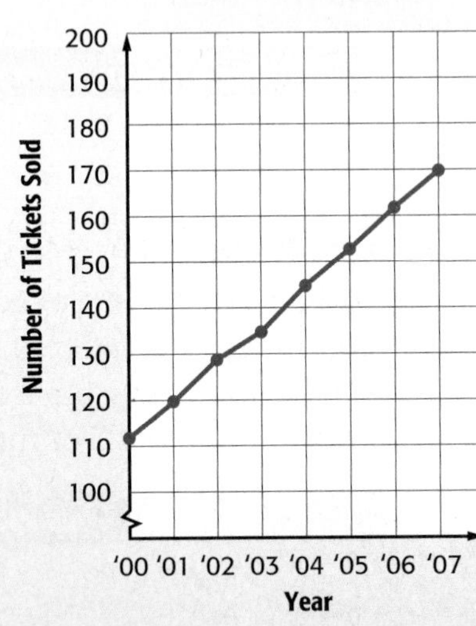

By extending the graph, it appears that about 195 tickets will be sold in 2010.

2-4 Stem-and-Leaf Plots (pp. 92–95)

Make a stem-and-leaf plot of each set of data.

17. Days left until vacation: 20, 8, 43, 39, 10, 47, 2, 27, 27, 39, 40

18. Length of last book read (pages): 224, 238, 235, 228, 126, 224, 252

19. **AGES** The ages of visitors to the Big Bend National Park one afternoon were 18, 35, 27, 56, 19, 22, 41, 28, 31, and 29. Make a stem-and-leaf plot of this set of data.

Example 4 Make a stem-and-leaf plot of the set of data on the number of U.S. states students in Mrs. Hamilton's class have visited: 3, 14, 22, 8, 11, 29, 6, 17, 31, 45, 10, 6, 7, 15, 21, 30, 32, 27, 9, and 21.

Number of U.S. States Visited

Stem	Leaf
0	3 6 6 7 8 9
1	0 1 4 5 7
2	1 1 2 7 9
3	0 1 2
4	5 *1 | 4 = 14 states*

Mixed Problem Solving
For mixed problem-solving practice, see page 707.

2-5 Line Plots (pp. 96–100)

BASKETBALL For Exercises 20–22, use the table of the number of points Patricia scored in each basketball game this season.

Patricia's Points Per Game				
5	8	10	8	6
5	14	9	8	6
8	10	10	8	5

20. Make a line plot of the data.
21. In how many games did Patricia score 8 points?
22. In how many games did Patricia score 10 or more points?

Example 5 The table shows the amount Jack earned for each lawn he mowed. Make a line plot.

Jack's Earnings ($)						
15	20	10	12	10	15	20
25	15	20	15	12	15	25

Draw a number line. Place an × above each value represented in the table.

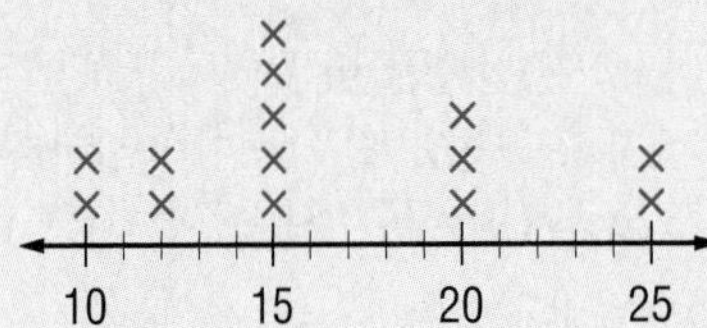

2-6 Mean (pp. 102–106)

Find the mean for each set of data.

23. Ski days remaining: 23, 34, 29, 36, 18, 22, 27
24. Hotel rooms: 103, 110, 98, 104, 110
25. **SPEED** The speeds of six cheetahs were recorded as 68, 72, 74, 72, 71, and 75 miles per hour. What was the mean speed of these animals?

Example 6 Find the mean for the set of data on miles driven: 117, 98, 104, 108, 104, 111.

$$\text{mean} = \frac{117 + 98 + 104 + 108 + 104 + 111}{6}$$
$$= 107 \text{ miles}$$

2-7 Median, Mode, and Range (pp. 108–113)

Find the median, mode, and range for each set of data.

26. Number of students: 21, 23, 27, 30
27. Ages of aunts: 36, 42, 48, 36, 82
28. **HOMEWORK** The minutes spent doing homework for one week were 30, 60, 77, 90, 88, 76, and 90. Find the median, mode, and range of these times.

Example 7 Find the median, mode, and range for the set of data on miles driven: 117, 98, 104, 108, 104, 111.

Order the data:
98, 104, 104, 108, 111, 117

median: $\frac{104 + 108}{2}$ or 106
mode: 104
range: 117 − 98 or 19

2-8 Selecting an Appropriate Display (pp. 114–118)

Select an appropriate type of display for data gathered about each situation.

29. the number of students who have traveled to Arizona, Louisiana, Kansas, Florida, Michigan, and Wisconsin
30. the change in the value of a house over a period of 30 years
31. the times, in seconds, that 40 teenagers each ran one lap around the track
32. the number of times each number turned up when a number cube was rolled ten times

Example 8 **Which display makes it easier to see how the distance of daily walks changed from Monday through Saturday?**

Distance of Daily Walks (mi)

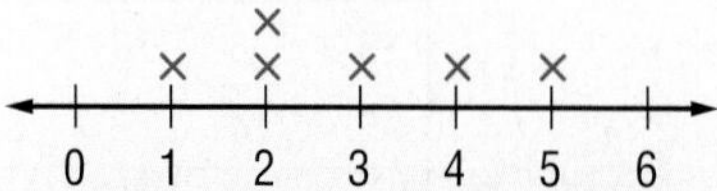

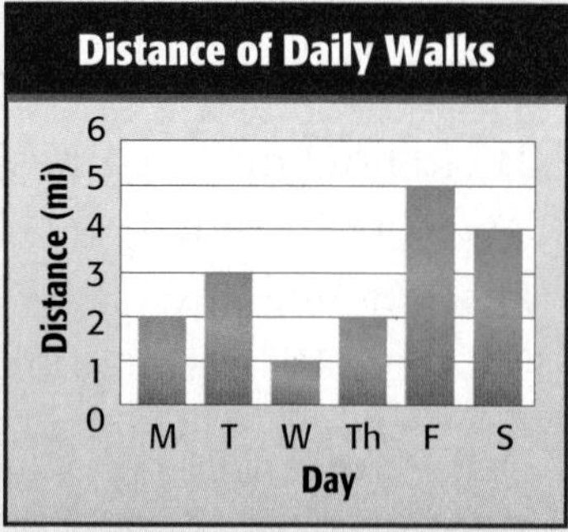

The bar graph shows the decrease and increase in the number of miles each day.

2-9 Integers and Graphing (pp. 121–125)

Write an integer to represent each piece of data.

33. The water retreated 20 feet.
34. An employee receives a holiday bonus of $500.
35. The temperature outside is 7° below zero.

Graph each integer on a number line.

36. -6 37. 9

38. **QUIZZES** On a five-problem quiz, Belinda received the following points on each problem: +5, −3, −2, +5, and −2. Make a line plot of the data.

Example 9 **Write an integer to represent the change in altitude of a plane after it has descended 15,000 feet.**

Since the altitude is decreasing, the integer would be −15,000.

Example 10 **Graph −2 on a number line.**

Draw a number line. Then draw a dot at the location that represents −2.

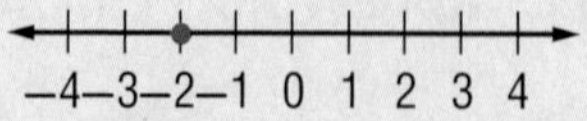

−4 −3 −2 −1 0 1 2 3 4

Chapter 2 Practice Test

Math Online glencoe.com
• Chapter Test

1. **MULTIPLE CHOICE** Mandy and her friends collected the following data during one week.

Which statement is supported by the graph?

A Gabriel watched TV three times as much as Mandy.
B Luke watched about 15 hours of TV.
C Gabriel watched the most TV.
D Ginger watched twice as much TV as Mandy.

PETS For Exercises 2 and 3, use the following information.

The weight of a kitten in ounces for each week since it was born are 4, 7, 9, 13, 17, and 20.

2. Create a line graph of this data.
3. Make a prediction of the kitten's weight at Week 7.

4. **MULTIPLE CHOICE** Julio collected the following data about the number of movies his classmates saw.

Number of Movies Seen at Theater								
0	4	3	2	0	4	5	2	1
4	6	1	3	7	4	8	10	0

Which measure of the data is represented by 10 movies?

F Mean
G Median
H Mode
J Range

FOOTBALL For Exercises 5–9, use data in the table below.

Number of Years the Leading Lifetime NFL Passers Played in NFL				
6	15	15	5	6
5	13	17	6	4
10	15	11	10	9
11	6	8	18	19

Source: *The World Almanac*

5. Make a stem-and-leaf plot of the data.
6. What is the greatest number of years a leading lifetime NFL passer has played?
7. Make a line plot of the data.
8. What is the difference between the greatest and least number of years a leading lifetime NFL passer has played?
9. Write two additional sentences that analyze the data.

Find the mean, median, mode, and range for each set of data.

10. Birthday money each year ($): 67, 68, 103, 65, 80, 54, 53
11. Bowling scores: 232, 200, 242, 242

Select an appropriate type of display for data gathered about each situation.

12. the number of digital cable subscribers each year since 2000
13. the number of pets owned by each student in a classroom

Write an integer to represent each piece of data.

14. He withdrew $75 from an ATM.
15. The snow level rose 4 inches.
16. Graph −3 and its opposite on a number line.

CHAPTER 2

Test Practice

Cumulative, Chapters 1–2

Math Online glencoe.com

• Test Practice

PART 1 Multiple Choice

Read each question. Then fill in the correct answer on the answer sheet provided by your teacher or on a sheet of paper.

1. A shop records the number of specialty shirts sold each month. What is the mean number of T-shirts sold?

T-Shirt Sales	
Month	**Number**
January	75
February	68
March	75
April	92
May	105

A 75

B 83

C 85

D 92

2. A grocery store collected data about ice cream sales over the last six months. Which statement is supported by the graph?

F More than one half of the pints of ice cream were sold in January and March.

G January through March had more sales than April through June.

H During the month of May, 45 pints of ice cream were sold.

J April through June had more sales than January through March.

3. The table shows the sales figures for five colors of automobiles sold at a dealership last year. Which graph most accurately displays the information in the table?

Automobile Sales	
Color	**Number**
Black	163
Blue	145
Red	129
Silver	212
White	205

A

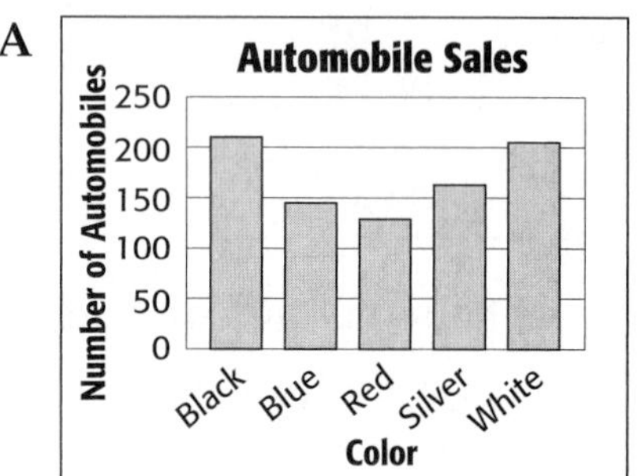

B

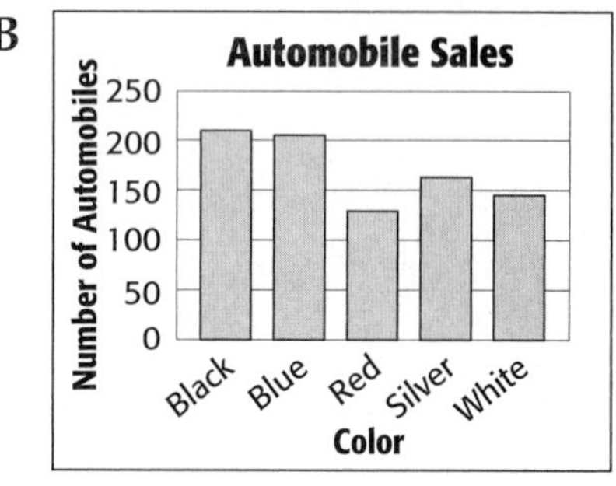

C

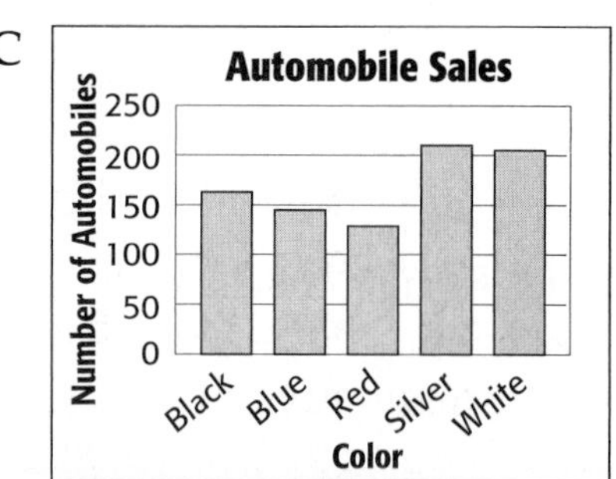

D

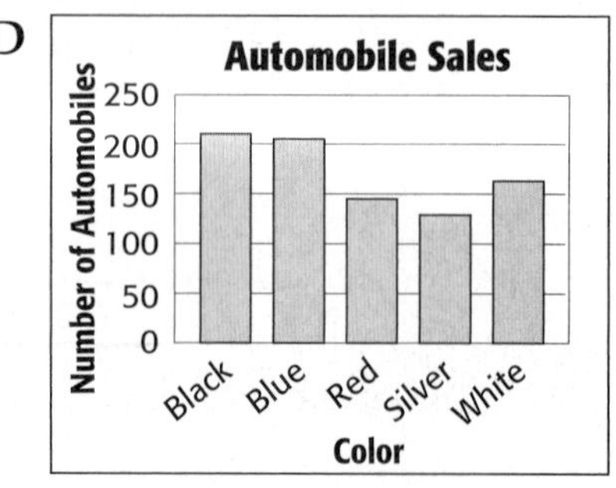

Preparing for Standardized Tests

For test-taking strategies and practice, see pages 718–735.

4. Jeremy's bill at the bakery was \$12.00. He bought 12 egg bagels at \$0.50 each. If an onion bagel costs \$0.75, how can he find how much he spent on onion bagels?
   - F Add \$0.50 and \$0.75.
   - G Subtract the product of 12 and \$0.50 from \$12.00.
   - H Multiply \$0.75 and 12.
   - J Divide 12 by \$0.50.

5. Each square tile in a kitchen measures 1 foot by 1 foot. The kitchen measures 16 feet by 12 feet. How many tiles cover the kitchen floor?
   - A 28
   - B 56
   - C 148
   - D 192

TEST-TAKING TIP

Question 5 Review any terms and formulas that you have learned before you take a test.

6. Find the median of the following ages of people attending a concert: 2, 7, 41, 25, 19, 22, 28, 32, and 24.
   - F 17
   - G 24
   - H 41
   - J 22

7. The weight limit for a small powerboat is 1,800 pounds. Which statement is best supported by this information?
   - A The boat can carry more than 25 adults.
   - B The boat can carry more than 30 children that weigh 75 pounds each.
   - C The boat can carry up to 9 people who each weigh at most 200 pounds.
   - D The boat can carry twice as many children as adults.

PART 2 Short Response/Grid In

Record your answers on the answer sheet provided by your teacher or on a sheet of paper.

8. Sherita bought 2 pairs of jeans for \$64 and 3 equally priced sweaters. She spent a total of \$136, not including tax. Find the price of each sweater in dollars.

9. Mrs. Elkin made 48 pancakes for her students. She has already served 16 students. If each student has exactly 2 pancakes, how many more students does she need to serve?

PART 3 Extended Response

Record your answers on the answer sheet provided by your teacher or on a sheet of paper. Show your work.

10. The graph shows the total number of credit cards owned by four different families in a neighborhood.

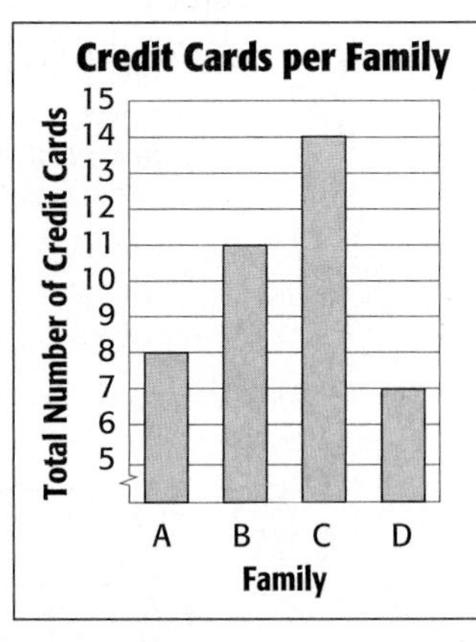

   a. The bar for Family C is three times the bar for Family A. Does this represent the data appropriately? Explain.

   b. How could the graph be changed to be more appropriate?

NEED EXTRA HELP?										
If You Missed Question...	1	2	3	4	5	6	7	8	9	10
Go to Lesson...	2-6	2-2	1-1	1-1	1-8	2-2	1-1	2-7	1-1	2-8

Unit 2

Number and Operations: Decimals and Fractions

Focus

Apply operations with decimals and fractions, including multiplication and division, to solve problems.

CHAPTER 3
Operations with Decimals

BIG Idea Understand, explain, and apply operations with decimals, including multiplication and division.

BIG Idea Multiply and divide decimals to solve problems.

CHAPTER 4
Fractions and Decimals

BIG Idea Understand the relationship between fractions and decimals.

CHAPTER 5
Operations with Fractions

BIG Idea Understand, explain, and apply operations with fractions, including multiplication and division.

BIG Idea Multiply and divide fractions to solve problems.

Problem Solving in Science

Real-World Unit Project

Space: It's Out of this World! Have you ever looked at the night sky and wondered just how large our solar system really is? How large is each planet and how much does each planet weigh? How fast does each planet travel around the Sun? How far is each planet from the Sun? You are about to launch into orbit on a quest to find the answers to these and many more questions about our solar system. You'll compare facts about each planet to Earth. Prepare to be amazed about the size of our solar system!

Math Online **Log on to glencoe.com to begin.**

Operations with Decimals

- Understand, explain, and apply operations with decimals, including multiplication and division.
- Multiply and divide decimals to solve problems.

Key Vocabulary

clustering (p. 151)
decimal (p. 138)
equivalent decimals (p. 143)
front-end estimation (p. 151)
standard form (p. 139)

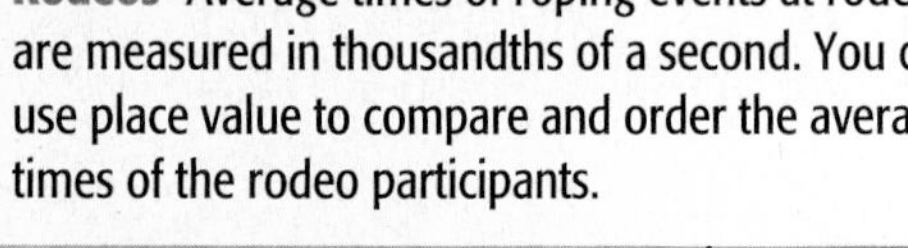

Real-World Link

Rodeos Average times of roping events at rodeos are measured in thousandths of a second. You can use place value to compare and order the average times of the rodeo participants.

Operations with Decimals Make this Foldable to help you organize your notes. Begin with two sheets of notebook paper.

1. **Fold** one sheet in half. Cut along fold from edges to margin.

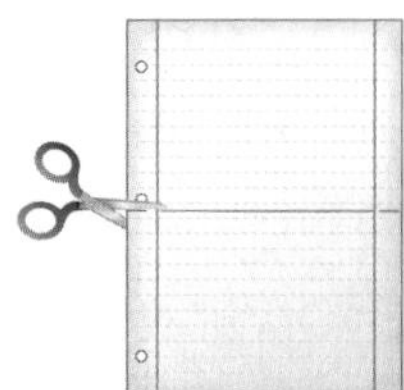

2. **Fold** the other sheet in half. Cut along fold between margins.

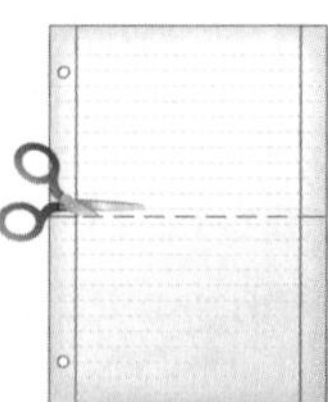

3. **Insert** first sheet through second sheet and along folds.

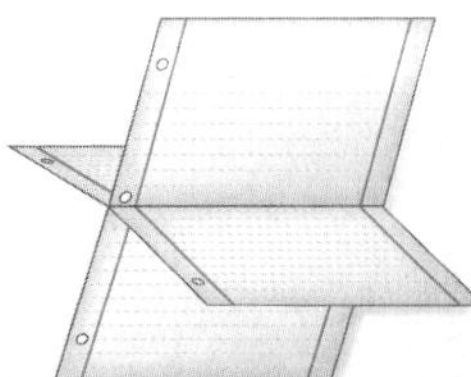

4. **Label** each side of each page with a lesson number and title.

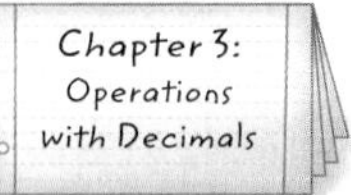

GET READY for Chapter 3

Diagnose Readiness You have two options for checking Prerequisite Skills.

Option 2

Math Online Take the Online Readiness Quiz at glencoe.com.

Option 1

Take the Quick Quiz below. Refer to the Quick Review for help.

QUICK Quiz

Multiply. (Page 744)

1. 17×28
2. 31×6
3. 109×14
4. 212×62
5. 228×19
6. 547×31
7. **SLEEP** The average adult gets 8 hours of sleep each night. How many total hours of sleep does the average adult get in one year (365 days)?

Divide. (Page 744)

8. $186 \div 3$
9. $171 \div 9$
10. $238 \div 14$
11. $832 \div 26$
12. $4{,}356 \div 36$
13. $1{,}728 \div 6$
14. **TRAVEL** Four friends drove from Chicago to Florida and spent $188 on gasoline. If they split the cost evenly, how much does each owe?

Replace each ● with < or > to make a true sentence. (Prior Grade)

15. 302,788 ● 203,788
16. 54,300 ● 543,000
17. 64,935 ● 61,935
18. 892,341 ● 892,431

QUICK Review

Example 1

Find 52 × 81.

$$\begin{array}{r} 52 \\ \times\ 81 \\ \hline 52 \\ +\ 4160 \\ \hline 4212 \end{array}$$

So, $52 \times 81 = 4{,}212$.

Example 2

Find 945 ÷ 15.

$$\begin{array}{r} 63 \\ 15\overline{)945} \\ -\ 90 \\ \hline 45 \\ -\ 45 \\ \hline 0 \end{array}$$

So, $945 \div 15 = 63$.

Example 3

Replace the ● in 71,238 ● 71,832 with < or > to make a true sentence.

Use place value.

71,238 Line up the digits.
71,832 Compare the hundreds place.
↑

Since $2 < 8$ in the hundreds place, $71{,}238 < 71{,}832$.

3-1 Representing Decimals

MAIN IDEA

Represent decimals in word form, standard form, and expanded form.

New Vocabulary

decimal
standard form
expanded form

Math Online

glencoe.com

- Concepts in Motion
- Extra Examples
- Personal Tutor
- Self-Check Quiz
- Reading in the Content Area

MINI Lab

The models below show some ways to represent the decimal 1.65.

Place-Value Chart

1,000	100	10	1	0.1	0.01	0.001
thousands	hundreds	tens	ones	tenths	hundredths	thousandths
0	0	0	1.	6	5	0

Money

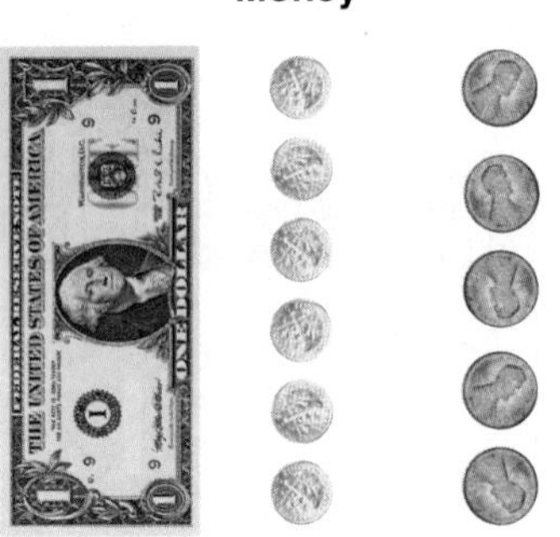

1 dollar 6 dimes 5 pennies

Decimal Model

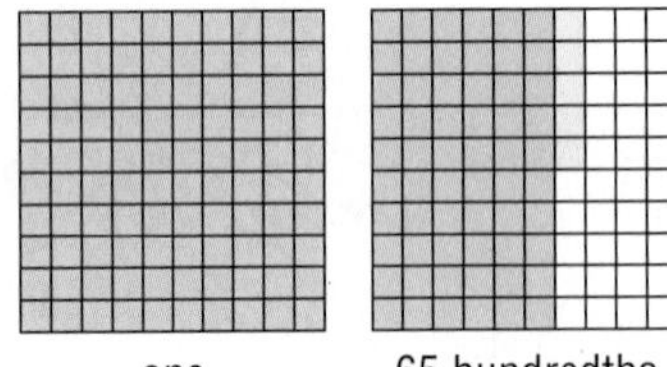

one 65 hundredths

Base-Ten Blocks

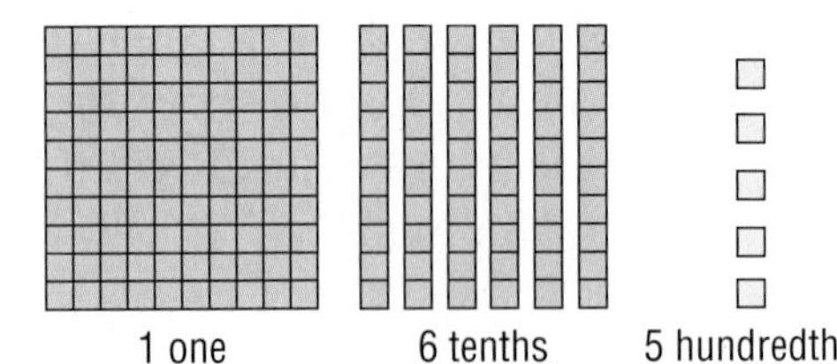

1 one 6 tenths 5 hundredths

Model each decimal using a place-value chart, money, a decimal model, and base-ten blocks.

1. 1.56 **2.** 0.85 **3.** 0.08 **4.** $2.25

Decimals, like whole numbers, are based on the number ten. In a place-value chart, the place to the right of the ones place has a value of one tenth. The next place has a value of one hundredth. Numbers that have digits in the tenths place and beyond are called **decimals.**

Place-Value Chart

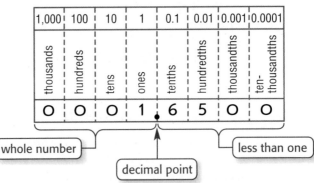

1,000	100	10	1	0.1	0.01	0.001	0.0001
thousands	hundreds	tens	ones	tenths	hundredths	thousandths	ten-thousandths
0	0	0	1.	6	5	0	0

whole number | decimal point | less than one

EXAMPLE Write a Decimal in Word Form

1 Write 17.542 in word form.

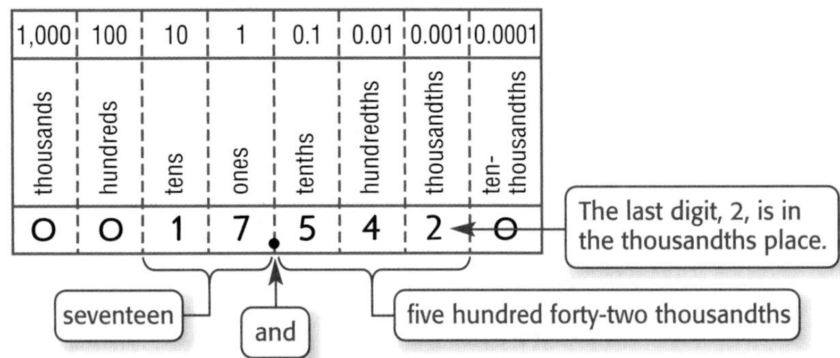

17.542 is seventeen and five hundred forty-two thousandths.

Reading Math

Decimal Point Use the word *and* only to read the decimal point. For example, read 0.235 as *two hundred thirty-five thousandths*. Read 235.035 as *two hundred thirty-five and thirty-five thousandths.*

CHECK Your Progress **Write each decimal in word form.**

a. 0.825 **b.** 16.08 **c.** 142.6

Standard form is the usual way to write a number. **Expanded form** is a sum of the products of each digit and its place value.

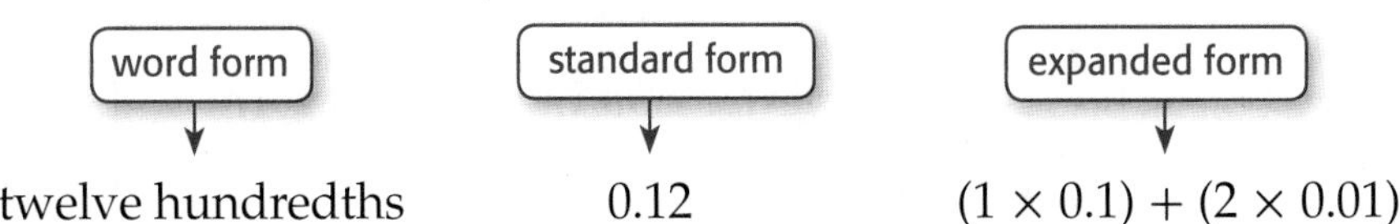

twelve hundredths 0.12 $(1 \times 0.1) + (2 \times 0.01)$

EXAMPLE Standard Form and Expanded Form

2 Write *thirty-five and ninety-six ten-thousandths* in standard form and in expanded form.

Place-Value Chart

1,000	100	10	1	0.1	0.01	0.001	0.0001
thousands	hundreds	tens	ones	tenths	hundredths	thousandths	ten-thousandths
0	0	3	5.	0	0	9	6

Standard form: 35.0096

Expanded form: $(3 \times 10) + (5 \times 1) + (0 \times 0.1) + (0 \times 0.01) + (9 \times 0.001) + (6 \times 0.0001)$

d. Write *three and eighty-five thousandths* in standard form and in expanded form.

CHECK Your Understanding

Example 1 (p. 139)

Write each decimal in word form.

1. 0.7
2. 0.08
3. 5.32
4. 0.022
5. 34.542
6. 8.6284

Example 2 (p. 139)

Write each decimal in standard form and in expanded form.

7. nine tenths
8. twelve thousandths
9. three and twenty-two hundredths
10. forty-nine and thirty-six ten-thousandths

Examples 1, 2 (p. 139)

11. **PACKAGING** A bag of dog food weighs 18.75 pounds. Write this number in two other forms.

Practice and Problem Solving

HOMEWORK HELP

For Exercises	See Examples
12–23, 32, 33	1
24–31	2

Write each decimal in word form.

12. 0.4
13. 0.9
14. 3.56
15. 1.03
16. 7.17
17. 4.94
18. 0.068
19. 0.387
20. 78.023
21. 20.054
22. 0.0036
23. 9.0769

Write each decimal in standard form and in expanded form.

24. five tenths
25. eleven and three tenths
26. two and five hundredths
27. thirty-four and sixteen hundredths
28. forty-one and sixty-two ten-thousandths
29. one hundred two ten-thousandths
30. eighty-three ten-thousandths
31. fifty-two and one hundredth
32. **HIKING** A state park has 19.8 miles of hiking and biking trails. Write this number in two other forms.
33. **MONEY** When writing a check, it is necessary to write the amount in both standard form and word form. Write $34.67 in words.
34. **ANALYZE TABLES** In the table, which numbers have their last digit in the hundredths place? Explain your reasoning. Write each number in expanded form.
35. How is 301.0019 written in word form?
36. Write $(5 \times 0.1) + (2 \times 0.01)$ in word form.
37. Write $(4 \times 0.001) + (8 \times 0.0001)$ in standard form.

All Time MLS Leaders in Goals Scored Per Game*

Player	Scoring Average
Stern John	0.8
Carlos Ruiz	0.695
Taylor Twellman	0.65
Mamadou Diallo	0.635
Raul Diaz Arce	0.55

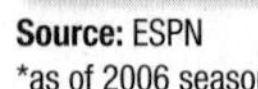

Source: ESPN
*as of 2006 season

EXTRA PRACTICE
See pages 678, 708.

H.O.T. Problems

CHALLENGE For Exercises 38 and 39, use the following information. The digits 3, 9, and 2 make up a decimal number.

38. What is the greatest possible decimal that is greater than 3, but less than 9?

39. What is the greatest possible decimal that is greater that 0, but less than 1?

40. **Which One Doesn't Belong?** Select the number that does not have the same value as the other three. Explain your reasoning.

thirty-four hundredths	$(3 \times 0.1) + (4 \times 0.01)$	three and four hundredths	0.34

41. **WRITING IN MATH** Explain how reading or hearing the word form of a decimal can help you write its standard form.

TEST PRACTICE

42. The world's smallest vegetable is the snow pea. It measures about 0.25 inch in diameter. Which phrase correctly represents this value?

    A twenty-five hundreds

    B twenty-five hundredths

    C twenty-five tenths

    D twenty-five thousandths

43. **SHORT RESPONSE** Write *two hundred eighty-four and twelve hundredths* in standard form.

44. Which of the following is another way to write the diameter of the tire in inches?

F $(3 \times 1) + (6 \times 0.1) + (1 \times 0.01)$

G $(3 \times 10) + (0 \times 1) + (1 \times 0.1) + (6 \times 0.01)$

H thirty and sixty-one tenths

J thirty and sixty-one hundredths

Spiral Review

45. **GEOGRAPHY** Jacksonville, Florida, is at sea level. Write this elevation as an integer. (Lesson 2-9)

46. **SURVEYS** Cheryl surveyed the students in her class to find their favorite type of music. What type of statistical display should Cheryl make to show the results? (Lesson 2-8)

GET READY for the Next Lesson

PREREQUISITE SKILL Choose the letter of the point that represents each decimal.

47. 6.3
48. 6.7
49. 6.2
50. 6.5
51. 7.2
52. 6.9

Comparing and Ordering Decimals

MAIN IDEA

Compare and order decimals.

New Vocabulary

inequality
equivalent decimals

Math Online

glencoe.com

- Extra Examples
- Personal Tutor
- Self-Check Quiz

GET READY for the Lesson

SUBWAYS Refer to the table.

World's Longest Subway Systems	
City	**Length (km)**
London	4.15
Moscow	3.4
New York City	3.71
Seoul	2.78
Tokyo	2.81

Source: *Jane's Urban Transport Systems*

1. Which city has the longest subway system? Explain.

Comparing decimals is similar to comparing whole numbers. You can use the symbols > or < to write an **inequality**. An inequality is a mathematical sentence indicating that two quantities are not equal. One quantity will be greater than or less than the other quantity.

EXAMPLE Compare Decimals

1 **SUBWAYS Refer to the table above. Use > or < to compare Tokyo's subway length with Seoul's.**

Use place value.

Tokyo: 2.81 First, line up the decimal points.

Seoul: 2.78 Then, starting at the left, find the first place the digits differ. Compare the digits.

Since $8 > 7$, $2.81 > 2.78$. Tokyo has a longer subway than Seoul.

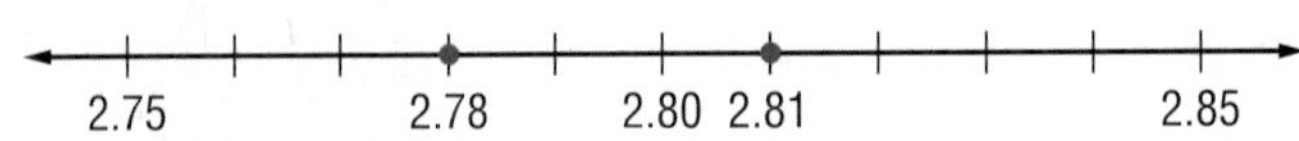

The number line shows that the answer is reasonable. Numbers to the right are greater than numbers to the left. So, $2.81 > 2.78$.

CHECK Your Progress

a. **SUBWAYS** Use >, <, or = to compare New York City's subway length with Moscow's.

Decimals that name the same number are called **equivalent decimals**. Examples are 0.8 and 0.80.

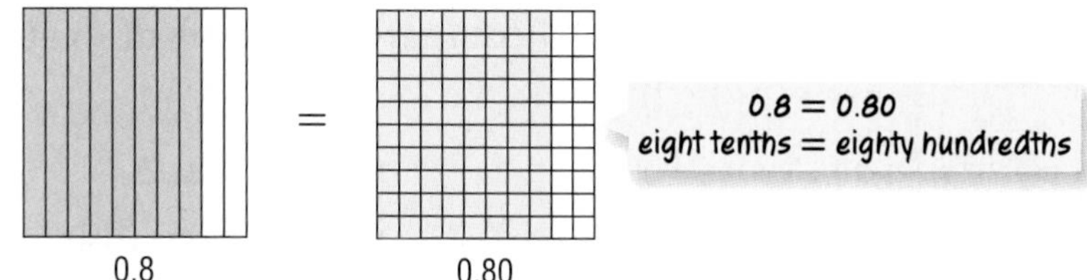

When you *annex*, or place zeros to the right of the last digit in a decimal, the value of the decimal does not change. Annexing zeros is useful when ordering a group of decimals.

EXAMPLE Order Decimals

2 Order 15, 14.95, 15.8, and 15.01 from least to greatest.

Study Tip

Check For Reasonableness You can check the reasonableness of the order by using a number line.

First, line up the decimal points.		Next, annex zeros so all numbers have the same final place value.	Finally, compare and order using place value.
15	→	15.00	14.95
14.95	→	14.95	15.00
15.8	→	15.80	15.01
15.01	→	15.01	15.80

The order from least to greatest is 14.95, 15, 15.01, and 15.8.

CHECK Your Progress

b. Order 35.06, 35.7, 35.5, and 35.849 from greatest to least.

CHECK Your Understanding

Example 1 (p. 142)

Use >, <, or = to compare each pair of decimals.

1. 0.4 ● 0.5
2. 0.38 ● 0.35
3. 2.7 ● 2.07
4. 25.5 ● 25.50
5. **POPULATION** Australia and Botswana are among the least populated countries in the world. In Australia, about 6.76 people live in each square mile, while 6.84 people live in each square mile in Botswana. Which country has the greater number of people per square mile?

Example 2 (p. 143)

6. **BASEBALL** The five highest career batting averages in Major League Baseball (MLB) are listed at the right. Order these averages from least to greatest.

Source: Major League Baseball

Practice and Problem Solving

HOMEWORK HELP	
For Exercises	See Examples
7–20	1
21–28	2

Use $>$, $<$, or $=$ to compare each pair of decimals.

7. 0.2 ● 2.0
8. 3.3 ● 3.30
9. 0.08 ● 0.8
10. 0.4 ● 0.004
11. 6.02 ● 6.20
12. 5.51 ● 5.15
13. 9.003 ● 9.030
14. 0.204 ● 0.214
15. 7.107 ● 7.011
16. 23.88 ● 23.880
17. 0.0624 ● 0.0264
18. 2.5634 ● 2.5364

19. **OLYMPICS** In the 2004 Summer Olympics, Carly Patterson had a total score of 38.387 in the all-around gymnastics event. Svetlana Khorkina had a total score of 38.211 in the same event. Who had a higher score in this event?

20. **FOOTBALL** Peyton Manning averaged 7.89 passing yards per attempt in 2006 and Carson Palmer averaged 7.76. Who had the higher average?

Order each set of decimals from least to greatest.

21. 16, 16.2, 16.02, 15.99
22. 44.5, 45.01, 44.11, 45
23. 5.545, 4.45, 4.9945, 5.6
24. 9.27, 9.6, 8.995, 9.0599

Order each set of decimals from greatest to least.

25. 2.1, 2.01, 2.11, 2.111
26. 7.66, 7.6, 7.666, 7.06
27. 32.32, 32.032, 32.302, 3.99
28. 57.68, 57.057, 5.75, 57.57

29. **INVENTORY** To keep track of the inventory at his store's warehouse, Akio must arrange items on shelves according to their stock numbers. Arrange the numbers in order from least to greatest.

Stock Number
321.53
321.539
321.5

30. **ANALYZE TABLES** The following table shows the amount of money Antoine spent on lunch each day this week. Order the amounts from least to greatest and then find the median amount he spent on lunch.

Day	Mon.	Tues.	Wed.	Thurs.	Fri.
Amount Spent ($)	3.31	3.45	3.18	3.43	3.29

H.O.T. Problems

31. **SELECT A TECHNIQUE** The average annual snowfall in Syracuse, New York, is 115.6 inches. Takeetna, Alaska, gets an average of 115.4 inches of snow per year. Which of the following techniques might you use to find which city, on average, gets more snowfall during a 10-year time period? Justify your selection(s). Then use the technique(s) to solve the problem.

mental math | number sense | estimation

32. **OPEN ENDED** Give an example of a decimal equivalent to 0.76.

33. **FIND THE ERROR** Ryan and Mateo are ordering 0.4, 0.5, and 0.49 from least to greatest. Who is correct? Explain your reasoning.

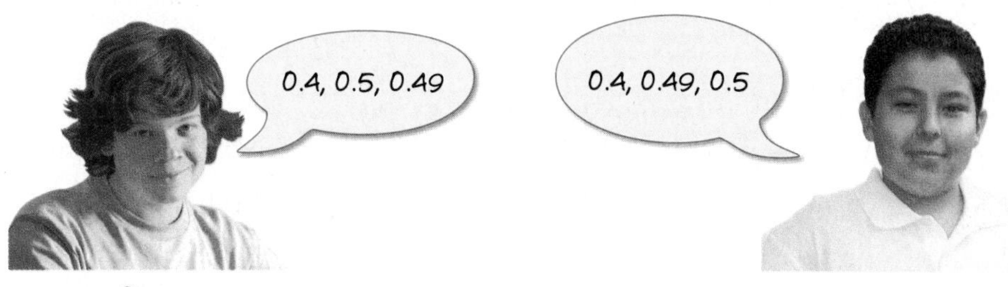

34. **CHALLENGE** Lindsay's cat weighs more than Marissa's cat, but less than Nate's cat. Kate's cat weighs 0.2 pound more than Nate's cat. The weights of each cat are 10.22, 10.2, 10.42, and 10.02 pounds. Identify each cat with its weight.

35. **WRITING IN MATH** Refer to the table in Exercise 36. Create a problem that involves comparing the times of two of the runners.

TEST PRACTICE

36. The table shows the finishing times for four runners in a 100-meter race.

Runner	Time (s)
Kara	14.31
Ariel	13.84
Mika	13.97
Nelia	13.79

In what order did the runners cross the finish line?

A Kara, Ariel, Mika, Nelia
B Nelia, Mika, Ariel, Kara
C Mika, Nelia, Ariel, Kara
D Nelia, Ariel, Mika, Kara

37. If Cheyenne correctly marked 1.005, 0.981, 0.899, and 0.93 on a number line, which number was closest to zero?

0 1 2

F 1.005
G 0.981
H 0.899
J 0.93

38. Which number is between 2.35 and 3.06?

A 2.315
B 2.571
C 3.084
D 3.628

Spiral Review

39. **TEMPERATURE** At the doctor, Clara's temperature was 101.5°F. Write this temperature in expanded form. (Lesson 3-1)

Graph each integer on a number line. (Lesson 2-9)

40. +3
41. −9
42. +2
43. −4

GET READY for the Next Lesson

PREREQUISITE SKILL Identify each underlined place-value position. (Page 738)

44. 14.0<u>6</u>
45. 3.<u>0</u>54
46. 0.42<u>7</u>8
47. 2.960<u>0</u>

3-3 Rounding Decimals

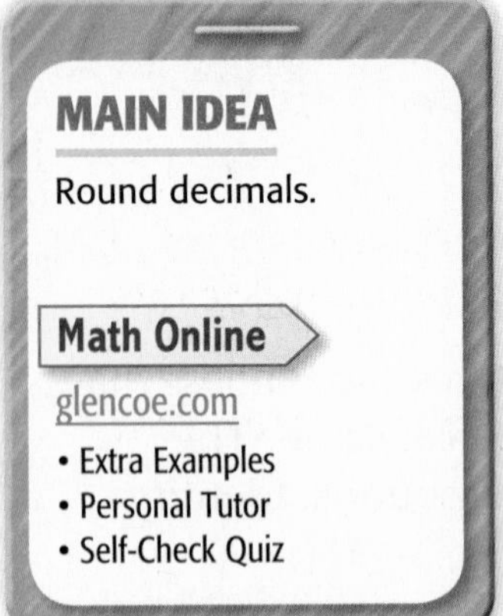

MAIN IDEA

Round decimals.

Math Online

glencoe.com

- Extra Examples
- Personal Tutor
- Self-Check Quiz

GET READY for the Lesson

MOVIES The prices of movie tickets from five different theaters are shown in the table.

1. Round each price to the nearest dollar.
2. How did you decide how to round each number?
3. Make a conjecture about how to round each price to the nearest dime.

Theater	Price ($)
Movie Max	$8.75
Star Theater	$7.95
Movie Mania	$6.25
Dollar Theater	$1.75
Cine-mart	$9.60

You can round decimals just as you round whole numbers.

Round Decimals

Key Concept

To round a decimal, first underline the digit to be rounded. Then look at the digit to the right of the place being rounded.

- If the digit is 4 or less, the underlined digit remains the same.
- If the digit is 5 or greater, add 1 to the underlined digit.
- After rounding, drop all digits after the underlined digit.

EXAMPLES Round Decimals

1 Round 1.324 to the nearest whole number.

Underline the digit to be rounded. → $\underline{1}.324$ ← Since 3 is less than 5, the digit 1 remains the same.

On the number line, 1.3 is closer to 1.0 than to 2.0. To the nearest whole number, 1.324 rounds to 1.

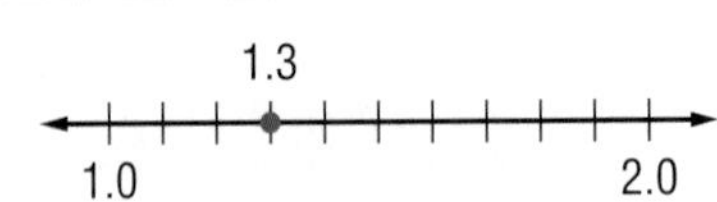

2 Round 99.96 to the nearest tenth.

Underline the digit to be rounded. → $99.\underline{9}6$ ← Since the digit is 6, add one to the underlined digit.

On the number line, 99.96 is closer to 100.0 than to 99.9. To the nearest tenth, 99.96 rounds to 100.0.

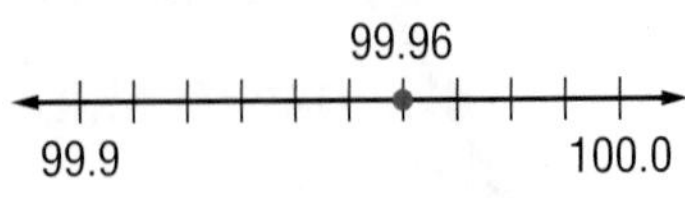

Round each decimal to the indicated place-value position.

a. 13.419; hundredths

b. 0.27838; ten-thousandths

3 PEANUTS Refer to the information at the left. To the nearest cent, how much did U.S. farmers receive for each pound of peanuts produced in 2006?

There are 100 cents in a dollar. So, rounding to the nearest cent means to round to the nearest hundredth.

Underline the digit in the hundredths place.

0.1<u>7</u>9

Then look at the digit to the right. The digit is greater than 5. So, add one to the underlined digit.

To the nearest cent, the average price is $0.18.

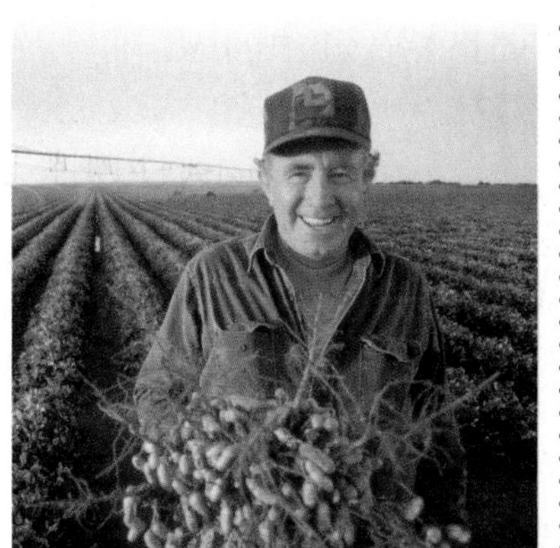

Real-World Link

On average, U.S. farmers received $0.179 for each pound of peanuts they produced in 2006.

Source: Farm Service Agency

CHECK Your Progress

c. **PASTA** A box of uncooked spaghetti costs $0.1369 per ounce. How much is this to the nearest cent?

d. **ANIMALS** An Arabian camel averages 3.45 meters tall. Round 3.45 to the nearest meter.

CHECK Your Understanding

Examples 1, 2 (p. 146)

Round each decimal to the indicated place-value position.

1. 0.329; tenths
2. 1.75; ones
3. 45.522; hundredths
4. 0.5888; thousandths
5. 7.67597; ten-thousandths
6. 34.59; tens

Example 3 (p. 147)

7. **ANALYZE TABLES** The 100-meter dash times for the Jackson Middle School boys' track team are shown in the table. To the nearest tenth, what is the time for each runner?

Runner	Time (s)
Jacob	11.92
Marquez	11.96
Alan	11.84
Tyrese	11.87

Example 3 (p. 147)

8. **GASOLINE** On May 15, 2007, the average price of a gallon of unleaded gasoline in North Carolina was $2.969. How much is this to the nearest cent?

Practice and Problem Solving

HOMEWORK HELP	
For Exercises	See Examples
9–16	1, 2
17–20	3

Round each decimal to the indicated place-value position.

9. 7.445; tenths
10. 7.999; tenths
11. 5.68; ones
12. 10.49; ones
13. 2.499; hundredths
14. 40.458; hundredths
15. 5.4572; thousandths
16. 45.0189; thousandths
17. **RACING** The winner of the 2007 Indianapolis 500, Dario Franchitti, had an average speed of 151.774 miles per hour. Round 151.774 miles per hour to the nearest ten miles per hour.
18. **CELL PHONES** In the U.S., there are 48.81 cell phones for every 100 people. Round 48.81 to the nearest whole number.
19. **MONEY MATTERS** The price of a 12-pack of soda is $4.39. How much is this to the nearest dollar?
20. **CURRENCY** Fifty Japanese yen are equal to $0.475441 U.S. dollars. Round this amount of U.S. dollars to the nearest cent.

CALCULATOR A calculator will often show the results of a calculation with a very long decimal. Round each of the numbers on the calculator displays to the nearest thousandth.

21. .2491666667
22. 1054.677828
23. 21.25103904

Round each decimal to the indicated place-value position.

24. 9.56303; hundredths
25. 988.08055; thousandths
26. 87.09; tens
27. 1,567.893; ten-thousandths
28. **CYCLING** The table shows the average winning speeds in the Tour de France from 2000–2005. Will it help to round these average speeds before listing them in order from least to greatest? Explain.

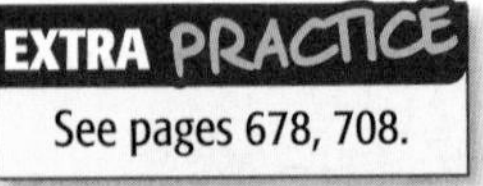

Tour de France Average Winning Speeds

Winner	Year	Average Speed (km/h)
Lance Armstrong	2005	41.654
Lance Armstrong	2004	40.553
Lance Armstrong	2003	40.94
Lance Armstrong	2002	39.93
Lance Armstrong	2001	40.02
Lance Armstrong	2000	38.57

Source: ESPN

H.O.T. Problems

29. **OPEN ENDED** Give an example of a decimal that when rounded to the nearest tenth is 15.0 and to the nearest hundredth is 15.00.
30. **Which One Doesn't Belong?** Identify the decimal that does not belong with the other three. Explain your reasoning.

11.23   11.26   11.19   11.24

31. **CHALLENGE** A number rounded to the nearest tenth is 6.1. The same number rounded to the nearest hundredth is 6.08 and rounded to the nearest thousandth is 6.083. Draw a conclusion as to what the original number could be.

32. **SELECT A TECHNIQUE** On four different days of walking on a treadmill, Mansi burned 149.6, 150.1, 150.4, and 149.8 Calories. Which of the following techniques might Mansi use to find the average number of Calories she burned to the nearest whole number over those four days? Justify your selection. Then use the technique to solve the problem.

mental math | number sense | estimation

33. **WRITING IN MATH** Use a model to show why 6.73 rounded to the nearest tenth is 6.7. Explain your reasoning.

TEST PRACTICE

34. The average annual precipitations for certain cities are given in the table.

City	Precipitation (in.)
Omaha, NE	30.22
Pittsburgh, PA	37.85
Kansas City, MO	37.98

What is the annual precipitation for Kansas City to the nearest tenth?

A 40.0 **C** 37.9

B 38.0 **D** 37.8

35. On July 28, 1976, a Lockheed SR–71A set the record for jet speed at 2,193.167 miles per hour. What is this speed rounded to the nearest mile per hour?

F 2,190

G 2,192

H 2,193

J 2,194

Spiral Review

Use >, <, or = to compare each pair of decimals. (Lesson 3-2)

36. 8.64 ● 8.065 37. 2.5038 ● 25.083 38. 12.004 ● 12.042

39. Write *thirty-two and five hundredths* in standard form. (Lesson 3-1)

40. **ZOOS** Admission to a zoo is $21 for adults and $14 for children. Define variables and write an expression to find the total cost of 2 adult tickets and 3 children's tickets. Then, find the value of the expression. (Lesson 1-4)

GET READY for the Next Lesson

PREREQUISITE SKILL Add or subtract. (Page 743)

41. 43 + 15 42. 68 + 37 43. 85 – 23 44. 52 – 29

3-4 Estimating Sums and Differences

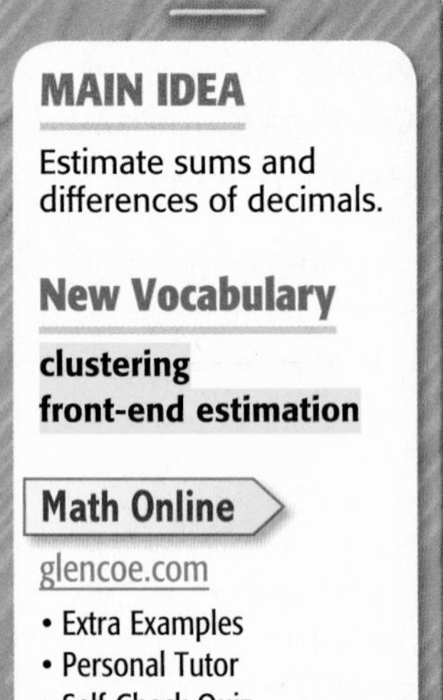

MAIN IDEA

Estimate sums and differences of decimals.

New Vocabulary

clustering
front-end estimation

Math Online

glencoe.com

- Extra Examples
- Personal Tutor
- Self-Check Quiz

GET READY for the Lesson

PARKS The table shows the five most visited national parks in the United States.

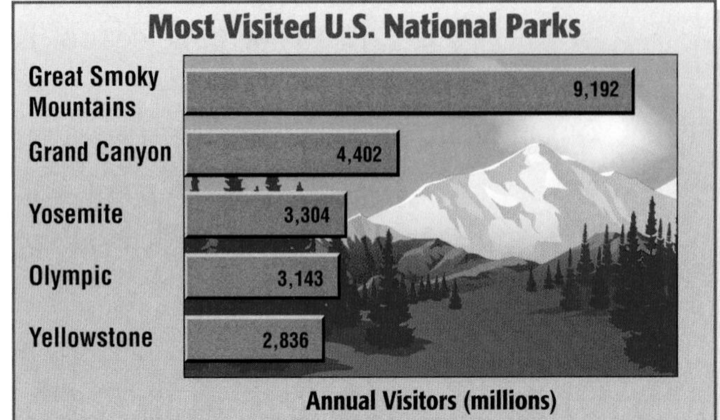

Source: National Park Service

1. Round the number of visitors to each park to the nearest million.
2. About how many more people visit the Great Smoky Mountains National Park each year than Yosemite National Park?

To estimate sums and differences of decimals, you can use the same methods you used for whole numbers.

EXAMPLES Use Estimation to Solve Problems

1 **Estimate the total number of annual visitors to Yellowstone and Olympic National Parks.**

Round each number to the nearest unit for easier adding.

2.836	→	3	2.836 rounds to 3.
+ 3.143	→	+ 3	3.143 rounds to 3.
		6	

About 6 million visitors visit these two parks annually.

2 **Estimate how many more annual visitors visit the Grand Canyon National Park than Yosemite.**

4.402	→	4	4.402 rounds to 4.
+ 3.304	→	− 3	3.304 rounds to 3.
		1	

About 1 million more visitors visit the Grand Canyon National Park.

CHECK Your Progress

a. Estimate the sum of 4.37 and 6.75 using rounding.

b. Estimate the difference of 42.18 and 17.25 using rounding.

When estimating a sum in which all of the addends are close to the same number, you can use **clustering.** Check to see if the addends are all close to one number. If so, round each addend to the same number. Then multiply.

TEST EXAMPLE

3 Julia is going on a hiking trip with her father. The table shows the prices of different items needed for the trip. Which is closest to the amount spent on the items?

Item	Price ($)
backpack	52.95
hiking boots	51.25
sleeping bag	48.75
food	45.50

A $100

B $175

C $200

D $250

Test-Taking Tip

Clustering Clustering is good for problems in which the addends are close together.

Read the Item

The addends are clustered around $50. Round each price to $50.

$52.95 → $50

$51.25 → $50

$48.75 → $50

$45.50 → $50

Solve the Item

Multiplication is repeated addition. So, a good estimate of the total cost of the hiking supplies is 4 × 50, or $200. The answer is C.

CHECK Your Progress

c. The table shows the number of miles Jaime ran last week. Estimate the total number of miles Jaime ran last week.

Day	Miles
Wednesday	5.1
Thursday	5.3
Friday	4.8
Saturday	5.0

F 10 miles

G 15 miles

H 20 miles

J 25 miles

Another type of estimation is front-end estimation. When you use **front-end estimation,** add or subtract the values of the digits in the front place or left-most place-value. Front-end estimation usually gives a sum that is less than the actual sum.

Study Tip

Estimation You can still use front-end estimation when the addends have a different number of digits. For example, to estimate 113 + 42, add the values in both the hundreds and tens place. So, an estimate of 113 + 42 is 110 + 40 or 150.

EXAMPLE Use Front-End Estimation

4 Estimate 34.6 + 55.3 using front-end estimation.

$$\begin{array}{rcr} 34.6 & \rightarrow & 30.0 \\ +\ 55.3 & \rightarrow & +\ 50.0 \\ \hline & & 80.0 \end{array}$$

Add the front digits.

Using front-end estimation, 34.6 + 55.3 is about 80.0.

CHECK Your Progress **Estimate using front-end estimation.**

d. 22.35 – 11.14

e. $47.92 – $21.62

Estimation Methods

Concept Summary

Rounding	Estimate by rounding each decimal to the nearest whole number that is easy for you to add or subtract mentally.
Clustering	Estimate by rounding a group of close addends to the same number. Then multiply.
Front-End Estimation	Estimate by adding or subtracting the values of the digits in the front place or front-most places.

CHECK Your Understanding

Example 1 (p. 150) **Estimate each sum using rounding.**

1. 0.36 + 0.83
2. $15.24 + $32.10

Example 2 (p. 150) **Estimate each difference using rounding.**

3. 4.44 – 2.79
4. 57.05 – 23.82

Example 3 (p. 151) **Estimate using clustering.**

5. 5.32 + 4.78 + 5.42
6. $0.95 + $0.79 + $1.02

7. **MULTIPLE CHOICE** The amount of time Omar spent on his homework each week last month is shown in the table.

Time Spent on Homework				
Week	1	2	3	4
Time (h)	11.24	9.47	12.36	10.38

Which is closest to the total time spent on homework?

A 30 hours **B** 35 hours **C** 40 hours **D** 50 hours

Example 4 (p. 152) **Estimate using front-end estimation.**

8. 109.4 + 513.8
9. $442.50 – $126.73

Practice and Problem Solving

HOMEWORK HELP	
For Exercises	See Examples
10–17	1, 2
18–23, 37, 38	3
24–29	4

Estimate using rounding.

10. 49.59 + 16.22
11. 33.15 + 86.85
12. 41.59 − 19.72
13. 62.61 − 13.05
14. 2.33 + 4.88 + 5.5
15. 9.05 + 1.42 + 6.79
16. **SHOPPING** Sandra bought a pair of shoes for $24.75 and a dress for $46.55. About how much did Sandra spend on the shoes and dress?
17. **MAGAZINES** Jackson and Lana each sold magazines. Jackson collected $432.17 and Lana collected $378.64. About how much more did Jackson collect than Lana?

Estimate using clustering.

18. 6.99 + 6.59 + 7.02 + 7.44
19. $3.33 + $3.45 + $2.78 + $2.99
20. 5.45 + 5.3948 + 4.7999
21. $55.49 + $54.99 + $55.33
22. 10.33 + 10.45 + 10.89 + 9.79
23. 99.8 + 100.2 + 99.5 + 100.4

Estimate using front-end estimation.

24. 75.45 − 15.23
25. 27.9 − 12.5
26. 28.65 + 71.53
27. 124.8 + 264.9
28. $315.65 + $130.42
29. $50.96 + $19.28
30. **MUSIC** The best-selling musician has sold 168.5 million albums. About how many more albums will need to be sold to reach 175 million?
31. **TUNNELS** The Flathead rail tunnel in Montana is 7.78 miles long. Colorado's Moffat rail tunnel is 6.21 miles long. About how much longer is the Flathead rail tunnel than the Moffat rail tunnel using rounding? If you use front-end estimation, would the estimate be the same? Why or why not?

Real-World Link . . . The current capacity for Fenway Park, home of the Boston Red Sox, is 36,108 for night games and 35,692 for day games.

Source: Major League Baseball

ANALYZE GRAPHS **For Exercises 32 and 33, use the graph.**

32. Use clustering to estimate the mean ticket cost for one game each of the St. Louis Cardinals, New York Yankees, and Chicago White Sox.
33. If the average price for a soda and hot dog at a New York Yankees game is $6.25, about how much would a family pay for four tickets, four sodas, and four hot dogs?

2007 Major League Baseball Ticket Prices

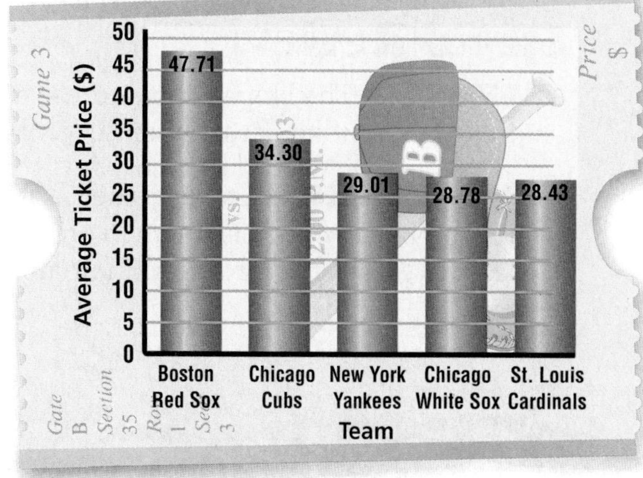

Source: Team Marketing Report

EXTRA PRACTICE
See pages 679, 708.

H.O.T. Problems

34. **NUMBER SENSE** How do you know that the sum of 7.4, 2.8, and 4.2 is less than 15?

35. **CHALLENGE** Donovan bought six identical hats for his friends. Based on rounding, the total estimate was $90. Decide what the maximum and minimum price of each hat could be.

36. **WRITING IN MATH** Explain the advantages and disadvantages of finding an approximate answer.

TEST PRACTICE

37. A school lunch menu is shown.

Monday	
Pizza $1.10	Soda $0.85
Salad $2.65	Milk $0.75
Taco $1.30	Fruit $1.15

Estimate how much money you will need to buy a slice of pizza, a taco, and a soda.

A a little less than $2

B a little more than $2

C a little more than $3

D a little less than $3

38. Refer to the table that shows the attendance in a recent year for the most popular theme parks in the United States.

Park	Attendance (millions)
Magic Kingdom	16.2
Disneyland	14.5
Epcot Center	9.9
Disney-MGM Studios	8.6
Disney's Animal Kingdom	8.2

Source: Coaster Grotto

Which is the best estimate for the total number of people that visited the parks?

F 55 million

G 58 million

H 60 million

J 65 million

Spiral Review

39. **GEMSTONES** The Hope Diamond has a weight of 45.52 carats. Round this amount to the nearest tenth. (Lesson 3-3)

40. **ANALYZE TABLES** The table at the right lists five common elements. List them in order from least to greatest according to their atomic masses. (Lesson 3-2)

Common Elements		
Element	Symbol	Atomic Mass
Argon	Ar	39.948
Calcium	Ca	40.078
Chlorine	Cl	35.453
Potassium	K	39.0983
Titanium	Ti	47.867

Source: *The Time Almanac*

GET READY for the Next Lesson

PREREQUISITE SKILL **Add or subtract.** (Page 743)

41. $\begin{array}{r} 278 \\ +\ 199 \\ \hline \end{array}$

42. $\begin{array}{r} 1,297 \\ +\ 86 \\ \hline \end{array}$

43. $\begin{array}{r} 700 \\ -\ 235 \\ \hline \end{array}$

44. $\begin{array}{r} 1,252 \\ -\ 79 \\ \hline \end{array}$

Explore 3-5

Math Lab
Adding and Subtracting Decimals Using Models

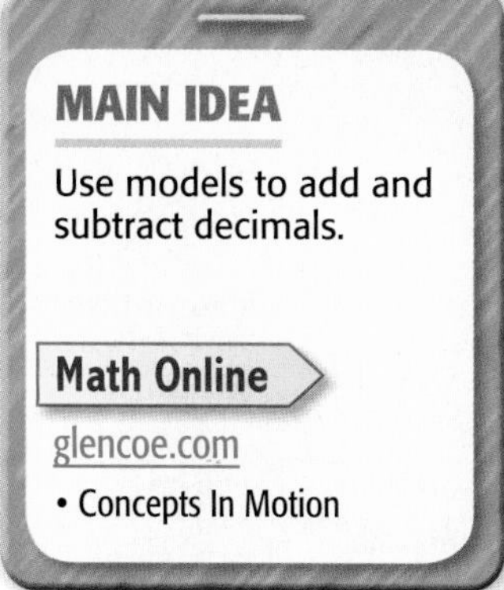

Decimal models can be used to add and subtract decimals.

Ones (1)	Tenths (0.1)	Hundredths (0.01)
One whole 10-by-10 grid represents 1 or 1.0.	Each row or column represents one tenth or 0.1.	Each square represents one hundredth or 0.01.

ACTIVITIES

1 Find 0.16 + 0.77 using decimal models.

STEP 1 Shade 0.16 green.

STEP 2 Shade 0.77 blue. The sum is the total shaded area.

So, $0.16 + 0.77 = 0.93$.

2 Find 0.52 − 0.08 using decimal models.

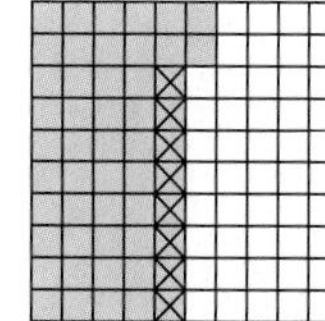

STEP 1 Shade 0.52 green.

STEP 2 Use ×s to cross out 0.08 from the shaded area. The difference is the amount of shaded area with no ×s.

So, $0.52 - 0.08 = 0.44$.

CHECK Your Progress

Find each sum or difference using decimal models.

a. $0.14 + 0.67$　　b. $0.35 + 0.42$　　c. $0.03 + 0.07$

d. $0.75 - 0.36$　　e. $0.68 - 0.27$　　f. $0.88 - 0.49$

Analyze the Results

1. Explain how you can use grid paper to model $0.8 - 0.37$.
2. **MAKE A CONJECTURE** Write a rule you can use to add or subtract decimals without using models.

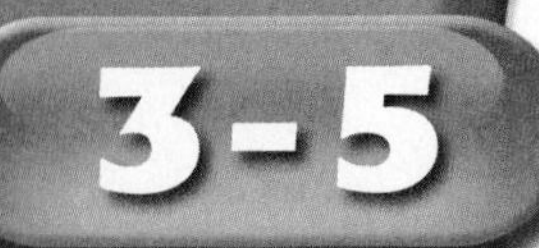

3-5 Adding and Subtracting Decimals

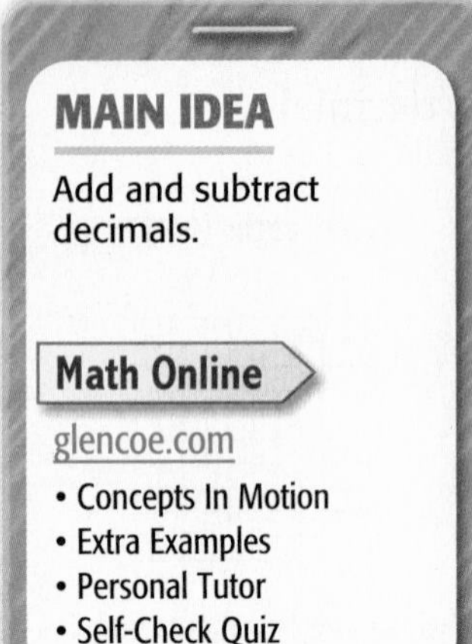

MAIN IDEA

Add and subtract decimals.

Math Online

glencoe.com

- Concepts In Motion
- Extra Examples
- Personal Tutor
- Self-Check Quiz

GET READY for the Lesson

SOFT DRINKS The table shows the top five consumers of carbonated soft drinks.

1. Estimate the sum of the top two countries.
2. Add the digits in the same place-value position for the top two countries.
3. Compare the estimate with the actual sum. Place the decimal point in the sum.
4. Make a conjecture about how to add decimals.

Carbonated Soft Drink Consumers

Country	Consumption Per Capita (gallons)
U.S.	51.7
Mexico	33.3
Norway	32.2
Ireland	32.0
Canada	30.9

Source: Top 10 of Everything

To add or subtract decimals, line up the decimal points. Then, add or subtract digits in the same place-value position.

EXAMPLES Add and Subtract Decimals

1 Find the sum of 23.1 and 5.8.

Estimate $23.1 + 5.8 \approx 23 + 6$ or 29

$$\begin{array}{r} 23.1 \\ +\ 5.8 \\ \hline 28.9 \end{array}$$

Line up the decimal points.

Add as with whole numbers.

The sum of 23.1 and 5.8 is 28.9.

Compare the answer to the estimate. Since 28.9 is close to 29, the answer is reasonable.

2 Find 5.774 − 2.371.

Estimate $5.774 - 2.371 \approx 6 - 2$ or 4

$$\begin{array}{r} 5.774 \\ -\ 2.371 \\ \hline 3.403 \end{array}$$

Line up the decimal points.

Subtract as with whole numbers.

So, $5.774 - 2.371 = 3.403$. **Check for Reasonableness** $3.403 \approx 4$ ✓

CHECK Your Progress

Find each sum or difference.

a. $54.7 + 21.4$ b. $14 + 23.5$ c. $17.3 + 33.5$

d. $9.543 - 3.67$ e. $18.4 - 12.9$ f. $\$50.62 - \$39.81$

Sometimes it is necessary to annex zeros before you subtract.

EXAMPLE Annex Zeros

3 Find 6 − 4.78.

Estimate $6 - 4.78 \approx 6 - 5$ or 1

$$\begin{array}{r} 6.00 \\ -\ 4.78 \\ \hline 1.22 \end{array}$$

Annex zeros so that both numbers have the same place value.

So, 6 − 4.78 = 1.22. **Check for Reasonableness** $1.22 \approx 1$ ✔

CHECK Your Progress

Find each difference.

g. 2 − 1.78 h. 14 − 9.09 i. 23 − 4.216

Real-World EXAMPLE

4 BONES The table shows the average length of the three longest bones in the human body. How much longer is the average femur than the average tibia?

Longest Bones in the Human Body	
Bone	**Length (in.)**
Femur (upper leg)	19.88
Tibia (inner lower leg)	16.94
Fibula (outer lower leg)	15.94

Source: *The Top 10 of Everything*

Estimate $19.88 - 16.94 \approx 20 - 17$ or 3

$$\begin{array}{r} 19.88 \\ -\ 16.94 \\ \hline 2.94 \end{array}$$

Line up the decimal points.

Subtract as with whole numbers.

So, the average femur is 2.94 inches longer than the average tibia. **Check for Reasonableness** $2.94 \approx 3$ ✔

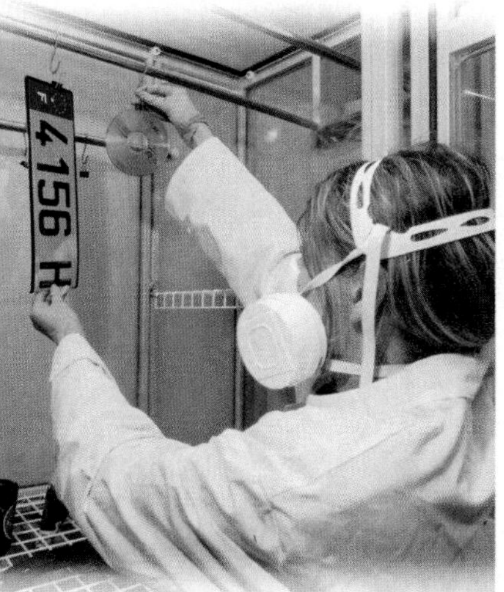

Real-World Career . . .
How Does a Forensic Scientist Use Math? Forensic scientists use math to analyze biological, chemical, or physical samples of evidence from a crime scene.

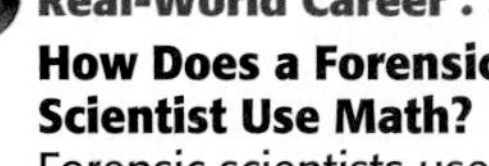

For more information, go to glencoe.com.

CHECK Your Progress

j. **SWIMMING** The table shows the top three times for the women's 100-meter butterfly event in a recent Summer Olympics. What is the difference between Petria Thomas' time and Inge de Bruijn's time?

Women's 100-Meter Butterfly		
Swimmer	**Country**	**Time (s)**
Petria Thomas	Australia	57.72
Otylia Jedrzejczak	Poland	57.84
Inge de Bruijn	Netherlands	57.99

Source: ESPN

You can also use decimals to evaluate algebraic expressions.

EXAMPLE Evaluate an Expression

5 ALGEBRA Evaluate $x + y$ if $x = 2.85$ and $y = 17.975$.

$x + y = 2.85 + 17.975$ Replace x with 2.85 and y with 17.975.

Estimate $2.85 + 17.975 \approx 3 + 18$ or 21

$$\begin{array}{r} 2.850 \\ +17.975 \\ \hline 20.825 \end{array}$$

2.850: Line up the decimal points. Annex a zero.

20.825: Add as with whole numbers.

The value is 20.825.

Check for Reasonableness $20.825 \approx 21$ ✔

CHECK Your Progress

Evaluate each expression if $a = 2.56$ and $b = 28.96$.

k. $3.23 + a$ l. $68.96 - b$ m. $b - a$

CHECK Your Understanding

Example 1 (p. 156)

Find each sum.

1. $5.5 + 3.2$
2. $72.4 + 12.5$
3. $9 + 29.34$

Example 2 (p. 156)

Find each difference.

4. $0.40 - 0.20$
5. $9.67 - 2.35$
6. $42.28 - 1.52$

Example 3 (p. 157)

Find each difference.

7. $8 - 5.78$
8. $15 - 6.24$
9. $36 - 7.3$

Example 4 (p. 157)

10. **ANALYZE TABLES** Use the table to find out how many more people there are per square mile in Iowa than in Colorado.

Population Density	
State	**People Per Square Mile**
Colorado	41.5
Iowa	52.4
Arkansas	51.3
Oklahoma	50.3

Source: U.S. Census Bureau

11. **MAGAZINES** In a recent year, *National Geographic* had an average paid circulation of 6.6 million magazines, and *Time* had an average paid circulation of 4.1 million magazines. What is the difference in circulation of these two magazines?

Example 5 (p. 158)

12. **ALGEBRA** Evaluate $s - t$ if $s = 8$ and $t = 4.25$.

Practice and Problem Solving

HOMEWORK HELP	
For Exercises	See Examples
13–18	1
19–24	2, 3
25, 26	4
27–30	5

Find each sum.

13. $7.2 + 9.5$
14. $4.9 + 3.0$
15. $1.34 + 2$
16. $0.796 + 13$
17. $54.5 + 48.51$
18. $15.63 + 24.36$

Find each difference.

19. $5.6 - 3.5$
20. $19.86 - 4.94$
21. $97 - 16.98$
22. $82 - 67.18$
23. $58.67 - 28.72$
24. $14.39 - 12.16$

25. **RODEO** The table shows the top three finishers in barrel racing at the Livestock Show and Rodeo. What is the time difference between first place and second place?

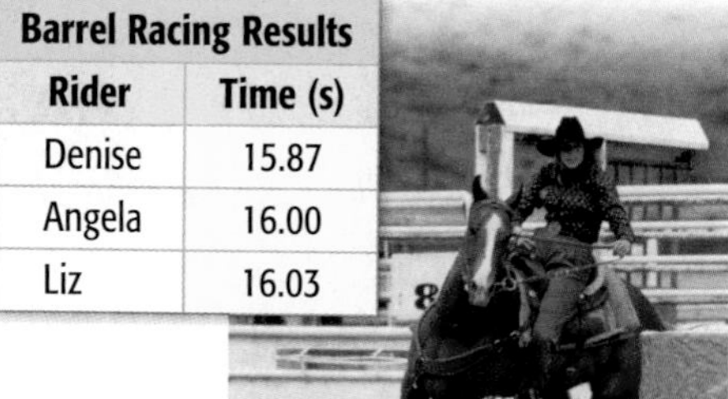

Barrel Racing Results	
Rider	**Time (s)**
Denise	15.87
Angela	16.00
Liz	16.03

26. **MONEY** You decide to buy a hat for \$10.95 and a T-shirt for \$14.20. How much change will you receive if you pay with a \$50 bill?

ALGEBRA Evaluate each expression if $a = 128.9$ and $d = 22.035$.

27. $a - 11.25$
28. $75 + a$
29. $a - d$
30. $d + a$

Use the order of operations to find the value of each expression.

31. $2 \cdot 6 + 0.073$
32. $3.4 + 5 \cdot 3$
33. $6 - 4.304 + 2.5$
34. $11.8 - 2^2$
35. $8 + 6.3 - 3.9$
36. $4^2 - 1.67$

37. **POPULATION** If the world's population is 6.3 billion people and grows by 2.6 billion by 2050, how many people will there be in 2050?

38. **ANALYZE TABLES** The table shows the top e-mail services in a recent year. How many more million people used the top three services than the bottom two? Do you believe that the services chosen from year to year would be the same? Explain your reasoning.

Top Five E-Mail Services	
Service	**Number of People (millions)**
Yahoo!	51.9
AOL	32.1
MSN	29.4
Google	7.3
BellSouth	2.5

Source: Nielsen Ratings

39. **FIND THE DATA** Refer to the Data File on pages 16–19. Choose some data and write a real-world problem in which you would add more than two decimals.

40. **SHOPPING** You have \$10 to buy art supplies. Can you buy construction paper that costs \$2.69, a glue stick that costs \$1.59, and markers that cost \$5.15? Explain your reasoning.

H.O.T. Problems

41. **CHALLENGE** Using each of the digits 1–8 only once, find two decimals that are each less than one and whose sum is the greatest possible value.

42. **REASONING** Find a counterexample for the following statement.

If two decimals each have their last nonzero digit in the hundredths place, their sum also has its last nonzero digit in the hundredths place.

43. **FIND THE ERROR** Noah and Yoko are finding 8.9 – 3.72. Who is correct? Explain your reasoning.

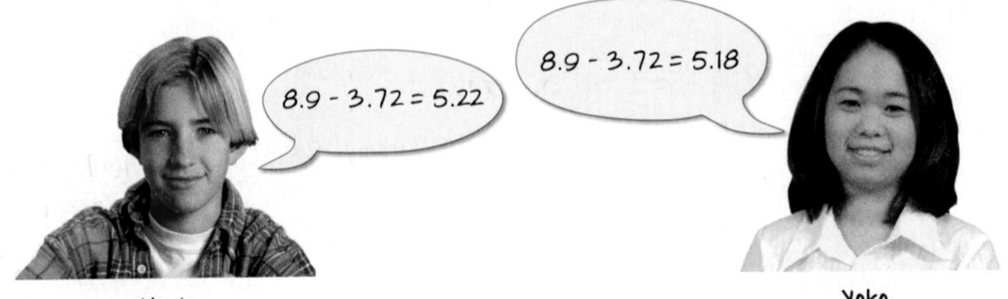

44. **WRITING IN MATH** Explain how you would find the difference of 3 and 2.89.

TEST PRACTICE

45. Jamal took $15.00 to spend at a sports card store. Baseball cards cost $1.75 per pack, and hockey cards cost $0.99 per pack. If Jamal buys 6 packs of baseball cards for $10.50, how can he determine how much money he has left to spend on hockey cards?

A Subtract $10.50 from $15.00

B Add $1.75 and $0.99

C Subtract $0.99 from $1.75

D Add $0.99 and $10.50

46. **SHORT RESPONSE** The table lists the average number of persons per square mile for several states.

State	Population
Florida	296.4
Indiana	169.5
Kentucky	101.7
North Carolina	165.2

Source: *The World Almanac 2007*

How many more people per square mile are in Florida than in Kentucky?

Spiral Review

Estimate. (Lesson 3-4)

47. 4.231 + 3.98

48. 3.945 + 1.92 + 3.55

49. 9.345 – 6.625

50. Round 28.561 to the nearest tenth. (Lesson 3-3)

GET READY for the Next Lesson

51. **PREREQUISITE SKILL** In a recent year, there were 45,033 beagles registered with the American Kennel Club. If there were 3 times as many labradors registered in the same year, find the number of registered labradors. (Page 744)

CHAPTER 3

Mid-Chapter Quiz

Lessons 3-1 through 3-5

Write each decimal in word form. (Lesson 3-1)

1. 0.6
2. 12.65
3. 3.0091
4. 0.25

Write each decimal in standard form and in expanded form. (Lesson 3-1)

5. four tenths
6. fifteen and seventy-two hundredths

7. **SKIING** Bianca's speed while cross-country skiing was 2.5 miles per hour. Write this number in two other forms. (Lesson 3-1)

Use $>$, $<$, or $=$ to compare each pair of decimals. (Lesson 3-2)

8. 0.06 ● 0.6
9. 8.04 ● 8.0004
10. 6.3232 ● 6.3202
11. 2.15 ● 2.150

12. **ANIMALS** The table shows the length of two of the world's smallest animals. Which animal is smaller? (Lesson 3-2)

Animal	Length (inches)
Brazilian Frog	0.33
Dwarf Goby Fish	0.30

Source: *The World Almanac for Kids*

13. Order 0.101, 0.0101, 0.011, 1.00001 from least to greatest. (Lesson 3-2)

14. **MULTIPLE CHOICE** Ruben recorded the lengths of his model airplanes in inches. Which list shows the lengths in order from greatest to least? (Lesson 3-2)

    A 7.2, 7.35, 8.01, 8.10
    B 7.35, 7.2, 8.01, 8.10
    C 8.01, 8.10, 7.2, 7.35
    D 8.10, 8.01, 7.35, 7.2

Round each decimal to the indicated place-value position. (Lesson 3-3)

15. 8.236; tenths
16. 10.0879; thousandths
17. 2.38141; ten-thousandths

18. **SPIDERS** The *Tegenaria atrica* spider can travel at a speed of 20.592 inches per second. What is the speed rounded to the nearest tenth? (Lesson 3-3)

Estimate using rounding, clustering, or front-end estimation. (Lesson 3-4)

19. 18.89 − 4.42
20. 42.33 + 13.48
21. 11.94 + 12.21 + 11.88 + 12.08

22. **MULTIPLE CHOICE** The table shows the weights of four packages that Martin is mailing to his cousin in New Jersey. Estimate the total weight of the packages. (Lesson 3-4)

Package	Weight (oz)
1	3.94
2	14.81
3	11.27
4	7.65

F 42 ounces
G 38 ounces
H 34 ounces
J 32 ounces

Find each sum or difference. (Lesson 3-5)

23. 67.13 + 31.7
24. 51.2 − 12.94

25. **RESTAURANTS** Andrea has a coupon for $1.75 off her next purchase of a deli sandwich. If the sandwich originally costs $5.65, how much will it cost with the coupon? (Lesson 3-5)

Explore 3-6

Math Lab
Multiplying Decimals by Whole Numbers

MAIN IDEA

Use models to multiply a decimal by a whole number.

You can use decimal models to multiply a decimal by a whole number. Recall that a 10-by-10 grid represents the number one.

ACTIVITY

Model 0.9 × 3 using decimal models.

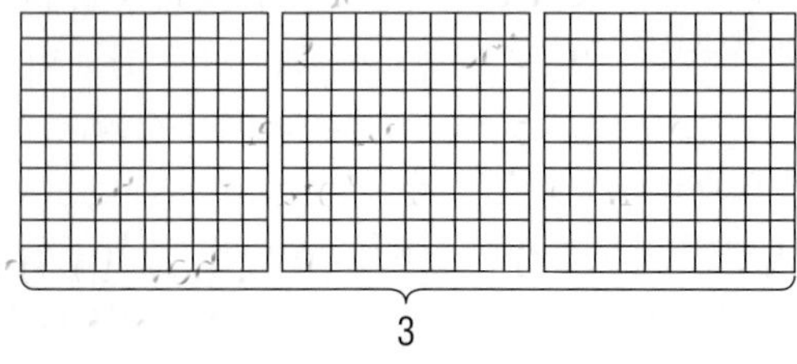

Draw three 10-by-10 decimal models to show the factor 3.

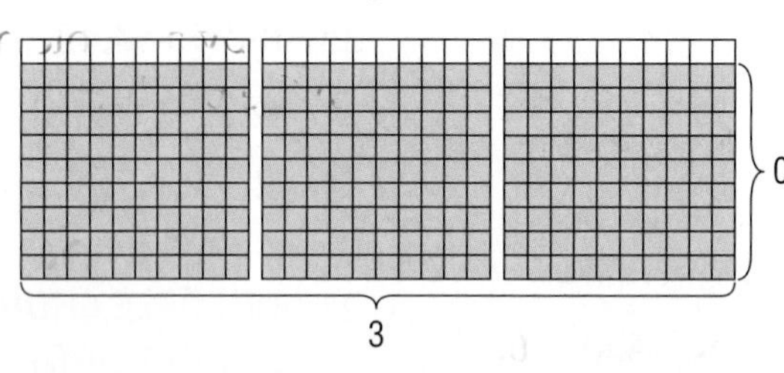

Shade nine rows of each decimal model to represent 0.9.

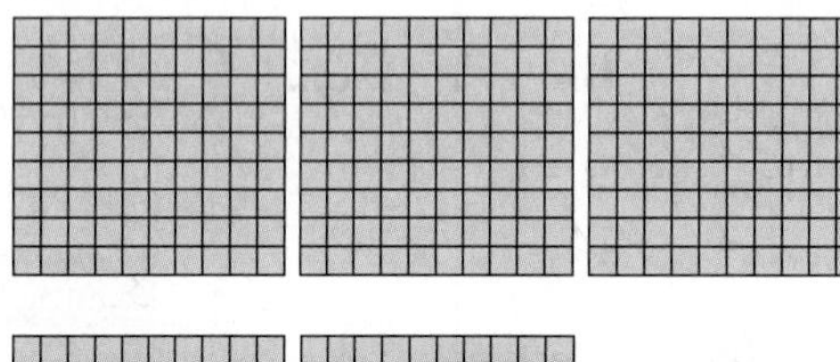

Cut off the shaded rows and rearrange them to form as many 10-by-10 grids as possible.

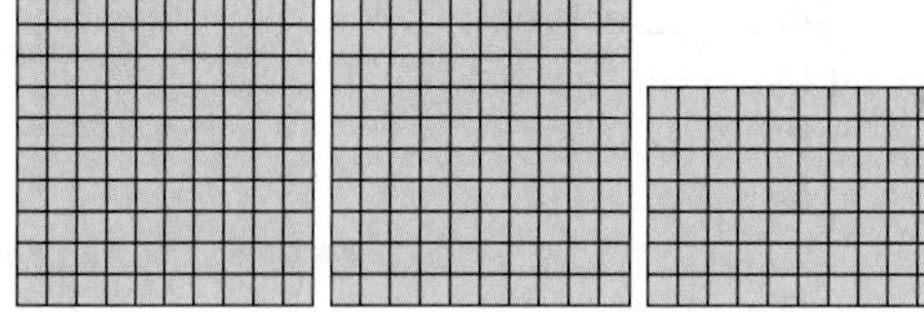

The product is *two and seven tenths.*

So, $0.9 \times 3 = 2.7$.

CHECK Your Progress

Use decimal models to show each product.

a. 3×0.5　　b. 2×0.7　　c. 0.8×4

Analyze the Results

1. **MAKE A CONJECTURE** Is the product of a whole number and a decimal greater or less than the whole number? Explain.
2. Test your conjecture on 7×0.3. Check your answer by making a model or by using a calculator.

3-6 Multiplying Decimals by Whole Numbers

MAIN IDEA

Estimate and find the product of decimals and whole numbers.

Math Online

glencoe.com

- Extra Examples
- Personal Tutor
- Self-Check Quiz

GET READY for the Lesson

PLANTS Bamboo can grow about 4.92 feet in height per day. The table shows different ways to find the total height a bamboo plant can grow in two days.

Growth of Bamboo over Two Days	
Add.	4.92 ft + 4.92 ft = 9.84 ft
Estimate.	4.92 is about 5. $2 \times 5 = 10$
Multiply.	2×4.92 ft = ■

1. Use the addition problem and the estimate to find 2×4.92.
2. Write an addition problem, an estimate, and a multiplication problem to find the total growth over 3 days, 4 days, and 5 days.
3. **MAKE A CONJECTURE** about how to find 5.35×4.

When multiplying a decimal by a whole number, use estimation to place the decimal point in the product. You can also count the number of decimal places.

EXAMPLES Multiply Decimals

1 **Find 14.2 × 6.**

METHOD 1 Use estimation.

Round 14.2 to 14.

$14.2 \times 6 \longrightarrow 14 \times 6$ or 84

$$\begin{array}{r} {}^{21} \\ 14.2 \\ \times\ 6 \\ \hline 85.2 \end{array}$$

Since the estimate is 84, place the decimal point after the 5.

METHOD 2 Count decimal places.

$$\begin{array}{r} {}^{21} \\ 14.2 \\ \times\ 6 \\ \hline 85.2 \end{array}$$

14.2: one decimal place

85.2: Count one decimal place from the right.

2 **Find 9 × 0.83.**

METHOD 1 Use estimation.

Round 0.83 to 1.

$9 \times 0.83 \longrightarrow 9 \times 1$ or 9

$$\begin{array}{r} {}^{2} \\ 0.83 \\ \times\ 9 \\ \hline 7.47 \end{array}$$

Since the estimate is 9, place the decimal point after the 7.

METHOD 2 Count decimal places.

$$\begin{array}{r} {}^{2} \\ 0.83 \\ \times\ 9 \\ \hline 7.47 \end{array}$$

0.83: two decimal places

7.47: Count two decimal places from the right.

CHOOSE Your Method Multiply.

a. 3.4×5 b. 11.4×8 c. 7×2.04

If there are not enough decimal places in the product, you need to annex zeros to the left.

EXAMPLES Annex Zeros in the Product

3 **Find 2×0.018.**

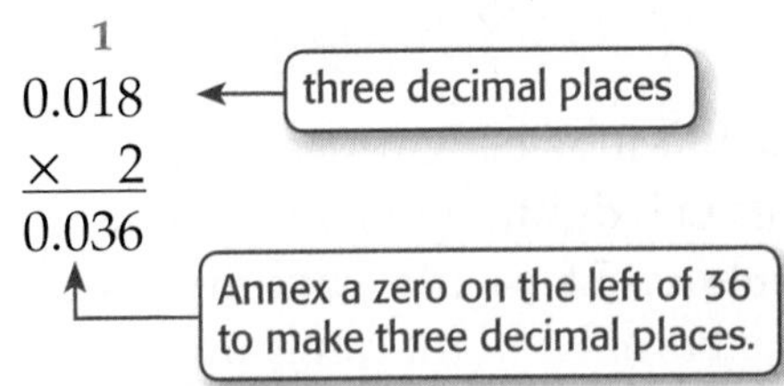

$$\begin{array}{r} \scriptstyle 1 \\ 0.018 \\ \times \quad 2 \\ \hline 0.036 \end{array}$$

three decimal places

Annex a zero on the left of 36 to make three decimal places.

Vocabulary Link

Annex

Everyday Use to add something

Math Use to annex a zero means to add a zero at the beginning or end of a decimal

4 **ALGEBRA Evaluate $4c$ if $c = 0.0027$.**

$4c = 4 \times 0.0027$ Replace c with 0.0027.

$$\begin{array}{r} \scriptstyle 2 \\ 0.0027 \\ \times \quad 4 \\ \hline 0.0108 \end{array}$$

four decimal places

Annex a zero to make four decimal places.

CHECK Your Progress **Multiply.**

d. 3×0.02

e. 0.12×8

f. 11×0.045

g. **ALGEBRA** Evaluate $7x$ if $x = 0.03$.

You can use paper and pencil or mental math to multiply a decimal by 10, 100, or 1,000.

EXAMPLE Multiply by 10, 100, or 1,000

5 **SCIENCE Find $5.7 \times 1{,}000$.**

METHOD 1 **Use paper and pencil.**

$$\begin{array}{r} 1{,}000 \\ \times \quad 5.7 \\ \hline 7000 \\ 50000 \\ \hline 5{,}700.0 \end{array}$$

one decimal place (5.7)

one decimal place (5,700.0)

METHOD 2 **Use mental math.**

Move the decimal point to the right the same number of zeros that are in 1,000, or 3 places.

$5.7 \times 1{,}000 = 5.700$ or 5,700.

Real-World Link The temperature of the surface of the Sun, in degrees Celsius, can be found by multiplying 5.7 by 1,000.

Source: *Solar Week*

CHOOSE Your Method

h. $7.9 \times 1{,}000$

i. 4.13×10

j. 2.3×100

CHECK Your Understanding

Examples 1, 2 (p. 163)

Multiply.

1. 2.7×6
2. 1.4×4
3. 0.52×3
4. $\$0.83 \times 6$

Examples 3, 4 (p. 164)

5. 5×0.09
6. 4×0.012
7. 0.065×18
8. 0.015×23
9. **ALGEBRA** Evaluate $14t$ if $t = 2.9$.

Example 5 (p. 164)

10. **MOON** The approximate radius of the Moon, in kilometers, can be found by multiplying 17.36 by 10. Find the Moon's radius.

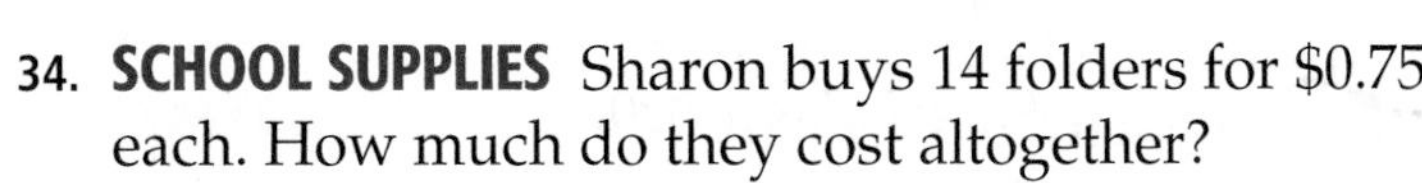

Practice and Problem Solving

HOMEWORK HELP

For Exercises	See Examples
11–18, 33, 34	1, 2
19–22	3
23, 24	4
25–32, 35, 36	5

Multiply.

11. 1.2×7
12. 1.7×5
13. 0.7×9
14. 0.9×4
15. 2×1.3
16. 2.4×8
17. 0.8×9
18. 3×0.5
19. 3×0.02
20. 7×0.012
21. 0.0036×19
22. 0.0198×75
23. **ALGEBRA** Evaluate $3.05n$ if $n = 27$.
24. **ALGEBRA** Evaluate $80.44w$ if $w = 2$.

Multiply.

25. 5.2×10
26. 4.8×100
27. $1.5 \times 1{,}000$
28. 9.3×100
29. 2.5×10
30. 1.26×100
31. $3.45 \times 1{,}000$
32. 2.17×10
33. **MEASUREMENT** Asher recently bought the poster shown at the right. What is its area?

34. **SCHOOL SUPPLIES** Sharon buys 14 folders for $0.75 each. How much do they cost altogether?
35. **MEASUREMENT** The height of Mount Everest, in meters, can be found by multiplying 8.85 by 1,000. Find the height of Mount Everest.
36. **TEMPERATURE** The hottest temperature recorded in the world, in degrees Fahrenheit, can be found by multiplying 13.46 by 10. Find this temperature.

Use the order of operations to find the value of each expression.

37. $2 \times 3.8 + 1.5$
38. $7 - 4 \times 0.8$
39. $3 \times 2.14 \times 10$
40. $5 + 2.6 \times 1{,}000$
41. $11 \times 7.85 + 33$
42. $19 + 0.4 \times 100$
43. **MEASUREMENT** The thickness of each type of coin is shown in the table. How much thicker is a stack of a dollar's worth of nickels than a dollar's worth of quarters?

Coin	Thickness (mm)
penny	1.55
nickel	1.95
dime	1.35
quarter	1.75

EXTRA PRACTICE
See pages 679, 708.

H.O.T. Problems

44. **OPEN ENDED** Create a problem about a real-world situation involving multiplication by a decimal factor. Then solve the problem.

45. **CHALLENGE** Discuss two different ways to find the value of the expression 5.4 × 1.17 × 100 that do not require you to first multiply 5.4 × 1.17.

46. **WRITING IN MATH** Summarize how to use mental math to multiply a decimal by a power of 10.

TEST PRACTICE

47. A recipe for a batch of cookies calls for one 5.75-ounce package of coconut. How many ounces of coconut are needed for 5 batches of cookies?

A 20.50 oz

B 25.25 oz

C 28.75 oz

D 29.75 oz

48. The table shows the admission prices to an amusement park.

Admission Prices	One-Day Pass	Two-Day Pass
Adult	$39.59	$43.99
Child (ages 3–9)	$30.59	$33.99

What is the total price of one-day passes for two adults and three children?

F $140.36

G $170.95

H $179.95

J $189.95

Spiral Review

For Exercises 49 and 50, refer to the table that shows the music sales in the United States in a recent year that were devoted to different types of music. (Lesson 3-5)

Total Music Sales	
Type of Music	Sales (millions of $)
rock	3,864.89
rap/hip-hop	1,631.84
R&B/urban	1,251.49
country	1,533.69
pop	993.83
other	2,785.18

Source: Recording Industry of America

49. What were the total sales for rock and country?

50. How much more money came from rap/hip-hop than pop?

51. **FUNDRAISING** During a fundraiser at her school, Careta sold $78.35 worth of candy. Diego sold $59.94 worth of candy. Use front-end estimation to find about how much more Careta sold. (Lesson 3-4)

Use >, <, or = to compare each pair of decimals. (Lesson 3-2)

52. 14.05 ● 14.5

53. 61.32 ● 61.23

54. 7.71 ● 7.17

GET READY for the Next Lesson

PREREQUISITE SKILL Find the value of each expression. (Page 744)

55. 43 × 25

56. 126 × 13

57. 18 × 165

Explore 3-7

Math Lab
Multiplying Decimals

MAIN IDEA

Use decimal models to multiply decimals.

In the lab on page 162, you used decimal models to multiply a decimal by a whole number. You can use similar models to multiply a decimal by a decimal.

ACTIVITY

1 Model 0.7 × 0.6 using decimal models.

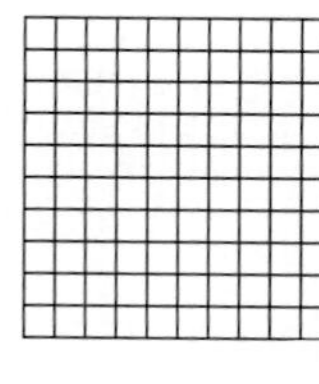

Draw a 10-by-10 decimal model. Recall that each small square represents 0.01.

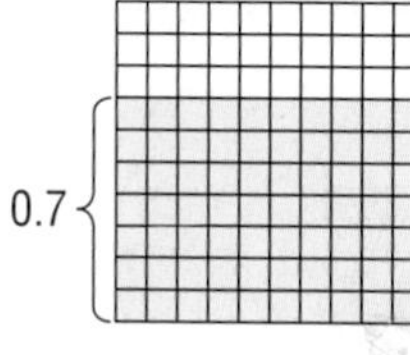

Shade seven rows of the model yellow to represent the first factor, 0.7.

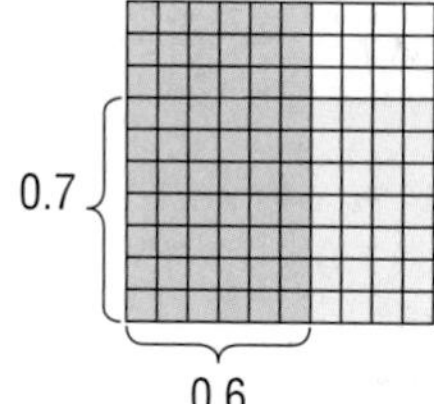

Shade six columns of the model blue to represent the second factor, 0.6.

There are *forty-two hundredths* in the region that is shaded green. So, $0.7 \times 0.6 = 0.42$.

CHECK Your Progress

Use decimal models to show each product.

a. 0.3×0.3 b. 0.4×0.9 c. 0.9×0.5

Analyze the Results

1. Tell how many decimal places are in each factor and in each product of Exercises a–c above.
2. **MAKE A CONJECTURE** Use the pattern you discovered in Exercise 1 to find 0.6×0.2. Check your conjecture with a model or a calculator.
3. Find two decimals with a product of 0.24.

ACTIVITY

2 **Model 0.8 × 2.9 using decimal models.**

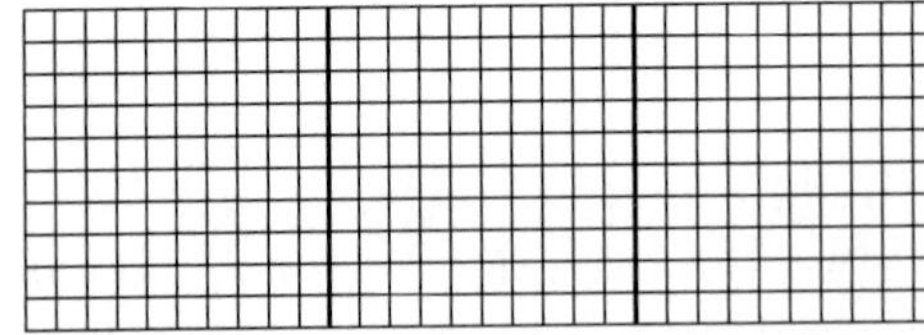

Draw three 10-by-10 decimal models.

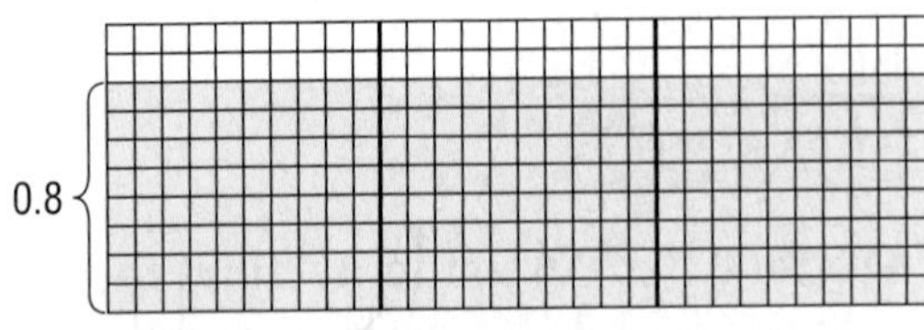

Shade 8 rows to represent 0.8.

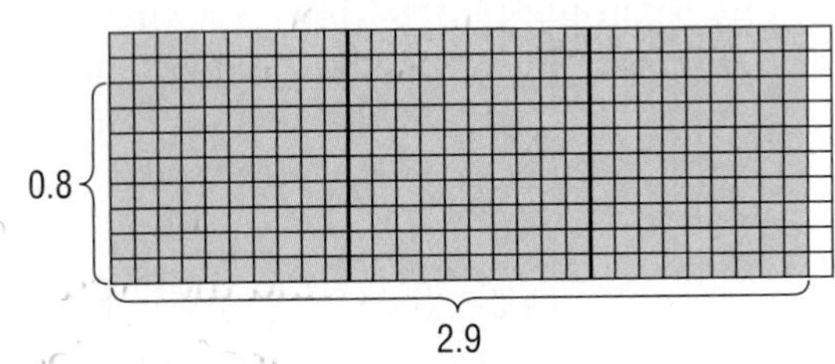

Shade 2 large squares and 9 columns of the next large square to represent 2.9.

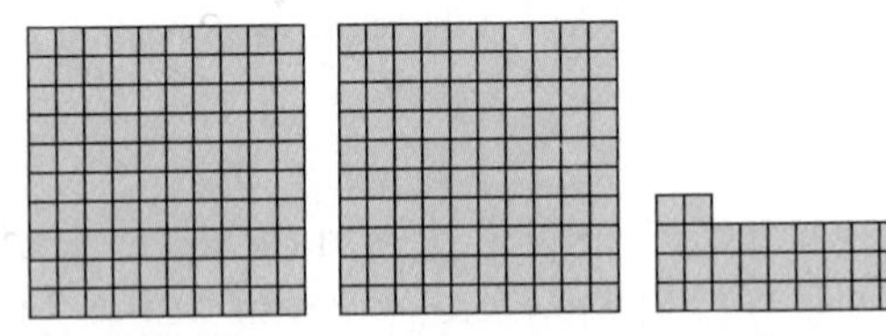

Cut off the squares that are shaded twice and rearrange them to form 10-by-10 grids.

There are *two and thirty-two hundredths* in the region that are shaded green. So, 0.8 × 2.9 = 2.32.

Study Tip

Arranging Squares Arrange the squares to form as many whole decimal models as possible. Then arrange the remaining squares into as many rows of 10 as possible to make counting them easier.

CHECK Your Progress

Use decimal models to show each product.

d. 1.5 × 0.7 **e.** 0.8 × 2.4 **f.** 1.3 × 0.3

Analyze the Results

4. **MAKE A CONJECTURE** How does the number of decimal places in the product relate to the number of decimal places in the factors?

5. Analyze each product.

First Factor		Second Factor		Product
0.9	×	0.6	=	0.54
1.0	×	0.6	=	0.60
1.5	×	0.6	=	0.90

   a. Explain why the first product is less than 0.6.
   b. Explain why the second product is equal to 0.6.
   c. Explain why the third product is greater than 0.6.

3-7 Multiplying Decimals

MAIN IDEA

Multiply decimals by decimals.

Math Online

glencoe.com

- Extra Examples
- Personal Tutor
- Self-Check Quiz

GET READY for the Lesson

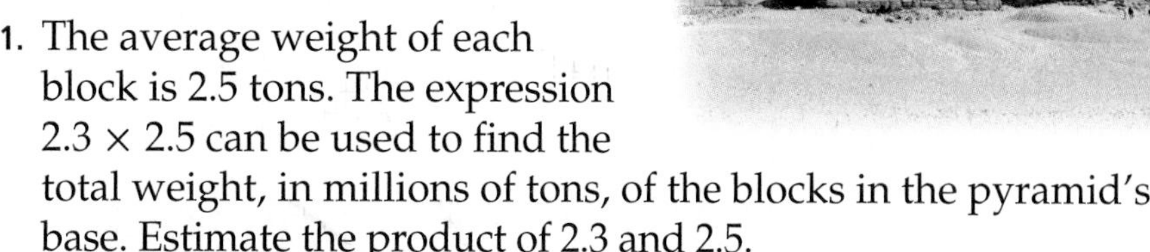

PYRAMIDS The largest of the Great Pyramids at Giza, in Egypt, contains 2.3 million blocks in its base.

1. The average weight of each block is 2.5 tons. The expression 2.3 × 2.5 can be used to find the total weight, in millions of tons, of the blocks in the pyramid's base. Estimate the product of 2.3 and 2.5.
2. Multiply 23 by 25.
3. **MAKE A CONJECTURE** about how you can use your answers in Exercises 2 and 3 to find the product of 2.3 and 2.5.
4. What is the total weight of the blocks in the pyramid's base?
5. Use your conjecture in Exercise 3 to find 1.7 × 5.4. Explain each step.

When multiplying a decimal by a decimal, multiply as with whole numbers. To place the decimal point, find the sum of the number of decimal places in each factor. The product has the same number of decimal places.

EXAMPLES Multiply Decimals

1 Find 4.2 × 6.7. **Estimate** 4.2 × 6.7 → 4 × 7 or 28

```
  4.2   ← one decimal place
× 6.7   ← one decimal place
  294
+ 252
28.14   ← two decimal places
```

The product is 28.14. Compared to the estimate, the product is reasonable.

2 Find 1.6 × 0.09. **Estimate** 1.6 × 0.09 → 2 × 0 or 0

```
   1.6  ← one decimal place
× 0.09  ← two decimal places
 0.144  ← three decimal places
```

The product is 0.144. Compared to the estimate, the product is reasonable.

CHECK Your Progress

a. 5.7 × 2.8 b. 4.12 × 0.07 c. 0.014 × 3.7

EXAMPLE Evaluate an Expression

3 ALGEBRA Evaluate $1.4x$ if $x = 0.067$.

$1.4x = 1.4 \times 0.067$ Replace x with 0.067.

$$\begin{array}{r} 0.067 \\ \times\ 1.4 \\ \hline 268 \\ +\ 67 \\ \hline 0.0938 \end{array}$$

0.067 ← three decimal places
× 1.4 ← one decimal place
0.0938 ← Annex a zero to make four decimal places.

CHECK Your Progress Evaluate each expression.

d. $0.04t$, if $t = 3.2$ e. $2.6b$, if $b = 2.05$ f. $1.33c$, if $c = 0.06$

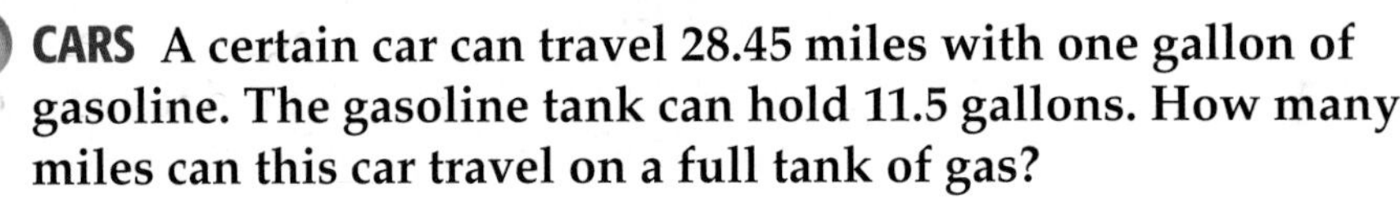

Real-World EXAMPLE

Real-World Link A car that can travel 20 miles on one gallon of gasoline will cost about $600 per year more, in gasoline costs alone, than a car that can travel 30 miles on one gallon of gasoline.

Source: Federal Trade Commission

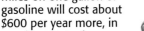

4 CARS A certain car can travel 28.45 miles with one gallon of gasoline. The gasoline tank can hold 11.5 gallons. How many miles can this car travel on a full tank of gas?

Estimate $28.45 \times 11.5 \rightarrow 30 \times 12$ or 360

$$\begin{array}{r} 28.45 \\ \times\ 11.5 \\ \hline 14225 \\ 2845 \\ +\ 2845 \\ \hline 327.175 \end{array}$$

28.45 ← two decimal places
× 11.5 ← one decimal place
327.175 ← The product has three decimal places.

The car could travel 327.175 miles.

CHECK Your Progress

g. **NUTRITION FACTS** A nutrition label indicates that one serving of apple crisp oatmeal has 2.5 grams of fat. How many grams of fat are there in 3.75 servings?

CHECK Your Understanding

Examples 1, 2 (p. 169)

Multiply.

1. 0.6×0.5 2. 1.4×2.56 3. 27.43×1.089
4. 0.3×2.4 5. 0.52×2.1 6. 0.45×0.053

Example 3 (p. 170)

ALGEBRA Evaluate each expression if $n = 1.35$.

7. $2.7n$ 8. $5.343 + 0.5n$ 9. $0.02n + 0.016$

Example 4 (p. 170)

10. **MEASUREMENT** A mile is approximately equal to 1.609 kilometers. How many kilometers is 2.5 miles?

Practice and Problem Solving

HOMEWORK HELP	
For Exercises	See Examples
11–22	1, 2
23–28	3
29–30	4

Multiply.

11. 0.7×0.4
12. 1.5×2.7
13. 0.4×3.7
14. 3.1×0.8
15. 0.98×7.3
16. 2.4×3.48
17. 6.2×0.03
18. 5.04×3.2
19. 14.7×11.36
20. 27.4×33.68
21. 0.28×0.08
22. 0.45×0.05

ALGEBRA Evaluate each expression if $x = 8.6$, $y = 0.54$, and $z = 1.18$.

23. $2.7x$
24. $6.34y$
25. $3.45x + 7.015$
26. $1.8y + 0.6z$
27. $9.1x - 4.7y$
28. $0.096 + 2.28y$

29. **ANIMALS** A giraffe can run up to 46.93 feet per second. How far could a giraffe run in 1.8 seconds?

30. **MEASUREMENT** Katelyn has a vegetable garden that measures 16.75 feet in length and 5.8 feet in width. Find the area of the garden.

Multiply.

31. 25.04×3.005
32. 1.03×1.005
33. 5.12×4.001

ALGEBRA Use the order of operations to evaluate each expression if $a = 1.3$, $b = 0.042$, and $c = 2.01$.

34. $ab + c$
35. $a \times 6.023 - c$
36. $3.25c + b$
37. abc

38. **MEASUREMENT** Find the area of the figure at the right. Justify your answer.

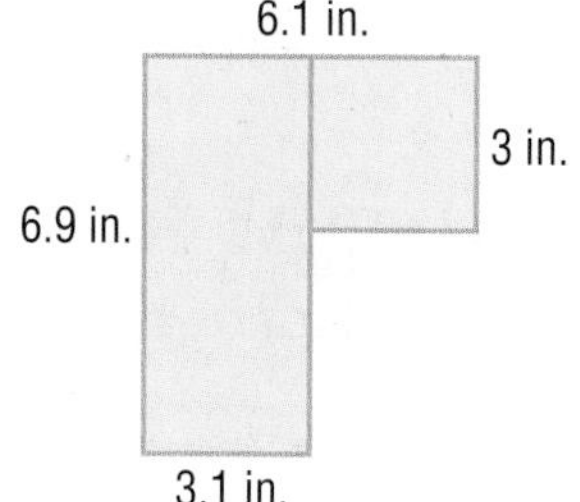

39. **ALGEBRA** Which of the three numbers 9.2, 9.5, or 9.7 is the correct solution of $2.65t = 25.705$?

40. **GROCERY SHOPPING** Pears cost \$0.98 per pound, and apples cost \$1.05 per pound. Mr. Bonilla bought 3.75 pounds of pears and 2.1 pounds of apples. How much did he pay for the pears and apples?

41. **FIND THE DATA** Refer to the Data File on pages 16–19. Choose some data and write a real-world problem in which you would multiply decimals.

For each statement below, find two decimals a and b that make the statement true. Then find two decimals a and b that make the statement false. Explain your reasoning.

EXTRA PRACTICE
See pages 680, 708.

42. If $a > 1$ and $b < 1$, then $ab < 1$.
43. If $ab < 1$, then $a < 1$ and $b < 1$.

H.O.T. Problems

CHALLENGE Evaluate each expression.

44. $0.3(3 - 0.5)$ 45. $0.16(7 - 2.8)$ 46. $1.06(2 + 0.58)$

47. **OPEN ENDED** Write a multiplication problem in which the product is between 0.05 and 0.75.

48. **NUMBER SENSE** Place the decimal point in the answer to make it correct. Explain your reasoning. $3.9853 \times 8.032856 = 32013341...$

49. **WRITING IN MATH** Describe two methods for determining where to place the decimal point in the product of two decimals.

TEST PRACTICE

50. What is the area of the rectangle?

1.4 cm
5.62 cm

A 14.04 cm^2
B 10.248 cm^2
C 8.992 cm^2
D 7.868 cm^2

51. Josefina took her grandmother to lunch. Josefina's lunch was \$6.70, her grandmother's lunch was \$7.25, and they split a dessert that cost \$3.50. If there was an 8.75% tax on the food, which procedure could be used to find the amount of tax Josefina paid for their lunch?

F Add the prices of the food items.
G Add the prices of the food items to the tax rate.
H Multiply the tax rate by the price of the most expensive food item.
J Multiply the tax rate by the sum of the prices of the food items.

Spiral Review

Multiply. (Lesson 3-6)

52. 45×0.27 53. 3.2×109 54. 27×0.45 55. 2.94×16

For Exercises 56 and 57, use the information below. The distance around Earth at the equator is about 24,889.78 miles. The distance around Earth through the North Pole and South Pole is about 24,805.94 miles. (Lesson 3-5)

56. How much greater is the distance at the equator than through the poles?

57. The mean distance around Earth is 24,847.86 miles. How much greater is the distance at the equator than the mean distance?

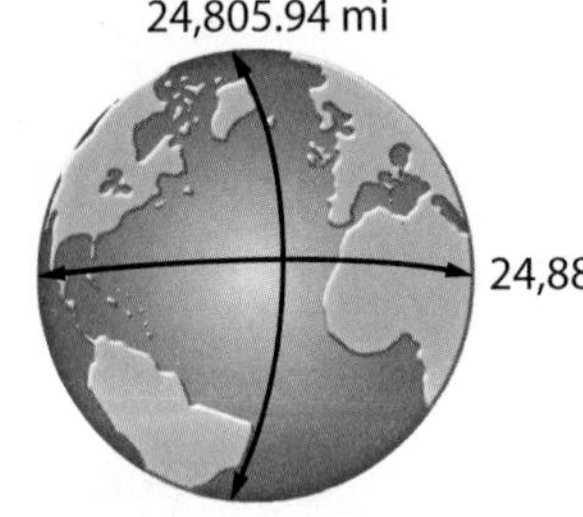

GET READY for the Next Lesson

PREREQUISITE SKILL Divide. (Page 744)

58. $21 \div 3$ 59. $81 \div 9$ 60. $56 \div 8$ 61. $63 \div 7$

3-8 Dividing Decimals by Whole Numbers

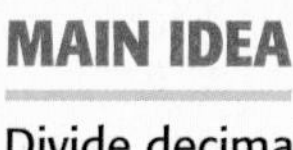

MAIN IDEA

Divide decimals by whole numbers.

Math Online

glencoe.com

- Extra Examples
- Personal Tutor
- Self-Check Quiz

MINI Lab

To find $3.6 \div 3$ using base-ten blocks, model 3.6 as 3 wholes and 6 tenths. Then separate into three equal groups.

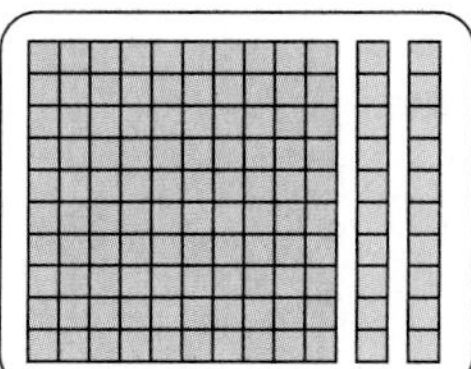 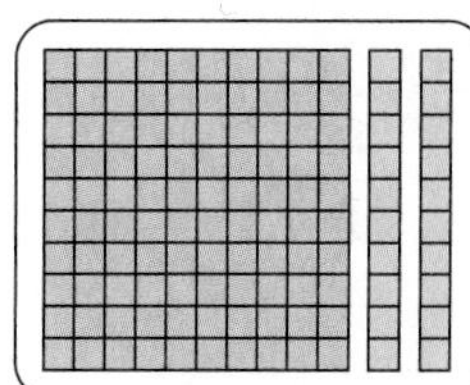 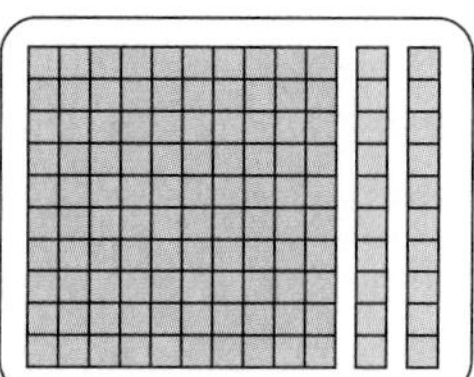

There is one whole and two tenths in each group.

So, $3.6 \div 3 = 1.2$.

Use base-ten blocks to show each quotient.

1. $3.4 \div 2$
2. $4.2 \div 3$
3. $5.6 \div 4$

Find each whole number quotient.

4. $34 \div 2$
5. $42 \div 3$
6. $56 \div 4$
7. Compare and contrast the quotients in Exercises 1–3 with the quotients in Exercises 4–6.
8. **MAKE A CONJECTURE** Write a rule for dividing a decimal by a whole number.

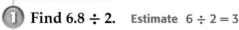

Dividing a decimal by a whole number is similar to dividing whole numbers.

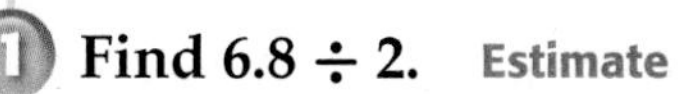

EXAMPLE Divide a Decimal by a 1-Digit Number

1 **Find 6.8 ÷ 2.** **Estimate** $6 \div 2 = 3$

$$\begin{array}{r} 3.4 \\ 2\overline{)6.8} \\ -6 \\ \hline 0\,8 \\ -8 \\ \hline 0 \end{array}$$

← Place the decimal point directly above the decimal point in the dividend.

$6.8 \div 2 = 3.4$ Compared to the estimate, the quotient is reasonable.

CHECK Your Progress

Divide.

a. $7.5 \div 3$

b. $3.5 \div 7$

c. $9.8 \div 2$

EXAMPLE Divide a Decimal by a 2-Digit Number

Study Tip

Checking Your Answer To check that the answer is correct, multiply the quotient by the divisor. In Example 2, $0.55 \times 14 = 7.7$.

2 **Find 7.7 ÷ 14.** **Estimate** $10 \div 10 = 1$

$$\begin{array}{r} 0.55 \\ 14\overline{)7.70} \\ -7\,0 \\ \hline 70 \\ -70 \\ \hline 0 \end{array}$$

0.55 ← Place the decimal point.

Annex a zero and continue dividing.

$7.7 \div 14 = 0.55$ Compared to the estimate, the quotient is reasonable.

CHECK Your Progress **Divide.**

d. $9.48 \div 15$ e. $3.49 \div 4$ f. $55.08 \div 17$

If the answer does not come out evenly, round the quotient to a specified place-value position.

TEST EXAMPLE

3 **SHORT RESPONSE** Michelle bought a dozen blueberry muffins for \$14.92. If each muffin costs the same amount, find the price of each muffin in dollars. **Estimate** \$15 ÷ 12

Read the Item

To find the price of one muffin, divide the total cost by the number of muffins. Round to the nearest cent, or hundredths place.

Solve the Item

$$\begin{array}{r} 1.243 \\ 12\overline{)14.92} \\ -12 \\ \hline 29 \\ -24 \\ \hline 52 \\ -48 \\ \hline 40 \\ -36 \\ \hline 4 \end{array}$$

1.243 Place the decimal point.

40 − 36: Divide until you place a digit in the thousandths place.

To the nearest cent, the cost in dollars is 1.24.

Test-Taking Tip

Dividing Decimals Be sure to place the decimal point in the quotient when dividing.

CHECK Your Progress

g. **SHORT RESPONSE** A bag of 12 bagels costs \$7.50. To the nearest cent, find the cost of each bagel.

CHECK Your Understanding

Examples 1, 2 (pp. 173–174)

Divide. Round to the nearest tenth if necessary.

1. 3.6 ÷ 4
2. 9.6 ÷ 2
3. 8.53 ÷ 6
4. 1087.9 ÷ 46
5. 12.32 ÷ 22
6. 69.904 ÷ 34

Example 3 (p. 174)

7. **TEST PRACTICE** A light-year, the distance that light travels in one year, is 5.88 trillion miles. How many trillion miles will light travel in one month?

Practice and Problem Solving

HOMEWORK HELP	
For Exercises	See Examples
8–13, 20, 21	1
14–19	2
31–32	3

Divide. Round to the nearest tenth if necessary.

8. 39.39 ÷ 3
9. 36.8 ÷ 2
10. 118.5 ÷ 5
11. 124.2 ÷ 9
12. 7.24 ÷ 7
13. 6.27 ÷ 4
14. 11.4 ÷ 19
15. 10.22 ÷ 14
16. 55.2 ÷ 46
17. 59.84 ÷ 32
18. 336.75 ÷ 31
19. 751.2 ÷ 25

20. **INSURANCE** Aurelia pays $414.72 per year for auto insurance. Suppose she makes 4 equal payments a year. How much does she pay every three months?

21. **BUILDINGS** Find the average height of the buildings shown in the table.

World's Tallest Buildings (thousands of feet)				
1.667	1.483	1.483	1.451	1.381

22. **MEASUREMENT** Mr. Jamison will stain the deck in his backyard. The deck has an area of 752.4 square feet. If the deck is 33 feet long, how wide is it?

23. **FOOD** The Student Council is raising money by selling bottled water at a band competition. The table shows the prices for different brands. Which brand is the best buy? Explain your reasoning.

Cost of Bottled Water (20-oz bottles)		
Brand A	6-pack	$3.45
Brand B	12-pack	$5.25
Brand C	24-pack	$10.99

24. **MEASUREMENT** The Verrazano-Narrows Bridge in New York City is 4.26 thousand feet long and is the seventh longest suspension bridge in the world. There are 3 feet in a yard. How long is the bridge in yards?

EXTRA PRACTICE
See pages 680, 708.

STATISTICS Find the mean for each set of data.

25. 22.6, 24.8, 25.4, 26.9
26. 1.43, 1.78, 2.45, 2.78, 3.25

H.O.T. Problems

27. **OPEN ENDED** Create a set of data for which the mean is 5.5.

28. **CHALLENGE Find each of the following quotients. Then find a pattern and explain how you can use this pattern to mentally divide 0.0096 by 3.**

844 ÷ 2   0.844 ÷ 2   84.4 ÷ 2   0.0844 ÷ 2   8.44 ÷ 2   0.00844 ÷ 2

29. **FIND THE ERROR** Felisa and Tabitha are finding 11.2 ÷ 14. Who is correct? Explain your reasoning.

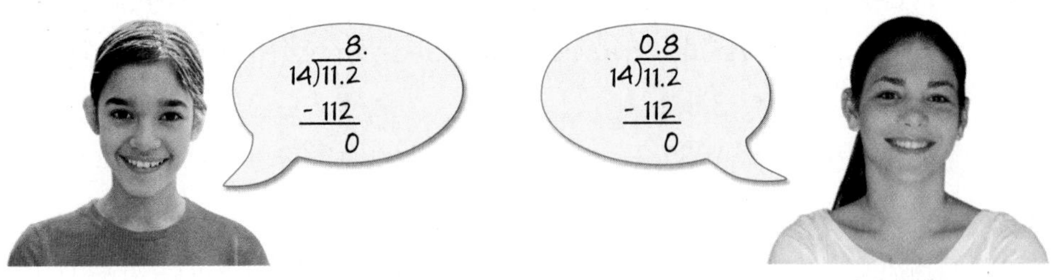

30. **WRITING IN MATH** Explain how you can use estimation to place the decimal point in the quotient 42.56 ÷ 22.

TEST PRACTICE

31. **SHORT RESPONSE** Tanner and three neighborhood friends are buying a basketball hoop that costs $249.84. If the cost is divided equally, how much will each person pay in dollars?

32. The table shows the number of subscribers to several Internet providers.

Internet Provider	Subscribers (millions)
Company A	2.45
Company B	3.12
Company C	2.8

What is the mean number of subscribers for these Internet providers?

A 2.9 million **C** 2.79 million

B 2.84 million **D** 2.52 million

Spiral Review

Multiply. (Lesson 3-7)

33. 2.4 × 5.7 34. 1.6 × 2.3 35. 0.32(8.1) 36. 2.68(0.84)

37. What is the product of 4.156 and 12? (Lesson 3-7)

For Exercises 38–40, write each power as a product of the same factor. Then find the value. (Lesson 1-3)

38. Carlos' great-grandmother is 3^4 years old.

39. James ran the 220-yard dash in 6^2 seconds.

40. Monique saved 5^3 dollars by the end of eight weeks.

GET READY for the Next Lesson

PREREQUISITE SKILL Divide. (Page 744 and Lesson 3-8)

41. 25 ÷ 5 42. 81 ÷ 9 43. 114.8 ÷ 14 44. 516.06 ÷ 18

Explore 3-9

Math Lab
Dividing by Decimals

MAIN IDEA

Use models to divide a decimal by a decimal.

The model below shows $18 \div 6$.

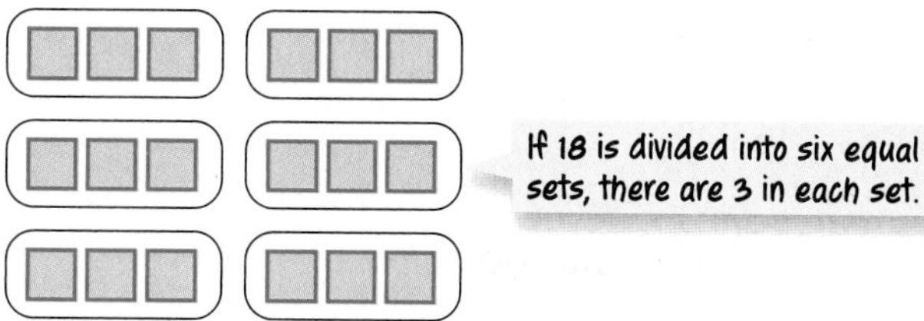

If 18 is divided into six equal sets, there are 3 in each set.

Dividing decimals is similar to dividing whole numbers. In the Activity below, 1.8 is the *dividend* and 0.6 is the *divisor*.

- Use base-ten blocks to model the dividend.
- Replace any ones block with tenths.
- Separate the tenths into groups represented by the divisor.
- The quotient is the number of groups.

ACTIVITY

1 Model $1.8 \div 0.6$.

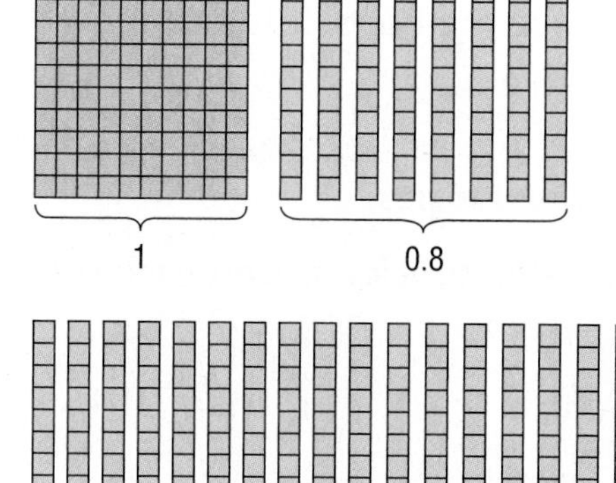

Place one and 8 tenths in front of you to show 1.8.

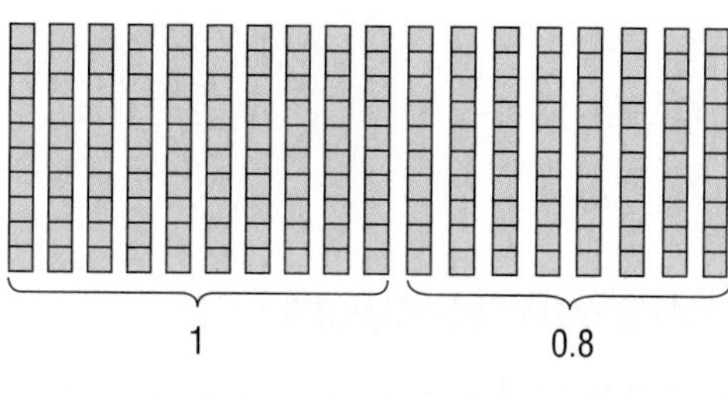

Replace the ones block with tenths. You should have a total of 18 tenths.

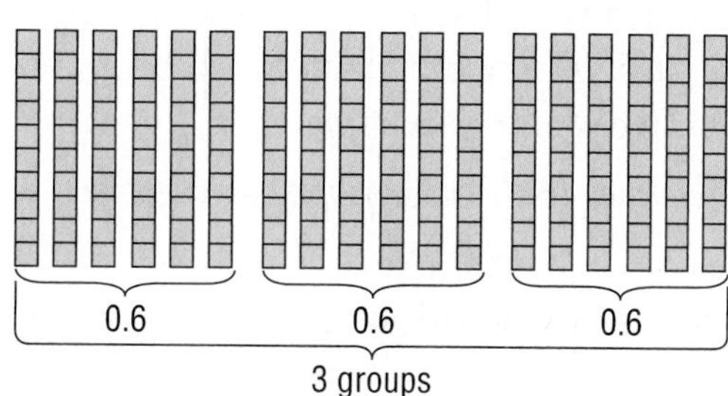

Separate the tenths into groups of six tenths to show dividing by 0.6.

There are three groups of six tenths in 1.8. So, $1.8 \div 0.6 = 3$.

You can use a similar model to divide by hundredths.

ACTIVITY

2 Model 0.2 ÷ 0.04.

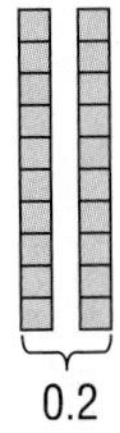

Model 0.2 with base-ten blocks.

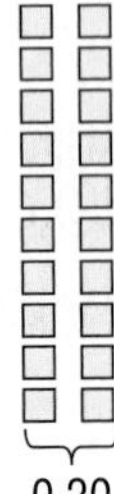

Replace the tenths with hundredths, since you are dividing by hundredths.

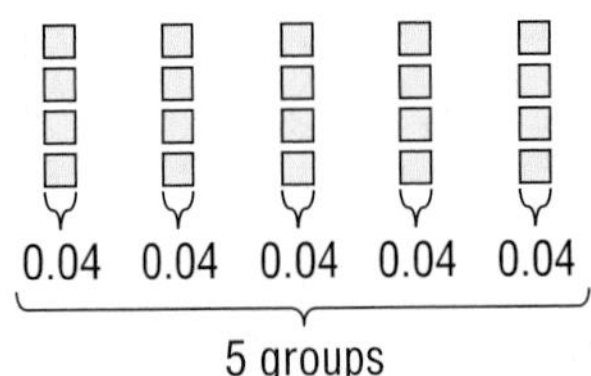

Separate the hundredths into groups of four hundredths to show dividing by 0.04.

There are five groups of four hundredths in 0.2.

So, 0.2 ÷ 0.04 = 5.

CHECK Your Progress

Use base-ten blocks to find each quotient.

a. 2.4 ÷ 0.6 **b.** 1.2 ÷ 0.4 **c.** 1.8 ÷ 0.6

d. 0.9 ÷ 0.09 **e.** 0.8 ÷ 0.04 **f.** 0.6 ÷ 0.05

ANALYZE THE RESULTS

1. Explain why the base-ten blocks representing the dividend must be replaced or separated into the smallest place value of the divisor.
2. Tell why the quotient 0.2 ÷ 0.04 is a whole number. What does the quotient represent?
3. Determine the missing divisor in the sentence 0.8 ÷ ■ = 20. Explain.
4. **MAKE A CONJECTURE** Tell whether 1.2 ÷ 0.03 is *less than, equal to,* or *greater than* 1.2. Explain your reasoning.

3-9 Dividing by Decimals

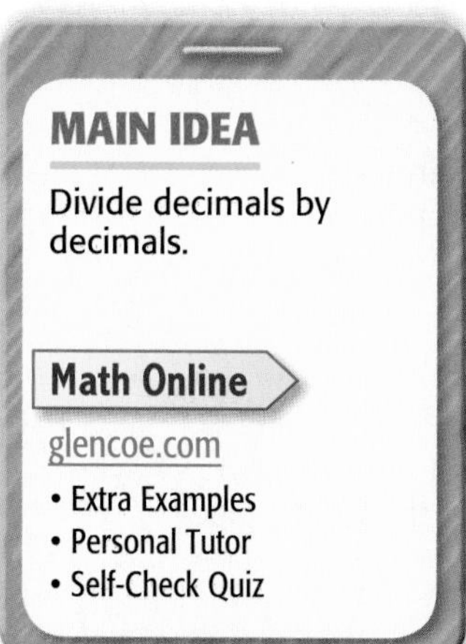

MAIN IDEA

Divide decimals by decimals.

Math Online

glencoe.com

- Extra Examples
- Personal Tutor
- Self-Check Quiz

MINI Lab

Use a calculator to copy and complete the table.

Division Problem	Quotient
36 ÷ 9	4
Set A	
36 ÷ 0.9	
36 ÷ 0.09	
36 ÷ 0.009	
Set B	
3.6 ÷ 9	
0.36 ÷ 9	
0.036 ÷ 9	
Set C	
3.6 ÷ 0.9	
0.36 ÷ 0.09	
0.036 ÷ 0.009	

1. Describe a pattern among the division problems and their quotients for each set.
2. Use the pattern in Set A to find 36 ÷ 0.0009 without a calculator.
3. Use the pattern in Set B to find 0.0036 ÷ 9 without a calculator.
4. Use the pattern in Set C to find 0.0036 ÷ 0.0009 without a calculator.
5. How could you find 0.042 ÷ 0.07 without a calculator?

When dividing by decimals, change the divisor into a whole number. To do this, multiply both the divisor and the dividend by the same power of 10. Then divide as with whole numbers.

EXAMPLE Divide by Decimals

1 Find 14.19 ÷ 2.2. **Estimate** 14 ÷ 2 = 7

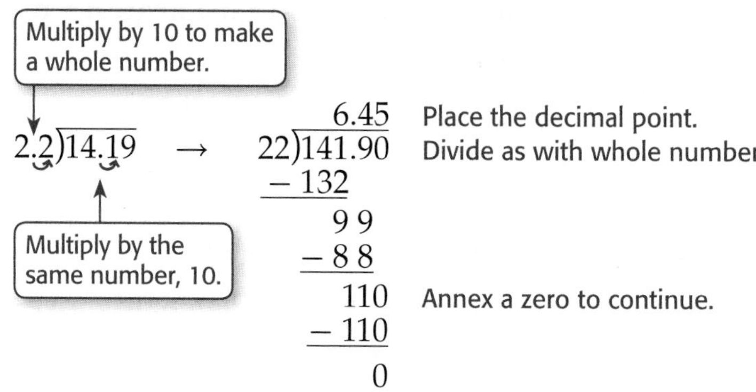

Multiply by 10 to make a whole number.
2.2)14.19 →
Multiply by the same number, 10.
6.45
22)141.90
− 132
9 9
− 8 8
110
− 110
0
Place the decimal point.
Divide as with whole numbers.
Annex a zero to continue.

14.19 divided by 2.2 is 6.45. Compare to the estimate.

Check 6.45 × 2.2 = 14.19 ✓

CHECK Your Progress Divide.

a. 54.4 ÷ 1.7 b. 8.424 ÷ 0.36 c. 0.0063 ÷ 0.007

EXAMPLES Zeros in the Quotient and Dividend

2 **Find 52 ÷ 0.4.**

0.4)52.0 — Multiply each by 10.

```
    130.    Place the decimal point.
  4)520.
   - 4
     12
   - 12
     00     Write a zero in the ones
            place of the quotient
            because 0 ÷ 4 = 0.
```

So, 52 ÷ 0.4 = 130.

Check 130 × 0.4 = 52 ✓

3 **Find 0.09 ÷ 1.8.**

1.8)0.09 — Multiply each by 10.

```
      0.05    Place the decimal point.
  18)0.90     18 does not go into 9, so
    - 0       write a 0 in the tenths place.
      09
    - 00
      90      Annex a 0 in the dividend
    - 90      and continue to divide.
       0
```

So, 0.09 ÷ 1.8 is 0.05.

Check 0.05 × 1.8 = 0.09 ✓

CHECK Your Progress **Divide.**

d. 5.6 ÷ 0.014 e. 62.4 ÷ 0.002 f. 0.4 ÷ 0.0025

EXAMPLE Round Quotients

4 **INTERNET How many times more Internet users are there in Japan than in Spain? Round to the nearest tenth.**

Find 86.3 ÷ 19.8.

Internet Users in 2006 (millions)	
U.S.	211.1
China	137.0
Japan	86.3
France	30.8
Canada	22.0
Spain	19.8

Source: Internet World Stats

```
                    4.35
19.8)86.3  →  198)863.
                 - 792
                    710
                  - 594
                  1,160
                  - 990
                    170
```

To the nearest tenth, 86.3 ÷ 19.8 = 4.4. So, there are about 4.4 times more Internet users in Japan than in Spain.

Study Tip

Rounding When rounding to the nearest tenth, you can stop dividing when there is a digit in the hundredths place.

CHECK Your Progress

g. **INTERNET** How many times more Internet users are there in the U.S. than in France? Round to the nearest tenth.

CHECK Your Understanding

Divide.

Example 1 (p. 179)

1. $3.69 \div 0.3$
2. $9.92 \div 0.8$
3. $0.45 \div 0.3$
4. $13.95 \div 3.1$

Examples 2, 3 (p. 180)

5. $0.6 \div 0.0024$
6. $0.462 \div 0.06$
7. $0.321 \div 0.4$
8. $2.943 \div 2.7$

Example 4 (p. 180)

9. **MEASUREMENT** Alicia bought 5.75 yards of fleece fabric to make blankets for a charity. She needs 1.85 yards of fabric for each blanket. How many blankets can Alicia make with the fabric she bought?

Practice and Problem Solving

HOMEWORK HELP

For Exercises	See Examples
10–13, 22–23	1
14–21	2, 3
24, 25	4

Divide.

10. $1.44 \div 0.4$
11. $0.68 \div 3.4$
12. $16.24 \div 0.14$
13. $2.07 \div 0.9$
14. $0.0338 \div 1.3$
15. $0.16728 \div 3.4$
16. $96.6 \div 0.42$
17. $1.08 \div 2.7$
18. $13.5 \div 0.03$
19. $8.4 \div 0.02$
20. $0.12 \div 0.15$
21. $0.242 \div 0.4$

22. **MEASUREMENT** A submarine sandwich 1.5 feet long is cut into 0.25-foot pieces. How many pieces will there be?

23. **MEASUREMENT** The average person's *stride length,* the distance covered by one step, is approximately 2.5 feet long. How many steps would the average person take to travel 50 feet?

Real-World Link The population of China is about 20% of the world's total population. So, one in every five people on Earth is a resident of China.

24. **POPULATION** The table shows the five most populated countries in the world. How many times more people live in China than in the United States? Round to the nearest tenth if necessary.

Most Populated Countries

Country	Approximate Population (billions)
China	1.322
India	1.13
United States	0.301
Indonesia	0.235
Brazil	0.19

Source: Central Intelligence Agency

25. **GEOGRAPHY** Alaska has the longest coastline in the United States, at about 6.64 thousand miles. Florida has about 1.35 thousand miles of coastline. How many times more coastline does Alaska have than Florida? Round to the nearest tenth if necessary.

26. **MEASUREMENT** Lake Superior, along the U.S.-Canadian border, has a maximum depth of 1.333 thousand feet. There are 5,280 feet in one mile. How deep is Lake Superior in miles? Round to the nearest hundredth if necessary.

ALGEBRA **Use the order of operations to evaluate each expression if $m = 88.2$, $n = 3$, and $p = 17.5$. Round to the nearest tenth if necessary.**

27. $\frac{m}{n}$
28. $\frac{mp}{n}$
29. $\frac{mn}{p}$
30. $\frac{m}{p}$
31. $\frac{p}{n}$
32. $\frac{m - p}{n}$
33. $\frac{p + n}{n}$
34. $\frac{m + n + p}{p}$

CARS **For Exercises 35 and 36, use the table that shows the most popular sports car colors in a recent year in North America.**

Most Popular Sports Car Colors	
Color	**Portion of Responses**
Silver	0.2
Gray	0.17
Blue	0.16
Black	0.14
White	0.1
Red	0.09
Green	0.06
Other	0.08

35. How many times more respondents chose silver than red? Round to the nearest tenth if necessary.

36. How many times more respondents chose either silver or black than red? Round to the nearest tenth if necessary.

37. **MEASUREMENT** The longest vehicle tunnel in the world is the Laerdal Tunnel in Norway with a length of 15.2 miles. How many vehicles could fit in the tunnel bumper to bumper if the average vehicle length is 0.004 mile?

See pages 680, 708.

38. **FIND THE DATA** Refer to the Data File on pages 16–19. Choose some data and write a real-world problem in which you would divide decimals.

H.O.T. Problems

39. **CHALLENGE** Find two positive decimals a and b that make the following statement true. Then find two positive decimals a and b that make the statement false.

If $a < 1$ and $b < 1$, then $a \div b < 1$.

40. **OPEN ENDED** Write a division problem with decimals in which it is necessary to annex one or more zeros to the dividend. Then solve the problem. Round to the nearest tenth if necessary.

41. **NUMBER SENSE** Use the number line below to determine if the quotient of $1.92 \div 0.5$ is closest to 2, 3, or 4. Do not calculate. Explain your reasoning.

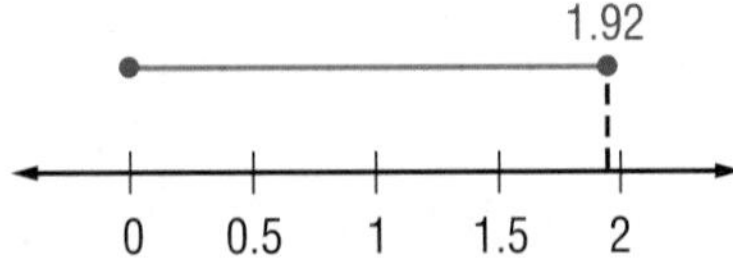

42. **Which One Doesn't Belong?** Identify the problem that does not have the same quotient as the other three. Explain your reasoning.

$49 \div 7$	$4.9 \div 7$	$0.49 \div 0.7$	$0.049 \div 0.07$

43. **WRITING IN MATH** Refer to the table in Exercise 24 on the world's most populated countries. Write and solve a problem in which you would divide decimals. Include instructions for rounding in your problem.

TEST PRACTICE

44. To the nearest tenth, how many times more people in the U.S. own dogs than own birds?

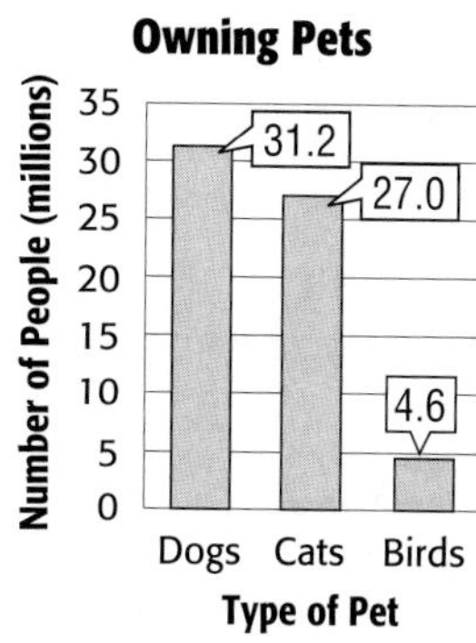

Source: *Statistical Abstract of the United States*

A 6.8
B 12.2
C 26.6
D 35.8

45. The table shows the approximate number of people in the world who speak either Spanish or French.

Language	Speakers (billions)
Spanish	0.425
French	0.129

To the nearest tenth, how many times more people speak Spanish than French?

F 0.054 billion
G 0.296 billion
H 0.304 billion
J 3.295 billion

Spiral Review

46. Find the quotient when 68.52 is divided by 12. (Lesson 3-8)

Multiply. (Lesson 3-7)

47. 19.2×2.45 **48.** 8.25×12.42 **49.** 9.016×51.9

Write an integer to represent each piece of data. (Lesson 2-9)

50. Miguel deposited $45 into his savings account.

51. Mrs. Bezant descended four flights of stairs.

52. The football team gained 16 yards.

53. Suki set her watch back by one hour.

54. GEOGRAPHY The four largest islands in the world are shown in the table. Find the mean and median number of square miles for these data. (Lesson 2-7)

World's Largest Islands

Island	Approximate Area (square miles)
Greenland	840,000
New Guinea	309,000
Borneo	287,300
Madagascar	227,000

GET READY for the Next Lesson

55. PREREQUISITE SKILL A number is multiplied by 8. Next, 4 is subtracted from the product. Then, 12 is added to the difference. If the result is 32, what is the number? Use the *guess and check* strategy. (Lesson 1-7)

3-10 Problem-Solving Investigation

MAIN IDEA: Determine reasonable answers to solve problems.

P.S.I. TEAM +

e-Mail: REASONABLE ANSWERS

STEPHANIE: I am burning a CD. I have picked out the first 5 songs. If the CD's capacity is 72 minutes, which is a more reasonable estimate for the number of minutes left on the CD: 40 minutes, 50 minutes, or 60 minutes?

Song	1	2	3	4	5
Length (min)	5.20	4.60	5.75	4.40	4.50

YOUR MISSION: Determine a reasonable estimate.

Understand	You know the lengths of the first 5 songs and the capacity of the CD. You need to determine a reasonable estimate for the remaining minutes on the CD.
Plan	Estimate the length of each song. Then add the estimated lengths. Finally, subtract that amount from 72, the capacity of the CD.
Solve	Song 1 → 5.20 → 5 Song 2 → 4.60 → 5 Song 3 → 5.75 → 6 Song 4 → 4.40 → 4 Song 5 → 4.50 → + 5 25 Since 72 − 25 = 47, a reasonable estimate for the number of minutes left is 50.
Check	Since 5.20 + 4.60 + 5.75 + 4.40 + 4.50 = 24.25 and 72 − 24.25 = 47.55, 50 minutes is a reasonable estimate.

Analyze The Strategy

1. Describe a situation where determining a reasonable answer would help you solve a problem.
2. **WRITING IN MATH** Write a problem that can be solved by determining a reasonable answer. Then tell the steps you would take to solve the problem.

Mixed Problem Solving

EXTRA PRACTICE
See pages 681, 708.

Determine reasonable answers for Exercises 3–5.

3. **CLOTHES** Annie wants to buy 2 pairs of capris for $34.99 each and 3 pairs of flip-flops for $7.99 each. Does she need to save $150, or is $100 enough?

4. **DONATIONS** Mario collected donations for the American Red Cross. He kept a record of the donations.

Donations	
Monday	$92.33
Tuesday	$107.08
Wednesday	$75.98
Thursday	$63.01
Friday	$111.64

Which is a more reasonable estimate for the amount of money Mario will collect next week if he doubles this week's donations: $700 or $800?

5. **PLAYGROUND** The length of a playground is 88.5 yards. Which is a more reasonable estimate for the length of the playground in feet: 240 or 270?

Use any strategy to solve Exercises 6–12. Some strategies are shown below.

PROBLEM-SOLVING STRATEGIES
• Make a table.
• Guess and check.

6. **CONCERT** In how many ways can 4 people stand in line at a concert if Terrez and Missy must stand next to each other?

7. **SHOPPING** An online store sells personalized magnets for $3.25 each and personalized keychains for $5.79 each. If Mrs. Anderson spent $56.78 on magnets and keychains, how many of each did she buy?

For Exercises 8 and 9, use the table below that shows the number of CD singles that were shipped for sale from 2001 to 2005.

Year	CD Singles Shipped (millions)
2001	17.3
2002	4.5
2003	8.3
2004	3.1
2005	2.8

8. Which year had about 3 times as many CD singles as in 2005?

9. Which year had about 5 million less CDs shipped than 2003?

10. **CHICKENS** The most eggs a chicken has ever laid in one day is 7. At this rate, how many eggs will a chicken lay in 8 years?

11. **NUMBERS** John wrote down two numbers. The product of the numbers is 48 and the difference between the two numbers is 8. What are the two numbers John wrote down?

12. **WHALES** The table below shows the weight of whales. Is the weight of a blue whale about 3 times, 4 times, or 5 times more than the weight of a gray whale?

Whale	Weight (tons)
Blue	151.0
Bowhead	95.0
Fin	69.9
Gray	38.5
Humpback	38.1

Source: *Top 10 of Everything*

CHAPTER 3

Study Guide and Review

Math Online glencoe.com

• STUDY TO GO
• Vocabulary Review

FOLDABLES® Study Organizer

GET READY to Study

Be sure the following Big Ideas are noted in your Foldable.

Chapter 3: Operations with Decimals

BIG Ideas

Estimation (Lesson 3-4)

- Rounding: Estimate by rounding each decimal to the nearest whole number that is easy for you to add or subtract mentally.
- Clustering: Estimate by rounding a group of close numbers to the same number.
- Front-End Estimation: Estimate by adding or subtracting the values of the digits in the front place or left-most place value.

Adding and Subtracting Decimals (Lesson 3-5)

- To add or subtract decimals, line up the decimal points. Then add or subtract as with whole numbers. Annex zeros if needed when subtracting.

Multiplying and Dividing Decimals (Lessons 3-8 and 3-9)

- To multiply decimals, multiply as with whole numbers. The product has the same number of decimal places as the sum of the number of decimal places in each factor.
- To divide a decimal by a whole number, place the decimal point directly above the decimal point in the dividend. Then divide as with whole numbers.
- To divide a decimal by a decimal, change the divisor into a whole number by multiplying both the dividend and the divisor by the same power of ten. Then divide.

Key Vocabulary

clustering (p. 151)
decimal (p. 138)
equivalent decimals (p. 143)
expanded form (p. 139)
front-end estimation (p. 151)
inequality (p. 142)
standard form (p. 139)

Vocabulary Check

State whether each sentence is *true* or *false*. If *false*, replace the underlined word or number to make a true sentence.

1. Standard form is a sum of the products of each digit and its place value.
2. The product of 0.423×100 is 42.3.
3. In 643.082, the digit 2 names the number two hundredths.
4. *Seven hundred and nine thousandths* written as a decimal is 0.079.
5. Estimation in which all of the decimals are close to the same number is called front-end estimation.
6. The product of 0.09×10 will have two decimal places.
7. The number 245 written in expanded form is two hundred forty-five.
8. The quotient of $4.5 \div 0.9$ is the same as the quotient of $45 \div 9$.
9. The symbol $>$ means less than.
10. Clustering can be used to estimate the sum of 119.3, 122.7, 118.9, 121.4, and 123.2 by rounding each number to 120 and multiplying 120 by 5.

Lesson-by-Lesson Review

3-1 Representing Decimals (pp. 138–141)

Write each decimal in standard form and in expanded form.

11. thirteen hundredths
12. six and five tenths
13. eighty-three and five thousandths
14. **NATURE** The largest sunflower head ever grown was *eighty-one and twenty-eight hundredths* centimeters across. Write this length in standard form.

Example 1 Write 21.62 in word form.

21.62 is twenty-one and sixty-two hundredths.

Example 2 Write three hundred forty-six thousandths in standard form and in expanded form.

Standard form: 0.346

Expanded form:

$(3 \times 0.1) + (4 \times 0.01) + (6 \times 0.001)$

3-2 Comparing and Ordering Decimals (pp. 142–145)

Use >, <, or = to compare each pair of decimals.

15. 0.35 ● 0.3
16. 6.024 ● 6.204
17. 0.10 ● 0.1
18. 8.34 ● 9.3

Order each set of decimals from least to greatest.

19. 9.501, 0.9051, 90.51, 0.0951
20. 7.403, 0.0743, 7.743, 74.43
21. **MONEY** The costs of four items are \$9, \$0.99, \$9.99, and \$19.99. Order these costs from least to greatest.

Example 3 Order 17.89, 0.17, 1.879, 10.789 from least to greatest.

17.89 ⟶ 17.890
0.17 ⟶ 0.170
1.879 ⟶ 1.879
10.789 ⟶ 10.789

Line up the decimal points and annex zeros so that each has the same number of decimal places.

Use place value to compare the decimals. The order from least to greatest is 0.17, 1.879, 10.789, and 17.89.

3-3 Rounding Decimals (pp. 146–149)

Round each decimal to the indicated place-value position.

22. 5.031; hundredths
23. 0.00042; ten-thousandths
24. 2.29; tenths
25. **AREA** The area of Hamilton County is 50.4 square miles. Round 50.4 to the nearest square mile.

Example 4 Round 8.0314 to the hundredths place.

8.0<u>3</u>14 Underline the digit to be rounded.

8.0314 Then look at the digit to the right.
↑ Since 1 is less than 5, the digit 3 stays the same.

So, 8.0314 rounds to 8.03.

3-4 Estimating Sums and Differences (pp. 150–154)

Estimate using rounding.

26. 37.82 + 14.24
27. \$72.18 − \$29.93
28. 6.8 + 4.2 + 3.5
29. 129.6 − 9.7

Estimate using clustering.

30. 12.045 + 11.81 + 12.3 + 11.56
31. \$6.45 + \$5.88 + \$5.61 + \$6.03

Estimate using front-end estimation.

32. $\begin{array}{r} 31.29 \\ +\ 58.07 \\ \hline \end{array}$
33. $\begin{array}{r} 93.65 \\ -\ 62.13 \\ \hline \end{array}$
34. $\begin{array}{r} 145.91 \\ +\ 131.65 \\ \hline \end{array}$
35. $\begin{array}{r} 87.25 \\ -\ 63.97 \\ \hline \end{array}$
36. **SHOPPING** Jodie buys a sweater for \$24.35, a bracelet for \$17.62, and a pair of earrings for \$11.19. If she uses front-end estimation to estimate the sum of her purchases, about how much does she spend?

Example 5 **Estimate 38.61 − 14.25 using rounding.**

$\begin{array}{rcr} 38.61 & \rightarrow & 39 \\ -\ 14.25 & \rightarrow & -\ 14 \\ \hline & & 25 \end{array}$ Round to the nearest whole number.

Example 6 **Estimate 8.12 + 7.65 + 8.31 + 8.08 using clustering.**

All addends of the sum are close to 8. So, an estimate is 4 × 8 or 32.

Example 7 **Estimate 24.6 + 35.1 using front-end estimation.**

24.6 + 35.1 Add the front digits to get 5.
An estimate is 50.

3-5 Adding and Subtracting Decimals (pp. 156–160)

Find each sum or difference.

37. $\begin{array}{r} 18.35 \\ +\ 23.61 \\ \hline \end{array}$
38. $\begin{array}{r} 148.93 \\ -\ 121.36 \\ \hline \end{array}$
39. $\begin{array}{r} 1.325 \\ +\ 0.081 \\ \hline \end{array}$
40. 248 − 131.28
41. **RELAY** The times for each leg of a 4 × 100-meter relay are 14.75, 14.49, 14.56, and 14.32 seconds. What was the total time of the relay team?
42. **MONEY** Coral has \$40 to buy a backpack. If the backpack costs \$35.99, how much money will she have left?

Example 8 **Find the sum of 48.23 and 11.65.**

Estimate 48.23 + 11.65 ≈ 48 + 12 = 60

$\begin{array}{r} 48.23 \\ +\ 11.65 \\ \hline 59.88 \end{array}$ Line up the decimals. Add as with whole numbers.

The sum is 59.88.

Check for Reasonableness 59.88 ≈ 60 ✔

Example 9 **Find the difference between 57.68 and 34.64.**

Estimate 58 − 35 ≈ 23

$\begin{array}{r} 57.68 \\ -\ 34.64 \\ \hline 23.04 \end{array}$ Line up the decimals. Subtract as with whole numbers.

The difference is 23.04.

Check for Reasonableness 23.04 ≈ 23 ✔

Mixed Problem Solving
For mixed problem-solving practice, see page 708.

3-6 Multiplying Decimals by Whole Numbers (pp. 163–166)

Multiply.

43. 1.4×6
44. 3×9.95
45. 0.082×17
46. 12.09×19
47. 5×0.048
48. 24.7×31
49. 16×6.65
50. 2.6×38
51. **GROCERIES** A loaf of bread costs $1.79. How much would five loaves of bread cost?
52. **ANIMALS** The average hamster weighs 0.3125 ounce. How much would 8 hamsters weigh altogether?

Example 10 Find 6.45×7.

Estimate $6.45 \times 7 \rightarrow 6 \times 7$ or 42

$$\begin{array}{r} {}^{3\,3}\\ 6.45 \\ \times\ 7 \\ \hline 45.15 \end{array}$$

6.45 × 7: There are two decimal places to the right of the decimal in 6.45.

45.15: Count the same number of places from right to left in the product.

3-7 Multiplying Decimals (pp. 169–172)

Multiply.

53. 0.6×1.3
54. 8.74×2.23
55. 0.04×5.1
56. 2.6×3.9
57. 0.002×50
58. 0.04×0.0063
59. **MEASUREMENT** A rectangular tomato garden measures 5.8 feet by 12.6 feet. What is the area of the garden?

Example 11 Find 38.76×4.2.

$$\begin{array}{rl} 38.76 & \leftarrow \text{two decimal places} \\ \times\ 4.2 & \leftarrow \text{one decimal place} \\ \hline 7752 & \\ +\ 15504 & \\ \hline 162.792 & \leftarrow \text{three decimal places} \end{array}$$

3-8 Dividing Decimals by Whole Numbers (pp. 173–176)

Divide.

60. $12.24 \div 36$
61. $203.84 \div 32$
62. $136.5 \div 35$
63. $37.1 \div 14$
64. $4.41 \div 5$
65. $26.96 \div 8$
66. **MONEY** In one year, Marcy made $214.68 in interest from her savings account. If she made the same amount of interest each month, how much did she make each month?

Example 12 Find the quotient of $16.1 \div 7$.

$$\begin{array}{r} 2.3 \\ 7\overline{)16.1} \\ -\,14 \\ \hline 2\,1 \\ -2\,1 \\ \hline 0 \end{array}$$

Place the decimal point.
Divide as with whole numbers.

Study Guide and Review

3-9 Dividing by Decimals (pp. 179–183)

Divide.

67. $0.96 \div 0.6$
68. $11.16 \div 6.2$
69. $0.276 \div 0.6$
70. $5.88 \div 0.4$
71. $18.45 \div 0.5$
72. $0.155 \div 0.25$
73. **MARATHONS** A marathon race is 26.2 miles long. David ran the marathon in 3.6 hours. On average, how many miles did he run per hour? Round to the nearest tenth.

Example 13 Find $11.48 \div 8.2$.

$8.2\overline{)11.48}$ Multiply the divisor and the dividend by 10 to move the decimal point one place to the right so that the divisor is a whole number.

$$\begin{array}{r} 1.4 \\ 82\overline{)114.8} \\ -82 \\ \hline 32\,8 \\ -32\,8 \\ \hline 0 \end{array}$$

Place the decimal point.
Divide as with whole numbers.

3-10 PSI: Reasonable Answers (pp. 184–185)

Determine reasonable answers for Exercises 74 and 75.

74. **HEIGHT** Evan is 5.75 feet tall. His sister, Cindy, is 0.8 times his height. Which is a reasonable height for Cindy: about 4 feet, 4.5 feet, or 6 feet? Explain your reasoning.
75. **MONEY** Derek has $23.80 in his pocket. He spent about 0.67 of this amount on a CD. Would $8, $16, or $20 be a reasonable price of the CD?

Example 14 There are 24 students in the Spanish club. If the number of students in the school is 19 times this amount, would about 400, 500, or 600 be a reasonable number of students in the school?

24×19 is about 25×20 or 500. So, 500 is a reasonable number of students in the school.

CHAPTER 3

Practice Test

Math Online glencoe.com

• Chapter Test

Write each decimal in word form.

1. 0.07
2. 8.051

Write each decimal in standard form and in expanded form.

3. six tenths
4. two and twenty-one thousandths

5. **SCIENCE** The mass of a particular chemical sample is given as 4.0023 grams. Write the mass in word form.

Use >, <, or = to compare each pair of decimals.

6. 2.03 ● 2.030
7. 7.960 ● 7.906

8. **MULTIPLE CHOICE** Dion recorded the daily high temperatures for Phoenix, Arizona, over five days in the table below.

Day	Temperature (°F)
Monday	109.8
Tuesday	108.9
Wednesday	111.08
Thursday	108.92
Friday	111.0

Which of the following shows the daily high temperatures in order from least to greatest?

A 108.9°F, 108.92°F, 109.8°F, 111.0°F, 111.08°F
B 108.92°F, 108.9°F, 109.8°F, 111.0°F, 111.08°F
C 108.9°F, 108.92°F, 109.8°F, 111.08°F, 111.0°F
D 108.92°F, 108.9°F, 109.8°F, 111.08°F, 111.0°F

Round each decimal to the indicated place-value position.

9. 27.35; tens
10. 3.4556; thousandths

Estimate each sum or difference using the indicated method.

11. 38.23 + 11.84; rounding
12. \$75.38 – \$22.04; front-end estimation
13. 6.72 + 7.09 + 6.6; clustering

Find each sum or difference.

14. 43.28 + 31.45
15. 392.802 – 173.521

Multiply.

16. 7.8 × 6
17. 0.92 × 4
18. 12 × 0.034
19. 4.56 × 9.7

20. **MULTIPLE CHOICE** Armando and his 3 friends ordered a 4-foot sub for \$25.99, 4 large drinks for \$1.79 each, and a salad for \$5.89. Which of the following represents the total cost, not including tax?

A \$134.68
B \$39.04
C \$37.25
D \$33.67

Divide. Round to the nearest tenth if necessary.

21. 7.2 ÷ 3
22. 0.45 ÷ 15
23. 36.08 ÷ 8.2
24. 10.79 ÷ 4.15

25. **ANIMALS** The greyhound can run as fast as 39.35 miles per hour. Without calculating, would about 12, 14, or 16 be a reasonable answer for the number of miles a greyhound could run at this rate in 0.4 hour? Explain your reasoning.

CHAPTER 3

Test Practice

Cumulative, Chapters 1–3

Math Online glencoe.com

• Test Practice

PART 1 Multiple Choice

Read each question. Then fill in the correct answer on the answer sheet provided by your teacher or on a sheet of paper.

1. Laura recorded the lengths in inches of a litter of newborn puppies. Which lists the lengths in order from least to greatest?

   A 8.42 in., 8.45 in., 8.9 in., 8.5 in., 8.64 in.

   B 8.42 in., 8.45 in., 8.5 in., 8.64 in., 8.9 in.

   C 8.9 in., 8.64 in., 8.5 in., 8.45 in., 8.42 in.

   D 8.42 in., 8.45 in., 8.64 in., 8.5 in., 8.9 in.

2. The table below shows Mr. Coughlin's monthly heating bills for November through February. He estimated that the heating cost a total of $800 over these four months. Which best describes his estimate?

Monthly Heating Bill	
Month	Bill ($)
November	196.43
December	214.89
January	204.58
February	222.76

   F More than the actual amount because he rounded to the nearest $10.

   G Less than the actual amount because he rounded to the nearest $10.

   H More than the actual amount because he rounded to the nearest $100.

   J Less than the actual amount because he rounded to the nearest $100.

3. Zack plans on buying 4 shirts. The cost of each shirt ranges from $19.99 to $35.99. What would be a reasonable total cost for the shirts?

   A $60 C $120

   B $70 D $160

4. On Monday, 75 adults and 250 children visited the science museum. On Tuesday, 65 adults and 200 children visited the museum. The cost of a ticket is $7.50 for an adult and $5.25 for a child. Read the problem-solving steps below. Arrange the steps in order to find how much money the museum took in on these two days. Which list shows the steps in the correct order?

   Step K: Add the two products together.

   Step L: Multiply the cost of an adult ticket by the number of adults.

   Step M: Write down the number of adults and the number of children.

   Step N: Multiply the cost of a child's ticket by the number of children.

   F L, K, M, N H M, N, K, L

   G L, M, N, K J M, N, L, K

5. The table shows the maximum speeds of winds in the U.S. for certain cities. What is the mean of the data?

Place	Maximum Wind Speed (mph)
Atlanta, GA	60
Houston, TX	51
Miami, FL	86
Mobile, AL	63
New York, NY	40

Source: *The World Almanac*

   A 46 mph C 60 mph

   B 58 mph D 86 mph

6. The number of hours that people studied for a Spanish test were 3, 2, 1, 0, 2, 1, 3, 5, 3, and 4. What is the mode of these hours?

   F 1 H 3

   G 2 J 5

Preparing for Standardized Tests

For test-taking strategies and practice, see pages 718–735.

7. Kenny recorded the heights of his tomato plants. Choose the group of numbers that lists the heights in order from least to greatest.

   A 3.28 ft, 3.29 ft, 3.06 ft, 3.41 ft

   B 4.15 ft, 4.10 ft, 4.10 ft, 4.01 ft

   C 3.23 ft, 3.30 ft, 3.35 ft, 3.53 ft

   D 2.89 ft, 2.98 ft, 2.99 ft, 2.88 ft

8. Danielle purchased 4 concert tickets. Each ticket was on sale for $5.95 off the original price. If the original price of each ticket was $29.95, which equation can be used to find t, the total price of the 4 tickets Danielle purchased?

   F $t = 4(5.95) - 4(29.95)$

   G $t = 29.95 - 5.95$

   H $t = 5.95 - 29.95$

   J $t = 4(29.95) - 4(5.95)$

TEST-TAKING TIP

Question 8 Read the question carefully to check that you answered the question that was asked. In question 8, you are asked to write an equation, not to find the answer.

9. A student arranged some books on the shelf using the Dewey Decimal System. Choose the group of book numbers that is listed in order from least to greatest.

   A 749, 749.01, 749.21, 749.11

   B 109.012, 109.021, 109.001, 109.3

   C 456.076, 465.076, 465.189, 465.2

   D 688.89, 687.9, 688.91, 688.95

PART 2 Short Response/Grid In

Record your answers on the answer sheet provided by your teacher or on a sheet of paper.

10. The temperature at 6:30 A.M. was 58.7°F. By 1:00 P.M., it was 92.6°F. Find the difference between the two temperatures in degrees Fahrenheit.

11. Before buying furniture, Sharon's mom had $7,420.60 in her checkbook. Afterward, the balance was $4,684.90. How much did Sharon's mom spend on her shopping trip?

PART 3 Extended Response

Record your answers on the answer sheet provided by your teacher or on a sheet of paper. Show your work.

12. Alexandra went to the mall on Saturday and bought the items in the table. Each item was on sale for the price shown.

Item	Original Price ($)	Sale Price ($)
bracelet	25.75	19.50
hat	19.95	15.00
movie	14.50	10.25
shirt	22.75	18.75

   a. How much did the items cost altogether?

   b. How much did Alexandra save?

   c. Explain how you determined how much she saved.

NEED EXTRA HELP?												
If You Missed Question...	1	2	3	4	5	6	7	8	9	10	11	12
Go to Lesson...	3-2	3-4	3-4	1-4	2-6	2-7	3-2	1-8	3-2	3-5	3-5	3-5

Fractions and Decimals

BIG Idea

- Understand the relationship between fractions and decimals.

Key Vocabulary

equivalent fractions (p. 204)

greatest common factor (p. 197)

least common multiple (p. 217)

simplest form (p. 217)

Real-World Link

Dairy Products In the United States, the consumption of dairy products in a recent year was $587\frac{1}{8}$ pounds per person, which can also be written as 587.125 pounds per person.

FOLDABLES® Study Organizer

Fractions and Decimals Make this Foldable to help you understand fractions and decimals. Begin with one sheet of $8\frac{1}{2}'' \times 11''$ paper.

1. **Fold** top of paper down and bottom of paper up as shown.

2. **Label** the top fold Fractions and the bottom fold Decimals.

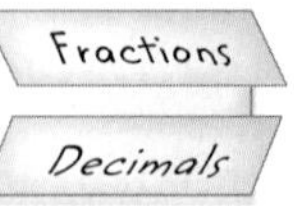

3. **Unfold** the paper and draw a number line in the middle of the paper.

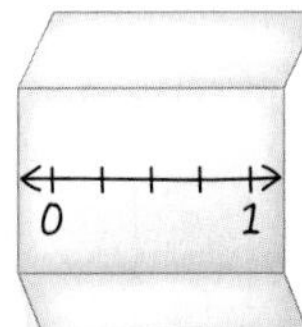

4. **Label** the fractions and decimals as shown.

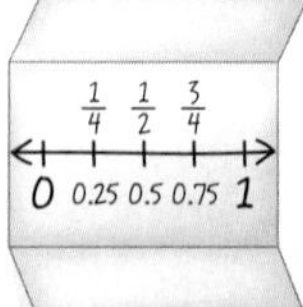

GET READY for Chapter 4

Diagnose Readiness You have two options for checking Prerequisite Skills.

Option 2

Math Online Take the Online Readiness Quiz at glencoe.com.

Option 1

Take the Quick Quiz below. Refer to the Quick Review for help.

QUICK Quiz

Tell whether each number is divisible by 2, 3, 4, 5, 6, 9, or 10. (page 736)

1. 67
2. 891
3. 145
4. 202
5. **GAMES** Is it possible to divide 78 marbles evenly among 6 players? Justify your response.

Find the prime factorization of each number. (Lesson 1-2)

6. 315
7. 264
8. 120
9. 28
10. **TRAVEL** Mary drove 225 miles in one day. Find the prime factorization of this number.

Write each decimal in standard form. (Lesson 3-1)

11. five and three tenths
12. seventy-four hundredths
13. two tenths
14. sixteen thousandths

QUICK Review

Example 1

Tell whether the number 756 is divisible by 2, 3, 5, 9, or 10.

2: Yes, the ones digit, 6, is divisible by 2.
3: Yes, the sum of the digits, 18, is divisible by 3.
5: No, the ones digit is neither 0 nor 5.
9: Yes, the sum of the digits is divisible by 9.
10: No, the ones digit is not 0.

Example 2

Find the prime factorization of 315.

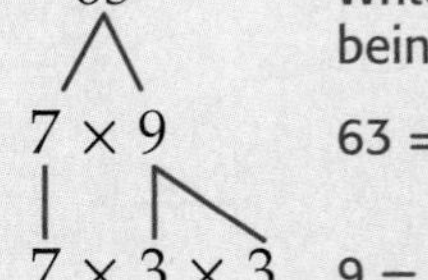

Write the number that is being factored at the top.

$63 = 7 \times 9$

$9 = 3 \times 3$

So, $63 = 3 \times 3 \times 7$ or $3^2 \times 7$.

Example 3

Write *twenty-seven and eighty-nine thousandths* in standard form.

10	1	0.1	0.01	0.001
tens	ones	tenths	hundredths	thousandths
2	7 .	0	8	9

The standard form is 27.089.

READING to SOLVE PROBLEMS

Make a Diagram

Making a diagram is a good strategy to use when you want to see how numbers or items are related. One kind of diagram is a Venn diagram. A *Venn diagram* uses overlapping circles to show the similarities and differences of two groups of items. Any item that is located where the circles overlap has a characteristic of both circles.

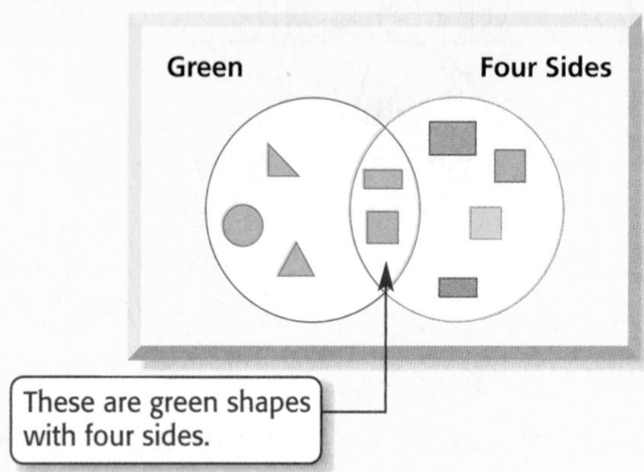

You can also make a Venn diagram using numbers. The Venn diagram below shows the factors of 28 in one circle and the factors of 36 in the second circle.

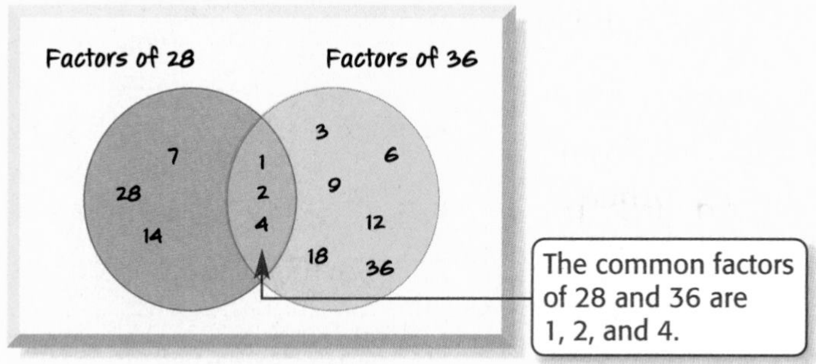

PRACTICE

Make a Venn diagram that shows the factors for each pair of numbers.

1. 8, 12
2. 20, 30
3. 25, 28
4. 15, 30
5. Organize the numbers 2, 5, 9, 27, 29, 35, and 43 into a Venn diagram. Use the headings *prime numbers* and *composite numbers*. What numbers are in the overlapping circles? Explain.

4-1 Greatest Common Factor

MAIN IDEA

Find the greatest common factor of two or more numbers.

New Vocabulary

Venn diagram
common factor
greatest common factor (GCF)

Math Online

glencoe.com

- Concepts in Motion
- Extra Examples
- Personal Tutor
- Self-Check Quiz

GET READY for the Lesson

SUMMER CAMP The Venn diagram below shows which activities each camper participated in on Monday. **Venn diagrams** use overlapping circles to show common elements.

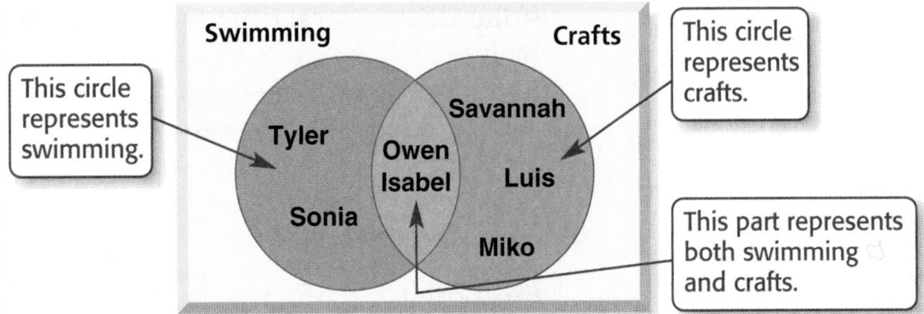

1. Who participated in swimming only?
2. Who participated in crafts only?
3. Who participated in both swimming and crafts?

Factors that are shared by two or more numbers are called **common factors**. The greatest of the common factors of two or more numbers is the **greatest common factor (GCF)** of the numbers.

To find common factors, you can make a list.

EXAMPLE Identify Common Factors

1 Identify the common factors of 16 and 24.

First, list the factors by pairs for each number. Then, circle the common factors.

Factors of 16
(1) × 16
(2) × (8)
(4) × 4

Factors of 24
(1) × 24
(2) × 12
3 × (8)
(4) × 6

The common factors are 1, 2, 4, and 8.

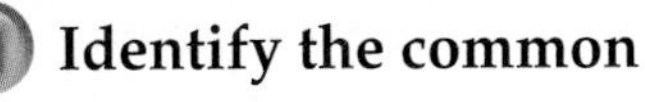

CHECK Your Progress

Identify the common factors of each set of numbers.

a. 25, 60

b. 18, 27, 36

EXAMPLE Find the GCF by Listing Factors

2 **Find the GCF of 60 and 54.**

First make an organized list of the factors for each number.

60: 1 × 60, 2 × 30, 3 × 20, 4 × 15, 5 × 12, 6 × 10 ⟶ 1, 2, 3, 4, 5, 6, 10, 12, 15, 20, 30, 60

54: 1 × 54, 2 × 27, 3 × 18, 6 × 9 ⟶ 1, 2, 3, 6, 9, 18, 27, 54

The common factors are 1, 2, 3, and 6, and the greatest of these is 6. So, the greatest common factor or GCF of 60 and 54 is 6.

Use a Venn diagram to show the factors. Notice that the factors 1, 2, 3, and 6 are the common factors of 60 and 54 and the GCF is 6.

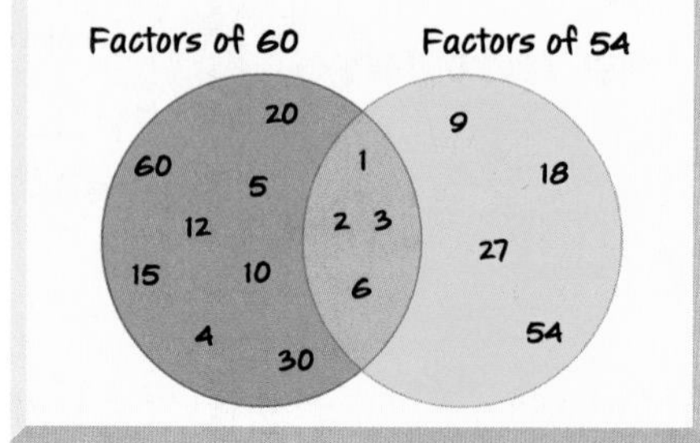

CHECK Your Progress

Find the GCF of each set of numbers.

c. 35, 60 d. 15, 45 e. 12, 19

EXAMPLE Find the GCF by Using Prime Factors

Review Vocabulary

prime number a whole number that has exactly two factors, 1 and the number itself; *Example:* 7 (Lesson 1-2)

prime factorization a composite number expressed as a product of prime numbers; *Example:* 12 = 2 × 2 × 3 (Lesson 1-2)

3 **Find the GCF of 18 and 30.**

METHOD 1 **Write the prime factorization.**

18 → 2 · 9 → (2) · (3) · 3

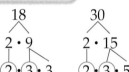

30 → 2 · 15 → (2) · (3) · 5

2 and 3 are common factors.

METHOD 2 **Divide by prime numbers.**

(2)) 18 30 Divide both 18 and 30 by 2.

(3)) 9 15 Divide the quotients by 3.

3 5

Using either method, the common prime factors are 2 and 3. So, the GCF of 18 and 30 is 2 × 3 or 6.

CHOOSE Your Method

Find the GCF of each set of numbers.

f. 12, 66 g. 36, 45 h. 32, 48

Real-World EXAMPLES

4 FOOD A bakery arranges three different types of muffins in a display case. There should be an equal number of muffins in each row in the case. What is the greatest possible number of muffins in each row?

Muffins	
Type	**Number**
blueberry	40
cinnamon raisin	24
chocolate chip	32

factors of 40: 1, 2, 4, 5, 8, 10, 20, 40

factors of 24: 1, 2, 3, 4, 6, 8, 12, 24

factors of 32: 1, 2, 4, 8, 16, 32

The GCF of 40, 24, and 32 is 8. So, the greatest number of muffins that could be placed in each row is 8.

5 How many rows of muffins are there if there are 8 in each row?

There is a total of $40 + 24 + 32$, or 96 muffins. So, the number of rows of muffins is $96 \div 8$, or 12.

CHECK Your Progress

HOBBIES Jerrica makes and sells beaded necklaces. She earned \$49 on Friday, \$42 on Saturday, and \$21 on Sunday selling necklaces at a local craft sale.

i. If Jerrica sold each necklace for the same amount, what is the most she could have charged per necklace?

j. How many necklaces did she sell?

CHECK Your Understanding

Example 1 (p. 197)

Identify the common factors of each set of numbers.

1. 11, 44
2. 12, 21, 30

Examples 2, 3 (p. 198)

Find the GCF of each set of numbers.

3. 8, 32
4. 24, 60
5. 3, 12, 18
6. 4, 10, 14

Examples 4, 5 (p. 199)

FOOD For Exercises 7 and 8, use the following information.

Oliver has 14 chocolate cookies and 21 iced cookies.

7. If Oliver gives each friend an equal number of each type of cookie, what is the greatest number of friends with whom he can share his cookies?
8. How many cookies did each friend receive?

Practice and Problem Solving

HOMEWORK HELP

For Exercises	See Examples
9–12	1
13–16	2
17–22	3
23, 25	4
24, 26	5

Identify the common factors of each set of numbers.

9. 45, 75
10. 36, 90
11. 6, 21, 30
12. 16, 24, 40

Find the GCF of each set of numbers.

13. 12, 18
14. 18, 42
15. 48, 60
16. 30, 72
17. 14, 35, 84
18. 9, 18, 42
19. 16, 52, 76
20. 12, 30, 72
21. 37, 64, 72
22. 35, 63, 84

SCRAPBOOKING For Exercises 23 and 24, use the following information.

Annika is placing photos in a scrapbook. She has eight large photos, twelve medium photos, and sixteen small photos. Each page will have only one size of photo. She also wants to place the same amount of photos on each page.

23. What is the greatest number of photos that could be on each page? Justify your response.
24. How many pages will she use in all? Justify your response.

SHOPPING For Exercises 25 and 26, use the following information.

A grocery store sells boxes of juice in equal-size packs. Carla bought 18 boxes, Rico bought 36 boxes, and Winston bought 45 boxes.

25. What is the greatest number of boxes in each pack?
26. How many packs did each person buy?

Find three numbers with a GCF that is the indicated value.

27. 6
28. 14
29. 15

30. **TOYS** The table shows the number of each type of toy in a store. The toys will be placed on shelves so that each shelf has the same number of each type of toy. How many shelves are needed for each type of toy so that it has the greatest number of toys?

Toy	Amount
dolls	45
footballs	105
small cars	75

31. **ARTWORK** The table shows the amount of money Ms. Ayala made over three days selling 4 × 6-inch prints at an arts festival. Each print cost the same amount. What is the most each print could have cost?

Ms. Ayala's Artwork	
Day	**Amount ($)**
Friday	60
Saturday	144
Sunday	96

EXTRA PRACTICE
See pages 681, 709.

32. **NUMBER SENSE** What is the GCF of all the numbers in the pattern 9, 18, 27, 36, ...? Explain your reasoning.

H.O.T. Problems

33. **REASONING** When is the GCF of two or more numbers equal to one of the numbers? Explain your reasoning.

CHALLENGE Determine whether each statement is *true* or *false*. If true, explain why. If false, give a counterexample.

34. The GCF of any two even numbers is always even.
35. The GCF of any two odd numbers is always odd.
36. The GCF of an odd number and an even number is always even.
37. **OPEN ENDED** Find three numbers with a GCF that is one of the numbers. The sum of the two lesser numbers must equal the greatest number.
38. **Which One Doesn't Belong?** Identify the number that does not have the same greatest common factor as the other three. Explain your reasoning.

16	8	24	20

39. **WRITING IN MATH** Which method would you prefer to use to find the GCF of 48, 64, and 144? Explain your reasoning.

TEST PRACTICE

40. **SHORT RESPONSE** Find the greatest common factor of 28, 42, and 70.

41. Which number is *not* a common factor of 24 and 36?

    A 2

    B 6

    C 12

    D 24

42. Jeremiah has 32 baseball cards and 48 football cards. He will share his collection with his brother so that they each have the same number of each type of card. What is the greatest number of baseball cards they will each have?

    F 4 cards

    G 8 cards

    H 12 cards

    J 16 cards

Spiral Review

43. **PLAYS** After five performances, the total attendance of a play was 39,963. Which is a more reasonable estimate for the number of people who attended each performance: 7,000 or 8,000? (Lesson 3-10)

44. **MONEY** Marcus bought several baseball caps. Each cap cost $16.40. If he spent a total of $114.80, how many caps did he buy? (Lesson 3-9)

Order each set of decimals from least to greatest. (Lesson 3-2)

45. 7, 9.85, 8.3, and 3.9

46. 12.1, 13.3, 11.49, and 12

GET READY for the Next Lesson

PREREQUISITE SKILL Tell whether both numbers in each number pair are divisible by 2, 3, 4, 5, 6, or 10. (Page 736)

47. 9, 24

48. 15, 25

49. 9, 10

50. 10, 30

Explore 4-2

Math Lab
Equivalent Fractions

MAIN IDEA

Use models to determine a procedure for generating equivalent fractions.

Math Online

glencoe.com

- Concepts in Motion

Fractions are often used to describe the relationship between part of a set of objects and the whole set.

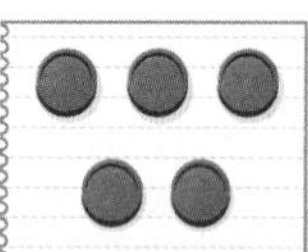

$\frac{3}{5}$ of the counters are red.

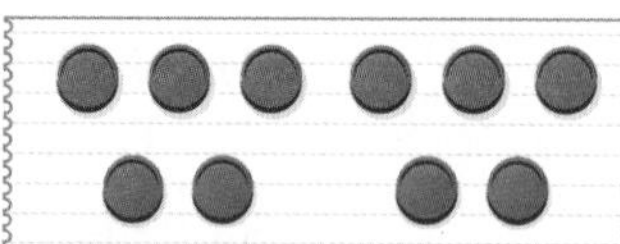

$\frac{6}{10}$ of the counters are red.

Fractions that share the same relationship between part and whole are said to be *equivalent*. In the models shown, 3 out of every 5 groups of counters are red. Therefore, $\frac{3}{5}$ and $\frac{6}{10}$ are equivalent fractions.

ACTIVITY

1 Use counters to generate a fraction equivalent to $\frac{2}{3}$.

STEP 1 Model $\frac{2}{3}$ by forming a group of counters in which 2 out of 3 are red.

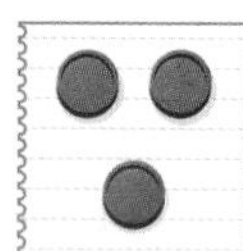

STEP 2 Combine two or more equal groups to form one larger group. The model shows 3 groups.

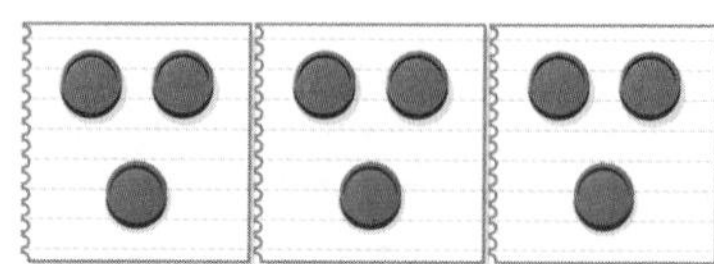

STEP 3 Name the fraction of the larger group that is red. Six out of 9 or $\frac{6}{9}$ of the larger group is red.

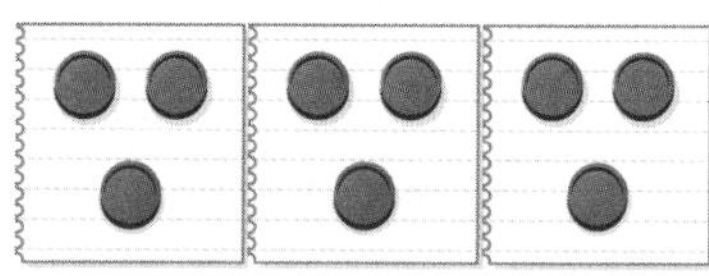

So, one fraction equivalent to $\frac{2}{3}$ is $\frac{6}{9}$.

CHECK Your Progress

Use counters to name three fractions equivalent to each fraction.

a. $\frac{3}{4}$ b. $\frac{1}{3}$ c. $\frac{2}{5}$ d. $\frac{5}{6}$

You can also generate equivalent fractions by separating a larger group into two or more smaller groups that share the same part to whole relationship. This process is called *simplifying* a fraction.

ACTIVITY

2 **Use counters to generate a simpler fraction that is equivalent to $\frac{6}{12}$.**

Study Tip

Equivalent fractions There may be more than one simpler fraction that is equivalent to a given fraction. For example, you could also separate the 12 counters into equal groups of 2 counters where 1 counter in each group is red. So, $\frac{6}{12}$ also equals $\frac{1}{2}$.

STEP 1 Model $\frac{6}{12}$ using counters.

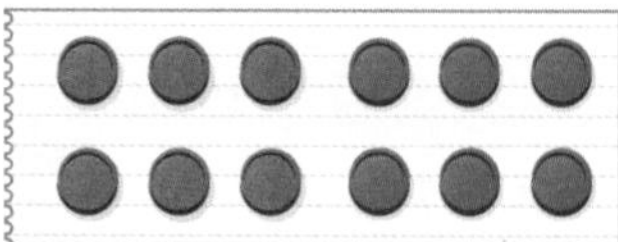

STEP 2 Separate the counters into equal groups so that the relationship between the red counters and total number of counters in each group is the same.

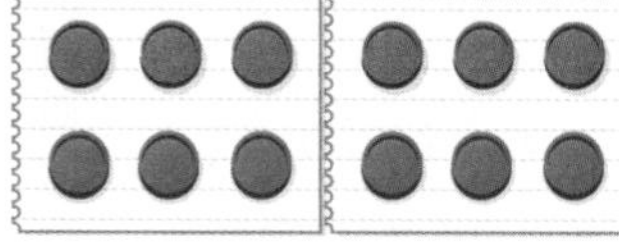

STEP 3 Name the fraction of each smaller group that is red. Three out of 6 or $\frac{3}{6}$ of each smaller group is red.

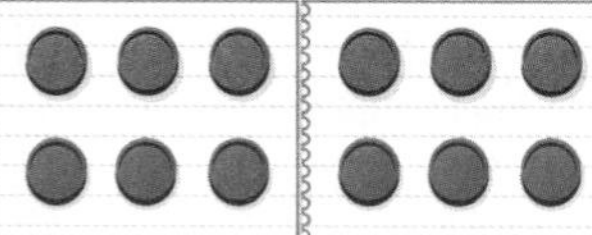

So, one simpler fraction equivalent to $\frac{6}{12}$ is $\frac{3}{6}$.

CHECK Your Progress

Use counters to name a simpler fraction that is equivalent to each fraction.

e. $\frac{10}{16}$ f. $\frac{6}{21}$ g. $\frac{8}{24}$ h. $\frac{24}{30}$

Analyze the Results

1. In Activity 1, an equivalent fraction is created by combining equal groups that have the same number of red counters and the same number of total counters. What operation does this model?

2. **MAKE A CONJECTURE** Use the operation you found in Exercise 1 to generate a fraction equivalent to $\frac{7}{8}$. Justify your answer.

3. In Activity 2, an equivalent fraction is created by separating a group of counters into equal groups that have the same number of red counters and the same number of total counters. What operation does this model?

4. **MAKE A CONJECTURE** Use the operation you found in Exercise 3 to generate a fraction equivalent to $\frac{30}{40}$. Justify your answer.

4-2 Simplifying Fractions

MAIN IDEA

Express fractions in simplest form.

New Vocabulary

equivalent fractions
simplest form

Math Online

glencoe.com

- Extra Examples
- Personal Tutor
- Self-Check Quiz

GET READY for the Lesson

ANIMALS The table shows the different types of kittens found at a local pet store.

1. How many kittens are at the pet store?
2. How many Siamese kittens are there?

Type of Kitten	Number
Siamese	4
Tortoise	3
Abyssinian	1
Persian	2
Angora	2

In the table above, you can compare the number of Siamese kittens to the total number of kittens by using a fraction.

$$\frac{4}{12} \begin{array}{l} \leftarrow \text{Siamese kittens} \\ \leftarrow \text{total number of kittens} \end{array}$$

Equivalent fractions are fractions that have the same value. The fractions $\frac{4}{12}$ and $\frac{1}{3}$ name the same part of the whole. So, the fractions are equivalent.

$\frac{4}{12}$ 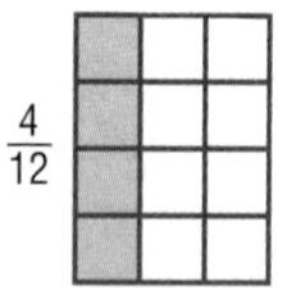

$\frac{1}{3}$ 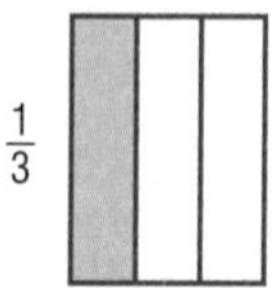

That is, $\frac{4}{12} = \frac{1}{3}$.

To find equivalent fractions, you can multiply or divide the numerator and denominator by the same nonzero number.

$$\frac{4}{12} = \frac{4 \div 4}{12 \div 4}$$
$$= \frac{1}{3}$$

So, 1 out of every 3 kittens at the pet store is Siamese.

EXAMPLES Write Equivalent Fractions

Replace each ■ with a number so the fractions are equivalent.

1. $\frac{5}{7} = \frac{■}{21}$

$\frac{5}{7} \overset{\times 3}{=} \frac{15}{21}$ Since $7 \times 3 = 21$, multiply the numerator and denominator by 3.

2 $\frac{12}{16} = \frac{6}{\blacksquare}$

$$\frac{12}{16} \overset{\div 2}{=} \frac{6}{8}$$

Since $12 \div 2 = 6$, divide the numerator and denominator by 2.

CHECK Your Progress

Replace each $\blacksquare$ with a number so the fractions are equivalent.

a. $\frac{3}{5} = \frac{\blacksquare}{20}$

b. $\frac{18}{24} = \frac{6}{\blacksquare}$

c. $\frac{\blacksquare}{7} = \frac{20}{35}$

A fraction is in **simplest form** when the GCF of the numerator and denominator is 1.

EXAMPLE Write Fractions in Simplest Form

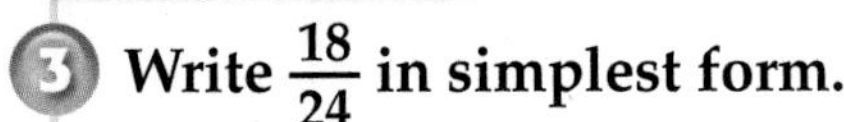

3 **Write $\frac{18}{24}$ in simplest form.**

METHOD 1 **Divide by common factors.**

$$\frac{18}{24} \overset{\div 2}{=} \frac{9}{12} \overset{\div 3}{=} \frac{3}{4}$$

A common factor of 18 and 24 is 2.
A common factor of 9 and 12 is 3.

METHOD 2 **Divide by the GCF.**

factors of 18: 1, 2, 3, 6, 9, 18

factors of 24: 1, 2, 3, 4, 6, 8, 12, 24

The GCF of 18 and 24 is 6.

$$\frac{18}{24} \overset{\div 6}{=} \frac{3}{4}$$

Divide the numerator and denominator by the GCF, 6.

Since the GCF of 3 and 4 is 1, the fraction $\frac{3}{4}$ is in simplest form.

Study Tip

Checking Solutions You can check the answer to Example 3 by multiplying the numerator and denominator by the GCF. The result should be the original fraction.

$\frac{3}{4} = \frac{3 \times 6}{4 \times 6} = \frac{18}{24}$

CHOOSE Your Method

Write each fraction in simplest form. If the fraction is already in simplest form, write *simplest form*.

d. $\frac{21}{24}$

e. $\frac{9}{15}$

f. $\frac{2}{3}$

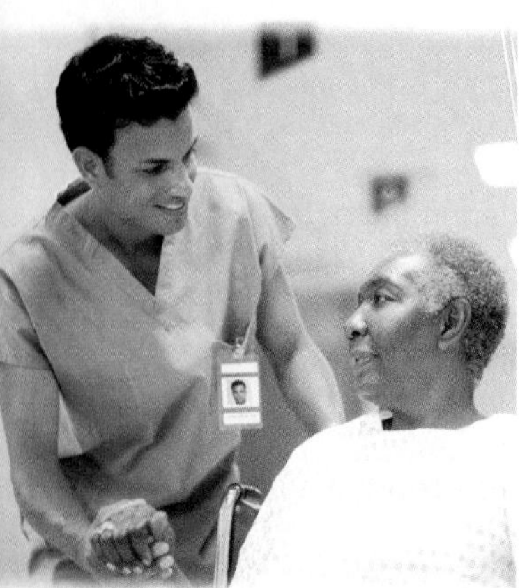

You can often use mental math to divide both the numerator and denominator by their GCF.

Real-World EXAMPLE

4 NURSES Approximately 36 out of 60 nurses work in hospitals. Express the fraction $\frac{36}{60}$ in simplest form.

The GCF of 36 and 60 is 12.

$\frac{\overset{3}{\cancel{36}}}{\underset{5}{\cancel{60}}} = \frac{3}{5}$ Mentally divide both the numerator and denominator by 12.

So, $\frac{3}{5}$ or 3 out of every 5 nurses work in hospitals.

Real-World Career
How Does a Nurse Use Math? Nurses use math to calculate correct doses of medicine for their patients.

Math Online
For more information, go to glencoe.com.

CHECK Your Progress

g. **BASKETBALL** In a recent NBA season, Kirk Hinrich of the Chicago Bulls started 66 of the 76 games he played. Express the fractional part of the games he started in simplest form.

h. **AIRPORTS** On Thursday, 40 out of a total of 192 flights were delayed due to weather. Express the fractional part of the delayed flights in simplest form.

CHECK Your Understanding

Examples 1, 2 (pp. 204–205)

Replace each ■ with a number so the fractions are equivalent.

1. $\frac{3}{8} = \frac{■}{24}$
2. $\frac{4}{5} = \frac{40}{■}$
3. $\frac{15}{25} = \frac{3}{■}$
4. $\frac{21}{28} = \frac{■}{4}$

Example 3 (p. 205)

Write each fraction in simplest form. If the fraction is already in simplest form, write *simplest form.*

5. $\frac{2}{10}$
6. $\frac{8}{25}$
7. $\frac{10}{38}$
8. $\frac{15}{45}$

Example 4 (p. 206)

9. **FOOD** The table shows the fraction of each type of baked good to be sold out of the total number of baked goods at the school bake sale. Express the fraction of baked goods that were muffins in simplest form.

School Bake Sale	
breads	$\frac{6}{50}$
cakes	$\frac{6}{20}$
cookies	$\frac{26}{100}$
muffins	$\frac{24}{100}$
pies	$\frac{4}{50}$

Practice and Problem Solving

HOMEWORK HELP	
For Exercises	See Examples
10–17	1, 2
18–25	3
26, 27	4

Replace each ■ with a number so the fractions are equivalent.

10. $\frac{1}{2} = \frac{\blacksquare}{8}$
11. $\frac{1}{3} = \frac{\blacksquare}{27}$
12. $\frac{\blacksquare}{5} = \frac{9}{15}$
13. $\frac{\blacksquare}{6} = \frac{20}{24}$
14. $\frac{7}{9} = \frac{14}{\blacksquare}$
15. $\frac{12}{16} = \frac{3}{\blacksquare}$
16. $\frac{30}{35} = \frac{\blacksquare}{7}$
17. $\frac{36}{45} = \frac{\blacksquare}{5}$

Write each fraction in simplest form. If the fraction is already in simplest form, write *simplest form.*

18. $\frac{6}{9}$
19. $\frac{4}{10}$
20. $\frac{10}{38}$
21. $\frac{27}{54}$
22. $\frac{19}{37}$
23. $\frac{32}{85}$
24. $\frac{28}{77}$
25. $\frac{15}{100}$

26. **BASEBALL** Brendan had a hit in 24 out of 36 times he batted. Express the fraction of times he hit safely in simplest form.

27. **MUSIC** In a typical symphony orchestra, 16 out of every 100 musicians are first and second violin players. Express the fraction of the orchestra that are violinists in simplest form.

28. **SURVEYS** The table shows the results of a survey about favorite movie theater snacks. Write a fraction in simplest form that compares the number of people who chose popcorn to the total number of people surveyed.

Favorite Movie Snack

Snack	Frequency
popcorn	24
hot dog	12
nachos	11
chocolate	8
licorice	5

Write two fractions that are equivalent to the given fraction.

29. $\frac{4}{10}$
30. $\frac{5}{12}$
31. $\frac{12}{20}$
32. $\frac{16}{44}$

33. **ANALYZE GRAPHS** The results of a survey of students' favorite hobbies are shown in the bar graph at the right. In simplest form, what fraction of the students chose video games as their favorite hobby?

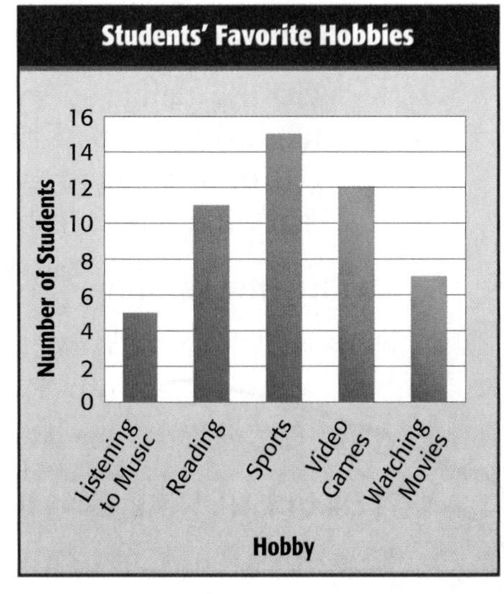

34. **FIND THE DATA** Refer to the Data File on pages 16–19. Choose some data and write a real-world problem in which you would write equivalent fractions.

EXTRA PRACTICE
See pages 681, 709.

H.O.T. Problems

35. **Which One Doesn't Belong?** Identify the fraction that does not belong with the other three. Explain your reasoning.

$\frac{6}{15}$ $\frac{10}{25}$ $\frac{4}{20}$ $\frac{22}{55}$

36. **CHALLENGE** Find a fraction equivalent to $\frac{3}{4}$. Its numerator and denominator, when added together, equal 84.

37. **WRITING IN MATH** Explain in your own words how to find a fraction that is equivalent to a given fraction.

TEST PRACTICE

38. Tyler has read $\frac{4}{5}$ of his novel for reading class. Which student has also read the same amount as Tyler?

Student	Amount Read
Abby	$\frac{1}{2}$
Gustavo	$\frac{12}{15}$
Tonisha	$\frac{4}{10}$
Lance	$\frac{16}{15}$

A Abby
B Gustavo
C Tonisha
D Lance

39. The fractions $\frac{2}{6}, \frac{3}{9}, \frac{4}{12}$, and $\frac{5}{15}$ are each equivalent to $\frac{1}{3}$. What is the relationship between the numerator and the denominator in each fraction that is equivalent to $\frac{1}{3}$?

F The numerator is three times the denominator.
G The numerator is three more than the denominator.
H The denominator is three times the numerator.
J The denominator is three more than the numerator.

Spiral Review

Find the GCF of each set of numbers. (Lesson 4-1)

40. 40, 36
41. 45, 75
42. 120, 150

43. **GASOLINE** Benita spent $38.40 at the gas station to fill up her car's gas tank. If she pumped 15 gallons of gasoline into her car, is about $2, $2.50, or $3 a more reasonable answer for the cost of each gallon of gasoline? (Lesson 3-10)

Identify the solution of each equation from the list given. (Lesson 1-8)

44. $45 - h = 38$; 6, 7, 8
45. $66 = z - 23$; 88, 89, 90

GET READY for the Next Lesson

PREREQUISITE SKILL **Divide. Include remainders in your answers.** (Page 726)

46. $8 \div 3$
47. $19 \div 6$
48. $52 \div 8$
49. $67 \div 9$

4-3 Mixed Numbers and Improper Fractions

MAIN IDEA

Write mixed numbers as improper fractions and vice versa.

New Vocabulary

mixed number
proper fraction
improper fraction

Math Online

glencoe.com

- Extra Examples
- Personal Tutor
- Self-Check Quiz
- Reading in the Content Area

MINI Lab

STEP 1 Shade one square self-stick note to represent the whole number 1.

STEP 2 Fold the shaded self-stick note into fourths.

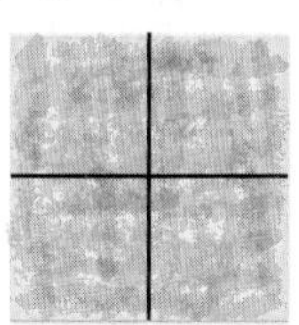

STEP 3 Fold a second square self-stick note into four equal parts to show fourths. Shade one part to represent $\frac{1}{4}$.

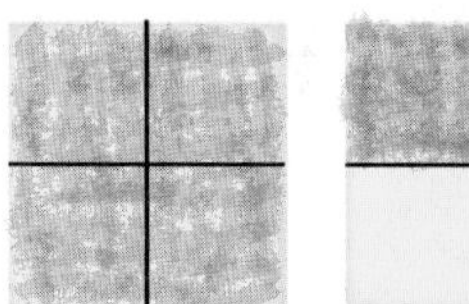

1. How many shaded $\frac{1}{4}$s are there?
2. What fraction is equivalent to $1\frac{1}{4}$?

Make a model to show each number.

3. the number of thirds in $2\frac{2}{3}$
4. the number of halves in $4\frac{1}{2}$

A number like $1\frac{1}{4}$ is an example of a mixed number. A **mixed number** indicates the sum of a whole number and a fraction.

$$1\frac{1}{4} = 1 + \frac{1}{4}$$

Notice that $1\frac{1}{4}$ and $\frac{5}{4}$ are graphed in the same position on the number line.

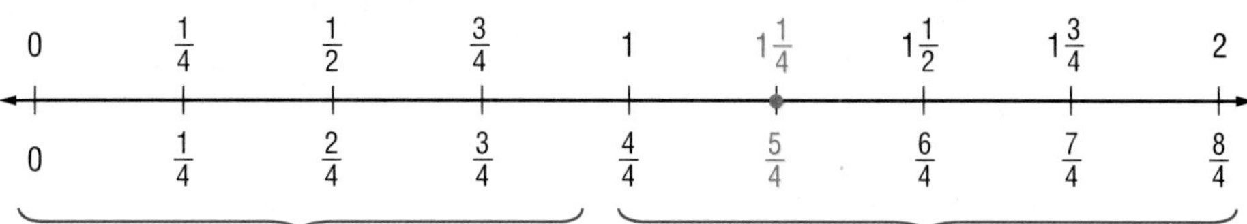

Mixed numbers and improper fractions have values that are greater than or equal to 1.

Real-World Link The distance around the crown of the Liberty Bell in Philadelphia, Pennsylvania, is $7\frac{1}{2}$ feet.

Source: Independence Hall Association in Philadelphia

You can write mixed numbers as equivalent improper fractions using mental math. Multiply the whole number and denominator. Then add the numerator.

EXAMPLE Mixed Numbers as Improper Fractions

1 LIBERTY BELL Use the information at the left. Write the distance around the crown of the Liberty Bell as an improper fraction.

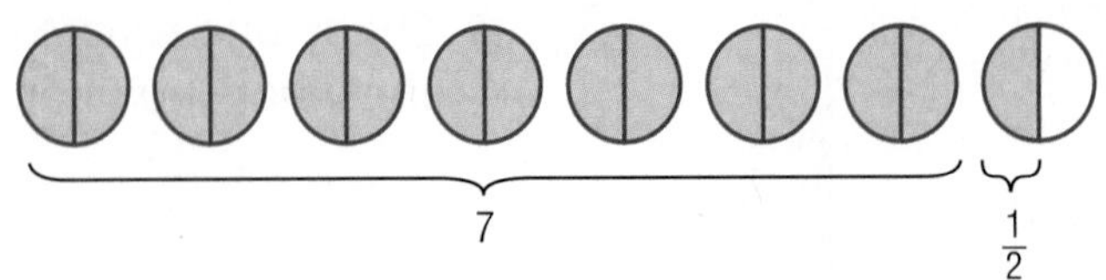

$$7\frac{1}{2} = \frac{(7 \times 2) + 1}{2}$$ There are seven wholes, each with two parts, plus one part.

$$= \frac{15}{2}$$

CHECK Your Progress

a. **SHIPS** The world's largest ship is the *Jahre Viking,* which measures 1,502 feet long. It can carry $4\frac{1}{5}$ million barrels of oil. Write $4\frac{1}{5}$ as an improper fraction.

Improper fractions can also be written as equivalent mixed numbers or whole numbers. Divide the numerator by the denominator and express the remainder as a fraction.

EXAMPLE Improper Fractions as Mixed Numbers

2 Write $\frac{23}{6}$ as a mixed number.

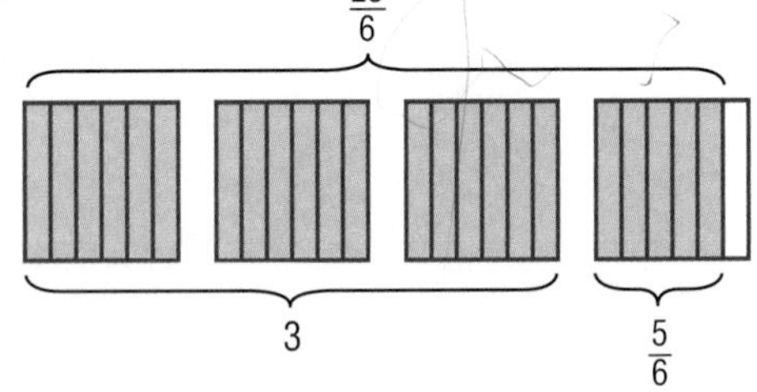

Divide 23 by 6.

$$\begin{array}{r} 3\frac{5}{6} \\ 6\overline{)23} \\ -18 \\ \hline 5 \end{array}$$ ← number of sixths left

So, $\frac{23}{6} = 3\frac{5}{6}$.

Reading Math

Fraction Bar Since a fraction represents division, $\frac{23}{6}$ means $23 \div 6$.

CHECK Your Progress

Write each improper fraction as a mixed number or a whole number.

b. $\frac{7}{3}$ c. $\frac{18}{5}$ d. $\frac{26}{2}$ e. $\frac{5}{5}$

CHECK Your Understanding

Example 1 (p. 210) **Write each mixed number as an improper fraction.**

1. $4\frac{1}{8}$
2. $2\frac{4}{5}$
3. $5\frac{2}{3}$

4. **BASEBALL** The width of a certain type of baseball bat is $2\frac{1}{4}$ inches. Write this width as an improper fraction.

Example 2 (p. 210) **Write each improper fraction as a mixed number or a whole number.**

5. $\frac{31}{6}$
6. $\frac{15}{4}$
7. $\frac{8}{8}$

Practice and Problem Solving

HOMEWORK HELP	
For Exercises	See Examples
8–17	1
18–25	2

Write each mixed number as an improper fraction.

8. $6\frac{1}{3}$
9. $8\frac{2}{3}$
10. $7\frac{4}{5}$
11. $1\frac{5}{8}$
12. $7\frac{1}{4}$
13. $5\frac{3}{4}$
14. $3\frac{5}{6}$
15. $4\frac{1}{6}$

16. **BOARD GAMES** The box for a popular board game is $10\frac{1}{2}$ inches wide. Write $10\frac{1}{2}$ as an improper fraction.

17. **RAIN FORESTS** The table shows the area of three tropical rain forests. Express the area of the Congo River Basin rain forest as an improper fraction.

Rain Forest	Area (square km)
Amazon	7 million
Congo River Basin	$1\frac{4}{5}$ million
Madagascar	110,000

Write each improper fraction as a mixed number or a whole number.

18. $\frac{16}{5}$
19. $\frac{27}{5}$
20. $\frac{9}{8}$
21. $\frac{19}{8}$
22. $\frac{15}{3}$
23. $\frac{28}{4}$
24. $\frac{10}{10}$
25. $\frac{9}{9}$

26. Express *six and three-fifths* as an improper fraction.

27. **ANIMALS** A nine-banded armadillo sleeps an average of $17\frac{2}{5}$ hours per day. Write $17\frac{2}{5}$ as an improper fraction.

28. **HEIGHTS** Find the height of each student listed in the table in terms of feet. Write as a mixed number in simplest form.

Student	Height (inches)
Emilio	65
Destiny	58
Hoshi	61
Jasmine	59

29. **TIME** Monifa spent 75 minutes at the park on Sunday. How many hours did Monifa spend at the park?

EXTRA PRACTICE
See pages 682, 709.

H.O.T. Problems

30. **OPEN ENDED** Select a mixed number that is between $6\frac{3}{5}$ and $\frac{36}{5}$.

31. **SELECT A TOOL** Which of the following tools might you use to write $4\frac{1}{6}$ as an improper fraction? Justify your selection(s). Then use the tool(s) to solve the problem.

draw a model | calculator | paper/pencil

32. **CHALLENGE** Write $2\frac{7}{4}$ and $3\frac{15}{15}$ in simplest form so that neither contains an improper fraction. Explain your reasoning.

33. **WRITING IN MATH** Explain how you know whether a fraction is less than, equal to, or greater than 1.

TEST PRACTICE

34. Which improper fraction is *not* equivalent to any of the mixed numbers in the table?

Cell Phone	Length (in.)
Julio's	$3\frac{1}{4}$
Morgan's	$2\frac{4}{5}$
Haylee's	$3\frac{3}{5}$

A $\frac{14}{5}$ B $\frac{13}{4}$ C $\frac{18}{5}$ D $\frac{14}{4}$

35. Serena bought 30 oranges. How many dozen oranges did she buy?

F $1\frac{3}{4}$

G $2\frac{1}{4}$

H $2\frac{1}{2}$

J $2\frac{2}{3}$

Spiral Review

Write each fraction in simplest form. (Lesson 4-2)

36. $\frac{35}{42}$ 37. $\frac{11}{12}$ 38. $\frac{5}{20}$

Find the GCF of each set of numbers. (Lesson 4-1)

39. 9, 39 40. 33, 88 41. 24, 48, 63

42. Order the decimals 27.025, 26.98, 27.13, 27.9, and 27.131 from least to greatest. (Lesson 3-2)

GET READY for the Next Lesson

43. **PREREQUISITE SKILL** Singer B had 18 more chart hits than Singer C. Singer A and Singer C had 227 chart hits combined. Determine a reasonable answer for the value of x. (Lesson 3-10)

Singer	Total Chart Hits
A	x
B	94
C	y
D	69

CHAPTER 4 Mid-Chapter Quiz

Lessons 4-1 through 4-3

Identify the common factors of each set of numbers. (Lesson 4-1)

1. 3, 9
2. 11, 33, 55

Find the GCF of each set of numbers. (Lesson 4-1)

3. 27, 45
4. 24, 40, 72

5. **MULTIPLE CHOICE** The table shows the number of shrimp ordered at a restaurant for three days.

Day	Shrimp
Monday	56
Tuesday	21
Wednesday	42

Each order contains the same number of shrimp. What is the greatest possible number of shrimp in each order? (Lesson 4-1)

A 8

B 7

C 6

D 3

Replace each ■ with a number so the fractions are equivalent. (Lesson 4-2)

6. $\frac{2}{9} = \frac{■}{45}$
7. $\frac{5}{12} = \frac{25}{■}$
8. $\frac{27}{36} = \frac{■}{4}$

9. **GRADES** On a quiz, Marta answered 4 out of 5 questions correctly. If each question is worth the same amount of points and the total number of points is twenty, what was Marta's score? (Lesson 4-2)

Write each fraction in simplest form. If the fraction is already in simplest form, write *simplest form*. (Lesson 4-2)

10. $\frac{15}{24}$
11. $\frac{12}{42}$
12. $\frac{9}{14}$

13. **RAINFALL** The world's driest city is Aswan, Egypt, which only receives an average of $\frac{32}{1,600}$ inches of rain each year. Write this fraction in simplest form. (Lesson 4-2)

Write each mixed number as an improper fraction. (Lesson 4-3)

14. $3\frac{5}{6}$
15. $7\frac{3}{5}$
16. $8\frac{4}{9}$

17. **MULTIPLE CHOICE** A local newspaper is reducing the width of its paper by $1\frac{3}{4}$ inches. What is this width as an improper fraction? (Lesson 4-3)

F $\frac{4}{3}$

G $\frac{8}{4}$

H $\frac{7}{3}$

J $\frac{7}{4}$

18. **BAKING** Express the amount of butter in the table as an improper fraction. (Lesson 4-3)

Ingredient	Amount
flour	$2\frac{3}{4}$ cups
butter	$1\frac{1}{3}$ cups
chocolate chips	$1\frac{1}{2}$ cups

Write each improper fraction as a mixed number or a whole number. (Lesson 4-3)

19. $\frac{37}{9}$
20. $\frac{69}{8}$
21. $\frac{42}{14}$

22. **WHALES** One of the world's heaviest whales is the Fin Whale, which weighs $\frac{248}{5}$ tons. Write this weight as a mixed number or a whole number. (Lesson 4-3)

Problem-Solving Investigation

MAIN IDEA: Solve problems by making an organized list.

P.S.I. TEAM +

e-Mail: MAKE AN ORGANIZED LIST

DELMAR: My three best friends, Bethany, Terrence, and Chris, are coming to my birthday party. I want all four of us to sit together on the same side of the table.

YOUR MISSION: Make an organized list to find how many ways they can sit together on the same side of the table.

Understand	You know that, counting Delmar, four people are sitting on one side of the table. You need to know the number of possible arrangements.
Plan	Make a list of all of the different possible arrangements. Use D for Delmar, B for Bethany, T for Terrence, and C for Chris.
Solve	Listing D first: DBTC, DBCT, DTBC, DTCB, DCBT, DCTB Listing B first: BDTC, BDCT, BTDC, BTCD, BCDT, BCTD Listing T first: TDBC, TDCB, TBDC, TBCD, TCDB, TCBD Listing C first: CDBT, CDTB, CBDT, CBTD, CTDB, CTBD There are 24 different ways the friends can sit along the same side of the table.
Check	Check the answer by seeing if each person is accounted for six times in the first, second, and third positions. ✔

Analyze The Strategy

1. Analyze the 24 possible arrangements. Do you agree or disagree with the possibilities? Explain your reasoning.
2. **WRITING IN MATH** Explain how making an organized list helps you to solve a problem.

Mixed Problem Solving

EXTRA PRACTICE
See pages 682, 709.

Use the *make an organized list* strategy to solve Exercises 3–6.

3. **JEANS** A store has the following options for jeans.

Length	Style	Color
short	straight leg	dark
medium	bootcut	light
long	flair	

How many combinations of length, style, and color are possible?

4. **NUMBER SENSE** How many different products are possible using the digits 2, 3, 6, and 8?

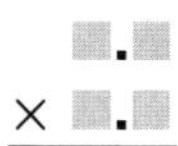

5. **PATTERNS** Where will the triangle with the circle be in the twentieth figure of this pattern?

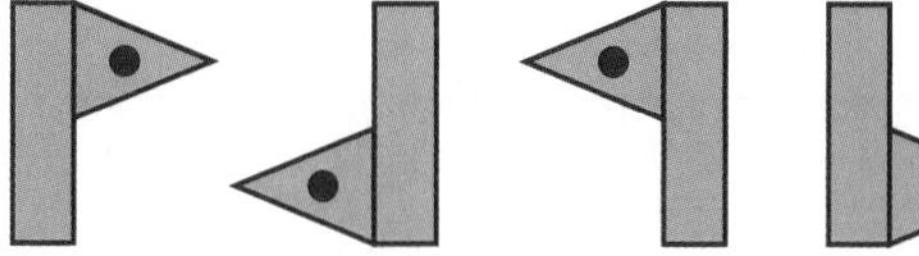

6. **MONEY** Joaquin has $0.75 to purchase a bottle of water from the vending machine. How many different combinations of change can he have if he only has nickels, dimes, and quarters? List the possibilities.

Use any strategy to solve Exercises 7–14. Some strategies are shown below.

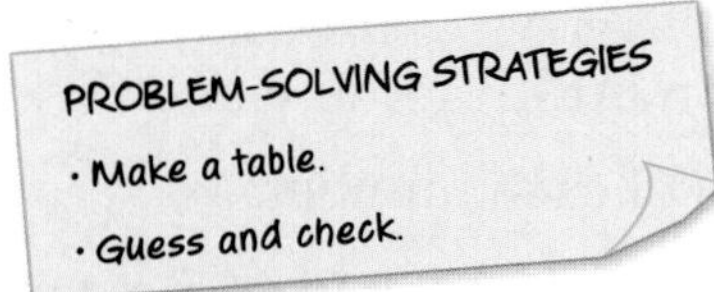
PROBLEM-SOLVING STRATEGIES
• Make a table.
• Guess and check.

7. **NUMBER SENSE** A whole number less than 10 is multiplied by 0.8. The product is then added to 14.4 and the result is 20. What is the number?

8. **FOOD** A grocery store deli sells turkey, roast beef, and ham sandwiches. In how many ways can the sandwiches be arranged in the display case?

9. **CODES** The letters A, B, C, and D are used to identify different types of dogs at a dog show. How many different identification codes for dogs are there if A is always the first letter?

10. **MALLS** The table shows the number of monthly trips to the mall for several sixth-grade students. How many students went to the mall six or more times in the month?

Students' Monthly Trips to the Mall					
5	10	0	1	11	4
12	4	3	6	8	5
8	9	6	2	13	2

11. **MONEY** You would like to buy four gifts that cost $15 each and one gift for $10.99. How much money will you have left if you start with $85.75?

12. **FOOD** Is $7 enough money to buy a loaf of bread for $0.98, one pound of cheese for $2.29, and one pound of luncheon meat for $3.29? Explain.

13. **HIKING** The number of miles Greg hiked in the first four days of a hiking trip are shown. At this rate, how many miles should he expect to hike at the end of the fifth day?

Day	Miles
1	2
2	3
3	5
4	8
5	■

14. **CARNIVALS** Lindsay and Marcello are setting up booths for the school carnival. There are six booths: ring toss, face-painting, snacks, tickets, balloon burst, and baseball toss. If the ticket booth and the snack booth must be first and second in line, respectively, in how many ways can the other booths be arranged?

4-5 Least Common Multiple

MAIN IDEA

Find the least common multiple of two or more numbers.

New Vocabulary

multiple
common multiples
least common multiple (LCM)

Math Online

glencoe.com

- Extra Examples
- Personal Tutor
- Self-Check Quiz

MINI Lab

STEP 1 Draw a number line from 0 to 15.

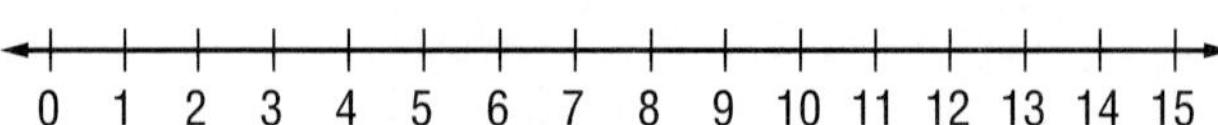

STEP 2 Find the products of 2 and each of the numbers 1, 2, 3, 4, 5, 6, and 7. Place a red tile above each of the products on the number line.

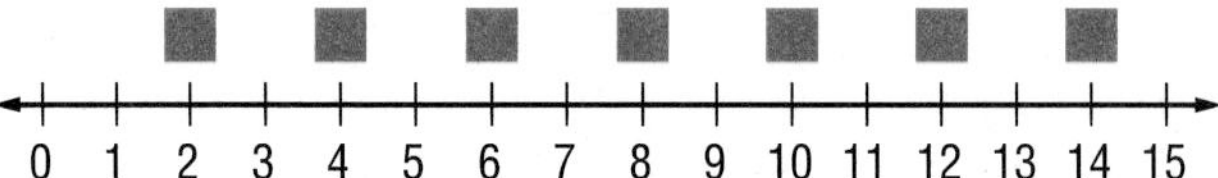

STEP 3 Find the products of 3 and each of the numbers 1, 2, 3, 4, and 5. Place a blue tile above each of the products on the same number line.

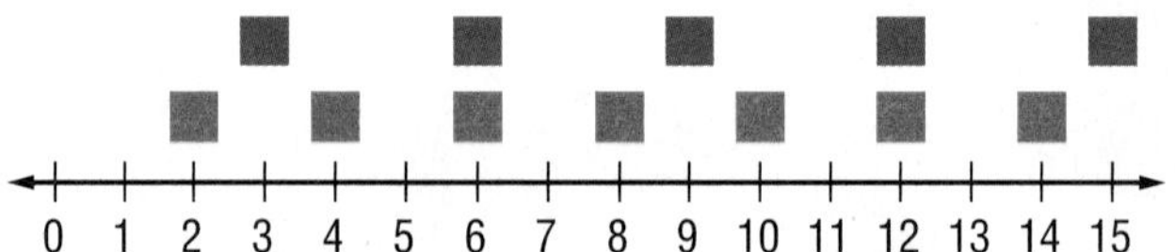

1. Which of the products of 2 are also products of 3?
2. Find the least number that is a product of both 2 and 3.

A **multiple** of a number is the product of the number and any whole number (0, 1, 2, 3, 4, ...). Multiples that are shared by two or more numbers are **common multiples**.

EXAMPLE Identify Common Multiples

1 Identify the first three common multiples of 4 and 8.

First, list the nonzero multiples of each number.

multiples of 4: 4, 8, 12, 16, 20, 24, … $1 \times 4, 2 \times 4, 3 \times 4, \ldots$

multiples of 8: 8, 16, 24, 32, 40, 48, … $1 \times 8, 2 \times 8, 3 \times 8, \ldots$

Notice that 8, 16, and 24 are multiples common to both 4 and 8. So, the first three common multiples of 4 and 8 are 8, 16, and 24.

CHECK Your Progress

Identify the first three common multiples of each set of numbers.

a. 2, 6

b. 4, 5, 10

The least number that is a multiple of two or more whole numbers is the **least common multiple (LCM)** of the numbers. In Example 1, the least common multiple of 4 and 8 is 8.

In addition to listing the multiples, you can also use prime factors to find the least common multiple.

EXAMPLE Find the LCM

2 Find the LCM of 15 and 40.

Write the prime factorization of each number. Identify all common prime factors.

$15 = 3 \times 5$

$40 = 2 \times 2 \times 2 \times 5$

Find the product of the prime factors using each common prime factor only once and any remaining factors.

The LCM is $2 \times 2 \times 2 \times 3 \times 5$ or 120.

CHECK Your Progress

Find the LCM of each set of numbers.

c. 4, 7

d. 3, 5, 7

Real-World Link There are more than 2,500 varieties of apples grown in the United States and more than 7,500 varieties of apples grown around the world.

Source: University of Illinois at Urbana-Champaign

Real-World EXAMPLE

3 FRUIT BASKETS Heritage Middle School is making fruit baskets for the community food bank. Apples are sold in bags of 10, oranges are sold in bags of 7, and there are 6 bananas in each bunch. How many of each should they buy so that they have an equal amount of each type of fruit in each basket?

Find the LCM using prime factors.

10: 2×5
7: 7
6: 2×3 Since 2 is a common prime factor, use it only once in the LCM.

They will have the same amount of each item when they buy $2 \times 5 \times 7 \times 3$, or 210 pieces of each kind of fruit.

CHECK Your Progress

e. **RADIO** A radio station is having a promotion in which every 12th caller receives a free CD and every 20th caller receives free movie passes. Which caller will be the first one to receive both prizes?

CHECK Your Understanding

Example 1 (p. 216) **Identify the first three common multiples of each set of numbers.**

1. 7, 14
2. 2, 8, 12

Example 2 (p. 217) **Find the LCM of each set of numbers.**

3. 6, 10
4. 2, 3, 13

Example 3 (p. 217)

5. **MEDICINE** Marco gets an allergy shot every 3 weeks. Percy gets an allergy shot every 5 weeks. If Marco and Percy meet while getting an allergy shot, how many weeks will it be before they see each other again?

Practice and Problem Solving

HOMEWORK HELP	
For Exercises	See Examples
6–11	1
12–17	2
18, 19	3

Identify the first three common multiples of each set of numbers.

6. 2, 10
7. 1, 7
8. 6, 9
9. 3, 8
10. 4, 8, 10
11. 3, 9, 18

Find the LCM of each set of numbers.

12. 3, 4
13. 7, 9
14. 16, 20
15. 15, 12
16. 15, 25, 75
17. 9, 12, 15

18. **MOON** A full moon occurs every 30 days. If the last full moon occurred on a Friday, how many days will pass before a full moon occurs again on a Friday?

19. **EVENTS** The cycles for two different events are shown in the table. Each of these events happened in the year 2000. What is the next year in which both will both happen?

Event	Cycle (yr)
Summer Olympics	4
United States Census	10

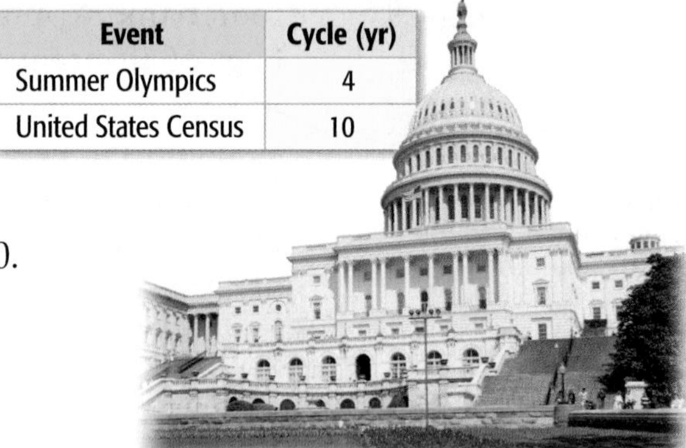

NUMBER SENSE For Exercises 20 and 21, use the following information.

The common multiples of x and 16 are 16, 32, 48, 64, 80, … .

The common multiples of y and z are 18, 36, 54, 72, 90, … .

20. Find four different possible values of x.
21. Find two different possible values each of y and z.

EXTRA PRACTICE
See pages 682, 709.

22. **PICTURES** For a yearbook picture, the marching band must line up in even rows. Describe the possible arrangements for the least number of people needed to be able to line up in rows of 5 or 6.

H.O.T. Problems

23. **FIND THE ERROR** D.J. and Trina are finding the LCM of 6 and 8. Who is correct? Explain your reasoning.

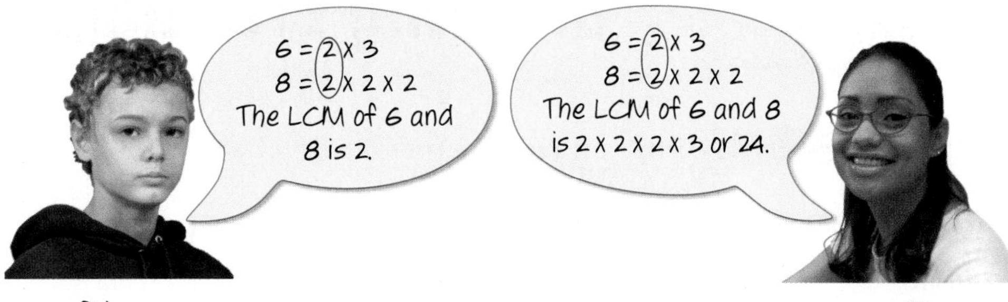

6 = 2 x 3
8 = 2 x 2 x 2
The LCM of 6 and 8 is 2.
6 = 2 x 3
8 = 2 x 2 x 2
The LCM of 6 and 8 is 2 x 2 x 2 x 3 or 24.
D.J.
Trina

24. **CHALLENGE** Is the statement below *sometimes, always,* or *never* true? Give at least two examples to support your reasoning.

The LCM of two numbers is the product of the two numbers.

25. **WRITING IN MATH** Create a problem about a real-world situation in which it would be helpful to find the least common multiple.

TEST PRACTICE

26. Micah is buying items for a birthday party. If he wants to have the same amount of each item, what is the least number of packages of cups he needs to buy?

Party Supplies	
Item	Number in Each Package
cups	6
plates	8

A 2 packages
B 3 packages
C 4 packages
D 5 packages

27. What is the least common multiple of 5, 9, and 15?

F 3
G 29
H 45
J 60

Spiral Review

28. **HOMEWORK** Tama needs to study for a math test, read a chapter in her novel, and write a social studies report tonight. How many different ways can Tama order these three activities? (Lesson 4-4)

29. **FOOD** Sabino bought a carton of 18 eggs for his dad at the grocery store. How many dozen eggs did Sabino buy? (Lesson 4-3)

Replace each ■ with a number so the fractions are equivalent. (Lesson 4-2)

30. $\frac{1}{5} = \frac{■}{25}$
31. $\frac{3}{17} = \frac{9}{■}$
32. $\frac{42}{48} = \frac{■}{8}$
33. $\frac{33}{55} = \frac{3}{■}$

GET READY for the Next Lesson

PREREQUISITE SKILL Choose the letter of the point that represents each fraction.

A B C
0 1

34. $\frac{1}{2}$
35. $\frac{3}{4}$
36. $\frac{1}{6}$

4-6 Comparing and Ordering Fractions

MAIN IDEA

Compare and order fractions.

New Vocabulary

least common denominator (LCD)

Math Online

glencoe.com

- Extra Examples
- Personal Tutor
- Self-Check Quiz

MINI Lab

Use a model to determine which fraction is greater, $\frac{3}{5}$ or $\frac{7}{10}$.

STEP 1 Draw a rectangle and shade $\frac{3}{5}$ of it.

$\frac{3}{5}$

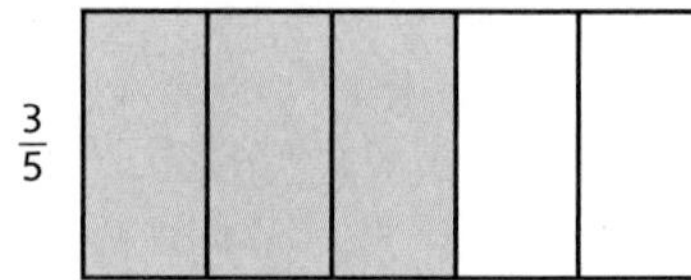

STEP 2 Draw another rectangle that is the same size and shade $\frac{7}{10}$ of it.

$\frac{7}{10}$

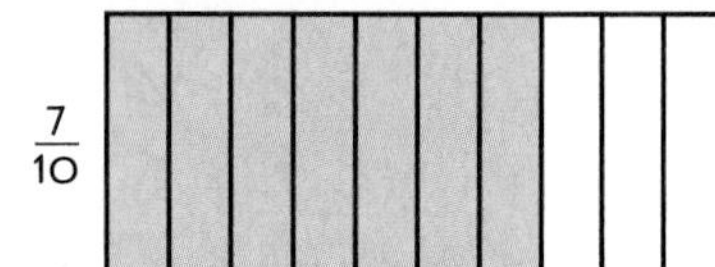

1. Which fraction is greater?

Use a model to determine which fraction is greater.

2. $\frac{1}{2}$ or $\frac{3}{7}$
3. $\frac{1}{6}$ or $\frac{2}{9}$
4. $\frac{3}{8}$ or $\frac{4}{7}$

To compare two fractions without using models, you can write them as fractions with the same denominator.

Compare Two Fractions — Key Concept

To compare two fractions you can follow these steps.

1. Find the **least common denominator (LCD)** of the fractions. That is, find the least common multiple of the denominators.
2. Write an equivalent fraction for each fraction using the LCD.
3. Compare the numerators.

EXAMPLES Compare Fractions and Mixed Numbers

Replace each ● with <, >, or = to make a true sentence.

1 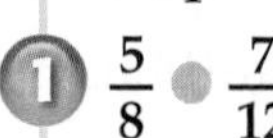

$\frac{5}{8}$ ● $\frac{7}{12}$

Step 1 The LCM of the denominators, 8 and 12, is 24. So, the LCD is 24.

Step 2 Write an equivalent fraction with a denominator of 24 for each fraction.

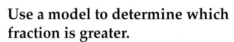

$\frac{5}{8} = \frac{15}{24}$ (×3) $\frac{7}{12} = \frac{14}{24}$ (×2)

Step 3 $\frac{15}{24} > \frac{14}{24}$, since 15 > 14. So, $\frac{5}{8} > \frac{7}{12}$.

Study Tip

Comparing Mixed Numbers When comparing mixed numbers like $5\frac{1}{8}$ and $3\frac{7}{10}$, it is not necessary to find a common denominator. Since $5 > 3$, $5\frac{1}{8} > 3\frac{7}{10}$.

2 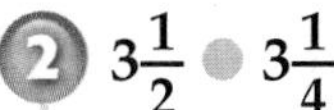 $3\frac{1}{2} \bullet 3\frac{1}{4}$

Since the whole numbers are the same, compare $\frac{1}{2}$ and $\frac{1}{4}$.

Step 1 The LCM of the denominators, 2 and 4, is 4. So, the LCD is 4.

Step 2 Write an equivalent fraction with a denominator of 4 for each fraction.

$$\frac{1}{2} \overset{\times 2}{=} \frac{2}{4} \qquad \frac{1}{4} \overset{\times 1}{=} \frac{1}{4}$$

Step 3 $\frac{2}{4} > \frac{1}{4}$, since $2 > 1$. So, $3\frac{1}{2} > 3\frac{1}{4}$.

Check Graph $3\frac{1}{2}$ and $3\frac{1}{4}$ on a number line. Since 4 is the LCD, separate the number line from 3 to 4 into four equal parts. Then graph $3\frac{2}{4}$ and $3\frac{1}{4}$.

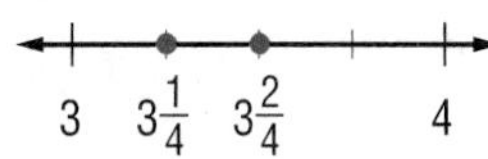

Since $3\frac{2}{4}$ is to the right of $3\frac{1}{4}$, the answer is correct.

Replace each ● with <, >, or = to make a true sentence.

a. $\frac{2}{3} \bullet \frac{4}{9}$ b. $\frac{5}{12} \bullet \frac{7}{8}$ c. $4\frac{1}{6} \bullet 4\frac{5}{18}$

You can use what you have learned about comparing fractions to order fractions.

EXAMPLE Order Fractions

3 **Order the fractions $\frac{1}{2}, \frac{9}{14}, \frac{3}{4}$, and $\frac{5}{7}$ from least to greatest.**

The LCD of the fractions is 28. So, rewrite each fraction with a denominator of 28.

$$\frac{1}{2} \overset{\times 14}{=} \frac{14}{28} \qquad \frac{3}{4} \overset{\times 7}{=} \frac{21}{28} \qquad \frac{9}{14} \overset{\times 2}{=} \frac{18}{28} \qquad \frac{5}{7} \overset{\times 4}{=} \frac{20}{28}$$

Since $\frac{14}{28} < \frac{18}{28} < \frac{20}{28} < \frac{21}{28}$, the order of the original fractions from least to greatest is $\frac{1}{2}, \frac{9}{14}, \frac{5}{7}, \frac{3}{4}$.

CHECK Your Progress

Order the fractions from least to greatest.

d. $\frac{1}{2}, \frac{5}{6}, \frac{2}{3}, \frac{3}{5}$ e. $\frac{4}{5}, \frac{3}{4}, \frac{2}{5}, \frac{1}{4}$ f. $4\frac{5}{6}, 4\frac{2}{3}, 4\frac{3}{5}, 4\frac{1}{5}$

TEST EXAMPLE

4 The fraction of Earth covered by each ocean is shown in the table. Which ocean covers the least amount of Earth?

A Arctic Ocean

B Atlantic Ocean

C Indian Ocean

D Pacific Ocean

Approximate Fraction of Earth Covered by Each Ocean	
Ocean	**Fraction**
Arctic	$\frac{1}{50}$
Atlantic	$\frac{1}{5}$
Indian	$\frac{7}{50}$
Pacific	$\frac{3}{10}$

Source: University of British Columbia Okanagan

Test-Taking Tip

Writing Equivalent Fractions Any common denominator can be used, but using the *least* common denominator usually makes the computation easier.

Read the Item You need to compare the fractions.

Solve the Item Rewrite the fractions with the LCD, 50.

$\frac{1}{50} = \frac{1}{50}$ (×1) $\quad \frac{1}{5} = \frac{10}{50}$ (×10) $\quad \frac{7}{50} = \frac{7}{50}$ (×1) $\quad \frac{3}{10} = \frac{15}{50}$ (×5)

So, $\frac{1}{50}$ is the least fraction, and the answer is A.

CHECK Your Progress

g. Each day, Kayla walks $\frac{1}{3}$ mile, Nora walks $\frac{1}{6}$ mile, and Mercedes walks $\frac{4}{5}$ mile. Which list shows these distances in order from least to greatest?

F $\frac{1}{3}$ mi, $\frac{1}{6}$ mi, $\frac{4}{5}$ mi

G $\frac{1}{6}$ mi, $\frac{4}{5}$ mi, $\frac{1}{3}$ mi

H $\frac{1}{3}$ mi, $\frac{4}{5}$ mi, $\frac{1}{6}$ mi

J $\frac{1}{6}$ mi, $\frac{1}{3}$ mi, $\frac{4}{5}$ mi

CHECK Your Understanding

Examples 1, 2 (pp. 220–221)

Replace each ● with <, >, or = to make a true sentence.

1. $\frac{3}{7}$ ● $\frac{1}{4}$
2. $\frac{5}{7}$ ● $\frac{15}{21}$
3. $8\frac{9}{16}$ ● $8\frac{5}{8}$

Example 3 (p. 221)

Order the fractions from least to greatest.

4. $\frac{4}{5}, \frac{1}{2}, \frac{9}{10}, \frac{3}{4}$
5. $6\frac{3}{8}, 6\frac{1}{4}, 6\frac{5}{6}, 6\frac{2}{3}$

Example 4 (p. 222)

6. **MULTIPLE CHOICE** In a survey about household odors, $\frac{7}{20}$ of the people said pet odors were the smelliest, $\frac{1}{10}$ voted for cooking odors, and $\frac{2}{5}$ chose garbage odors. Which odor did people choose the most?

A pet odors

B cooking odors

C garbage odors

D cannot tell from data

Practice and Problem Solving

HOMEWORK HELP	
For Exercises	See Examples
7–16	1, 2
17–22	3
33–35	4

Replace each ● with <, >, or = to make a true statement.

7. $\frac{1}{3}$ ● $\frac{3}{5}$
8. $\frac{7}{8}$ ● $\frac{5}{6}$
9. $5\frac{6}{9}$ ● $5\frac{2}{3}$
10. $7\frac{3}{4}$ ● $7\frac{9}{16}$
11. $\frac{7}{12}$ ● $\frac{1}{2}$
12. $\frac{14}{18}$ ● $\frac{7}{9}$
13. $2\frac{4}{5}$ ● $2\frac{13}{15}$
14. $10\frac{5}{8}$ ● $10\frac{20}{32}$
15. **MEASUREMENT** Which is shorter, $\frac{5}{8}$ of a foot or $\frac{3}{4}$ of a foot?
16. **MEASUREMENT** Which is greater, $\frac{2}{3}$ of a dozen or $\frac{3}{4}$ of a dozen?

Order the fractions from least to greatest.

17. $\frac{1}{2}, \frac{2}{3}, \frac{1}{4}, \frac{5}{6}$
18. $\frac{2}{3}, \frac{2}{9}, \frac{5}{6}, \frac{11}{18}$
19. $9\frac{1}{6}, 9\frac{2}{5}, 9\frac{3}{7}, 9\frac{3}{5}$
20. $3\frac{5}{8}, 3\frac{3}{4}, 3\frac{1}{2}, 3\frac{9}{16}$
21. **INSECTS** Ling collected four small insects for his science class. The insects measured $\frac{3}{8}$ inch, $\frac{5}{16}$ inch, $\frac{3}{4}$ inch, and $\frac{1}{2}$ inch. Which insect is the longest?
22. **NECKLACES** Kate has three different beads that she is using to make a necklace. The first bead is $\frac{5}{6}$ inch long, the second bead is $\frac{7}{8}$ inch long, and the third bead is $\frac{3}{16}$ inch long. Which of these beads is the longest?

Replace each ● with <, >, or = to make a true statement.

23. $\frac{3}{5}$ ● $\frac{3}{20}$
24. $5\frac{1}{3}$ ● $6\frac{1}{3}$
25. $\frac{15}{24}$ ● $1\frac{5}{8}$
26. $\frac{18}{4}$ ● $3\frac{1}{2}$
27. **ANALYZE TABLES** The world's five largest deserts are shown in the table. Order the areas from least to greatest.
28. **BIKING** Tobias, Marcus, and Dominic rode their bicycles to the arcade. Tobias rode $\frac{12}{5}$ miles, Marcus rode $2\frac{1}{3}$ miles, and Dominic rode $\frac{9}{4}$ miles. Which person rode closest to 2 miles? Explain your reasoning.

Desert	Area (millions of square miles)
Sahara	$\frac{7}{2}$
Kalahari	$\frac{2}{10}$
Gobi	$\frac{2}{5}$
Australian	$1\frac{4}{10}$
Arabian	$\frac{1}{2}$

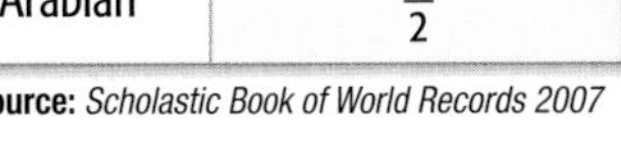

Source: *Scholastic Book of World Records 2007*

29. **FIND THE DATA** Refer to the Data File on pages 16–19. Choose some data and write a real-world problem in which you would compare two fractions.
30. **OPEN ENDED** Specify three fractions with different denominators that have an LCD of 24. Then arrange the fractions in order from least to greatest.

EXTRA PRACTICE
See pages 683, 709.

H.O.T. Problems

31. **CHALLENGE** Order $\frac{3}{8}$, $\frac{3}{7}$, and $\frac{3}{9}$ from least to greatest without writing equivalent fractions with a common denominator. Explain your strategy.

32. **WRITING IN MATH** Explain how to determine if $\frac{1}{6}$ is less than, greater than, or equal to $\frac{7}{9}$ without using the least common denominator.

TEST PRACTICE

33. Which statement about the mixed number $2\frac{3}{4}$ is true?

   A $2\frac{3}{4} > 2\frac{2}{3}$

   B $3 < 2\frac{3}{4}$

   C $2\frac{3}{4} < 2\frac{2}{3}$

   D $2\frac{1}{4} > 2\frac{3}{4}$

34. A plumber needs to drill a hole that is just slightly larger than $\frac{3}{16}$ inch in diameter. Which measure is the smallest but still larger than $\frac{3}{16}$ inch?

   F $\frac{3}{32}$ inch

   G $\frac{5}{16}$ inch

   H $\frac{13}{64}$ inch

   J $\frac{17}{32}$ inch

35. The table shows the fraction of Internet users that have done each activity online.

Activity	Fraction of Internet Users
Search for information	$\frac{9}{10}$
Check the weather	$\frac{19}{25}$
Download music	$\frac{1}{4}$
Write or read a blog	$\frac{9}{25}$

Source: Pew Internet and American Life Project

Which activity was reported most often?

   A downloading music

   B checking the weather

   C searching for information

   D writing or reading a blog

Spiral Review

36. **DECORATING** Sahale decorated his house with three strands of party lights. The red strand blinks every 4 seconds, the green strand blinks every 6 seconds, and the white strand blinks every 10 seconds. How many seconds will go by until the three strands blink at the same time? (Lesson 4-5)

37. Express $5\frac{3}{8}$ as an improper fraction. (Lesson 4-3)

38. **MONEY** Lydia has four quarters, three dimes, two nickels, two one-dollar bills, and one five-dollar bill. Write a fraction in simplest form that compares the number of bills to the total number of pieces of money she has. (Lesson 4-2)

GET READY for the Next Lesson

PREREQUISITE SKILL Write each decimal in standard form. (Lesson 3-1)

39. seven tenths

40. four and six tenths

41. eighty-nine hundredths

42. twenty-five thousandths

4-7 Writing Decimals as Fractions

MAIN IDEA

Write decimals as fractions or mixed numbers in simplest form.

New Vocabulary

rational number

Math Online

glencoe.com

- Extra Examples
- Personal Tutor
- Self-Check Quiz

GET READY for the Lesson

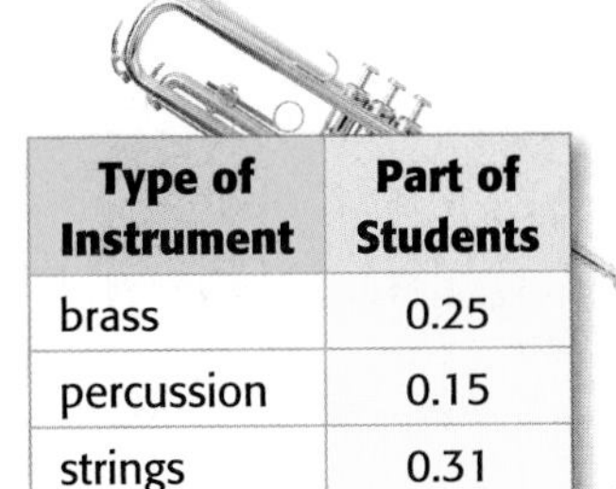

MUSIC The table shows the part of students in the school orchestra that play each type of musical instrument.

Type of Instrument	Part of Students
brass	0.25
percussion	0.15
strings	0.31
woodwind	0.29

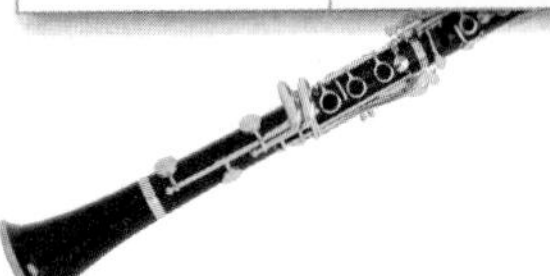

1. Write the word form of the decimal that represents the part of those surveyed who play a stringed instrument.
2. Write this decimal as a fraction.
3. Repeat Exercises 1 and 2 with each of the other decimals.

Decimals like 0.25, 0.15, 0.31, and 0.29 can be written as fractions with denominators of 10, 100, 1,000, and so on. Any number that can be written as a fraction is a **rational number**.

Write Decimals as Fractions — Key Concept

To write a decimal as a fraction, you can follow these steps.

1. Identify the place value of the last decimal place.
2. Write the decimal as a fraction using the place value as the denominator. If necessary, simplify the fraction.

EXAMPLES Write Decimals as Fractions

Write each decimal as a fraction in simplest form.

1 **0.6**

The place-value chart shows that the place value of the last decimal place is tenths. So, 0.6 means six tenths.

1,000	100	10	1	0.1	0.01	0.001	0.0001
thousands	hundreds	tens	ones	tenths	hundredths	thousandths	ten-thousandths
0	0	0	0.	6	0	0	0

$0.6 = \frac{6}{10}$ Say *six tenths.*

$= \frac{\overset{3}{\cancel{6}}}{\underset{5}{\cancel{10}}}$ Simplify. Divide the numerator and denominator by the GCF, 2.

$= \frac{3}{5}$

Study Tip

Mental Math Here are some commonly used decimal fraction equivalencies:

$0.1 = \frac{1}{10}$

$0.2 = \frac{1}{5}$

$0.25 = \frac{1}{4}$

$0.5 = \frac{1}{2}$

$0.75 = \frac{3}{4}$

It is helpful to memorize these.

2 **0.45**

$0.45 = \frac{45}{100}$ Say *forty-five hundredths.*

$= \frac{\overset{9}{\cancel{45}}}{\underset{20}{\cancel{100}}}$ Simplify. Divide by the GCF, 5.

$= \frac{9}{20}$

1,000	100	10	1	0.1	0.01	0.001	0.0001
thousands	hundreds	tens	ones	tenths	hundredths	thousandths	ten-thousandths
0	0	0	0.	4	5	0	0

3 **0.375**

$0.375 = \frac{375}{1,000}$ Say *three hundred seventy-five thousandths.*

$= \frac{\overset{3}{\cancel{375}}}{\underset{8}{\cancel{1,000}}}$ Simplify. Divide by the GCF, 125.

$= \frac{3}{8}$

1,000	100	10	1	0.1	0.01	0.001	0.0001
thousands	hundreds	tens	ones	tenths	hundredths	thousandths	ten-thousandths
0	0	0	0.	3	7	5	0

CHECK Your Progress

Write each decimal as a fraction in simplest form.

a. 0.8 **b.** 0.28 **c.** 0.125

Decimals like 3.25, 26.82, and 125.54 can be written as mixed numbers in simplest form.

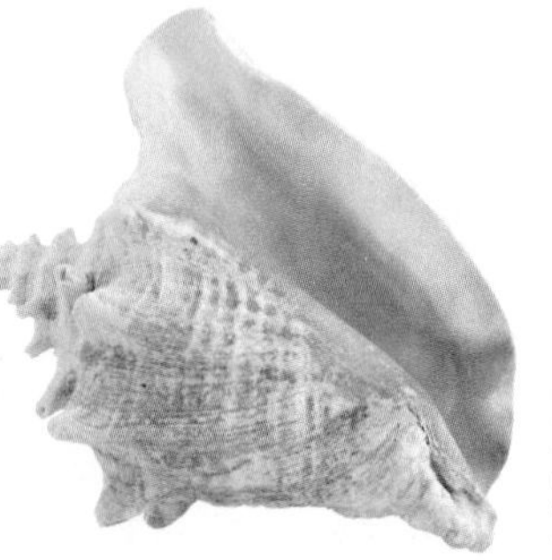

Real-World Link The Queen Conch is a mollusk that produces the beautiful shell shown above. A Queen Conch can live for 20–25 years in captivity.

Source: Conch Heritage Network

EXAMPLE Write Decimals as Mixed Numbers

4 **SHELLS** **The table shows the average length of several kinds of seashells. Express the average length of the conch shell as a mixed number in simplest form.**

Length of Seashells

Shell	Average Length (in.)
Conch	9.85
Nautilus	6.5
Scallop	2.75
Tulip	8.0

Source: Sanibel Seashell Industries

$9.85 = 9\frac{85}{100}$ Say *nine and eighty-five hundredths.*

$= 9\frac{\overset{17}{\cancel{85}}}{\underset{20}{\cancel{100}}}$ Simplify.

$= 9\frac{17}{20}$

CHECK Your Progress

d. **MILK** It takes approximately 4.65 quarts of milk to make a pound of cheese. Express this amount as a mixed number in simplest form.

Check Your Understanding

Examples 1–4 (pp. 225–226)

Write each decimal as a fraction or mixed number in simplest form.

1. 0.4 **2.** 0.5 **3.** 0.64 **4.** 0.75

5. 0.525 **6.** 0.375 **7.** 2.75 **8.** 5.12

Example 4 (p. 226)

9. **CARS** Mr. Ravenhead's car averages 23.75 miles per gallon of gasoline. Express this amount as a mixed number in simplest form.

Practice and Problem Solving

HOMEWORK HELP

For Exercises	See Examples
10–13	1
14–17, 22, 23	2
18–21	3
24–33	4

Write each decimal as a fraction in simplest form.

10. 0.3 **11.** 0.7 **12.** 0.2 **13.** 0.5

14. 0.33 **15.** 0.21 **16.** 0.65 **17.** 0.82

18. 0.875 **19.** 0.425 **20.** 0.018 **21.** 0.004

22. **STOCKS** Last week a share of stock gained a total of 0.64 point. Express this gain as a fraction in simplest form.

23. **DISTANCE** Evita lives 0.85 mile from her school. Write this distance as a fraction in simplest form.

Write each decimal as a mixed number in simplest form.

24. 8.9 **25.** 12.1 **26.** 14.06 **27.** 17.03

28. 9.35 **29.** 42.96 **30.** 7.425 **31.** 50.605

SANDWICHES For Exercises 32–34, refer to the table that shows the ingredients in an Italian sandwich at Johnny's Deli.

Ingredient	Amount (lb)
meat	0.35
vegetables	0.15
secret sauce	0.05
bread	0.05

32. What fraction of a pound is each ingredient?

33. How much more meat is in the sandwich than vegetables? Write the amount as a fraction in simplest form.

34. What is the total weight of the Italian sandwich? Write the amount as a fraction in simplest form.

35. **LADYBUGS** The average length of a ladybug can range from 0.08 to 0.4 inch. Find two lengths that are within the given span. Write them as fractions in simplest form.

EXTRA PRACTICE

See pages 683, 709.

36. **FENCES** William bought 20 yards of fencing. He used 5.9 yards to surround one flower garden and 10.3 yards to surround another garden. Write the amount remaining as a fraction in simplest form.

H.O.T. Problems

37. **CHALLENGE** Decide whether the following statement is *sometimes*, *always*, or *never* true. Explain your reasoning.

    Any decimal that ends with a digit in the thousandths place can be written as a fraction with a denominator that is divisible by both 2 and 5.

38. **FIND THE ERROR** Eduardo and Laura are writing 4.28 as a mixed number. Who is correct? Explain your reasoning.

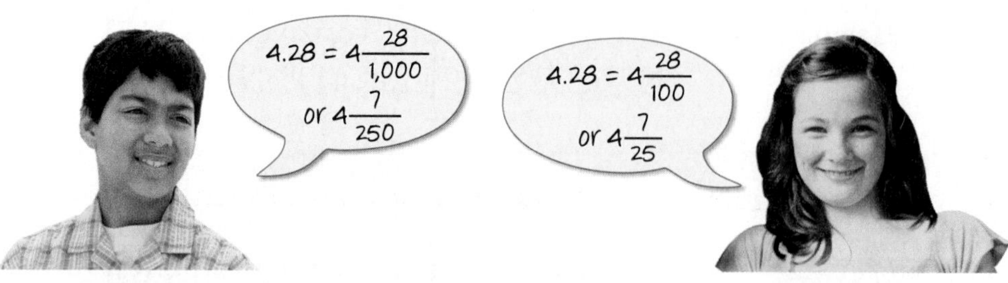

$4.28 = 4\frac{28}{1,000}$ or $4\frac{7}{250}$
$4.28 = 4\frac{28}{100}$ or $4\frac{7}{25}$
Eduardo
Laura

39. **WRITING IN MATH** Explain how to express 0.36 as a fraction.

TEST PRACTICE

40. Rafael shaded 0.25 of the design.

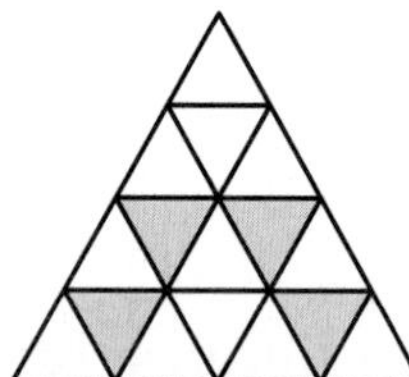

Which fraction in simplest form represents the shaded part of the design?

A $\frac{1}{2}$ B $\frac{25}{100}$ C $\frac{4}{16}$ D $\frac{1}{4}$

41. Which of the following statements is *not* true?

F $0.6 = \frac{3}{5}$

G $0.125 = \frac{1}{8}$

H $2.015 = 2\frac{1}{200}$

J $10.38 = 10\frac{19}{50}$

Spiral Review

Replace each ● with <, >, or = to make a true sentence. (Lesson 4-6)

42. $\frac{1}{3}$ ● $\frac{2}{7}$

43. $7\frac{5}{9}$ ● $7\frac{6}{11}$

44. $\frac{3}{5}$ ● $\frac{12}{20}$

45. $8\frac{4}{15}$ ● $9\frac{8}{27}$

46. Find the LCM of 15, 20, and 25. (Lesson 4-5)

47. **SWEATERS** A store sells sweaters in 5 different styles and 4 different colors. How many combinations of style and color are available? (Lesson 4-4)

GET READY for the Next Lesson

PREREQUISITE SKILL Divide. (Page 726)

48. $45 \div 5$

49. $72 \div 4$

50. $112 \div 8$

51. $84 \div 4$

4-8 Writing Fractions as Decimals

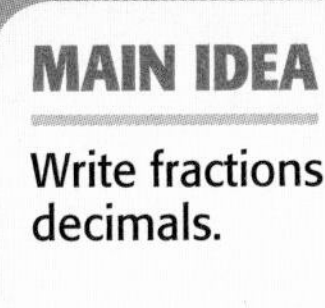

MAIN IDEA

Write fractions as decimals.

Math Online

glencoe.com

- Extra Examples
- Personal Tutor
- Self-Check Quiz

GET READY for the Lesson

BIRTH ORDER The table shows the responses to a survey of birth order.

1. Write the decimal for $\frac{3}{10}$.
2. Write the fraction equivalent to $\frac{1}{2}$ with a denominator of 10.
3. Write the decimal for the fraction you found in Exercise 2.

What is Your Birth Order?	Response
oldest child	$\frac{1}{20}$
middle child	$\frac{1}{2}$
youngest child	$\frac{3}{10}$
only child	$\frac{3}{20}$

Fractions with denominators of 10, 100, or 1,000 can be written as a decimal using place value. For fractions with denominators that are *factors* of 10, 100, or 1,000, you can write equivalent fractions with these denominators.

EXAMPLES Write Fractions as Decimals

1 Write $\frac{2}{5}$ as a decimal.

Since 5 is a factor of 10, write an equivalent fraction with a denominator of 10.

$\frac{2}{5} = \frac{4}{10}$ (×2) Since 5 × 2 = 10, multiply the numerator and denominator by 2.

$= 0.4$ Read 0.4 as four tenths.

2 Write $\frac{3}{4}$ as a decimal.

Since 4 is a factor of 100, write an equivalent fraction with a denominator of 100.

$\frac{3}{4} = \frac{75}{100}$ (×25) Since 4 × 25 = 100, multiply the numerator and denominator by 25.

$= 0.75$ Read 0.75 as seventy-five hundredths.

CHECK Your Progress

Write each fraction as a decimal.

a. $\frac{3}{5}$ b. $\frac{14}{25}$ c. $\frac{102}{250}$

Any fraction can be written as a decimal by dividing the numerator by the denominator.

EXAMPLE Fractions as Decimals

3 **Write $\frac{7}{8}$ as a decimal.**

METHOD 1 **Use paper and pencil.**

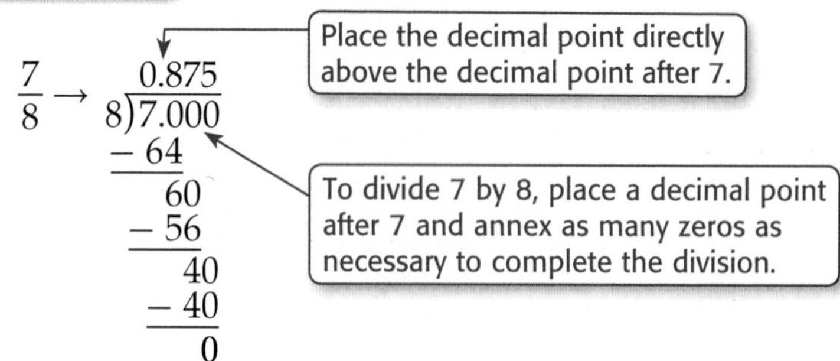

$$\frac{7}{8} \rightarrow 8\overline{)7.000} = 0.875$$

$$\begin{array}{r} 0.875 \\ 8\overline{)7.000} \\ -64 \\ \hline 60 \\ -56 \\ \hline 40 \\ -40 \\ \hline 0 \end{array}$$

Place the decimal point directly above the decimal point after 7.

To divide 7 by 8, place a decimal point after 7 and annex as many zeros as necessary to complete the division.

METHOD 2 **Use a calculator.**

7 [÷] 8 [ENTER] 0.875

Therefore, $\frac{7}{8} = 0.875$.

CHOOSE Your Method **Write each fraction as a decimal.**

d. $\frac{1}{8}$ e. $\frac{1}{2}$ f. $\frac{5}{4}$

Real-World EXAMPLE Mixed Numbers as Decimals

4 **INTERNET Use the information at the left to write the number of Internet users per 100 people as a decimal.**

$55\frac{7}{50} = 55 + \frac{7}{50}$	Definition of a mixed number
$= 55 + \frac{14}{100}$	Since $50 \times 2 = 100$, multiply the numerator and the denominator by 2.
$= 55 + 0.14$ or 55.14	Read 55.14 as *fifty-five and fourteen hundredths.*

The number of Internet users per 100 people is 55.14.

Check Use a calculator. 55 [+] 7 [÷] 50 [ENTER] 55.14 ✔

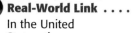

Real-World Link
In the United States, there are $55\frac{7}{50}$ Internet users per 100 people.
Source: *Top 10 of Everything*

CHECK Your Progress

g. **POPULATION** In Nevada, there are $20\frac{2}{5}$ people per square mile. Express this fraction as a decimal.

CHECK Your Understanding

Examples 1–3 (pp. 229–230)

Write each fraction or mixed number as a decimal.

1. $\frac{9}{10}$
2. $\frac{2}{5}$
3. $\frac{7}{2}$
4. $\frac{1}{8}$
5. $\frac{9}{25}$
6. $\frac{5}{16}$

Example 4 (p. 230)

7. $3\frac{7}{10}$
8. $6\frac{4}{25}$
9. $4\frac{9}{40}$

10. **ANIMALS** The Siberian tiger can grow up to $10\frac{4}{5}$ feet long. Express this length as a decimal.

Practice and Problem Solving

HOMEWORK HELP	
For Exercises	See Examples
11–14	1, 2
15–18	3
19–24	4

Write each fraction or mixed number as a decimal.

11. $\frac{1}{20}$
12. $\frac{19}{25}$
13. $\frac{77}{200}$
14. $\frac{311}{500}$
15. $\frac{5}{8}$
16. $\frac{12}{75}$
17. $\frac{9}{16}$
18. $\frac{5}{32}$
19. $6\frac{1}{16}$
20. $8\frac{21}{40}$
21. $12\frac{43}{80}$
22. $9\frac{9}{32}$

23. **GAMES** A handheld video game is $5\frac{13}{16}$ inches long. Express this length as a decimal.

24. **SCHOOL** Rancho Middle School has an average of $23\frac{3}{8}$ students per teacher. Write this fraction as a decimal.

Replace each ● with <, >, or = to make a true sentence.

25. $\frac{3}{4}$ ● 0.8
26. $\frac{17}{40}$ ● 0.4
27. 0.72 ● $\frac{3}{4}$

28. **GEOMETRY** The length s of a side of a square can be found using the formula $s = \frac{1}{4}P$, where P is the perimeter. Express $\frac{1}{4}$ as a decimal.

29. **TRACK** Paloma can run the 100-meter dash in $16\frac{1}{5}$ seconds. Savannah's best time is 19.8 seconds. How much faster is Paloma than Savannah in the 100-meter dash?

30. **MEASUREMENT** The table shows the wingspans of different birds. Using decimals, name the bird that has the smallest minimum wingspan and the bird that has the greatest maximum wingspan.

Bird	Minimum Wingspan (ft)	Maximum Wingspan (ft)
Whooping Crane	$7\frac{1}{6}$	$7\frac{1}{4}$
Bald Eagle	$6\frac{7}{12}$	$7\frac{1}{2}$
Black-Footed Albatross	$6\frac{3}{10}$	$7\frac{1}{12}$

Source: Cornell Lab of Ornithology

EXTRA PRACTICE
See pages 683, 709.

H.O.T. Problems

CHALLENGE **Express each fraction as a decimal.**

31. $\frac{1}{3}$ 32. $\frac{2}{3}$ 33. $\frac{4}{9}$

34. **REASONING** Explain why the decimals in Exercises 31–33 are called *repeating decimals.*

35. **CHALLENGE** Write a fraction that can be expressed as a repeating decimal when two digits repeat.

36. **OPEN ENDED** Write a fraction with a decimal value between $\frac{1}{2}$ and $\frac{3}{4}$. Write both the fraction and the equivalent decimal.

37. **WRITING IN MATH** Summarize the two methods for expressing fractions as decimals. Describe when it is appropriate to use each method in your summary.

TEST PRACTICE

38. Which decimal represents the shaded portion of the figure below?

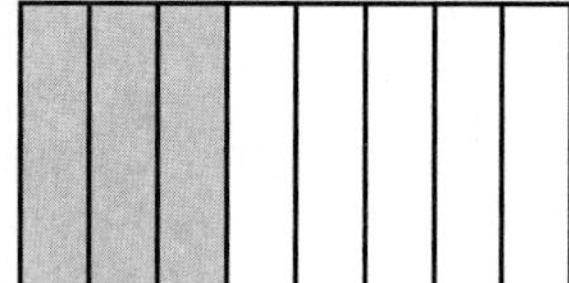

A 0.25

B 0.333

C 0.375

D 0.4

39. The formula $d = v + \frac{1}{20}v^2$ can be used to find the distance d required to stop a certain model car traveling at v miles per hour. Which of the following best represents $\frac{1}{20}$?

F 0.05

G 0.21

H 0.4

J 1.2

Spiral Review

Write each decimal as a fraction or mixed number in simplest form. (Lesson 4-7)

40. 0.25 41. 0.73 42. 8.118 43. 11.14

44. Which fraction is greater, $\frac{13}{40}$ or $\frac{3}{7}$? (Lesson 4-6)

45. **FOOD** Twenty out of two dozen cupcakes are chocolate cupcakes. Write this amount as a fraction in simplest form. (*Hint:* 1 dozen = 12) (Lesson 4-2)

GET READY for the Next Lesson

PREREQUISITE SKILL **Graph each number on the same number line.**

46. 2.25 47. 1.5 48. 0.5 49. 3.75

Algebra: Ordered Pairs and Functions

MAIN IDEA

Use ordered pairs to locate points and organize data.

New Vocabulary

coordinate plane
origin
***x*-axis**
***y*-axis**
ordered pair
***x*-coordinate**
***y*-coordinate**
graph

Math Online

glencoe.com

- Concepts in Motion
- Extra Examples
- Personal Tutor
- Self-Check Quiz

GET READY for the Lesson

MAPS A street map is shown.

1. How is the map labeled?
2. Location C5 is closest to the end of which street?
3. Identify where Cedar Court and Juniper Lane intersect on the map.

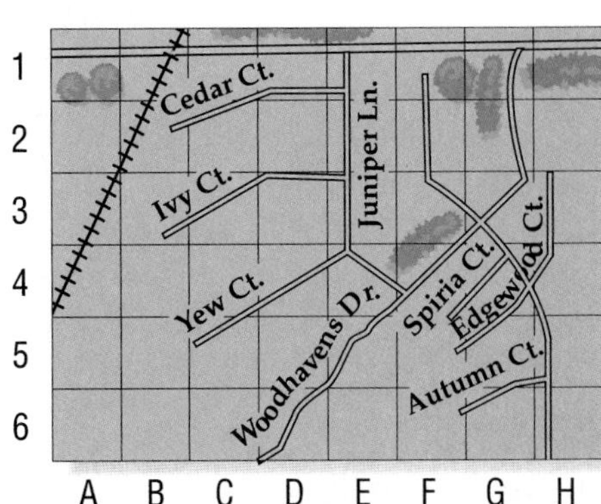

In mathematics, points are located on a coordinate plane.

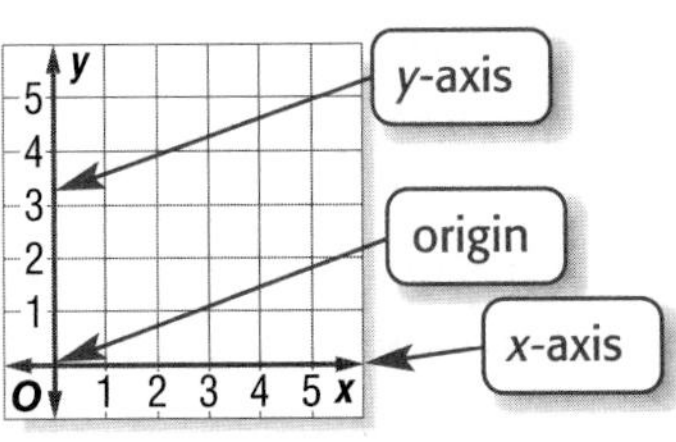

The **coordinate plane** is formed when two number lines intersect at their zero points. This point is called the **origin**. The horizontal number line is called the ***x*-axis**, and the vertical number line is called the ***y*-axis**.

You can use an **ordered pair** to name any point on the coordinate plane. The first number in an ordered pair is the ***x*-coordinate**, and the second number is the ***y*-coordinate**.

The *x*-coordinate corresponds to a number on the *x*-axis. — (3, 6) — The *y*-coordinate corresponds to a number on the *y*-axis.

EXAMPLE Naming Points Using Ordered Pairs

1 Write the ordered pair that names point *L*.

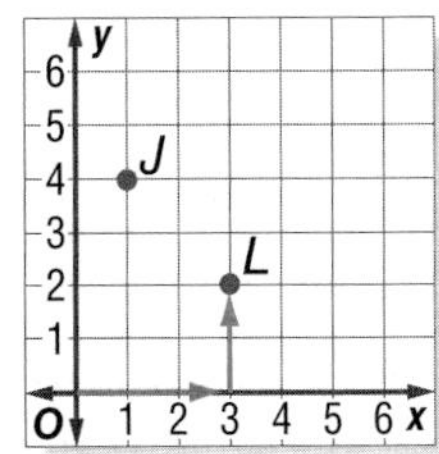

Step 1 Start at the origin. Move right along the *x*-axis until you are under point *L*. The *x*-coordinate of the ordered pair is 3.

Step 2 Now move up until you reach point *L*. The *y*-coordinate is 2.

So, point *L* is named by the ordered pair (3, 2).

CHECK Your Progress

a. Write the ordered pair that names point *J*.

Study Tip

Coordinate Plane A coordinate plane is also called a coordinate grid.

You can also graph a point on a coordinate plane. To **graph** a point means to place a dot at the point named by an ordered pair.

EXAMPLES Graphing Ordered Pairs

2 **Graph the point $M(5, 6)$.**

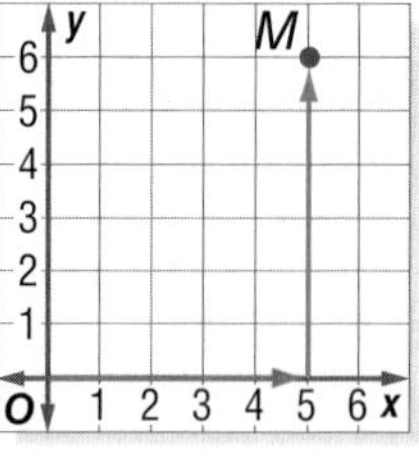

- Start at the origin.
- Move 5 units to the right on the x-axis.
- Then move 6 units up to locate the point.
- Draw a dot and label the dot M.

3 **Graph the point $N\left(2\frac{1}{2}, 2\right)$.**

- Start at the origin.
- The value $2\frac{1}{2}$ is halfway between 2 and 3. So, on the x-axis, move halfway between 2 and 3.
- Then move 2 units up to locate the point.
- Draw a dot and label the dot N.

Study Tip

Alternative Method To graph a fraction or a decimal, you can also draw the coordinate plane so that the axes are separated into halves, thirds, and so on.

4 **Graph the point $P(1, 1.25)$.**

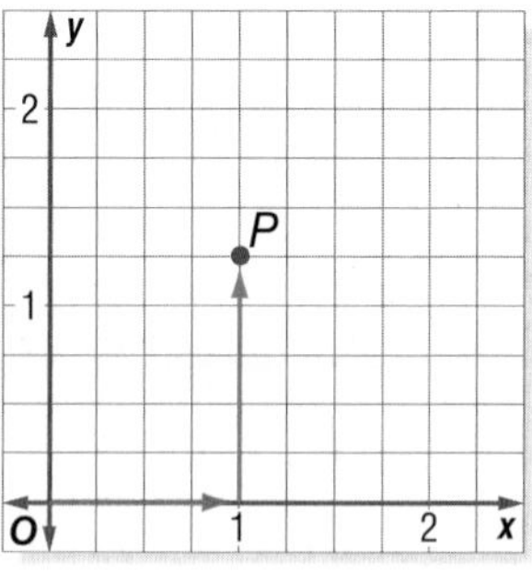

- Start at the origin.
- Move 1 unit to the right on the x-axis.
- The value 1.25 is one-fourth of the way between 1 and 2. So, move one-fourth of the way between 1 and 2.
- Draw a dot and label the dot P.

CHECK Your Progress

Graph and label each point on a coordinate plane.

b. $X(8, 0)$ **c.** $Y\left(2, 5\frac{1}{4}\right)$ **d.** $Z(3.75, 6)$

Ordered pairs can also be used to graph functions.

Real-World EXAMPLES

5 **FUNDRAISER The cheerleaders at South Middle School are selling bracelets for a fundraiser. The costs of 1, 2, 3, and 4 bracelets are shown in the table. List this information as ordered pairs (number of bracelets, cost).**

Bracelet Costs

Number of Bracelets	Cost ($)
1	3
2	6
3	9
4	12

The ordered pairs are (1, 3), (2, 6), (3, 9), and (4, 12).

6 **Graph the ordered pairs in Example 5. Then describe the graph.**

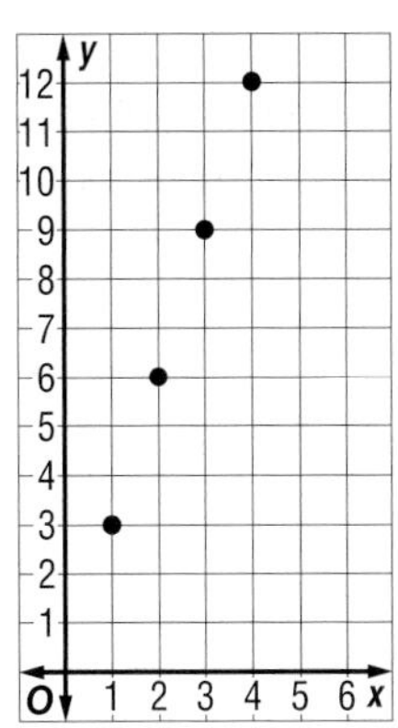

The points appear to fall on a line.

CHECK Your Progress

e. **BABYSITTING** Gloria earns $5.50 each hour babysitting. She made a table showing her earnings for 0, 1, 2, and 3 hours of babysitting. List this information as ordered pairs (time, earnings).

Time (h)	Earnings ($)
0	0
1	5.5
2	11
3	16.5

f. Graph the ordered pairs. Then describe the graph.

CHECK Your Understanding

Example 1 (p. 233) **Use the coordinate plane at the right to name the ordered pair for each point.**

1. G
2. D
3. C
4. F

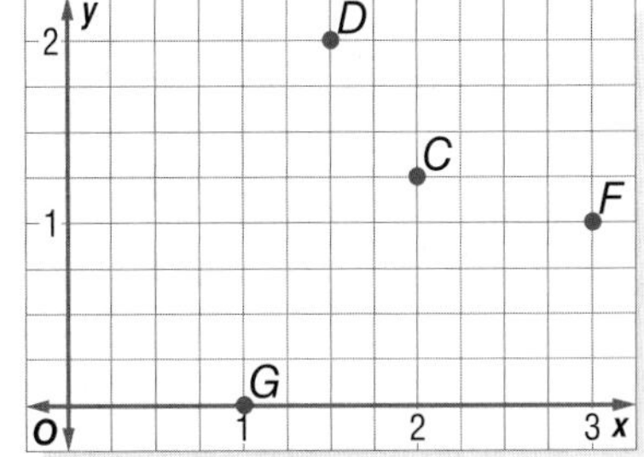

Examples 2–4 (p. 234) **Graph and label each point on a coordinate plane.**

5. $A(3, 7)$
6. $B(0, 4)$
7. $C(1.5, 6)$
8. $D\left(2, 4\frac{3}{4}\right)$

BASKETBALL For Exercises 9 and 10, use the following information.

In basketball, each shot made from outside the 3-point line scores 3 points. The table at the right shows this relationship.

3-Point Shots Made	Total Points
0	0
1	3
2	6
3	9

Examples 5, 6 (pp. 234–235)

9. List this information as ordered pairs (3-point shots made, total number of points).
10. Graph the ordered pairs. Then describe the graph.

Practice and Problem Solving

HOMEWORK HELP	
For Exercises	See Examples
11–20	1
21–28	2–4
29–32	5, 6

Use the coordinate plane at the right to name the ordered pair for each point.

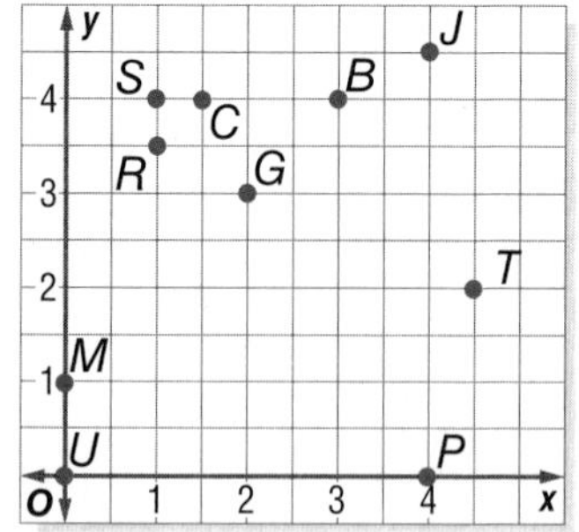

11. P
12. G
13. B
14. M
15. S
16. U
17. T
18. R
19. J
20. C

Graph and label each point on a coordinate plane.

21. $L(6, 1)$
22. $M(3, 0)$
23. $N\left(3\frac{1}{2}, 5\right)$
24. $Q\left(0, 5\frac{1}{4}\right)$
25. $R(1.5, 7)$
26. $P(4, 1.75)$
27. $A\left(2\frac{3}{4}, 2\right)$
28. $B\left(5, 4\frac{1}{4}\right)$

MEASUREMENT For Exercises 29 and 30, use the following information.

The table gives the amount of fencing needed to create square pens with side lengths 5, $5\frac{1}{4}$, $5\frac{1}{2}$, and $5\frac{3}{4}$ feet.

Side Length (ft)	5	$5\frac{1}{4}$	$5\frac{1}{2}$	$5\frac{3}{4}$
Amount of Fencing (ft)	20	21	22	23

29. List this information as ordered pairs (side length, amount of fencing).
30. Graph the ordered pairs. Then describe the graph.

READING For Exercises 31 and 32, use the following information.

It took Kevin 4 minutes to read one page in his book. The table shows the total time it took him to read 0, 1, 2, and 3 pages of the book.

Number of Pages	Total Time (min)
0	0
1	4
2	8
3	12

31. List this information as ordered pairs (number of pages, total time).
32. Graph the ordered pairs. Then describe the graph.

33. **MAPS** Your house is located at (4, 1), which is 4 blocks east and 1 block north of the map's center, (0, 0). If you walk two blocks east and one block north from your house to your friend's house, what are the coordinates of your friend's house?

34. **GEOMETRY** Three of the corners of a square drawn on a coordinate plane are located at (3.5, 8), (3.5, 3), and (8.5, 8). What is the ordered pair of the fourth corner?

H.O.T. Problems

35. **OPEN ENDED** Give an example of an ordered pair that represents a point located on the x-axis.

36. **CHALLENGE** Give the coordinates of the point located halfway between (2, 1) and (2, 4).

37. **WRITING IN MATH** Explain why the ordered pair (3, 2) is graphed at a different location than the ordered pair (2, 3).

TEST PRACTICE

38. Which ordered pair represents a point located inside both the square and the circle?

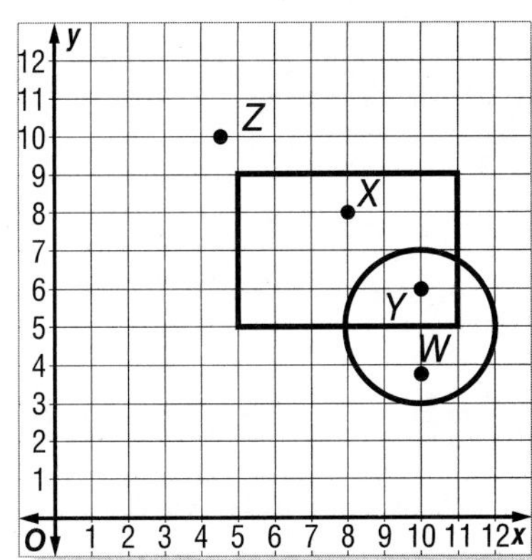

A (9, 4)

B (10, 6)

C (8, 7)

D (11, 8)

39. What point on the grid below corresponds to the coordinate pair $\left(9, 3\frac{1}{2}\right)$?

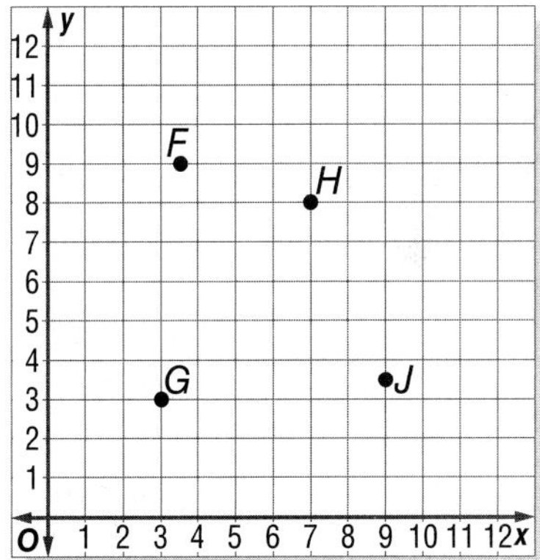

F Point F

G Point G

H Point H

J Point J

Spiral Review

40. **AREA** The formula $A = \frac{1}{2}(b_1 + b_2)h$ can be used to find the area A of a trapezoid given the length of the bases b_1 and b_2 and the height h. Express $\frac{1}{2}$ as a decimal. (Lesson 4-8)

Write each decimal as a fraction or mixed number in simplest form. (Lesson 4-7)

41. 1.34

42. 0.052

43. 13.008

44. **ROLLER COASTERS** If train A and train B, on a side-by-side roller coaster track, both leave the starting point at 9:00 A.M., at what time will they next leave the starting point together? (Lesson 4-5)

Roller Coaster Schedule	
Train	**Departs**
A	every 8 minutes
B	every 6 minutes

45. **BUSINESS** The manager of a shoe store wants to post a display of the number of shoes sold by each sales associate over each of the past 6 months. What type of statistical display would be most appropriate for this situation? (Lesson 2-8)

CHAPTER 4

Study Guide and Review

Math Online glencoe.com

- STUDY TO GO
- Vocabulary Review

GET READY to Study

Be sure the following Big Ideas are noted in your Foldable.

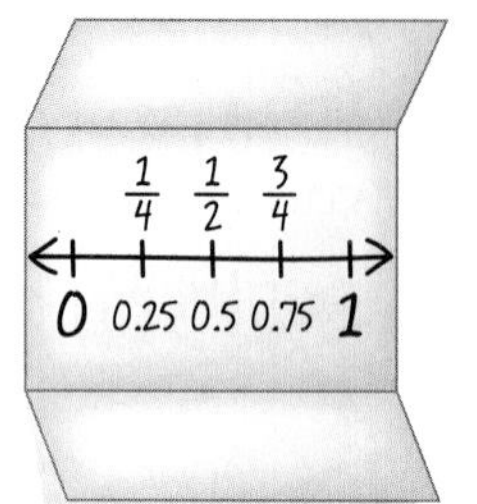

BIG Ideas

Simplest Form (Lesson 4-2)

To write a fraction in simplest form, either

- Divide the numerator and denominator by common factors until the only common factor is 1, or
- Divide the numerator and denominator by the GCF.

Comparing Fractions (Lesson 4-6)

To compare two fractions, follow these steps.

Step 1 Find the least common denominator (LCD) of the fractions. That is, find the least common multiple of the denominators.

Step 2 Write an equivalent fraction for each fraction using the LCD.

Step 3 Compare the numerators.

Writing Decimals as Fractions (Lesson 4-7)

To write a decimal as a fraction, follow these steps.

Step 1 Identify the place value of the last decimal place.

Step 2 Write the decimal as a fraction using the place value as the denominator.

Step 3 If necessary, simplify the fraction.

Key Vocabulary

common multiples (p. 216)
equivalent fractions (p. 204)
greatest common factor (GCF) (p. 197)
improper fraction (p. 209)
least common denominator (LCD) (p. 220)
least common multiple (LCM) (p. 217)
mixed number (p. 209)
multiple (p. 216)
simplest form (p. 205)
Venn diagram (p. 197)

Vocabulary Check

State whether each sentence is *true* or *false*. If *false*, replace the underlined word or number to make a true sentence.

1. Fractions that have the same value are called <u>equivalent fractions</u>.
2. The <u>LCM</u> of 2 and 8 is 2.
3. To write a fraction in simplest form, divide the numerator and denominator by the <u>GCF</u>.
4. The least common multiple of the denominators of two fractions is called the <u>greatest common factor</u>.
5. The LCM of 2 and 4 is <u>less</u> than the GCF of 2 and 4.
6. Multiples that are shared by two or more numbers are <u>mixed numbers</u>.
7. A fraction is in simplest form when the <u>LCD</u> of the numerator and denominator is 1.
8. An improper fraction is <u>less than</u> 1.
9. $\frac{4}{5}$ and $\frac{12}{15}$ are <u>common multiples</u>.
10. The <u>GCF</u> of 5 and 3 is 1.

Lesson-by-Lesson Review

4-1 Greatest Common Factor (pp. 197–201)

Find the GCF of each set of numbers.

11. 15, 18
12. 30, 36
13. 28, 70
14. 26, 52, 65
15. **SCHOOL** For a class field trip, 45 boys and 72 girls will be placed into several groups. Each group will have the same number of boys and girls. What is the greatest number of groups that can be formed?

Example 1 **Find the GCF of 36 and 54.**

To find the GCF, you can use prime factors.

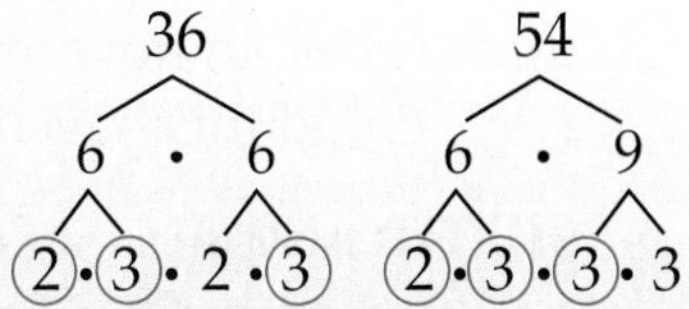

The common prime factors are 2, 3, and 3. The GCF of 36 and 54 is $2 \times 3 \times 3$ or 18.

4-2 Simplifying Fractions (pp. 204–208)

Replace each ■ with a number so the fractions are equivalent.

16. $\frac{2}{3} = \frac{■}{24}$
17. $\frac{5}{8} = \frac{35}{■}$
18. $\frac{■}{6} = \frac{12}{24}$
19. $\frac{7}{■} = \frac{63}{81}$

Write each fraction in simplest form. If the fraction is already in simplest form, write *simplest form*.

20. $\frac{21}{24}$
21. $\frac{15}{80}$
22. **MUFFINS** Out of a dozen muffins, nine were blueberry muffins. Write this fraction in simplest form.

Example 2 **Replace the ■ with a number so that $\frac{4}{9}$ and $\frac{■}{27}$ are equivalent.**

Since $9 \times 3 = 27$, multiply the numerator and denominator by 3.

$$\frac{4}{9} \overset{\times 3}{=} \frac{12}{27}$$

4-3 Mixed Numbers and Improper Fractions (pp. 209–212)

Write each mixed number as an improper fraction.

23. $3\frac{1}{4}$
24. $5\frac{3}{8}$

Write each improper fraction as a mixed number or a whole number.

25. $\frac{23}{4}$
26. $9\frac{5}{5}$
27. **CRAFTS** Beth cut a piece of fabric into four $\frac{1}{3}$-foot-long pieces. How long was the original piece of string?

Example 3 **Write $4\frac{2}{5}$ as an improper fraction.**

$$4\frac{2}{5} = \frac{(4 \times 5) + 2}{5} = \frac{22}{5}$$

Example 4 **Write $\frac{49}{6}$ as a mixed number.**

Divide 49 by 6.

$$\begin{array}{r} 8\frac{1}{6} \\ 6\overline{)49} \\ -48 \\ \hline 1 \end{array}$$

So, $\frac{49}{6} = 8\frac{1}{6}$.

Study Guide and Review

4-4 PSI: Make an Organized List (pp. 214–215)

Solve. Use the *make an organized list* strategy.

28. **COINS** When Rosa tossed a coin four times, she noticed that tails came up three times. In how many different ways could this have happened?

29. **STUFFED ANIMALS** Shaunae has 4 stuffed animals: a tiger, teddy bear, bunny, and frog. In how many ways can she arrange these animals on her shelf?

30. **CHILDREN** A couple has 4 children, 2 of whom are boys. How many different birth orders are possible if the boys were born in consecutive years?

Example 5 **A true-false test contains three questions. How many different ways are there to complete the test?**

Make a list of all possible arrangements. Use T for a *true* answer to a question and F for a *false* answer.

TTT — answering *false* for none of the three questions
TTF, TFT, FTT — answering *false* for one question
TFF, FTF, FFT — answering *false* for two questions
FFF — answering *false* for all three questions

There are 8 ways to complete the test.

4-5 Least Common Multiple (pp. 216–219)

Find the LCM of each set of numbers.

31. 10, 25
32. 28, 35
33. Find the LCM of 8, 12, and 16.
34. What is the LCM of 12, 15, and 20?
35. **CRAFTS** Diana is making craft puppies out of clothespins, which are sold 40 per bag. One bag of plastic eyes is enough for exactly 25 puppies. How many bags of each should she buy so there will be no leftover clothespins or plastic eyes?
36. **LAWNS** Mr. Kwan mows his lawn every 2 days. Mr. Kwan's neighbor mows his lawn every 5 days. If both men mowed their lawn today, how many days will it be until they both mow their lawn on the same day?

Example 6 **Find the LCM of 8 and 18.**

multiples of 8: 8, 16, 24, 32, 40, 48, 56, 64, 72, …
multiples of 18: 18, 36, 54, 72, …

So, the LCM of 8 and 18 is 72.

Example 7 **Find the LCM of 9 and 24.**

Write the prime factorization of each number.

$9 = 3 \times 3$
$24 = 2 \times 2 \times 2 \times 3$

3 is a common prime factor

So, the LCM of 9 and 24 is $2 \times 2 \times 2 \times 3 \times 3$ or 72.

Mixed Problem Solving
For mixed problem-solving practice, see page 709.

4-6 Comparing and Ordering Fractions (pp. 220–224)

Replace each ● with <, >, or = to make a true sentence.

37. $\frac{2}{5}$ ● $\frac{4}{9}$

38. $2\frac{12}{15}$ ● $2\frac{4}{5}$

39. $7\frac{3}{8}$ ● $7\frac{4}{10}$

40. $\frac{7}{12}$ ● $\frac{5}{9}$

Order the fractions from least to greatest.

41. $\frac{2}{3}, \frac{3}{4}, \frac{1}{2}, \frac{5}{9}$

42. $3\frac{7}{12}, \frac{5}{8}, 3\frac{5}{6}, \frac{3}{4}$

43. **MONEY** Which is more, $\frac{3}{4}$ of a dollar or $\frac{3}{5}$ of a dollar?

Example 8 Replace ● with <, >, or = to make $\frac{2}{5}$ ● $\frac{3}{8}$ true.

First, find the LCD. The LCM of 5 and 8 is 40. So, the LCD is 40.

Next, rewrite both fractions with a denominator of 40.

$$\frac{2}{5} = \frac{16}{40} \text{ (}\times 8\text{) and } \frac{3}{8} = \frac{15}{40} \text{ (}\times 5\text{)}$$

Since 16 > 15, $\frac{16}{40} > \frac{15}{40}$. So, $\frac{2}{5} > \frac{3}{8}$.

4-7 Writing Decimals as Fractions (pp. 225–228)

Write each decimal as a fraction or mixed number in simplest form.

44. 0.9
45. 0.35
46. 0.72
47. 0.125
48. 3.006
49. 9.315
50. 2.64
51. 0.048

52. **MEATBALLS** Peter bought 5.65 pounds of hamburger to make meatballs for a family reunion. Write 5.65 as a mixed number in simplest form.

Example 9 Write 0.85 as a fraction in simplest form.

$0.85 = \frac{85}{100}$ Say *eighty-five hundredths.*

$= \frac{\cancel{85}^{17}}{\cancel{100}_{20}}$ Simplify. Divide the numerator and denominator by the GCF, 5.

$= \frac{17}{20}$

Example 10 Write 7.4 as a mixed number in simplest form.

$7.4 = 7\frac{4}{10}$ Say *seven and four tenths.*

$= 7\frac{\cancel{4}^{2}}{\cancel{10}_{5}}$ Simplify.

$= 7\frac{2}{5}$

Study Guide and Review

4-8 Writing Fractions as Decimals (pp. 229–232)

Write each fraction or mixed number as a decimal.

53. $\frac{7}{8}$

54. $\frac{9}{15}$

55. $\frac{21}{25}$

56. $4\frac{2}{16}$

57. $12\frac{3}{4}$

58. $8\frac{9}{16}$

59. **HOMEWORK** Jonah spent $\frac{3}{4}$ of an hour on his math homework. Write this time as a decimal.

Example 11 Write $\frac{5}{8}$ as a decimal.

$$\begin{array}{r} 0.625 \\ 8\overline{)5.000} \\ -\underline{48} \\ 20 \\ -\underline{16} \\ 40 \\ -\underline{40} \\ 0 \end{array}$$

Divide 5 by 8.

So, $\frac{5}{8} = 0.625$.

4-9 Algebra: Ordered Pairs and Functions (pp. 233–237)

Use the coordinate plane at the right to name the ordered pair for each point.

60. B

61. C

62. D

63. E

Graph and label each point on a coordinate plane.

64. $X(5, 0)$

65. $Y(4.75, 6)$

66. $Z\left(2, 8\frac{1}{2}\right)$

67. **MEASUREMENT** The table gives the ages and heights, in feet, of five students in Mr. Cole's science class.

Age	11	12	11.5	12.5
Height	5	5.5	5.25	5.75

List this information as ordered pairs. Graph the ordered pairs. Then describe the graph.

Example 12 Use the coordinate plane below to name the ordered pair for point A.

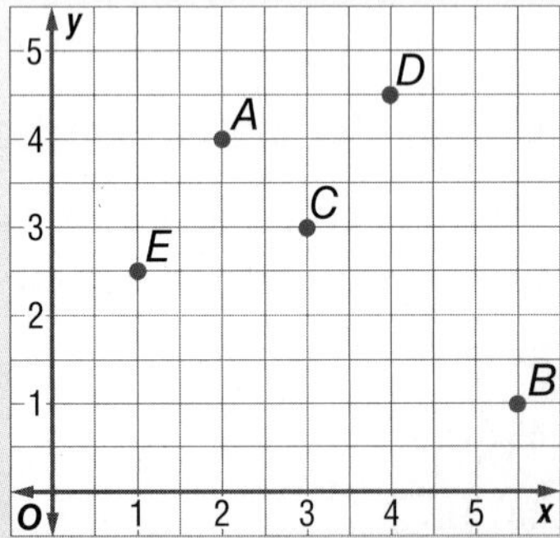

Point A is named by the ordered pair (2, 4).

Example 13 Graph the point $M\left(3, 2\frac{1}{4}\right)$.

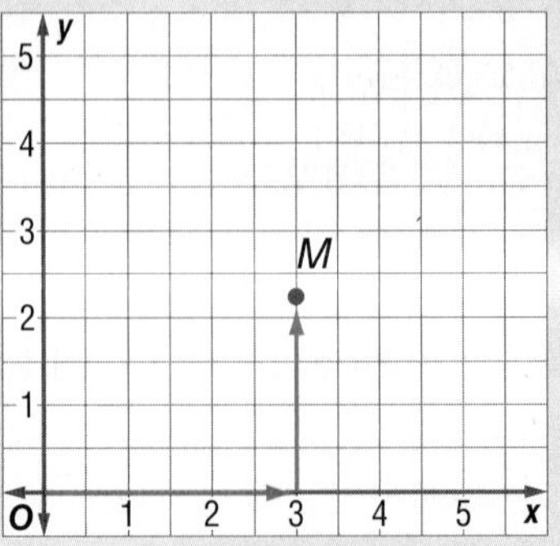

CHAPTER 4

Practice Test

Math Online glencoe.com
• Chapter Test

1. **MULTIPLE CHOICE** Find the GCF of 24, 48, and 84.

   A 24 C 8
   B 12 D 6

Replace each ■ with a number so the fractions are equivalent.

2. $\frac{12}{18} = \frac{■}{6}$ 3. $\frac{7}{9} = \frac{35}{■}$

4. **DVDs** Danny has 8 action DVDs, 4 comedy DVDs, and 2 drama DVDs. Write a fraction in simplest form that compares the number of comedy DVDs to the total number of DVDs.

Write each mixed number as an improper fraction.

5. $2\frac{5}{7}$ 6. $4\frac{2}{3}$ 7. $1\frac{4}{7}$

8. **PHYSICS** The speed of sound is about $\frac{3{,}806}{5}$ miles per hour. Write this speed as a mixed number.

9. **MOVIES** In how many different ways can four friends sit next to each other in one row of a movie theater?

10. **MULTIPLE CHOICE** At the gym, Hilary swims every 6 days, runs every 4 days, and cycles every 16 days. If she did all three activities today, in how many days will she do all three activities again on the same day?

    F 24 days H 48 days
    G 26 days J 64 days

Find the LCM of each set of numbers.

11. 6, 15 12. 4, 9, 18

Replace each ● with <, >, or = to make a true sentence.

13. $\frac{4}{7} ● \frac{3}{5}$ 14. $6\frac{1}{4} ● 6\frac{4}{18}$ 15. $\frac{2}{9} ● \frac{6}{27}$

16. Order the fractions $1\frac{5}{6}, 1\frac{3}{4}, 1\frac{2}{3}$, and $1\frac{7}{9}$ from least to greatest.

17. **MONEY** $\frac{19}{20}$ of all bills that are printed by the U.S. Treasury Department are used to replace worn-out money. Write this fraction as a decimal.

Write each decimal as a fraction or mixed number in simplest form.

18. 0.84 19. 7.015 20. 1.3

21. **SAVINGS** The table shows the amount of money Andrew saved in November.

Week	Total Saved ($)
1	6
2	12
3	18
4	24

List this information as ordered pairs. Then graph the ordered pairs on a coordinate plane.

Use the coordinate plane to name the ordered pair for each point.

22. *A*
23. *B*
24. *C*
25. *D*

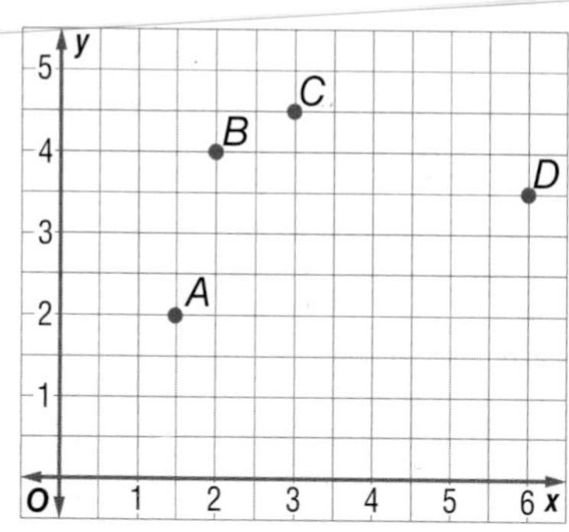

CHAPTER 4

Test Practice

Cumulative, Chapters 1–4

Math Online glencoe.com

• Test Practice

PART 1 Multiple Choice

Read each question. Then fill in the correct answer on the answer sheet provided by your teacher or on a sheet of paper.

1. Find the greatest common factor of 16, 24, and 40.

   A 2
   C 8
   B 4
   D 40

TEST-TAKING TIP

Question 1 Use the answer choices to help find a solution. To find the GCF, divide 16, 24 and 40 by each possible choice. The greatest value that divides evenly into all three numbers is the solution.

2. The formula $C = \frac{5}{9}(F - 32)$ can be used to convert a temperature from degrees Fahrenheit to degrees Celsius. Which of the following is closest in value to $\frac{5}{9}$?

   F 5.9
   H 1.8
   G 4
   J 0.56

3. The ages of people eating at a restaurant were 12, 7, 31, 15, 9, 12, 18, 22, and 14. What is the mean of these ages?

   A 7
   C 31
   B 15.6
   D 12.9

4. Brandi recorded the monthly rainfall for Portland, Oregon. Which list shows the monthly rainfall in order from greatest to least?

   F 4.03 in., 4.14 in., 4.30 in., 4.31 in., 4.51 in.
   G 4.51 in., 4.31 in., 4.30 in., 4.03 in., 4.14 in.
   H 4.51 in., 4.31 in., 4.30 in., 4.14 in., 4.03 in.
   J 4.51 in., 4.14 in., 4.30 in., 4.31 in., 4.03 in.

5. Of the 200 people Melanie surveyed about their favorite flavor of ice cream, 64 said chocolate, 36 said vanilla, 48 said chocolate chip, and 52 said peanut butter chip. Which circle graph best displays the data?

   A Ice Cream Flavor

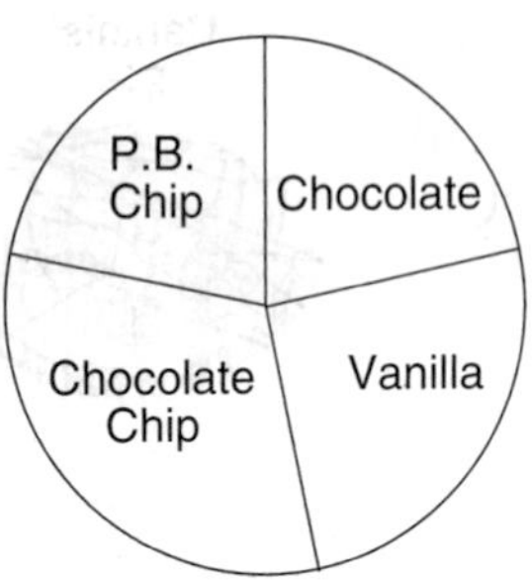

   B Ice Cream Flavor

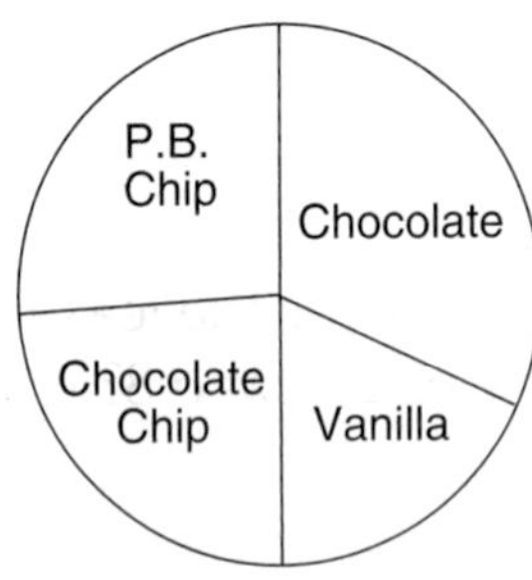

   C Ice Cream Flavor

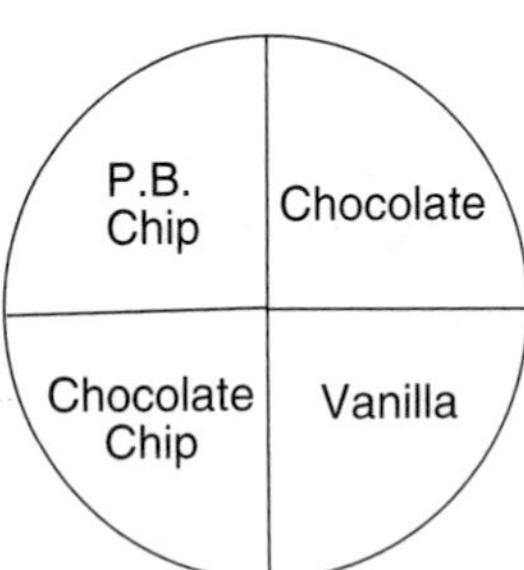

   D Ice Cream Flavor

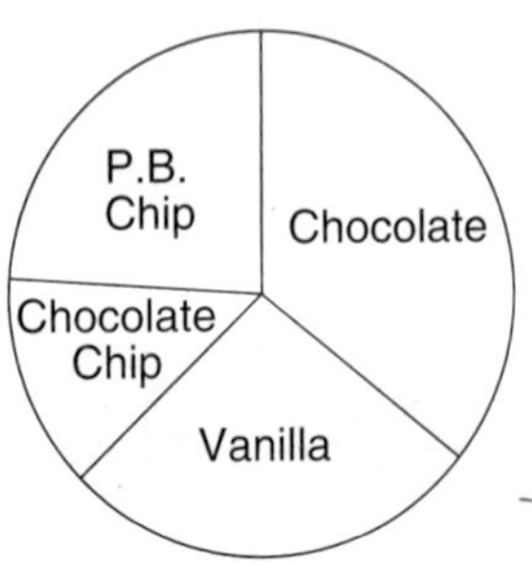

Preparing for Standardized Tests

For test-taking strategies and practice, see pages 718–735.

6. The Sonoma family and the Canini family each brought a pie to the picnic. Only a portion of each pie was eaten. The pictures below show how much of the pies were left. What portion of the pies was eaten altogether?

Sonomas' Pie

Caninis' Pie

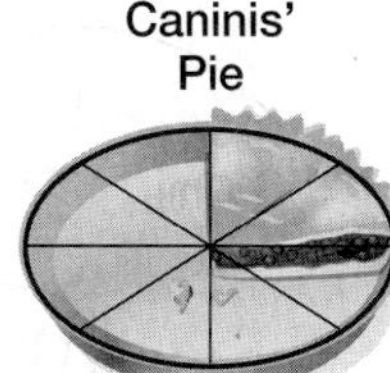

F $\frac{5}{8}$

G $1\frac{1}{4}$

H $1\frac{3}{8}$

J $1\frac{3}{4}$

7. Which of the following is the least common multiple of 4, 6, and 8?

A 12

B 16

C 24

D 48

8. Jill and 3 friends bought 4 movie tickets for $24, 4 large drinks for $4.25 each, and a jumbo popcorn for $5.30. If they split the cost evenly, which equation can be used to find *c*, the amount each person should pay, not including tax?

F $c = 24.00 + 4.25 + 5.30 \div 4$

G $c = 24.00 + 4 \times 4.25 + (5.30 \div 4)$

H $c = (24.00 + 4 \times 4.25 + 5.30) \div 4$

J $c = (24.00 + 4.25 + 5.30) \div 4$

PART 2 Short Response/Grid In

Record your answers on the answer sheet provided by your teacher or on a sheet of paper.

9. Agnes spent 12 minutes making her bed, 17 minutes dusting, 15 minutes vacuuming, and 24 minutes putting away laundry. How much total time in minutes did Agnes spend on cleaning her room?

10. Several families in a neighborhood were asked how many gallons of milk they buy each week. The results are shown below. What is the mode of the data?

1, 3, 2, 2, 1, 1, 1, 3, 2, 1, 1, 1, 2, 2, 1, 3, 1, 1

PART 3 Extended Response

Record your answers on the answer sheet provided by your teacher or on a sheet of paper. Show your work.

11. Copy the models below. Both models have the same area.

Model A

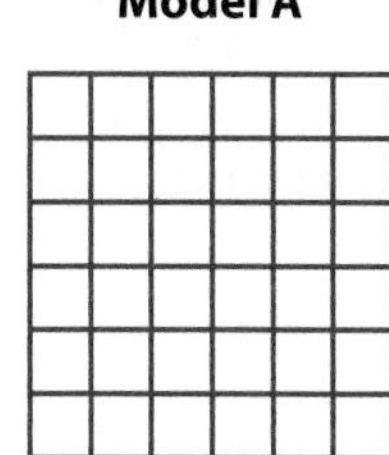

Model B

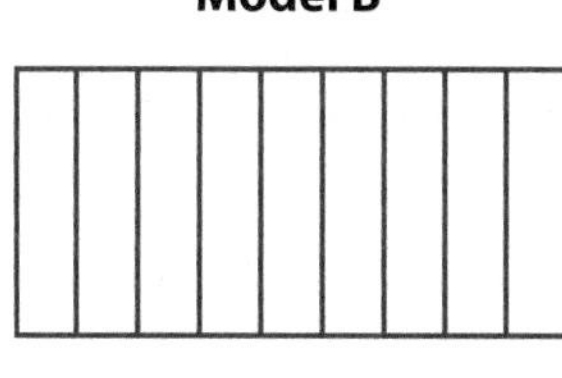

a. Shade 0.25 of Model A.

b. Shade $\frac{1}{3}$ of Model B.

c. Which model has the greater fraction of shaded area? Explain your answer.

NEED EXTRA HELP?											
If You Missed Question...	1	2	3	4	5	6	7	8	9	10	11
Go to Lesson...	4-1	4-7	2-6	3-2	1-1	1-1	4-5	1-4	1-1	2-7	4-2

Operations with Fractions

BIG Ideas

- Understand, explain, and apply operations with fractions, including multiplication and division.
- Multiply and divide fractions to solve problems.

Key Vocabulary

like fractions (p. 256)

unlike fractions (p. 263)

Real-World Link

Animals The state animal of New York is the beaver, whose average length is $3\frac{1}{2}$ feet long. The beaver's tail averages $1\frac{3}{5}$ feet long.

Operations with Fractions Make this Foldable to help you organize your notes. Begin with two sheets of plain 11″ × 17″ paper, four index cards, and glue.

1. **Fold** one sheet in half widthwise.

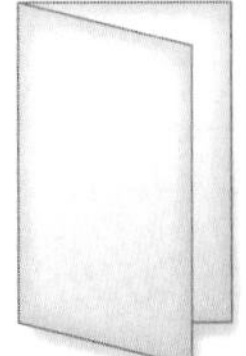

2. **Open** and fold the bottom to form a pocket. Glue edges.

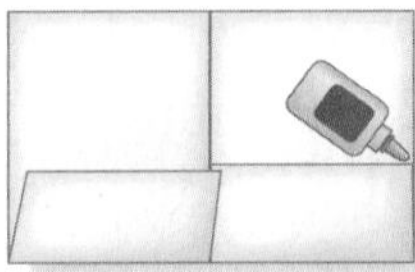

3. **Repeat** steps 1 and 2. Glue the back of one piece to the front of the other to form a booklet.

4. **Label** each left-hand pocket *What I Know* and each right-hand pocket *What I Need to Know.* Place an index card in each pocket.

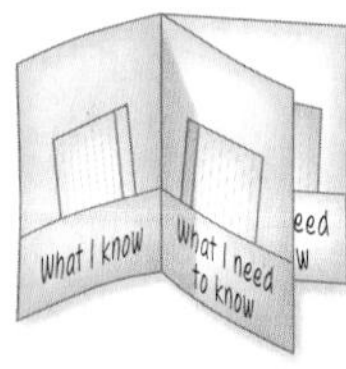

GET READY for Chapter 5

Diagnose Readiness You have two options for checking Prerequisite Skills.

Option 2

Math Online Take the Online Readiness Quiz at glencoe.com.

Option 1

Take the Quick Quiz below. Refer to the Quick Review for help.

QUICK Quiz

Estimate using rounding. (Prior Grade)

1. $1.2 + 6.6$
2. $9.6 - 2.3$
3. $8.25 - 4.8$
4. $5.85 + 7.1$
5. **MONEY** Braden spent $17.88 on a hat and $4.22 on lunch. About how much did he spend altogether?

Write each fraction in simplest form. (Lesson 4-2)

6. $\frac{3}{18}$
7. $\frac{21}{28}$
8. $\frac{16}{40}$
9. $\frac{6}{38}$
10. **HOMEWORK** Sandra finished 21 out of 39 problems. Write the fraction, in simplest form, of homework that she completed.

Write each improper fraction as a mixed number. (Lesson 4-2)

11. $\frac{11}{10}$
12. $\frac{14}{5}$
13. $\frac{7}{5}$
14. $\frac{15}{9}$

QUICK Review

Example 1

Estimate 8.74 − 2.15 using rounding.

Round 8.74 to 9 and 2.15 to 2.

$9 - 2 = 7$

So, 8.74 − 2.15 is *about* 7.

Example 2

Write $\frac{24}{36}$ in simplest form.

$$\frac{24 \div 12}{36 \div 12} = \frac{2}{3}$$

Divide the numerator and denominator by the GCF, 12.

Since the GCF of 2 and 3 is 1, the fraction $\frac{2}{3}$ is in simplest form.

Example 3

Write $\frac{19}{7}$ as a mixed number.

Divide 19 by 7.

$$\begin{array}{r} 2\frac{5}{7} \\ 7\overline{)19} \\ -14 \\ \hline 5 \end{array}$$

Use the remainder as the numerator of the fraction.

So, $\frac{19}{7} = 2\frac{5}{7}$.

Explore 5-1

Math Lab
Rounding Fractions

MAIN IDEA

Use models to round fractions to the nearest half.

In Lesson 3-3, you learned to round decimals. You can use a similar method to round fractions.

ACTIVITY

Draw and shade a model to represent each fraction. Then use the model to round each fraction to the nearest half.

1. $\frac{4}{20}$

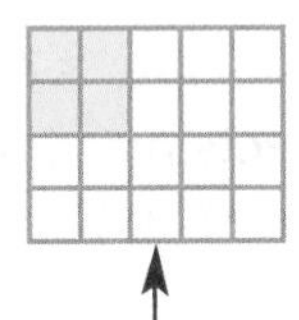

Shade 4 out of 20.

Very few sections are shaded. So, $\frac{4}{20}$ rounds to 0.

2. $\frac{4}{10}$

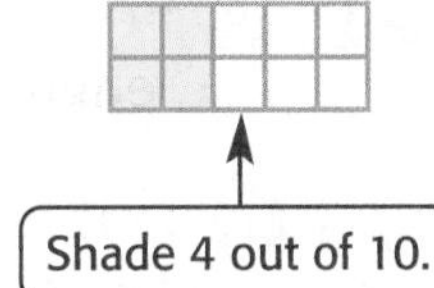

Shade 4 out of 10.

About one half of the sections are shaded. So, $\frac{4}{10}$ rounds to $\frac{1}{2}$.

3. $\frac{4}{5}$

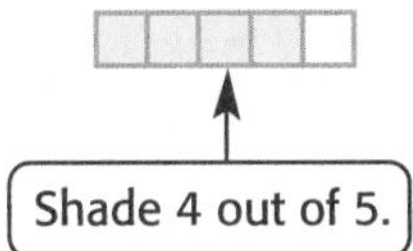

Shade 4 out of 5.

Almost all of the sections are shaded. So, $\frac{4}{5}$ rounds to 1.

CHECK Your Progress

Draw and shade a model to represent each fraction. Then use the model to round each fraction to the nearest half.

a. $\frac{13}{20}$ b. $\frac{7}{8}$ c. $\frac{9}{10}$ d. $\frac{1}{5}$ e. $\frac{11}{15}$

f. $\frac{2}{25}$ g. $\frac{6}{10}$ h. $\frac{17}{20}$ i. $\frac{1}{8}$ j. $\frac{7}{16}$

Analyze the Results

1. Sort the fractions in Exercises a–j into three groups: those that round to 0, those that round to $\frac{1}{2}$, and those that round to 1.
2. **MAKE A CONJECTURE** Compare the numerators and denominators of the fractions in each group. Explain how to round any fraction to the nearest half without using a model.
3. Test your conjecture by repeating the activity and Exercise 1 using the fractions $\frac{3}{5}$, $\frac{3}{17}$, $\frac{16}{20}$, $\frac{2}{13}$, $\frac{5}{24}$, $\frac{7}{15}$, $\frac{7}{9}$, and $\frac{9}{11}$.

Rounding Fractions and Mixed Numbers

MAIN IDEA

Round fractions and mixed numbers.

Math Online

glencoe.com

- Extra Examples
- Personal Tutor
- Self-Check Quiz
- Reading in the Content Area

MINI Lab

Using a ruler, measure the thickness of your textbook.

1. What is the thickness of your book?
2. Looking at the ruler, is the thickness of the book at the right closer to 1 inch, $1\frac{1}{2}$ inches, or 2 inches?

STEP 1 Pick several objects from your classroom. Measure the lengths of the objects to the nearest eighth of an inch.

STEP 2 Sort the different measurements into different categories: those that round up to the next greater whole number, those that round to a half inch, and those that round down to the smaller whole number.

3. Compare the numerators and denominators of the fractions in each group. How do they compare?
4. Write a rule about how to round to the nearest half inch.

It is often helpful to be able to round fractions and mixed numbers to the nearest half in real-world situations. To round fractions and mixed numbers to the nearest half, you can use the following guidelines.

Rounding to the Nearest Half

Key Concept

Round Up	Round to $\frac{1}{2}$	Round Down
If the numerator is almost as large as the denominator, round the number up to the next whole number.	If the numerator is about half of the denominator, round the fraction to $\frac{1}{2}$.	If the numerator is much smaller than the denominator, round the number down to the previous whole number.
Example	**Example**	**Example**
$\frac{7}{8}$ rounds to 1. (7 is almost as large as 8.)	$2\frac{3}{8}$ rounds to $2\frac{1}{2}$. (3 is about half of 8.)	$\frac{1}{8}$ rounds to 0. (1 is much smaller than 8.)

Study Tip

Common Fractions
$\frac{1}{3}$ and $\frac{2}{3}$ each round to $\frac{1}{2}$.
$\frac{1}{4}$ and $\frac{3}{4}$ may be rounded up or down.

EXAMPLE Round to the Nearest Half

1 **Round $3\frac{1}{6}$ to the nearest half.**

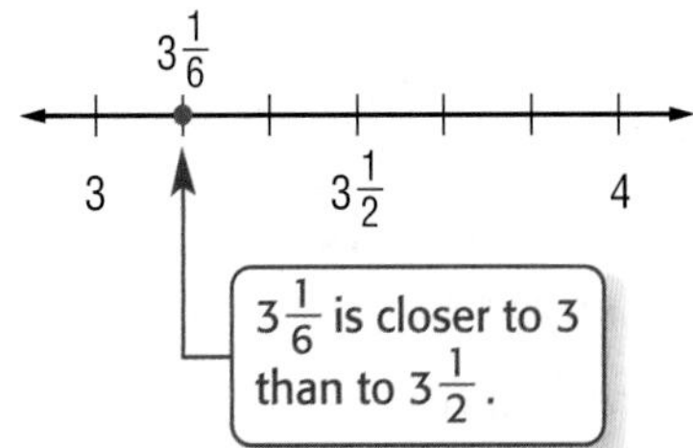

$3\frac{1}{6}$
3
$3\frac{1}{2}$
4
$3\frac{1}{6}$ is closer to 3 than to $3\frac{1}{2}$.

The numerator of $\frac{1}{6}$ is much smaller than the denominator. So, $3\frac{1}{6}$ rounds to 3.

CHECK Your Progress

Round each number to the nearest half.

a. $8\frac{1}{12}$ b. $2\frac{9}{10}$ c. $\frac{2}{9}$

d. $\frac{5}{12}$ e. $1\frac{2}{5}$ f. $4\frac{3}{7}$

EXAMPLE Measure to the Nearest Half

2 **Find the length of the leaf to the nearest half inch.**

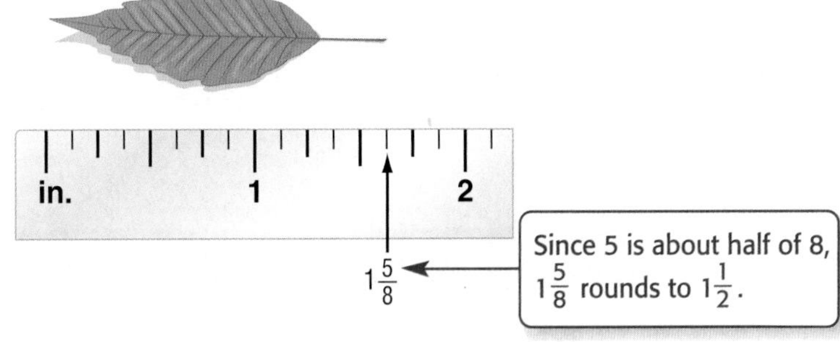

in.
1
2
$1\frac{5}{8}$
Since 5 is about half of 8, $1\frac{5}{8}$ rounds to $1\frac{1}{2}$.

To the nearest half inch, the leaf is $1\frac{1}{2}$ inches.

CHECK Your Progress

g. **Find the width of the bracelet to the nearest half inch.**

Real-World Career...
How Does a Veterinarian Use Math?
A veterinarian uses math to calculate the proper dosages of medication for different-sized animals.

Math Online
For more information, go to glencoe.com.

Sometimes you should round a number down when it is better for a measure to be too small than too large. Other times you should round up despite what the rule says.

Real-World EXAMPLE

3 ANIMALS A pet store sells pet collars in different lengths. Jonathan measured the distance around his puppy's neck to be $11\frac{1}{4}$ inches. Should he buy the 11-inch collar or the 12-inch collar?

Even though $11\frac{1}{4}$ rounds down to 11, the puppy's neck is too large for the 11-inch collar. Jonathan should buy the 12-inch collar.

CHECK Your Progress

h. **FURNITURE** The Turners are buying a sofa for their basement. The width of their basement door is $29\frac{3}{4}$ inches. Should they round $29\frac{3}{4}$ inches up or down to guarantee that the sofa fits through the door? Explain your reasoning.

CHECK Your Understanding

Example 1 (p. 250) **Round each number to the nearest half.**

1. $\frac{7}{8}$
2. $3\frac{1}{10}$
3. $\frac{3}{8}$
4. $6\frac{2}{3}$
5. $\frac{1}{5}$

Example 2 (p. 250) **Find the length of each item to the nearest half inch.**

6. 

7. 

Example 3 (p. 251)

8. **DRAWINGS** To carry her drawings home from school this year, Sara wants to make her drawings small enough to fit into an $8\frac{1}{2}$-inch-wide binder pocket. When deciding on the width of the drawings she will make, should she round $8\frac{1}{2}$ inches up or down? Explain your reasoning.

9. **GARDENING** Based on the area of his flowerbed, a gardener calculates that he needs to dilute $4\frac{3}{8}$ gallons of fertilizer with water. Should he round $4\frac{3}{8}$ gallons up or down when deciding on the amount of fertilizer to purchase? Explain your reasoning.

Practice and Problem Solving

HOMEWORK HELP	
For Exercises	See Examples
10–19	1
20–23	2
24, 25	3

Round each number to the nearest half.

10. $\frac{5}{6}$
11. $2\frac{4}{5}$
12. $4\frac{2}{9}$
13. $9\frac{1}{6}$
14. $3\frac{2}{9}$
15. $3\frac{1}{12}$
16. $\frac{1}{3}$
17. $5\frac{3}{10}$
18. $\frac{7}{12}$
19. $3\frac{2}{3}$

Find the length of each item to the nearest half inch.

20. 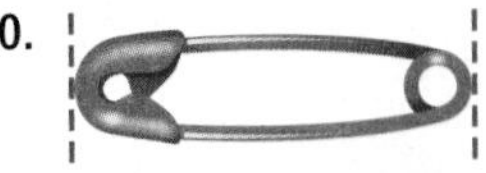

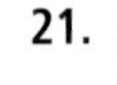
21. 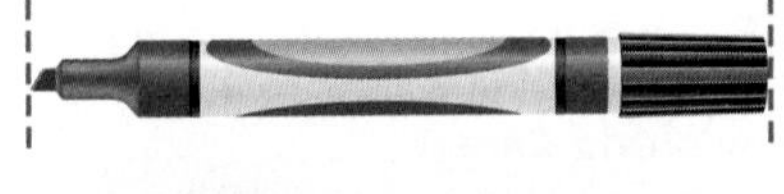

22. 

23. 

24. **DECORATING** The Santiagos are buying blinds to fit in a window opening that is $24\frac{3}{4}$ inches wide. Should they round $24\frac{3}{4}$ inches up or down when deciding on the size of blinds to purchase? Explain your reasoning.

25. **PACKAGES** Martin is mailing a gift that is $14\frac{3}{8}$ inches tall. He can choose from several shipping boxes. Should he round $14\frac{3}{8}$ inches up or down when selecting a shipping box? Explain your reasoning.

Round each number to the nearest half.

26. $\frac{13}{16}$
27. $6\frac{5}{16}$
28. $9\frac{7}{24}$
29. $4\frac{19}{32}$

30. **SHELVES** Your bedroom has an $8\frac{1}{4}$-foot ceiling. To the nearest half foot, what is the tallest bookcase that can fit in your bedroom?

31. **CRAFTS** Marina is making birthday cards. She is using envelopes that are $6\frac{3}{4}$ inches by $4\frac{5}{8}$ inches. To the nearest half inch, how large can she make her cards?

Use rounding to order each set of numbers from least to greatest.

32. $\frac{7}{8}, \frac{2}{11}, \frac{4}{7}$
33. $3\frac{5}{9}, 3\frac{3}{14}, 3\frac{6}{7}$
34. $7\frac{6}{11}, 7\frac{9}{10}, 7\frac{1}{7}$

35. **ANALYZE GRAPHS** Several students were asked to name their favorite free-time activity. Are more than half the students represented by any one category? Explain your reasoning.

EXTRA PRACTICE
See pages 684, 710.

H.O.T. Problems

CHALLENGE Round each number to the nearest fourth. Explain your reasoning.

36. $\frac{3}{16}$

37. $\frac{79}{100}$

38. $\frac{21}{40}$

39. **Which One Doesn't Belong?** Identify the number that does not belong with the other three. Explain your reasoning.

$3\frac{7}{8}$ | $4\frac{4}{5}$ | $4\frac{2}{7}$ | $3\frac{8}{9}$

40. **OPEN ENDED** Select three mixed numbers with different denominators that each round up to $7\frac{1}{2}$.

41. **WRITING IN MATH** Explain how to decide when to round a fraction to 0, $\frac{1}{2}$, or 1 when rounding to the nearest half.

TEST PRACTICE

42. What is the length of the worm to the nearest half inch?

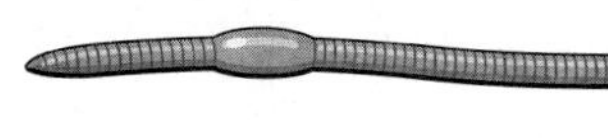

A $1\frac{1}{5}$

B 2

C $2\frac{1}{2}$

D 3

43. The pages in Brooke's scrapbook are $9\frac{3}{4}$ inches by $10\frac{3}{8}$ inches. To the nearest half inch, what is the largest photograph she can place on a page?

F 9 inches by 10 inches

G $9\frac{1}{2}$ inches by 10 inches

H 9 inches by $10\frac{1}{2}$ inches

J $9\frac{1}{2}$ inches by $10\frac{1}{2}$ inches

Spiral Review

44. Graph and label each point in the table at the right on a coordinate plane. (Lesson 4-9)

Point	*x*	*y*
A	3	1
B	2.5	4
C	4.25	0
D	1	2

Write each fraction or mixed number as a decimal. (Lesson 4-8)

45. $\frac{1}{8}$

46. $4\frac{4}{5}$

47. $\frac{2}{5}$

48. $2\frac{3}{16}$

49. **ANTARCTICA** The average depth of ice covering Antarctica is about 1.12 miles. Write this depth as a mixed number in simplest form. (Lesson 4-7)

GET READY for the Next Lesson

50. **PREREQUISITE SKILL** Six friends will split the cost of 2 large pizzas. If each pizza costs $14.99, is $4, $5, or $6 a reasonable answer for the amount that each friend will pay? (Lesson 3-10)

Problem-Solving Investigation

MAIN IDEA: Solve problems by acting them out.

P.S.I. TEAM +

e-Mail: ACT IT OUT

BETHANY: Tonya, Liseli, Meghan, and I want to ride the new ride Teradactyl at the amusement park. Each car on the ride has two rows with two seats in each row.

YOUR MISSION: Act it out to find how many different ways the four friends can sit in a car on the ride so that Tonya and Meghan sit next to each other.

Understand	You know that each car on the ride has two rows of seats. There are two seats in each row. Tonya and Meghan want to sit next to each other.
Plan	You can arrange student desks to model the amusement park ride. Place four desks in two rows with two desks in each row. Have four students act out possible arrangements and record each one. Use B for Bethany, T for Tonya, L for Liseli, and M for Meghan.
Solve	Meghan and Tonya can either sit in the front row or the back row. There are 8 possible ways for the friends to sit in the car on the ride. M T / L B; M T / B L; T M / L B; T M / B L; L B / T M; L B / M T; B L / T M; B L / M T
Check	In each row, there are four ways for the friends to sit. So, having 8 possible ways makes sense.

Analyze The Strategy

1. Explain how this strategy could help determine the reasonableness of your answer after the calculations were completed.
2. **WRITING IN MATH** Write a problem that could be solved by using the *act it out* strategy. Then explain how you would act it out.

Mixed Problem Solving

EXTRA PRACTICE
See pages 684, 710.

Use the *act it out* strategy to solve Exercises 3–5.

3. **RESTAURANTS** A restaurant serves a chicken entrée and a fish entrée. Each entrée comes with a choice of coffee, tea, lemonade, or water. How many entrée-beverage choices are possible? List them.

4. **RUNNING** Mike, Juliana, Tyrone, and Elisa are entered in a 4-person relay race. In how many orders can they run the relay, if Mike must run last? List them.

5. **TEAMS** Twenty-four students will be divided into four equal-size teams. Each student will count off, beginning with the number 1 as the first team. If Nate is the eleventh student to count off, to which team number will he be assigned?

Use any strategy to solve Exercises 6–14. Some strategies are shown below.

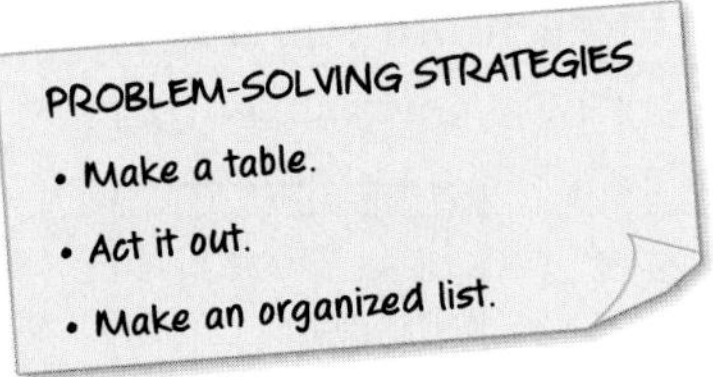
PROBLEM-SOLVING STRATEGIES
• Make a table.
• Act it out.
• Make an organized list.

6. **SEATING** Six students are sitting at a lunch table. Two more students arrive, and at the same time three students leave. How many students are at the table now?

7. **MONEY** Tetuso bought a clock radio for $9 less than the regular price. If he paid $32, what was the regular price?

8. **TESTS** A list of test scores is shown.

English Test Scores						
68	77	99	86	73	75	100
86	70	97	93	80	91	72
85	98	79	77	65	89	71

How many more students scored 71 to 80 than 91 to 100?

9. **INTERNET** Cesar needs to visit three Web sites for a homework assignment. In how many orders can he visit the Web sites?

10. **SCHOOL** The birth months of the students in Miss Miller's geography class are shown below. How many more students were born in June than in August?

Birth Months		
June	July	April
March	July	June
October	May	August
June	April	October
May	October	April
September	December	January

11. **ANIMALS** Corey has one cat and one hamster. His cat weighs 9.75 pounds and his hamster weighs 1.8 pounds. About how many times more does Corey's cat weigh than his hamster?

12. **PATTERNS** What number is missing in the pattern below?

…, 234, 345, ■, 567, …

13. **DVDs** The table shows the cost of DVD rentals at a video store. Darren purchases a 3-night rental and gets a new release rental for half price. How much money will he have left over if he had $20 to spend originally?

DVD Rentals

Type of Rental	Cost
New Release	$4.50
1-Night	$3.75
2-Night	$4.25
3-Night	$5.25

14. **SCHOOL** Every 8 minutes, Daniela can study twelve Spanish vocabulary words. How many Spanish vocabulary words can she study in 1 hour and 20 minutes?

5-3

Adding and Subtracting Fractions with Like Denominators

MAIN IDEA

Add and subtract fractions with like denominators.

New Vocabulary

like fractions

Math Online

glencoe.com

- Concepts in Motion
- Extra Examples
- Personal Tutor
- Self-Check Quiz

MINI Lab

You can use grid paper to model adding fractions such as $\frac{4}{18}$ and $\frac{3}{18}$.

STEP 1 On grid paper, draw a rectangle like the one shown. Since the grid has 18 squares, each square represents $\frac{1}{18}$.

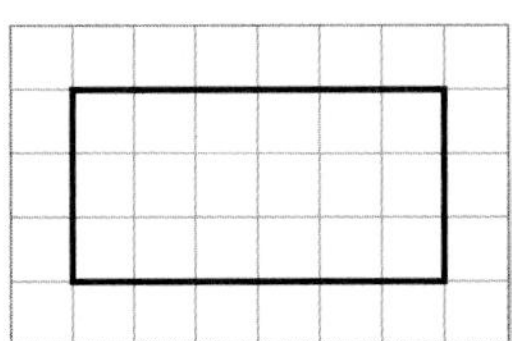

STEP 2 With a marker, color four squares to represent $\frac{4}{18}$. With a different marker, color three more squares to represent $\frac{3}{18}$.

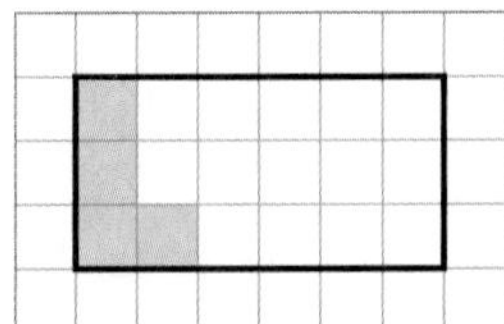

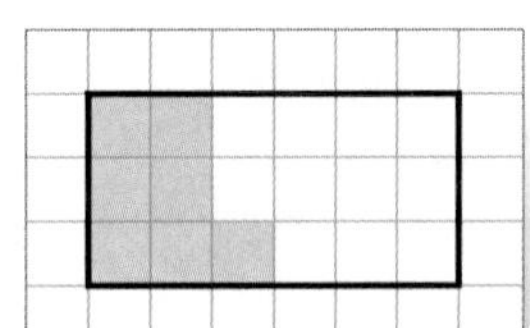

STEP 3 Seven of the 18 squares are colored. So, the sum of $\frac{4}{18}$ and $\frac{3}{18}$ is $\frac{7}{18}$.

Find each sum using grid paper.

1. $\frac{4}{12} + \frac{3}{12}$
2. $\frac{1}{6} + \frac{1}{6}$
3. $\frac{3}{10} + \frac{5}{10}$
4. What patterns do you notice with the numerators?
5. What patterns do you notice with the denominators?
6. Explain how you could find the sum $\frac{3}{8} + \frac{1}{8}$ without using grid paper.

Fractions with the same denominator are called **like fractions**. When you add and subtract like fractions, the denominator names the units being added or subtracted.

4 eighteenths	plus	3 eighteenths	equals	7 eighteenths
$\frac{4}{18}$	$+$	$\frac{3}{18}$	$=$	$\frac{7}{18}$

Add Like Fractions

Key Concept

Words To add fractions with the same denominators, add the numerators. Use the same denominator in the sum. For example, *2 fifths plus 1 fifth equals 3 fifths.*

Examples

Model

$\frac{2}{5} + \frac{1}{5}$

$\frac{3}{5}$

Numbers

$$\frac{2}{5} + \frac{1}{5} = \frac{2+1}{5}$$
$$= \frac{3}{5}$$

EXAMPLE Add Like Fractions

1 **Find the sum of $\frac{4}{5}$ and $\frac{3}{5}$.**

Estimate $1 + \frac{1}{2} = 1\frac{1}{2}$

$\frac{4}{5} + \frac{3}{5} = \frac{4+3}{5}$ Add the numerators.

$= \frac{7}{5}$ Simplify.

$= 1\frac{2}{5}$ Write as a mixed number.

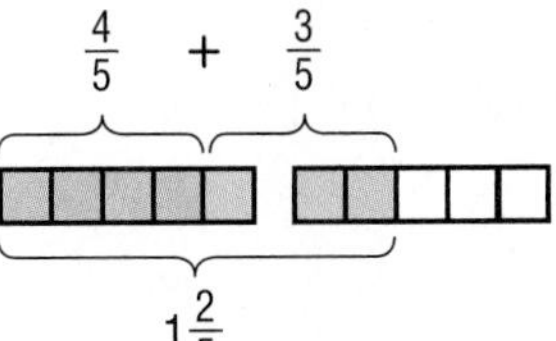

Check for Reasonableness Compare $1\frac{2}{5}$ to the estimate. $1\frac{2}{5} \approx 1\frac{1}{2}$ ✔

CHECK Your Progress

Add. Write in simplest form.

a. $\frac{1}{6} + \frac{5}{6}$ b. $\frac{4}{7} + \frac{6}{7}$ c. $\frac{1}{9} + \frac{5}{9}$

Review Vocabulary

simplest form the form of a fraction when the GCF of the numerator and denominator is 1; *Example:* $\frac{3}{4}$ (Lesson 4-2)

The rule for subtracting fractions is similar to the rule for adding fractions.

Subtract Like Fractions

Key Concept

Words To subtract fractions with the same denominators, subtract the numerators. Use the same denominator in the difference. For example, *3 fifths minus 1 fifth equals 2 fifths.*

Examples

Model

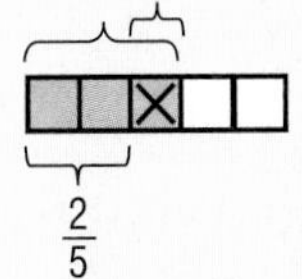

Numbers

$$\frac{3}{5} - \frac{1}{5} = \frac{3-1}{5}$$
$$= \frac{2}{5}$$

EXAMPLE Subtract Like Fractions

2 **Find $\frac{7}{8} - \frac{5}{8}$. Write in simplest form.**

$\frac{7}{8} - \frac{5}{8} = \frac{7-5}{8}$ Subtract the numerators.

$= \frac{2}{8}$ or $\frac{1}{4}$ Simplify.

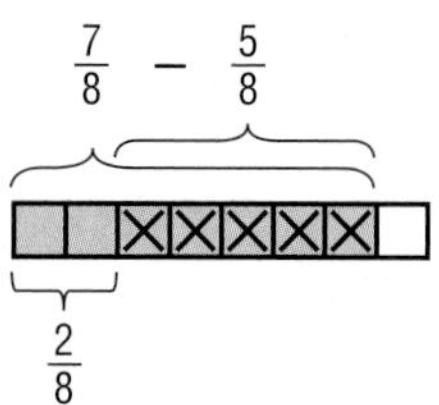

Check *7 eighths minus 5 eighths equals 2 eighths.* ✔

CHECK Your Progress **Subtract. Write in simplest form.**

d. $\frac{5}{9} - \frac{2}{9}$ e. $\frac{11}{12} - \frac{5}{12}$ f. $\frac{7}{10} - \frac{3}{10}$

Real-World EXAMPLE

3 **POPULATION** About $\frac{6}{100}$ of the population of the United States lives in Florida. Another $\frac{4}{100}$ lives in Ohio. How much more of the U.S. population lives in Florida than in Ohio?

$\frac{6}{100} - \frac{4}{100} = \frac{6-4}{100}$ Subtract the numerators.

$= \frac{2}{100}$ or $\frac{1}{50}$ Simplify.

About $\frac{1}{50}$ more of the U.S. population lives in Florida than in Ohio.

Check *6 hundredths minus 4 hundredths equals 2 hundredths.* ✔

Real-World Link Ohio is the seventh largest state in the United States ranked by population. In 2006, its estimated population was 11,478,006.

Source: U.S. Census Bureau

CHECK Your Progress

g. **JUICE** Two-fifths quart of pineapple juice was added to a bowl containing $\frac{3}{5}$ quart of orange juice. How many total quarts of pineapple juice and orange juice are in the bowl?

CHECK Your Understanding

Examples 1, 2 (pp. 257–258)

Add or subtract. Write in simplest form.

1. $\frac{3}{5} + \frac{1}{5}$ 2. $\frac{2}{7} + \frac{1}{7}$ 3. $\frac{3}{4} + \frac{3}{4}$

4. $\frac{3}{8} - \frac{1}{8}$ 5. $\frac{4}{5} - \frac{1}{5}$ 6. $\frac{6}{7} - \frac{2}{7}$

Example 3 (p. 258)

7. **PRESIDENTS** As of 2007, $\frac{8}{42}$ of the U.S. presidents were born in Virginia and $\frac{7}{42}$ were born in Ohio. What fraction of the U.S. presidents were born in either Virginia or Ohio? Write in simplest form.

Practice and Problem Solving

HOMEWORK HELP	
For Exercises	See Examples
8–13	1
14–19	2
20–23	3

Add or subtract. Write in simplest form.

8. $\frac{4}{5} + \frac{3}{5}$
9. $\frac{5}{7} + \frac{6}{7}$
10. $\frac{3}{8} + \frac{7}{8}$
11. $\frac{1}{9} + \frac{5}{9}$
12. $\frac{5}{6} + \frac{5}{6}$
13. $\frac{15}{16} + \frac{7}{16}$
14. $\frac{9}{10} - \frac{3}{10}$
15. $\frac{5}{8} - \frac{3}{8}$
16. $\frac{5}{14} - \frac{1}{14}$
17. $\frac{5}{9} - \frac{2}{9}$
18. $\frac{7}{12} - \frac{2}{12}$
19. $\frac{15}{18} - \frac{13}{18}$

20. **GRADES** In Mr. Navarro's first period class, $\frac{17}{28}$ of the students got an A on their math test. In his second period class, $\frac{11}{28}$ of the students got an A. How many more of the students got an A in Mr. Navarro's first period class than his second period class?

21. **COOKING** A recipe for Michigan blueberry pancakes calls for $\frac{3}{4}$ cup flour, $\frac{1}{4}$ milk, and $\frac{1}{4}$ cup blueberries. How much more flour is needed than milk?

ANALYZE TABLES For Exercises 22 and 23, use the table and the information below.

The table shows the Instant Messenger abbreviations that students use the most at Hillside Middle School.

Instant Messenger Abbreviations	
L8R (Later)	$\frac{48}{100}$
LOL (Laughing out loud)	$\frac{26}{100}$
BRB (Be right back)	$\frac{19}{100}$
CUL8R (See you later)	$\frac{7}{100}$

22. What fraction of these students uses LOL or CUL8R when using Instant Messenger?

23. What fraction of these students uses L8R or BRB when using Instant Messenger?

Use the order of operations to add or subtract. Write in simplest form.

24. $\frac{4}{5} + \frac{1}{5} + \frac{3}{5}$
25. $\frac{7}{8} + \frac{5}{8} - \frac{1}{8}$
26. $\frac{13}{14} - \frac{5}{14} + \frac{6}{14}$

Write an addition or subtraction expression for each model. Then add or subtract.

27. 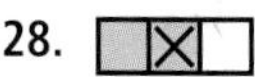

28. [model]

29. **ANALYZE GRAPHS** The graph shows the location of volcanic eruptions in 2006. What fraction represents the volcanic eruptions for both North and South America? How much larger is the section for Asia and South Pacific than for Europe?

Worldwide Volcano Eruptions, 2006

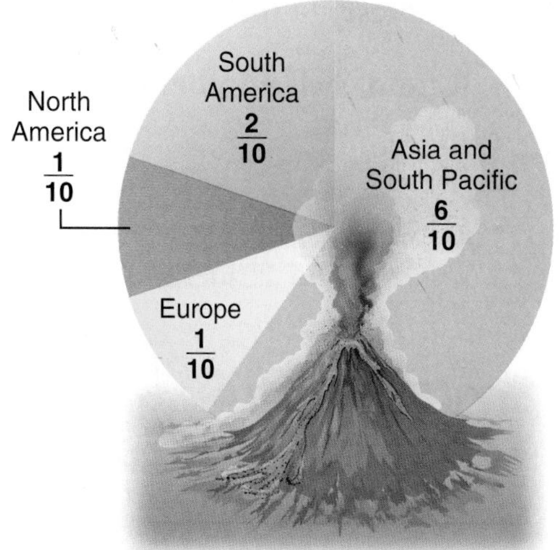

30. **MEASUREMENT** How much longer than $\frac{5}{16}$ inch is $\frac{13}{16}$ inch?

EXTRA PRACTICE
See pages 685, 710.

Draw a model for each expression. Then add or subtract.

31. $\frac{3}{11} + \frac{6}{11}$

32. $\frac{3}{4} - \frac{1}{4}$

33. $\frac{4}{9} + \frac{7}{9}$

H.O.T. Problems

34. **OPEN ENDED** Select two like fractions with a difference of $\frac{1}{3}$ and with denominators that are *not* 3. Justify your selection.

35. **CHALLENGE** Simplify the following expression.

$$\frac{14}{15} + \frac{13}{15} - \frac{12}{15} + \frac{11}{15} - \frac{10}{15} + \dots - \frac{4}{15} + \frac{3}{15} - \frac{2}{15} + \frac{1}{15}$$

36. **WRITING IN MATH** Write a simple rule for adding and subtracting like fractions.

TEST PRACTICE

37. A group of friends bought two large pizzas and ate only part of each pizza. The pictures show how much of the pizzas were left.

First Pizza

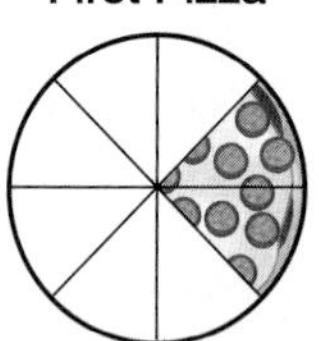

Second Pizza

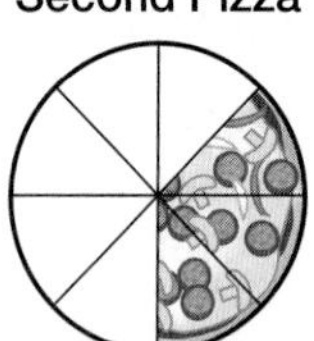

How many pizzas did they eat?

A $\frac{3}{8}$ B $\frac{5}{8}$ C $1\frac{1}{4}$ D $1\frac{3}{8}$

38. At a school carnival, homemade pies were cut into 8 equal-sized pieces. Eric sold 13 pieces, Elena sold 7 pieces, and Tanya sold 10 pieces. Which expression can be used to find the total number of pies sold by Eric, Elena, and Tanya?

F $13 + 7 + 10$

G $8(13 + 7 + 10)$

H $\frac{13}{8} \times \frac{7}{8} \times \frac{10}{8}$

J $\frac{13}{8} + \frac{7}{8} + \frac{10}{8}$

Spiral Review

39. **SCHOOL** Three students need to give their presentations in science class. How many different ways can the teacher arrange the presentations? (Lesson 5-2)

Round each number to the nearest half. (Lesson 5-1)

40. $3\frac{2}{5}$

41. $\frac{1}{12}$

42. $6\frac{4}{7}$

43. **GAMES** Find the area of a rectangular game board that is 25 inches long and 11 inches wide. (Lesson 1-9)

GET READY for the Next Lesson

PREREQUISITE SKILL Find the LCD for each pair of fractions. (Lesson 4-5)

44. $\frac{3}{4}$ and $\frac{5}{8}$

45. $\frac{2}{3}$ and $\frac{1}{2}$

46. $\frac{3}{10}$ and $\frac{3}{4}$

47. $\frac{4}{5}$ and $\frac{2}{9}$

Explore 5-4

Math Lab
Unlike Denominators

MAIN IDEA

Use models to add and subtract fractions with unlike denominators.

Math Online

glencoe.com

- Concepts in Motion

In this lab, you will use fraction strips to add and subtract fractions with *unlike* denominators.

ACTIVITY

1 Use fraction strips to find $\frac{1}{2} + \frac{1}{5}$.

STEP 1 Model each fraction.

$\frac{1}{2}$ $\frac{1}{5}$

STEP 2 To add, line up the end of the shaded part of the first strip with the beginning of the second strip.

$\frac{1}{2}$ $+\frac{1}{5}$

STEP 3 Test different fraction strips below the model, lining up each with the beginning of the first strip. Do the marks line up? If not, try another strip.

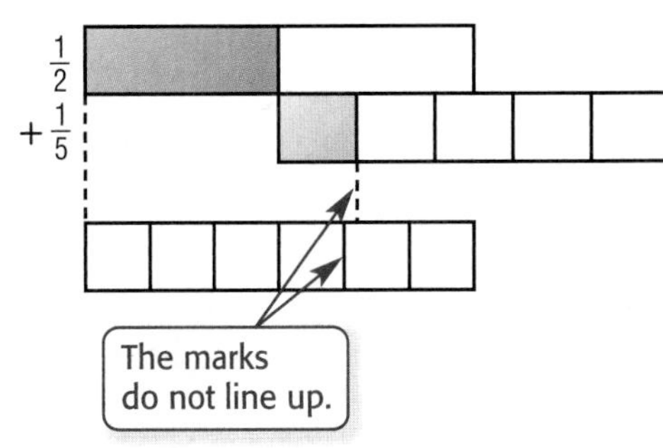

STEP 4 Once the correct strip is found, shade the sections between the beginning of the strip to the point where they line up.

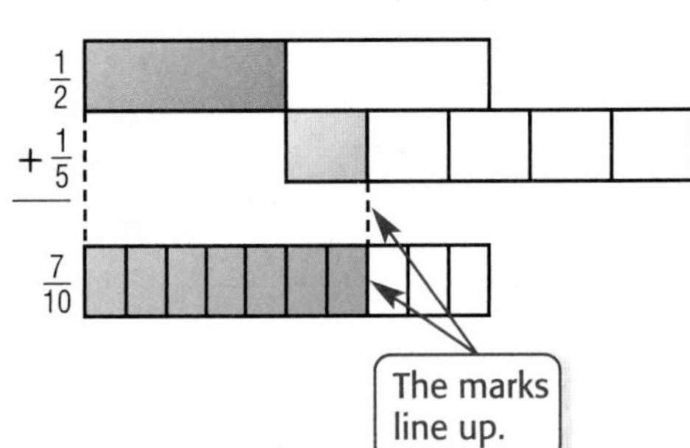

So, $\frac{1}{2} + \frac{1}{5} = \frac{7}{10}$.

CHECK Your Progress **Use fraction strips to add.**

a. $\frac{1}{10} + \frac{2}{5}$

b. $\frac{1}{6} + \frac{1}{2}$

c. $\frac{1}{2} + \frac{3}{4}$

ACTIVITY

2 Use fraction strips to find $\frac{7}{8} - \frac{3}{4}$.

STEP 1 Model each fraction.

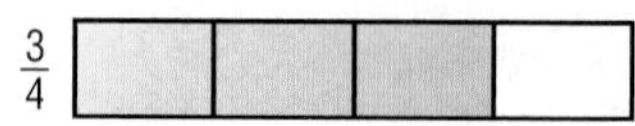

STEP 2 To subtract, line up the ends of the shaded parts of each strip.

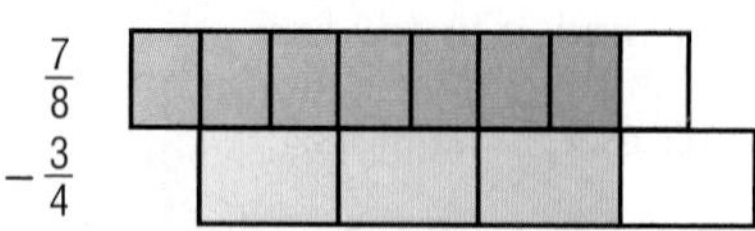

STEP 3 Test different fraction strips below the model, checking to see if the marks line up. Then shade the sections between the beginning of the strip and the point where they line up.

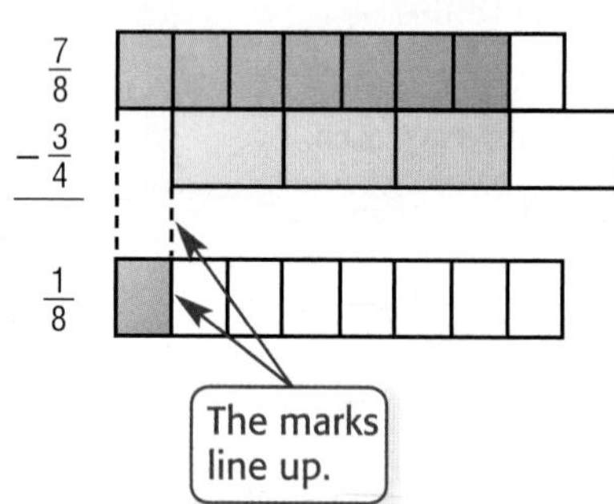

So, $\frac{7}{8} - \frac{3}{4} = \frac{1}{8}$.

CHECK Your Progress

Use fraction strips to subtract.

d. $\frac{3}{8} - \frac{1}{4}$ e. $\frac{8}{9} - \frac{1}{3}$ f. $\frac{2}{3} - \frac{1}{4}$

ANALYZE THE RESULTS

Use the models from Activities 1 and 2 to complete the following.

1. $\frac{1}{2} + \frac{1}{5} = \frac{\blacksquare}{10} + \frac{\blacksquare}{10}$
2. $\frac{7}{8} - \frac{3}{4} = \frac{\blacksquare}{8} - \frac{\blacksquare}{8}$

Write an addition or subtraction expression for each model. Then add or subtract.

3. 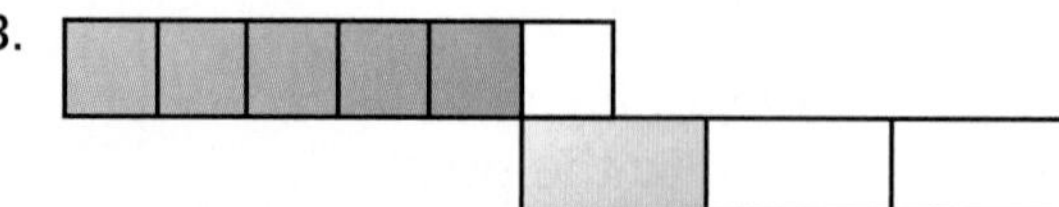

4. 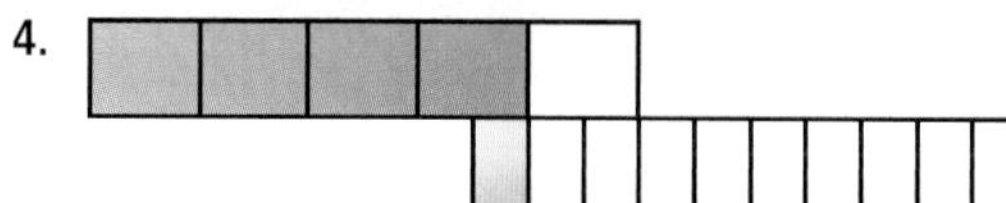

5. **MAKE A CONJECTURE** What is the relationship between the number of separations on the answer fraction strip and the denominators of the fractions added or subtracted?

5-4 Adding and Subtracting Fractions with Unlike Denominators

MAIN IDEA

Add and subtract fractions with unlike denominators.

New Vocabulary

unlike fractions

Math Online

glencoe.com

- Extra Examples
- Personal Tutor
- Self-Check Quiz

GET READY for the Lesson

MEASUREMENT The table shows the fractions of one hour for different minutes.

1. Write each fraction in simplest form.
2. What fraction of one hour is equal to the sum of 15 minutes and 20 minutes? Write in simplest form.
3. Explain why $\frac{1}{6}$ hour + $\frac{1}{3}$ hour = $\frac{1}{2}$ hour.
4. Explain why $\frac{1}{12}$ hour + $\frac{1}{2}$ hour = $\frac{7}{12}$ hour.

Number of Minutes	Fraction of One Hour
1	$\frac{1}{60}$
5	$\frac{5}{60}$
10	$\frac{10}{60}$
15	$\frac{15}{60}$
20	$\frac{20}{60}$
30	$\frac{30}{60}$
45	$\frac{45}{60}$

Before you can add two **unlike fractions**, or fractions with different denominators, one or both of the fractions must be renamed so that they have a common denominator.

Add or Subtract Unlike Fractions — Key Concept

To add or subtract fractions with different denominators,

- Rename the fractions using the least common denominator (LCD).
- Add or subtract as with like fractions.
- If necessary, simplify the sum or difference.

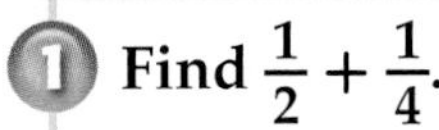

EXAMPLE Add Unlike Fractions

1 Find $\frac{1}{2} + \frac{1}{4}$.

METHOD 1 **Use a model.**

$\frac{1}{2}$

$+\frac{1}{4}$

$\frac{3}{4}$

Review Vocabulary

least common denominator (LCD) the least common multiple (LCM) of the denominators of two or more fractions; *Example:* the LCD of $\frac{1}{2}$ and $\frac{1}{4}$ is 4. (Lesson 4-5)

METHOD 2 **Use the LCD.**

The least common denominator of $\frac{1}{2}$ and $\frac{1}{4}$ is 4.

Write the problem.		Rename using the LCD, 4.				Add the fractions.
$\frac{1}{2}$	$\longrightarrow$	$\frac{1 \times 2}{2 \times 2}$	$=$	$\frac{2}{4}$	$\longrightarrow$	$\frac{2}{4}$
$+\frac{1}{4}$	$\longrightarrow$	$+\frac{1 \times 1}{4 \times 1}$	$=$	$+\frac{1}{4}$	$\longrightarrow$	$+\frac{1}{4}$
						$\frac{3}{4}$

CHOOSE Your Method

Add. Write in simplest form.

a. $\frac{1}{6} + \frac{2}{3}$ b. $\frac{9}{10} + \frac{1}{2}$ c. $\frac{1}{4} + \frac{3}{8}$

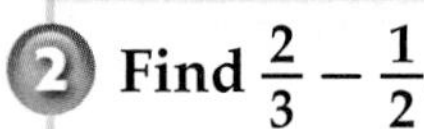

EXAMPLE Subtract Unlike Fractions

2 **Find $\frac{2}{3} - \frac{1}{2}$.**

METHOD 1 **Use a model.**

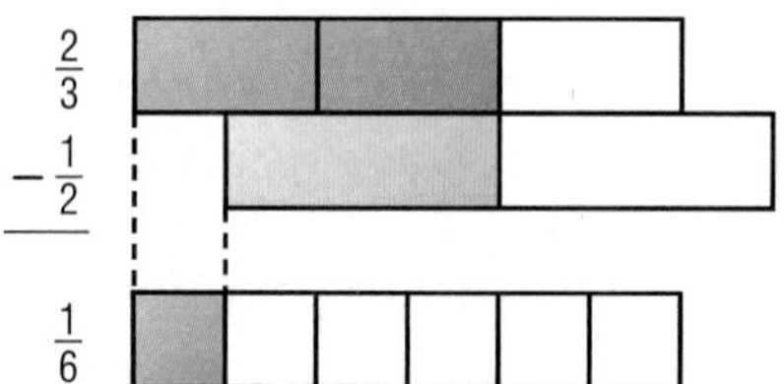

METHOD 2 **Use the LCD.**

The least common denominator of $\frac{2}{3}$ and $\frac{1}{2}$ is 6.

Write the problem.		Rename using the LCD, 6.				Subtract the fractions.
$\frac{2}{3}$	$\longrightarrow$	$\frac{2 \times 2}{3 \times 2}$	$=$	$\frac{4}{6}$	$\longrightarrow$	$\frac{4}{6}$
$-\frac{1}{2}$	$\longrightarrow$	$-\frac{1 \times 3}{2 \times 3}$	$=$	$-\frac{3}{6}$	$\longrightarrow$	$-\frac{3}{6}$
						$\frac{1}{6}$

Study Tip

Check for Reasonableness Estimate the difference in Example 2.

$\frac{2}{3} - \frac{1}{2} \approx \frac{1}{2} - \frac{1}{2}$ or 0.

Compare $\frac{1}{6}$ to the estimate. $\frac{1}{6} \approx 0$. So, the answer is reasonable.

CHOOSE Your Method

Subtract. Write in simplest form.

d. $\frac{5}{8} - \frac{1}{4}$ e. $\frac{3}{4} - \frac{1}{3}$ f. $\frac{1}{2} - \frac{2}{5}$

Real-World EXAMPLE

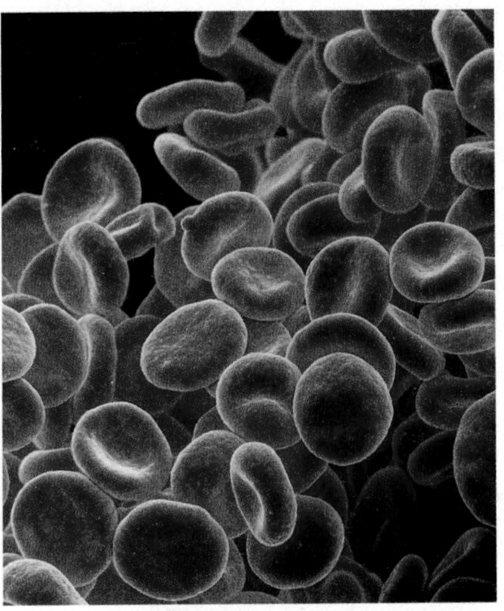

Real-World Link Two to three drops of blood contain about one billion red blood cells.

3 HEALTH Use the table to find the fraction of the population that has type A or type B blood.

Blood Type Frequencies				
ABO Type	O	A	B	AB
Fraction	$\frac{11}{25}$	$\frac{21}{50}$	$\frac{1}{10}$	$\frac{1}{25}$

Source: Palomar College

Find $\frac{21}{50} + \frac{1}{10}$.

The least common denominator of $\frac{21}{50}$ and $\frac{1}{10}$ is 50.

Write the problem.		Rename using the LCD, 50.				Add the fractions.
$\frac{21}{50}$	$\longrightarrow$	$\frac{21 \times 1}{50 \times 1}$	$=$	$\frac{21}{50}$	$\longrightarrow$	$\frac{21}{50}$
$+\frac{1}{10}$	$\longrightarrow$	$+\frac{1 \times 5}{10 \times 5}$	$=$	$+\frac{5}{50}$	$\longrightarrow$	$+\frac{5}{50}$
						$\frac{26}{50}$ or $\frac{13}{25}$

So, $\frac{13}{25}$ of the population has type A or type B blood.

CHECK Your Progress

g. **SURVEY** The table shows the results of an online survey of over 36,000 youth. How much greater was the part of youth that said their favorite way to be "artsy" was by drawing than by acting?

What is your favorite way to be artsy?
Drawing $\frac{8}{25}$
Acting $\frac{7}{50}$
Making music $\frac{7}{50}$
Taking pictures $\frac{11}{100}$
Writing $\frac{3}{50}$

Source: PBS Kids

EXAMPLE Evaluate an Expression with Fractions

4 ALGEBRA Evaluate $a - b$ if $a = \frac{3}{4}$ and $b = \frac{1}{6}$.

$a - b = \frac{3}{4} - \frac{1}{6}$	Replace a with $\frac{3}{4}$ and b with $\frac{1}{6}$.
$= \frac{3 \times 3}{4 \times 3} - \frac{1 \times 2}{6 \times 2}$	Rename $\frac{3}{4}$ and $\frac{1}{6}$ using the LCD, 12.
$= \frac{9}{12} - \frac{2}{12}$	Simplify.
$= \frac{7}{12}$	Subtract the numerators.

Study Tip

Look Back You can review **evaluating expressions** in Lesson 1-5.

CHECK Your Progress

h. **ALGEBRA** Evaluate $c + d$ if $c = \frac{2}{5}$ and $d = \frac{3}{10}$.

CHECK Your Understanding

Examples 1, 2 (pp. 263–264)

Add or subtract. Write in simplest form.

1. $\frac{2}{3} + \frac{2}{9}$
2. $\frac{1}{4} + \frac{5}{8}$
3. $\frac{2}{3} - \frac{1}{2}$
4. $\frac{3}{5} - \frac{1}{2}$
5. $\frac{3}{10} + \frac{1}{5}$
6. $\frac{2}{3} + \frac{1}{4}$
7. $\frac{3}{4} - \frac{1}{8}$
8. $\frac{5}{7} - \frac{1}{2}$

Example 3 (p. 265)

9. **TOOLS** A certain drill set includes drill bits ranging from $\frac{1}{16}$ inch to $\frac{1}{4}$ inch. What is the range of drill bits in this set?

Example 4 (p. 265)

ALGEBRA Evaluate each expression.

10. $x + y$ if $x = \frac{5}{6}$ and $y = \frac{7}{12}$
11. $r - s$ if $r = \frac{7}{10}$ and $s = \frac{1}{4}$

Practice and Problem Solving

HOMEWORK HELP	
For Exercises	See Examples
12–27	1, 2
28, 29	3
30, 31	4

Add or subtract. Write in simplest form.

12. $\frac{3}{8} + \frac{1}{4}$
13. $\frac{2}{5} + \frac{1}{2}$
14. $\frac{9}{10} - \frac{1}{2}$
15. $\frac{5}{8} - \frac{1}{4}$
16. $\frac{1}{6} + \frac{3}{4}$
17. $\frac{1}{4} + \frac{2}{3}$
18. $\frac{5}{6} - \frac{7}{10}$
19. $\frac{3}{4} - \frac{2}{5}$
20. $\frac{8}{9} + \frac{1}{2}$
21. $\frac{5}{7} + \frac{1}{2}$
22. $\frac{9}{10} - \frac{2}{5}$
23. $\frac{7}{8} - \frac{3}{4}$
24. $\frac{7}{8} + \frac{3}{4}$
25. $\frac{7}{12} + \frac{2}{3}$
26. $\frac{3}{4} - \frac{2}{7}$
27. $\frac{9}{11} - \frac{1}{2}$

ANALYZE TABLES For Exercises 28 and 29, use the table showing the fraction of total coupon book sales of four students in a class.

28. What is the difference between Jabar's and Corey's fraction of total sales?
29. What part of the total sales did Billy and Domanick have altogether?

Coupon Book Sales

Student	Fraction of Total Sales
Corey	$\frac{1}{12}$
Billy	$\frac{3}{40}$
Domanick	$\frac{1}{3}$
Jabar	$\frac{2}{15}$

ALGEBRA Evaluate each expression.

30. $a + b$ if $a = \frac{7}{10}$ and $b = \frac{5}{6}$
31. $x - y$ if $x = \frac{4}{5}$ and $y = \frac{1}{2}$

Use the order of operations to add or subtract. Write in simplest form.

32. $\frac{9}{10} + \frac{2}{3} - \frac{11}{15}$

33. $\frac{7}{12} + \frac{5}{8} + \frac{5}{6}$

34. $\frac{15}{16} - \frac{1}{3} - \frac{1}{12}$

Write an addition or subtraction sentence for each model.

35. 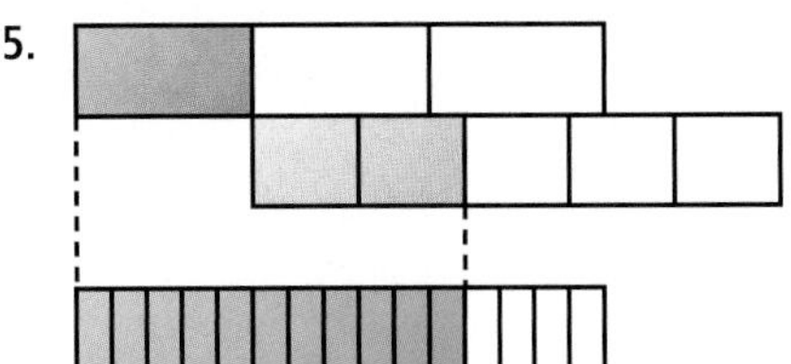

36. 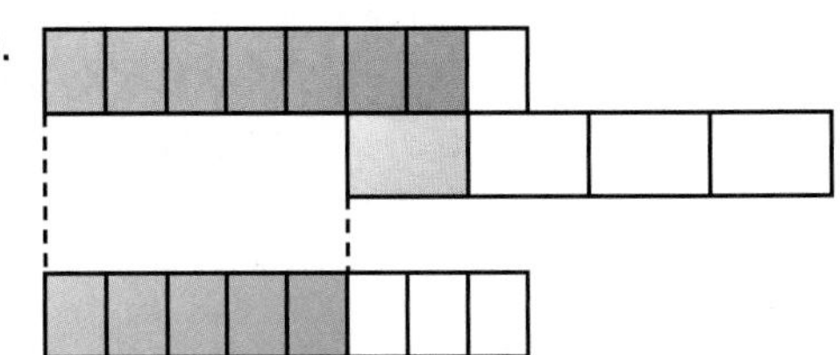

Use fraction strips to model each expression. Then add or subtract.

37. $\frac{1}{3} + \frac{1}{6}$

38. $\frac{5}{8} - \frac{1}{2}$

39. $\frac{5}{6} + \frac{2}{3}$

40. **GARDENING** Suppose an herb plant grew $\frac{9}{16}$ inch the first week and $\frac{7}{8}$ inch the second week. How much more did the herb plant grow the second week? Justify your solution.

ANALYZE TABLES For Exercises 41–43, use the table.

41. What portion of the Earth's landmass is Asia and Africa?

42. How much more is the landmass of North America than South America?

43. What portion of Earth's landmass is Antarctica, Europe, Australia, and Oceania?

Continent or Island Group	Portion of Earth's Landmass
Antarctica, Europe, Australia, and Oceania	■
Asia	$\frac{3}{10}$
Africa	$\frac{1}{5}$
North America	$\frac{1}{6}$
South America	$\frac{1}{8}$

Source: *Oxford Atlas of the World*

44. **STUDYING** Nikki knows that studying each night is better than cramming for a test. Thus, she makes a habit of studying every night for $\frac{3}{5}$ of an hour on math and $\frac{3}{4}$ of an hour on English. Which subject does she spend more time studying and by how much?

EXTRA PRACTICE
See pages 685, 710.

H.O.T. Problems

45. **OPEN ENDED** Create and use a model to represent the sum of two fractions with unlike denominators.

46. **FIND THE ERROR** Simona and Kenji are finding $\frac{5}{8} + \frac{1}{4}$. Who is correct? Explain your reasoning.

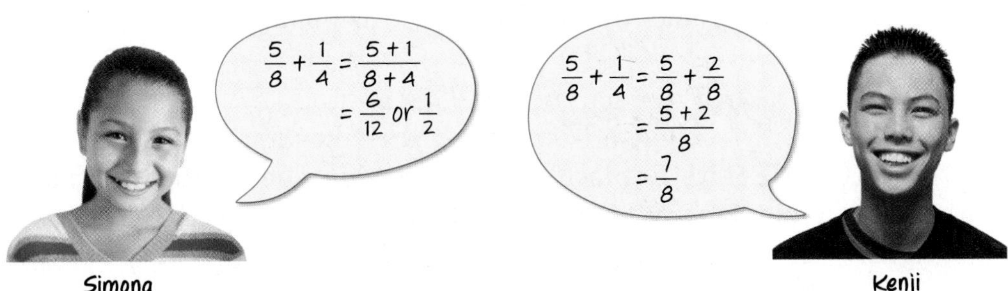

CHALLENGE **Decide whether each sentence is *sometimes, always,* or *never* true. Explain your reasoning.**

47. The sum of two fractions that are less than 1 is less than 1.

48. The difference of two fractions is less than both fractions.

49. **WRITING IN MATH** Write a problem about a real-world situation in which you would subtract $\frac{4}{5}$ and $\frac{3}{4}$.

50. Hernando made a drawing of his bedroom. The length of his drawing is $\frac{3}{4}$ foot, and the width is $\frac{1}{3}$ foot less than the length. Find the width of the drawing.

A $\frac{1}{4}$ ft

B $\frac{5}{12}$ ft

C $\frac{7}{12}$ ft

D $1\frac{1}{12}$ ft

51. On a camping trip, Rebecca hiked $\frac{5}{8}$ mile to a cave, and then $\frac{1}{4}$ mile inside the cave. Each strip below represents 1 mile. Which strip is shaded to show the total number of miles, one way, Rebecca hiked?

F

G

H

J

Spiral Review

Add or subtract. Write in simplest form. (Lesson 5-3)

52. $\frac{7}{10} + \frac{1}{10}$

53. $\frac{3}{8} - \frac{1}{8}$

54. $\frac{5}{18} + \frac{7}{18}$

55. $\frac{11}{20} - \frac{3}{20}$

56. **PAPER FOLDING** Paloma folded a piece of paper in half vertically. Then she folded it in half horizontally. If she repeats this process one more time and opens up the piece of paper, how many regions will be separated by the fold lines? Use the *act it out* strategy. (Lesson 5-2)

BASKETBALL **For Exercises 57 and 58, use the stem-and-leaf plot that shows the number of points the basketball team scored each game this season.** (Lesson 2-4)

Basketball Team Points

Stem	Leaf
3	9
4	3 5 5 7 8 9
5	0 0 2 4 7 7 7
6	0 2

4|3 = 43 points

57. What is the fewest number of points the team scored?

58. How many games did the team score 57 points?

GET READY for the Next Lesson

PREREQUISITE SKILL **Replace each ■ with a number so that the fractions are equivalent.** (Lesson 4-2)

59. $\frac{3}{4} = \frac{■}{12}$

60. $\frac{1}{8} = \frac{■}{24}$

61. $\frac{1}{3} = \frac{■}{12}$

62. $\frac{5}{6} = \frac{■}{18}$

READING to SOLVE PROBLEMS

Meaning of Subtraction

You know that one meaning of subtraction is *to take away*. But there are other meanings too. Look for these meanings when you're solving a word problem.

To take away

Chad found $\frac{5}{8}$ of a pizza in the refigerator. He ate $\frac{1}{8}$ of the original pizza. How much of the original pizza is left?

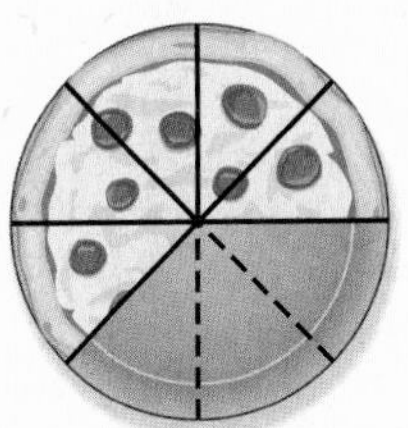

To find a missing addend

Heather made a desktop by gluing a sheet of oak veneer to a sheet of $\frac{3}{4}$-inch plywood. The total thickness of the desktop is $\frac{13}{16}$ inch. What was the thickness of the oak veneer?

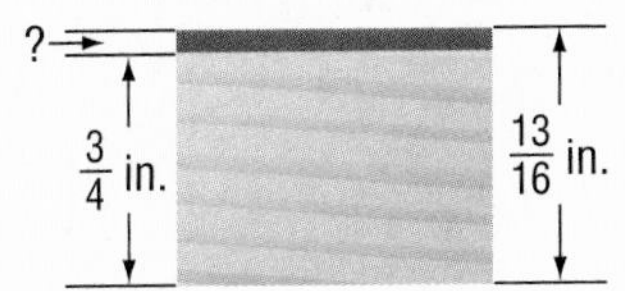

To compare the size of two sets

Yesterday, it rained $\frac{7}{8}$ inch. Today, it rained $\frac{1}{4}$ inch. How much more did it rain yesterday than today?

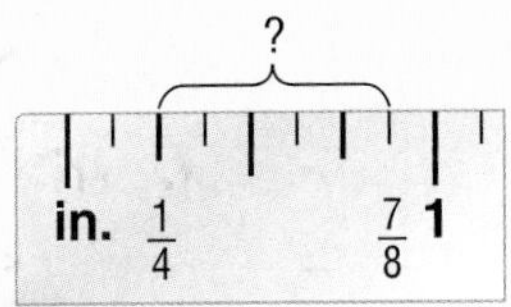

PRACTICE

1. Solve each problem above.

Identify the meaning of subtraction shown in each problem. Then solve the problem.

2. Marcus opened a carton of milk and drank $\frac{1}{4}$ of it. How much of the carton of milk is left?
3. How much bigger is a $\frac{15}{16}$-inch wrench than a $\frac{3}{8}$-inch wrench?
4. Part of a hiking trail is $\frac{3}{4}$ mile long. When you pass the $\frac{1}{8}$-mile marker, how much farther is it until the end of the trail?
5. A cornbread recipe calls for $\frac{3}{4}$ cup of cornmeal. Ali has only $\frac{1}{4}$ cup. How much more cornmeal does she need?

5-5 Adding and Subtracting Mixed Numbers

MAIN IDEA

Add and subtract mixed numbers.

Math Online

glencoe.com

- Extra Examples
- Personal Tutor
- Self-Check Quiz

MINI Lab

You can use paper plates to add and subtract mixed numbers.

STEP 1 Cut a paper plate into fourths and another plate into halves.

STEP 2 Use one whole plate and three fourths of a plate to show the mixed number $1\frac{3}{4}$.

STEP 3 Use two whole plates and one half of a plate to show $2\frac{1}{2}$.

STEP 4 Make as many whole paper plates as you can.

1. How many whole paper plates can you make?
2. What fraction is represented by the leftover pieces?

Use paper plate models to find each sum or difference.

3. $1\frac{3}{4} + 2\frac{1}{2}$
4. $2\frac{3}{4} - 1\frac{1}{4}$
5. $1\frac{2}{3} + 2\frac{1}{6}$

The Mini Lab suggests the following rule.

Add and Subtract Mixed Numbers — Key Concept

- Add or subtract the fractions.
- Then add or subtract the whole numbers.
- Rename and simplify if necessary.

EXAMPLES Add or Subtract Mixed Numbers

1 **Find $4\frac{5}{6} - 2\frac{1}{6}$.** **Estimate** $5 - 2 = 3$

Subtract the fractions. Subtract the whole numbers.

$$\begin{array}{r} 4\frac{5}{6} \\ -\ 2\frac{1}{6} \\ \hline \frac{4}{6} \end{array} \quad \rightarrow \quad \begin{array}{r} 4\frac{5}{6} \\ -\ 2\frac{1}{6} \\ \hline 2\frac{4}{6} \text{ or } 2\frac{2}{3} \end{array}$$

Check for Reasonableness $2\frac{2}{3} \approx 3$ ✓

2 **Find $5\frac{1}{4} + 10\frac{2}{3}$.** **Estimate** $5 + 11 = 16$

Write the problem.		Rename the fractions using the LCD, 12.				Add the fractions. Then add the whole numbers.
$5\frac{1}{4}$	$\rightarrow$	$\frac{1 \times 3}{4 \times 3}$	$\rightarrow$	$5\frac{3}{12}$	$\rightarrow$	$5\frac{3}{12}$
$+\ 10\frac{2}{3}$	$\rightarrow$	$\frac{2 \times 4}{3 \times 4}$	$\rightarrow$	$+\ 10\frac{8}{12}$	$\rightarrow$	$+\ 10\frac{8}{12}$
						$15\frac{11}{12}$

Check for Reasonableness $15\frac{11}{12} \approx 16$ ✔

CHECK Your Progress

Add or subtract. Write in simplest form.

a. $5\frac{2}{8} + 3\frac{1}{8}$ b. $5\frac{1}{2} - 2\frac{1}{3}$ c. $6\frac{2}{5} + 3\frac{1}{2}$

EXAMPLES Rename Numbers to Subtract

Study Tip

Compensation You can calculate $5 - 2\frac{7}{8}$ mentally.

Think: $2\frac{7}{8} + \frac{1}{8} = 3$

$5 + \frac{1}{8} = 5\frac{1}{8}$

Since $5\frac{1}{8} - 3 = 2\frac{1}{8}$, $5 - 2\frac{7}{8} = 2\frac{1}{8}$.

3 **Find $5 - 2\frac{7}{8}$.** **Estimate** $5 - 3 = 2$

5	$\rightarrow$	$4\frac{8}{8}$	Rename 5 as $4\frac{8}{8}$.
$-\ 2\frac{7}{8}$	$\rightarrow$	$-\ 2\frac{7}{8}$	
		$2\frac{1}{8}$	Subtract.

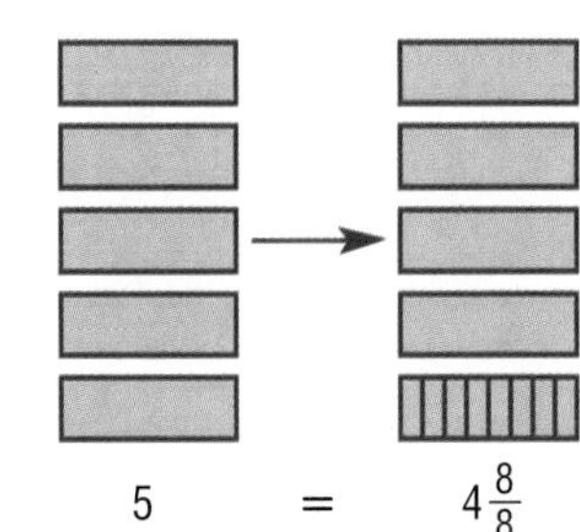

Check for Reasonableness $2\frac{1}{8} \approx 2$ ✔

4 **Find $12\frac{1}{8} - 9\frac{1}{4}$.** **Estimate** $12 - 9 = 3$

Step 1

$12\frac{1}{8}$	$\rightarrow$	$12\frac{1}{8}$
$-\ 9\frac{1}{4}$	$\rightarrow$	$-\ 9\frac{2}{8}$

Rename $\frac{1}{8}$ and $\frac{1}{4}$ using their LCD, 8.

Step 2

$12\frac{1}{8}$	$\rightarrow$	$11\frac{9}{8}$
$-\ 9\frac{2}{8}$	$\rightarrow$	$-\ 9\frac{2}{8}$
		$2\frac{7}{8}$

Rename $12\frac{1}{8}$ as $11\frac{8}{8} + \frac{1}{8}$ or $11\frac{9}{8}$.

Check for Reasonableness $2\frac{7}{8} \approx 3$ ✔

CHECK Your Progress

d. $5 - 3\frac{1}{2}$ e. $7 - 2\frac{1}{4}$ f. $2 - 1\frac{6}{7}$

g. $11\frac{1}{2} - 7\frac{1}{8}$ h. $6\frac{2}{5} - 3\frac{3}{5}$ i. $8\frac{7}{10} - 6\frac{3}{4}$

TEST EXAMPLE

5 **Refer to the table. How much longer is the NBA basketball court than the Olympic basketball court?**

Sport	Length of Court (ft)	Width of Court (ft)
Olympic Basketball	$91\frac{5}{6}$	$49\frac{1}{6}$
NBA Basketball	94	50

A $3\frac{1}{6}$ feet **C** $2\frac{1}{6}$ feet

B $2\frac{5}{6}$ feet **D** $1\frac{1}{6}$ feet

Test-Taking Tip

Eliminating Choices By estimating $94 - 91\frac{5}{6}$, you know the difference must be greater than 2 feet and less than 3 feet. So, you can eliminate choices A and D.

Read the Item You need to find $94 - 91\frac{5}{6}$.

Solve the Item

$$\begin{array}{r} 94 \\ -\ 91\frac{5}{6} \\ \hline \end{array} \rightarrow \begin{array}{r} 93\frac{6}{6} \\ -\ 91\frac{5}{6} \\ \hline 2\frac{1}{6} \end{array}$$

Rename 94 as $93\frac{6}{6}$.

The NBA court is $2\frac{1}{6}$ feet longer than the Olympic court. The answer is C.

CHECK Your Progress

d. A recipe for pumpkin bread calls for $3\frac{1}{4}$ cups flour, and a recipe for cornbread calls for $1\frac{1}{3}$ cups flour. How much more flour is needed for pumpkin bread than cornbread?

F $2\frac{11}{12}$ c **G** $2\frac{7}{12}$ c **H** $2\frac{1}{12}$ c **J** $1\frac{11}{12}$ c

CHECK Your Understanding

Examples 1–4 (pp. 270–271)

Add or subtract. Write in simplest form.

1. $\begin{array}{r} 5\frac{3}{4} \\ -\ 1\frac{1}{4} \\ \hline \end{array}$

2. $\begin{array}{r} 2\frac{3}{8} \\ +\ 4\frac{1}{8} \\ \hline \end{array}$

3. $\begin{array}{r} 14\frac{3}{5} \\ -\ 6\frac{3}{10} \\ \hline \end{array}$

4. $6\frac{9}{10} + 8\frac{1}{4}$

5. $3\frac{2}{3} - 2\frac{4}{5}$

6. $4\frac{1}{3} - 1\frac{3}{4}$

Example 5 (p. 272)

7. **MULTIPLE CHOICE** A g-force is a unit of measurement for an object being accelerated. A roller coaster has a g-force of $4\frac{3}{5}$. A second roller coaster has a g-force of $3\frac{1}{2}$. How much greater is the g-force of the first roller coaster than the second?

A $\frac{9}{10}$ **B** $1\frac{1}{10}$ **C** $1\frac{1}{5}$ **D** $2\frac{1}{5}$

Practice and Problem Solving

HOMEWORK HELP	
For Exercises	See Examples
8–15	1, 2
16–23	3, 4
24	

Add or subtract. Write in simplest form.

8. $\begin{array}{r} 3\frac{5}{6} \\ +\ 4\frac{1}{6} \\ \hline \end{array}$

9. $\begin{array}{r} 4\frac{5}{12} \\ +\ 6\frac{7}{12} \\ \hline \end{array}$

10. $\begin{array}{r} 4\frac{5}{8} \\ -\ 2\frac{3}{8} \\ \hline \end{array}$

11. $\begin{array}{r} 9\frac{4}{5} \\ -\ 4\frac{2}{5} \\ \hline \end{array}$

12. $6\frac{3}{5} + \frac{4}{5}$

13. $3\frac{3}{8} + 6\frac{5}{8}$

14. $7\frac{7}{9} - 4\frac{1}{3}$

15. $6\frac{6}{7} - 4\frac{5}{14}$

16. $7 - 5\frac{1}{2}$

17. $9 - 3\frac{3}{5}$

18. $4\frac{1}{4} - 2\frac{3}{4}$

19. $9\frac{3}{8} - 6\frac{5}{8}$

20. $12\frac{1}{5} - 5\frac{3}{10}$

21. $8\frac{1}{3} - 1\frac{5}{6}$

22. $14\frac{3}{8} - 5\frac{3}{4}$

23. $10\frac{5}{9} - 3\frac{2}{3}$

24. **DELI** Caroline bought $2\frac{1}{4}$ pounds of turkey and $1\frac{2}{3}$ pounds of roast beef. How much more turkey than roast beef did Caroline buy?

25. **PAINTING** Pamela is going to paint three different rooms. She will need $2\frac{1}{2}$ gallons of paint for the first room, $4\frac{1}{3}$ gallons of paint for the second room, and $3\frac{3}{4}$ gallons of paint for the third room. How much paint does Pamela need for all three rooms?

26. **ANALYZE TABLES** Sei ("say") whales can reach different sizes based on their location. Find the difference between the longest and shortest sei whales according to their location. Justify your solution.

Sei Whale Lengths

Location	Length (feet)
Southern Hemisphere	$65\frac{3}{5}$
North Pacific	61
North Atlantic	$56\frac{4}{5}$

Source: Sea World

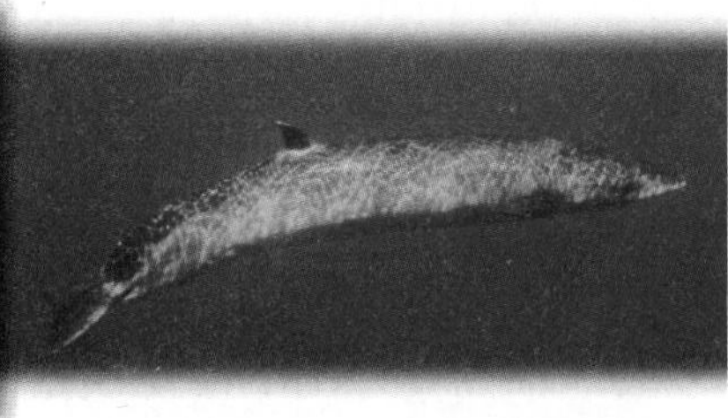

27. **DISTANCE** Neil lives $3\frac{1}{2}$ blocks from Dario's house. Dario lives $2\frac{1}{4}$ blocks from the library, and the video store is $1\frac{1}{8}$ blocks from the library. How far will Neil travel if he walks from his home to Dario's house, the library, and then the video store?

Write an addition or subtraction expression for each model. Then add or subtract.

28.  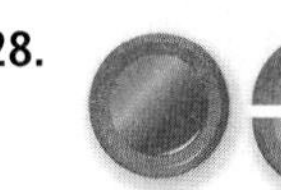 + 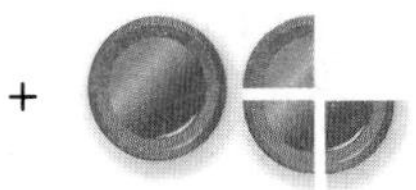

29. 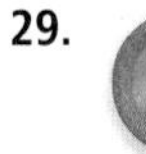  – 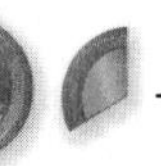 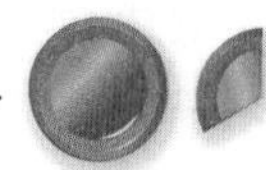

EXTRA PRACTICE
See pages 685, 710.

H.O.T. Problems

30. **FIND THE ERROR** Karen and Daniel are finding $7\frac{1}{2} - 4$. Who is correct? Explain your reasoning.

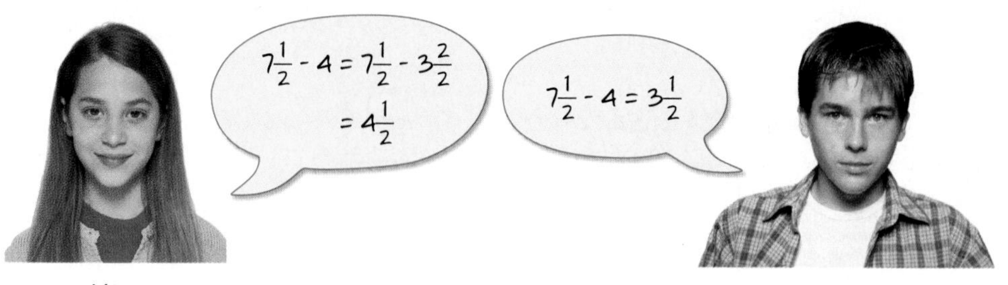

31. **CHALLENGE** Use the digits 1, 1, 2, 2, 3, and 4 to create two mixed numbers with a sum of $4\frac{1}{4}$.

32. **WRITING IN MATH** Describe a method of renaming $5\frac{3}{7}$ as $4\frac{10}{7}$ that involves mental math. Explain why your method works.

TEST PRACTICE

33. Mrs. Matthews bought $2\frac{2}{3}$ pounds of fish, $4\frac{1}{2}$ pounds of chicken, and $3\frac{1}{4}$ pounds of beef. How many pounds did she buy altogether?

A $10\frac{5}{12}$ lb

B $10\frac{1}{3}$ lb

C 10 lb

D $9\frac{3}{4}$ lb

34. Trey's hamster weighs $14\frac{1}{8}$ ounces and Gina's hamster weighs $12\frac{2}{3}$ ounces. How much more does Trey's hamster weigh than Gina's?

F $2\frac{11}{24}$ oz

G $1\frac{1}{2}$ oz

H $1\frac{11}{24}$ oz

J $1\frac{1}{4}$ oz

Spiral Review

Add or subtract. Write in simplest form. (Lessons 5-3 and 5-4)

35. $\frac{1}{3} + \frac{1}{3}$

36. $\frac{9}{10} - \frac{3}{10}$

37. $\frac{4}{5} - \frac{3}{4}$

38. $\frac{7}{9} + \frac{5}{12}$

39. **VIDEO GAMES** Corey bought a video game that cost $37.85 and paid with $40. Is $2, $3, or $4 a reasonable amount for how much change he received? (Lesson 3-10)

GET READY for the Next Lesson

PREREQUISITE SKILL Round each number to the nearest half. (Lesson 5-1)

40. $1\frac{2}{5}$

41. $7\frac{4}{9}$

42. $5\frac{3}{8}$

43. $2\frac{5}{6}$

44. $2\frac{1}{12}$

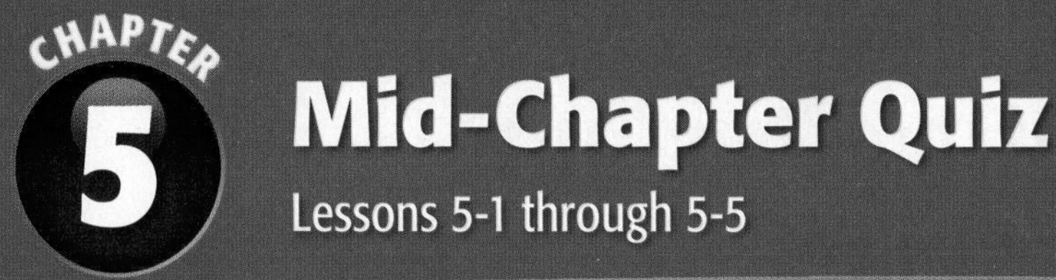

Mid-Chapter Quiz

Lessons 5-1 through 5-5

Round each number to the nearest half. (Lesson 5-1)

1. $\frac{7}{8}$
2. $3\frac{2}{7}$
3. $6\frac{3}{4}$

4. **STICKERS** Find the length of the sticker to the nearest half inch. (Lesson 5-1)

5. **SCHOOL** It takes Monica $1\frac{3}{4}$ minutes to walk to the bus stop. Should she leave her house $1\frac{1}{2}$ minutes or 2 minutes before the bus arrives? (Lesson 5-1)

6. **MAZES** In a corn maze, you begin by walking north. You turn at the next right and then at the next left. In which direction are you facing now? Use the *act it out* strategy. (Lesson 5-2)

7. **ART** Tia is making a sign with her name to hang in her bedroom. She wants each letter of her name to be a different color. How many different ways can she write her name using red, green, and yellow markers? Use the *act it out* strategy. (Lesson 5-2)

Add or subtract. Write in simplest form. (Lesson 5-3)

8. $\frac{5}{9} + \frac{7}{9}$
9. $\frac{9}{11} - \frac{5}{11}$
10. $\frac{1}{6} + \frac{5}{6}$

11. **MEASUREMENT** How much longer is a section of rope measuring $\frac{11}{16}$ inch than a section of rope measuring $\frac{7}{16}$ inch? Write in simplest form. (Lesson 5-3)

Add or subtract. Write in simplest form. (Lesson 5-4)

12. $\frac{5}{8} + \frac{3}{4}$
13. $\frac{2}{3} - \frac{1}{2}$
14. $\frac{3}{5} + \frac{5}{6}$

15. **MULTIPLE CHOICE** On Tuesday, Trent spent $\frac{11}{20}$ hour on the Internet. On Wednesday, he spent $\frac{8}{15}$ hour on the Internet. How much more time did Trent spend on the Internet on Tuesday than on Wednesday? (Lesson 5-4)

    A $\frac{1}{60}$ hour
    C $\frac{1}{15}$ hour
    B $\frac{1}{20}$ hour
    D $\frac{1}{12}$ hour

Add or subtract. Write in simplest form. (Lesson 5-5)

16. $1\frac{5}{12} + 4\frac{4}{12}$
17. $5\frac{1}{8} - 3\frac{1}{2}$
18. $8\frac{1}{6} + 7\frac{3}{4}$

19. **CRAFTS** Tiffany cut $1\frac{9}{32}$ inches from each side of a square piece of scrapbook paper. If the scrapbook paper now measures $5\frac{1}{4}$ inches on each side, what was its original side length? (Lesson 5-5)

20. **MULTIPLE CHOICE** To win horse racing's Triple Crown, a horse must win all three races shown. How much longer is the longest race than the shortest? (Lesson 5-5)

Race	Length (mi)
Kentucky Derby	$1\frac{1}{4}$
Preakness Stakes	$1\frac{3}{16}$
Belmont Stakes	$1\frac{1}{2}$

    F $\frac{1}{4}$ mi
    H $\frac{1}{2}$ mi
    G $\frac{5}{16}$ mi
    J $1\frac{1}{16}$ mi

5-6 Estimating Products of Fractions

MAIN IDEA

Estimate products of fractions using compatible numbers and rounding.

New Vocabulary

compatible numbers

Math Online

glencoe.com

- Extra Examples
- Personal Tutor
- Self-Check Quiz

GET READY for the Lesson

NATURE A wildlife preserve has 16 tigers, of which about $\frac{1}{3}$ are male. Use 16 counters to represent the 16 tigers.

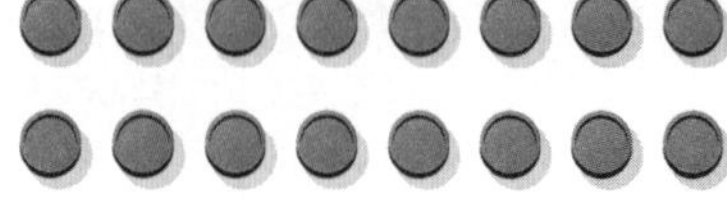

1. Can you separate the counters into three groups so that each group has the same number of counters? Explain.
2. What multiple of 3 is closest to 16?
3. About how many tigers in the preserve are male? Explain.

One way to estimate products involving fractions is to use **compatible numbers**, or numbers that are easy to divide mentally.

EXAMPLES Estimate Using Compatible Numbers

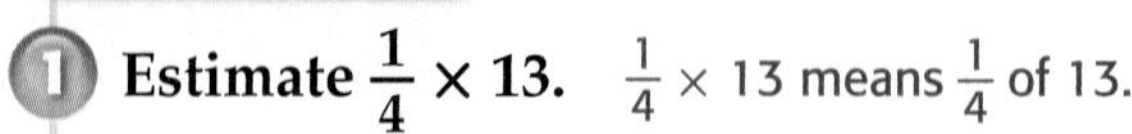

1 Estimate $\frac{1}{4} \times 13$. $\frac{1}{4} \times 13$ means $\frac{1}{4}$ of 13.

Find a multiple of 4 close to 13.

$\frac{1}{4} \times 13 \approx \frac{1}{4} \times 12$ 12 and 4 are compatible numbers since $12 \div 4 = 3$.

≈ 3 $12 \div 4 = 3$.

So, $\frac{1}{4} \times 13$ is *about* 3.

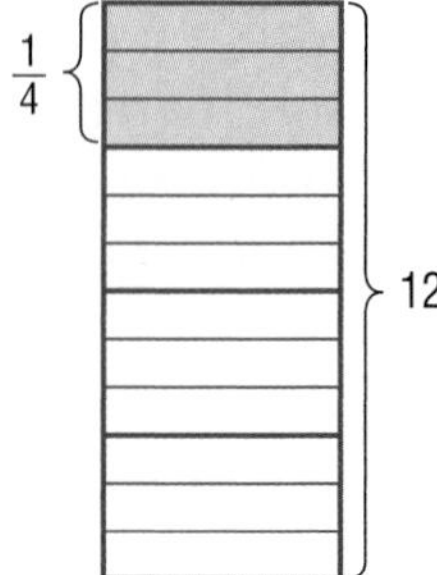

2 Estimate $\frac{2}{5}$ of 11.

$\frac{1}{5} \times 11 \approx \frac{1}{5} \times 10$ Use 10 since 10 and 5 are compatible numbers.

≈ 2 $10 \div 5 = 2$

If $\frac{1}{5}$ of 10 is 2, then $\frac{2}{5}$ of 10 is 2×2, or 4.

So, $\frac{2}{5} \times 11$ is *about* 4.

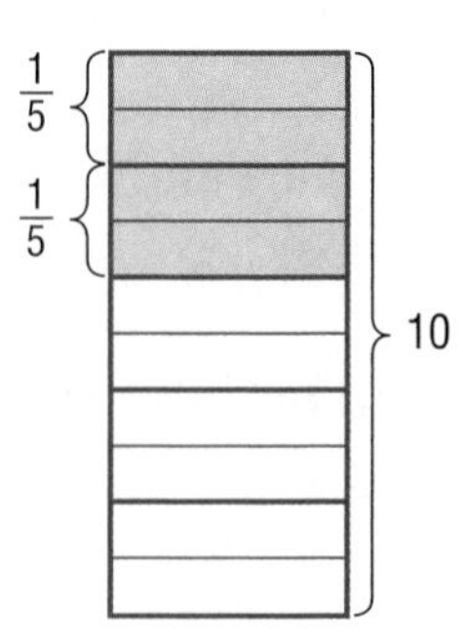

CHECK Your Progress

Estimate each product.

a. $\frac{1}{5} \times 16$

b. $\frac{5}{6} \times 13$

c. $\frac{3}{4}$ of 23

EXAMPLE Estimate by Rounding to 0, $\frac{1}{2}$, or 1

3 **Estimate $\frac{1}{3} \times \frac{7}{8}$.**

$\frac{1}{3} \times \frac{7}{8} \rightarrow \frac{1}{2} \times 1$

$\frac{1}{2} \times 1 = \frac{1}{2}$

So, $\frac{1}{3} \times \frac{7}{8}$ is *about* $\frac{1}{2}$.

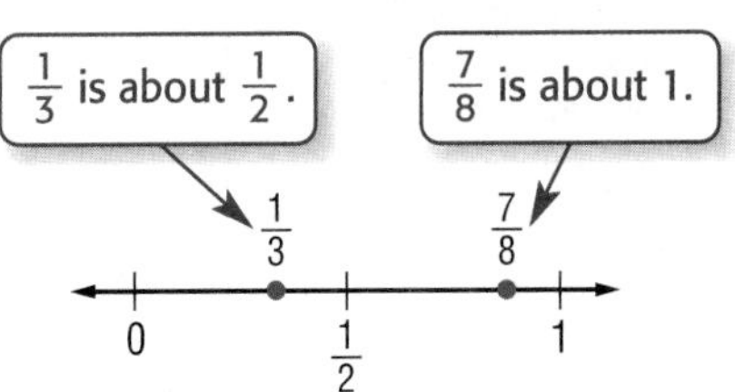

CHECK Your Progress

Estimate each product.

d. $\frac{5}{8} \times \frac{9}{10}$

e. $\frac{5}{6} \times \frac{9}{10}$

f. $\frac{5}{6}$ of $\frac{1}{9}$

EXAMPLE Estimate With Mixed Numbers

> **Study Tip**
> **Look Back** You can review **rounding fractions** in Lesson 5-1.

4 **MEASUREMENT Estimate the area of the flower bed.**

Round each mixed number to the nearest whole number.

$14\frac{7}{8} \times 6\frac{1}{3} \quad \rightarrow \quad 15 \times 6 = 90$

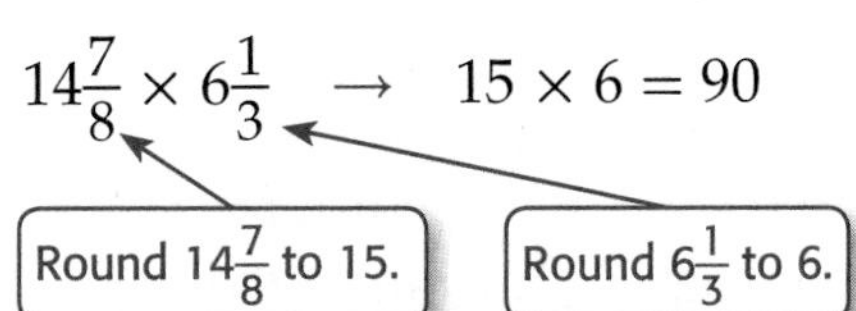

So, the area is *about* 90 square feet.

CHECK Your Progress

g. **MEASUREMENT** A border is made up of $32\frac{2}{3}$ bricks that are $1\frac{1}{6}$ feet long. About how long is the border?

CHECK Your Understanding

Examples 1–4 (pp. 276–277)

Estimate each product.

1. $\frac{1}{8} \times 15$
2. $\frac{3}{4} \times 21$
3. $\frac{2}{5}$ of 26
4. $\frac{1}{10}$ of 68
5. $\frac{1}{4} \times \frac{8}{9}$
6. $\frac{5}{8} \times \frac{1}{9}$
7. $6\frac{2}{3} \times 4\frac{1}{5}$
8. $\frac{9}{10} \times 10\frac{3}{4}$

Example 4 (p. 277)

9. **MEASUREMENT** Hakeem's front porch measures $9\frac{3}{4}$ feet by 4 feet. Estimate the area of his front porch.
10. **MEASUREMENT** A kitchen measures $24\frac{1}{6}$ feet by $9\frac{2}{3}$ feet. Estimate the area of the kitchen.

Practice and Problem Solving

HOMEWORK HELP

For Exercises	See Examples
11–20	1, 2
21–28	3
29, 30	4

Estimate each product.

11. $\frac{1}{4} \times 21$
12. $\frac{1}{5} \times 26$
13. $\frac{1}{3}$ of 41
14. $\frac{1}{6}$ of 17
15. $\frac{5}{7}$ of 22
16. $\frac{2}{9}$ of 88
17. $\frac{2}{3} \times 10$
18. $\frac{3}{8} \times 4$
19. **PIZZA** Cyrus is inviting 11 friends over for pizza. He would like to have enough pizza so each friend can have $\frac{1}{4}$ of a pizza. About how many pizzas should he order?
20. **BOOKS** Tara would like to finish $\frac{2}{5}$ of her book by next Friday. If the book has 203 pages, about how many pages does she need to read?

Estimate each product.

21. $\frac{5}{7} \times \frac{1}{9}$
22. $\frac{1}{10} \times \frac{7}{8}$
23. $\frac{11}{12} \times \frac{3}{8}$
24. $\frac{2}{5} \times \frac{9}{10}$
25. $4\frac{1}{3} \times 2\frac{3}{4}$
26. $6\frac{4}{5} \times 4\frac{1}{9}$
27. $5\frac{1}{8} \times 9\frac{1}{12}$
28. $2\frac{9}{10} \times 8\frac{5}{6}$

Estimate the area of each rectangle.

29. 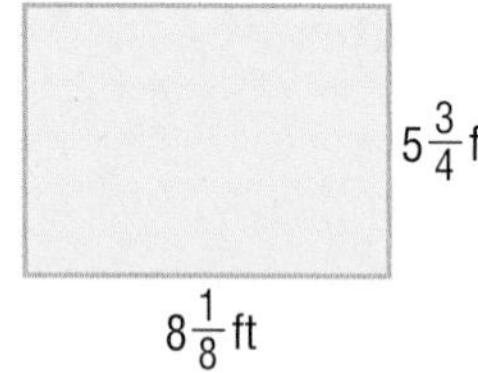

30. 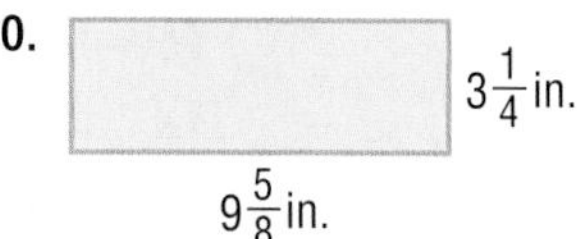

31. **VACATION** The circle graph shows when people pack for a vacation. Suppose 58 people were surveyed. About how many people pack the day they leave?
32. **MEASUREMENT** A wall measures $8\frac{1}{2}$ feet by $12\frac{3}{4}$ feet. If a gallon of paint covers about 150 square feet, will one gallon of paint be enough to cover the wall? Explain.

When Vacationers Pack

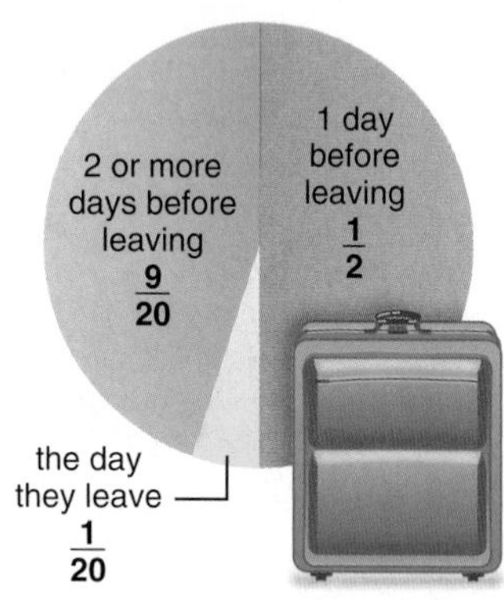

Source: Carlson Wagonlit Travel Survey

BAKING **For Exercises 33 and 34, use the shown that Angelina is using to make the cake.**

33. A cup of walnuts weighs about 8 ounces. About how many ounces are called for in the recipe?
34. If Angelina wants to make 3 cakes, about how many cups of milk will she need?
35. **WEATHER** Seattle, Washington, received rain on $\frac{7}{10}$ of the days in a recent month. If this pattern continues, about how many days would it rain in 90 days?

See pages 686, 710.

H.O.T. Problems

36. **SELECT A TECHNIQUE** Which of the following techniques could you use to easily determine whether an answer is reasonable to the multiplication of $4\frac{10}{11}$ by $7\frac{1}{13}$? Justify your response.

mental math | number sense | estimation

37. **CHALLENGE** Determine which point on the number line could be the graph of the product of the numbers graphed at *C* and *D*. Explain your reasoning.

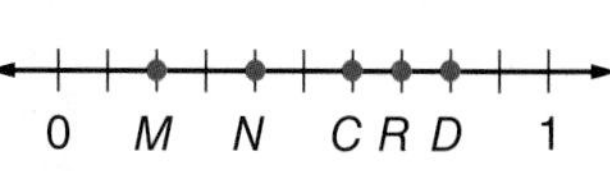

38. **WRITING IN MATH** Write a real-world problem that can be solved by estimating $\frac{3}{5} \times 21$. Then solve using compatible numbers.

TEST PRACTICE

39. Which is the best estimate of the area of the rectangle?

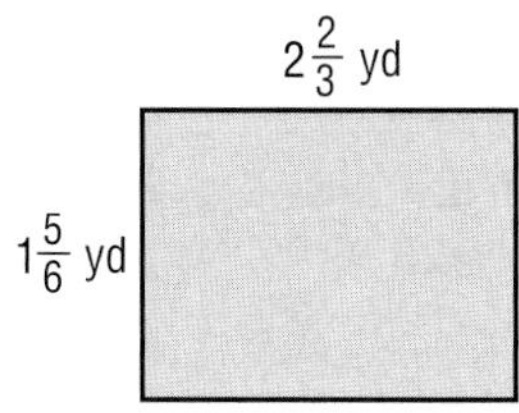

A 2 yd²
B 3 yd²
C 4 yd²
D 6 yd²

40. A total of 133 sixth-grade students went to a local museum. Of these, between one half and three fourths packed their lunch. Which of the following ranges could represent the number of students who packed their lunch?

F Less than 65
G Between 65 and 100
H Between 100 and 130
J More than 130

Spiral Review

41. **COOKING** A recipe for enchiladas calls for $1\frac{1}{4}$ pounds of ground beef and $\frac{1}{3}$ pound of cheddar cheese. How much more ground beef is needed than cheese? (Lesson 5-5)

Add or subtract. Write in simplest form. (Lesson 5-4)

42. $\frac{2}{3} + \frac{4}{5}$
43. $\frac{8}{9} - \frac{1}{3}$
44. $\frac{5}{6} + \frac{5}{12}$
45. $\frac{9}{10} - \frac{1}{2}$

46. **FOOD** Three people equally share 7.5 ounces of juice. How much juice does each receive? (Lesson 3-8)

47. **MEASUREMENT** Find the area of a rectangle with a length of 17 yards and a width of 42 yards. (Lesson 1-9)

GET READY for the Next Lesson

PREREQUISITE SKILL Find the GCF of each set of numbers. (Lesson 4-1)

48. 6, 9
49. 10, 4
50. 15, 9
51. 24, 16

Explore 5-7

Math Lab
Multiplying Fractions

MAIN IDEA

Multiply fractions using models.

In Explore 3-7, you used decimal models to multiply decimals. You can use a similar model to multiply fractions.

ACTIVITY

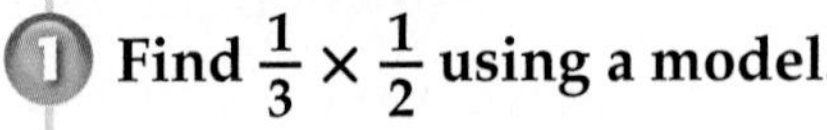

1 Find $\frac{1}{3} \times \frac{1}{2}$ using a model.

To find $\frac{1}{3} \times \frac{1}{2}$, find $\frac{1}{3}$ of $\frac{1}{2}$.

Begin with a square to represent 1.

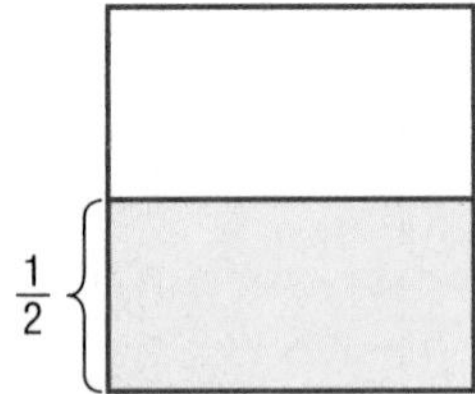

Shade $\frac{1}{2}$ of the square yellow.

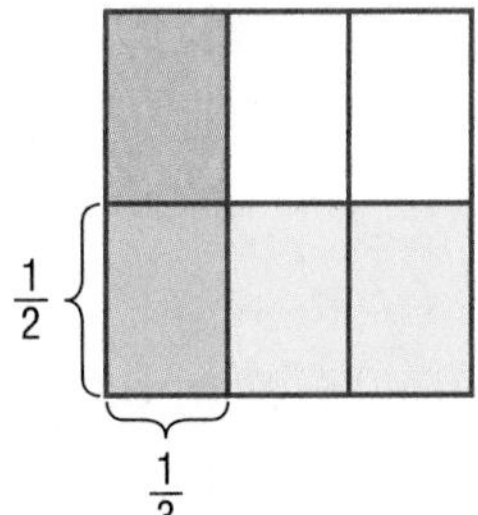

Shade $\frac{1}{3}$ of the square blue. The part that was shaded both yellow and blue appears green.

One sixth of the square is shaded green. So, $\frac{1}{3} \times \frac{1}{2} = \frac{1}{6}$.

CHECK Your Progress

Find each product using a model.

a. $\frac{1}{4} \times \frac{1}{2}$

b. $\frac{1}{3} \times \frac{1}{4}$

c. $\frac{1}{2} \times \frac{1}{5}$

Analyze the Results

1. Describe how you would change the model to find $\frac{1}{2} \times \frac{1}{3}$. Is the product the same as $\frac{1}{3} \times \frac{1}{2}$? Explain.

ACTIVITY

2 **Find $\frac{3}{5} \times \frac{2}{3}$ using a model. Write in simplest form.**

To find $\frac{3}{5} \times \frac{2}{3}$, find $\frac{3}{5}$ of $\frac{2}{3}$.

Begin with a square to represent 1.

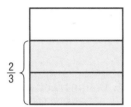

Shade $\frac{2}{3}$ of the square yellow.

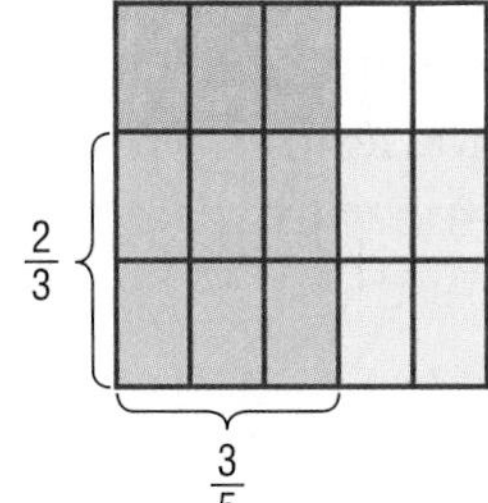

Shade $\frac{3}{5}$ of the square blue.

Six out of 15 parts are shaded green. So, $\frac{3}{5} \times \frac{2}{3} = \frac{6}{15}$ or $\frac{2}{5}$.

Study Tip

Multiplying Fractions
Finding $\frac{3}{5} \times \frac{2}{3}$ is the same as finding $\frac{2}{3} \times \frac{3}{5}$. So, you could also begin by shading $\frac{3}{5}$ of the square yellow and then $\frac{2}{3}$ of the square blue.

CHECK Your Progress

Find each product using a model. Then write in simplest form.

d. $\frac{3}{4} \times \frac{2}{3}$ **e.** $\frac{2}{5} \times \frac{5}{6}$ **f.** $\frac{4}{5} \times \frac{3}{8}$

ANALYZE THE RESULTS

2. Draw a model to show that $\frac{2}{3} \times \frac{5}{6} = \frac{10}{18}$. Then explain how the model shows that $\frac{10}{18}$ simplifies to $\frac{5}{9}$.
3. Explain the relationship between the numerators of the problem and the numerator of the product. What do you notice about the denominators of the problem and the denominator of the product?
4. **MAKE A CONJECTURE** Write a rule you can use to multiply fractions.

5-7 Multiplying Fractions

MAIN IDEA

Multiply fractions.

Math Online

glencoe.com

- Concepts in Motion
- Extra Examples
- Personal Tutor
- Self-Check Quiz

GET READY for the Lesson

REPTILES A chameleon's body is about $\frac{1}{2}$ the length of its tongue. A certain chameleon has a tongue that is $\frac{2}{3}$ foot long. The overlapping region represents the length of the chameleon, which is $\frac{1}{2}$ of $\frac{2}{3}$ or $\frac{1}{2} \times \frac{2}{3}$.

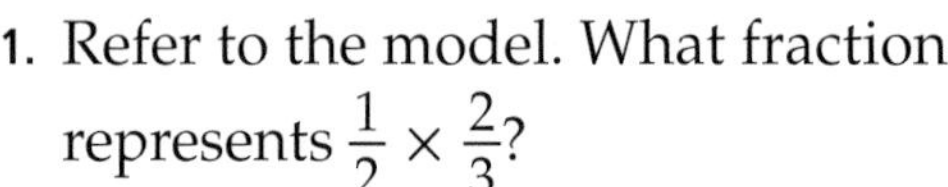

1. Refer to the model. What fraction represents $\frac{1}{2} \times \frac{2}{3}$?

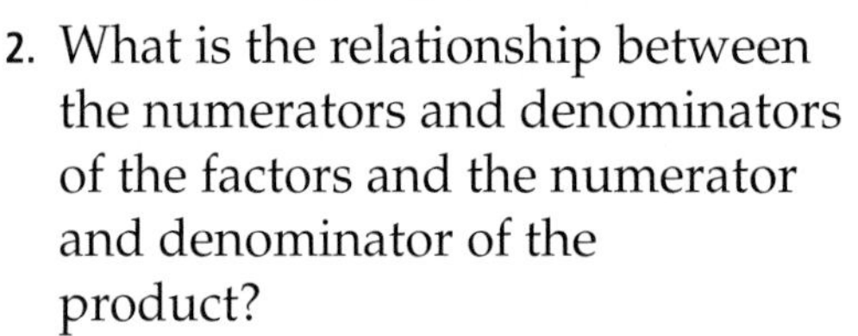

2. What is the relationship between the numerators and denominators of the factors and the numerator and denominator of the product?

Tongue is $\frac{2}{3}$ foot.

Body is $\frac{1}{2}$ of $\frac{2}{3}$.

Multiply Fractions — Key Concept

Words Multiply the numerators and multiply the denominators.

Examples

Numbers $\frac{2}{5} \times \frac{1}{2} = \frac{2 \times 1}{5 \times 2}$

Algebra $\frac{a}{b} \times \frac{c}{d} = \frac{a \times c}{b \times d}$, where b and d are not 0.

EXAMPLE Multiply Fractions

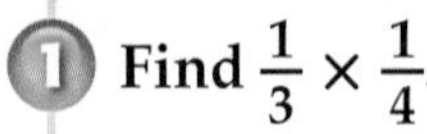

1 Find $\frac{1}{3} \times \frac{1}{4}$.

$\frac{1}{3} \times \frac{1}{4} = \frac{1 \times 1}{3 \times 4}$ Multiply the numerators. Multiply the denominators.

$= \frac{1}{12}$ Simplify.

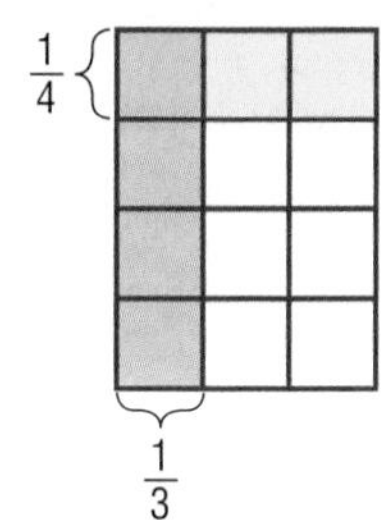

CHECK Your Progress

Multiply. Write in simplest form.

a. $\frac{1}{2} \times \frac{3}{5}$

b. $\frac{1}{3} \times \frac{3}{4}$

c. $\frac{2}{3} \times \frac{5}{6}$

To multiply a fraction and a whole number, first write the whole number as a fraction.

EXAMPLE Multiply Fractions and Whole Numbers

Find $\frac{3}{5} \times 4$. **Estimate** $\frac{1}{2} \times 4 = 2$

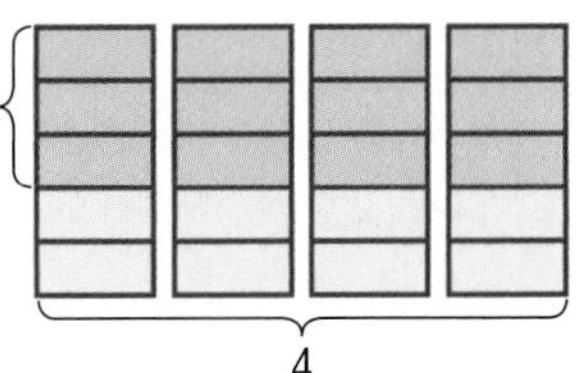

$\frac{3}{5} \times 4 = \frac{3}{5} \times \frac{4}{1}$ Write 4 as $\frac{4}{1}$.

$= \frac{3 \times 4}{5 \times 1}$ Multiply.

$= \frac{12}{5}$ or $2\frac{2}{5}$ Simplify. Compare to the estimate.

CHECK Your Progress

d. $\frac{2}{3} \times 6$

e. $\frac{3}{4} \times 5$

f. $3 \times \frac{1}{2}$

Review Vocabulary

factor two or more numbers that are multiplied together to form a product; *Example:* 1, 2, 3, and 6 are all factors of 6 (Lesson 1-2)

If the numerators and the denominators have a common factor, you can simplify *before* you multiply.

EXAMPLE Simplify Before Multiplying

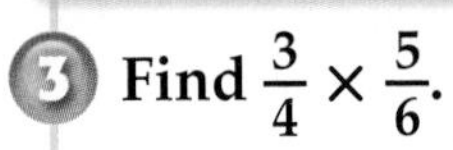

Find $\frac{3}{4} \times \frac{5}{6}$. **Estimate** $\frac{1}{2} \times 1 = \frac{1}{2}$

$\frac{3}{4} \times \frac{5}{6} = \frac{\overset{1}{\cancel{3}} \times 5}{4 \times \underset{2}{\cancel{6}}}$ Divide both the numerator and the denominator by 3.

$= \frac{5}{8}$ Simplify. Compare to the estimate.

CHECK Your Progress

g. $\frac{3}{4} \times \frac{4}{9}$

h. $\frac{5}{6} \times \frac{9}{10}$

i. $\frac{3}{5} \times 10$

EXAMPLE Evaluate Expressions

4 ALGEBRA Evaluate ab if $a = \frac{2}{3}$ and $b = \frac{3}{8}$.

$ab = \frac{2}{3} \times \frac{3}{8}$ Replace a with $\frac{2}{3}$ and b with $\frac{3}{8}$.

$= \frac{\overset{1}{\cancel{2}} \times \overset{1}{\cancel{3}}}{\underset{1}{\cancel{3}} \times \underset{4}{\cancel{8}}}$ The GCF of 2 and 8 is 2. The GCF of 3 and 3 is 3. Divide both the numerator and the denominator by 2 and then by 3.

$= \frac{1}{4}$ Simplify.

Study Tip

Mental Math You can multiply some fractions mentally. For example, $\frac{1}{3}$ of $\frac{3}{8} = \frac{1}{8}$. So, $\frac{2}{3}$ of $\frac{3}{8} = \frac{2}{8}$ or $\frac{1}{4}$.

CHECK Your Progress

j. Evaluate $\frac{3}{4}c$ if $c = \frac{2}{5}$.

k. Evaluate $5a$ if $a = \frac{3}{10}$.

CHECK Your Understanding

Examples 1–3 (pp. 282–283)

Multiply. Write in simplest form.

1. $\frac{1}{8} \times \frac{1}{2}$
2. $\frac{2}{3} \times \frac{4}{5}$
3. $\frac{4}{5} \times 10$
4. $\frac{3}{4} \times 12$
5. $\frac{3}{10} \times \frac{5}{6}$
6. $\frac{3}{5} \times \frac{5}{6}$

Example 2 (p. 283)

7. **FROGS** The male Cuban tree frog is about $\frac{2}{5}$ the size of the female Cuban tree frog. The average size of the female Cuban tree frog is shown at the right. What is the size of the male Cuban tree frog?

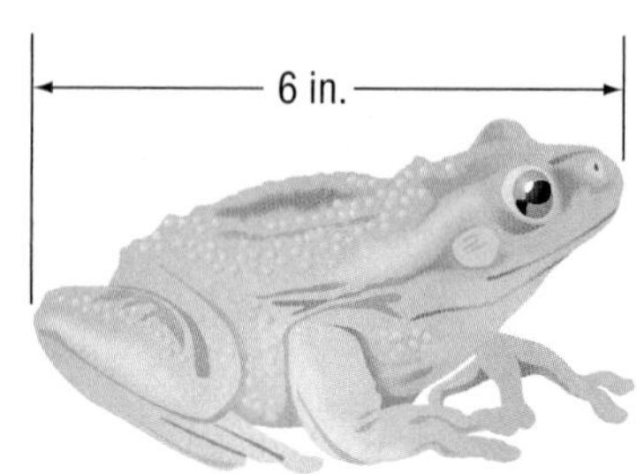

Example 4 (p. 283)

8. **ALGEBRA** Evaluate xy if $x = \frac{1}{4}$ and $y = \frac{5}{6}$.

Practice and Problem Solving

HOMEWORK HELP

For Exercises	See Examples
9–12	1
13–16, 25–28	2
17–20	3
21–24	4

Multiply. Write in simplest form.

9. $\frac{1}{3} \times \frac{2}{5}$
10. $\frac{1}{8} \times \frac{3}{4}$
11. $\frac{3}{4} \times \frac{5}{8}$
12. $\frac{2}{5} \times \frac{3}{7}$
13. $\frac{3}{4} \times 2$
14. $\frac{2}{3} \times 4$
15. $\frac{5}{6} \times 15$
16. $\frac{3}{8} \times 11$
17. $\frac{2}{3} \times \frac{1}{4}$
18. $\frac{3}{5} \times \frac{5}{7}$
19. $\frac{4}{9} \times \frac{3}{8}$
20. $\frac{2}{5} \times \frac{5}{6}$

ALGEBRA Evaluate each expression if $a = \frac{3}{5}$, $b = \frac{1}{2}$, and $c = \frac{1}{3}$.

21. ab
22. bc
23. $\frac{1}{3}a$
24. $\frac{6}{7}c$

25. **ANIMALS** A sloth spends about $\frac{4}{5}$ of its life asleep. If a sloth lives to be 28 years old, how many years did it spend asleep?

26. **RIVERS** The Mississippi River is the second longest river in the United States, second only to the Missouri River. The Mississippi River is about $\frac{23}{25}$ the length of the Missouri River. If the Missouri River is 2,540 miles long, how long is the Mississippi River?

27. **WEATHER** In a recent year, the weather was partly cloudy $\frac{2}{5}$ of the days. Assuming there are 365 days in a year, how many days were partly cloudy?

28. **PIZZA** Alvin ate $\frac{5}{8}$ of a pizza. If there were 16 slices of pizza, how many slices did Alvin eat?

Multiply.

29. $\frac{1}{2} \times \frac{1}{3} \times \frac{1}{4}$

30. $\frac{2}{3} \times \frac{3}{4} \times \frac{2}{3}$

31. $\frac{1}{2} \times \frac{2}{5} \times \frac{15}{16}$

32. $\frac{2}{3} \times \frac{9}{10} \times \frac{5}{9}$

ALGEBRA Use the order of operations to evaluate each expression if $x = \frac{4}{5}$, $y = \frac{3}{7}$, and $z = \frac{7}{10}$.

33. $\frac{2}{3}xz$

34. xyz

35. $\frac{3}{4}x + z$

36. $\frac{7}{8}y + \frac{5}{7}z$

37. **GEOGRAPHY** Michigan's area is 96,810 square miles. Water makes up about $\frac{2}{5}$ of the area of the state. About how many square miles of water does Michigan have?

38. **HEALTH** About $\frac{1}{15}$ of a pint of blood is pumped through the human body with every heartbeat. If the average human heart beats 72 times per minute, how many quarts of blood are pumped through the human body each minute? $\left(\textit{Hint}: 1 \text{ pint} = \frac{1}{2} \text{ quart}\right)$

39. **FRANCE** In a poll of the students in Lily's French class, $\frac{1}{6}$ have been to France. Of these, 4 have been to Paris. Would 18, 26, or 30 be a reasonable number of students in Lily's French class? Explain.

40. **RECYCLING** In a community survey, $\frac{13}{20}$ of residents claim to recycle. Of these, $\frac{1}{4}$ recycle only paper products. If there are 720 residents, how many residents recycle only paper products?

EXTRA PRACTICE
See pages 686, 710.

H.O.T. Problems

41. **OPEN ENDED** Create a model to explain why $\frac{2}{3} \times \frac{1}{2} = \frac{1}{3}$.

REASONING State whether each statement is *true* or *false*. If the statement is *false*, provide a counterexample.

42. The product of two fractions that are each between 0 and 1 is also between 0 and 1.

43. The product of a mixed number between 4 and 5 and a fraction between 0 and 1 is less than 4.

44. The product of two mixed numbers that are each between 4 and 5 is between 16 and 25.

45. **NUMBER SENSE** If the product of two positive fractions, a and b, is $\frac{15}{56}$, find three pairs of possible values for a and b.

46. **CHALLENGE** Is the product of two positive fractions, that are each less than 1, also less than 1? Explain.

47. **WRITING IN MATH** Explain why $\frac{a}{b} \times \frac{b}{c} \times \frac{c}{d} \times \frac{d}{e}$ is equal to $\frac{a}{e}$.

TEST PRACTICE

48. In a recent survey, $\frac{5}{8}$ of pet owners stated that they allow their pet to go outside. Of these, $\frac{1}{3}$ allow their pet outside without supervision. Which expression gives the fraction of the pet owners surveyed that allow their pet outside without supervision?

    A $\frac{5}{8} + \frac{1}{3}$

    B $\frac{5}{8} - \frac{1}{3}$

    C $\frac{5}{8} \times \frac{1}{3}$

    D $\frac{5}{8} \div \frac{1}{3}$

49. There are 150 students in the band and 90 students in the chorus. One half of the band members and $\frac{4}{5}$ of the chorus members participated in a charity concert. How many more band members than chorus members participated in the concert?

    F 3

    G 18

    H 27

    J 72

Spiral Review

Estimate each product. (Lesson 5-6)

50. $\frac{1}{6}$ of 29

51. $1\frac{8}{9} \times 5\frac{1}{6}$

52. $\frac{1}{7} \times \frac{35}{6}$

53. $\frac{4}{9} \times \frac{8}{9}$

54. **MEASUREMENT** How much longer is $\frac{7}{8}$ of a mile than $\frac{5}{6}$ of a mile? (Lesson 5-5)

55. **MAGAZINES** Samuel receives a car magazine once every four weeks, a music magazine once every six weeks, and a movie magazine once every nine weeks. If he received all three magazines this week, in how many weeks will he receive all three magazines again? (Lesson 4-5)

56. **SCHEDULING** Fatima is scheduling the five courses shown at the right. Social studies is only offered first or second period. In how many different ways can the classes be scheduled if she has science class after art class and if English must be the first period of the day? (Lesson 4-4)

Courses
Science
Social Studies
Mathematics
Art
English

57. **RESTAURANTS** Marcus and four friends went to dinner at a local restaurant. The total cost of each friend's bill was $14.78, $15.24, $14.87, $15.42, and $14.75. Write these bills in order from least to greatest. (Lesson 3-2)

GET READY for the Next Lesson

PREREQUISITE SKILL **Write each mixed number as an improper fraction.** (Lesson 4-3)

58. $3\frac{1}{4}$

59. $5\frac{2}{3}$

60. $2\frac{5}{7}$

61. $6\frac{5}{8}$

5-8 Multiplying Mixed Numbers

MAIN IDEA

Multiply mixed numbers.

Math Online

glencoe.com

- Extra Examples
- Personal Tutor
- Self-Check Quiz

GET READY for the Lesson

ANATOMY The Atlantic Giant Squid has an eyeball that is about 12 times as large as the average human eyeball. If the average human eyeball is $1\frac{1}{4}$ inches across, how large is the Atlantic Giant Squid's eyeball?

1. Write a multiplication expression that shows the size of the Atlantic Giant Squid's eyeball.
2. Use repeated addition to find $12 \times 1\frac{1}{4}$. (*Hint:* $12 \times 1\frac{1}{4}$ means there are 12 groups of $1\frac{1}{4}$.)
3. Write the multiplication expression from Exercise 1 using improper fractions.
4. Multiply the improper fractions from Exercise 3. How large is the Atlantic Giant Squid's eyeball?

Multiplying mixed numbers is similar to multiplying fractions.

Multiply Mixed Numbers — Key Concept

To multiply mixed numbers, write the mixed numbers as improper fractions and then multiply as with fractions.

EXAMPLE Multiply a Fraction and a Mixed Number

1 **Find $\frac{1}{4} \times 4\frac{4}{5}$.** **Estimate** Use compatible numbers $\longrightarrow \frac{1}{4} \times 4 = 1$

$\frac{1}{4} \times 4\frac{4}{5} = \frac{1}{4} \times \frac{24}{5}$ Write $4\frac{4}{5}$ as $\frac{24}{5}$.

$= \frac{1 \times \overset{6}{\cancel{24}}}{\underset{1}{\cancel{4}} \times 5}$ Divide 24 and 4 by their GCF, 4.

$= \frac{6}{5}$ or $1\frac{1}{5}$ Simplify. Compare to the estimate.

CHECK Your Progress

Multiply. Write in simplest form.

a. $\frac{2}{3} \times 2\frac{1}{2}$

b. $\frac{3}{8} \times 3\frac{1}{3}$

c. $3\frac{1}{2} \times \frac{1}{3}$

EXAMPLE Multiply Mixed Numbers

Real-World Link The Hoover Dam, located on the Arizona-Nevada border, contains enough concrete to pave a highway, 16 feet wide, from San Francisco to New York City.

Source: U.S. Department of the Interior

2 **DAMS Hoover Dam contains $4\frac{1}{2}$ million cubic yards of concrete. The Grand Coulee Dam, in Washington state, contains $2\frac{2}{3}$ times as much concrete. How much concrete does it contain?**

Estimate $4 \times 3 = 12$

$4\frac{1}{2} \times 2\frac{2}{3} = \frac{9}{2} \times \frac{8}{3}$ Write the mixed numbers as improper fractions.

$= \frac{\overset{3}{\cancel{9}}}{\underset{1}{\cancel{2}}} \times \frac{\overset{4}{\cancel{8}}}{\underset{1}{\cancel{3}}}$ Divide 9 and 3 by their GCF, 3. Then divide 8 and 2 by their GCF, 2.

$= \frac{3}{1} \times \frac{4}{1}$ Multiply the numerators and multiply the denominators.

$= \frac{12}{1}$ or 12 Simplify.

There are 12 million cubic yards of concrete in the Grand Coulee Dam.

CHECK Your Progress

d. **MEASUREMENT** Mr. Wilkins is laying bricks to make a rectangular patio. The area he is covering with bricks is $15\frac{1}{2}$ feet by $9\frac{3}{4}$ feet. What is the area of the patio?

EXAMPLE Evaluate Expressions

3 **ALGEBRA If $c = 1\frac{7}{8}$ and $d = 3\frac{1}{3}$, what is the value of cd?**

$cd = 1\frac{7}{8} \times 3\frac{1}{3}$ Replace c with $1\frac{7}{8}$ and d with $3\frac{1}{3}$.

$= \frac{\overset{5}{\cancel{15}}}{\underset{4}{\cancel{8}}} \times \frac{\overset{5}{\cancel{10}}}{\underset{1}{\cancel{3}}}$ Divide the numerator and denominator by 3 and by 2.

$= \frac{25}{4}$ or $6\frac{1}{4}$ Simplify.

CHECK Your Progress

e. **ALGEBRA** If $a = 3\frac{1}{5}$ and $b = 2\frac{3}{4}$, what is the value of ab?

CHECK Your Understanding

Examples 1, 2 (pp. 287–288)

Multiply. Write in simplest form.

1. $\frac{1}{2} \times 2\frac{3}{8}$
2. $1\frac{1}{2} \times \frac{2}{3}$
3. $1\frac{3}{4} \times 2\frac{4}{5}$

Example 2 (p. 288)

4. **COOKING** A waffle recipe calls for $2\frac{1}{4}$ cups of flour. If Chun wants to make $1\frac{1}{2}$ times the recipe, how much flour does he need?

Example 3 (p. 288)

5. **ALGEBRA** If $x = \frac{9}{10}$ and $y = 1\frac{1}{3}$, find xy.

Practice and Problem Solving

HOMEWORK HELP

For Exercises	See Examples
6–11, 23	1
12–17, 22	2
18–21	3

Multiply. Write in simplest form.

6. $\frac{1}{2} \times 2\frac{1}{3}$
7. $\frac{3}{4} \times 2\frac{5}{6}$
8. $1\frac{7}{8} \times \frac{4}{5}$
9. $1\frac{4}{5} \times \frac{5}{6}$
10. $\frac{7}{8} \times 3\frac{1}{4}$
11. $\frac{3}{10} \times 2\frac{5}{6}$
12. $1\frac{1}{3} \times 1\frac{1}{4}$
13. $3\frac{1}{5} \times 3\frac{1}{6}$
14. $3\frac{3}{4} \times 2\frac{2}{5}$
15. $4\frac{1}{2} \times 2\frac{5}{6}$
16. $6\frac{2}{3} \times 3\frac{3}{10}$
17. $3\frac{3}{5} \times 5\frac{5}{12}$

ALGEBRA **Evaluate each expression if $a = \frac{2}{3}$, $b = 3\frac{1}{2}$, and $c = 1\frac{3}{4}$.**

18. ab
19. $\frac{1}{2}c$
20. bc
21. $\frac{1}{8}a$

22. **MEASUREMENT** A reproduction of Claude Monet's *Water-Lilies* has dimensions $34\frac{1}{2}$ inches by $36\frac{1}{2}$ inches. Find the area of the painting.

23. **ANIMALS** A three-toed sloth can travel at a speed of $\frac{3}{20}$ mile per hour. At this rate, how far can a three-toed sloth travel in $2\frac{1}{2}$ hours?

Multiply. Write in simplest form.

24. $\frac{3}{4} \times 2\frac{1}{2} \times \frac{4}{5}$
25. $1\frac{1}{2} \times \frac{2}{3} \times \frac{3}{5}$
26. $3\frac{2}{5} \times 4\frac{1}{2} \times 2\frac{2}{3}$
27. $\frac{1}{7} \times 5\frac{5}{6} \times 1\frac{1}{4}$

28. **RUNNING** Use the formula $d = rt$ to find the distance d a long-distance runner can run at a rate r of $9\frac{1}{2}$ miles per hour for time t of $1\frac{3}{4}$ hours.

ASTRONOMY **For Exercises 29–32, use the table and the following information.**

Earth is about $92\frac{9}{10}$ million miles from the Sun.

Planet	Approximate Number of Times as Far from the Sun as Earth
Venus	$\frac{3}{4}$
Mars	$1\frac{1}{2}$
Jupiter	$5\frac{1}{4}$
Saturn	$9\frac{1}{2}$

Source: *World Almanac for Kids*

29. How far is Venus from the Sun?
30. How far is Mars from the Sun?
31. How far is Jupiter from the Sun?
32. How far is Saturn from the Sun?
33. **FIND THE DATA** Refer to the Data File on pages 16–19. Choose some data and write a real-world problem in which you would multiply mixed numbers.

EXTRA PRACTICE
See pages 686, 710.

ALGEBRA **Evaluate each expression if $g = 5\frac{3}{4}$, $k = 2\frac{1}{3}$, and $h = 1\frac{7}{8}$.**

34. $gk + h$
35. gkh
36. $gh - k$

H.O.T. Problems

37. **OPEN ENDED** Find two positive mixed numbers, each greater than 1 and less than 2, with a product greater than 1 and less than 2.

38. **NUMBER SENSE** Without multiplying, determine whether the product $2\frac{1}{2} \times \frac{2}{3}$ is located on the number line at point *A*, *B*, or *C*. Explain your reasoning.

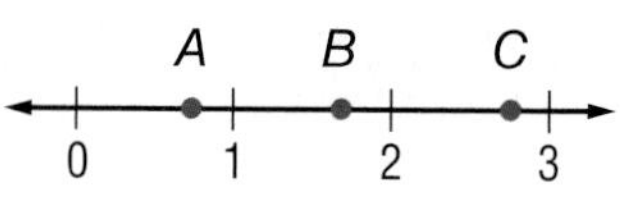

39. **CHALLENGE** Determine if the product of two positive mixed numbers is *always*, *sometimes*, or *never* less than 1. Explain your reasoning.

40. **WRITING IN MATH** Summarize how to multiply mixed numbers.

TEST PRACTICE

41. The table shows some ingredients needed to make lasagna.

mozzarella cheese	chopped onion	tomato sauce
$3\frac{1}{2}$ cups	$\frac{1}{4}$ cup	$2\frac{2}{3}$ cups

If you make four times the recipe, how many cups of tomato sauce are needed?

A $9\frac{3}{4}$ c

B $10\frac{1}{2}$ c

C $10\frac{2}{3}$ c

D $5\frac{1}{3}$ c

42. Dario buys a bag of lawn fertilizer that weighs $35\frac{3}{4}$ pounds. He wants to use $\frac{1}{2}$ of the bag on his front lawn. How many pounds of fertilizer will he use on the front lawn?

F $13\frac{5}{8}$ lb

G $17\frac{7}{8}$ lb

H $35\frac{1}{4}$ lb

J $36\frac{1}{4}$ lb

Spiral Review

Multiply. Write in simplest form. (Lesson 5-7)

43. $\frac{5}{7} \times \frac{3}{4}$

44. $\frac{2}{3} \times \frac{1}{6}$

45. $\frac{3}{8} \times \frac{2}{5}$

46. $\frac{1}{2} \times \frac{4}{7}$

47. **RECREATION** There are about 300 million people who visit a national park in the United States each year. If about $\frac{2}{5}$ come from overseas, about how many visitors come from abroad? (Lesson 5-6)

48. **FLOWERS** The table shows the number of each type of flower a florist is using to arrange flower bouquets. Each bouquet will have the same number of each type of flower. What is the greatest number of tulips that can be in each bouquet? (Lesson 4-1)

Flower	Number
roses	32
tulips	24
daisies	40

GET READY for the Next Lesson

PREREQUISITE SKILL Multiply. Write in simplest form. (Lesson 5-7)

49. $\frac{1}{4} \times \frac{3}{8}$

50. $\frac{2}{7} \times \frac{3}{4}$

51. $\frac{1}{2} \times \frac{1}{6}$

52. $\frac{2}{5} \times \frac{5}{6}$

Explore 5-9

Math Lab
Dividing Fractions

MAIN IDEA

Divide fractions using models.

There are 8 small prizes that are given away 2 at a time. How many people will get prizes?

1. How many 2s are in 8? Write as a division expression.

Suppose there are two granola bars divided equally among 8 people. What part of a granola bar will each person get?

2. What part of 8 is in 2? Write as a division expression.

ACTIVITY

1 Find $1 \div \frac{1}{5}$ using a model.

STEP 1 Make a model of the dividend, 1.

THINK How many $\frac{1}{5}$s are in 1?

STEP 2 Rename 1 as $\frac{5}{5}$ so the numbers have common denominators. So, the problem is $\frac{5}{5} \div \frac{1}{5}$. Redraw the model to show $\frac{5}{5}$.

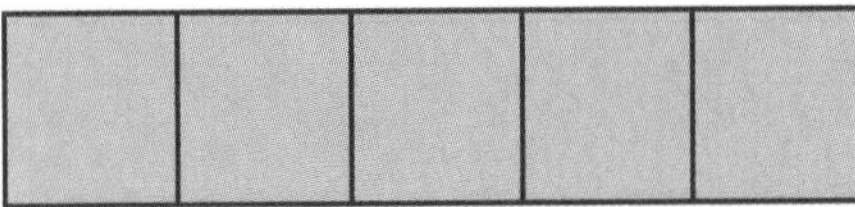

How many $\frac{1}{5}$s are in $\frac{5}{5}$?

STEP 3 Circle groups that are the size of the divisor, $\frac{1}{5}$.

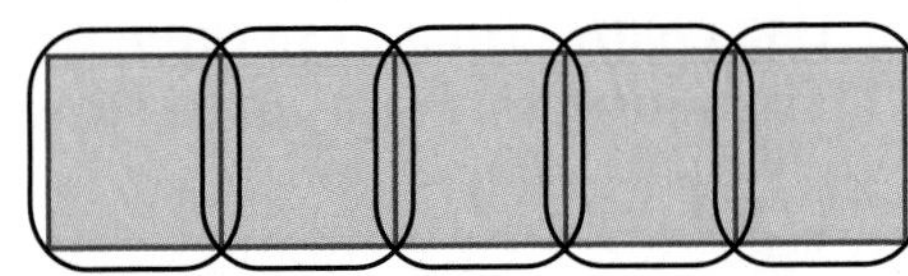

There are five $\frac{1}{5}$s in $\frac{5}{5}$.

So, $1 \div \frac{1}{5} = 5$.

CHECK Your Progress

Find each quotient using a model.

a. $2 \div \frac{1}{5}$ b. $3 \div \frac{1}{3}$ c. $3 \div \frac{2}{3}$ d. $2 \div \frac{3}{4}$

A model can also be used to find the quotient of two fractions.

ACTIVITY

2 **Find $\frac{3}{4} \div \frac{3}{8}$ using a model.**

STEP 1 Rename $\frac{3}{4}$ as $\frac{6}{8}$ so the fractions have common denominators. So, the problem is $\frac{6}{8} \div \frac{3}{8}$. Draw a model of the dividend, $\frac{6}{8}$.

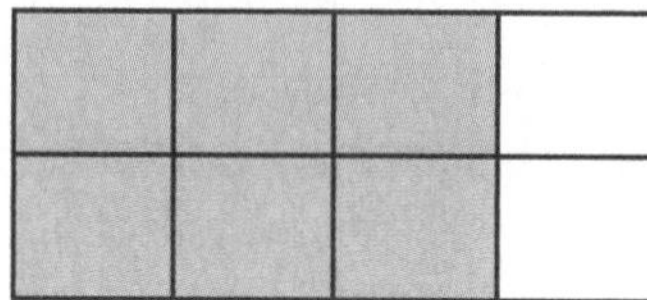

THINK How many $\frac{3}{8}$s are in $\frac{6}{8}$?

STEP 2 Circle groups that are the size of the divisor, $\frac{3}{8}$.

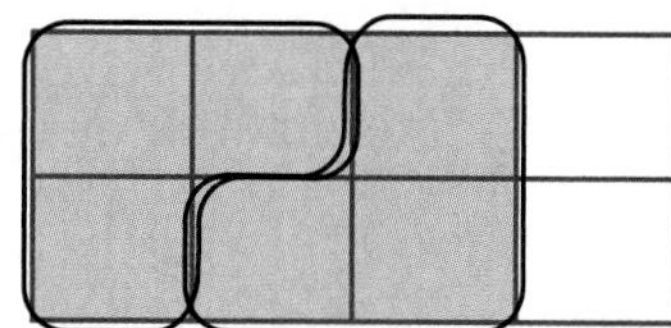

There are two $\frac{3}{8}$s in $\frac{6}{8}$.

So, $\frac{3}{4} \div \frac{3}{8} = 2$.

CHECK Your Progress

Find each quotient using a model.

e. $\frac{4}{10} \div \frac{1}{5}$ f. $\frac{3}{4} \div \frac{1}{2}$ g. $\frac{4}{5} \div \frac{1}{5}$ h. $\frac{1}{6} \div \frac{1}{3}$

Analyze the Results

Use *greater than, less than,* or *equal to* to complete each sentence. Then give an example to support your answer.

1. When the dividend is equal to the divisor, the quotient is ■ 1.
2. When the dividend is greater than the divisor, the quotient is ■ 1.
3. When the dividend is less than the divisor, the quotient is ■ 1.

4. **MAKE A CONJECTURE** You know that multiplication is commutative because the product of 3 × 4 is the same as 4 × 3. Is division commutative? Give examples to explain your answer.

5-9 Dividing Fractions

MAIN IDEA

Divide fractions.

New Vocabulary

reciprocal

Math Online

glencoe.com

- Extra Examples
- Personal Tutor
- Self-Check Quiz

MINI Lab

James and his friend Ethan ordered three one-foot submarine sandwiches. They estimate that $\frac{1}{2}$ of a sandwich will serve one person.

1. How many $\frac{1}{2}$-sandwich servings are there?
2. The model shows $3 \div \frac{1}{2}$. What is $3 \div \frac{1}{2}$?

Draw a model to find each quotient.

3. $3 \div \frac{1}{4}$
4. $2 \div \frac{1}{6}$
5. $4 \div \frac{1}{2}$

Dividing by $\frac{1}{2}$ gives the same result as multiplying by 2. The numbers $\frac{1}{2}$ and 2 have a special relationship. Their product is 1. Any two numbers with a product of 1 are called **reciprocals**.

$$3 \div \frac{1}{2} = 6 \qquad 3 \times 2 = 6$$

reciprocals: $\frac{1}{2}$ and 2; same result: 6

EXAMPLES Find Reciprocals

1 Find the reciprocal of 5.

Since $5 \times \frac{1}{5} = 1$, the reciprocal of 5 is $\frac{1}{5}$.

2 Find the reciprocal of $\frac{2}{3}$.

Since $\frac{2}{3} \times \frac{3}{2} = 1$, the reciprocal of $\frac{2}{3}$ is $\frac{3}{2}$.

CHECK Your Progress **Find the reciprocal of each number.**

a. 11

b. $\frac{3}{5}$

c. $\frac{1}{3}$

You can use reciprocals to divide fractions.

Divide Fractions — Key Concept

Words To divide by a fraction, multiply by its reciprocal.

Examples

Numbers $\frac{1}{2} \div \frac{2}{3} = \frac{1}{2} \times \frac{3}{2}$

Algebra $\frac{a}{b} \div \frac{c}{d} = \frac{a}{b} \times \frac{d}{c}$, where b, c, and $d \neq 0$

EXAMPLES Divide by a Fraction

Study Tip

Mental Math To find the reciprocal of a fraction, *invert* the fraction. That is, switch the numerator and denominator.

3 **Find $\frac{1}{8} \div \frac{3}{4}$.**

$\frac{1}{8} \div \frac{3}{4} = \frac{1}{8} \times \frac{4}{3}$ Multiply by the reciprocal, $\frac{4}{3}$.

$= \frac{1 \times \overset{1}{\cancel{4}}}{\underset{2}{\cancel{8}} \times 3}$ Divide 8 and 4 by the GCF, 4.

$= \frac{1}{6}$ Multiply numerators. Multiply denominators.

4 **Find $3 \div \frac{1}{2}$.**

$3 \div \frac{1}{2} = \frac{3}{1} \times \frac{2}{1}$ Multiply by the reciprocal of $\frac{1}{2}$.

$= \frac{6}{1}$ or 6 Simplify.

Your Progress

Divide. Write in simplest form.

d. $\frac{1}{4} \div \frac{3}{8}$

e. $\frac{2}{3} \div \frac{3}{8}$

f. $4 \div \frac{3}{4}$

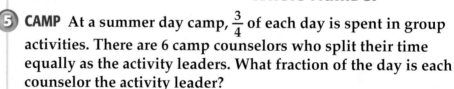

Real-World EXAMPLE Divide by a Whole Number

5 **CAMP** **At a summer day camp, $\frac{3}{4}$ of each day is spent in group activities. There are 6 camp counselors who split their time equally as the activity leaders. What fraction of the day is each counselor the activity leader?**

Divide $\frac{3}{4}$ into 6 equal parts.

$\frac{3}{4} \div 6 = \frac{3}{4} \times \frac{1}{6}$ Multiply by the reciprocal.

$= \frac{\overset{1}{\cancel{3}}}{4} \times \frac{1}{\underset{2}{\cancel{6}}}$ Divide 3 and 6 by the GCF, 3.

$= \frac{1}{8}$ Simplify.

Each camp counselor spends $\frac{1}{8}$ of his or her day at camp as the activity leader.

Real-World Link
In the United States, about 50% of campers and staff return to the same summer camp.
Source: American Camping Association

Your Progress

g. **MEASUREMENT** A neighborhood garden that is $\frac{2}{3}$ of an acre is to be divided into 4 equal-size areas. What is the size of each area?

CHECK Your Understanding

Examples 1, 2 (p. 293)

Find the reciprocal of each number.

1. $\frac{2}{3}$
2. $\frac{1}{7}$
3. $\frac{2}{5}$
4. 4

Examples 3–5 (p. 294)

Divide. Write in simplest form.

5. $\frac{1}{4} \div \frac{1}{2}$
6. $\frac{5}{6} \div \frac{1}{3}$
7. $2 \div \frac{1}{3}$
8. $5 \div \frac{2}{7}$
9. $\frac{4}{5} \div 2$
10. $\frac{5}{6} \div 3$

Example 5 (p. 294)

11. **HORSES** The average adult horse needs $\frac{2}{5}$ bale of hay each day to meet dietary requirements. A horse farm has 44 bales of hay. How many horses can be fed with 44 bales of hay in one day?

Practice and Problem Solving

HOMEWORK HELP	
For Exercises	See Examples
12–17	1, 2
18–21, 33	3
22–25, 30	4
26–29, 31, 32	5

Find the reciprocal of each number.

12. $\frac{1}{4}$
13. $\frac{1}{10}$
14. $\frac{2}{5}$
15. $\frac{7}{9}$
16. 8
17. 1

Divide. Write in simplest form.

18. $\frac{1}{8} \div \frac{1}{2}$
19. $\frac{1}{2} \div \frac{2}{3}$
20. $\frac{3}{4} \div \frac{2}{3}$
21. $\frac{3}{4} \div \frac{9}{10}$
22. $3 \div \frac{3}{4}$
23. $2 \div \frac{3}{5}$
24. $5 \div \frac{3}{4}$
25. $8 \div \frac{4}{7}$
26. $\frac{3}{5} \div 6$
27. $\frac{5}{6} \div 5$
28. $\frac{5}{8} \div 2$
29. $\frac{8}{9} \div 4$

30. **FOOD** Rafael had $\frac{3}{4}$ of a pumpkin pie left. He divided this into six equal-size slices. What fraction of the original pie was each slice?

31. **MEASUREMENT** Jamar has a piece of plywood that is $\frac{8}{9}$ yard long. He wants to cut this into 3 equal-size pieces to use as small shelves in his bedroom. What will be the length of each of these shelves?

32. **TIME** Chelsea devoted $\frac{3}{8}$ of her day to run errands, exercise, visit with her friends, and go shopping. If she devotes an equal amount of time to each of these four activities, what fraction of her day is spent on each activity?

33. **MEASUREMENT** A piece of string is to be cut into equal-size pieces. If the length of the string is $\frac{11}{12}$ foot long and each piece is to be $\frac{1}{24}$ foot long, how many pieces can be cut?

CRAFTS For Exercises 34 and 35, refer to the following information.

To tie-dye one T-shirt, $\frac{3}{8}$ of a cup of dye is needed. The table shows the number of cups of each color of dye in Mr. Nielson's art class.

34. Mr. Nielson notices that he is running out of orange dye. How many T-shirts can be made using only orange dye?

35. Mr. Nielson has three classes. For each class, he wants to use the same amount of red dye. How many T-shirts can be made using only the color red for each class?

Amount of Dye	
Color	**Number of Cups**
red	12
orange	$\frac{3}{4}$
yellow	2
green	$2\frac{5}{6}$
blue	8
purple	$5\frac{1}{2}$
black	6

ALGEBRA Use the order of operations to evaluate each expression if $a = \frac{2}{3}$, $b = \frac{3}{4}$, and $c = \frac{1}{2}$.

36. $a \div b$

37. $b \div c - a$

38. $a \div c$

39. $c \div b + a$

40. **DATA ANALYSIS** The expression $\left(5\frac{1}{2} + 5\frac{3}{4} + 5\frac{7}{8}\right) \div 3$ gives the mean of the numbers $5\frac{1}{2}$, $5\frac{3}{4}$, and $5\frac{7}{8}$. Use the order of operations to find the mean.

41. **FIND THE DATA** Refer to the Data File on pages 16–19. Choose some data and write a real-world problem in which you would divide fractions.

H.O.T. Problems

42. **OPEN ENDED** Find two positive fractions with a quotient of $\frac{5}{6}$.

43. **FIND THE ERROR** Raul and Tom are solving $\frac{8}{9} \div 4$. Who is correct? Explain your reasoning.

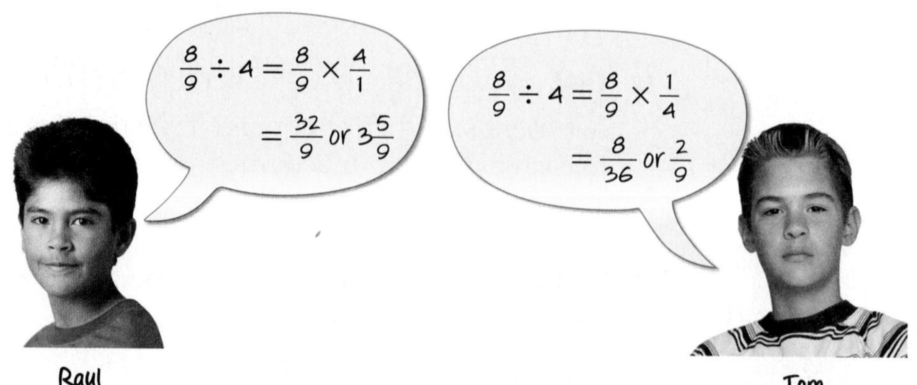

$\frac{8}{9} \div 4 = \frac{8}{9} \times \frac{4}{1}$
$= \frac{32}{9}$ or $3\frac{5}{9}$
$\frac{8}{9} \div 4 = \frac{8}{9} \times \frac{1}{4}$
$= \frac{8}{36}$ or $\frac{2}{9}$
Raul
Tom

CHALLENGE For Exercises 44 and 45, simplify each expression. Then, write one or two sentences describing each result.

44. $\frac{a}{b} \div \frac{a}{c}$

45. $\frac{a}{b} \div \frac{c}{b}$

46. **WRITING IN MATH** Create two real-world problems that involve the fraction $\frac{1}{2}$ and the whole number 3. One problem should involve multiplication, and the other should involve division.

TEST PRACTICE

47. In cooking, 1 drop is equal to $\frac{1}{6}$ of a dash. If a recipe calls for $\frac{2}{3}$ of a dash, which expression would give the number of drops that are needed?

A $\frac{1}{6} + \frac{2}{3}$

B $\frac{1}{6} - \frac{2}{3}$

C $\frac{1}{6} \times \frac{2}{3}$

D $\frac{2}{3} \div \frac{1}{6}$

48. Which of the following numbers, when divided by $\frac{1}{2}$, gives a result *less than* $\frac{1}{2}$?

F $\frac{2}{8}$

G $\frac{7}{12}$

H $\frac{2}{3}$

J $\frac{5}{24}$

Spiral Review

Multiply. Write in simplest form. (Lesson 5-8)

49. $2\frac{2}{5} \times 3\frac{1}{3}$

50. $1\frac{5}{6} \times 2\frac{3}{4}$

51. $3\frac{3}{7} \times 2\frac{3}{8}$

52. $4\frac{4}{9} \times 5\frac{1}{4}$

53. **VOLUNTEERING** According to a survey, 9 in 10 teens volunteer at least once a year. Of these, about $\frac{1}{3}$ help clean up their communities. What fraction of teens volunteer by helping clean up their communities? (Lesson 5-7)

54. **SCHOOL** Nathan, Angelina, and Carlos are each being considered for president, vice president, and secretary of Student Council. In how many ways can the three positions be filled by these students? Use the *make an organized list* strategy. (Lesson 4-4)

GET READY for the Next Lesson

PREREQUISITE SKILL Write each mixed number as an improper fraction. Then find the reciprocal of each. (Lesson 4-3)

55. $1\frac{2}{3}$

56. $1\frac{5}{9}$

57. $4\frac{1}{2}$

58. $3\frac{3}{4}$

59. $6\frac{4}{5}$

5-10 Dividing Mixed Numbers

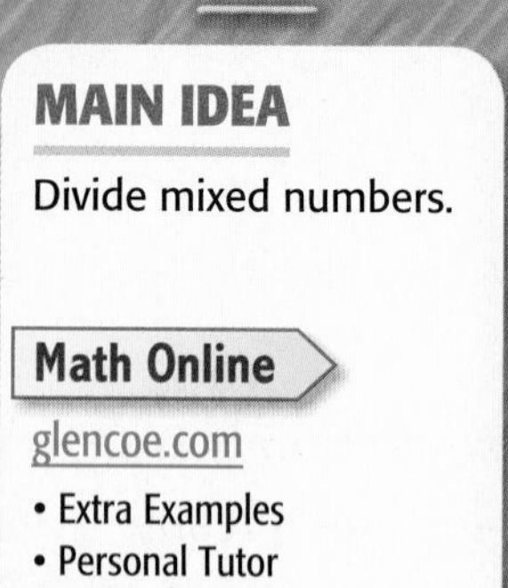

MAIN IDEA

Divide mixed numbers.

Math Online

glencoe.com

- Extra Examples
- Personal Tutor
- Self-Check Quiz

GET READY for the Lesson

DEPTH The deepest point in Earth's oceans is the Mariana Trench, which is located $6\frac{4}{5}$ miles beneath the ocean's surface. The average depth of Earth's oceans is $2\frac{1}{2}$ miles. By contrast, the highest elevation of Earth is Mt. Everest, which is about $5\frac{1}{2}$ miles high.

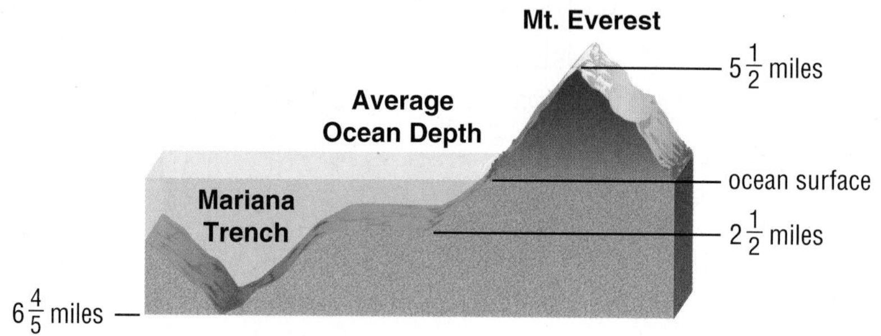

1. Write a division expression to find how many times as tall is Mt. Everest than the depth of the average ocean.
2. Write a division expression to find how many times as deep is the Mariana Trench than the average ocean on Earth.

Dividing mixed numbers is similar to dividing fractions. To divide mixed numbers, write the mixed numbers as improper fractions and then divide as with fractions.

EXAMPLE Divide by a Mixed Number

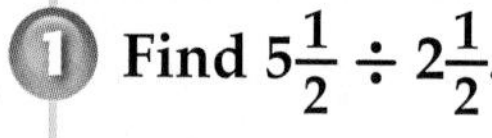

1 **Find $5\frac{1}{2} \div 2\frac{1}{2}$.**

Estimate $6 \div 3 = 2$

$5\frac{1}{2} \div 2\frac{1}{2} = \frac{11}{2} \div \frac{5}{2}$ Write mixed numbers as improper fractions.

$= \frac{11}{2} \times \frac{2}{5}$ Multiply by the reciprocal.

$= \frac{11}{\cancel{2}_1} \times \frac{\cancel{2}^1}{5}$ Divide 2 and 2 by the GCF, 2.

$= \frac{11}{5}$ or $2\frac{1}{5}$ Simplify.

Check for Reasonableness $2\frac{1}{5} \approx 2$ ✔

CHECK Your Progress **Divide. Write in simplest form.**

a. $4\frac{1}{5} \div 2\frac{1}{3}$

b. $8 \div 2\frac{1}{2}$

c. $1\frac{5}{9} \div 2\frac{1}{3}$

EXAMPLE Evaluate Expressions

2) ALBEBRA Find $m \div n$ if $m = 1\frac{3}{4}$ and $n = \frac{2}{5}$.

$m \div n = 1\frac{3}{4} \div \frac{2}{5}$ Replace m with $1\frac{3}{4}$ and n with $\frac{2}{5}$.

$= \frac{7}{4} \div \frac{2}{5}$ Write the mixed number as an improper fraction.

$= \frac{7}{4} \times \frac{5}{2}$ Multiply by the reciprocal.

$= \frac{35}{8}$ or $4\frac{3}{8}$ Simplify.

CHECK Your Progress

d. **ALGEBRA** Find $c \div d$ if $c = 2\frac{3}{8}$ and $d = 1\frac{1}{4}$.

Real-World Link
At birth, a Giant Panda is about the size of a stick of butter. The average adult male weighs about $1\frac{1}{5}$ times as much as the average adult female.

Source: San Diego Zoo

Real-World EXAMPLE

3) PANDAS Refer to the information at the left. If the average weight of a male Giant Panda is 330 pounds, how much does the average female Giant Panda weigh? **Estimate** $300 \div 1 = 300$

$330 \div 1\frac{1}{5} = \frac{330}{1} \div \frac{6}{5}$ Write the mixed number as an improper fraction.

$= \frac{330}{1} \times \frac{5}{6}$ Multiply by the reciprocal.

$= \frac{\overset{55}{\cancel{330}}}{1} \div \frac{5}{\underset{1}{\cancel{6}}}$ Divide 330 and 6 by their GCF, 6.

$= \frac{275}{1}$ or 275 Simplify.

So, the average female Giant Panda weighs about 275 pounds.

Check for Reasonableness $275 \approx 300$ ✔

CHECK Your Progress

e. **FUNDRAISING** The soccer team has $16\frac{1}{2}$ boxes of wrapping paper to sell to raise money for team T-shirts. If selling the wrapping paper is split equally among the 12 players, how many boxes should each player sell?

CHECK Your Understanding

Example 1 (p. 298)

Divide. Write in simplest form.

1. $3\frac{1}{2} \div 2$
2. $8 \div 1\frac{1}{3}$
3. $3\frac{1}{5} \div \frac{2}{7}$

Example 2 (p. 299)

4. **ALGEBRA** What is the value of $c \div d$ if $c = \frac{3}{8}$ and $d = 1\frac{1}{2}$?

Example 3 (p. 299)

5. **BAKING** Jay is cutting a roll of biscuit dough into slices that are $\frac{3}{8}$ inch thick. If the roll is $10\frac{1}{2}$ inches long, how many slices can he cut?

Practice and Problem Solving

HOMEWORK HELP

For Exercises	See Examples
6–17	1
18–23	2
24–27	3

Divide. Write in simplest form.

6. $5\frac{1}{2} \div 2$
7. $4\frac{1}{6} \div 10$
8. $3 \div 4\frac{1}{2}$
9. $6 \div 2\frac{1}{4}$
10. $6\frac{1}{2} \div \frac{3}{4}$
11. $7\frac{4}{5} \div \frac{1}{5}$
12. $6\frac{1}{2} \div 3\frac{1}{4}$
13. $8\frac{3}{4} \div 2\frac{1}{6}$
14. $3\frac{3}{5} \div 1\frac{4}{5}$
15. $3\frac{3}{4} \div 5\frac{5}{8}$
16. $4\frac{2}{3} \div 2\frac{2}{9}$
17. $6\frac{3}{5} \div 2\frac{3}{4}$

ALGEBRA **Evaluate each expression if $a = 4\frac{4}{5}$, $b = \frac{2}{3}$, $c = 6$, and $d = 1\frac{1}{2}$.**

18. $12 \div a$
19. $b \div 1\frac{2}{9}$
20. $a \div b$
21. $a \div c$
22. $c \div d$
23. $c \div (ab)$

24. **SCIENCE** A human has 46 chromosomes. This is $5\frac{3}{4}$ times the number of chromosomes of a fruit fly. How many chromosomes does a fruit fly have?

25. **MEASUREMENT** Gisele is making a scrapbook in which the pages are $13\frac{1}{2}$ inches long. She wants to place square photos across a page. Each photo is $2\frac{1}{4}$ inches long. How many photos can she place on a page? Assume there is no spacing between photos.

26. **CASHEWS** How many $\frac{3}{8}$-pound bags of cashews can be made from $6\frac{3}{8}$ pounds of cashews?

27. **BORDERS** The length of a kitchen wall is $24\frac{2}{3}$ feet long. A border will be placed along the wall of the kitchen. If the border comes in strips that are each $1\frac{3}{4}$ feet long, how many strips of border are needed?

28. **EGGS** The table gives the official U.S. weight, in minimum ounces per dozen, of eggs. How many times as large is the minimum size per dozen of jumbo eggs than the minimum size per dozen of small eggs?

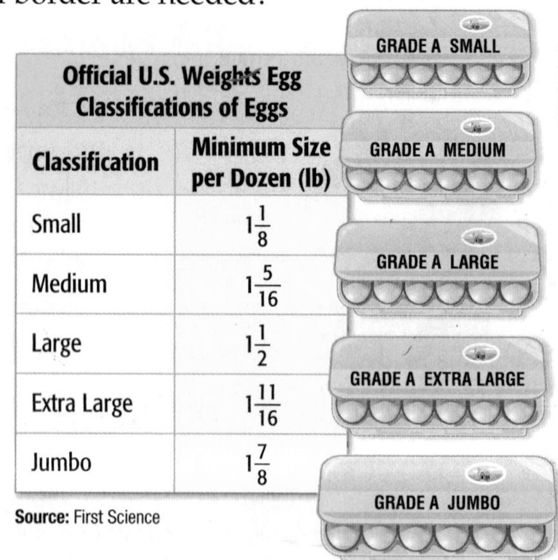

Official U.S. Weights Egg Classifications of Eggs

Classification	Minimum Size per Dozen (lb)
Small	$1\frac{1}{8}$
Medium	$1\frac{5}{16}$
Large	$1\frac{1}{2}$
Extra Large	$1\frac{11}{16}$
Jumbo	$1\frac{7}{8}$

Source: First Science

HURRICANES **For Exercises 29 and 30, use the following information.**

Suppose a hurricane traveled 130 miles from a point in the Atlantic Ocean to the Florida coastline in $6\frac{1}{2}$ hours.

29. How many miles per hour did the hurricane travel?

30. How far would the hurricane travel in $1\frac{1}{2}$ hours at the same speed?

EXTRA PRACTICE

See pages 687, 710.

H.O.T. Problems

31. **OPEN ENDED** Find two mixed numbers with a quotient of $2\frac{2}{3}$.

32. **Which One Doesn't Belong?** Select the expression that has a quotient greater than 1. Explain your reasoning.

$4\frac{2}{3} \div 5\frac{1}{4}$ $\quad$ $3\frac{1}{8} \div 2\frac{2}{5}$ $\quad$ $1\frac{6}{7} \div 2\frac{1}{3}$ $\quad$ $5\frac{3}{4} \div 7\frac{3}{8}$

33. **CHALLENGE** Without dividing, explain whether $5\frac{1}{6} \div 3\frac{5}{8}$ is greater than or less than $5\frac{1}{6} \div 2\frac{2}{5}$.

34. **WRITING IN MATH** In your own words, explain how to find the quotient of 12 and $2\frac{2}{3}$.

TEST PRACTICE

35. The largest meteorite crater is in Winslow, Arizona, with a depth of about $\frac{2}{50}$ mile and a distance across of about $\frac{4}{5}$ mile. How many times greater is the distance across the meteorite than its depth?

A about 20
B about $15\frac{1}{2}$
C about $5\frac{1}{2}$
D about 5

36. Lola used $1\frac{1}{2}$ cups of dried apricots to make $\frac{5}{6}$ of her trail mix. How many more cups of dried apricots does she need to finish making her trail mix?

F $\frac{3}{10}$ c
G $\frac{1}{2}$ c
H $\frac{5}{9}$ c
J $\frac{2}{3}$ c

Spiral Review

37. **MEASUREMENT** If a quart is $\frac{1}{4}$ of a gallon and a pint is $\frac{1}{8}$ of a gallon, how much of a quart is a pint? (Lesson 5-9)

38. **TRAINS** The fastest recorded train is the TGV in France, with a speed of about 320 miles per hour. How far would this train go in $2\frac{1}{2}$ hours? (Lesson 5-8)

Multiply. Write in simplest form. (Lesson 3-8)

39. $\frac{4}{5} \times 1\frac{3}{4}$ $\quad$ 40. $2\frac{5}{8} \times \frac{2}{7}$ $\quad$ 41. $1\frac{1}{8} \times 5\frac{1}{3}$

Problem Solving in Science

Space: It's Out of This World! It's time to complete your project. Use the data and information you have gathered about the solar system to prepare a Web page or a poster. Be sure to include several graphs displaying the information you have collected.

Math Online **Unit Project at** glencoe.com

CHAPTER 5 Study Guide and Review

Math Online glencoe.com

- STUDY TO GO
- Vocabulary Review

GET READY to Study

Be sure the following Big Ideas are noted in your Foldable.

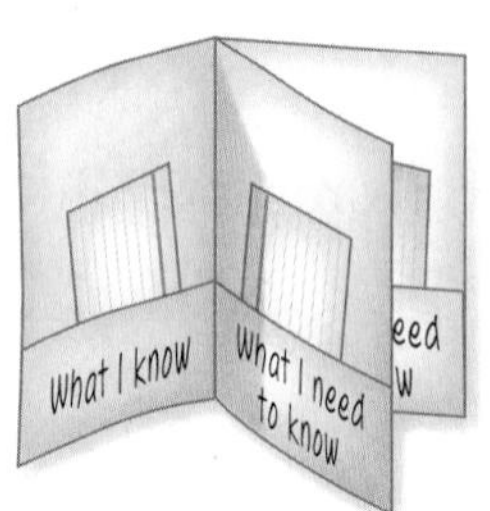

BIG Ideas

Fractions with Like Denominators (Lesson 5-3)

To add or subtract fractions with the same denominators follow these steps:

1. Add or subtract the numerators.
2. Use the same denominator in the sum or difference.
3. If necessary, simplify the sum or difference.

Fractions with Unlike Denominators (Lesson 5-4)

To add or subtract fractions with different denominators, follow these steps:

1. Rename the fractions using the least common denominator (LCD).
2. Add or subtract as with like fractions.
3. If necessary, simplify the sum or difference.

Multiplying Fractions (Lessons 5-6 to 5-8)

- To multiply fractions, multiply the numerators and multiply the denominators. Write the result in simplest form.
- To multiply mixed numbers, write the mixed numbers as improper fractions. Then multiply as with fractions. Write the result in simplest form.

Dividing Fractions (Lessons 5-9 and 5-10)

- To divide by a fraction, multiply by its reciprocal. Write the result in simplest form.
- To divide mixed numbers, write the mixed numbers as improper fractions. Then divide as with fractions. Write the result in simplest form.

Key Vocabulary

compatible numbers (p. 276)
like fractions (p. 256)
reciprocal (p. 293)
unlike fractions (p. 263)

Vocabulary Check

State whether each sentence is *true* or *false*. If *false*, replace the underlined word or number to make a true sentence.

1. When painting a wall that is $9\frac{3}{5}$- by $8\frac{1}{4}$-feet, it would make sense to round the number of gallons of paint that is needed <u>up</u> to the nearest gallon.
2. To the nearest half, $5\frac{1}{5}$ rounds to $\underline{5\frac{1}{2}}$.
3. When adding or subtracting <u>like fractions</u>, use the same denominator in the sum.
4. The product of $\frac{5}{6} \times \frac{2}{7}$ has a denominator of <u>13</u> when simplified.
5. Sometimes it is necessary to rename the fraction part of a mixed number as an <u>improper fraction</u> in order to subtract.
6. The mixed number $\underline{9\frac{1}{4}}$ can be renamed as $8\frac{5}{4}$.
7. To add or subtract fractions with unlike denominators, first rename the fractions using the <u>GCF</u>.
8. When dividing fractions, multiply by the reciprocal of the <u>first</u> fraction.
9. The LCD of $\frac{1}{8}$ and $\frac{3}{10}$ is <u>80</u>.

Lesson-by-Lesson Review

5-1 Rounding Fractions and Mixed Numbers (pp. 249–253)

Round each number to the nearest half.

10. $\frac{4}{5}$
11. $4\frac{1}{3}$
12. $6\frac{6}{14}$
13. $\frac{11}{20}$
14. $2\frac{2}{11}$
15. $9\frac{4}{9}$
16. **CELL PHONES** Maria wants to buy a carrying case for her $3\frac{1}{5}$-inch cell phone. Cases come in 3-inch, $3\frac{1}{2}$-inch, or 4-inch lengths. Which case size would provide the most appropriate fit for Maria's cell phone?

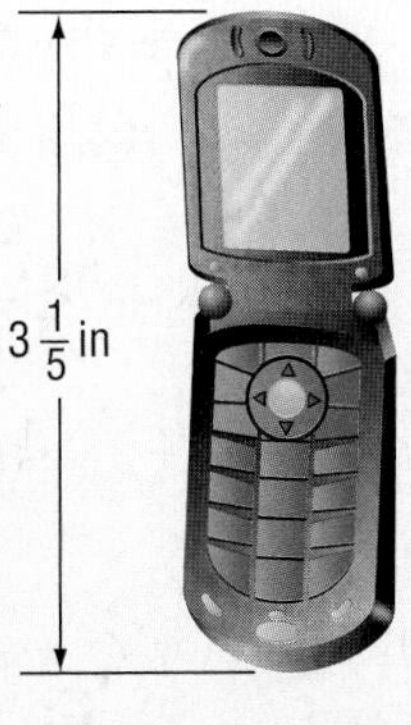

Example 1 Round $\frac{5}{8}$ to the nearest half.

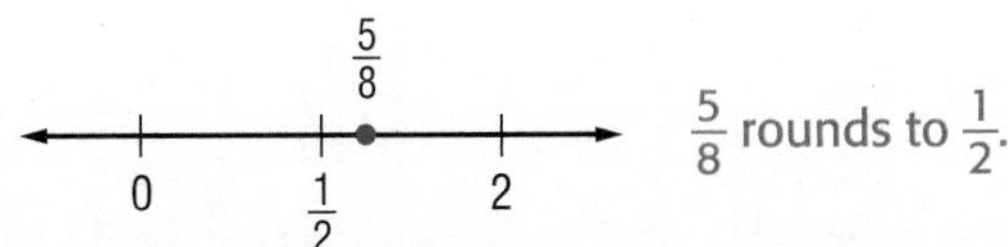

$\frac{5}{8}$ rounds to $\frac{1}{2}$.

Example 2 Round $2\frac{4}{5}$ to the nearest half.

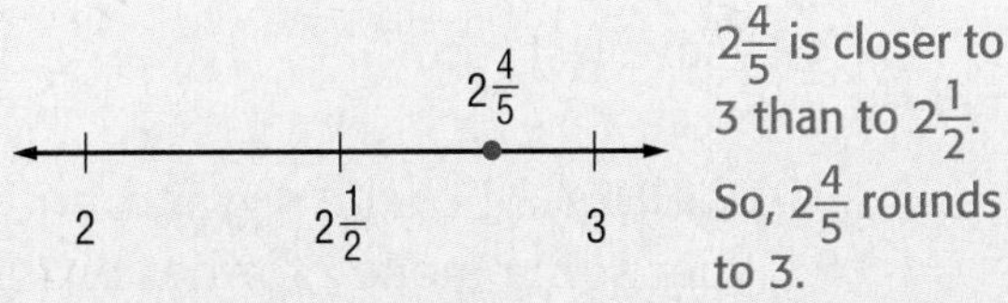

$2\frac{4}{5}$ is closer to 3 than to $2\frac{1}{2}$. So, $2\frac{4}{5}$ rounds to 3.

5-2 PSI: Act It Out (pp. 254–255)

Solve. Use the *act it out* strategy.

17. **SEATING** In how many different ways can four students be seated in a row of four seats?
18. **WALKING** Kenneth walks at the rate of 1 foot every five seconds while Ebony walks at the rate of 2 feet every five seconds. If Kenneth has a head start of 3 feet, after how many seconds will they be at the same spot?
19. **GAMES** Eight friends are seated in a circle. Juana walks around the circle and lightly taps every third person on the shoulder. How many times does she need to walk around the circle in order to have tapped each person on the shoulder at least once?

Example 3 **Drew wants to center the word *manatee* on a square sheet of paper that is 10 inches long. If the dimensions of the word are $7\frac{1}{2}$- by 2-inches, what should the top and side margins of the word be?**

Cut a $7\frac{1}{2}$- by 2-inch rectangular piece of paper with the word *manatee* written on it. Place it roughly in the center of a square sheet of paper 10 inches long. Measure the top and side margins to see if they are equal and adjust if necessary. The top margin should be 4 inches. The side margin should be $1\frac{1}{4}$ inches.

Study Guide and Review

5-3 Adding and Subtracting Fractions with Like Denominators (pp. 256–260)

Add or subtract. Write in simplest form.

20. $\frac{5}{8} + \frac{1}{8}$
21. $\frac{7}{12} + \frac{1}{12}$
22. $\frac{7}{10} + \frac{3}{10}$
23. $\frac{6}{7} - \frac{2}{7}$
24. $\frac{11}{12} - \frac{7}{12}$
25. $\frac{7}{9} + \frac{4}{9}$

26. **MEASUREMENT** How much longer is $\frac{17}{20}$ hour than $\frac{13}{20}$ hour? Write in simplest form.

27. **MONEY** Michelle's grandmother gave her some money for her birthday. Michelle saved $\frac{5}{8}$ of this amount toward the purchase of a new MP3 player and $\frac{1}{8}$ of this amount toward the purchase of a new bicycle. If she spent the rest, what fraction did she save for these two items? Write in simplest form.

Example 4

Find $\frac{3}{8} + \frac{1}{8}$. **Estimate** $\frac{1}{2} + 0 = \frac{1}{2}$

$\frac{3}{8} + \frac{1}{8} = \frac{3+1}{8}$ Add the numerators.

$= \frac{4}{8}$ or $\frac{1}{2}$ Simplify.

Example 5

Find $\frac{7}{12} - \frac{5}{12}$. **Estimate** $\frac{1}{2} - \frac{1}{2} = 0$

$\frac{7}{12} - \frac{5}{12} = \frac{7-5}{12}$ Subtract the numerators.

$= \frac{2}{12}$ or $\frac{1}{6}$ Simplify.

5-4 Adding and Subtracting Fractions with Unlike Denominators (pp. 263–268)

Add or subtract. Write in simplest form.

28. $\frac{1}{2} + \frac{2}{3}$
29. $\frac{5}{8} + \frac{1}{4}$
30. $\frac{7}{9} - \frac{1}{12}$
31. $\frac{9}{10} - \frac{1}{4}$
32. $\frac{7}{9} - \frac{1}{6}$
33. $\frac{4}{5} + \frac{2}{10}$

34. **RUNNING** Teresa ran $\frac{5}{6}$ mile while Yolanda ran $\frac{1}{4}$ mile. By what fraction did Teresa run more than Yolanda?

Example 6

Find $\frac{3}{8} + \frac{2}{3}$. **Estimate** $\frac{1}{2} + \frac{1}{2} = 1$

The LCD of $\frac{3}{8}$ and $\frac{2}{3}$ is 24.

$$\frac{3}{8} \rightarrow \frac{3 \times 3}{8 \times 3} \rightarrow \frac{9}{24}$$

$$+\frac{2}{3} \rightarrow \frac{2 \times 8}{3 \times 8} \rightarrow +\frac{16}{24}$$

$$\frac{25}{24} \text{ or } 1\frac{1}{24}$$

Mixed Problem Solving
For mixed problem-solving practice, see page 710.

5-5 Adding and Subtracting Mixed Numbers (pp. 270–274)

Add or subtract. Write in simplest form.

35. $3\frac{2}{5} + 1\frac{3}{5}$
36. $9\frac{7}{8} - 5\frac{3}{8}$
37. $7\frac{5}{6} + 9\frac{3}{4}$
38. $4\frac{3}{7} - 2\frac{5}{14}$
39. **ANIMALS** The average length of a giraffe's tongue is $1\frac{1}{2}$ feet long. The average length of a human's tongue is $\frac{1}{3}$ foot long. How much longer is the average giraffe's tongue than the average human's tongue?

Subtract. Write in simplest form.

40. $5 - 3\frac{2}{3}$
41. $6\frac{3}{8} - 3\frac{5}{6}$
42. $12\frac{2}{5} - 9\frac{2}{3}$
43. $8\frac{5}{8} - 1\frac{3}{4}$
44. **BAKING** A recipe for pumpkin bread calls for $2\frac{2}{3}$ cups of pumpkin and $3\frac{1}{4}$ cups of flour. How many more cups of flour are needed than pumpkin?

Example 7

Find $6\frac{5}{8} - 2\frac{2}{5}$. Estimate $7 - 2 = 5$

$$\begin{array}{rcccr} 6\frac{5}{8} & \rightarrow & \frac{5 \times 5}{8 \times 5} & \rightarrow & 6\frac{25}{40} \\ -\,2\frac{2}{5} & \rightarrow & \frac{2 \times 8}{5 \times 8} & \rightarrow & -\,2\frac{16}{40} \\ \hline & & & & 4\frac{9}{40} \end{array}$$

Example 8

Find $3\frac{1}{5} - 1\frac{4}{5}$. Estimate $3 - 2 = 1$

$$\begin{array}{rcr} 3\frac{1}{5} & \rightarrow & 2\frac{6}{5} \\ -\,1\frac{4}{5} & \rightarrow & -\,1\frac{4}{5} \\ \hline & & 1\frac{2}{5} \end{array}$$

Rename $3\frac{1}{5}$ as $2\frac{6}{5}$.

5-6 Estimating Products of Fractions (pp. 276–279)

Estimate each product.

45. $\frac{1}{5} \times 21$
46. $10 \times 2\frac{3}{4}$
47. $\frac{5}{6} \times 13$
48. $7\frac{3}{4} \times \frac{1}{4}$
49. $4\frac{5}{6} \times 8\frac{3}{10}$
50. $\frac{3}{7} \times \frac{11}{12}$
51. **AMUSEMENT PARKS** The average wait time to ride the Super Coaster is 55 minutes. If Joy and her friends have waited $\frac{5}{6}$ of that time, estimate how long they have waited.

Example 9 **Estimate $\frac{1}{7} \times 41$.**

$\frac{1}{7} \times 41 \longrightarrow \frac{1}{7} \times 42$ 42 and 7 are compatible numbers since $42 \div 7 = 6$.

$\frac{1}{7} \times 42 = 6$ $\frac{1}{7}$ of 42 is 6.

So, $\frac{1}{7} \times 41$ is *about* 6.

5-7 Multiplying Fractions (pp. 282–286)

Multiply. Write in simplest form.

52. $\frac{1}{3} \times \frac{1}{4}$ 53. $\frac{7}{8} \times \frac{4}{21}$ 54. $\frac{5}{6} \times 9$

55. **SCHOOL** Half of Mr. Carson's class play a sport. Of these, two thirds are male. What fraction of the class is male and play a sport?

Example 10 Find $\frac{3}{10} \times \frac{4}{9}$.

$\frac{3}{10} \times \frac{4}{9} = \frac{\overset{1}{\cancel{3}} \cdot \overset{2}{\cancel{4}}}{\underset{5}{\cancel{10}} \cdot \underset{3}{\cancel{9}}}$ Divide the numerator and denominator by the GCF.

$= \frac{2}{15}$ Simplify.

5-8 Multiplying Mixed Numbers (pp. 287–290)

Multiply. Write in simplest form.

56. $\frac{2}{3} \times 4\frac{1}{2}$ 57. $6\frac{5}{8} \times 4$ 58. $2\frac{1}{4} \times 6\frac{2}{3}$

59. **ART** Find the area of the painting.

Example 11 Find $3\frac{1}{2} \times 4\frac{2}{3}$.

$3\frac{1}{2} \times 4\frac{2}{3} = \frac{7}{2} \times \frac{14}{3}$ Write the numbers as improper fractions.

$= \frac{7}{\underset{1}{\cancel{2}}} \times \frac{\overset{7}{\cancel{14}}}{3}$ Divide 2 and 14 by their GCF, 2.

$= \frac{49}{3}$ or $16\frac{1}{3}$ Simplify.

5-9 Dividing Fractions (pp. 293–297)

Divide. Write in simplest form.

60. $\frac{2}{3} \div \frac{4}{5}$ 61. $\frac{1}{8} \div \frac{3}{4}$ 62. $5 \div \frac{4}{9}$

63. **COOKING** Ashanti uses $\frac{3}{4}$ cup of oats to make cookies. This is $\frac{1}{3}$ the amount called for in the recipe. How many cups of oats are called for in the recipe?

Example 12 Find $\frac{3}{8} \div \frac{2}{3}$.

$\frac{3}{8} \div \frac{2}{3} = \frac{3}{8} \times \frac{3}{2}$ Multiply by the reciprocal of $\frac{2}{3}$.

$= \frac{9}{16}$ Multiply the numerators and multiply the denominators.

5-10 Dividing Mixed Numbers (pp. 298–301)

Divide. Write in simplest form.

64. $2\frac{4}{5} \div 5\frac{3}{5}$ 65. $8 \div 2\frac{1}{2}$

66. **ICE CREAM** To make $4\frac{1}{2}$ gallons of ice cream, it takes $6\frac{3}{10}$ gallons of milk. How many gallons of milk does it take to make one gallon of ice cream?

Example 13 Find $5\frac{1}{2} \div 1\frac{5}{6}$.

$5\frac{1}{2} \div 1\frac{5}{6} = \frac{11}{2} \div \frac{11}{6}$ Rewrite as improper fractions.

$= \frac{11}{2} \times \frac{6}{11}$ Multiply by the reciprocal.

$= \frac{\overset{1}{\cancel{11}}}{\underset{1}{\cancel{2}}} \times \frac{\overset{3}{\cancel{6}}}{\underset{1}{\cancel{11}}}$ Divide by the GCF.

$= \frac{3}{1}$ or 3 Simplify.

Practice Test

Math Online glencoe.com
• Chapter Test

Round each number to the nearest half.

1. $4\frac{7}{8}$ 2. $1\frac{10}{18}$ 3. $11\frac{1}{17}$

4. **TRACK** For a 3-person relay race, a coach can choose from 4 of his top runners. How many different 3-person teams can he choose? Use the *act it out* strategy.

5. **MULTIPLE CHOICE** The table shows the amount of rainfall over a one-week period in May. It did not rain on any other days of the week. Find the total amount of rainfall for the week.

Day of Week	Rainfall (in.)
Monday	$1\frac{1}{4}$
Thursday	$\frac{5}{8}$
Saturday	$1\frac{5}{16}$

A $2\frac{3}{16}$ in.

B $2\frac{5}{16}$ in.

C $3\frac{3}{16}$ in.

D $3\frac{5}{16}$ in.

Add or subtract. Write in simplest form.

6. $\frac{2}{9} + \frac{5}{9}$ 7. $\frac{11}{12} - \frac{3}{8}$ 8. $\frac{2}{5} + \frac{2}{4}$

9. **CAKES** At a party, if $\frac{1}{3}$ of one sheet cake and $\frac{1}{6}$ of another sheet cake remain uneaten, what fraction of a whole sheet cake remains uneaten?

Add or subtract. Write in simplest form.

10. $2\frac{1}{5} + 4\frac{2}{5}$

11. $6\frac{5}{8} - 4\frac{1}{2}$

12. $11\frac{1}{2} - 7\frac{3}{5}$

13. **MULTIPLE CHOICE** If you use $1\frac{1}{4}$ pounds of a 3-pound package of ground beef and freeze the rest, how much ground beef do you freeze?

F $2\frac{3}{4}$ lb H $1\frac{1}{4}$ lb

G $1\frac{3}{4}$ lb J $\frac{3}{4}$ lb

Estimate each product.

14. $\frac{1}{3} \times 22$ 15. $3\frac{2}{3} \times 5\frac{1}{9}$

16. $\frac{7}{8} \times 39$ 17. $6\frac{4}{5} \times 8\frac{1}{7}$

Multiply. Write in simplest form.

18. $\frac{3}{5} \times \frac{2}{9}$

19. $\frac{3}{8} \times 2\frac{2}{3}$

20. $7\frac{7}{8} \times 5\frac{1}{3}$

21. **GEOMETRY** To find the area of a parallelogram, use the formula $A = bh$, where b is the length of the base and h is the height. Find the area of the parallelogram.

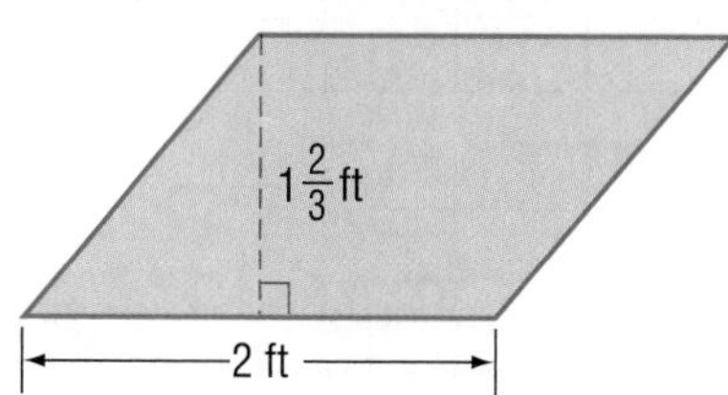

Divide. Write in simplest form.

22. $\frac{1}{8} \div \frac{3}{4}$

23. $\frac{2}{5} \div 4$

24. $5\frac{3}{4} \div 1\frac{1}{2}$

25. **ALGEBRA** Evaluate $x \div y$ if $x = 7\frac{2}{3}$ and $y = 1\frac{4}{5}$. Write in simplest form.

CHAPTER 5

Test Practice

Cumulative, Chapters 1–5

Math Online glencoe.com

• Test Practice

PART 1 Multiple Choice

Read each question. Then fill in the correct answer on the answer sheet provided by your teacher or on a sheet of paper.

1. Ted is going to make three different picture frames. He will need $3\frac{1}{4}$ feet of wood for the first frame, $1\frac{2}{3}$ feet of wood for the second frame, and $2\frac{1}{2}$ feet of wood for the third frame. How much wood does Ted need for all three picture frames?

   A $6\frac{3}{4}$ feet
   B $7\frac{7}{8}$ feet
   C $7\frac{5}{12}$ feet
   D $8\frac{1}{2}$ feet

2. Trevor plans to buy rope for two projects. One project requires $\frac{5}{8}$ yard of rope, and the other requires $\frac{1}{4}$ yard of rope. Each strip below represents 1 yard of rope. Which strip is shaded to show the total amount of rope that Trevor needs for both projects?

   F 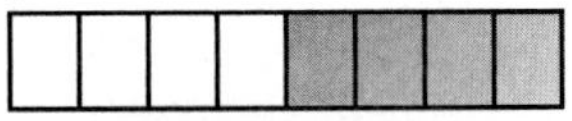

   G 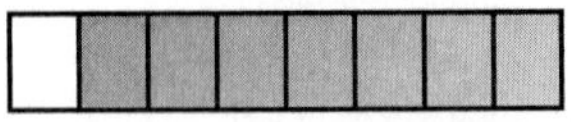

   H 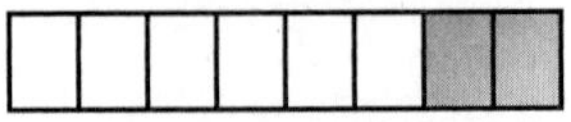

   J 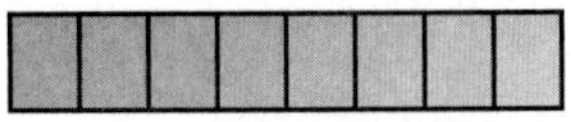

3. Two-thirds of a blueberry pie is left in the refrigerator. If the pie is cut in 6 equal-size slices, what fraction of the original pie is each slice?

   A $\frac{1}{9}$
   B $\frac{1}{6}$
   C $\frac{1}{4}$
   D $\frac{1}{3}$

4. Annmarie bought two DVDs that were originally priced at $19 each. Each DVD was on sale for $5 off the original price of the DVD. Which equation could be used to find *c*, the total sale price of the 2 DVDs?

   F $c = 2(19) - 2(5)$
   G $c = 2(19) - 5$
   H $c = 19 - 2(5)$
   J $c = 19 - 5$

5. Keith shipped a present to his mom in a box that has a length of 9.5 inches. The width of the box is 3.4 inches less than the length. What is the width of the box?

   A 12.9 in.
   B 6.9 in.
   C 6.1 in.
   D 5.1 in.

TEST-TAKING TIP

Question 5 You can use number sense to eliminate possible answer choices. You know that the width is *less* than the length. So, eliminate answer choice A.

6. At Medina Middle School there are 53 homerooms. If 955 students attend Medina Middle School, about how many students are in each homeroom?

   F 17
   G 18
   H 19
   J 20

7. Estimate the amount of money needed to pay for the groceries shown in the table.

Shopping List	
Cereal	$2.89
Meats	$7.75
Lettuce	$1.29
Detergent	$5.89

   A $21
   B $20
   C $19
   D $18

Preparing for Standardized Tests

For test-taking strategies and practice, see pages 718–735.

8. Jordan ran $2\frac{3}{4}$ miles on Monday. On Wednesday, he ran twice as many miles than he did on Monday. On Friday, Jordan ran $1\frac{1}{2}$ times as many miles than he did on Wednesday. How many miles did he run on Friday?

   F $8\frac{3}{4}$ miles

   G $8\frac{1}{4}$ miles

   H $6\frac{1}{2}$ miles

   J $5\frac{3}{4}$ miles

9. Evelyn must buy plastic bowls and plastic spoons for an ice cream party. Bowls are sold in packages of 16 and spoons in packages of 24. What is the least number of packages of bowls and spoons that Evelyn can buy to have an equal number of bowls and spoons?

   A 5 packages of bowls and 3 packages of spoons

   B 2 packages of bowls and 3 packages of spoons

   C 3 packages of bowls and 2 packages of spoons

   D 4 packages of bowls and 4 packages of spoons

10. Drew read 120 pages of his book, which was $\frac{3}{5}$ of the book. What decimal represents the fraction that he has read?

    F 0.12 H 0.60

    G 0.35 J 0.80

PART 2 Short Response/Grid In

Record your answers on the answer sheet provided by your teacher or on a sheet of paper.

11. Mr. Thompson recorded these geography quiz scores: 23, 21, 19, 25, 24, 15, 18, 19, and 23. What is the median of these quiz scores?

12. Lamar recorded his times for several 100-meter dash trials in seconds. Which was the fastest time in seconds?

    8.9 s, 8.64 s, 8.45 s, 8.5 s, 8.42 s

PART 3 Extended Response

Record your answers on the answer sheet provided by your teacher or on a sheet of paper. Show your work.

13. Misha planted a tomato garden with the following dimensions.

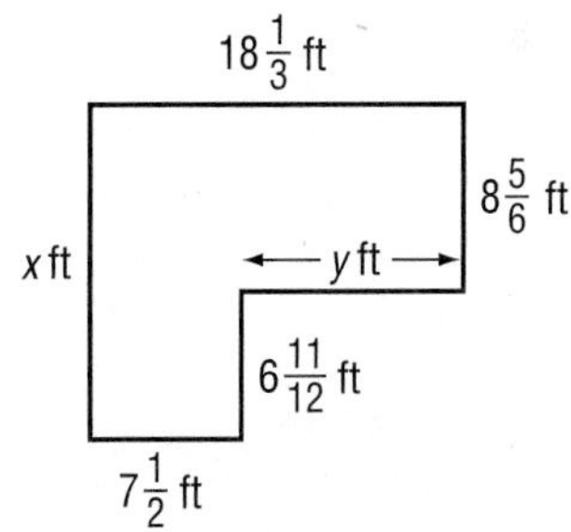

   a. Find the value of x as a fraction in simplest form.

   b. Find the value of y as a fraction in simplest form.

   c. Misha wishes to enclose the garden with a fence. How many feet of fencing would be needed to enclose the entire garden? Write as a fraction in simplest form.

NEED EXTRA HELP?													
If You Missed Question...	1	2	3	4	5	6	7	8	9	10	11	12	13
Go to Lesson...	5-5	5-4	1-3	1-8	3-5	3-3	3-4	4-6	4-5	4-8	2-7	3-2	5-5

Unit 3

Patterns, Relationships, and Algebraic Thinking

Focus

Connect ratios and rates to multiplication and division, use ratios to solve problems involving percent and probability, and write mathematical expressions and equations.

CHAPTER 6
Ratio, Proportion, and Functions

BIG Idea Solve problems involving ratios and rates.

BIG Idea Write mathematical expressions and equations.

CHAPTER 7
Percent and Probability

BIG Idea Solve problems involving percent and probability.

BIG Idea Solve problems involving ratios.

Problem Solving in Social Studies

Real-World Unit Project

The Nifty Fifty States Is New Jersey more crowded than Colorado? Are there more people per square mile in Ohio than in Michigan? Which states have a higher percentage of youth than older adults? Which states have a higher ratio of women compared to men? You have been selected to compile data on the population profiles of each of the fifty states. You'll be asked to write and compare ratios and find percents. Then you'll *state* your findings in a report about the nifty fifty states. Are you in the right *state* of mind? Let's begin!

Math Online Log on to glencoe.com to begin.

Ratio, Proportion, and Functions

BIG Ideas

- Solve problems involving ratios and rates.
- Write mathematical expressions and equations.

Key Vocabulary

proportion (p. 329)
proportional (p. 329)
ratio (p. 314)

Real-World Link

Dunes The Sleeping Bear Dunes National Lakeshore runs for 35 miles along the Michigan coastline. There are several types of park passes available including a \$5 per person fee valid for 7 days, a \$10 per vehicle fee valid for 7 days, and a \$20 per vehicle annual fee.

FOLDABLES® Study Organizer

Ratio, Proportion, and Functions Make this Foldable to help you organize your notes. Begin with a piece of graph paper.

1. **Fold** one sheet of graph paper in thirds lengthwise.

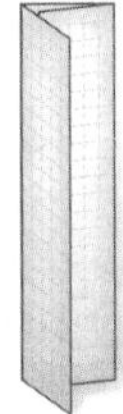

2. **Unfold** lengthwise and fold one-fourth down widthwise. Cut to make three tabs as shown.

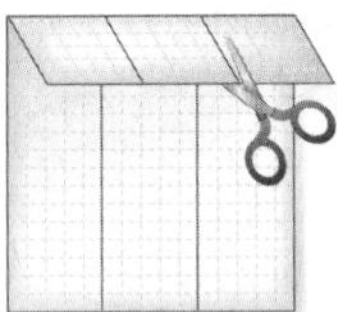

3. **Unfold** the tabs. Label the paper as shown.

Definition & Notes | Definition & Notes | Definition & Notes
Examples | Examples | Examples

4. **Refold** the tabs and label as shown.

Ratio | Proportion | Function
Examples | Examples | Examples

GET READY for Chapter 6

Diagnose Readiness You have two options for checking Prerequisite Skills.

Option 2

Math Online Take the Online Readiness Quiz at glencoe.com.

Option 1

Take the Quick Quiz below. Refer to the Quick Review for help.

QUICK Quiz

Write each fraction in simplest form. (Lesson 4-2)

1. $\frac{32}{48}$
2. $\frac{7}{28}$
3. $\frac{15}{25}$
4. $\frac{30}{35}$
5. **TRAVEL** An airplane has flown 260 miles out of a total trip of 500 miles. What fraction, in simplest form, of the trip has been completed?

Solve each equation. (Lesson 1-8)

6. $16m = 48$
7. $5x = 40$
8. $15p = 150$
9. $3n = 15$
10. $7y = 56$
11. $12z = 72$
12. $8h = 96$
13. $10e = 90$

Find the next three values in each pattern. (Lesson 1-1)

14. 4, 7, 10, 13, ...
15. 62, 66, 70, 74, ...
16. 1.8, 2.4, 3.0, 3.6, ...
17. **MUSIC** Mario played the drums for 30 minutes on Tuesday, 45 minutes on Wednesday, and 60 minutes on Thursday. At this rate, how many minutes will he play on Friday?

QUICK Review

Example 1

Write $\frac{40}{64}$ in simplest form.

$$\frac{40}{64} \underset{\div 8}{\overset{\div 8}{=}} \frac{5}{8}$$

Divide the numerator and denominator by the GCF, 8.

Since the GCF of 5 and 8 is 1, the fraction $\frac{5}{8}$ is in simplest form.

Example 2

Solve $14k = 84$ mentally.

$14k = 84$ — **THINK** 14 times what number equals 84?

$14 \cdot 6 = 84$

$84 = 84$

The solution is 6.

Example 3

Find the next three values in the pattern 5, 16, 27, 38, ...

Look for a pattern. Each number is obtained by adding 11 to the previous number.

5, 16, 27, 38, ...
(+11, +11, +11)

The next three numbers are 49, 60, and 71.

6-1 Ratios and Rates

MAIN IDEA

Express ratios and rates in fraction form.

New Vocabulary

ratio
rate
unit rate

Math Online

glencoe.com

- Extra Examples
- Personal Tutor
- Self-Check Quiz

MINI Lab

Consider the set of paper clips shown.

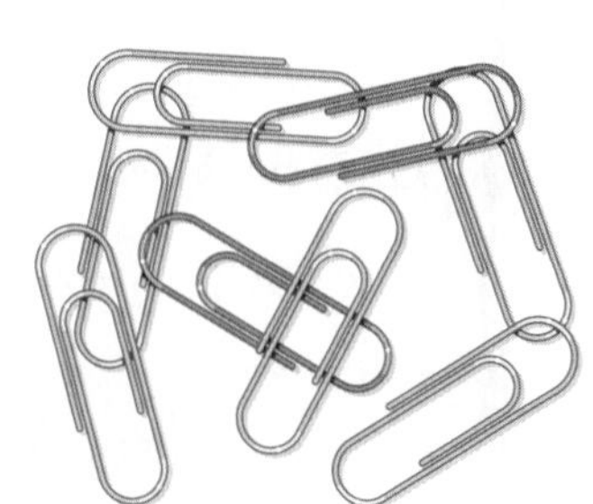

1. Compare the number of blue paper clips to the number of red paper clips using the word *more* and then using the word *times*.
2. Compare the number of red paper clips to the number of blue paper clips using the word *less* and then using a fraction.

There are many different ways to compare amounts or *quantities*. A **ratio** is a comparison of two quantities by division. A ratio of 2 red paper clips to 6 blue paper clips can be written in three ways.

$$2 \text{ to } 6 \quad 2{:}6 \quad \frac{2}{6}$$

As with fractions, ratios are often expressed in simplest form.

EXAMPLE Write a Ratio in Simplest Form

1 Write the ratio in simplest form that compares the number of red paper clips to the number of blue paper clips in the Mini Lab. Then explain its meaning.

red paper clips ⟶ 2
blue paper clips ⟶ 6

$$\frac{2}{6} = \frac{1}{3}$$ (÷2 on top and bottom) ← The GCF of 2 and 6 is 2.

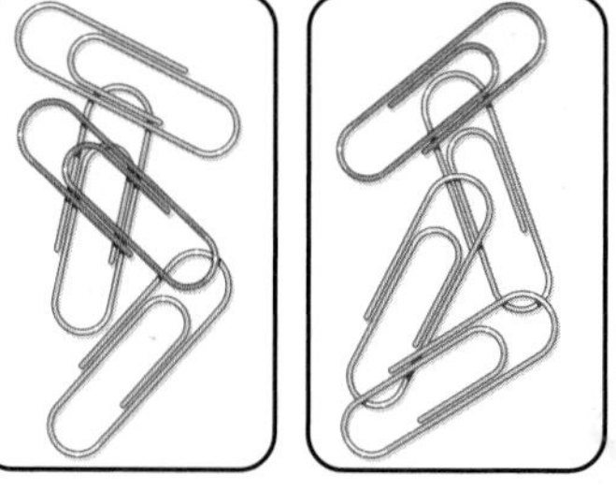

The ratio of red to blue paper clips is $\frac{1}{3}$, 1 to 3, or 1:3. This means that for every 1 red paper clip there are 3 blue paper clips.

CHECK Your Progress

a. Write the ratio in simplest form that compares the number of suns to the number of moons. Then explain its meaning.

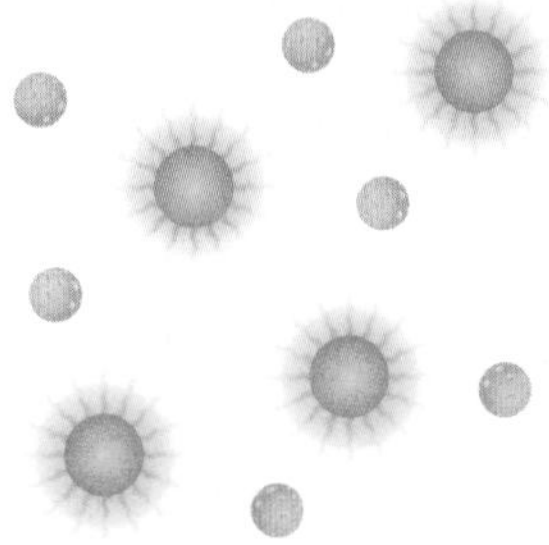

Ratios can also be used to compare a part to a whole.

EXAMPLE Use Ratios to Compare Parts to a Whole

2 SURVEYS Several students were asked to name their favorite flavor of gum. Write the ratio that compares the number of students who chose fruit to the total number of students who responded.

Favorite Flavor of Gum	
Flavor	**Number of Responses**
Peppermint	9
Cinnamon	8
Fruit	3
Spearmint	1

Three students preferred fruit out of a total of 9 + 8 + 3 + 1 or 21 responses.

fruit flavor responses →
total responses →

$$\frac{3}{21} = \frac{1}{7} \quad (\div 3)$$

The GCF of 3 and 21 is 3.

The ratio of the number of students who chose fruit to the total number of responses is $\frac{1}{7}$, 1 to 7, or 1:7. Analyzing the ratio tells us that one out of every 7 students preferred fruit-flavored gum.

CHECK Your Progress

b. **PETS** A pet store sold the animals listed in the table in one week. What was the ratio of cats to pets sold that week? Then explain its meaning.

Pet	Number Sold
Birds	10
Dogs	9
Cats	8
Gerbils	7
Lizards	2

A **rate** is a ratio comparing two quantities with different kinds of units.

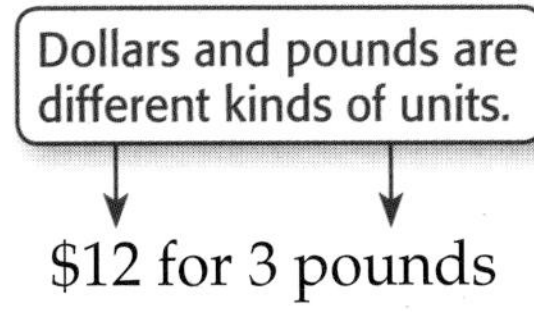

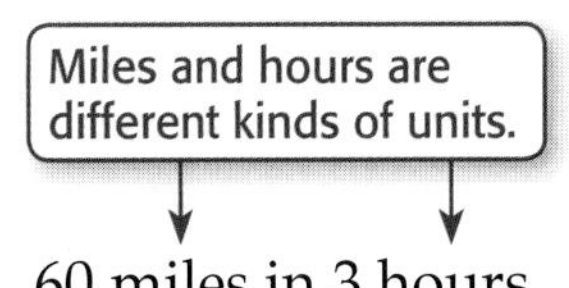

$12 for 3 pounds

60 miles in 3 hours

Study Tip

Unit Rates Some common unit rates are miles per hour, miles per gallon, price per pound, and dollars per hour.

The rate for one unit of a given quantity is called a **unit rate**.

The model shows that the dollars divided by the number of pounds is the number of dollars for 1 pound.

A unit rate of $4 for 1 pound can be read as $4 per pound.

Real-World Link The world's fastest insect is the Australian Dragonfly. It would take an Australian Dragonfly 4 hours to fly 144 miles.

When written as a fraction, a unit rate has a denominator of 1. Therefore, to write a rate as a unit rate, divide the numerator and denominator of the rate by the denominator. A unit rate can also be called a *rate of change*.

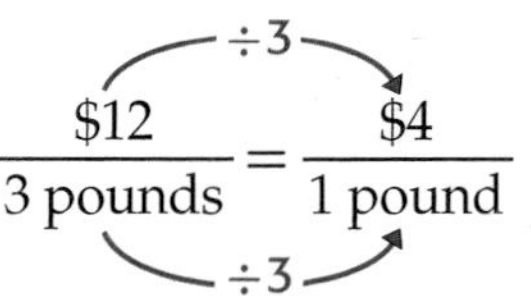

$$\frac{\$12}{3 \text{ pounds}} = \frac{\$4}{1 \text{ pound}}$$

EXAMPLE Find a Unit Rate

3 DRAGONFLIES Use the information at the left to find how many miles an Australian Dragonfly can fly per hour.

Write the rate that compares the number of miles to the number of hours. Then divide to find the unit rate.

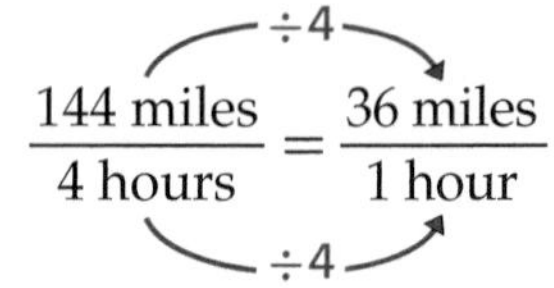

$$\frac{144 \text{ miles}}{4 \text{ hours}} = \frac{36 \text{ miles}}{1 \text{ hour}}$$

So, an Australian Dragonfly can fly about 36 miles per hour.

CHECK Your Progress

c. **WATER PARK** For Carolina's birthday, her mom took her and 4 friends to a water park. Carolina's mom paid $40 for 5 student tickets. What was the price for one student ticket?

Example 1 (p. 314)

For Exercises 1–3, write each ratio as a fraction in simplest form. Then explain its meaning.

1. 

pens to pencils

2. 

pennies:dimes

3. **MOVIES** A theater is showing 8 romantic comedies and 12 action thrillers. What is the ratio of action thrillers to romantic comedies?

Example 2 (p. 315)

4. **FRUIT** Last month, Amber ate 9 apples, 5 bananas, 4 peaches, and 7 oranges. Find the ratio of bananas to the total number of pieces of fruit Amber ate last month. Then explain its meaning.

Example 3 (p. 316)

Write each rate as a unit rate.

5. $9 for 3 cases of soda

6. 25 meters in 2 seconds

7. **HEALTH** Shina's heart beats 410 times in 5 minutes. At this rate, how many times does Shina's heart beat per minute?

Practice and Problem Solving

HOMEWORK HELP	
For Exercises	See Examples
8–13	1
14–17	2
18–23	3

For Exercises 8–13, write each ratio as a fraction in simplest form. Then explain its meaning.

8. 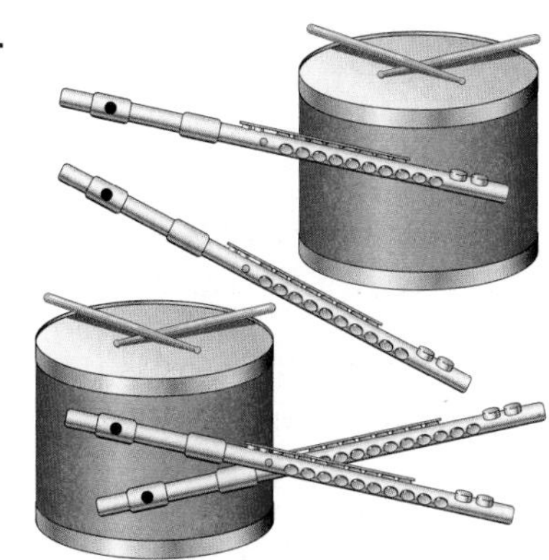

flutes:drums

9. 

sandwiches to milk cartons

10. **SCHOOL** A class has 6 boys and 15 girls. What is the ratio of boys to girls?

11. **CARS** Audrey counted 6 motorcycles and 27 cars at the restaurant parking lot. Find the ratio of motorcycles to cars.

12. **JEWELRY** The jewelry store is having a sale on 25 emerald rings and 15 ruby rings. Find the ratio of ruby rings to emerald rings.

13. **ANIMALS** An animal shelter has 36 kittens and 12 puppies available for adoption. What is the ratio of puppies to kittens?

14. **ANALYZE TABLES** For reading class, Salvador is keeping track of the types of books he has read so far this year. Find the ratio of mystery books to the total number of books Salvador has read. Then explain its meaning.

Type	Number of Books
Mystery	10
Nonfiction	7
Science Fiction/ Fantasy	5
Western	2

15. **ANALYZE TABLES** Last week, a wireless phone company sold the cell phone covers listed in the table. Find the ratio of black cell phone covers to the total number of cell phone covers sold last week. Then explain its meaning.

Color	Number of Cell Phone Covers
Green	5
Silver	6
Red	3
Black	4

16. **CLOTHES** For a trip, Ramona packed 6 blouses, 5 pairs of shorts, 3 pairs of jeans, and 1 skirt. Find the ratio of pairs of jeans to the total number of pieces of clothing Ramona packed. Then explain its meaning.

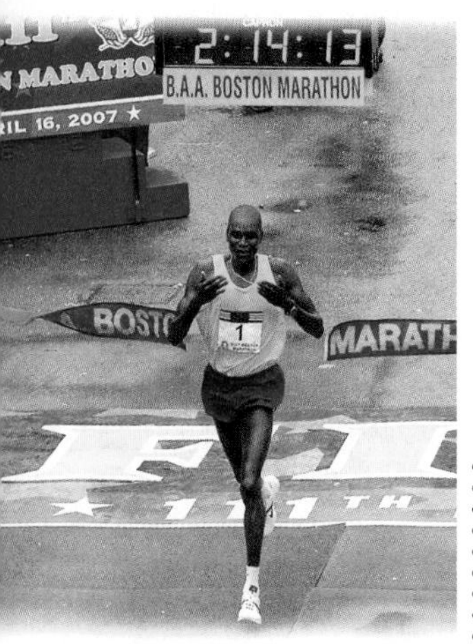

Real-World Link The 2007 Boston Marathon winning time was 2:14:13 for the men's division.

Source: Boston Athletic Association

17. **FOOD DRIVE** On the first day of the food drive, Mrs. Teasley's classes brought in 6 cans of fruit, 4 cans of beans, 7 boxes of noodles, and 4 cans of soup. Find the ratio of cans of fruit to the total number of food items collected. Then explain its meaning.

Write each rate as a unit rate.

18. 180 words in 3 minutes
19. $36 for 4 tickets
20. $4 for 8 bottles of water
21. $3 for a dozen eggs

22. **MARATHON** A marathon is approximately 26 miles. If Joshua ran the marathon in 4 hours at a constant rate, how far did he run per hour?

23. **RECYCLING** 340 trees are saved by recycling 20 tons of paper. How many trees are saved from 1 ton of recycled paper?

ANALYZE GRAPHS For Exercises 24 and 25, use the graphic. Write each ratio in simplest form. Then explain its meaning.

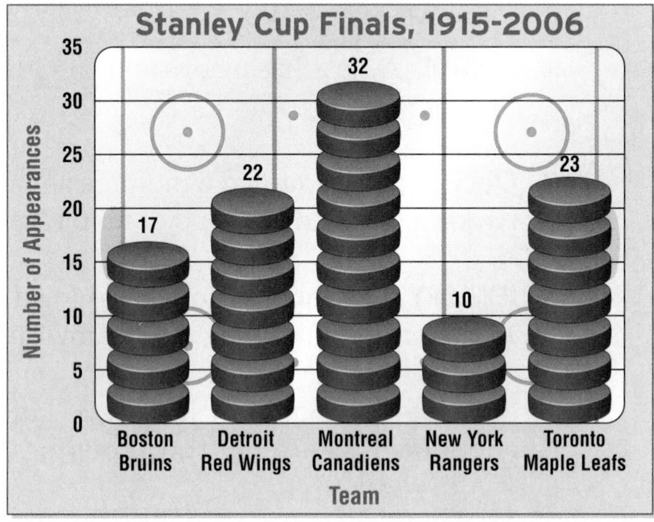

Source: National Hockey League

24. Write the ratio that compares the appearances made by the Rangers to the appearances made by the Red Wings.

25. Write the ratio that compares the appearances made by the Maple Leafs to the appearances made by the Bruins.

26. **FUNDRAISING** The 24 students in Mr. Brown's homeroom sold 72 magazine subscriptions. The 28 students in Mrs. Garcia's homeroom sold 98 magazine subscriptions. Whose homeroom sold more magazine subscriptions per student? Explain your reasoning.

See pages 687, 711.

27. **PACKAGING** A 6-pack of bottled water is on sale for $3. The same bottled water is also available in a 24-pack for $10. Which is less expensive per bottle: the 6-pack or the 24-pack? Explain your reasoning.

H.O.T. Problems

28. **OPEN ENDED** Create three different drawings showing a number of circles and triangles in which the ratio of circles to triangles is 2:3.

29. **CHALLENGE** Student Council sold 8 tickets to the spring dance in 15 minutes. At this rate, how many tickets will they sell per hour?

30. **FIND THE ERROR** Halley and Alma are writing the rate \$108 in 6 weeks as a unit rate. Who is correct? Explain your reasoning.

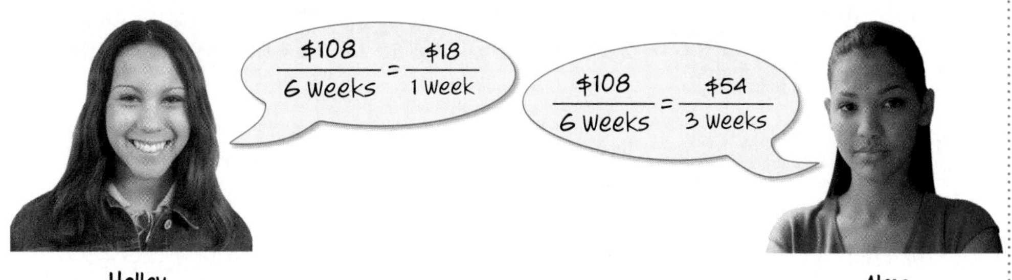

31. **WRITING IN MATH** What is the difference between a ratio and a rate? Give two examples of each.

TEST PRACTICE

32. While working out at the gym, Rodrigo spends 25 minutes on a treadmill and 35 minutes lifting weights. What is the ratio of the time Rodrigo spends on the treadmill to the time spent lifting weights?

A 2 to 3

B 5 to 7

C 4 to 5

D 1 to 7

33. The table shows the age ranges of the guests at Margo's birthday party. Which ratio accurately compares the number of guests ages 15 to 40 to the total number of guests at the party?

Age Range	Number of Guests
Under 15	11
15–40	6
41–65	3
Over 65	2

F 1:2 **H** 1:11

G 3:22 **J** 3:11

Spiral Review

Divide. Write in simplest form. (Lessons 5-9 and 5-10)

34. $\frac{3}{4} \div \frac{6}{7}$ 35. $\frac{1}{8} \div \frac{1}{6}$ 36. $3\frac{8}{9} \div 1\frac{2}{3}$ 37. $5\frac{5}{8} \div 2\frac{1}{2}$

38. **BAKING** Viho needs $2\frac{1}{4}$ cups of flour for cookies, $1\frac{2}{3}$ cups for almond bars, and $3\frac{1}{2}$ cups for cinnamon rolls. How much flour does he need in all? (Lesson 5-5)

39. **DECORATING** Janie is arranging a bookshelf, a chair, and a dresser along one wall of her bedroom. Use the *make a list* strategy to find the number of ways Janie can arrange the furniture. (Lesson 4-4)

GET READY for the Next Lesson

PREREQUISITE SKILL Write each fraction in simplest form. (Lesson 4-2)

40. $\frac{6}{10}$ 41. $\frac{15}{18}$ 42. $\frac{3}{12}$ 43. $\frac{25}{35}$

Extend 6-1

Math Lab
Ratios and Tangrams

MAIN IDEA

Explore ratios and the relationship between ratio and area.

Math Online

glencoe.com

- Concepts in Motion

A tangram is a puzzle that is made by cutting a square into seven geometric figures. The puzzle can be formed into many different figures.

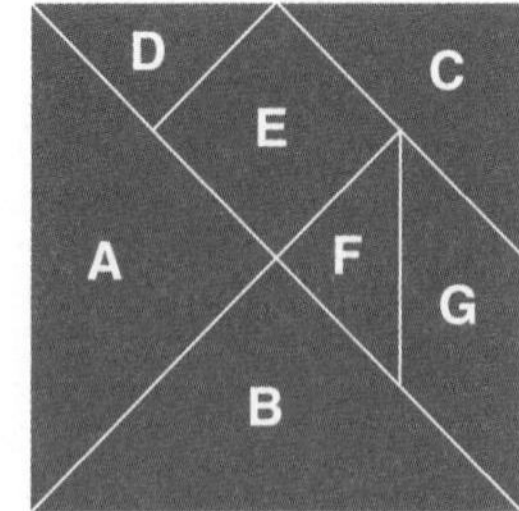

D
C
E
A
F
G
B

In this lab, you will use a tangram to explore ratios and the relationship between ratio and area.

STEP 1 Begin with one sheet of patty paper. Fold the top left corner to the bottom right corner. Unfold and cut along the fold so that two large triangles are formed.

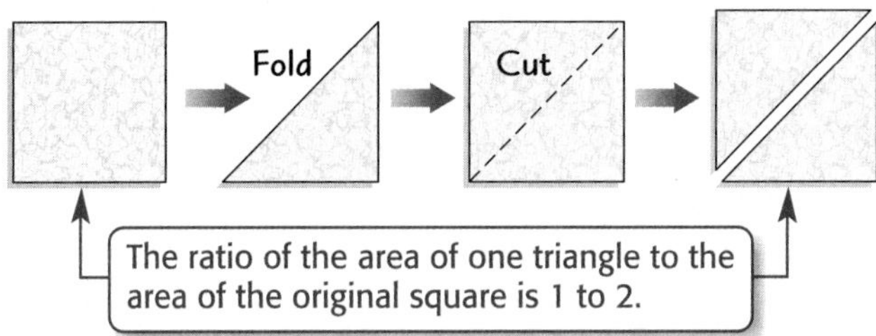

Fold
Cut
The ratio of the area of one triangle to the area of the original square is 1 to 2.

STEP 2 Use one of the cut triangles. Fold the bottom left corner to the bottom right corner. Unfold and cut along the fold. Label the triangles A and B.

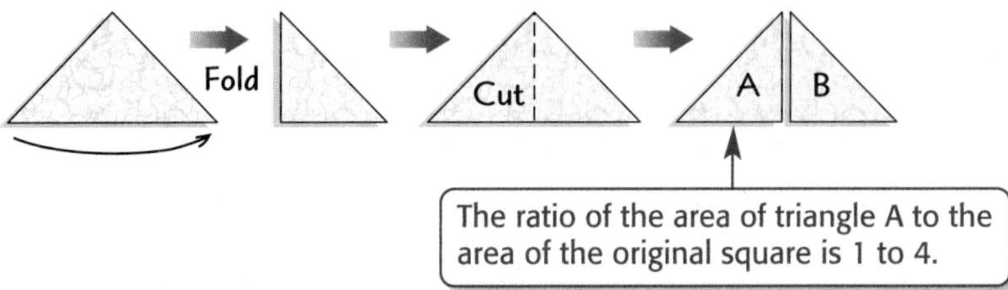

Fold
Cut
A
B
The ratio of the area of triangle A to the area of the original square is 1 to 4.

STEP 3 Use the other large triangle from step 1. Fold the bottom left corner to the bottom right corner. Make a crease and unfold. Next, fold the top down along the crease as shown. Make a crease and cut along the second crease line. Cut out the small triangle and label it C.

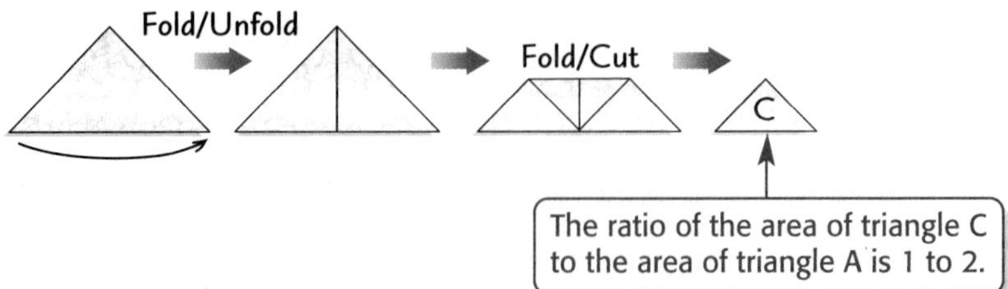

Fold/Unfold
Fold/Cut
C
The ratio of the area of triangle C to the area of triangle A is 1 to 2.

STEP 4 Use the remaining piece. Fold it in half from left to right. Cut along the fold. Using the left figure, fold the bottom left corner to the bottom right corner. Cut along the fold and label the triangle D and the square E.

Study Tip

Ratios of Areas If the ratio of the area of triangle D to the area of square E is 1 to 2, then the area of triangle D is *half* the area of square E.

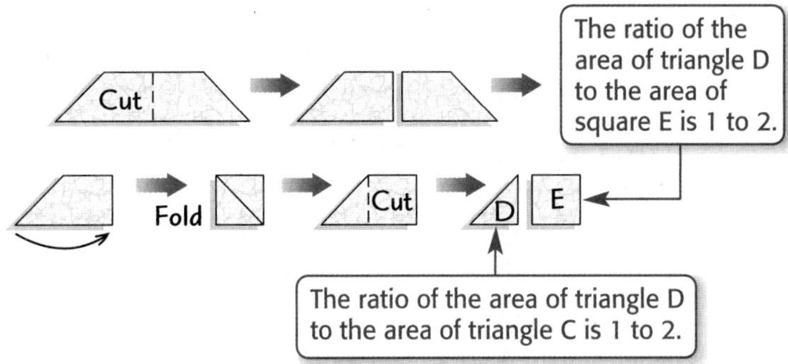

Cut
Fold
Cut
D
E
The ratio of the area of triangle D to the area of square E is 1 to 2.
The ratio of the area of triangle D to the area of triangle C is 1 to 2.

STEP 5 Use the remaining piece. Fold the bottom left corner to the top right corner. Cut along the fold. Label the triangle F and the other figure G.

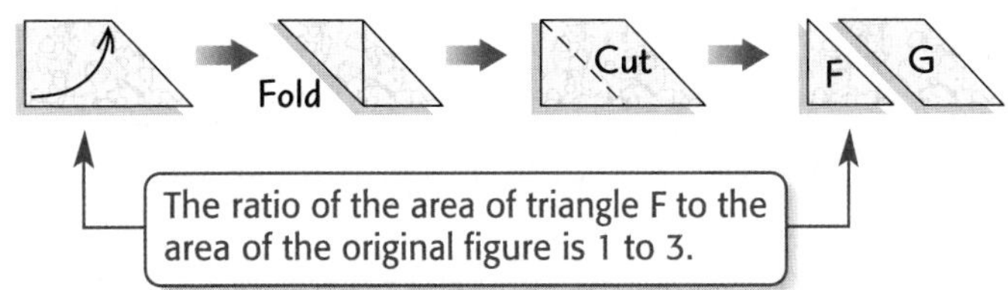

Fold
Cut
F
G
The ratio of the area of triangle F to the area of the original figure is 1 to 3.

Analyze the Results

1. Suppose the area of triangle B is 1 square unit. Find the area of each triangle.
   a. triangle C
   b. triangle F
2. Explain how the area of each of these triangles compares to the area of triangle B.
3. Explain why the ratio of the area of triangle C to the original large square is 1 to 8.
4. Tell why the area of square E is equal to the area of figure G.
5. Find the ratio of the area of triangle F to the original large square. Explain your reasoning.
6. Complete the table. Write the fraction that compares the area of each figure to the original square. What do you notice about the denominators?

Figure	A	B	C	D	E	F	G
Fractional Part of the Large Square							

6-2 Ratio Tables

MAIN IDEA

Use ratio tables to represent and solve problems involving equivalent ratios.

New Vocabulary

ratio table
equivalent ratio
scaling

Math Online

glencoe.com

- Extra Examples
- Personal Tutor
- Self-Check Quiz

GET READY for the Lesson

JUICE One can of frozen orange juice concentrate is mixed with 3 cans of water to make one batch of orange juice.

1. How many cans of juice and how many cans of water would you need to make 2 batches that have the same taste? 3 batches? Draw a picture to support your answers.
2. Find the ratio in simplest form of juice to water needed for 1, 2, and 3 batches of juice. What do you notice?

The quantities found in the activity above can be organized into a table. This table is called a **ratio table** because the columns are filled with pairs of numbers that have the same ratio.

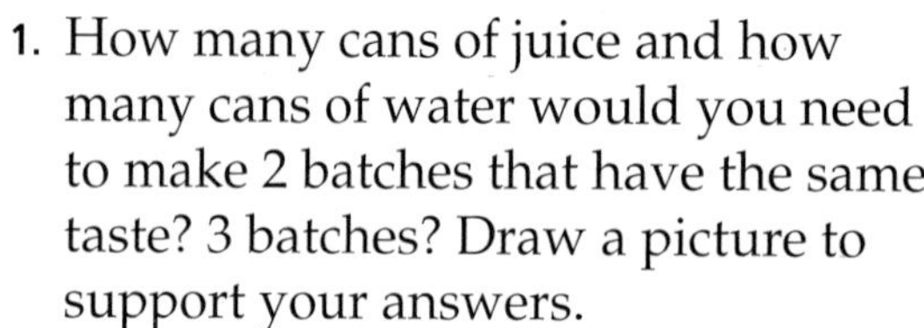

Cans of Concentrate	1	2	3
Cans of Water	3	6	9

The ratios $\frac{1}{3}$, $\frac{2}{6}$, and $\frac{3}{9}$ are equivalent since each simplifies to a ratio of $\frac{1}{3}$.

Equivalent ratios express the same relationship between two quantities. You can use a ratio table to find equivalent ratios or rates.

EXAMPLE Equivalent Ratios of Larger Quantities

1 **ICING To make yellow icing, 6 drops of yellow food coloring are added to 1 cup of white icing. Use the ratio table to find how much yellow to add to 5 cups of white icing to get the same shade.**

Cups of Icing	1				5
Drops of Yellow	6				■

METHOD 1 Find a pattern and extend it.

For 2 cups of icing, you would need a total of 6 + 6 or 12 drops.

+1 +1 +1 +1

Cups of Icing	1	2	3	4	5
Drops of Yellow	6	12	18	24	30

+6 +6 +6 +6

Continue this pattern until you reach 5 cups.

METHOD 2 **Multiply each quantity by the same number.**

$\times 5$

Cups of Icing	1	5
Drops of Yellow	6	**30**

$\times 5$

Since $1 \times 5 = 5$, multiply each quantity by 5.

So, add 30 drops of yellow food coloring to 5 cups of icing.

Study Tip

Check for Accuracy To check your answer for Example 1, check to see if the ratio of the two new quantities is equivalent to the ratio of the original quantities.

$\frac{5}{30} = \frac{5 \div 5}{30 \div 5}$ or $\frac{1}{6}$ ✓

CHOOSE Your Method

a. **NURSING** A patient receives 1 liter of IV fluids every 8 hours. At that rate, use the ratio table to find how many hours it will take to receive 4 liters of IV fluids.

IV Fluids (L)	1	4
Time (h)	8	■

You can also divide each quantity in a ratio by the same number to produce an equivalent ratio involving smaller quantities.

EXAMPLE Equivalent Ratios of Smaller Quantities

2 **HOT DOGS In a recent year, Takeru Kobayashi won a hot dog eating contest by eating nearly 54 hot dogs in 12 minutes. If he ate at a constant rate, use the ratio table to determine about how many hot dogs he ate every 2 minutes.**

Hot Dogs	54		■
Time (min)	12		2

$\div 2$ $\quad$ $\div 3$

Hot Dogs	54	**27**	**9**
Time (min)	12	**6**	2

$\div 2$ $\quad$ $\div 3$

Divide each quantity by one or more common factors until you reach a quantity of 2 minutes.

So, Kobayashi ate about 9 hot dogs every 2 minutes.

Real-World Link

In 2007, Joey Chestnut beat Takeru Kobayashi by consuming 66 hot dogs in 12 minutes. Takeru Kobayashi ate 63 hot dogs in 12 minutes.

Source: Nathan's Famous International July Fourth Hot Dog Eating Contest

CHECK Your Progress

b. **JAM** To make cranberry jam, you need 12 cups of sugar for every 16 cups of cranberries. Use the ratio table to find the amount of sugar needed for 4 cups of cranberries.

Sugar (c)	12		■
Cranberries (c)	16		4

Multiplying or dividing two related quantities by the same number is called **scaling**. Sometimes you may need to *scale back* and then *scale forward* to find an equivalent ratio.

EXAMPLE Use Scaling

3 GROCERIES Cans of corn are on sale at 10 for $4. Use the ratio table to find the cost of 15 cans.

Cans of Corn	10		15
Cost in Dollars	4		■

There is no whole number by which you can multiply 10 to get 15. So, scale back to 5 and then scale forward to 15.

÷2 ×3

Cans of Corn	10	**5**	15
Cost in Dollars	4	**2**	**6**

÷2 ×3

Divide each quantity by a common factor, 2.

Then, since 5 × 3 = 15, multiply each quantity by 3.

So, 15 cans of corn would cost $6.

CHECK Your Progress

c. **HEIGHT** A child's height measures 105 centimeters. If 25 centimeters is about 10 inches, use the ratio table to estimate the height in inches.

Height (cm)	25		105
Height (in.)	10		■

Real-World EXAMPLE Use a Ratio Table

Real-World Link The U.S. dollar and the euro account for approximately half of all currency exchanged in the world.

4 MONEY On her vacation, Leya exchanged $50 American and received $90 Canadian. Use a ratio table to find how many Canadian dollars she would receive for $20 American.

Set up a ratio table.

Canadian Dollars	90		■
American Dollars	50		20

Label the rows with the two quantities being compared. Then fill in what is given.

Use scaling to find the desired quantity.

÷10 ×4

Canadian Dollars	90	**9**	**36**
American Dollars	50	**5**	20

÷10 ×4

Divide each quantity by a common factor, 10.

Then, since 5 × 4 = 20, multiply each quantity by 4.

Leya would receive $36 Canadian for $20 American.

CHECK Your Progress

d. **AUTOMOBILES** Landon owns a hybrid SUV that can travel 400 miles on a 15-gallon tank of gas. Use a ratio table to determine how many miles he can travel on 6 gallons.

CHECK Your Understanding

For Exercises 1–3, use the ratio tables given to solve each problem.

Example 1 (pp. 322–323)

1. **MONEY** Santiago receives an allowance of $7 every week. How much total does he receive after 4 weeks?

Allowance	7			■
Number of Weeks	1			4

Example 2 (p. 323)

2. **EXERCISE** Tonya runs 8 kilometers in 60 minutes. At this rate, how long would it take her to run 2 kilometers?

Distance Run (km)	8		2
Time (min)	60		■

Example 3 (p. 324)

3. **BEVERAGES** A certain 12-ounce soft drink contains about 10 teaspoons of sugar. If you drink 18 ounces of this soft drink, how many teaspoons of sugar have you consumed?

Ounces of Soft Drink	12		18
Teaspoons of Sugar	10		■

Example 4 (p. 324)

4. **FOOD** Lamika buys 12 packs of juice boxes that are on sale and pays a total of $48. Use a ratio table to determine how much Lamika will pay to buy 8 more packs of juice boxes at the same store.

Practice and Problem Solving

HOMEWORK HELP

For Exercises	See Examples
5, 6	1
7, 8	2
9, 10	3
11, 12	4

For Exercises 5–10, use the ratio tables given to solve each problem.

5. **PIES** To make 5 apple pies, you need about 2 pounds of apples. How many pounds of apples do you need to make 20 apple pies?

Number of Pies	5		20
Pounds of Apples	2		■

6. **FIELD TRIP** A zoo requires that 1 adult accompany every 7 students that visit the zoo. How many adults must accompany 28 students?

Number of Adults	1			■
Number of Students	7			28

7. **MONEY** Before leaving to visit Mexico, Levant traded 270 American dollars and received 3,000 Mexican pesos. When he returned from Mexico, he had 100 pesos left. How much will he receive when he exchanges these pesos for dollars?

American Dollars	270		■
Mexican Pesos	3,000		100

8. **JEWELRY** Valentina purchased 200 beads for $48 to make necklaces. If she needs to buy 25 more beads, how much will she pay if she is charged the same rate?

Number of Beads	200		25
Cost in Dollars	48		■

9. **KNITTING** Four balls of wool will make 8 knitted caps. How many balls of wool will Malcolm need if he wants to make 6 caps?

Balls of Wool	4		■
Number of Caps	8		6

10. **BIRDS** If a hummingbird were to get all of its food from a feeder, then a 16-ounce nectar feeder could feed about 80 hummingbirds a day. How many hummingbirds would you expect to be able to feed with a 12-ounce feeder?

Real-World Link. To make a 10-Calorie solution of nectar for a hummingbird feeder, mix one part sugar with four parts water.

Ounces of Nectar	16		12
Number of Birds Fed	80		■

11. **BIKING** On a bike trip across the United States, Jason notes that he covers about 190 miles every 4 days. If he continues at this rate, use a ratio table to determine about how many miles he could bike in 6 days.

12. **PHOTOGRAPHY** When a photo is reduced or enlarged, its length to width ratio usually remains the same. Aurelia wants to enlarge a 4-inch by 6-inch photo so that it has a height of 15 inches. Use a ratio table to determine the new width of the photo.

13. **PETS** Before administering medicine, a veterinarian needs to know the animal's weight in kilograms. If 20 pounds is about 9 kilograms and a dog weighs 30 pounds, use a ratio table to find the dog's weight in kilograms. Explain your reasoning.

14. **TRAVEL** On a typical day, flights at a local airport arrive at a rate of 10 every 15 minutes. At this rate, how many flights would you expect to arrive in 1 hour?

RECIPES For Exercises 15–17, use the following information.

A punch recipe that serves 24 people calls for 4 liters of lemon-lime soda, 2 pints of sherbet, and 6 cups of ice.

15. Create a ratio table to represent this situation.

16. How much of each ingredient would you need to make an identical recipe that serves 12 people? 36 people?

17. How much of each ingredient would you need to make an identical recipe that serves 18 people? Explain your reasoning.

See pages 688, 711.

H.O.T. Problems

18. **CHALLENGE** Use the ratio table to determine how many people 13 subs would serve. Explain your reasoning.

Number of Subs	3	5	8	13
People Served	12	20	32	■

19. **NUMBER SENSE** There are 10 girls and 8 boys in Mr. Augello's class. If 5 more girls and 5 more boys join the class, will the ratio of girls to boys remain the same? Justify your answer using a ratio table.

20. **WRITING IN MATH** Explain two different methods that can be used to find the missing value in the ratio table.

Pages Read	60		80
Number of Days	9		■

TEST PRACTICE

21. Paul buys 5 DVDs for $60. At this rate, how much would he pay for 3 DVDs?

   A $10

   B $30

   C $36

   D $58

22. **SHORT RESPONSE** Beth walks 2 blocks in 15 minutes. How many blocks would Beth walk if she walked at the same rate for an hour?

23. Jay Len is making biscuits using the recipe below.

Whole Wheat Biscuits

2 c	Whole wheat flour
4 tsp	Baking powder
$\frac{1}{2}$ tsp	Salt
2 tbsp	Shortening
1 c	Milk
1	Small egg

Makes 20 biscuits

How many cups of flour will he need to make 30 biscuits?

   F $1\frac{1}{2}$ cups

   G 3 cups

   H 10 cups

   J 15 cups

Spiral Review

24. **HORSES** A horse ranch has 6 mustangs and 18 Arabians. Write the ratio of mustangs to Arabians as a fraction in simplest form. Then explain its meaning. (Lesson 6-1)

Divide. Write in simplest form. (Lesson 5-10)

25. $5\frac{3}{4} \div 3\frac{3}{4}$

26. $5 \div 2\frac{1}{3}$

27. $2\frac{1}{2} \div 1\frac{4}{5}$

28. List the next five common multiples after the LCM of 6 and 9. (Lesson 4-5)

GET READY for the Next Lesson

PREREQUISITE SKILL **Write each rate as a unit rate.** (Lesson 6-1)

29. $24 for 3 hats

30. 130 miles in 2 hours

31. 145 students for 5 teachers

Extend 6-2

Graphing Calculator Lab
Ratio Tables

MAIN IDEA

Use technology to compare output/input ratios for functions.

Math Online

glencoe.com

- Other Calculator Keystrokes

You can use the CellSheet application on a TI-83/84 Plus graphing calculator to compare the output/input ratios of real-world functions.

ACTIVITY

1 **MOVIES** **The total cost of purchasing 1, 2, 3, 4, and 5 DVDs for $19 each is found by multiplying the number of DVDs purchased by 19. Create a table to model this situation. Include a column that calculates the ratio of cost to DVDs.**

STEP 1 Access CellSheet by pressing [APPS] 8 [ENTER] [ENTER].

STEP 2 Enter the heading DVDS in cell A1 by pressing 2nd [ALPHA] ["] [D] [V] [D] [S] [ENTER]. Similarly, enter the heading COST in cell B1 and RATIO in cell C1.

STEP 3 Enter the numbers 1 through 5 into cells A2 through A6, respectively. Then, insert the formula =A2*19 in cell B3 by pressing [STO▸] [ALPHA] [A] 2 [×] 19.

STEP 4 Calculate the cost for each number of DVDs by copying the formula to cells B4 through B6. Move to cell B3 and press [F3] to copy, [F1] [▼] [▼] [▼] [▼] to select the range of cells, and [F4] to paste.

STEP 5 Use a similar procedure to insert, copy, and paste the formula B2/A2 in cells C3 through C6.

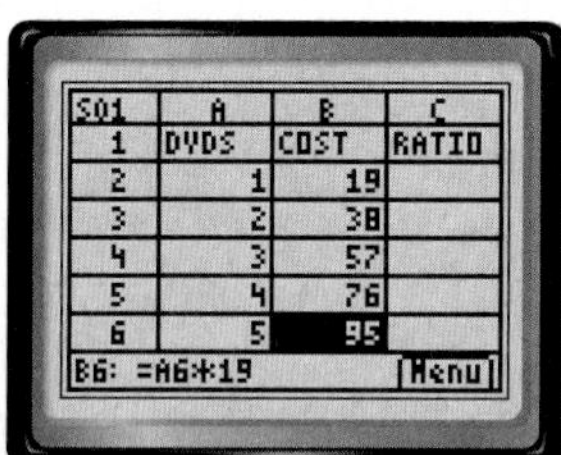

Analyze the Results

1. Does the 2-column table of values for DVDs and Cost form a ratio table? Explain your reasoning.
2. **CLOTHING** A store offers $5 off any purchase over $10. Create a graphing calculator table that models the total cost of purchasing $11 through $14 in clothing. Include a ratio column of cost:amount.
3. Does the 2-column table of values for the amount and cost form a ratio table? Explain your reasoning.

6-3 Proportions

MAIN IDEA

Determine if two quantities are proportional.

New Vocabulary

proportional
proportion

Math Online

glencoe.com

- Extra Examples
- Personal Tutor
- Self-Check Quiz
- Reading in the Content Area

GET READY for the Lesson

PHOTOGRAPHY Leon spent \$2 to make 10 prints from his digital camera. Later, he went back to the same store and spent \$6 to make 30 prints.

Number of Prints	Cost (\$)
10	2
30	6

1. Express the relationship between the total cost and number of prints he made for each situation as a rate in fraction form.
2. Compare the relationship between the numerators of each rate you wrote in Exercise 1. Compare the relationship between the denominators of these rates.
3. Are the rates you wrote in Exercise 1 equivalent? Explain.

In the situations above, there are two related quantities: the number of prints and the cost for these prints. Notice that both quantities change, but in the same way.

Number of Prints	10	30
Cost (\$)	2	6

×3 (number of prints), ×3 (cost)

As the number of prints triples, the cost also triples.

By comparing these quantities as rates in simplest form, you can see that the relationship between the two quantities stays the same.

$$\frac{10 \text{ prints}}{\$2} = \frac{5 \text{ prints}}{\$1} \quad \text{and} \quad \frac{30 \text{ prints}}{\$6} = \frac{5 \text{ prints}}{\$1}$$

(÷2 numerator and denominator; ÷6 numerator and denominator)

Two quantities are **proportional** if they have a constant ratio or rate. In the example above, the cost for making prints is proportional to the number of prints because each quantity has a constant rate of \$1 for 5 prints.

A proportional relationship is often expressed by writing a proportion.

Proportion — Key Concept

Words A **proportion** is an equation stating that two ratios or rates are equivalent.

Examples $\frac{2}{5} = \frac{6}{15}$ $\quad \frac{\$2}{10 \text{ prints}} = \frac{\$6}{30 \text{ prints}}$

There are different ways to determine if the relationship between two quantities is proportional. One way is by examining unit rates.

EXAMPLES Use Unit Rates

Determine if the quantities in each pair of rates are proportional. Explain your reasoning and express each proportional relationship as a proportion.

1 20 miles in 5 hours; 45 miles in 9 hours

Write each rate as a fraction. Then find its unit rate.

$$\frac{20 \text{ miles}}{5 \text{ hours}} = \frac{4 \text{ miles}}{1 \text{ hour}} \quad (\div 5) \qquad \frac{45 \text{ miles}}{9 \text{ hours}} = \frac{5 \text{ miles}}{1 \text{ hour}} \quad (\div 9)$$

Since the rates do not have the same unit rate, they are not equivalent. So, the number of miles is not proportional to the number of hours.

Study Tip

Unit Rates *The unit rate in Example 2,* $\frac{\$7}{1 \text{ T-shirt}}$ *or \$7 per T-shirt, is called the unit price since it gives the cost per unit.*

2 3 T-shirts for \$21; 5 T-shirts for \$35

$$\frac{\$21}{3 \text{ T-shirts}} = \frac{\$7}{1 \text{ T-shirt}} \quad (\div 3) \qquad \frac{\$35}{5 \text{ T-shirts}} = \frac{\$7}{1 \text{ T-shirt}} \quad (\div 5)$$

Since the rates have the same unit rate, they are equivalent. The cost is proportional to the number of T-shirts. So, $\frac{\$21}{3 \text{ T-shirts}} = \frac{\$35}{5 \text{ T-shirts}}$.

3 READING Felisa read the first 60 pages of a book in 3 days. She read the last 90 pages in 6 days. Are these reading rates proportional? Explain your reasoning.

$$\frac{60 \text{ pages}}{3 \text{ days}} = \frac{20 \text{ pages}}{1 \text{ day}} \quad (\div 3) \qquad \frac{90 \text{ pages}}{6 \text{ days}} = \frac{15 \text{ pages}}{1 \text{ day}} \quad (\div 6)$$

Since the rates do not have the same unit rate, they are not equivalent. So, Felisa's reading rates were not proportional.

CHECK Your Progress

a. **JEWELRY** Marcia made 10 bracelets for 5 friends. Jen made 12 bracelets for 4 friends. Are these rates proportional? Explain.

b. **FUNDRAISING** Club A raised \$168 by washing 42 cars. Club B raised \$152 by washing 38 cars. Are these fundraising rates proportional? Explain your reasoning.

If a unit rate is not easily found, check to see if the rates are equivalent. If they are, then the quantities are proportional.

Real-World Link
In 2004, in the U.S. DVD sales totaled \$15.5 billion.
Source: *USA Today*

EXAMPLES Use Equivalent Fractions

Determine if the quantities in each pair of ratios or rates are proportional. Explain your reasoning and express each proportional relationship as a proportion.

4 **3 free throws made out of 7 attempts; 9 free throws made out of 14 attempts**

Write each ratio as a fraction.

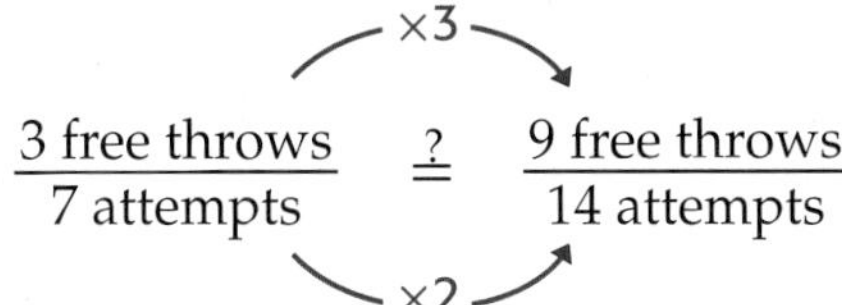

×3
3 free throws / 7 attempts ≟ 9 free throws / 14 attempts
×2

The numerator and the denominator are not multiplied by the same number. So, the fractions are not equivalent.

The number of free throws made is not proportional to the number of attempts.

5 **6 DVDs for \$90; 3 DVDs for \$45**

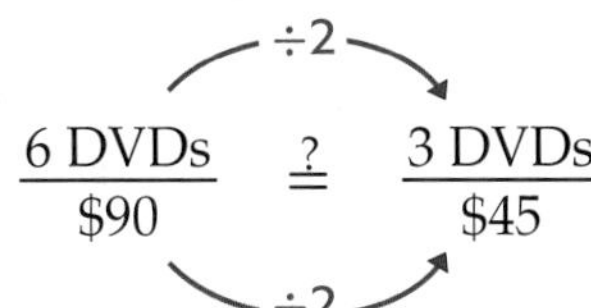

The numerator and the denominator are divided by the same number. So, the fractions are equivalent.

The number of DVDs is proportional to the cost.

CHECK Your Progress

c. 5 packs of baseball cards for \$15; 10 packs of baseball cards for \$30

d. 12 girls out of 16 students; 4 girls out of 8 students

CHECK Your Understanding

Determine if the quantities in each pair of ratios or rates are proportional. Explain your reasoning and express each proportional relationship as a proportion.

Examples 1, 2 (p. 330)

1. \$24 saved after 3 weeks; \$52 saved after 7 weeks
2. 270 Calories in 3 servings; 450 Calories in 5 servings

Examples 4, 5 (p. 331)

3. 3 hours worked for \$12; 9 hours worked for \$36
4. 16 breaths in 60 seconds; 14 breaths in 15 seconds

Example 3 (p. 330)

5. **FITNESS** Micah can do 75 push-ups in 3 minutes. Eduardo can do 130 push-ups in 5 minutes. Are these rates proportional? Explain.

Practice and Problem Solving

HOMEWORK HELP	
For Exercises	See Examples
6–9	1, 2
10–13	4, 5
14, 15	3

Determine if the quantities in each pair of ratios or rates are proportional. Explain your reasoning and express each proportional relationship as a proportion.

6. $12 for 3 paperback books; $28 for 7 paperback books
7. 16 points scored in 4 games; 48 points scored in 8 games
8. 96 words typed in 3 minutes; 160 words typed in 5 minutes
9. $3 for 6 bagels; $9 for 24 bagels
10. 288 miles driven on 12 gallons of fuel; 240 miles driven on 10 gallons of fuel
11. 15 computers for 45 students; 45 computers for 135 students
12. 12 minutes to drive 30 laps; 48 minutes to drive 120 laps
13. 16 out of 28 students own pets; 240 out of 560 students own pets

14. **PHOTOGRAPHY** Jade enlarged the photograph at the right to a poster. The size of the poster is 60 inches by 100 inches. Is the size of the poster proportional to the size of the photograph? Explain your reasoning.

15. **SURVEY** One school survey showed that 3 out of 5 students buy their lunch. Another survey showed that 12 out of 19 students buy their lunch. Are these results proportional? Explain.

BASEBALL For Exercises 16–18, refer to the table below. Determine if each pair of players made proportionally the same number of hits. Explain.

16. Mark Teixeira and Eric Bruntlett
17. Hideki Matsui and Mark Teixeira
18. Brad Eldred and Ramon Santiago

2007 Spring Batting Statistics

Player	Team	At Bats	Hits
Mark Teixeira	Texas Rangers	48	12
Brad Eldred	Pittsburgh Pirates	66	20
Hideki Matsui	New York Yankees	60	16
Eric Bruntlett	Houston Astros	56	14
Ramon Santiago	Detroit Tigers	33	10

Source: Major League Baseball

19. **ANALYZE TABLES** Of the players listed in the table above, did the player who made the most hits have the best record? Explain.

20. **TESTS** On a math test, it took Kiera 30 minutes to do 6 problems. Heath finished 18 problems in 40 minutes. Did the students work proportionally at the same rate? Explain your reasoning.

21. **SAVINGS** Rosalinda saved $35 in 5 weeks. Her sister saved $56 in 56 days. Did each sister save proportionally the same amount of money? Explain.

EXTRA PRACTICE
See pages 688, 711.

H.O.T. Problems

CHALLENGE For Exercises 22–25, use the following information to verify each proportion. Justify your response.

To verify a proportion, you can use cross products. If the product of the *means* equals the product of the *extremes*, then the two ratios form a proportion. In a proportion, the top left and bottom right numbers are the extremes. The top right and bottom left numbers are the means. In the proportion in Exercise 22, the numbers 5 and 9 are the means and the numbers 3 and 15 are the extremes.

22. $\frac{3}{5} = \frac{9}{15}$ **23.** $\frac{2}{7} = \frac{5}{21}$ **24.** $\frac{1}{8} = \frac{3}{28}$ **25.** $\frac{4}{9} = \frac{12}{27}$

26. **WRITING IN MATH** Cecil's Pizza offers two large pizzas for $15 and four large pizzas for $28. Describe and use three different ways to determine if the pair of ratios is proportional.

TEST PRACTICE

27. The ratio of girls to boys in the junior high band is 3 to 4. Which of these shows possible numbers of the girls and boys in the band?

A 30 girls, 44 boys

B 27 girls, 36 boys

C 22 girls, 28 boys

D 36 girls, 50 boys

28. Which of the following shows an equivalent way to show the cost of the tomatoes?

HOME-GROWN VEGETABLES	
Cucumbers	6 for $2
Peppers	12 for $9
Tomatoes	6 for $4

F 15 for $10 **H** 12 for $9

G 20 for $15 **J** 8 for $6

Spiral Review

29. SHOPPING Walter purchased 2 CDs for $26. Use a ratio table to find how much he would pay for 6 CDs. (Lesson 6-2)

30. MEASUREMENT A caterpillar is 4 centimeters long, and a Monarch butterfly is 10 centimeters long. Write the ratio of the caterpillar's length to the butterfly's length as a fraction in simplest form. (Lesson 6-1)

Find the prime factorization of each number. (Lesson 1-2)

31. 15 **32.** 94 **33.** 102 **34.** 126

GET READY for the Next Lesson

PREREQUISITE SKILL Write each rate as a unit rate. (Lesson 6-1)

35. 56 wins in 8 years **36.** $12 for 5 hot dogs **37.** $21 for 3 hours

6-4 Algebra: Solving Proportions

MAIN IDEA

Solve proportions.

Math Online

glencoe.com

- Extra Examples
- Personal Tutor
- Self-Check Quiz

GET READY for the Lesson

SHOES A department store is selling flip flops for \$5 a pair.

1. How many pairs of flip flops can you buy with \$20? \$25?
2. Write a proportion to express the relationship between the cost of 3 pairs of flip flops and the cost *c* of 7 pairs of flip flops.
3. How much will it cost to buy 6 pairs of flip flops?

Number of Pairs	Price (\$)
1	5
2	10
3	15

When you find an unknown value in a proportion, you are *solving the proportion*. As you discovered in Lesson 6-3, there are different methods to determine if a relationship is proportional. You can use these same methods to solve a proportion.

EXAMPLES Solve Using Equivalent Fractions

Solve each proportion.

1 $\frac{4}{7} = \frac{m}{35}$

Find a value for *m* so the fractions are equivalent.

$\frac{4}{7} = \frac{m}{35}$ (×5 on numerator and denominator) Since 7 × 5 = 35, multiply the numerator and denominator by 5.

$\frac{4}{7} = \frac{20}{35}$ Since $4 \times 5 = 20$, $m = 20$.

2 $\frac{12}{15} = \frac{4}{y}$

$\frac{12}{15} = \frac{4}{y}$ (÷3 on numerator and denominator) Since 12 ÷ 3 = 4, divide the numerator and denominator by 3.

$\frac{12}{15} = \frac{4}{5}$ Since $15 \div 3 = 5$, $y = 5$.

Study Tip

Check Your Answer Write each ratio in simplest form. If they simplify to the same ratio, they are equivalent.

3 $\frac{x}{16} = \frac{7}{8}$

$\frac{x}{16} = \frac{7}{8}$ (÷2 on numerator and denominator) Since 16 ÷ 2 = 8, divide the numerator and denominator by 2.

$\frac{14}{16} = \frac{7}{8}$ THINK What number divided by 2 is 7? The answer is 14.

So, $x = 14$.

CHECK Your Progress

Solve each proportion.

a. $\frac{2}{3} = \frac{n}{9}$ b. $\frac{30}{54} = \frac{z}{9}$ c. $\frac{5}{8} = \frac{40}{x}$

Proportions can also be used to make predictions.

EXAMPLE Make Predictions in Proportional Situations

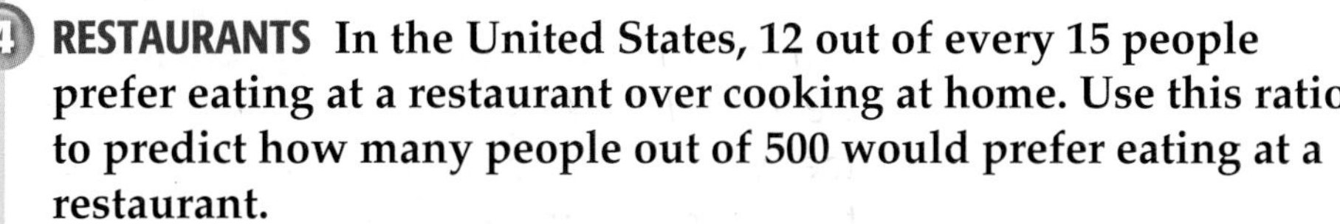

4 **RESTAURANTS In the United States, 12 out of every 15 people prefer eating at a restaurant over cooking at home. Use this ratio to predict how many people out of 500 would prefer eating at a restaurant.**

Write and solve a proportion. Let p represent the number of people who can be expected to prefer eating at a restaurant.

$$\frac{12}{15} = \frac{p}{500}$$

prefer restaurants → 12, total people → 15; p ← prefer restaurants, 500 ← total people

The denominators 15 and 500 are not easily related by multiplication. So, simplify the ratio 12 out of 15. Then solve using equivalent fractions.

$$\frac{12}{15} = \frac{4}{5} = \frac{400}{500}$$

(÷3, then ×100) Since 5 × 100 = 500, multiply the numerator and denominator by 100.

So, about 400 out of 500 people can be expected to prefer eating at a restaurant.

Real-World Career... How Does a Chef Use Math? Chefs use ratios and proportions to double, triple, or quadruple the quantities of ingredients needed in a recipe.

Math Online For more information, go to glencoe.com.

CHECK Your Progress

d. **ICE CREAM** There are 810 Calories in 3 scoops of vanilla ice cream. About how many Calories are there in 7 scoops of ice cream?

e. **BED TIME** If 15 out of 25 students go to bed before 10 P.M., predict how many go to bed before 10 P.M. in a school of 1,000 students.

You can also examine unit rates to solve a proportion.

EXAMPLE Solve Using Unit Rates

5 DRIVING The Millers drove 105 miles on 3 gallons of gas. At this rate, how many miles can they drive on 10 gallons of gas?

Step 1 Set up the proportion. Let a represent the number of miles that can be driven on 10 gallons of gas.

$$\frac{105 \text{ miles}}{3 \text{ gallons of gas}} = \frac{a \text{ miles}}{10 \text{ gallons of gas}}$$

Step 2 Find the unit rate.

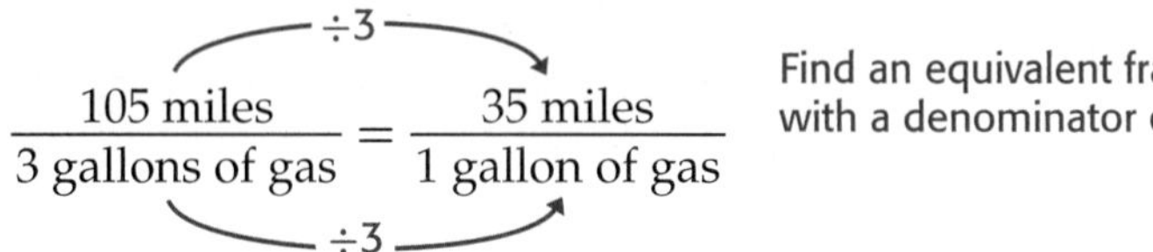

$$\frac{105 \text{ miles}}{3 \text{ gallons of gas}} \overset{\div 3}{=} \frac{35 \text{ miles}}{1 \text{ gallon of gas}}$$

Find an equivalent fraction with a denominator of 1.

Step 3 Rewrite the proportion using the unit rate and solve using equivalent fractions.

$$\frac{105 \text{ miles}}{3 \text{ gallons of gas}} \overset{\div 3}{=} \frac{35 \text{ miles}}{1 \text{ gallon of gas}} \overset{\times 10}{=} \frac{350 \text{ miles}}{10 \text{ gallons of gas}}$$

So, the value of a is 350. At the given rate, the Millers can drive 350 miles on 10 gallons of gas.

CHECK Your Progress

f. **NUTRITION** Three average-size apples contain 180 Calories. How many average-size apples contain 300 Calories?

CHECK Your Understanding

Examples 1–3 (pp. 334–335)

Solve each proportion.

1. $\frac{3}{4} = \frac{x}{20}$
2. $\frac{5}{4} = \frac{a}{36}$
3. $\frac{18}{20} = \frac{9}{n}$

Example 4 (p. 335)

4. **PETS** Out of 30 students surveyed, 17 have a dog. Based on these results, predict how many of the 300 students in the school have a dog.
5. **LOCKERS** If one out of 12 students at a school share a locker, predict how many share a locker in a school of 456 students.

Example 5 (p. 336)

6. **PARTIES** If 84 cookies will serve 28 students, how many cookies are needed for 30 students?
7. **SUNGLASSES** Chet spent $24 on two pairs of sunglasses. At this rate, how much would 6 pairs of sunglasses cost?

Practice and Problem Solving

HOMEWORK HELP	
For Exercises	See Examples
8–15	1–3
16, 17	4
18–21	5

Solve each proportion.

8. $\frac{2}{5} = \frac{w}{15}$
9. $\frac{3}{4} = \frac{z}{28}$
10. $\frac{7}{d} = \frac{35}{10}$
11. $\frac{4}{x} = \frac{16}{28}$
12. $\frac{p}{3} = \frac{25}{15}$
13. $\frac{h}{8} = \frac{6}{16}$
14. $\frac{6}{7} = \frac{18}{c}$
15. $\frac{21}{35} = \frac{3}{r}$

16. **NEWSPAPER** A recent survey reported that out of 50 teenagers, 9 said they get most of their news from a newspaper. At this rate, how many out of 300 teenagers would you expect to get their news from a newspaper?

17. **HORSES** A Clydesdale drinks about 120 gallons of water every 4 days. At this rate, about how many gallons of water does a Clydesdale drink in 28 days?

18. **DVDs** Nata spent $28 on 2 DVDs. At this rate, how much would 5 DVDs cost?

19. **LUNCH** Four students spent $12 on school lunch. At this rate, find the amount 10 students would spend on the same school lunch.

20. **BASEBALL** If 15 baseballs weigh 75 ounces, how many baseballs weigh 15 ounces?

21. **HEALTH** In 10 minutes, a heart can beat 700 times. At this rate, in how many minutes will a heart beat 140 times?

Real-World Link
A Clydesdale consumes about 20 quarts of feed and 40 to 50 pounds of hay each day.

Solve each proportion.

22. $\frac{11}{13} = \frac{x}{91}$
23. $\frac{96}{128} = \frac{12}{c}$
24. $\frac{5}{12} = \frac{x}{6}$

25. **SCHOOL** Suppose 8 out of every 20 students are absent from school less than five days a year. Predict how many students would be absent from school less than five days a year in a school system of 40,000 students.

26. **ANALYZE TABLES** The table shows which school subjects are favored by a group of students. Write and solve a proportion that could be used to predict the number of students out of 400 that would pick science as their favorite subject.

Favorite School Subject

Subject	Number of Responses
Math	6
Science	3
English	4
History	7

27. **YOGA** Liliana takes 4 breaths per 10 seconds during yoga. At this rate, about how many breaths would Liliana take in 2 minutes of yoga?

28. **CONTESTS** For a store contest, 4 out of every 65 people who visit a store will receive a free DVD. If 455 people have visited the store, how many DVDs have been given away?

29. **ANALYZE TABLES** There were 340,000 cattle placed on feed in Texas in February 2005. Write a proportion that could be used to find how many of these cattle were between 700 and 799 pounds. How many of the 340,000 cattle placed on feed were between 700 and 799 pounds?

Cattle Placed on Feed in Texas, February 2005

Weight Group	Fraction of Total Cattle
Less than 600 pounds	$\frac{1}{5}$
600–699 pounds	$\frac{11}{50}$
700–799 pounds	$\frac{2}{5}$
800 pounds	$\frac{9}{50}$

Source: National Agriculture Statistics Service

30. **FIND THE DATA** Refer to the Data File on pages 16–19. Choose some data and write a real-world problem in which you would write and solve a proportion.

H.O.T. Problems

31. **OPEN ENDED** One rate of a proportion is $\frac{9}{n}$. Select two other rates to form the proportion, one that can be solved using equivalent fractions and the other that can be solved with unit rates. Then solve the proportion using each method.

32. **FIND THE ERROR** James and Akiko are setting up a proportion to solve the following problem. Who set up their proportion correctly? Explain your reasoning.

Angelina's mom teaches at a preschool. There is 1 teacher for every 12 students at the preschool. There are 276 students at the preschool. How many teachers are there at the preschool?

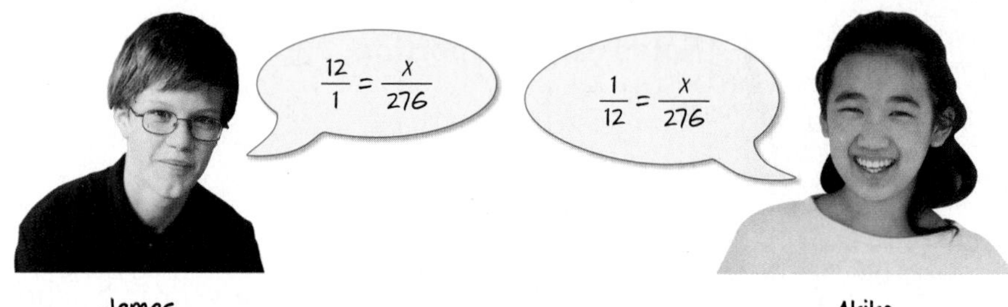

33. **REASONING** Tell whether the following statement is *sometimes*, *always*, or *never* true for numbers greater than zero. Explain.

In a proportion, if the numerator of the first ratio is greater than the denominator of the first ratio, then the numerator of the second ratio is greater than the denominator of the second ratio.

34. **CHALLENGE** Suppose 25 out of 175 people said they like to play disc golf and 5 out of every 12 of the players have a personalized flying disc. At the same rate, in a group of 252 people, predict how many you would expect to have a personalized flying disc.

35. **WRITING IN MATH** Jonah can run 3 laps in 24 minutes. At the same rate, about how many laps can Jonah run in 50 minutes? Explain your reasoning.

TEST PRACTICE

36. A spinner is divided into equal sections. There are 6 green sections and 4 yellow sections. Damon spins the spinner 30 times. Which proportion can be used to find y, the number of times that the spinner can be expected to land on a yellow section?

    A $\frac{y}{30} = \frac{4}{6}$

    B $\frac{y}{30} = \frac{6}{10}$

    C $\frac{y}{30} = \frac{4}{10}$

    D $\frac{y}{30} = \frac{6}{4}$

37. **SHORT RESPONSE** Student Council sells bottled water at the cheerleading competition. They sold 3 cases in 20 minutes. If they continue selling bottled water at this rate, how many cases of bottled water would they sell in 3 hours?

38. The ratio of green pepper plants to red pepper plants in Adeline's garden is about 3 to 5. If there are 20 red pepper plants, about how many green pepper plants are there?

    F 35 H 12

    G 16 J 6

Spiral Review

Determine if the quantities in each pair of rates are proportional. Explain your reasoning and express each proportional relationship as a proportion. (Lesson 6-3)

39. \$36 for 4 baseball hats; \$56 for 7 baseball hats

40. 12 posters for 36 students; 21 posters for 63 students

41. **TUTORS** A tutor charges \$30 for 2 hours. Use the ratio table to determine how much she charges for 5 hours. (Lesson 6-2)

Cost	30		■
Number of Hours	2		5

42. Order $\frac{1}{4}, \frac{7}{40}, \frac{1}{5}, \frac{3}{20}$ from least to greatest. (Lesson 4-8)

WEATHER For Exercises 43 and 44, refer to the table at the right. (Lesson 2-2)

43. Make a line graph of the data.

44. Describe the change in the daily high temperature from Tuesday to Friday.

5-Day Forecast	
Day	High Temperature (°F)
Monday	76
Tuesday	81
Wednesday	78
Thursday	62
Friday	53

GET READY for the Next Lesson

45. **PREREQUISITE SKILL** The Sears Tower in Chicago has 30 stories more than the Aon Centre in Chicago. Together, both buildings have 190 stories. Find the number of stories in each building. Use the *guess and check* strategy. (Lesson 1-7)

CHAPTER 6 Mid-Chapter Quiz

Lessons 6-1 through 6-4

1. **CLASSES** Tyson's math class has 12 boys and 8 girls. What is the ratio of boys to girls? (Lesson 6-1)

2. **SALES** At a bake sale, 15 cookies and 40 brownies were sold. What is the ratio of cookies sold to brownies sold? (Lesson 6-1)

Write each rate as a unit rate. (Lesson 6-1)

3. 171 miles in 3 hours
4. $15 for 3 pounds

5. **MULTIPLE CHOICE** A hockey team made four of their 10 attempted goals. Which ratio compares the number of goals made to the number of goals attempted? (Lesson 6-1)

   A $\frac{4}{5}$

   B $\frac{3}{5}$

   C $\frac{5}{2}$

   D $\frac{2}{5}$

For Exercises 6 and 7, use the ratio tables given to solve each problem. (Lesson 6-2)

6. **MONEY** Peyton spends $15 on lunch every week. At this rate, how much money will he spend in 5 weeks?

Number of Weeks	1				5
Money Spent ($)	15				■

7. **DISHES** Charlee washes 10 dishes in 8 minutes. At this rate, how long will it take her to wash 25 dishes?

Number of Dishes	10		25
Number of Minutes	8		■

Determine if the quantities in each pair of ratios or rates are proportional. Explain your reasoning and express each proportional relationship as a proportion. (Lesson 6-3)

8. $4 for 12 doughnuts; $9 for 36 doughnuts
9. 24 pages read in 8 minutes; 72 pages read in 24 minutes
10. 48 out of 64 students own cell phones; 192 out of 258 students own cell phones

11. **MULTIPLE CHOICE** The ratio of brown tiles to tan tiles in a mosaic is 2 to 3. Which of these shows the possible numbers of brown tiles and tan tiles in the mosaic? (Lesson 6-3)

    F 16 brown tiles, 24 tan tiles

    G 14 brown tiles, 20 tan tiles

    H 12 brown tiles, 19 tan tiles

    J 8 brown tiles, 9 tan tiles

Solve each proportion. (Lesson 6-4)

12. $\frac{x}{6} = \frac{12}{18}$

13. $\frac{8}{20} = \frac{30}{x}$

14. $\frac{3}{d} = \frac{9}{15}$

15. $\frac{24}{72} = \frac{x}{6}$

16. **MULTIPLE CHOICE** Christina made 4 bracelets in 36 minutes. At this rate, how many bracelets would she make in 108 minutes? (Lesson 6-4)

    A 8

    B 9

    C 12

    D 16

17. **ACTIVITIES** Suppose 8 out of 24 students in a classroom participate in after-school activities. At the same rate, predict how many students in a school of 960 can be expected to participate in after-school activities. (Lesson 6-4)

6-5 Problem-Solving Investigation

MAIN IDEA: Solve problems by looking for a pattern.

P.S.I. TEAM +

e-Mail: LOOK FOR A PATTERN

LISETE: I'm building a model of a set of stairs using cubes. I used 4 cubes to make the first step, 8 cubes for the second step, and 12 cubes for the third step.

YOUR MISSION: Use the look for a pattern strategy to find how many cubes will be used to make the eighth step.

Understand	You know how many cubes were used to make the first three steps. You need to find how many cubes are needed to make the eighth step.
Plan	Look for a pattern to find the total number of cubes needed.
Solve	Use a table to find the pattern.
Check	Draw a sketch of all eight steps. Count the total number of cubes. Since there are 32 cubes altogether, the answer is correct. ✓

Step Number	Number of Cubes
1	4
2	8
3	12
⋮	⋮
8	■

The number of cubes is 4 times the step number. So, the number of cubes needed for the eighth step is equal to 8×4, or 32 cubes.

Analyze The Strategy

1. Explain when you would use the *look for a pattern* strategy to solve a problem.
2. WRITING IN MATH Write a problem that can be solved by looking for a pattern. Then write the steps you would take to find the solution.

Mixed Problem Solving

EXTRA PRACTICE
See pages 689, 711.

Use the *look for a pattern* strategy to solve Exercises 3–5.

3. **MONEY** Every year, Miguel receives $20 for his birthday, plus $1 for each year of his age. Lauren receives $10 for her birthday and $2 for each year of her age. In 2009, Miguel is 10, and Lauren is 8. In what year will they both receive the same amount of money?

4. **GEOMETRY** Draw the next two figures in the pattern below.

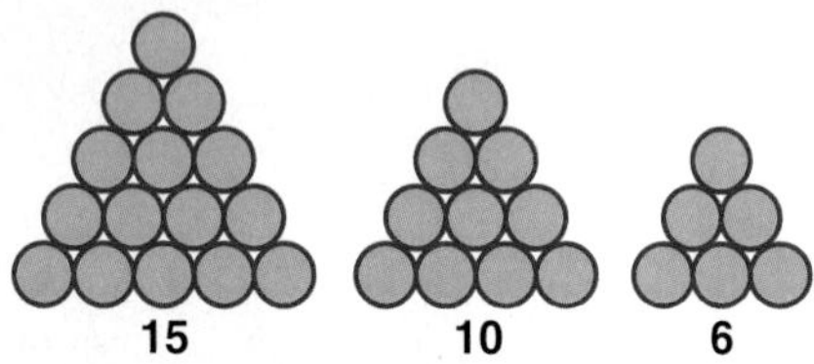

5. **NUMBER SENSE** Describe the pattern below. Then find the next three numbers.

   3, 6, 10, 15, 21, ■, ■, ■

Use any strategy to solve Exercises 6–14. Some strategies are shown below.

PROBLEM-SOLVING STRATEGIES
· Guess and check.
· Look for a pattern.
· Act it out.

6. **FOOD** Which is more, $\frac{3}{8}$ of a pizza or $\frac{1}{3}$ of a pizza?

7. **MONEY** The admission for a fair is $6 for adults, $4 for children, and $3 for senior citizens. Twelve people paid a total of $50 for admission. If 8 children attended, how many adults and senior citizens attended?

8. **FOOD** About how much more money is spent on strawberry and grape jelly than the other types of jelly each year?

Yearly Jelly Sales (thousands)	
strawberry and grape	$366.2
all others	$291.5

Source: Nielson Marketing Research

9. **NUMBER THEORY** The triangle below is known as *Pascal's Triangle*. If the pattern continues, what will the numbers in the next row be from left to right?

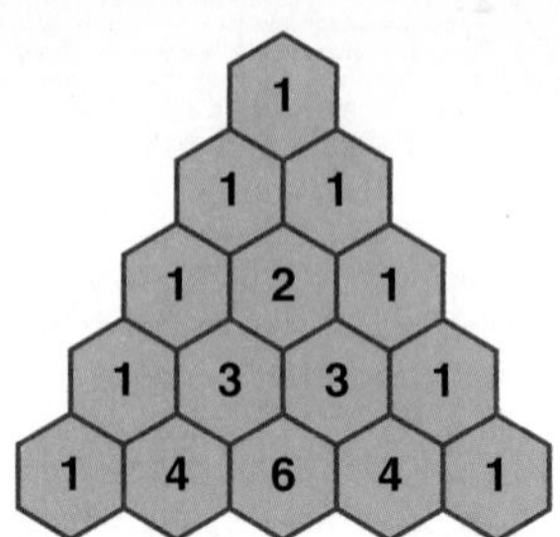

10. **LANGUAGE ARTS** On Monday, 86 science fiction books were sold at a book sale. This is 8 more than twice the amount sold on Thursday. How many science fiction books were sold on Thursday?

11. **PATTERNS** Find the number of toothpicks needed to create figure 8 in the pattern below.

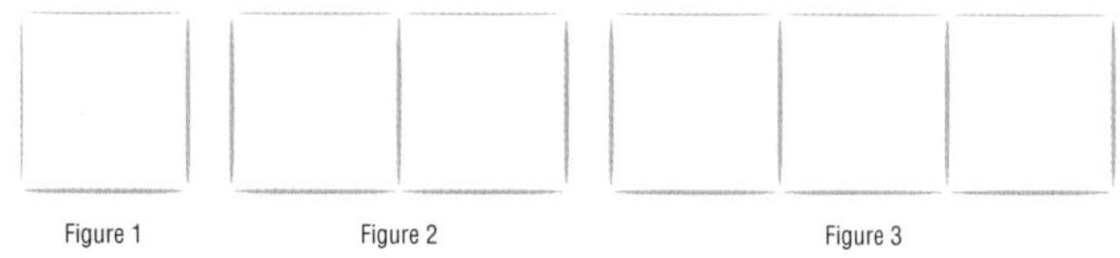

12. **TICKETS** The total price, including a service fee, of a concert ticket is $66.45. If the service fee was $4.95, what was the original cost of the concert ticket?

13. **TRAVEL** Mr. Ishikawa left Kansas City, Missouri, at 3:00 P.M. and arrived in Tulsa, Oklahoma, at 8:00 P.M., driving a distance of approximately 240 miles. During his trip, he took a one-hour dinner break. What was Mr. Ishikawa's average speed in miles per hour?

14. **FIELD TRIPS** Mrs. Samuelson had $350 to spend on a field trip for herself and 18 students. The rate of admission was $13 per person and lunch cost about $5 per person. How much money was left after the trip?

6-6 Sequences and Expressions

MAIN IDEA

Extend and describe arithmetic sequences using algebraic expressions.

New Vocabulary

sequence
term
arithmetic sequence

Math Online

glencoe.com

- Extra Examples
- Personal Tutor
- Self-Check Quiz

GET READY for the Lesson

PIZZA The table shows the number of slices of pizza for different numbers of large pizzas.

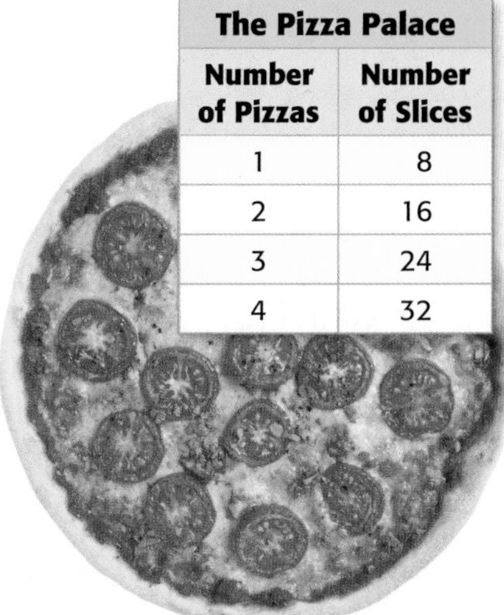

The Pizza Palace	
Number of Pizzas	**Number of Slices**
1	8
2	16
3	24
4	32

1. Find the rate of slices to the number of pizzas for each row in the table.
2. Is the number of pizzas proportional to the number of slices? Explain your reasoning.
3. Make an ordered list of the number of slices and describe the pattern between consecutive numbers in this list.
4. What relationship appears to exist between the pattern found in Exercise 3 and the rates found in Exercise 1?

The number of slices in the table above is an example of a sequence. A **sequence** is a list of numbers in a specific order. Each number in the list at the right is called a **term** of the sequence.

sequence

8, 16, 24, 32, ...

term

This sequence is an **arithmetic sequence** because each term is found by adding the same number to the previous term.

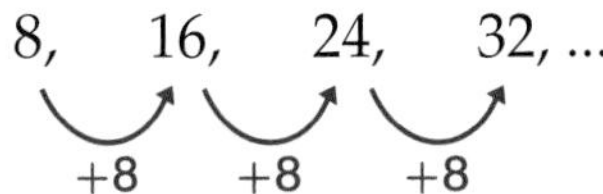

There are several ways of showing a sequence. In addition to being shown as a list, a sequence can also be shown in a table. The table gives both the position of each term in the list and the value of the term.

List

8, 16, 24, 32, ...

Table

Position	1	2	3	4
Value of Term	8	16	24	32

You can also write an algebraic expression to describe a sequence. The value of each term can be described as a function of its position in the sequence.

EXAMPLES Describe Sequences

Use words and symbols to describe the value of each term as a function of its position. Then find the value of the tenth term in the sequence.

1

Position	1	2	3	4	n
Value of Term	3	6	9	12	■

Notice that the value of each term is 3 times its position number. So, the value of the term in position n is $3n$.

Position	Multiply by 3	Value of Term
1	1×3	3
2	2×3	6
3	3×3	9
4	4×3	12
n	$n \times 3$	$3n$

Study Tip

Look Back You can review **evaluating algebraic expressions** in Lesson 1-5.

Now find the value of the tenth term.

$3n = 3 \cdot 10$ Replace n with 10.

$= 30$ Multiply.

The value of the tenth term in the sequence is 30.

Position	6	7	8	9	n
Value of Term	2	3	4	5	■

The value of each term is 4 less than its position number. So, the value of the term in position n is $n - 4$.

Position	Subtract 4	Value of Term
6	$6 - 4$	2
7	$7 - 4$	3
8	$8 - 4$	4
9	$9 - 4$	5
n	$n - 4$	$n - 4$

Study Tip

Arithmetic Sequences Some arithmetic sequences are proportional. In Example 2, $\frac{6}{2} \neq \frac{7}{3} \neq \frac{8}{4} \neq \frac{9}{5}$. However, all arithmetic sequences can be described by an algebraic expression.

Now find the value of the tenth term.

$n - 4 = 10 - 4$ Replace n with 10.

$= 6$ Subtract.

The value of the tenth term in the sequence is 6.

CHECK Your Progress

Use words and symbols to describe the value of each term as a function of its position. Then find the value of the eighth term in the sequence.

a.

Position	10	11	12	13	n
Value of Term	7	8	9	10	■

b.

Position	2	3	4	5	n
Value of Term	12	18	24	30	■

Real-World EXAMPLE Make a Table

3 **MEASUREMENT There are 12 inches in 1 foot. Make a table and write an algebraic expression relating the number of feet to the number of inches. Then find Becca's height in feet if she is 60 inches tall.**

Inches	Feet
12	1
24	2
36	3
48	4
n	$n \div 12$

Notice that the number of inches divided by 12 gives the number of feet. So, to find Becca's height, use the expression $n \div 12$.

$n \div 12 = 60 \div 12$ Replace n with 60.

$= 5$ Divide.

So, Becca is 5 feet tall.

CHECK Your Progress

c. **RUNNING** There are 3 feet in 1 yard. Make a table and write an algebraic expression relating the number of feet to the number of yards. Then find how many feet Summer ran if she ran 400 yards.

TEST EXAMPLE

4 **Which expression was used to create the table at the right?**

Position	Value of Term
1	5
2	7
3	9
n	■

A $n + 3$ **C** $2n + 3$

B $2n$ **D** $n - 3$

Test-Taking Tip

Eliminate Answer Choices As you examine a multiple-choice test item, eliminate answer choices you know to be incorrect.

Read the Item To find the expression, determine the function.

Solve the Item Notice that the values 5, 7, 9, ... increase by 2, so the rule contains $2n$. Therefore, choices A and D can be eliminated.

If the rule were simply $2n$, then the value for position 1 would be 2×1 or 2. But this value is 5. So, choice B can be eliminated.

This leaves choice C. Test a few values.

Row 1: $2n + 3 = 2(1) + 3 = 2 + 3$ or 5
Row 3: $2n + 3 = 2(3) + 3 = 6 + 3$ or 9

So, the answer is C.

CHECK Your Progress

d. **Which expression was used to create the table at the right?**

Position	Value of Term
1	4
2	9
3	14
n	■

F $5n - 1$ **H** $n + 1$

G $5n$ **J** $n + 5$

CHECK Your Understanding

Examples 1, 2 (p. 344)

Use words and symbols to describe the value of each term as a function of its position. Then find the value of the fifteenth term in the sequence.

1.

Position	1	2	3	4	n
Value of Term	2	4	6	8	■

2.

Position	3	4	5	6	n
Value of Term	10	11	12	13	■

Example 3 (p. 345)

3. **MEASUREMENT** There are 16 ounces in 1 pound. Make a table and write an algebraic expression relating the number of ounces to the number of pounds. Then find the number of ounces of potatoes Mr. Padilla bought if he bought a ten-pound bag of potatoes.

Example 4 (p. 345)

4. **MULTIPLE CHOICE** The table at the right shows the fee for overdue books at a library, based on the number of weeks the book is overdue. Which expression was used to find the fee for a book that is n weeks overdue?

Weeks Overdue	Fee ($)
1	3
2	5
3	7
4	9
n	■

A $n + 1$ C $n - 1$

B $2n$ D $2n + 1$

Practice and Problem Solving

HOMEWORK HELP	
For Exercises	See Examples
5–8	1, 2
9, 10	3
11, 12, 28, 29	4

Use words and symbols to describe the value of each term as a function of its position. Then find the value of the twelfth term in the sequence.

5.

Position	3	4	5	6	n
Value of Term	12	13	14	15	■

6.

Position	6	7	8	9	n
Value of Term	2	3	4	5	■

7.

Position	1	2	3	4	n
Value of Term	5	10	15	20	■

8.

Position	2	3	4	5	n
Value of Term	24	36	48	60	■

9. **MEASUREMENT** There are 60 minutes in 1 hour. Make a table and write an algebraic expression relating the number of hours to the number of minutes. Then find the duration of the movies in hours if Hannah and her friends watched two movies that together were 240 minutes long.

10. **MEASUREMENT** There are 12 months in 1 year. Make a table and write an algebraic expression relating the number of months to the number of years. Then find Andre's age in months if he is 12 years old.

Real-World Link . . . About 15 years ago, there were only 3 rock gyms in the U.S. Today, there are over 700 rock gyms.

Source: Health A to Z

ANALYZE TABLES Use the table at the right and the following information for Exercises 11 and 12.

The table shows the amount it costs to rock climb at an indoor rock climbing facility, based on the number of hours.

Time (h)	Amount ($)
1	13
2	21
3	29
4	37
n	■

11. How does the cost change with each additional hour?

12. What is the rule to find the amount charged to rock climb for n hours?

Determine how the next term in each sequence can be found. Then find the next two terms in the sequence.

13. 1, 4, 7, 10, ...
14. 4, 16, 28, 40, ...
15. 2.3, 3.2, 4.1, 5.0, ...
16. 1.5, 3.9, 6.3, 8.7, ...
17. $1\frac{1}{2}, 3, 4\frac{1}{2}, 6, \ldots$
18. $2\frac{1}{4}, 2\frac{3}{4}, 3\frac{1}{4}, 3\frac{3}{4}, \ldots$

Find the missing number in each sequence.

19. $7, ■, 16, 20\frac{1}{2}, \ldots$
20. 14.6, ■, 24, 28.7
21. $30, ■, 19, 13\frac{1}{2}, \ldots$
22. 43.8, 36.7, ■, 22.5, ...

23. **MEASUREMENT** There are 24 hours in 1 day. Make a table and write an algebraic expression relating the number of hours to the number of days. Then find the number of hours in 1 week.

24. **GEOMETRY** Assume the pattern below continues. Write an algebraic expression to find the number of squares in Figure 8. Then find the number of squares in Figure 8.

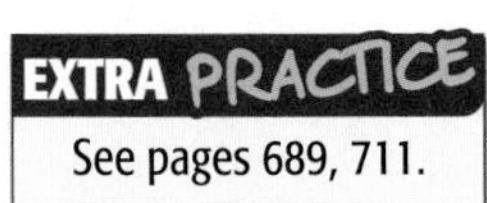

EXTRA PRACTICE See pages 689, 711.

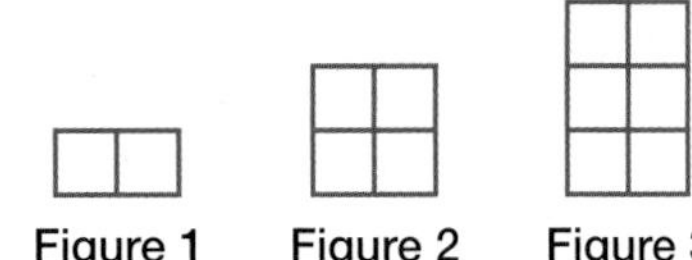

H.O.T. Problems

25. **OPEN ENDED** Create a sequence in which $1\frac{1}{4}$ is added to each number.

26. **SELECT A TECHNIQUE** Gene charges a base fee of $5 for each lawn he mows plus $2 for the number of hours it takes to complete the job. Which of the following techniques might Gene use to determine an expression he can use to represent the total charge for mowing a lawn based on the number of hours? Justify your selection(s). Then use the technique(s) to solve the problem.

mental math | number sense | estimation

27. **ANALYZE TABLES** Use words and symbols to generalize the relationship of each term as a function of its position. Then determine the value of the term when $n = 100$.

Position	1	2	3	4	5	n
Value of Term	1	4	9	16	25	■

28. **WRITING IN MATH** Write a problem about a real-world situation in which you would use a sequence to describe a pattern.

TEST PRACTICE

29. What is the rule to find the value of the missing term in the sequence below?

Position, x	Value of Term
1	1
2	5
3	9
4	13
5	17
x	■

A $x + 4$

B $4x - 3$

C $4x$

D $x - 3$

30. The table shows Samantha's age and Ling's age over four consecutive years.

Samantha's Age, x (years)	Ling's Age, y (years)
9	6
10	7
11	8
12	9

Which expression represents Ling's age in terms of Samantha's age?

F $y - 3$

G $3x$

H $3y$

J $x - 3$

Spiral Review

31. **ART** Mr. Torres is hanging his students' drawings in rows on one wall of his classroom. He places 1 drawing on the top row, 3 on the second row, and 5 on the third row. If this pattern continues, how many drawings will be on the seventh row? (Lesson 6-5)

Solve each proportion. (Lesson 6-4)

32. $\frac{3}{4} = \frac{a}{28}$

33. $\frac{5}{x} = \frac{45}{63}$

34. $\frac{24}{38} = \frac{12}{m}$

35. $\frac{x}{75} = \frac{5}{25}$

GET READY for the Next Lesson

PREREQUISITE SKILL Determine what number should be added to the first number to get the second number. (Lesson 5-5)

36. $4\frac{1}{4}, 4\frac{1}{2}$

37. $8\frac{1}{2}, 10$

38. $9, 12\frac{1}{2}$

39. $1\frac{2}{3}, 2\frac{1}{3}$

6-7 Proportions and Equations

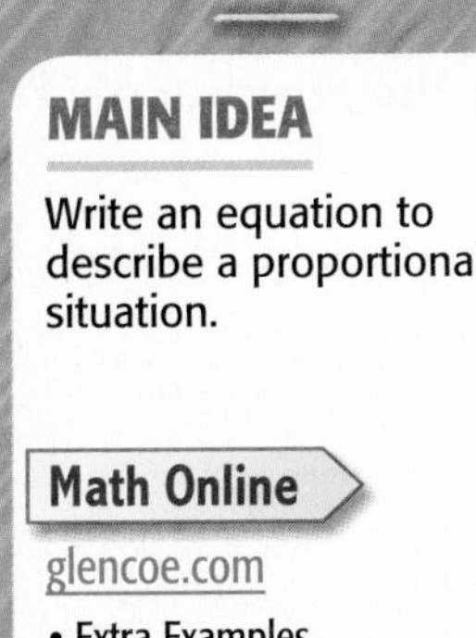

MAIN IDEA

Write an equation to describe a proportional situation.

Math Online

glencoe.com

- Extra Examples
- Personal Tutor
- Self-Check Quiz

GET READY for the Lesson

BABYSITTING The table at the right shows the amount of money Carli earns based on the number of hours she babysits.

Hours Babysitting	Earnings ($)
1	5
2	10
3	15
4	20

1. Write a sentence that describes the relationship between the number of hours she babysits and her earnings.
2. Is the relationship proportional? Explain.
3. What is the rule to find the amount Carli earns for babysitting h hours?
4. If e represents the amount Carli earns, what equation can you use to represent this situation?

In Lesson 6-6, you used algebraic expressions to determine how to find the value of a term in a sequence. Similarly, you can use an equation to represent a function.

EXAMPLE Write an Equation for a Function

1 Write an equation to represent the function displayed in the table.

Input, x	1	2	3	4	5
Output, y	9	18	27	36	45

Examine how the value of each input and output changes. Each output y is equal to 9 times the input x. So, the equation that represents the function is $y = 9x$.

Input, x	Multiply by 9	Output, y
1	1×9	9
2	2×9	18
3	3×9	27
4	4×9	36
5	5×9	45

CHECK Your Progress

a. Write an equation to represent the function displayed in the table.

Input, x	1	2	3	4	5
Output, y	16	32	48	64	80

You can often use an equation to represent a real-world situation.

FUNDRAISING The cheerleading squad is holding a car wash to raise money. They are charging $7 for each car they wash.

2 Make a table to show the relationship between the number of cars washed *c* and the total amount earned *t*.

The total earned (output) is equal to $7 times the number of cars washed (input).

Cars Washed, *c*	Multiply by 7	Total Earned ($), *t*
1	1×7	7
2	2×7	14
3	3×7	21
4	4×7	28

3 Write an equation to find the total amount earned *t* for washing *c* cars.

Study the table from Example 2. The total earned equals $7 times the number of cars washed.

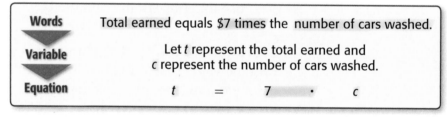

Words	Total earned equals $7 times the number of cars washed.
Variable	Let *t* represent the total earned and *c* represent the number of cars washed.
Equation	$t = 7 \cdot c$

So, the equation is $t = 7c$.

Real-World Link. Bald eagles can dive at speeds up to 200 miles per hour.
Source: San Diego Zoo

4 How much will the cheerleading squad earn if they wash 25 cars?

$t = 7c$ Write the equation.

$t = 7 \cdot 25$ Replace *c* with 25.

$t = 175$ Multiply.

The cheerleading squad will earn $175 for washing 25 cars.

CHECK Your Progress

BALD EAGLE While in normal flight, a bald eagle flies at an average speed of 30 miles per hour.

b. Make a table to show the relationship between the total distance *d* that a bald eagle can travel in *h* hours.

c. Write an equation to find the total distance *d* that a bald eagle can travel in *h* hours while in normal flight.

d. How many miles can a bald eagle travel in 2 hours?

5 **JET SKIING The cost of renting a jet ski at a local marina is shown in the table. Write a sentence and an equation to describe the data. Then find the total cost of renting a jet ski for 6 hours.**

Number of Hours, h	Total Cost ($), t
1	10
2	20
3	30
4	40

The cost of renting a jet ski is $10 for each hour. The total cost t is $10 times the number of hours h. Therefore, $t = 10h$. Use this equation to find the total cost t of renting a jet ski for 6 hours.

$t = 10h$ Write the equation.

$t = 10 \cdot 6$ or 60 Replace h with 6. Multiply.

The total cost of renting a jet ski for 6 hours is $60.

e. **FITNESS** A fitness center charges the amount shown in the table for using the facility. Write a sentence and an equation to describe the data. Then find the total cost of using the fitness center for 12 months.

Number of Months, m	Total Fee ($), t
1	25
2	50
3	75
4	100

CHECK Your Understanding

Example 1 (p. 349)

Write an equation to represent the function displayed in each table.

1.

Input, x	0	1	2	3	4
Output, y	0	4	8	12	16

2.

Input, x	1	2	3	4	5
Output, y	8	16	24	32	40

Examples 2–4 (p. 350)

LUNCH Use the following information for Exercises 3–5.

The school cafeteria sells lunch passes that allow a student to purchase any number of lunches in advance for $3 a lunch.

3. Make a table to show the relationship between the number of lunches n and the total cost t.
4. Write an equation to find t, the total cost in dollars for a lunch card with n lunches.
5. If Lolita buys a lunch pass for 20 lunches, how much will it cost?

Example 5 (p. 351)

6. **CARNIVAL** The general admission to a local carnival is shown in the table. Write a sentence and an equation to describe the data. Then find the total cost of admission for 7 people.

Number of People, n	Total Admission ($), t
1	4
2	8
3	12
4	16

Practice and Problem Solving

HOMEWORK HELP	
For Exercises	See Examples
7–10	1
11, 14	2
12, 15	3
13, 16	4
17, 18	5

Write an equation to represent the function displayed in each table.

7.

Input, x	1	2	3	4	5
Output, y	6	12	18	24	30

8.

Input, x	0	1	2	3	4
Output, y	0	11	22	33	44

9.

Input, x	0	1	2	3	4
Output, y	0	15	30	45	60

10.

Input, x	1	2	3	4	5
Output, y	10	20	30	40	50

VIDEO GAMES Use the following information for Exercises 11–13.

In a video game, each player earns 15 points for each coin they collect.

11. Make a table to show the relationship between the number of coins collected c and the total points p.
12. Write an equation to find p, the total points for collecting c coins.
13. How many points will a player earn if she collects 21 coins?

ELEPHANTS Use the following information for Exercises 14–16.

An African elephant eats at a rate of 400 pounds of vegetation each day.

14. Make a table to show the relationship between the number of pounds v an African elephant eats in days d.
15. Write an equation to find v, the number of pounds of vegetation an African elephant eats in d days.
16. How many pounds of vegetation does an African elephant eat in 5 days?

Real-World Link. An African elephant spends up to 16 hours each day searching for and feeding on vegetation in the form of grasses, tree limbs, tubers, fruits, vines, and shrubs.

17. **ENTERTAINMENT** The disc jockey hired for the spring dance charges the amount shown in the table. Write a sentence and an equation to describe the data. At this rate, how much will it cost to hire the disc jockey for 5 hours?

Number of Hours, h	Total Charge (\$), t
1	35
2	70
3	105

18. **FOOD** A catering service provides the table shown as a guide to help its customers decide how many pans of lasagna to order for events. Write a sentence and an equation to describe the data. At this rate, how many people would 8 pans of lasagna serve?

Number of Pans, p	Total Number Served, n
1	24
2	48
3	72

19. **RESEARCH** Use the Internet or another source to find the average amount of food that another animal eats per day. Then write an equation to find f, the amount of food the animal eats in d days.

Write an equation to represent the function displayed in each table.

20.

Input, x	2	4	6	8	10
Output, y	1	2	3	4	5

21.

Input, x	3	6	9	12	15
Output, y	1	2	3	4	5

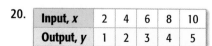

22. **FIND THE DATA** Refer to the Data File on pages 16–19. Choose some data and write a real-world problem in which you would write an equation using two variables to represent a problem situation.

23. **WEATHER** Write an equation to find the total precipitation t in inches for Burbank in m months. How much precipitation does Burbank receive in 4 months? Compare this to the total precipitation in 4 months for Coronado.

City	Average Annual Precipitation (in.)
Burbank	12
Coronado	9
Pasadena	20

Source: Weatherbase

EXTRA PRACTICE
See pages 689, 711.

H.O.T. Problems

24. **OPEN ENDED** Write about a real-world situation that can be represented by the equation $y = 5x$. Be sure to explain what the variables represent in the situation.

25. **CHALLENGE** Write an equation to represent the function in the table.

Input, x	6	8	10	12	14	16
Output, y	0	1	2	3	4	5

26. **WRITING IN MATH** Choose an exercise from this lesson, and explain why the relationship is proportional.

TEST PRACTICE

27. The table shows admission prices at a local zoo based on the number of guests.

Number of Guests, x	Total Admission ($), y
1	7
2	14
3	21
4	28

Which equation can be used to find y, the total admission for x guests?

A $x = 7y$
B $y = 7 + x$
C $y = 7x$
D $x = 7 + y$

28. If the cost of snorkeling is $5 for the equipment plus an additional $7 for each hour that you snorkel, which equation represents c, the cost in dollars for snorkeling for h hours?

F $c = 7h + 5$
G $c = 5h + 7$
H $c = 7(h + 5)$
J $c = 5(h + 7)$

Spiral Review

29. Find the next two terms of the sequence 3, 11, 19, 27, … . (Lesson 6-6)

30. **MONEY** Maxine withdraws the same amount of money from her savings account each month. In March, her account balance was $100. Her next monthly balances were $85, $70, and $55. At this rate, what will her account balance be in September? (Lesson 6-5)

Extend 6-7

Graphing Calculator Lab
Graphing Proportional Relationships

MAIN IDEA

Graph proportional relationships.

Math Online

glencoe.com

- Other Calculator Keystrokes

You can use a TI-83/84 Plus graphing calculator to graph an equation that represents a proportional relationship.

ACTIVITY

CORN DOGS Suppose Booth A at a fair sells corn dogs for $2 each while Booth B charges $3 for each corn dog. Write and graph an equation representing the total cost y for selling x corn dogs.

STEP 1 Write an equation to show the relationship between the number of corn dogs sold x and total cost y at each booth.

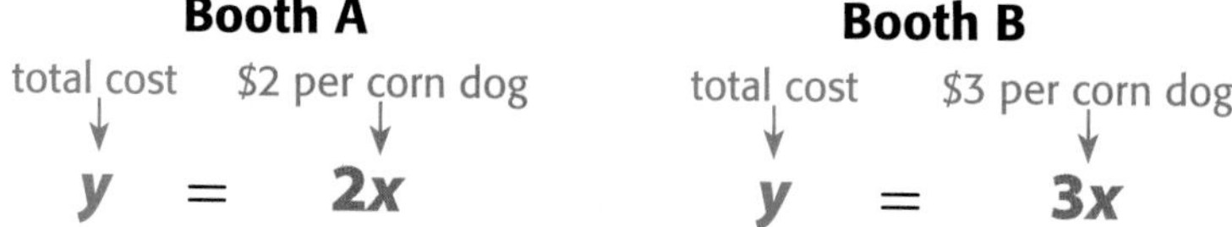

Booth A

total cost → y = $2x$ ← $2 per corn dog

Booth B

total cost → y = $3x$ ← $3 per corn dog

STEP 2 Press [Y=] on your calculator and enter the expression $2x$ into Y_1 and $3x$ into Y_2.

STEP 3 Next, change the view of your graph by pressing [WINDOW] and entering the values shown below.

STEP 4 Finally, graph the equations by pressing [Graph].

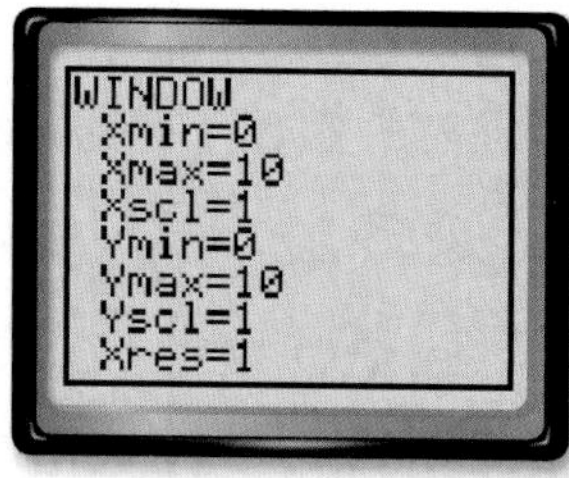

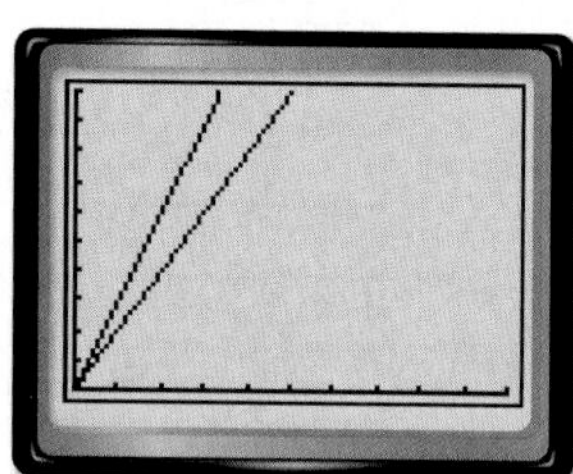

Analyze the Results

1. **MAKE A CONJECTURE** Which equation is represented by the steeper line? Explain. To check your conjecture, press [Trace] and then use [↑] and [↓] keys to switch between the two lines.

2. Press [2nd] [CALC] 1 and enter a value of 3. Switch between the two lines. What do the x- and y-values displayed represent?

3. **MAKE A CONJECTURE** Suppose Booth C sells corn dogs for $2.50 each. How should the graph of this equation appear in relationship to the graphs of the other two equations? Explain. Check your answer by entering the appropriate equation into Y_3, pressing [Graph], and switching between the equations.

Chapter 6 Study Guide and Review

Math Online glencoe.com
- STUDY TO GO
- Vocabulary Review

FOLDABLES® Study Organizer — GET READY to Study

Be sure the following Big Ideas are noted in your Foldable.

Ratio	Proportion	Function
Examples	Examples	Examples

BIG Ideas

Ratios (Lessons 6-1 and 6-2)
- Ratios are a comparison of two quantities by division.
- Equivalent ratios express the same relationship between two quantities.
- To find an equivalent ratio, multiply or divide each quantity in the ratio by the same number.

Rates (Lesson 6-1)
- Rates are a ratio comparing two quantities with different kinds of units.
- A unit rate is a rate for one unit of a given quantity.

Proportions (Lessons 6-3 and 6-4)
- Proportions are equations stating that two ratios are equivalent.
- To solve a proportion, use either an equivalent fraction or a unit rate.

Sequences (Lesson 6-6)
- Sequences are a list of numbers in a specific order.
- Each number listed in a sequence is called a term.
- Each term of an arithmetic sequence is found by adding the same number to the previous term.
- Algebraic expressions can be used to describe a sequence.

Equations (Lesson 6-7)
- An equation can be used to represent a function.

Key Vocabulary

arithmetic sequence (p. 343)
equivalent ratio (p. 322)
proportion (p. 329)
proportional (p. 329)
rate (p. 315)
rate of change (p. 343)
ratio (p. 314)
ratio table (p. 322)
scaling (p. 323)
sequence (p. 343)
term (p. 343)
unit rate (p. 315)

Vocabulary Check

State whether each sentence is *true* or *false*. If *false*, replace the underlined word or number to make a true sentence.

1. A ratio is a comparison of two numbers by <u>multiplication</u>.
2. A <u>rate</u> is the ratio of two measurements that have different units.
3. Two quantities are said to be <u>proportional</u> if they have a constant ratio or rate.
4. A proportion is an equation stating that two ratios or rates are <u>not equal</u>.
5. <u>Adding or subtracting</u> two related quantities by the same number is called scaling.
6. A <u>sequence</u> is a list of numbers in a specific order.
7. In an arithmetic sequence, each term is found by <u>multiplying</u> the same number.
8. In a ratio table, the columns are filled with pairs of numbers that have the same <u>term</u>.

Lesson-by-Lesson Review

6-1 Ratios and Rates (pp. 314–319)

Write each ratio or rate as a fraction in simplest form.

9. 12 blue marbles out of 20 marbles
10. 9 goldfish out of 36 fish
11. 18 boys out of 21 students

Write each rate as a unit rate.

12. 3 inches of rain in 6 hours
13. 189 pounds of garbage in 12 weeks
14. 78 erasers in 3 packages
15. **DVDs** Rick has 12 action, 15 comedy, and 9 drama DVDs. Find the ratio of action DVDs to the total number of DVDs. Then explain its meaning.

Example 1 **Write the ratio 30 sixth graders out of 45 students as a fraction in simplest form.**

$$\frac{30}{45} = \frac{2}{3} \quad (\div 15)$$

The GCF of 30 and 45 is 15.

Example 2 **Write the rate 150 miles in 4 hours as a unit rate.**

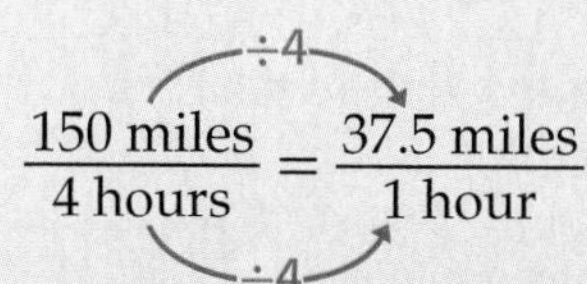

÷4
150 miles / 4 hours = 37.5 miles / 1 hour
÷4

Divide the numerator and the denominator by 4 to get the denominator of 1.

6-2 Ratio Tables (pp. 322–327)

For Exercises 16 and 17, use the ratio tables given to solve each problem.

16. **MONEY** Arthur bought 5 notebooks for $3. How much will he spend on 10 notebooks?

Number of Notebooks	5	10
Money Spent ($)	3	■

17. **TRUCKS** In a parking lot, 3 out of 8 vehicles were trucks. If there were 128 vehicles, how many were trucks?

Number of Trucks	3		■
Number of Vehicles	8		128

18. **TICKETS** Roman spent $306 for 17 baseball tickets. Use a ratio table to determine how much each ticket costs.

Example 3 **Boston received 6 inches of rain in 24 hours. If it rained at a constant rate, use a ratio table to determine how much rain Boston would receive in 48 hours.**

Set up a ratio table. Label the rows with the two quantities being compared. Then fill in what is given.

Inches of Rain	6	■
Number of Hours	24	48

Use scaling to find the desired quantity.

×2

Inches of Rain	6	12
Number of Hours	24	48

×2

So, Boston would receive 12 inches of rain in 48 hours.

Mixed Problem Solving
For mixed problem-solving practice, see page 711.

6-3 Proportions (pp. 329–333)

Determine if the quantities in each pair of ratios or rates are proportional. Explain your reasoning and express each proportional relationship as a proportion.

19. 220 ft to 25 in.; 88 ft to 10 in.
20. 14 girls to 21 boys; 21 girls to 34 boys
21. **JEWELRY** Stacey made 8 necklaces in 48 minutes. Nick made 4 necklaces in 24 minutes. Are these rates proportional? Explain your reasoning.

Example 4 **Are the ratios 18 weeks to 3 years and 54 weeks to 10 years proportional? Explain your reasoning.**

Find the unit ratio for each ratio.

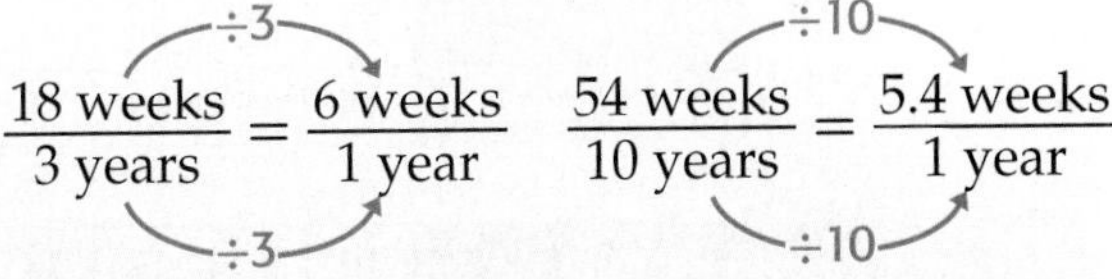

$$\frac{18 \text{ weeks}}{3 \text{ years}} = \frac{6 \text{ weeks}}{1 \text{ year}} \quad (\div 3) \qquad \frac{54 \text{ weeks}}{10 \text{ years}} = \frac{5.4 \text{ weeks}}{1 \text{ year}} \quad (\div 10)$$

Since the ratios do not share the same unit ratio, they are not equivalent. Therefore, they are not proportional.

6-4 Algebra: Solving Proportions (pp. 334–339)

Solve each proportion.

22. $\frac{7}{11} = \frac{m}{33}$
23. $\frac{3}{20} = \frac{15}{k}$
24. $\frac{g}{20} = \frac{9}{60}$
25. $\frac{25}{h} = \frac{5}{7}$
26. **DOGS** In the sixth grade, 12 out of 27 students have a dog. If there are 162 students, predict how many have a dog.

Example 5 **Solve the proportion $\frac{9}{12} = \frac{g}{4}$.**

$\frac{9}{12} = \frac{3}{4}$ (÷3) Since 12 ÷ 3 = 4, divide the numerator and denominator by 3.

So, $g = 3$.

6-5 PSI: Look for a Pattern (pp. 341–342)

Solve by looking for a pattern.

27. **LAPS** Cheyenne ran 1 lap on day 1. She ran 2 laps on day 2, 4 laps on day 3, and 8 laps on day 4. If this pattern continues, how many laps will she run on day 7?
28. **BASEBALL** The table shows the scores at a baseball game. If the scoring pattern continues, how many runs will the Reds score in the 7th inning?

Inning	1	2	3	4	5	6	7
Cubs	0	1	2	4	0	1	3
Reds	4	3	2	0	4	3	■

Example 6 **A display of cans is stacked in the shape of a pyramid. There are 2 cans in the top row, 4 in the second row, 6 in the third row, and so on. The display contains 7 rows. How many cans are in the display?**

Make a table to see the pattern.

Row	1	2	3	4	5	6	7
Number of Cans	2	4	6	8	10	12	14

+2 +2 +2 +2 +2 +2

There are 2 + 4 + 6 + 8 + 10 + 12 + 14 or 56 cans in the display.

CHAPTER 6 Study Guide and Review

6-6 Sequences and Expressions (pp. 343–348)

Use words and symbols to describe the value of each term as a function of its position. Then find the value of the sixteenth term in the sequence.

29.

Position	22	23	24	25	n
Value of Term	15	16	17	18	■

30.

Position	10	11	12	13	n
Value of Term	28	31	34	37	■

31. **BUSINESS** Jennifer earns $12 for every dog she walks. Write an algebraic expression to find the amount of money she would earn for walking n number of dogs. Then find the amount of money she would earn for walking 45.

Example 7 **Use words and symbols to describe the value of each term as a function of its position. Then find the value of the ninth term in the sequence.**

Position	5	6	7	8	n
Value of Term	20	24	28	32	■

Study the relationship between each position and the value of its term.

Position	Multiply by 4	Value of Term
5	5 × **4**	20
6	6 × **4**	24
7	7 × **4**	28
8	8 × **4**	32
n	n × **4**	**4*n***

Notice that the value of each term is 4 times its position number.

$4n = 4 \cdot 9$ Replace n with 9.

$= 36$ Multiply.

The value of ninth term in the sequence is 36.

6-7 Proportions and Equations (pp. 349–353)

Write an equation to represent the function displayed in each table.

32.

Input, x	1	2	3	4	5
Output, y	6	12	18	24	30

33.

Input, x	1	2	3	4	5
Output, y	7	14	21	28	35

34. **BOWLING** A bowling alley charges $4 per game. Write an equation to find the total cost in dollars t for the number of games bowled n. Then find the total cost for 8 games bowled.

Example 8 **Write an equation to represent the function displayed in the table.**

Examine how the value of each input and output changes.

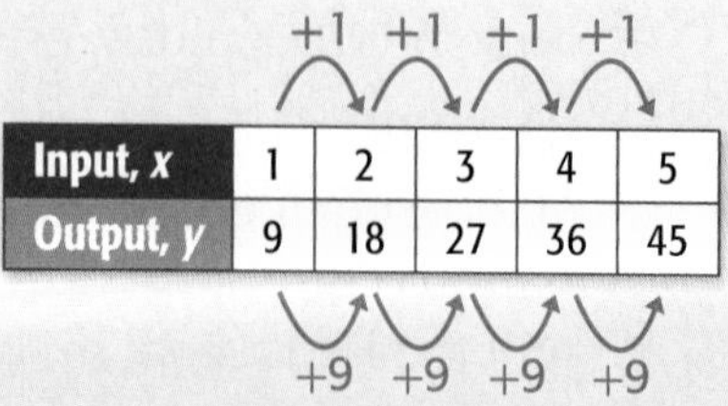

Input, x	1	2	3	4	5
Output, y	9	18	27	36	45

Each output is equal to 9 times the input. So, the equation that represents the function is $y = 9x$.

CHAPTER 6 Practice Test

Math Online glencoe.com

• Chapter Test

Write each ratio as a fraction in simplest form.

1. 12 red blocks out of 20 blocks
2. 24 potato chips out of 144 chips
3. 65 rotten apples out of 520 apples
4. **WORD PROCESSING** The world record for the fastest typing speed is 212 words per minute. How many words per second is this? Round to the nearest tenth.

Write each rate as a unit rate.

5. \$2 for 36 erasers
6. 180 pages in 90 minutes
7. **MULTIPLE CHOICE** Candace buys 12 cans of orange juice for \$6. At this rate, how much would she pay for 48 cans of orange juice?

   A \$20

   B \$22

   C \$24

   D \$30

Determine if the quantities in each pair of ratios or rates are proportional. Explain your reasoning and express each proportional relationship as a proportion.

8. 32 pencils for \$8; 16 pencils for \$4
9. 72 out of 90 students have siblings; 362 out of 450 students have siblings
10. 524 Calories for 4 servings; 786 Calories for 6 servings

Solve each proportion.

11. $\frac{4}{6} = \frac{x}{12}$
12. $\frac{10}{p} = \frac{2}{8}$
13. $\frac{n}{13} = \frac{8}{52}$
14. $\frac{7}{13} = \frac{a}{52}$
15. **SEASONS** If 7 of the 28 students in a class prefer the winter months, predict how many would prefer the winter months in a school of 400 students.
16. **DISCOUNT** Ellie is using the following table to help her calculate the discount on baseball caps. Mr. Gomez would like to order 8 baseball caps. How much of a discount should Ellie give him?

Baseball caps	1	2	3
Discount (\$)	2	3	4

17. **MULTIPLE CHOICE** Which expression was used to create the table?

Position, x	Value of Term
3	11
4	14
5	17
6	20
7	23
x	■

   F $x + 8$

   G $2x + 3$

   H $x - 8$

   J $3x + 2$

READING For Exercises 18–20, use the following information.

Darnell reads an average of 2 hours each day.

18. Make a table to show the relationship between the number of h hours Darnell spends reading in d days.
19. Write an equation to find h, the number of hours Darnell spends reading in d days.
20. On average, how many hours will Darnell spend reading in 12 days?

Test Practice
Cumulative, Chapters 1–6

Math Online glencoe.com
• Test Practice

PART 1 Multiple Choice

Read each question. Then fill in the correct answer on the answer sheet provided by your teacher or on a sheet of paper.

1. Dante made a scale model of a tree. The actual tree is 32 feet tall, and the height of the model is 2 feet. How many actual feet does one foot on the model represent?

   A 2 feet

   B 8 feet

   C 16 feet

   D 32 feet

2. The ratio of cats to dogs seen by a vet in a day is about 2 to 5. If a vet saw 40 animals in one day, about how many were dogs?

   F 5

   G 12

   H 29

   J 40

3. Which statement about the mixed number $1\frac{5}{8}$ is true?

   A $1\frac{5}{8} > \frac{1}{2}$

   B $1\frac{5}{8} > 2$

   C $1 > 1\frac{5}{8}$

   D $1\frac{5}{8} < \frac{1}{4}$

4. Mr. Lee's car odometer read 55,085.4 miles. After his trip, it read 56,002.8 miles. Choose the best estimate for the number of miles the car was driven on his trip.

   F 100 miles

   G 150 miles

   H 1,000 miles

   J 2,000 miles

5. David's time in a race was 9 minutes. Armando's time was two minutes less than Rachel's time, and Rachel's time was 4 minutes more than David's time. Which table could be used to find Armando's time?

   A

Name	Time (min)
Armando	9 − 4 + 2
Rachel	4 − 2
David	9

   B

Name	Time (min)
Armando	4 − 2
Rachel	9 + 4
David	9

   C

Name	Time (min)
Armando	9 + 4 − 2
Rachel	9 + 4
David	9

   D

Name	Time (min)
Armando	4 + 2
Rachel	9 + 4
David	9

6. On Monday, Ted ran $3\frac{1}{2}$ miles. Then on Wednesday he ran $1\frac{2}{3}$ miles and on Friday he ran $2\frac{1}{4}$ miles. Find the total number of miles Ted ran.

   F $6\frac{1}{2}$ miles

   G $7\frac{5}{12}$ miles

   H $7\frac{1}{2}$ miles

   J $8\frac{1}{12}$ miles

7. At a concert, backstage passes were awarded to the person sitting in the seat numbered with the least common multiple of 9, 12, and 18. Find the number of the prizewinning seat.

   A 3

   B 12

   C 36

   D 45

Preparing for Standardized Tests

For test-taking strategies and practice, see pages 718–735.

8. Juanita was making a cake. She added $\frac{1}{4}$ cup of brown sugar, $1\frac{1}{3}$ cups of flour, and $\frac{3}{4}$ cup of white sugar. Which procedure can you use to find the total number of cups Juanita used?

   F Divide the sum of the whole numbers by the sum of the fractions, using a common denominator when necessary.

   G Find the difference between the sum of the whole numbers and the sum of the fractions, using a common denominator when necessary.

   H Multiply the sum of the whole numbers by the sum of the fractions, using a common denominator when necessary.

   J Add the sum of the whole numbers and the sum of the fractions, using a common denominator when necessary.

9. At a sports camp, there is 1 counselor for every 12 campers. If there are 156 campers attending the camp, which proportion can be used to find x, the number of counselors?

   A $\frac{x}{12} = \frac{1}{156}$

   B $\frac{12}{1} = \frac{x}{156}$

   C $\frac{1}{12} = \frac{x}{156}$

   D $\frac{x}{1} = \frac{12}{156}$

TEST-TAKING TIP

Question 9 When setting up a proportion, make sure the numerators and denominators in each ratio have the same units, respectively.

PART 2 Short Response/Grid In

Record your answers on the answer sheet provided by your teacher or on a sheet of paper.

10. Which is the prime factorization of 252 using exponents?

11. When Shawnel held her cat and stepped onto a scale, the scale's display read 107.2 pounds. If Shawnel weighs 97.9 pounds, how many pounds does her cat weigh?

12. Find the greatest common factor of 15, 20, and 25.

PART 3 Extended Response

Record your answers on the answer sheet provided by your teacher or on a sheet of paper. Show your work.

13. Samantha's class sorted books in the library. The class sorted 45 books in 90 minutes.

    a. Write a proportion to find how long it would take to sort 120 books.

    b. How many hours did it take the class to sort 120 books?

    c. Suppose their rate slowed to 30 books in 90 minutes. How long would it take the class to sort the 120 books?

NEED EXTRA HELP?													
If You Missed Question...	1	2	3	4	5	6	7	8	9	10	11	12	13
Go to Lesson...	6-1	6-4	4-6	3-4	5-5	5-5	4-5	1-4	6-4	1-3	3-5	4-1	6-4

Percent and Probability

BIG Ideas

- Solve problems involving ratio, percent, and probability.

Key Vocabulary

percent (p. 365)
probability (p. 381)
sample space (p. 389)
tree diagram (p. 390)

Real-World Link

Soccer Suppose a soccer player made 6 goals in his last 15 attempts. You can use probability to predict the number of goals he will score in his next 50 attempts.

Percent and Probability Make this Foldable to help you understand percents and probability. Begin with a piece of 11″ × 17″ paper.

1. **Fold** a 2″ tab along the long side of the paper.

2. **Unfold** the paper and fold in thirds widthwise.

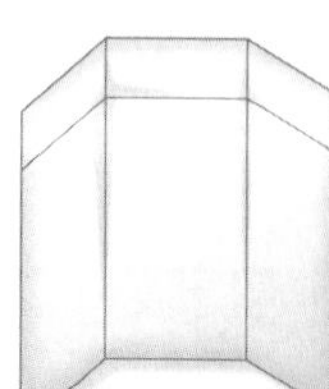

3. **Draw** lines along the folds and label the head of each column as shown. Label the front of the folded table with the chapter title.

Fraction	Percent	Decimal
$\frac{1}{2}$ →	50% →	0.5

GET READY for Chapter 7

Diagnose Readiness You have two options for checking Prerequisite Skills.

Option 2

Math Online Take the Online Readiness Quiz at glencoe.com.

Option 1

Take the Quick Quiz below. Refer to the Quick Review for help.

QUICK Quiz

Write each fraction in simplest form. If the fraction is already in simplest form, write *simplest form*. (Lesson 4-2)

1. $\frac{25}{100}$
2. $\frac{17}{100}$
3. $\frac{30}{100}$
4. $\frac{15}{100}$

Solve each proportion. (Lesson 6-4)

5. $\frac{1}{a} = \frac{3}{9}$
6. $\frac{7}{16} = \frac{h}{48}$
7. $\frac{5}{8} = \frac{30}{y}$
8. $\frac{t}{35} = \frac{6}{7}$
9. $\frac{s}{18} = \frac{2}{3}$
10. $\frac{36}{p} = \frac{2}{3}$
11. **BAKING** If baking 4 apple pies requires 2 pounds of apples, how many pounds of apples are needed to bake 12 apple pies?

Round each number to the nearest ten. (Prior Grade)

12. 42
13. 5
14. 68
15. 74
16. 18
17. 9
18. **COMPUTERS** Rebekah has saved $77 toward new speakers for her computer. To the nearest ten, how much has she saved?

QUICK Review

Example 1

Simplify $\frac{3}{15}$.

$\frac{3}{15} = \frac{1}{5}$ (÷3) Divide the numerator and denominator by the GCF, 3.

Since the GCF of 1 and 5 is 1, the fraction $\frac{1}{5}$ is in simplest form.

Example 2

Solve $\frac{9}{16} = \frac{n}{32}$.

$\frac{9}{16} = \frac{n}{32}$ Write the proportion.

$\frac{9}{16} = \frac{n}{32}$ (×2) Since 16 × 2 = 32, multiply 9 by 2.

So, $n = 18$.

Example 3

Round 86 to the nearest ten.

Look at the number in the ones place, 6.

Since 6 is 5 or greater, add one to the number in the tens place.

86 rounded to the tens place is 90.

Explore 7-1

Math Lab
Modeling Percents

MAIN IDEA

Use models to illustrate the meaning of percents.

In Lesson 3-1, you learned that a 10 × 10 grid can be used to represent *hundredths*. The word *percent* (%) means *out of one hundred*, so you can also use a 10 × 10 grid to model percents.

ACTIVITY

1 Model 18%.

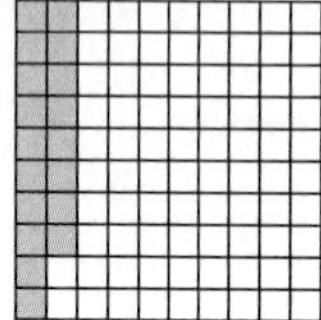

18% means 18 out of 100.

So, shade 18 of the 100 squares on the decimal model.

CHECK Your Progress **Model each percent.**

a. 30% b. 8% c. 42% d. 75%

ACTIVITIES

Identify each percent that is modeled.

2

There are 40 out of 100 squares shaded.

So, the model shows 40%.

3

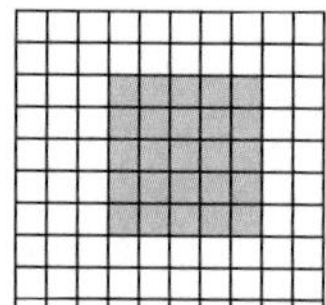

There are 25 out of 100 squares shaded.

So, the model shows 25%.

CHECK Your Progress **Identify each percent modeled.**

e.

f.

g.

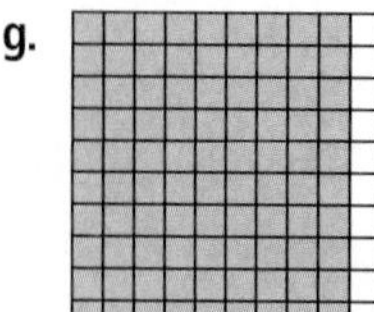

ANALYZE THE RESULTS

1. Identify the fraction of each model in Exercises a–g that is shaded.
2. **MAKE A CONJECTURE** How can you write a percent as a fraction? How can you write a fraction with a denominator of 100 as a percent?

7-1 Percents and Fractions

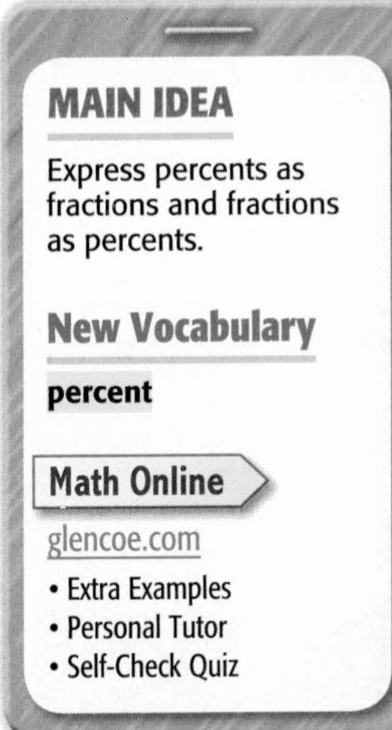

MAIN IDEA
Express percents as fractions and fractions as percents.
New Vocabulary
percent
Math Online
glencoe.com
• Extra Examples
• Personal Tutor
• Self-Check Quiz

GET READY for the Lesson

FOOD Kimiko asked 100 students in the cafeteria what their favorite fruit bar flavor was: cherry, grape, strawberry, or blueberry. The results are shown in the bar graph.

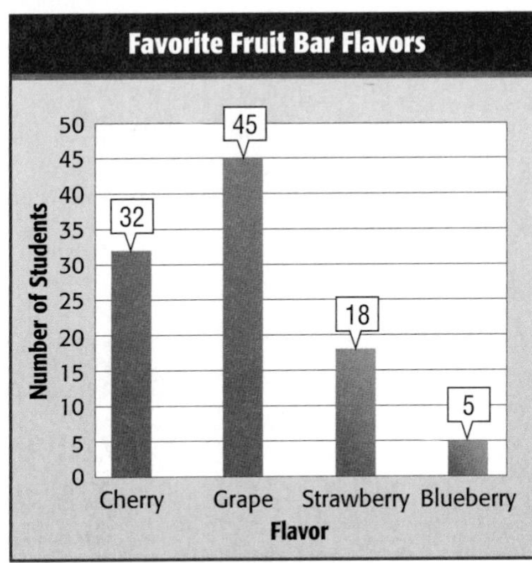

1. What ratio compares the number of students who prefer grape fruit bars to the total number of students?
2. Draw a decimal model to represent this ratio.
3. What fraction represents this ratio?

Ratios such as 32 out of 100, 45 out of 100, 18 out of 100, and 5 out of 100, can be written as percents.

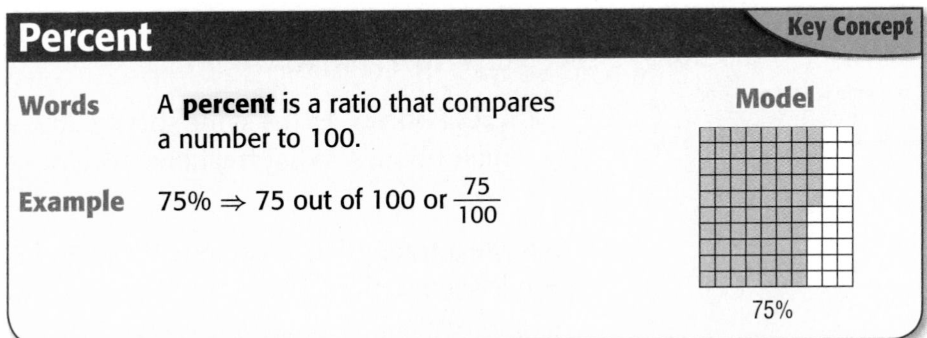

Percent — Key Concept

Words A **percent** is a ratio that compares a number to 100.

Example 75% ⇒ 75 out of 100 or $\frac{75}{100}$

Model

75%

EXAMPLES Write a Percent as a Fraction

1 Write 50% as a fraction in simplest form.

50% means *50 out of 100.*

$50\% = \frac{50}{100}$ Definition of percent

$= \frac{\cancel{50}^{1}}{\cancel{100}_{2}}$ or $\frac{1}{2}$ Simplify. Divide the numerator and the denominator by the GCF, 50.

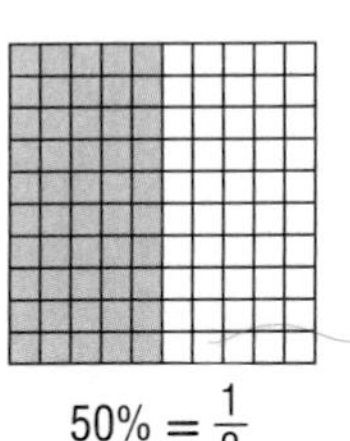

$50\% = \frac{1}{2}$

Study Tip

Percents A percent can be greater than 100%. Since percent means *hundredths*, or *per 100*, a percent like 125% means 125 hundredths, or 125 per 100.

2 **Write 125% as a mixed number in simplest form.**

125% means *125 for every 100*.

$125\% = \frac{125}{100}$ Definition of percent

$= 1\frac{25}{100}$ Write as a mixed number.

$= 1\frac{\cancel{25}^{1}}{\cancel{100}_{4}}$ or $1\frac{1}{4}$ Divide the numerator and denominator by the GCF, 25.

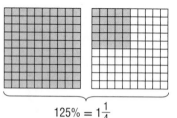

$125\% = 1\frac{1}{4}$

CHECK Your Progress

Write each percent as a fraction or mixed number in simplest form.

a. 10%

b. 97%

c. 135%

Real-World EXAMPLE

3 **CELL PHONES** **In a recent survey, 35% of cell phone owners said they use the text messaging feature on their cell phones. What fraction of cell phone owners is this?**

$35\% = \frac{35}{100}$ Definition of percent

$= \frac{7}{20}$ Simplify.

So, $\frac{7}{20}$ of cell phone owners use text messaging features on their phones.

Study Tip

Check for Reasonableness You can conclude that $\frac{7}{20}$ is a reasonable answer because 35% is a little less than 50%, and $\frac{7}{20}$ is a little less than $\frac{10}{20}$ or $\frac{1}{2}$.

CHECK Your Progress

d. **CELL PHONES** In the same survey, 28% said they take pictures with their phones. What fraction of cell phone owners is this?

To write a fraction as a percent, write and solve a proportion. One ratio is the fraction. The other is an unknown compared to 100.

EXAMPLES Write a Fraction as a Percent

4 **Write $\frac{9}{20}$ as a percent.**

$\frac{9}{20} = \frac{n}{100}$ Write a proportion.

$\frac{9}{20} = \frac{45}{100}$ (×5) Since 20 × 5 = 100, multiply 9 by 5 to find n.

So, $\frac{9}{20} = \frac{45}{100}$ or 45%.

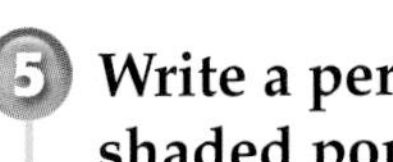

5 **Write a percent to represent the shaded portion of the model at the right.**

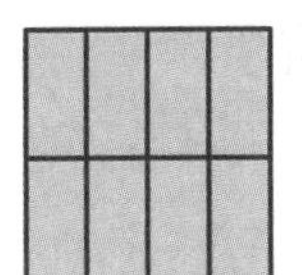
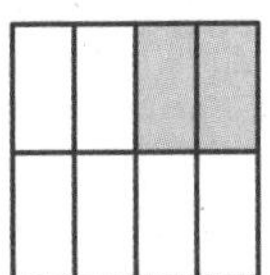

Study Tip

Models If a model shows two figures, assume that each figure represents one whole.

The portion shaded is $1\frac{2}{8}$ or $1\frac{1}{4}$.

$1\frac{1}{4} = \frac{5}{4}$ Write $1\frac{1}{4}$ as an improper fraction.

$\frac{5}{4} = \frac{n}{100}$ Write a proportion.

$\frac{5}{4} = \frac{125}{100}$ (×25 numerator and denominator) Since 4 × 25 = 100, multiply 5 by 25 to find *n*.

So, $\frac{125}{100}$ or 125% of the model is shaded.

CHECK Your Progress

Write each fraction, mixed number, or shaded portion of each model as a percent.

e. $\frac{3}{5}$

f. $2\frac{9}{10}$

g.

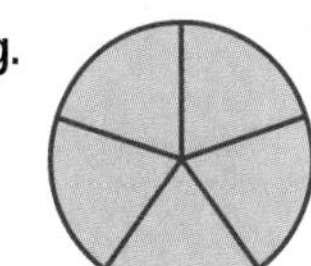
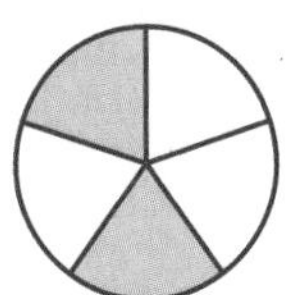

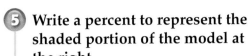

CHECK Your Understanding

Examples 1, 2 (pp. 365–366)

Write each percent as a fraction or mixed number in simplest form.

1. 15%
2. 80%
3. 180%

Example 3 (p. 366)

4. **SCHOOLS** About 18% of Kentucky's public schools are middle schools. What fraction of Kentucky public schools is this?

Example 4 (p. 366)

Write each fraction or mixed number as a percent.

5. $\frac{1}{4}$
6. $\frac{2}{5}$
7. $2\frac{1}{4}$

Example 5 (p. 367)

Write a percent to represent the shaded portion of each model.

8.

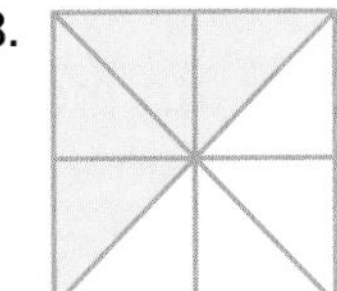

9.

10.

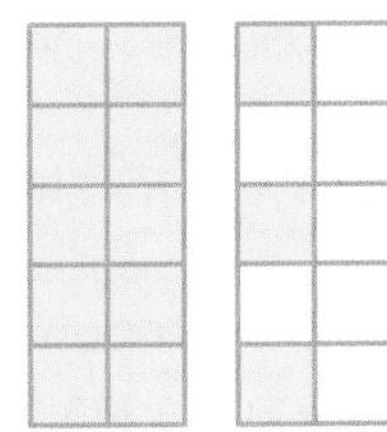

11.

Practice and Problem Solving

HOMEWORK HELP	
For Exercises	See Examples
12–17	1, 2
18, 19	3
20–27	4
28–33	5

Write each percent as a fraction or mixed number in simplest form.

12. 14%
13. 47%
14. 2%
15. 20%
16. 185%
17. 280%

18. **E-MAIL** In a recent year, 22% of e-mail users said they spend less time using e-mail because of spam. What fraction of e-mail users is this?

19. **SOCCER** In a recent season, the Dallas Burn won or tied about 54% of their games. What fraction of their games did they win or tie?

Write each fraction or mixed number as a percent.

20. $\frac{3}{10}$
21. $\frac{7}{20}$
22. $1\frac{1}{4}$
23. $1\frac{2}{5}$
24. $\frac{1}{100}$
25. $\frac{5}{100}$

26. **PETS** About $\frac{7}{10}$ of a cat's day is spent dozing. About what percent of a cat's day is spent dozing?

27. **FOOD** About $\frac{23}{25}$ of a watermelon is water. About what percent is this?

Write a percent to represent the shaded portion of each model.

28. 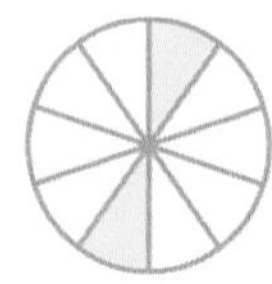

29. 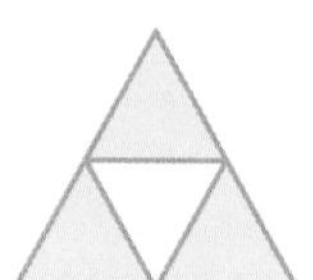

30. 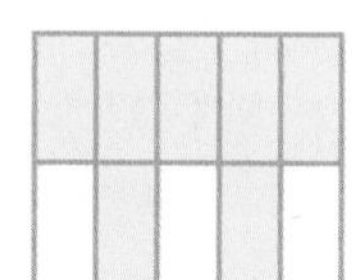

31. 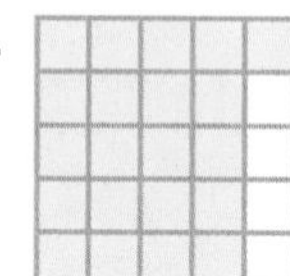

32. 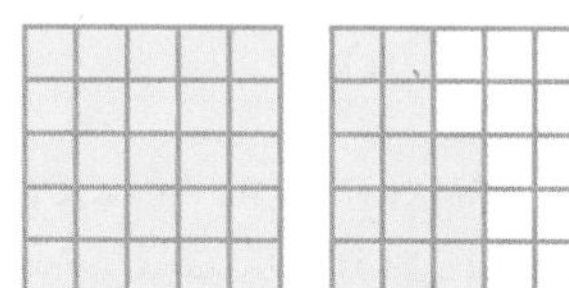

33. 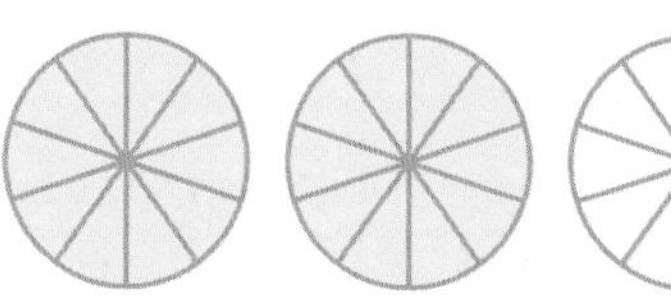

34. **CLOTHES** Use the table to determine what percent of Khaliah's pants are jeans and what percent are khakis. What is the relationship between these two percents?

Khaliah's Pants	
Jeans	Khakis

35. **INTERNET** A survey showed that 82% of youth most often use the Internet at home. What fraction of youth surveyed most often use the Internet somewhere else?

36. **ANALYZE TABLES** The table shows the approximate fraction of each coastline as compared to the entire coastline of the United States. Write each fraction as a percent. About what percent of coastline is the Pacific coast? Then order the coastlines from least to greatest.

Coast	Atlantic	Gulf	Arctic	Pacific
Fractions	$\frac{1}{5}$	$\frac{1}{10}$	$\frac{2}{25}$	■

EXTRA PRACTICE
See pages 690, 712.

H.O.T. Problems

37. **OPEN ENDED** Write three fractions that can be written as percents between 50% and 75%. Justify your solution.

38. **CHALLENGE** Write $\frac{1}{200}$ as a percent.

39. **Which One Doesn't Belong?** Identify the number that does not belong with the other three. Explain your reasoning.

$\frac{9}{20}$	$\frac{45}{100}$	45%	$\frac{8}{45}$

40. **WRITING IN MATH** Determine whether the following statement is *true* or *false*. Explain your reasoning. If false, provide a counterexample.

When writing a number greater than 1 as a percent, one ratio of the proportion should be an unknown number compared to 1,000.

TEST PRACTICE

41. On Friday, 65% of the students at Plainview Middle School bought a hot lunch in the cafeteria. What fractional part of the school did *not* buy a hot lunch in the cafeteria?

A $\frac{1}{65}$

B $\frac{13}{20}$

C $\frac{7}{20}$

D $\frac{6}{5}$

42. Marita used black and white tiles to create the mosaic below.

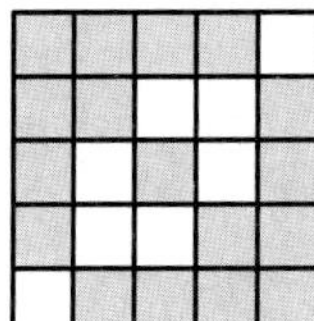

What percent of her mosaic consists of white tiles?

F 8%

G 17%

H 32%

J 68%

Spiral Review

BABYSITTING Use the following information for Exercises 43–44. (Lesson 6-7)

Vonzell earns $7 per hour for babysitting twin boys.

43. Write an equation to represent the total amount t that Vonzell earns for babysitting these boys for h hours.

44. How much will she earn if she babysits them for 6 hours?

Describe how the next term in each sequence can be found. Then find the next two terms. (Lesson 6-6)

45. 5, 8, 11, 14, ...

46. $\frac{1}{2}, \frac{3}{4}, 1, 1\frac{1}{4}, \ldots$

GET READY for the Next Lesson

PREREQUISITE SKILL Write each fraction in simplest form. (Lesson 4-2)

47. $\frac{26}{100}$

48. $\frac{50}{100}$

49. $\frac{10}{100}$

50. $\frac{75}{100}$

7-2 Circle Graphs

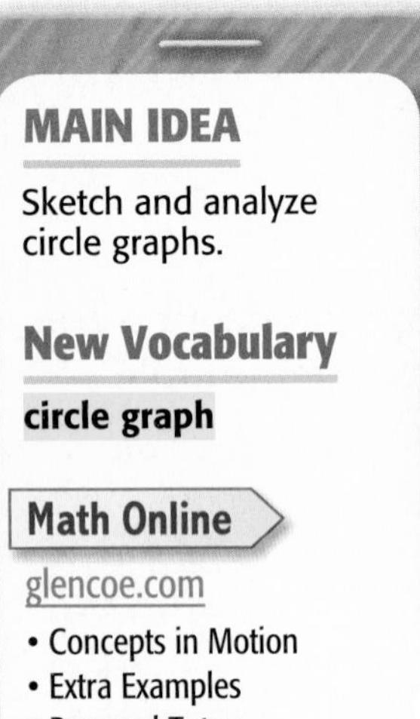

MAIN IDEA

Sketch and analyze circle graphs.

New Vocabulary

circle graph

Math Online

glencoe.com

- Concepts in Motion
- Extra Examples
- Personal Tutor
- Self-Check Quiz

MINI Lab

The table below shows the results of a survey about which sport people prefer to watch. These data can be displayed in a circle graph.

STEP 1 For each category, let 1 centimeter equal 1%. On a piece of adding machine tape, mark the length, in centimeters, that represents each percent. Label each section.

Sport	Percent
auto racing	4
baseball	11
basketball	12
football	43
none	12
other	18

Source: The Gallup Poll

auto racing 4%	baseball 11%	basketball 12%	football 43%	none 12%	other 18%

STEP 2 Tape the ends together to form a circle.

STEP 3 Tape one end of a piece of string to the center of the circle. Tape the other end to the point where two sections meet. Repeat with four more pieces of string.

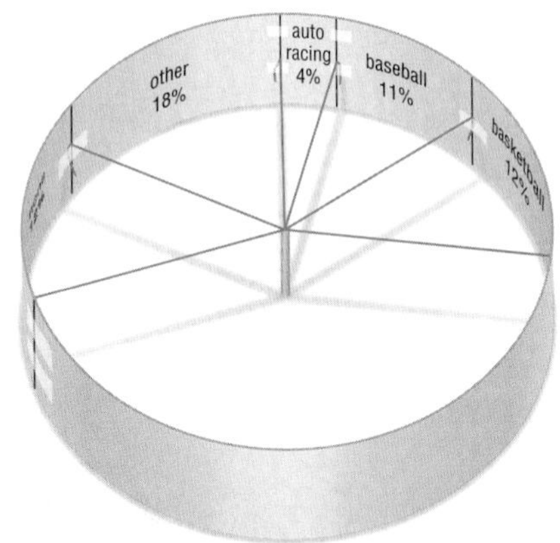

1. Make a bar graph of the data.
2. Which graph represents the data better, a circle graph or a bar graph? Explain.

A **circle graph** is used to compare data that are parts of a whole.

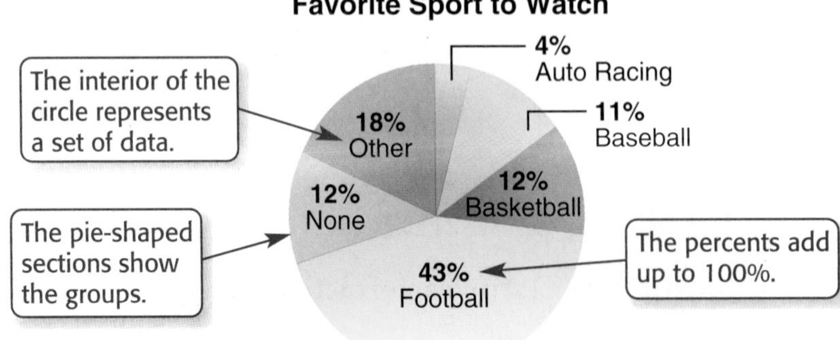

You can use number sense to sketch circle graphs.

EXAMPLE Sketch Circle Graphs

1 TRANSPORTATION The students at Adams Middle School were asked how they get to school in the morning. The results are shown at the right. Sketch a circle graph to display the data.

Transportation to School	
Mode of Transportation	Percent
school bus	54
walk	27
bike	9.5
carpool	9.5

Study Tip

Circle Graphs Before constructing a circle graph, check that the percents add up to 100%.

- Write a fraction to estimate each percent.

  $54\% \approx 50\%$ and $50\% = \frac{50}{100}$ or $\frac{1}{2}$

  $27\% \approx 25\%$ and $25\% = \frac{25}{100}$ or $\frac{1}{4}$

  $9.5\% \approx 10\%$ and $10\% = \frac{10}{100}$ or $\frac{1}{10}$

- Use a compass to draw a circle with at least a 1-inch radius.

- Since 54% is a little more than 50% or $\frac{1}{2}$, shade a little more than $\frac{1}{2}$ of the circle for "School Bus." Since 27% is a little more than 25% or $\frac{1}{4}$, shade a little more than $\frac{1}{4}$ of the circle for "Walk."

  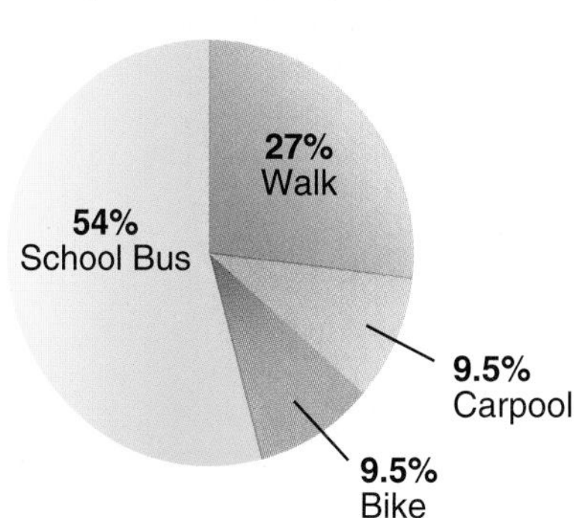

  Since the last two sections are equal, take the remaining portion of the circle and divide it into two equal parts.

- Label each section of the circle graph. Give the graph a title.

CHECK Your Progress

a. **VACATION** The table shows how people responded to a question about the importance of sunny weather in a vacation location. Sketch a circle graph to display the data.

Sunshine While on Vacation	
Response	**Percent**
very important	45
important	30
not very important	15
not at all important	10

Source: Opinion Research Corp.

You can analyze data displayed in a circle graph.

EXAMPLES Analyze Circle Graphs

POPCORN Sandi surveyed her class to see which flavor of popcorn they like the best. The circle graph at the right shows the results of her survey.

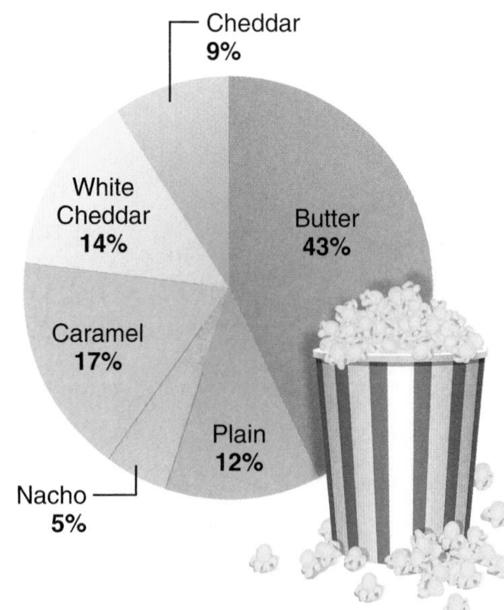

2 Which flavor of popcorn did the students like the most?

The largest section of the graph is the section that represents butter flavored popcorn. So, most students said that butter popcorn is their favorite flavor of popcorn.

> **Study Tip**
>
> **Check for Reasonableness** For Example 3, you can also compare the size of the sections. Since the "Cheddar" and "Nacho" sections are about the same size, the answer seems reasonable.

3 Which two sections represent the flavors chosen least often?

By comparing the percents, the sections labeled "Cheddar" and "Nacho" are the smallest. So, cheddar and nacho flavored popcorn were chosen least often.

4 Which three sections altogether make up about the same percent as butter flavored popcorn?

Together, the sections representing white cheddar, caramel, and plain are the same size as the section representing butter popcorn.

CHECK Your Progress

LANDFILLS The circle graph at the right shows the items that fill landfills in the United States.

b. Which item is found most in U.S. landfills?

c. Which item(s) is found the least in U.S. landfills?

d. Which two sections represent items that are found about the same amount in U.S. landfills?

e. How does the amount of food and yard waste compare to the amount of rubber and leather in U.S. landfills?

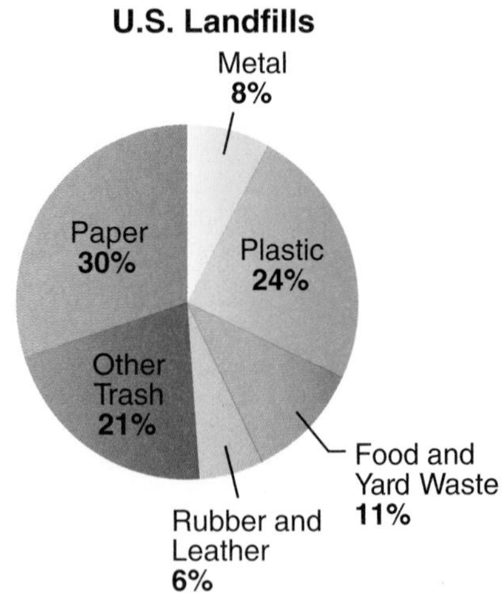

Source: *The World Almanac*

Example 1 (p. 371)

1. **PARTIES** In a national online survey, 13-year-olds were asked if they like having a surprise party thrown for them. The table shows the results. Sketch a circle graph to display the data.

Do You Like Surprise Parties?	
Yes	72%
No	25%
No Answer	3%

MUSIC For Exercises 2–4, use the graph at the right that shows the result of a student survey.

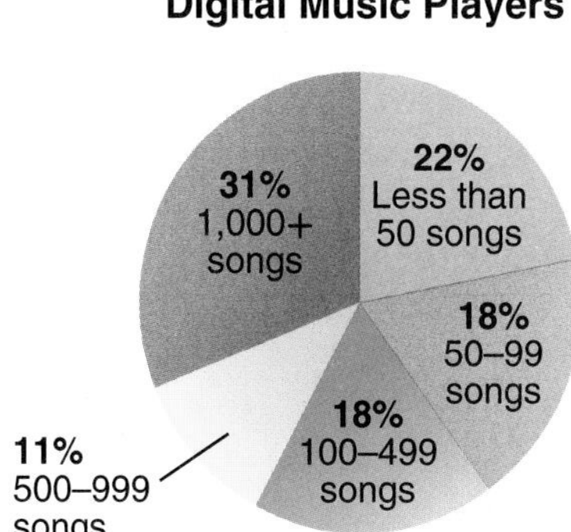

Examples 2, 3 (p. 372)

2. Which category has the least percent of students?
3. What percent of students said they have 500 or more songs?

Example 4 (p. 372)

4. How does the percent of students with 100–499 songs compare with the number of students that have 50–99 songs?

Practice and Problem Solving

HOMEWORK HELP	
For Exercises	See Examples
5, 6	1
7–12	2–4

5. **ELECTIONS** The table shows the results of an election for class president. Sketch a circle graph to display the data.

Class President Ballots	
Melissa	31%
Lacey	25%
Troy	25%
Omar	19%

6. **MUSIC** In Mrs. Castro's class, 75% of the students like rock music, 10% favor country music, 5% prefer classical, and 10% chose another type of music. Sketch a circle graph to display the data.

LIFE SCIENCE For Exercises 7–9, use the graph below that shows the elements found in the human body.

7. Which element makes up the greatest percent of the body?
8. How does the amount of nitrogen in the body compare to the amount of calcium in the body?
9. What can you determine about the amount of oxygen and carbon in the human body?

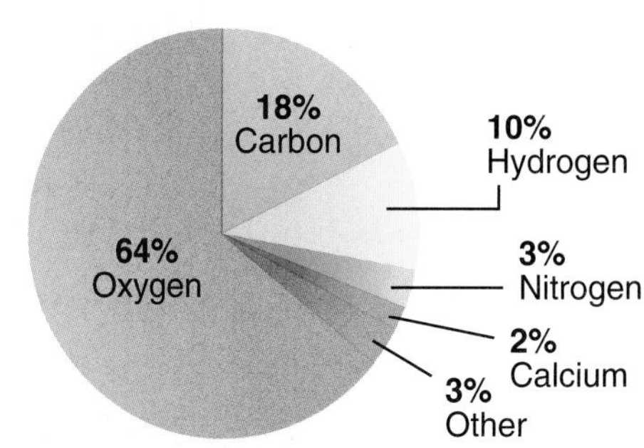

LAKES For Exercises 10–12, use the graph that shows the approximate portion of the entire Great Lakes each lake covers.

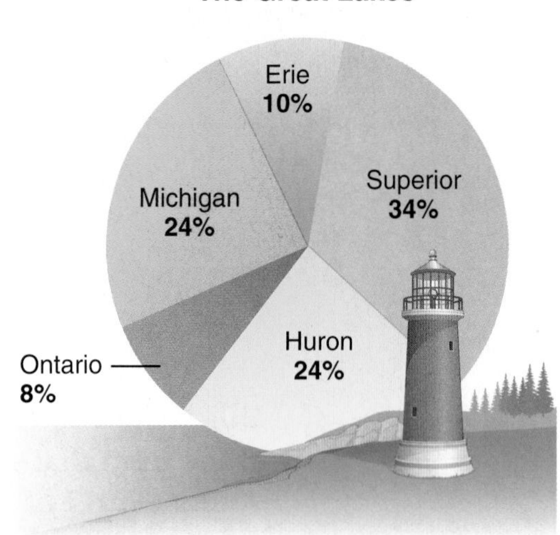

10. Which Great Lake is the smallest?
11. What percent of the entire Great Lakes are the two smallest lakes?
12. How does the size of Lake Superior compare to the size of Lake Ontario?

13. Refer to the graph in Example 1 on page 371. How does the number of students that walk to school compare to the number of students that ride the school bus?

14. **MOVIES** A group of 100 students was asked about their movie-going experience pet peeves. The table shows their responses. Sketch a circle graph to compare the students' responses. What percent of the students chose bad manners or overpriced food as their pet peeves?

Movie Pet Peeves

Response	Number of Students
bad manners	49
preshow commercials	24
overpriced food	12
babies/noisy children	12
other	3

EXTRA PRACTICE
See pages 690, 712.

H.O.T. Problems

15. **COLLECT THE DATA** Record your activities for one 24-hour period. Sketch a circle graph to display your data. Then write a few sentences that analyze the data?

16. **CHALLENGE** Can data that show how many DVDs were purchased in 2009 be represented in a circle graph? If it cannot, what should be included to be able to make a circle graph of the data?

17. **SELECT A TECHNIQUE** Carter wants to sketch a circle graph of the data in the table. Which of the following techniques might Carter use to determine how big to make each section of the circle graph? Justify your selection(s). Then use the technique(s) to sketch the circle graph.

How Students Travel to Clint Middle School	
bicycle	5%
bus	47%
carpool	25%
walk	22%

mental math | number sense | estimation

18. **WRITING IN MATH** Write a problem that can be solved using one of the circle graphs in this lesson. Be sure to include a statement about the graph to which you are referring.

TEST PRACTICE

19. A group of adults was asked to give a reason why they honor their moms. Fifty percent of the adults said, "she survived raising me," 22% said, "she was a great role model," 19% said, "she has become my best friend," and 9% said, "she always had dinner on the table and clean clothes in the closet." Which circle graph best displays the data?

A
Why My Mom is Great
Dinner/ Clothes
Best Friend
Survived Raising Me
Role Model

C

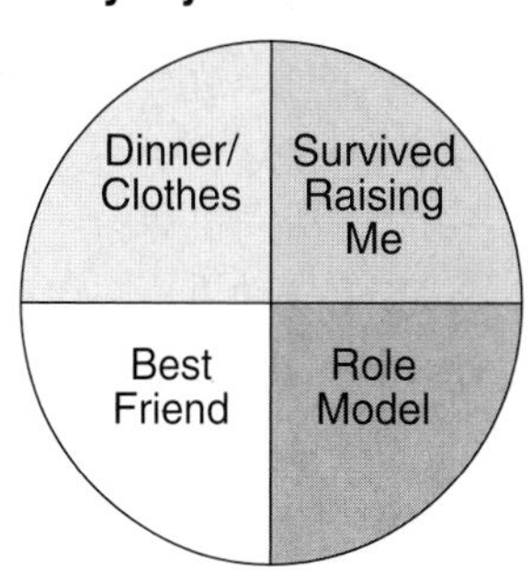

B
Why My Mom is Great
Dinner/ Clothes
Best Friend
Role Model
Survived Raising Me

D

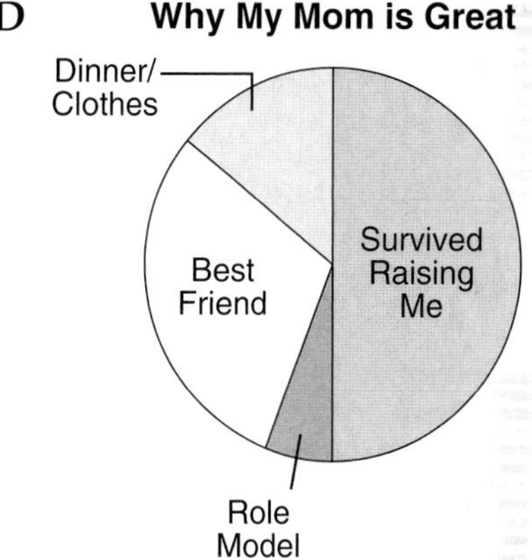

Spiral Review

Write each fraction or mixed number as a percent. (Lesson 7-1)

20. $\frac{43}{100}$ 21. $\frac{9}{10}$ 22. $1\frac{1}{5}$

MOVIES Use the following information for Exercises 23 and 24. (Lesson 6-7)

A video store charges $3 to rent a DVD.

23. Write an equation to represent the total cost *c* for renting *d* DVDs.
24. How much will it cost to rent 4 DVDs?

25. Order 3.8, 3.05, 0.39, and 3.5 from greatest to least. (Lesson 3-2)

GET READY for the Next Lesson

PREREQUISITE SKILL Write each fraction as a decimal. (Lesson 4-8)

26. $\frac{65}{100}$ 27. $\frac{1}{8}$ 28. $\frac{15}{100}$ 29. $\frac{1}{5}$

READING to SOLVE PROBLEMS

Draw a Picture

Have you heard the expression *a picture is worth a thousand words*? Sometimes drawing a picture can help you better understand the numbers you find in a word problem. For example, a *number map* can show how numbers are related to each other. Start by placing a number in the center of the map.

Below is a number map that shows various meanings of the decimal 0.5. Notice that you can add both mathematical meanings and everyday meanings to the number map.

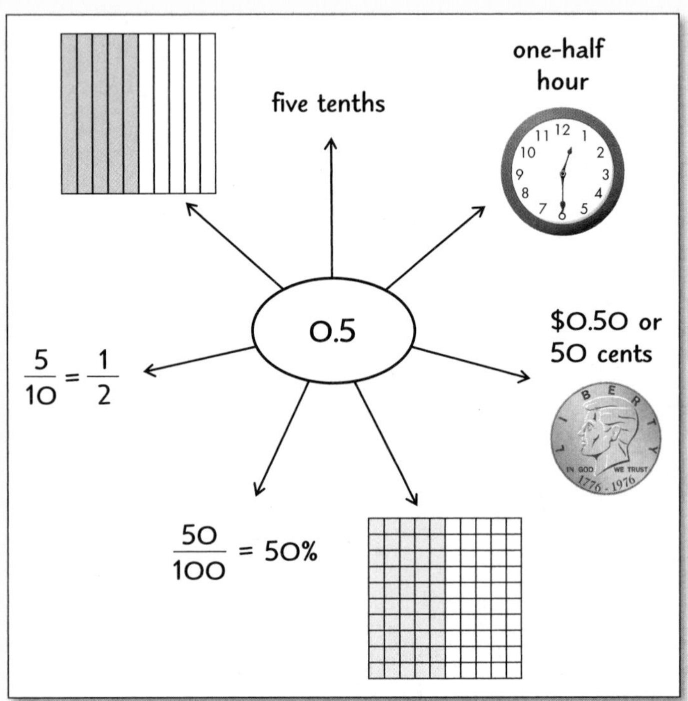

PRACTICE

Make a number map for each number. (*Hint:* For whole numbers, think of factors, prime factors, divisibility, place value, and so on.)

1. 0.75
2. 0.1
3. 0.01
4. 1.25
5. 2.5
6. 25
7. 45
8. 60
9. 100
10. Refer to Exercise 1. Explain how each mathematical or everyday meaning on the number map relates to the decimal 0.75.

7-3 Percents and Decimals

MAIN IDEA

Express percents as decimals and decimals as percents.

Math Online

glencoe.com

- Extra Examples
- Personal Tutor
- Self-Check Quiz

GET READY for the Lesson

SCHOOL The circle graph shows the favorite subjects of students in a recent survey.

1. What percent does the entire circle graph represent?
2. What fraction represents the section of the graph labeled math?
3. Write the fraction from Exercise 2 as a decimal.

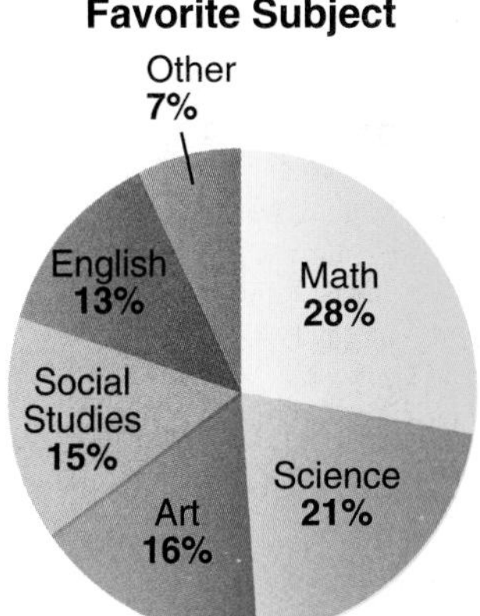

Source: *Time for Kids Almanac 2005*

Percents can be written as decimals. To write a percent as a decimal, rewrite the percent as a fraction with a denominator of 100. Then write the fraction as a decimal.

EXAMPLES Write a Percent as a Decimal

Write each percent as a decimal.

1 **56%**

$56\% = \frac{56}{100}$ Rewrite the percent as a fraction with a denominator of 100.

$= 0.56$ Write *56 hundredths* as a decimal.

2 **8%**

$8\% = \frac{8}{100}$ Rewrite the percent as a fraction with a denominator of 100.

$= 0.08$ Write *8 hundredths* as a decimal.

3 **120%**

$120\% = \frac{120}{100}$ Rewrite the percent as a fraction with a denominator of 100.

$= 1\frac{20}{100}$ Write as a mixed number.

$= 1.20$ or 1.2 Write *1 and 20 hundredths* as a decimal.

Study Tip

Mental Math To write a percent as a decimal, move the decimal point two places to the left and remove the % sign. This is the same as dividing by 100.

120% = 120%

= 1.20 or 1.2

CHECK Your Progress

Write each percent as a decimal.

a. 32% b. 6% c. 190%

You can also write a decimal as a percent. To write a decimal as a percent, write the decimal as a fraction with a denominator of 100. Then write the fraction as a percent.

EXAMPLES Write a Decimal as a Percent

Study Tip

Mental Math To write a decimal as a percent, move the decimal point two places to the right and add a % sign. This is the same as multiplying by 100.
$0.38 = 0.38$
$= 38\%$

Write each decimal as a percent.

4 **0.38**

$0.38 = \frac{38}{100}$ Write *38 hundredths* as a fraction.

$= 38\%$ Write the fraction as a percent.

5 **1.45**

$1.45 = 1\frac{45}{100}$ Write *1 and 45 hundredths* as a mixed number.

$= \frac{145}{100}$ Write the mixed number as an improper fraction.

$= 145\%$ Write the fraction as a percent.

CHECK Your Progress

Write each decimal as a percent.

d. 0.47 **e.** 1.75 **f.** 0.52

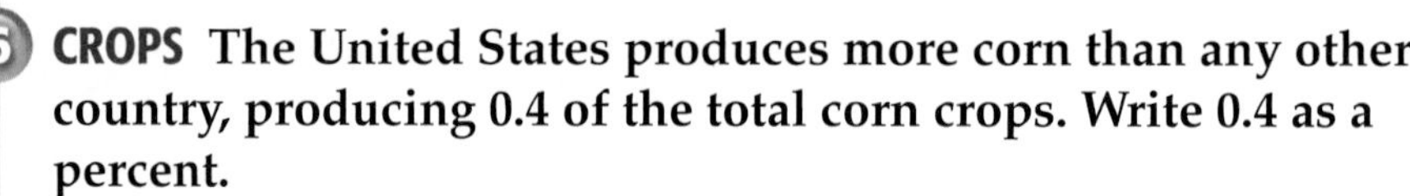

6 **CROPS The United States produces more corn than any other country, producing 0.4 of the total corn crops. Write 0.4 as a percent.**

$0.4 = \frac{4}{10}$ Write *4 tenths* as a fraction.

$= \frac{4 \times 10}{10 \times 10}$ Multiply the numerator and denominator by 10 so that the denominator is 100.

$= \frac{40}{100}$ Simplify.

$= 40\%$ Write the fraction as a percent.

Real-World Link Of all the corn grown in the U.S., more than 50% is used for livestock feed, 25% is exported, and 15% is sold for food.

Source: *Scholastic Book of World Records*

CHECK Your Progress

g. **ANIMALS** The komodo dragon, which can weigh as much as 300 pounds, can eat up to 0.8 of its body weight during one meal. What percent is equivalent to 0.8?

h. **ATTENDANCE** About 0.01 of the students at Central Middle School were born outside of the United States. What percent is equivalent to 0.01?

CHECK Your Understanding

Examples 1–3 (p. 377)

Write each percent as a decimal.

1. 27%
2. 15%
3. 4%
4. 9%
5. 115%
6. 136%

Examples 4, 5 (p. 378)

Write each decimal as a percent.

7. 0.32
8. 0.15
9. 0.91
10. 1.25
11. 2.91
12. 4.63

Example 6 (p. 378)

13. **BIOLOGY** About 0.7 of the human body is water. What percent is equivalent to 0.7?

Practice and Problem Solving

HOMEWORK HELP	
For Exercises	See Examples
14–23	1–3
24–31	4, 5
32, 33	6

Express each percent as a decimal.

14. 17%
15. 35%
16. 2%
17. 3%
18. 125%
19. 104%
20. 11%
21. 95%

22. **MONEY** A bank offers an interest rate of 4% on a savings account. Write 4% as a decimal.

23. **FOOD** When making a peanut butter and jelly sandwich, 96% of people put the peanut butter on first. Write 96% as a decimal.

Express each decimal as a percent.

24. 0.22
25. 0.99
26. 1.75
27. 3.55
28. 0.5
29. 0.6
30. 0.16
31. 0.87

32. **DIGITAL CAMERAS** In a recent year, the number of homes with digital cameras grew 0.44 from the previous year. Write 0.44 as a percent.

33. **PODCASTS** In 2006, 0.12 of Americans downloaded a podcast from the Internet. What percent is equivalent to 0.12?

34. **MONEY** The formula $I = prt$ gives the simple interest I earned on an account where an amount p is deposited at an interest rate r for a certain number of years t. Use the table to order the accounts from least to greatest interest earned after 5 years.

Accounts at First Savings Bank		
Account	p ($)	r (%)
A	350	4
B	500	3.5
C	280	4.25

See pages 690, 712.

Replace each ● with <, >, or = to make a true sentence.

35. 18% ● 0.2
36. 0.5 ● 5%
37. 2.3 ● 23%

H.O.T. Problems

38. **OPEN ENDED** Write a decimal between 0.5 and 0.75. Then write it as a fraction in simplest form and as a percent.

39. **CHALLENGE** Order 15.37%, 1.537, 0.01537, and 10.37% from least to greatest.

40. **CHALLENGE** How would you write $43\frac{3}{4}\%$ as a decimal?

41. **WRITING IN MATH** Write a problem about a real-world situation in which you would either write a percent as a decimal or write a decimal as a percent.

TEST PRACTICE

42. Each square below is divided into sections of equal size. Which square has 75% of its total area shaded?

A 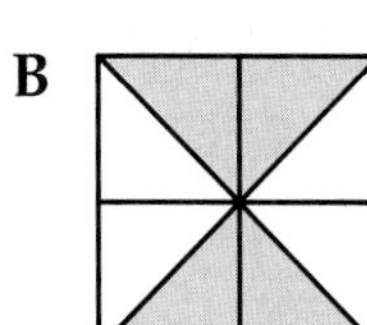

C 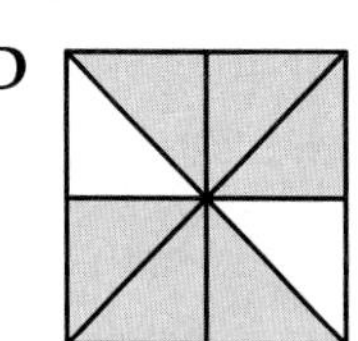

B

D

43. **SHORT RESPONSE** Tamika is buying the baseball hat shown below.

What decimal represents 25%?

Spiral Review

44. **NEWSPAPERS** The circle graph shows how many adults read the daily newspaper by age. Which age group has the largest percent of readers? (Lesson 7-2)

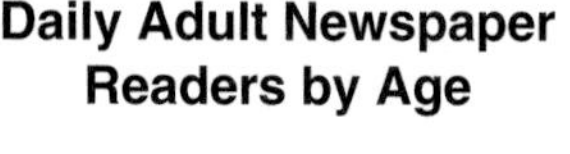

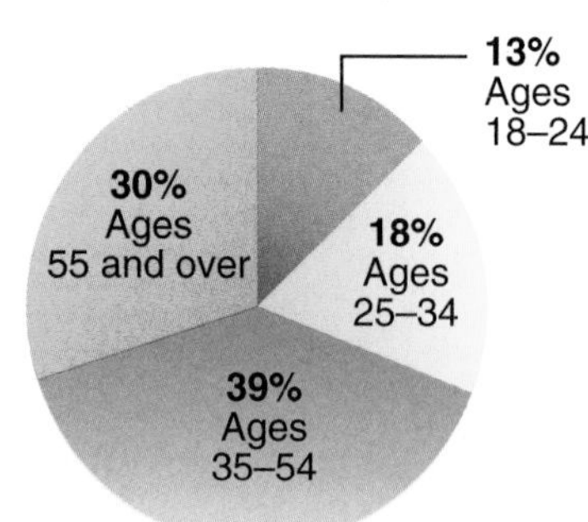

Source: Newspaper Association of America

Write each percent as a fraction or mixed number in simplest form. (Lesson 7-1)

45. 24%

46. 38%

47. 125%

48. 35%

49. **AGE** The equation $17 - v = 5$ gives Virginia's age v in years. Solve the equation mentally. (Lesson 1-8)

GET READY for the Next Lesson

PREREQUISITE SKILL **Write each fraction in simplest form.** (Lesson 4-2)

50. $\frac{5}{45}$

51. $\frac{15}{40}$

52. $\frac{21}{30}$

53. $\frac{9}{21}$

7-4 Probability

MAIN IDEA

Find and interpret the probability of a simple event.

New Vocabulary

outcomes
simple event
probability
random
complementary events

Math Online

glencoe.com

- Extra Examples
- Personal Tutor
- Self-Check Quiz
- Reading in the Content Area

GET READY for the Lesson

FLOWERS A flower shop sells carnations in several different colors. Morgan is selecting a carnation for her mom from the colors shown. She decides to close her eyes and pick a carnation.

1. Write a ratio that compares the number of yellow carnations to the total number of carnations.
2. What percent of the carnations are yellow?
3. Does Morgan have a good chance of selecting a yellow carnation?
4. What would happen to her chances of picking a yellow carnation if a green, lilac, orange, dark purple, and teal carnation were added to the flowers shown?
5. What happens to her chances if there is only one yellow carnation and one pink carnation?

It is equally likely to select any one of the five carnations. The five carnations represent the possible **outcomes**. A **simple event** is one outcome or a collection of outcomes. For example, selecting a yellow carnation is a simple event.

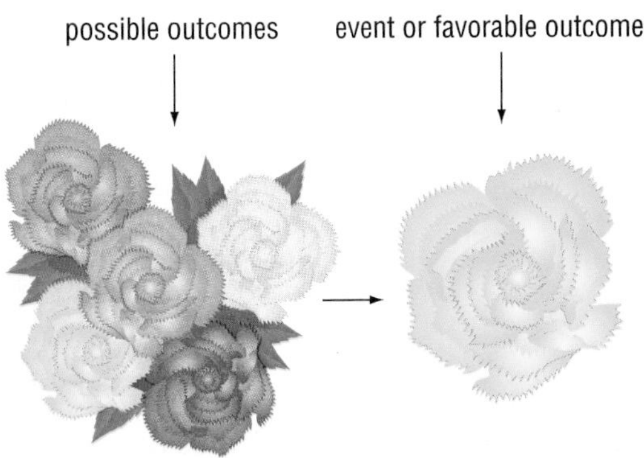

Probability is the chance that some event will occur. You can use a ratio to find probability.

Probability — Key Concept

Words A The probability of an event is a ratio that compares the number of favorable outcomes to the number of possible outcomes.

Symbols $P(\text{event}) = \dfrac{\text{number of favorable outcomes}}{\text{number of possible outcomes}}$

The probability that an event will occur is a number from 0 to 1, including 0 and 1. The closer a probability is to 1, the more likely it is that an event will happen.

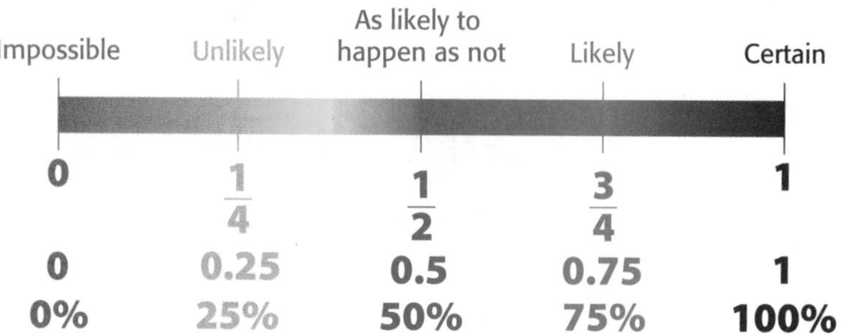

Outcomes occur at **random** if each outcome is equally likely to occur.

EXAMPLES Find Probability

There are six equally likely outcomes if a number cube with sides labeled 1 through 6 is rolled.

1 Find the probability of rolling a 6.

Reading Math

Probability The notation $P(6)$ is read *the probability of rolling a 6.*

There is only one 6 on the number cube.

$$P(6) = \frac{\text{number of favorable outcomes}}{\text{number of possible outcomes}}$$

$$= \frac{1}{6}$$

The probability of rolling a 6 is $\frac{1}{6}$.

2 Find the probability of rolling a 2, 3, or 4.

The word *or* indicates that the number of favorable outcomes needs to include the numbers 2, 3, and 4.

$$P(2, 3, \text{ or } 4) = \frac{\text{number of favorable outcomes}}{\text{number of possible outcomes}}$$

$$= \frac{3}{6} \text{ or } \frac{1}{2} \quad \text{Simplify.}$$

The probability of rolling a 2, 3, or 4 is $\frac{1}{2}$.

CHECK Your Progress

The spinner at the right is spun once. Find the probability of each event. Write each answer as a fraction.

a. $P(\text{F})$

b. $P(\text{D or G})$

c. $P(\text{vowel})$

Study Tip

Angles on Spinners You know that the outcomes of spinning each letter at the right are equally likely since the angles formed are the same size.

Vocabulary Link

Complement

Everyday Use the quantity required to make something complete

Complementary Events

Math Use two events that are the only ones that can happen

If you toss a coin, it can either land on heads or *not* land on heads. These two events are complementary events. **Complementary events** are two events in which either one or the other must happen, but they cannot happen at the same time. The sum of the probability of an event and its complement is 1 or 100%.

EXAMPLE Find Probability of the Complement

3 Find the probability of *not* rolling a 6 in Example 1.

The probability of *not* rolling a 6 and the probability of rolling a 6 are complementary. So, the sum of the probabilities is 1.

$P(6) + P(not\ 6) = 1$

$\frac{1}{6} + P(not\ 6) = 1$ Replace $P(6)$ with $\frac{1}{6}$.

$\frac{1}{6} + \frac{5}{6} = 1$ **THINK** $\frac{1}{6}$ plus what number equals 1?

So, the probability of *not* rolling a 6 is $\frac{5}{6}$.

Study Tip

Look Back You can review solving **equations** in Lesson 1-8.

CHECK Your Progress

A bag contains 5 blue, 8 red, and 7 green marbles. A marble is selected at random. Find the probability of each event.

d. $P(not\ \text{red})$

e. $P(not\ \text{blue or green})$

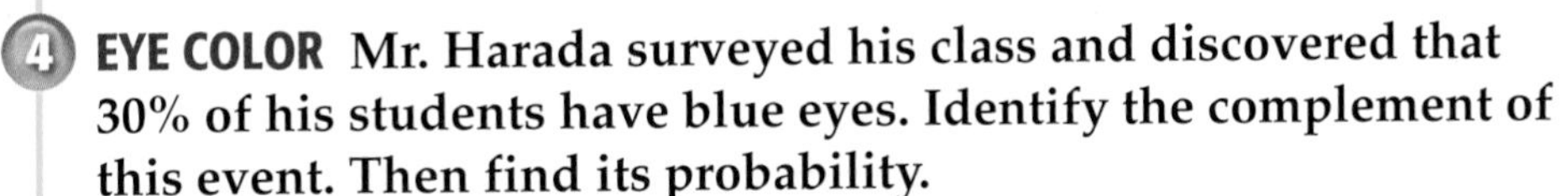

Real-World EXAMPLE

4 EYE COLOR Mr. Harada surveyed his class and discovered that 30% of his students have blue eyes. Identify the complement of this event. Then find its probability.

The complement of having blue eyes is *not* having blue eyes. The sum of the probabilities is 100%.

$P(\text{blue eyes}) + P(not\ \text{blue eyes}) = 100\%$

$30\% + P(not\ \text{blue eyes}) = 100\%$ Replace $P(\text{blue eyes})$ with 30%.

$30\% + 70\% = 100\%$ **THINK** 30% plus what number equals 100%?

So, the probability that a student does *not* have blue eyes is 70%.

CHECK Your Progress

MOVIES Wahlid surveyed his classmates about their favorite type of movies. Identify the complement of each event. Then find the probability of the complement.

Type of Movie	Percent of Students
comedy	46
romance	22
adventure	18
thriller	14

f. adventure

g. romance or thriller

CHECK Your Understanding

Examples 1–3 (pp. 382–383)

A letter tile is chosen randomly. Find the probability of each event. Write each answer as a fraction.

V D I L S I B I E

1. P(D)
2. P(I)
3. P(B or E)
4. P(S, V, or L)
5. P(*not* a vowel)
6. P(*not* a consonant)

Example 4 (p. 383)

7. **GAMES** The probability of choosing a "Go Back 1 Space" card in a board game is 25%. Describe the complement of this event and find its probability. Write the answer as a fraction, decimal, and percent.

Practice and Problem Solving

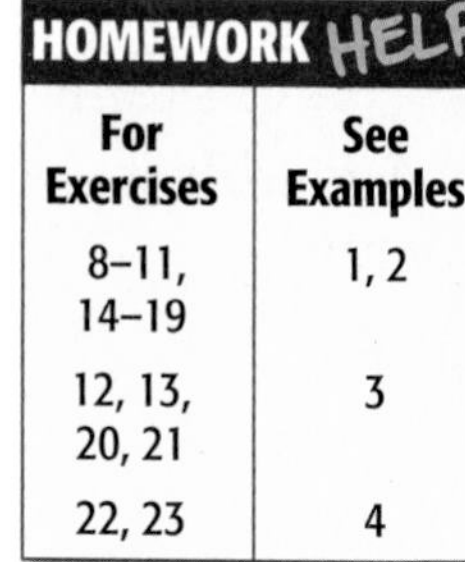

HOMEWORK HELP

For Exercises	See Examples
8–11, 14–19	1, 2
12, 13, 20, 21	3
22, 23	4

The spinner shown is spun once. Find the probability of each event. Write each answer as a fraction.

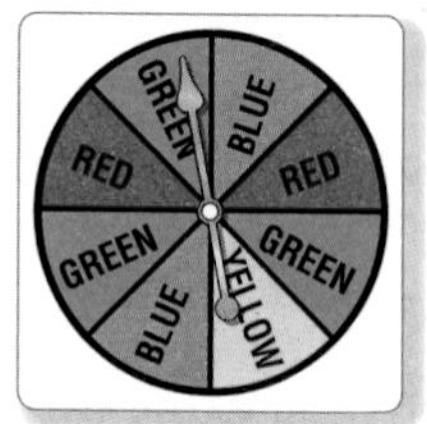

8. P(blue)
9. P(orange)
10. P(red or yellow)
11. P(red, yellow, or green)
12. P(*not* brown)
13. P(*not* green)

Ten cards numbered 1 through 10 are mixed together and then one card is drawn. Find the probability of each event. Write each answer as a fraction.

14. P(8)
15. P(7 or 9)
16. P(less than 5)
17. P(greater than 3)
18. P(odd)
19. P(even)
20. P(*not* a multiple of 4)
21. P(*not* 5, 6, 7, or 8)

ANALYZE TABLES **For Exercises 22 and 23, use the table on air travel at selected airports.**

22. Suppose a flight that arrived at El Centro is selected at random. What is the probability that the flight did not arrive on time?
23. Suppose a flight that arrived at Islip is selected at random. What is the probability that the flight did arrive on time?

Air Travel

Airport	Arrivals (Percent on-time)
El Centro (CA)	80
Baltimore (MD)	82
Charleston (SC)	77
Islip (NY)	83
Milwaukee (WI)	76

Source: U.S. Department of Transportation

One jelly bean is picked, without looking, from the jar shown. Write a sentence stating how likely it is for each event to happen. Justify your answer.

24. black
25. purple
26. purple, red, or yellow
27. green

28. **SCHOOL** Of the students at Grant Middle School, 63% are girls. The school newspaper is randomly selecting a student to be interviewed. Describe the complement of selecting a girl and find the probability of the complement. Write the answer as a fraction, decimal, and percent.

GEOMETRY For Exercises 29–31, use the spinners shown and the information below.

The probability of landing in a certain section on a spinner can be found by considering the size of the angle formed by that section. In spinner A, the angle formed by the blue section is one-fourth of the angle formed by the entire circle. So, $P(\text{blue}) = \frac{1}{4}$, 0.25, or 25%.

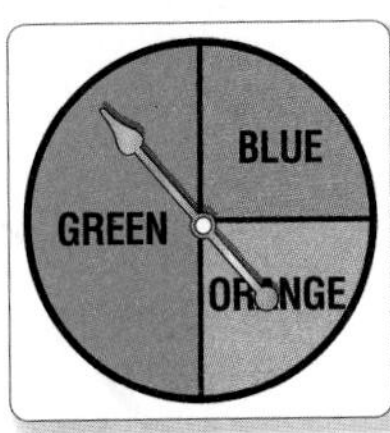

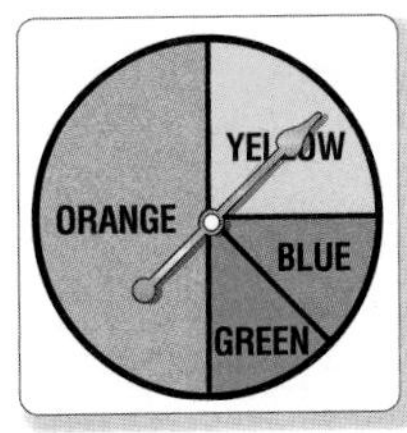

29. Determine *P*(green) for each spinner. Write the probabilities as fractions, decimals, and percents. Justify your response.

30. Determine *P*(*not* orange) for each spinner. Write the probabilities as fractions, decimals, and percents. Justify your response.

See pages 691, 712.

31. For each of the three spinners, are the outcomes of spinning each color equally likely? Explain your reasoning.

H.O.T. Problems

32. **FIND THE ERROR** Alyssa and Luisa are finding the probability of rolling a 3 on a number cube. Who is correct? Explain your reasoning.

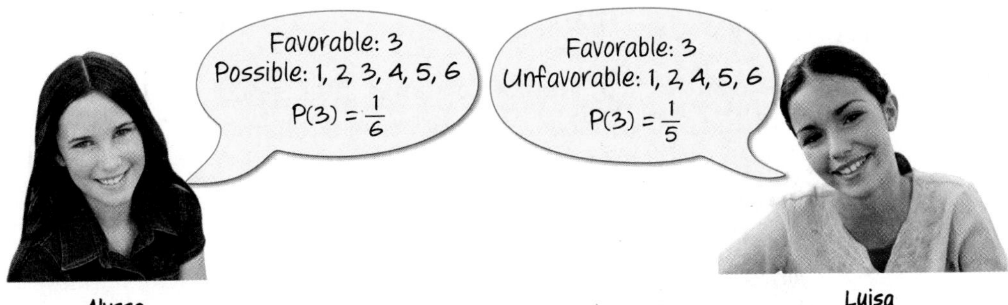

CHALLENGE Another way to describe the chance of an event occurring is with odds. The *odds* in favor of an event is the ratio that compares the number of ways the event can occur to the ways that the event *cannot* occur.

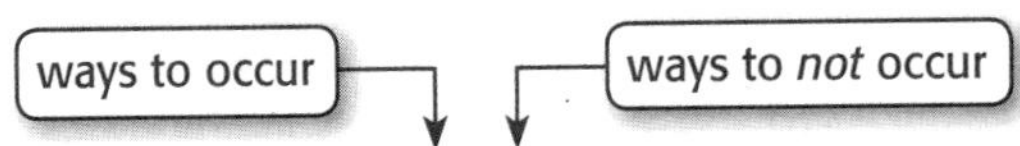

odds of rolling a 3 or a 4 on a number cube → 2 : 4 or 1 : 2

Find the odds of each outcome if a number cube is rolled.

33. a 2, 3, 5, or 6

34. a number less than 3

35. an even number

36. a number greater than 5

37. **CHALLENGE** A spinner for a board game has more than three sections, all of equal size, and the probability of the spinner stopping on blue is 0.5. Design two possible spinners for the game. Explain why each spinner drawn makes sense.

38. **WRITING IN MATH** Explain the relationship between the probability of an event and its complement. Give an example.

TEST PRACTICE

39. Joel has a bowl containing the mints shown in the table.

Color	Number
Red	5
Orange	3
Yellow	1
Green	6

If he randomly chooses one mint from the bowl, what is the probability that the mint will be orange?

A $\frac{1}{5}$

B $\frac{2}{3}$

C $\frac{11}{15}$

D $\frac{4}{5}$

40. A miniature golf course has a bucket with 7 yellow golf balls, 6 green golf balls, 3 blue golf balls, and 8 red golf balls. If Tamika draws a golf ball at random from the bucket, what is the probability that she will *not* draw a green golf ball?

F $\frac{1}{4}$

G $\frac{1}{3}$

H $\frac{2}{3}$

J $\frac{3}{4}$

Spiral Review

41. **FARMS** About 93% of Nebraska's land area is occupied by 48,500 farms and ranches. Write 93% as a decimal. (Lesson 7-3)

42. **VIDEO GAMES** The table shows the time several students spend playing video games. Sketch a circle graph to display the data. (Lesson 7-2)

Time Spent Playing Video Games	
Time (h)	**Percent**
0–1	35
1–2	10
2–3	25
3 or more	30

Add or subtract. Write in simplest form. (Lesson 5-5)

43. $4\frac{3}{8} + 7\frac{1}{8}$

44. $1\frac{2}{3} + 5\frac{3}{4}$

45. $4\frac{3}{5} - 1\frac{1}{2}$

46. $8\frac{1}{6} - 2\frac{2}{3}$

GET READY for the Next Lesson

PREREQUISITE SKILL List all possible outcomes for each situation.

47. tossing a coin

48. rolling a number cube

49. selecting a month of the year

50. choosing a color of the American flag

Extend 7-4

Probability Lab
Experimental and Theoretical Probability

MAIN IDEA

Compare experimental probability with theoretical probability.

New Vocabulary

theoretical probability
experimental probability

Theoretical probability is based on what *should* happen under perfect conditions. These are the probabilities you found in Lesson 7-4. **Experimental probability** is based on what *actually* happens in an experiment. In this lab, you will investigate the relationship between these two types of probability.

STEP 1 Place 3 blue cubes and 5 red cubes in a paper bag.

STEP 2 Without looking, draw a cube out of the bag. If the cube is blue, record a B in a table like the one shown. If the cube is red, record an R.

Trials	Outcome
1	R
2	B
3	R
⋮	
30	

STEP 3 Replace the cube and repeat steps 1 and 2 for a total of 30 trials.

Analyze the Results

1. To find the experimental probability of selecting a blue cube, write the ratio of the number of times a blue cube was selected to the number of trials. What is the experimental probability of selecting a blue cube?
2. What is the theoretical probability of selecting a blue cube? How does this probability compare to the experimental probability found in Exercise 1? Explain any differences.
3. Compare your results to the results of other groups in your class. Why do you think the experimental probabilities usually vary when an experiment is repeated?
4. Find the experimental probability for the entire class's trials. How do the experimental and theoretical probability compare?
5. **MAKE A CONJECTURE** Explain why the experimental probability obtained in Exercise 4 may be closer in value to the theoretical probability than the experimental probability in Exercise 1.
6. **COLLECT THE DATA** Work with a partner. Have your partner place a different number of red and blue cubes totaling 10 into the bag. Use experimental probability to guess the correct number of red and blue cubes in the bag. Explain your reasoning.

Reading Math

Trials A *trial* is a single part of a well-defined experiment. In this lab, a trial is the selection of a cube from the bag.

CHAPTER 7 Mid-Chapter Quiz

Lessons 7-1 through 7-4

Write each percent as a fraction in simplest form. (Lesson 7-1)

1. 39% 2. 18% 3. 175%

4. **MULTIPLE CHOICE** On Tuesday, 48% of the students at West Middle School rode the bus to school. What fractional part of the school did *not* ride the bus to school? (Lesson 7-1)

A $\frac{13}{25}$

B $\frac{1}{48}$

C $\frac{5}{4}$

D $\frac{12}{15}$

Write each fraction or mixed number as a percent. (Lesson 7-1)

5. $\frac{8}{20}$ 6. $1\frac{1}{2}$ 7. $\frac{3}{100}$

FOOD For Exercises 8 and 9, use the graph below. (Lesson 7-2)

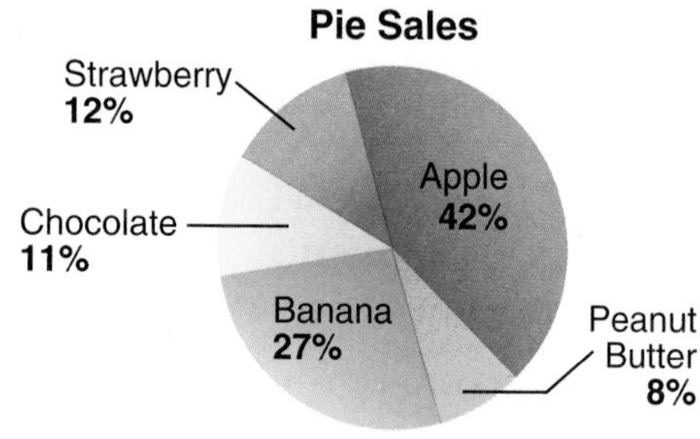

8. What part of the total sales is either strawberry or peanut butter?
9. Which two types of pie have about the same amount of sales?

10. **SPORTS** In Ms. Thorne's class, 60% of the students like soccer, 25% favor football, 5% prefer tennis, and 10% choose another type of sport. Sketch a circle graph to display the data. (Lesson 7-2)

Write each percent as a decimal. (Lesson 7-3)

11. 73% 12. 145% 13. 9%

14. **PLAYS** Twyla has memorized 85% of her lines for the school play. What decimal is equivalent to 85%? (Lesson 7-3)

Write each decimal as a percent. (Lesson 7-3)

15. 0.22 16. 6.75 17. 0.1

18. **MUSIC** The number of chorus students increased by a factor of 1.2 from the previous year. Write 1.2 as a percent. (Lesson 7-3)

19. **MULTIPLE CHOICE** Each circle is divided into sections of equal size. Which circle has 25% of its total area shaded? (Lesson 7-3)

F 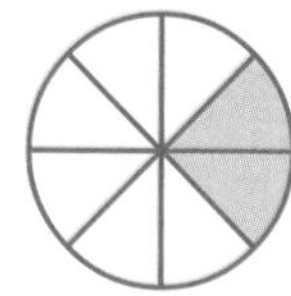

H 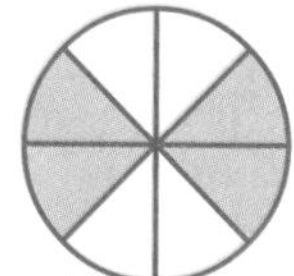

G 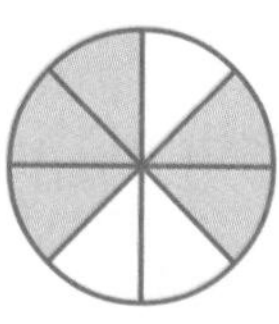

J 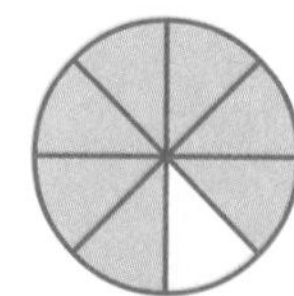

The table shows the types of shirts on a store shelf. Find the probability of choosing each type of shirt. (Lesson 7-4)

Type of Shirt	Number on Shelf
polo	7
T-shirt	6
sweater	4
tank top	3

20. P(polo)
21. P(T-shirt or tank top)
22. P(*not* sweater)
23. P(*not* polo or tank top)

24. **SCHOOL** There is a 60% chance that Suzette will be the next group leader in her reading class. What is the probability that she will *not* be the next group leader? Write the answer as a fraction, decimal, and percent. (Lesson 7-4)

7-5 Sample Spaces

MAIN IDEA

Construct sample spaces using tree diagrams or lists.

New Vocabulary

Fundamental Counting Principle
sample space
tree diagram

Math Online

glencoe.com

- Extra Examples
- Personal Tutor
- Self-Check Quiz

GET READY for the Lesson

MOVIES A movie theater's concession stand sign is shown.

1. List all the possible ways to choose a soft drink, a popcorn, and a candy.
2. How do you know you have accounted for all possible combinations?

SOFT DRINK
Jumbo, Large, Medium

POPCORN
Giant, Large, Small

CANDY
Licorice, Chocolate

The set of all possible outcomes is called the **sample space**. The sample space for choosing a marble and the sample space for picking a card shown are listed at the right.

You can make a list to determine the sample space.

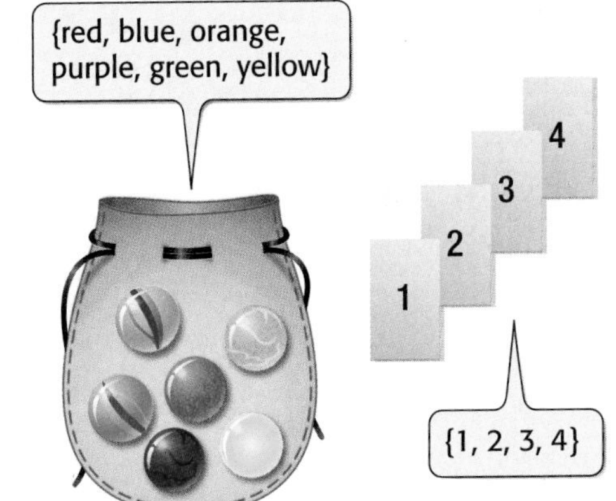

EXAMPLE Use a List to Find Sample Space

1 **ASSEMBLIES The three students chosen to represent Mr. Balderick's class in a school assembly are shown. All three of them need to sit in a row on the stage. In how many different ways can they sit in a row?**

Students
Adrienne
Carlos
Greg

Make an organized list. Use A for Adrienne, C for Carlos, and G for Greg. Use each letter exactly once.

ACG AGC CAG CGA GAC GCA

There are 6 ways the students can sit on the stage.

CHECK Your Progress

a. **FOOD** How many chicken and sauce combinations are possible if you can choose from crispy or grilled chicken with ranch or barbeque sauce? Make an organized list to show the sample space.

A tree diagram can also be used to show a sample space. A **tree diagram** is a diagram that shows all possible outcomes of an event.

EXAMPLE Use a Tree Diagram to Find Sample Space

2 **Use a tree diagram to find how many ice cream cones are possible from a choice of a waffle cone or sugar cone and a choice of chocolate, mint, or peanut butter ice cream.**

List each type of cone first.

Cone	Flavor	Outcome
waffle (W)	chocolate (C)	WC
	mint (M)	WM
	peanut butter (P)	WP
sugar (S)	chocolate (C)	SC
	mint (M)	SM
	peanut butter (P)	SP

There are six possible ice cream cones.

Reading Math

Outcomes The outcome WC means waffle cone and a chocolate ice cream.

CHECK Your Progress

b. Use a tree diagram to find how many different words can be made using the words *fast, slow, old,* and *young* and the suffixes *-er* and *-est.*

Another way to find the sample space is to use the **Fundamental Counting Principle**. This principle states that if there are m outcomes for a first choice and n outcomes for a second choice, then the total number of possible outcomes can be found by finding $m \times n$.

Real-World EXAMPLE Use Fundamental Counting Principle

3 **PIZZA Casey's Pizza House offers the choices shown in the table. Use the Fundamental Counting Principle to find the total number of possible outcomes of a 1-topping pizza.**

Casey's Pizza House	
Crust	**Toppings**
hand-tossed	green pepper
thick	ham
thin	mushroom
	onion
	pepperoni
	sausage

number of outcomes for crust choice		number of outcomes for topping choice		total number of outcomes	
3	$\cdot$	6	=	18	Fundamental Counting Principle

There are 18 different outcomes.

Check Draw a tree diagram to show the sample space.

CHECK Your Progress

c. A number cube is rolled and a spinner with four equal sections marked A, B, C, and D is spun. Use the Fundamental Counting Principle to find the total number of possible outcomes.

CHECK Your Understanding

Example 1 (p. 389)

1. **LIBRARY** How many ways can Ramiro, Garth, and Lakita line up to check out library books? Make an organized list to show the sample space.

Example 2 (p. 390)

2. Use a tree diagram to find how many different backpacks can be made if the backpack comes in nylon or leather and red, green, or black.

Example 3 (p. 390)

3. How many possible outcomes are there if a number cube with sides labeled 1–6 is rolled, and a letter is chosen from the bag shown?

Practice and Problem Solving

HOMEWORK HELP	
For Exercises	See Examples
4–7	1
8–11	2
12–17	3

For Exercises 4–7, make an organized list to show the sample space for each situation.

4. **AMUSEMENT PARK** The names of three roller coasters at Six Flags over Georgia are shown. In how many different ways can Felipe and his friend ride each of the three roller coasters, one time on each roller coaster?

Roller Coasters at Six Flags Over Georgia
Superman-Ultimate Flight
Deja Vu
The Georgia Cyclone

Source: Six Flags

5. **MUSIC** In how many ways can Kame listen to 4 CDs assuming he listens to each CD once?

6. **BOOKS** Refer to the table at the right. Ms. Collins plans on buying one of the books listed for her nephew. She can also choose from yellow or green gift bags. How many book and gift bag combinations are possible?

Best-Selling Children's Hardcover Books of All Time
1. *The Poky Little Puppy* (1942)
2. *The Tale of Peter Rabbit* (1902)
3. *Tootle* (1945)
4. *Green Eggs and Ham* (1960)

Source: *Publishers Weekly*

7. **RESEARCH** Use the Internet or another source to find the fifth book on the all-time best-selling list of children's hardcover books. How many book and gift bag combinations are possible if Ms. Collins can also choose from the fifth best-selling book?

Real-World Link Over 14.8 million copies of *The Poky Little Puppy* have been sold.

Source: American Library Association

Draw a tree diagram to show the sample space for each situation. Then tell how many outcomes are possible.

8. jeans or khakis and a yellow, white, or blue shirt
9. sesame, raisin, or onion bagel with butter, jelly, cream cheese, or peanut butter
10. spin a spinner with 4 equal sections and roll a number cube
11. select a letter from the word FUN, toss a coin, and spin a spinner with 2 sections

For Exercises 12–17, use the Fundamental Counting Principle to find the total number of possible outcomes for each of the following.

12. rolling a number cube and spinning a spinner with eight equal sections
13. tossing a coin and selecting one letter from the word *outcome*
14. selecting one sweatshirt from a choice of five sweatshirts and one pair of pants from a choice of four pairs of pants
15. selecting one entrée from a choice of nine entrées and one dessert from a choice of three desserts
16. selecting either Mark, Padma, or Terrence to be captain and either Flora, Miguel, or Derek to be co-captain
17. selecting one month of the year and one day of the week
18. **DELI** Use a tree diagram to find how many different sandwiches can be made from a choice of white or multigrain bread, a choice of ham, turkey, or roast beef, and a choice of American or provolone cheese.
19. **SCHOOL** A science quiz has one multiple-choice question with answer choices A, B, and C, and two true/false questions. Draw a tree diagram that shows all of the ways a student can answer the questions. Then find the probability of answering all three questions correctly by guessing.

For Exercises 20–22, use the required clothing list at the right for Camp Wood Springs.

1 pair of green shorts
1 pair of white shorts
1 pair of jeans
1 green Wood Springs T-shirt
1 white Wood Springs T-shirt
1 white Wood Springs sweatshirt
1 pair of green socks
1 pair of white socks

20. How many different outfits are possible?
21. What is P(shorts, T-shirt, white socks)?
22. Find the probability of selecting an outfit consisting of shorts, a T-shirt, and socks with no green articles of clothing.

H.O.T. Problems

23. **REASONING** The names of 5 students, Kayla, Jeremy, Chi-Wei, Martin, and Sunil, are written on 5 pieces of paper and placed in a hat. Without looking, three names will be selected from the hat. Find the sample space of each situation below. Then explain how the situations are different and how the sample space is affected.
    a. the number of three-student groups that are possible
    b. the number of different ways that three students are chosen such that the first student is captain, the second student is co-captain, and the third student is the group secretary
24. **CHALLENGE** While playing a game, the spinner shown is spun and a coin is tossed. If the spinner lands on 2 or 4 and the coin lands on tails, Nguyen gets a point. If any other number is spun and the coin lands on heads, Gretchen gets a point. Is this a fair game? Explain.

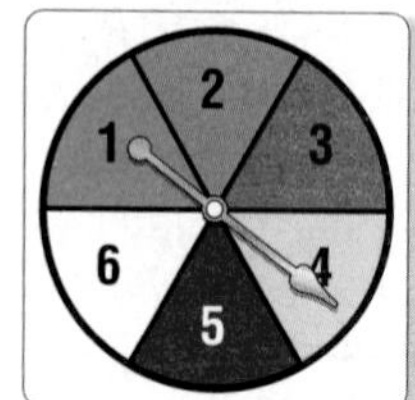

25. **WRITING IN MATH** Describe a situation in which there are 12 possible outcomes.

TEST PRACTICE

26. Claire is deciding between a red shirt and a blue shirt. The shirt also comes in small, medium, and large sizes. Which diagram shows all of the possible combinations of shirt color and size?

A Red: Small, Medium, Large; Blue: Small, Large

C Red: Small, Medium, Large; Blue: Small, Medium, Large

B / **D**

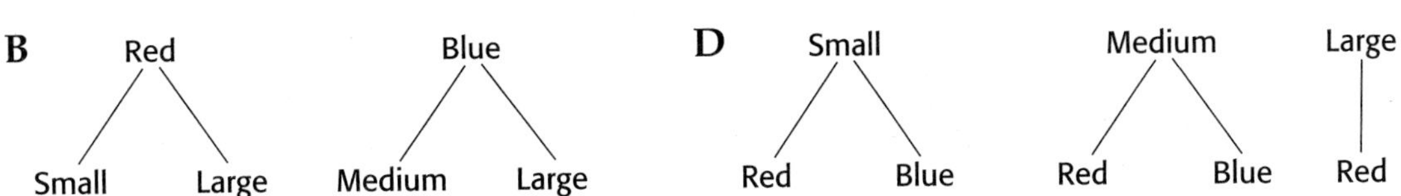

27. Joey's Pizza Parlor offers 3 kinds of toppings and 3 sizes of pizza. Which table shows all the possible 1-topping pizzas?

F

Size	Toppings
Small	Pepperoni
Medium	Pepperoni
Large	Pepperoni
Small	Cheese
Medium	Cheese
Large	Cheese

H

Size	Toppings
Small	Pepperoni
Medium	Cheese
Large	Veggie

G

Size	Toppings
Small	Pepperoni
Small	Pepperoni
Small	Pepperoni
Medium	Cheese
Medium	Cheese
Medium	Cheese
Large	Veggie
Large	Veggie
Large	Veggie

J

Size	Toppings
Small	Pepperoni
Small	Cheese
Small	Veggie
Medium	Pepperoni
Medium	Cheese
Medium	Veggie
Large	Pepperoni
Large	Cheese
Large	Veggie

Spiral Review

A drawer of silverware contains 6 forks, 5 knives, and 3 spoons. One piece of silverware is selected at random. Find the probability of each event. (Lesson 7-4)

28. $P(\text{fork})$

29. $P(\text{knife or spoon})$

30. $P(\text{fork or spoon})$

31. **SNOWBOARDING** At a popular ski resort, 35% of all people who buy tickets are snowboarders. What decimal is equivalent to 35%? (Lesson 7-3)

GET READY for the Next Lesson

PREREQUISITE SKILL Solve each proportion. (Lesson 6-4)

32. $\frac{2}{3} = \frac{8}{x}$

33. $\frac{k}{9} = \frac{10}{45}$

34. $\frac{5}{c} = \frac{30}{96}$

35. $\frac{15}{35} = \frac{3}{d}$

7-6 Making Predictions

MAIN IDEA

Predict the actions of a larger group using a sample.

New Vocabulary

survey
population
sample

Math Online

glencoe.com

- Extra Examples
- Personal Tutor
- Self-Check Quiz

MINI Lab

In this activity, you will predict the number of students in your school who have green eyes.

Eye Color	
Color	Students
blue	
brown	
green	
hazel	

STEP 1 Have one student in each group copy the table shown.

STEP 2 Count the number of students with each eye color in your group. Record the results.

STEP 3 Predict the number of students in your school with green eyes.

STEP 4 Combine your results with the other groups in your class. Make a prediction based on the class data.

1. When working in a group, how did your group predict the number of students in your school with green eyes?
2. Compare your group's prediction with the class prediction. Which do you think is more accurate? Explain.

A **survey** is a method of collecting information. The group being studied is the **population**. Sometimes, the population is very large. To save time and money, part of the group, called a **sample**, is surveyed.

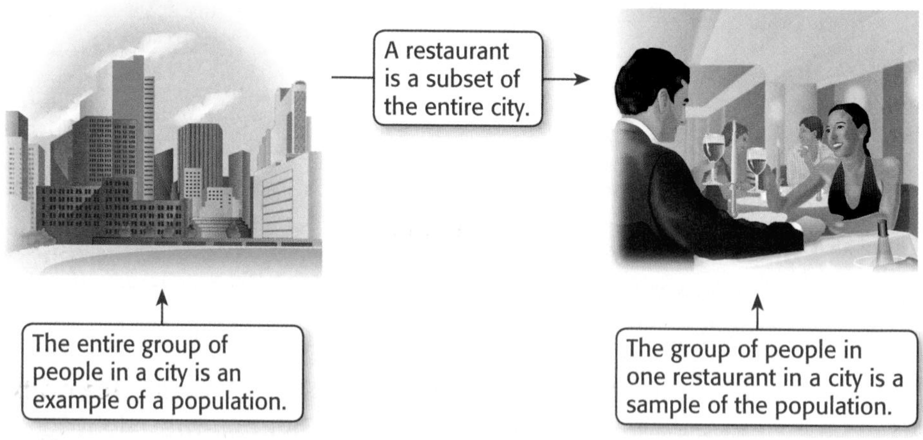
A restaurant is a subset of the entire city.
The entire group of people in a city is an example of a population.
The group of people in one restaurant in a city is a sample of the population.

A good sample is:

- selected at random, or without preference,
- representative of the population, and
- large enough to provide accurate data.

The responses of a good sample are proportional to the responses of the population. So, you can use the results of a survey or past actions to predict the actions of a larger group.

Real-World Career. . . How Does a Photographer Use Math? A photographer uses math when he or she is adjusting settings on a camera based on distance, lighting, and setting.

Math Online

For more information, go to glencoe.com.

EXAMPLES Make Predictions Using Proportions

PHOTOS The students in Mr. Blackwell's class brought photos from their summer break. The table shows how many students brought each type of photo.

Summer Break Photos	
Location	Students
beach	6
campground	4
home	7
theme park	11

1 What was the probability that a student brought a photo taken at a theme park?

$$P(\text{theme park}) = \frac{\text{number of students with theme park photos}}{\text{number of students with a photo}}$$

$$= \frac{11}{28}$$

So, the probability that a student brought a photo taken at a theme park was $\frac{11}{28}$.

2 There are 560 students at the school where Mr. Blackwell teaches. Predict how many students would bring in a photo taken at a theme park.

Let s represent the number of students who would bring in a photo taken at a theme park.

$\frac{11}{28} = \frac{s}{560}$ Write a proportion.

$\frac{11}{28} \underset{\times 20}{\overset{\times 20}{=}} \frac{s}{560}$ Since $28 \times 20 = 560$, multiply 11 by 20 to find s.

$\frac{11}{28} = \frac{220}{560}$ $s = 220$

Of the 560 students, you can expect about 220 to bring a photo from a theme park.

Study Tip

Look Back You can review **solving proportions** in Lesson 6-4.

CHECK Your Progress

INTERNET A survey at a school found that 6 out of every 10 students have an Internet weblog.

a. What is the probability that a student at the school has a weblog?

b. If there are about 250 students at the school, about how many have a weblog?

CHECK Your Understanding

CAREERS For Exercises 1 and 2, use the following information and the table shown.

Every tenth student entering Hamilton Middle School is asked what career field he or she may pursue when they finish school.

Career Field	Students
entertainment	17
education	14
medicine	11
public service	6
sports	2

Example 1 (p. 395)

1. Find the probability that a student will choose public service as a career field.

Example 2 (p. 395)

2. Predict how many students out of 400 will enter the education field.

Practice and Problem Solving

HOMEWORK HELP

For Exercises	See Examples
3, 5	1
4, 6–10	2

MAGAZINES For Exercises 3 and 4, use the following information.

Three out of every 10 students ages 6–14 have a magazine subscription.

3. Find the probability that Annabelle has a magazine subscription.
4. Suppose there are 30 students in Annabelle's class. About how many will have a magazine subscription?

VIDEO GAMES For Exercises 5 and 6, use the following information.

Luther won 12 of the last 20 video games he played.

5. Find the probability of Luther winning the next game he plays.
6. Suppose Luther plays a total of 60 games with his friends over the next month. Predict how many of these games Luther will win.

SPORTS For Exercises 7–10, use the table to predict the number of students out of 500 that would participate in each sport.

Sport	Students
baseball/softball	6
basketball	5
football	9
gymnastics	2
tennis	3

7. football
8. tennis
9. gymnastics
10. basketball

Real-World Link... In a recent year, 28.8 percent of the population of the U.S. volunteered at some point.

Source: U.S. Department of Labor

VOLUNTEERING For Exercises 11–13, use the graph.

11. About 2.8 million kids ages 10–14 live in California. Predict the number of kids who volunteer a few times a year.
12. North Carolina has about 600,000 kids ages 10–14. Predict the number of kids in this age group who volunteer once a week.
13. About 300,000 kids ages 10–14 live in South Carolina. Predict the number of kids in this age group who volunteer once a year.

How Often Kids Volunteer

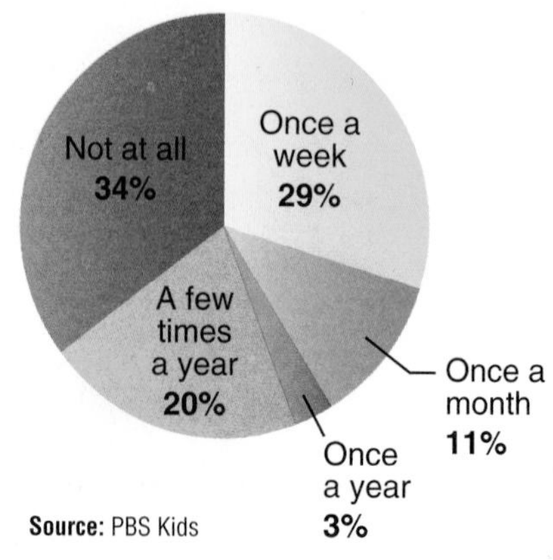

Source: PBS Kids

14. **BOOKS** The school librarian recorded the types of books students checked out on a typical day. If there are 605 students enrolled at the school, predict the number of students that prefer humor books. Compare this to the number of students at the school who prefer nonfiction.

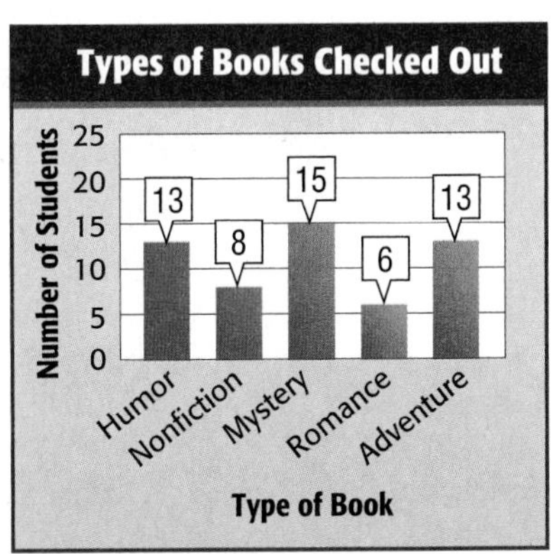

15. **BASKETBALL** The probability of Jaden making a free throw is 15%. Predict the number of free throws that he can expect to make if he attempts 40 free throws.

H.O.T. Problems

16. **FIND THE ERROR** A survey of a sixth-grade class showed that 4 out of every 10 students are taking a trip during spring break. There are 150 students in the sixth grade. Xavier and Elisa are trying to determine how many of the sixth-grade students can be expected to take a trip during spring break. Who is correct? Explain your reasoning.

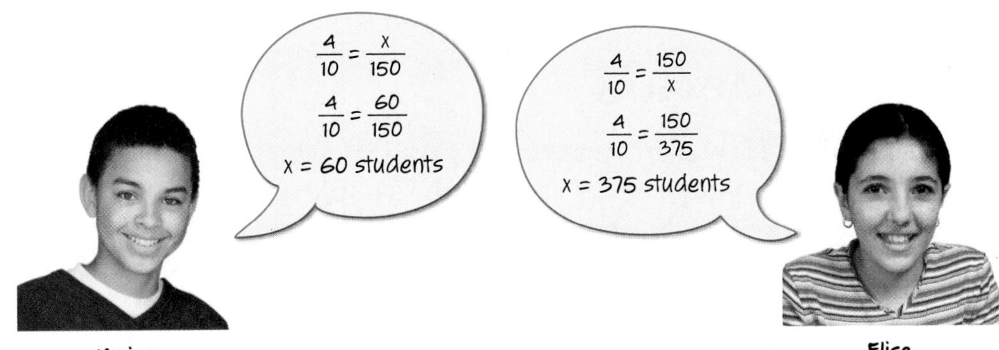

17. **CHALLENGE** One letter tile is drawn from the bag and replaced 300 times. Predict how many times a consonant will *not* be picked.

18. **OPEN ENDED** Give an example of a situation in which you would make a prediction.

19. **SELECT A TOOL** Nolan is going to listen to a CD on random mode. There are 14 songs on the CD, and 4 of them are Nolan's favorites. Which of the following tools can Nolan use to find the probability that the first song played will be one of his favorites? Justify your selection(s). Then use the tool(s) to solve the problem.

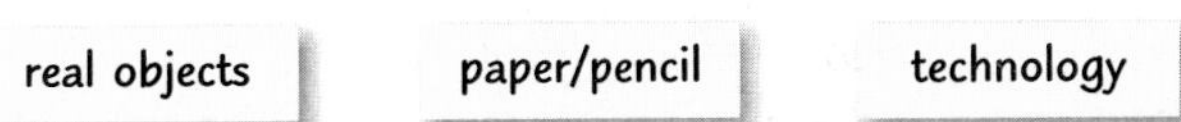

20. **WRITING IN MATH** Three out of four of Mitch's sixth-grade friends say that they will not attend the school dance. Based on this information, Mitch predicts that only 25 of the 100 sixth graders at his school will attend the dance. Is this a valid prediction? Explain your reasoning.

TEST PRACTICE

21. At the school carnival, Jesse won the balloon dart game 1 out of every 5 times he played. If he plays the game 15 more times, about how many times can he expect to win?

    A 3

    B 4

    C 5

    D 15

22. **SHORT RESPONSE** If 7 out of 30 students are going on the ski trip, predict the number of students out of 150 that are going on the ski trip.

23. The table shows the results of a survey of sixth-grade students in the lunch line.

Favorite Drink	
Drink	**Students**
Chocolate Milk	15
Soda	12
Milk	6
Water	2

If there are 245 sixth graders in the school, how many can be expected to prefer chocolate milk?

F 45 **H** 90

G 84 **J** 105

Spiral Review

24. **VIDEOS** How many ways can a person watch 3 different videos? Make an organized list to show the sample space. (Lesson 7-5)

Juanita randomly turns to a page in a 15-page booklet. Find the probability of each event. (Lesson 7-4)

25. P(odd page)
26. P(even page)
27. P(page that is a composite number)
28. P(3, 5, or 7)

29. **INSECTS** A mosquito's proboscis, the part that sucks blood, is the first $\frac{1}{3}$ of its body's length. The rest of the mosquito is made up of the head, thorax, and abdomen. How much of a mosquito is the head, thorax, and abdomen? (Lesson 5-3)

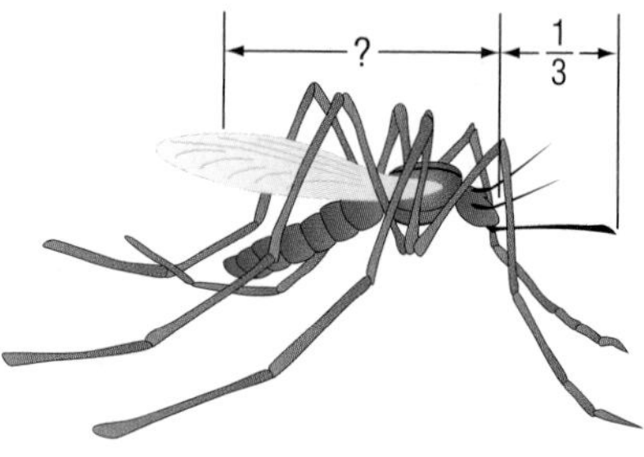

Write each improper fraction as a mixed number. (Lesson 4-3)

30. $\frac{16}{9}$
31. $\frac{22}{9}$
32. $\frac{35}{6}$
33. $\frac{50}{6}$

34. **MONEY** About how much more is \$74.50 than \$29.95? (Lesson 3-4)

GET READY for the Next Lesson

35. **PREREQUISITE SKILL** Chandler collected \$4 from each of his 28 classmates to buy a gift for their teacher. Is \$150, \$180, or \$200 a more reasonable estimate for how much money was collected? (Lesson 3-10)

Problem-Solving Investigation

MAIN IDEA: Solve problems by solving a simpler problem.

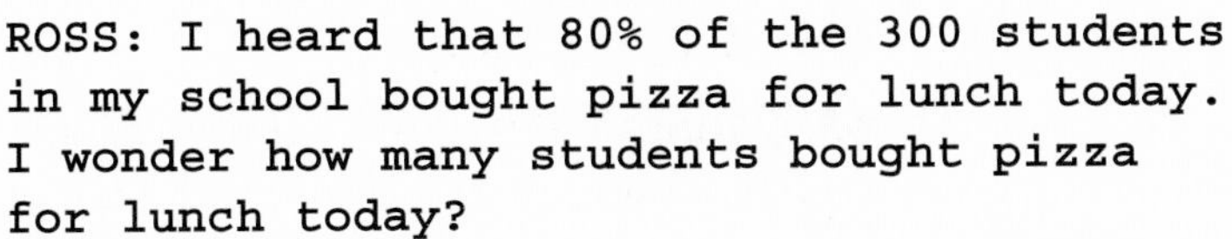

e-mail: SOLVE A SIMPLER PROBLEM

ROSS: I heard that 80% of the 300 students in my school bought pizza for lunch today. I wonder how many students bought pizza for lunch today?

THE MISSION: Solve a simpler problem to find how many students bought pizza.

Understand	You know the number of students in the school and that 80% of them bought pizza. You need to find the number of students that bought pizza.
Plan	Solve a simpler problem by finding 10% of the total students and then use the result to find 80% of the total students.
Solve	Since $80\% = \frac{80}{100}$ or $\frac{8}{10}$, 8 out of every 10 students bought pizza. There are $300 \div 10$ or 30 groups with ten students in each group. Multiply 30 by 8. So, 240 students bought pizza for lunch.
Check	You know that 80% is close to 75%, which is $\frac{3}{4}$. Since $\frac{1}{4}$ of 300 is 75, $\frac{3}{4}$ of 300 is 225. So, 240 is a reasonable answer. ✔

Analyze The Strategy

1. Explain when you would use the *solve a simpler problem* strategy.
2. Explain why the students found it simpler to work with 10%.
3. **WRITING IN MATH** Write a problem that can be solved by working a simpler problem. Then write the steps to find the solution.

Mixed Problem Solving

Use the *solve a simpler problem* strategy to solve Exercises 4–6.

4. **MONEY** Heidi's dad wants to leave an 18% tip for a $24.60 restaurant bill. About how much money should he leave?

5. **MOVIES** Ebony estimates that she watches 500 movies per year. About how many movies does she watch per week?

6. **CANDY** A candy factory can make 1,200 individually wrapped pieces of candy in one minute. About how many pieces do they make per second?

Use any strategy to solve Exercises 7–16. Some strategies are shown below.

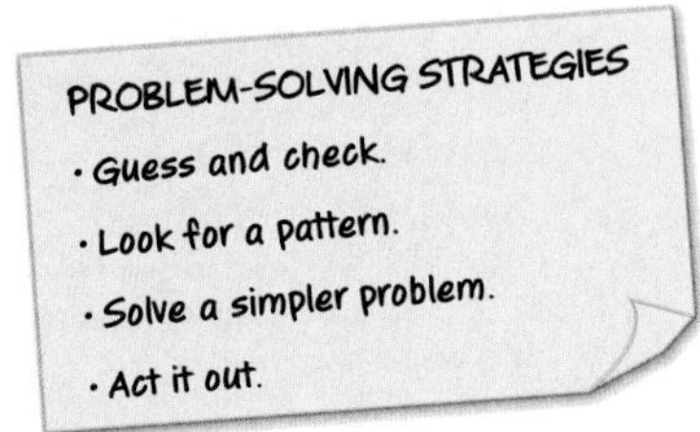

7. **WATCHES** Yuma's watch beeps every hour. How many times will it beep in one week?

8. **BORDER** Part of a strip of border for a bulletin board is shown.

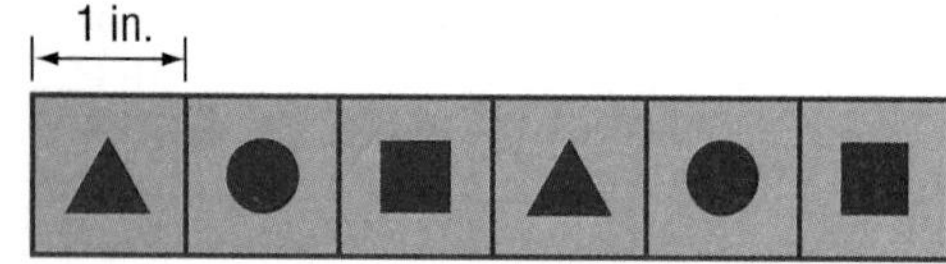

All the sections of the border are the same width. If the first shape on the strip is a triangle and the strip is 74 inches long, what is the last shape on the strip?

9. **EXERCISE** To train for a marathon, you plan to run one mile the first week and double the number of miles each week for 6 weeks. How many miles will you run the sixth week?

10. **AREA** Find the area of the figure.

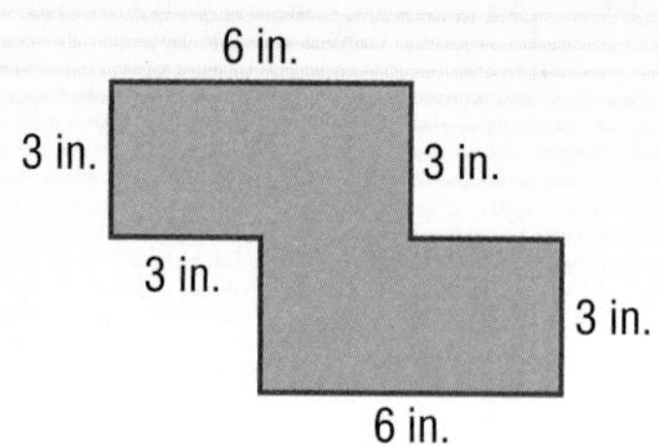

11. **PARTY FAVORS** Mia has a string of ribbon 72 inches long. She needs to cut the ribbon into 2-inch pieces to tie around party favors. If it takes 1 second to make each cut, how long will it take to cut all of the ribbon into 2-inch pieces?

12. **HANDSHAKES** If a total of 10 handshakes were exchanged at a party and each person shook hands exactly once with each of the others, how many people were at the party?

13. **PIE** The number of each type of pie in Rose's bakery is shown in the bar graph. How many times as many strawberry pies are there as peach pies?

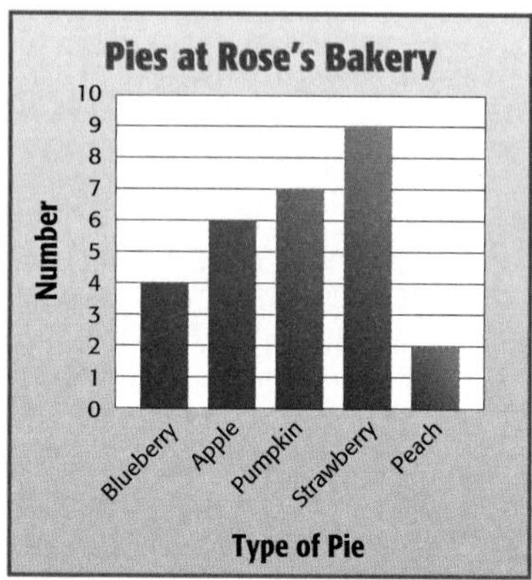

14. **GEOGRAPHY** Water covers 1,911 square miles of the state of South Carolina. If this is only 5.9% of the area of the state, about how many total square miles is South Carolina?

15. **PATTERNS** Describe the pattern then find the missing number.

4, 12, ■, 108, 324

16. **STICKERS** Julie has 32 stickers. She plans to give each of 4 friends an equal number of the stickers. How many stickers will each person receive?

7-8 Estimating with Percents

MAIN IDEA

Estimate the percent of a number.

Math Online

glencoe.com

- Concepts in Motion
- Extra Examples
- Personal Tutor
- Self-Check Quiz

MINI Lab

You can use a model and find a fractional part of a number. The model below shows how to find $\frac{1}{4}$ of 20.

STEP 1 Model 20 on a piece of grid paper.

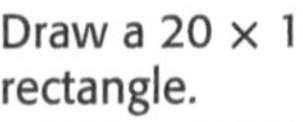

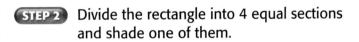

STEP 2 Divide the rectangle into 4 equal sections and shade one of them.

Each section contains 5 grid squares.

So, $\frac{1}{4}$ of 20 is 5.

Use grid paper to find the fractional portion of each number.

1. $\frac{1}{2}$ of 10
2. $\frac{1}{5}$ of 10
3. $\frac{2}{5}$ of 20
4. $\frac{5}{6}$ of 36
5. **MAKE A CONJECTURE** How can you find a fractional part of a number without drawing a model on grid paper?

Estimating with percents will provide a reasonable solution to many real-world problems. The table below shows some commonly used percents and their fraction equivalents.

Percent–Fraction Equivalents — Key Concept

$20\% = \frac{1}{5}$	$50\% = \frac{1}{2}$	$80\% = \frac{4}{5}$	$25\% = \frac{1}{4}$	$33\frac{1}{3}\% = \frac{1}{3}$
$30\% = \frac{3}{10}$	$60\% = \frac{3}{5}$	$90\% = \frac{9}{10}$	$75\% = \frac{3}{4}$	$66\frac{2}{3}\% = \frac{2}{3}$
$40\% = \frac{2}{5}$	$70\% = \frac{7}{10}$	$100\% = 1$		

EXAMPLES Estimate the Percent of a Number

1 Estimate 52% of 298.

52% is close to 50% or $\frac{1}{2}$. Round 298 to 300.

$\frac{1}{2}$ of 300 is 150. $\frac{1}{2}$ or *half* means to divide by 2.

So, 52% of 298 is about 150.

2 **Estimate 60% of 27.**

60% is $\frac{3}{5}$.

Round 27 to 25 since it is divisible by 5.

$\frac{1}{5}$ of 25 is 5. $\frac{1}{5}$ or *1 fifth* means divide by 5.

So, $\frac{3}{5}$ of 25 is 3×5 or 15.

Thus, 60% of 27 is about 15.

CHECK Your Progress

Estimate each percent.

a. 48% of 76 b. 18% of 42 c. 73% of 41

Real-World EXAMPLE

Real-World Link The largest male polar bear ever found weighed 2,209 pounds and was 12 feet in length.

Source: Sea World

3 **POLAR BEARS Polar bears can eat as much as 10% of their body weight in less than one hour. If an adult male polar bear weighs 715 pounds, about how much food can he eat in one hour?**

To determine how much food a polar bear can eat in one hour, you need to estimate 10% of 715.

METHOD 1 **Use a proportion.**

$10\% = \frac{1}{10}$ and $715 \approx 700$

$\frac{1}{10} = \frac{x}{700}$ Write the proportion.

$\frac{1}{10} = \frac{x}{700}$ (×70 on numerator and denominator) Since $10 \times 70 = 700$, multiply 1 by 70.

$x = 70$

METHOD 2 **Use mental math.**

$10\% = \frac{1}{10}$ and $715 \approx 700$

$\frac{1}{10}$ of 700 is 70.

So, a polar bear can eat about 70 pounds of food in one hour.

CHOOSE Your Method

d. **HIKING** A group of friends went on a hiking trip. They planned to hike a total of 38 miles. They want to complete 25% of the hike by the end of the first day. About how many miles should they hike the first day?

TEST EXAMPLE

4 Yutaka surveyed the students in his health class regarding their favorite juice. Which is the most likely number of students out of 353 that would prefer orange juice?

Favorite Juice	Percent of Students
Mixed Fruit	34
Apple	29
Orange	22
Grape	15

A 20 **C** 175

B 70 **D** 350

Test-Taking Tip

Eliminate Choices When taking a multiple-choice test, eliminate the choices you know to be incorrect. The percent of students that would prefer orange juice is clearly less than 50%. So, you can eliminate choices C and D.

Read the Item

You need to estimate the number of students out of 353 that would prefer orange juice. 22% of the students chose orange juice.

Solve the Item

22% is about 20% or $\frac{1}{5}$. Round 353 to 350.

$\frac{1}{5}$ of 350 is 70.

So, about 70 students would prefer orange juice. The answer is B.

CHECK Your Progress

e. According to a recent survey, about 80% of sixth graders eat lunch in their school cafeteria. About how many of Westside Middle School's 126 sixth graders eat lunch in the cafeteria?

F 50 **G** 80 **H** 100 **J** 125

CHECK Your Understanding

Examples 1, 2 (pp. 401–402)

Estimate each percent.

1. 19% of $53
2. 34% of 62
3. 47% of $118
4. 38% of $50
5. 59% of 16
6. 75% of 33

Example 3 (p. 402)

7. **PURSES** A purse that originally cost $29.99 is on sale for 50% off. About how much is the sale price of the purse?

Example 4 (p. 403)

8. **MULTIPLE CHOICE** Ayana surveyed several classmates about their favorite winter activity. Predict the number of students out of 164 that prefer snowboarding.

A 30 **C** 75

B 54 **D** 100

Favorite Winter Activity

Practice and Problem Solving

HOMEWORK HELP	
For Exercises	See Examples
9–18	1, 2
19, 20	3
31–34	4

Estimate each percent.

9. 21% of 96
10. 53% of 59
11. 19% of 72
12. 35% of 147
13. 26% of 125
14. 42% of 16
15. 79% of 82
16. 67% of 296
17. 89% of 195
18. Estimate seventy-four percent of forty-five.
19. **RIDES** Trevon spent 8 hours and 15 minutes at an amusement park yesterday. He spent 75% of the time at the park riding rides. About how much time did he spend riding rides?
20. **ANIMALS** Penguins spend as much as 75% of their lives in the sea. An Emperor Penguin living in the wild has a life span of about 18 years. About how many years does a wild Emperor Penguin spend in the sea?
21. **POPULATION** About 42% of Alaska's population lives in the city of Anchorage. If Alaska has a total population of 648,818, about how many people live in Anchorage?
22. **BASKETBALL** During the basketball season, Michael made 37 baskets out of 71 attempts. About what percent of his shots did he miss? Explain.
23. **CIRCLE GRAPHS** A group of students were asked how they most often communicate with their grandparents. Sketch a circle graph of the results shown in the table.

Communicating with Grandparents	
Phone	43%
E-mail	32%
Letters	19%
Instant Messages	6%

24. **FIND THE DATA** Refer to the Data File on pages 16–19. Choose some data and write a real-world problem in which you would estimate the percent of a number.

Estimate the percent that is shaded in each figure.

25. 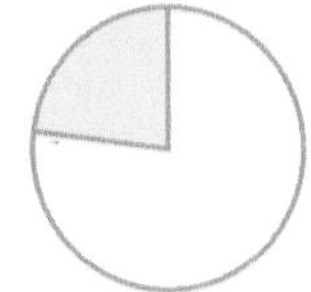
26. 
27. 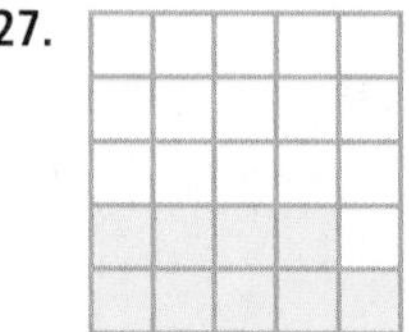

EXTRA PRACTICE
See pages 692, 712.

H.O.T. Problems

28. **CHALLENGE** Order 10% of 20, 20% of 20, and $\frac{1}{5}$% of 20 from least to greatest.
29. **NUMBER SENSE** Rachel wants to buy a shirt regularly priced at $32. It is on sale for 40% off. Rachel estimates that she will save $\frac{2}{5}$ of $30 or $12. Will the actual amount be more or less than $12? Explain your reasoning.
30. **WRITING IN MATH** A classmate is trying to estimate 42% of $122. Explain how your classmate should solve the problem.

TEST PRACTICE

31. Refer to the graph. If 4,134 people were surveyed, which equation can be used to estimate the number of 18–24-year-olds that own a portable MP3 player?

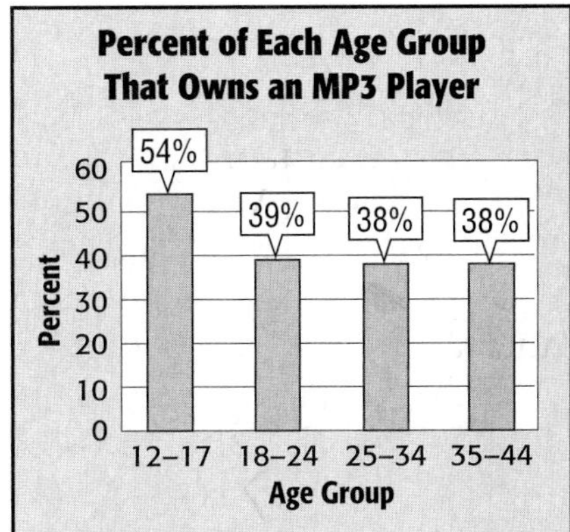

Source: Edison Media research

A $\frac{1}{2}$ of 4,000 = 2,000

B $\frac{2}{5}$ of 4,000 = 1,600

C $\frac{1}{3}$ of 4,000 = 1,300

D $\frac{1}{5}$ of 4,000 = 800

32. **SHORT RESPONSE** In a recent survey of teens, 21% said their friends like to read and talk about books. About how many teens out of 1,095 would say their friends read and talk about books?

33. After a group of 24 parts were tested, 5 were found to be defective. About what percent of the parts tested were defective?

F 5%

G 20%

H 25%

J 33%

34. The volleyball team played 20 games this season. If they won 75% of their games, about how many games did they win?

A 7 C 15

B 10 D 20

Spiral Review

35. **GYMS** Every 12th person entering the gym on Saturday will receive a free T-shirt. How many T-shirts will be given away if 190 people go to the gym on Saturday? (Lesson 7-7)

36. **GLASSES** In Mr. Cardona's second period class, 9 out of the 20 students wear glasses. If Mr. Cardona has 100 students, predict the number of his students who wear glasses. (Lesson 7-6)

Express each decimal as a percent. (Lesson 7-3)

37. 0.45 38. 0.02 39. 0.362 40. 0.058

Problem Solving in Social Studies

Real-World Unit Project

The Nifty Fifty States It's time to complete your project. Use the data you've gathered about the fifty United States to prepare a Web page, poster, or electronic presentation. Be sure to include statistical displays and/or a spreadsheet with your project.

Math Online **Unit Project at** glencoe.com

CHAPTER 7 Study Guide and Review

Math Online glencoe.com
- STUDY TO GO
- Vocabulary Review

GET READY to Study

Be sure the following Big Ideas are noted in your Foldable.

Fraction	Percent	Decimal
$\frac{1}{2}$ →	50% →	0.5

BIG Ideas

Percent (Lesson 7-1)

- A percent is a ratio that compares a number to 100.

Percent Conversions (Lessons 7-1 and 7-3)

- To write a percent as a fraction, write the percent as a fraction with a denominator of 100. Then simplify.
- To write a percent as a decimal, rewrite the percent as a fraction with a denominator of 100. Then write the fraction as a decimal.
- To write a decimal as a percent, write the decimal as a fraction with a denominator of 100. Then write the fraction as a percent.

Probability (Lesson 7-4)

- The probability of an event is a ratio that compares the number of favorable outcomes to the number of possible outcomes.

Percent-Fraction Equivalents (Lesson 7-8)

$20\% = \frac{1}{5}$	$50\% = \frac{1}{2}$	$80\% = \frac{4}{5}$
$25\% = \frac{1}{4}$	$33\frac{1}{3}\% = \frac{1}{3}$	$30\% = \frac{3}{10}$
$60\% = \frac{3}{5}$	$90\% = \frac{9}{10}$	$75\% = \frac{3}{4}$
$66\frac{2}{3}\% = \frac{2}{3}$	$40\% = \frac{2}{5}$	$70\% = \frac{7}{10}$
$100\% = 1$		

Key Vocabulary

circle graph (p. 370)
complementary events (p. 383)
Fundamental Counting Principle (p. 390)
percent (p. 365)
population (p. 394)
probability (p. 381)
random (p. 382)
sample (p. 394)
sample space (p. 389)
simple event (p. 381)
survey (p. 394)
tree diagram (p. 390)

Vocabulary Check

State whether each sentence is *true* or *false*. If *false*, replace the underlined word or number to make a true sentence.

1. An organized list that is used to show all of the possible outcomes is called a <u>survey</u>.
2. A percent is a ratio that compares a number to <u>10</u>.
3. A ratio that compares the number of favorable outcomes to the number of possible outcomes is the <u>probability</u> of an event.
4. If an event is certain to occur, the probability of that event is <u>0</u>.
5. Two events in which either one or the other must happen, but cannot happen at the same time, are called <u>random</u>.
6. In tossing a coin, the notation <u>*P*(tails)</u> denotes the probability that the coin will land on tails.

Lesson-by-Lesson Review

7-1 Percents and Fractions (pp. 365–369)

Write each percent as a fraction or mixed number in simplest form.

7. 3%
8. 48%
9. 120%

Write each fraction as a percent.

10. $\frac{7}{8}$
11. $1\frac{3}{5}$
12. $\frac{19}{100}$

13. **DESERTS** One third of Earth's land surface is covered by desert and about 13% of the world's population live in a desert area. Write the percent of the world's population that live in a desert area as a fraction in simplest form.

Example 1 **Write 24% as a fraction in simplest form.**

$24\% = \frac{24}{100}$ Definition of percent

$= \frac{6}{25}$ Simplify. Divide numerator and denominator by the GCF, 4.

Example 2 **Write $\frac{3}{5}$ as a percent.**

$\frac{3}{5} = \frac{n}{100}$ Write a proportion.

$\frac{3}{5} = \frac{60}{100}$ (×20) Since 5 × 20 = 100, multiply 3 by 20 to find *n*.

So, $\frac{3}{5} = 60\%$.

7-2 Circle Graphs (pp. 370–375)

14. **MOVIES** A group of adults were asked to name their favorite type of movie. The results are shown in the table. Sketch a circle graph of the data.

Type of Movie	Percent
Comedy	40
Action	25
Drama	25
Romance	10

For Exercises 15–17, refer to the circle graph from Exercise 14.

15. Which type of movie did the adults say is their favorite?
16. Which two sections represent the responses by the same amount of adults?
17. How does the number of adults who say their favorite type of movie is comedy compare to the number of adults who say their favorite type of movie is romance?

Example 3 **In Mr. Finn's class, 52% of the students have one sibling, 24% have two or more siblings, and 24% have no siblings. Sketch a circle graph of the data.**

Use a compass to draw a circle with at least a 1-inch radius.

Since 52% is a little more than 50% or $\frac{1}{2}$, shade a little more than $\frac{1}{2}$ of the circle for "One Sibling." Since the last two sections are equal, take the remaining part of the circle and divide it into two equal parts.

Label each section of the graph. Then give the graph a title.

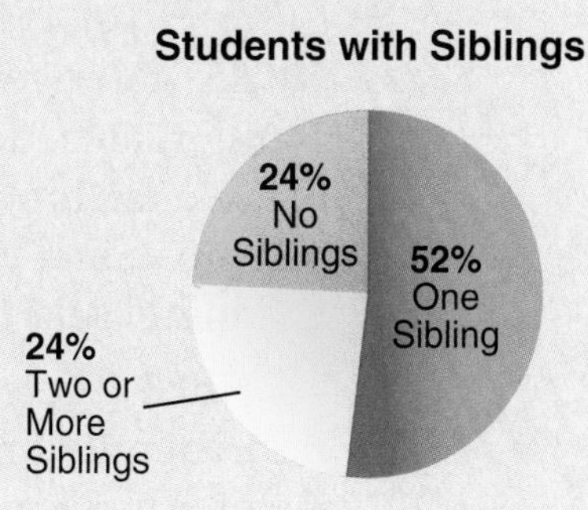

Study Guide and Review

7-3 Percents and Decimals (pp. 377–380)

Write each percent as a decimal.

18. 2%
19. 38%
20. 140%
21. 90%

Write each decimal as a percent.

22. 0.03
23. 1.3
24. 1.75
25. 0.51
26. **BREAD** A slice of bread is 30% water. Write this as a decimal.

Example 4 **Write 46% as a decimal.**

$46\% = \frac{46}{100}$ Rewrite the percent as a fraction with a denominator of 100.

$= 0.46$ Write *46 hundredths* as a decimal.

Example 5 **Write 0.85 as a percent.**

$0.85 = \frac{85}{100}$ Write *85 hundredths* as a fraction.

$= 85\%$ Write the fraction as a percent.

7-4 Probability (pp. 381–386)

One marble is pulled from the bag without looking. Find the probability of each event. Write each answer as a fraction.

27. P(purple)
28. P(*not* red)
29. P(yellow or green)
30. P(purple or red)

The letters in the word ALMANAC are each written on a piece of paper and placed in a bag. Find the probability of each event. Write each answer as a fraction.

31. P(M or N)
32. P(vowel)
33. P(*not* A)
34. P(A through M)
35. **SOCKS** A drawer contains 14 unorganized socks of which 4 are black, 2 are brown, 2 are blue, and the rest are white. Find the probability of randomly selecting a white sock.
36. **WEATHER** There is a 78% chance of rain. Identify the complement of this event. Then find its probability.

Example 6 **The spinner shown is spun once. Find the probability of landing on an even number.**

There are 10 equally likely outcomes on the spinner. Five of the numbers are even numbers. Those numbers are 2, 4, 6, 8, and 10.

$P(2, 4, 6, 8, \text{or } 10) = \frac{5}{10} \text{ or } \frac{1}{2}$

So, $P(2, 4, 6, 8, \text{or } 10)$ is $\frac{1}{2}$, 0.5, or 50%.

Example 7 **Refer to Example 6. Identify the complement of landing on 2. Then find the probability.**

The complement of land on 2 is *not* landing on 2. The sum of the probabilities is 1.

$P(2) + P(\textit{not } 2) = 1$

$\frac{1}{10} + P(\textit{not } 2) = 1$ $P(2) = \frac{1}{10}$

$\frac{1}{10} + \frac{9}{10} = 1$ Use mental math.

So, $P(\textit{not } 2)$ is $\frac{9}{10}$.

Mixed Problem Solving
For mixed problem-solving practice, see page 712.

7-5 Sample Spaces (pp. 389–393)

For Exercises 37 and 38, make an organized list to show the sample space for the situation. Then tell how many outcomes are possible.

37. a choice of black or blue jeans in classic fit, stretch, or bootcut style
38. a choice of comedy, action, horror, or science fiction DVD in widescreen or full screen format

Draw a tree diagram to show the sample space for each situation. Then tell how many outcomes are possible.

39. apple, peach, or cherry pie with milk, juice, or tea
40. toss a dime, quarter, and penny

A coin is tossed, and a number cube is rolled.

41. How many outcomes are possible?
42. Find P(tails, *not* odd).

Example 8 **Suppose you have a choice of a plain (P) or frosted (F) doughnut with cream (C), jelly (J), or custard (S) filling. How many different doughnuts are possible?**

Use a tree diagram.

Doughnut	Filling	Outcome
plain (P)	cream (C)	PC
	jelly (J)	PJ
	custard (S)	PS
frosted (F)	cream (C)	FC
	jelly (J)	FJ
	custard (S)	FS

There are 6 possible doughnuts.

7-6 Making Predictions (pp. 394–398)

PIZZA **For Exercises 43 and 44, use the following information.**

Out of 32 students, 8 prefer only cheese on their pizza.

43. What is the probability that a student in this group likes only cheese on a pizza? Write the answer as a fraction, decimal, and percent.
44. If there were 240 students, how many would you expect to like only cheese on their pizza?

Example 9 **If 9 out of 25 people prefer rock music, how many people out of 1,000 would you expect to prefer rock music?**

Let p represent the number of people who prefer rock music.

$\frac{9}{25} = \frac{p}{1,000}$ Write the proportion.

$\frac{9}{25} \overset{\times 40}{=} \frac{360}{1,000}$ Since $25 \times 40 = 1,000$, multiply 9 by 40.

So, of the 1,000 people, you would expect 360 to prefer rock music.

7-7 PSI: Solve a Simpler Problem (pp. 399–400)

Solve. Use the *solve a simpler problem* strategy.

45. **TABLES** The cafeteria has 45 square tables that can be pushed together to form one long table for the lacrosse team's banquet. Each square table can seat one person on each side. How many people can be seated at the banquet table?

46. **RAINFALL** In one year, Orlando received 62.51 inches of rain. In September, the city received 25% of that rainfall. About how much rain did Orlando receive in September?

Example 10 In May, Marisol had earned \$240 from babysitting. She wants to save 70% of this amount to put toward a new laptop. How much money does she need to save?

Solve a simpler problem.

$10\% \text{ of } \$240 = \frac{1}{10} \text{ of } \$240$ $\quad 10\% = \frac{1}{10}$

$= \$24$ $\quad \frac{1}{10}$ means divide by 10.

Since 70% is 7 • 10%, multiply \$24 by 7.

So, Marisol needs to save 7 • \$24 or \$168.

7-8 Estimating with Percents (pp. 401–405)

Estimate each percent.

47. 40% of 78
48. 73% of 20
49. 25% of 122
50. 19% of 99
51. 48% of 48
52. 41% of 243

53. **POPULATION** According to the 2005 Census estimate, about 23.5% of Kentucky residents were under 18 years old. If the population of Kentucky was estimated at 4,173,405, about how many were under 18 years old?

54. **SAVINGS** Sofia wants to save 30% of her paycheck. If her paycheck is \$347.89, what would be a reasonable amount for her to save?

55. **SCHOOL UNIFORMS** According to a recent national survey, about 83% of teens oppose school uniforms. Predict the number of teens out of 2,979 that would *not* oppose school uniforms.

Example 11 Estimate 33% of 60.

33% is close to $33\frac{1}{3}\%$ or $\frac{1}{3}$.

$\frac{1}{3}$ of 60 is 20.

So, 33% of 60 is about 20.

Example 12 When the library surveyed the entire school regarding their favorite type of magazine, about 28% of the students preferred sports magazines. Predict the number of students out of 1,510 that would prefer sports magazines.

28% of the students preferred sports.

28% is about 30% or $\frac{3}{10}$.

Round 1,510 to 1,500.

$\frac{1}{10}$ of 1,500 is 150. So, $\frac{3}{10}$ of 1,500 is 3 • 150 or 450.

So, about 450 students would prefer sports magazines.

CHAPTER 7 Practice Test

Math Online glencoe.com
• Chapter Test

Write each percent as a fraction or mixed number in simplest form.

1. 42%
2. 110%
3. 18%

Write each fraction as a percent.

4. $\frac{2}{5}$
5. $\frac{11}{20}$
6. $1\frac{1}{2}$

7. **MULTIPLE CHOICE** Eighty-five percent of the students dressed up for spirit week. What fractional part of the student body did *not* dress up for spirit week?

   A $\frac{17}{20}$
   C $\frac{1}{85}$
   B $\frac{3}{20}$
   D $\frac{1}{5}$

SEASONS **For Exercises 8 and 9, refer to the circle graph.**

Favorite Season

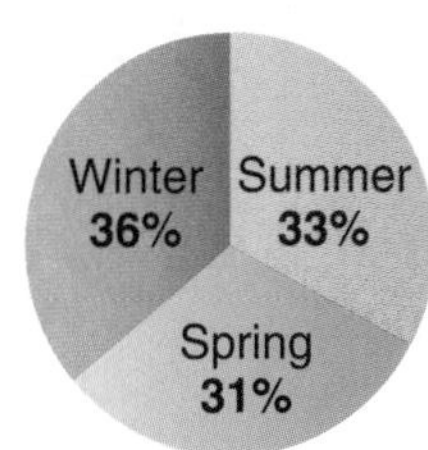

8. Which season is most frequently named as favorite?
9. How does winter compare to summer as a favorite?

Express each decimal as a percent.

10. 0.3
11. 0.87
12. 1.49

A set of 20 cards is numbered 1–20. One card is chosen without looking. Find each probability. Write each probability as a fraction, a decimal, and a percent.

13. $P(8)$
14. $P(3 \text{ or } 10)$
15. $P(\text{prime})$
16. $P(\textit{not} \text{ odd})$

FOOD **For Exercises 17 and 18, use the following information.**

A food cart offers a choice of iced tea or soda and nachos, popcorn, or pretzels.

17. Draw a tree diagram that shows all of the choices for a beverage and a snack.
18. Find the probability that the next customer who orders a beverage and a snack will choose iced tea and popcorn.

VACATION **For Exercises 19 and 20, use the table below and the following information.**

Julio asked every second sixth-grade student what they enjoy doing most on an extended break from school.

Activity	Students
playing outside	31
shopping	24
traveling	16
playing video games	15
sports	14

19. Find the probability a student enjoys playing outside most.
20. If there are 280 students in the sixth grade, how many can be expected to enjoy playing outside most?
21. **MULTIPLE CHOICE** Jose was outside practicing free throws. He has made 10 out of 12 shots. If he shoots 24 more free throws, about how many of these free throws can he expect to make?

    F 30
    H 24
    G 28
    J 20

Estimate each percent.

22. 19% of 51
23. 49% of 26
24. 77% of 51
25. 69% of 203

CHAPTER 7

Test Practice
Cumulative, Chapters 1–7

Math Online glencoe.com
• Test Practice

PART 1 Multiple Choice

Read each question. Then fill in the correct answer on the answer sheet provided by your teacher or on a sheet of paper.

1. The cost of renting a speed boat is \$25 plus an additional fee of \$12 for each hour that the boat is rented. Which equation can be used to find c, the cost in dollars for the rental for h hours?

   A $c = 12h + 25$ **C** $c = 12(h + 25)$

   B $c = 25h + 12$ **D** $c = 25(h + 12)$

2. A jar contains 4 oatmeal cookies, 9 chocolate chip cookies, 3 sugar cookies, and 4 peanut butter cookies. If you draw a cookie at random from the jar, what is the probability that you will *not* draw a sugar cookie?

   F $\frac{2}{5}$ **H** $\frac{3}{20}$

   G $\frac{3}{4}$ **J** $\frac{17}{20}$

3. In a piggy bank there are 9 dimes, 13 pennies, 6 nickels, and 7 quarters. If a coin is selected at random from the piggy bank, what is the probability that a quarter will be drawn?

   A $\frac{4}{5}$ **C** $\frac{28}{35}$

   B $\frac{5}{7}$ **D** $\frac{1}{5}$

4. There were 6 buses and 150 people signed up for the field trip. What is the ratio of people to buses?

   F 1:25 **H** 6:150

   G 25:1 **J** 150:6

TEST-TAKING TIP

Question 4 Be sure to simplify your ratio.

5. Mrs. Patterson has 28 yards of fabric. She needs a certain amount of fabric for a bedspread and bedskirt. Mrs. Patterson wants to find the number of yards of fabric she will have left for curtains.

   Look at the problem-solving steps shown below. Arrange the steps in the correct order to find the number of yards of fabric Mrs. Patterson will have left for curtains.

   Step R: Find the sum of the yards needed for the bedspread and the bedskirt.

   Step S: Find the difference between 28 and the sum of the yards needed for the bedspread and the bedskirt.

   Step T: Identify the number of yards needed for the bedspread and the bedskirt.

   Which list shows the steps in the correct order?

   A R, S, T **C** T, S, R

   B T, R, S **D** S, R, T

6. Which statement best describes the data shown in the line graph?

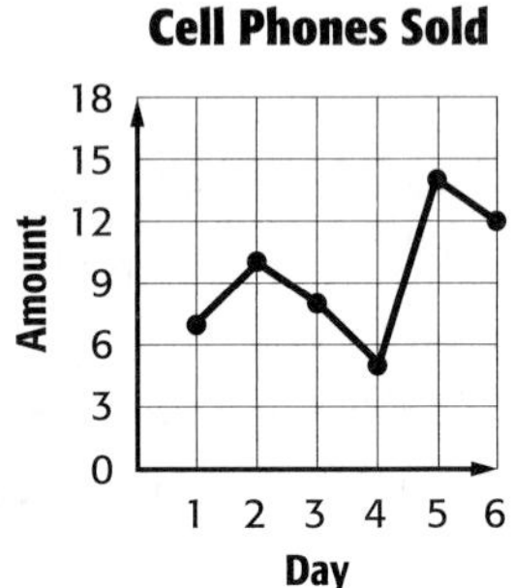

   F The greatest decrease in the number of cell phones sold was from Day 5 to Day 6.

   G The number of cell phones sold on Day 7 can be expected to be 20.

   H The greatest increase in the number of cell phones sold was from Day 4 to Day 5.

   J The number of cell phones sold increased each day.

Preparing for Standardized Tests
For test-taking strategies and practice, see pages 718–735.

7. Guillermo went to the hardware store. He bought paint for $28.75, paint brushes for $13.50, and masking tape for $2.95. If the total cost was $48.57, which procedure could be used to find the amount of tax Guillermo paid?

   A Divide the total cost by the sum of the prices of the items.

   B Multiply the total cost by the sum of the prices of the items.

   C Add the total cost and the sum of the prices of the items.

   D Subtract the sum of the prices of the items from the total cost.

8. Sanford's Shoe Store received a shipment of shoes for its newest location. The manager determined that 35% of the shoes were athletic shoes. What fraction of the shoes were athletic shoes?

   F $\frac{7}{20}$ **H** $\frac{3}{8}$

   G $\frac{1}{6}$ **J** $\frac{13}{20}$

9. A class of 26 students ordered 27 doughnuts for $1.25 each, 5 gallons of orange juice for $1.99 each, and a package of napkins for $1.25. If the class agreed to split the cost evenly, which equation can be used to find t, the amount each student should pay?

   A $t = 27(1.25) + 5(1.99) + 1.25$

   B $t = (27 \times 1.25 + 5 \times 1.99 + 1.25) \div 26$

   C $t = (27 \times 1.25 + 5 \times 1.99 + 1.25) \div 27$

   D $t = 26 \times 1.25 + 5 \times 1.99 + 1.25 \div 26$

PART 2 Short Response/Grid In

Record your answers on the answer sheet provided by your teacher or on a sheet of paper.

10. Cindy planted 4 flowers in 9 minutes. About how many flowers can Cindy plant in 36 minutes?

11. A bakery offers 5 kinds of muffins and 4 kinds of coffee. How many possible combinations of one muffin and one coffee are available at the bakery?

PART 3 Extended Response

Record your answers on the answer sheet provided by your teacher or on a sheet of paper. Show your work.

12. Serena's Boutique is having a sale. If you pick one item from each category, you get all three for a total of $25.

Polo Shirt	Hat	Socks
teal	blue	striped
mauve	green	polka dots
white	red	checkered

   a. What are the possible combinations you can buy to get the sale price? Show these combinations in a tree diagram.

   b. If the mauve shirt were removed from the choices, how many fewer combinations would there be?

   c. If you choose a combination at random, what is the probability that it will contain a white shirt, a blue or green hat, and striped socks?

NEED EXTRA HELP?												
If You Missed Question...	1	2	3	4	5	6	7	8	9	10	11	12
Go to Lesson...	6-7	7-4	7-4	6-1	1-1	2-3	1-1	7-1	1-8	6-4	7-5	7-5

Unit 4

Measurement and Geometry

Focus

Solve application problems involving estimation and measurement.

CHAPTER 8
Systems of Measurement

BIG Idea Solve application problems involving estimation and measurement of length, weight, volume, time, and temperature.

CHAPTER 9
Geometry: Angles and Polygons

BIG Idea Use geometric vocabulary to describe angles, polygons, and circles.

BIG Idea Use ratios to solve problems involving similarity.

CHAPTER 10
Measurement: Perimeter, Area, and Volume

BIG Idea Relate properties of two- and three-dimensional shapes to find perimeter, area, and volume of figures.

Problem Solving in Industrial Education

Real-World Unit Project

A New Zoo Have you ever wondered how a zoo is designed? How much space is needed for each kind of animal? How much more space do elephants need than reptiles? In this project, you will design a new zoo. You will select the types of animals you want to have in your zoo and determine the appropriate dimensions for their living quarters. You'll determine the area of each section in your zoo as well as the zoo's total area. You'll create feeding schedules and determine appropriate temperature conditions for each kind of animal. Are you ready to embark on your journey to the wild side? Let's begin!

Math Online **Log on to glencoe.com to begin.**

Systems of Measurement

BIG Idea

- Solve application problems involving estimation and measurement of length, weight, volume, time, and temperature.

Key Vocabulary

capacity (p. 424)
mass (p. 437)
metric system (p. 432)
temperature (p. 455)

Real-World Link

Hiking The Appalachian Trail twists throughout the Appalachian Mountains. About 229 miles of the trail run through eastern Pennsylvania. This is the length of about 4,000 football fields placed end to end.

Systems of Measurement Make this Foldable to help you organize your notes on metric and customary units. Begin with a sheet of 11″ × 17″ paper.

1. **Fold** the paper in half along the length. Then fold in thirds along the width.

2. **Unfold** and cut along the two top folds to make three strips. Cut off the first strip.

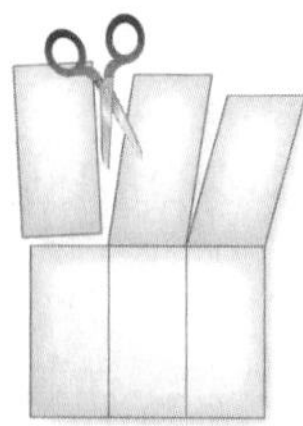

3. **Refold** the two top strips. Then fold the entire booklet in thirds along the length.

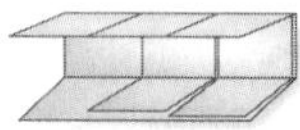

4. **Unfold** and draw lines along the folds. Label as shown.

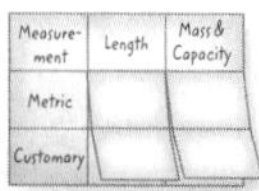

GET READY for Chapter 8

Diagnose Readiness You have two options for checking Prerequisite Skills.

Option 2

Math Online Take the Online Readiness Quiz at glencoe.com.

Option 1

Take the Quick Quiz below. Refer to the Quick Review for help.

QUICK Quiz

Add. (Lesson 3-5)

1. 8.73 + 11.96
2. 54.26 + 21.85
3. 3.04 + 9.92
4. 76.38 + 44.15
5. 7.9 + 8.62
6. 15.37 + 9.325
7. **EXERCISE** Aisha walked 3.5 miles on Monday, 2.75 miles on Wednesday, and 3.25 miles on Friday. How many miles did Aisha walk on those three days?

Subtract. (Lesson 3-5)

8. 17.46 − 3.29
9. 68.05 − 24.38
10. 9.85 − 2.74
11. 8.4 − 3.26
12. 73.91 − 50.68
13. 27 − 8.62
14. **DINNER** Joey's dinner cost $14.88 including tax. If he paid with $20, how much change did he receive?

Multiply. (Prior Grade)

15. 38 × 100
16. 5,264 × 10
17. 675 × 10
18. 89 × 1,000
19. 718 × 100
20. 249 × 100
21. **TICKETS** A ticket to a popular concert sells for $56. If 1,000 tickets were sold, how much money was spent for tickets?

QUICK Review

Example 1

Find 46.2 + 8.08.

Line up the decimal points.

$$\begin{array}{r} \scriptstyle 1 \\ 46.20 \\ +\ 8.08 \\ \hline 54.28 \end{array}$$

Annex a zero.

Example 2

Find 52.08 − 12.96.

Line up the decimal points.

$$\begin{array}{r} \scriptstyle 4\,11\,10 \\ \not{5}\not{2}.\not{0}8 \\ -\ 12.96 \\ \hline 39.12 \end{array}$$

Since 9 is larger than 0, rename 0 as 10. Rename the 2 in the ones place as 11 and the 5 in the tens place as 4. Then subtract.

Example 3

Find 45 × 100.

$$\begin{array}{r} 100 \\ \times\ 45 \\ \hline 500 \\ +\ 4000 \\ \hline 4{,}500 \end{array}$$

So, 45 × 100 = 4,500.

Length in the Customary System

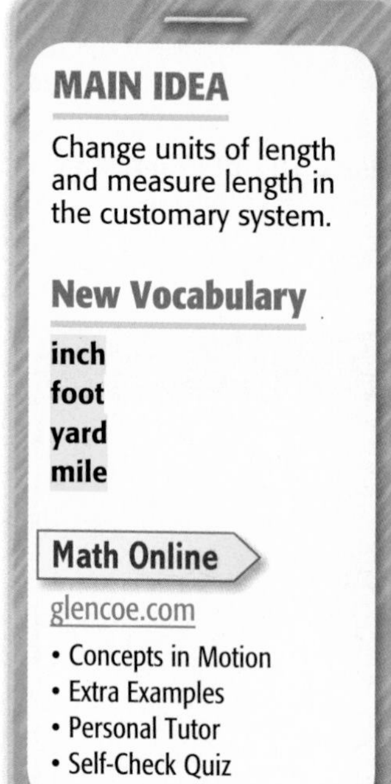

MAIN IDEA

Change units of length and measure length in the customary system.

New Vocabulary

inch
foot
yard
mile

Math Online

glencoe.com

- Concepts in Motion
- Extra Examples
- Personal Tutor
- Self-Check Quiz

MINI Lab

STEP 1 Find three objects in your classroom that are best measured in inches.

STEP 2 Estimate the length of each object to the nearest inch.

STEP 3 Measure each object and record your measurements in a table like the one shown at the right.

Object	Estimated Length	Actual Length

STEP 4 Repeat Steps 1–3 with different objects using feet and then yards.

1. Compare the objects measured in inches with those measured in feet and yards. How are they different?
2. Would you measure the distance from one city to another in yards? Explain.

The most commonly used customary units of length are shown below.

Customary Units of Length — Key Concept

Unit	Model
1 **inch** (in.)	width of a quarter
1 **foot** (ft) = 12 in.	length of a large adult foot
1 **yard** (yd) = 3 ft	length from nose to fingertip
1 **mile** (mi) = 1,760 yd	10 city blocks

You can use a customary ruler to measure length in inches or parts of an inch.

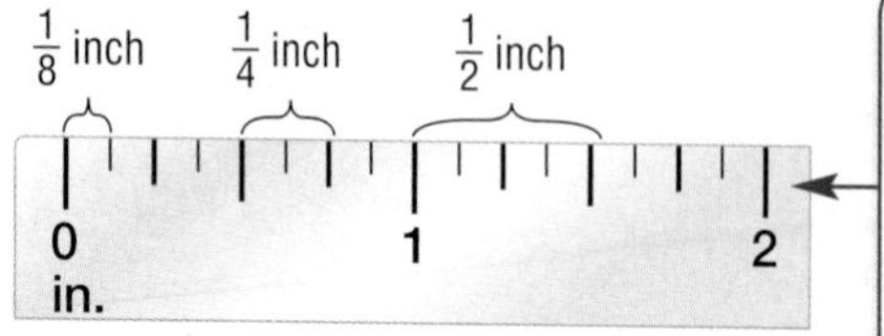

The smallest mark on this ruler represents $\frac{1}{8}$ inch. The next larger mark represents $\frac{1}{4}$ inch, and the next larger mark represents $\frac{1}{2}$ inch. The longest mark on a ruler represents an inch.

EXAMPLE Draw a Line Segment

1 Draw a line segment measuring $2\frac{7}{8}$ inches.

Draw a line segment from 0 to $2\frac{7}{8}$.

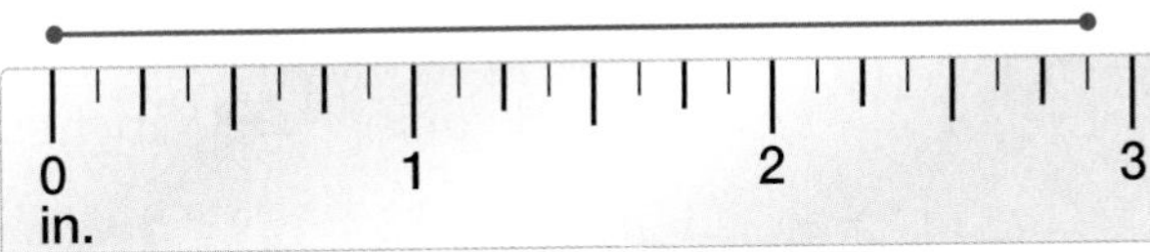

 **CHECK Your Progress**

a. Draw a line segment measuring $1\frac{3}{4}$ inches.

Real-World EXAMPLE Measure Length

2 ELECTRONICS Measure the MP3 player's length to the nearest eighth inch.

The MP3 player is between $1\frac{3}{4}$ inches and $1\frac{7}{8}$ inches. It is closer to $1\frac{3}{4}$ inches.

The length of the MP3 player is about $1\frac{3}{4}$ inches.

> **Study Tip**
>
> **Measuring** The $\frac{3}{4}$-inch mark is the same as $\frac{6}{8}$.

 **CHECK Your Progress**

b. **BOOKS** Measure the width of the cover of this textbook to the nearest eighth inch.

EXAMPLE Change Larger Units to Smaller Units

3 3 ft = ■ in.

METHOD 1 Use a ratio table.

You know that 1 foot is equal to 12 inches. Set up the ratio table with the measures you know.

×3

Feet	1	3
Inches	12	**36**

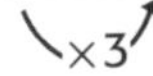

Since 1 × 3 = 3, multiply each quantity by 3.

So, 3 feet = 36 inches.

> **Study Tip**
>
> **Look Back** You can review **ratio tables** in Lesson 6-2.

METHOD 2 **Select an appropriate operation.**

Since 1 foot = 12 inches, multiply 3 by 12.

$3 \times 12 = 36$

So, 3 feet = 36 inches.

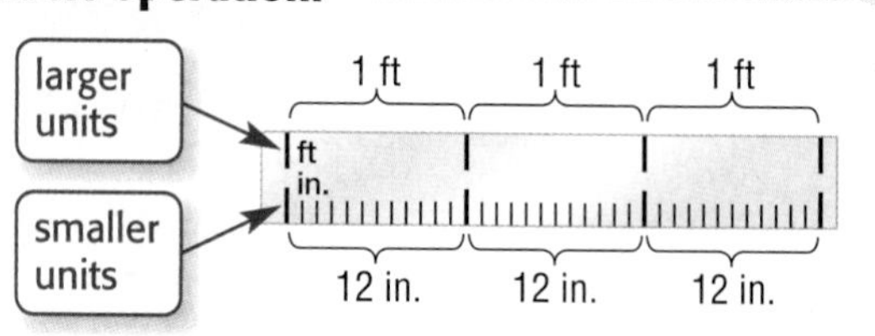

CHOOSE Your Method

c. 5 ft = ■ in. d. 3 yd = ■ ft e. 2 mi = ■ yd

Study Tip

Measurement When changing from larger units to smaller units, there will be a greater number of smaller units than larger units. So, multiply. When changing from smaller units to larger units, there will be fewer larger units than smaller units. So, divide.

EXAMPLE Change Smaller Units to Larger Units

4 **20 ft = ■ yd**

Since 3 feet = 1 yard, divide 20 by 3.

$20 \div 3 = 6\frac{2}{3}$

So, 20 feet = $6\frac{2}{3}$ yards.

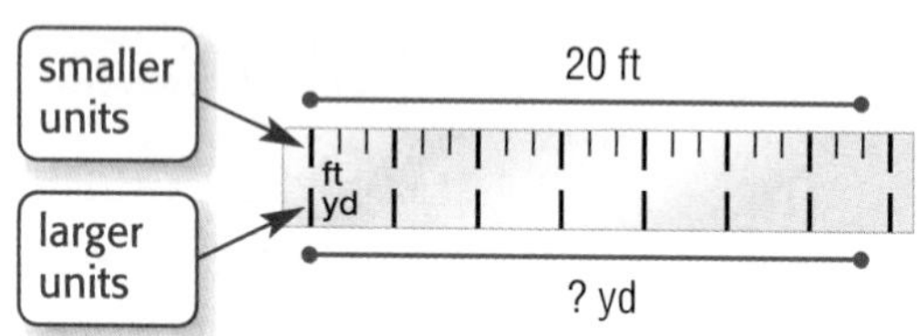

CHECK Your Progress

f. 36 ft = ■ yd g. 54 in. = ■ ft h. 2,640 yd = ■ mi

TEST EXAMPLE

5 A bookcase is 59 inches tall. The distance between the top of the bookcase and the ceiling is about 4 feet. Which is closest to the distance between the floor and the ceiling?

A 4 ft **B** 5 ft **C** 8 ft **D** 9 ft

Test-Taking Tip

Eliminate Choices Eliminate any answer choices with unreasonably small or large measurements.

Read the Item

You need to find the distance from the floor to the ceiling.

Solve the Item

The bookcase is about 60 inches. This is $60 \div 12$, or 5 feet tall. So, the distance between the floor and the ceiling is $5 + 4$, or 9 feet. The answer is D.

CHECK Your Progress

i. Kylee hiked 118 feet and then another 7 yards before resting. Which is closest to the distance Kylee hiked before resting?

F 7 yd **G** 15 yd **H** 45 yd **J** 47 yd

CHECK Your Understanding

Example 1 (p. 419)

Draw a line segment of each length.

1. $1\frac{1}{4}$ in.
2. $\frac{5}{8}$ in.

Example 2 (p. 419)

Measure the length of each object to the nearest eighth inch.

3. 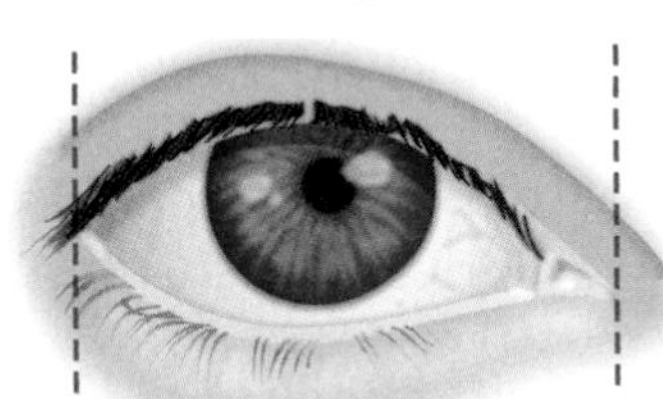

4. 

Examples 3, 4 (pp. 419–420)

Complete.

5. 4 yd = ■ ft
6. 4 mi = ■ yd
7. 72 in. = ■ yd
8. 54 ft = ■ yd

Example 5 (p. 420)

9. **MULTIPLE CHOICE** Brianna's brother is about 25 inches shorter than she is. If Brianna is 5 feet tall, which is closest to her brother's height in feet?

A 2 ft
B 3 ft
C 4 ft
D 5 ft

Practice and Problem Solving

HOMEWORK HELP

For Exercises	See Examples
10–13	1
14–19	2
20–23, 28	3
24–27, 29	4
48, 49	5

Draw a line segment of each length.

10. $2\frac{1}{2}$ in.
11. $3\frac{1}{4}$ in.
12. $\frac{3}{4}$ in.
13. $1\frac{3}{8}$ in.

Measure the length of each line segment or object to the nearest eighth inch.

14. 

15. 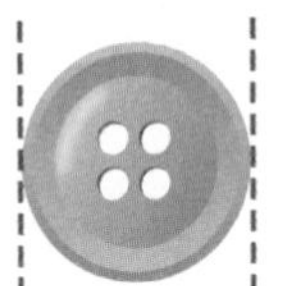

16. 

17. 

18. •————•

19. •——————————•

Complete.

20. 5 yd = ■ in.
21. 6 yd = ■ ft
22. 6 ft = ■ in.
23. 3 mi = ■ ft
24. 48 in. = ■ ft
25. 10 ft = ■ yd
26. 6,160 yd = ■ mi
27. 510 in. = ■ ft

28. **SPACE SCIENCE** The largest telescope in the world is powerful enough to identify a penny that is 5 miles away. How many yards is this?

Real-World Career **How Does a Roller Coaster Designer Use Math?** A roller coaster designer uses different combinations of angles and lengths to build roller coasters that can obtain top speeds and maximum forces.

Math Online For more information, go to glencoe.com.

29. **ROLLER COASTERS** Kingda Ka at Six Flags Great Adventure in Jackson, New Jersey, is the tallest roller coaster in the United States. It has a height of 456 feet. What is this height in yards?

Determine the greater measurement. Explain your reasoning.

30. $1\frac{1}{2}$ yards or 48 inches
31. 54 inches or $4\frac{1}{3}$ feet

32. **ANIMALS** The length with tail of a bighorn sheep ranges from 50 inches to 62 inches long. What is the range of this length in feet?

33. **BACKPACKS** Kathy estimates that her backpack is 30 inches long. Is this a reasonable estimate? Why or why not?

Determine whether you would measure each length or distance in inches, feet, yards, or miles. Explain your reasoning.

34. length of a computer monitor
35. distance from your home to school
36. distance from home plate to the pitchers mound on a baseball field

ESTIMATION **Estimate the length of each object. Then measure to find the actual length.**

37. the length of your bedroom to the nearest foot
38. the width of your student ID card to the nearest eighth inch
39. the height of your dresser to the nearest foot
40. the height of a classroom wall to the nearest yard
41. the length of a new pencil to the nearest half inch

EXTRA PRACTICE See pages 692, 713.

42. **FIND THE DATA** Refer to the Data File on pages 16–19. Choose some data and write a real-world problem in which you would convert a customary measurement of length.

H.O.T. Problems

43. **REASONING** Explain the math error in the comic.

44. **OPEN ENDED** Draw a segment that measures between $1\frac{1}{2}$ inches and $2\frac{1}{4}$ inches long. State the measure of the segment to the nearest fourth inch. Then state the measure to the nearest eighth inch.

45. **CHALLENGE** How many sixteenths of an inch are in one foot? How many half inches are in one yard?

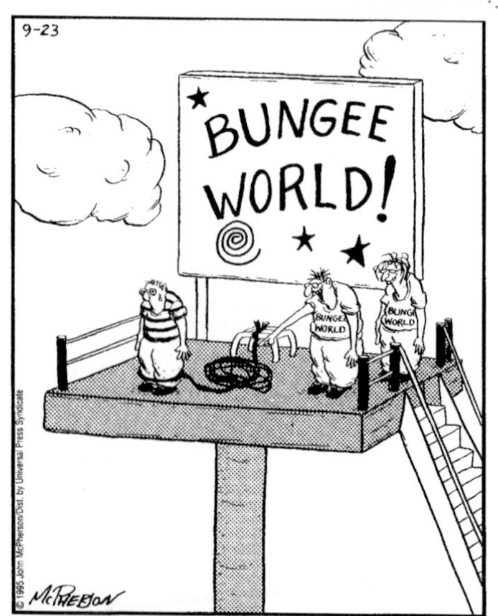

"Okee-doke! Let's just double-check. We're 130 feet up and we've got 45 yards of bungee cord, that's uh ... 90 feet. Allow for 30 feet of stretching, that gives us a total of ... 120 feet. Perfect!"

46. **FIND THE ERROR** Liseli and Huang are changing 168 inches to feet. Who is correct? Explain your reasoning.

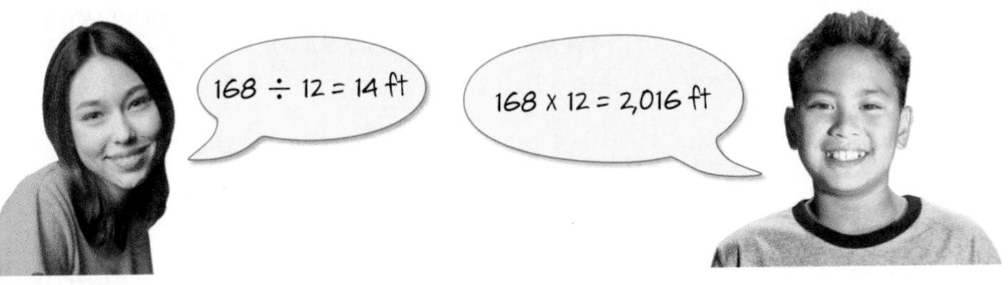

168 ÷ 12 = 14 ft
168 x 12 = 2,016 ft
Liseli
Huang

47. **WRITING IN MATH** Suppose your friend says that 24 feet is equal to 2 inches. Is this reasonable? Explain.

TEST PRACTICE

48. The diagram below shows the dimensions of a football field.

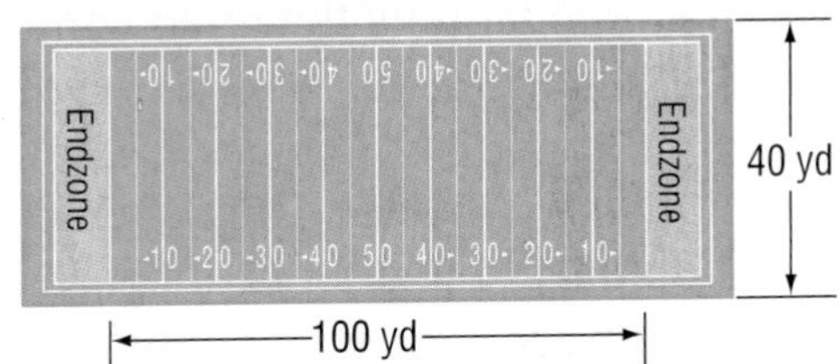

What is the width of the field expressed in feet?

A $3\frac{1}{4}$ ft

B $13\frac{1}{4}$ ft

C 120 ft

D 480 ft

49. Mr. Cortez's car is about 71 inches wide. His garage door is 9 feet wide. How much wider is the garage door than Mr. Cortez's car?

F 1 ft

G 2 ft

H 3 ft

J 4 ft

Spiral Review

Estimate each percent. (Lesson 7-8)

50. 23% of 97
51. 34% of 117
52. 44% of 39
53. 78% of 83

54. **TENNIS** Christina hit the ball over the net 3 out of her last 5 attempts. Find the probability of Christina hitting the ball over the net on her next attempt. Suppose Christina attempts 15 hits. About how many hits over the net will she make? (Lesson 7-6)

55. **SALES** What type of display would be most appropriate to show the change in the number of magazines Wade sold over each of the last 5 days? (Lesson 2-8)

GET READY for the Next Lesson

PREREQUISITE SKILL Multiply or divide. (Page 744)

56. 4×8
57. 16×5
58. $5{,}000 \div 2{,}000$
59. $400 \div 8$

8-2 Capacity and Weight in the Customary System

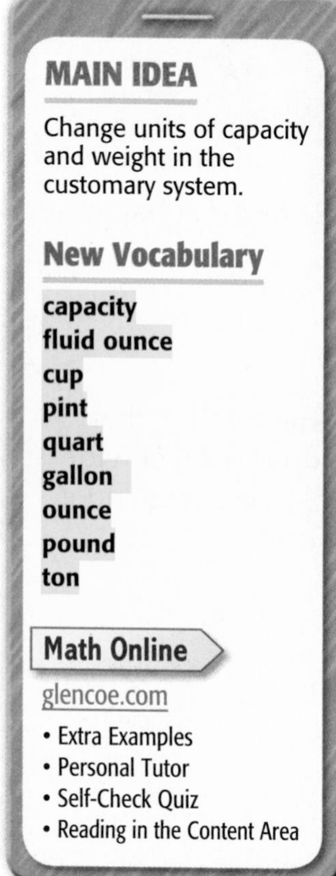

MAIN IDEA

Change units of capacity and weight in the customary system.

New Vocabulary

capacity
fluid ounce
cup
pint
quart
gallon
ounce
pound
ton

Math Online

glencoe.com

- Extra Examples
- Personal Tutor
- Self-Check Quiz
- Reading in the Content Area

MINI Lab

Several different milk containers are shown below.

STEP 1 Fill the pint container with water. Then pour the water into the quart container. Repeat until the quart container is full. Record the number of pints needed to fill the quart.

STEP 2 Fill the quart container with water. Then pour the water into the gallon container. Repeat until the gallon container is full. Record the number of quarts needed to fill the gallon.

Complete.

1. 1 quart = ■ pints
2. 2 quarts = ■ pints
3. 1 gallon = ■ quarts
4. 1 gallon = ■ pints
5. What fractional part of 1 gallon would fit in 1 quart?
6. How many gallons are equal to 16 quarts? Explain.

Capacity refers to the amount that can be held in a container. The most commonly used customary units of capacity are shown.

Customary Units of Capacity — Key Concept

Unit	Model
1 **fluid ounce** (fl oz)	2 tablespoons of water
1 **cup** (c) = 8 fl oz	coffee cup
1 **pint** (pt) = 2 c	small ice cream container
1 **quart** (qt) = 2 pt	large liquid measuring cup
1 **gallon** (gal) = 4 qt	large plastic jug of milk

As with units of length, you can use a ratio table to change between units of capacity.

Vocabulary Link

Capacity

Everyday Use the maximum amount that can be contained, as in a theater filled to capacity

Math Use amount that can be held in a container

EXAMPLES Change Units of Capacity

Complete.

1 **3 qt = ■ pt**

METHOD 1 **Use a ratio table.**

You know that there are 2 pints in 1 quart. Set up the ratio table with the measures you know. Then multiply each quantity by the same number.

×3

Quarts	1	3
Pints	2	6

×3

Since $1 \times 3 = 3$, multiply 2 by 3.

METHOD 2 **Select an appropriate operation.**

You are changing a larger unit to a smaller unit.
Since 1 quart = 2 pints, multiply 3 by 2.

$3 \times 2 = 6$

So, 3 quarts = 6 pints.

2 **64 fl oz = ■ pt**

First, find the number of cups in 64 fluid ounces.

Since 8 fluid ounces = 1 cup, divide 64 by 8.

$64 \div 8 = 8$

So, 64 fluid ounces = 8 cups. Next, find the number of pints in 8 cups.

Since 2 cups = 1 pint, divide 8 by 2.

$8 \div 2 = 4$

So, 64 fluid ounces = 4 pints.

Study Tip

Check For Example 2, since 8 fluid ounces = 1 cup and 2 cups = 1 pint, you need to divide twice. $64 \div 8 = 8$ and $8 \div 2 = 4$. So, 64 fluid ounces = 4 pints.

CHOOSE Your Method

a. 4 pt = ■ c b. 32 fl oz = ■ c c. 3 gal = ■ qt

The most commonly used customary units of weight are shown.

Customary Units of Weight (Key Concept)

Unit	Model
1 **ounce** (oz)	pencil
1 **pound** (lb) = 16 oz	package of notebook paper
1 **ton** (T) = 2,000 lb	small passenger car

Real-World Link The statue of Abraham Lincoln at the Lincoln Memorial in Washington, D.C., weighs approximately 350,000 pounds.

Source: National Park Service

EXAMPLES Change Units of Weight

3 SCULPTURES Use the information at the left to determine the weight of the statue of Abraham Lincoln in tons.

350,000 lb = ■ T — **THINK** 2,000 pounds = 1 ton

350,000 ÷ 2,000 = 175 — Divide 350,000 by 2,000.

So, the weight of the statue is 175 tons.

4 RAISINS How many 4-ounce snack boxes can be made with a 3-pound bag of raisins?

First, find the total number of ounces in 3 pounds.

3 × 16 = 48 — Multiply by 16.

Next, find how many sets of 4 ounces are in 48 ounces.

48 oz ÷ 4 oz = 12

So, 12 snack boxes can be made with 3 pounds of raisins.

d. **CONSTRUCTION** At a construction site, 3 tons of rocks were hauled away. How many pounds is this?

e. **PETS** Justin's dog eats 20 ounces of dry dog food each day. If Justin buys a 40-pound bag of dog food, how many days will it last?

CHECK Your Understanding

Examples 1, 2 (p. 425)

Complete.

1. 7 pt = ■ c
2. 24 qt = ■ gal
3. 16 pt = ■ gal
4. 5 c = ■ fl oz
5. 16 pt = ■ qt
6. 8 c = ■ pt

Example 3 (p. 426)

7. **MAMMALS** The heaviest land mammal, the African elephant, can weigh more than 7 tons. How many pounds is this?

8. **AIRCRAFTS** The maximum takeoff weight of an F-15E Strike Eagle is 81,000 pounds. How many tons is this?

Example 4 (p. 426)

9. **FOOD** Miguela bought a 10-pound bag of potatoes. How many people can be served 8 ounces of potatoes?

10. **BREAKFAST** Roman uses 1 cup of milk for his cereal every morning. How many times will he be able to have cereal with milk with 1 quart of milk?

Practice and Problem Solving

HOMEWORK HELP	
For Exercises	See Examples
11–22	1, 2
23, 24	3
25, 26	4

Complete.

11. 5 qt = ■ pt
12. 8 gal = ■ qt
13. 24 fl oz = ■ c
14. 32 qt = ■ gal
15. 6 pt = ■ c
16. 13 qt = ■ gal
17. 9 gal = ■ pt
18. 24 fl oz = ■ pt
19. 1,500 lb = ■ T
20. 112 oz = ■ lb
21. 84 oz = ■ lb
22. 4 T = ■ lb

23. **MAMMALS** The heaviest marine mammal, the blue whale, can weigh more than 143 tons. How many pounds is this?

24. **ICE CREAM** In the United States, the annual consumption of ice cream is 24 pints per person. How many gallons of ice cream is this per person?

25. **RECIPES** Juanita needs 6 ounces of chocolate chips for a cookie recipe. A large bag of chocolate chips weighs 2 pounds. About how many batches of cookies can she make?

26. **MAPLE SYRUP** Vermont produces about 430,000 gallons of maple syrup each year. How many 2-quart containers of maple syrup can be made from 430,000 gallons of syrup?

Complete.

27. 1 lb 8 oz = ■ oz
28. 2 pt 1 c = ■ c
29. 3 gal 2 qt = ■ qt

Choose the better estimate for each measure.

30. cups or quarts?

31. fluid ounces or pints?

32. ounces or pounds?

33. pounds or tons?

Find the greater quantity. Explain your reasoning.

34. 14 cups or 5 pints
35. 4 pints or 60 fluid ounces

Estimate. Then check the reasonableness of the estimate.

36. The number of cups of juice in a 12-ounce can.

37. The number of pints in a 9-quart bottle of laundry detergent.

38. **TRIATHLON** During the Ironman Triathlon World Championships, about 250,000 cups of water are given away. Each cup contains 8 fluid ounces. About how many gallons of water are given away?

39. **FIND THE DATA** Refer to the Data File on pages 16–19. Choose some data and write a real-world problem in which you would convert a customary measurement of capacity or weight.

40. **RECIPES** Ellen has 12 quart jars and 24 pint jars to fill with strawberry jam. If her recipe makes 5 gallons of jam, will she have enough jars? Explain.

COOKING For Exercises 41–43, use the following information.

Ron uses the ingredients at the right in one of his favorite dishes.

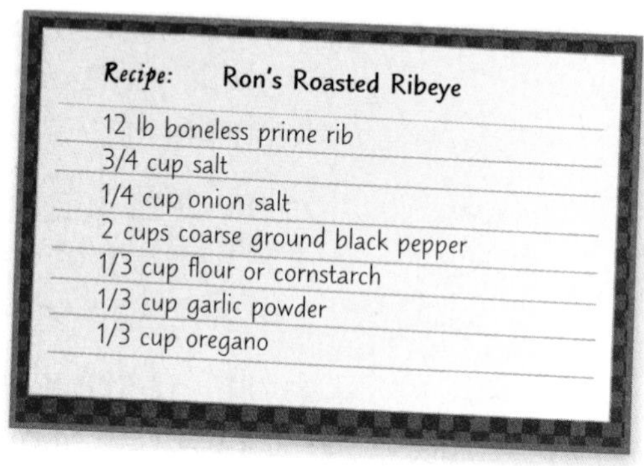

Recipe: Ron's Roasted Ribeye
12 lb boneless prime rib
3/4 cup salt
1/4 cup onion salt
2 cups coarse ground black pepper
1/3 cup flour or cornstarch
1/3 cup garlic powder
1/3 cup oregano

41. How many ounces of prime rib are needed?

42. There are 16 tablespoons in 1 cup. How many tablespoons of coarse ground black pepper are needed?

43. How can you find how many tablespoons of garlic powder are needed?

H.O.T. Problems

44. **OPEN ENDED** Without looking at their labels, estimate the weight or capacity of three packaged food items in your kitchen. Then compare your estimate to the actual weight or capacity.

45. **SELECT A TECHNIQUE** A homemade ice cream recipe calls for 2 pints of heavy cream. At the grocery store, Antonia finds that heavy cream is sold in 10-ounce containers and 24-ounce containers. Which of the following techniques might Antonia use to determine if buying one 10-ounce container and one 24-ounce container will be enough for the ice cream recipe? Justify your selection(s). Then use the technique(s) to solve the problem.

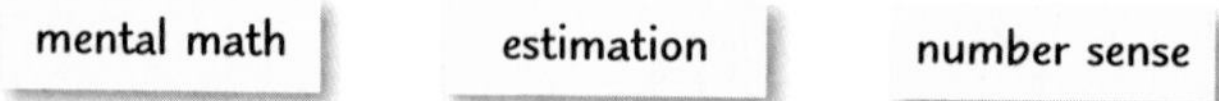

46. **CHALLENGE** Create a function table that shows the number of fluid ounces in 1, 2, 3, and 4 cups. Graph the ordered pairs (cups, fluid ounces) on a coordinate plane. Then describe the graph.

47. **WRITING IN MATH** Determine whether 1 cup of sand and 1 cup of cotton balls would have the same capacity, the same weight, both, or neither. Explain your reasoning.

TEST PRACTICE

48. A store advertises a 32-ounce container of juice for $0.99. What is the capacity of the container in cups?

A 1 cup

B 2 cups

C 4 cups

D 8 cups

49. **SHORT RESPONSE** A can of green beans weighs 13 ounces. How many pounds does a case of 24 cans weigh?

50. Which table represents the relationship between pounds and ounces?

F

Pounds	Ounces
1	16
2	32
3	48
4	64

H

Pounds	Ounces
1	8
2	16
3	24
4	32

G

Pounds	Ounces
16	1
32	2
48	3
64	4

J

Pounds	Ounces
1	16
2	24
3	32
4	40

Spiral Review

51. **HEIGHT** Marissa is 5 feet 6 inches tall. How tall is she in inches? (Lesson 8-1)

52. **GAMES** An air hockey table that normally sells for $158.99 is on sale for 75% of the regular price. What would be a reasonable amount for the sale price? (Lesson 7-8)

BASKETBALL For Exercises 53 and 54, use the following information.

In the first five basketball games, Jamil made 9 out of 12 free-throw attempts. (Lesson 7-6)

53. Find the probability of Jamil making his next free-throw attempt.

54. Suppose Jamil attempts 40 free throws throughout the season. About how many free throws will you expect him to make? Justify your reasoning.

55. **PARKS** The table shows the acreage of the largest national parks in the United States. To the nearest tenth of a million, what is the acreage of each of the parks? (Lesson 3-3)

The Largest U.S. National Parks

Park	Acreage (millions)
Wrangell-St. Elias, Alaska	13.27
Gates of the Arctic, Alaska	8.47
Denali, Alaska	6.07
Katmai, Alaska	4.73
Death Valley, California	3.37

Source: *Scholastic Book of World Records*

GET READY for the Next Lesson

PREREQUISITE SKILL Estimate each measure. (Lesson 8-1)

56. the width of a quarter

57. the width of a doorway

58. the width of your palm

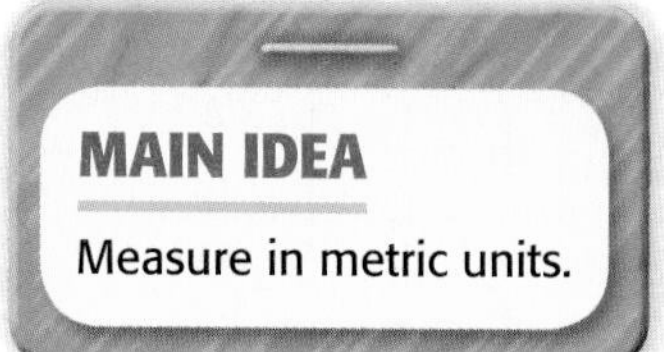

Measurement Lab
The Metric System

MAIN IDEA

Measure in metric units.

The basic unit of length in the metric system is the *meter*. All other metric units of length are defined in terms of the meter.

The most commonly used metric units of length are shown in the table.

Metric Unit	Symbol	Meaning
millimeter	mm	thousandth
centimeter	cm	hundredth
meter	m	one
kilometer	km	thousand

Units on a metric ruler or tape measure are divided into tenths. The ruler below is labeled using *centimeters*.

The bracelet below is about 12.4 centimeters long.

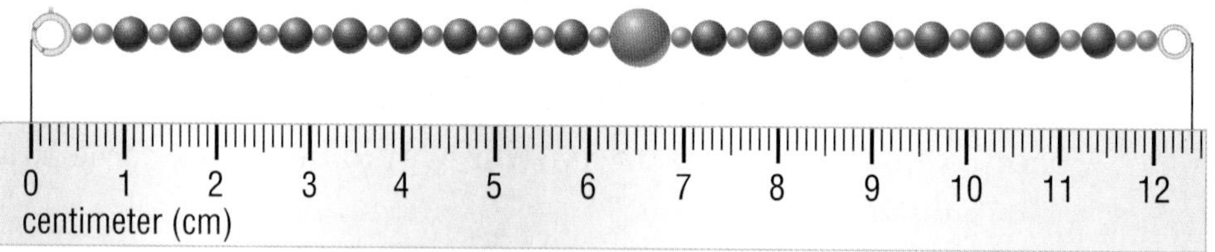

To read *millimeters*, count each individual unit or mark on the metric ruler.

There are ten millimeter marks for each centimeter mark. The bracelet is about 124 millimeters long.

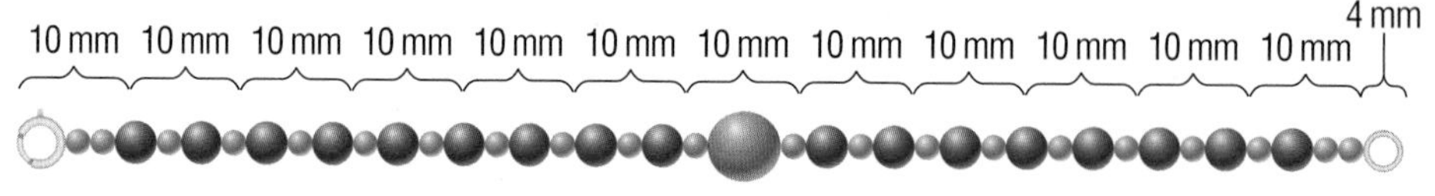

124 mm = 12.4 cm

There are 100 centimeters in one meter. Since there are 10 millimeters in one centimeter, there are 10 × 100 or 1,000 millimeters in one meter.

The bracelet is $\frac{124}{1,000}$ of a meter or 0.124 meter long.

124 mm = 12.4 cm

12.4 cm = 0.124 m

ACTIVITY

Use metric units of length to measure various items.

STEP 1 Copy the table.

Object	Measure		
	mm	cm	m
length of pencil			
length of sheet of paper			
length of your hand			
width of your little finger			
length of table or desk			
length of chalkboard eraser			
width of door			
height of door			
distance from doorknob to the floor			
length of classroom			

Use a metric ruler or tape measure to measure the objects listed in the table. Complete the table.

Study Tip

Appropriate Tools To measure longer objects such as the height of a door or length of a classroom, use a tape measure.

ANALYZE THE RESULTS

1. Tell which unit of measure is most appropriate for each item. How did you decide which unit was most appropriate?

2. **LOOK FOR A PATTERN** Examine the pattern between the numbers in each column. How are the numbers in the first and second columns related? in the first and third columns? in the second and third columns?

3. **MAKE A CONJECTURE** If you know the length of an object measured in millimeters, explain how you could find its length measured in centimeters.

4. **MAKE A CONJECTURE** If you know the length of an object measured in meters, explain how you could find its length measured in centimeters.

5. Select three objects around your classroom that would be best measured in meters, three objects that would be best measured in centimeters, and three objects that would be best measured in millimeters. Explain your choices.

6. Write the name of a common object that you think has a length that corresponds to each length. Explain your choices.

   a. 5 centimeters
   b. 3 meters
   c. 1 meter
   d. 75 centimeters

8-3 Length in the Metric System

MAIN IDEA

Use metric units of length.

New Vocabulary

meter
metric system
millimeter
centimeter
kilometer

Math Online

glencoe.com

- Extra Examples
- Personal Tutor
- Self-Check Quiz

GET READY for the Lesson

WATERFALLS The table shows the tallest waterfalls in the world.

World's Tallest Waterfalls	
Waterfall	**Height (m)**
Angel Falls	979
Olu'upena	900
Tres Hermanas	914
Tugela Falls	948

1. What unit of measure is used?
2. What is the height of the tallest waterfall?
3. Use the Internet or another source to find the meaning of *meter*. Then write a sentence explaining how a meter compares to a yard.

A **meter** is the basic unit of length in the metric system. The **metric system** is a decimal system of weights and measures. The most commonly used metric units of length are shown below.

Metric Units of Length — Key Concept

Unit	Model	Benchmark
1 **millimeter** (mm)	thickness of a dime	1 mm ≈ 0.04 inch
1 **centimeter** (cm)	half the width of a penny	1 cm ≈ 0.4 inch
1 meter (m)	width of a doorway	1 m ≈ 1.1 yards
1 **kilometer** (km)	six city blocks	1 km ≈ 0.6 mile

The segment at the right is 1 centimeter or 10 millimeters long. This is about $\frac{3}{8}$ inch in customary units.

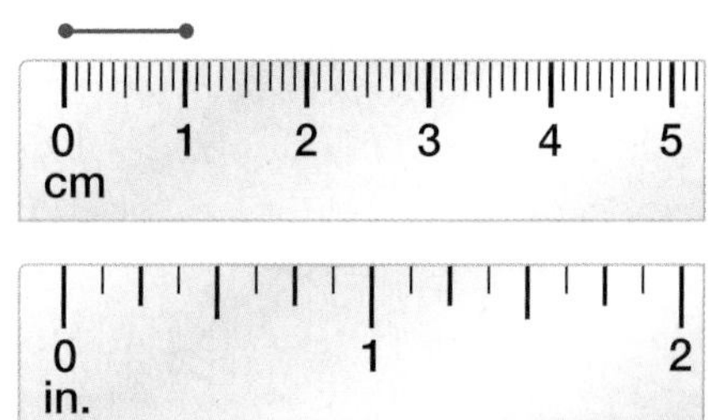

EXAMPLES Use Metric Units of Length

1 **Write the metric unit of length that you would use to measure the thickness of your pencil's eraser.**

The width of a pencil eraser is greater than the thickness of a dime, but less than half the width of a penny. So, the millimeter is an appropriate unit of measure.

Write the metric unit of length that you would use to measure each of the following.

2 **height of your school**

Since the height of your school is much greater than half the width of a penny, but much less than six city blocks, the meter is an appropriate unit of measure.

3 **length of the Missouri River**

Since the length of the Missouri River is much greater than 6 city blocks, it would be measured in kilometers.

4 **width of a skateboard**

Since the width of a skateboard is much greater than half the width of a penny and less than the width of a doorway, the centimeter is an appropriate unit of measure.

CHECK Your Progress

a. thickness of a nickel

b. height of a cereal box

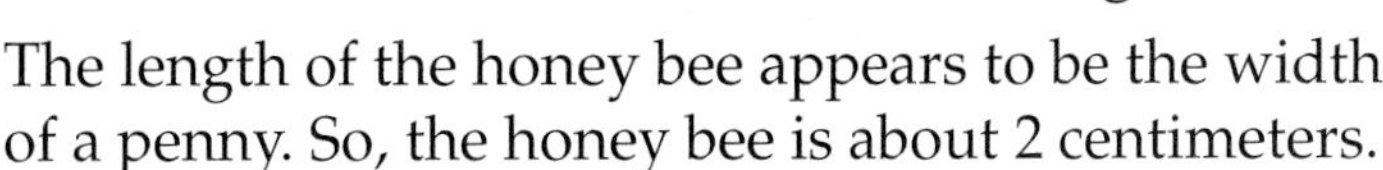

Real-World EXAMPLE **Estimate and Measure Length**

5 **INSECTS Estimate the metric length of the honey bee. Then measure to find the actual length.**

The length of the honey bee appears to be the width of a penny. So, the honey bee is about 2 centimeters.

Use a ruler to measure the actual length of the honey bee.

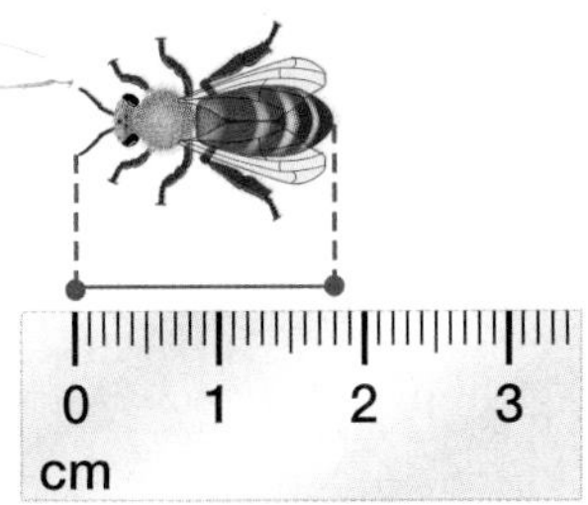

The honey bee is 18 millimeters long or 1.8 centimeters long.

Real-World Link It takes the life work of about 300 bees to make one pound of honey.

Source: Woodland Park Zoo

CHECK Your Progress

c. **FOOD** Estimate the length of the blueberry. Then measure to find the actual length.

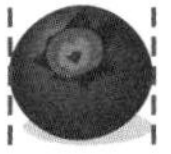

Examples 1–4 (pp. 432–433)

Write the metric unit of length that you would use to measure each of the following.

1. thickness of a calculator
2. distance from home to school
3. height of a tree
4. width of a computer screen

Example 5 (p. 433)

Estimate the metric length of each figure. Then measure to find the actual length.

5. 

6. 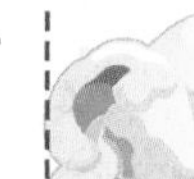

Practice and Problem Solving

HOMEWORK HELP	
For Exercises	See Examples
7–14	1–4
15–20	5

Write the metric unit of length that you would use to measure each of the following.

7. thickness of a note pad
8. thickness of a watchband
9. length of a trombone
10. width of a dollar bill
11. length of a bracelet
12. length of the Mississippi River
13. distance from Knoxville, Tennessee, to Asheville, North Carolina
14. distance from home plate to first base on a baseball field

Estimate the metric length of each figure. Then measure to find the actual length.

15.  

16.  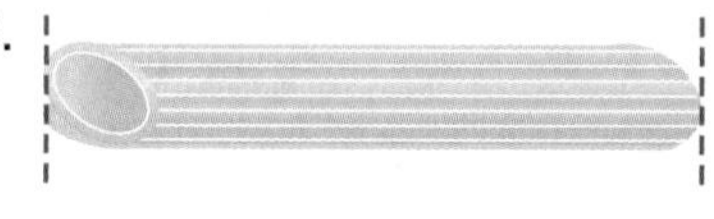

17. 

18. 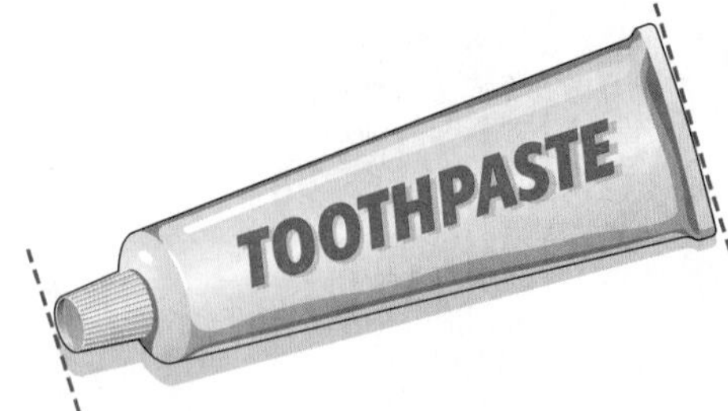

19. 

20.

21. **SKYSCRAPER** Which metric unit of length would be the best to use to describe the height of the Marriott Hotel in Detroit, Michigan?

22. **RESEARCH** Use the Internet or another source to find the height of the Marriott Hotel in Detroit, Michigan. Is the height given in the metric unit you selected in Exercise 21? If not, in what unit is the measurement given?

Real-World Link The Marriott Hotel at the Renaissance Center is the tallest building in Detroit, Michigan. It was built in 1977 and has 73 stories.

Source: *The World Almanac*

Estimate the metric length of each of the following. Then measure to find the actual length.

23. student ID card
24. chalkboard
25. eraser on end of a pencil
26. width of a cell phone

27. **FLOOR PLANS** Estimate the metric length and width of your bedroom or classroom. Then use a meterstick to check your measurement.

28. **MAPS** Estimate the distance in centimeters between Ocean City and Atlantic City on the map. Check your measurement with a ruler.

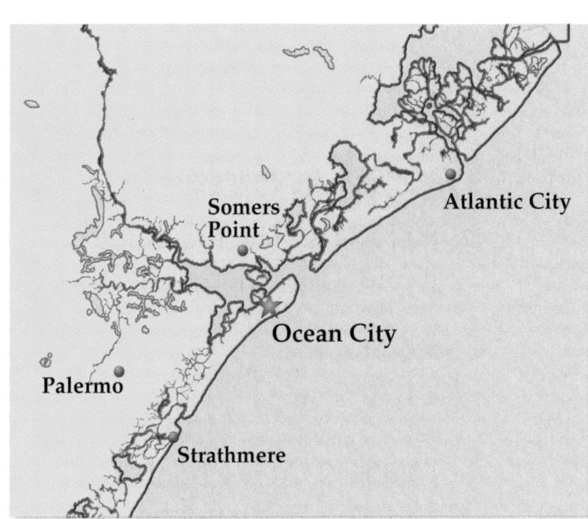

29. Which customary unit of length is comparable to a meter?

30. Is a mile or a foot closer in length to a kilometer?

Find the greater length. Explain your reasoning.

31. 15 millimeters or 3 centimeters
32. 3 feet or 1 meter
33. 1 mile or 2 kilometers
34. 5 centimeters or 1 inch

35. **FARMING** If you were to build a fence around a cattle pasture, would you need to be accurate to the nearest kilometer, to the nearest meter, or to the nearest centimeter? Explain your reasoning.

36. **COLLECT THE DATA** Choose three classmates or three members of your family. Which metric unit of length would you use to measure each person's height? Estimate the combined height of all three people. Then measure to check the reasonableness of your estimate.

See pages 693, 713.

H.O.T. Problems

37. **OPEN ENDED** Give three examples of objects that are larger than the thickness of a dime, but smaller than the width of a doorway. What unit of measure would you use for the objects you chose?

38. **CHALLENGE** Order 4.8 mm, 4.8 m, 4.8 cm, 0.48 m, and 0.048 km from greatest to least measurement.

39. **WRITING IN MATH** Identify the four most commonly used metric units of length and describe an object having each length. Use objects that are different from those given in the lesson.

TEST PRACTICE

40. What is the best estimate for the length of the paper clip?

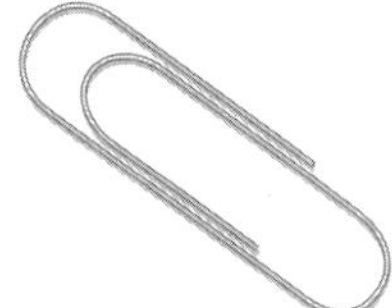

A 3 mm

B 3 cm

C 0.3 m

D 0.3 km

41. Which metric unit would you use to measure the distance an athlete jumps in a long jump competition?

F millimeter

G centimeter

H meter

J kilometer

Spiral Review

42. **FOOTBALL** The football team drank 20 gallons of water during one game. How many quarts is this? (Lesson 8-2)

Complete. (Lesson 8-1)

43. 4 ft = ■ in.

44. 5280 yd = ■ mi

45. 144 in. = ■ yd

46. **FAMILIES** At a school, there are 108 students in the 6th grade. Of these, 18 students do not have any siblings. In Mr. Romain's class, 8 of the 26 students have no siblings. Is the number of students without siblings in Mr. Romain's class proportional to the number of students without siblings in the 6th grade? Explain. (Lesson 6-3)

Add or subtract. Write in simplest form. (Lesson 5-3)

47. $\frac{1}{5} + \frac{2}{5}$

48. $\frac{3}{8} + \frac{2}{8}$

49. $\frac{6}{7} - \frac{3}{7}$

50. $\frac{9}{10} - \frac{3}{10}$

51. **MOVIES** The table shows the top grossing animated films of all time. To the nearest tenth of a million, how much did each film earn? (Lesson 3-3)

52. **MONEY** Write an integer that represents a direct deposit of $200 into a savings account. (Lesson 2-9)

Top Grossing Animated Movies of All Time

Movie	Earnings (millions of dollars)
Shrek 2	436.47
Finding Nemo	339.71
The Lion King	328.54
Shrek	267.67
The Incredibles	261.44

GET READY for the Next Lesson

PREREQUISITE SKILL Name an item sold in a grocery store that is measured using each type of unit.

53. fluid ounce

54. pound

55. quart

Mass and Capacity in the Metric System

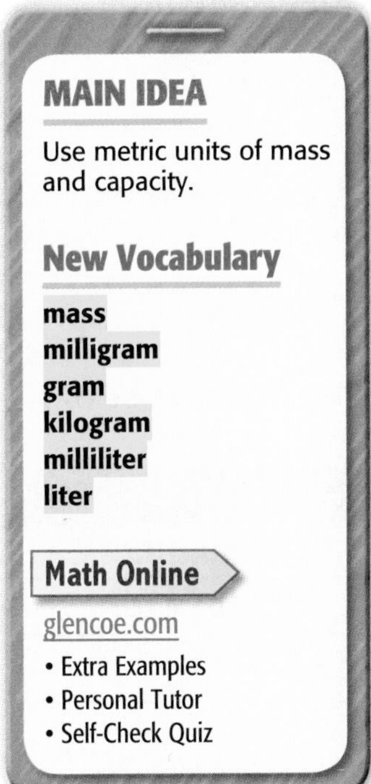

MAIN IDEA
Use metric units of mass and capacity.
New Vocabulary
mass
milligram
gram
kilogram
milliliter
liter
Math Online
glencoe.com
• Extra Examples
• Personal Tutor
• Self-Check Quiz

MINI Lab

The gram and kilogram are two units used to measure mass in the metric system. A paper clip has a mass of one gram. A novel has a mass of one kilogram.

STEP 1 Find two objects in your classroom that you think have a mass of about 1 gram.

STEP 2 Place one of the objects on one side of a balance scale and a small paper clip on the other side of the balance scale. Then, replace the first object with the second object.

1. Which object's mass was closest to one gram?
2. Repeat Steps 1 and 2 with objects that have a mass of one kilogram. Which object's mass was closest to one kilogram?

The **mass** of an object is the amount of material it contains. The most commonly used metric units of mass are shown below.

Metric Units of Mass — Key Concept

Unit	Model	Benchmark
1 **milligram** (mg)	grain of salt	1 mg ≈ 0.00004 oz
1 **gram** (g)	small paper clip	1 g ≈ 0.04 oz
1 **kilogram** (kg)	six medium apples	1 kg ≈ 2 lb

EXAMPLES Use Metric Units of Mass

Write the metric unit of mass that you would use to measure each of the following. Then estimate the mass.

1 sheet of notebook paper

A sheet of paper has a mass greater than a small paper clip, but less than six medium apples. So, the gram is the appropriate unit.

Estimate A sheet of paper has slightly more mass than a paper clip.

One estimate for the mass of a sheet of paper is about 6 grams.

2 **bag of potatoes**

A bag of potatoes has a mass greater than six apples. So, the kilogram is the appropriate unit.

Estimate A bag of potatoes contains about 15 potatoes.

One estimate for the mass of a bag of potatoes is about 2 or 3 kilograms.

CHECK Your Progress

Write the metric unit of mass that you would use to measure each of the following. Then estimate the mass.

a. tennis ball b. horse c. aspirin

The most commonly used metric units of capacity are shown below.

Metric Units of Capacity — Key Concept

Unit	Model	Benchmark
1 **milliliter** (mL)	eyedropper	1 mL ≈ 0.03 fl oz
1 **liter** (L)	small pitcher	1 L ≈ 1 qt

There are 1,000 milliliters in a liter. You can use this information to estimate capacity.

EXAMPLES Use Metric Units of Capacity

Write the metric unit of capacity that you would use to measure each of the following. Then estimate the capacity.

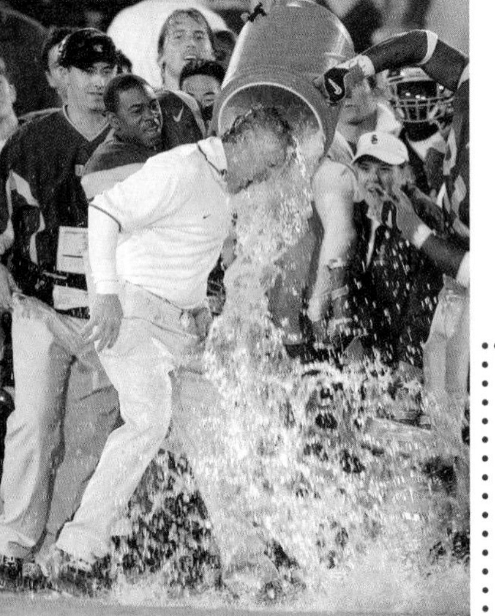

Real-World Link Large water coolers that hold about 40 liters are kept on the sidelines during a football game. After a big win, many teams continue the tradition of dumping the contents of the cooler over their coach.

3 **small water cooler**

A small water cooler has a capacity greater than a small pitcher. So, the liter is the appropriate unit.

Estimate A water cooler will hold about 12 small pitchers of water.

One estimate for the capacity of a water cooler is about 12 liters.

4 **glass of juice**

A glass of juice is greater than an eyedropper and less than a small pitcher. So, the milliliter is the appropriate unit.

Estimate There are 1,000 milliliters in a liter. A small pitcher can fill about 4 glasses.

One estimate for the capacity of a glass of juice is about 1,000 ÷ 4 or 250 milliliters.

CHECK Your Progress

d. cooler of lemonade e. raindrop

One kilogram is equal to 1,000 grams. You can use this information to compare metric units.

Real-World EXAMPLE Compare Metric Units

5 LIFE SCIENCE The table shows the average mass of several human organs. Is the combined mass of the lungs more or less than one kilogram?

Human Organs	Average Mass (g)
Skin	10,886
Right Lung	580
Left Lung	510
Male Heart	315
Female Heart	265
Thyroid	35

Source: *Top 10 of Everything*

Find the total mass.

right lung	580 g
left lung	+ 510 g
total	1,090 g

Since 1 kilogram = 1,000 grams and 1,090 grams is more than 1,000 grams, the combined mass of the lungs is more than one kilogram.

CHECK Your Progress

f. **RECIPES** The table shows the liquid ingredients of a fruit punch recipe. Does the recipe call for more or less than a liter of pineapple juice and ginger ale? Explain.

Fruit Punch Liquid Ingredients	Amount (mL)
pineapple juice	510
water	769
ginger ale	375

CHECK Your Understanding

Examples 1–4 (pp. 437–438)

Write the metric unit of mass or capacity that you would use to measure each of the following. Then estimate the mass or capacity.

1. nickel
2. bucket of water
3. laptop computer
4. juice in a lemon
5. light bulb
6. one-gallon paint can

Example 5 (p. 439)

ANIMALS For Exercises 7–9, use the table below that shows the average amount of food that selected animals eat per day.

7. Is the total daily amount eaten by the animals listed in the table more or less than 250 kg?
8. Write the amounts listed in the table in order from least to greatest.
9. Is the total amount eaten by a flamingo over four days more or less than one kilogram? Explain.

Animal	Average Daily Consumption
Bald eagle	400 g
Elephant	200 kg
Flamingo	270 g
Giant panda	12 kg
Gorilla	32 kg
Prairie dog	190 g

Practice and Problem Solving

HOMEWORK HELP

For Exercises	See Examples
10–21	1–4
22, 23	5

Write the metric unit of mass or capacity that you would use to measure each of the following. Then estimate the mass or capacity.

10. granola bar
11. grape
12. large watermelon
13. cow
14. large bowl of punch
15. bathtub
16. chipmunk
17. shoe
18. grain of sugar
19. postage stamp
20. 10 drops of food coloring
21. ink in a ballpoint pen

ANALYZE TABLES **For Exercises 22 and 23, use the table at the right that shows the mass of ducks.**

22. Is the combined mass of a cinnamon teal, cape teal, and marbled teal more or less than one kilogram?
23. Which birds from the table will have a combined mass closest to one kilogram? Explain your reasoning.

Duck Mass

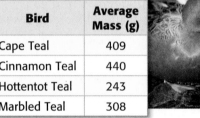

Bird	Average Mass (g)
Cape Teal	409
Cinnamon Teal	440
Hottentot Teal	243
Marbled Teal	308

Source: Sea World

24. **MINTS** You can purchase your favorite mints in 295-gram or 1.2-kilogram packages. Which has a greater mass? Explain.
25. **AIR FRESHENER** Liquid air freshener comes in 1.36-liter bottles and 243-milliliter bottles. Which container has less mass? Explain.

ANALYZE TABLES **For Exercises 26 and 27, use the following information and the table at the right.**

A kiloliter is equal to 1,000 liters and is about the amount needed to fill 5 bathtubs.

Average Annual Consumption of Bottled Water, 2006

Country	Per Person Consumption (L)
U.S.	97.5
Canada	61.4
Germany	140.1
Spain	142.9
Mexico	136.2

Source: Nestlé Waters Press

26. Is the amount of bottled water consumed by all countries in the table more or less than half a kiloliter?
27. About how many bathtubs could be filled with the amount of bottled water consumed by 15 people in Mexico in 2006?
28. **VITAMINS** One orange provides you with 70 milligrams of Vitamin C. One stalk of raw broccoli contains about 200 milligrams of Vitamin C. About how many oranges would it take to have the same amount of Vitamin C as the broccoli?

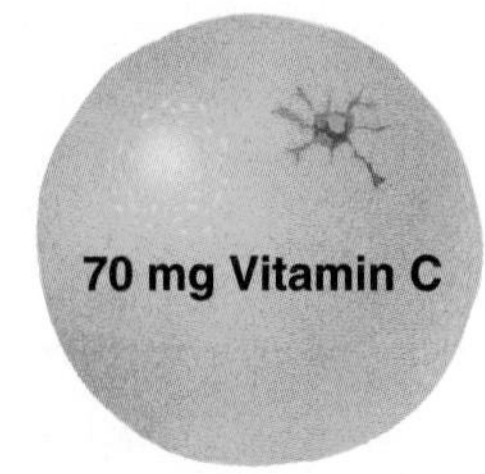

See pages 693, 713.

H.O.T. Problems

29. **OPEN ENDED** Locate and identify an item found at your home that has a capacity of about one liter.

30. **NUMBER SENSE** The mass of a dime is recorded as 4. What metric unit was used to measure the mass? Explain your reasoning.

31. **CHALLENGE** Determine whether the following statement is *true* or *false*. If false, give a counterexample.

    Any two items filled to the same capacity will also have the same mass.

32. **WRITING IN MATH** Write a problem about a real-world situation in which you would have to decide which metric unit to use to measure the mass or capacity of an item.

TEST PRACTICE

33. The capacity of a glass of iced tea would best be measured in what metric unit?

    A milliliters

    B liters

    C milligrams

    D grams

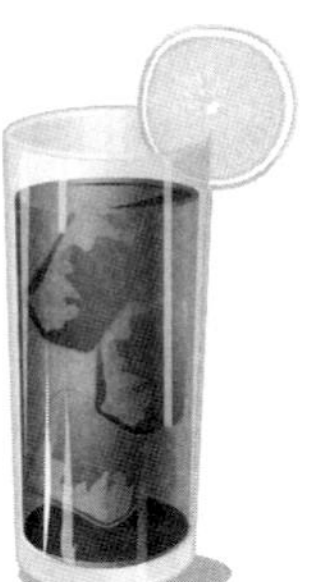

34. Which of the following items on Allie's grocery list has a mass of about 2 kilograms?

    F bag of marshmallows

    G can of green beans

    H loaf of bread

    J bag of flour

Spiral Review

Write the metric unit of length that you would use to measure each of the following. (Lesson 8-3)

35. length of a hand

36. thickness of a folder

37. **MEASUREMENT** How many ounces are in 5 pounds? (Lesson 8-2)

WEATHER For Exercises 38 and 39, use the following information. (Lesson 7-5)

A morning radio announcer reports that the chance of rain today is 85%.

38. What is the probability that it will *not* rain?

39. Should you carry an umbrella? Justify your answer.

GET READY for the Next Lesson

40. **PREREQUISITE SKILL** Leon has 45 baseball cards. He is collecting 5 more cards each month. Alicia has 30 baseball cards, and she is collecting 10 more each month. How many months will it be before Alicia has more cards than Leon? Use the *look for a pattern* strategy. (Lesson 6-5)

8-5 Problem-Solving Investigation

MAIN IDEA: Solve problems using benchmarks.

P.S.I. TEAM +

e-Mail: USE BENCHMARKS

TONYA: I want to paint a foursquare court that has a perimeter of about 12 yards. I know that my "foot" is about one foot long. So, if I walk heel-to-toe for 3 steps, the distance covered will be about three feet or one yard.

YOUR MISSION: Use a benchmark to make a square with a perimeter of 12 yards with no standard measuring tools.

Understand	You want to make a foursquare court that has a perimeter of 12 yards. Since the figure is a square, each side must be 3 yards or 9 feet. You know that you can step heel-to-toe 9 times to equal 3 yards.
Plan	A benchmark is a measurement by which other items can be measured. Mark your starting point. Step heel to toe 9 times in one direction, mark your spot, and then make a 90° turn to step again. Do this until you have completed the square. 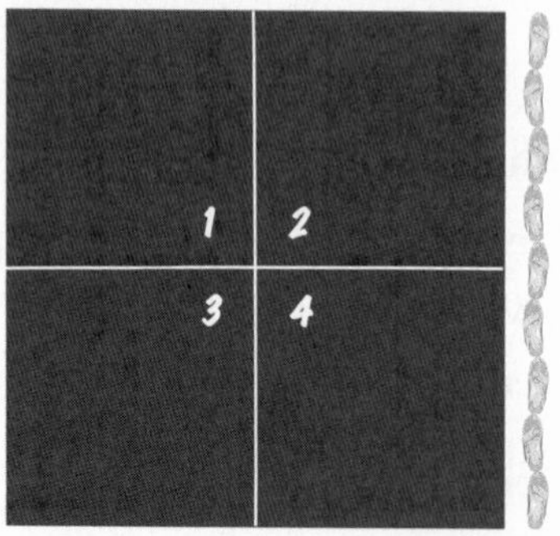
Solve	It will take a total of 36 heel-to-toe steps to mark off the 12-yard perimeter of the foursquare court.
Check	Since 36 feet is equal to 12 yards, this result is reasonable.

Analyze The Strategy

1. **WRITING IN MATH** Explain why using 3 heel-to-toe steps is a good benchmark to use for measuring the perimeter in yards.
2. Describe how you would determine a method for dividing the large square into 4 smaller squares.

Mixed Problem Solving

EXTRA PRACTICE
See pages 694, 713.

Use a benchmark to solve Exercises 3 and 4.

3. **KITCHEN** André wants to buy a new refrigerator for his kitchen. The refrigerator will need to fit underneath some cabinets, but he doesn't know the exact measurement. He has some string and he knows that the width of a door is about one meter. Describe a way André could estimate the distance in meters.

4. **RIBBON** Marie has a spool of ribbon to tie around party favors. She knows that her hand span is 20 centimeters. She needs to know the total length of the spool of ribbon in order to determine how many equal-size pieces she can cut. Describe a way Marie could estimate the length of the ribbon.

Use any strategy to solve Exercises 5–12. Some strategies are shown below.

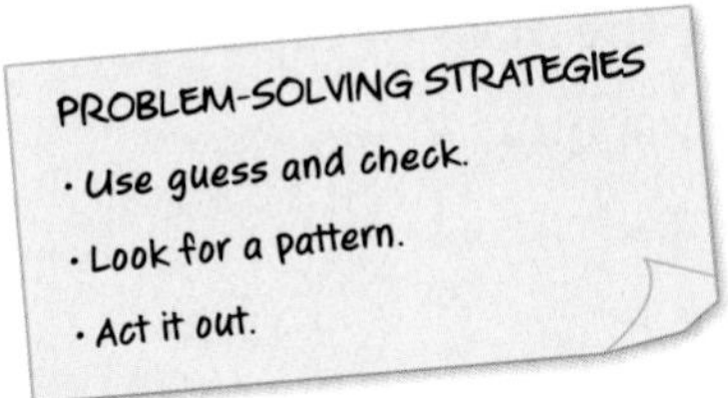
PROBLEM-SOLVING STRATEGIES
• Use guess and check.
• Look for a pattern.
• Act it out.

5. **SPELLING** Demarco is keeping track of the number of words he gets wrong on each of his spelling tests. Which is greater, the mean or the median number of words he misspells?

Test Number	Amount Misspelled
1	2
2	3
3	2
4	1
5	1
6	2
7	5
8	2

6. **PATTERNS** What is the missing number in the pattern?

   ..., 0.07, 0.1, ■, 0.7, 1.0, ...

7. **HEIGHT** The students in Mrs. Delgado's class want to find the probability that a student picked at random is taller than 200 centimeters. They know that the doorway is 3 meters high. Describe a way the students can find who is taller than 200 centimeters.

8. **NUMBER SENSE** A number is multiplied by 6, and then 13 is added to the product. The result is 79. What is the number?

9. **PATTERNS** Draw the seventeenth figure in the pattern.

10. **MUSIC** Randy purchased a saxophone for $317.89, including tax. How much change should he receive from $350?

11. **SOFTBALL** The Grayson Middle School softball team won three times as many games as they lost. If they lost 5 games, how many games did they play?

12. **FOOTBALL** The table shows the passing leaders of Super Bowl games. How much greater was the distance that Donovan McNabb passed the ball in the 2005 Super Bowl than Peyton Manning in the 2007 Super Bowl in feet?

Passing Leaders		
Player	**Year**	**Yards**
Donovan McNabb	2005	357
Matt Hasselbeck	2006	273
Peyton Manning	2007	247

CHAPTER 8 Mid-Chapter Quiz

Lessons 8-1 through 8-5

Complete. (Lesson 8-1)

1. 10,560 ft = ■ mi
2. ■ in. = 2 yd

Find the length of each line segment or object to the nearest eighth inch. (Lesson 8-1)

3. 

4. 

5. **HEIGHT** Scott is 78 inches tall. The height of a ceiling is 9 feet. How many feet are between him and the ceiling when he is standing? (Lesson 8-1)

Complete. (Lesson 8-2)

6. 22 pt = ■ qt
7. ■ qt = 14 gal
8. 32 oz = ■ lb
9. ■ fl oz = 5 c
10. 9 pt = ■ c
11. ■ gal = 48 pt

12. **MULTIPLE CHOICE** How many 8-ounce servings are in 2 gallons of juice? (Lesson 8-2)

   A 16 **C** 32

   B 24 **D** 48

Write the metric unit of length that you would use to measure each of the following. (Lesson 8-3)

13. length of a textbook
14. distance between two cities
15. thickness of a pencil
16. length of a classroom

Estimate the metric length of each of the following. Then measure to find the actual length. (Lesson 8-3)

17. 

18.

19. **VOLLEYBALL COURTS** Which metric unit would be the best to use to describe the length and width of a volleyball court? (Lesson 8-3)

Write the metric unit of mass or capacity that you would use to measure each of the following. Then estimate the mass or capacity. (Lesson 8-4)

20. washing machine
21. can of soup
22. tank of gas
23. shoelace
24. packet of sugar

25. **MULTIPLE CHOICE** The mass of a cell phone would best be measured in what metric units? (Lesson 8-4)

   F milliliters **H** grams

   G liters **J** milligrams

26. **ELECTRONICS** Josefina is buying a stereo and wants to place it on a shelf on her entertainment center. She needs to know if there is enough space to place the stereo on the shelf. If Josefina only has a piece of string that is 5 inches long, describe a way she could estimate the height of the shelf. (Lesson 8-5)

Changing Metric Units

MAIN IDEA

Change units within the metric system.

Math Online

glencoe.com

- Concepts in Motion
- Extra Examples
- Personal Tutor
- Self-Check Quiz

GET READY for the Lesson

FOOD The table shows the estimated consumption of baked beans per person for several countries.

Baked Beans Consumed per Person		
Country	**g**	**kg**
Ireland	5,600	5.6
United Kingdom	4,800	4.8
United States	2,000	2
Canada	1,200	1.2

Source: *Top 10 of Everything*

1. How many grams of baked beans are consumed per person in the United States?
2. How many kilograms of baked beans are consumed per person in the United States?
3. Describe the relationship between the quantities you found in Exercises 1 and 2.
4. Compare the number of grams and kilograms of baked beans consumed by the other countries in the table. Make a conjecture about how to convert from grams to kilograms.

To change from one unit to another in the metric system, you multiply or divide by powers of 10. The chart below shows the relationship between the units in the metric system and the powers of 10.

1,000	100	10	1	0.1	0.01	0.001
thousands	hundreds	tens	ones	tenths	hundredths	thousandths
kilo	hecto	deka	base unit	deci	centi	milli

Each place value is 10 times the place value to its right.

In Lesson 8-1, you learned the following methods for changing customary units of measure.

- To change from a larger unit to a smaller unit, multiply.
- To change from a smaller unit to a larger unit, divide.

You can use the same methods for changing metric units of measure.

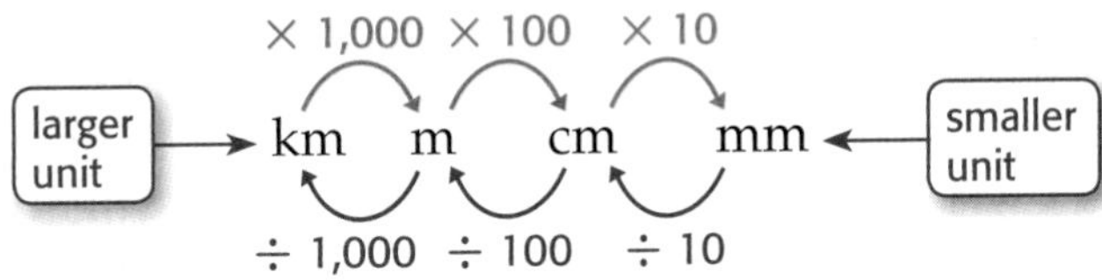

EXAMPLES Change Metric Units

> **Study Tip**
> **Check for Reasonableness** Since a millimeter is a smaller unit than a centimeter, the number of millimeters needed to equal 26 centimeters should be greater than 26. Since 260 > 26, the answer seems reasonable.

Complete.

1 **■ mm = 26 cm**

Since 1 centimeter = 10 millimeters, multiply 26 by 10.

$26 \times 10 = 260$

So, 260 mm = 26 cm.

2 **135 g = ■ kg**

Since 1,000 grams = 1 kilogram, divide 135 by 1,000.

$135 \div 1{,}000 = 0.135$

So, 135 g = 0.135 kg.

CHECK Your Progress

a. 513 mL = ■ L **b.** 5 cm = ■ mm **c.** ■ mg = 82 g

Real-World Link In some areas, carpoolers can drive in a lane called the High-Occupancy Vehicle (HOV) lane. There are currently about 1,300 miles (2,100 kilometers) of HOV lanes across the country.

Real-World EXAMPLE

3 **CARPOOLING Angela picks up Niko and Howard on the way to school. How many kilometers does she travel in all?**

Angela
500 m
Niko
4 km
Howard
11 km
School

First, change 500 meters to kilometers. You can use a proportion.

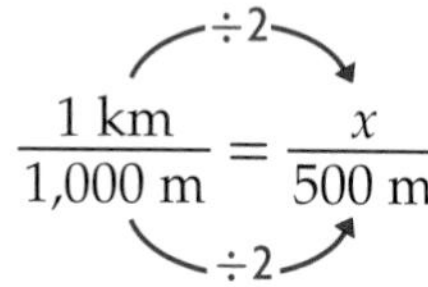

$$\frac{1 \text{ km}}{1{,}000 \text{ m}} = \frac{x}{500 \text{ m}}$$

$x = 1 \div 2$ or 0.5 kilometer

Add to find the total number of kilometers Angela travels.

$0.5 + 4 + 11 = 15.5$ kilometers

So, Angela travels 15.5 kilometers.

CHECK Your Progress

d. **WATER** A person should drink about 1.9 liters of water daily. Miko drank 1,650 milliliters one morning. How much more water should Miko drink during the day?

CHECK Your Understanding

Examples 1, 2 (p. 446)

Complete.

1. 95 g = ■ mg
2. 5 L = ■ mL
3. ■ mm = 38 cm
4. ■ L = 75 mL
5. 205 mg = ■ g
6. 85 mm = ■ cm

Example 3 (p. 446)

7. **TRAVEL** Booker's family drove 42 kilometers from Brownsville to Harlingen and then another 2,300 meters to his aunt's house. What is the total number of kilometers Booker's family drove?

Practice and Problem Solving

HOMEWORK HELP

For Exercises	See Examples
8–19	1, 2
20, 21	3

Complete.

8. ■ L = 95 mL
9. ■ g = 1,900 mg
10. 52 mm = ■ cm
11. 354 cm = ■ m
12. ■ mg = 6 g
13. ■ mL = 238 L
14. 4 m = ■ mm
15. 18 L = ■ mL
16. ■ L = 136 mL
17. ■ g = 7 mg
18. 1,300 g = ■ kg
19. 450 m = ■ km

20. **ANIMALS** If a rhinoceros has a mass of 3,600 kilograms and a pygmy mouse has a mass of 8 grams, how much more mass is the rhinoceros than the pygmy mouse?

21. **TRACK** A running track at a college is 200 meters long. Isabel wants to run 1 kilometer on this track. How many laps will she have to run?

Complete.

22. 500 mg = ■ kg
23. 250 mm = ■ km
24. 200,000 mL = ■ kL
25. 3 km = ■ cm

Order each set of measurements from least to greatest.

26. 4.2 kg, 420 g, 400,000 mg
27. 560 mm, 55 cm, 5.6 km
28. 630 mg, 63 g, 6.3 kg
29. 8.2 km, 8,500 mm, 80 m

30. **BRIDGES** The table shows the length of the three longest suspension bridges in the United States. If Perez biked over the main span of the Golden Gate Bridge and back, about how many kilometers did he bike?

U.S. Suspension Bridges

Bridge	Location	Length of Main Span (m)
Verrazano Narrows	New York	1,298
Golden Gate	San Francisco	1,280
Mackinac Straits	Michigan	1,158

Source: *Top 10 of Everything*

31. **FITNESS** Danielle walked 0.75 kilometer each day for five days. How many meters did she walk in all?

32. **FIND THE DATA** Refer to the Data File on pages 16–19. Choose some data and write a real-world problem in which you would need to change metric units.

33. **TRACK** At a track meet, Andres raced in the 5,000-meter run, 10,000-meter run, and 400-meter hurdles. How many total kilometers did Andres race at the track meet?

34. **RESEARCH** Use the Internet or another source to find other metric prefixes for very large and very small units of measure. List three of each type and explain their meaning.

BASEBALL For Exercises 35–37, use the table at the right.

Distance of Mickey Mantle's Five Longest Home Runs
192 m
201 m
19,600 cm
22,400 cm
198,000 mm

35. List the home runs in order from greatest to least.

36. How much longer was Mickey Mantle's longest home run than his fifth longest home run?

See pages 694, 713.

37. Find the mean, median, and mode of the home runs. Which of these measures best represents the data?

H.O.T. Problems

38. **OPEN ENDED** Choose a metric measure between 1 and 100. Then write two measures equivalent to that measure.

39. **CHALLENGE** If Tyra has x milligrams of food for her parrot, write an algebraic expression for the amount of kilograms of parrot food she has.

40. **SELECT A TOOL** Rachelle takes a large jug of lemonade to her brother's soccer games. She sells cups of the lemonade to the fans. The jug contains 10 liters of lemonade, and each cup will hold 400 milliliters. Which of the following tools might Rachelle use to determine how many cups to bring with her? Justify your selection(s). Then use the tool(s) to solve the problem.

real objects | paper/pencil | calculator

41. **FIND THE ERROR** Raul and Shayla are changing 470 milliliters to liters. Who is correct? Explain your reasoning.

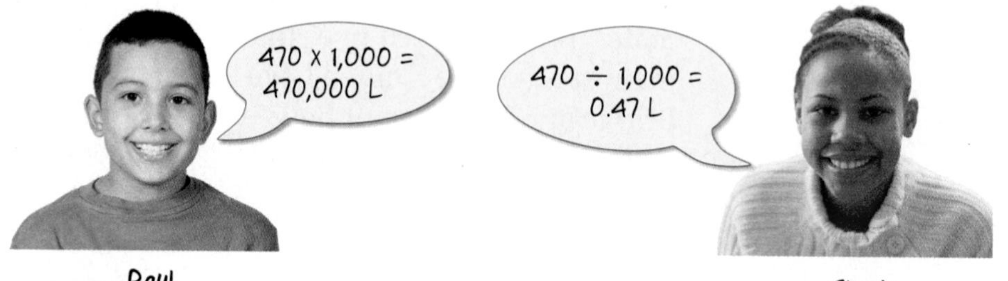

42. **WRITING IN MATH** Explain the steps you would use to change 7 kiloliters to milliliters.

TEST PRACTICE

43. **SHORT RESPONSE** Nestor bought a 3-meter telephone cord. What is the length of the cord in millimeters?

44. The mass of Ethan's dog is 25,900 grams. What is the mass of his dog in kilograms?
    - A 2.59 kg
    - B 25.9 kg
    - C 259 kg
    - D 2,590 kg

45. It is recommended that children have an intake of about 7,000 milligrams of calcium per week. Which proportion can be used to find x, the number of grams of calcium children should have per week?
    - F $\frac{7{,}000 \text{ mg}}{x \text{ g}} = \frac{1{,}000 \text{ mg}}{1{,}000 \text{ g}}$
    - G $\frac{7{,}000 \text{ mg}}{x \text{ g}} = \frac{1 \text{ mg}}{1{,}000 \text{ g}}$
    - H $\frac{7{,}000 \text{ mg}}{x \text{ g}} = \frac{1{,}000 \text{ mg}}{1 \text{ g}}$
    - J $\frac{1{,}000 \text{ mg}}{1 \text{ g}} = \frac{7{,}000 \text{ mg}}{1{,}000 \text{ g}}$

Spiral Review

46. **DRIVEWAYS** Before seal coating his driveway, Tito needs to know its length and width. He knows that it is about 3 meters wide. Describe a way he could estimate the length and width of his driveway without using a metric ruler. (Lesson 8-5)

47. **MEASUREMENT** Which is the better estimate for the capacity of a glass of water, 360 liters or 360 milliliters? (Lesson 8-4)

48. **MEASUREMENT** Estimate the metric length of the battery. Then measure to find the actual length. (Lesson 8-3)

Write each mixed number as an improper fraction. (Lesson 4-3)

49. $1\frac{7}{8}$

50. $7\frac{3}{8}$

51. $6\frac{6}{7}$

52. $3\frac{2}{5}$

53. **ARCHITECTURE** Use front-end estimation to find the difference in the ceiling heights between the kitchen and the bedroom. (Lesson 3-4)

Room	Kitchen	Bedroom
Ceiling Height (ft)	12.35	8.59

54. If the input values of a function are 0, 1, and 6 and the corresponding outputs are 4, 5, and 10, what is the function rule? (Lesson 1-6)

GET READY for the Next Lesson

PREREQUISITE SKILL **Add or subtract.** (Lesson 3-5)

55. $3.26 + 4.86$

56. $9.32 - 4.78$

57. $27.48 + 78.92$

58. $7.18 - 2.31$

8-7 Measures of Time

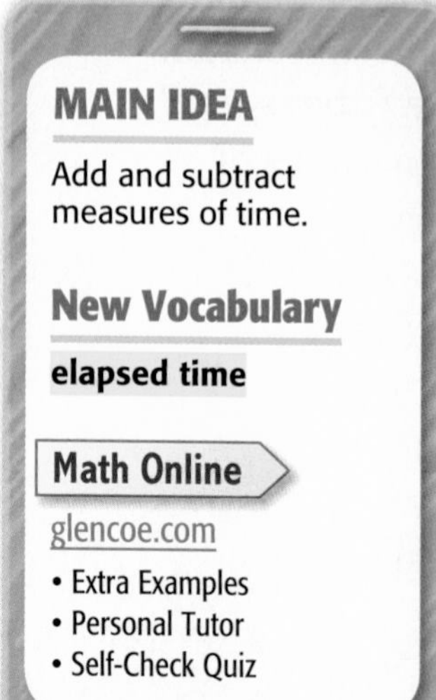

MAIN IDEA
Add and subtract measures of time.
New Vocabulary
elapsed time
Math Online
glencoe.com
• Extra Examples
• Personal Tutor
• Self-Check Quiz

GET READY for the Lesson

VACATION Rachel is going on vacation with her family to Orlando, Florida. They are flying from Des Moines, Iowa. They have to catch a connecting flight in Atlanta, Georgia, before arriving in Orlando.

Departure	Arrival City	Flight Time
Des Moines, IA	Atlanta, GA	2 h 11 min
Atlanta, GA	Orlando, FL	1 h 22 min

1. How long was the flight from Des Moines to Atlanta? Atlanta to Orlando?
2. What is the sum of the minutes? of the hours?
3. How long is the total flight time from Des Moines to Orlando?

The most commonly used units of time are shown below.

Units of Time — Key Concept

Unit	Model
1 second (s)	time needed to say 1,001
1 minute (min) = 60 seconds	time for 2 average TV commercials
1 hour (h) = 60 minutes	time for 2 weekly TV sitcoms

To add or subtract measures of time, use the following steps.

Step 1 Add or subtract the seconds.
Step 2 Add or subtract the minutes.
Step 3 Add or subtract the hours.

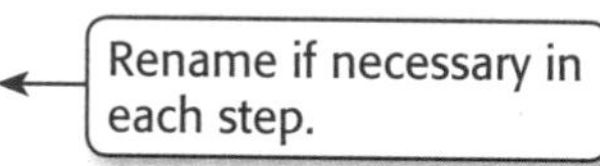

EXAMPLES Add and Subtract Units of Time

1 Find the sum of 4 h 20 min and 2 h 50 min.

Estimate 4 h 20 min + 2 h 50 min ≈ 4 h + 3 h or 7 h

4 h 20 min + 2 h 50 min	Add minutes first, then hours.
6 h 70 min	70 minutes is greater than 60 minutes or 1 hour.
6 h (1 h 10 min)	Rename 70 minutes as 1 hour and 70 – 60 or 10 min.
7 h 10 min	Add hours.

Check for Reasonableness 7 h 10 min ≈ 7 h ✓

2 **Find the difference of 8 h 20 min 35 s and 3 h 45 min 30 s.**

Estimate 8 h 20 min 35 s − 3 h 45 min 30 s ≈ 8 h − 4 h or 4 h

8 h 20 min 35 s − 3 h 45 min 30 s 5 s	Subtract the seconds first. Notice that you cannot subtract 45 minutes from 20 minutes.
(7 h 60 min) 20 min 35 s − 3 h 45 min 30 s 5 s	Rename 8 h as 7 h + 1 h or 7 h 60 min.
7 h 80 min 35 s − 3 h 45 min 30 s 5 s	Add the minutes.
7 h 80 min 35 s − 3 h 45 min 30 s 4 h 35 min 5 s	Subtract the minutes, then the hours.

Check for Reasonableness 4 h 35 min 5 s ≈ 4 h ✓

CHECK Your Progress Add or subtract.

a. 5 h 55 min + 6 h 17 min

b. 11 h 25 min 20 s − 4 h 5 min 35 s

c. 9 h 35 s + 2 h 59 min 49 s

Real-World Link Jupiter has the shortest synodic day, the time from sunrise to sunrise, at 9 h 55 min 33 s.

Source: *The World Almanac*

Real-World EXAMPLE

3 **ASTRONOMY** **The table shows the rotation of several planets relative to the Sun. How much longer is a day on Mars than a day on Saturn?**

Planet	Synodic Day
Earth	24 h 0 min 0 s
Mars	24 h 39 min 35 s
Neptune	16 h 6 min 37 s
Saturn	10 h 39 min 23 s

Source: *The World Almanac*

Estimate

24 h 39 min 35 s − 10 h 39 min 23 s ≈ 25 h − 11 h or 14 h

24 h 39 min 35 s − 10 h 39 min 23 s 14 h 0 min 12 s	Subtract seconds first, then minutes, and finally the hours.

The day on Mars is 14 hours 12 seconds longer than on Saturn.

Check for Reasonableness 14 h 12 s ≈ 14 h ✓

CHECK Your Progress

d. **HOMEWORK** Shiro spent 1 hour and 25 minutes working on a social studies project and 40 minutes on math homework. How much time did Shiro spend on homework?

Sometimes you need to determine the **elapsed time**, which is how much time has passed from beginning to end.

Real-World EXAMPLE Elapsed Time

Elapsed Time If the starting and ending times are both A.M. or both P.M., you do not need to first find the elapsed time before noon. Simply subtract the starting time from the ending time.

4 **MOVIES A movie begins at 11:15 A.M. and ends at 1:17 P.M. How long is the movie?**

You need to find how much time has elapsed.

11:15 A.M. to 12:00 noon is 45 minutes.
12:00 noon to 1:17 is 1 hour 17 minutes.

Add the elapsed time before noon and the elapsed time after noon to find the total elapsed time.

$$\begin{array}{r} 45 \text{ min} \\ +\ 1 \text{ h } 17 \text{ min} \\ \hline 1 \text{ h } 62 \text{ min} \end{array} = 2 \text{ h } 2 \text{ min}$$ Rename 62 min as 1 h 2 min.

The length of the movie is 2 hours 2 minutes.

CHECK Your Progress

e. **APPOINTMENTS** Lucita left school at 8:25 A.M. for an orthodontics appointment and returned at 10:50 A.M. How long was she gone from school?

CHECK Your Understanding

Examples 1, 2 (pp. 450–451)

Add or subtract.

1. $\begin{array}{r} 4 \text{ h } 23 \text{ min} \\ +\ 6 \text{ h } 52 \text{ min} \\ \hline \end{array}$

2. $\begin{array}{r} 5 \text{ h } 15 \text{ min } 10 \text{ s} \\ -\ 2 \text{ h } 30 \text{ min } 45 \text{ s} \\ \hline \end{array}$

3. $\begin{array}{r} 8 \text{ h } \qquad\quad 35 \text{ s} \\ +\ 7 \text{ h } 29 \text{ min } 54 \text{ s} \\ \hline \end{array}$

Example 3 (p. 451)

4. **TELEVISION** The table shows the average daily television viewing time for all viewers. How much more time did a viewer spend watching television in 2005 than in 1950?

Average Television Viewing Time per Week

Year	Time
1950	4 h 35 min
2005	8 h 11 min

Source: Nielsen Media Research

Example 4 (p. 452)

5. **SCHOOL** Anoki left his house at 6:45 A.M. to go to school and returned at 2:55 P.M. How much time elapsed between the time Anoki left his house and returned home from school?

Practice and Problem Solving

HOMEWORK HELP

For Exercises	See Examples
6–11	1, 2
12, 13	3
14–19	4

Add or subtract.

6. $\begin{array}{r} 15\text{ h }45\text{ min} \\ +\ 20\text{ h }30\text{ min} \\ \hline \end{array}$

7. $\begin{array}{r} 35\text{ min }25\text{ s} \\ +\ 24\text{ min }40\text{ s} \\ \hline \end{array}$

8. $\begin{array}{r} 6\text{ h }29\text{ min }28\text{ s} \\ -\ 2\text{ h }48\text{ min }14\text{ s} \\ \hline \end{array}$

9. $\begin{array}{r} 2\text{ h }57\text{ min }19\text{ s} \\ -\ 1\text{ h }23\text{ min }42\text{ s} \\ \hline \end{array}$

10. $\begin{array}{r} 12\text{ h }21\text{ min }45\text{ s} \\ +\ 8\text{ h }45\text{ min }16\text{ s} \\ \hline \end{array}$

11. $\begin{array}{r} 5\text{ h }\qquad\quad 28\text{ s} \\ +\ 3\text{ h }8\text{ min }40\text{ s} \\ \hline \end{array}$

12. **ASTRONOMY** Refer to the table in Example 3. How much longer is a day on Earth than a day on Neptune?

13. **MUSIC** Benjamin spent 90 minutes practicing the piano and then 1 hour and 20 minutes listening to the radio. How much time did Benjamin spend practicing the piano and listening to the radio?

Find the elapsed time.

14. 7:28 A.M. to 10:07 A.M.
15. 5:30 P.M. to 9:56 P.M.
16. 6:25 A.M. to 4:45 P.M.
17. 9:30 P.M. to 3:39 A.M.

18. **BASKETBALL** The school basketball game started at 6:30 P.M. and ended at 8:22 P.M. How long was the game?

19. **SHOPPING** Berta, Crystal, and Taylor went shopping and then to a movie. If they left Berta's house at 9:30 A.M. and got back at 6:40 P.M., how long were they gone?

Add or subtract.

20. $\begin{array}{r} 8\text{ h }\qquad\quad 41\text{ s} \\ 3\text{ h }11\text{ min }\ 8\text{ s} \\ +\quad 58\text{ min }10\text{ s} \\ \hline \end{array}$

21. $\begin{array}{r} 8\text{ h }25\text{ s} \\ +\qquad 50\text{ s} \\ \hline \end{array}$

22. $\begin{array}{r} 5\text{ h}\qquad\qquad\quad \\ -\ 1\text{ h }15\text{ min }12\text{ s} \\ \hline \end{array}$

23. **ANALYZE TABLES** The table shows the start and end times for three exercise classes. If Alyssa only has time in her schedule for a class lasting for 1 hour and 15 minutes, which class should she choose?

Class	Start Time	End Time
Jazzercise	3:00 P.M.	4:45 P.M.
Spinning	2:40 P.M.	3:20 P.M.
Aerobics	4:25 P.M.	5:40 P.M.

24. **COOKING** Suppose Mr. James puts a meat loaf in the oven at 11:49 A.M. It needs to bake for 1 hour and 35 minutes. At what time should he take the meat loaf out of the oven?

EXTRA PRACTICE
See pages 694, 713.

25. **SCHOOL** You have 6 classes and a lunch period that each last 45 minutes. Between each period is a 7-minute break. If your day begins at 8:00 A.M., what time will your day end?

H.O.T. Problems

26. **CHALLENGE** Kimmie and her family went out of town for several days. If they left Friday at 2:45 P.M. and returned on Wednesday at 11:00 A.M., find the elapsed time.

27. **REASONING** Determine if a stopwatch is *always, sometimes*, or *never* a good way to measure the length of a movie. Explain your reasoning.

28. **OPEN ENDED** Identify a starting time in the morning and an ending time in the afternoon where the elapsed time is 3 hours 45 minutes.

29. **Which One Doesn't Belong?** Identify the time that is not the same as the others. Explain your reasoning.

3 h 42 min 24 s	2 h 102 min 24 s	3 h 40 min 65 s	3 h 41 min 84 s

30. **WRITING IN MATH** Write a problem about an activity you do on a regular basis in which you need to figure the elapsed time.

TEST PRACTICE

31. Denzell left for an amusement park at 9:15 A.M. and returned home at 6:05 P.M. About how many hours elapsed between the time he left and the time he returned home from the amusement park?

A 3 h

B 8 h

C 9 h

D 10 h

32. **SHORT RESPONSE** Evan spent 2 hours and 35 minutes doing research for a paper. Then he spent 1 hour and 25 minutes writing the paper. How many hours did Evan spend on these two activities?

33. The table shows the time Heather spends on each activity in the morning. About how much time in all does it take for Heather to get ready and arrive at work?

Activity	Time
Get ready in the morning	35 minutes
Walk to the bus stop	12 minutes
Bus ride to work	48 minutes
Walk from the bus stop to work	8 minutes

F 1 hour 43 minutes

G 1 hour 3 minutes

H 1.43 hours

J 1.03 hours

Spiral Review

Complete. (Lesson 8-6)

34. ■ L = 450 mL

35. 65 m = ■ cm

36. 8,800 g = ■ kg

37. **MEASUREMENT** To measure the water in a large birdbath, which metric unit of capacity would you use? (Lesson 8-4)

GET READY for the Next Lesson

PREREQUISITE SKILL **Add or subtract.** (Page 743)

38. 364 + 132

39. 55 + 249

40. 189 – 162

41. 204 – 79

8-8 Measures of Temperature

MAIN IDEA

Choose and estimate reasonable temperatures.

New Vocabulary

temperature
degree
Celsius (°C)
Fahrenheit (°F)

Math Online

glencoe.com

- Extra Examples
- Personal Tutor
- Self-Check Quiz

MINI Lab

Use a thermometer that has both a Fahrenheit (°F) and a Celsius (°C) scale to measure the temperature of the items listed.

Item	Temperature	
	(°F)	(°C)
temperature of glass of ice water		
temperature of cold water from faucet		
temperature of hot water from faucet		

1. Copy the table and record your findings.

Use your findings to predict the temperature of each item.

2. cold glass of milk
3. hot cup of coffee
4. frozen dessert

Temperature is the measure of hotness or coldness of an object or environment. It is measured as **degrees** on a temperature scale.

In the metric system, temperature is measured in degrees **Celsius (°C)**. Water freezes at 0°C and boils at 100°C.

In the customary system, temperature is measured in degrees **Fahrenheit (°F)**. Water freezes at 32°F and boils at 212°F.

The thermometers at the right show common temperatures in degrees Celsius and degrees Fahrenheit.

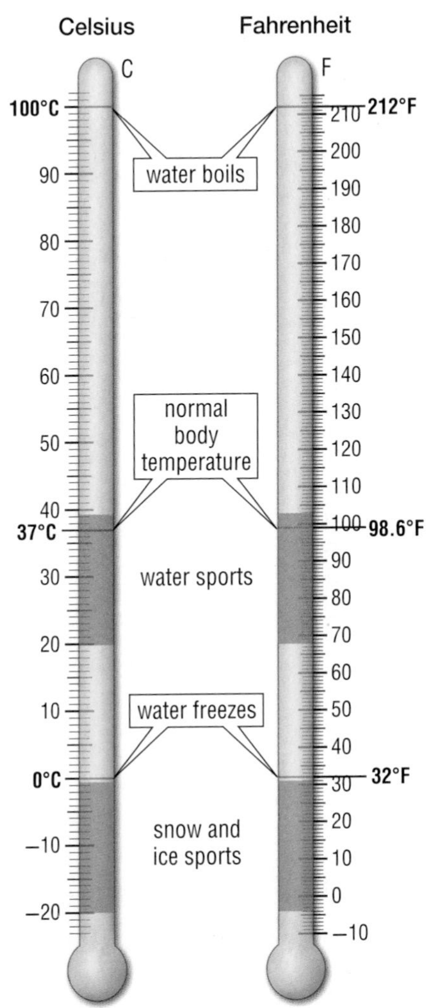

EXAMPLES Choose Reasonable Temperatures

Choose the more reasonable temperature for each.

1 water in a warm bath: 75°F or 105°F

Normal body temperature is 98.6°F, so you would want a warm bath to be warmer than your body temperature. So, 105°F is a more reasonable temperature.

2 inside a classroom: 21°C or 84°C

On the Celsius scale, water boils at 100°C. So, 84°C would be too hot for the temperature inside of a classroom. The more reasonable temperature is 21°C.

3 glass of iced tea: 5°C or −25°C

Water freezes at 0°C, so a glass of iced tea at 5°C would be cool but not freezing yet, nor below freezing at −25°C. So, 5°C is a more reasonable temperature.

Real-World Link Water is the only natural substance found in all three states: liquid, solid (ice), and gas (steam).

Source: U.S. Geological Survey

CHECK Your Progress

Choose the more reasonable temperature for each.

a. hot soup: 40°C or 90°C

b. slice of warm apple pie: 60°F or 110°F

EXAMPLES Give Reasonable Temperatures

Give a reasonable estimate of the temperature in degrees Fahrenheit and degrees Celsius for each situation.

4 outdoor swimming

Swimming is typically an activity for a warm summer day. So, a reasonable temperature is 90°F and 30°C.

5 temperature in a refrigerator

The temperature in a refrigerator should be colder than room temperature so that food will not spoil but not cold enough for food to freeze. So, a reasonable temperature is 35°F and 2°C.

6 temperature of hail

Hail is made of ice. So, the temperature should be a little less than freezing. A reasonable temperature is 30°F and −1°C.

CHECK Your Progress

Give a reasonable estimate of the temperature in degrees Fahrenheit and degrees Celsius for each activity.

c. snowboarding

d. playing baseball

Check Your Understanding

Examples 1–3 (p. 456)

Choose the more reasonable temperature for each.

1. cake in oven: 200°F or 350°F
2. ice cream: −10°C or 10°C
3. person with a fever: 81°F or 101°F
4. hot chocolate: 60°C or 30°C

Examples 4–6 (p. 456)

Give a reasonable estimate of the temperature in degrees Fahrenheit and degrees Celsius for each activity.

5. attending a football game
6. snow skiing
7. planting flowers
8. hiking
9. **FISHING** Kendrick plans on going ice fishing at Bitterroot Lake, Montana, this weekend. What is a reasonable temperature Kendrick can expect while ice fishing?

Practice and Problem Solving

HOMEWORK HELP	
For Exercises	See Examples
10–17	1–3
18–23	4–6

Choose the more reasonable temperature for each.

10. inside a restaurant: 31°F or 71°F
11. ice water: 5°C or 25°C
12. walk-in freezer: 19°F or 39°F
13. hot grill: 120°F or 200°F
14. baking pie in oven: 70°C or 170°C
15. frozen vegetables: −10°C or 10°C
16. **HOT TUBS** The Pecks purchased a hot tub. Should they set the hot tub heater thermostat at 39°C or 80°C? Explain your reasoning.
17. **HOCKEY** At a hockey game, Liana states that the temperature on the ice is 40°F. Her friend thinks 20°F is more reasonable. Who is correct? Explain your reasoning.

Give a reasonable estimate of the temperature in degrees Fahrenheit and degrees Celsius for each activity.

18. jogging
19. taking a hot shower
20. horseback riding
21. ice skating
22. going to an amusement park
23. sunbathing on the beach
24. **SUMMER** It is a warm summer day. If the temperature reads 30 degrees, is this 30°C or 30°F?
25. **COOKING** Use the table at the right. Makayla began cooking chicken that had a temperature of 38°F. What is a reasonable amount the temperature of the chicken will need to rise in order to be safe to eat?

Safe Meat Cooking Temperatures

Meat	Temperature (°F)
Beef, medium well	150–155
Chicken	165–175
Turkey	165–175
Pork	150

Source: What's Cooking America

26. **ELEVATION** Air temperature decreases about 6°C for every increase in the elevation of 1,000 meters. If the temperature outside starts out at 30°C, make a table of values for the temperature at elevations of 1,000, 2,000, 3,000, 4,000, 5,000, and 6,000 meters. What is the difference in the temperature at 3,000 meters and 6,000 meters? (Assume the starting elevation is 0 meters.)

EXTRA PRACTICE
See pages 695, 713.

H.O.T. Problems

CHALLENGE The expression $\frac{5(F - 32)}{9}$, where F is the temperature in degrees Fahrenheit, can be used to convert temperatures to degrees Celsius. Convert each temperature to degrees Celsius.

27. 77°F
28. 41°F
29. 194°F

30. **WRITING IN MATH** A local newspaper used the display at the right to illustrate the weather for the upcoming week. Write and solve a real-world problem using the temperatures in the display.

5-Day Forecast				
Mon	Tue	Wed	Thur	Fri
Partly cloudy	Showers	Showers	Scattered showers	Sunny
Hi: 85°F Lo: 58°F	Hi: 82°F Lo: 55°F	Hi: 78°F Lo: 49°F	Hi: 83°F Lo: 55°F	Hi: 88°F Lo: 60°F

TEST PRACTICE

31. Which of the following is a reasonable temperature for the activity shown in the illustration below?

A 25°C
B 35°F
C 80°C
D 150°F

32. On a cold winter morning, Cara walked outside to get the newspaper. The temperature outside was 5°F. What is a reasonable estimate for the difference between this temperature and Cara's normal body temperature?

F 80°F
G 90°F
H 100°F
J 110°F

Spiral Review

33. **MARATHONS** The first winner of the Boston Marathon in 1897 had a winning time of 2 hours 55 minutes 10 seconds. In 2007, the winner of the Boston Marathon had a winning time of 2 hours 14 minutes 13 seconds. How much faster was the winning time in 2007 than in 1897? (Lesson 8-7)

Complete. (Lesson 8-6)

34. 6,000 L = ■ kL
35. 84 mm = ■ cm
36. ■ g = 3,700 mg

Extend 8-8

Measurement Lab
Using Appropriate Units and Tools

MAIN IDEA

Select appropriate units and tools to measure objects or activities.

An *attribute* is a characteristic of an object. For example, one attribute of a quarter is that it is made of metal. Some attributes of an object can be measured. For example, you can measure how much a quarter weighs and its thickness. In this lab, you will measure the attributes of several objects or activities.

ACTIVITY

1

STEP 1 Select an object in your classroom such as a desk, book, backpack, or trash can.

STEP 2 Make a list of all the measurable attributes of your object. Choose from among length, weight or mass, or capacity.

STEP 3 Select an appropriate tool from among those provided by your teacher and measure each attribute. Record each measure using appropriate units in a table like the one below.

Object	Attribute(s)	Tool	Measurement

STEP 4 Choose a different object with at least one attribute that requires the use of a different tool to measure. Then repeat Steps 1 through 3.

Analyze the Results

1. Express each attribute of the object you measured using different units. For example, if you measured the length of the object in centimeters, write this length in meters.
2. Write a real-world problem in which one of your measurements is needed to solve the problem. For example, if you measured the time it takes one student to go through the lunch line, a problem might be to estimate the time it would take for each student in your class to go through the lunch line.

ACTIVITY

2 **STEP 1** Select a classroom activity such as sharpening your pencil or walking from the door to your desk.

STEP 2 Make a list of all the measurable attributes of your activity. Choose from among length, time, or temperature.

STEP 3 Select an appropriate tool from among those provided by your teacher and measure each attribute. Record each measure using appropriate units.

STEP 4 Choose a different activity with at least one attribute that requires the use of a different tool to measure. Then repeat Steps 1 through 3.

Analyze the Results

3. Express each attribute of the activity you measured using different units. For example, if you measured the time it took to do an activity in minutes, write this time using seconds.
4. Write a real-world problem in which one of your measurements is needed to solve the problem. For example, if you measured the time it took to sharpen one new pencil, a problem might be to estimate the time it would take for each student in your class to sharpen a new pencil before taking a test.

Suppose you were going to organize the following events for a field day at your school. What tools would you need to set up the event and determine a winner?

5. 50-meter dash: Who can run the fastest?
6. Softball throw: Who can throw the farthest?
7. Water relay: Which team can fill up their bucket the fastest using a leaking cup?
8. Indicate an ideal outdoor temperature, in Fahrenheit and Celsius degrees, for the field day event described in Exercise 7. Explain your reasoning.

CHAPTER 8 Study Guide and Review

Math Online glencoe.com
• STUDY TO GO
• Vocabulary Review

FOLDABLES® Study Organizer — GET READY to Study

Be sure the following Big Ideas are noted in your Foldable.

Measurement	Length	Mass & Capacity
Metric		
Customary		

BIG Ideas

Customary Units of Length (Lesson 8-1)
- 1 inch (in.)
- 1 foot (ft) = 12 in.
- 1 yard (yd) = 3 ft
- 1 mile (mi) = 1,760 yd

Customary Units of Capacity (Lesson 8-2)
- 1 fluid ounce (fl oz)
- 1 cup (c) = 8 fl oz
- 1 pint (pt) = 2 c
- 1 quart (qt) = 2 pt
- 1 gallon (gal) = 4 qt

Customary Units of Weight (Lesson 8-2)
- 1 ounce (oz)
- 1 pound (lb) = 16 oz
- 1 ton (T) = 2,000 lb

Metric Units of Length (Lesson 8-3)
- 1 millimeter (mm)
- 1 centimeter (cm)
- 1 meter (m)
- 1 kilometer (km)

Metric Units of Mass and Capacity (Lesson 8-4)
- 1 milligram (mg)
- 1 milliliter (mL)
- 1 gram (g)
- 1 liter (L)
- 1 kilogram (kg)

Changing Metric Units (Lesson 8-6)
- To change from a larger unit to a smaller unit, multiply by powers of 10. To change from a smaller unit to a larger unit, divide by powers of 10.

Measures of Time (Lesson 8-7)
- Elapsed time is how much time has passed from beginning to end.

Measures of Temperature (Lesson 8-8)
- Temperature is measured in degrees Fahrenheit (°F) and in degrees Celsius (°C).

Key Vocabulary

capacity (p. 424)
Celsius (°C) (p. 455)
centimeter (p. 432)
cup (p. 424)
degree (p. 455)
elapsed time (p. 452)
Fahrenheit (°F) (p. 455)
fluid ounce (p. 424)
foot (p. 418)
gallon (p. 424)
gram (p. 437)
inch (p. 418)
kilogram (p. 437)
kilometer (p. 432)
liter (p. 438)
mass (p. 437)
meter (p. 432)
metric system (p. 432)
mile (p. 418)
milligram (p. 437)
milliliter (p. 438)
millimeter (p. 432)
ounce (p. 425)
pint (p. 424)
pound (p. 425)
quart (p. 424)
temperature (p. 455)
ton (p. 425)
yard (p. 418)

Vocabulary Check

Choose the correct term or number to complete each sentence.

1. A centimeter equals (one tenth, one hundredth) of a meter.
2. You should (multiply, divide) to change from larger to smaller units.
3. One paper clip has a mass of about one (gram, kilogram).
4. One cup is equal to (16, 8) fluid ounces.
5. The basic unit of capacity in the metric system is the (liter, gram).
6. To convert from 15 yards to feet, you should (multiply, divide) by 3.
7. One centimeter is (longer, shorter) than 1 millimeter.

CHAPTER 8

Study Guide and Review

Lesson-by-Lesson Review

8-1 Length in the Customary System (pp. 418–423)

Complete.

8. 2 mi = ▪ ft
9. ▪ in. = 5 ft
10. 9 yd = ▪ ft
11. 72 in. = ▪ yd

Draw a line segment of each length.

12. $2\frac{7}{8}$ in.
13. $1\frac{1}{2}$ in.
14. $3\frac{1}{4}$ in.

15. **BASKETBALL** The length of an NBA basketball court is 94 feet. The length of a high school basketball court is 28 yards. What is the difference in feet between the two courts?

Example 1 **Complete 36 ft = ▪ yd.**

$36 \div 3 = 12$ Since 1 yard = 3 feet, divide 36 by 3.

So, 36 ft = 12 yd.

Example 2 **Draw a line segment measuring $1\frac{3}{8}$ inches.**

Draw a line segment from 0 to $1\frac{3}{8}$.

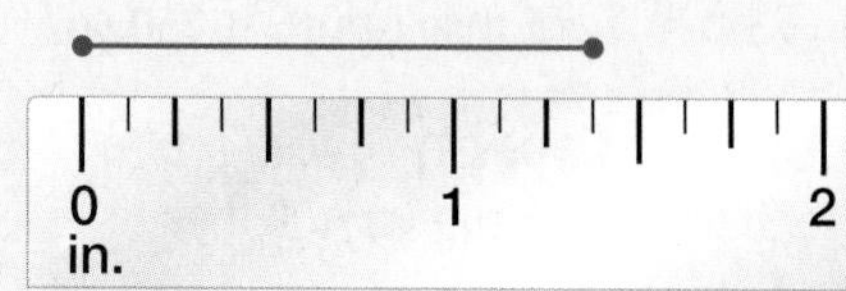

8-2 Capacity and Weight in the Customary System (pp. 424–429)

Complete.

16. 5 T = ▪ lb
17. ▪ qt = 44 c
18. ▪ lb = 12 oz
19. ▪ pt = 8 qt
20. 64 fl oz = ▪ c
21. 3 gal = ▪ qt

22. **TRUCKS** A pickup truck weighs about $1\frac{1}{2}$ tons. How many pounds is this?

Example 3 **Complete 5 qt = ▪ pt.**

$5 \times 2 = 10$ Since 1 quart = 2 pints, multiply 5 by 2 to change a larger unit to a smaller unit.

So, 5 quarts = 10 pints.

8-3 Length in the Metric System (pp. 432–436)

Write the metric unit of length that you would use to measure each of the following.

23. height of your school
24. the length of the state of Kentucky
25. thickness of slice of bread
26. distance across school gym
27. length of your arm

Example 4 **Write the metric unit of length that you would use to measure the height of the monkey bars on the school playground.**

Use the Key Concept box on page 432. The height of the monkey bars is larger than half the width of a penny and smaller than six city blocks. So, use the meter.

Mixed Problem Solving
For mixed problem-solving practice, see page 713.

8-4 Mass and Capacity in the Metric System (pp. 437–441)

Write the metric unit of mass or capacity that you would use to measure each of the following. Then estimate the mass or capacity.

28. an apple
29. a pitcher of lemonade
30. a snowflake
31. an automobile
32. a can of soda
33. **SPORTS DRINK** You can buy your favorite sports drink in 2.5-liter bottles or 550-milliliter bottles. Which bottle has less sports drink?

Example 5 Write the metric unit of mass that you would use to measure a cell phone. Then estimate the mass.

The mass of a cell phone is greater than a paper clip, but less than a textbook. So, the gram is the appropriate unit.

Estimate There are 1,000 grams in a kilogram. A cell phone is much heavier than a paper clip, but not nearly as heavy as a textbook.

One estimate for the mass of a cell phone is about 500 grams.

8-5 PSI: Use Benchmarks (pp. 442–443)

Use a benchmark to solve each problem.

34. **SCARVES** Cho is making scarves. Each scarf must be 12 inches long. She only has a piece of string that is 3 inches long to measure. How can she measure the scarves?
35. **PUNCH** Arturo needs to add 3 quarts of orange juice to the punch. He only has a gallon jug to measure the orange juice. Describe a way he can measure the orange juice.

Example 6 Leo needs to add 2 pints of cream to make ice cream, but all he has is a $\frac{1}{2}$ cup to measure out the amount of cream. How can he measure the cream? Use the benchmark strategy.

There are two $\frac{1}{2}$ cups in one cup and 2 cups in one pint. Leo needs 2 pints or 4 cups of cream. So, he should measure out eight $\frac{1}{2}$ cups of cream.

8-6 Changing Metric Units (pp. 445–449)

Complete.

36. 300 mL = ■ L
37. ■ g = 1 mg
38. ■ m = 75 km
39. 5 kg = ■ g
40. 345 cm = ■ m
41. ■ m = 23 mm
42. 5,200 L = ■ kL
43. 35 m = ■ cm
44. **FOOD** A recipe calls for 3 liters of chicken broth. How many milliliters is this?

Example 7 Complete 9 g = ■ mg.

Since 1 gram = 1,000 milligrams, multiply 9 by 1,000.

$9 \times 1{,}000 = 9{,}000$

So, 9 g = 9,000 mg.

CHAPTER 8

Study Guide and Review

8-7 Measures of Time (pp. 450–454)

Add or subtract.

45. 5 h 20 min + 2 h 16 min

46. 7 h 45 min − 4 h 32 min

47. 9 h 7 min − 8 h 7 min 8 s

48. 2 h 35 min + 6 h 41 min

49. 7 h 20 min + 2 h 48 min 10 s

50. 6 h 50 min 40 s − 3 h 35 min 20 s

51. **FRENCH** Luanda's French lesson started at 6:45 P.M. and ended at 7:30 P.M. How long was her lesson?

52. **BUS** Meredith takes a bus from Lexington, Kentucky, at 8:45 A.M. and travels to Charlotte, North Carolina. If the entire trip takes 8 hours and 22 minutes, what time will she arrive in Charlotte?

Example 8 **Add.**

3 h 50 min
+ 2 h 15 min
5 h 65 min

Rename 65 min as 1 h 5 min

5 h + 1 h 5 min = 6 h 5 min

Example 9 **Subtract.**

5 h 10 min 53 s
− 2 h 29 min 30 s

Subtract the seconds first.

5 h 10 min 53 s
− 2 h 29 min 30 s
23 s

You cannot subtract 29 min from 10 min, so rename 5 h 10 min as 4 h 70 min.

4 h 70 min 53 s
− 2 h 29 min 30 s
2 h 41 min 23 s

8-8 Measures of Temperature (pp. 455–458)

Choose the more reasonable temperature for each.

53. boiling water: 100°F or 212°F

54. inside your bedroom: 45°F or 74°F

55. hot apple cider: 58°C or 28°C

56. **FROZEN DINNERS** Theodore took a frozen dinner out of the freezer. Give a reasonable estimate of the temperature of the frozen dinner in degrees Fahrenheit.

Example 10 **Choose the more reasonable temperature for an ice pop: −5°C or 35°C.**

Water freezes at 0°C, and an ice pop would need to be frozen. 35°C is too warm for a frozen ice pop.

So, a more reasonable temperature for an ice pop is −5°C.

CHAPTER 8

Practice Test

Math Online glencoe.com

• Chapter Test

Complete.

1. 48 in. = ■ ft
2. 2 yd = ■ in.
3. ■ mm = 7 cm
4. ■ fl oz = 3 c
5. 48 c = ■ gal
6. ■ yd = 8 mi
7. 328 mL = ■ L
8. ■ pt = 6 qt
9. 150 g = ■ kg
10. ■ km = 57 m
11. 1,000 mg = ■ g
12. 8 L = ■ mL

13. **PUNCH** Student Council orders 4 gallons of punch for an after-school meeting. How many cups of punch is this?

14. **MULTIPLE CHOICE** Determine which box of cereal is the best buy by finding the price to the nearest ounce.

Cereal	
Amount	**Price**
2 lb	\$4.20
1 lb	\$2.05
25 oz	\$2.75
20 oz	\$2.50

A 2 lb
B 1 lb
C 25 oz
D 20 oz

15. **FOOD** Estimate the metric length of the almond. Then measure to find the actual length.

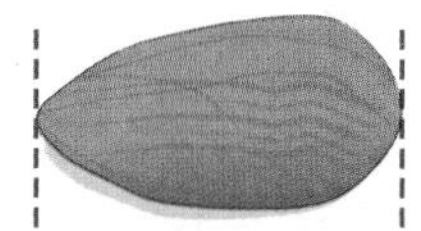

Write the metric unit of length that you would use to measure each of the following.

16. length of a skateboard
17. height of a giraffe

18. **FISH TANKS** Kelly needs to fill her 10-gallon fish tank. She only has a quart container to measure out the water. Describe how she could fill the fish tank using only the quart container.

Write the metric unit of mass or capacity that you would use to measure each of the following. Then estimate the mass or capacity.

19. five \$1 bills
20. a bucket of water

Add or subtract.

21. 19 min 30 s
 − 12 min 40 s

22. 7 h 20 min
 + 2 h 48 min 10 s

23. **MULTIPLE CHOICE** The table shows the time Jerry spent doing work in his yard. About how much time does it take for Jerry to do all of these activities?

Activity	Time
Mowing the lawn	45 minutes
Raking leaves	18 minutes
Edging	9 minutes
Trimming the bushes	15 minutes

F 1.3 hours
G 1.15 hours
H 1 hour 30 minutes
J 1 hour 45 minutes

Choose the more reasonable temperature for each item.

24. Is a glass of apple juice more likely to be −8°C or 18°C?
25. Is the inside of a sauna more likely to be 55°F or 86°F?

CHAPTER 8

Test Practice

Cumulative, Chapters 1–8

Math Online glencoe.com

• Test Practice

PART 1 Multiple Choice

Read each question. Then fill in the correct answer on the answer sheet provided by your teacher or on a sheet of paper.

1. The length of a table is 2 meters. What is the length of the table in centimeters?

   A 2,000 cm C 20 cm

   B 200 cm D 2 cm

2. Mrs. Baker has 25 students in her class. If each student needs 4 file cards, which equation can be used to find s, the total number of file cards needed?

   F $s = 25 \div 4$ H $s = 25 - 4$

   G $s = 25 \times 4$ J $s = 25 + 4$

3. Edmundo had 1 penny, 1 nickel, 1 dime, and 1 quarter in a bag. He picked 2 coins at random from the bag. Which diagram shows all the possible unique coin combinations of the 2 coins that Edmundo picked?

   A 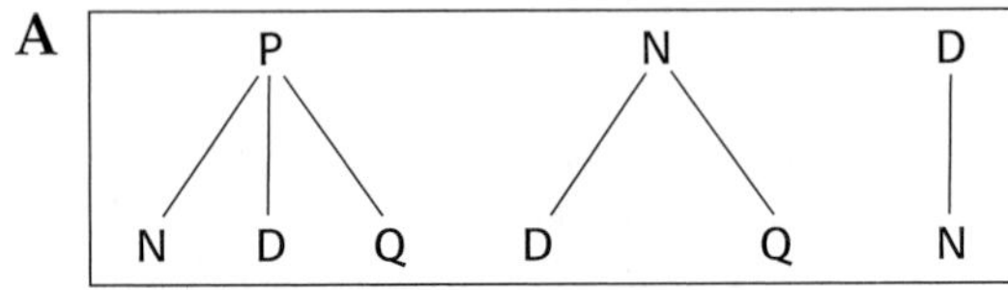

   B 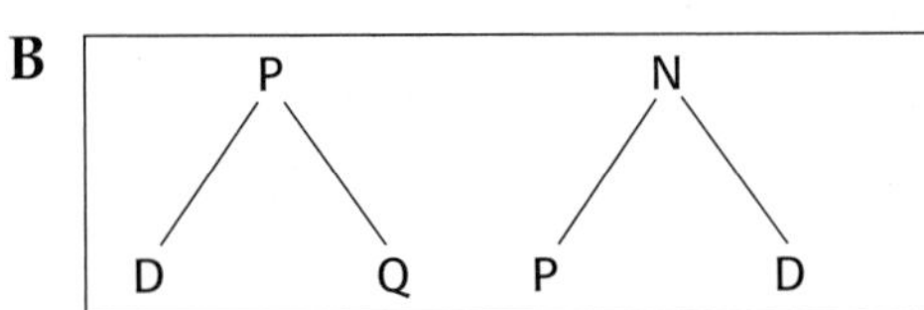

   C 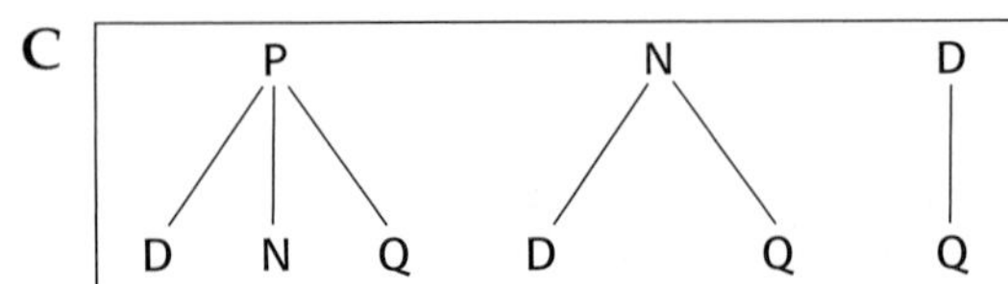

   D 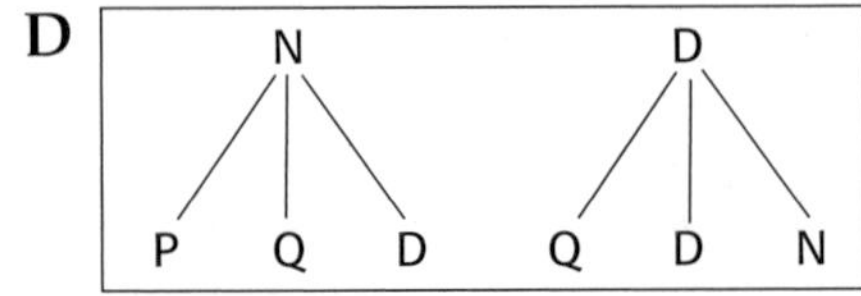

4. Bob's Boot Shop sold 60% of its stock of winter boots before the first snow of the year. What fraction of the stock of winter boots has *not* yet been sold?

   F $\frac{3}{5}$ H $\frac{1}{4}$

   G $\frac{2}{5}$ J $\frac{1}{40}$

TEST-TAKING TIP

Question 4 Be sure to read the question carefully. The question asks for the fraction that has *not* been sold.

5. A bag of apples weighs 2,450 grams. What is this weight in kilograms?

   A 0.00245 kg

   B 24.5 kg

   C 2.45 kg

   D 0.245 kg

6. The Music Shop records the number of CDs sold each month. What is the median number of CDs sold?

Month	Number of CDs Sold
January	50
February	35
March	42
April	110
May	97

   F 32 H 97

   G 50 J 110

7. Alex needs to take 250 milliliters of a particular type of medicine. How many liters of medicine is this?

   A 2.5 L

   B 25 L

   C 0.25 L

   D 2,500 L

Preparing for Standardized Tests
For test-taking strategies and practice, see pages 718–735.

8. Daniel went shopping for shoes. He bought one pair of shoes for \$32.50, another pair for \$29.99, and a third pair for \$49.50. Which procedure could be used to find the average price of each pair?
   - **F** Multiply the sum of the prices of the shoes by the total number of shoes purchased.
   - **G** Add the sum of the prices of the shoes to the total number of shoes purchased.
   - **H** Divide the sum of the prices of the shoes by the total number of shoes purchased.
   - **J** Subtract the sum of the prices of the shoes from the total number of shoes purchased.

9. The formula $V = \frac{1}{3}Bh$ can be used to find the volume of a pyramid. Which of the following best represents $\frac{1}{3}$?
   - **A** 0.33
   - **B** 0.67
   - **C** 3
   - **D** 3.3

10. The cost of renting a car is \$50 plus an additional \$.10 for each mile driven. Which equation can be used to find t, the cost in dollars of the rental for m miles?
    - **F** $t = 0.10m + 25$
    - **G** $t = 50 + 0.10$
    - **H** $t = 50(m + 0.10)$
    - **J** $t = 50 + 0.10m$

PART 2 Short Response/Grid In

Record your answers on the answer sheet provided by your teacher or on a sheet of paper.

11. Francesca has a bag containing 2 purple, 5 orange, 7 blue, and 6 red marbles. If she randomly chooses one marble from the bag, what is the probability that the marble will be purple? Write your answer as a decimal.

PART 3 Extended Response

Record your answers on the answer sheet provided by your teacher or on a sheet of paper. Show your work.

12. Mya works on Saturdays, and then attends a study group. She needs to be home before her younger brother. She walks from work to the study group in 5 minutes. She studies for 40 minutes. Then, she takes 5 minutes to walk to the bus stop. The bus ride is 30 minutes long. It takes her 10 minutes to walk home from where the bus drops her off.
    - a. If Mya is supposed to be home at 5:00 P.M., what is the latest time at which she can leave work?
    - b. Explain the strategy you used to answer part a.
    - c. If Mya had to be home by 3:00 P.M. and it takes her 5 minutes longer to get home, what time should she leave work?

NEED EXTRA HELP?												
If You Missed Question...	1	2	3	4	5	6	7	8	9	10	11	12
Go to Lesson...	8-6	6-7	7-5	7-1	8-6	2-7	8-6	2-6	4-8	6-7	7-4	8-7

Geometry: Angles and Polygons

- Use geometric vocabulary to describe angles, polygons, and circles.
- Use ratios to solve problems involving similarity.

Key Vocabulary

angle (p. 470)
degree (p. 470)
quadrilateral (p. 494)
vertex (p. 470)

Real-World Link

Roller Coasters Riders of the Greezed Lightnin' roller coaster in Kentucky experience a 138-foot-long drop at a 70° angle.

Geometry: Angles and Polygons Make this Foldable to help organize information about angles and polygons. Begin with ten half-sheets of notebook paper.

1. **Fold** a sheet in half lengthwise. Then cut a 1″ tab along the left edge through one thickness.

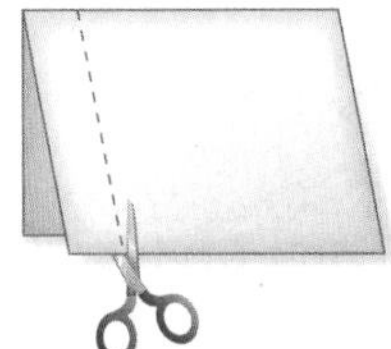

2. **Glue** the 1″ tab down. Write the word *Geometry* on this tab and the lesson title on the front tab.

3. **Write** *Definitions* and *Examples* under the tab.

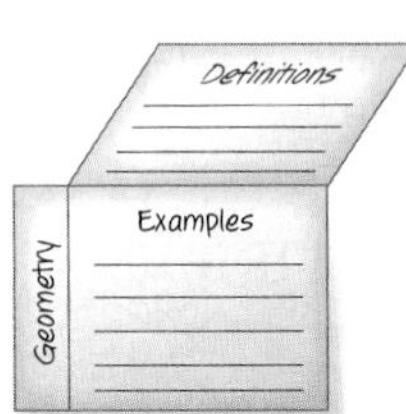

4. **Repeat** Steps 1–3 for each lesson using the remaining paper. Staple them to form a booklet.

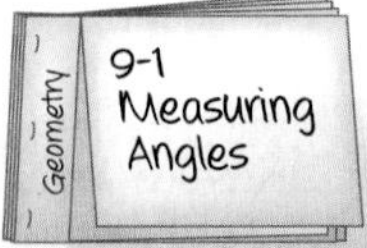

GET READY for Chapter 9

Diagnose Readiness You have two options for checking Prerequisite Skills.

Option 2

Math Online Take the Online Readiness Quiz at glencoe.com.

Option 1

Take the Quick Quiz below. Refer to the Quick Review for help.

QUICK Quiz

Solve each equation. (Lesson 1-8)

1. $x + 44 = 90$
2. $68 + x = 90$
3. $x + 122 = 180$
4. $87 + x = 180$
5. **BASKETBALL** In the first two season games, Lee scored a total of 40 points. If he scored 21 points in the second game, how many points did he score in the first game?

Solve each equation. (Lesson 1-8)

6. $77 + 44 + x = 180$
7. $90 + x + 32 = 180$
8. $53 + x + 108 + 82 = 360$
9. $29 + 38 + 112 + x = 360$
10. **HOTELS** A hotel room for four people costs $360. If three people each pay $85, how much will the fourth person have to pay?

Tell whether each pair of figures has the same size and shape. (Prior Grade)

11. 

12. 

QUICK Review

Example 1

Solve $54 + x = 180$.

$54 + x = 180$ **THINK** What number added to 54 equals 180?

$54 + 126 = 180$ You know that $54 + 126 = 180$.

The solution is 126.

Example 2

Solve $61 + x + 22 = 180$.

$61 + x + 22 = 180$ Add 61 and 22.

$83 + x = 180$ **THINK** What number added to 83 equals 180?

$83 + 97 = 180$ You know that $83 + 97 = 180$.

The solution is 97.

Example 3

Tell whether the triangles below have the same size and shape.

Yes, the triangles have the same size and same shape.

9-1 Measuring Angles

MAIN IDEA

Measure and classify angles.

New Vocabulary

- angle
- side
- vertex
- degree
- right angle
- acute angle
- obtuse angle
- straight angle

Math Online

glencoe.com

- Extra Examples
- Personal Tutor
- Self-Check Quiz
- Reading in the Content Area

GET READY for the Lesson

CELL PHONES The circle graph shows the charges on Corinne's most recent cell phone bill.

Corinne's Cell Phone Bill

Sending Pictures
Downloading Ringtones
Text Messages
Monthly Minutes Included in Plan
Overage Minutes

1. Did Corinne spend more money text messaging or on overage minutes? downloading ringtones or sending pictures? Explain.
2. The percents 43%, 28%, 17%, 8%, and 4% correspond to the sections in the graph. Explain how you would match each percent with its corresponding section.

Each section of the circle graph above shows an angle. **Angles** have two **sides** that share a common endpoint called the **vertex** of the angle. An angle is often named by the label at the vertex. Angle C can be written as $\angle C$.

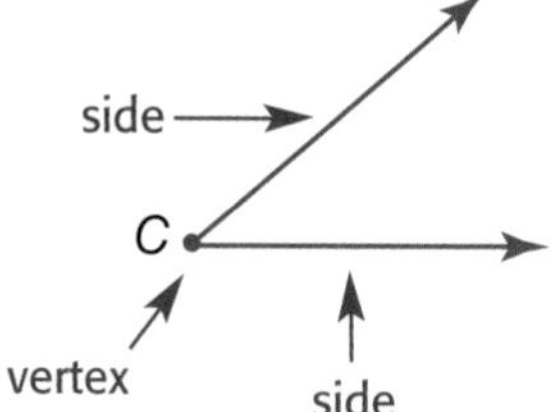
side
C
vertex
side

The most common unit of measure for angles is the **degree**. A circle can be separated into 360 equal-sized parts. Each part would make up a one-degree (1°) angle.

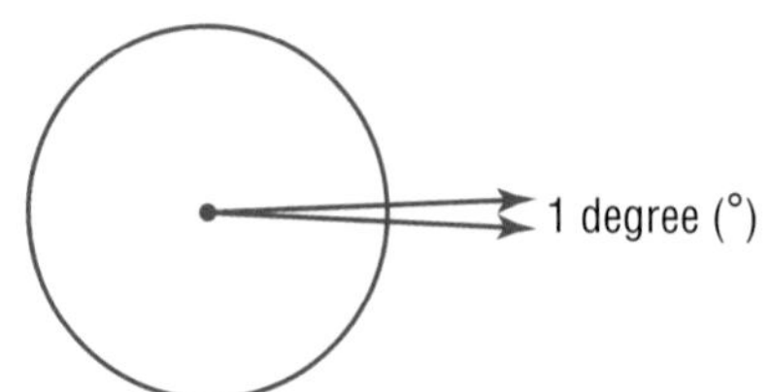
1 degree (°)

EXAMPLES Measure Angles

Use a protractor to find the measure of each angle.

1

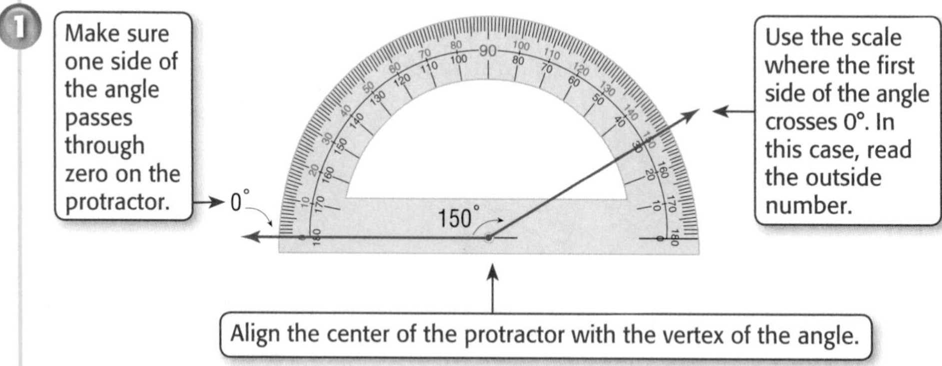
Make sure one side of the angle passes through zero on the protractor.
0°
150°
Use the scale where the first side of the angle crosses 0°. In this case, read the outside number.
Align the center of the protractor with the vertex of the angle.

The angle measures 150°.

Reading Math

Degrees The measurement 0° is read *zero degrees*.

2

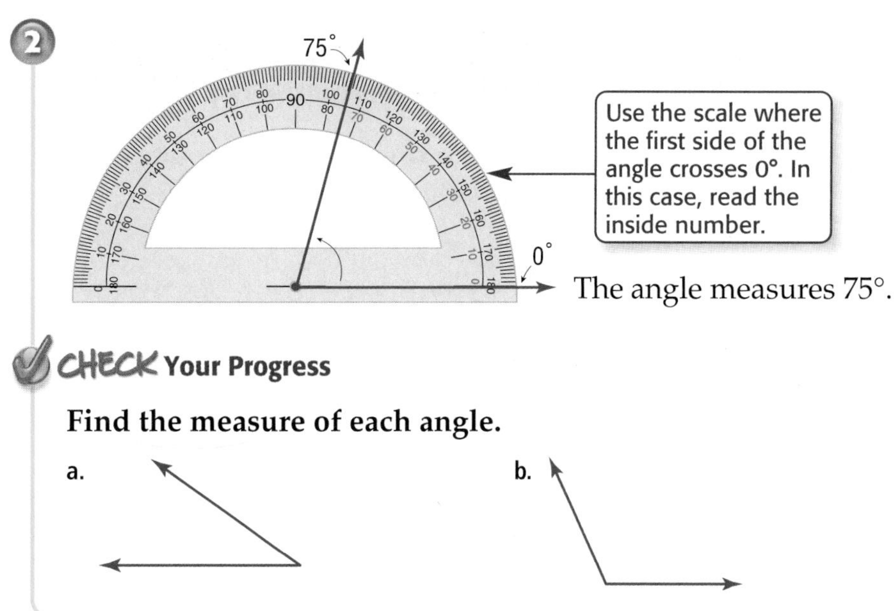

Use the scale where the first side of the angle crosses 0°. In this case, read the inside number.

The angle measures 75°.

CHECK Your Progress

Find the measure of each angle.

a. b.

Angles can be classified according to their measures.

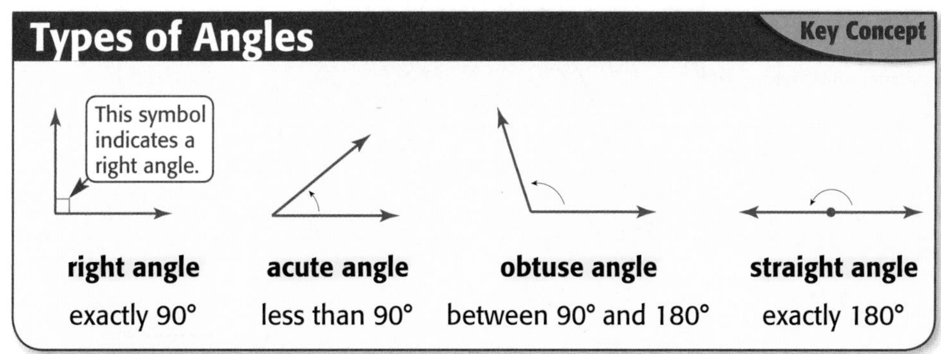

Types of Angles
Key Concept
This symbol indicates a right angle.
right angle
exactly 90°
acute angle
less than 90°
obtuse angle
between 90° and 180°
straight angle
exactly 180°

EXAMPLES Classify Angles

Classify each angle as *acute*, *obtuse*, *right*, or *straight*.

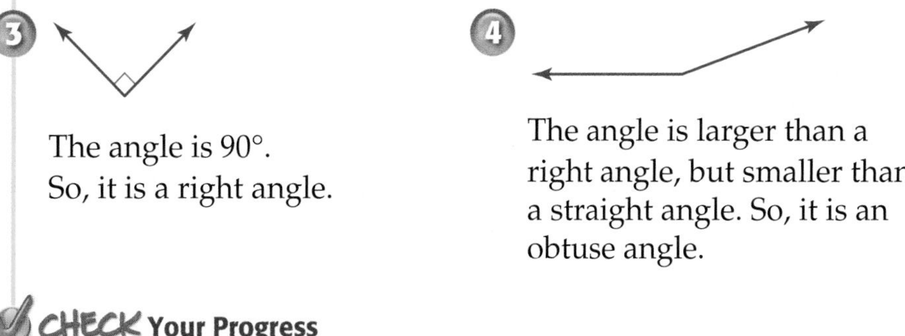

3 The angle is 90°. So, it is a right angle.

4 The angle is larger than a right angle, but smaller than a straight angle. So, it is an obtuse angle.

CHECK Your Progress

Classify each angle as *acute*, *obtuse*, *right*, or *straight*.

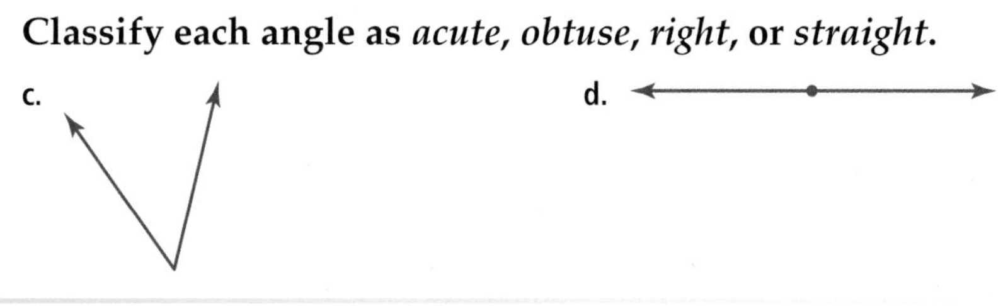

CHECK Your Understanding

Examples 1–4 (pp. 470–471)

Use a protractor to find the measure of each angle. Then classify each angle as *acute, obtuse, right,* or *straight.*

1. 

2.

3. 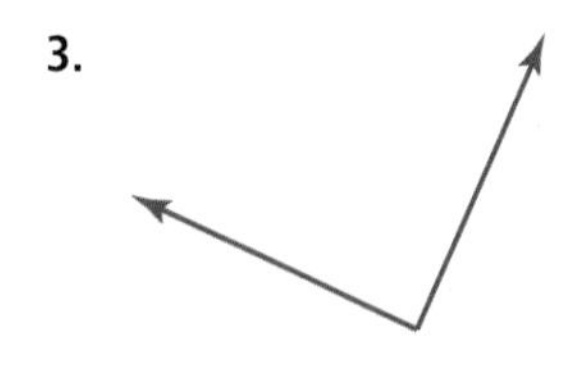

Examples 3, 4 (p. 471)

4. **HOCKEY** The *lie* of a hockey stick is the angle between the blade and the shaft. Classify this angle.

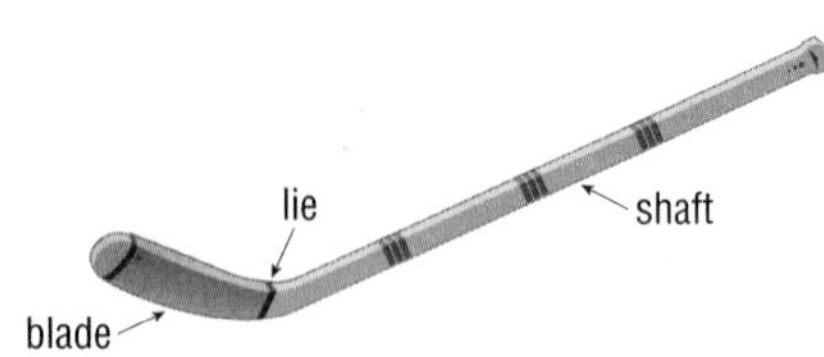

Practice and Problem Solving

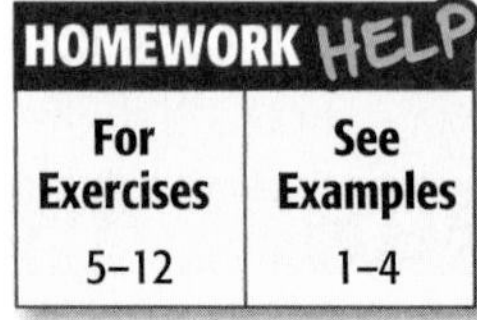

HOMEWORK HELP

For Exercises	See Examples
5–12	1–4

Use a protractor to find the measure of each angle. Then classify each angle as *acute, obtuse, right,* or *straight.*

5. 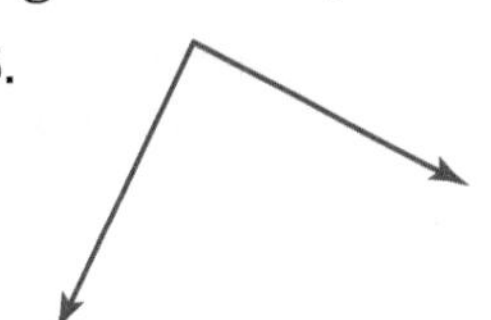

6. 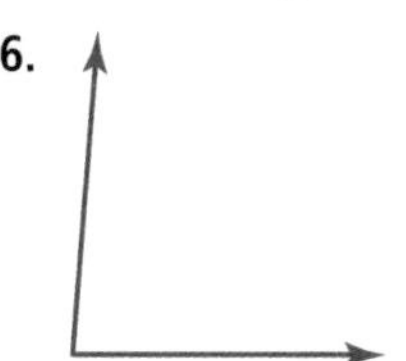

7. 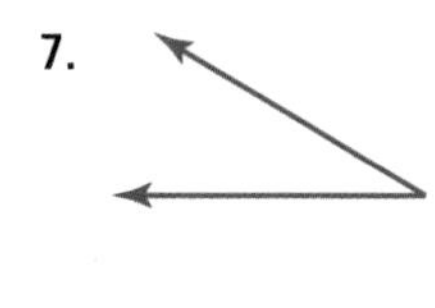

8.

9. 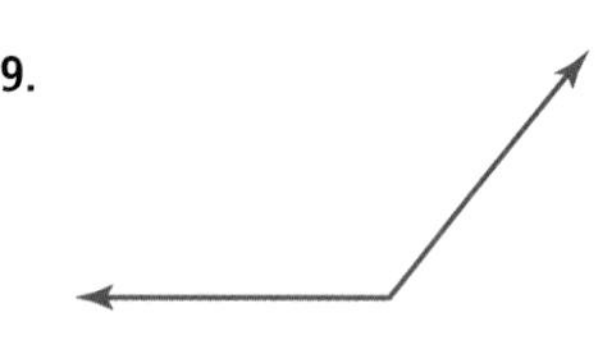

10. 

CLOCKS **Classify the angles shown in the time on each clock.**

11. 

12. 

Use a protractor to find the measure of the indicated angle.

13. 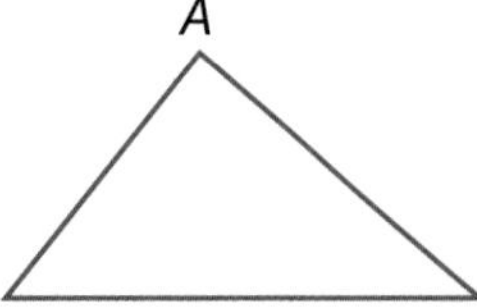

14. 

15. 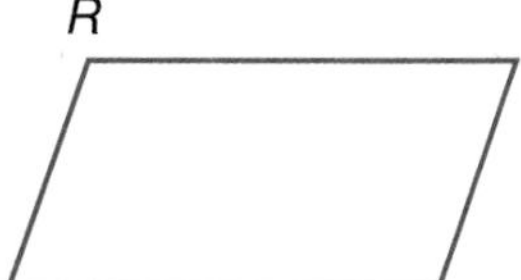

16. **FLAGS** What is the measure of $\angle F$ on the Ohio flag shown at the right?

17. **AIRPLANES** What type of angle does the path of an airplane make with the ground when it is landing?

EXTRA PRACTICE
See pages 695, 714.

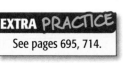

H.O.T. Problems

18. **OPEN ENDED** Select three objects in your classroom or at home. Measure the angles found in the object, and then classify each angle.

19. **FIND THE ERROR** Lorena and Nathan are measuring angles. Who is correct? Explain your reasoning.

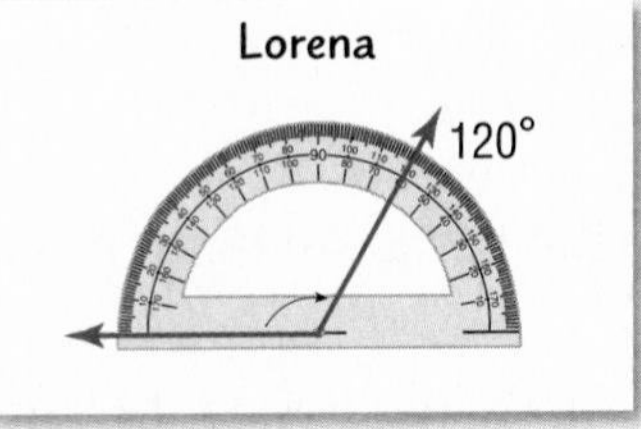

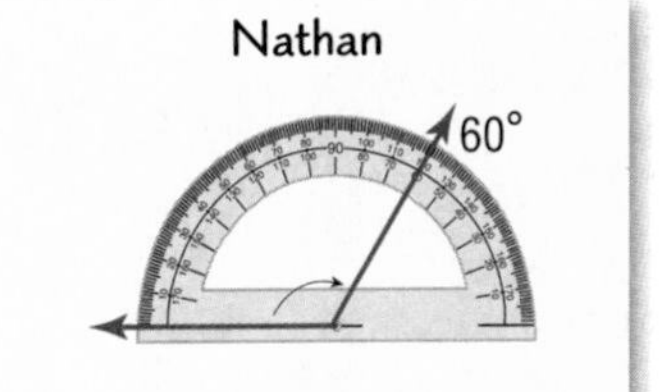

20. **CHALLENGE** Measure ∠Z to the nearest degree. Then describe the method you used to find the measure.

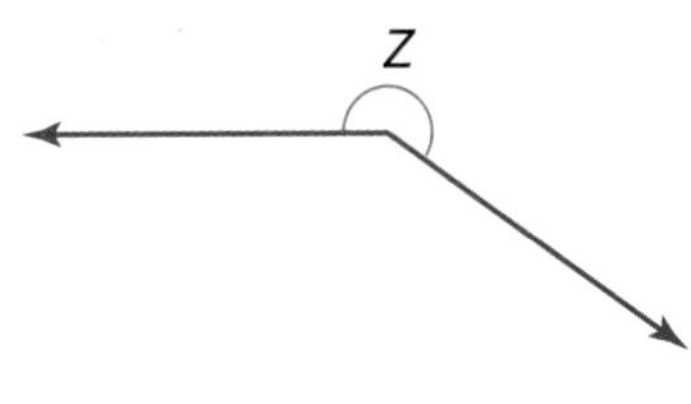

21. **WRITING IN MATH** Without measuring, explain in your own words how you can classify an angle as acute, obtuse, or right.

TEST PRACTICE

22. Find the measure of ∠FGK to the nearest degree.

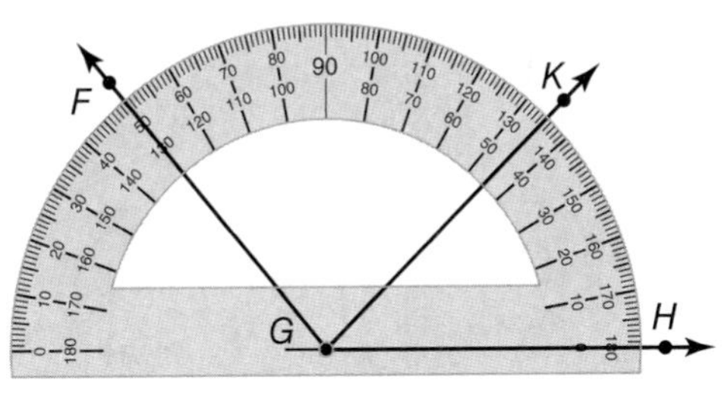

A 130° **C** 85°
B 95° **D** 45°

23. What type of angle is at each vertex of the regular hexagon below?

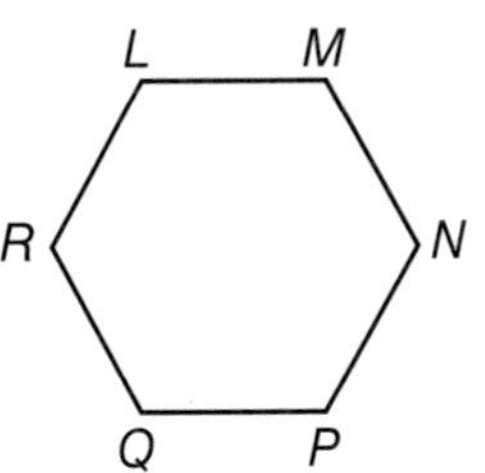

F acute **H** right
G obtuse **J** straight

Spiral Review

24. **SCIENCE** The melting point of sulfur is 246°F. Its boiling point is 833°F. How much does the temperature need to increase for sulfur to go from melting to boiling? (Lesson 8-8)

25. Find the sum of 13 hours 45 minutes and 27 hours 50 minutes. (Lesson 8-7)

GET READY for the Next Lesson

PREREQUISITE SKILL Use a ruler to draw a diagram that shows how the hands on a clock appear at each time. (Page 745)

26. 12:10 27. 9:00 28. 2:30 29. 3:45

9-2 Estimating and Drawing Angles

MAIN IDEA

Estimate measures of angles and draw angles.

Math Online

glencoe.com

- Concepts in Motion
- Extra Examples
- Personal Tutor
- Self-Check Quiz

MINI Lab

To estimate the measure of an angle, use angles that measure 45°, 90°, and 180°.

STEP 1 Fold a paper plate in half to find the center of the plate.

STEP 2 Cut wedges as shown. Then measure and label each angle.

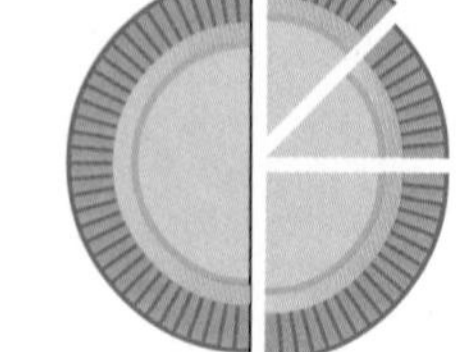

1. Use the wedges to estimate the measure of each angle shown.

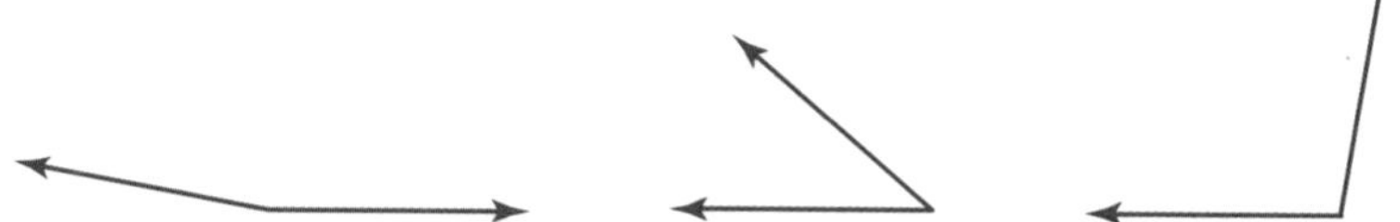

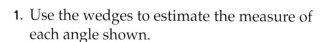

2. How did the wedges help you to estimate each angle?
3. Explain how the 90° and 45° wedges can be used to estimate the angle at the right. What is a reasonable estimate for the angle?
4. How would you estimate the measure of any angle without using the wedges?

To estimate the measure of an angle, compare it to an angle whose measure you know.

EXAMPLE Estimate Angle Measures

1 Estimate the measure of the angle.

The angle is a little less than a 90° angle. So, a reasonable estimate is about 80°.

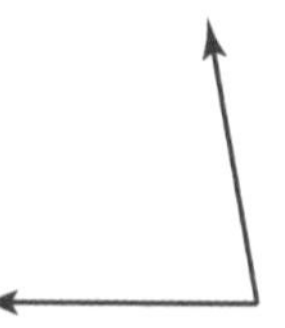

CHECK Your Progress

Estimate the measure of each angle.

a. 

b. 

Reading Math

straightedge A straightedge is an object, like a ruler, used to draw straight lines.

EXAMPLE Draw an Angle

2 **Use a protractor and a straightedge to draw a 74° angle.**

Step 1 Draw one side of the angle. Then mark the vertex.

Step 2 Place the center point of the protractor on the vertex. Align the mark labeled 0 on the protractor with the line. Find 74° on the correct scale and make a dot.

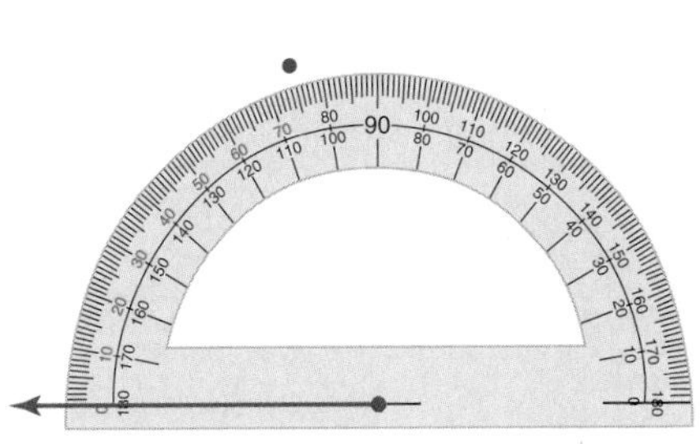

Step 3 Remove the protractor and use a straightedge to draw the side that connects the vertex and the dot.

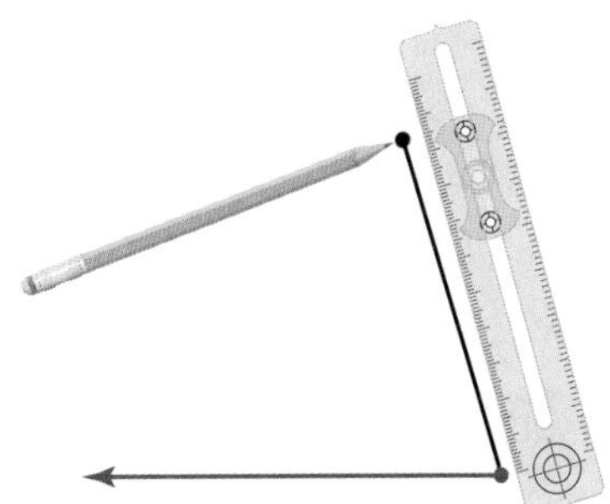

Study Tip

Check for Reasonableness You can check whether you have used the correct scale by comparing your angle with an estimate of its size.

CHECK Your Progress

Use a protractor and a straightedge to draw angles having the following measurements.

c. 68° d. 105° e. 85°

CHECK Your Understanding

Example 1 (p. 474) **Estimate the measure of each angle.**

1. 

2. 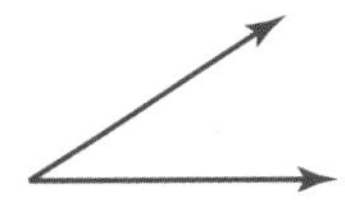

3.

BICYCLE For Exercises 4 and 5, refer to the bicycle diagram.

4. Estimate the measure of the head angle.

5. Estimate the measure of the seat angle.

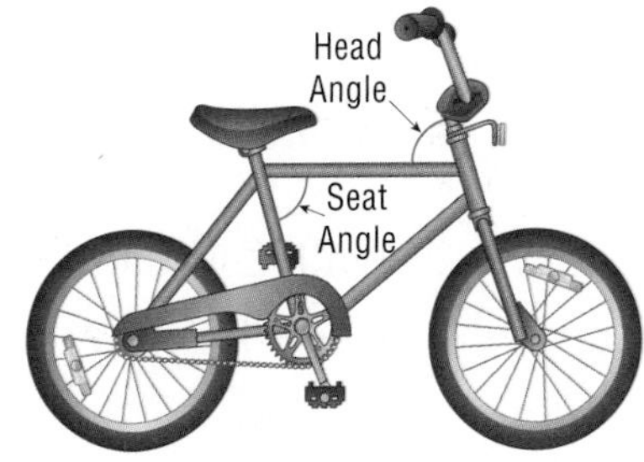

Example 2 (p. 475) **Use a protractor and a straightedge to draw angles having the following measurements.**

6. 25° 7. 140° 8. 60°

Practice and Problem Solving

HOMEWORK HELP	
For Exercises	See Examples
9–14, 23, 24	1
15–22	2

Estimate the measure of each angle.

9. 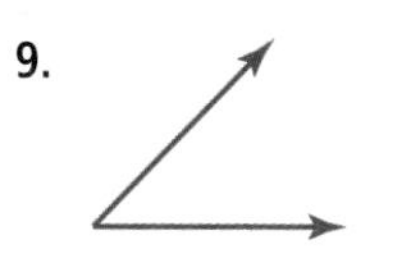

10. 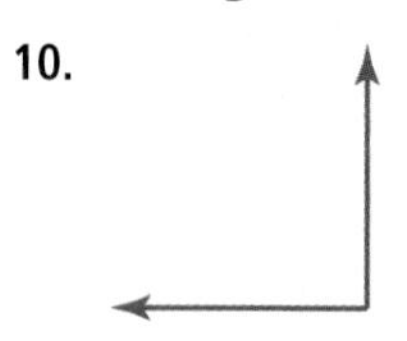

11. 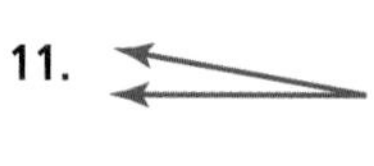

12. 

13. 

14. 

Use a protractor and a straightedge to draw angles having the following measurements.

15. 75°
16. 50°
17. 45°
18. 20°
19. 115°
20. 175°
21. 133°
22. 79°

23. **TIME** Estimate the measure of the angle formed by the hands on the clock shown.

24. **GOLF** Estimate the measure of the golfer's spine angle in the photo at the left.

Spine angle

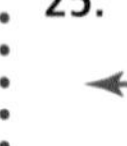

Real-World Link While swinging a golf club, it is important to keep your spine angle constant throughout your entire swing.

Source: *Golf Tips Magazine*

Estimate the measure of each angle. Explain your reasoning.

25. 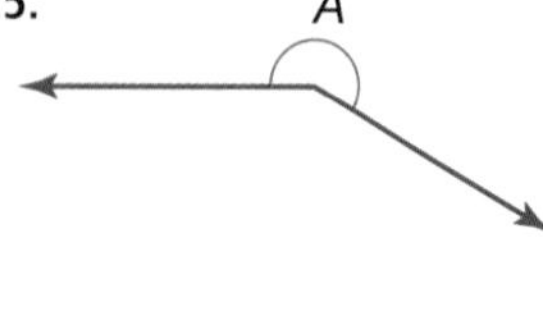

26. 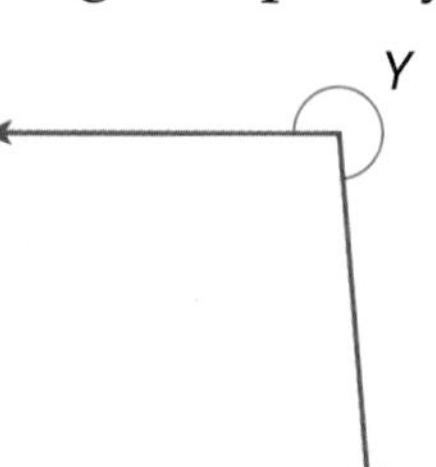

27. 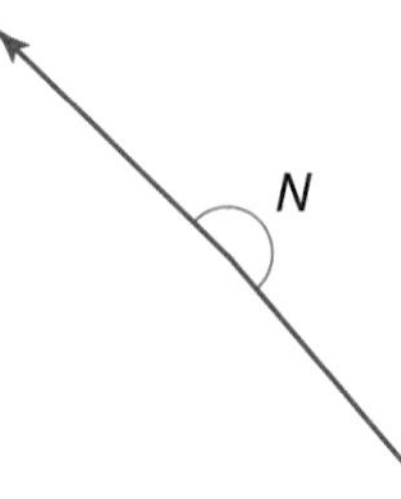

28. **RESEARCH** Use the Internet or another source to find a photo of a humpback whale. Draw an example of the angle formed by the two tail fins. Then give a reasonable estimate for the measure of this angle.

29. **LADDERS** To be considered *safe,* a ladder should be leaned at an angle of about fifteen degrees formed by the top of the ladder and the vertical wall. Estimate the measure of the angles formed by each ladder below and determine which ladders would be considered *safe.*

a. 
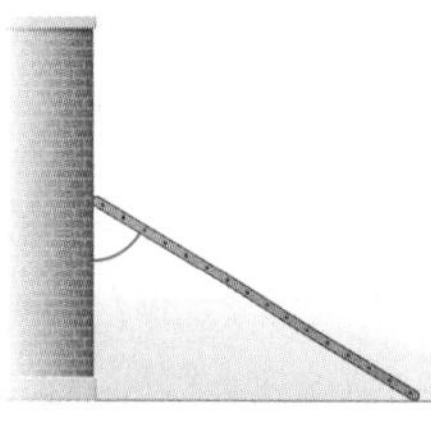

b. 

c. 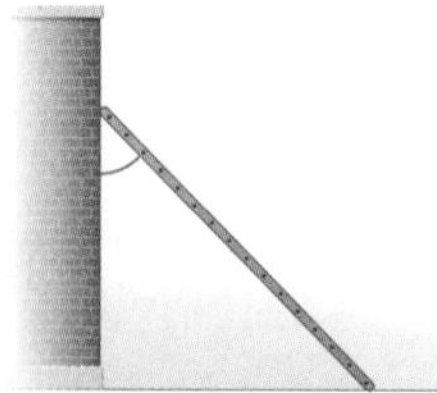

TREES For Exercises 30–32, use the following information and the diagram at the right.

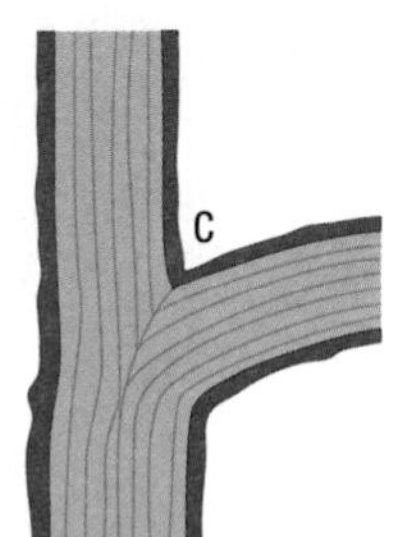

When pruning apple trees, if the angle formed by a branch growing off a main trunk is less than 60°, the branch should be pruned, or trimmed, as the branch will become weak.

Estimate the measure of each angle labeled *C*. Then determine whether the branch should be pruned.

30. 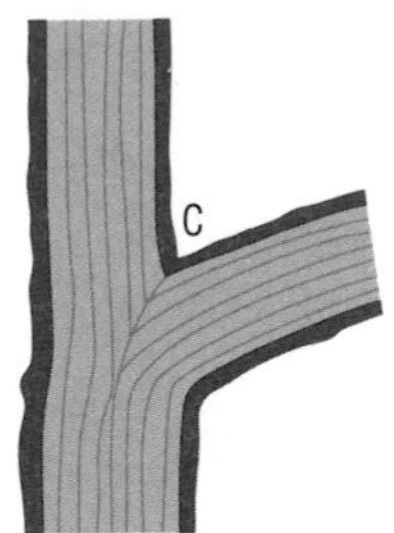

31. 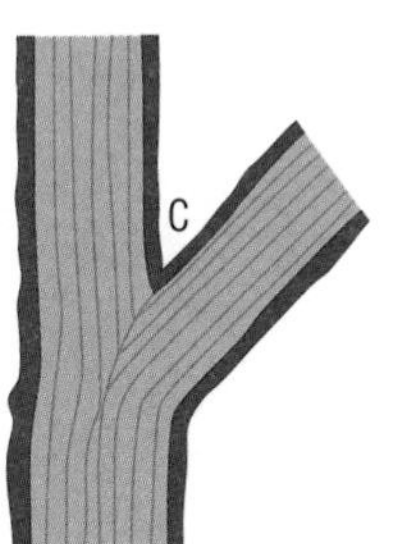

32. 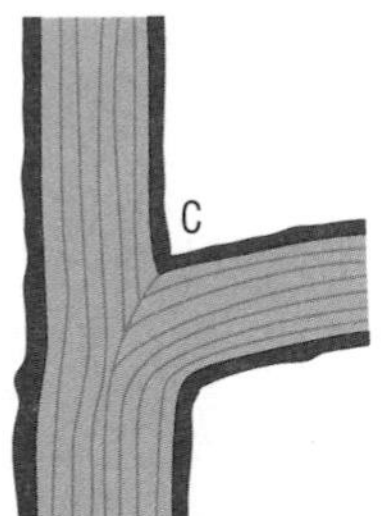

33. **GYROSCOPES** A gyroscope can be used to show how different objects spin on an axis (like a bicycle tire or a basketball on someone's finger). When a gyroscope is spinning level on a desk, it is spinning at a 90° angle to the desk. Draw a diagram of a gyroscope spinning at a 30° angle, a 50° angle, a 110° angle, and a 135° angle.

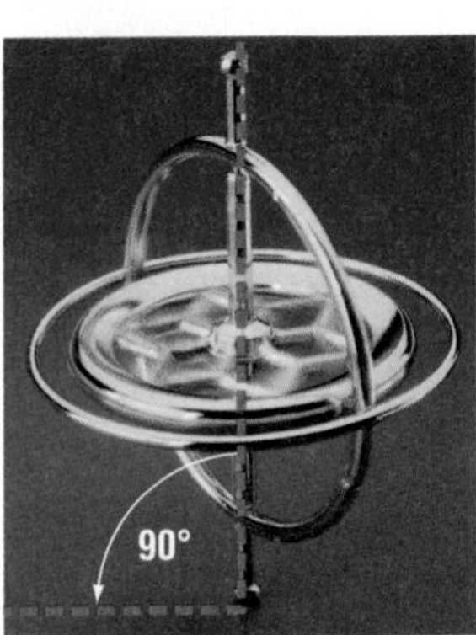

See pages 695, 714.

34. **TRIANGLES** Use a protractor and straightedge to draw a triangle with angle measures of 70°, 60°, and 50°. Label each angle with its measure.

H.O.T. Problems

35. **CHALLENGE** Estimate the measure of each angle in the figure below. Then analyze any relationships you observe in these angle measures.

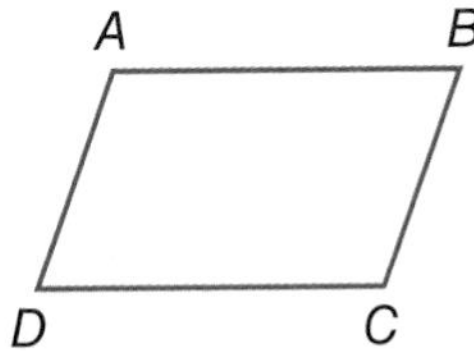

36. **REASONING** Mr. Morales is a physical therapist. At each visit with a patient recovering from knee surgery, he determines the angle at which the patient can bend his or her knee. Do you think Mr. Morales should use estimation to track his patient's progress? Explain your reasoning.

37. **OPEN ENDED** Choose a capital letter that is made up of straight line segments, at least two of which meet to form an acute angle. Draw this letter using a straightedge. Label the angle ∠1. Then estimate the measure of this angle.

38. **WRITING IN MATH** Describe a situation in which drawing a diagram with approximate angle measures would be appropriate and useful.

TEST PRACTICE

39. Which angle has an approximate measure of 50°?

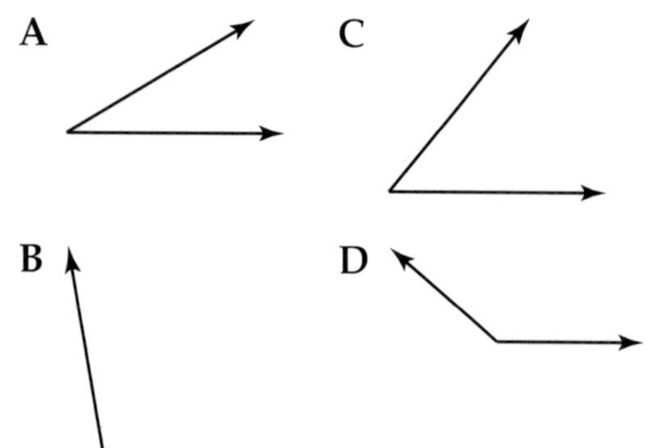

40. Below is the shape of a kite that Jermain made.

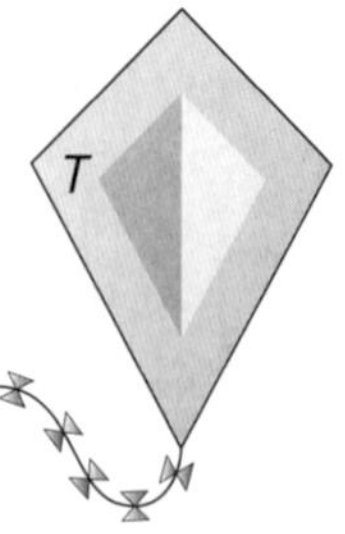

Estimate the measure of $\angle T$.

F 45° H 100°

G 80° J 140°

Spiral Review

Use a protractor to find the measure of each angle. Then classify each angle as *acute, obtuse, right,* or *straight.* (Lesson 9-1)

41. 42. 43.

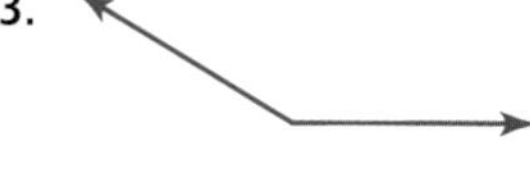

Choose the more reasonable temperature for each. (Lesson 8-8)

44. air temperature on a snowy day: 35°C or 0°C
45. bread just out of the oven: 70°C or 140°C
46. ice cream sandwich: 15°F or 50°F

OCEANS For Exercises 47 and 48, refer to the graph at the right. (Lesson 3-5)

47. What is the combined area of the two smallest oceans?
48. How many millions of miles greater is the Pacific Ocean than the Atlantic Ocean?

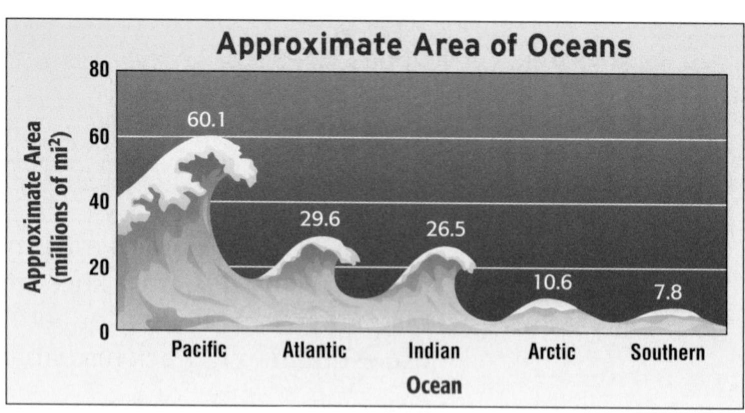

Source: *Time Almanac for Kids*

GET READY for the Next Lesson

PREREQUISITE SKILL Solve each equation mentally. (Lesson 1-8)

49. $x + 45 = 180$ 50. $25 + x = 90$ 51. $130 + x = 180$ 52. $x + 50 = 90$

9-3 Angle Relationships

MAIN IDEA

Classify and apply angle relationships.

New Vocabulary

vertical angles
congruent angles
supplementary angles
complementary angles

Math Online

glencoe.com

- Extra Examples
- Personal Tutor
- Self-Check Quiz

MINI Lab

STEP 1 Copy the figure shown on dot paper.

STEP 2 Use a protractor to find the measure of each angle.

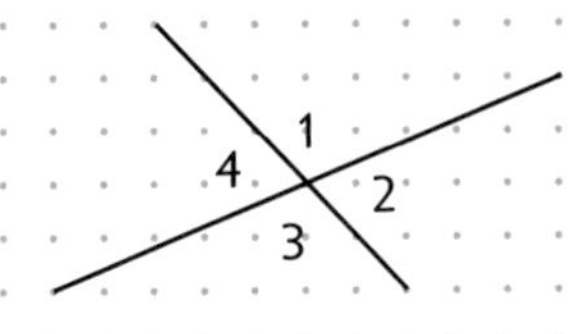

1. What do you notice about the measures of $\angle 1$ and $\angle 3$? $\angle 2$ and $\angle 4$?
2. **MAKE A CONJECTURE** Describe the relationship between opposite angles formed by intersecting lines.
3. Find the sum of the measures of $\angle 3$ and $\angle 4$ and of $\angle 2$ and $\angle 3$.
4. What type of angle is formed by $\angle 3$ and $\angle 4$? $\angle 2$ and $\angle 3$?
5. **MAKE A CONJECTURE** Describe the relationship between the angles that form a straight angle.

When two lines intersect, they form two pairs of opposite angles called **vertical angles**. Vertical angles have the same measure. Angles with the same measure are **congruent angles**.

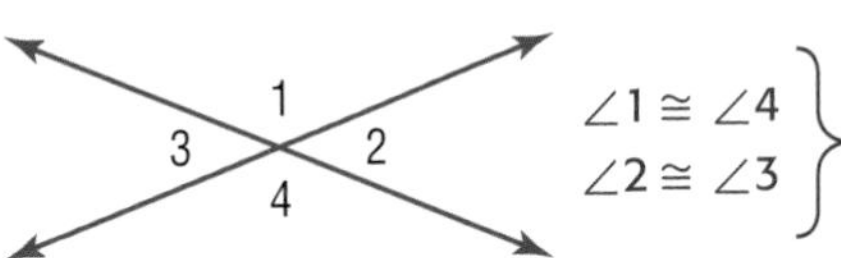

$\angle 1 \cong \angle 4$
$\angle 2 \cong \angle 3$

The symbol $\cong$ is used to show that the angles are congruent.

EXAMPLE Find a Missing Angle Measure

1 Find the value of x in the figure.

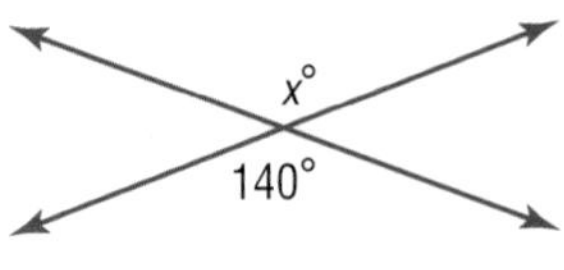

The angle labeled $x°$ and the angle labeled 140° are vertical angles. Therefore, they are congruent.

So, the value of x is 140.

CHECK Your Progress

Find the value of x in each figure.

a. 

b. 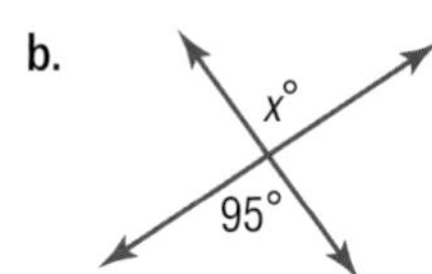

Pairs of angles can also have other relationships. In the Mini Lab, you found pairs of angles that have a sum of 180°. Two angles are **supplementary** if the sum of their measures is 180°. Two angles are **complementary** if the sum of their measures is 90°.

Reading Math

Notation The notation $m\angle 1$ is read as *the measure of angle 1.*

Pairs of Angles

Key Concept

Words Two angles that have a sum of 180° are supplementary angles.

Models

$m\angle 1 = 120°, m\angle 2 = 60°, m\angle 1 + m\angle 2 = 180°$

Words Two angles that have a sum of 90° are complementary angles.

Models

$m\angle 1 = 30°, m\angle 2 = 60°, m\angle 1 + m\angle 2 = 90°$

You can use the definitions of complementary and supplementary to classify angles.

EXAMPLES Classify Pairs of Angles

Classify each pair of angles as *complementary, supplementary,* or *neither.*

Study Tip

Angle Relationships Angles do not have to share the same vertex in order to be classified as complementary or supplementary.

2

45° 45°

$45° + 45° = 90°$

Since the sum of their measures is 90°, the angles are complementary.

3

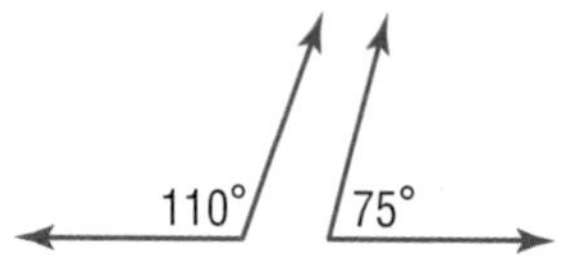

$110° + 75° = 185°$

Since the sum of their measures is not 90° or 180°, the angles are neither complementary nor supplementary.

CHECK Your Progress

Classify each pair of angles as *complementary, supplementary,* or *neither.*

c. 45° 135°

d. 10° 80°

EXAMPLES Find Missing Angle Measures

Find the value of x in each figure.

4

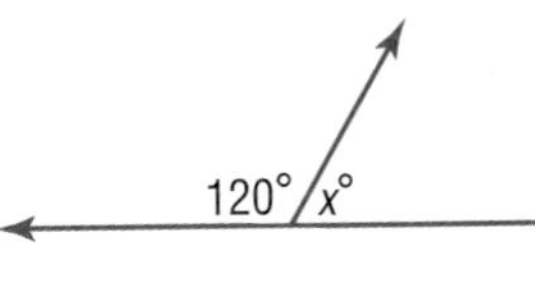

Since the angles form a straight line, they are supplementary.

$120° + x° = 180°$ Definition of supplementary angles

$120° + 60° = 180°$ **THINK** What measure added to 120° equals 180°?

So, the value of x is 60.

> **Study Tip**
>
> **Check for Reasonableness** The angle measuring $x°$ is acute, so x should be less than 90. Since $60 < 90$, the answer is reasonable. ✓

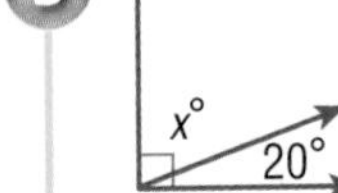

Since the angles form a right angle, they are complementary.

$x° + 20° = 90°$ Definition of complementary angles

$70° + 20° = 90°$ **THINK** What measure added to 20° equals 90°?

So, the value of x is 70.

CHECK Your Progress

e.

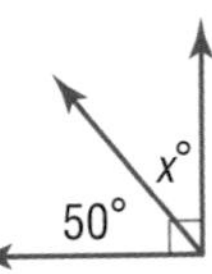

f.

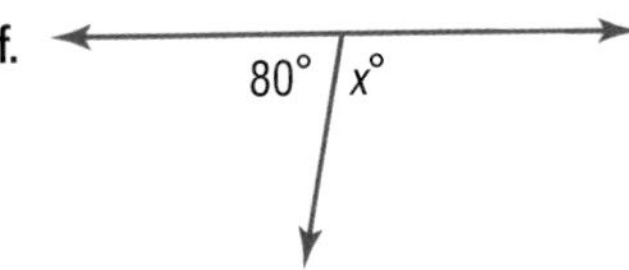

CHECK Your Understanding

Examples 2, 3 (p. 480)

Classify each pair of angles as *complementary, supplementary,* or *neither.*

1.

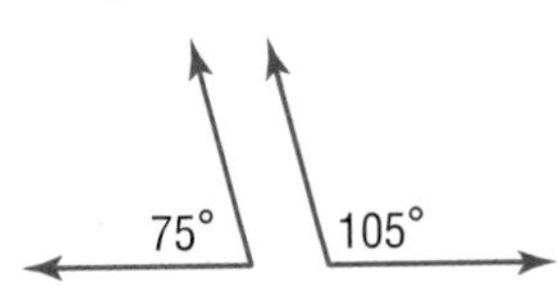

2.

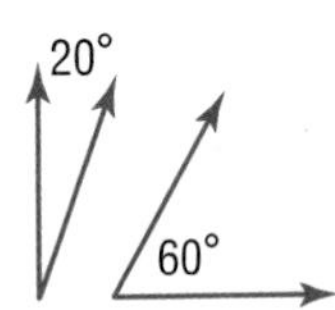

3.

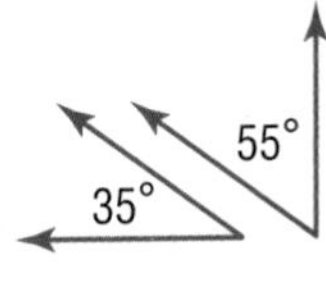

Examples 1, 4, 5 (pp. 479, 481)

Find the value of x in each figure.

4.

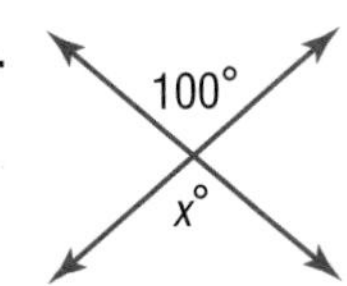

5.

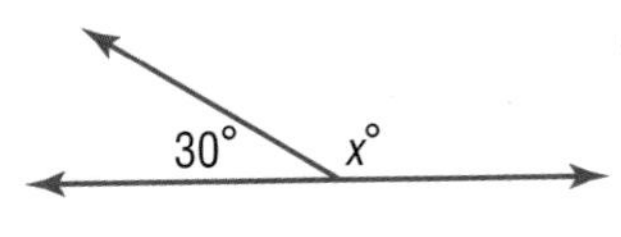

6.

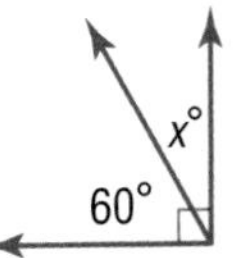

Example 4 (p. 481)

7. **TREES** What is the value of x in the maple leaf at the right?

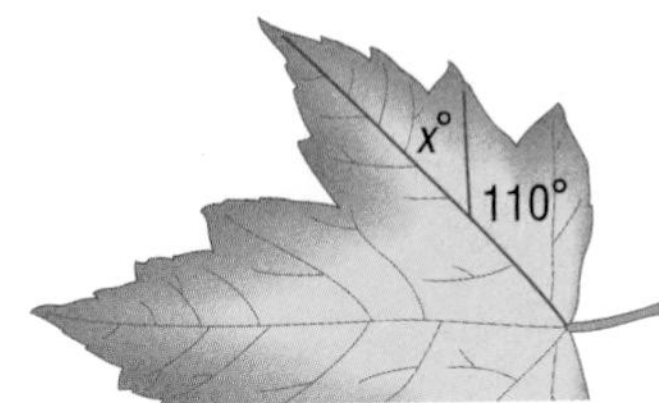

Practice and Problem Solving

HOMEWORK HELP	
For Exercises	**See Examples**
8–13	2, 3
14–19	1, 4, 5
20, 21	4

Classify each pair of angles as *complementary, supplementary,* or *neither.*

8.
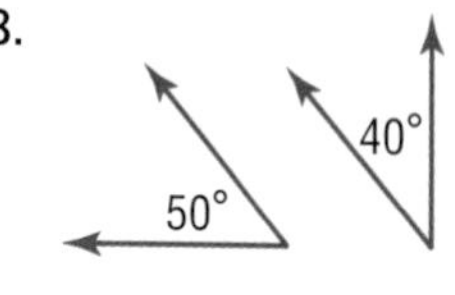

9.
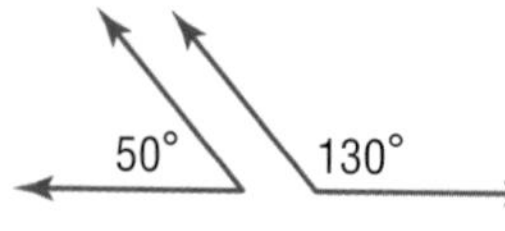

10.
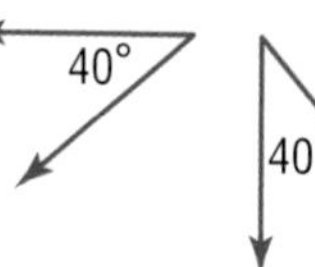

11.
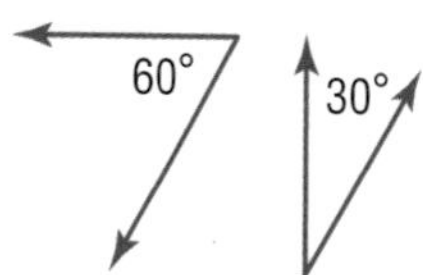

12.
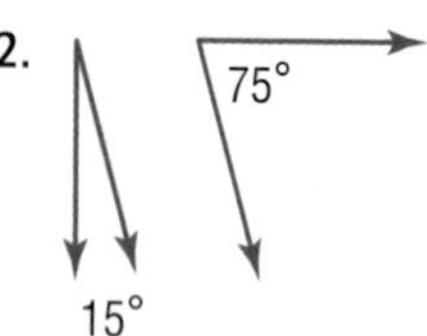

13.
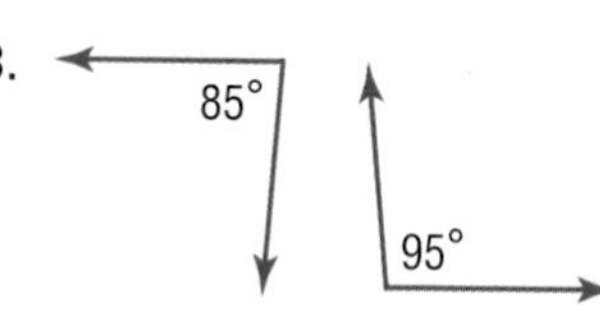

Find the value of x in each figure.

14.
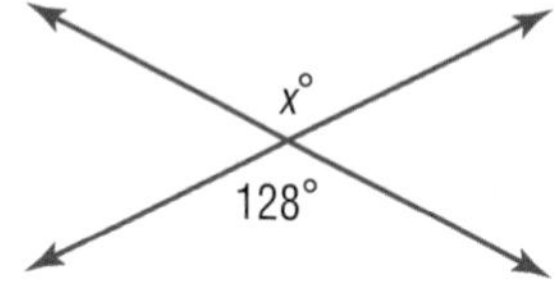

15.
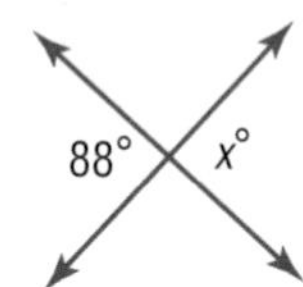

16.
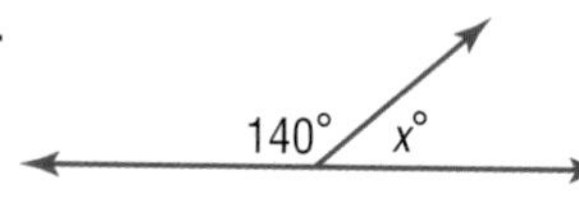

17.
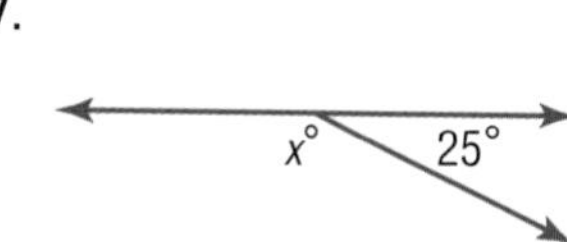

18.
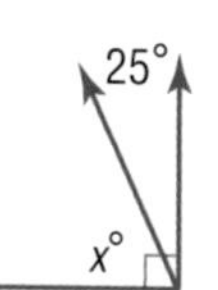

19.
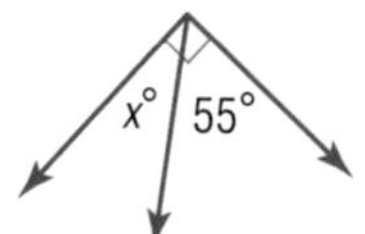

20. **HORSES** What is the value of x in the hurdle shown at the right?

Real-World Career...
How Does a Bridge Engineer Use Math? Bridge engineers use geometry when they design buildings and bridges.

21. **BRIDGES** A *truss bridge* is made of many short straight beams as shown in the diagram below. Create a problem that can be solved by referring to the angles labeled 1–4 in the diagram.

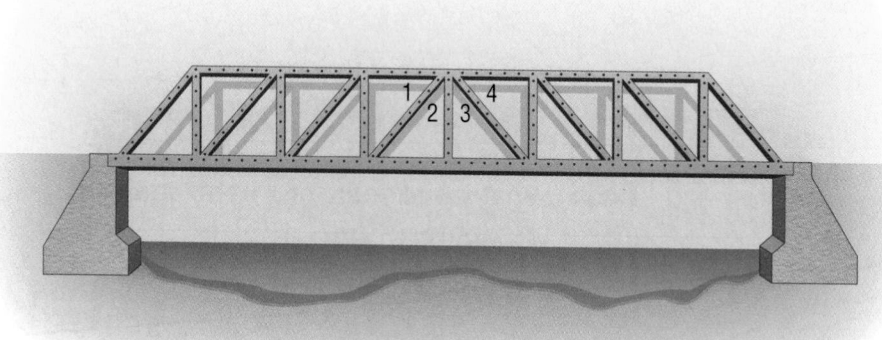

Math Online
For more information, go to glencoe.com.

22. Angles G and H are complementary. Find $m\angle H$ if $m\angle G = 40°$.

23. Angles J and K are supplementary. Find $m\angle J$ if $m\angle K = 65°$.

ARCHITECTURE For Exercises 24–28, refer to the photo of the John Hancock Center in Chicago, Illinois, at the right.

Classify each pair of angles.

24. $\angle 1$ and $\angle 2$

25. $\angle 2$ and $\angle 4$

26. $\angle 3$ and $\angle 4$

27. $\angle 1$ and $\angle 3$

28. If the $m\angle 3$ is 46°, what is the measure of $\angle 2$?

Determine whether each statement is *sometimes, always,* or *never* true. Explain your reasoning.

29. Vertical angles have the same angle measure.

30. Two right angles are complementary.

31. Two obtuse angles are supplementary.

32. Two vertical angles are complementary.

For Exercises 33–35, refer to the figure below.

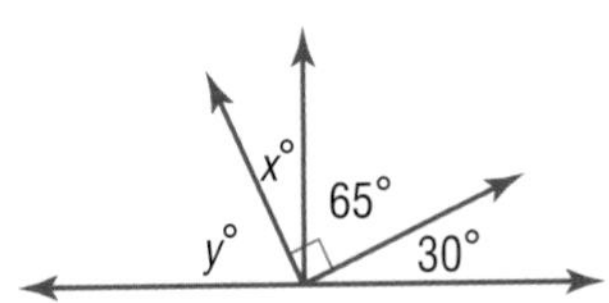

33. What is the value of x?

34. What is the value of y?

35. How many pairs of complementary angles are in the figure? Describe each pair.

H.O.T. Problems

36. **OPEN ENDED** Draw and label an angle in which you can find both a complementary angle and supplementary angle to the angle drawn. Classify the type of angle that must be drawn. Explain your reasoning.

37. **REASONING Answer each of the following questions.**
    a. What kind of angle is the supplement of an acute angle?
    b. What kind of angle is the supplement of a right angle?
    c. Can two acute angles be supplementary? Justify your answer.

38. **CHALLENGE** Refer to the figure at the right. If $m\angle 1 = m\angle 2$ and $m\angle 3 = m\angle 4$, what can you conclude about the sum of $m\angle 1$ and $m\angle 3$? Justify your answer.

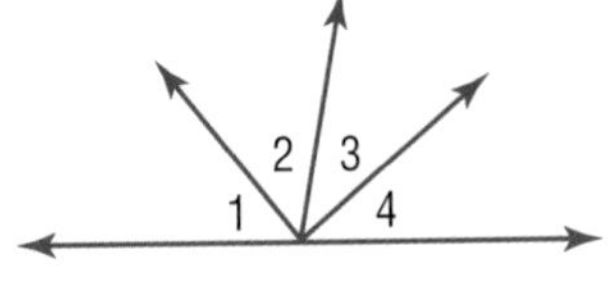

39. **WRITING IN MATH** Two angles are supplementary to the same angle. What is true about the measures of the two angles? Explain your reasoning.

TEST PRACTICE

40. If $\angle A$ and $\angle B$ are complementary and the measure of $\angle A$ is 60°, what is the measure of $\angle B$?

A 30°

B 60°

C 90°

D 120°

41. **SHORT RESPONSE** What is the value of x in the figure below?

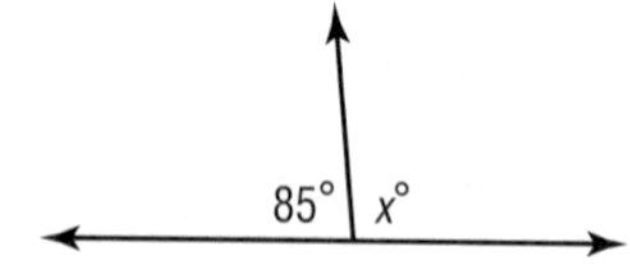

42. Which two angle pairs are NOT supplementary?

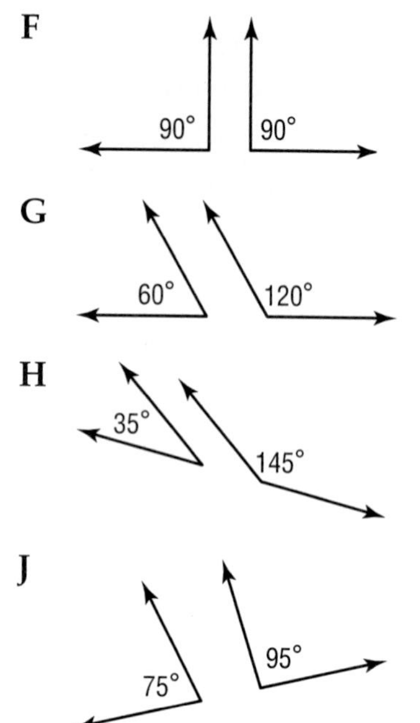

Spiral Review

Use a protractor and a straightedge to draw angles having the following measurements. (Lesson 9-2)

43. 75°

44. 25°

45. 110°

Use a protractor to find the measure of each angle. Then classify each angle as *acute, obtuse, right,* or *straight.* (Lesson 9-1)

46. 

47. 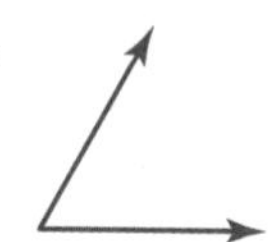

48. 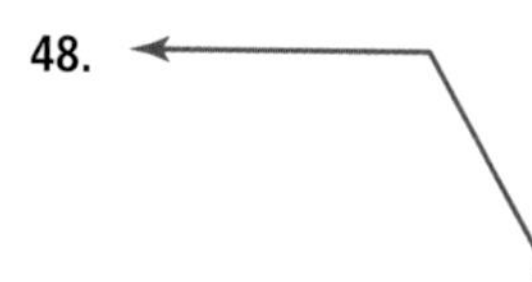

BICYCLING Use the following information for Exercises 49 and 50. (Lesson 6-7)

Devon rides his bike at an average rate of 12 miles per hour.

49. Write an equation to find d, the distance Devon bikes in h hours.

50. How many miles can Devon bike in 3 hours?

ALGEBRA Evaluate each expression. (Lesson 5-4)

51. $m + n$ if $m = \frac{11}{16}$ and $n = \frac{1}{4}$

52. $c - d$ if $c = \frac{7}{8}$ and $d = \frac{3}{5}$

GET READY for the Next Lesson

PREREQUISITE SKILL Find the value of each expression. (Lesson 1-4)

53. $180 - (45 + 60)$

54. $180 - (70 + 70)$

55. $180 - (37 + 83)$

Explore 9-4

Geometry Lab
Angles in Triangles

MAIN IDEA

Explore the relationship among the angles of a triangle.

Math Online

glencoe.com

- Concepts in Motion

Triangle means *three angles*. In this lab, you will explore how the three angles of a triangle are related.

ACTIVITY

STEP 1 Draw a triangle similar to the one shown below on notebook or construction paper.

STEP 2 Label the corners 1, 2, and 3. Then tear each corner off.

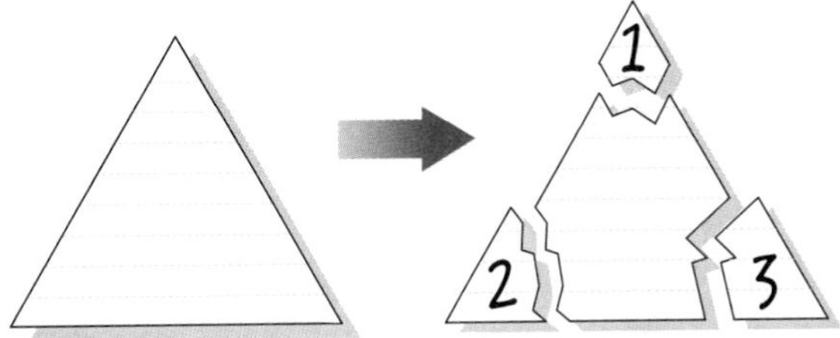

STEP 3 Rearrange the torn pieces so that the corners all meet at one point as shown.

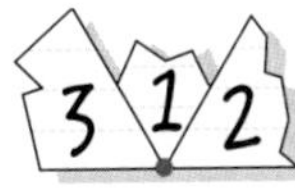

STEP 4 Repeat steps 1 and 2 with two differently shaped triangles.

ANALYZE THE RESULTS

1. What does each torn corner represent?
2. The point where these three corners meet is the vertex of another angle as shown. Classify this angle as right, acute, obtuse, or straight. Explain.

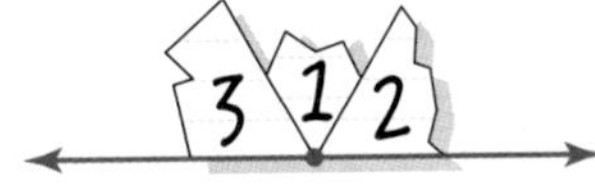

3. What is the measure of this angle?
4. **MAKE A CONJECTURE** What is the sum of the measures of angles 1, 2, and 3 for each of your triangles? Verify your conjecture by measuring each angle using a protractor. Then find the sum of these measures for each triangle.
5. **MAKE A CONJECTURE** What is the sum of the measures of the angles of any triangle?

9-4 Triangles

MAIN IDEA

Classify triangles and find missing angle measures in triangles.

New Vocabulary

acute triangle
right triangle
obtuse triangle
line segment
congruent segments
scalene triangle
isosceles triangle
equilateral triangle

Math Online

glencoe.com

- Extra Examples
- Personal Tutor
- Self-Check Quiz

MINI Lab

STEP 1 Draw the triangle shown at the right on dot paper. Then cut it out.

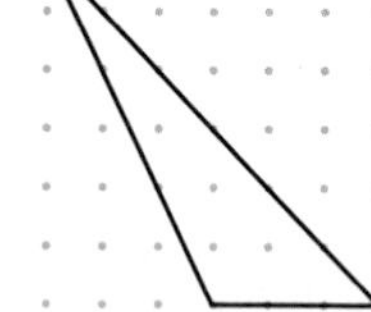

STEP 2 Measure each angle of the triangle and label each angle with its measure.

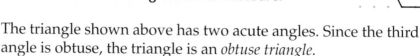

The triangle shown above has two acute angles. Since the third angle is obtuse, the triangle is an *obtuse triangle.*

1. Repeat the activity with nine other triangles.
2. Sort your triangles into three groups based on the third angle measures. Name the groups *acute, right,* and *obtuse.*

All triangles have at least two acute angles. As you discovered in the Mini Lab above, a triangle can be classified according to the angle measure of its third angle.

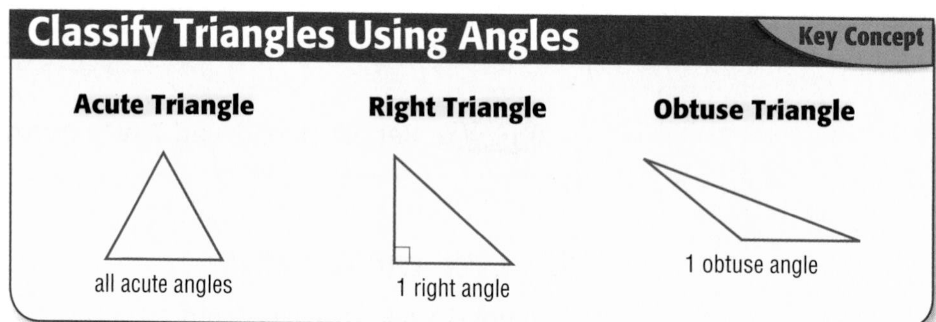

Classify Triangles Using Angles
Key Concept
Acute Triangle
all acute angles
Right Triangle
1 right angle
Obtuse Triangle
1 obtuse angle

EXAMPLES Classify a Triangle by Its Angles

Classify each triangle as *acute, right,* or *obtuse.*

1

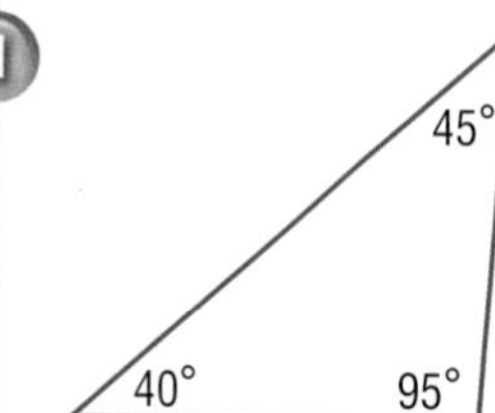

The 95° angle is obtuse.
So, the triangle is an obtuse triangle.

2

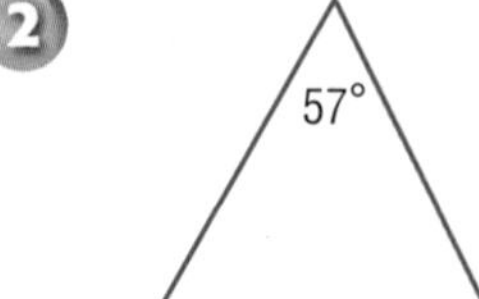

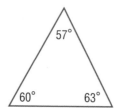

All the angles are acute.
So, the triangle is an acute triangle.

Classify each triangle as *acute, right,* or *obtuse*.

a. 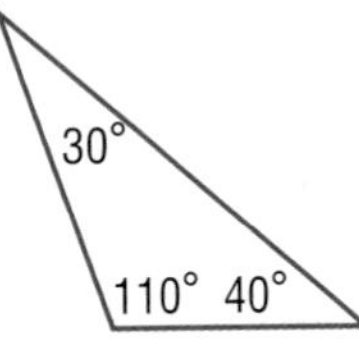

b. 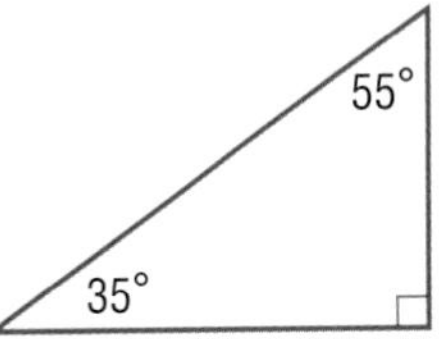

In the Lab on page 485, you discovered the following relationship.

Sum of Angle Measures in a Triangle

Key Concept

Words The sum of the measures of the angles in a triangle is 180°.

Model 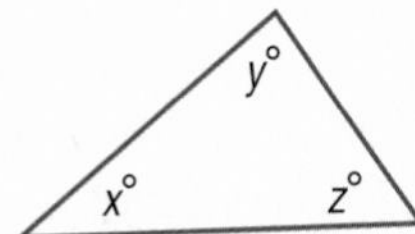

Symbols $x + y + z = 180°$

You can find a missing angle measure by using the fact that the sum of the measures of the angles is 180°.

Real-World EXAMPLE Find Angle Measures

3 FLAGS Find the value of x in the Antigua and Barbuda flag.

The three angles marked are the angles of a triangle. Since the sum of the angle measures in a triangle is 180°, $x + 55 + 90 = 180$. Use mental math to solve the equation.

$x + 55 + 90 = 180$ Write the equation.

$x + 145 = 180$ Add 55 and 90.
THINK What measure added to 145 equals 180?

$35 + 145 = 180$ You know that 35 + 145 = 180.

So, the value of x is 35.

Study Tip

Alternative Method If you know two of the angle measures in a triangle, you can find the third angle measure by subtracting the two known measures from 180°. The value of x is $180 - 55 - 90$ or 35.

CHECK Your Progress

Find the value of x in each triangle below.

c. 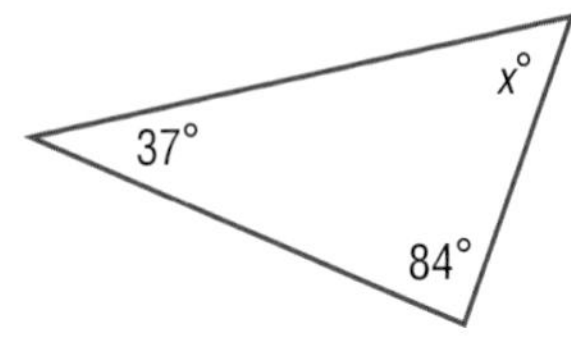

d. 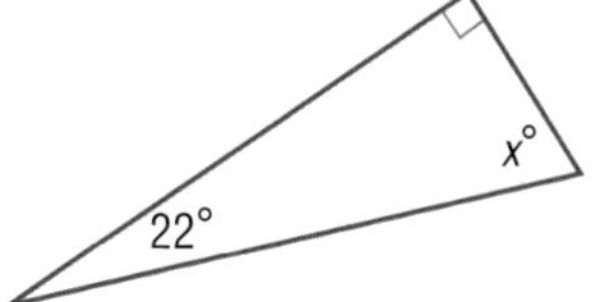

You can also classify triangles by their sides. Each side of a triangle is a **line segment**, or a straight path between two points. Line segments that have the same length are called **congruent segments**. On a figure, congruent sides are indicated by the tick marks on the sides of the figure.

Reading Math

Segments $\overline{AB}$ is read as *segment AB.* So the sides of the triangle below are $\overline{AB}$, $\overline{BC}$, and $\overline{CD}$.

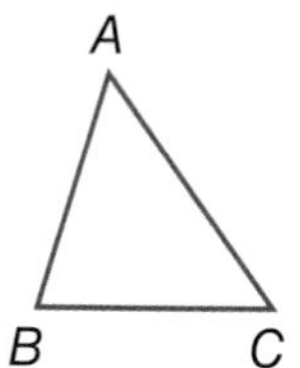

Classify Triangles Using Sides

Key Concept

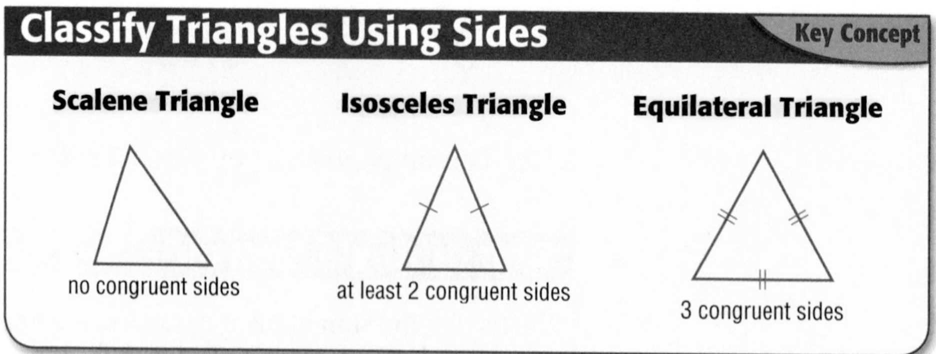

Scalene Triangle
no congruent sides
Isosceles Triangle
at least 2 congruent sides
Equilateral Triangle
3 congruent sides

Since an isosceles triangle is defined as having *at least* two congruent sides, all equilateral triangles are also isosceles.

EXAMPLES Classify a Triangle by Its Sides

Classify each triangle as *scalene, isosceles,* or *equilateral.*

4

Only two of the sides are congruent.
So, the triangle is an isosceles triangle.

5

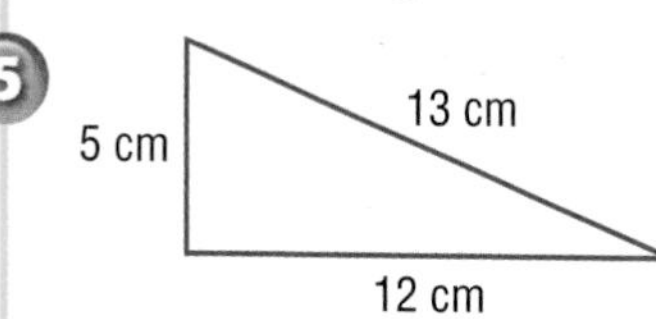

None of the sides are congruent.
So, the triangle is a scalene triangle.

Real-World Link In music, a "triangle" is generally an equilateral triangle but earlier forms of the instrument were often isosceles triangles. The side lengths of each modern triangle are usually 4 inches, 6 inches, or 9 inches.

Source: Heritage Ethnic Music

CHECK Your Progress

Classify each triangle as *scalene, isosceles,* or *equilateral.*

CHECK Your Understanding

Examples 1, 2 (pp. 486–487)

Classify each triangle as *acute, right,* or *obtuse.*

1. 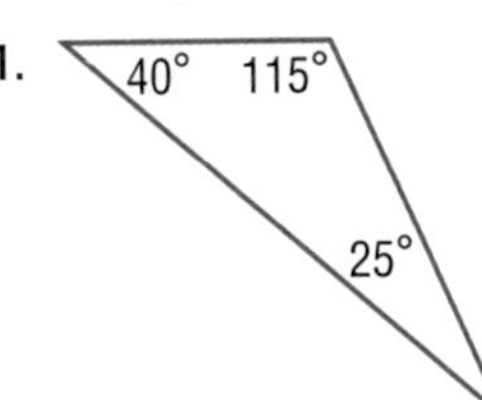

2.

Example 3 (p. 487)

Find the value of x in each triangle.

3. 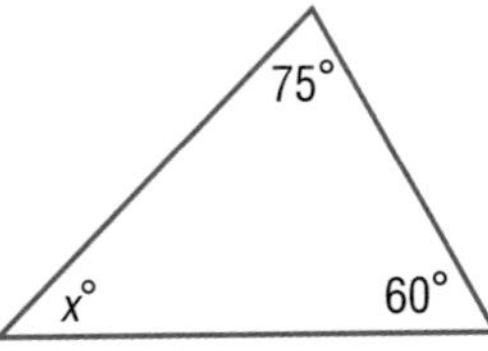

4. 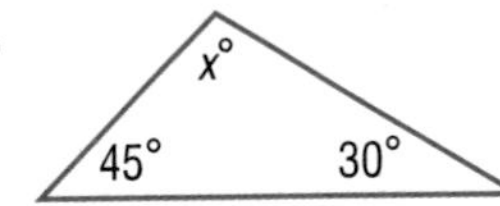

5. **SAILING** What is the value of x in the sail of the sailboat at the right?

Examples 4, 5 (p. 488)

Classify each triangle as *scalene, isosceles,* or *equilateral.*

6. 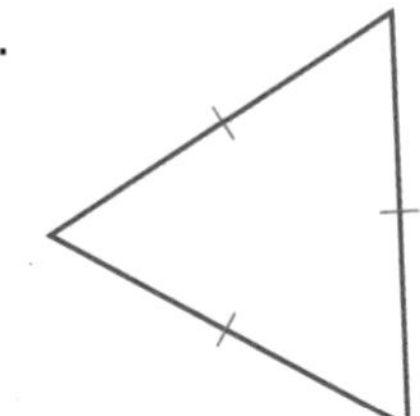

7. 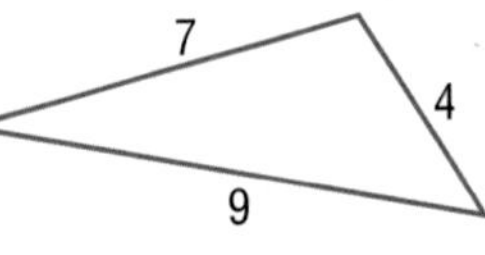

Practice and Problem Solving

HOMEWORK HELP	
For Exercises	See Examples
8–13	1, 2
14–21	3
22–26	4, 5

Classify each triangle drawn or having the given angle measures as *acute, right,* or *obtuse.*

8. 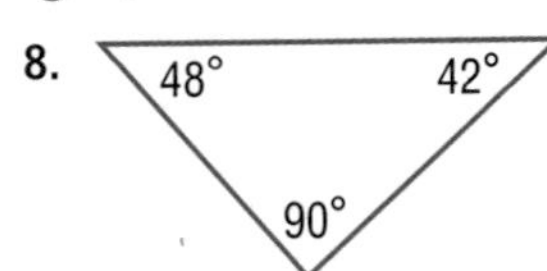

9. 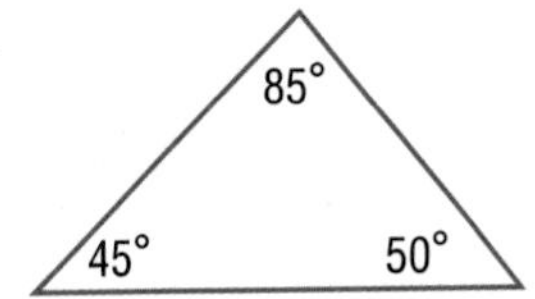

10. 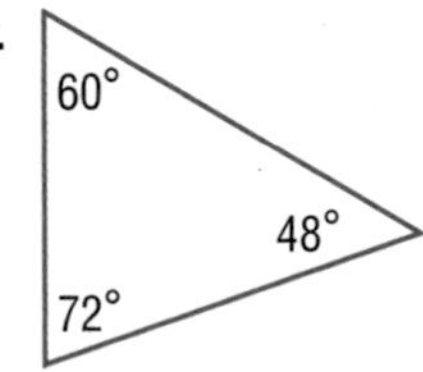

11. $100°, 45°, 35°$
12. $90°, 75°, 15°$
13. $114°, 33°, 33°$

Find the value of x in each triangle drawn or having the given angle measures.

14. 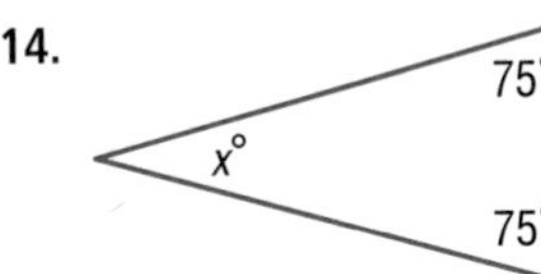

15. 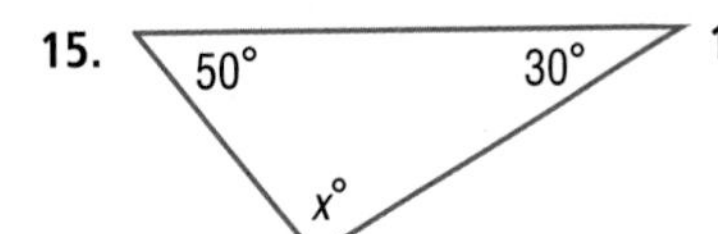

16.

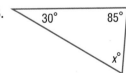

17. $70°, 60°, x°$
18. $x°, 60°, 25°$
19. $x°, 35°, 25°$

20. **SKYSCRAPERS** The diagram below shows the view of the top of Fountain Place in Dallas. What is the value of x?

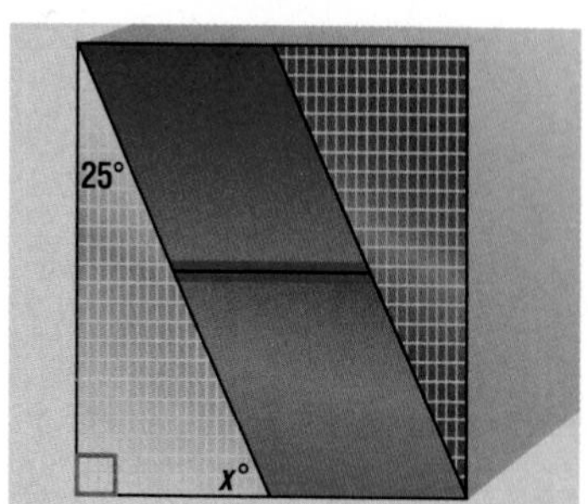

21. **PARKS** An A-frame picnic shelter at George Rogers Clark Historic Park in Ohio is shown below. What is the value of x?

Real-World Link Fountain Place is a 60-story skyscraper in downtown Dallas. It is a unique building with a 10-sided exterior structure.

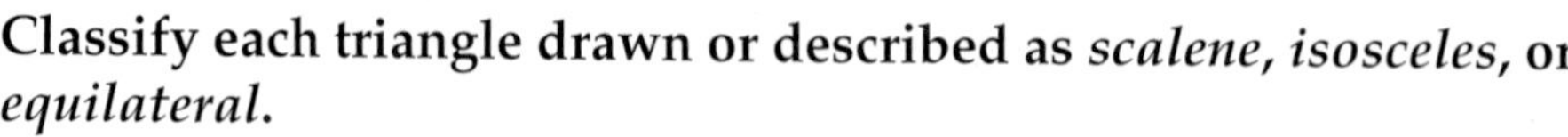

Classify each triangle drawn or described as *scalene, isosceles,* or *equilateral.*

22. 

23. 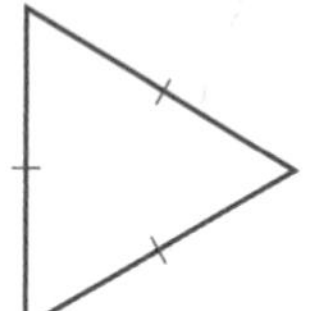

24. 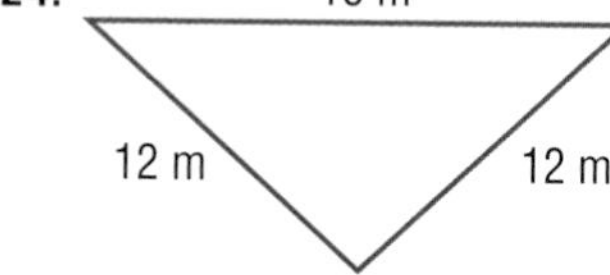

25. sides: 9 in., 11 in., 13 in.

26. sides: 5 cm, 6 cm, 5 cm

27. What is the measure of the third angle of a triangle if one angle measures 25° and the second angle measure 50°?

28. What is the measure of the third angle of a right triangle if one of the angles measures 30°?

29. What is the relationship between the two acute angles of a right triangle?

30. **ART** The boat below was created using *origami,* a paper-folding technique. Measure the angles of the triangles labeled 1, 2, and 3. Then classify the triangles as *acute, right,* or *obtuse.*

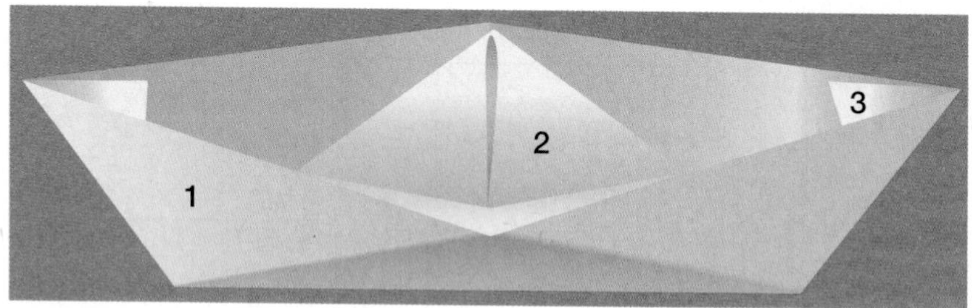

Measure the angles and sides of each triangle. Classify each triangle as *acute, right,* or *obtuse.* Then classify each triangle as *scalene, isosceles,* or *equilateral.*

31. 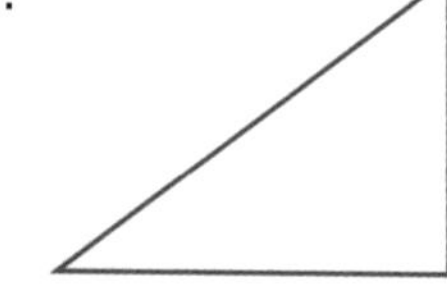

32. 

33. 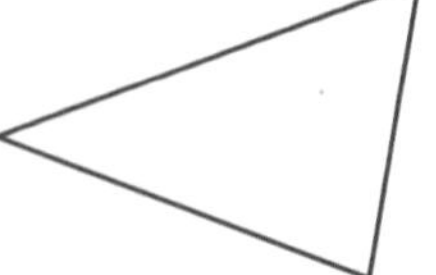

EXTRA PRACTICE
See pages 696, 714.

H.O.T. Problems

34. **OPEN ENDED** Draw an obtuse scalene triangle using a ruler and protractor. Label each side and angle with its measure.

35. **CHALLENGE** Apply what you know about the sum of the measures of the angles of a triangle to find the values of x and y in the figure below.

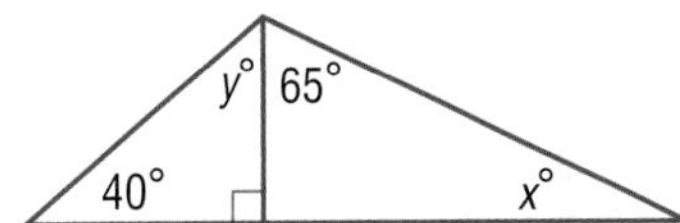

36. **WRITING IN MATH** Explain why a triangle must always have at least two acute angles. Include drawings in your explanation.

TEST PRACTICE

37. A triangle has angles measuring 25° and 60°. What is the measure of the triangle's third angle?

    A 15°

    B 85°

    C 95°

    D 115°

38. **SHORT RESPONSE** Triangle ABC is isosceles. If the measure of $\angle B$ is 48° and the measures of $\angle A$ and $\angle C$ are equal, what is the measure of $\angle A$ in degrees?

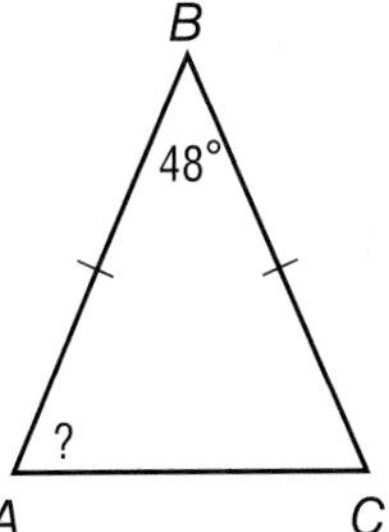

Spiral Review

39. Angles A and B are complementary. Find $m\angle A$ if $m\angle B = 35°$. (Lesson 9-3)

Use a protractor and a straightedge to draw angles having the following measurements. (Lesson 9-2)

40. 85°

41. 20°

42. 125°

43. **VEHICLES** The Thrust SuperSonic Car is the fastest vehicle in the world and can reach a speed of 763 miles per hour. It is 54 feet long. What is the length of this vehicle in yards? (Lesson 8-1)

44. **SCHOOL** Twenty-seven percent of the students in Ms. Malan's class are female. Identify the complement of selecting a female student at random from the class. Then find its probability. (Lesson 7-4)

GET READY for the Next Lesson

PREREQUISITE SKILL Draw an example of each figure.

45. rectangle

46. parallelogram

47. triangle

CHAPTER 9

Mid-Chapter Quiz

Lessons 9-1 through 9-4

Use a protractor to find the measure of each angle. Then classify each angle as *acute, obtuse, right,* or *straight.* (Lesson 9-1)

1. 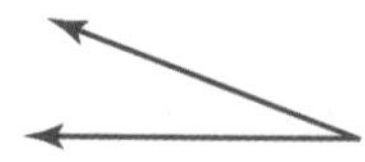

2. 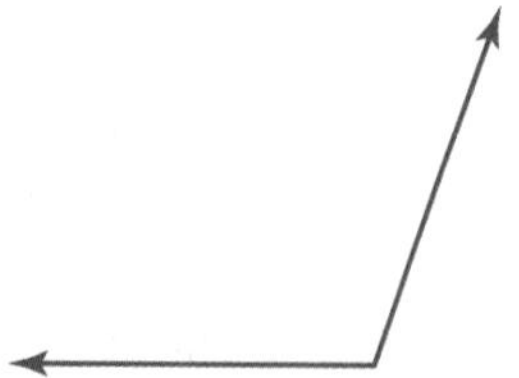

3. **MULTIPLE CHOICE** Find the measure of $\angle HJE$ to the nearest degree. (Lesson 9-1)

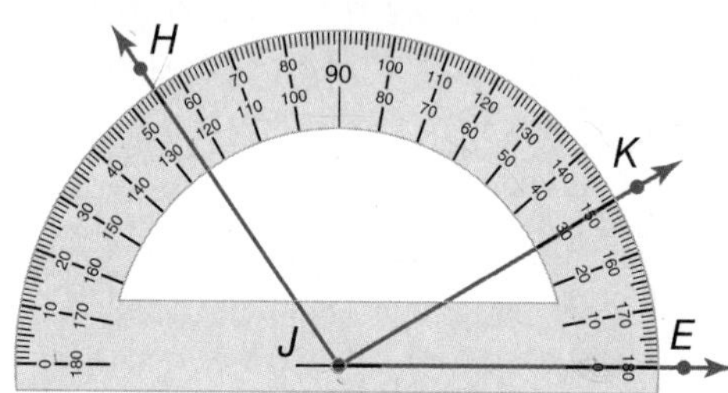

A 180°
B 125°
C 90°
D 30°

4. Estimate the measure of $\angle R$ on the leaf rake shown. (Lesson 9-2)

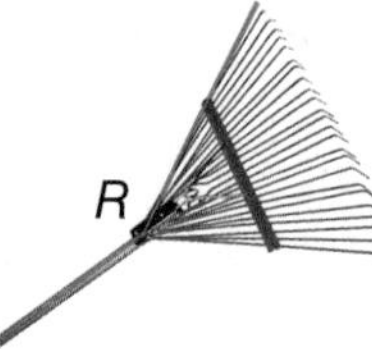

5. **MULTIPLE CHOICE** Which angle measures between 45° and 90°? (Lesson 9-2)

F 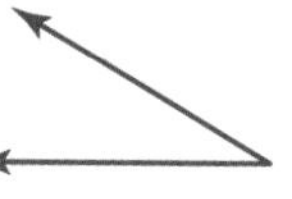

H 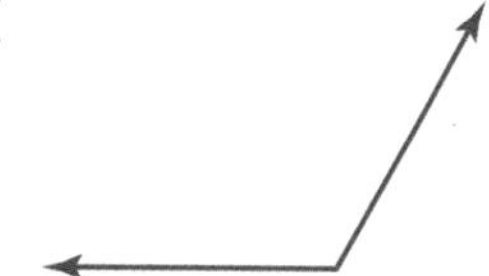

G 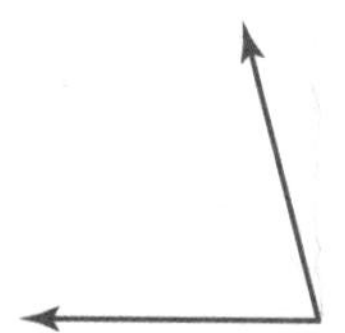

J 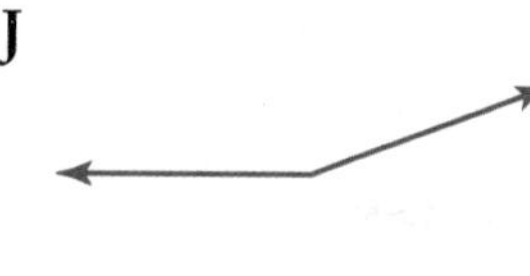

Use a protractor to draw angles having the following measurements. (Lesson 9-2)

6. 35°
7. 110°
8. 80°

Classify each pair of angles as *complementary, supplementary,* or *neither.* (Lesson 9-3)

9. 10.

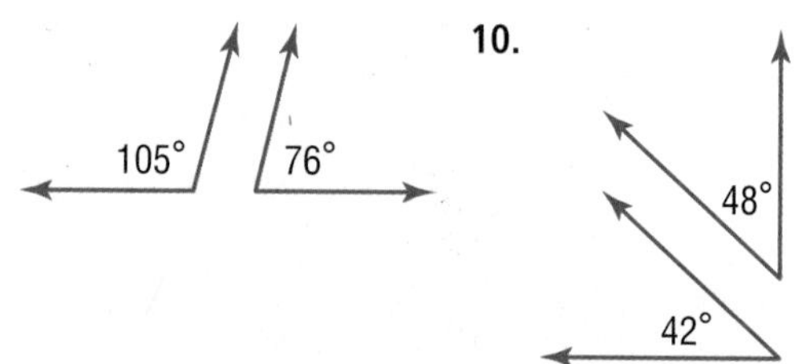

11. **MULTIPLE CHOICE** If $\angle P$ and $\angle Q$ are supplementary and the measure of $\angle P$ is 41°, what is the measure of $\angle Q$? (Lesson 9-3)

A 49°
B 59°
C 139°
D 149°

12. Find the value of x. (Lesson 9-3)

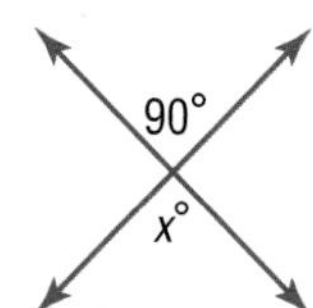

Classify each triangle as *acute, right,* or *obtuse.* (Lesson 9-4)

13.

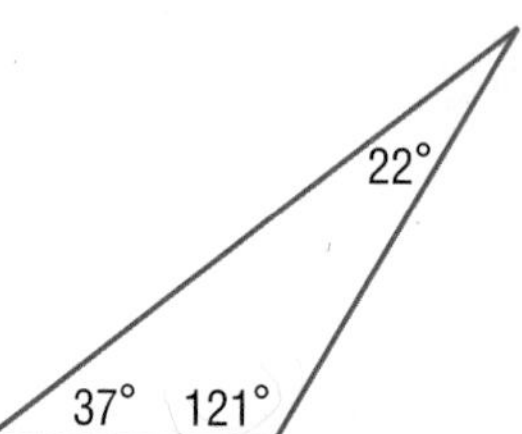

14.

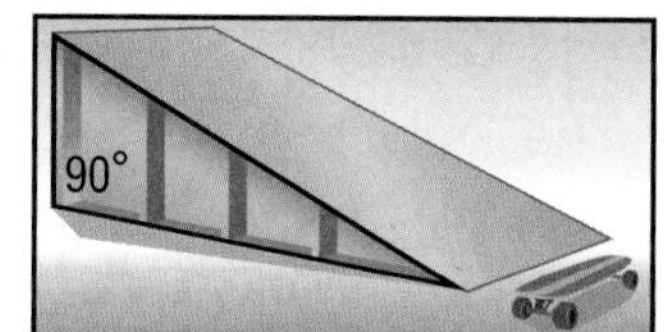

15. Find the value of x. (Lesson 9-4)

Explore 9-5

Geometry Lab
Angles in Quadrilaterals

MAIN IDEA

Explore the relationship among the angles of different quadrilaterals.

Quadrilateral means *four sides*. Four-sided figures also have four angles. In this lab, you will explore how the angles of different quadrilaterals are related.

ACTIVITY

STEP 1 Draw the quadrilaterals shown on grid paper.

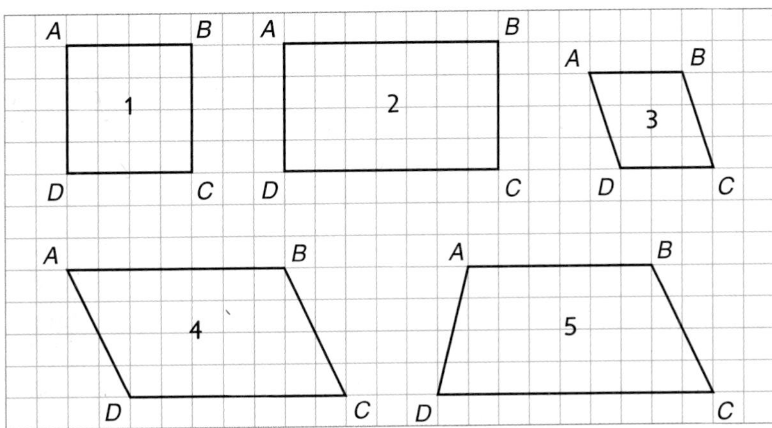

STEP 2 Use a protractor to measure the angles of each figure. Record your results in a table like the one shown.

Quadrilateral	$m\angle A$	$m\angle B$	$m\angle C$	$m\angle D$	Sum of Angles
1					
2					
3					
4					
5					

Analyze the Results

1. Describe any patterns you see in the angle measurements of Quadrilateral 1 and Quadrilateral 2.
2. Describe any patterns you see in the angle measurements of Quadrilaterals 1–4.
3. **MAKE A CONJECTURE** Are any of the patterns found in Quadrilaterals 1–4 present in Quadrilateral 5? If not, make a conjecture as to what makes Quadrilateral 5 different from Quadrilaterals 1–4.

9-5 Quadrilaterals

MAIN IDEA

Classify quadrilaterals and find missing angle measures in quadrilaterals.

New Vocabulary

quadrilateral
rectangle
square
parallelogram
rhombus
trapezoid

Math Online

glencoe.com

- Extra Examples
- Personal Tutor
- Self-Check Quiz

MINI Lab

The figure below is a **quadrilateral**, since it has four sides and four angles.

STEP 1 Draw a quadrilateral.

STEP 2 Pick one vertex and draw the diagonal to the opposite vertex.

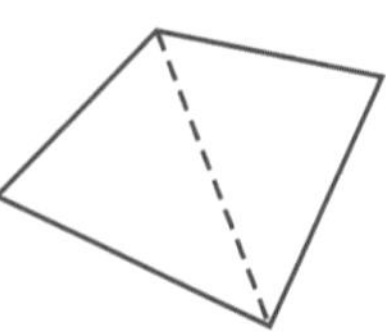

1. Name the shape of the figures formed when you drew the diagonal. How many figures were formed?
2. **MAKE A CONJECTURE** Use the relationship among the angle measures in a triangle to find the sum of the angle measures in a quadrilateral. Explain.
3. Find the measure of each angle of your quadrilateral. Compare the sum of these measures to the sum you found in Exercise 2.

The angles of a quadrilateral have a special relationship.

Angles of a Quadrilateral — Key Concept

Words The sum of the measures of the angles of a quadrilateral is 360°.

Model

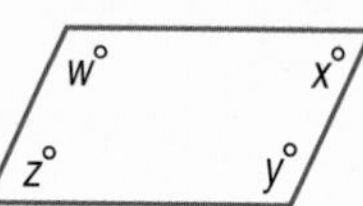

Symbols $w + x + y + z = 360$

EXAMPLE Find Angle Measures

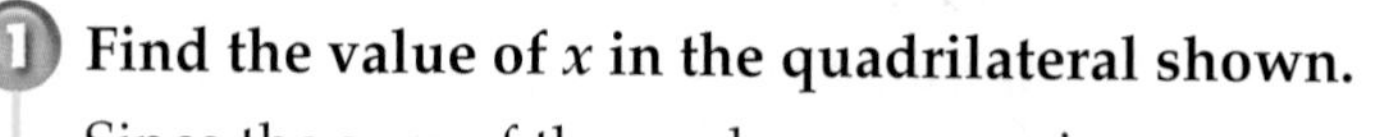

1 Find the value of x in the quadrilateral shown.

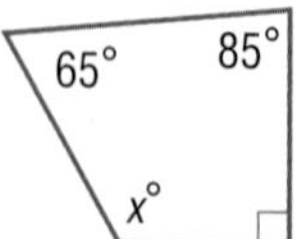

Since the sum of the angle measures in a quadrilateral is 360°, $x + 65 + 85 + 90 = 360$.

$x + 65 + 85 + 90 = 360$	Write the equation.
$x + 240 = 360$	Add 65, 85, and 90. **THINK** What measure added to 240 equals 360?
$120 + 240 = 360$	You know that $120 + 240 = 360$.

So, the value of x is 120.

Find the value of x in each quadrilateral.

a. 146° 116° 40° $x°$

b. 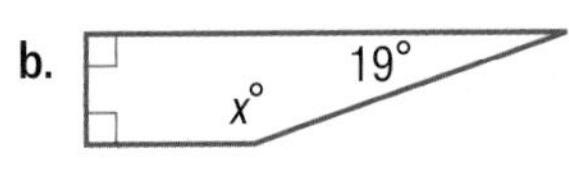

The table shows the characteristics of 5 special quadrilaterals.

Classifying Quadrilaterals

Key Concept

Quadrilateral	Figure	Characteristics
Rectangle		• Opposite sides congruent • All angles are right angles • Opposite sides parallel
Square		• All sides congruent • All angles are right angles • Opposite sides parallel
Parallelogram		• Opposite sides congruent • Opposite sides parallel • Opposite angles congruent
Rhombus		• All sides congruent • Opposite sides parallel • Opposite angles congruent
Trapezoid		• Exactly one pair of opposite sides parallel

Reading Math

Congruent Angles The red arcs show congruent angles.

Study Tip

Parallel Parallel means that if you extend the lengths of the sides, the opposite sides will never intersect or meet.
Perpendicular Lines that meet to form a right angle are perpendicular lines.

Real-World EXAMPLE Classify Quadrilaterals

2 QUILTS Classify the quadrilaterals labeled 1 and 2 in the quilt piece.

Figure 1 is a square. Figure 2 is a rhombus.

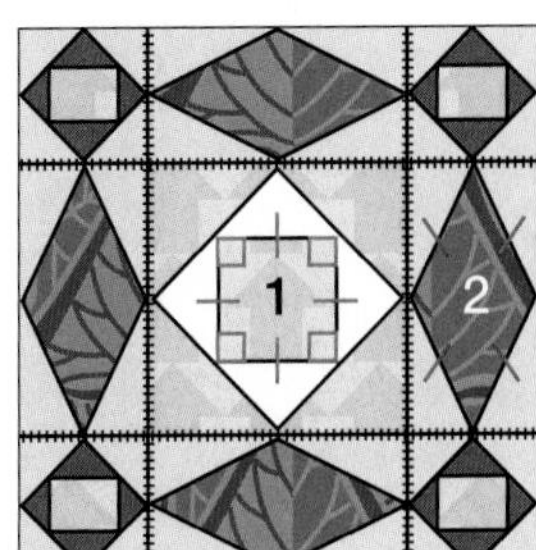

c. **LOGOS** Classify the quadrilaterals used in the logo below.

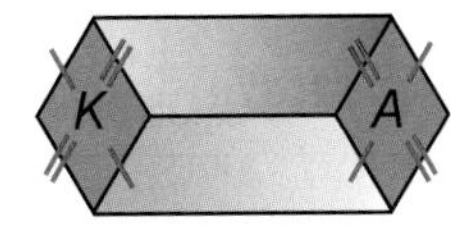

Test-Taking Tip

Check for Reasonableness After you solve the problem, look back at the quadrilateral to determine if your answer is a reasonable estimate for the measure of the angle.

TEST EXAMPLE

3 **SHORT RESPONSE** What is the value of x in the parallelogram at the right?

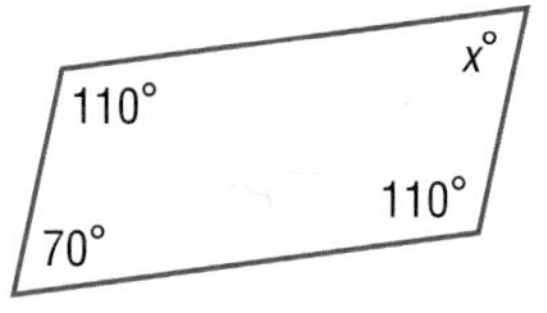

Read the Item

You need to find the value of x.

Solve the Item

You know that in a parallelogram, opposite angles are congruent. Since the angle opposite the missing measure has a measure of 70°, $x = 70$.

Check You know the angles in a quadrilateral add up to 360°. Since $70° + 110° + 70° + 110° = 360°$, the answer is reasonable. ✔

CHECK Your Progress

d. **SHORT RESPONSE** A rhombus is shown at the right. Find the measure in degrees of $\angle P$.

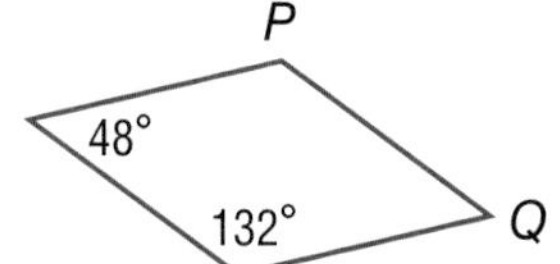

e. **SHORT RESPONSE** Refer to the rhombus above. Find the measure in degrees of $\angle Q$.

CHECK Your Understanding

Example 1 (pp. 494–495)

Find the value of x in each quadrilateral.

1. 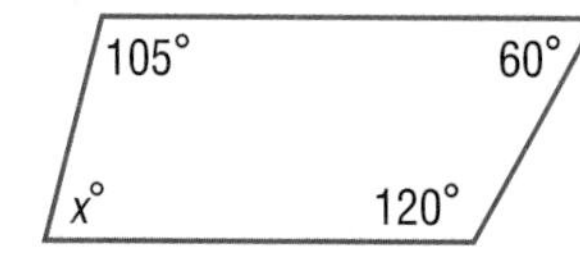

2. 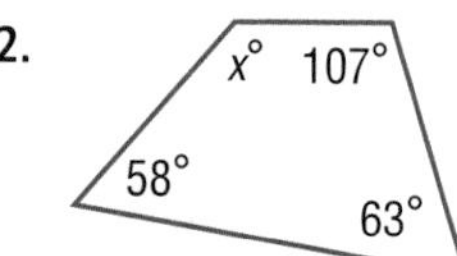

Example 2 (p. 495)

3. **SIGNS** Classify each quadrilateral.

Example 3 (p. 496)

4. **SHORT RESPONSE** Find the value of x in the parallelogram at the right.

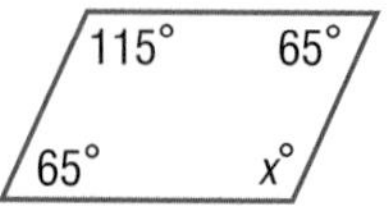

Practice and Problem Solving

HOMEWORK HELP	
For Exercises	See Examples
5–10	1
11–18	2
33, 34	3

Find the value of *x* in each quadrilateral.

5.
80° 120° x° 65°

6. 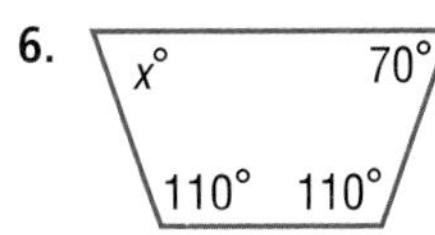

7. 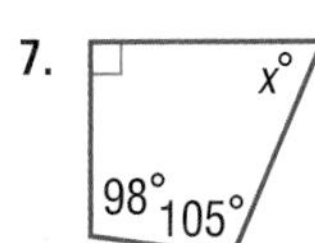

8. 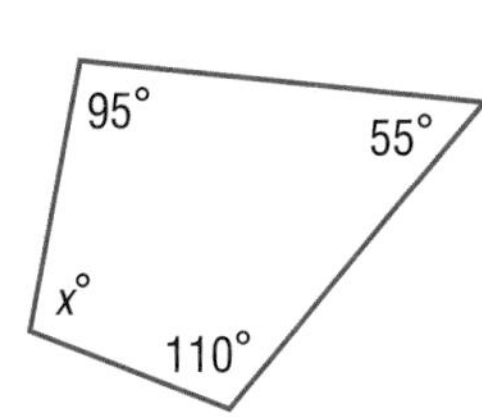

9. 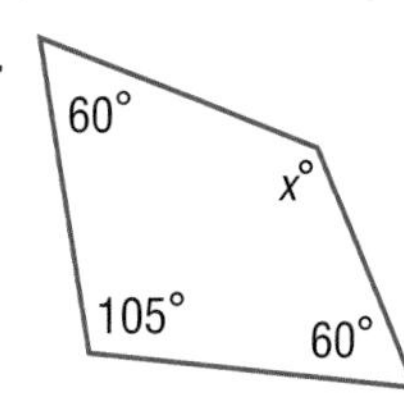

10. 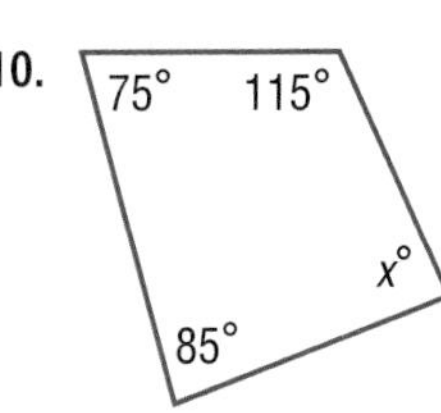

Classify each quadrilateral.

11. 

12. 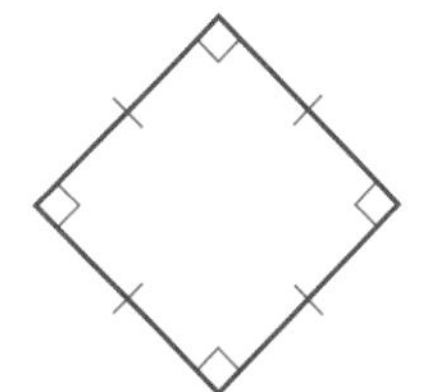

13. 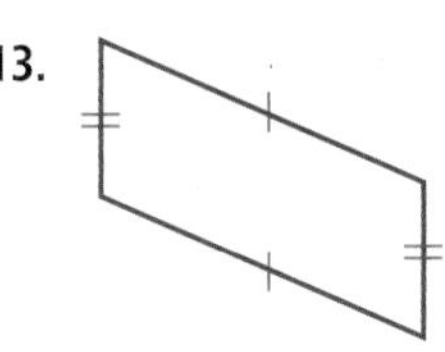

14. 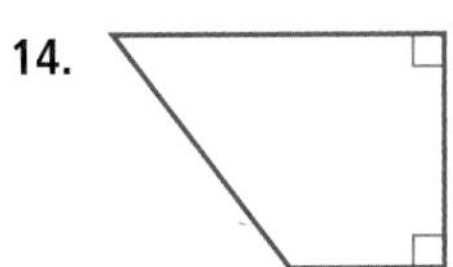

15. 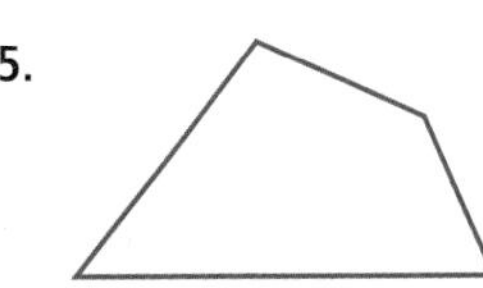

16. 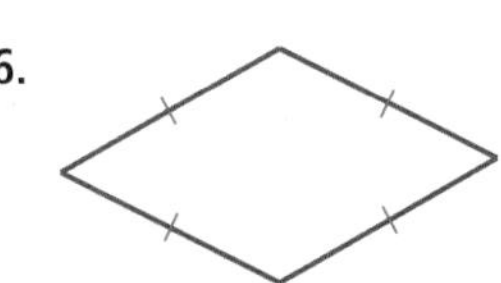

17. **FLAGS** Many aircraft display the shape of the American flag slightly distorted to indicate motion. Classify each quadrilateral.

18. **SIGNS** Classify each quadrilateral.

Real-World Link A tangram is an ancient Chinese puzzle consisting of 7 geometric shapes.

19. **TANGRAM** Triangles and quadrilaterals are polygons. A *polygon* is a simple closed figure formed by three or more sides. A *regular polygon* has all sides and all angles congruent. Refer to the seven tangram pieces shown at the left. Classify the polygons numbered 3 and 5. Then use a ruler and protractor to identify any regular polygons.

Find the value of *x* in each quadrilateral.

20. 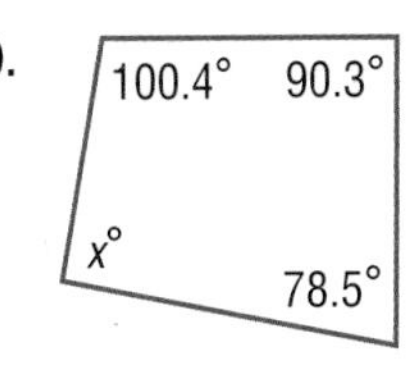

21. 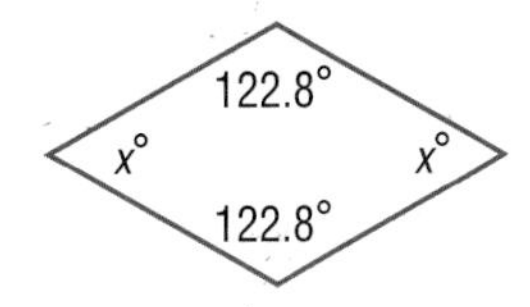

22.
2x° 2x° 2x° 2x°

23. **SORTING** Grace sorted a set of quadrilaterals into two categories according to a certain rule. The shapes that followed the rule were put in Set A, and the shapes that did not follow the rule were put in Set B.

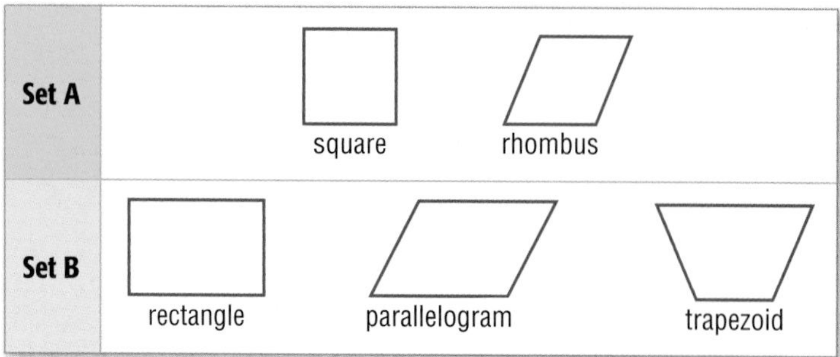

What rule did Grace use to sort the quadrilaterals?

24. **RESEARCH** Use the Internet or another source to look up the meaning of the term *isosceles trapezoid*. Explain how an isosceles trapezoid is related to an isosceles triangle. Then draw an example of an isosceles trapezoid.

H.O.T. Problems

25. **OPEN ENDED** Describe two different real-world items that are shaped as quadrilaterals. Then classify those quadrilaterals.

26. **NUMBER SENSE** Three of the angle measures of a quadrilateral are congruent. Without calculating, determine if the measure of the fourth angle in each of the following situations is greater than, less than, or equal to 90°. Explain your reasoning.
    - **a.** The three congruent angles each measure 89°.
    - **b.** The three congruent angles each measure 90°.
    - **c.** The three congruent angles each measure 91°.

CHALLENGE Determine whether each statement is *sometimes, always,* or *never* true. Explain your reasoning.

27. A rhombus is a square.
28. A quadrilateral is a parallelogram.
29. A rectangle is a square.
30. A square is a rectangle.

31. **CHALLENGE** Refer to Exercise 19 for the definitions of *polygon* and *regular polygon*. Draw two regular polygons. The first regular polygon should be a triangle. The second regular polygon should be a quadrilateral. Measure the angles of your regular polygons. What are the angle measures of a regular triangle and a regular quadrilateral? Then, classify each of your regular polygons, choosing the most specific terms.

32. **WRITING IN MATH** Make a diagram that shows the relationship between each of the following shapes: rectangle, parallelogram, square, rhombus, quadrilateral, and trapezoid. Then write a few sentences that explain your diagram.

TEST PRACTICE

33. The drawing below shows the shape of Hinto's patio.

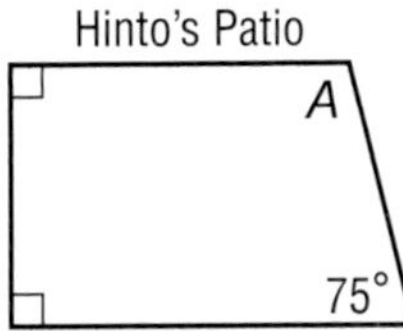

Find the measure of ∠*A*.

A 75°

B 105°

C 165°

D 195°

34. A parallelogram is shown below. Find the measure of ∠*M* to the nearest degree.

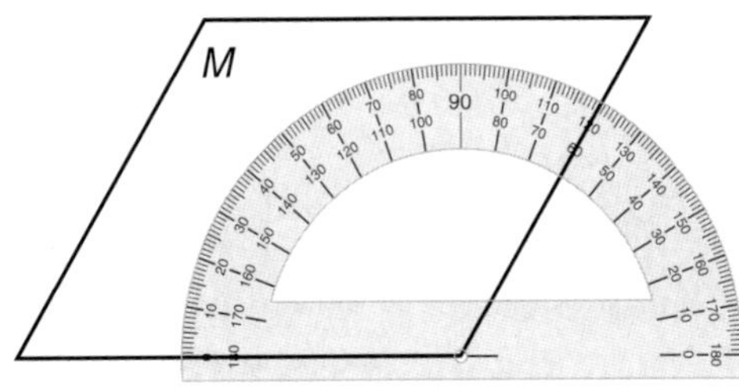

F 30° **H** 120°

G 60° **J** 150°

Spiral Review

Find the value of *x* in each triangle. (Lesson 9-4)

35. 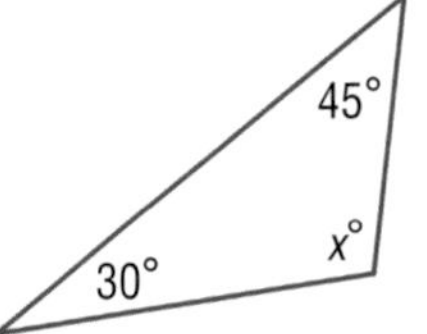

36. 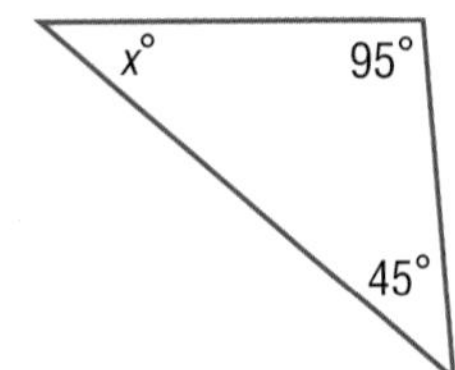

37. 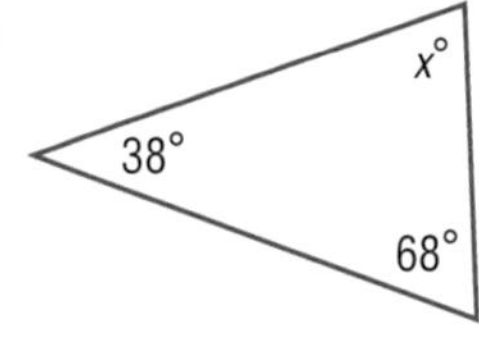

Classify each pair of angles as *complementary, supplementary,* or *neither.* (Lesson 9-3)

38. 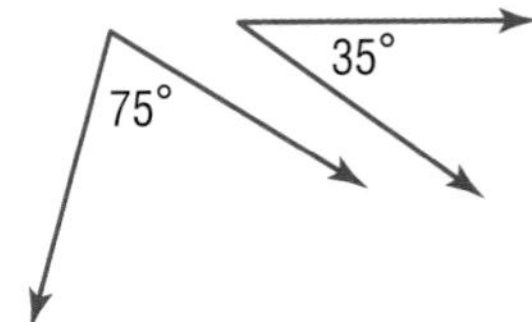

39. 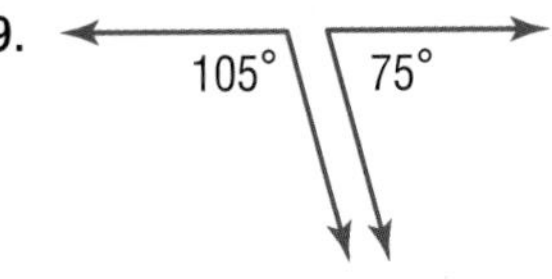

40. 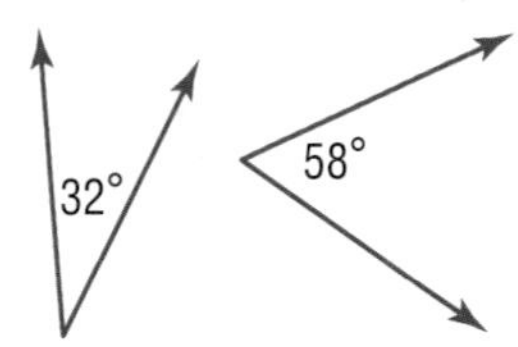

41. **AIRPLANES** An airplane is flying at 34,848 feet from the ground. How many miles from the ground is the airplane? (Lesson 8-1)

42. **BAKING** Mallory needs $4\frac{3}{4}$ cups of flour for a bread recipe. Write $4\frac{3}{4}$ as an improper fraction. (Lesson 4-3)

GET READY for the Next Lesson

PREREQUISITE SKILL Tell whether each pair of figures appear to have the same size and shape.

43. 

44. 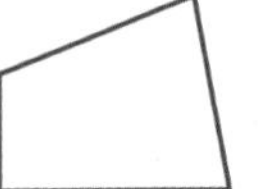

45. 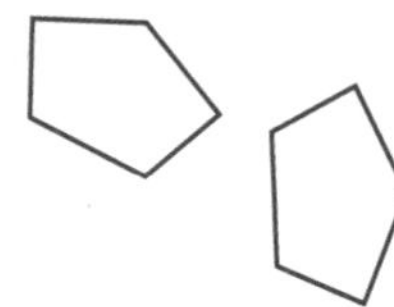

9-6 Problem-Solving Investigation

MAIN IDEA: Solve problems by drawing a diagram.

P.S.I. TEAM +

e-MAIL: DRAW A DIAGRAM

MANUEL: I want to tell people about a party I'm having, so I will tell Ross and Cara and have each of them tell two friends, and so on.

YOUR MISSION: Draw a diagram to find how many people will be invited to the party in three minutes. Two friends will tell another two friends each minute.

Understand	You know that Manuel tells Ross and Cara about the party, and then each friend tells two other friends each minute. You need to find the total number of people who will be invited to the party in three minutes.
Plan	Draw a diagram.
Solve	Manuel Ross Cara — 1 minute 1 2 1 2 — 2 minutes 1 2 1 2 1 2 1 2 — 3 minutes So, after 3 minutes, a total of 14 people will be invited to the party.
Check	Check the diagram to make sure that it meets all of the requirements. Since the diagram is correct, the answer is correct. ✔

Analyze The Strategy

1. Explain why you think Manuel drew a diagram to solve the problem.
2. WRITING IN MATH Write and solve a problem that can be solved by drawing a diagram.

Mixed Problem Solving

EXTRA PRACTICE
See pages 697, 714.

Solve Exercises 3–5. Use the *draw a diagram* strategy.

3. **DRIVING** The downtown section of a city is rectangular, 4 blocks by 3 blocks. How many ways are there to drive from one corner of downtown to the opposite corner, if you must make exactly two turns?

4. **FLOWERS** Keyana is planting flowers around the outside edge of her square garden. There will be 8 flowers on each side of the garden. What is the least number of flowers she needs to plant?

5. **BROWNIES** Colin is arranging brownies on a rectangular sheet 9 inches by 12 inches. How many brownies will fit if each is 1 inch square and needs to be placed 2 inches apart from another brownie?

Use any strategy to solve Exercises 6–14. Some strategies are shown below.

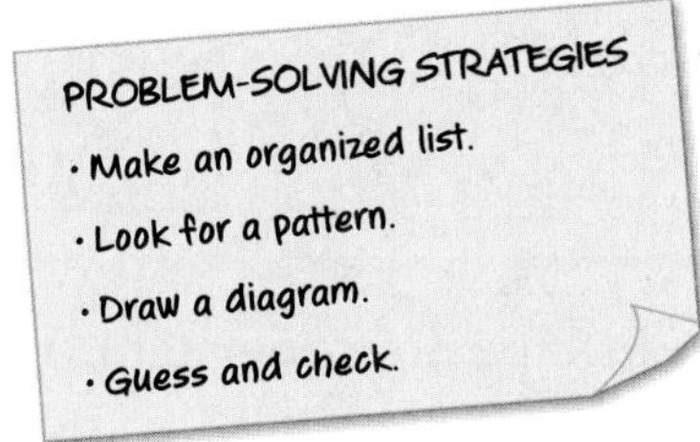
PROBLEM-SOLVING STRATEGIES
• Make an organized list.
• Look for a pattern.
• Draw a diagram.
• Guess and check.

6. **MONEY** Rodrigo paid for his bottle of juice with $1.25 in coins. If he used all 11 coins in his pocket, what is a possible combination of the coins that he used?

7. **PATTERNS** Draw the next two figures in the pattern.

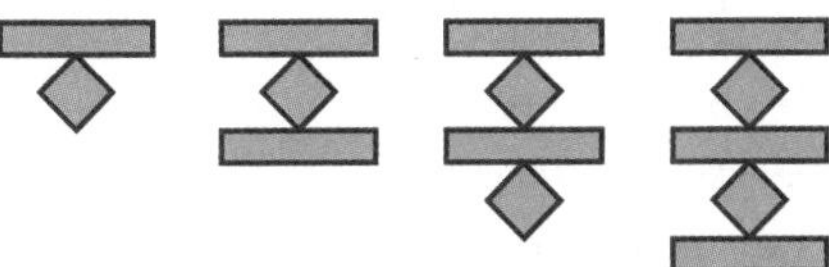

8. **PORTRAITS** The Lee family is having their family portrait taken. If Mr. and Mrs. Lee want to sit in the front row and have their 3 children standing behind them, in how many different ways can the Lee family pose for their portrait?

9. **AGE** William's uncle is three times as old as William. In 12 years, his uncle will be twice his age. How old is William now?

10. **GIFTS** At a family gift exchange, everyone gave everyone else exactly one gift. If there were a total of 30 gifts given, how many people were at the gift exchange?

11. **GEOMETRY** How many times greater is the length of the longest side in the triangle below than the length of the shortest side?

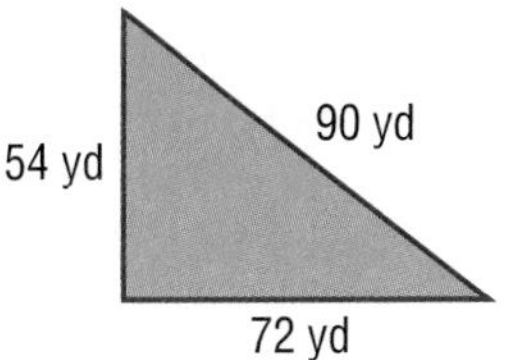
54 yd
90 yd
72 yd

12. **FRUIT** Use the table that shows the prices of different amounts of mixed fruit at the grocery store.

Pounds	Cost ($)
2	4.50
4	9.00
6	13.50
8	18.00

How much will 13 pounds of fruit cost?

13. **DANCES** For a school dance, there are 5 columns arranged in the shape of a pentagon. One large streamer is hung from each column to every other column. How many streamers are there in all?

14. **TRAVEL** Jeffrey drove a total of 285 miles to visit his sister. He drove 55 miles per hour for the first 165 miles and then 60 miles per hour for the rest of the trip. How many hours did it take him to complete the trip?

9-7 Similar and Congruent Figures

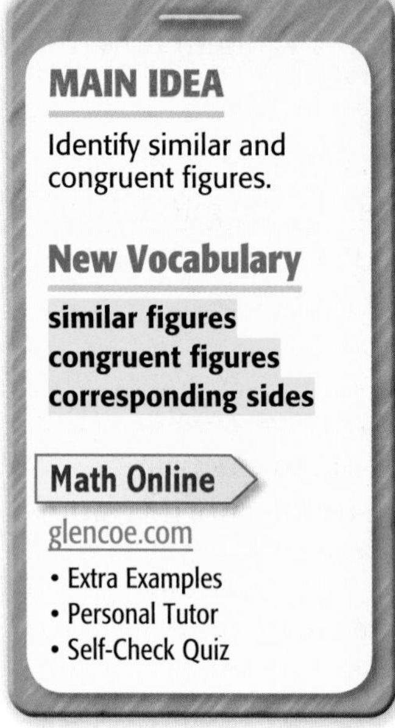

MAIN IDEA

Identify similar and congruent figures.

New Vocabulary

similar figures
congruent figures
corresponding sides

Math Online

glencoe.com

- Extra Examples
- Personal Tutor
- Self-Check Quiz

GET READY for the Lesson

FLAGS The flags of four U.S. states are displayed at the right.

1. How many different-sized stars are displayed among the four flags?
2. Compare the size and the shape of these stars.

Alaska

Ohio

Arizona

Tennessee

Figures that have the same shape but not necessarily the same size are called **similar figures**.

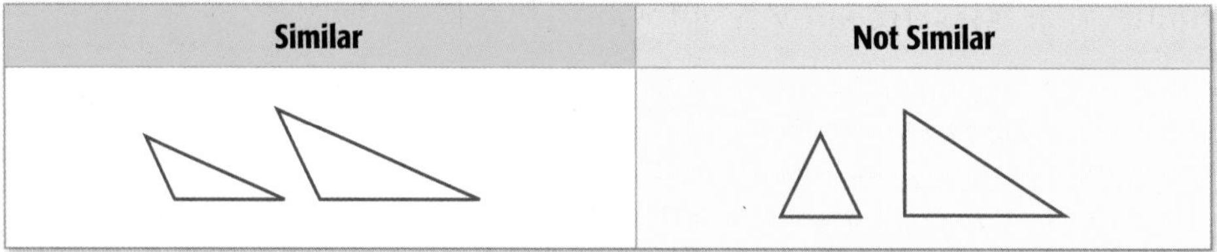

Figures that have the same size and shape are **congruent figures**.

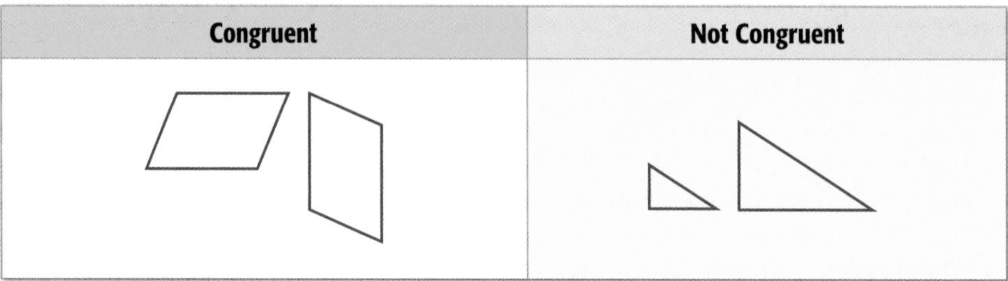

EXAMPLES Identify Similar and Congruent Figures

Tell whether each pair of figures is *congruent*, *similar*, or *neither*.

1

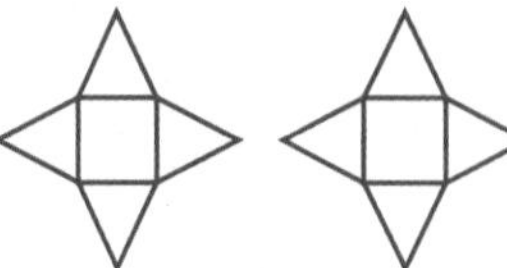

The figures have the same size and shape. They are congruent.

2

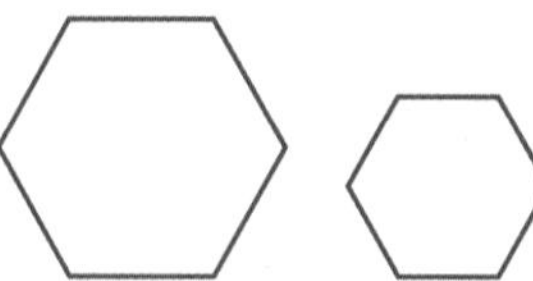

The figures have the same shape but not the same size. They are similar.

Vocabulary Link

Similar

Everyday Use nearly, but not exactly, the same or alike

Math Use figures that have the same shape but not necessarily the same size

Tell whether each pair of figures is *congruent, similar,* or *neither.*

a. 

b. 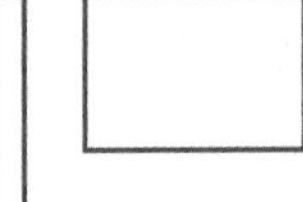

c. 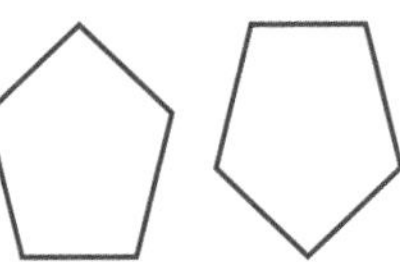

The sides of congruent or similar figures that match are called **corresponding sides**. In congruent figures, these sides are congruent. In similar figures, these matching sides are proportional.

The triangles below are similar with corresponding sides $\overline{FG}$ and $\overline{XY}$, $\overline{GH}$ and $\overline{YZ}$, and $\overline{HF}$ and $\overline{ZX}$.

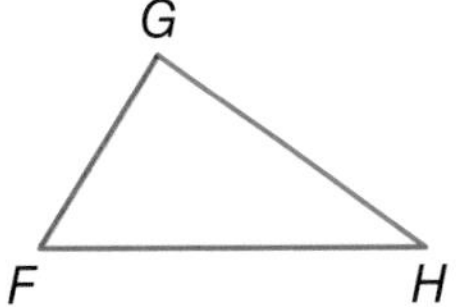

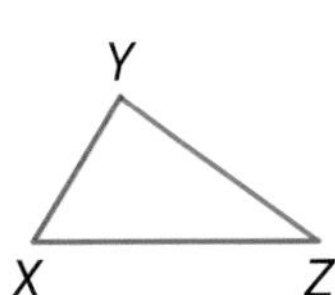

Real-World EXAMPLE Identify Corresponding Sides

Reading Math

Identifying Polygons Uppercase letters are used to name polygons. The larger photograph is referred to as rectangle *ABCD*. The smaller photograph is referred to as rectangle *PQRS*.

3 PHOTOGRAPHY **The photographs below are similar rectangles. What side of rectangle *ABCD* corresponds to $\overline{QR}$?**

Corresponding sides represent the same side of similar figures. So, $\overline{BC}$ corresponds to $\overline{QR}$.

d. **PHOTOGRAPHY** Determine which side corresponds to $\overline{DC}$ in the photographs in Example 3.

In a rectangle, all angles are right angles and therefore congruent. So, to determine if two rectangles are similar, you need to only see if corresponding sides are proportional. Recall from Lesson 6-3 that quantities are said to be *proportional* if they have a constant ratio.

EXAMPLE Identify Similar Figures

4 Which rectangle(s) below is similar to rectangle *ABCD*?

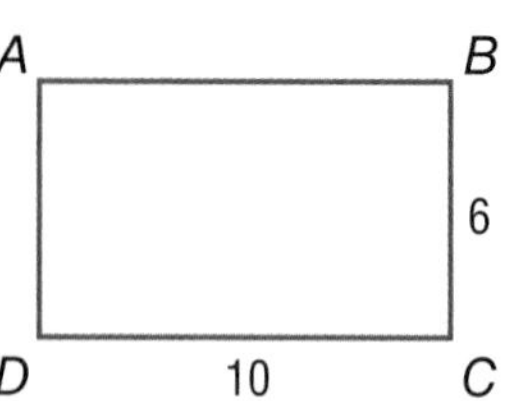

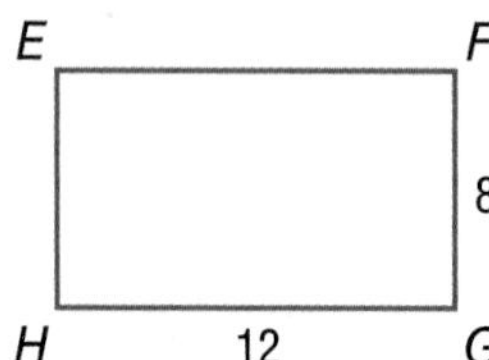

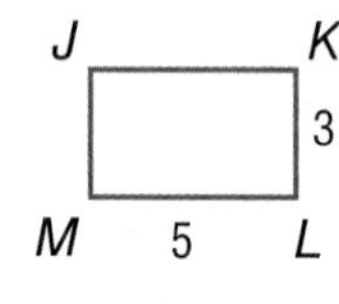

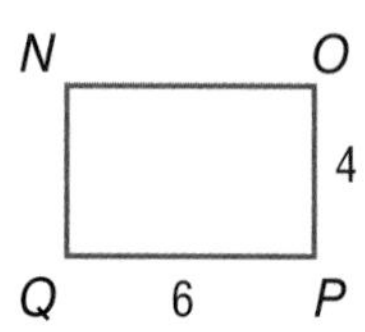

Examine the ratios of corresponding sides to see if they have a constant ratio.

Rectangle *EFGH*	Rectangle *JKLM*	Rectangle *NOPQ*
$\frac{BC}{FG} = \frac{6}{8}$ or $\frac{3}{4}$	$\frac{BC}{KL} = \frac{6}{3}$ or $\frac{2}{1}$	$\frac{BC}{OP} = \frac{6}{4}$ or $\frac{3}{2}$
$\frac{CD}{GH} = \frac{10}{12}$ or $\frac{5}{6}$	$\frac{CD}{LM} = \frac{10}{5}$ or $\frac{2}{1}$	$\frac{CD}{PQ} = \frac{10}{6}$ or $\frac{5}{3}$
Not similar	Similar	Not similar

So, rectangle *JKLM* is similar to rectangle *ABCD*.

Reading Math

Segments *BC* is read as *the measure of side BC.*

Study Tip

Right Triangles Two right triangles are similar if the ratios of two pairs of corresponding sides are proportional.

CHECK Your Progress

e. Which triangle(s) below is similar to triangle *DEF*? Assume corresponding angles are congruent.

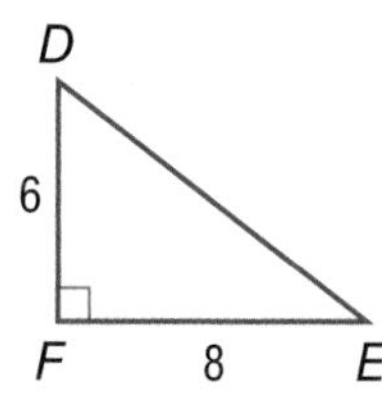

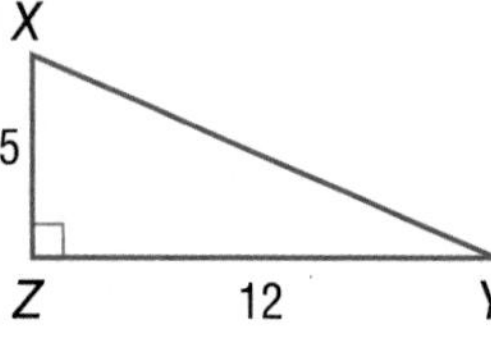

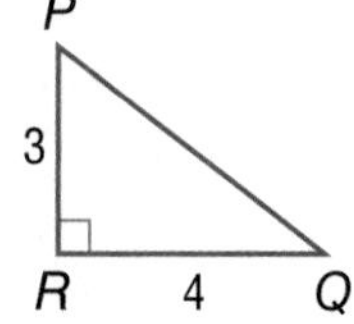

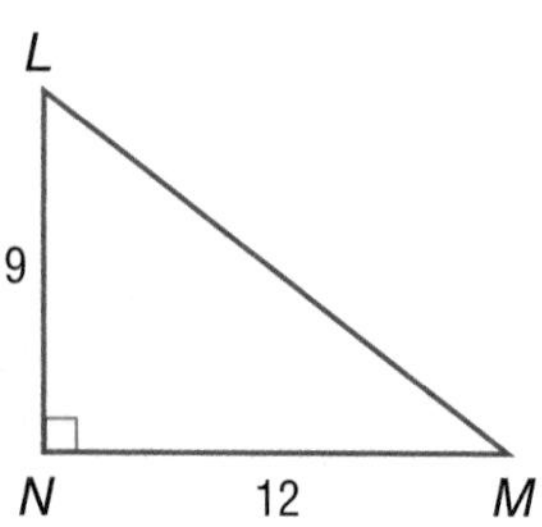

Examples 1, 2 (pp. 502–503)

Tell whether each pair of figures is *congruent, similar,* or *neither.*

1. 

2. 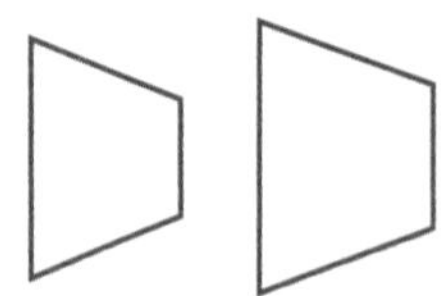

3. 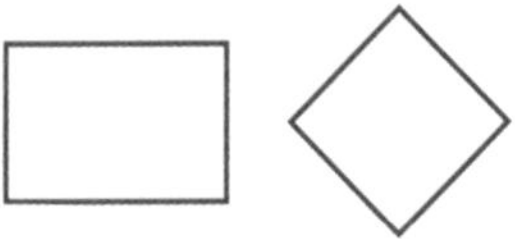

Example 3 (p. 503)

BILLIARDS For Exercises 4 and 5, refer to the similar billiards tables below.

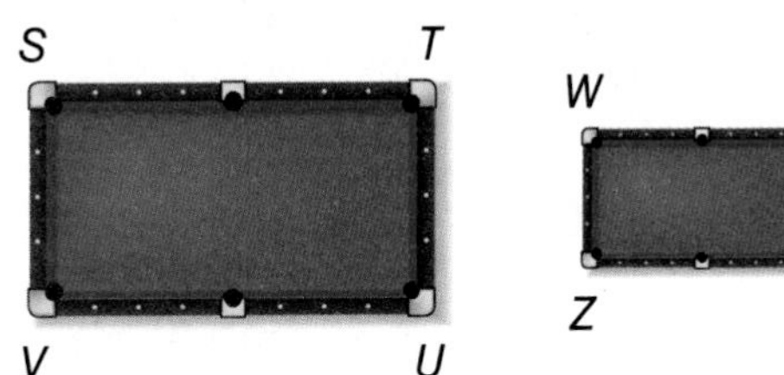

4. What side of rectangle *WXYZ* corresponds to $\overline{TU}$?
5. What side of rectangle *STUV* corresponds to $\overline{WX}$?

Example 4 (p. 504)

State whether each rectangle is similar to rectangle *WXYZ*.

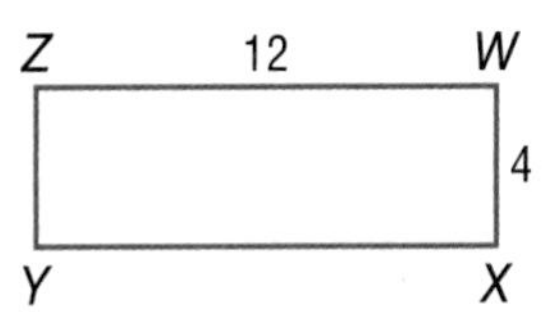

6. 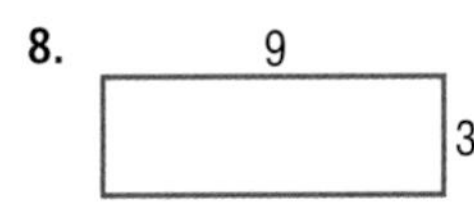

7.

8.

Practice and Problem Solving

HOMEWORK HELP	
For Exercises	See Examples
9–14	1, 2
15, 16	3
17–24	4

Tell whether each pair of figures is *congruent, similar,* or *neither.*

9. 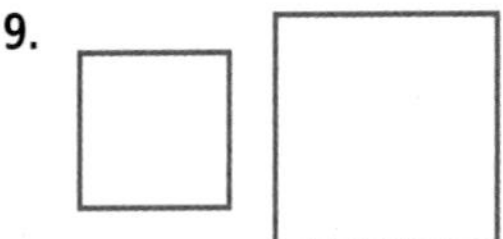

10. 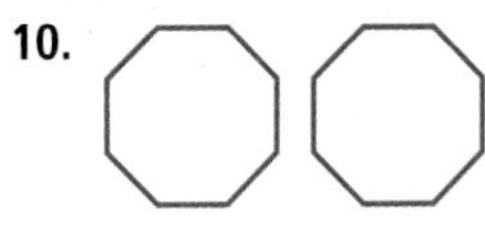

11. 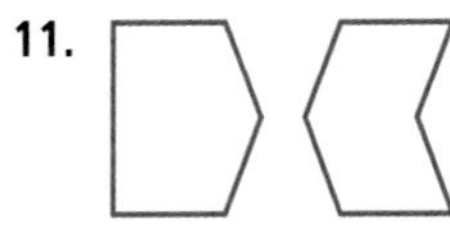

12.

13. 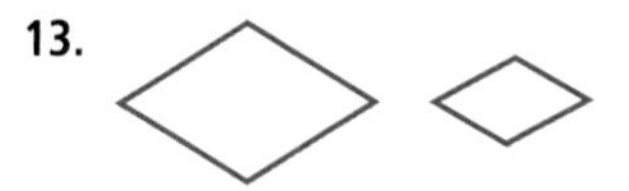

14. 

QUILTING For Exercises 15 and 16, refer to the quilt at the right.

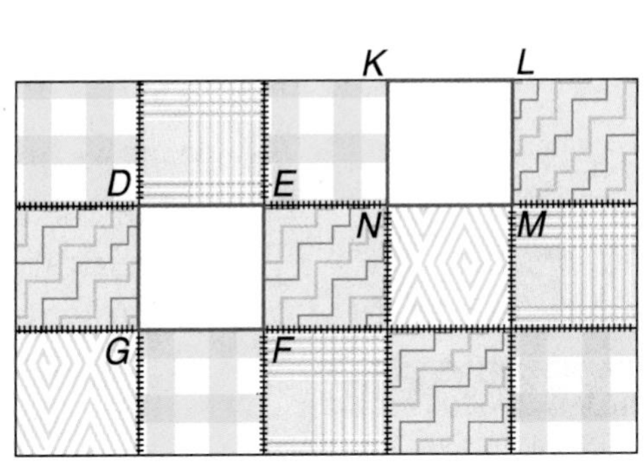

The quilt is made up of many congruent quadrilaterals.

15. What side of quadrilateral *DEFG* corresponds to $\overline{KL}$?

16. What side of quadrilateral *KLMN* corresponds to $\overline{EF}$?

TELESCOPE For Exercises 17 and 18, refer to the photo of the Hobby-Eberly Telescope Dome and the following information.

The triangles that form the outside of the telescope dome are similar triangles.

17. What side of triangle *LMN* corresponds to $\overline{AC}$?

18. What side of triangle *ABC* corresponds to $\overline{MN}$?

State whether each triangle is similar to triangle *ABC*.

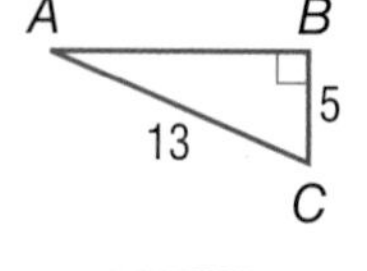

19. 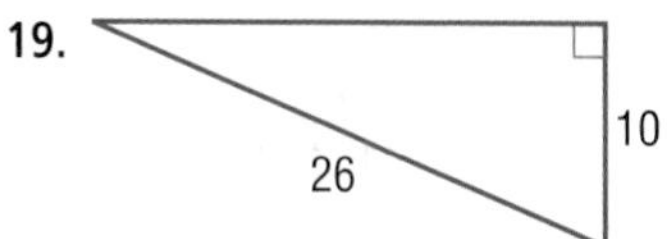

20. 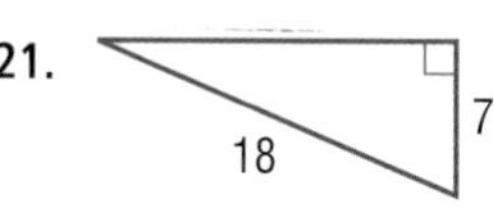

21.

State whether each rectangle is similar to rectangle *JKLM*.

22. 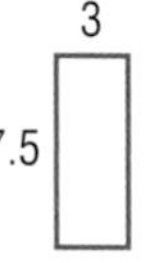

23. 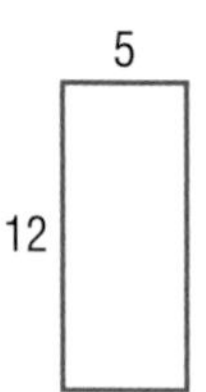

24. 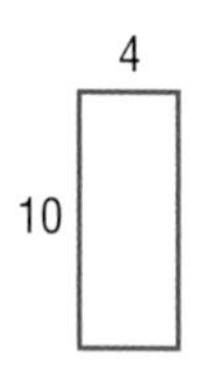

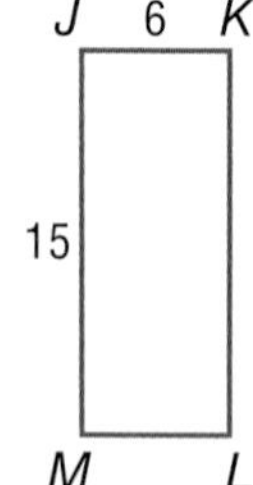

25. **STATE FAIR** A plot of land will be separated into ten congruent rectangles with rope as shown for the judging of cows. Find the total length of rope needed. Then find the total length of rope needed to section the same plot of land into five congruent rectangles.

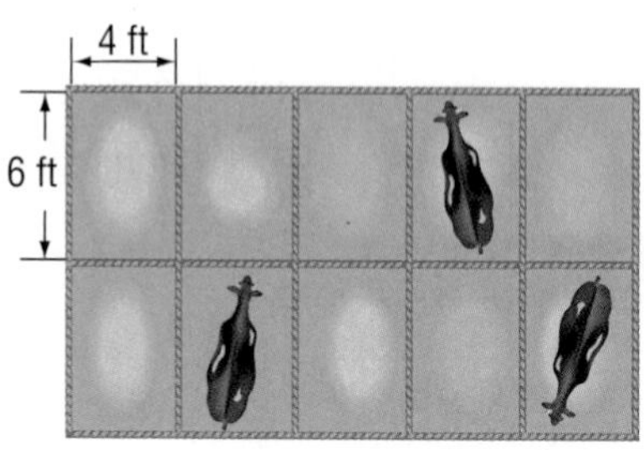

Each pair of figures is similar. Find the value of *x*.

26. 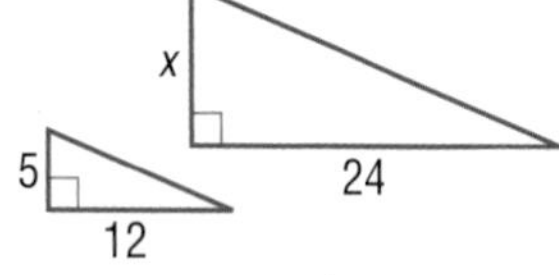

27. 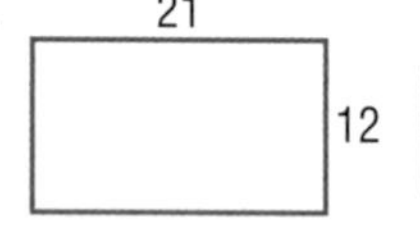

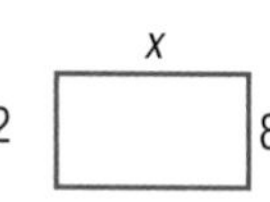

28. **JEWELRY** Flora made the earrings shown out of colored wire. The earrings are similar. Write a ratio that compares a side length of the larger earring to that of the smaller earring. Then, compare this to the ratio of the total length of wire needed to make the larger earring to that of the smaller earring.

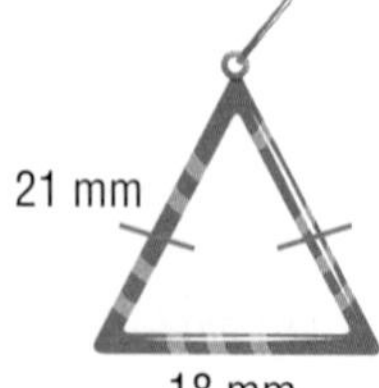

EXTRA PRACTICE
See pages 697, 714.

H.O.T. Problems

29. **OPEN ENDED** Draw a pair of similar triangles and a pair of congruent quadrilaterals. Label the vertices of the figures. Using these labels, generalize a statement about each pair of figures.

CHALLENGE For Exercises 30 and 31, use the figure at the right.

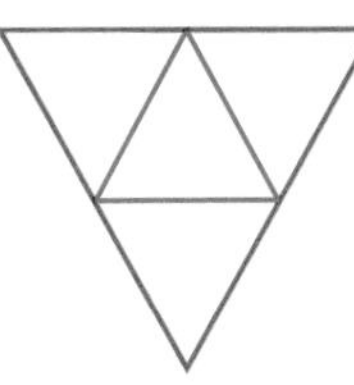

30. Measure the side lengths and the angles of the blue and red triangles. How do the triangles compare to one another?

31. Find the perimeter of each triangle. How does the ratio of the perimeters compare to the ratio of the side lengths?

32. **WRITING IN MATH** Explain why the statement that two squares are *always* similar is true.

TEST PRACTICE

33. Mrs. Daisy's garden contains pathways which form congruent triangles as shown.

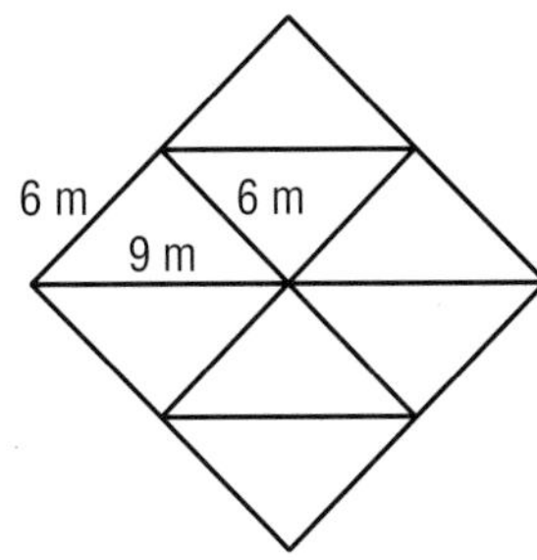

How many meters of pathways are there in her garden?

A 66 m **C** 108 m

B 96 m **D** 180 m

34. Which of the following is NOT true about the triangles below, given they are similar?

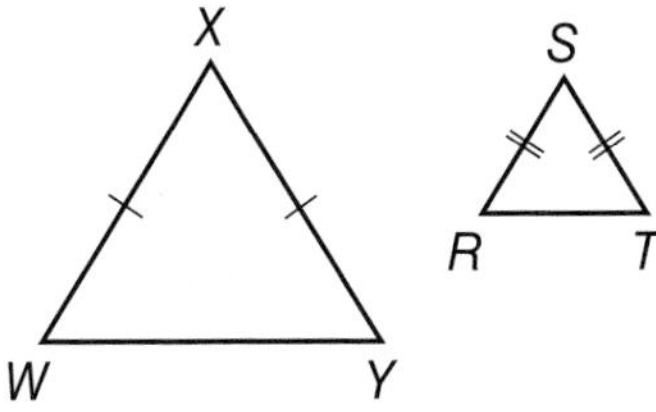

F $\overline{XY}$ is congruent to $\overline{XW}$.

G Triangle *WXY* has the same shape as triangle *RST*.

H $\overline{RS}$ is congruent to $\overline{WX}$.

J Triangle *WXY* and triangle *RST* are both isosceles triangles.

Spiral Review

35. **SATELLITES** A main satellite sends a signal to each of two smaller satellites. If each of those two satellites sends a signal to each other and a signal back to the main satellite, draw a diagram to determine the number of signals sent. (Lesson 9-6)

Find the value of *x* in each quadrilateral. (Lesson 9-5)

36. 105° x° 75° 70°

37. 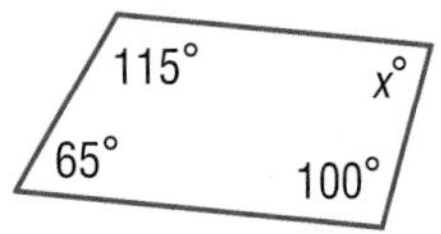

38. 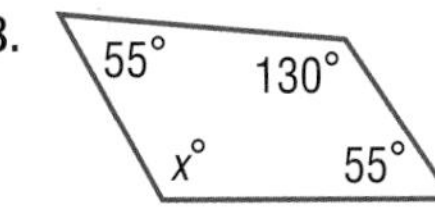

39. **WEATHER** The probability that it will snow tomorrow is forecasted at 60%. Describe the complement of this event and find its probability. (Lesson 7-4)

Extend 9-7

Geometry Lab
Tessellations

MAIN IDEA

Create tessellations using pattern blocks.

New Vocabulary

tessellation

A pattern formed by repeating figures that fit together without gaps or overlaps is a **tessellation**. Tessellations are formed using slides, flips, or turns of congruent figures.

ACTIVITY

STEP 1 Select the three pattern blocks shown.

STEP 2 Choose one of the blocks and trace it on your paper. Choose a second block that will fit next to the first without any gaps or overlaps and trace it.

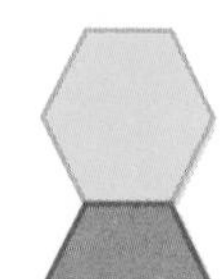

STEP 3 Trace the third pattern block into the tessellation.

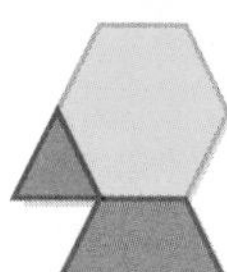

STEP 4 Continue the tessellation by expanding the pattern.

Create a tessellation using the pattern blocks shown.

a. 

b. 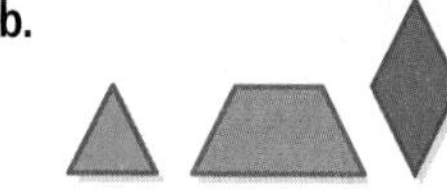

c. 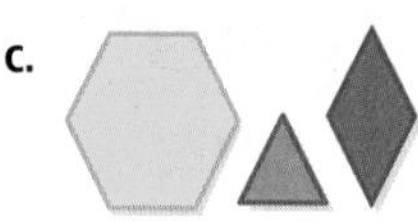

Analyze the Results

1. Tell if a tessellation can be created using a square and an equilateral triangle. Justify your answer with a drawing.
2. **MAKE A CONJECTURE** What is the sum of the measures of the angles where the vertices of the figures meet? Is this true for all tessellations?
3. Name two figures that cannot be used to create a tessellation. Use a drawing to justify your answer.

CHAPTER 9

Study Guide and Review

Math Online glencoe.com
- STUDY TO GO
- Vocabulary Review

FOLDABLES® Study Organizer GET READY to Study

Be sure the following Big Ideas are noted in your Foldable.

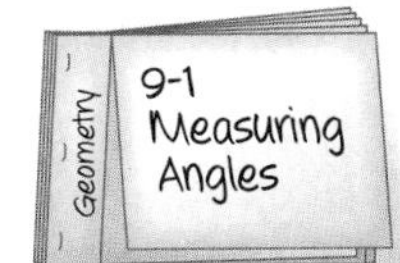

BIG Ideas

Types of Angles (Lesson 9-1)
- An acute angle measures less than 90°.
- An obtuse angle measures between 90° and 180°.
- A right angle measures exactly 90°.
- A straight angle measures exactly 180°.

Classifying Triangles (Lesson 9-4)
- An acute triangle has all acute angles.
- A right triangle has one right angle.
- An obtuse triangle has one obtuse angle.

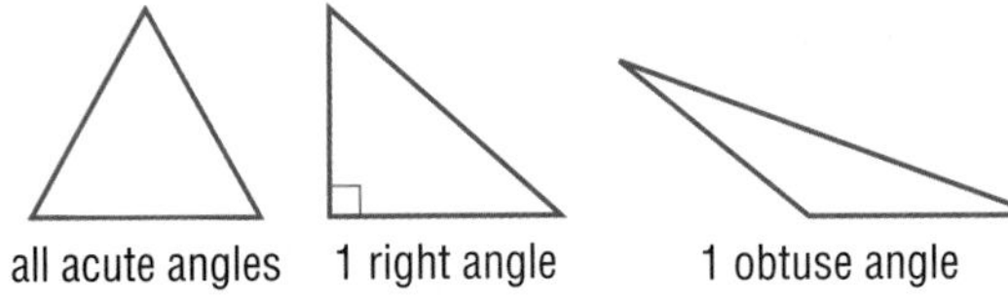

Sum of Angle Measures (Lessons 9-4 and 9-5)
- The sum of the angle measures in a triangle is 180°.
- The sum of the measures of the angles in a quadrilateral is 360°.

Similar and Congruent Figures (Lesson 9-7)
- Figures that have the same shape but not necessarily the same size are similar figures.
- Figures that have the same shape and size are congruent figures.

Key Vocabulary

acute angle (p. 471)
acute triangle (p. 486)
angle (p. 470)
complementary angles (p. 480)
congruent angles (p. 479)
congruent figures (p. 502)
congruent segments (p. 488)
corresponding sides (p. 503)
degree (p. 470)
equilateral triangle (p. 488)
isosceles triangle (p. 488)
line segment (p. 488)
obtuse angle (p. 471)
obtuse triangle (p. 486)
parallelogram (p. 495)
quadrilateral (p. 494)
rectangle (p. 495)
rhombus (p. 495)
right angle (p. 471)
right triangle (p. 486)
scalene triangle (p. 488)
side (p. 470)
similar figures (p. 502)
square (p. 495)
straight angle (p. 471)
supplementary angles (p. 480)
tessellation (p. 508)
trapezoid (p. 495)
vertex (p. 470)
vertical angles (p. 479)

Vocabulary Check

State whether each sentence is *true* or *false*. If *false*, replace the underlined word or number to make a true sentence.

1. When two lines intersect, they form two pairs of opposite angles called <u>supplementary angles</u>.
2. A four-sided figure with exactly one pair of opposite sides parallel is a <u>trapezoid</u>.
3. All triangles have at least two <u>obtuse</u> angles.
4. The sum of the angle measures in a quadrilateral is <u>360°</u>.
5. Congruent figures are also <u>similar</u> figures.

CHAPTER 9

Study Guide and Review

Lesson-by-Lesson Review

9-1 Measuring Angles (pp. 470–473)

Use a protractor to find the measure of each angle. Then classify each angle as *acute, obtuse, right,* or *straight*.

6. 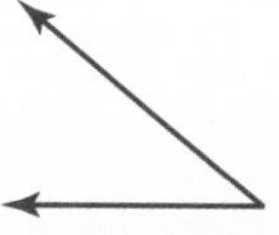

7. 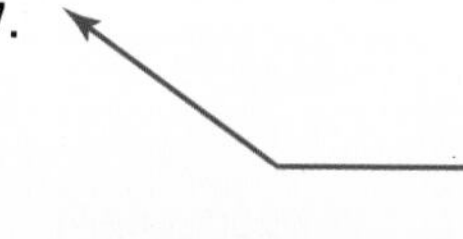

8. **ROLLER COASTERS** Use a protractor to measure the angle formed at the top of the roller coaster. Then classify it as *right, acute, obtuse,* or *straight*.

Example 1 Use a protractor to find the measure of the angle. Then classify the angle as *acute, obtuse, right,* or *straight*.

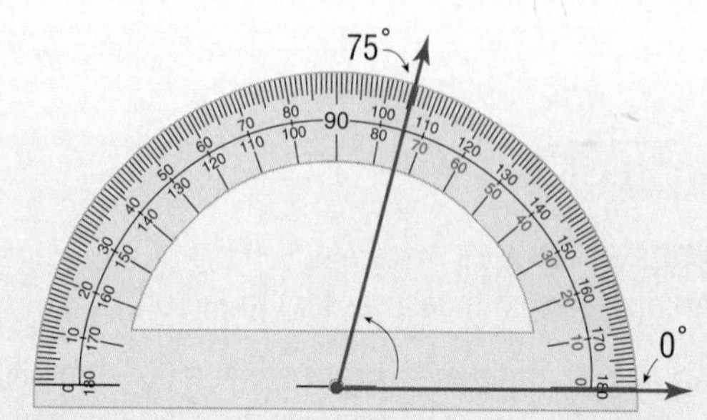

The angle measures 75°. Since it is less than 90°, it is an acute angle.

9-2 Estimating and Drawing Angles (pp. 474–478)

Use a protractor and a straightedge to draw angles having the following measurements.

9. 36°
10. 127°
11. 180°
12. 90°

Estimate the measure of each angle.

13. 14. 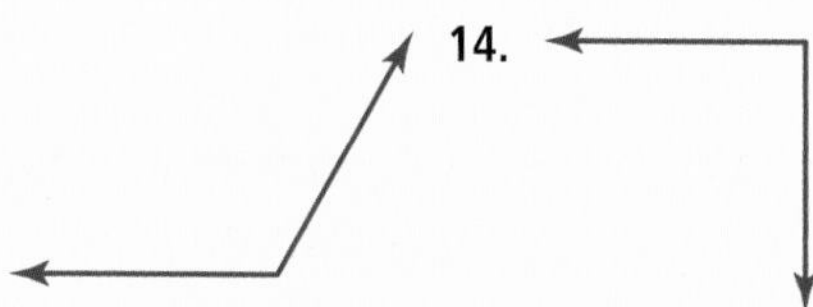

15. **AIRPLANES** Estimate the measure of the angle shown below.

Example 2 Use a protractor and a straightedge to draw a 47° angle.

Draw one side of the angle. Align the center of the protractor and the 0° with the line. Find 47°. Make a mark.

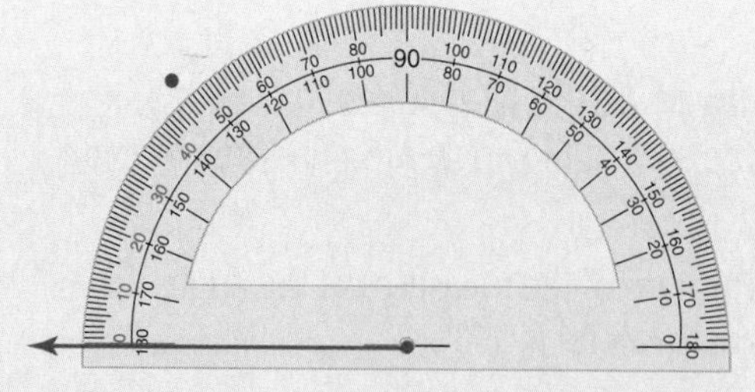

Draw the other side of the angle.

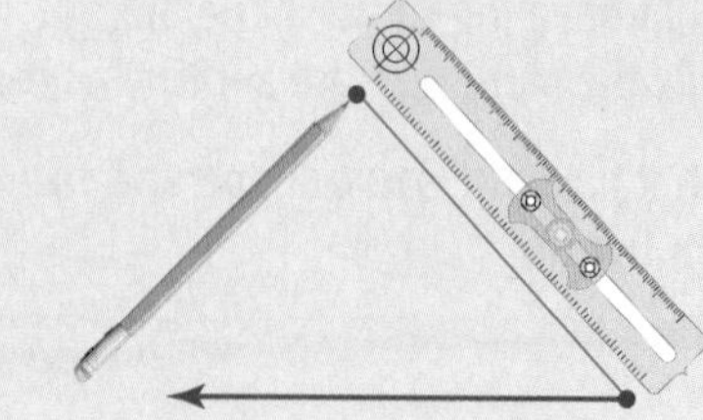

9-3 Angle Relationships (pp. 479–484)

Classify each pair of angles as *complementary, supplementary,* or *neither.*

16.

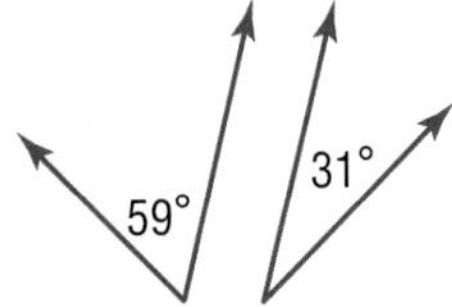

17.

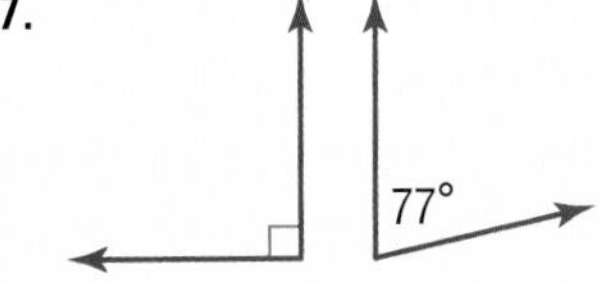

18.

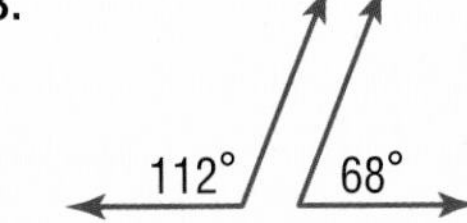

Find the value of x in each figure.

19.

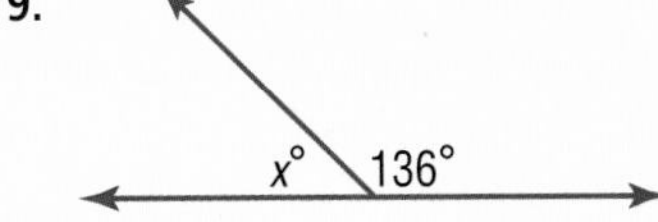

20.

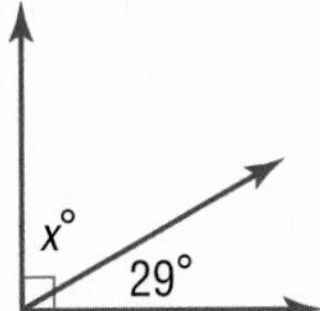

21. **TRAFFIC** Find the value of x in the intersection shown.

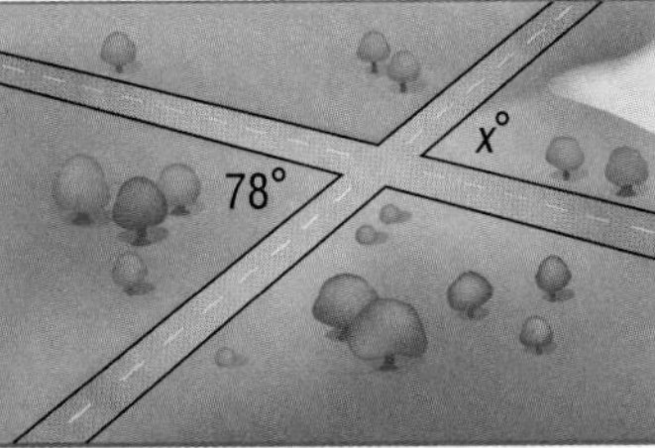

Classify each pair of angles as *complementary, supplementary,* or *neither.*

Example 3

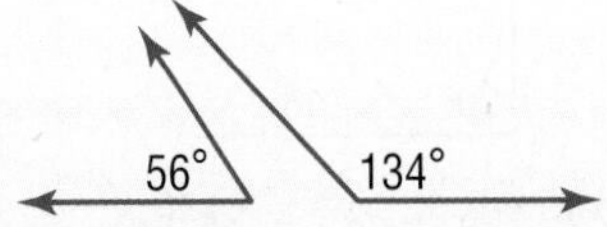

$56° + 134° = 190°$

The angles are neither complementary nor supplementary.

Example 4

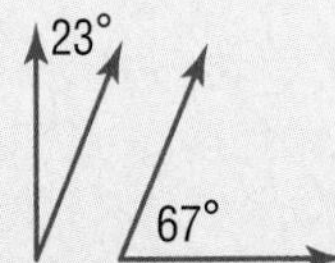

$23° + 67° = 90°$

The angles are complementary.

Example 5 Find the value of x in the figure.

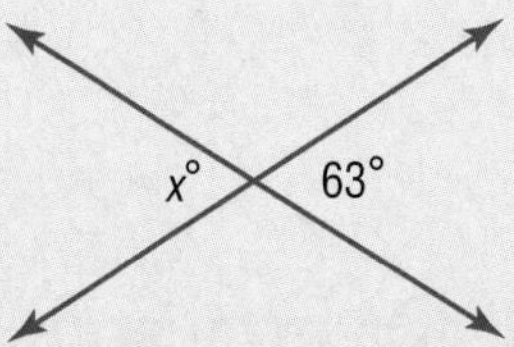

The angle labeled $x°$ and the angle labeled 63° are vertical angles. Therefore, they are congruent.

So, the value of x is 63.

9-4 Triangles (pp. 486–491)

Classify each triangle as *acute, right,* or *obtuse.*

22. 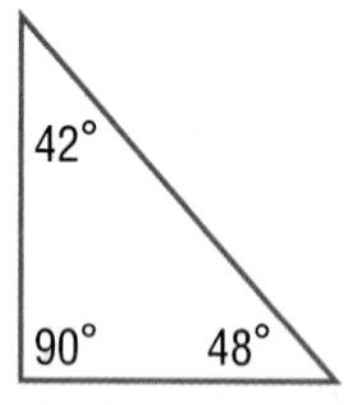

23. 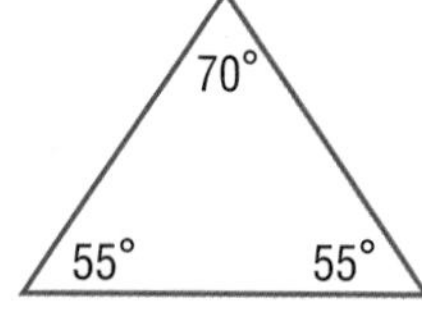

24. 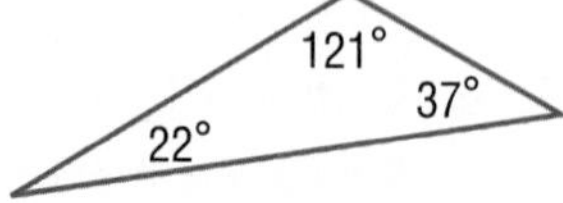

Find the value of x in each triangle.

25. 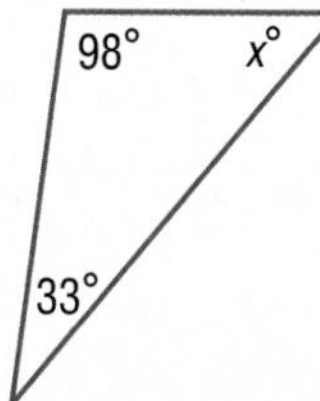

26. 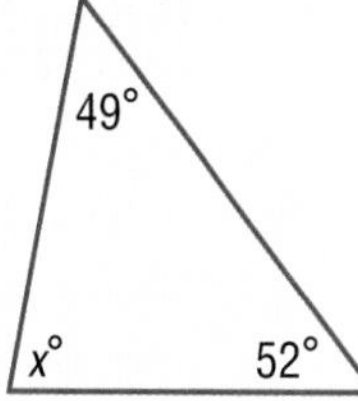

Classify each triangle as *scalene, isosceles,* or *equilateral.*

27. 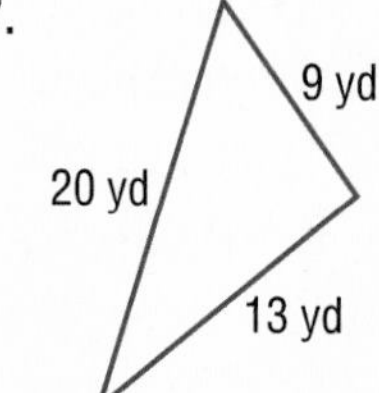

28. 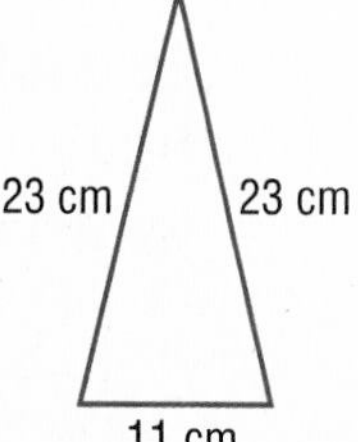

29. **ARCHITECTURE** The top of the R.R. Donnelly Building in Chicago, Illinois, contains a triangle. Classify the triangle as *scalene, isosceles,* or *equilateral.*

Example 6 **Classify the triangle shown as *acute, right,* or *obtuse.***

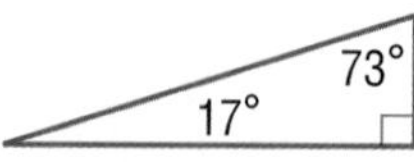

Since the triangle has one right angle, it is a right triangle.

Example 7 **Find the value of x in the triangle shown.**

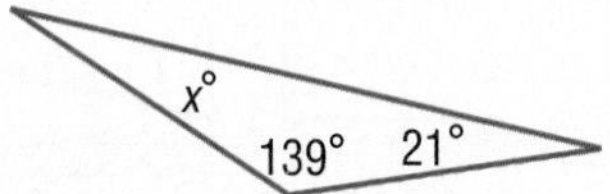

The sum of the angle measures in a triangle is 180°. So, $x + 139 + 21 = 180$.

$x + 21 + 139 = 180$	Write the equation.
$x + 160 = 180$	Add 21 and 139.
$20 + 160 = 180$	You know that $20 + 160 = 180$.

So, the value of x is 20.

Example 8 **Classify the triangle shown as *scalene, isosceles,* or *equilateral.***

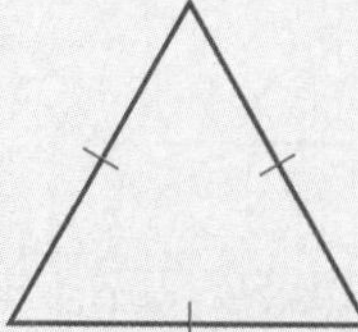

Since the triangle has all three congruent sides, it is an equilateral triangle. It is also an isosceles triangle since it has at least two congruent sides.

Mixed Problem Solving
For mixed problem-solving practice, see page 714.

9-5 Quadrilaterals (pp. 494–499)

Find the value of x in each quadrilateral.

30. 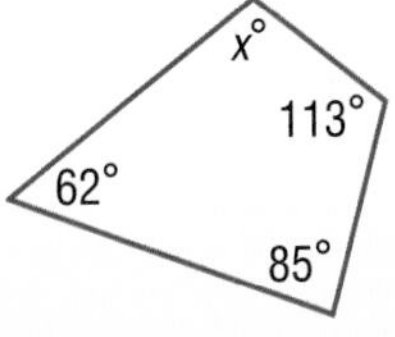

31. 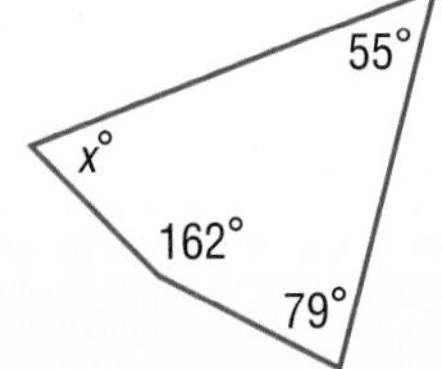

Classify each quadrilateral.

32. 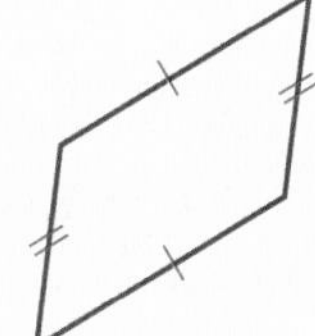

33. 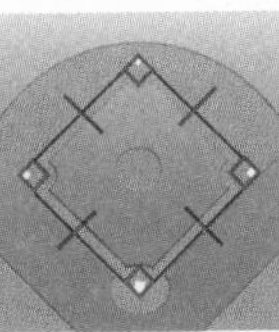

34. **TABLES** Identify the quadrilateral outlined.

Example 9 **Find the value of x in the quadrilateral shown.**

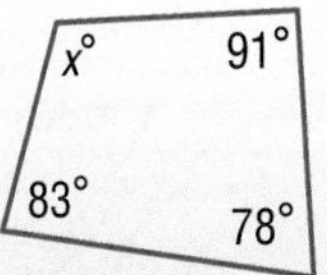

The sum of the angle measures in a quadrilateral is 360°.

$x + 91 + 78 + 83 = 360$

$x + 252 = 360$ Add 91, 78, and 83.

$108 + 252 = 360$ You know that $108 + 252 = 360$.

So, the value of x is 108.

Example 10 **Classify the quadrilateral shown below.**

The quadrilateral has exactly one pair of parallel sides, so it is a trapezoid.

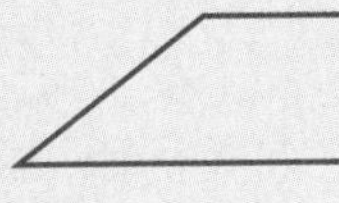

9-6 PSI: Draw a Diagram (pp. 500–501)

Solve by drawing a diagram.

35. **ART** Gina is painting a design. She will paint six dots in a circle. Each dot is to be connected to every other dot by a straight line. How many straight lines will Gina need to draw?

36. **WINDOWS** A contractor is designing one side of an office building 48 feet in length. The windows are each 5 feet long, and will be placed 6 feet apart and 5 feet from the end of the building wall. How many windows can the contractor design?

Example 11 **Five friends are seated in a circle at a restaurant. How many ways can any two friends split a meal?**

Draw five dots to represent the friends. Connect all dots to represent all possible choices for splitting a meal.

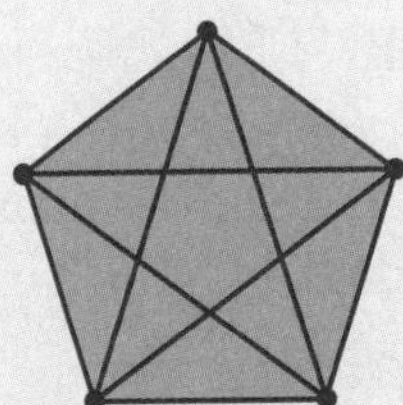

Count the number of lines that were drawn, 10. So, there are 10 ways for any two friends to split a meal.

9-7 Similar and Congruent Figures (pp. 502–507)

37. Tell whether the pair of figures shown below is *congruent, similar,* or *neither.*

The triangles shown are congruent.

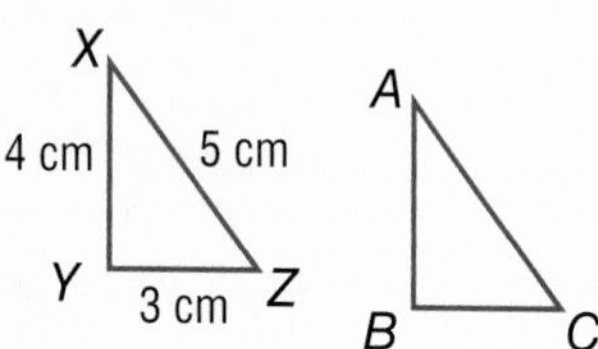

38. What side of triangle XYZ corresponds to $\overline{AC}$?

39. Which rectangle below is similar to rectangle $MNOP$?

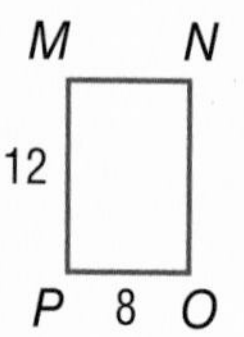

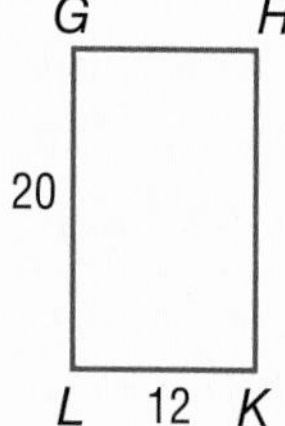

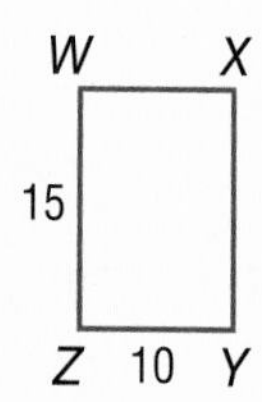

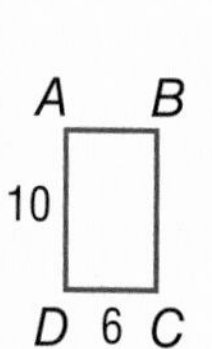

40. **ART** Elisa created the design below out of ribbon for an art display. If the design consists of 10 equilateral triangles, how much ribbon was used?

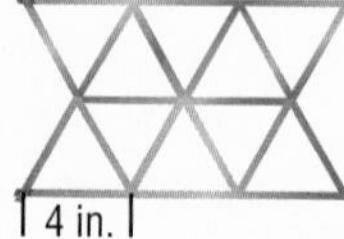

Example 12 **Tell whether the pair of figures shown below is *congruent, similar,* or *neither.***

The figures are congruent since they have the same shape and same size.

Example 13 **The figures shown are congruent. What side of quadrilateral *RSTU* corresponds to $\overline{MN}$?**

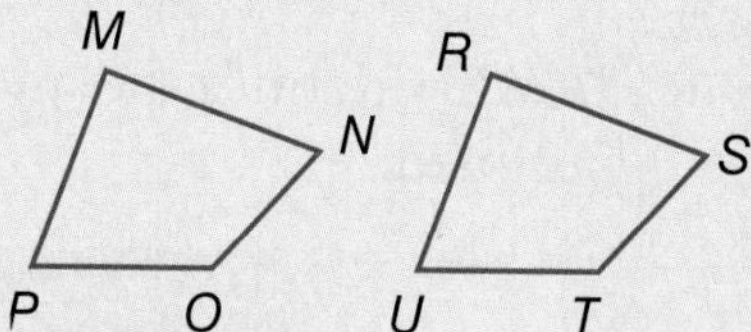

The parts of congruent figures that match are corresponding parts.
In the figures, $\overline{RS}$ corresponds to $\overline{MN}$.

Example 14 **State whether triangle *DEF* is similar to triangle *SRT*.**

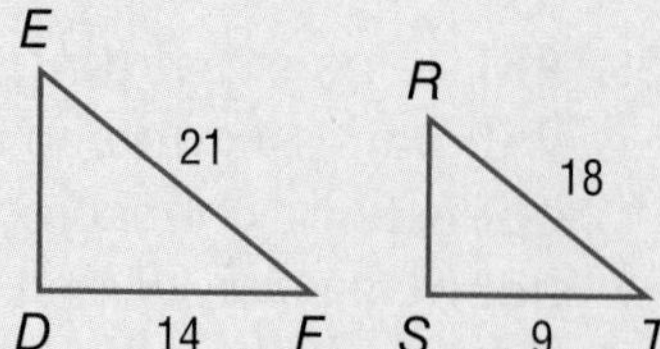

Examine the ratios of the corresponding sides to see if they form a constant ratio.

Triangle *DEF*	**Triangle *SRT***
$\frac{EF}{RT} = \frac{21}{18}$ or $\frac{7}{6}$	$\frac{DF}{ST} = \frac{14}{9}$

So, the triangles are not similar.

Practice Test

Math Online glencoe.com
• Chapter Test

Use a protractor to measure each angle and classify as *acute, obtuse, right,* or *straight*.

1. 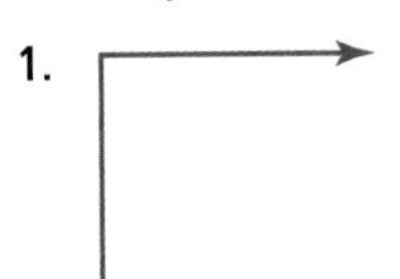

2.  

3. 

4. **HILLS** Estimate the measure of x.

Classify each pair of angles as *complementary, supplementary,* or *neither*.

5.  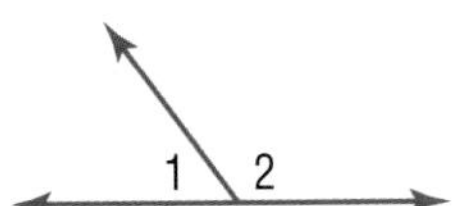

6. 

7. Find the value of x in the triangle at the right.

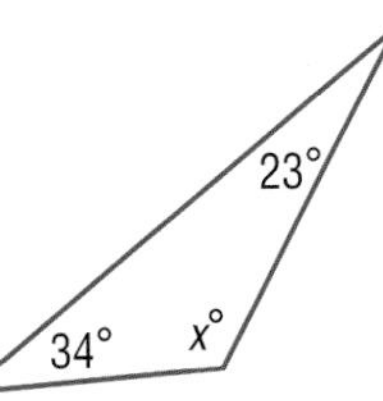

Classify each triangle as *scalene, isosceles,* or *equilateral*.

8. 

9. 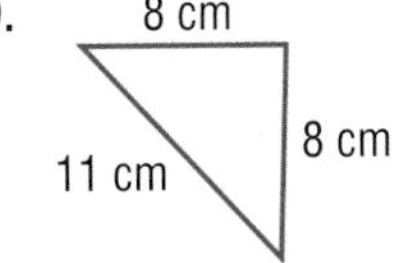

10. **MULTIPLE CHOICE** Find the measure of $\angle R$ in the trapezoid below.

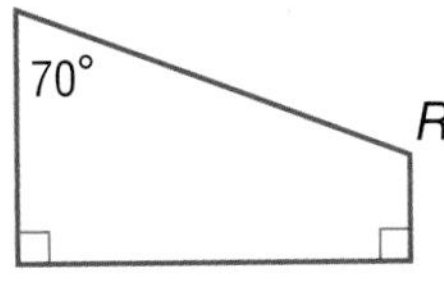

A 110° C 90°
B 100° D 20°

11. **PATIOS** Classify each quadrilateral outlined in the brick patio.

12. **MULTIPLE CHOICE** Refer to the shapes.

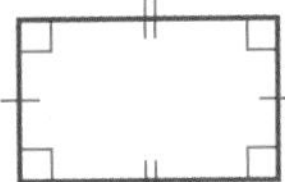 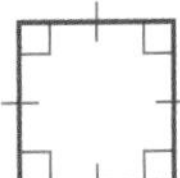 

Which statement is NOT true?

F Each shape is a quadrilateral.
G Each shape is a polygon.
H Each shape is a parallelogram.
J The sum of the angle measures of each shape is 360°.

13. **SPORTS** To block off the boundaries for a game of Capture the Flag, Jonas plans to use 5 orange cones on each side of the rectangular field. This includes one cone at each corner. How many cones are needed?

14. The quadrilaterals shown are similar. What side of quadrilateral $ABCD$ corresponds to $\overline{GH}$?

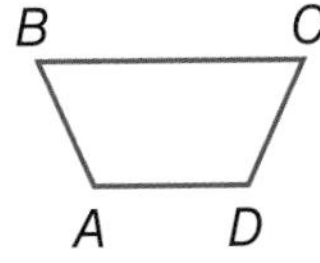

Tell whether each pair of figures is *congruent, similar,* or *neither*.

15. 

16. 

CHAPTER 9 Test Practice
Cumulative, Chapters 1–9

Math Online glencoe.com
• Test Practice

PART 1 Multiple Choice

Read each question. Then fill in the correct answer on the answer sheet provided by your teacher or on a sheet of paper.

1. At a concession stand, each pizza was cut into 8 equal-sized pieces. Dave sold 7 pieces, Sally sold 5 pieces, and Lori sold 6 pieces. Find the total portion of pizza sold by these 3 people.

   A 18

   B $2\frac{3}{4}$

   C $2\frac{1}{4}$

   D $2\frac{9}{10}$

2. Which fraction is *not* equivalent to $\frac{2}{3}$?

   F $\frac{4}{6}$ **H** $\frac{9}{12}$

   G $\frac{6}{9}$ **J** $\frac{10}{15}$

3. Find the measure of $\angle P$ of the parallelogram shown to the nearest degree.

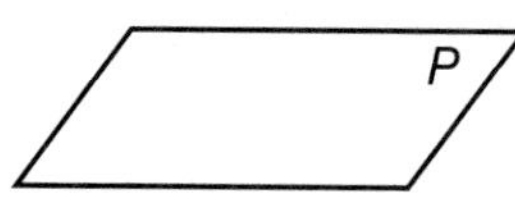

   A 127° **C** 28°

   B 96° **D** 53°

4. The table shows the scores at a gymnastics competition. Who had the highest overall score?

Gymnast	Overall Score
Amber	8.95
Camille	8.82
Jacy	8.73
Luisa	8.99

   F Amber **H** Jacy

   G Camille **J** Luisa

5. Sarah will select two sweaters from 1 red sweater, 1 yellow sweater, 1 green sweater, and 1 blue sweater. Which diagram shows all the possible outcomes?

   A

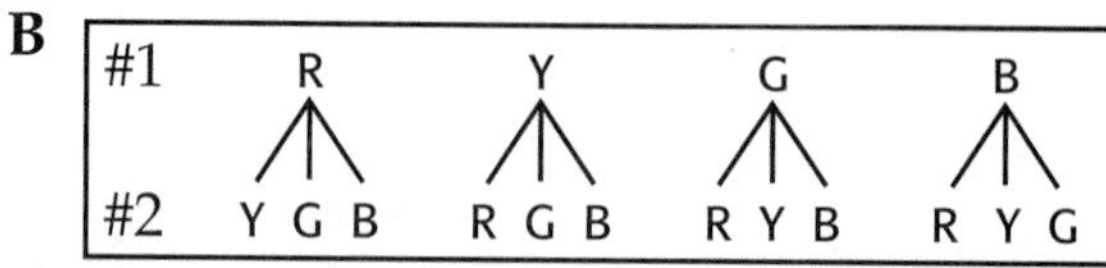

   B

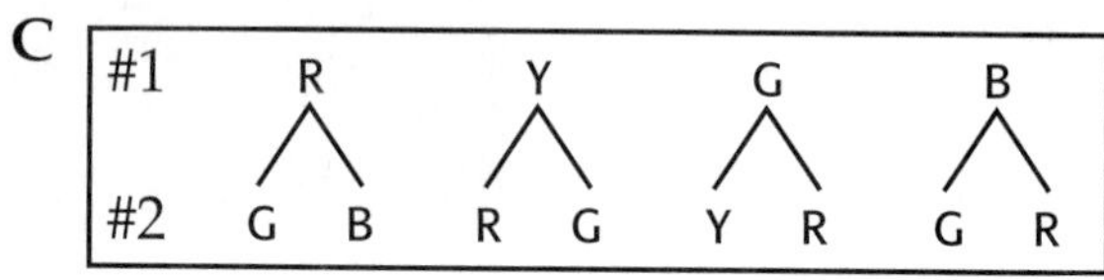

   C

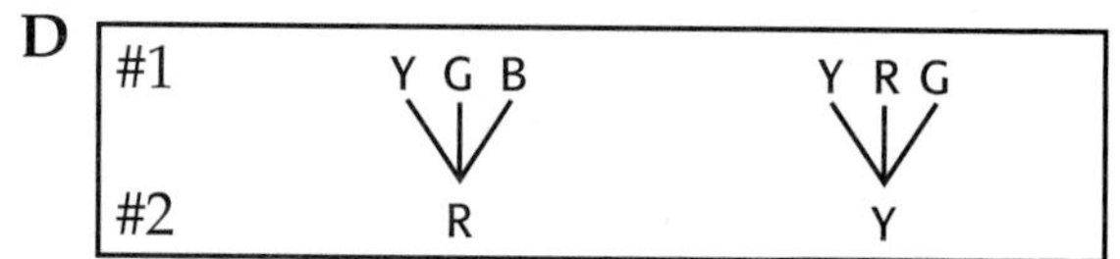

   D

#1 Y G B Y R G

#2 R Y

6. Each day after school, Bernardo runs for 10 minutes. If he runs an average of 900 feet per minute, how can you find the number of feet he runs per day?

   F Divide his average running speed by the number of minutes he runs each day.

   G Multiply his average running speed by the number of minutes he runs each day.

   H Subtract the number of minutes he runs each day from his average running speed.

   J Add his average running speed to the number of minutes he runs each day.

7. The length of Rashid's bedroom is 432 centimeters. What is the length of his bedroom in meters?

   A 43.2 m **C** 0.432 m

   B 4.32 m **D** 0.0432 m

Preparing for Standardized Tests
For test-taking strategies and practice, see pages 718–735.

8. A rectangular garden has a width of 5.3 feet and length that is 1.8 feet more than the width. What is the length of the garden?

   F 2.9 ft **H** 7.1 ft
   G 3.5 ft **J** 8.4 ft

TEST-TAKING TIP

Question 8 This problem does not include a drawing. Make one to help you quickly see how to solve the problem.

9. Find the measure of $\angle H$ below.

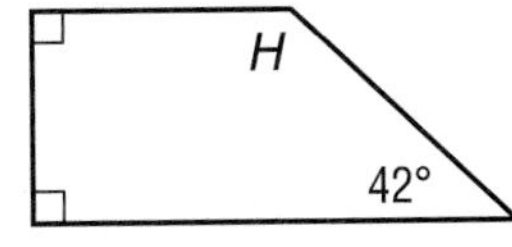

   A 26° **C** 180°
   B 138° **D** 212°

10. In which of the figures below does $\angle M$ appear to be an acute angle?

F
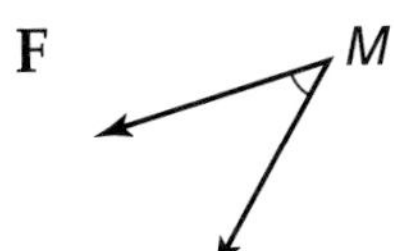

G
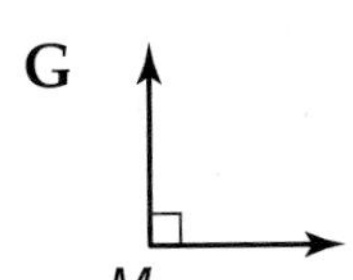

H
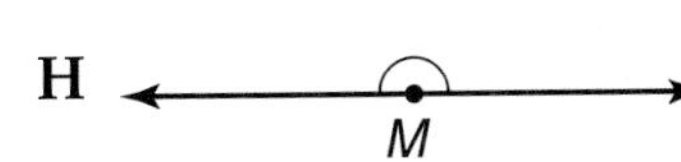

J
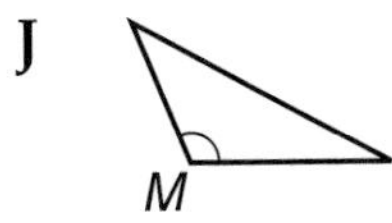

PART 2 Short Response/Grid In

Record your answers on the answer sheet provided by your teacher or on a sheet of paper.

11. The ratio of boys to girls at a rock concert was about 7 to 6. If there were 1,092 boys at the rock concert, about how many girls were at the concert?

PART 3 Extended Response

Record your answers on the answer sheet provided by your teacher or on a sheet of paper. Show your work.

12. Alexander is designing a logo for his father's electronics store. The logo will consist of different geometric figures.

    a. The first figure meets the following criteria.

    - a polygon
    - has an obtuse angle
    - has all sides congruent

    Make a possible classification and sketch of the figure.

    b. The second figure meets the following criteria.

    - a polygon similar to the first figure
    - positioned inside the first figure

    Make a possible classification and sketch of the figure.

NEED EXTRA HELP?												
If You Missed Question...	1	2	3	4	5	6	7	8	9	10	11	12
Go to Lesson...	5-4	4-2	9-1	3-2	7-5	1-1	8-6	3-5	9-5	9-1	6-4	9-1

Measurement: Perimeter, Area, and Volume

- Relate properties of two- and three-dimensional shapes to find perimeter, area, and volume of figures.

Key Vocabulary

perimeter (p. 522)
circle (p. 528)
circumference (p. 528)
volume (p. 548)

Real-World Link

Seashores Cape Hatteras National Seashore in North Carolina stretches across 31,263 acres, or about 49 square miles.

Measurement: Perimeter, Area, and Volume Make this Foldable to help you organize your notes. Begin with a sheet of 11″ × 17″ paper and six index cards.

1. **Fold** lengthwise about 3″ from the bottom.

2. **Fold** the paper in thirds.

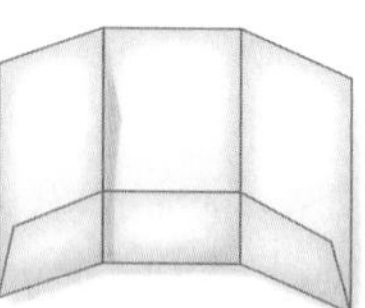

3. **Open** and staple the edges on either side to form three pockets.

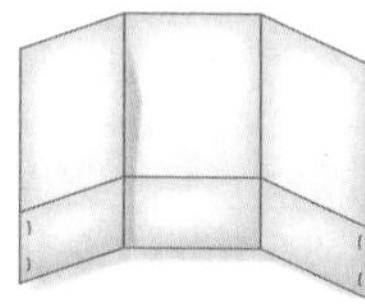

4. **Label** the pockets as shown. Place two index cards in each pocket.

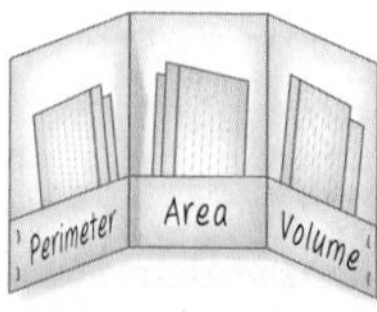

GET READY for Chapter 10

Diagnose Readiness You have two options for checking Prerequisite Skills.

Option 2

Math Online Take the Online Readiness Quiz at glencoe.com.

Option 1

Take the Quick Quiz below. Refer to the Quick Review for help.

QUICK Quiz

Evaluate each expression. (Lesson 1-4)

1. 4(9)
2. 4(17)
3. 2(8) + 2(5)
4. 2(16) + 2(11)
5. **SHOPPING** Lou bought two pairs of pants and two shirts. If each pair of pants cost \$22 and each shirt cost \$13, how much did Lou spend?

Use the π button on your calculator to evaluate each expression. Round to the nearest tenth. (Prior Grade)

6. $\pi \times 7$
7. $\pi \times 12$
8. $2 \times \pi \times 8$
9. $2 \times \pi \times 13$

Evaluate each expression. (Lesson 1-4)

10. $16 \cdot 7$
11. $23 \cdot 5$
12. $\frac{8 \times 9}{2}$
13. $\frac{14 \times 11}{2}$
14. $10 \times 12 \times 8$
15. $33 \times 7 \times 5$
16. (2)(3)(5) + (2)(3)(9) + (2)(5)(9)
17. (2)(8)(4) + (2)(8)(6) + (2)(4)(6)
18. **BAKE SALE** Lucinda bought four packages of muffin mix. Each mix makes 12 muffins. If Lucinda sells each muffin for \$2, what is the most she could earn?

QUICK Review

Example 1

Evaluate 3(15) − 8.

$3(15) - 8 = 45 - 8$ Multiply.
$= 37$ Subtract.

Example 2

Evaluate 2(31) + 2(9).

$2(31) + 2(9) = 62 + 18$ Multiply.
$= 80$ Add.

Example 3

Use the π button on your calculator to evaluate $2 \times \pi \times 3$. Round to the nearest tenth.

$2 \times \pi \times 3 = 6 \times \pi$ Multiply 2 by 3.
$= 18.8$ Multiply 6 by π.

Example 4

Evaluate $\frac{8 \times 4}{2}$.

$\frac{8 \times 4}{2} = \frac{32}{2}$ Multiply 8 by 4.
$= 16$ Divide 32 by 2.

Example 5

Evaluate (2)(9)(3) + (2)(9)(4) + (2)(3)(4).

(2)(9)(3) + (2)(9)(4) + (2)(3)(4)
$= 54 + 72 + 24$ Multiply.
$= 150$ Add.

Explore 10-1

Measurement Lab
Area and Perimeter

MAIN IDEA

Explore changes in area and perimeter of rectangles and squares.

If you increase the side lengths of a rectangle or square proportionally, how are the area and the *perimeter*, or distance around the rectangle, affected? In this lab, you will investigate relationships between the areas and perimeters of original figures and those of the similar figures.

ACTIVITY

1 **STEP 1** On centimeter grid paper, draw and label a rectangle with a length of 6 centimeters and a width of 2 centimeters.

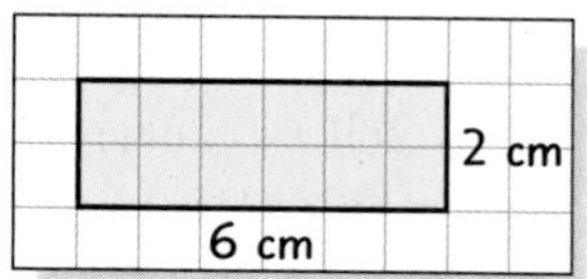

STEP 2 Find the area and perimeter of this original rectangle. Then record the information in a table like the one shown.

Rectangle	Length (cm)	Width (cm)	Area (sq cm)	Perimeter (cm)
original	6	2		
A	12	4		
B	18	6		
C	24	8		

STEP 3 Repeat Steps 1 and 2 for rectangles A, B, and C, whose dimensions are shown in the table.

Analyze the Results

1. Describe how the dimensions of rectangles A, B, and C are different than the original rectangle.
2. Describe how the area of the original rectangle changed when the length and width were both doubled.
3. Describe how the perimeter of the original rectangle changed when the length and width were both doubled.
4. How did the area and the perimeter of the original rectangle change when the length and width were both tripled? quadrupled?
5. Compare the areas and perimeters of rectangle A and a rectangle with dimensions of 3 centimeters and 4 centimeters.

6. Draw a rectangle with a length and width that are half those of the original rectangle. Describe how the area and perimeter changes.

7. **MAKE A CONJECTURE** How are the perimeter and area of a rectangle affected if the length and the width are changed proportionally?

ACTIVITY

2 **STEP 1** On centimeter grid paper, draw and label a square with a length of 4 centimeters.

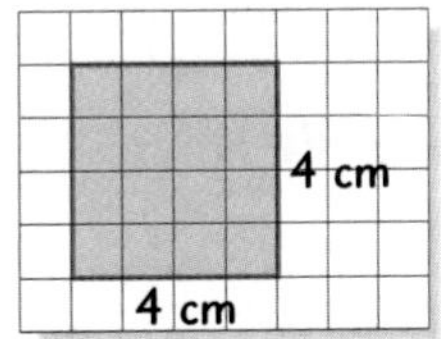

STEP 2 Find the area and perimeter of this original square. Then record the information in a table like the one shown.

Square	Side Length (cm)	Area (sq cm)	Perimeter (cm)
original	4		
A	5		
B	6		
C	7		

STEP 3 Repeat Steps 1 and 2 for squares A, B, and C, whose dimensions are shown in the table.

Study Tip

Look Back You can review **area of squares** in Lesson 1-9.

Analyze the Results

8. Describe how the dimensions of squares A, B, and C are different from the original square.
9. Describe how the perimeter of the original square changed when the side lengths increased by one centimeter.
10. Compare the ratios $\frac{\text{perimeter}}{\text{side length}}$ in the table above.
11. Suppose the perimeter of a square is 60 centimeters. Explain how you can use the ratio in Exercise 10 to find the length of its side. Then find its side length.
12. **WRITE A FORMULA** If P represents the perimeter of a square, write an equation that describes the relationship between the square's side length s and perimeter P. Compare this equation to the formula for a square's area.
13. **MAKE A CONJECTURE** Suppose you double the side lengths of the orginal square. Use what you learned in Activity 1 to predict the area and perimeter of the new square. Explain your reasoning.

10-1 Perimeter

MAIN IDEA

Find the perimeters of squares and rectangles.

New Vocabulary

perimeter

Math Online

glencoe.com

- Extra Examples
- Personal Tutor
- Self-Check Quiz

MINI Lab

STEP 1 Use a centimeter ruler to measure the side length of each square shown below. Round each to the nearest centimeter.

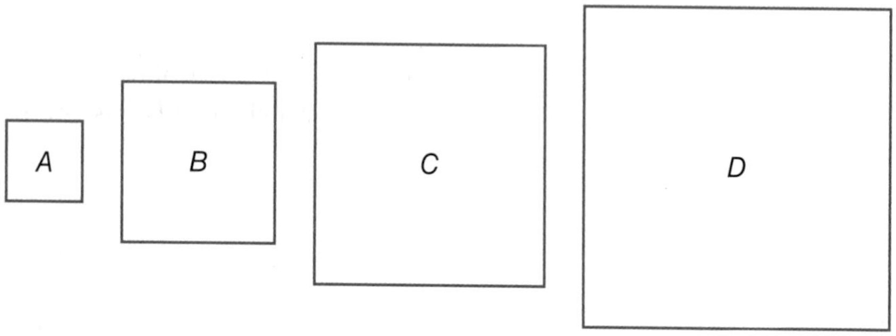

STEP 2 Copy and complete the table below. Find the distance around each square by adding the measures of its sides.

Square	Side Length	Distance Around
A		
B		
C		
D		

1. Write the ratio $\frac{\text{distance around}}{\text{side length}}$ in simplest form for squares A through D. What do you notice about these ratios?
2. **MAKE A CONJECTURE** Write an expression for the distance around a square that has a side length of x centimeters.

The distance around any closed figure is called its **perimeter**. As you discovered in the Mini Lab above, you can multiply the measure of any side of a square by 4 to find its perimeter.

Perimeter of a Square — Key Concept

Words The perimeter P of a square is four times the measure of any of its sides s.

Symbols $P = 4s$

Model

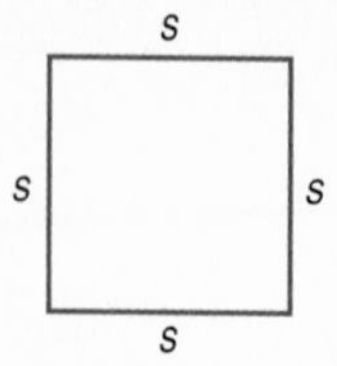

Real-World Link In 1964, indoor volleyball became an Olympic sport. Beach volleyball followed many years later in 1996.

Real-World EXAMPLE Perimeter of a Square

1 **VOLLEYBALL In volleyball, the shape of the court on one side of the net is a square with a side length of 9 meters. What is the perimeter of one half of a volleyball court?**

$P = 4s$ Perimeter of a square

$P = 4(9)$ Replace s with 9.

$P = 36$ Multiply.

The perimeter of one half of a volleyball court is 36 meters.

CHECK Your Progress

a. **SOFTBALL** The infield of a softball field is a square that measures 60 feet on each side. What is the perimeter of the infield?

Perimeter of a Rectangle

Key Concept

Words The perimeter P of a rectangle is the sum of the lengths and widths. It is also two times the length ℓ plus two times the width w.

Symbols $P = \ell + w + \ell + w$

$P = 2\ell + 2w$

Model

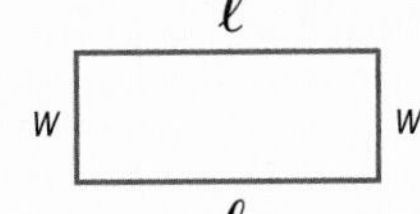

EXAMPLE Perimeter of a Rectangle

2 **Find the perimeter of the rectangle.**

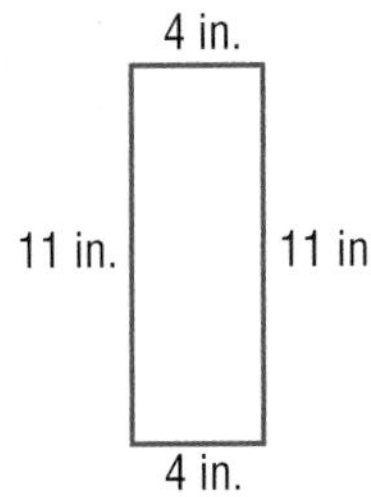

$P = 2\ell + 2w$ Write the formula.

$P = 2(11) + 2(4)$ Replace ℓ with 11 and w with 4.

$P = 22 + 8$ Multiply.

$P = 30$ Add.

The perimeter is 30 inches.

Study Tip

Check You can check your answer in Example 2 by finding the sum of the lengths and widths of the rectangle. $P = 11 + 4 + 11 + 4$ or 30 inches. So, the answer is correct.

CHECK Your Progress

Find the perimeter of each rectangle.

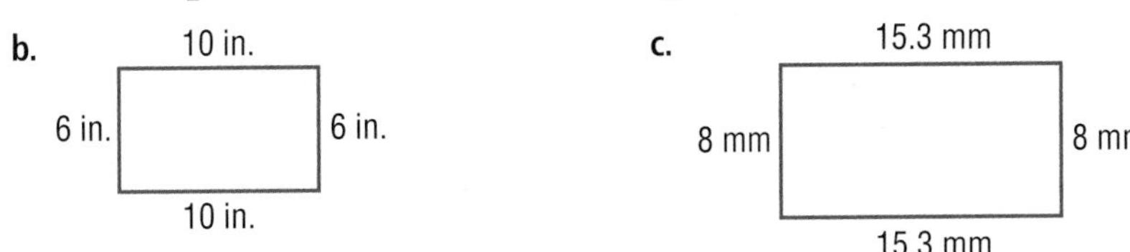

CHECK Your Understanding

Example 1 (p. 523)

1. **CHESS** The game of chess is played on a square-shaped board. What is the perimeter of the chess board shown?

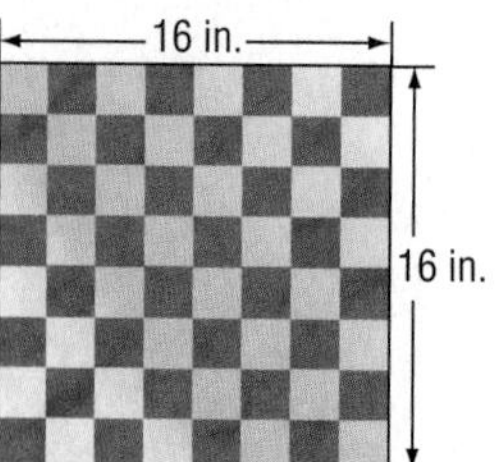

Example 2 (p. 523)

Find the perimeter of each rectangle.

2. 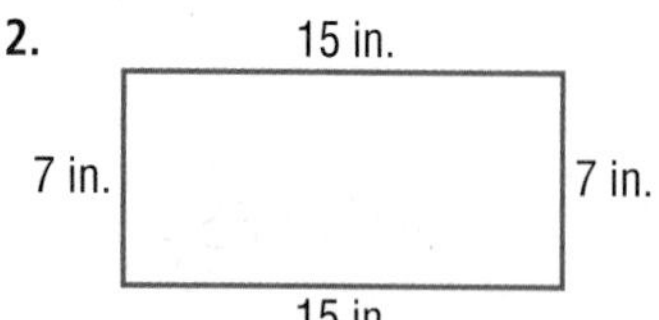

3. 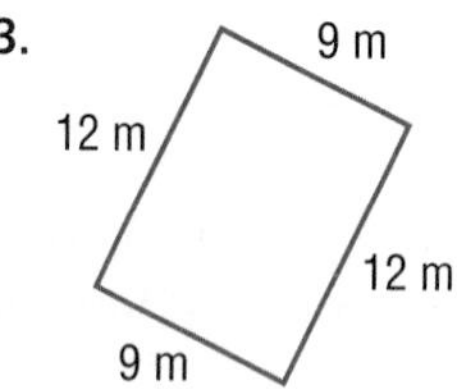

4. 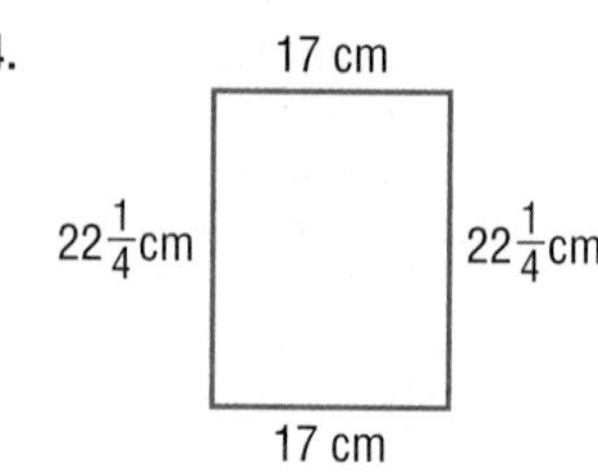

Practice and Problem Solving

HOMEWORK HELP	
For Exercises	See Examples
5–8	1
9–12	2

5. **SIGNS** A typical *Do Not Enter* sign is 750 millimeters on each side. What is the perimeter of the sign?

6. **COUNTY** Gray County is an approximate square with each side measuring 30 miles. What is the approximate perimeter of Gray County?

Find the perimeter of each square or rectangle.

7. 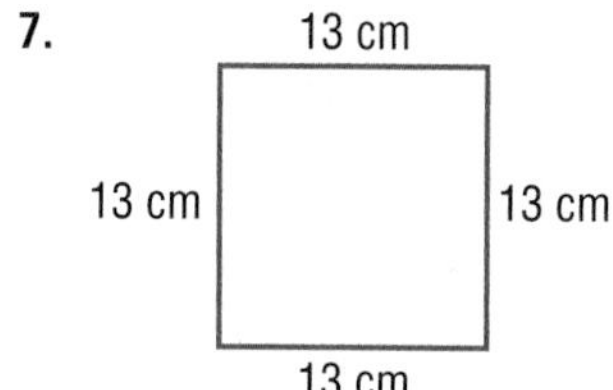

8. 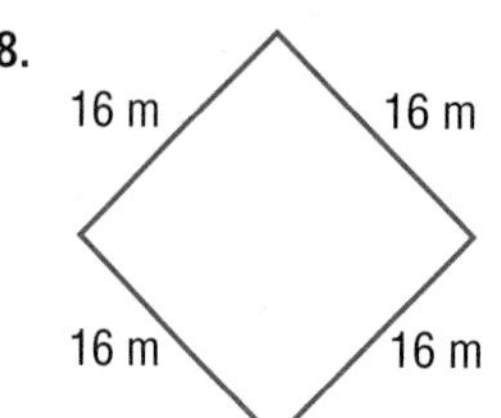

9. 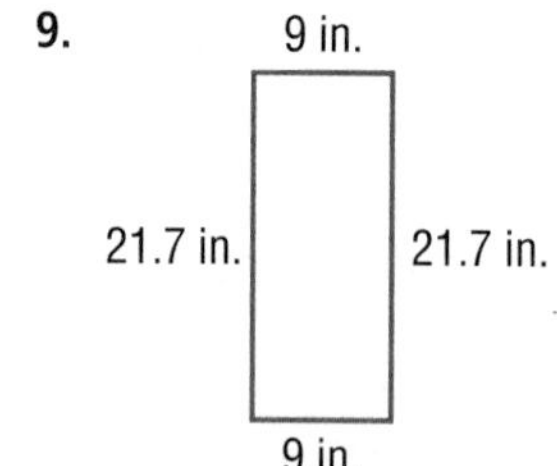

10. 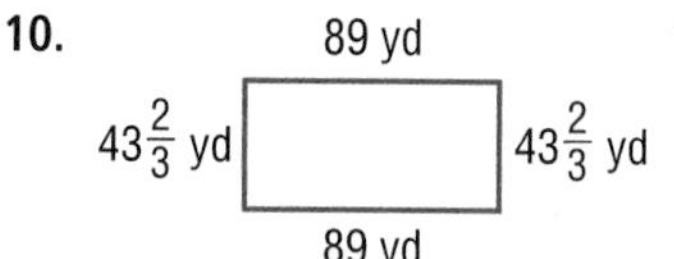

11. 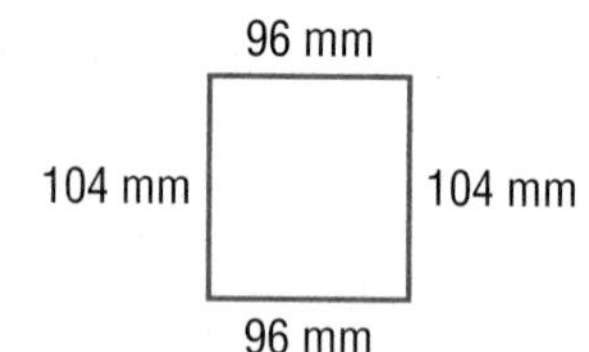

12. 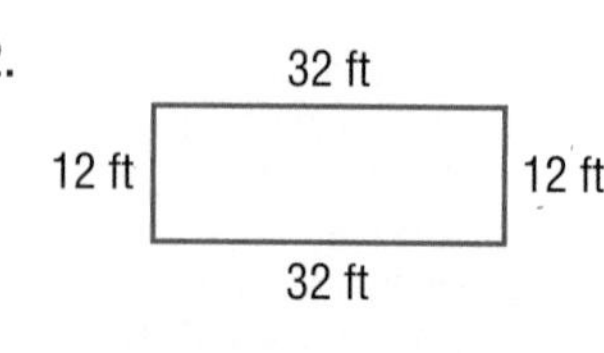

13. **FRAMES** Nadia has a square picture frame that will hold a 5-inch by 5-inch photo. The picture frame has a border that is 1 inch thick all the way around. How much larger is the perimeter of the frame than the perimeter of the picture?

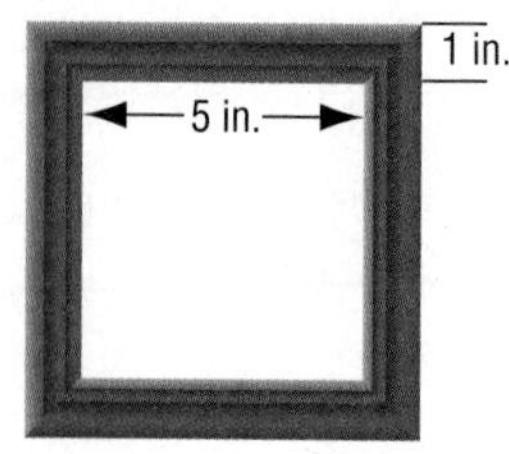

Find the perimeter of each figure.

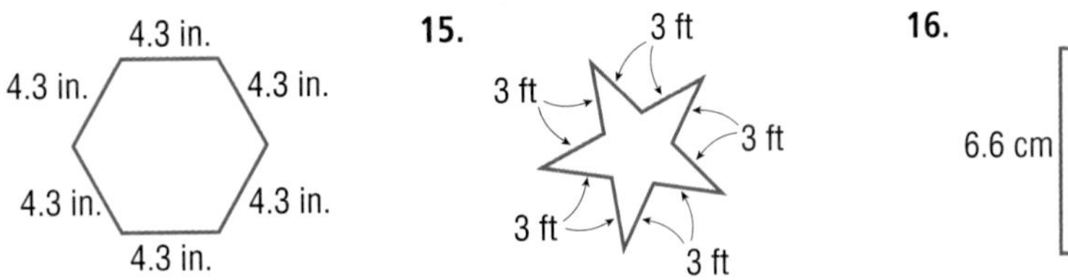

17. **ROWING** The blades of the oars shown are quadrilaterals. What is the perimeter of quadrilateral *ABCD*?

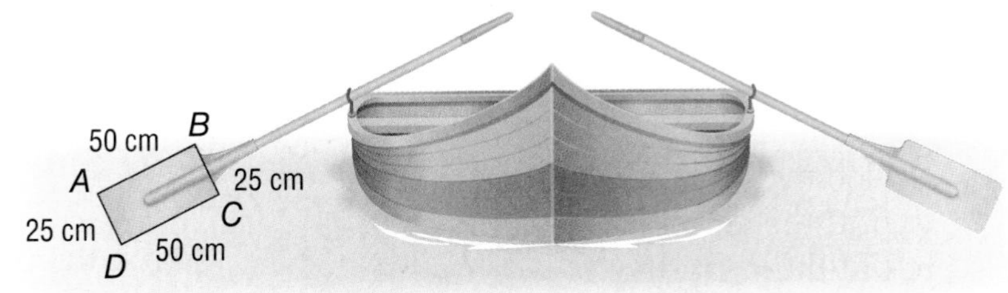

Real-World Link... Rowing teams keep the oars on each side of the boat parallel to optimize speed and performance.

18. **BASKETBALL** A basketball court measures 26 meters by 14 meters. Ten meters of seating is added to each side of the court. Find the perimeter of the rectangle enclosed by the court and the seating area.

19. **SEWING** A lace border will be sewn on square pillows. The amount of lace needed for one pillow is $58\frac{1}{2}$ inches. What is the length of the pillow?

For Exercises 20 and 21, find the value of y given the perimeter P. How many segments y units long are needed for the perimeter of each figure?

20.

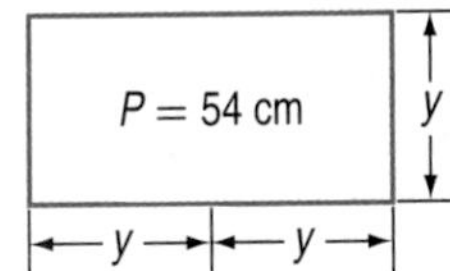

21.

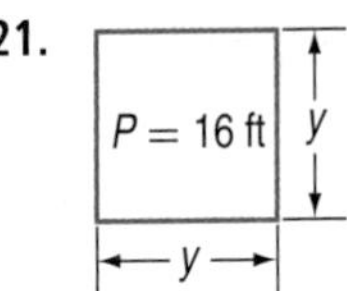

H.O.T. Problems

22. **OPEN ENDED** Draw a quadrilateral that has a perimeter of 20 centimeters.

23. **REASONING** Are two rectangles with equal perimeters always congruent? Explain your reasoning.

24. **FIND THE ERROR** Daquan and Jasmine are finding the perimeter of a rectangle that is 14 inches by 12 inches. Who is correct? Explain.

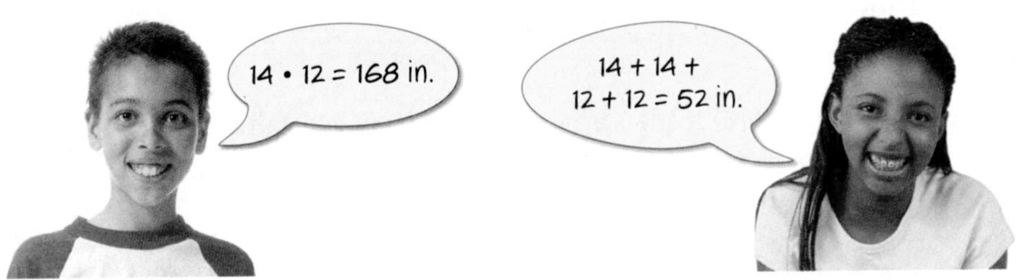

25. **CHALLENGE** Find and compare the perimeters of the rectangles whose dimensions are listed in the table. Then create another set of at least three rectangles that share a similar relationship.

Length (ft)	Width (ft)
6	1
5	2
4	3

26. **WRITING IN MATH** Compare and contrast the formulas for the perimeter of squares and rectangles.

TEST PRACTICE

27. Mr. Johnson is building a bottomless square sandbox using cedar wood.

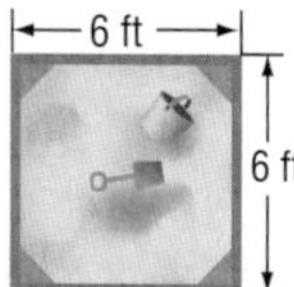

Which method can Mr. Johnson use to find the amount of cedar needed to build the sandbox?

A Multiply the length of a side by 2.

B Multiply the length of a side by 4.

C Square the length of a side.

D Multiply the length of each side by 2 and add the result.

28. **SHORT RESPONSE** Francisco cut a rectangle out of construction paper for a geometry project.

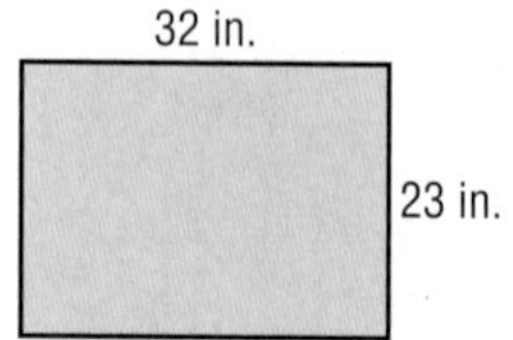

Find the perimeter of the rectangle in inches.

Spiral Review

Tell whether each pair of figures is *congruent, similar,* or *neither.* (Lesson 9-7)

29.  

30. 

31. 

32. **TRAVEL** Chayton lives in Glacier and works in Alpine. There is no direct route from Glacier to Alpine, so Chayton drives through either Elm or Perth. How many different ways can he drive to work? Use the *draw a diagram* strategy. (Lesson 9-6)

Estimate each percent. (Lesson 7-8)

33. 31% of 157

34. 74% of 45

35. 33% of 92

36. **BUSINESS** The table shows the choices available when ordering a pie from the Taste-n-Tell Bakery. How many different pies are available? (Lesson 7-5)

Taste-n-Tell Bakery	
flavor	apple cherry peach
crust	single double
size	medium large

GET READY for the Next Lesson

PREREQUISITE SKILL Round each number to the nearest tenth. (Lesson 3-3)

37. 43.363

38. 9.8767

39. 37.6219

40. 42.961

Explore 10-2

Measurement Lab
Circumference

MAIN IDEA

Describe the relationship between the diameter and circumference of a circle.

In this investigation, you will discover the relationship between the distance around a circle (circumference), and the distance across a circle through its center (diameter).

ACTIVITY

STEP 1 Make a table like the one shown.

Object	C	d	$\frac{C}{d}$

STEP 2 Cut a piece of string the length of the distance C around a circular object such as a jar lid. Use a centimeter ruler to measure the length of the string to the nearest tenth of a centimeter.

STEP 3 Measure the distance d across the lid. Record this measurement in the table.

STEP 4 Use a calculator to find the ratio of the distance around each circle to the distance across the circle.

STEP 5 Repeat steps 2 though 4 for several other circular objects.

Analyze the Results

1. **MAKE A CONJECTURE** If you know the diameter of a circle, how can you find the distance around the circle?
2. **MAKE A PREDICTION** What would be the approximate distance around a circle that is 4 inches across?
3. **MAKE A CONJECTURE** How can you find the distance around a circle if you know the distance from the center of the circle to the edge of the circle?

10-2 Circles and Circumference

MAIN IDEA

Estimate and find the circumference of circles.

New Vocabulary

circle
center
chord
diameter
circumference
radius

Math Online

glencoe.com

- Extra Examples
- Personal Tutor
- Self-Check Quiz

GET READY for the Lesson

DREAMCATCHERS The table shows the approximate distance around (circumference), the distance across through the center (diameter), and the distance from the center to the edge (radius) of several dreamcatchers.

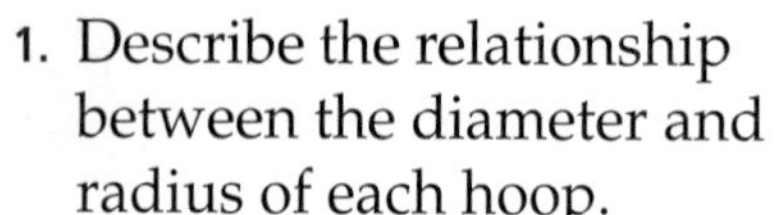

1. Describe the relationship between the diameter and radius of each hoop.
2. Describe the relationship between the circumference and diameter of each hoop.

Circumference (in.)	Diameter (in.)	Radius (in.)
9.4	3	1.5
37.7	12	6
62.8	20	10

A **circle** is the set of all points in a plane that are the same distance from a point called the **center**. A **chord** is any segment with both endpoints on the circle.

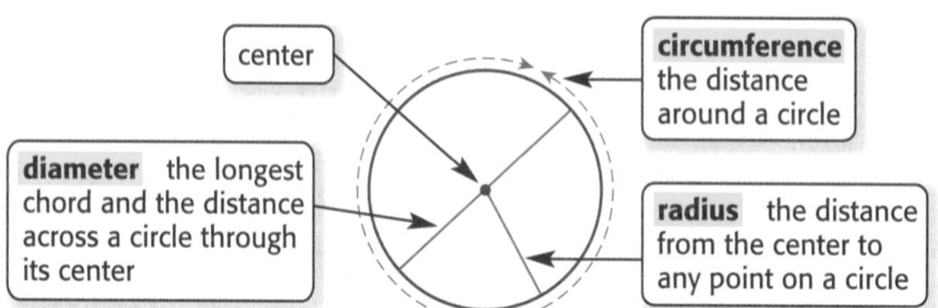
center
circumference the distance around a circle
diameter the longest chord and the distance across a circle through its center
radius the distance from the center to any point on a circle

Radius and Diameter — Key Concept

Words The diameter d of a circle is twice its radius r. The radius r of a circle is half of its diameter d.

Symbols $d = 2r$ $\qquad r = \frac{d}{2}$

EXAMPLES Find the Radius and Diameter

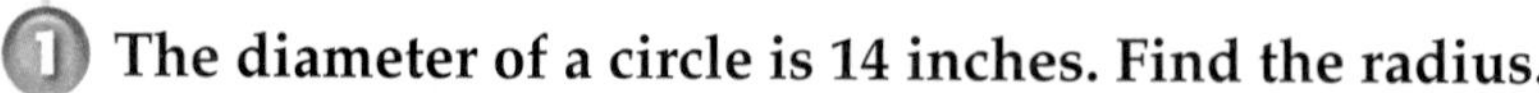

1 The diameter of a circle is 14 inches. Find the radius.

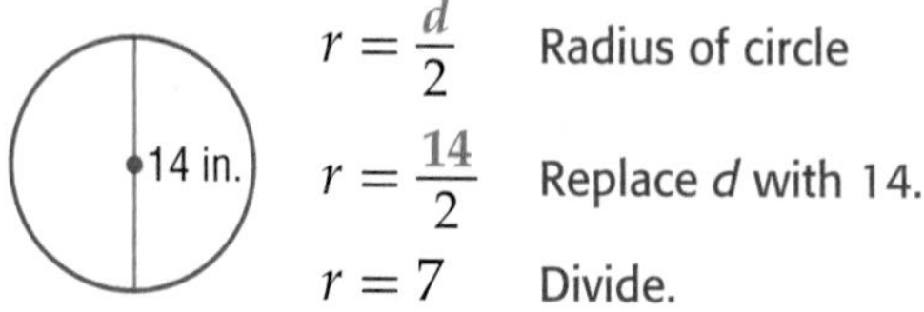
14 in.

$r = \frac{d}{2}$ Radius of circle

$r = \frac{14}{2}$ Replace d with 14.

$r = 7$ Divide.

The radius is 7 inches.

2 **The radius of a circle is 8 feet. Find the diameter.**

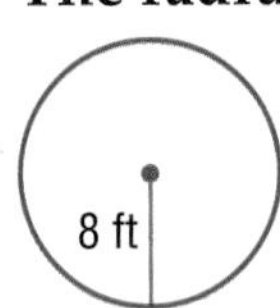

$d = 2r$	Diameter of circle
$d = 2 \cdot 8$	Replace r with 8.
$d = 16$	Multiply.

The diameter is 16 feet.

 **CHECK Your Progress**

Find the radius or diameter of each circle with the given dimension.

a. $d = 23$ cm **b.** $r = 3$ in. **c.** $d = 16$ yd

Study Tip

Pi The exact value of pi never ends, but it is often approximated as 3 or 3.14.

The circumference of a circle is always a little more than three times its diameter. The exact number of times is represented by the Greek letter π (pi). The exact value of π is 3.1415926....

Circumference — Key Concept

Words The circumference of a circle is equal to π times its diameter or π times twice its radius.

Symbols $C = \pi d$ or $C = 2\pi r$

Model

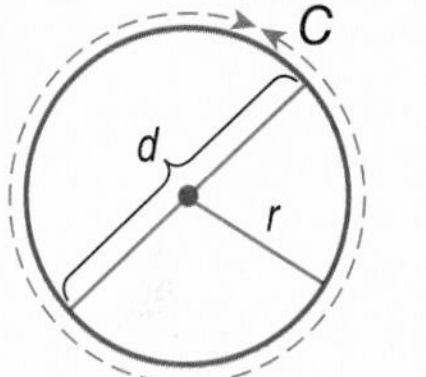

You can estimate the circumference of a circle by rounding the value of π to 3.

EXAMPLES Estimate the Circumference

Estimate the circumference of each circle.

Symbols The symbol ≈ means *approximately equal to.*

3 The diameter of the circle is 9 centimeters.

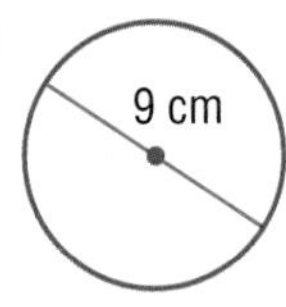

$C = \pi d$	Circumference of circle
$C \approx 3 \cdot 9$	Replace π with 3 and d with 9.
$C \approx 27$	Multiply.

The circumference is about 27 centimeters.

4 The radius of the circle is 6 inches.

$C = 2\pi r$	Circumference of circle
$C \approx 2 \cdot 3 \cdot 6$	Replace π with 3 and r with 6.
$C \approx 36$	Multiply.

The circumference is about 36 inches.

d. $d = 7$ in. **e.** $r = 5$ ft **f.** $r = 12$ mm

EXAMPLE Find Circumference

5 **Find the circumference of a circle with a diameter of 4 inches. Round to the nearest tenth.**

METHOD 1 **Use 3.14.**

$C = \pi d$

$C \approx (3.14)(4)$

$C \approx 12.56$

METHOD 2 **Use a calculator.**

$C = \pi d$

$C = \pi \cdot 4$

$C \approx$ [2nd] [π] [×] 4 [ENTER]

≈ 12.56637061

To the nearest tenth, the circumference is 12.6 inches.

Study Tip

Check for Reasonableness In Example 5, since $3 \times 4 = 12$ and 12.6 is close to 12, the answer is reasonable.

CHOOSE Your Method

g. Find the circumference of a circle with a diameter of 15 meters. Round to the nearest tenth.

TEST EXAMPLE

6 A bicycle wheel has spokes for support. Each spoke extends from the center of the wheel to the rim. Which method can be used to find the circumference of the bicycle wheel?

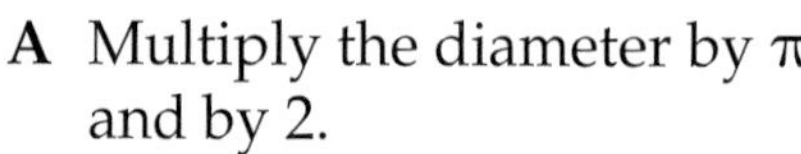

A Multiply the diameter by π and by 2.

B Divide the diameter by π.

C Multiply the radius by π.

D Multiply the radius by π and by 2.

Test-Taking Tip

Formulas Many standardized tests have a list of formulas you may need to solve problems. However, it is always a good idea to familiarize yourself with the formulas before taking the test.

Read the Item You need to find the circumference of the bicycle tire. You know the radius of the wheel.

Solve the Item Use the formula for the circumference of a circle, $C = 2\pi r$. The formula states that the circumference of a circle is equal to 2 times π times the radius. So, the answer is D.

CHECK Your Progress

h. An above-ground circular swimming pool is 18 feet in diameter. How does the pool's diameter d compare to its circumference C?

F $d \approx \frac{1}{2}C$

G $d \approx 2C$

H $d \approx 3C$

J $d \approx \frac{1}{3}C$

Examples 1, 2 (pp. 528–529)

Find the radius or diameter of each circle with the given dimension.

1. $d = 3$ m
2. $r = 14$ ft
3. $d = 20$ in.

Examples 3, 4 (p. 529)

Estimate the circumference of each circle.

4. 

5. 

6. 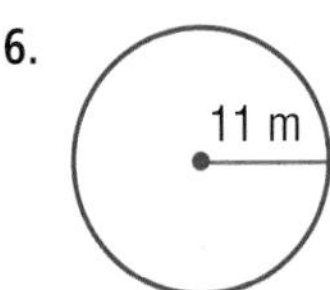

Example 5 (p. 530)

Find the circumference of each circle. Round to the nearest tenth.

7. 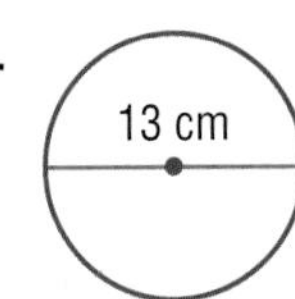

8. 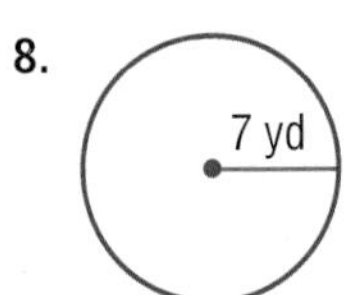

9. 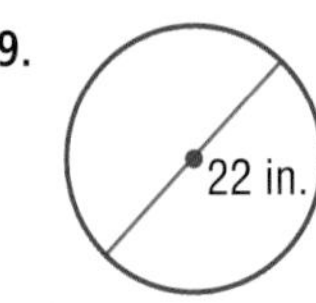

Example 6 (p. 530)

10. **MULTIPLE CHOICE** Paul knows the circumference of Earth around the Equator but would like to find the radius. Which method can Paul use to find the radius of Earth?

    A Multiply the circumference by the diameter.

    B Divide the circumference by π and then divide by 2.

    C Multiply the circumference by π.

    D Divide the circumference by π and then multiply by 2.

Practice and Problem Solving

HOMEWORK HELP

For Exercises	See Examples
11–14	1, 2
15–20	3, 4
21–28	5
38, 39	6

Find the radius or diameter of each circle with the given dimensions.

11. $d = 5$ mm
12. $d = 24$ ft
13. $r = 17$ cm
14. $r = 36$ in.

Estimate the circumference of each circle.

15. 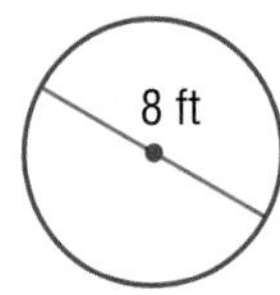

16. 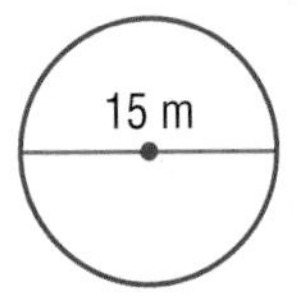

17. 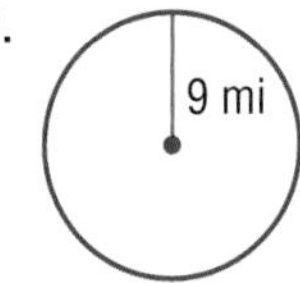

18. $r = 15$ yd
19. $d = 13$ ft
20. $d = 27$ cm

Find the circumference of each circle. Round to the nearest tenth.

21. 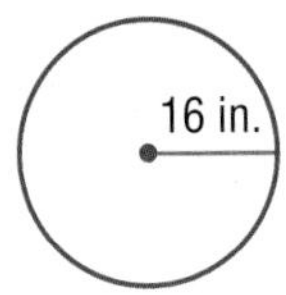

22. 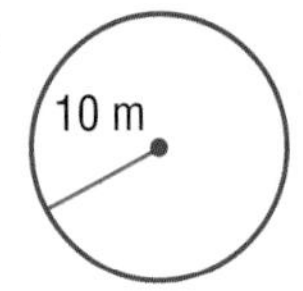

23. 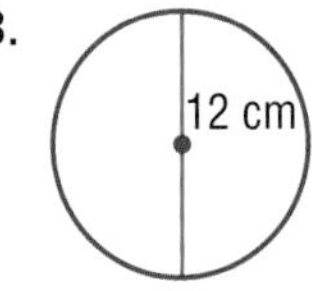

24. $d = 28$ ft
25. $r = 21$ mm
26. $r = 35$ in.

Real-World Link In California and Oregon, many shield volcanoes have diameters of three or four miles.

Source: U.S. Geological Survey

27. **MUSIC** The diameter of a music CD is 12 centimeters. Find the circumference of a CD to the nearest tenth.

28. **VOLCANOES** The Belknap shield volcano is located in Oregon. The volcano is circular and has a diameter of 5 miles. What is the circumference of this volcano to the nearest tenth?

29. **TREES** The largest tree in the world by volume is The General Sherman Tree in Sequoia National Park. The diameter at the base is 36 feet. If a person with outstretched arms can reach 6 feet, how many people would it take to reach around the base of the tree?

30. **WALKING** At a local park, Dawn can choose between two circular paths to walk. One path has a diameter of 120 yards, and the other has a radius of 45 yards. How much farther can Dawn walk on the longer path than the shorter path if she walks around the path once?

31. **ESTIMATION** Without calculating, determine if the circumference of a circle with a radius of 4 feet will be greater or less than 24 feet. Explain your reasoning.

32. **FIND THE DATA** Refer to the Data File on pages 16–19. Choose some data and write a real-world problem in which you would estimate the circumference of a circular object.

EXTRA PRACTICE
See pages 698, 715.

33. **ESTIMATION** Catalina is giving pillar candles as favors at her birthday party. She wants to glue a piece of ribbon around each candle. The diameter of each candle is 4 inches. She has 8 candles and 2 yards of ribbon. Does she have enough ribbon? Explain.

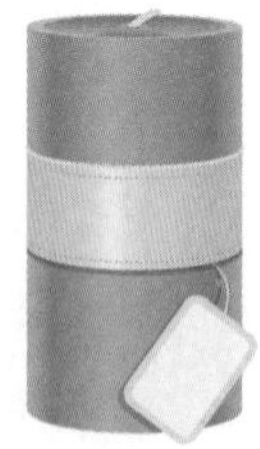

H.O.T. Problems

34. **OPEN ENDED** Draw and label a circle that has a diameter more than 5 inches, but less than 10 inches. Estimate its circumference and then find its circumference. Then compare your estimate to the value you found on your calculator.

35. **FIND THE ERROR** Bena and Orlando are using a calculator to find the circumference of a circle with a radius of 7 inches. Who entered the correct keystrokes to find the circumference? Explain your reasoning.

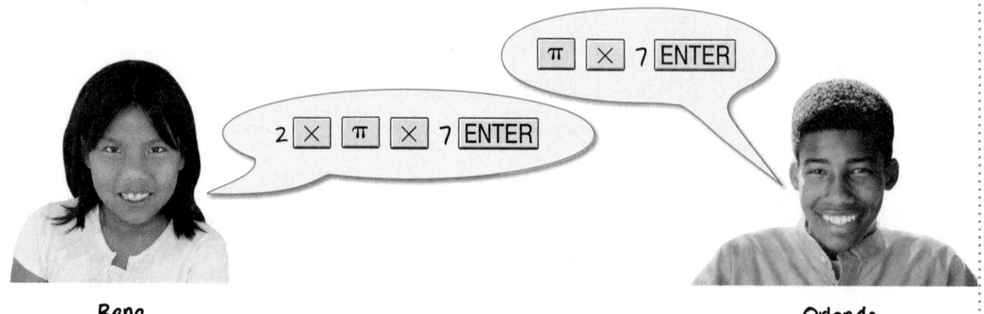

36. **CHALLENGE** Analyze how the circumference of a circle would change if the diameter was doubled. Provide an example to support your explanation.

37. **WRITING IN MATH** Explain how you could estimate the diameter of a circle with a circumference of 15.7 meters.

TEST PRACTICE

38. A circle with center at point *O* is shown below.

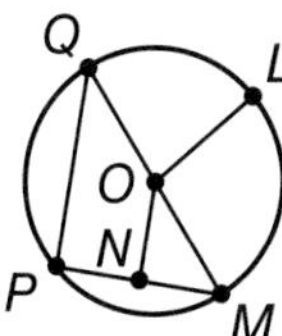

Which line segment is half the length of diameter *QM*?

A Segment *ON*

B Segment *PM*

C Segment *QP*

D Segment *OL*

39. The circumference of the Ferris wheel at the county fair is stated in the local newspaper. Which method can you use to find the diameter of the Ferris wheel?

F Multiply the circumference by π.

G Multiply the circumference by 2 and divide by the radius.

H Divide the circumference by π.

J Divide the circumference by the radius and multiply by 2.

Spiral Review

Find the perimeter of each rectangle with the dimensions given. (Lesson 10-1)

40. 4 inches by 7 inches

41. 15 feet by 17 feet

42. 17 yards by 24 yards

43. 25 miles by 15 miles

State whether each triangle is similar to triangle *XYZ*. (Lesson 9-7)

44.

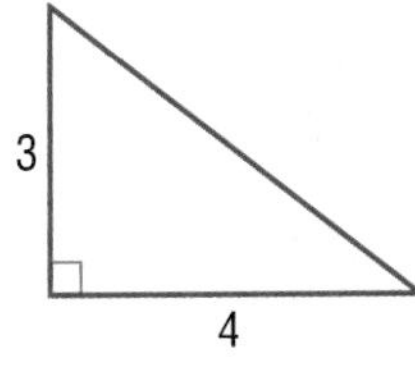

45.

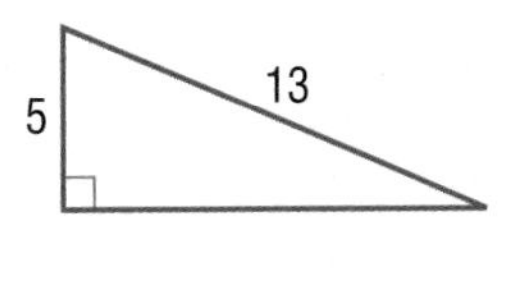

46.

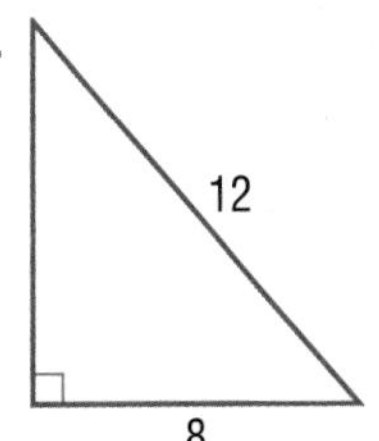

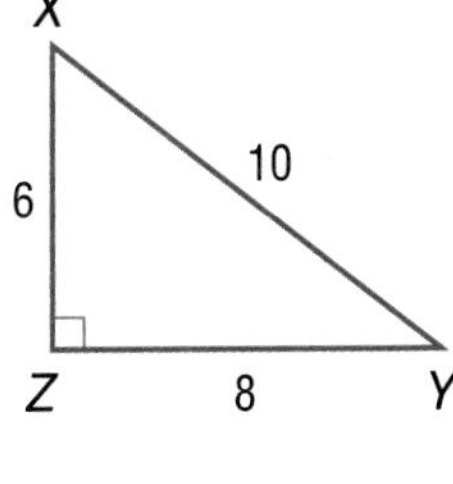

47. **GAMES** At the county fair, Alejandra tosses a beanbag onto an alphabet board. It is equally likely that the bag will land on any letter. Find the probability that the beanbag will land on one of the letters in her name. (Lesson 7-4)

48. **BABYSITTING** Arianna started babysitting at 5:30 P.M. The children's parents were home at 9:15 P.M. How long did Arianna babysit? (Lesson 8-7)

GET READY for the Next Lesson

PREREQUISITE SKILL Multiply. (Page 744)

49. 6×17

50. 11×13

51. 20×9

52. 18×27

10-3 Area of Parallelograms

MAIN IDEA
Find the areas of parallelograms.
New Vocabulary
base
height
Math Online
glencoe.com
• Extra Examples
• Personal Tutor
• Self-Check Quiz

MINI Lab

STEP 1 Draw and then cut out a rectangle as shown.

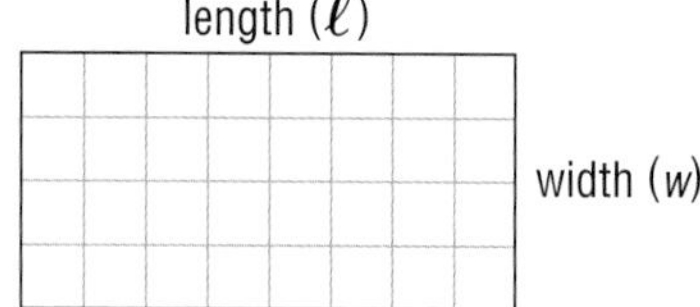

STEP 2 Cut a triangle from one side of the rectangle and move it to the other side to form a parallelogram.

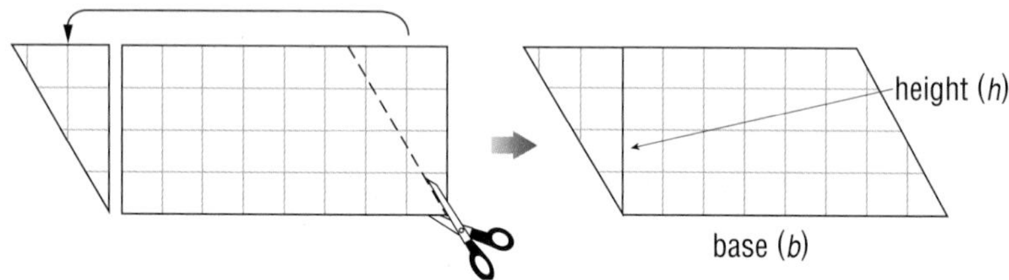

STEP 3 Repeat Steps 1 and 2 with two other rectangles of different dimensions on grid paper.

STEP 4 Copy and complete the table below using the three rectangles and three corresponding parallelograms you created.

	Length (ℓ)	Width (w)		Base (b)	Height (h)
Rectangle 1			Parallelogram 1		
Rectangle 2			Parallelogram 2		
Rectangle 3			Parallelogram 3		

1. How does a parallelogram relate to a rectangle?
2. What part of the parallelogram corresponds to the length of the rectangle?
3. What part corresponds to the rectangle's width?
4. **MAKE A CONJECTURE** What is the formula for the area of a parallelogram?

In the Mini Lab, you discovered how the area of a parallelogram is related to the area of a rectangle.

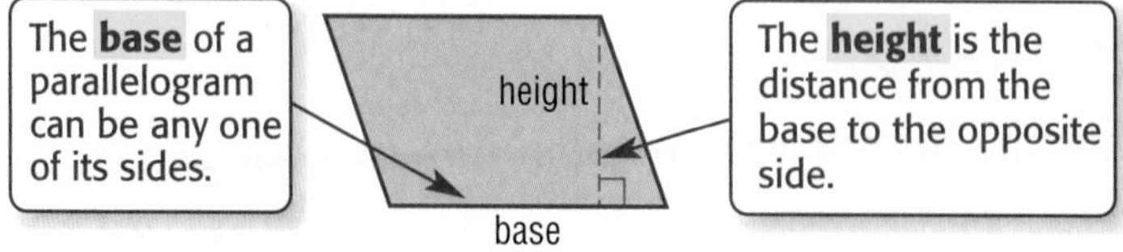

The **base** of a parallelogram can be any one of its sides.
height
base
The **height** is the distance from the base to the opposite side.

To find the area of a parallelogram, multiply the measures of the base and the height.

Area of a Parallelogram

Key Concept

Words The area A of a parallelogram is the product of any base b and its height h.

Model

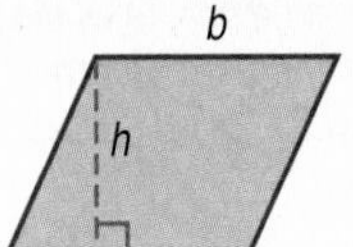

Symbols $A = bh$

EXAMPLES Find Areas of Parallelograms

Find the area of each parallelogram.

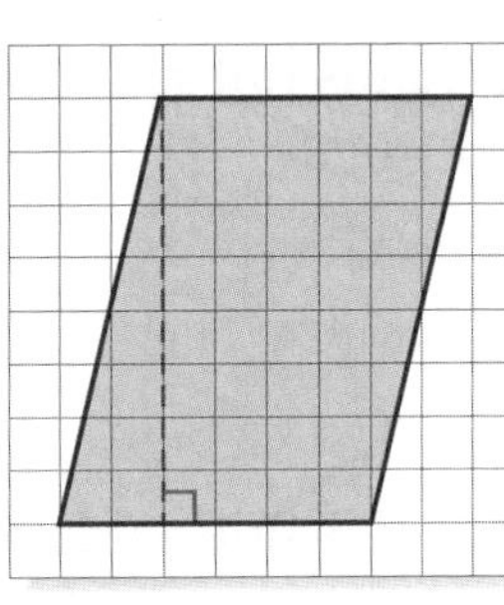

The base is 6 units, and the height is 8 units.

$A = bh$ Area of parallelogram

$A = 6 \cdot 8$ Replace b with 6 and h with 8.

$A = 48$ Multiply.

The area is 48 square units or 48 units^2.

Reading Math

Area Measurement An area measurement can be written using abbreviations and an exponent of 2. For example:
square units = units^2
square inches = in^2
square feet = ft^2
square meters = m^2

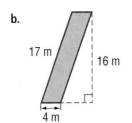

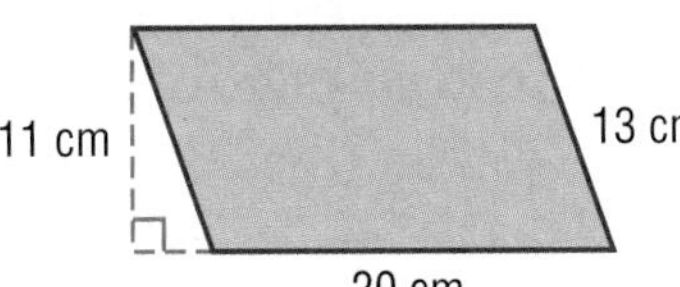

Estimate $A \approx 20 \cdot 10$ or 200 cm^2

$A = bh$ Area of parallelogram

$A = 20 \cdot 11$ Replace b with 20 and h with 11.

$A = 220$ Multiply.

The area is 220 square centimeters or 220 cm^2.

Check for Reasonableness Compare 220 with the estimate. $220 \approx 200$ ✔

CHECK Your Progress

a.

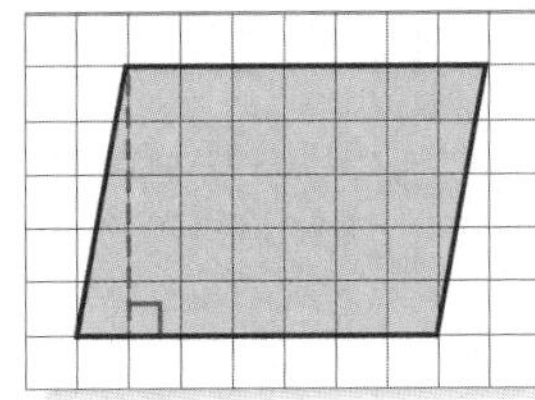

b.

17 m
16 m
4 m

Study Tip

Height of Parallelograms For the parallelogram formed by the area shaded black in Example 3, its height, 12 inches, is labeled *outside* the parallelogram.

Real-World EXAMPLE

3 FLAGS Romilla is doing a research project on the nation of Trinidad and Tobago. Part of the project is to paint a replica of the nation's flag. Find the area of the flag that is not red.

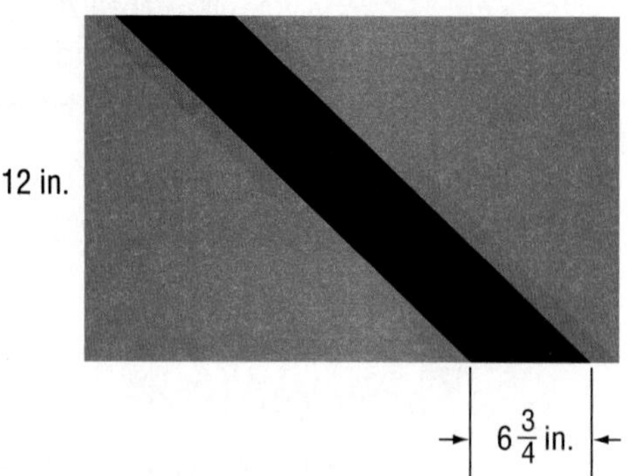

The area of the flag that is not red is shaped like a parallelogram, so use the formula $A = bh$.

$A = bh$	Area of parallelogram
$A = 6\frac{3}{4} \cdot 12$	Replace b with $6\frac{3}{4}$ and h with 12.
$A = 81$	$6\frac{3}{4} \cdot 12 = \frac{27}{4} \cdot 12$, or 81

The area of the flag that is not red is 81 square inches.

CHECK Your Progress

c. **ART** Guadalupe and her dad made parallelogram-shaped picture frames to display some of her artwork. Find the area of the artwork that will be visible in each picture frame.

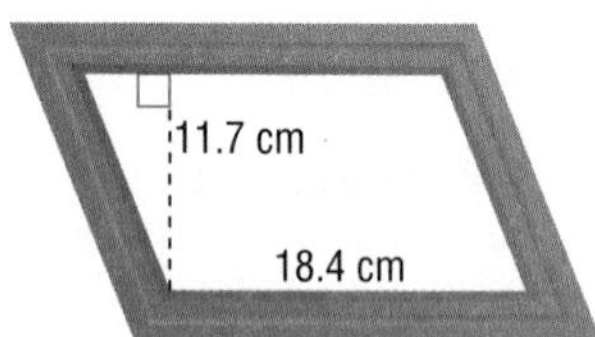

CHECK Your Understanding

Examples 1, 2 (p. 535)

Find the area of each parallelogram.

1.

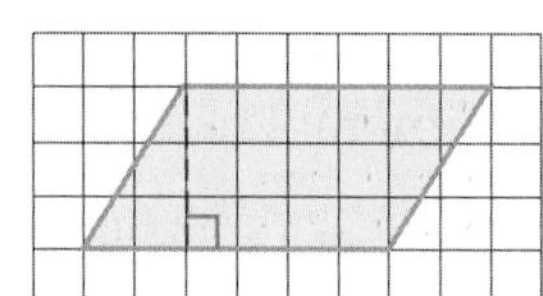

2.

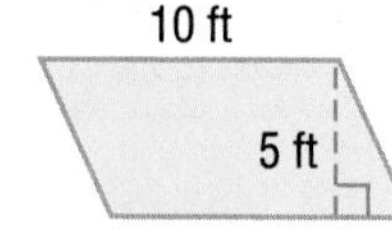

3.

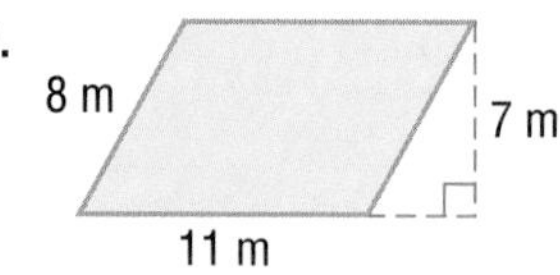

Example 3 (p. 536)

4. Find the area of a parallelogram with base 15 yards and height $21\frac{2}{3}$ yards.

5. **TANGRAMS** The size of the parallelogram piece in a set of tangrams is shown at the right. Find the area of the piece.

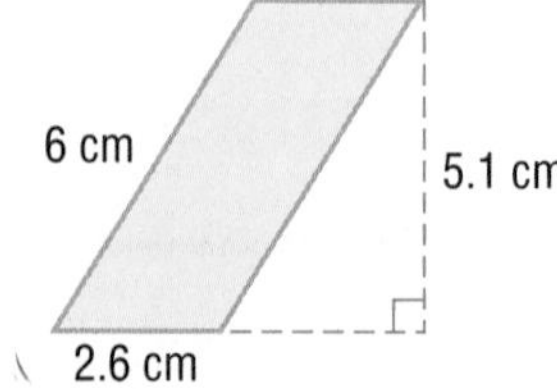

Practice and Problem Solving

HOMEWORK HELP

For Exercises	See Examples
6–11	1, 2
12–15	3

Find the area of each parallelogram.

6. 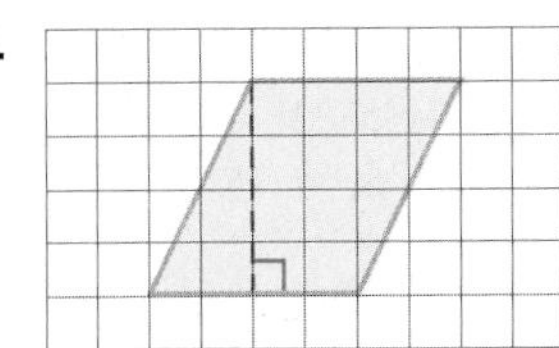

7. 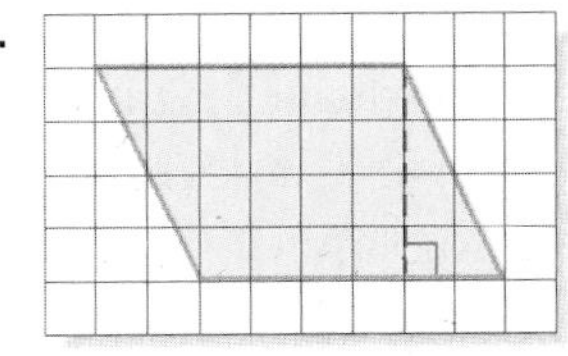

8. 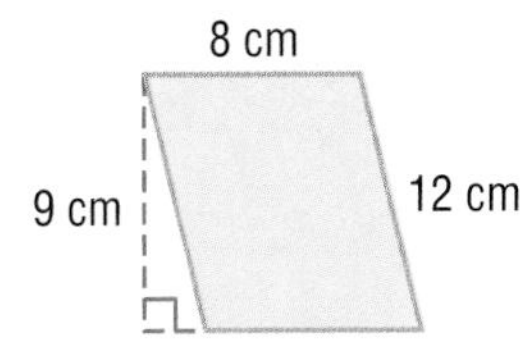

9. 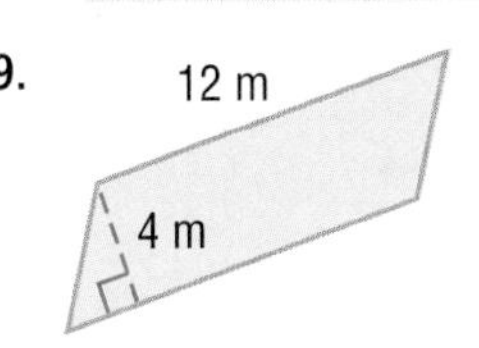

10. 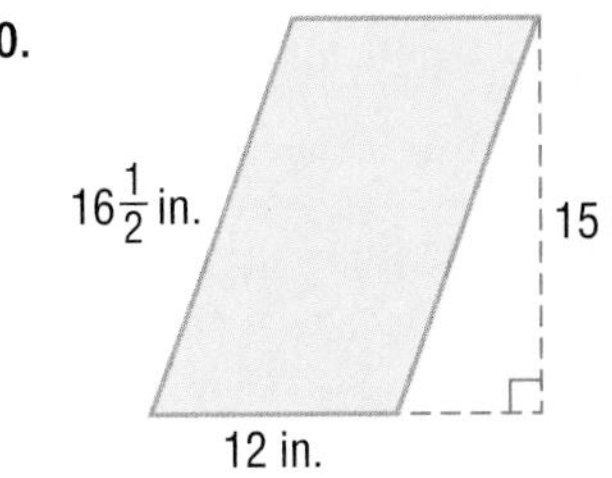

11. 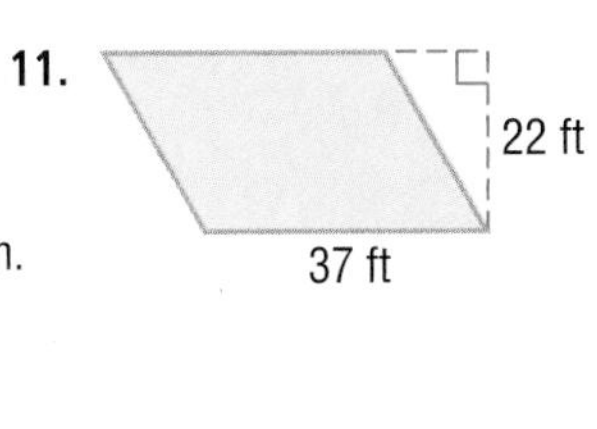

12. Find the area of a parallelogram with base 24 feet and height $2\frac{1}{4}$ feet.

13. Find the area of a parallelogram with base 6.75 meters and height 4.8 meters.

14. **PARKING** Find the area of the parking space below.

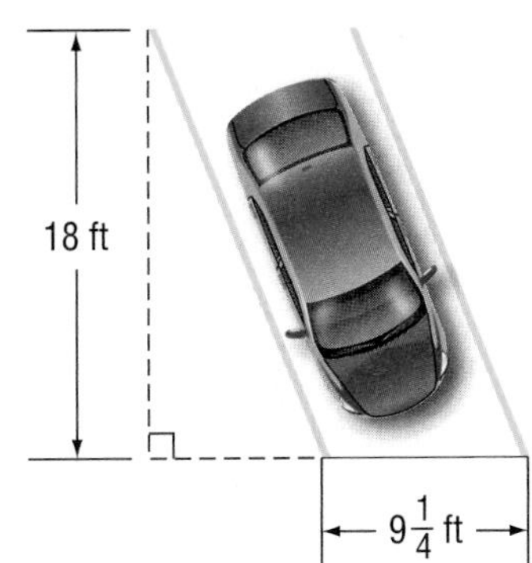

15. **MAPS** What is the area of the region shown on the map?

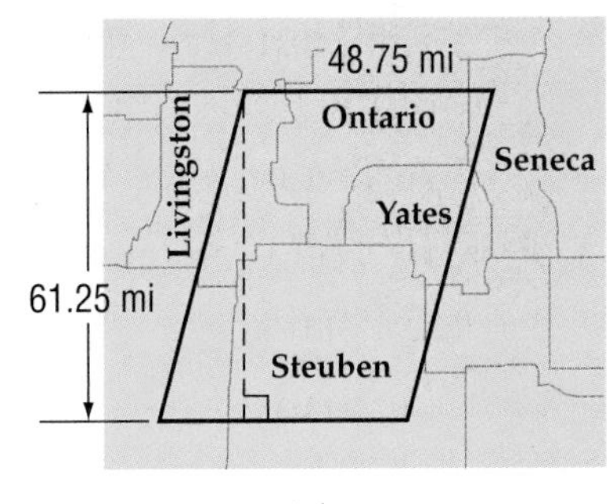

Find the area of the shaded region in each figure.

16. 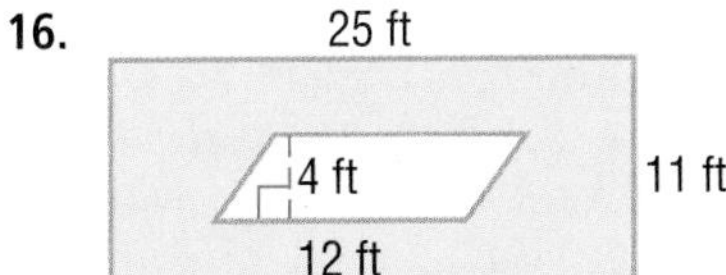

17. 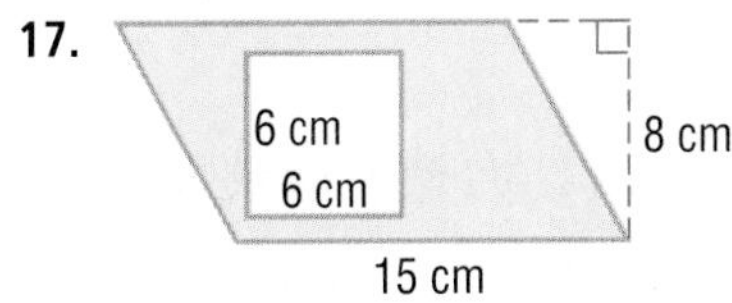

18. **BUILDINGS** The base of a building is shaped like a parallelogram. The first floor has an area of 20,000 square feet. If the base of this parallelogram is 250 feet, can its height be 70 feet? Explain.

19. **ANALYZE TABLES** An architect designed three different parallelogram-shaped brick patios. Find the missing dimensions in the table.

Patio	Base (ft)	Height (ft)	Area (ft²)
1	$15\frac{3}{4}$	■	147
2	■	$11\frac{1}{4}$	$140\frac{5}{8}$
3	$10\frac{1}{4}$	■	$151\frac{3}{16}$

EXTRA PRACTICE
See pages 698, 715.

H.O.T. Problems

20. **REASONING** Refer to parallelogram *KLMN* at the right. If the area of parallelogram *KLMN* is 35 square inches, what is the area of triangle *KLN*?

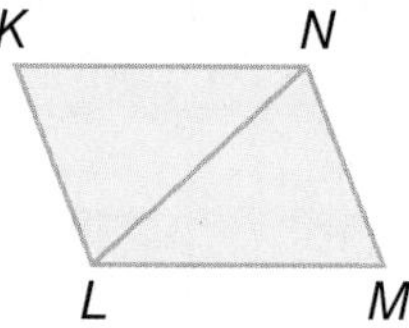

21. **OPEN ENDED** On grid paper, draw three different parallelograms that each have an area of 24 units and a height of 4 units. Compare and contrast the parallelograms.

22. **CHALLENGE** If $x = 5$ and $y < x$, which figure has the greater area? Explain your reasoning.

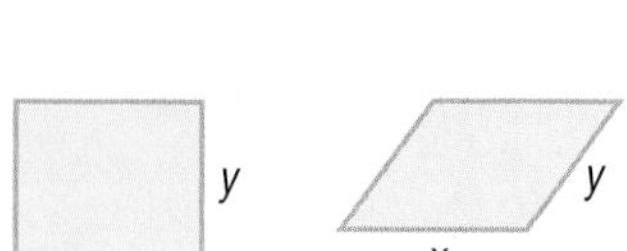

23. **WRITING IN MATH** Explain how the formula for the area of a parallelogram is related to the formula for the area of a rectangle.

TEST PRACTICE

24. Robert used a piece of poster board shaped like a parallelogram to make a sign for his campaign as class president. The base of the poster board is 52 inches, and the area is 1,872 square inches. Find the height of the poster board.

A 884 in.

B 176 in.

C 42 in.

D 36 in.

25. A family has a flower garden in the shape of a parallelogram in their backyard. They planted grass in the rest of the yard. What is the area of the backyard that is planted with grass?

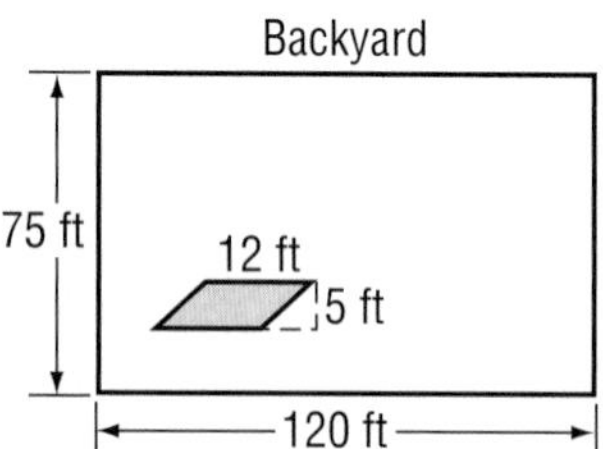

F 390 sq ft

G 8,940 sq ft

H 9,060 sq ft

J 9,144 sq ft

Spiral Review

Estimate the circumference of each circle. (Lesson 10-2)

26. $d = 15$ in.

27. $r = 19$ m

28. $d = 6$ ft

29. **MONUMENTS** The Lincoln Memorial is a rectangular structure whose base is 188 feet by 118 feet. What is the perimeter of the base of the Lincoln Memorial? (Lesson 10-1)

GET READY for the Next Lesson

PREREQUISITE SKILL Find the value of each expression. (Lesson 1-4)

30. $\frac{6 \times 3}{2}$

31. $\frac{5 \times 12}{2}$

32. $\frac{7 \times 8}{2}$

33. $\frac{14 \times 12}{2}$

Explore 10-4

Measurement Lab
Area of Triangles

MAIN IDEA

Discover the formula for the area of a triangle using the properties of parallelograms and a table of values.

In this investigation, you will discover the formula for the area of a triangle using the properties of parallelograms and a table of values.

ACTIVITY

STEP 1 Copy the table shown.

Parallelogram	Base, b	Height, h	Area of Parallelogram	Area of Each Triangle
A	4	6		
B	2	5		
C	3	4		
D	5	3		
E	7	5		

STEP 2 Draw Parallelogram A on grid paper using the dimensions given in the table.

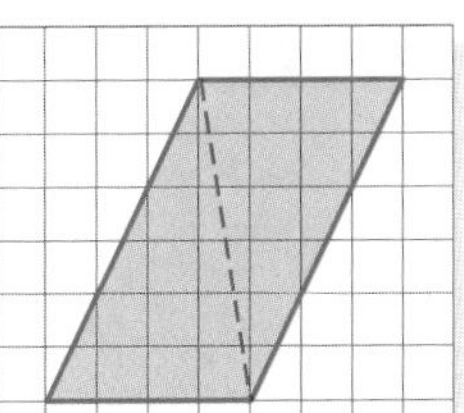

STEP 3 Draw a diagonal as shown.

STEP 4 Cut out the parallelogram. Then calculate its area. Record this measure in the table.

STEP 5 Cut along the diagonal to form two triangles.

Analyze the Results

1. Compare the base and height of each triangle to the base and height of the original parallelogram. What do you notice?
2. Compare the two triangles formed. How are they related?
3. What is the area of each triangle? Record your answer in the table.
4. Repeat Steps 2 through 5 for Parallelograms B through E. Calculate the area of each triangle formed and record your results in the table.
5. **LOOK FOR A PATTERN** What patterns do you notice in the rows of the table?
6. **MAKE A CONJECTURE** Write a formula that relates the area A of a triangle to the length of its base b and height h.

10-4 Area of Triangles

MAIN IDEA

Find the areas of triangles.

Math Online

glencoe.com

- Concepts in Motion
- Extra Examples
- Personal Tutor
- Self-Check Quiz

GET READY for the Lesson

BIOSPHERE The structure of the different sections in the Biosphere 2 complex in Tucson, Arizona, are made of interlocking triangles that are all the same size.

1. Compare the two outlined triangles.
2. What figure is formed by the two triangles?
3. **MAKE A CONJECTURE** Describe the relationship that exists between the area of one triangle and the area of the parallelogram.

A parallelogram can be formed by two congruent triangles. Since congruent triangles have the same area, the area of a triangle is one half the area of the parallelogram.

The base of a triangle can be any one of its sides. The height is the shortest distance from a base to the opposite vertex.

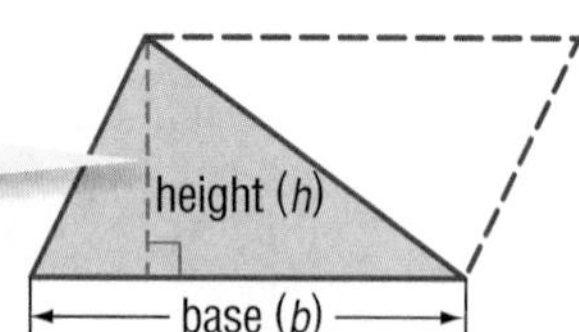

Area of a Triangle — Key Concept

Words The area A of a triangle is one half the product of the base b and its height h.

Symbols $A = \frac{1}{2}bh$ or $A = \frac{bh}{2}$

Model

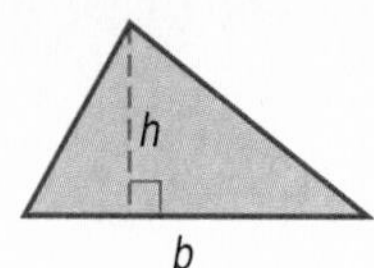

EXAMPLES Find the Area of a Triangle

Find the area of each triangle.

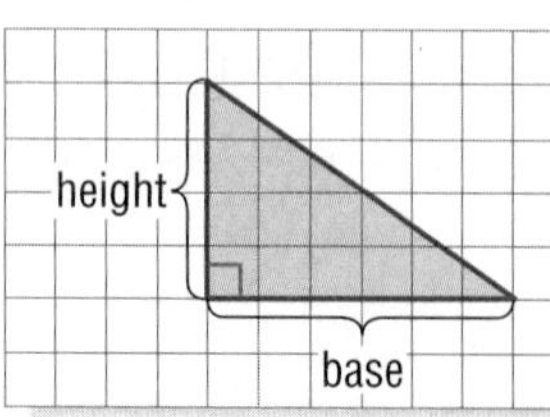

By counting, you find that the measure of the base is 6 units and the height is 4 units.

Study Tip

Mental Math You can use mental math to multiply $\frac{1}{2}(6)(4)$. Think: Half of 6 is 3, and 3×4 is 12.

$A = \frac{1}{2}bh$	Area of a triangle
$A = \frac{1}{2}(6)(4)$	Replace *b* with 6 and *h* with 4.
$A = \frac{1}{2}(24)$	Multiply.
$A = 12$	Multiply.

The area of the triangle is 12 square units.

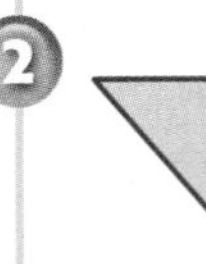

2

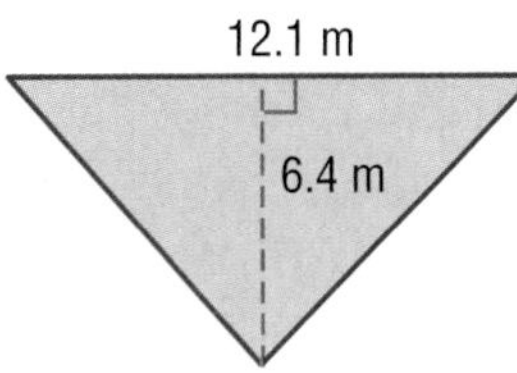

$A = \frac{1}{2}bh$	Area of a triangle
$A = \frac{1}{2}(12.1)(6.4)$	Replace *b* with 17 and *h* with 9.
$A = \frac{1}{2}(77.44)$	Multiply.
$A = 38.72$	Divide. $\frac{1}{2}(77.44) = 77.44 \div 2$, or 38.72

The area of the triangle is 38.72 square meters.

Study Tip

Check for Reasonableness To estimate the area of the triangle in Example 2, round the base to 12 meters and the height to 6 meters. The area is then $\frac{12 \times 6}{2}$ or 36 square meters. Since 38.72 is close to 36, the answer is reasonable.

CHECK Your Progress

a.

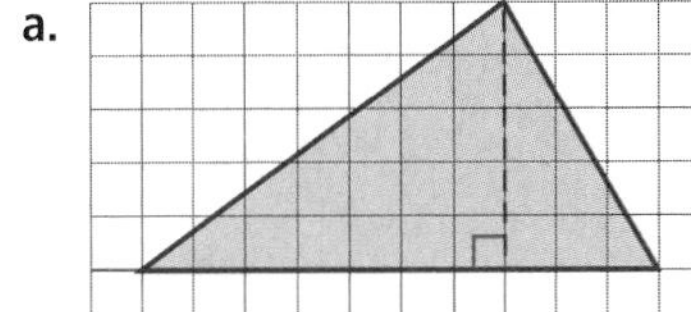

b.

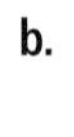

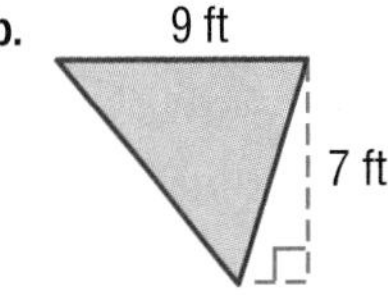

Real-World EXAMPLE

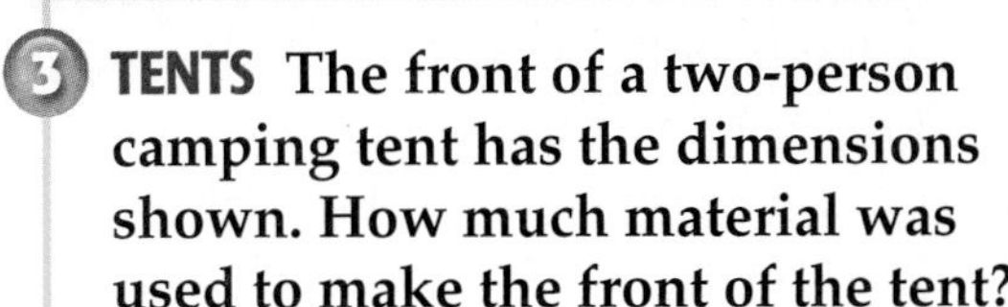

3 **TENTS The front of a two-person camping tent has the dimensions shown. How much material was used to make the front of the tent?**

$A = \frac{1}{2}bh$	Area of a triangle
$A = \frac{1}{2}(5)(3)$	Replace *b* with 5 and *h* with 3.
$A = \frac{1}{2}(15)$ or 7.5	Multiply.

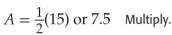

The front of the tent has an area of 7.5 square feet.

CHECK Your Progress

c. **SNACKS** A triangular cracker has a height of 4 centimeters and a base of 5 centimeters. Find the area of the cracker.

CHECK Your Understanding

Examples 1, 2 (pp. 540–541)

Find the area of each triangle.

1. 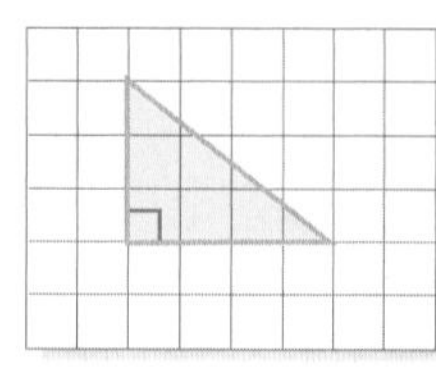

2. 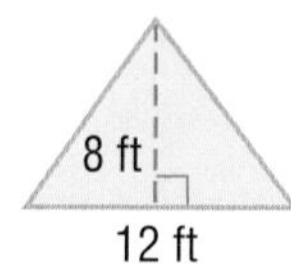

3. 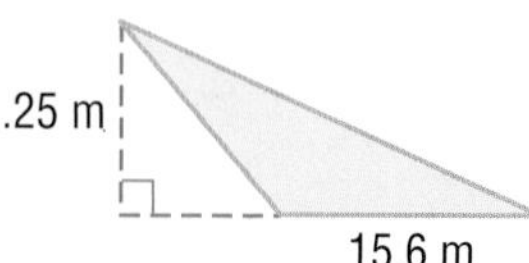

Example 3 (p. 541)

4. **CRAFTS** Consuela made a triangular paper box as shown. What is the area of the top of the box?

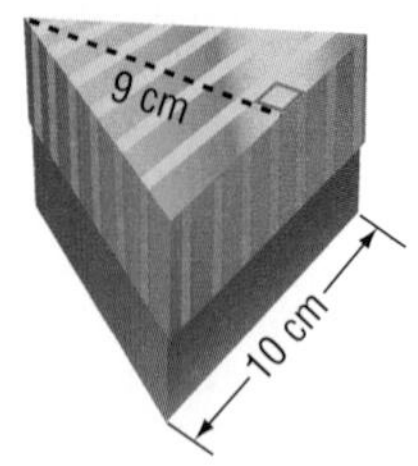

Practice and Problem Solving

HOMEWORK HELP	
For Exercises	See Examples
5, 6	1
7–12	2
13, 14	3

Find the area of each triangle.

5. 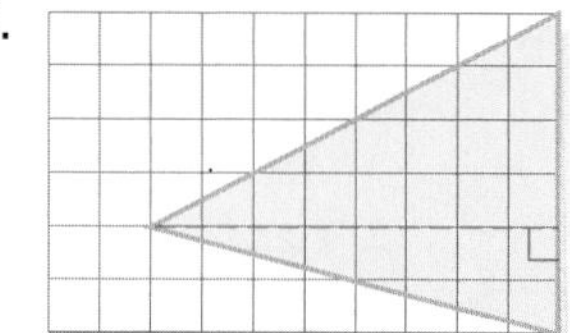

6. 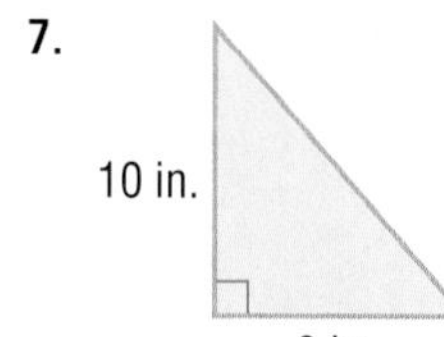

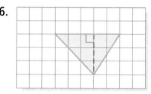

7. 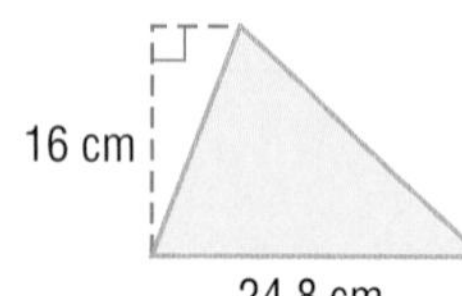

8. 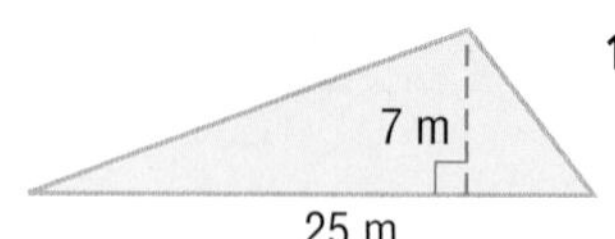

9. 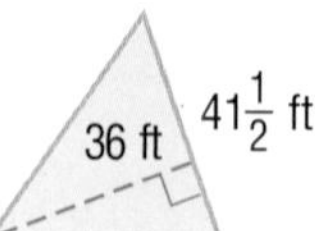

10. 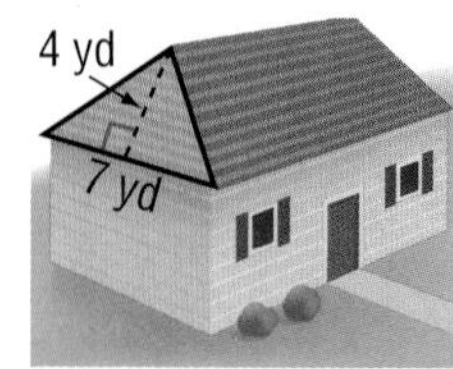

11. height: 14 in., base: 35 in.

12. height: 27 cm, base: 19 cm

13. **ROOFING** Ansley is going to help his father shingle the roof of their house. What is the area of the triangular portion of one end of the roof to be shingled?

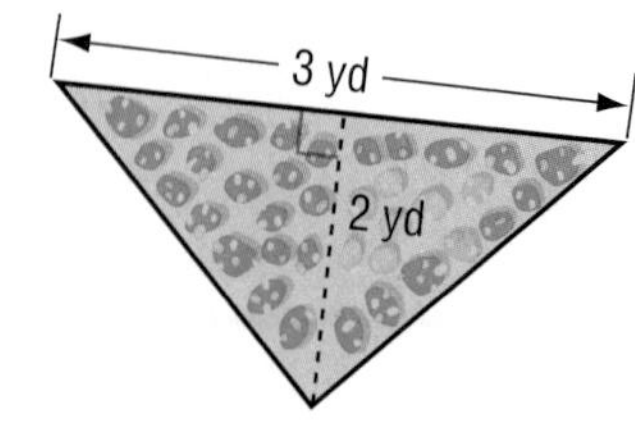

14. **ARCHITECTURE** An architect plans on designing a building on a triangular plot of land. If the base of the triangle is 100.8 feet and the height is 96.3 feet, find the available floor area the architect has to design the building.

15. **FLOWER BEDS** A flower bed in a parking lot is shaped like a triangle as shown. Find the area of the flower bed in square feet. If one bag of topsoil covers 10 square feet, how many bags are needed to cover this flower bed?

16. **ALGEBRA** The table at the right shows the areas of a triangle where the base of the triangle stays the same but the height changes. Write an algebraic expression that can be used to find the area of a triangle that has a base of 5 units and a height of n units.

Area of Triangles

Base (units)	Height (units)	Area (units²)
5	2	5
5	4	10
5	6	15
5	8	20
5	n	■

17. **REASONING** Which is smaller, a triangle with an area of 1 square foot or a triangle with an area of 64 square inches?

18. **FLAGS** What is the area of the triangle on the flag of the Philippines at the right?

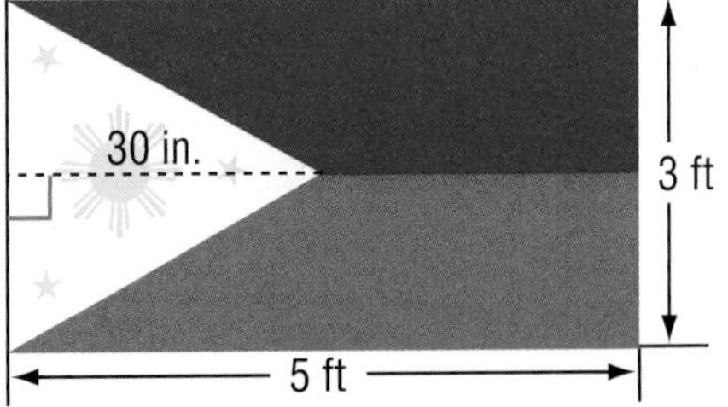

COMPOSITE FIGURES Find the perimeter and area of each figure.

19.
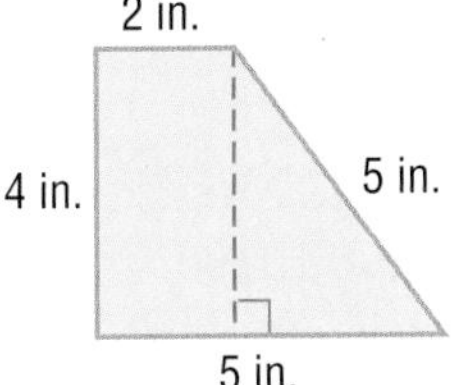

20.
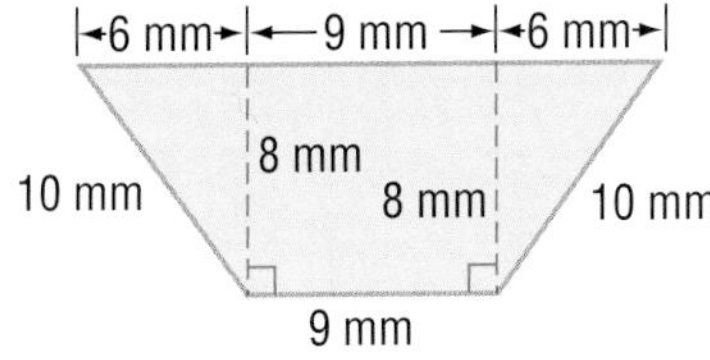

EXTRA PRACTICE
See pages 699, 715.

H.O.T. Problems

21. **FIND THE ERROR** Dolores and Demetrius are finding the base of the triangle shown. Its area is 148.5 square meters. Who is correct? Explain.

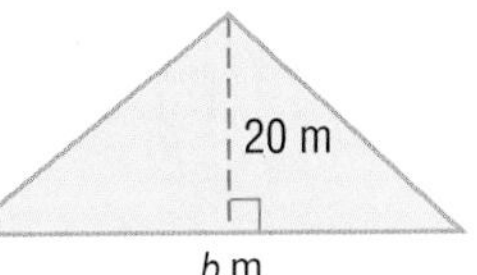

Dolores: $100 = \frac{1}{2}(b)(20)$, $100 = 10b$, $10 = b$

Demetrius: $100 = (b)20$, $100 = 20b$, $5 = b$

CHALLENGE For Exercises 22–25, use the information below.

All the triangles and squares in the quilt pattern shown are congruent.

12 in.

12 in.

22. Find the measure of the base and height of one of the triangles.
23. Calculate the area of one triangle and then find the area of all the triangles.
24. Calculate the area of one of the smaller squares and then find the area of all of the smaller squares.
25. What is the total area of the figure? Is your answer reasonable?

26. **REASONING** If two triangles have an area of 24 square feet, do they always have the same base and height? Use a model to explain your answer.

27. **WRITING IN MATH** Draw a triangle and label its base and height. Draw another triangle that has the same base, but a height twice that of the first triangle. Find the area of each triangle. Then write a ratio that expresses the area of the first triangle to the area of the second triangle.

TEST PRACTICE

28. The table below shows the areas of a triangle where the height of the triangle stays the same but the base changes.

Areas of Triangles		
Height (units)	Base (units)	Area (square units)
7	2	7
7	3	$10\frac{1}{2}$
7	4	14
7	5	$17\frac{1}{2}$
7	x	?

Which expression can be used to find the area of a triangle that has a height of 7 units and a base of x units?

A $7x$

B $\frac{7x}{2}$

C $\frac{7}{2}$

D $\frac{x}{2}$

29. Norma cut a triangle out of construction paper for an art project.

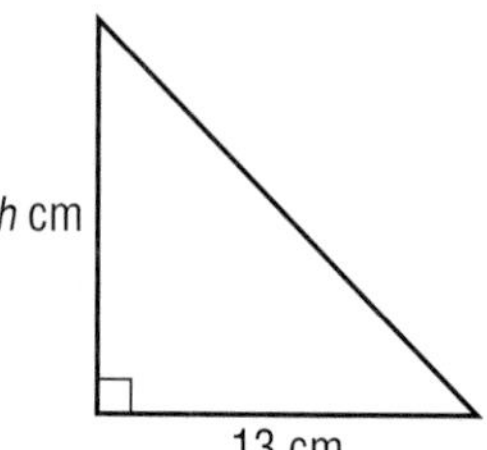

The area of the triangle is 84.5 square centimeters. What is the height of the triangle?

F 6.5 cm

G 13 cm

H 26 cm

J 169 cm

Spiral Review

30. Find the area of a parallelogram with base 15 inches and height 10 inches. (Lesson 10-3)

31. Find the circumference of a circle with a radius of 5 meters. Round to the nearest tenth. (Lesson 10-2)

32. **IDENTIFICATION** Measure the length and width of a student ID card or library card to the nearest eighth inch. Then find the perimeter of the card. (Lessons 1-9 and 10-1)

GET READY for the Next Lesson

33. **PREREQUISITE SKILL** A bookstore arranges its best-seller books in the front window. In how many ways can four best-seller books be arranged in a row? Use the *act it out* strategy. (Lesson 5-3)

CHAPTER 10

Mid-Chapter Quiz

Lessons 10-1 through 10-4

Find the perimeter of each figure. (Lesson 10-1)

1. 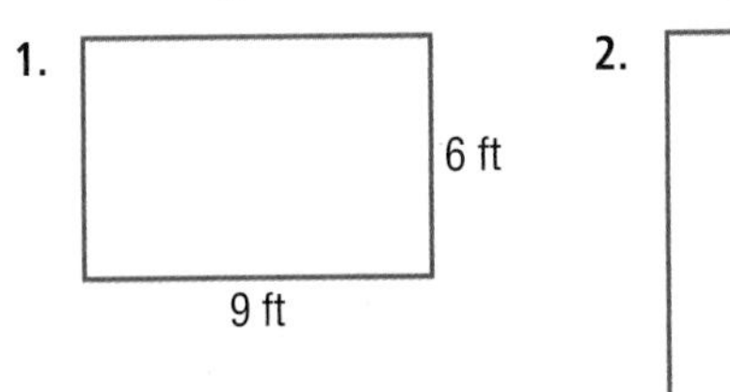

2. 7.6 cm, 3.4 cm

3. **FIELDS** How many feet of fencing is needed to fence a rectangular field 126 feet by 84 feet? (Lesson 10-1)

Find the radius or diameter of each circle with the given dimensions. (Lesson 10-2)

4. $d = 7$ in.
5. $r = 32$ ft
6. $r = 16$ yd
7. $d = 18$ cm

Estimate the circumference of each circle. (Lesson 10-2)

8. 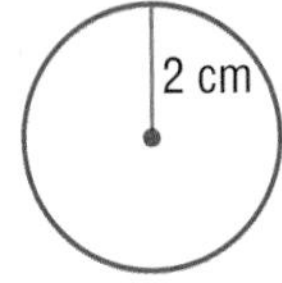

9. 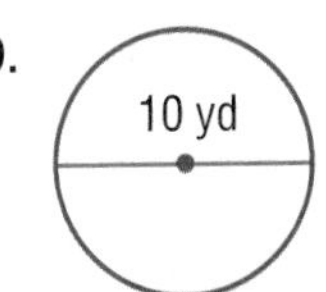

10. **POOLS** Find the circumference of a circular pool with a diameter of 3.7 feet. Round to the nearest tenth. (Lesson 10-2)

11. **MULTIPLE CHOICE** Ernesto knows the circumference of a DVD but would like to find the diameter. Which method can Ernesto use to find the diameter of the DVD? (Lesson 10-2)

    A Multiply the circumference of the DVD by its radius.

    B Divide the circumference of the DVD by π and then divide by 2.

    C Divide the circumference of the DVD by π.

    D Multiply the circumference of the DVD by 2.

Find the area of each parallelogram. (Lesson 10-3)

12. 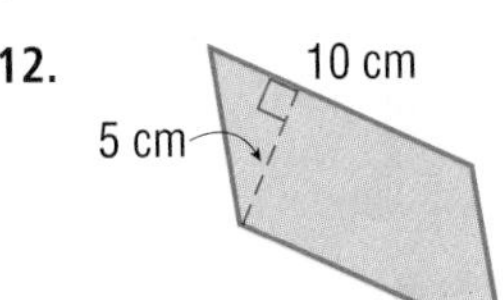

13. 6 ft, $8\frac{1}{2}$ ft, 8 ft

14. Find the area of a parallelogram with base $5\frac{1}{4}$ feet and height $7\frac{1}{2}$ feet. (Lesson 10-3)

15. **MULTIPLE CHOICE** Which expression can be used to find the area of a triangle that has a height of 9 units and a base of n units? (Lesson 10-4)

    F $9n$

    G $\frac{9n}{2}$

    H $\frac{9}{2}$

    J $\frac{n}{2}$

Find the area of each triangle. (Lesson 10-4)

16.

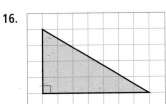

17. 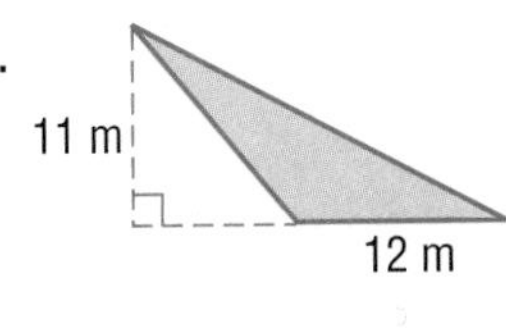

18. **PENNANTS** A pennant for a baseball team is a triangular flag with a base of 12 inches and a height of 30 inches. What is the area of the pennant? (Lesson 10-4)

10-5 Problem-Solving Investigation

MAIN IDEA: Solve problems by making a model.

P.S.I. TEAM +

e-MAIL: MAKE A MODEL

D.J.: I'm helping set up 7 rows of chairs for a school assembly. There are eight chairs in the first row. Each row after that has two more chairs than the previous row. If I have 100 chairs, can I set up enough rows?

YOUR MISSION: Make a model to find whether D.J. has enough chairs to set up all 7 rows.

Understand	You know that each row has two more chairs than the previous row. The first row has 8 chairs and there are 7 rows. You need to determine if 100 chairs are enough.
Plan	Make a model to see if there are enough chairs.
Solve	Use counters to show the layout of the chairs. Row 1 — 8 chairs Row 2 — 10 chairs Row 3 — 12 chairs Row 4 — 14 chairs Row 5 — 16 chairs Row 6 — 18 chairs Row 7 — 20 chairs Add the number of chairs in each row: $8 + 10 + 12 + 14 + 16 + 18 + 20 = 98$ Since $98 < 100$, there are enough chairs.
Check	The average number of chairs in the first and last row is $\frac{8 + 20}{2} = \frac{28}{2}$ or 14. Since there are 7 rows and $7 \times 14 = 98$, the answer is reasonable. ✔

Analyze The Strategy

1. Tell how making a model helped D.J. solve the problem.

2. **WRITING IN MATH** Write a problem that can be solved by making a model.

Mixed Problem Solving

Use the *make a model* strategy to solve Exercises 3–5.

3. **GEOMETRY** For a school assignment, Santiago has to give three different possibilities for the dimensions of a rectangle that has a perimeter of 28 feet and an area greater than 30 square feet. One of the models he made is shown below. What are two other possibilities for the dimensions of the rectangle?

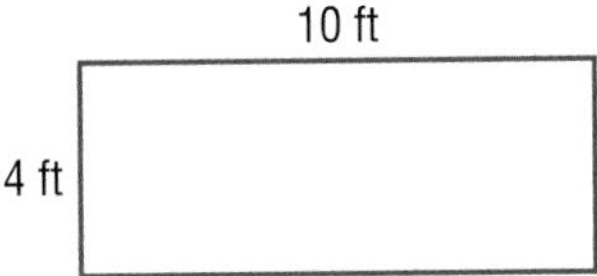

4. **DESIGN** A designer wants to arrange 12 square glass bricks into a rectangular shape with the least perimeter possible. How many blocks will be in each row?

5. **PAPER** Timothy took a piece of notebook paper and cut it in half. Then he placed the 2 pieces on top of each other and cut them in half again to have 4 pieces of paper. If he could keep cutting the paper, how many pieces of paper would he have after 6 cuts?

Use any strategy to solve Exercises 6–13. Some strategies are shown below.

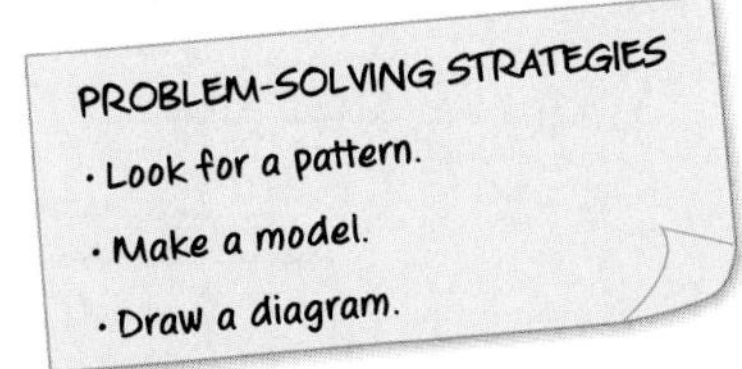

6. **SKATES** Of 50 students surveyed, 22 have a skateboard, and 18 have shoes with wheels. Of those, 6 students have both. How many students have neither a skateboard nor shoes with wheels?

7. **BOOSTERS** In 2008, 25 parents participated in the band booster organization at King Middle School. Participation increased to 40 parents in 2009 and 55 parents in 2010. If the trend continues, about how many parents can be expected to participate in the band booster organization in 2011?

8. **E-MAIL** Meghan sends four friends an e-mail. Each friend then forwards the e-mail to another four friends, and so on. If four friends forward the e-mail to another four friends each hour, how long will it take for 84 friends to receive the e-mail?

9. **ART** Rhonda folded a piece of notebook paper in half twice. Then she punched a hole through all layers. How many holes will there be when she unfolds the paper?

10. **GEOMETRY** The base and height of each triangle are half their length than in the previous triangle. What will be the area of the fourth triangle?

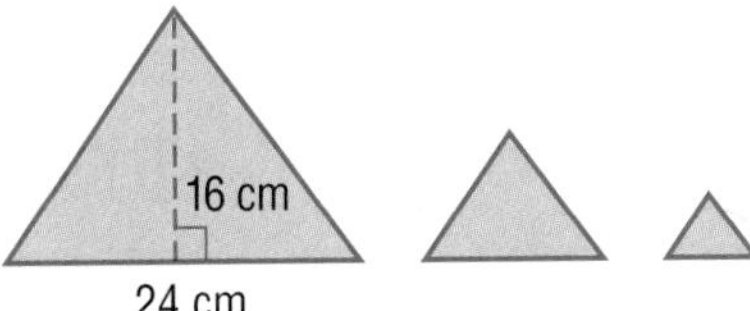

11. **WATER PARKS** What is the total price for two adult and three children one-day passes to a local water park?

	One-day Pass	Two-day Pass
Adults	$40	$45
Child	$30	$35

12. **LOANS** Willow's father purchased a new car. His loan, including interest, is $12,720. How much are his monthly payments if he has 12 payments per year for 5 years?

13. **SOCCER** Refer to the graph. How many more boys signed up for soccer in 2010 than 2008?

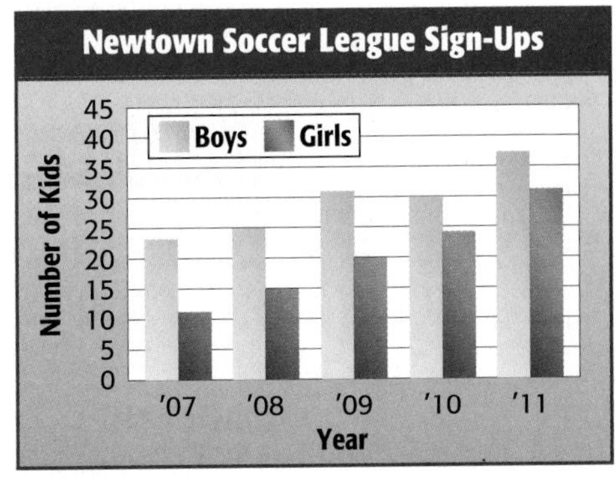

10-6 Volume of Rectangular Prisms

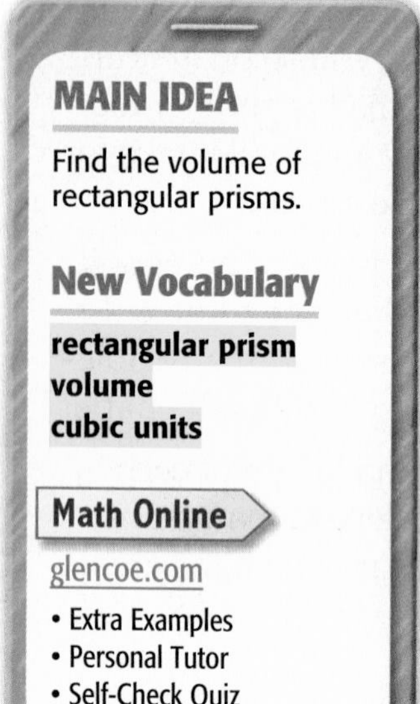

MAIN IDEA

Find the volume of rectangular prisms.

New Vocabulary

rectangular prism
volume
cubic units

Math Online

glencoe.com

- Extra Examples
- Personal Tutor
- Self-Check Quiz

MINI Lab

The figures at the right are *prisms.*

STEP 1 Copy the table below.

Prism	Number of Cubes	Height of Prism	Length of Base	Width of Base	Area of Base
A					
B					
C					
D					
E					

STEP 2 Using centimeter cubes, build five different prisms. For each prism, record the dimensions and the number of cubes used.

1. Examine the rows of the table. What patterns do you notice?
2. **MAKE A CONJECTURE** Describe the relationship between the number of cubes needed and the dimensions of the prism.

A **rectangular prism** is a three-dimensional figure with two parallel bases that are congruent rectangles.

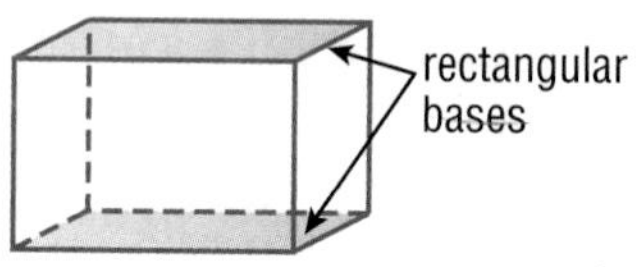

Volume is the amount of space inside a three-dimensional figure. Volume is measured in **cubic units**. Decomposing the prism tells you the number of cubes of a given size it will take to fill the prism.

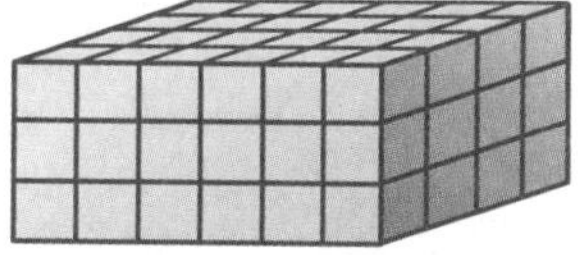

The volume of a rectangular prism is related to its dimensions.

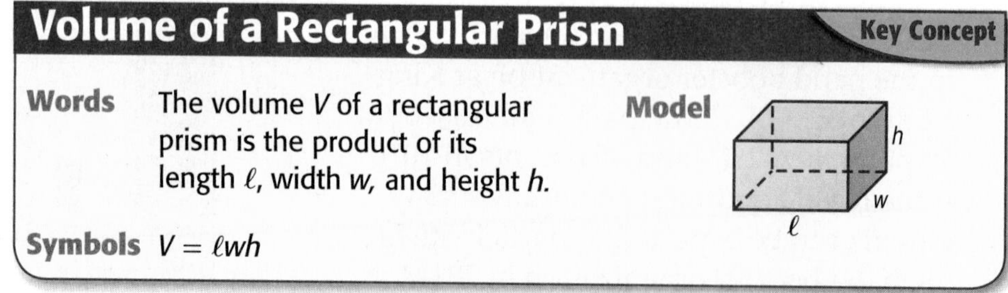

Volume of a Rectangular Prism

Key Concept

Words The volume V of a rectangular prism is the product of its length ℓ, width w, and height h.

Model

Symbols $V = \ell wh$

Reading Math

Volume Measurement A volume measurement can be written using abbreviations and an exponent of 3.
For example:
cubic units = units^3
cubic inches = in^3
cubic feet = ft^3
cubic meters = m^3

Another method to decompose a rectangular prism is to find the area of the base (B) and multiply it by the height (h).

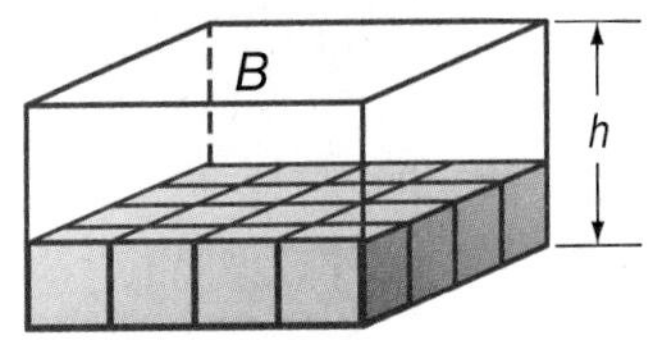

$V = Bh$

number of rows of cubes needed to fill the prism (h)

area of the base, or the number of cubes needed to cover the base (B)

EXAMPLE Find the Volume of a Rectangular Prism

1 **Find the volume of the rectangular prism.**

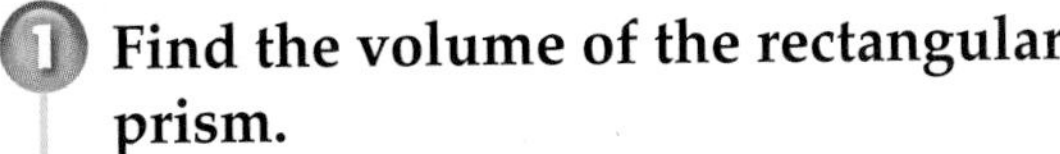

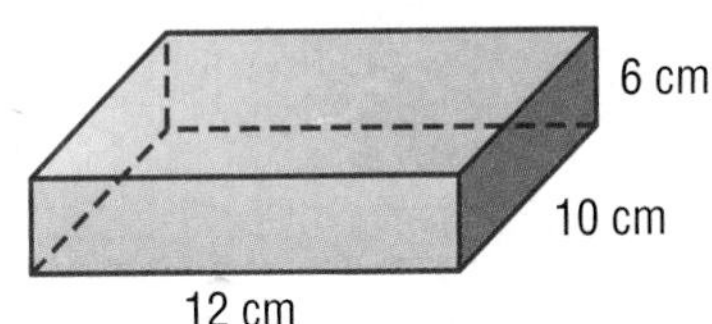

Estimate

$V \approx 10 \text{ cm} \times 10 \text{ cm} \times 6 \text{ cm}$ or 600 cm^3

In the figure, the length is 12 centimeters, the width is 10 centimeters, and the height is 6 centimeters.

METHOD 1 **Use $V = \ell wh$.**

$V = \ell wh$	Volume of rectangular prism
$V = 12 \times 10 \times 6$	Replace ℓ with 12, w with 10, and h with 6.
$V = 720$	Multiply.

METHOD 2 **Use $V = Bh$.**

B, or the area of the base, is 10×12 or 120 square centimeters.

$V = Bh$	Volume of rectangular prism
$V = 120 \times 6$	Replace B with 120 and h with 6.
$V = 720$	Multiply.

The volume is 720 cubic centimeters.

Check for Reasonableness Since we underestimated, the answer should be greater than the estimate. $720 > 600$ ✓

Study Tip

Decomposing Figures You can think of the volume of the prism as consisting of six congruent slices. Each slice contains the area of the base, 120 cm^2, multiplied by a height of 1 cm.

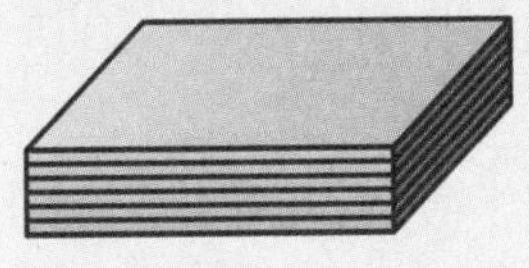

CHOOSE Your Method

Find the volume of each prism.

a. 5 in. by 5 in. by 5 in. cube

5 in.
5 in.
5 in.

b.

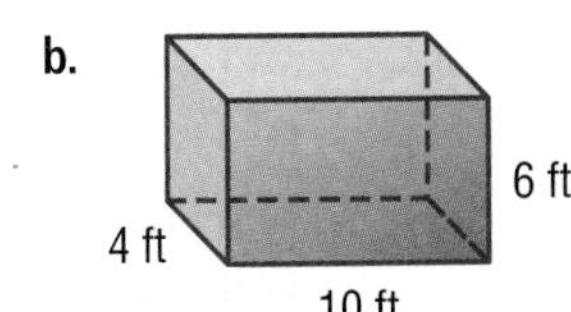

Real-World Career.
How Does a Packaging Manager Use Math?
A packaging manager coordinates and oversees the entire production process of a multi-shift packaging operation.

Math Online
For more information, go to glencoe.com.

Real-World EXAMPLE

2 **PACKAGING** **A cereal box has the dimensions shown. What is the volume of the cereal box?**

8 in.
$12\frac{1}{2}$ in.
$3\frac{1}{4}$ in.

Estimate $10 \times 3 \times 10 = 300$

Find the volume.

$V = \ell wh$

$V = 8 \times 3\frac{1}{4} \times 12\frac{1}{2}$ Replace ℓ with 8, w with $3\frac{1}{4}$, and h with $12\frac{1}{2}$.

$V = \frac{\overset{1}{\cancel{8}}}{1} \times \frac{13}{\underset{1}{\cancel{4}}} \times \frac{25}{\underset{1}{\cancel{2}}}$ Write as improper fractions. Then divide by the GCFs.

$V = \frac{325}{1}$ or 325 Multiply.

The volume of the cereal box is 325 cubic inches.

Check for Reasonableness $325 \approx 300$ ✓

CHECK Your Progress

c. **CONTAINERS** A storage container measures 4 inches long, 5 inches high, and $8\frac{1}{2}$ inches wide. Find the volume of the storage container.

CHECK Your Understanding

Find the volume of each prism.

Example 1 (p. 549)

1. 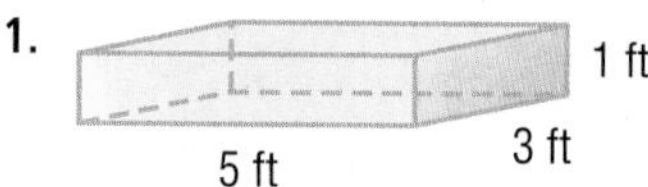

2. 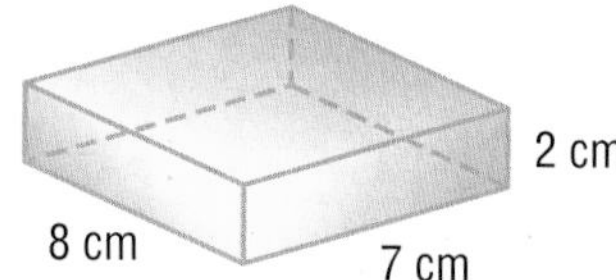

3. 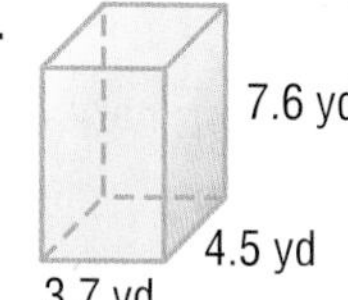

4. 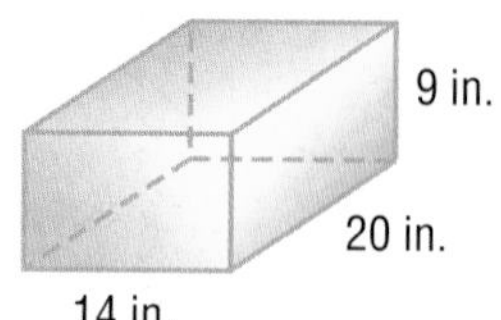

Example 2 (p. 550)

5. **SINKS** A rectangular kitchen sink is 25.25 inches long, 19.75 inches wide, and 10 inches deep. Find the amount of water that can be contained in the sink.

6. **FISHING** A fishing tackle box is 13 inches long, 6 inches wide, and 2.5 inches high. What is the volume of the tackle box?

Practice and Problem Solving

HOMEWORK HELP	
For Exercises	See Examples
7–12	1
13, 14	2

Find the volume of each prism.

7. 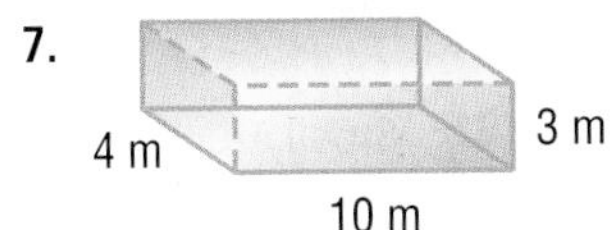

8. 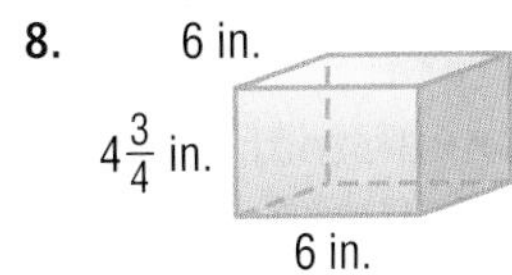

9. 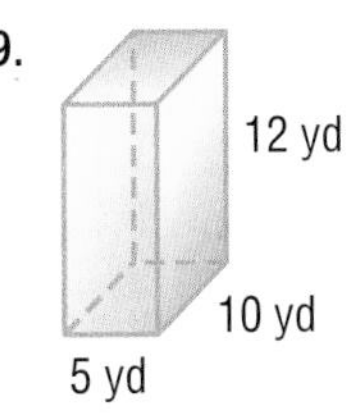

10. 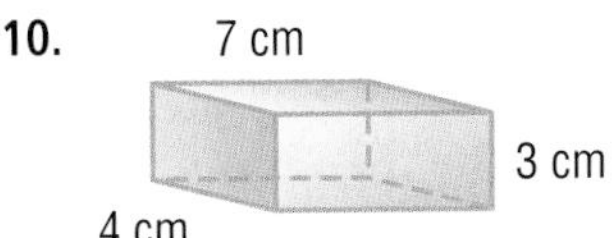

11. 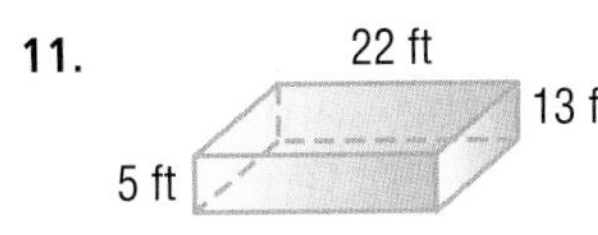

12. 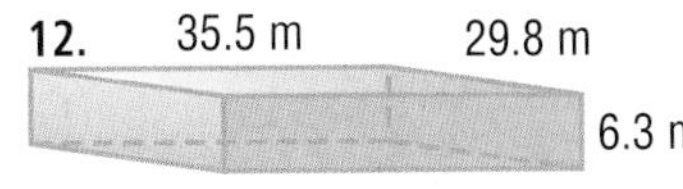

13. **PETS** Find the volume of the pet carrier shown at the right.

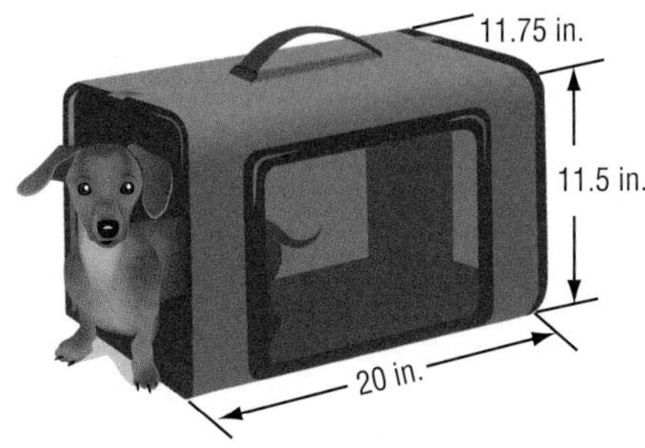

14. **CANYONS** The Palo Duro Canyon is 120 miles long, as much as 20 miles wide, and has a maximum depth of more than 0.15 mile. What is the approximate volume of this canyon?

Real-World Link Palo Duro Canyon State Park in Canyon, Texas, opened on July 4, 1934, and contains 18,438 acres.

Source: Palo Duro Canyon

15. Find the length of a rectangular prism having a volume of 2,830.5 cubic meters, width of 18.5 meters, and height of 9 meters.

16. What is the width of a rectangular prism with a length of 13 feet, volume of 11,232 cubic feet, and height of 36 feet?

Replace each ● with <, >, or = to make a true sentence.

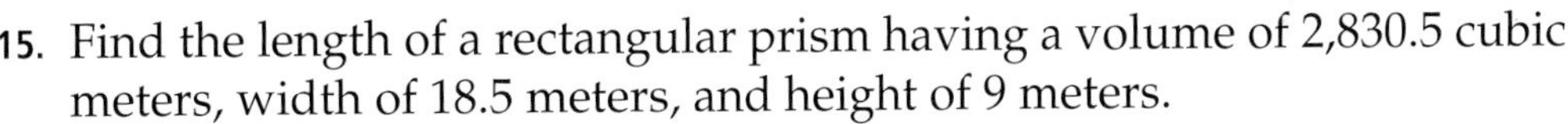

17. 1 ft^3 ● 1 yd^3
18. 5 m^3 ● 5 yd^3
19. 27 ft^3 ● 1 yd^3

SAND ART For Exercises 20 and 21, use the following information.

The glass container shown is filled to a height of 2.25 inches.

20. How much sand is currently in the container?

21. How much more sand could the container hold before it overflows?

22. **NUMBER SENSE** The volume of a cube is 64 cubic feet. What is the height of the cube?

23. **REASONING** Which has the greater volume: a prism with a length of 5 inches, a width of 4 inches, and a height of 10 inches or a prism with a length of 10 inches, a width of 5 inches, and a height of 4 inches? Justify your selection.

ANALYZE TABLES **For Exercises 24–26, use the table at the right.**

Inside Dimensions of Moving Trucks			
Truck	Length (ft)	Width (ft)	Height (ft)
Van	10	$6\frac{1}{2}$	6
Small Truck	$11\frac{1}{3}$	$7\frac{5}{12}$	$6\frac{3}{4}$
2-Bedroom Moving Truck	$14\frac{1}{12}$	$7\frac{7}{12}$	$7\frac{1}{6}$
3-Bedroom Moving Truck	$20\frac{5}{6}$	$7\frac{1}{2}$	$8\frac{1}{12}$
Mega Moving Truck	$22\frac{1}{4}$	$7\frac{7}{12}$	$8\frac{5}{12}$

24. What is the approximate volume of the small truck?
25. The Davis family is moving, and they estimate that they will need a truck with about 1,300 cubic feet. Which truck would be best for them to rent?
26. About how many cubic feet greater is the volume of the Mega Moving Truck than the 2-bedroom moving truck?
27. **ESTIMATION** Jeffrey estimates that the volume of a rectangular prism with a length of 5.8 centimeters, a width of 3 centimeters, and a height of 12.2 centimeters is less than 180 cubic centimeters. Is he correct? Explain.

See pages 699, 715.

28. **REASONING** The volume of a rectangular prism is 16 cubic feet. The height of the prism is 4 feet and the base of the prism is a square. What is the length of one side of the base?

H.O.T. Problems

29. **Which One Doesn't Belong?** Identify the rectangular prism that does not belong with the other three. Explain your reasoning.

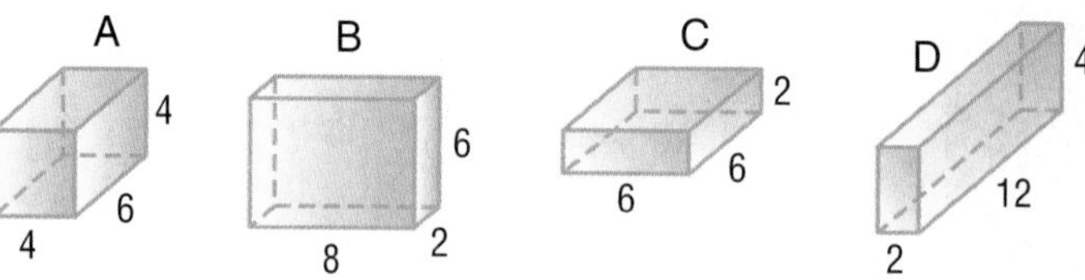

30. **OPEN ENDED** Draw and label a rectangular prism that has a volume between 200 and 400 cubic inches. Then give an example of a real-world object that is this approximate size.
31. **SELECT A TOOL** Basilio is filling his new fish tank with water. The dimensions of the fish tank are 36 inches by 13 inches by 16 inches. Basilio knows that 1 gallon equals 231 cubic inches. Which of the following tools might Basilio use to determine about how many gallons of water he needs to fill the fish tank? Justify your selection(s). Then use the tool(s) to solve the problem.

calculator | centimeter cubes | paper/pencil

32. **CHALLENGE** Refer to the prism at the right. If all the dimensions of the prism doubled, would the volume double? Explain your reasoning.

33. **WRITING IN MATH** Explain why cubic units are used to measure volume instead of linear units or square units.

TEST PRACTICE

34. Justin used the shoebox to create a home for the toad he caught.

Find the volume of the shoebox.

A 222 in^3

B 864 in^3

C 1,620 in^3

D 1,710 in^3

35. A cereal company is creating a new size box in which to package cereal. The box has a width of 27 centimeters, a length of 7 centimeters, and a volume of 6,426 cubic centimeters. Find the height of the cereal box.

F 34 cm

G 38 cm

H 42 cm

J 46 cm

Spiral Review

36. **TOYS** Tiffany is using wooden cube blocks to make rectangular prisms. If she has exactly 8 wooden cube blocks, make a model to find the length, width, and height of two possible rectangular prisms. (Lesson 10-5)

37. What is the area of a triangle with base 52 feet and height 38 feet? (Lesson 10-4)

Find the value of x in each quadrilateral. (Lesson 9-5)

38. 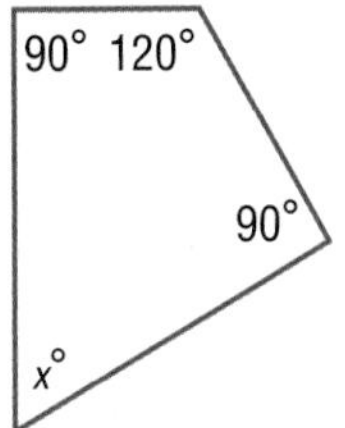

39. 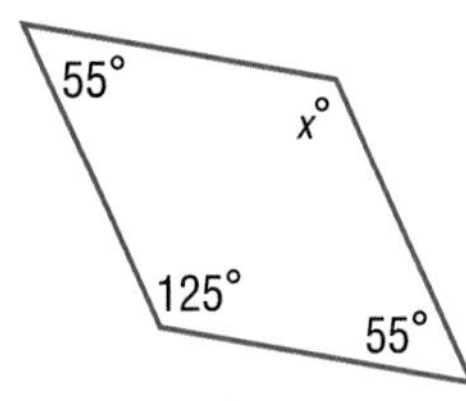

Complete. (Lesson 8-6)

40. ■ cm = 47 mm

41. 3,500 g = ■ kg

42. **CLOTHES** How many outfits can you make with two different colored sweatshirts and four types of jeans? Make an organized list to show the sample space. (Lesson 7-5)

GET READY for the Next Lesson

PREREQUISITE SKILL Find the area of each rectangle. (Lesson 1-9)

43. 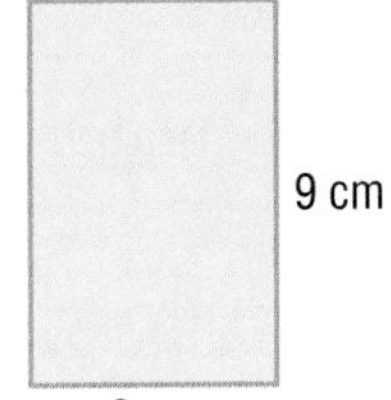

44. 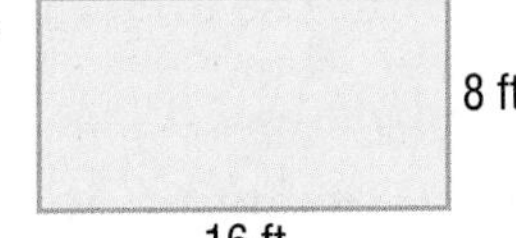

45. 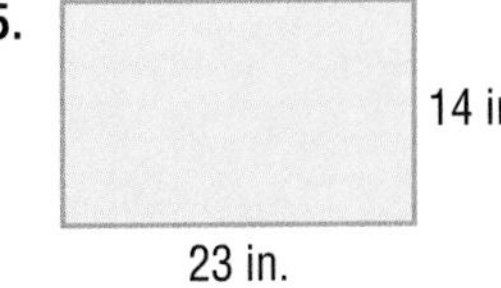

Explore 10-7

Geometry Lab
Using a Net to Build a Cube

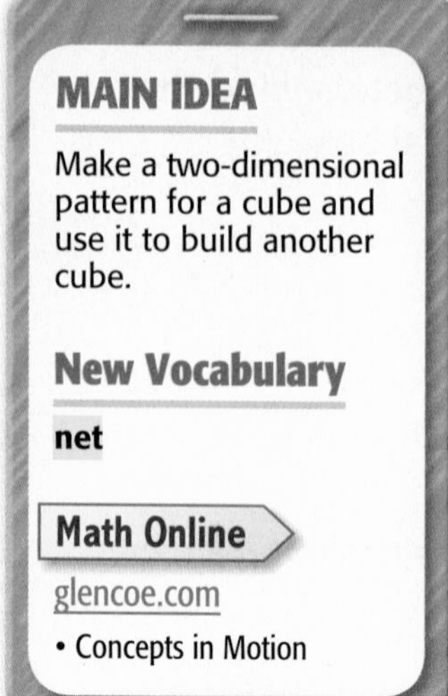

MAIN IDEA

Make a two-dimensional pattern for a cube and use it to build another cube.

New Vocabulary

net

Math Online

glencoe.com

- Concepts in Motion

In this lab, you will make a two-dimensional pattern of a cube called a **net** and use it to build the three-dimensional figure.

ACTIVITY

STEP 1 Place the cube on paper as shown. Trace the base of the cube, which is a square.

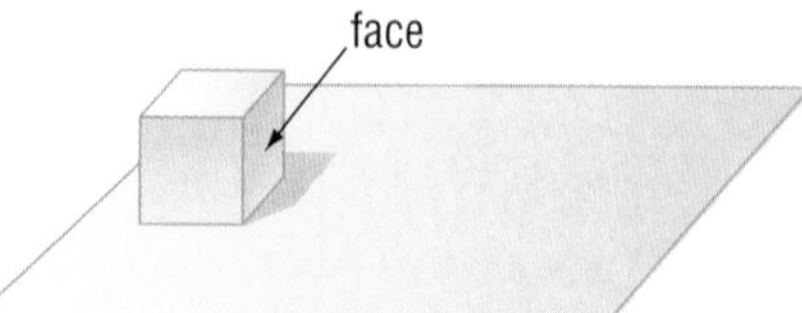

STEP 2 Roll the cube onto another side. Continue tracing each side to make the figure shown. This two-dimensional figure is called a net.

STEP 3 Cut out the net. Then build the cube.

STEP 4 Make a net like the one shown. Cut out the net and try to build a cube.

ANALYZE THE RESULTS

1. Explain whether both nets formed a cube. If not, describe why the net or nets did not cover the cube.
2. Draw three other nets that will form a cube and three other nets that will not form a cube. Describe a pattern in the nets that do form a cube.
3. Measure the edges of the cube in the activity above. Use this measure to find the area of one side of the cube.
4. **MAKE A CONJECTURE** Write an expression for the total area of all the surfaces of a cube with edge length *s*.
5. Draw a net for a rectangular prism. Explain the difference between this net and the nets that formed a cube.

Surface Area of Rectangular Prisms

MAIN IDEA

Find the surface areas of rectangular prisms.

New Vocabulary

surface area

Math Online

glencoe.com

- Extra Examples
- Personal Tutor
- Self-Check Quiz
- Reading in the Content Area

MINI Lab

STEP 1 Draw and cut out a net of the prism.

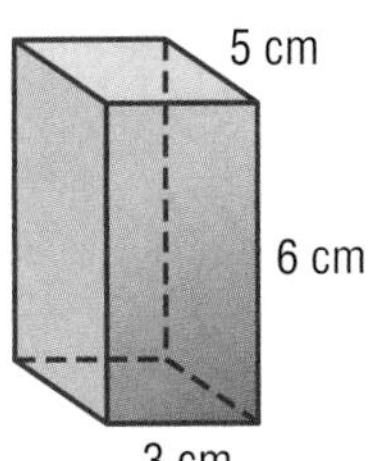

STEP 2 Fold along the dashed lines. Tape the edges.

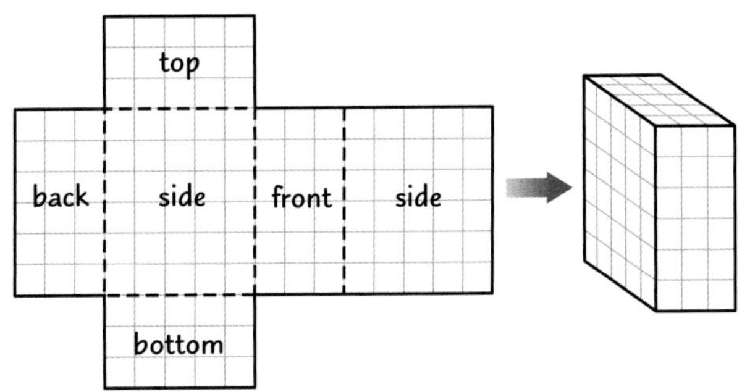

1. Find the area of each face of the prism.
2. What is the sum of the areas of the faces of the prism?

The sum of the areas of all the faces of a prism is called the **surface area** of the prism.

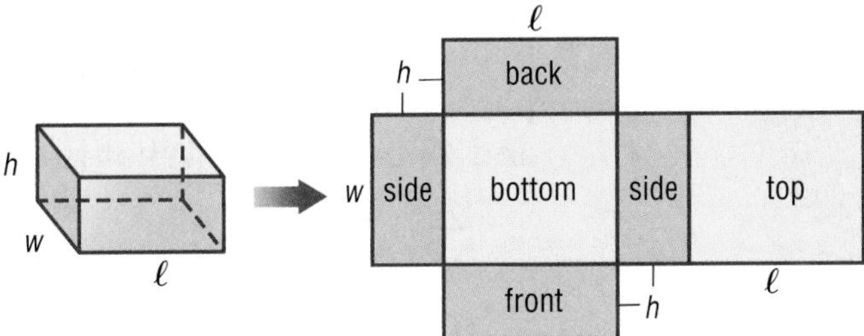

top and bottom	$\ell w + \ell w = 2\ell w$
front and back	$\ell h + \ell h = 2\ell h$
two sides	$wh + wh = 2wh$
sum of the areas	$2\ell w + 2\ell h + 2wh$

Surface Area of a Rectangular Prism — Key Concept

Words The surface area S of a rectangular prism with length ℓ, width w, and height h is the sum of the areas of the faces.

Model

h
w
ℓ

Symbols $S = 2\ell w + 2\ell h + 2wh$

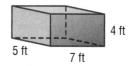

EXAMPLE Find the Surface Area of a Rectangular Prism

1 **Find the surface area of the rectangular prism.**

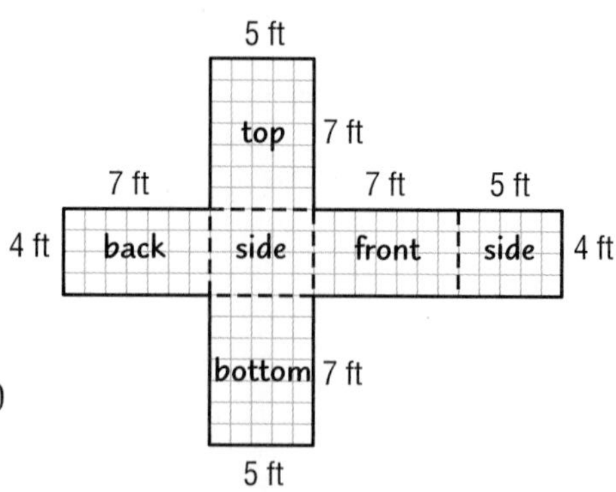

Find the area of each face.

top and bottom:
$2\ell w = 2(7)(5)$ or 70

front and back:
$2\ell h = 2(7)(4)$ or 56

two sides:
$2wh = 2(5)(4)$ or 40

Add to find the surface area.

The surface area is $70 + 56 + 40$ or 166 square feet.

CHECK Your Progress

a. Find the surface area of the rectangular prism.

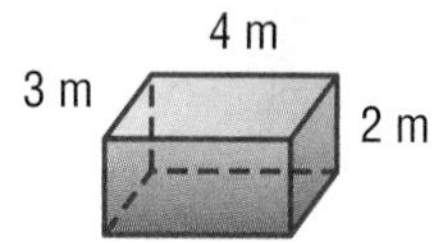

Surface area can be applied to many real-world situations.

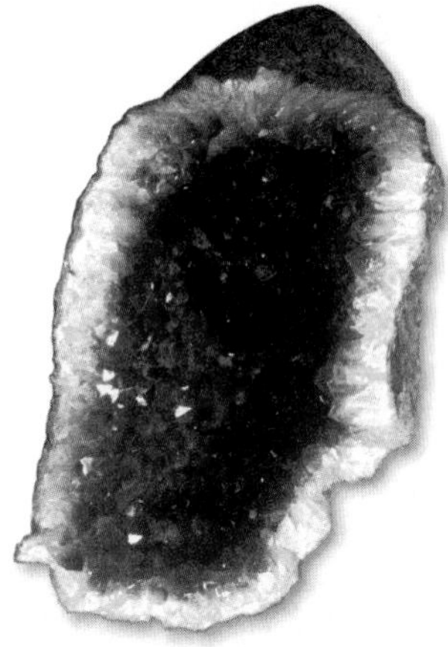

Real-World Link A *geode* is a hollow rock that is lined on the inside with crystal. The largest geode ever found is 30 feet deep and large enough for people to walk through.

Real-World EXAMPLE

2 **GEOLOGY A geode is shaped like a rectangular prism. It is packed in a box that measures 7 inches long, 3 inches wide, and 16 inches tall. What is the surface area of the box?**

$S = 2\ell w + 2\ell h + 2wh$	Surface area of a prism
$S = 2(7)(3) + 2(7)(16) + 2(3)(16)$	$\ell = 7, w = 3, h = 16$
$S = 14(3) + 14(16) + 6(16)$	Multiply.
$S = 42 + 224 + 96$	Multiply.
$S = 362$	Add.

The surface area of the box is 362 square inches.

CHECK Your Progress

b. **PAINTING** Nadine is going to paint her younger sister's toy chest, including the bottom. What is the approximate surface area that she will paint?

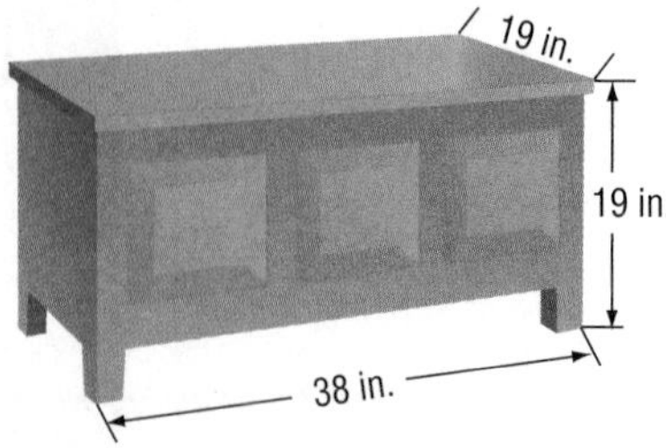

CHECK Your Understanding

Example 1 (p. 556)

Find the surface area of each rectangular prism.

1.

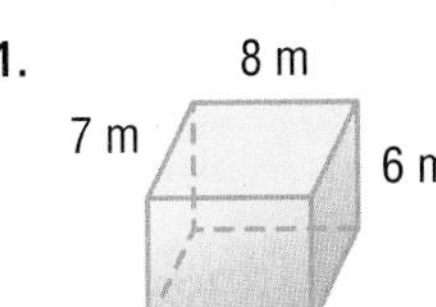

2.

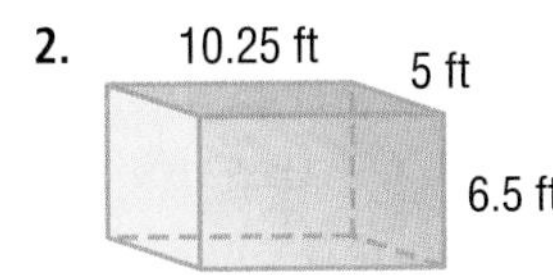

3.

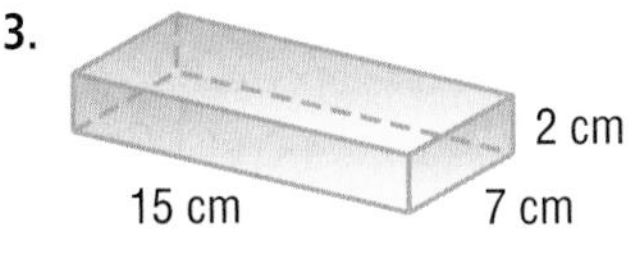

Example 2 (p. 556)

4. **VIDEO GAMES** A game box for video games is shaped like a rectangular prism. What is the surface area of the game box?

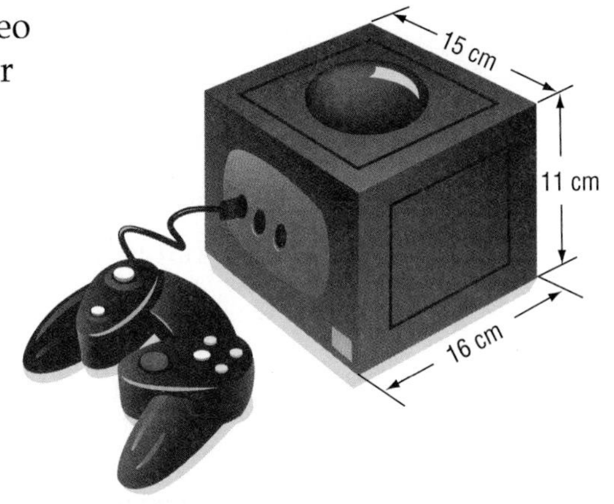

Practice and Problem Solving

HOMEWORK HELP

For Exercises	See Examples
5–10	1
11, 12	2

Find the surface area of each rectangular prism.

5.

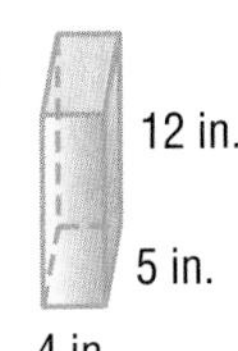

6. 5 ft, 7 ft, 3 ft

7.

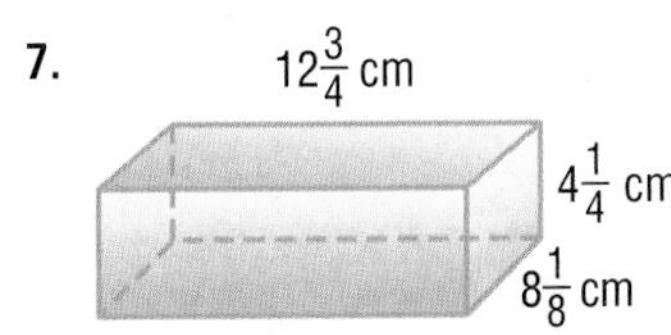

8.

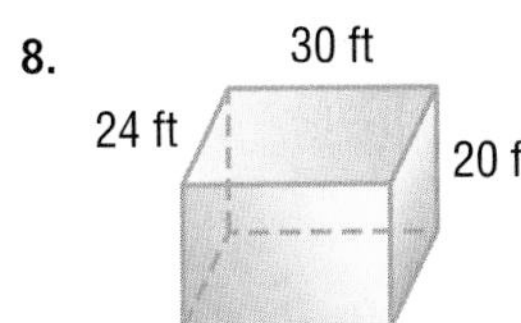

9.

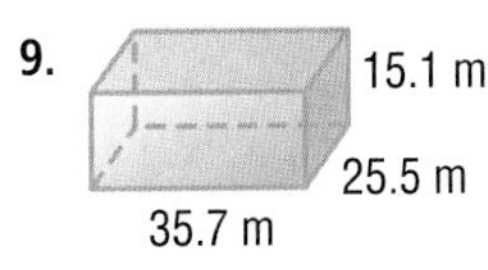

10.

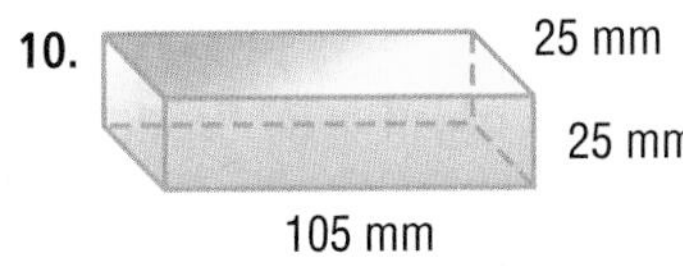

11. **DISPLAYS** Tomás keeps his diecast car in a glass display case as shown. What is the surface area of the glass?

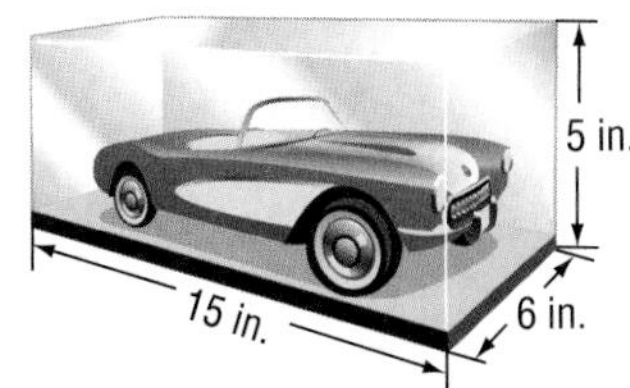

12. **CAKES** A full sheet cake is typically 18 inches by 24 inches by 2 inches. What is the minimum surface area of a rectangular box that will contain the cake?

13. **ESTIMATION** Stella estimates that the surface area of a rectangular prism with a length of 13.2 feet, a width of 6 feet, and a height of 8 feet is about 460 square feet. Is her estimate reasonable? Explain your reasoning.

Classify each measure as *length, area, surface area,* or *volume*. Explain your reasoning. Include an appropriate unit of measure.

14. the amount of water in a lake
15. the amount of land available to build a house
16. the amount of wrapping paper needed to cover a box
17. the number of tiles needed to tile a bathroom floor
18. the amount of tin foil needed to cover a sandwich
19. the amount of cereal that will fit in a box
20. the height of a tree

BIRDS For Exercises 21–23, use the following information.

Julia is making a bird nesting box for her backyard.

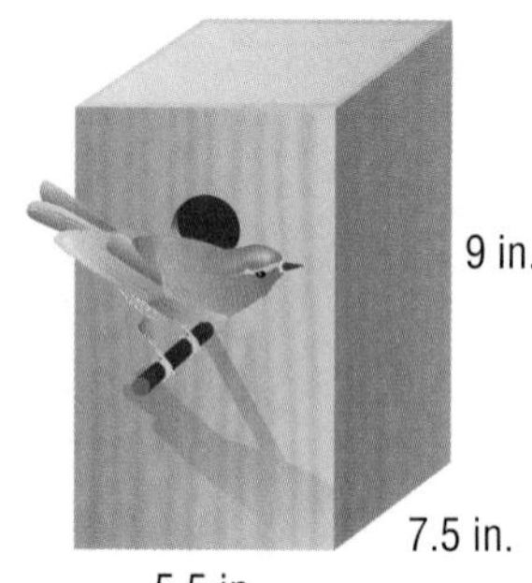

21. What is the surface area of the nesting box?
22. What is the surface area if the depth is doubled?
23. What is the surface area if the depth is half as great?

24. **SHIPPING** Find the surface area of each shipping package. Which package has the greater surface area? Does the same package have a greater volume? Explain.

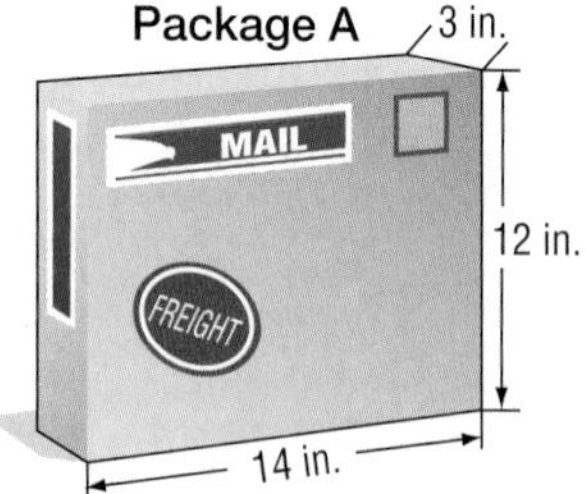

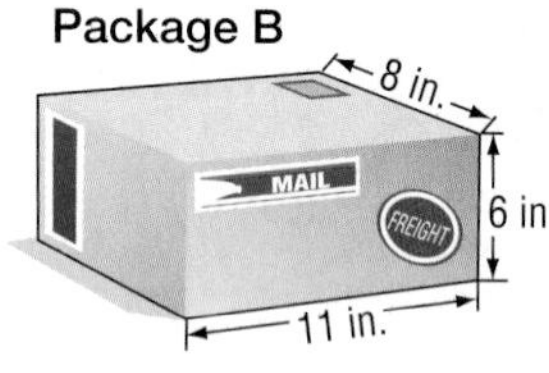

EXTRA PRACTICE
See pages 700, 715.

H.O.T. Problems

25. **OPEN ENDED** Draw and label a rectangular prism that has a surface area of 208 square feet.

26. **REASONING** Determine whether the following statement is *sometimes, always,* or *never* true. Explain your reasoning.

    If all the dimensions of a cube are doubled, the surface area is four times greater.

CHALLENGE For Exercises 27 and 28, use the figure shown. All of the triangular faces are congruent.

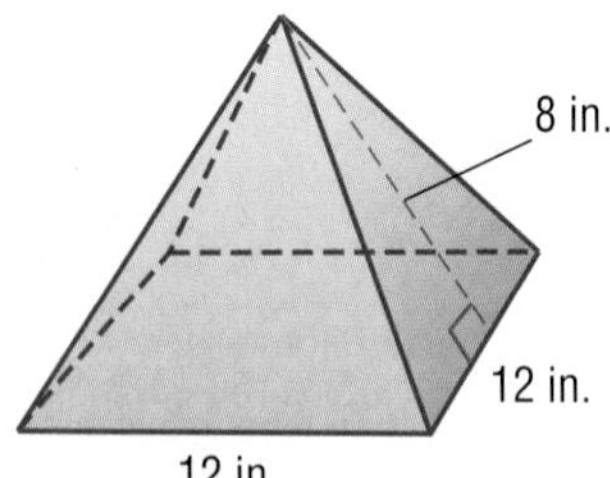

27. What is the area of one of the triangular faces? the square face?
28. Use what you know about finding the surface area of a rectangular prism to find the surface area of the square pyramid.

29. **WRITING IN MATH** Write a problem about a real-world situation in which you would need to find the surface area of a rectangular prism.

TEST PRACTICE

30. Which net can be used to make and find the surface area of a cube?

A
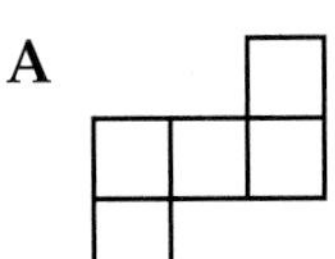

B
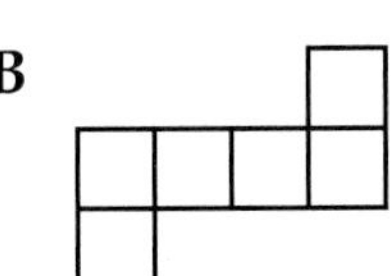

C
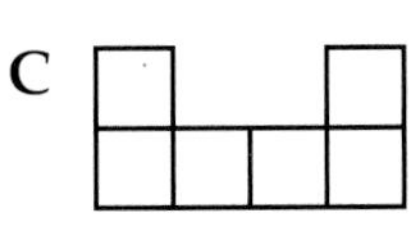

D
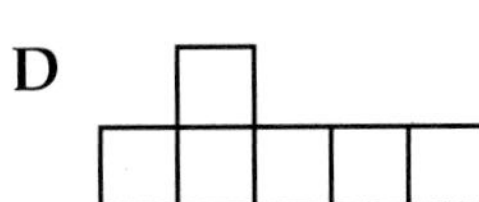

31. Horacio is going to paint a shoebox to use for storage of his trading cards. The shoebox is 23 inches long, 10 inches wide, and 8 inches high. Find the surface area of the shoebox.

F 246 in^2

G 828 in^2

H 988 in^2

J 1,840 in^2

Spiral Review

32. Find the volume of a rectangular prism with sides measuring 5 feet, 8 feet, and 12 feet. (Lesson 10-6)

33. **MINIATURE GOLF** Find the perimeter and area of the miniature golf hole shown. (Lessons 1-9 and 10-6)

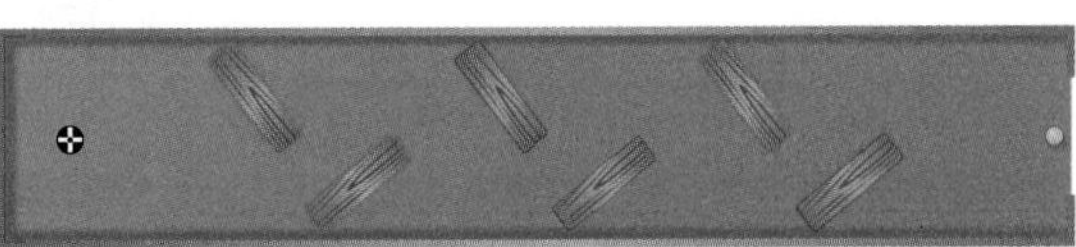

34. **MONEY** Measure the length of a dollar bill to the nearest sixteenth inch. (Lesson 8-1)

35. **PHONES** How many ways can A.J. call three of his friends? Make an organized list to show the sample space. (Lesson 7-5)

Write each decimal as a percent. (Lesson 7-3)

36. 0.44
37. 5.35
38. 0.6
39. 2.1

Problem Solving in Industrial Education

Real-World Unit Project

A New Zoo It's time to complete your project. Use the data you have collected about your selected zoo animals to create a set of drawings or blueprints for your zoo. Be sure to include all dimensions, areas, surface areas, and volumes of each section of your zoo as well as feeding schedules and temperature conditions.

Math Online **Unit Project at** glencoe.com

Extend 10-7

Measurement Lab
Selecting Formulas and Units

MAIN IDEA

Select appropriate units, tools, and formulas to measure objects.

Recall from Lesson 8-8 that an *attribute* is a characteristic of an object. Some attributes, like length and width, can be measured directly on the object. These measures are called *direct measures*. Others, like perimeter, circumference, area, and volume, can be calculated from direct measures. These are *calculated measures*.

ACTIVITY

STEP 1 Copy the table below.

Object	Attribute	Formula Needed	Direct Measure(s)	Calculated Measure(s)
shoebox				
chalkboard				
desktop				
cereal box				
clock face				
bulletin board				
basketball				

STEP 2 Choose an attribute for each object that involves a calculated measure. Then determine what attributes of the object you must measure directly in order to calculate this measure. Record this information in the table.

STEP 3 Indicate what formula you need to use in order to calculate each measure.

STEP 4 Select a measuring tool from among those provided by your teacher, and find the direct measure(s) for each object using the smallest unit on your measuring tool. Record each measure in the table. Be sure to include appropriate units.

Analyze the Results

1. Which object did you find most difficult to measure directly? How did you solve this problem?

2. **WRITING IN MATH** Write a real-world problem that could be solved using one of the objects and the measure you calculated.

Study Guide and Review

Math Online glencoe.com
• STUDY TO GO
• Vocabulary Review

GET READY to Study

Be sure the following Big Ideas are noted in your Foldable.

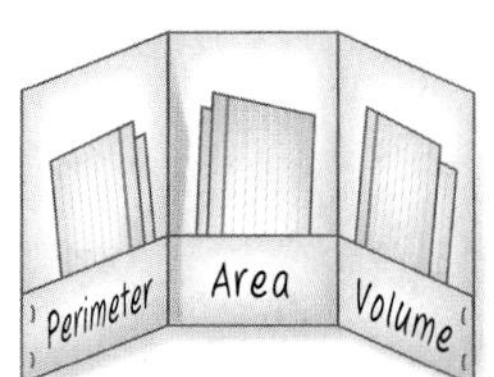

BIG Ideas

Perimeter (Lesson 10-1)

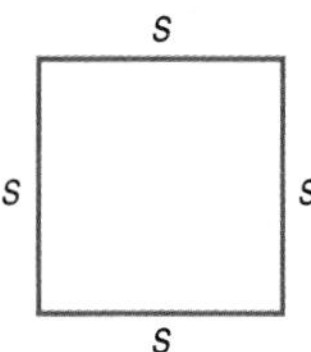

$P = 4s$

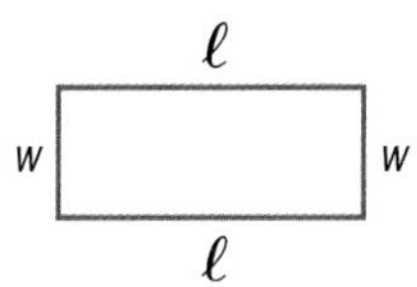

$P = 2\ell + 2w$

Circles and Circumference (Lesson 10-2)

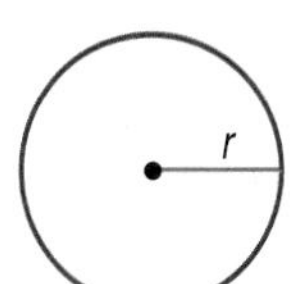

$d = 2r$
$C = \pi d$ or $C = 2\pi r$

Area (Lessons 10-3 and 10-4)

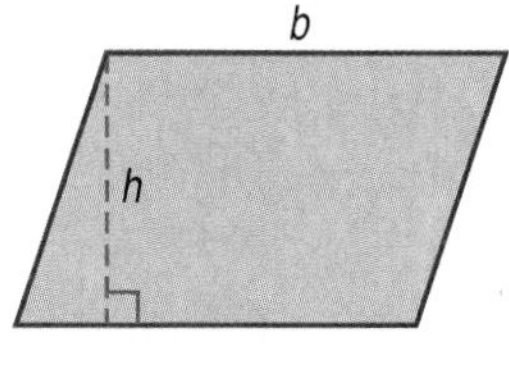

$A = bh$

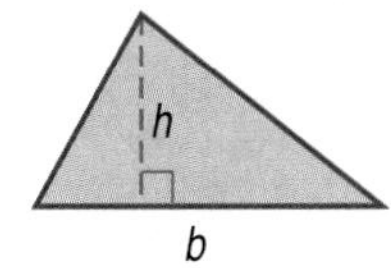

$A = \frac{bh}{2}$ or $A = \frac{1}{2}bh$

Volume and Surface Area (Lessons 10-6 and 10-7)

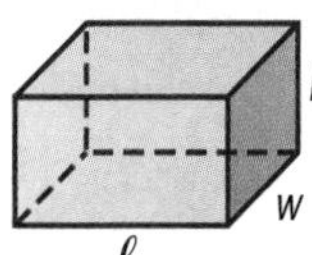

$V = \ell wh$
$S = 2\ell w + 2\ell h + 2wh$

Key Vocabulary

base (p. 534)
center (p. 528)
chord (p. 528)
circle (p. 528)
circumference (p. 528)
diameter (p. 528)
height (p. 534)
net (p. 554)
perimeter (p. 522)
radius (p. 528)
rectangular prism (p. 548)
surface area (p. 555)
volume (p. 548)

Vocabulary Check

Choose the correct term to complete each sentence.

1. The amount of space that a three-dimensional figure contains is called its (area, volume).
2. The shortest distance from the base to the opposite side of a parallelogram is called the (height, center).
3. The distance around any closed figure is called its (surface area, perimeter).
4. In estimating the circumference of a circle, round the value of π to (3, 4).
5. Cubic units are used when calculating (surface area, volume).
6. The distance around a circle is called the (diameter, circumference).
7. The longest chord of a circle is the (radius, diameter).

Lesson-by-Lesson Review

10-1 Perimeter (pp. 522–526)

Find the perimeter of each figure.

8. 7 yd, 7 yd, 7 yd, 7 yd

9. 28 ft, 17 ft, 17 ft, 28 ft

10. **WALLPAPER** How many feet of wallpaper border are needed for a bedroom wall that is 11 feet long and 9 feet wide?

Example 1 **Find the perimeter of the rectangle.**

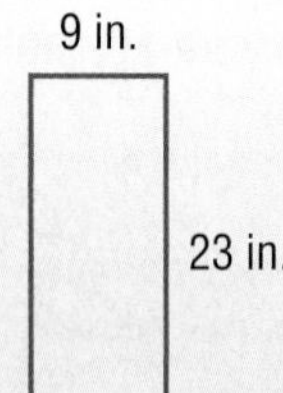

$P = 2\ell + 2w$

$P = 2(23) + 2(9)$

$P = 46 + 18$

$P = 64$ in.

The perimeter is 64 inches.

10-2 Circles and Circumference (pp. 528–533)

Find the radius or diameter of each circle with the given dimensions.

11. $d = 58$ cm
12. $r = 27$ in.
13. $r = 9$ ft
14. $d = 32$ yd

Estimate the circumference of each circle.

15. 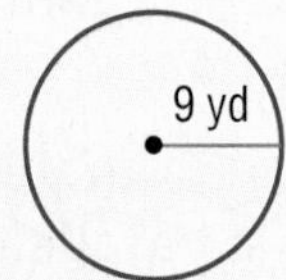

16. 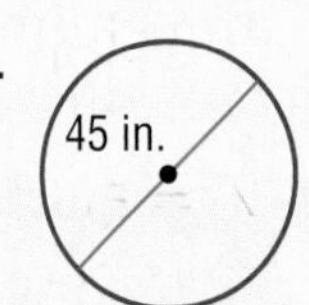

Find the circumference of each circle. Round to the nearest tenth.

17. 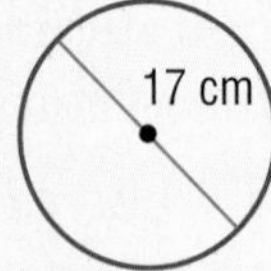

18. 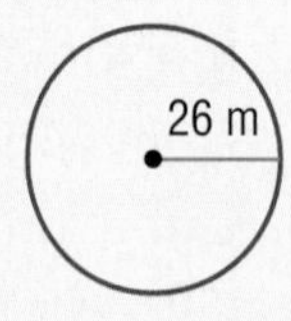

19. **RIDES** The plans for a carousel call for a circular floor with a diameter of 40 feet. Find the circumference of the floor.

Example 2 **Find the radius of a circle with diameter 68 yards.**

The radius of a circle is half its diameter. So, the radius of a circle with a diameter of 68 yards is $\frac{1}{2}$ of 68 yards, or 34 yards.

Example 3 **Estimate the circumference of a circle with radius 8 feet.**

The circumference of a circle is π times twice its radius. The circumference is about $3 \times 2 \times 8$, or 48 feet.

Example 4 **Find the circumference of the circle at the right. Round to the nearest tenth.**

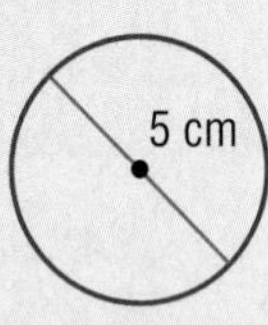

The circumference of a circle is π times its diameter. The circumference is $\pi \times 5$ centimeters, or 15.7 centimeters.

Mixed Problem Solving
For mixed problem-solving practice, see page 715.

10-3 Area of Parallelograms (pp. 534–538)

Find the area of each parallelogram.

20.

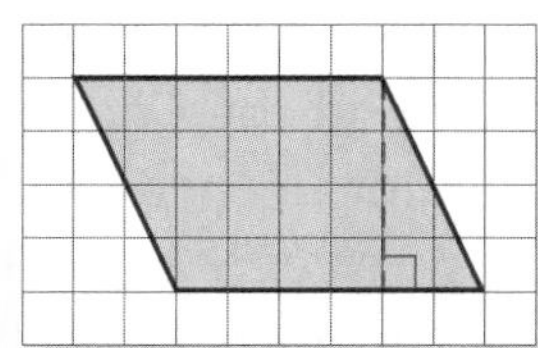

21.

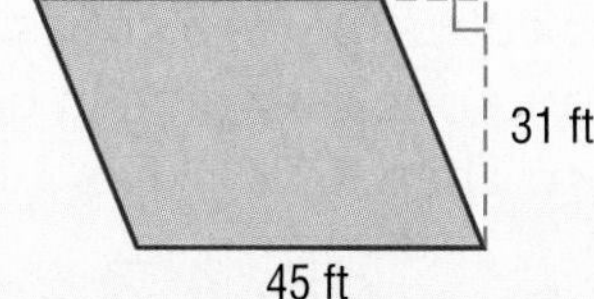

22. 23.

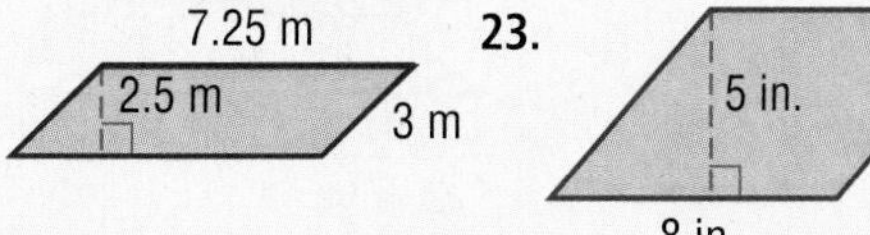

24. **DECKS** Find the area of a deck if it is a parallelogram with base $8\frac{1}{4}$ feet and height 6 feet.

Example 5 Find the area of the parallelogram.

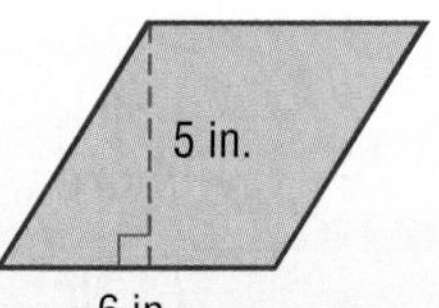

$A = bh$

$A = 6 \cdot 5$

$A = 30 \text{ in}^2$

Example 6 Find the area of a parallelogram with base 4.3 meters and height 11.2 meters.

$A = bh$

$A = 4.3 \cdot 11.2$

$A = 48.16$

The area is 48.16 square meters.

10-4 Area of Triangles (pp. 540–544)

Find the area of each triangle.

25.

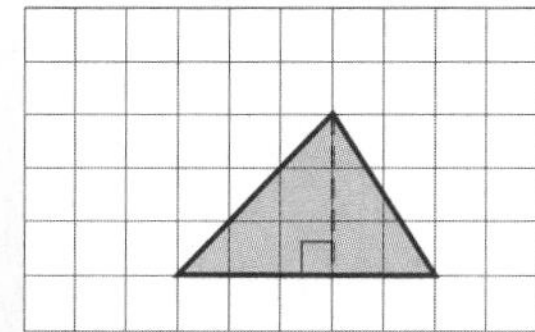

26.

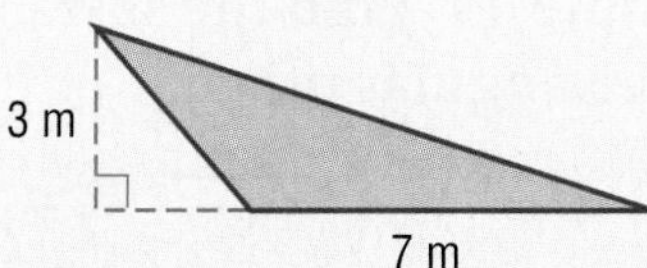

27.

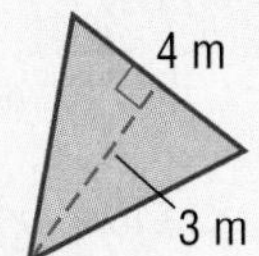

28.

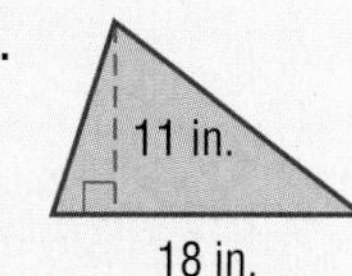

29. **FLAGS** How much material is needed to make a triangular flag with base $2\frac{1}{4}$ feet and height $8\frac{1}{2}$ feet?

Example 7 Find the area of the triangle.

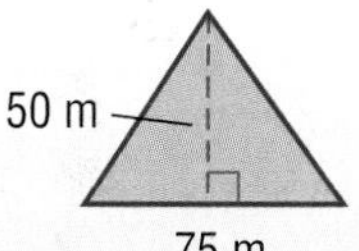

$A = \frac{1}{2}bh$

$A = \frac{1}{2}(75 \cdot 50)$

$A = 1{,}875 \text{ m}^2$

Example 8 Find the area of a triangular garden with base 8 feet and height 7 feet.

$A = \frac{1}{2}bh$

$A = \frac{1}{2}(8)(7)$

$A = \frac{1}{2}(56)$

$A = 28$

The area is 28 square feet.

10-5 PSI: Make a Model (pp. 546–547)

Solve. Use the *make a model* strategy.

30. **CANS** A grocer is stacking cans of tomato soup into a pyramid-shaped display. The bottom layer has 8 cans. There is one less can in each layer and there are 6 layers. How many cans are in the display?

31. **BRICKS** A brick layer wants to arrange 16 bricks into a rectangular shape with the greatest perimeter possible. How many bricks will be in each row?

Example 9 **A cheerleading squad formed a pyramid. There were 5 cheerleaders on the bottom and one less cheerleader in each row. How many rows were in the pyramid, if there are 12 cheerleaders?**

Using 12 cubes, place 5 cubes on the bottom and one less cube in each layer as shown. There are 3 rows.

10-6 Volume of Rectangular Prisms (pp. 548–553)

Find the volume of each figure.

32. 33.

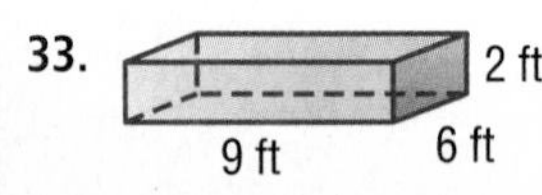

34. **BUILDINGS** What is the volume of an office building with length 168 yards, width 115 yards, and height 96 yards?

Example 10 **Find the volume of the figure.**

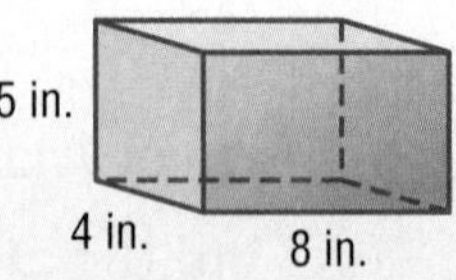

$V = \ell wh$

$V = 8 \times 4 \times 5$

$V = 160$

The volume is 160 cubic inches.

10-7 Surface Area of Rectangular Prisms (pp. 555–559)

Find the surface area of each rectangular prism.

35.

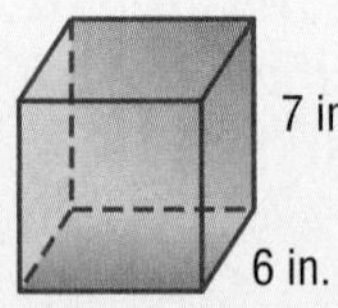

36.

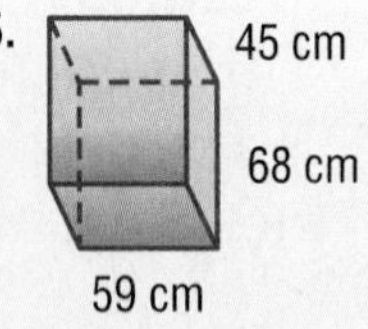

37. **TISSUE BOXES** How much cardboard covers the outside of a tissue box if the dimensions of the box are to be 4 inches by 3 inches by 5 inches?

Example 11 **Find the surface area of the rectangular prism.**

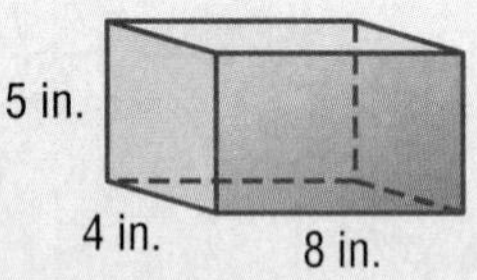

top and bottom: $2(8 \times 4)$ or 64

front and back: $2(8 \times 5)$ or 80

two sides: $2(4 \times 5)$ or 40

The surface area is $64 + 80 + 40$ or 184 square feet.

CHAPTER 10 Practice Test

Math Online glencoe.com
• Chapter Test

Find the perimeter of each figure.

1. 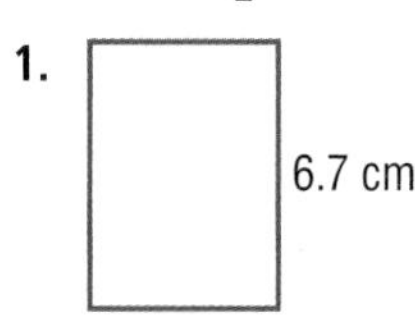

2. 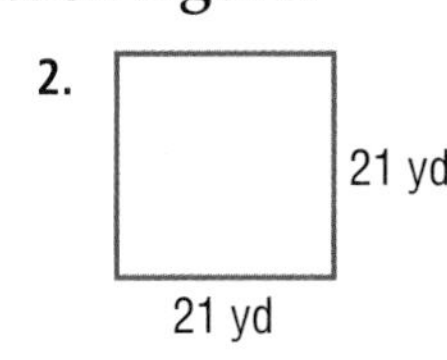

Find the radius or diameter of each circle with the given dimensions.

3. $r = 9$ in.

4. $d = 46$ mm

5. **MULTIPLE CHOICE** The drawing shows two circles that have the same center.

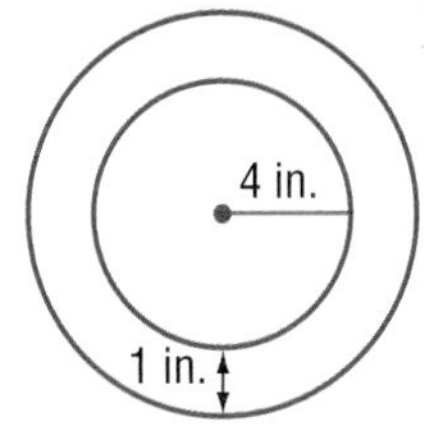

Which expression can be used to find the approximate circumference of the outer circle in inches?

A $\pi(4 + 1)$

B $\frac{1}{2}(4 + 1)$

C $2\pi(4 + 1)$

D $2(4 + 1)$

Find the area of each parallelogram or triangle.

6. 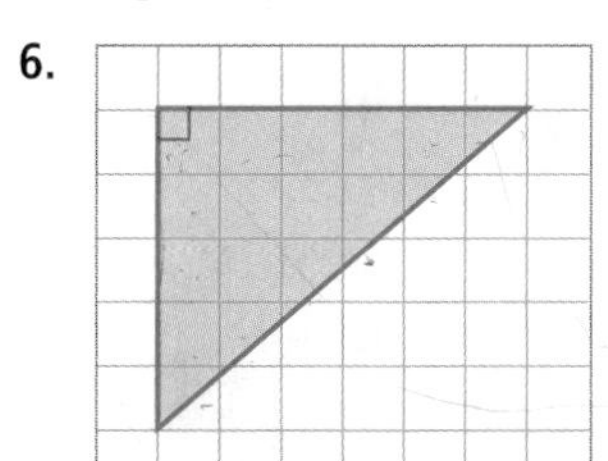

7. 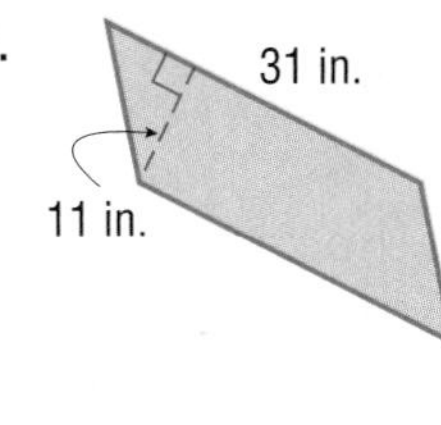

8. **REASONING** Which has the greater area, a triangle with a base of 8 meters and a height of 12 meters, or a triangle with a base of 4 meters and a height of 16 meters? Justify your response.

9. **GARDENING** A triangular garden has a base of 7 meters and a height of 6 meters. If one bag of fertilizer covers 25 square meters, how many bags of fertilizer are needed to fertilize the garden?

10. **GEOMETRY** A rectangular prism is made using exactly 12 cubes. Find a possible length, width, and height of the prism. Use the *make a model* strategy.

Find the volume of each figure.

11. 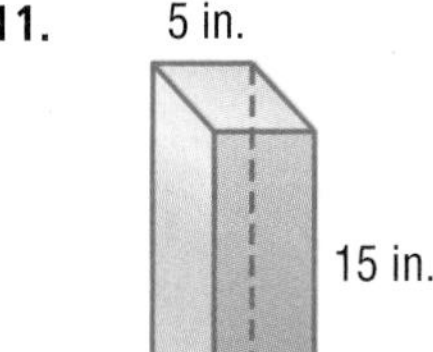

12. 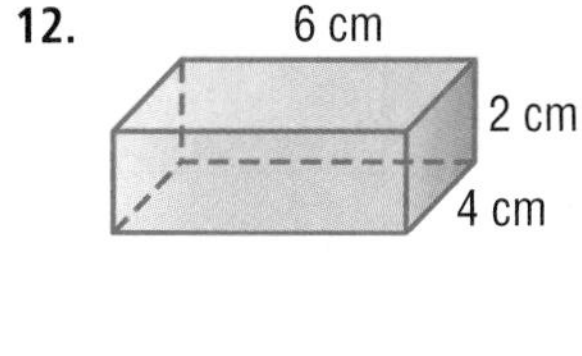

13. **POOLS** A rectangular pool is 21 feet long by 18 feet wide. Find the number of cubic feet of water required to fill the pool so that the water is 9 feet deep.

14. **MULTIPLE CHOICE** Which expression gives the surface area of a rectangular prism with length 5 units, width 8 units, and height 3 units?

F $(2)(5^2) + (2)(8^2) + (2)(3^2)$

G $2(5)(8) + 2(5)(3) + 2(8)(3)$

H $2(5)(8)(3)$

J $(2)(5)(8 + 3)$

Find the surface area of each rectangular prism.

15. 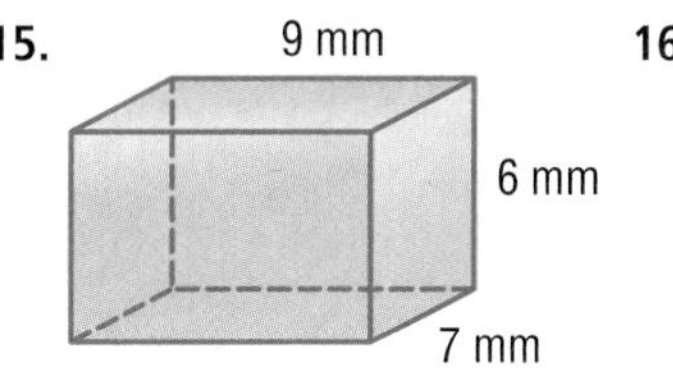

16. 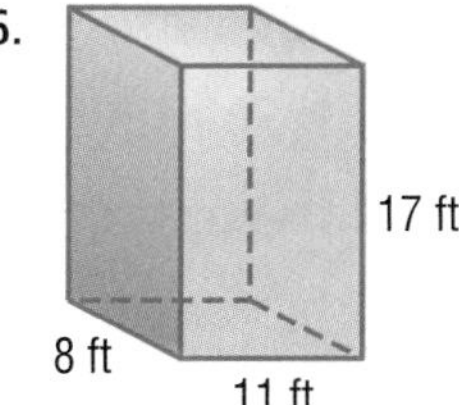

CHAPTER 10 Test Practice
Cumulative, Chapters 1–10

Math Online glencoe.com
• Test Practice

PART 1 Multiple Choice

Read each question. Then fill in the correct answer on the answer sheet provided by your teacher or on a sheet of paper.

1. The table below shows the areas of a triangle where the height of the triangle stays the same, but the base changes.

Area of Triangles		
Height (units)	Base (units)	Area (square units)
4	3	6
4	4	8
4	5	10
4	6	12
4	n	■

Which expression can be used to find the area of a triangle that has a height of 4 units and a base of n units?

A $\frac{n}{4}$ C $\frac{4}{2n}$

B $\frac{4n}{2}$ D $4n$

TEST-TAKING TIP

Question 1 Many standardized tests list any geometry formulas you will need to solve problems. However, it is a good idea to familiarize yourself with the formulas before the test.

2. Annalese is making necklaces for her friends. She has determined that it takes her about 28 minutes to make each necklace. About how long will it take Annalese to make 7 necklaces?

F 1 h 96 min

G 2 h 40 min

H 3 h 16 min

J 4 h

3. Julie has a circular garden in her front yard with a diameter of 8 feet. How does the diameter d compare to the circumference C of the garden?

A $d \approx \frac{1}{3}C$ C $d \approx 2C$

B $d \approx \frac{1}{2}C$ D $d \approx 3C$

4. An angle of an isosceles triangle measures 40°. The other two angles in the triangle are congruent. Which method can be used to find the measure of each congruent angle?

F Multiply 40 by 2. Then add 180.

G Subtract 40 from 180. Then divide by 2.

H Add 40 to 180. Then divide by 3.

J Divide 50 by 2. Then subtract from 180.

5. Mrs. Bixler designed a quilt by outlining equilateral triangles with ribbon as shown below. How much ribbon did Mrs. Bixler use to complete her quilt?

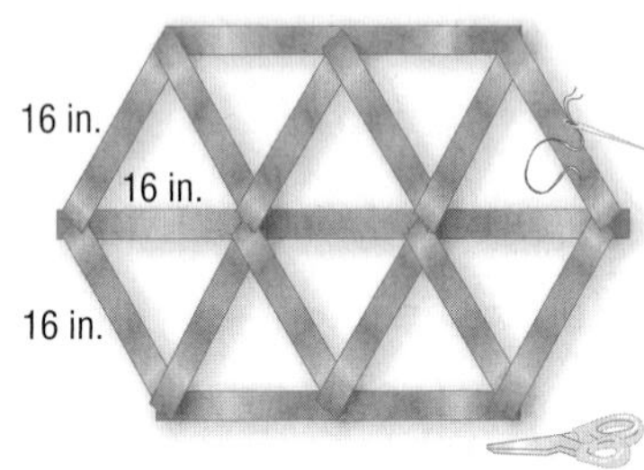

A 125 in. C 304 in.

B 264 in. D 320 in.

6. In the spreadsheet below, a formula applied to the values in columns A and B results in the values in column C. What is the formula?

F $C = A - B$

G $C = A - 2B$

H $C = A + B$

J $C = A + 2B$

	A	B	C
1	4	0	4
2	5	1	3
3	6	2	2
4	7	3	1

Preparing for Standardized Tests
For test-taking strategies and practice, see pages 718–735.

7. In Mrs. Baumgartner's classroom library, the ratio of fiction to non-fiction books is 3 to 4. Which of the following shows possible numbers of fiction to non-fiction books in Mrs. Baumgartner's library?

   A 132 fiction, 172 non-fiction

   B 165 fiction, 228 non-fiction

   C 168 fiction, 224 non-fiction

   D 186 fiction, 242 non-fiction

8. The owner of an ice skating rink recorded the number of paying customers for one week. The table below shows the results. About how many customers paid during the week?

Day	Customers
Monday	42
Tuesday	38
Wednesday	56
Thursday	62
Friday	81
Saturday	112
Sunday	143

   F 600

   G 580

   H 550

   J 500

9. Jermil left home at 2:55 P.M. for field hockey practice. He returned home from practice at 5:05 P.M. About how long was Jermil gone?

   A 2 h

   B 3 h

   C 4 h

   D 5 h

PART 2 Short Response/Grid In

Record your answers on the answer sheet provided by your teacher or on a sheet of paper.

10. The side lengths and perimeters of regular polygons are shown in the table below. Which geometric figure is represented by the information in the table?

Side Length (inches)	Perimeter (inches)
3	12
5	20
8	32
10	40

11. Mario used a square baking pan to make a cake. The length of each side of the pan was 16 inches. Find the area of the pan in square inches.

PART 3 Extended Response

Record your answers on the answer sheet provided by your teacher or on a sheet of paper. Show your work.

12. Leora is gift wrapping a box that measures 15 inches long, 9 inches wide, and 3 inches high.

    a. Find the surface area and the volume of the box.

    b. What is the effect on the surface area and the volume if each dimension is doubled?

    c. What is the effect if only one dimension is doubled? Does it matter which dimension is doubled? Explain.

NEED EXTRA HELP?												
If You Missed Question...	1	2	3	4	5	6	7	8	9	10	11	12
Go to Lesson...	10-4	8-7	10-2	9-4	9-4	6-7	6-1	1-1	8-7	10-1	1-9	10-6

Unit 5

Number, Operations, and Algebraic Thinking

Focus

Apply integers, fractions, and decimals to solve problems and simple one-step equations.

CHAPTER 11
Integers and Transformations

BIG Idea Add, subtract, multiply, and divide with integers to solve problems.

BIG Idea Apply understanding of the coordinate plane and operations with integers to translations, reflections, and rotations.

CHAPTER 12
Algebra: Properties and Equations

BIG Idea Write, evaluate, and use algebraic expressions to solve problems.

BIG Idea Solve simple one-step equations.

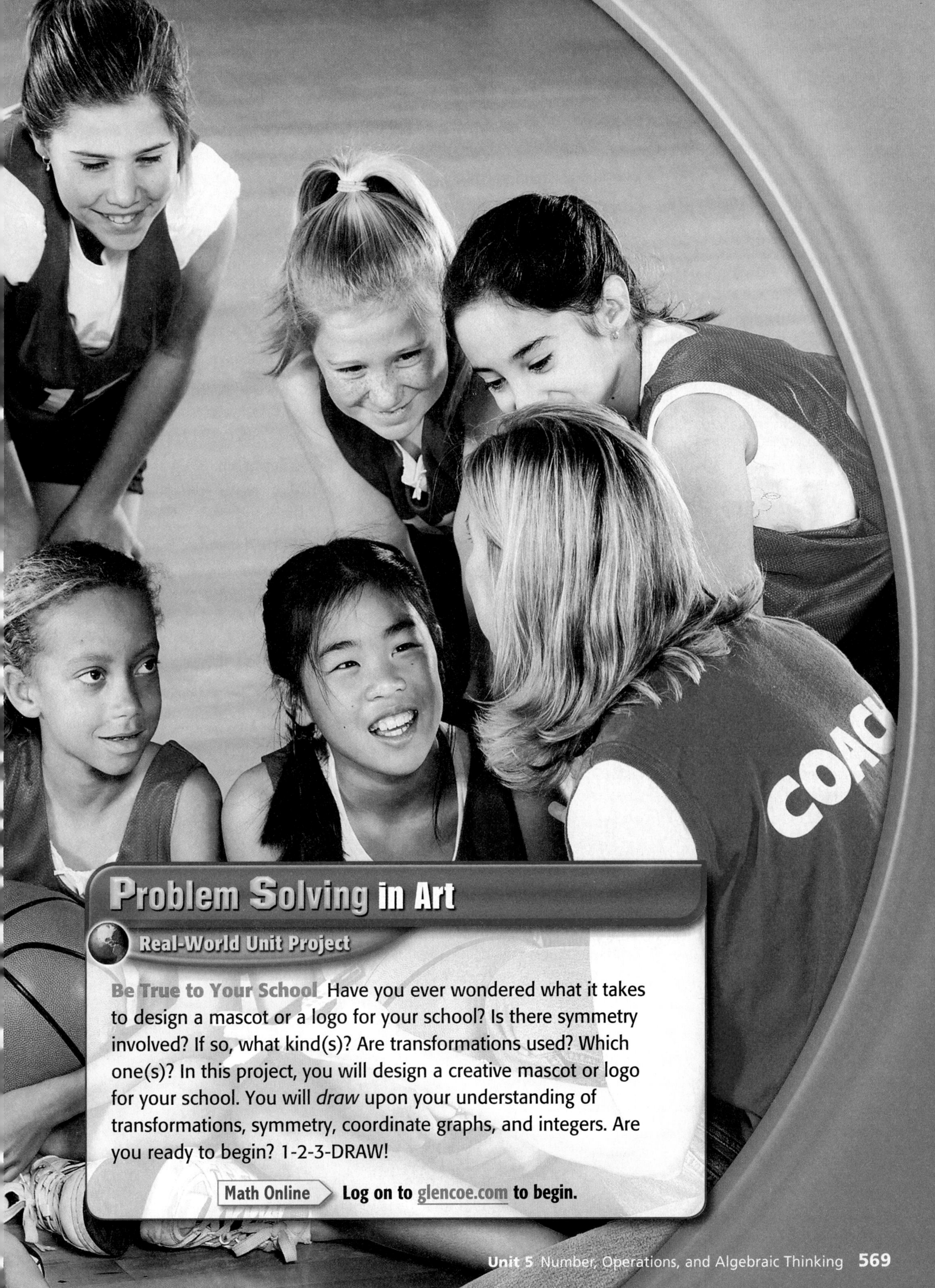

Problem Solving in Art

Real-World Unit Project

Be True to Your School Have you ever wondered what it takes to design a mascot or a logo for your school? Is there symmetry involved? If so, what kind(s)? Are transformations used? Which one(s)? In this project, you will design a creative mascot or logo for your school. You will *draw* upon your understanding of transformations, symmetry, coordinate graphs, and integers. Are you ready to begin? 1-2-3-DRAW!

Math Online **Log on to glencoe.com to begin.**

Integers and Transformations

BIG Ideas

- Add, subtract, multiply, and divide with integers to solve problems.
- Apply understanding of the coordinate plane and operations with integers to translations, reflections, and rotations.

Key Vocabulary

reflection (p. 610)
rotation (p. 615)
transformation (p. 604)
translation (p. 604)
quadrants (p. 599)

Real-World Link

Buildings The total height of the Empire State Building is 1,454 feet, including the lightning rod. The foundation is 55 feet below ground, and the lobby is 47 feet above ground.

Integers and Transformations Make this Foldable to help you organize your notes. Begin with eleven sheets of notebook paper.

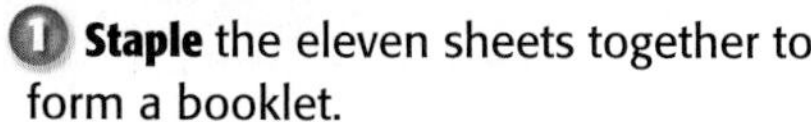

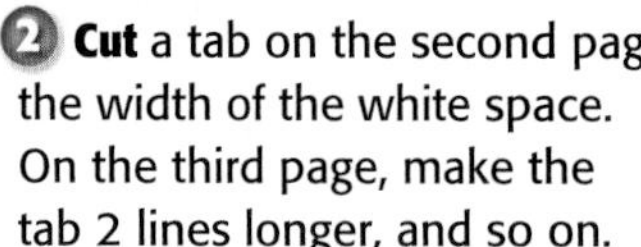

1. **Staple** the eleven sheets together to form a booklet.

2. **Cut** a tab on the second page the width of the white space. On the third page, make the tab 2 lines longer, and so on.

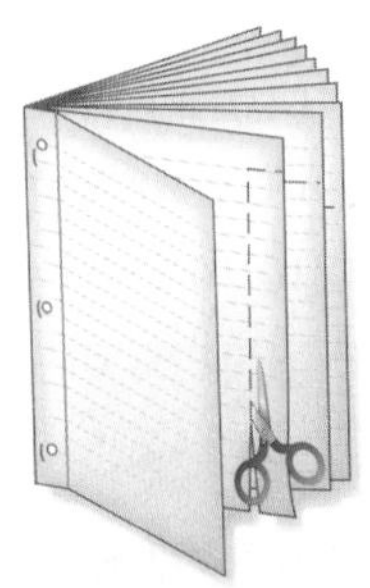

3. **Write** the chapter title on the cover and label each tab with the lesson number.

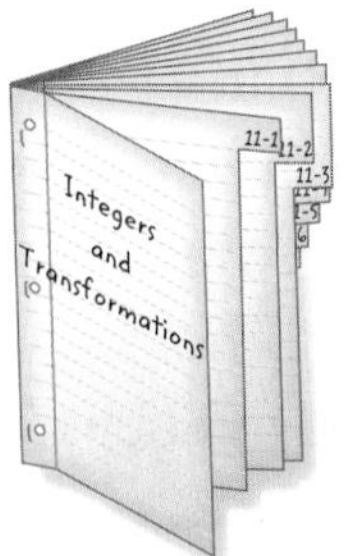

GET READY for Chapter 11

Diagnose Readiness You have two options for checking Prerequisite Skills.

Option 2

Math Online Take the Online Readiness Quiz at glencoe.com.

Option 1

Take the Quick Quiz below. Refer to the Quick Review for help.

QUICK Quiz

Add. (Prior Grade)

1. $12 + 15$
2. $3 + 4$
3. $5 + 7$
4. $16 + 9$
5. **SPORTS** Kalib scored 15 points. Andrea scored 21 points. How many points did they score altogether?

Subtract. (Prior Grade)

6. $14 - 6$
7. $9 - 4$
8. $11 - 5$
9. $8 - 3$
10. **EXERCISE** Stan exercises 3 days a week. How many days during the week does he *not* exercise?

Multiply. (Prior Grade)

11. 7×6
12. 10×2
13. 5×9
14. 8×3
15. 4×4
16. 6×8
17. **SCHOOL** Marty studied 3 hours a day for five days. How many hours total did he study?

Divide. (Prior Grade)

18. $32 \div 4$
19. $63 \div 7$
20. $21 \div 3$
21. $18 \div 9$
22. $72 \div 9$
23. $45 \div 3$

QUICK Review

Example 1

Find 23 + 8.

$$\begin{array}{r} {\scriptstyle 1} \\ 23 \\ +\ 8 \\ \hline 31 \end{array}$$

Example 2

Find 26 − 9.

$$\begin{array}{r} {\scriptstyle 1\ 16} \\ \not{2}\not{6} \\ -\ 9 \\ \hline 17 \end{array}$$

Since 9 is larger than 6, rename 6 as 16. Rename the 2 in the tens place as 1. Then subtract.

Example 3

Find 9 × 8.

$9 \times 8 = 72$ Multiplication fact

or

$9 \times 8 = 9 + 9 + 9 + 9 + 9 + 9 + 9 + 9$
$= 72$

Example 4

Find 56 ÷ 7.

$56 \div 7 = 8$ Division fact

11-1 Ordering Integers

MAIN IDEA

Compare and order integers.

Math Online

glencoe.com

- Extra Examples
- Personal Tutor
- Self-Check Quiz
- Reading in the Content Area

GET READY for the Lesson

SNACKS Lynn, Chi, and Todd are reviewing their accounts at the school's Snack Emporium. Lynn has $7 left in her account, Chi has $10 left in his account, and Todd owes the Snack Emporium $3.

1. Write an integer to represent the amount of money that each person has in his or her account at the Snack Emporium.
2. Order the integers from least to greatest.
3. Who has the least money in his or her Snack Emporium account?

In Lesson 2-9, you learned that an integer is any number from the set {…−4, −3, −2, −1, 0, 1, 2, 3, 4…}. You can use a number line to compare and order integers.

EXAMPLE Compare Integers

Replace the ● with < or > to make a true sentence.

1 **12 ● −4**

Graph 12 and −4 on a number line. Then compare.

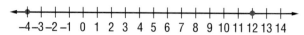

−4 −3 −2 −1 0 1 2 3 4 5 6 7 8 9 10 11 12 13 14

Since 12 is to the right of −4, 12 > −4.

CHECK Your Progress

a. −3 ● −5 **b.** −5 ● 0 **c.** 6 ● −1 **d.** 2 ● −2

EXAMPLE Order Integers

2 **Order −9, 6, −3, and 0 from least to greatest.**

Graph the numbers on a number line.

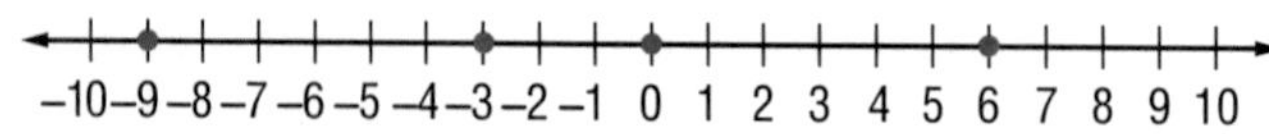

The order from least to greatest is −9, −3, 0, and 6.

CHECK Your Progress

e. Order −4, 3, 11, and −25 from greatest to least.

f. Order −18, 30, 2, −6, and 3 from least to greatest.

Study Tip

Ordering Integers When ordering integers from greatest to least, write the integers as they appear from right to left.

Real-World EXAMPLE

3 **ELEVATION The table shows the lowest elevation for several continents. Order the elevations from least to greatest.**

Continent	Lowest Elevation (m)
Africa	−156
Asia	−418
Australia	−12
Europe	−28
North America	−86
South America	−105

Source: *The World Factbook*

First, graph each integer. Then, write the integers as they appear on the number line from left to right.

−500 −450 −400 −350 −300 −250 −200 −150 −100 −50 0

The elevations from least to greatest are −418, −156, −105, −86, −28, and −12.

CHECK Your Progress

g. **GO-KARTS** The table shows the results of a go-kart race. Negative values indicate seconds less than the average time, and positive values indicate seconds greater than the average time. Arrange the racers from the least amount of time to the greatest.

Name	Times
Sareeta	−6
Bárbara	12
Tamara	−3

CHECK Your Understanding

Example 1 (p. 572)

Replace each ● with < or > to make a true sentence.

1. 17 ● 31
2. −6 ● −10
3. 9 ● −8
4. −83 ● −38

Example 2 (p. 572)

Order each set of integers from least to greatest.

5. 9, −5, −13, −8, 1
6. 22, 4, 14, −2, 5

Order each set of integers from greatest to least.

7. −54, 7, −8, −14, 9, −33
8. −17, −16, 12, 24, −7

Example 3 (p. 573)

9. **RUNNING** The number line shows the position of different runners in relationship to Yolanda. Which runner is ahead of Annika but behind Yolanda? Write an integer to represent her position.

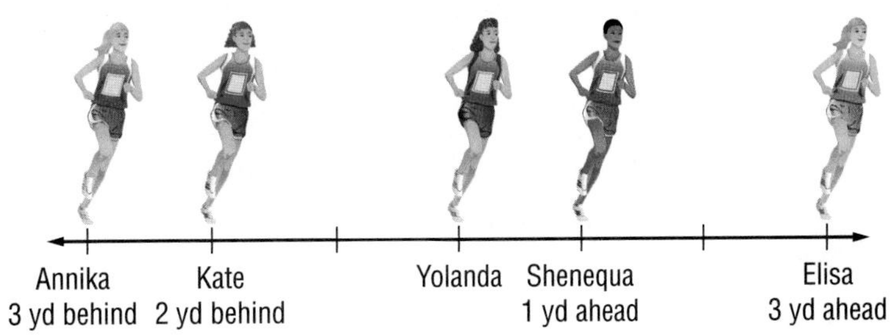

Practice and Problem Solving

HOMEWORK HELP	
For Exercises	See Examples
10–15	1
16–20	2
21, 22	3

Replace each ● with < or > to make a true sentence.

10. −2 ● −4
11. −2 ● 4
12. 1 ● −3
13. −6 ● 3
14. 5 ● 0
15. −3 ● 2

Order each set of integers from least to greatest.

16. 15, 17, 21, 6, 3
17. 14, 1, 6, 23, 14, 5
18. −55, 143, 18, −79, 44, 101
19. −221, 63, 54, −89, −71, −10
20. Order 5, 33, 24, 17, and 6 from greatest to least.
21. **TRAINS** Gary, Sindhu, and Beth are all waiting for their trains to arrive. Gary's train leaves at 5 minutes before noon, Sindhu's leaves at 25 minutes after noon, and Beth's leaves 5 minutes before Sindhu's train. Order the three by who will leave first.
22. **CELL PHONES** The table indicates Xavier's cell phone use over the last four months. Positive values indicate the number of minutes he went over his allotted time, and negative values indicate the number of minutes he was under. Arrange the months from least to most minutes used.

Month	Time (min)
February	−156
March	12
April	0
May	−45

Real-World Link The Andromeda Galaxy, about 2.5 million light-years away, is the farthest object that is visible to the unaided human eye.

Source: *Astronomy, A Self-Teaching Guide*

LIGHT For Exercises 23–25, refer to the table and the following information.

The apparent magnitude of an object measures how bright the object appears to the human eye. A negative magnitude identifies a brighter object than a positive magnitude.

Object	Approximate Apparent Magnitude
100-Watt Bulb	−19
Alpha Centauri	4
Andromeda Galaxy	0
Full Moon	−13
Sun	−27
Venus	−5

Source: *Astronomy, A Self-Teaching Guide*

23. Which object appears the brightest to the human eye?
24. Order the objects from the brightest to the faintest.
25. Find the median apparent magnitude of this data set.

GOLF For Exercises 26 and 27, use the bar graph and the information below.

The bar graph gives the scores of four golfers (A, B, C, and D). The numbers indicate scores above and below par.

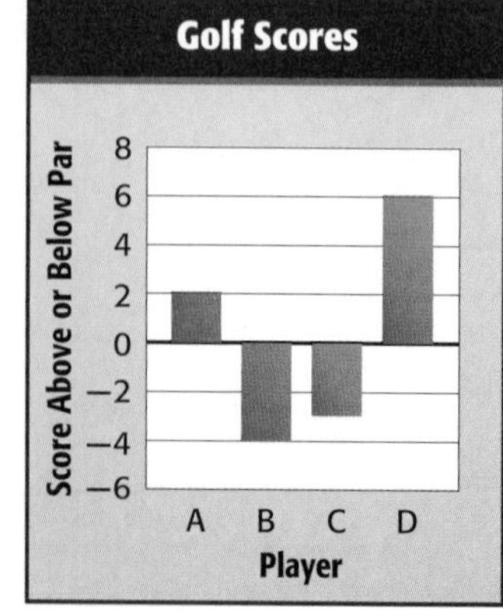

26. Order the scores on a number line.
27. Which player had the worst score? Explain your answer.

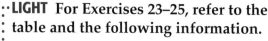

EXTRA PRACTICE
See pages 700, 716.

H.O.T. Problems

28. **OPEN ENDED** Give a set of five integers, two positive and three negative, for which the mean, median, and mode are all −3.

29. **NUMBER SENSE** Explain why −11 is less than −7.

30. **CHALLENGE** Order the fractions $-\frac{1}{2}, \frac{5}{2}, -\frac{12}{4}, \frac{1}{6}$, and $\frac{7}{8}$ from least to greatest.

31. **WRITING IN MATH** In your own words, explain how to list integers from greatest to least.

TEST PRACTICE

32. The table shows the temperatures for a four-day period.

Temperature (°F)	
Monday	−7
Tuesday	8
Wednesday	−2
Thursday	−1

Which list shows the temperatures from least to greatest?

A 8, −2, −1, −7

B 8, −1, −2, −7

C −7, −2, −1, 8

D −7, −1, −2, 8

33. Verónica (V) was 12 minutes early to class, Deshawn (D) was right on time, and Kendis (K) was 3 minutes late. Which time line represents the students' arrival to class?

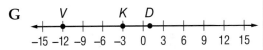

F V at −12, D at 0, K at 3
−15 −12 −9 −6 −3 0 3 6 9 12 15

G V at −12, K at −3, D at 0
−15 −12 −9 −6 −3 0 3 6 9 12 15

H

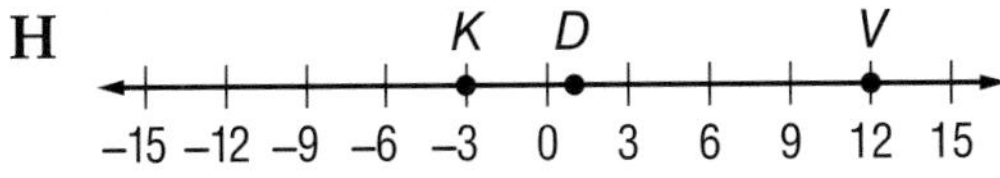

J D at 0, K at 3, V at 12
−15 −12 −9 −6 −3 0 3 6 9 12 15

Spiral Review

34. **GEOMETRY** Find the surface area of the rectangular prism at the right. (Lesson 10-7)

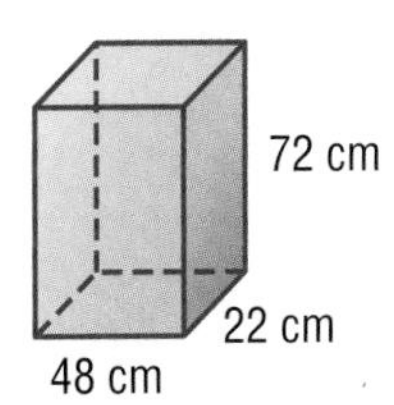

35. **ROOMS** A living room is in the shape of a rectangular prism. If the dimensions of the living room are 15.5 feet by 20 feet by 8 feet, how many cubic feet of space does the room occupy? (Lesson 10-6)

Add or subtract. Write in simplest form. (Lesson 5-6)

36. $2\frac{1}{4} + \frac{3}{4}$

37. $5\frac{2}{3} - 3\frac{1}{3}$

38. $2\frac{2}{5} - \frac{3}{5}$

39. $3\frac{3}{8} + 1\frac{6}{8}$

GET READY for the Next Lesson

PREREQUISITE SKILL **Add or subtract.** (Page 743)

40. 6 + 4

41. 6 − 4

42. 10 + 3

43. 10 − 3

Explore 11-2

Algebra Lab
Zero Pairs

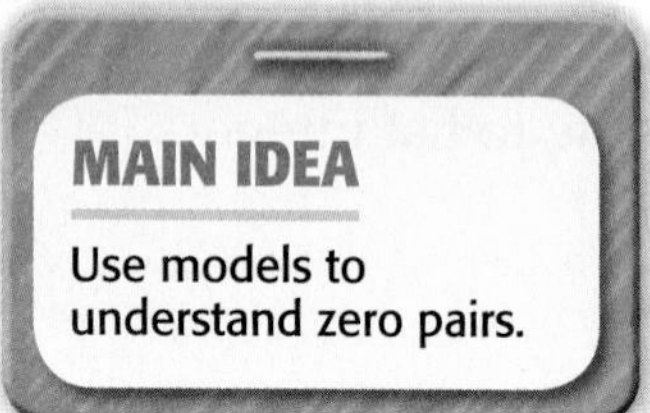

MAIN IDEA
Use models to understand zero pairs.

Counters can be used to help you understand integers. A yellow counter ⊕ represents the integer +1. A red counter ⊖ represents the integer −1. When one yellow counter is paired with one red counter, the result is zero. This pair of counters is called a **zero pair**.

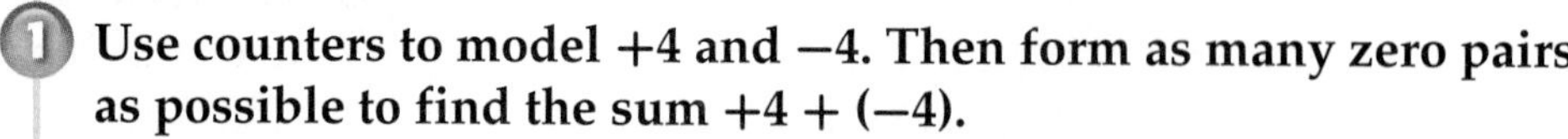

ACTIVITY

1 Use counters to model +4 and −4. Then form as many zero pairs as possible to find the sum +4 + (−4).

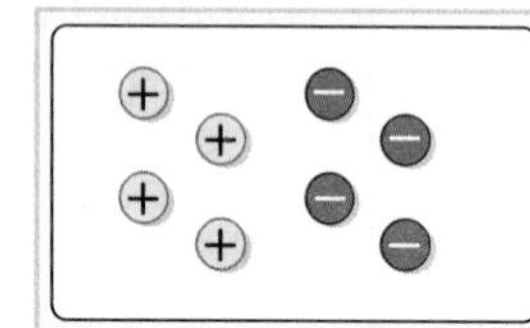

Place four yellow counters on the mat to represent +4. Then place four red counters on the mat to represent −4.

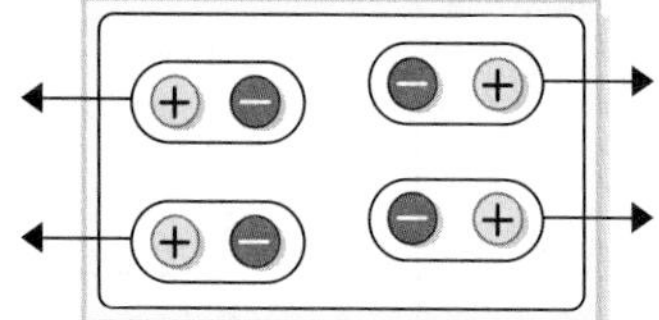

Pair the positive and negative counters. Then remove all zero pairs.

There are no counters on the mat. So, +4 + (−4) = 0.

CHECK Your Progress

Use counters to model each pair of integers. Then form zero pairs to find the sum of the integers.

a. +3, −3 b. +5, −5 c. −7, +7

Analyze the Results

1. What is the value of a zero pair? Explain your reasoning.
2. Suppose there are 5 zero pairs on an integer mat. What is the value of these zero pairs? Explain.
3. Explain the effect of removing a zero pair from the mat. What effect does this have on the remaining counters?
4. Integers like +4 and −4 are called *opposites*. What is the sum of any pair of opposites?
5. Write a sentence describing how zero pairs are used to find the sum of any pair of opposites.
6. **MAKE A CONJECTURE** How do you think you could find +5 + (−2) using counters?

11-2 Adding Integers

MAIN IDEA

Add integers.

Math Online

glencoe.com

- Extra Examples
- Personal Tutor
- Self-Check Quiz

GET READY for the Lesson

GAMES One of Marco's video games involves moving back and forth between colored squares to reach a space on the opposite side of the board. On his first three plays, he moved five squares to the right, 3 squares back to the left, and 1 square right.

1. If he started on the purple square with the star, how many squares to the right is he at the end?

To add integers, you can use counters or a number line.

EXAMPLES Add Integers with the Same Sign

1 **Find +3 + (+2).**

METHOD 1 **Use counters.**

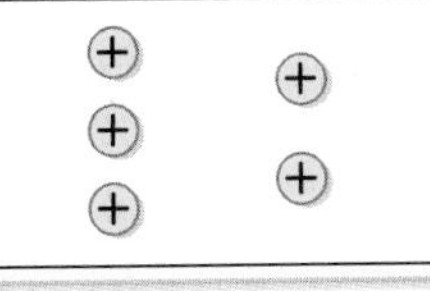

METHOD 2 **Use a number line.**

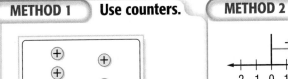

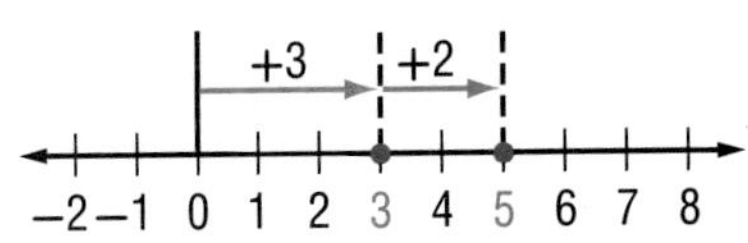

So, $+3 + (+2) = +5$ or 5.

2 **Find −2 + (−4).**

METHOD 1 **Use counters.**

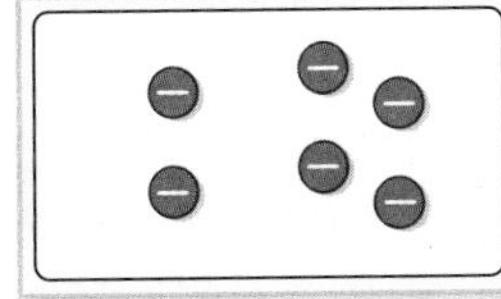

METHOD 2 **Use a number line.**

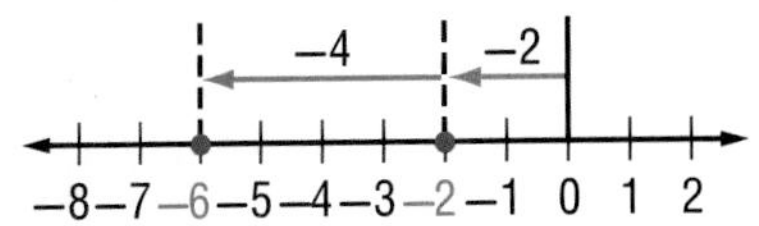

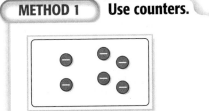

So, $-2 + (-4) = -6$.

CHOOSE Your Method

Add. Use counters or a number line if necessary.

a. $+1 + (+4)$ b. $-3 + (-4)$ c. $-7 + (-4)$

To add two integers with different signs, it is necessary to remove any zero pairs. A *zero pair* is a pair of counters that includes one positive counter and one negative counter.

Reading Math

Positive Integers A number without a sign is assumed to be positive.

EXAMPLE Add Integers with Different Signs

3 **Find $-8 + 6$.**

METHOD 1 **Use counters.**

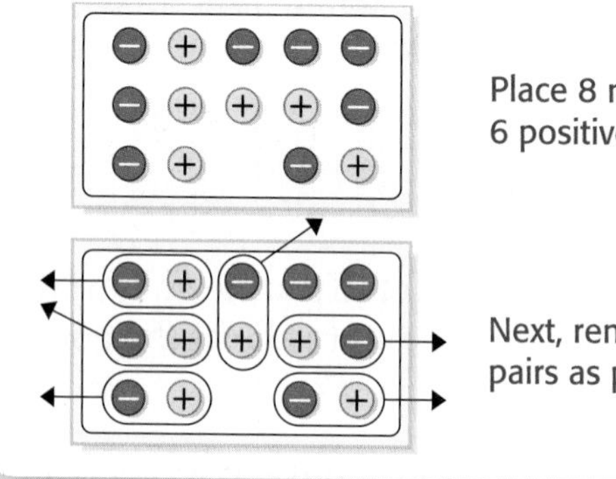

Place 8 negative counters and 6 positive counters on the mat.

Next, remove as many zero pairs as possible.

METHOD 2 **Use a number line.**

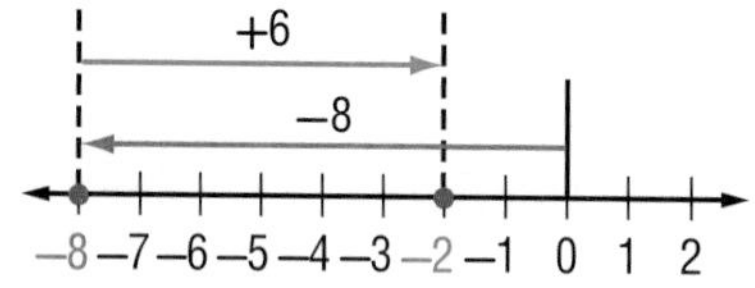

Start at 0. Move 8 units to the left to show -8. From there, move 6 units right to show $+6$.

So, $-8 + 6 = -2$.

CHOOSE Your Method

Add. Use counters or a number line if necessary.

d. $-6 + 3$ e. $4 + (-4)$ f. $+7 + (-3)$

The following rules are often helpful when adding integers.

Add Integers — Key Concept

Words The sum of two positive integers is always positive. The sum of two negative integers is always negative.

Examples $5 + 1 = 6$ $-5 + (-1) = -6$

Words The sum of a positive integer and a negative integer is sometimes positive, sometimes negative, and sometimes zero.

Examples $5 + (-1) = 4$ $-5 + 1 = -4$ $-5 + 5 = 0$

Check Your Understanding

Examples 1–3 (pp. 577–578)

Add. Use counters or a number line if necessary.

1. $+3 + (+1)$
2. $4 + (+2)$
3. $-3 + (-5)$
4. $-6 + (-4)$
5. $+2 + (-5)$
6. $-4 + 9$

Example 3 (p. 578)

7. **MONEY** The deposits and withdrawals from Max's checking account are shown at the right. If Max originally had \$75 in his account, how much does he have now?

Check Number	Amount (\$)
881	+20
1246	−35
1247	−10
882	+40

Practice and Problem Solving

HOMEWORK HELP	
For Exercises	See Examples
8–13	1
14–19, 26	2
20–25, 27	3

Add. Use counters or a number line if necessary.

8. $+2 + (+1)$
9. $+5 + (+1)$
10. $6 + (+2)$
11. $3 + (+4)$
12. $+8 + 3$
13. $+9 + 4$
14. $-4 + (-1)$
15. $-5 + (-4)$
16. $-2 + (-4)$
17. $-3 + (-3)$
18. $-2 + (-3)$
19. $-6 + (-10)$
20. $-7 + (+5)$
21. $-3 + (+3)$
22. $-2 + 6$
23. $-12 + 7$
24. $15 + (-6)$
25. $8 + (-18)$

For Exercises 26 and 27, write an addition problem that represents the situation. Then solve.

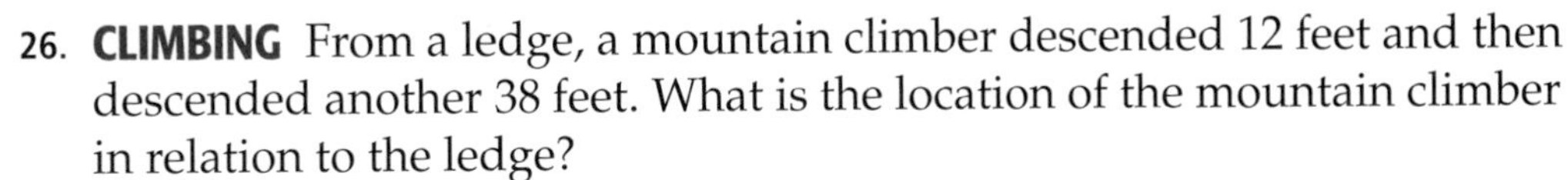

26. **CLIMBING** From a ledge, a mountain climber descended 12 feet and then descended another 38 feet. What is the location of the mountain climber in relation to the ledge?

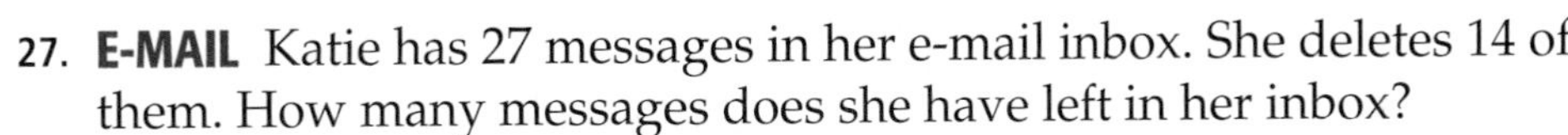

27. **E-MAIL** Katie has 27 messages in her e-mail inbox. She deletes 14 of them. How many messages does she have left in her inbox?

28. **ALGEBRA** Evaluate $a + b$ if $a = 7$ and $b = -3$.

29. **ALGEBRA** Evaluate $a + b$ if $a = -9$ and $b = 4$.

Real-World Link... The temperature change described in Exercise 34 was caused by a *chinook* wind. Chinook winds are extremely fast and can cause dramatic changes in temperature.

Source: National Weather Service

Add.

30. $12 + 13 + (-4)$
31. $5 + (-8) + (-7) + 11$
32. $5 + 2 + (-8) + (-3) + 9$
33. $3 + (-5) + 7 + 0 + (-2)$

34. **WEATHER** The record for the fastest temperature change in the United States occurred in 1980 in Great Falls, Montana, where the temperature rose 47°F in just seven minutes. If the original temperature was −32°F, what was the temperature 7 minutes later?

35. **MONEY** Sophia's checking account statement shows that her current balance is −\$247. She deposits \$225 into her checking account. What is Sophia's new checking account balance?

36. **GOLF** Jeff's golf score was 5 strokes less than Mike's golf score. Mike's score was 2 strokes more than Sam's. If Sam's score relative to par was +2, what was Jeff's golf score?

37. **TIME ZONES** The map and table below show how the time in several of the world's cities is related to the time in London. If it is 1 P.M. in London, what time is it in New York City and Jakarta? If it is 11 A.M. in Shanghai, what time is it in Paris?

City	Time Ahead of London (h)
Cape Town	2
Jakarta	−10
New York City	−5
Paris	1
Shanghai	8
Vancouver	−8

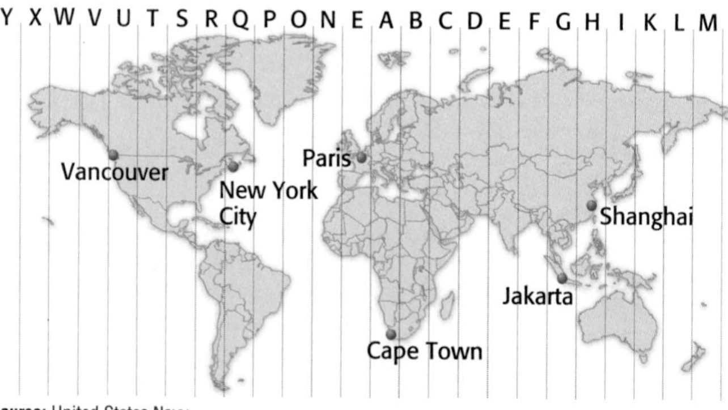

Source: United States Navy

38. **FIND THE DATA** Refer to the Data File on pages 16–19. Choose some data and write a real-world problem in which you would add integers.

39. **CLIMBING** Lakin is painting his grandmother's house. He climbs 10 feet up the ladder. Then he climbs down 8 feet and then ascends another 7 feet. What is Lakin's location on the ladder?

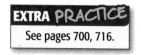

EXTRA PRACTICE
See pages 700, 716.

H.O.T. Problems

40. **OPEN ENDED** Write two different addition sentences with sums that are each −10.

41. **FIND THE ERROR** Monifa and Ramón are finding 4 + (−6). Who is correct? Explain your reasoning.

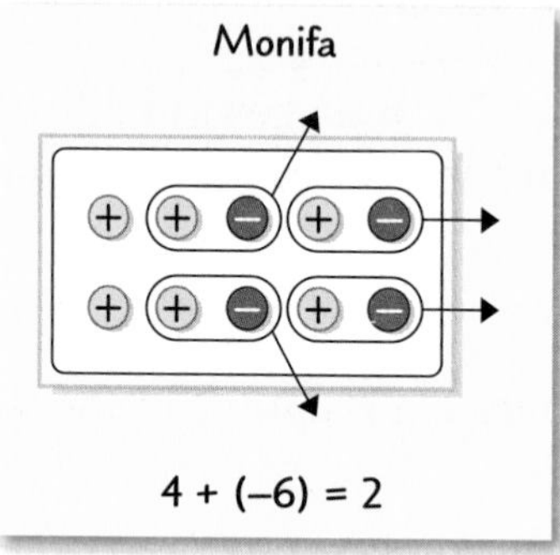

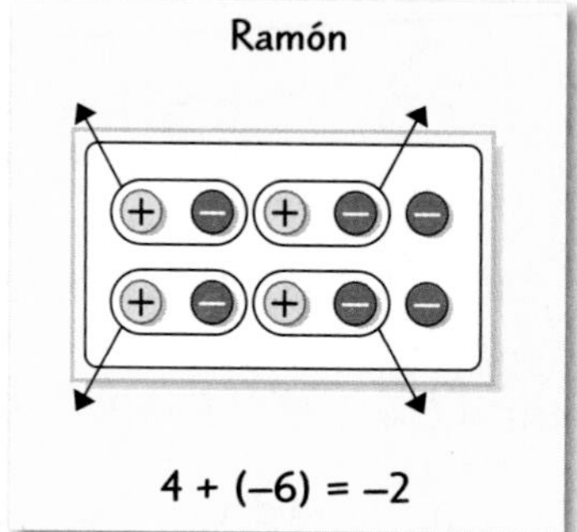

NUMBER SENSE Tell whether each sum is *positive, negative,* or *zero* without adding.

42. $-8 + (-8)$
43. $-2 + 2$
44. $6 + (+6)$
45. $-5 + 2$
46. $-3 + 8$
47. $-2 + (-6)$

CHALLENGE For Exercises 48 and 49, find possible integer values for x and y that make each statement true.

48. $x + y = -x$
49. $x + (-y) = y + y$

50. **WRITING IN MATH** Explain how a number line can be used to model adding positive and negative integers. Relate this to a real-world situation.

TEST PRACTICE

51. Which expression is represented by the model?

A $0 + (-4)$

B $-10 + 4$

C $-14 + 10$

D $-10 + 14$

52. Alyssa owes her brother \$13. If she pays back \$8 of this amount and then borrows an additional \$4, which addition sentence represents this situation?

F $13 + 8 + 4$

G $-13 + (-8) + (-4)$

H $-13 + 8 + (-4)$

J $13 + 8 + (-4)$

Spiral Review

53. **TESTS** Jacqui's score on the biology test was the average score for her class. Marquez's score was three points above the average, Melanie's score was 10 points below the average, and Len's score was 5 points above the average. Write the scores as integers and order them from least to greatest. (Lesson 11-1)

54. **PACKAGING** How much cardboard is needed to make a box in the shape of a rectangular prism with length 5 inches, width 7 inches, and height 3 inches? (Lesson 10-7)

Estimate the circumference of each circle. (Lesson 10-2)

55. 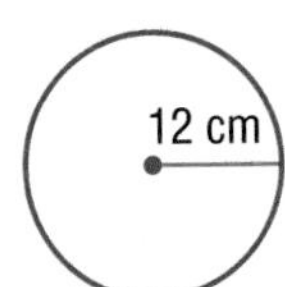

56. 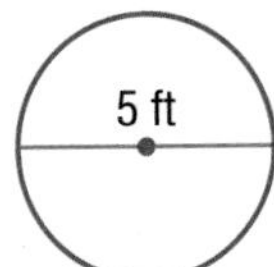

57. 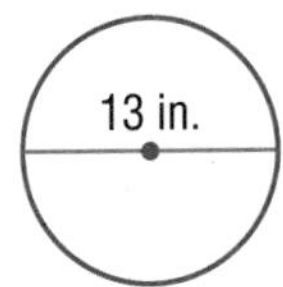

58. **TIME** It took Brady 24 minutes to answer 8 questions on his math test. If he continues at this rate, how long will it take him to answer all 48 questions? (Lesson 6-4)

59. **RAPPELLING** LaTasha is going on a rappelling trip with some friends. They're going to rappel down a 50-foot cliff, a 180-foot cliff, a 90-foot tower, and a 200-foot canyon. What is the average height they are going to rappel? (Lesson 2-6)

GET READY for the Next Lesson

PREREQUISITE SKILL Subtract. (Page 743)

60. $5 - 3$

61. $6 - 4$

62. $9 - 5$

63. $10 - 3$

11-3 Subtracting Integers

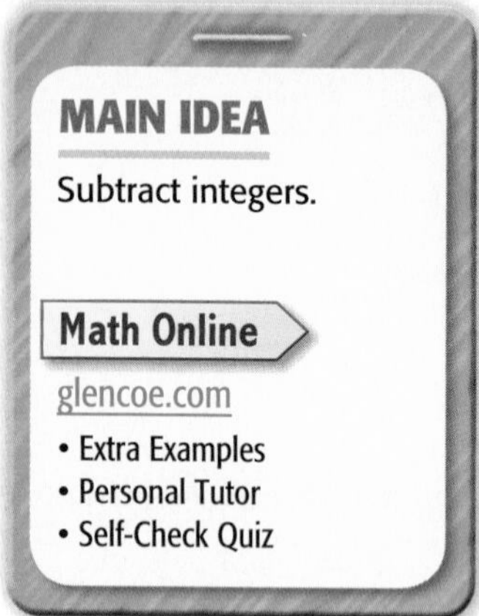

MINI Lab

The number lines below model the subtraction problems 8 – 2 and –3 – 4.

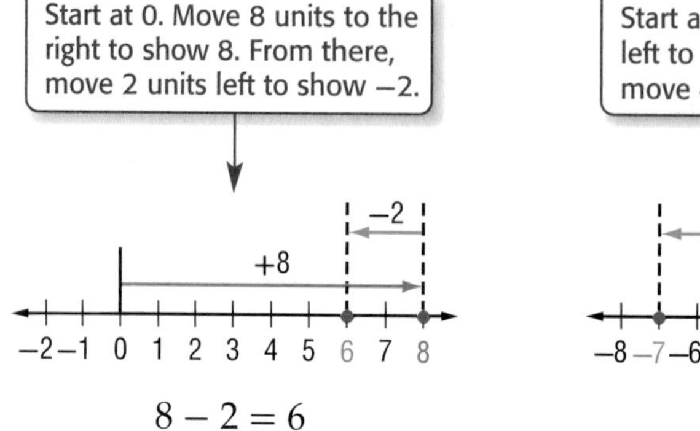

$8 - 2 = 6$

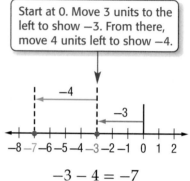

Start at 0. Move 3 units to the left to show –3. From there, move 4 units left to show –4.

–4 –3

–8 –7 –6 –5 –4 –3 –2 –1 0 1 2

$-3 - 4 = -7$

1. Model 8 + (–2) using a number line.
2. Compare this model to the model for 8 – 2. How is 8 – 2 related to 8 + (–2)?
3. Use a number line to model –3 + (–4).
4. Compare this model to the model for –3 – 4. How is –3 – 4 related to –3 + (–4)?

The Mini Lab shows that when you subtract a number, the result is the same as adding the opposite of the number.

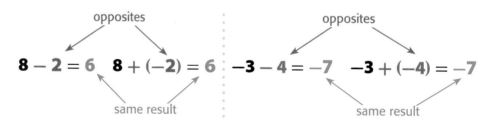

opposites | opposites

$8 - 2 = 6$ $8 + (-2) = 6$ | $-3 - 4 = -7$ $-3 + (-4) = -7$

same result | same result

To subtract integers, you can use counters or the following rule.

Subtract Integers — Key Concept

Words To subtract an integer, add its opposite.

Examples
$5 - 2 = 5 + (-2)$
$-3 - 4 = -3 + (-4)$
$-1 - (-2) = -1 + 2$

EXAMPLE Subtract Positive Integers

1 **Find 3 − 1.**

METHOD 1 Use counters.

Place 3 positive counters on the mat to show +3. Then, remove 1 counter.

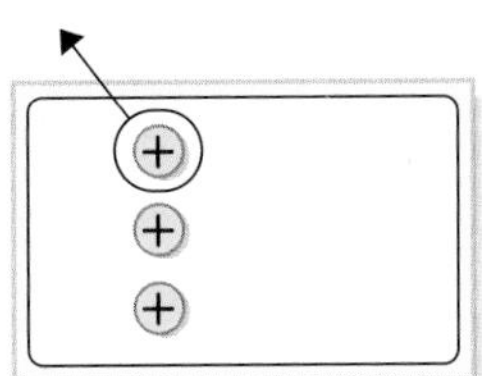

METHOD 2 Add the opposite.

To subtract 1, add −1.

$3 - 1 = 3 + (-1)$
$= 2$

So, $3 - 1 = 2$.

Study Tip

Check by Adding In Example 1, you can check $3 - 1 = 2$ by adding. $2 + 1 = 3$ ✓ In Example 2, you can check $-5 - (-3) = -2$ by adding. $-2 + (-3) = -5$ ✓

CHOOSE Your Method **Subtract. Use counters if necessary.**

a. $6 - 4$ b. $+5 - 2$ c. $9 - 6$

EXAMPLE Subtract Negative Integers

2 **Find −5 − (−3).**

METHOD 1 Use counters.

Place 5 negative counters on the mat to show −5. Then, remove 3 counters.

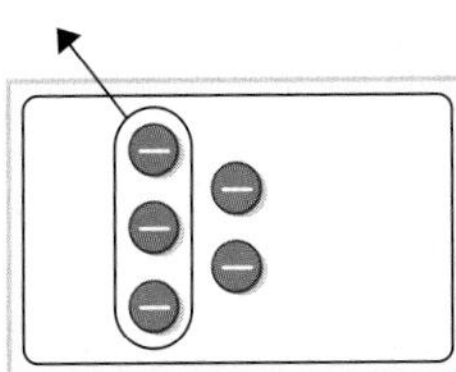

METHOD 2 Add the opposite.

To subtract −3, add 3.

$-5 - (-3) = -5 + 3$
$= -2$

So, $-5 - (-3) = -2$.

Check Use a number line to find $-5 + 3$.

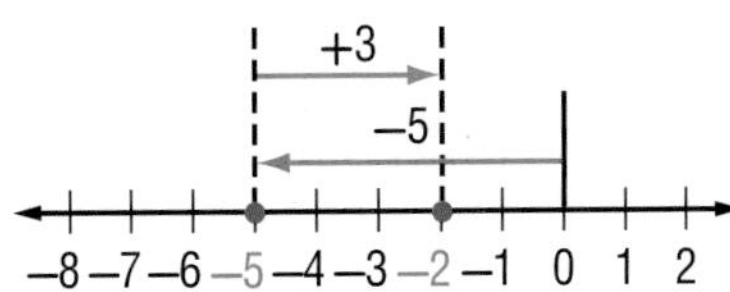

CHOOSE Your Method **Subtract. Use counters if necessary.**

d. $-8 - (-2)$ e. $-6 - (-1)$ f. $-5 - (-4)$

Sometimes you need to add zero pairs before you can subtract.

EXAMPLE Subtract Integers Using Zero Pairs

3 **Find $-2 - 3$.**

METHOD 1 **Use counters.**

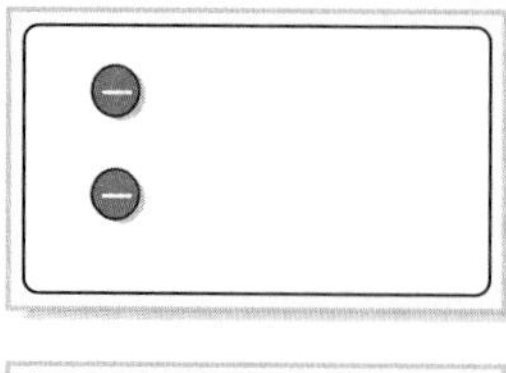

Place 2 negative counters on the mat to show -2.

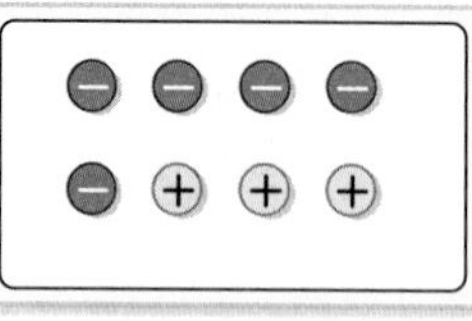

Since there are no positive counters, add 3 zero pairs.

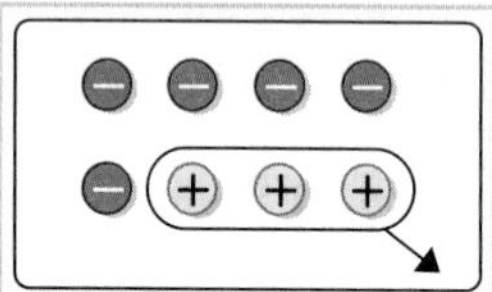

Now, remove 3 positive counters. The result is 5 negative counters.

Study Tip

Zero Pairs Recall that when you add zero pairs, the values of the integers on the mat do not change.

METHOD 2 **Add the opposite.**

$-2 - 3 = -2 + (-3)$ To subtract 3, add -3.

$= -5$

So, $-2 - 3 = -5$.

Study Tip

Use a Number Line The number line shows that $-2 - 3 = -5$.

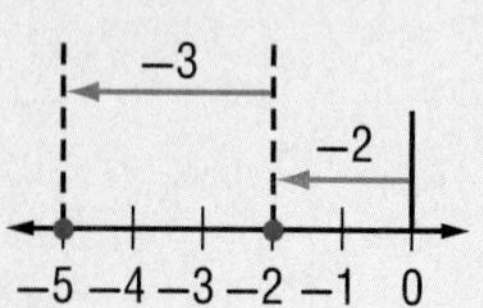

CHOOSE Your Method **Subtract. Use counters if necessary.**

g. $-8 - 3$

h. $-2 - 5$

i. $-9 - 5$

Real-World EXAMPLE

4 **GAMES Matt and Isabel are contestants on a game show. Currently, Isabel has 13 points and Matt has -7 points. How many more points does Isabel have than Matt?**

Subtract Matt's score from Isabel's.

$13 - (-7) = 13 + 7$ To subtract -7, add 7.

$= 20$ Simplify.

Isabel has 20 points more than Matt.

CHECK Your Progress

j. **TRACK** Jamie ran a mile in 15 seconds more than her average time. Yesterday she ran a mile in 13 seconds less than her average time. What is the difference between these times?

CHECK Your Understanding

Examples 1–3 (pp. 583–584)

Subtract. Use counters if necessary.

1. $7 - 5$
2. $+4 - 1$
3. $-9 - (-4)$
4. $-6 - (-6)$
5. $-1 - 5$
6. $-2 - (-3)$

Example 4 (p. 584)

7. **ALLOWANCE** Rodney receives $20 every month for his allowance. He owes his brother $13. After Rodney pays back his brother, how much of his allowance will he have left?

Practice and Problem Solving

HOMEWORK HELP

For Exercises	See Examples
8–13	1
14–19	2
20–25	3
26, 27	4

Subtract. Use counters if necessary.

8. $8 - 3$
9. $6 - 5$
10. $11 - 7$
11. $15 - 8$
12. $+6 - 2$
13. $+8 - 1$
14. $-7 - (-5)$
15. $-8 - (-4)$
16. $-10 - (-5)$
17. $-9 - (-6)$
18. $-9 - (-9)$
19. $-2 - (-2)$
20. $-7 - 9$
21. $-6 - 2$
22. $-4 - 2$
23. $-5 - 4$
24. $-12 - 3$
25. $-15 - 5$

26. **TIME** Time zones all over the world base their standard time on their distance from Greenwich, England. Located on the prime meridian, any time zone east of Greenwich is behind GMT (Greenwich Mean Time). Any time zone west of Greenwich is ahead. The table shows certain cities' time in relation to GMT. What is the difference in time between Los Angeles and Paris?

City	Time (h)
New York	−5
Tokyo	+8
Los Angeles	−5
Paris	+1

27. **STATISTICS** For a class project, Mykia recorded the outside temperature for five nights. Find the range of his data: −3°F, 7°F, −5°F, 9°F, and 12°F.

28. **ALGEBRA** Evaluate $a - b$ if $a = 5$ and $b = 7$.

29. **ALGEBRA** Find the value of $m - n$ if $m = -3$ and $n = 4$.

30. **BANKING** Colleen's bank subtracts a positive number from her account balance when she makes a withdrawal and adds a positive number when she deposits money. The table shows her account activity over the last month. If she started the month with $500, how much does she have now?

Week	Deposits ($)	Withdrawals ($)
1		45
2		75
3	200	
4		115

EXTRA PRACTICE

See pages 701, 716.

H.O.T. Problems

31. **OPEN ENDED** Find two negative integers, a and b, where $a - b$ is positive.

32. **Which One Doesn't Belong?** Identify the expression that does not belong with the other three. Explain your reasoning.

7 – 3	–3 + 7	–3 – (–7)	–7 + 3

CHALLENGE For Exercises 33–36, find the value of each expression if the value of $x - y = 2$ and $x + y = 8$. Explain.

33. $x + (-y)$

34. $(x + y) - (x - y)$

35. $x - (-y)$

36. $(x - y) - (x + y)$

37. **WRITING IN MATH** Is the difference between two integers with different signs always negative? Explain.

TEST PRACTICE

38. Which expression is represented by the model?

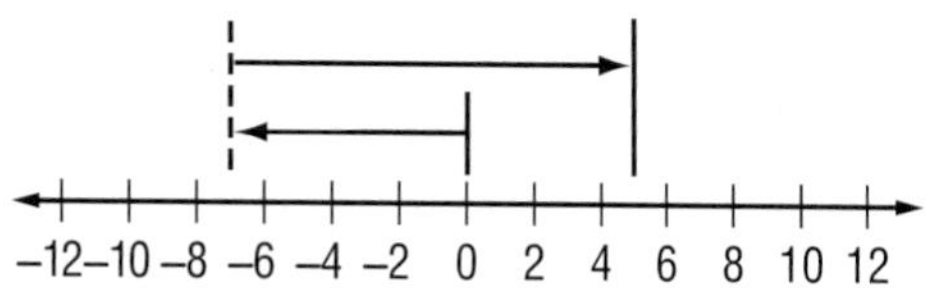

A $0 - (-5)$

B $-6 + 5$

C $-7 - (-12)$

D $0 - (-12)$

39. In a video game, Juliana had −11 points. On her next turn, she lost an additional 4 points. Which integer represents her score now?

F 15 points

G 7 points

H −7 points

J −15 points

Spiral Review

Add. Use counters or a number line if necessary. (Lesson 11-2)

40. $3 + (-4)$

41. $4 + (-4)$

42. $-2 + (-3)$

43. Order −6, 2, 0, −2, and 4 on the same number line. (Lesson 11-1)

Write each fraction as a percent. (Lesson 7-1)

44. $\frac{4}{5}$

45. $\frac{3}{2}$

46. $\frac{1}{20}$

47. $\frac{7}{5}$

48. **PAINTBALL** Cristiano and four of his friends are going to play paintball this weekend. The total cost of their trip will be $118.75. If they split the total evenly, how much will the trip cost each person? (Lesson 3-8)

49. **SHOES** The shoe sizes of the boy's basketball team are 11, 10, 15, 9, 17, 12, and 10. What is the average shoe size of the boy's basketball team? (Lesson 2-6)

GET READY for the Next Lesson

PREREQUISITE SKILL Multiply. (Page 744)

50. 5×6

51. 8×7

52. 9×4

53. 8×9

11-4 Multiplying Integers

MAIN IDEA

Multiply integers.

Math Online

glencoe.com

- Concepts in Motion
- Extra Examples
- Personal Tutor
- Self-Check Quiz

MINI Lab

The models show 3×2 and $3 \times (-2)$.

For 3×2, place 3 sets of 2 positive counters on the mat.

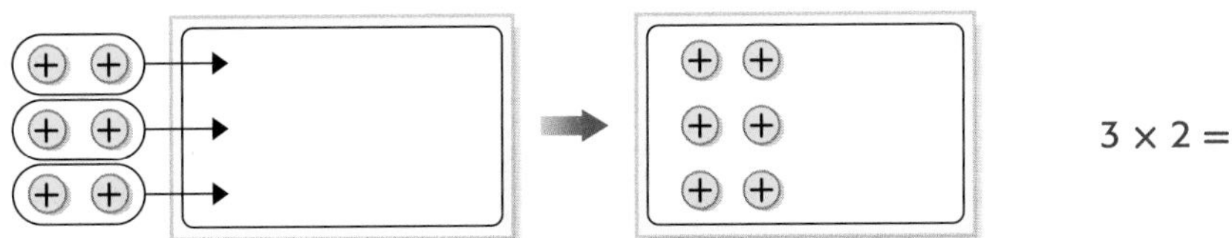

$3 \times 2 = 6$

For $3 \times (-2)$, place 3 sets of 2 negative counters on the mat.

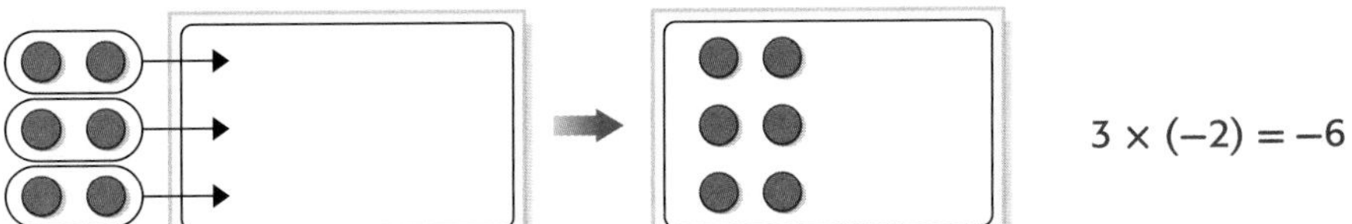

$3 \times (-2) = -6$

1. Use counters to find $4 \times (-3)$ and $5 \times (-2)$.
2. **MAKE A CONJECTURE** What is the sign of the product of a positive and negative integer?

To find the sign of products like -3×2 and $-3 \times (-2)$, you can use patterns.

$3 \times 2 = 6$	$3 \times (-2) = -6$
$2 \times 2 = 4$	$2 \times (-2) = -4$
$1 \times 2 = 2$	$1 \times (-2) = -2$
$0 \times 2 = 0$	$0 \times (-2) = 0$
$-1 \times 2 = -2$	$-1 \times (-2) = 2$
$-2 \times 2 = -4$	$-2 \times (-2) = 4$
$-3 \times 2 = ■$	$-3 \times (-2) = ■$

By extending the number pattern, you find that $-3 \times 2 = -6$.

By extending the number pattern, you find that $-3 \times (-2) = 6$.

Multiply Integers

Key Concept

Words The product of two integers with different signs is negative.

Examples $3 \times (-2) = -6$ $\quad -3 \times 2 = -6$

Words The product of two integers with the same sign is positive.

Examples $3 \times 2 = 6$ $\quad -3 \times (-2) = 6$

EXAMPLES Multiply Integers with Different Signs

Multiply.

1 $4 \times (-2)$

$4 \times (-2) = -8$ The integers have different signs. The product is negative.

2 -8×3

$-8 \times 3 = -24$ The integers have different signs. The product is negative.

CHECK Your Progress

a. $6 \times (-3)$ b. $3(-1)$ c. -4×5 d. $-2(7)$

EXAMPLES Multiply Integers with Same Signs

Multiply.

3 4×8

$4 \times 8 = 32$ The integers have the same sign. The product is positive.

4 $-5 \times (-6)$

$-5 \times (-6) = 30$ The integers have the same sign. The product is positive.

CHECK Your Progress

e. 3×3 f. $6(8)$ g. $-5 \times (-3)$ h. $-4(-3)$

Real-World EXAMPLE

5 **BATHTUBS A bathtub drains at the rate of 4 gallons per minute. What will be the change in the volume of the bathtub after 6 minutes?**

To find the change in the volume of the bathtub after 6 minutes, you can multiply 6 by the amount of change per minute, -4 gallons.

$6 \times (-4) = -24$

So, after 6 minutes, the change in the volume of the bathtub will be -24 gallons. This means the bathtub will have drained 24 gallons of water.

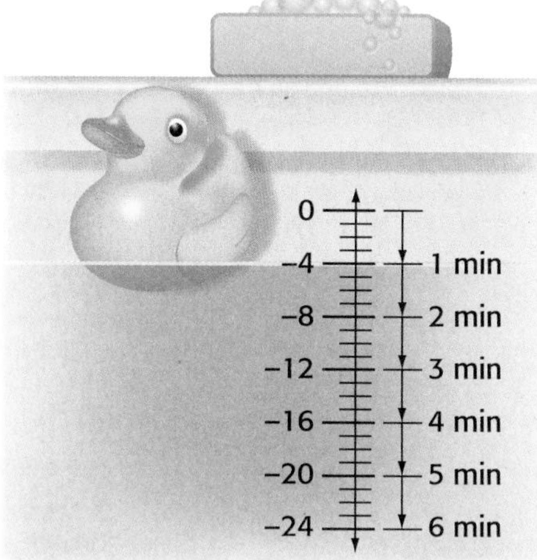

CHECK Your Progress

i. **MONEY** Justine made three $20 withdrawals from her savings account in the past month. She did not make any deposits. How much money does Justine have in her savings account now, in relation to how much she had at the beginning of the month?

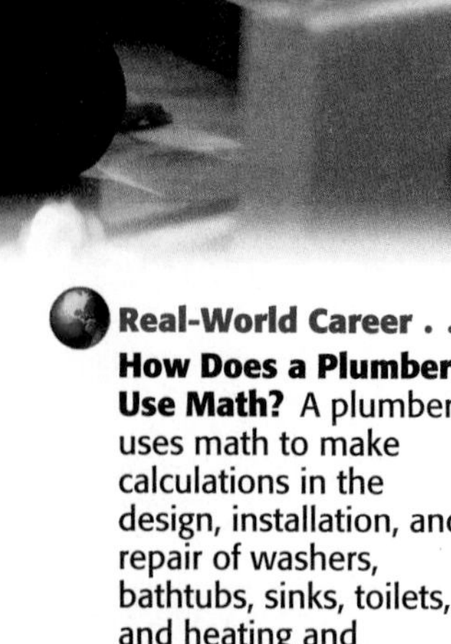

Real-World Career . . .
How Does a Plumber Use Math? A plumber uses math to make calculations in the design, installation, and repair of washers, bathtubs, sinks, toilets, and heating and cooling systems.

Math Online

For more information, go to glencoe.com.

CHECK Your Understanding

Examples 1–4 (p. 588)

Multiply.

1. $4 \times (-7)$
2. $8(-7)$
3. 4×4
4. $9(3)$
5. $-1 \times (-7)$
6. $-7(-6)$

Example 5 (p. 588)

7. **ANCHORS** An anchor dropped from a boat descends 8 meters every minute. What will be the change in the location of the anchor after 9 minutes?

Practice and Problem Solving

HOMEWORK HELP	
For Exercises	See Examples
8–15	1, 2
16–23	3, 4
24, 25	5

Multiply.

8. $6 \times (-6)$
9. $9 \times (-1)$
10. $7(-3)$
11. $2(-10)$
12. -7×5
13. -2×9
14. $-5(6)$
15. $-6(9)$
16. 8×7
17. $9(4)$
18. $-5 \times (-8)$
19. $-9 \times (-7)$
20. 12×7
21. 9×5
22. $-6(-10)$
23. $-1(-9)$

24. **MONEY** Jackson withdraws \$30 from his bank account every week for lunch and spending money. What integer represents the change in value of his bank account after 8 weeks if he makes no additional deposits or withdrawals?

25. **SUBMARINES** A submarine descends from the surface of the water at the rate of 220 feet per minute. What integer represents the change in the submarine's location related to the surface of the water after 40 minutes?

26. **ALGEBRA** Evaluate st if $s = -4$ and $t = 9$.

27. **ALGEBRA** Find the value of ab if $a = -12$ and $b = -5$.

PATTERNS For Exercises 28 and 29, find the next two numbers in the pattern. Then describe the pattern.

28. 2, −4, 8, −16, ...
29. −2, −6, −18, −54, ...

Multiply.

30. $3(-4 - 7)$
31. $-2(3)(-4)$
32. $-4[5 + (-9)]$

EXTRA PRACTICE See pages 701, 716.

33. **WEATHER** On Wednesday morning, the temperature dropped 4°F every hour for 5 hours. If the temperature was 11°F before it started dropping, what was the temperature after 5 hours?

H.O.T. Problems

34. **Which One Doesn't Belong?** Identify the expression that does not belong with the other three. Explain your reasoning.

-9(3)	-4(-8)	6 x 5	-1 x (-7)

35. **OPEN ENDED** Write three different pairs of integers that each have the product of −18.

REASONING **For Exercises 36–39, decide whether each statement is true or false for any positive integers *a* and *c* and any negative integers *b* and *d*. Explain.**

36. $a \times b$ is positive

37. $b \times d$ is negative

38. $a \times c$ is positive

39. $b \times c$ is negative

40. **CHALLENGE** What is the sign of the product of three negative numbers? What is the sign of the product of four negative numbers? Justify your reasoning.

41. **WRITING IN MATH** Write a problem about a real-world situation in which you would multiply integers.

TEST PRACTICE

42. A rock climber descended from the top of a rock at a rate of 4 meters per minute. Where will the rock climber be in relation to the top of the rock after 8 minutes?

A −8 m

B −12 m

C −32 m

D −48 m

43. Which expression has a product of −36?

F $-4 \times (-9)$

G 4×9

H $-12 \times (-3)$

J $3 \times (-12)$

Spiral Review

Subtract. Use counters if necessary. (Lesson 11-3)

44. $9 - 2$

45. $+3 - 1$

46. $-5 - (-8)$

47. $-7 - 6$

Add. Use counters or a number line if necessary. (Lesson 11-2)

48. $-7 + 2$

49. $+4 + (-3)$

50. $-2 + (-2)$

51. $7 + (-8)$

52. **HEALTH** The table shows the approximate number of heartbeats per second for humans and housecats. About how many more times does a housecat's heart beat in one hour than a human? (Lesson 7-7)

Heart Rates

Animal	Number of Beats per Second
Housecat	2.0
Human	1.2

53. How many meters are in 82 centimeters? (Lesson 8-6)

54. Estimate the product of $\frac{3}{4}$ and 37. (Lesson 5-6)

GET READY for the Next Lesson

55. **PREREQUISITE SKILL** Parker has a piece of ribbon measuring $8\frac{3}{4}$ yards. How many pieces of ribbon each measuring $1\frac{3}{4}$ yards can be cut from the large piece of ribbon? Use the *act it out* strategy. (Lesson 5-2)

CHAPTER 11

Mid-Chapter Quiz

Lessons 11-1 through 11-4

Replace each ● with <, >, or = to make a true sentence. (Lesson 11-1)

1. -7 ● -3
2. 4 ● -2
3. -8 ● 6
4. 78 ● 76

5. Order $-5, -7, 4, -3$, and -2 from greatest to least. (Lesson 11-1)

6. Order $0, -9, 6, -1$, and 5 from least to greatest. (Lesson 11-1)

7. **GOLF** In golf, the lowest score wins. The table shows the scores relative to par of the winners of the U.S. Open from 2000 to 2007. What was the lowest winning score? (Lesson 11-1)

Year	Winner	Score
2007	Angel Cabrera	+5
2006	Geoff Ogilvy	+5
2005	Michael Campbell	even
2004	Retief Goosen	−4
2003	Jim Furyk	−8
2002	Tiger Woods	−3
2001	Retief Goosen	−4
2000	Tiger Woods	−12

Add. Use counters or a number line if necessary. (Lesson 11-2)

8. $+8 + (-3)$
9. $-6 + 2$
10. $-4 + (-7)$

11. **ALGEBRA** Evaluate $x + y$ if $x = 7$ and $y = -12$. (Lesson 11-2)

12. **MULTIPLE CHOICE** A mole is in a burrow 12 inches below ground. It digs down 2 more inches. Which addition sentence represents the situation? (Lesson 11-2)

    A $12 + (-2)$

    B $-12 + 2$

    C $-12 + (-2)$

    D $12 + 2$

Subtract. Use counters if necessary. (Lesson 11-3)

13. $9 - (+3)$
14. $-3 - 5$
15. $8 - (-2)$
16. $-4 - (-7)$

17. **ALGEBRA** Evaluate $g - h$ if $g = -2$ and $h = -6$. (Lesson 11-3)

18. **MULTIPLE CHOICE** In Antarctica the temperature can be −8°F with a wind chill of −32°F. Find the difference between the temperature and the wind chill temperature. (Lesson 11-3)

    F -40

    G -24

    H 24

    J 40

Multiply. (Lesson 11-4)

19. $-4 \times (-7)$
20. $6(-9)$
21. 8×2
22. $-7(10)$

23. **ELEVATORS** An elevator descends at a rate of 15 feet per second. Write an integer to represent the change in the position of the elevator after 20 seconds. (Lesson 11-4)

24. **DISTANCE** The table shows the number of yards a cyclist traveled. If the cyclist continues at the same rate, how many yards will she have traveled after four minutes? (Lesson 11-4)

Time (min)	Distance (yd)
1	360
2	720
3	1,080
4	■

25. **SCIENCE** For each kilometer above Earth's surface, the temperature decreases 7°C. If the temperature at Earth's surface is 0°C, what will the temperature be 2 kilometers above the surface? (Lesson 11-4)

11-5 Problem-Solving Investigation

MAIN IDEA: Solve problems by working backward.

P.S.I. TEAM +

e-Mail: WORK BACKWARD

LOURDES: The baseball team sold twice as many adult tickets to Friday's game as student tickets. The number of student tickets sold was eighteen more than the number of seats that were not sold. Twenty-one seats were not sold.

YOUR MISSION: Work backward to find the total number of tickets that could be sold in all.

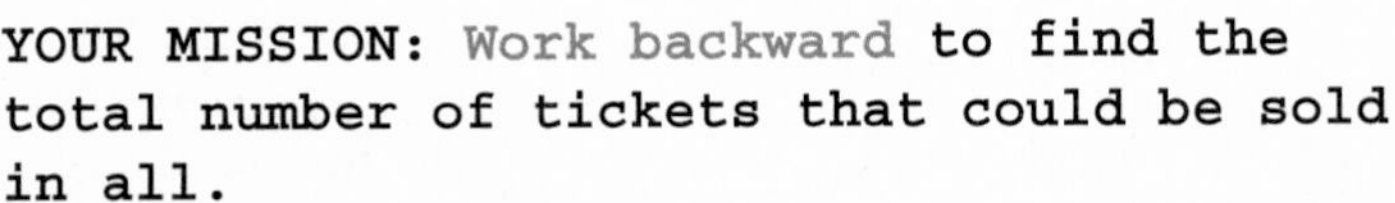

Understand	You know twice as many adult tickets were sold as student tickets. You know that 18 more student tickets were sold than the number of seats that were not sold. You know that 21 seats were not sold. You need to find how many total tickets could be sold.
Plan	Begin with the number of seats that were not sold. By working backward, find the number of student tickets and then the number of adult tickets. The sum of all of these will give the total number of tickets that could be sold.
Solve	21 number of seats that were not sold $21 + 18 = 39$ number of student tickets $39 \times 2 = 78$ number of adult tickets The sum of 21, 39, and 78 is 138. So, a total of 138 tickets could be sold.
Check	Begin with 138 tickets. $138 - 78 - 39 = 21$. Since 78 is twice as much as 39, and 39 is 18 more than 21, the answer is correct. ✔

Analyze The Strategy

1. What is the best way to check your solution when using the *work backward* strategy?
2. WRITING IN MATH Explain when you would use the *work backward* strategy to solve a problem.

Mixed Problem Solving

For Exercises 3–5, solve using the *work backward* strategy.

3. **EXERCISE** David swims and lifts weights for a total of 35 minutes at the gym before school. He spends 10 minutes getting ready for school after he works out. It takes him 15 minutes to get to the gym from his house and 20 minutes to get to school from the gym. If he needs to be at school at 8:15 A.M., what time should he leave his house in the morning to get to the gym?

4. **AGE** Marcia is five years younger than her sister, Lynn. Lynn is half as old as her aunt Jen. If Jen is 28, how old is Marcia?

5. **NUMBER SENSE** A number is multiplied by 4, and then 6 is added to the product. The result is 18. What is the number?

Use any strategy to solve Exercises 6–14. Some strategies are shown below.

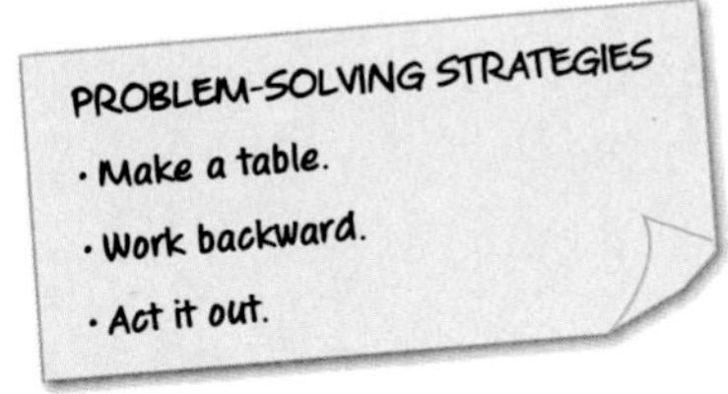

PROBLEM-SOLVING STRATEGIES
• Make a table.
• Work backward.
• Act it out.

6. **ROLLER COASTERS** The list shows how many times each of 20 students rode a roller coaster at an amusement park one day.

Number of Times Rode Roller Coaster									
5	10	0	12	8	7	2	6	4	1
0	14	3	11	5	9	13	8	6	3

How many more students rode a roller coaster 5 to 9 times than 10 to 14 times?

7. **CAMERAS** Francesca saved 13 new pictures on her digital camera's memory card and deleted 32 pictures. If there are now 108 pictures, how many pictures had she saved originally?

8. **NUMBER SENSE** What is the least positive number that you can divide by 7 and get a remainder of 4, divide by 8 and get a remainder of 0, and divide by 9 and get a remainder of 4?

9. **CABLE** In 1977, there were 12,168,450 cable television subscribers. By 2003, there were 73,365,880 subscribers. How many more cable television subscribers were there in 2003 than in 1977?

10. **COMETS** Halley's comet is visible from Earth approximately every 76 years. If the next scheduled appearance is in 2062, when was the last visit by Halley's comet?

11. **RECIPES** A fruit punch recipe calls for 8 cups of orange juice. If there are 4 cups in one quart, how many quarts of juice are needed for the recipe?

12. **NUMBER SENSE** A number is subtracted from 20. If the difference is divided by 3 and the result is 4, what is the number?

13. **COMPUTERS** A megabyte is equal to 1,000 kilobytes. A kilobyte is equal to 1,024 bytes. If a byte is equal to 8 bits, how many bits are in a megabyte?

14. **PARKING** The diagram shows the number of parking spaces available in several different parking lots on a college campus. Find the mean number of parking spaces available in each lot.

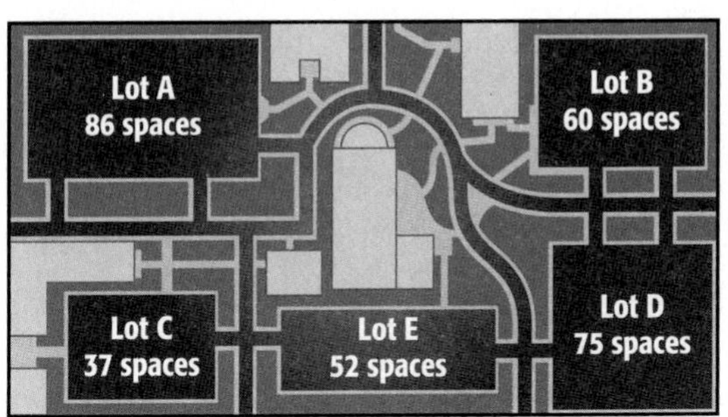

11-6 Dividing Integers

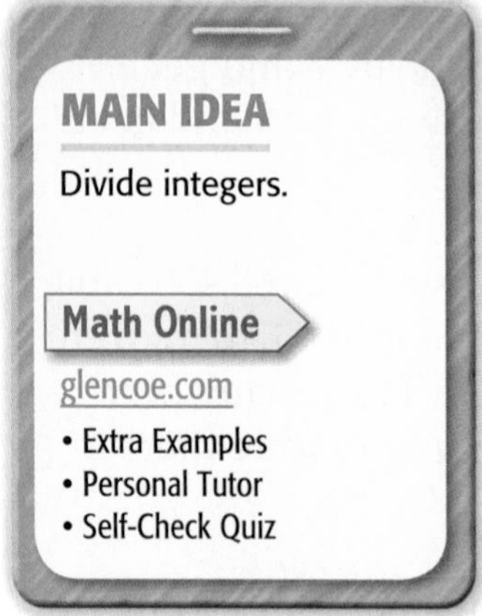

MAIN IDEA

Divide integers.

Math Online

glencoe.com

- Extra Examples
- Personal Tutor
- Self-Check Quiz

MINI Lab

You can use counters to model $-12 \div 3$.

STEP 1 Place 12 negative counters on the mat to represent -12.

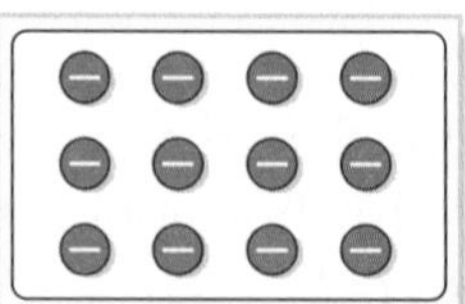

STEP 2 Separate the 12 negative counters into three equal-size groups. There are 3 groups of 4 negative counters.

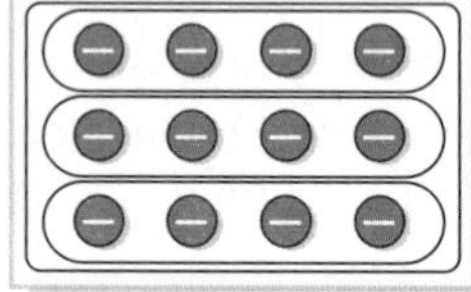

So, $-12 \div 3 = -4$.

1. Explain how you would model $-9 \div 3$.
2. What would you do differently to model $8 \div 2$?

Division means to separate the number into equal-size groups.

EXAMPLES Divide Integers

Divide.

1 $-8 \div 4$

Separate 8 negative counters into 4 equal-size groups.

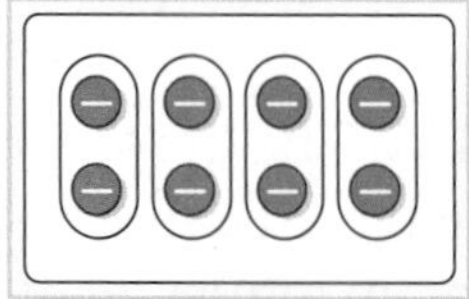

There are 4 groups of 2 negative counters.

So, $-8 \div 4 = -2$.

2 $15 \div 3$

Separate 15 positive counters into 3 equal-size groups.

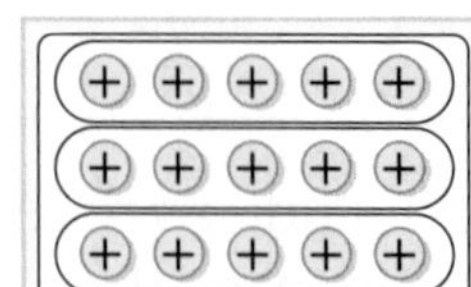

There are 3 groups of 5 positive counters.

So, $15 \div 3 = 5$.

CHECK Your Progress **Divide. Use counters if necessary.**

a. $-4 \div 2$ b. $-20 \div 4$ c. $18 \div 3$

EXAMPLES Divide Integers

3 **Find −10 ÷ 2.**

Since −5 × 2 = −10, it follows that −10 ÷ 2 = −5. ← negative quotient

4 **Find 14 ÷ (−7).**

Since −2 × (−7) = 14, it follows that 14 ÷ (−7) = −2. ← negative quotient

5 **Find −24 ÷ (−6).**

Since 4 × (−6) = −24, it follows that −24 ÷ (−6) = 4. ← positive quotient

CHECK Your Progress **Divide. Work backward if necessary.**

d. −16 ÷ 4 e. 32 ÷ (−8) f. −21 ÷ (−3)

Divide Integers — Key Concept

Words The quotient of two integers with different signs is negative.

Examples 8 ÷ (−2) = −4 −8 ÷ 2 = −4

Words The quotient of two integers with the same sign is positive.

Examples 8 ÷ 2 = 4 −8 ÷ (−2) = 4

TEST EXAMPLE

6 The elevator in the Sears Tower in Chicago, Illinois, descends about 420 meters in 60 seconds. Which integer represents the change in the height of the elevator each second?

A 7 B 6 C −6 D −7

Test-Taking Tip

Key Words Look for words that indicate negative numbers. To descend means to go down, so a descent would be represented by a negative integer.

Read the Item You need to find the number of meters per second. Represent the total descent as the integer −420.

Solve the Item Since −420 ÷ 60 = −7, the answer is D.

CHECK Your Progress

g. **DOGS** Over 5 years, the number of registered Beagles declined by 4,525. If the decline in numbers was the same each year, which integer represents the change per year?

F −905 G −560 H 560 J 905

CHECK Your Understanding

Examples 1–5 (pp. 594–595)

Divide.

1. $-6 \div 2$
2. $-12 \div 3$
3. $15 \div 3$
4. $-25 \div 5$
5. $72 \div (-9)$
6. $-36 \div (-4)$

Example 6 (p. 595)

7. **MULTIPLE CHOICE** In 2006, there were 83 endangered mammals in the U.S. In the same year, the number of endangered plants in the U.S. was 747. How many times greater was the number of endangered plants than mammals in 2006?

A -9 **C** 9

B -7 **D** 13

Practice and Problem Solving

HOMEWORK HELP	
For Exercises	See Examples
8–19	1–5
20, 21	6

Divide.

8. $-8 \div 2$
9. $-12 \div 4$
10. $-18 \div 6$
11. $-32 \div 4$
12. $21 \div 7$
13. $35 \div 7$
14. $-40 \div 8$
15. $-45 \div 5$
16. $63 \div (-9)$
17. $81 \div (-9)$
18. $-48 \div (-6)$
19. $-54 \div (-6)$

20. **GAMES** Anando lost a total of 28 points over the last 2 rounds of a game. If he lost the same number of points each round, what integer represents the change in his score each round?

21. **CAVE EXPLORING** A cave explorer starts at the entrance to a cave and descends 195 meters into the cave in 3 hours. If the explorer traveled an equal distance each hour, what integer gives the distance and direction traveled each hour relative to the entrance of the cave?

22. **ALGEBRA** Find the value of $c \div d$ if $c = -22$ and $d = 11$.

23. **ALGEBRA** What value of m makes $48 \div m = -16$ true?

Find the value of each expression.

24. $\dfrac{-3 + (-7)}{2}$

25. $\dfrac{[4 + (-6)] \times (-1 + 7)}{-3}$

26. **WEATHER** The table shows the record low temperatures for certain locations in Arizona. What is the average record low temperature for these cities?

City	Temp (°F)	Year
Flagstaff	−30	1937
Glendale	20	1971
Grand Canyon	−23	1985
Mesa	15	1950
Tucson	8	1954

Source: National Weather Service

Real-World Link The highest temperature recorded for Arizona was 128°F in June, 1994, at Lake Havasu City.

Source: *The World Almanac*

27. **FIND THE DATA** Refer to the Data File on pages 16–19. Choose some data and write a real-world problem in which you would divide integers.

HOT AIR BALLOONS **For Exercises 28 and 29, the table shows the change in altitude over time for several hot air balloons at a ballooning festival.**

Balloon	Change in Altitude (ft)	Time (min)	Balloon	Change in Altitude (ft)	Time (min)
Air Adam	−3,600	120	Flying Felicia	800	50
Benji's Balloon	−1,200	60	Jersey Jen	−2,700	135
Captain Kate	480	30	Magnificent Meg	−1,500	60
Daring Diego	−350	14	Soaring Susana	720	30

28. Find the change in altitude in feet per second for each balloon listed in the table.

29. For the balloons listed in the table that descended, find the median change in altitude in feet per minute.

30. **WETLANDS** A study revealed that 6,540,000 acres of coastal wetlands have been lost over the past 50 years due to draining, dredging, landfills, and spoil disposal. If the loss continues at the same rate, how many acres will be lost in the next 10 years?

EXTRA PRACTICE
See pages 702, 716.

H.O.T. Problems

31. **OPEN ENDED** Write a division sentence whose quotient is −9. Then describe a real-world situation that this division sentence could represent.

32. **FIND THE ERROR** Bena and Suki are finding −56 ÷ 8. Who is correct? Explain your reasoning.

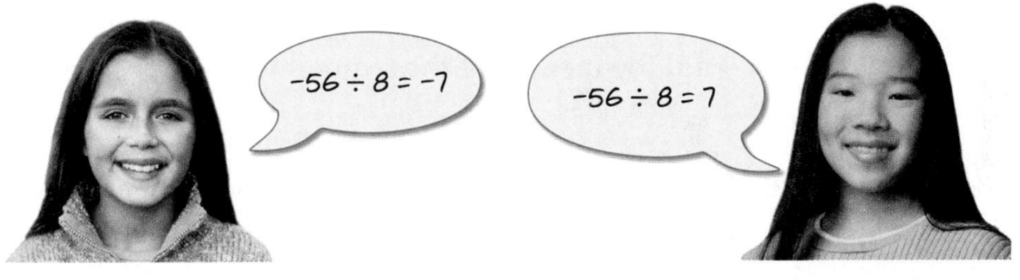

33. **SELECT A TOOL** Melissa wants to show her brother how to divide −15 by 3. Which of the following tools might Melissa use to show this division problem? Justify your selection(s). Then use the tool(s) to demonstrate this division problem.

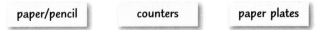

paper/pencil | counters | paper plates

34. **NUMBER SENSE** A number is divided by −6 and then divided again by −2. This gives the same result as the original number being divided by which integer?

35. **CHALLENGE** If $x \div y$ is negative, what is the sign of xy? Explain.

36. **WRITING IN MATH** Without calculating, explain how you know whether the quotient 432 ÷ (−18) is greater than or less than 0.

TEST PRACTICE

37. During the past week, Mrs. Hirosho recorded the following amounts in her checkbook: \$20, –\$53, –\$62, and –\$27. Which expression can be used to find the average of these amounts?

    A $[20 + (-53) + (-62) + (-27)] \div 4$

    B $(20 + 53 + 62 + 27) \div 4$

    C $20 + 53 + 62 + 27 \div 4$

    D $20 + (-53) + (-62) + (-27) \div 4$

38. The table shows the number of points each student earned on the first math test. Each question on the test was worth an equal number of points.

Student	Points
Charlie	76
Serefina	84
Kaneesha	96

If Charlie answered 19 questions correctly, how many questions did Serefina answer correctly?

F –4   **H** 22

G –3   **J** 21

Spiral Review

39. **MONEY** Brianna and 5 of her friends bought a pack of six fruit juices after their lacrosse game. If the pack costs \$3.29, how much does each person owe to the nearest cent if the cost is divided equally? Use the *work backward* strategy. (Lesson 11-5)

40. Find the product of –3 and –7. (Lesson 11-4)

41. **WEATHER** Find the median of the temperatures –9°F, 7°F, –4°F, –1°F, and 11°F. (Lessons 2-7 and 11-1)

42. **AQUARIUMS** A shark petting tank is 20 feet long, 8 feet wide, and 3 feet deep. What is the surface area if the top of the tank is open? (Lesson 10-7)

Add or subtract. (Lesson 8-7)

43. 17 min 45 s
– 9 min 24 s

44. 2 h 25 min
+ 9 h 53 min

45. 25 min 17 s
– 12 min 38 s

Write each percent as a fraction in simplest form. (Lesson 7-1)

46. 30%   47. 28%   48. 145%   49. 85%

GET READY for the Next Lesson

PREREQUISITE SKILL Draw a number line from –10 to 10. Then graph each point on the number line. (Lesson 11-1)

50. 3   51. 0   52. –2   53. –6

11-7 The Coordinate Plane

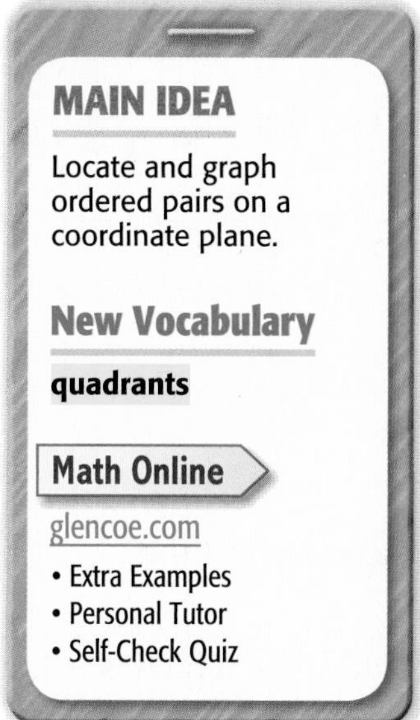

MAIN IDEA

Locate and graph ordered pairs on a coordinate plane.

New Vocabulary

quadrants

Math Online

glencoe.com

- Extra Examples
- Personal Tutor
- Self-Check Quiz

GET READY for the Lesson

MAPS The map shows the layout of a small town. The locations of buildings are described in respect to the town hall. Each unit on the grid represents one block.

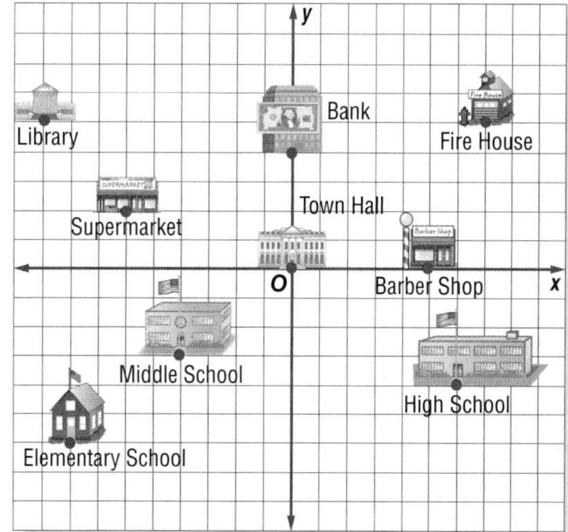

1. Describe the location of the barber shop in relation to the town hall.
2. What building is located 7 blocks east and 5 blocks north of the town hall?
3. Violeta is at the library. Describe how many blocks and in what direction she should travel to get to the supermarket.
4. Let north and east directions be represented by positive integers. Let west and south directions be represented by negative integers. Describe the location of the high school as an ordered pair using integers.
5. Describe the location of the bank as an ordered pair using integers.

Recall that a coordinate plane is formed when the x-axis and y-axis intersect at their zero points. The x-axis and y-axis extend forever in each direction and separate the coordinate plane into four regions called **quadrants**. Each quadrant has an area of infinite size. In Lesson 4-9, you graphed and named ordered pairs in Quadrant I.

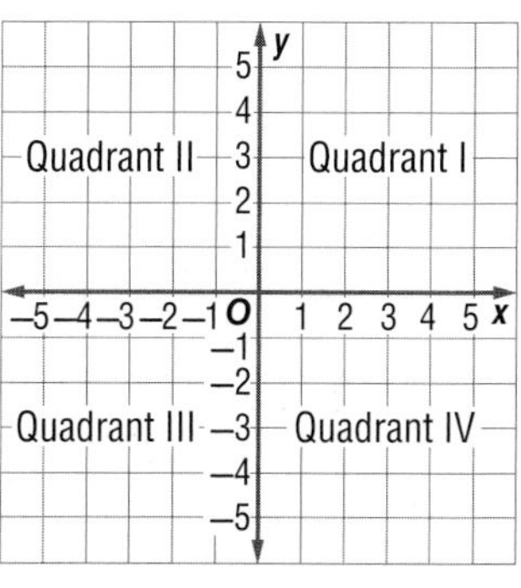

EXAMPLES Identify Ordered Pairs

Identify the ordered pair that names each point. Then identify the quadrant in which each point is located.

1 **point *B***

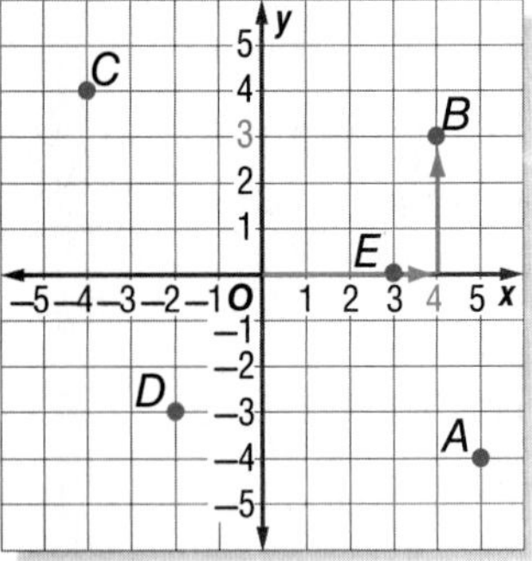

Step 1 Start at the origin. Move right on the x-axis to find the x-coordinate of point B, which is 4.

Step 2 Move up the y-axis to find the y-coordinate, which is 3.

Point B is named by (4, 3). Point B is in the first quadrant.

2 **point *D***

Step 1 Start at the origin. Move left on the x-axis to find the x-coordinate of point D, which is -2.

Step 2 Move down the y-axis to find the y-coordinate, which is -3.

Point D is named by $(-2, -3)$. Point D is in the third quadrant.

CHECK Your Progress

Write the ordered pair that names each point. Then identify the quadrant in which each point is located.

a. *A* **b.** *C* **c.** *E*

Study Tip

Ordered Pairs *A point located on the x-axis will have a y-coordinate of 0. A point located on the y-axis will have an x-coordinate of 0. Points located on an axis are not in any quadrant.*

To graph an ordered pair, draw a dot at the point that corresponds to the coordinates.

EXAMPLE Graph Ordered Pairs

3 **Graph point M at $(-3, 5)$.**

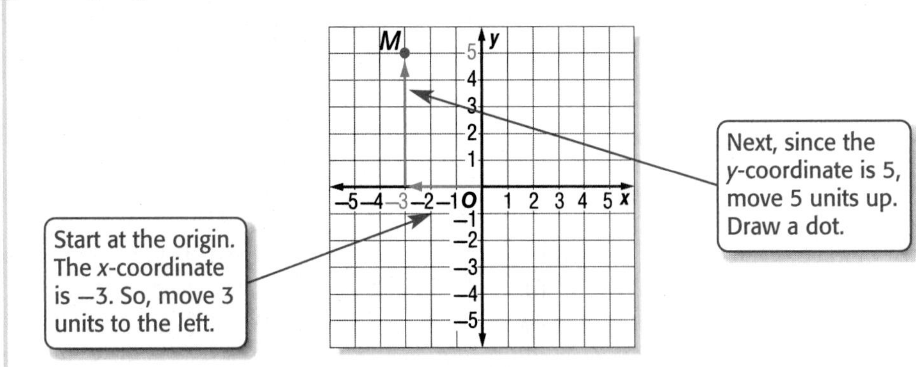

M
y
x
O
Start at the origin. The x-coordinate is −3. So, move 3 units to the left.
Next, since the y-coordinate is 5, move 5 units up. Draw a dot.

CHECK Your Progress **Graph and label each point.**

d. $N(2, 4)$ **e.** $P(0, 4)$ **f.** $Q(5, 1)$

CHECK Your Understanding

Examples 1, 2 (p. 600)

For Exercises 1–6, use the coordinate plane at the right. Identify the point for each ordered pair.

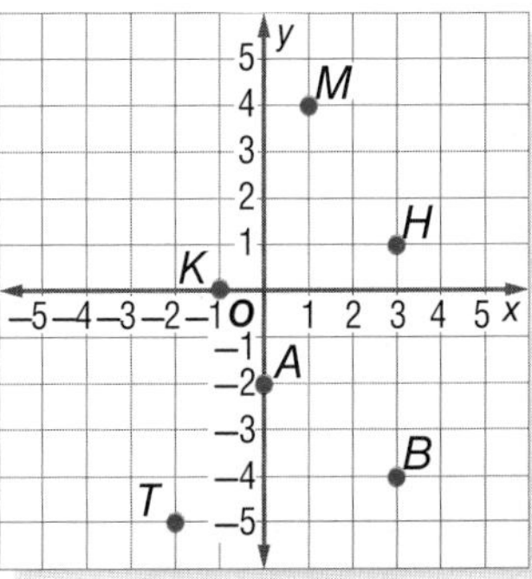

1. (3, 1)
2. (−1, 0)
3. (3, −4)

Write the ordered pair that names each point. Then identify the quadrant where each point is located.

4. M
5. A
6. T

SCHOOL **For Exercises 7 and 8, refer to the diagram below of a school.**

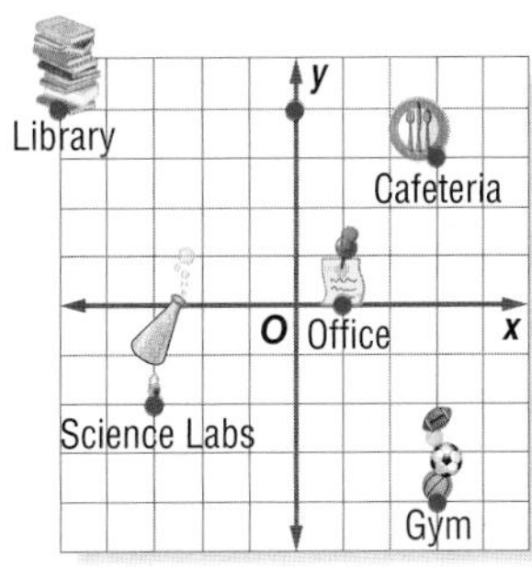

7. What part of the school is located at (3, −4)?
8. What ordered pair represents the location of the science labs?

Example 3 (p. 600)

Graph and label each point on a coordinate plane.

9. $D(2, 1)$
10. $K(-3, 3)$
11. $N(0, -1)$

Practice and Problem Solving

HOMEWORK HELP

For Exercises	See Examples
12–23, 30–33	1, 2
24–29	3

Use the coordinate plane at the right. Identify the point for each ordered pair.

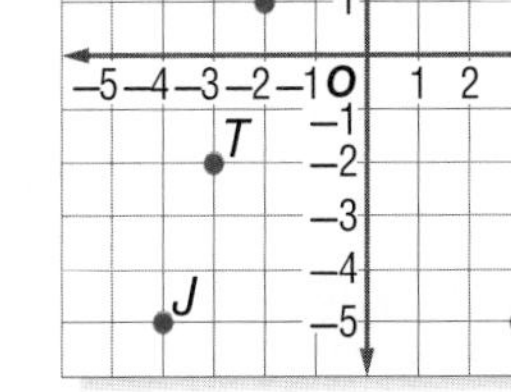

12. (2, 2)
13. (1, 3)
14. (−4, 2)
15. (−2, 1)
16. (−3, −2)
17. (−4, −5)

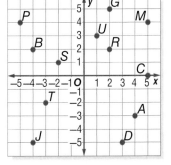

Write the ordered pair that names each point. Then identify the quadrant where each point is located.

18. G
19. C
20. A
21. D
22. P
23. M

Graph and label each point on a coordinate plane.

24. $N(1, 2)$
25. $T(0, 0)$
26. $B(-3, 4)$
27. $F(5, -2)$
28. $H(-4, -1)$
29. $K(-2, -5)$

PARKS For Exercises 30–33, refer to the map of Wonderland Park.

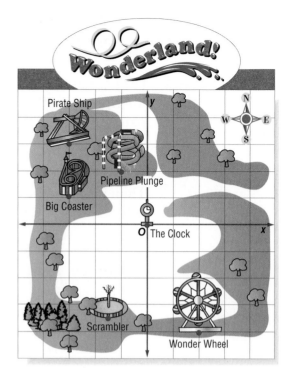

30. In which quadrant is the Wonder Wheel located?
31. What attraction is located at $(-3, 3)$?
32. What attraction is located at $(-1.5, -3.5)$?
33. What is located closest to the origin?

Graph and label each point on a coordinate plane.

34. $J\left(2\frac{1}{2}, -2\frac{1}{2}\right)$
35. $A\left(4\frac{3}{4}, -1\frac{1}{4}\right)$
36. $D(-1.5, 2.5)$

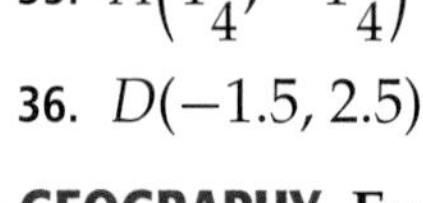

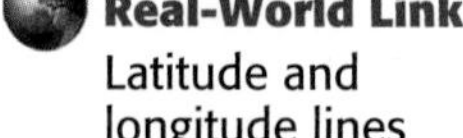

Real-World Link Latitude and longitude lines make up a coordinate system for Earth. Latitude lines go north and south, and longitude lines go east and west.

GEOGRAPHY For Exercises 37–40, refer to the map of South America shown.

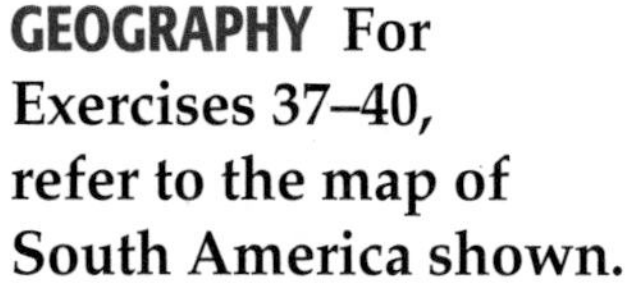

37. What country is located at (10°S latitude, 60°W longitude)?
38. Use latitude and longitude to name a location in the country Argentina as an ordered pair. (*Hint:* Refer to Exercise 37 for the format of the ordered pair.)
39. What line on a coordinate plane is similar to the line labeled 0° on the map?
40. Begin at (20°S latitude, 40°W longitude). Travel 20° west and 10° south. What country will you be in?

EXTRA PRACTICE See pages 702, 716.

H.O.T. Problems

CHALLENGE Without graphing, identify the quadrant(s) for which each of the following statements is true for any point (x, y). Justify your response.

41. The product of the x- and y-coordinates is positive.
42. The quotient of the x- and y-coordinates is negative.

43. **OPEN ENDED** Give the coordinates of three points that form a straight line when connected.

44. **WRITING IN MATH** Explain what all points located on the x-axis have in common. Then explain what points located on the y-axis have in common.

TEST PRACTICE

45. Which of the following coordinates lie within the circle graphed below?

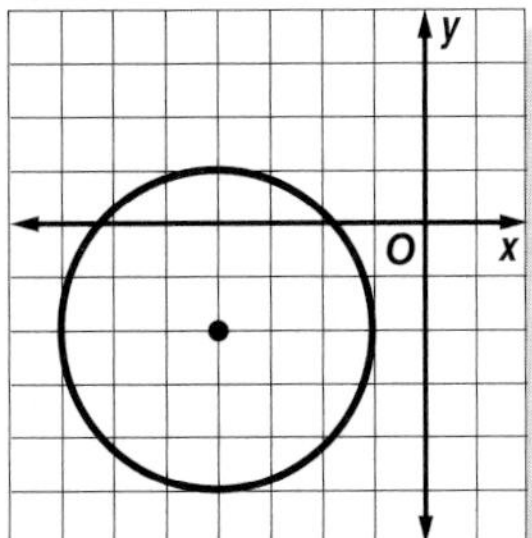

A $(-2, 3)$

B $(-3, -4)$

C $(-1, 2)$

D $(-3, 4)$

46. What point on the grid below corresponds to the coordinate pair $(-3, 5)$?

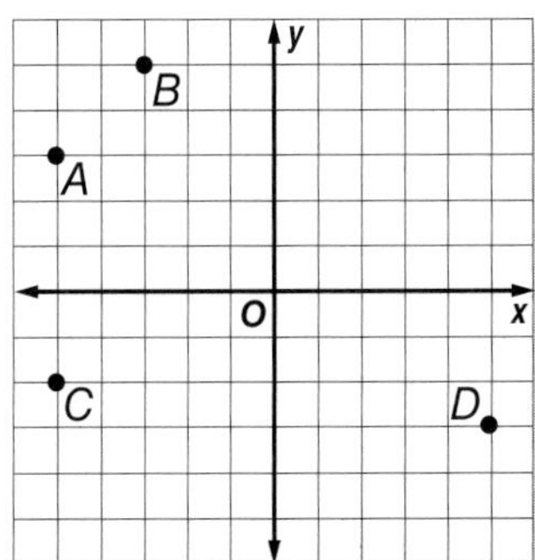

F Point A **H** Point C

G Point B **J** Point D

Spiral Review

47. Find the value of $x \div y$ if $x = -15$ and $y = -3$. (Lesson 11-6)

48. **SAVINGS** Pia saved \$55 in one week toward the purchase of a new bicycle. She continued to save at this rate for a total of six weeks and did not make any withdrawals. Write an integer to represent the change in her savings account at the end of the six weeks. (Lesson 11-4)

Add or subtract. (Lessons 11-2 and 11-3)

49. $7 + (-3)$ 50. $-19 - 3$ 51. $-22 + (-14)$ 52. $-6 - (-1)$

Find the value of x in each triangle. (Lesson 9-4)

53. 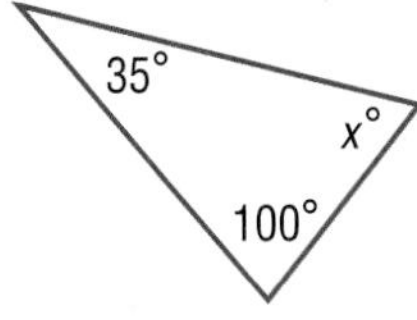

54. 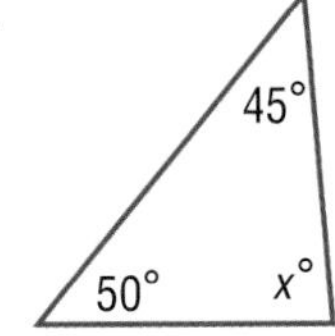

55. 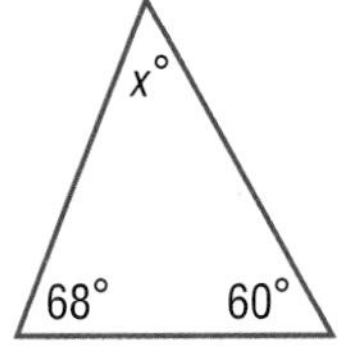

GET READY for the Next Lesson

PREREQUISITE SKILL Subtract. (Lesson 11-3)

56. $2 - 3$ 57. $4 - 7$ 58. $-6 - 8$ 59. $-2 - 9$

11-8 Translations

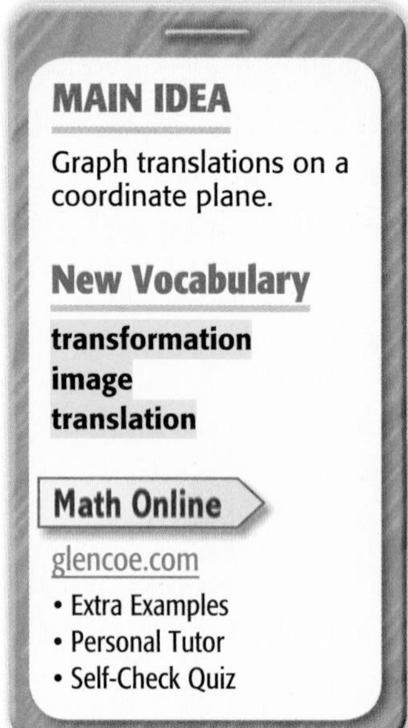

GET READY for the Lesson

GAMES Jose and Ronatta are playing the game shown at the right. Jose is using the blue game piece and Ronatta is using the yellow game piece.

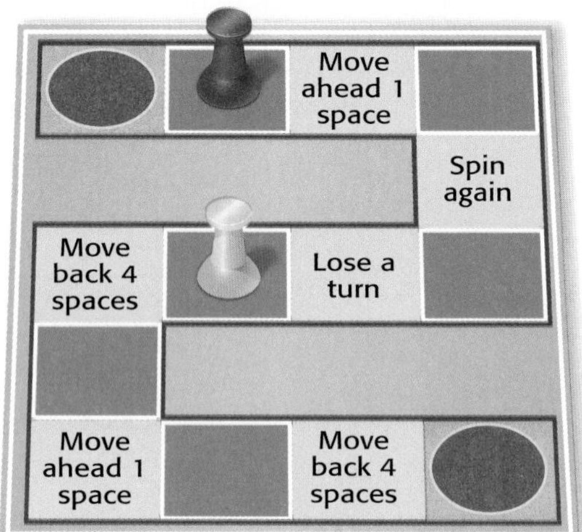

1. On Jose's turn, he spins a 4. Describe where his piece will be after moving.
2. Did the size or shape of the game piece change after moving? Explain.

A **transformation** is a movement of a geometric figure. The resulting figure is called an **image**. Sliding a figure without turning it is called a **translation**. A translation does not change the size or shape of a figure.

EXAMPLE Graph a Translation

1 Translate square *QRST* 6 units left. Graph square *Q′R′S′T′*.

Step 1 Move each vertex of the square 6 units left. Label the new vertices *Q′*, *R′*, *S′*, and *T′*.

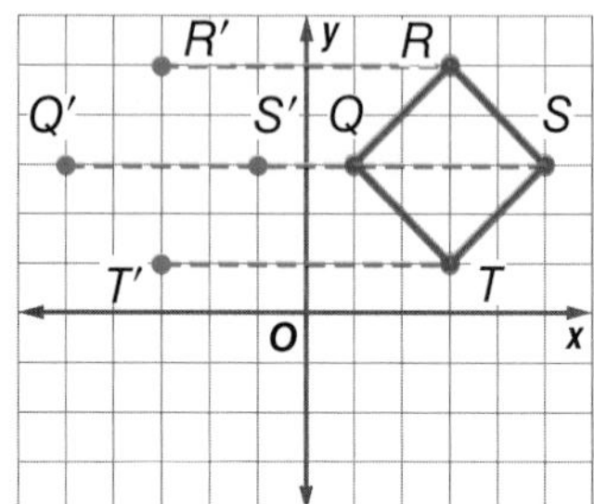

Step 2 Connect the new vertices to draw the square. The coordinates of the vertices of the new square are *Q′*(−5, 3), *R′*(−3, 5), *S′*(−1, 3), and *T′*(−3, 1).

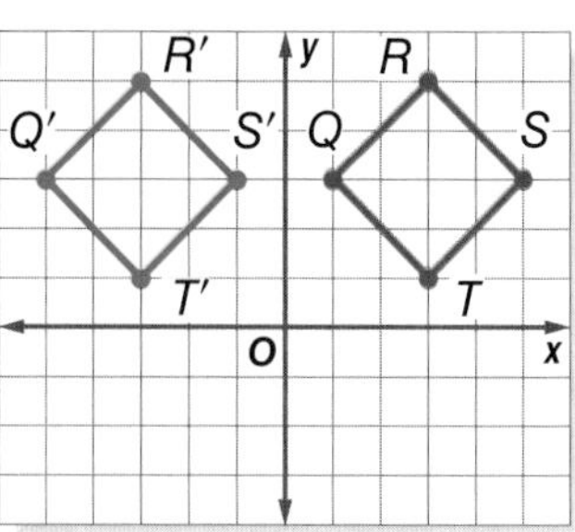

CHECK Your Progress

a. Translate square *QRST* 5 units down. Graph square *Q′R′S′T′*.

EXAMPLE Graph a Translation

2 **Translate triangle *XYZ* 2 units right and 4 units down. Graph triangle *X′Y′Z′*.**

Step 1 Move each vertex of the triangle 2 units right and 4 units down. Label the new vertices X', Y', and Z'.

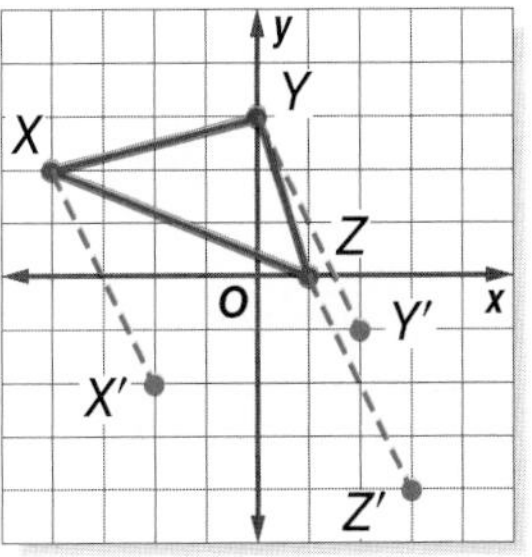

Step 2 Connect the new vertices to draw the triangle. The coordinates of the vertices of the new triangle are $X'(-2, -2)$, $Y'(2, -1)$, and $Z'(3, -4)$.

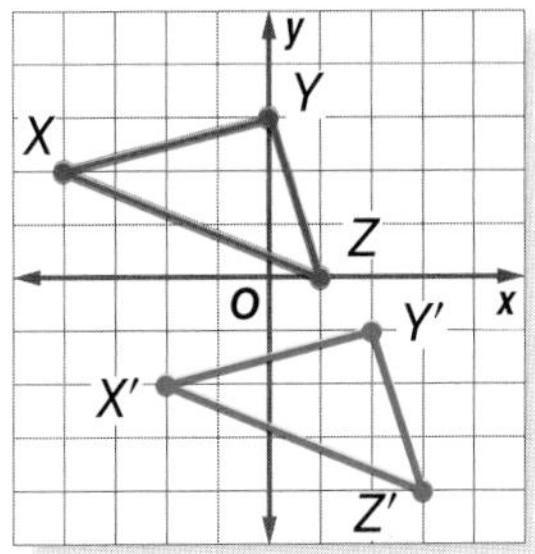

Study Tip

Congruent Figures You can check to make sure that your translated image is correct by verifying that the two images are congruent figures. If the images are congruent, then the translation is probably correct.

CHECK Your Progress

b. Translate triangle *XYZ* 3 units down and 4 units right. Graph triangle *X′Y′Z′*.

Translation

Key Concept

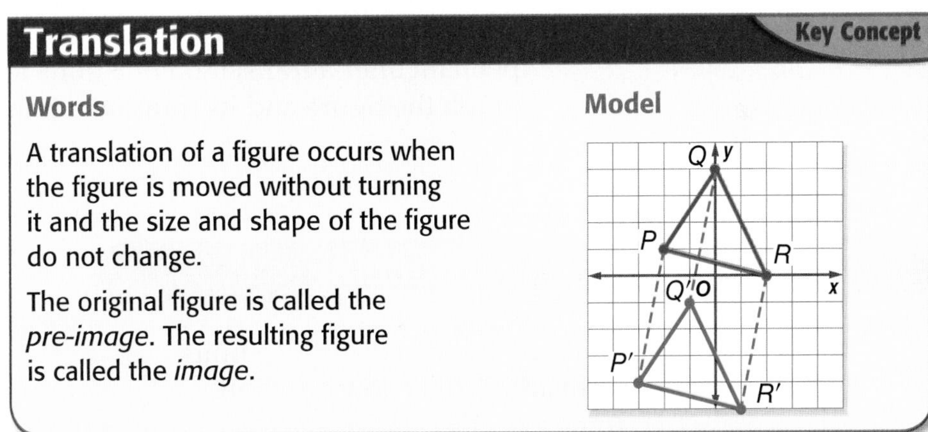

Words

A translation of a figure occurs when the figure is moved without turning it and the size and shape of the figure do not change.

The original figure is called the *pre-image*. The resulting figure is called the *image*.

Model

Reading Math

Prime Symbols Use prime symbols for the vertices in a transformed image.
$P \rightarrow P'$
$Q \rightarrow Q'$
$R \rightarrow R'$

You can use what you have learned about adding and subtracting integers to help you find the coordinates of a translated image. A translation is a vertical shift, a horizontal shift, or a combination of vertical and horizontal shifts.

If an image is translated right or left, you add or subtract the value from the *x*-coordinate. If an image is translated up or down, you add or subtract the value from the *y*-coordinate.

Real-World EXAMPLE Find Coordinates of a Translation

3 VIDEO GAMES **A video game animator wants to move a car in a game 5 units left and 4 units up. If the car had original coordinates at $A(1, 2)$, $B(5, 2)$, $C(5, -1)$, and $D(1, -1)$, find the new vertices of the car after the translation. Then graph the figure and its translated image.**

The vertices of the car after the translation can be found by subtracting 5 from the *x*-coordinates and adding 4 to the *y*-coordinates.

$ABCD$	$(x - 5, y + 4)$	$A'B'C'D'$
$A(1, 2)$	$(1 - 5, 2 + 4)$	$A'(-4, 6)$
$B(5, 2)$	$(5 - 5, 2 + 4)$	$B'(0, 6)$
$C(5, -1)$	$(5 - 5, -1 + 4)$	$C'(0, 3)$
$D(1, -1)$	$(1 - 5, -1 + 4)$	$D'(-4, 3)$

Use the vertices of the car to graph the car at $ABCD$ and $A'B'C'D'$.

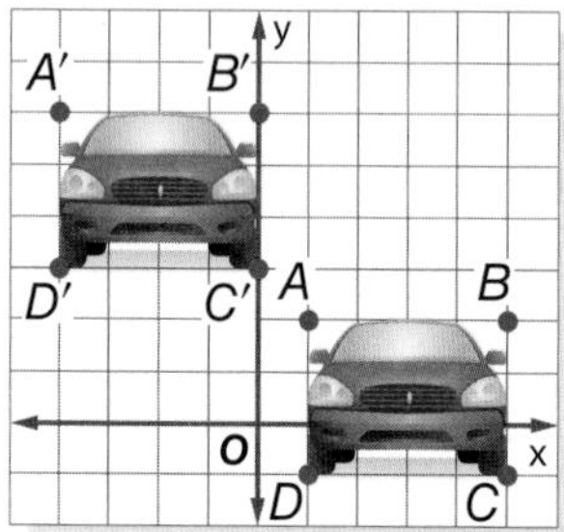

Real-World Link Video game animators use formulas to change the coordinates of an object in the game to a different location.

CHECK Your Progress

c. **PENNANTS** A triangular baseball pennant hanging on a wall has vertices $L(-1, 1)$, $M(-1, 5)$, and $N(4, 3)$. Find the vertices of the pennant after a translation of 3 units left and 4 units down. Then graph the figure and its translated image.

CHECK Your Understanding

Example 1, 2 (pp. 604–605)

1. Translate trapezoid $ABCD$ 5 units right and 3 units down. Graph trapezoid $A'B'C'D'$.

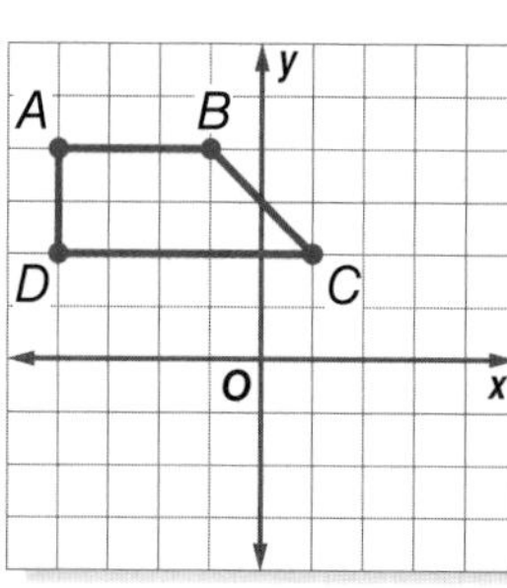

Example 3 (p. 606)

SAILS **A sail on a boat has vertices $F(-1, 2)$, $G(2, 3)$, and $H(3, 1)$. Find the vertices of the sail after each translation. Then graph the figure and its translated image.**

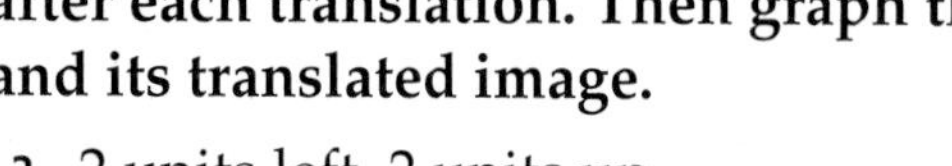

2. 2 units left, 2 units up
3. 3 units right, 1 unit down
4. 4 units left

Practice and Problem Solving

HOMEWORK HELP	
For Exercises	See Examples
5–8	1
9–16	2

5. Translate triangle TUV 6 units left and 4 units down. Graph triangle $T'U'V'$.

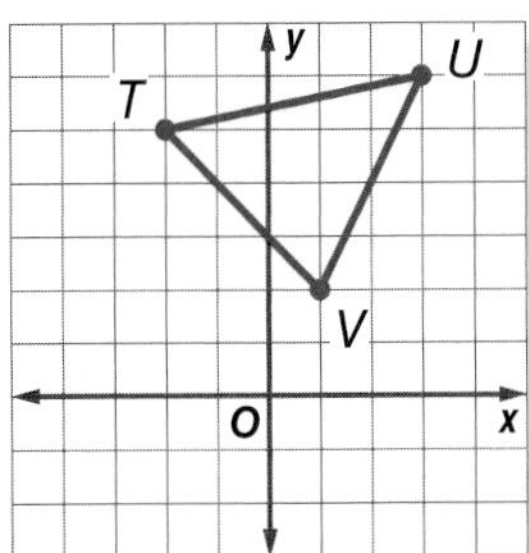

6. Translate quadrilateral $WXYZ$ 7 units right and 3 units up. Graph quadrilateral $W'X'Y'Z'$.

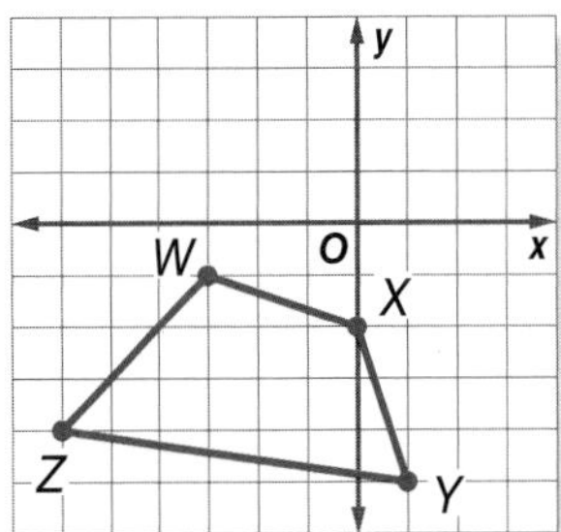

7. Translate triangle PQR 6 units right and 4 units down. Graph triangle $P'Q'R'$.

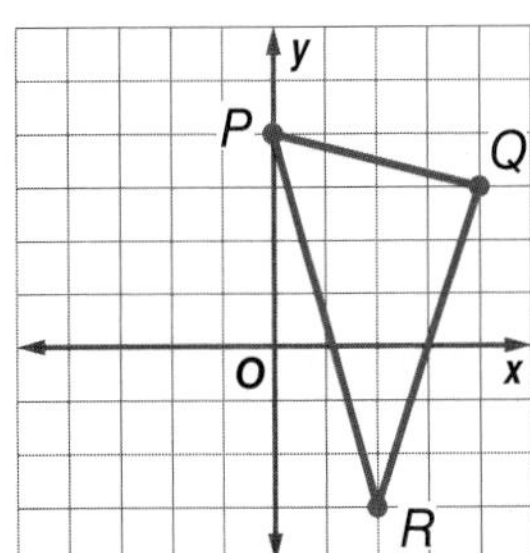

8. Translate quadrilateral $JKLM$ 6 units left and 5 units up. Graph quadrilateral $J'K'L'M'$.

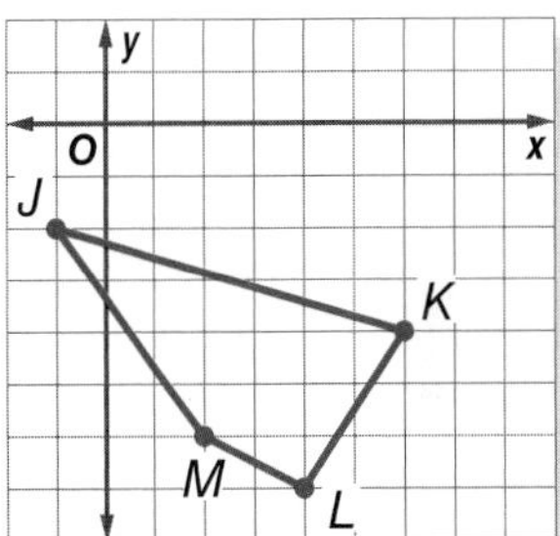

CANDLES A decorative candle on a table has vertices $R(-5, -4)$, $S(-1, -2)$, and $T(1, -5)$. Find the vertices of the candle after each translation. Then graph the figure and its translated image.

9. 3 units right
10. 2 units right, 4 units up
11. 5 units left, 3 units down
12. 1 unit left, 6 units up

RUGS A rug has vertices $D(-2, 4)$, $E(1, 3)$, $F(1, -2)$, and $G(-2, -1)$. Find the vertices of the rug after each translation. Then graph the figure and its translated image.

13. 4 units right, 2 units down
14. 4 units left, 5 units up
15. 6 units down
16. 2 units left, 4 units down

17. **VIDEO GAMES** The detective in a popular video game walks through a maze searching for clues. The detective begins her search at the coordinates $(1, -4)$. There is a clue located 6 units left of the origin and 8 units down. The detective moves 2 units up, 3 units left, 4 units down, and 6 units left. State the detective's final coordinates. Will she discover the clue?

18. **SANDBOXES** After moving the sandbox at the local park, the coordinates of its corners are at $(2, 5)$, $(2, -2)$, $(9, 5)$, and $(9, -2)$. If the sandbox was moved 3 units right and 4 units up, what were the original coordinates of the sandbox?

For Exercises 19–22, use the layout of Tamika's bedroom to find the new coordinates of each object after the translation given.

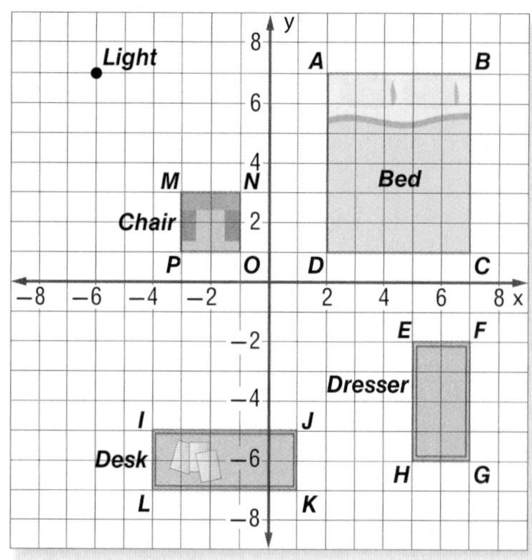

Real-World Link
People in the Ukraine developed a technique called Pysanka for decorating eggs. The largest Pysanka egg was created using 524 star patterns and 2,206 triangular pieces.

19. Bed: 9 units left, 1 unit down
20. Dresser: 6 units up
21. Desk: 5 units right, 2 units up
22. Chair: 3 units left, 5 units down
23. After making all the translations in Exercises 19–22, draw the new layout of Tamika's bedroom on a coordinate plane.
24. **TESSELLATIONS** A tessellation is a pattern formed by repeating figures that fit together without gaps or overlaps. Use the information at the left to describe how tessellations and translations were used to create the pattern on the egg.

GEOMETRY For Exercises 25–28, find the missing coordinates of each figure described below. Then graph the figure and its image after each translation.

25. Square $ABCD$ has vertices $A(1, -2)$, $B(6, -2)$, $C(1, 3)$, and $D(■, ■)$ and is translated 2 units left and 5 units up.
26. Rectangle $FGHJ$ has vertices $F(-6, 4)$, $G(-3, 4)$, $H(-3, 9)$, and $J(■, ■)$ and is translated 4 units right.
27. Parallelogram $QRST$ has vertices $Q(-4, 0)$, $R(-1, 0)$, $S(-2, 3)$, and $T(■, ■)$ and is translated 5 units left.
28. Right isosceles triangle XYZ has vertices $X(4, -1)$, $Y(4, -4)$, and $Z(■, ■)$ and is translated 3 units right and 1 unit down.

GEOMETRY For Exercises 29 and 30, refer to the following information.

Right triangle LMN has vertices $L(3, 2)$, $M(3, -3)$, and $N(■, ■)$, and is translated three units left and one unit down.

29. Without graphing, find the vertical distance between vertices L' and M'. Explain your method.
30. If the horizontal distance between vertices M and N is the integer x and $x < 4$, how many coordinates are possible for vertex N'? List them.

EXTRA PRACTICE
See pages 702, 716.

H.O.T. Problems

CHALLENGE A translation can also be described using an ordered pair. The ordered pair (−4, 3) means a translation of 4 units left and 3 units up. If triangle *ABC* has vertices at *A*(−3, 5), *B*(1, −1), and *C*(−4, −2), give the coordinates of the vertices of triangle *A′B′C′* after each translation.

31. (5, 2)
32. (−6, −3)
33. (2, −4)
34. (−5, 1)

35. **WRITING IN MATH** Describe how you would translate rectangle *QRST* 7 units right and 4 units down.

TEST PRACTICE

36. Find the coordinates of *W′* of quadrilateral *WXYZ* after a translation 5 units left and 4 units down.

A (−2, 10)
B (−3, 1)
C (6, 0)
D (7, 9)

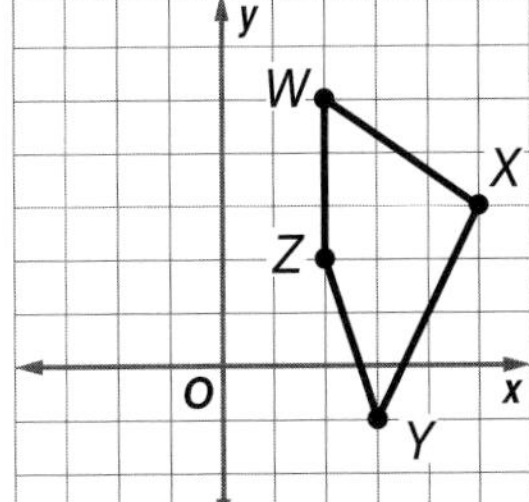

37. Which graph shows a translation of the triangle?

F
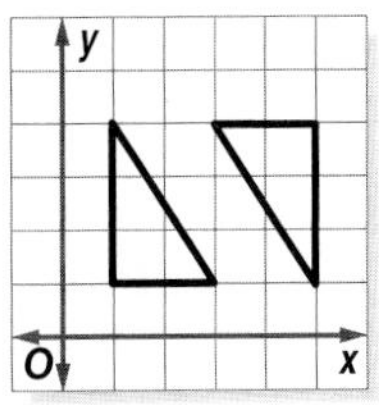

H
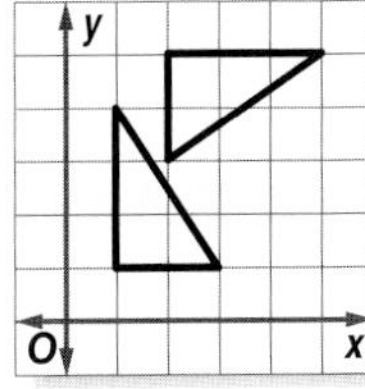

G
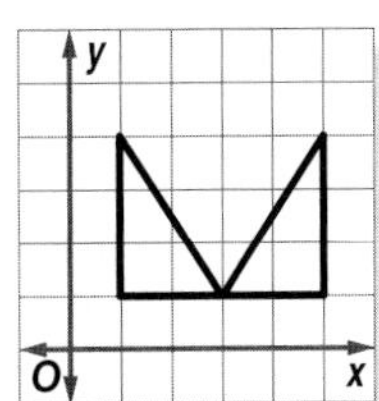

J
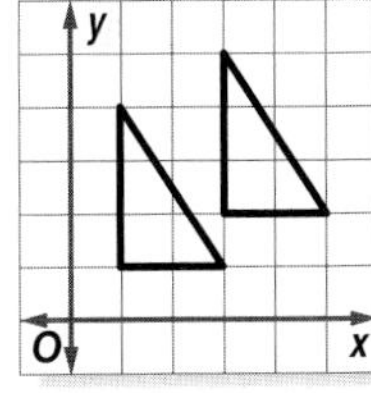

Spiral Review

For Exercises 38–41, use the coordinate plane at the right. Identify the point for each ordered pair. (Lesson 11-7)

38. (4, 1)
39. (−3, 3)
40. (−4, 0)
41. (2, −4)

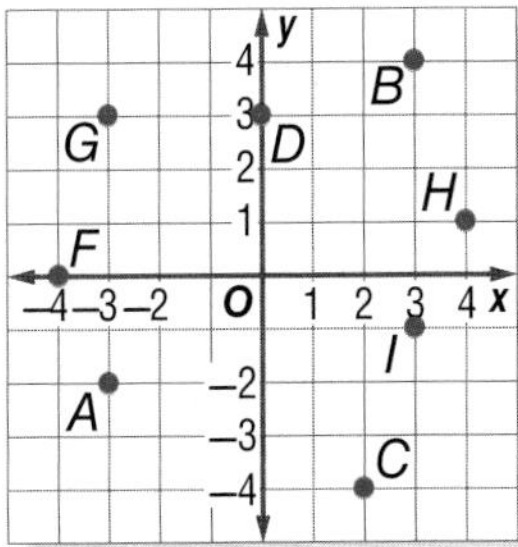

42. **ALGEBRA** Find the value of $a \div b$ if $a = -8$ and $b = 4$. (Lesson 11-6)

GET READY for the Next Lesson

PREREQUISITE SKILL Determine whether each letter could be folded in half so that one side matches the other. Write *yes* or *no*.

43. A
44. F
45. M
46. L

11-9 Reflections

MAIN IDEA

Graph reflections on a coordinate plane.

New Vocabulary

reflection

Math Online

glencoe.com

- Extra Examples
- Personal Tutor
- Self-Check Quiz

GET READY for the Lesson

BUTTERFLIES The monarch butterfly is found throughout the United States.

1. What do you notice about each wing of the butterfly?
2. If the butterfly were to fold its wings together, would the markings on the butterfly's wings line up? Explain.

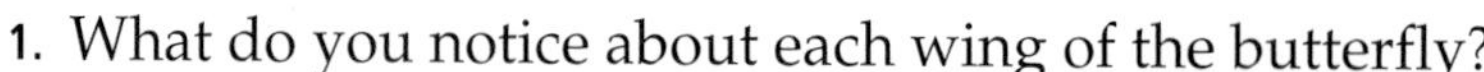

A **reflection** is the mirror image that is created when a figure is flipped over a line. A reflection is another type of geometric transformation. If you were to draw a line down the middle of the butterfly's body, one wing is a reflection of the other.

EXAMPLE Reflect a Figure Over the *x*-Axis

1 **Triangle *LMN* has vertices *L*(3, −2), *M*(4, −4), and *N*(−2, −3). Graph the figure and its reflected image over the *x*-axis. Then find the coordinates of the reflected image.**

Step 1 Graph triangle *MLH* on a coordinate plane. Then, count the number of units between each vertex and the *x*-axis.

L is 2 units from the axis.
M is 4 units from the axis.
N is 3 units from the axis.

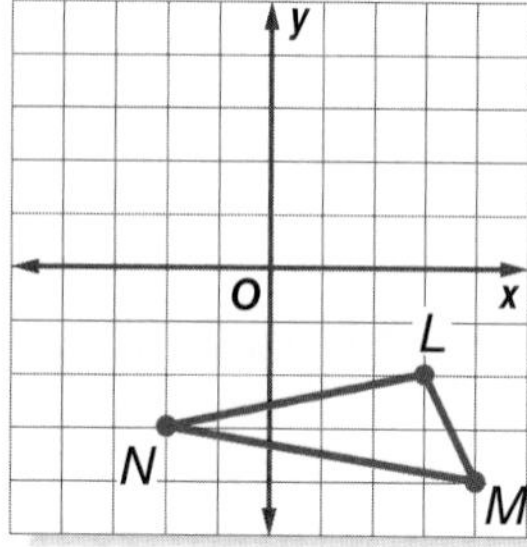

Step 2 Make a point for each vertex the same distance away from the *x*-axis on the opposite side of the *x*-axis and connect the new points to form the image of triangle *L′M′N′*. The coordinates are *L*(3, 2), *M*(4, 4), and *N*(−2, 3).

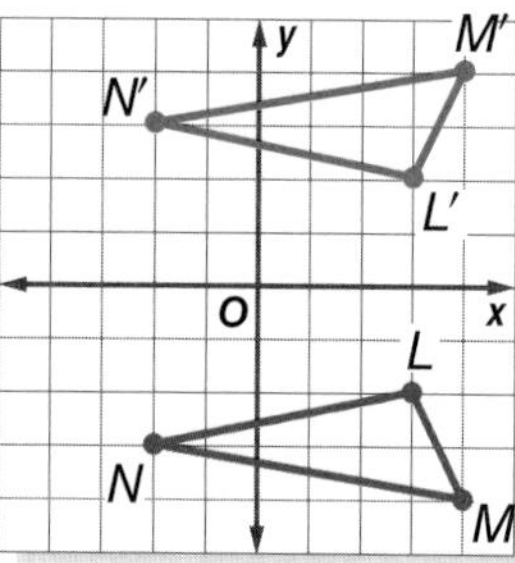

CHECK Your Progress

a. Parallelogram *FGHJ* has vertices *F*(−4, 1), *G*(−2, 3), *H*(3, 3), and *J*(1, 1). Graph the figure and its reflected image over the *x*-axis. Then find the coordinates of the reflected image.

EXAMPLE Reflect a Figure Over the y-Axis

2 Quadrilateral $WXYZ$ has vertices $W(-3, 2)$, $X(-1, 3)$, $Y(-2, 3)$ and $Z(-4, -1)$. Graph the figure and its reflected image over the y-axis. Then find the coordinates of the reflected image.

Step 1 Graph quadrilateral $WXYZ$ on a coordinate plane then count the number of units between each vertex and the y-axis.

W is 3 units from the axis.
X is 1 unit from the axis.
Y is 2 units from the axis.
Z is 4 units from the axis.

Step 2 Make a point for each vertex the same distance away from the y-axis on the opposite side of the y-axis and connect the new points to form the image of quadrilateral $W'X'Y'Z'$.

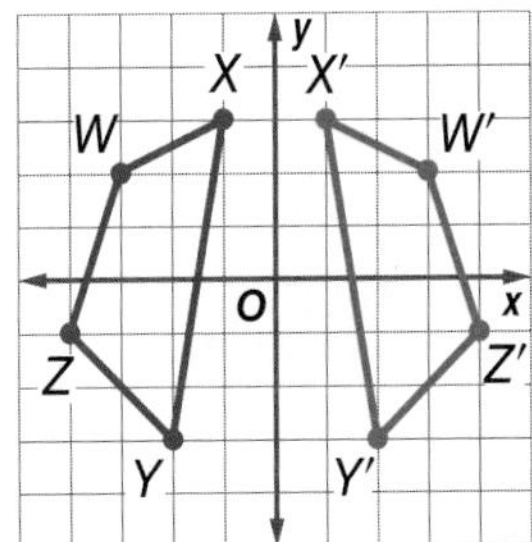

Study Tip

Line of Reflection The line that an image is reflected over is called the *line of reflection*. In this case, the line of reflection is the y-axis.

CHECK Your Progress

b. Triangle DEF has vertices $D(5, 3)$, $E(2, 1)$, and $F(2, -4)$. Graph the figure and its reflected image over the y-axis. Then find the coordinates of the reflected image.

Reflection

Key Concept

Words

A reflection of a figure occurs when the image is flipped over a line to create a mirror image of the figure.

Models

Reflection over the x-axis

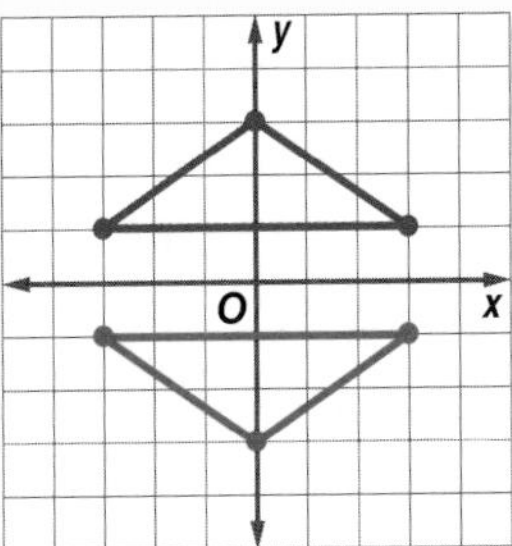

Reflection over the y-axis

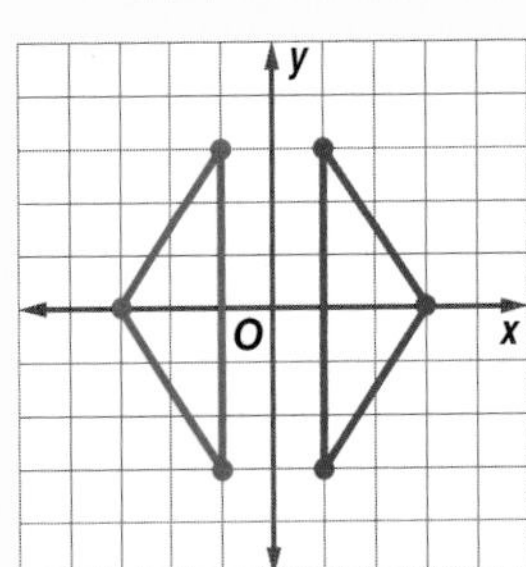

CHECK Your Understanding

Example 1 (p. 610) **Graph each figure and its reflection over the x-axis. Then find the coordinates of the reflected image.**

1. Quadrilateral $QRST$ with vertices $Q(-2, 4)$, $R(4, 1)$, $S(-1, 0)$, and $T(-3, 2)$.
2. Triangle XYZ with vertices $X(2, 3)$, $Y(6, 5)$, and $Z(4, 1)$.

Example 2 (p. 611) **Graph each figure and its reflection over the y-axis. Then find the coordinates of the reflected image.**

3. Triangle FGH with vertices $F(1, 1)$, $G(4, 3)$, and $H(5, -4)$.
4. Trapezoid $LMNP$ with vertices $L(-4, 2)$, $M(-1, 4)$, $N(-1, -2)$, and $P(-4, -2)$.

Practice and Problem Solving

HOMEWORK HELP

For Exercises	See Examples
5–7	1
8–10	2

Graph each figure and its reflection over the x-axis. Then find the coordinates of the reflected image.

5. Triangle PQR with vertices $P(-4, 4)$, $Q(2, 0)$, and $R(-3, 1)$.
6. Quadrilateral $WXYZ$ with vertices $W(-4, 2)$, $X(1, 4)$, $Y(2, 2)$, and $Z(-3, 0)$.
7. Find the vertices of quadrilateral $A'B'C'D'$ after a reflection over the x-axis.

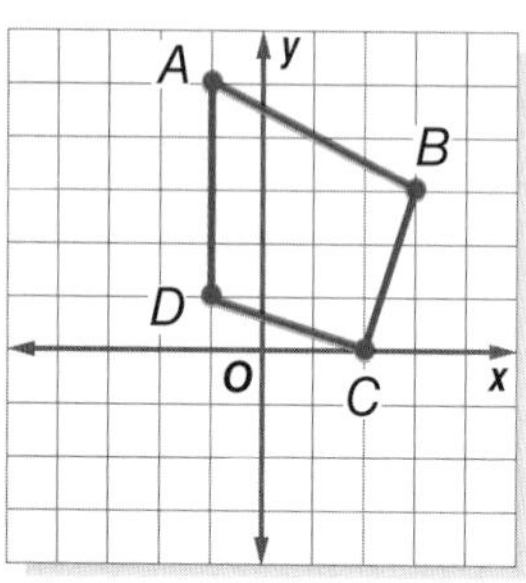

Graph each figure and its reflection over the y-axis. Then find the coordinates of the reflected image.

8. Triangle BCD with vertices $B(0, 1)$, $C(4, 0)$, and $D(2, 4)$.
9. Parallelogram $PQRS$ with vertices $P(3, -2)$, $Q(3, 4)$, $R(5, 6)$, and $S(5, 0)$.
10. Find the vertices of quadrilateral $J'K'L'M'$ after a reflection over the y-axis.

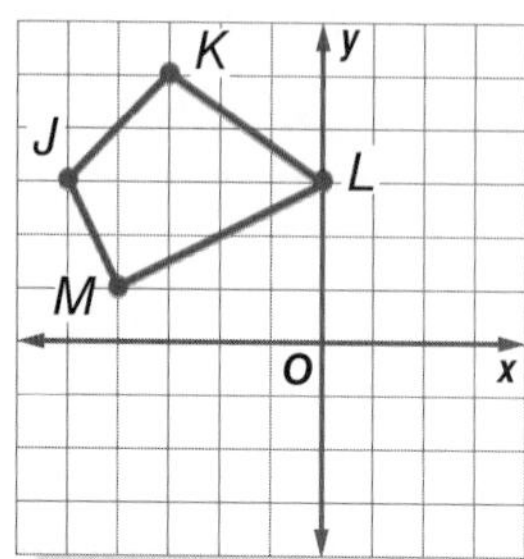

DESIGNS **Use the design at the right to answer Exercises 11 and 12.**

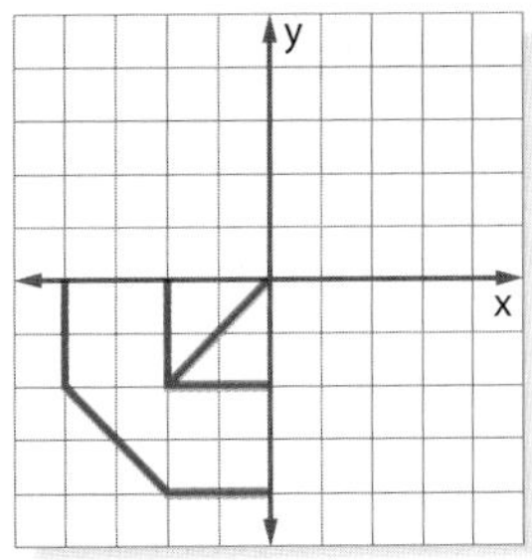

11. Copy the design on a coordinate plane. Graph the reflection of the design over the y-axis.
12. Graph the reflection of the image you created in Exercise 11 over the x-axis.

For Exercises 13–15, use the following information.

Triangle ABC has vertices $A(1, 3)$, $B(3, 5)$, and $C(4, 1)$.

13. Reflect the triangle over the y-axis and then over the x-axis. What are the coordinates of triangle $A'B'C'$? Graph triangle ABC and triangle $A'B'C'$ on the same coordinate plane.
14. Reflect triangle ABC over the x-axis and then over the y-axis. Graph both triangles. Compare the result from Exercise 13 with this result.
15. What are the coordinates of triangle $A'B'C'$ from Exercise 13 after it is translated three units up and 4 units right?

16. **ART** Navajo art is known for its use of reflections. A traditional Navajo blanket generally has an image reflected one or more times over a line. The right half of a Navajo blanket is shown. Copy the design onto a piece of paper. Then draw the design after it has been reflected over a vertical line.

Real-World Link

The Navajo Nation covers 27,000 square miles in Arizona, Utah, and New Mexico. It is the largest area for Native Americans in the United States.

Source: Navajo Nation

For Exercises 17–19, use the graph shown at the right.

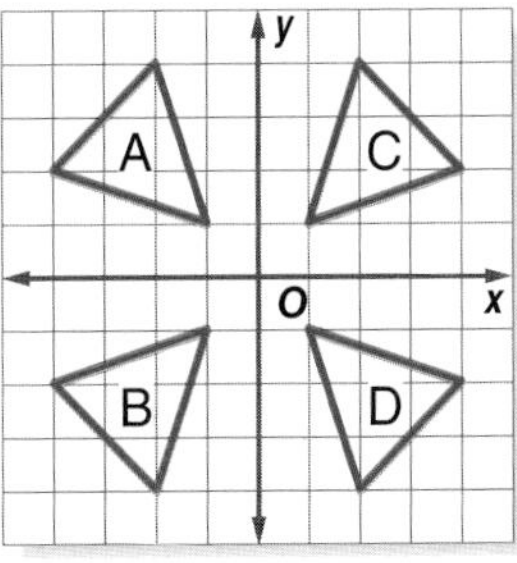

17. Which pair(s) of figures is a reflection over the x-axis?
18. Which pair(s) of figures is a reflection over the y-axis?
19. If figure B was reflected over the x-axis and then over the y-axis, which figure(s) would the resulting image look like?

GEOMETRY **For Exercises 20–22, use the figure at the right and the following information.**

A figure has *line symmetry* when it can be folded in half across a line, called the *line of symmetry*, so that the two halves match exactly. The figure at the right has two lines of symmetry.

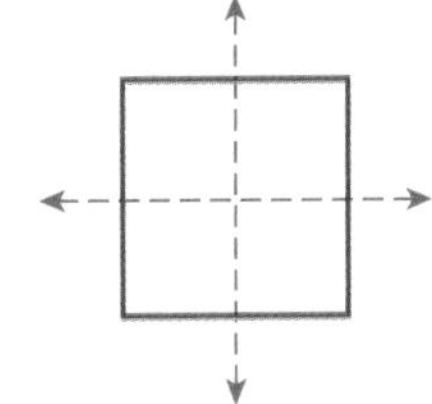

20. Draw a figure with one line of symmetry.
21. Draw a figure with three lines of symmetry.
22. Draw a figure with no lines of symmetry.

23. **RESEARCH** Refer to Exercises 20–22. Use the Internet or another source to find an example of line symmetry in nature.

H.O.T. Problems

24. **OPEN ENDED** Give three examples of capital letters that if reflected over a horizontal line would appear the same.

25. **CHALLENGE** A piece of square paper is folded in half twice. The image at the right shows what it looks like after some of the paper is cut away. If the piece of paper is opened up all the way, what will the resulting image look like?

26. **WRITING IN MATH** When you find the coordinates of an image after a reflection over the x-axis or the y-axis, what do you notice about the coordinates of the new image in relation to the coordinates of the original image?

TEST PRACTICE

27. Quadrilateral $ABCD$ has vertices at $A(-3, 5)$, $B(-2, 2)$, $C(-3, -2)$, and $D(-5, 1)$. What are the coordinates of C' after a reflection over the y-axis?

   A $(-3, 2)$

   B $(3, -2)$

   C $(3, 2)$

   D $(-3, -2)$

28. The graph below shows triangle $X'Y'Z'$ after a reflection over the x-axis. What are the original coordinates of the point Y?

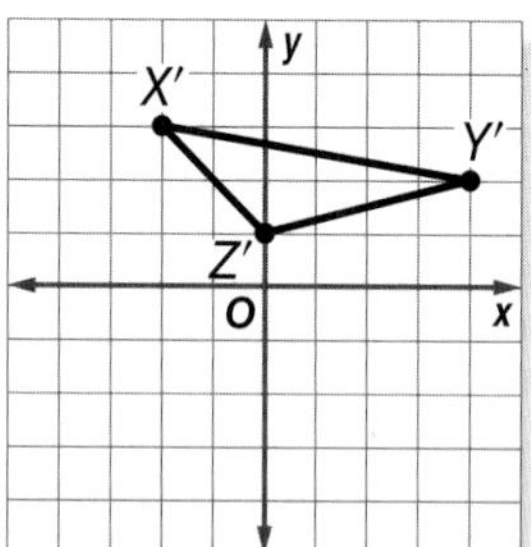

   F $(4, -2)$ **H** $(-4, 2)$

   G $(2, 4)$ **J** $(-4, -2)$

Spiral Review

Trapezoid $LMNO$ has vertices $L(-4, -1)$, $M(-1, 2)$, $N(2, 2)$, and $O(2, -1)$. Find the vertices of trapezoid $L'M'N'O'$ after each translation. Then graph the figure and its translated image. (Lesson 11-8)

29. 3 units right, 4 units down

30. 2 units left, 2 units up

Graph and label each point on a coordinate plane. (Lesson 11-7)

31. $A(4, 0)$

32. $B(-2, -5)$

33. $C(3, 4)$

34. $D(2, -1)$

GET READY for the Next Lesson

PREREQUISITE SKILL Find the opposite of each integer. (Lesson 2-9)

35. 6

36. −7

37. 3

38. −2

11-10 Rotations

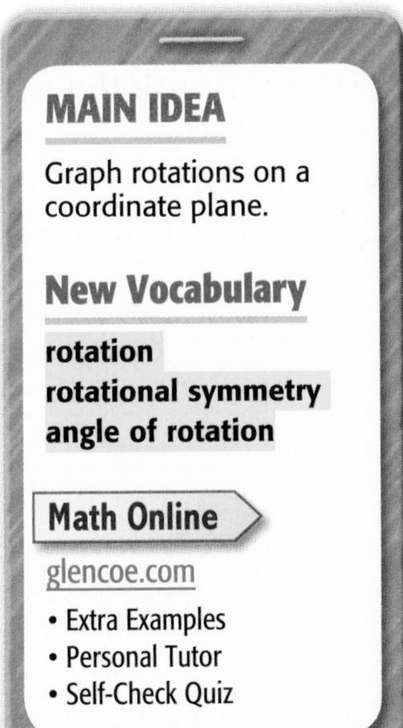

MAIN IDEA

Graph rotations on a coordinate plane.

New Vocabulary

rotation
rotational symmetry
angle of rotation

Math Online

glencoe.com

- Extra Examples
- Personal Tutor
- Self-Check Quiz

MINI Lab

STEP 1 Draw and label triangle ABC with vertices $A(-4, 1)$, $B(-4, 5)$, and $C(-1, 1)$.

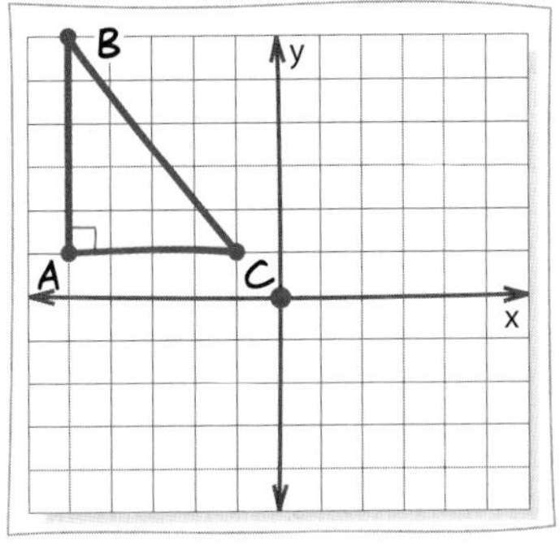

STEP 2 Attach a piece of tracing paper to the coordinate plane with a fastener. Then trace the triangle and the x- and y-axis.

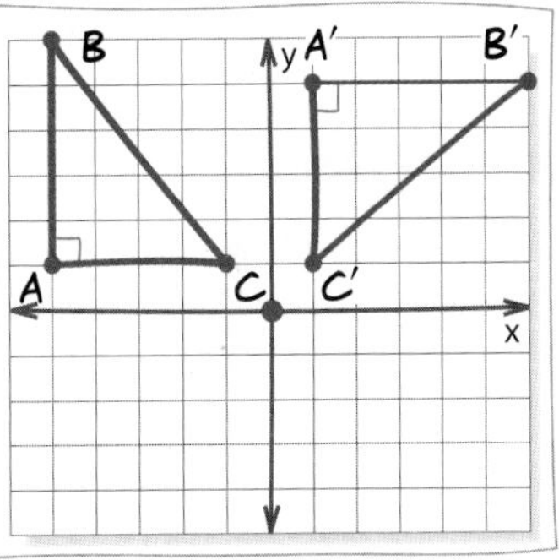

STEP 3 Turn the tracing paper clockwise so that the original y-axis is on top of the original x-axis.

1. Describe the transformation that occurred from triangle ABC to triangle $A'B'C'$.
2. What are the coordinates of triangle $A'B'C'$?

The type of transformation above is a rotation. A **rotation** occurs when a figure is rotated around a point. It can also be called a turn. A rotation does not change the size or shape of the figure. In the Mini Lab, the triangle was rotated 90° clockwise. The rotations shown below are clockwise around the origin.

90° Rotation

180° Rotation

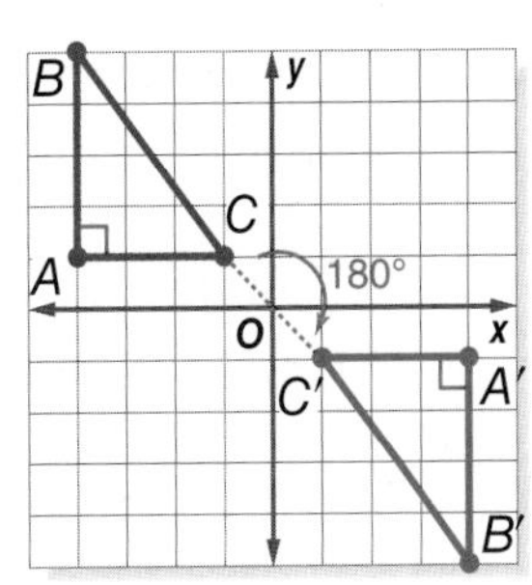

270° Rotation

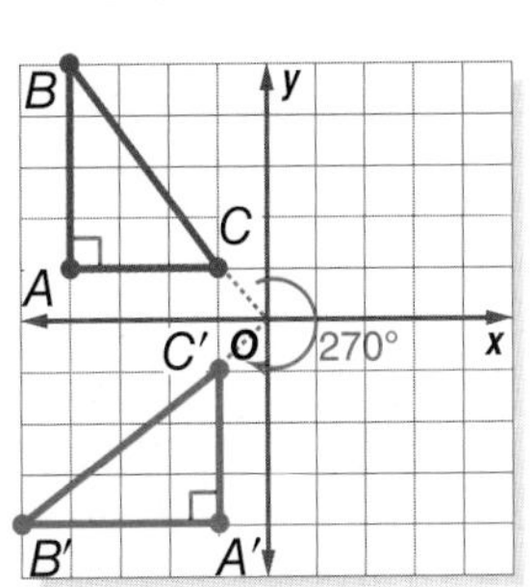

A rotation of a figure is based on a circle, so it can be anywhere from 0° to 360°.

EXAMPLE Rotate a Figure Clockwise

1 **Triangle *RST* has vertices *R*(1, 3), *S*(4, 4), and *T*(2, 1). Graph the figure and its image after a clockwise rotation of 90° about the origin. Then give the coordinates of the vertices for triangle *R′S′T′*.**

Step 1 Graph triangle *RST* on a coordinate plane.

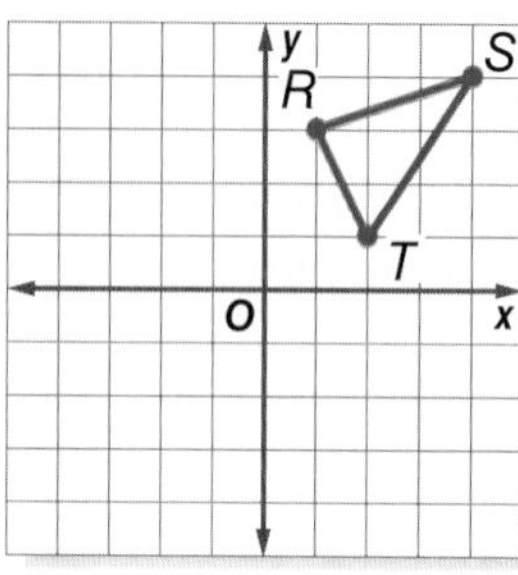

Reading Math

Segment Notation
$\overline{RO}$ means the line segment connecting points *R* and *O*.

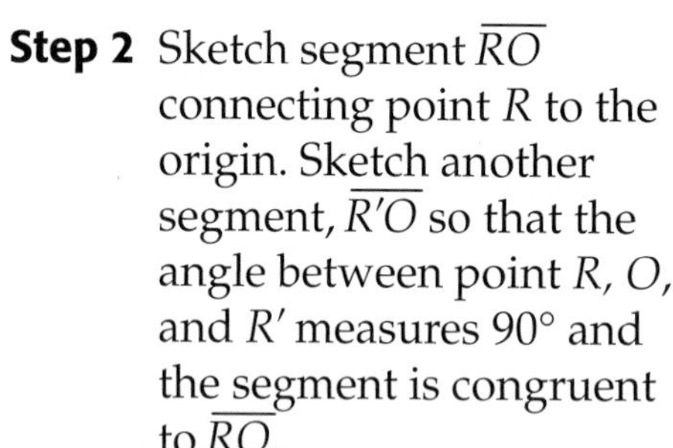

Step 2 Sketch segment $\overline{RO}$ connecting point *R* to the origin. Sketch another segment, $\overline{R'O}$ so that the angle between point *R*, *O*, and *R′* measures 90° and the segment is congruent to $\overline{RO}$.

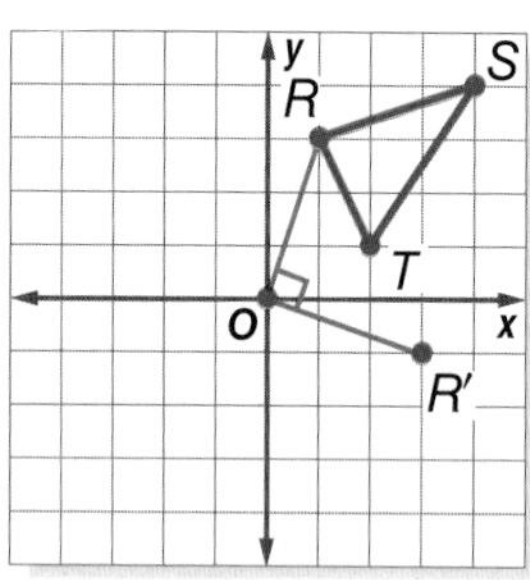

Reading Math

Clockwise To rotate the figure the same way the hands on a clock rotate.

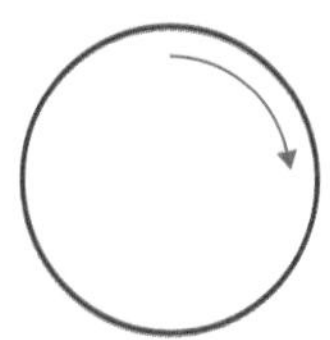

Counterclockwise
To rotate the figure the opposite direction of the way the hands on a clock rotate.

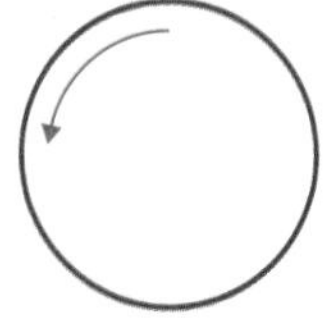

Step 3 Repeat Step 2 for points *S* and *T*. Then connect the vertices to form triangle *R′S′T′*.

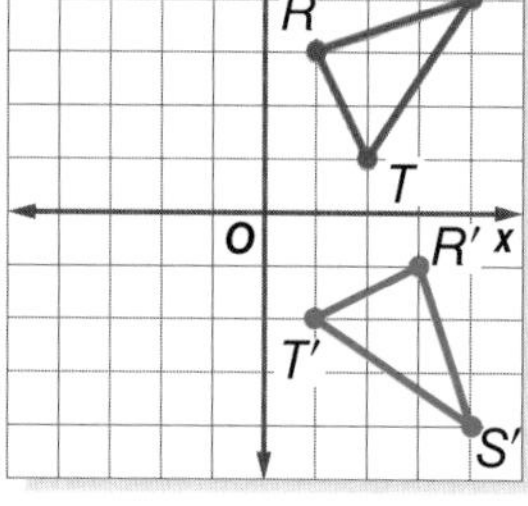

So, the coordinates of the vertices of triangle *R′S′T′* are *R′*(3, –1), *S′*(4, –4), and *T′*(1, –2).

CHECK Your Progress

a. Triangle *XYZ* has vertices *X*(–5, 4), *Y*(–1, 2), and *Z*(–3, 1). Graph the figure and its image after a counterclockwise rotation about the origin of 180°. Then give the coordinates of the vertices for triangle *X′Y′Z′*.

A figure can have **rotational symmetry** if the figure can be rotated a certain number of degrees about its center and still look like the original. The **angle of rotation** is the degree measure of the angle through which the figure is rotated.

Real-World EXAMPLE Determine Rotational Symmetry

2 SNOW Determine whether the snowflake has rotational symmetry. Write *yes* or *no*. If *yes*, name its angle(s) of rotation.

Since the snowflake can be rotated and still look like it does in its original position, the snowflake has rotational symmetry.

The snowflake will match itself after being rotated 60°, 120°, 180°, 240°, 300°, and 360°.

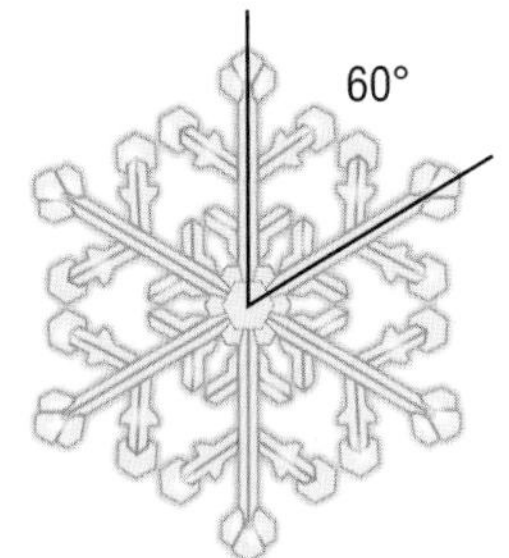

CHECK Your Progress

b. **LOGOS** Determine whether the Boston Bruins logo shown has rotational symmetry. Write *yes* or *no*. If *yes*, name its angle(s) of rotation.

CHECK Your Understanding

Example 1 (p. 616) **Graph triangle *ABC* and its image after each rotation. Then give the coordinates of the vertices for triangle *A′B′C′*.**

1. 90° counterclockwise
2. 180° clockwise
3. 270° counterclockwise
4. 270° clockwise

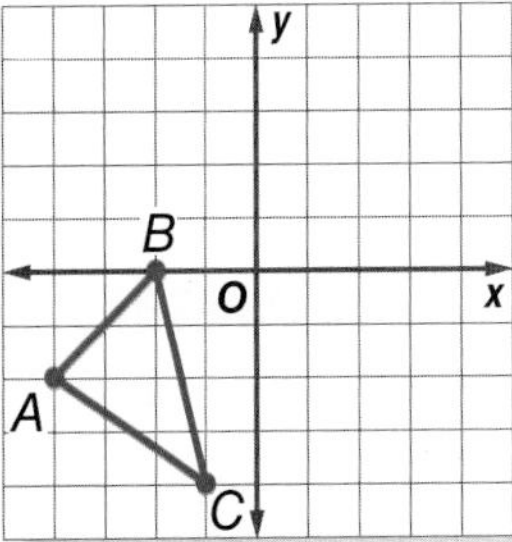

Example 2 (p. 617) **Determine whether each figure has rotational symmetry. Write *yes* or *no*. If *yes*, name its angle(s) of rotation.**

5.

6.

Practice and Problem Solving

HOMEWORK HELP	
For Exercises	See Examples
7–14	1
15–18	2

Triangle *PQR* has vertices *P*(1, −5), *Q*(2, −1), and *R*(5, −4). Graph the figure and its image after each rotation. Then give the coordinates of the vertices for triangle *P′Q′R′*.

7. 270° clockwise
8. 180° counterclockwise
9. 90° counterclockwise
10. 90° clockwise

Quadrilateral *FGHJ* has vertices *F*(1, 1), *G*(2, 5), *H*(5, 3), and *J*(4, 0). Graph the figure and its image after each rotation. Then give the coordinates of the vertices for quadrilateral *F′G′H′J′*.

11. 90° clockwise
12. 270° clockwise
13. 270° counterclockwise
14. 180° clockwise

Determine whether each figure has rotational symmetry. Write *yes* or *no*. If *yes*, name its angle(s) of rotation.

15. 

16. 

17. 

18. **DESIGNS** Rotate the design shown 90° clockwise, 180° clockwise, and 270° clockwise. Draw the completed image after all the rotations.

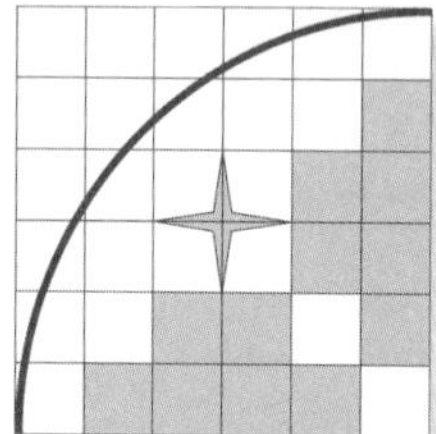

GEOMETRY For Exercises 19–21, use the information on line symmetry on page 611 and the following information.

19. Draw a figure that has line symmetry but does not have rotational symmetry.
20. Draw a figure that has both line symmetry and rotational symmetry.
21. Is it possible for a figure to have rotational symmetry, but not line symmetry? Justify your response with a drawing or an explanation.

EXTRA PRACTICE
See pages 703, 716.

22. **GEOMETRY** The right isosceles triangle *PQR* has vertices *P*(3, 3), *Q*(3, 1), and *R*(■, ■) and is rotated 90° counterclockwise around the origin. Find the missing vertex of the triangle. Then graph the triangle and its image.

H.O.T. Problems

23. **FIND THE ERROR** Elena and Samuel are graphing triangle *ABC* with vertices at *A*(0, 0), *B*(2, 3), and *C*(4, 1) and its image after a rotation 90° counterclockwise about the origin. Who is correct? Explain.

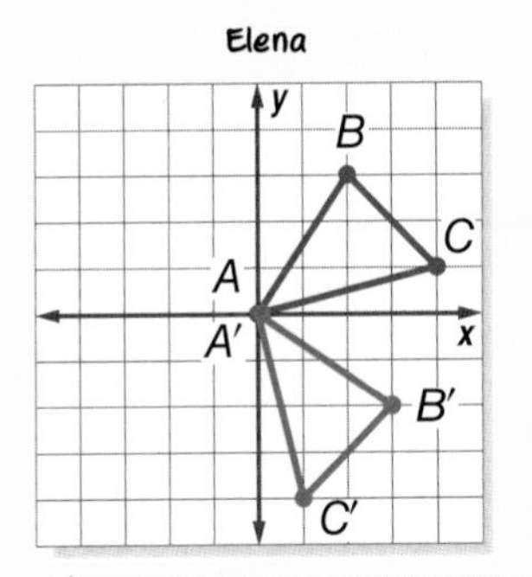

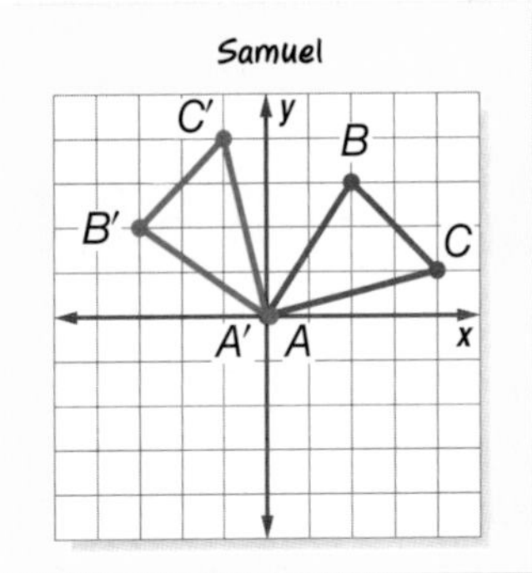

CHALLENGE Triangle *JKL* has vertices *J*(−4, −1), *K*(−1, −2), and *L*(−5, −5). Graph the figure and its image after each rotation about the origin. Then give the coordinates of the vertices for triangle *J′K′L′*.

24. 540° clockwise
25. 450° counterclockwise
26. 720° counterclockwise
27. 630° counterclockwise

28. **WRITING IN MATH** Describe what information is needed to rotate a figure.

TEST PRACTICE

29. Which figure shows the letter F after a rotation of 270° clockwise?

A ⅃ **B** ⅃ **C** ⅃ **D** F

30. Triangle *XYZ* has vertices *X*(2, −2), *Y*(5, 0), and *Z*(3, −4). What are the coordinates of point *Y′* after a rotation of 180°?

F (0, −5)
G (−5, 0)
H (0, 5)
J (5, 0)

Spiral Review

Graph quadrilateral *ABCD* and its resulting image after each transformation. (Lessons 11-8 and 11-9)

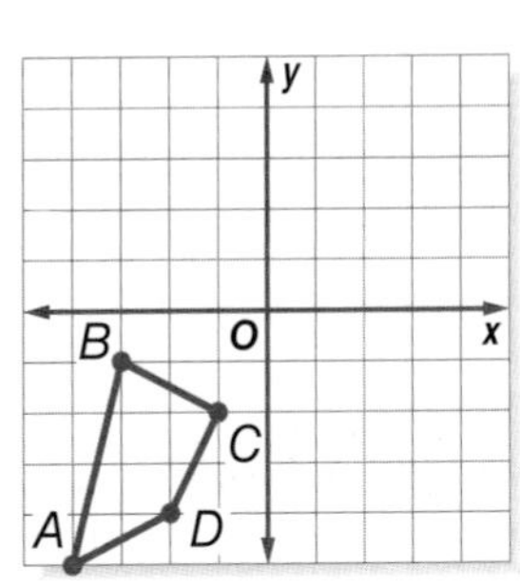

31. reflection over the *y*-axis
32. translation 2 units right, 3 units up
33. reflection over the *x*-axis
34. translation 3 units left, 4 units up
35. reflection over the *x*-axis, then over the *y*-axis

CHAPTER 11 Study Guide and Review

Math Online glencoe.com
• STUDY TO GO
• Vocabulary Review

FOLDABLES® Study Organizer — GET READY to Study

Be sure the following Big Ideas are noted in your Foldable.

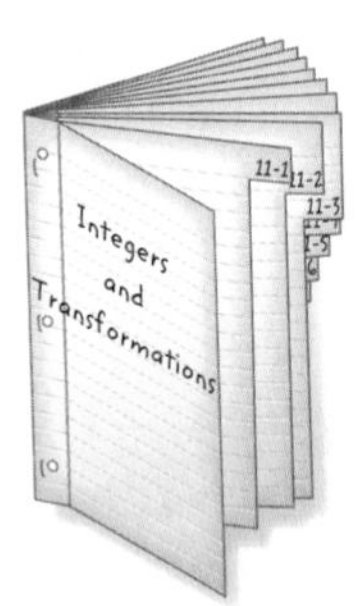

BIG Ideas

Integers (Lessons 11-1 to 11-4, 11-6)

- **Addition and Subtraction**
  The sum of two positive integers is always positive.

  The sum of two negative integers is always negative.

  To subtract an integer, add its opposite.

- **Multiplication and Division**
  The product or quotient of two integers with different signs is negative.

  The product or quotient of two integers with the same sign is positive.

Coordinate Plane (Lesson 11-7)

- The *x*-axis and *y*-axis separate the coordinate plane into four regions called quadrants.

Transformations (Lessons 11-8 to 11-10)

- **Translations**
  A translation of a figure occurs when the figure is moved horizontally, vertically, or both.

- **Reflections**
  A reflection of a figure occurs when the figure is flipped over a line to create a mirror image of the figure.

- **Rotations**
  A rotation occurs when a figure is rotated, or turned, about a point, such as the origin.

Key Vocabulary

angle of rotation (p. 617)
image (p. 604)
quadrants (p. 599)
reflection (p. 610)
rotation (p. 615)
rotational symmetry (p. 617)
translation (p. 604)
transformation (p. 604)
zero pair (p. 576)

Vocabulary Check

State whether each sentence is *true* or *false*. If *false*, replace the underlined word or number to make a true sentence.

1. An <u>integer</u> is any number from the set $\{\ldots, -3, -2, -1, 0, 1, 2, 3, \ldots\}$.
2. The numerical factor of a term that contains a variable is called a <u>quadrant</u>.
3. Any integer that is <u>less</u> than zero is a positive integer.
4. The point $(3, -2)$ translated <u>up</u> 4 units becomes $(7, -2)$.
5. The sign of the sum of two positive integers is <u>negative</u>.
6. The point $(-1, 5)$ reflected across the <u>y-axis</u> becomes $(-1, -5)$.
7. The four regions into which the two axes of a coordinate plane are separated are called <u>integers</u>.
8. To subtract an integer, add its <u>opposite</u>.
9. The sign of the quotient of a positive integer and a negative integer is <u>positive</u>.
10. The <u>y-coordinate</u> is the first number in an ordered pair.

Lesson-by-Lesson Review

11-1 Ordering Integers (pp. 572–575)

Replace each ● with <, >, or = to make a true sentence.

11. −4 ● 0
12. 6 ● 2
13. −1 ● 1
14. 7 ● −6
15. **MONEY** Order $8, −$7, −$5, and −$12 from greatest to least.

Example 1 Replace the ● in −5 ● −2 with <, >, or = to make a true sentence.

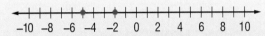

−10 −8 −6 −4 −2 0 2 4 6 8 10

Since −5 is to the left of −2, −5 < −2.

11-2 Adding Integers (pp. 577–581)

Add. Use counters or a number line if necessary.

16. −8 + (−5)
17. −4 + (−2)
18. 8 + (−3)
19. 6 + (−9)
20. −10 + 4
21. −9 + 5
22. **FOOTBALL** In training for the football season, Antwon ran forward 20 yards, backward 10 yards, and forward 20 yards. Write an integer that represents the change in his location after the starting point.

Example 2 Find −3 + 2.

Use a number line.

Start at 0. Move 3 units to the left to show −3. From there, move 2 units right to show +2.

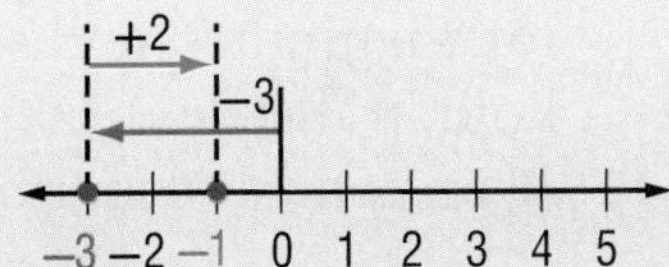

So, −3 + 2 = −1.

11-3 Subtracting Integers (pp. 582–586)

Subtract. Use counters if necessary.

23. −4 − (−9)
24. −12 − (−8)
25. −8 − (−3)
26. 6 − 4
27. 15 − 6
28. −3 − (−8)
29. **STAIRS** From the first floor, Chelsea climbed 48 stairs and then descended 60 stairs. Write an integer that describes her change in location related to the first floor in the number of stairs.

Example 3 Find −2 − 3.

Use counters.

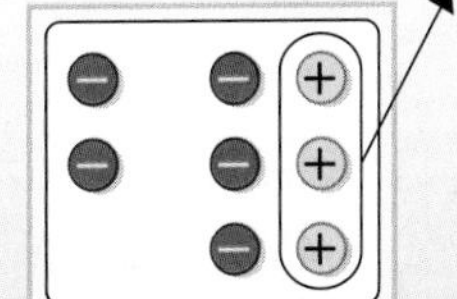

First, place 2 negative counters on the mat. Then add 3 zero pairs. Remove +3.

So, −2 − 3 = −5.

Chapter 11 Study Guide and Review

11-4 Multiplying Integers (pp. 587–590)

Multiply.

30. -3×5
31. $6(-4)$
32. $-2 \times (-8)$
33. $7(5)$
34. $8(-7)$
35. -6×9
36. **MONEY** Sherita invested in the stock market. Over three months, she lost an average of \$2 each month. Write an integer that represents the amount of money she lost.

Example 4 **Find -4×3.**

$-4 \times 3 = -12$ The integers have different signs. The product is negative.

Example 5 **Find $-8 \times (-3)$.**

$-8 \times (-3) = 24$ The integers have the same sign. The product is positive.

11-5 PSI: Work Backward (pp. 592–593)

Solve. Use the *work backward* strategy.

37. **TIME** After school, Lynn watched TV for half an hour, played basketball for 45 minutes, and studied for an hour. If it is now 7:00 P.M., when did Lynn come home from school?
38. **HEIGHT** Fernando is 2 inches taller than Jason. Jason is 1.5 inches shorter than Deirdre and 1 inch taller than Nicole. Hao, who is 5 feet 10 inches tall, is 2.5 inches taller than Fernando. How tall is each student?

Example 6 **A number is divided by 2. Then 5 is added to the quotient. After subtracting 7, the result is 28. What is the number?**

Start with the final value and perform the opposite operation with each resulting value until you arrive at the starting value.

$28 + 7 = 35$ Undo subtracting 7.

$35 - 5 = 30$ Undo adding 5.

$30 \cdot 2 = 60$ Undo dividing by 2.

The number is 60.

11-6 Dividing Integers (pp. 594–598)

Divide.

39. $8 \div (-2)$
40. $-56 \div -8$
41. $-81 \div -9$
42. $-36 \div (-3)$
43. $24 \div (-8)$
44. $-72 \div 6$
45. $-21 \div (-3)$
46. $42 \div (-7)$
47. **ALGEBRA** Evaluate $k \div j$ if $k = -28$ and $j = 7$.

Example 7 **Find $-6 \div 3$.**

$-2 \times 3 = -6$. So, $-6 \div 3 = -2$.

Example 8 **Find $-20 \div (-5)$.**

$4 \times -5 = -20$. So, $-20 \div (-5) = 4$.

Mixed Problem Solving
For mixed problem-solving practice, see page 716.

11-7 The Coordinate Plane (pp. 599–603)

CITY For Exercises 48 and 49, refer to the diagram of a city.

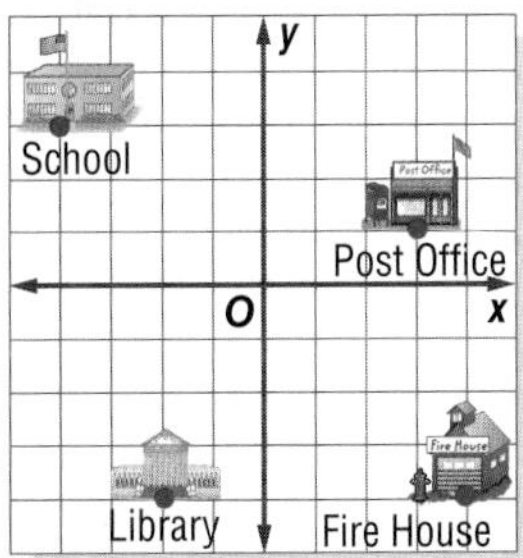

48. Which building is located at (−2, −4)?

49. What ordered pair represents the location of the school?

Graph and label each point on a coordinate plane.

50. $E(2, -1)$
51. $F(-3, -2)$
52. $G(4, 1)$
53. $H(-2, 4)$

Example 9 **Graph $M(-3, -4)$.**

Start at 0. Since the x-coordinate is −3, move 3 units left. Then, since the y-coordinate is −4, move 4 units down.

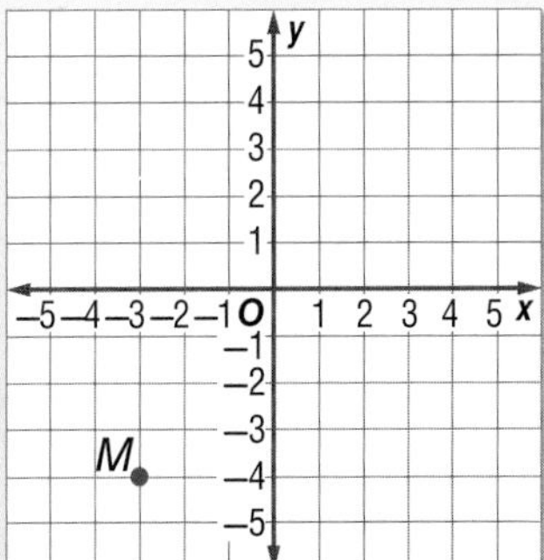

11-8 Translations (pp. 604–609)

Refer to Example 10. Find the vertices of $\triangle R'S'T'$ after each translation. Then graph $\triangle R'S'T'$.

54. 1 unit right, 2 units down

55. 3 units left

56. 4 units left, 1 unit up

57. **GAMES** Maria is playing a board game. Her game piece is located at (3, −4). On her next move, she moves her game piece 2 units left and 1 unit up. What are the coordinates of her game piece after the move?

Example 10 **Find the vertices of $\triangle R'S'T'$ after a translation 5 units left and 4 units down. Then graph $\triangle R'S'T'$.**

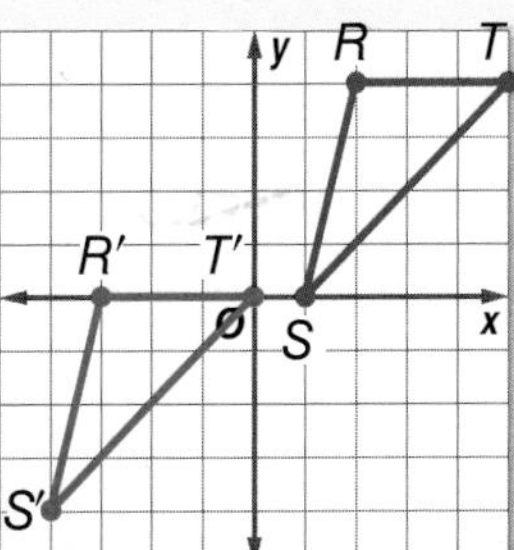

Subtract 5 from each x-coordinate and subtract 4 from each y-coordinate.

The vertices of $\triangle R'S'T'$ are $R'(-3, 0)$, $S'(-4, -4)$, and $T'(0, 0)$.

11-9 Reflections (pp. 610–614)

Graph each figure and its reflection over the indicated axis. Then find the coordinates of the reflected image.

58. $\triangle ABC$ with vertices $A(-1, 3)$, $B(2, -2)$, $C(4, 0)$; y-axis
59. Quadrilateral $GHKJ$ with vertices $G(0, -3)$, $H(-1, -2)$, $K(3, 0)$, $J(4, -1)$; x-axis
60. $\triangle LMN$ with vertices $L(-2, -1)$, $M(1, 2)$, $N(-5, 4)$; x-axis
61. **ALPHABET** Copy the figure at the right. Draw the image of the letter Z after a reflection across the vertical line.

Z

Example 11 Find the vertices of $\triangle P'Q'R'$ after a reflection across the y-axis. Then graph $\triangle P'Q'R'$.

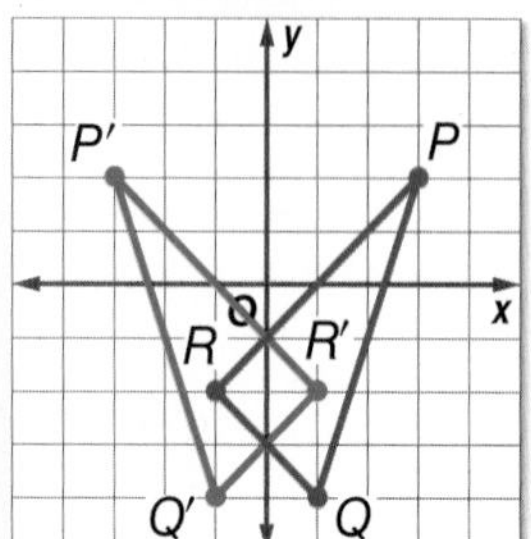

Point P is 3 units from the y-axis.
Point R is 1 unit from the y-axis.
Point Q is 1 unit from the y-axis.
The vertices of $\triangle P'Q'R'$ are $P'(-3, 2)$, $Q'(-1, -4)$, and $R'(1, -2)$.

11-10 Rotations (pp. 615–619)

Refer to Example 12. Graph $\triangle X'Y'Z'$ after each rotation about the origin. Then find the coordinates of $\triangle X'Y'Z'$.

62. 270° clockwise
63. 180° counterclockwise
64. 360° counterclockwise
65. **PAPER FOLDING** Magdalene created the design below out of construction paper. Determine whether the design has rotational symmetry. Write *yes* or *no*. If yes, name its angle(s) of rotation.

Example 12 Graph $\triangle X'Y'Z'$ after a 90° clockwise rotation around the origin. Then find the coordinates of $\triangle X'Y'Z'$.

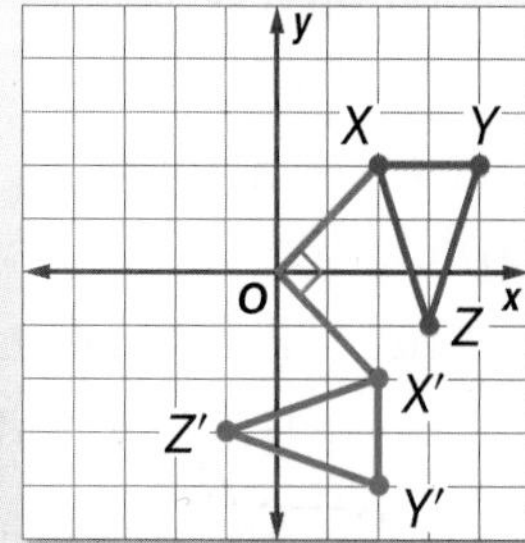

Point X becomes $X'(2, -2)$.
Point Y becomes $Y'(2, -4)$.
Point Z becomes $Z'(-1, -3)$.

CHAPTER 11 Practice Test

Math Online glencoe.com

• Chapter Test

Order each set of integers from least to greatest.

1. $-9, 1, 4, -1, -4$
2. $-5, 3, -8, 1, 0$
3. $11, -13, -8, 6, 12$
4. $2, 0, -2, 4, -1$

Add or subtract. Use counters or a number line if necessary.

5. $-5 + (-7)$
6. $-13 + 10$
7. $-4 - (-9)$
8. $6 - (-5)$
9. $2 + (-8)$
10. $-11 - 3$

11. **ANIMALS** A whale is 8 meters below the ocean's surface. It descended 16 meters. What is the current position of the whale in relation to the ocean's surface?

Multiply or divide.

12. $4(-9)$
13. $24 \div (-4)$
14. $-63 \div (-9)$
15. $-2(-7)$
16. $-3(8)$
17. $-15 \div (-5)$

18. **JUICE** Ciera poured 150 milliliters of juice into a glass from a juice carton. She then poured 175 milliliters into a second glass from the same carton. What was the final change in the volume of juice from the carton?

19. **MONEY** Soledad spent \$7.98 on a book. Then she spent \$12.44 on dinner and \$3.87 on ice cream. About how much money did she spend altogether?

Graph and label each point on a coordinate grid.

20. $C(-4, -2)$
21. $D(5, 3)$

Name the ordered pair for each point and identify its quadrant.

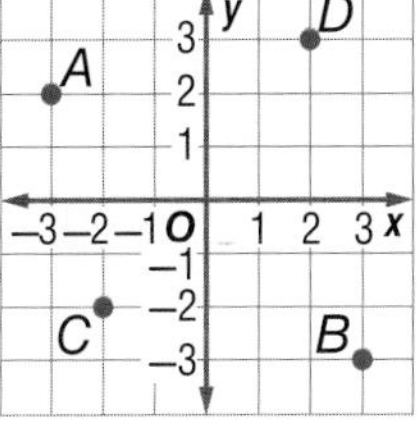

22. A
23. B
24. C
25. D

26. **MULTIPLE CHOICE** Which line contains the ordered pair $(-2, 1)$?

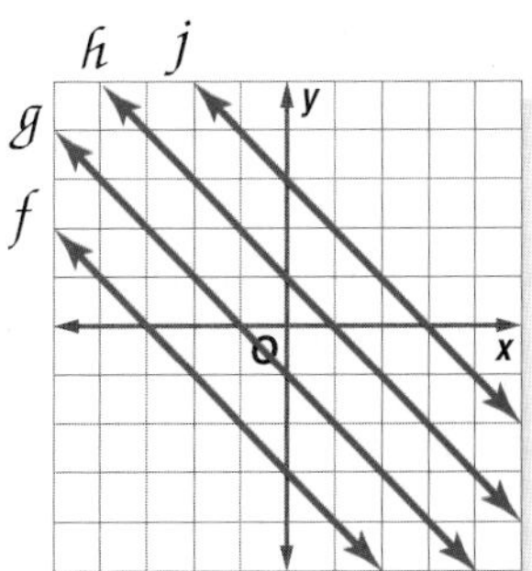

A Line f
C Line h
B Line g
D Line j

27. **MULTIPLE CHOICE** A golfer's score relative to par on the first hole was -2. Suppose he could continue to score at this rate. What would his score relative to par be after eighteen holes?

F -36
H 36
G -20
J 16

For Exercises 28–33, refer to the figure at the right.

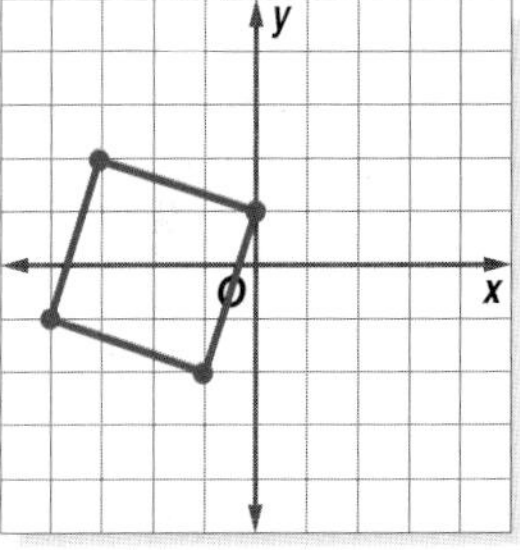

28. Graph the figure after a translation 3 units right and 1 unit down.
29. What are the coordinates of the vertices after the translation?
30. Graph the original figure after a reflection over the y-axis.
31. What are the coordinates of the vertices after the reflection?
32. Graph the original figure after a rotation 90° clockwise about the origin.
33. What are the coordinates of the vertices after the rotation?

CHAPTER 11 Test Practice
Cumulative, Chapters 1–11

Math Online glencoe.com
• Test Practice

PART 1 Multiple Choice

Read each question. Then fill in the correct answer on the answer sheet provided by your teacher or on a sheet of paper.

1. The temperature at noon was 8 degrees below zero Celsius. At five o'clock it was 12 below zero Celsius. Which integer represents the temperature at noon in degrees Celsius?

   A −12 C 8
   B −8 D 12

TEST-TAKING TIP

Question 1 Look for words that indicate negative numbers. A temperature below zero should be represented by a negative integer.

2. What point on the grid below corresponds to the coordinate pair $\left(3, 7\frac{1}{2}\right)$?

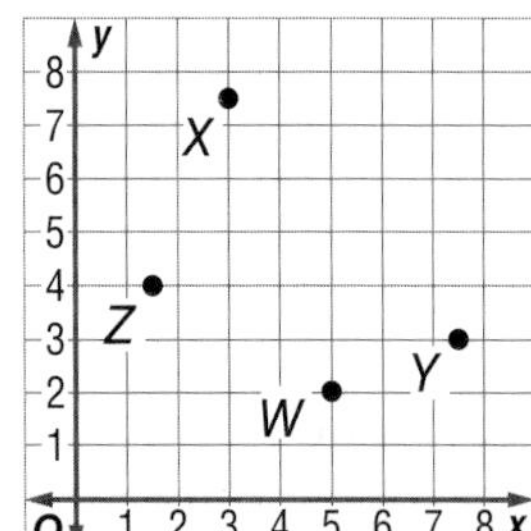

   F Point *W* H Point *Y*
   G Point *X* J Point *Z*

3. It costs $0.25 to run a dryer at the laundromat for one-quarter of an hour. How many dollars does it cost to run the dryer for 2 hours?

   A 4 C 2
   B 3 D 1

4. Of the sixth-grade students surveyed, 25% say that math is their favorite subject, 23% say English, 37% say science, and 15% say music. Which graph best represents these data?

F

Music, Math, Science, English

H

Music, Math, Science, English

G

Music, Math, Science, English

J

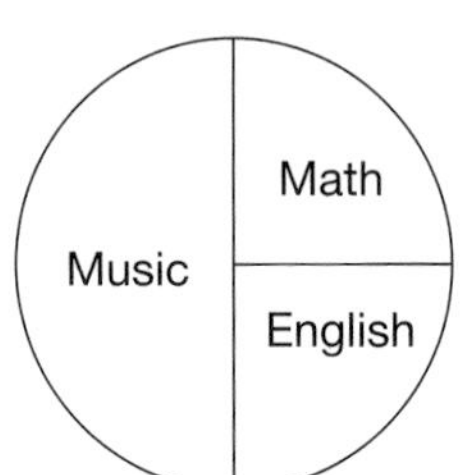

5. The circumference of a circle is 37.68 centimeters. Find the approximate length of the circle's radius.

   A 12 centimeters C 5 centimeters
   B 6 centimeters D 3 centimeters

6. The clock below shows the position of the hands at 3:10 P.M. What kind of angle is ∠*s*?

   F Straight H Obtuse
   G Acute J Right

7. There are 16 cars and 64 passengers scheduled to go to a concert. Which ratio accurately compares the number of passengers to the number of cars?

   A 2 to 8 C 8 to 21
   B 4 to 1 D 16 to 64

Preparing for Standardized Tests
For test-taking strategies and practice, see pages 718–735.

8. Mary's temperature on Sunday night was 101°F. On Monday morning, her temperature was 103°F. Which integer represents Mary's temperature on Sunday night in degrees Fahrenheit?

   F -103

   G -101

   H 101

   J 103

9. Which point best represents the location of the ordered pair $\left(1\frac{1}{4}, 2\right)$?

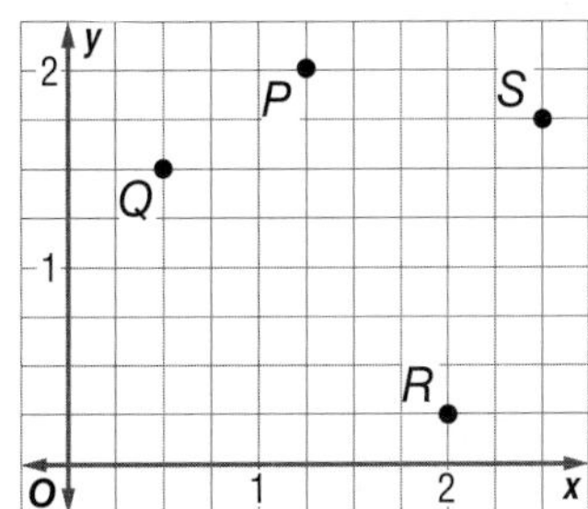

   A Point *P*

   B Point *R*

   C Point *Q*

   D Point *S*

10. You can drive your car 19.56 miles with one gallon of gasoline. How many miles can you drive with 11.96 gallons of gasoline?

   F 11.96 miles

   H 19.56 miles

   G 31.52 miles

   J 233.94 miles

PART 2 Short Response/Grid In

Record your answers on the answer sheet provided by your teacher or on a sheet of paper.

11. Find the measure, in degrees, of ∠2 in rectangle *WXYZ*.

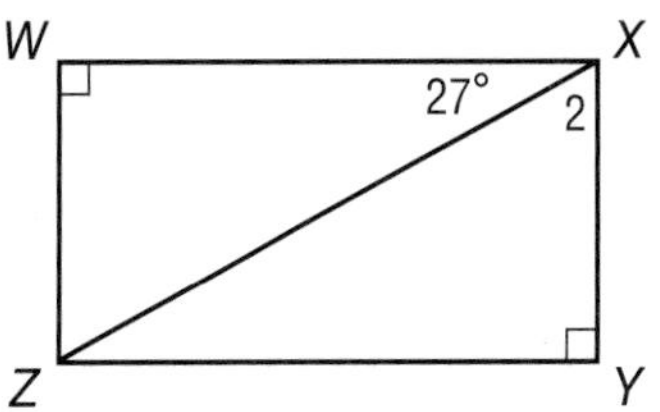

PART 3 Extended Response

Record your answers on the answer sheet provided by your teacher or on a sheet of paper. Show your work.

12. Refer to the figure at the right.

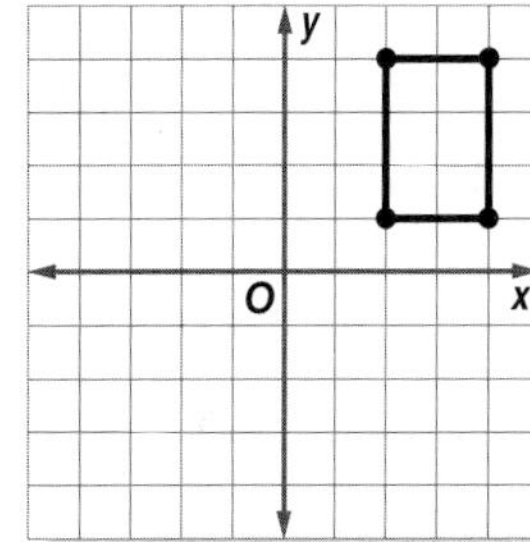

   a. Translate the figure 4 units left. What are the coordinates of the vertices after the translation? How are the new coordinates related to the original figure?

   b. Reflect the original figure across the *x*-axis. What are the coordinates of the vertices after the reflection? How are the new coordinates related to the original figure?

NEED EXTRA HELP?												
If You Missed Question...	1	2	3	4	5	6	7	8	9	10	11	12
Go to Lesson...	2-9	4-9	5-9	7-2	10-2	9-1	6-1	2-9	4-9	6-4	9-5	11-7

Algebra: Properties and Equations

BIG Ideas

- Write, evaluate, and use algebraic expressions to solve problems.
- Solve simple one-step equations.

Key Vocabulary

inverse operations (p. 644)
Distributive Property (p. 632)
equivalent expressions (p. 636)
like terms (p. 637)

Real-World Link

Turtles The state reptile of South Carolina is the loggerhead sea turtle. The average adult weighs about 136 kilograms. The heaviest turtle observed weighed 227 kilograms. You can solve the equation $x + 136 = 227$ to find the difference x in kilograms between these weights.

Algebra: Properties and Equations Make this Foldable to help you organize information about properties and equations. Begin with seven sheets of notebook paper.

1. **Staple** the seven sheets together to form a booklet.

2. **Cut** a tab on the second page the width of the white space. On the third page, make the tab 2 lines longer, and so on.

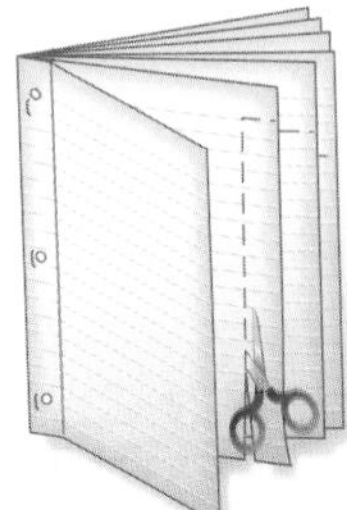

3. **Write** the chapter title on the cover and label each tab with the lesson number.

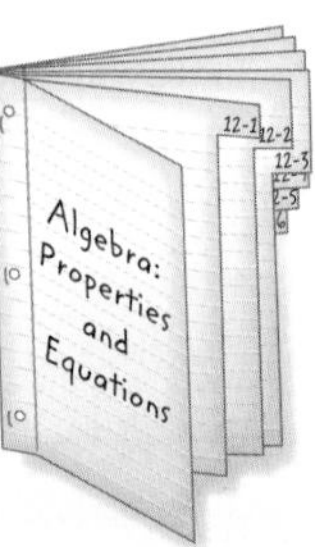

GET READY for Chapter 12

Diagnose Readiness You have two options for checking Prerequisite Skills.

Option 2

Math Online Take the Online Readiness Quiz at glencoe.com.

Option 1

Take the Quick Quiz below. Refer to the Quick Review for help.

QUICK Quiz

Evaluate each expression. (Lessons 1-5, 11-2, 11-3, and 11-4)

1. $4a + 2b$ if $a = 3$ and $b = 7$
2. $-3x - 2x$ if $x = 6$
3. $5y + 11z + y$ if $y = 1$ and $z = -2$
4. $6w + 2w - 3w$ if $w = -8$
5. **TICKETS** Jeffrey bought 3 student tickets to see a movie and 2 children's tickets. If the price of a student ticket is \$6.50 and the price of a child's ticket is \$4.75, what was the total cost?

Add or subtract. (Lessons 11-2 and 11-3)

6. $3 + (-11)$
7. $-4 + 19$
8. $5 - (-2)$
9. $-9 - (-8)$
10. **GAMES** Marjorie earned 7 points during the first round of a game and lost 12 points during the second round. How many points does she have now?

Multiply. (Lesson 11-4)

11. $5(-9)$
12. $-6(-12)$
13. -3×14
14. $-8(-7)$
15. **MONEY** Morgan withdraws \$5 from her checking account every week. Write an integer that represents the change in her account value after 7 weeks.

QUICK Review

Example 1

Evaluate $7p + 3p$ if $p = -1$.

$7p + 3p = 7(-1) + 3(-1)$ Replace p with -1.

$= -7 + (-3)$ Multiply.

$= -10$ Add.

Example 2

Find $-1 + (-4)$.

Use a number line to find the sum.

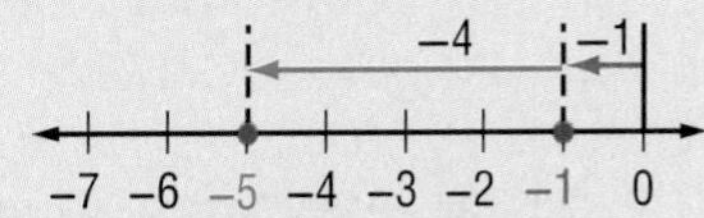

So, $-1 + (-4) = -5$.

Example 3

Find $-2(-13)$.

The two integers have the same sign. The product is positive.

So, $-2(-13) = 26$.

Explore 12-1

Algebra Lab
The Distributive Property

MAIN IDEA

Model the Distributive Property.

To find the area of a rectangle, multiply the length and width. To find the area of a rectangle formed by two smaller rectangles, you can use either one of two methods.

ACTIVITY

1 **Find the area of the blue and yellow rectangles.**

METHOD 1 **Add the lengths. Then multiply.**

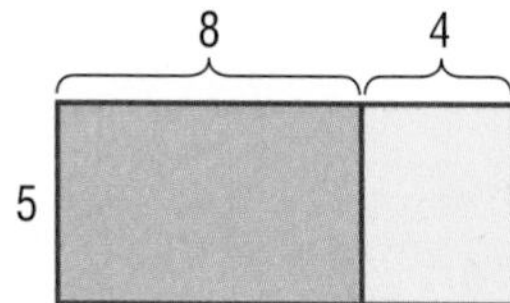

$5(8 + 4) = 5(12)$ Add.

$= 60$ Simplify.

METHOD 2 **Find each area. Then add.**

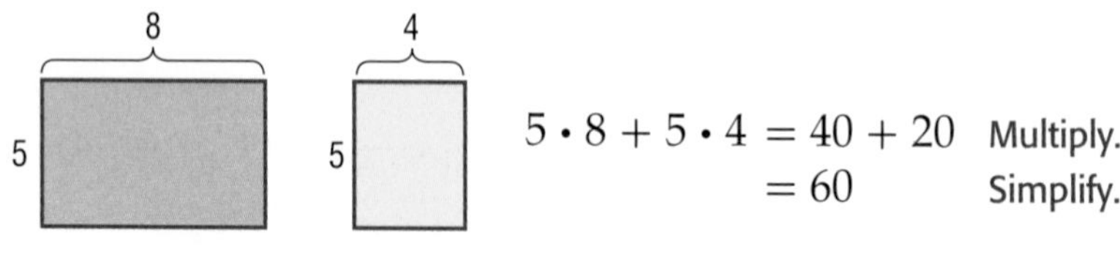

$5 \cdot 8 + 5 \cdot 4 = 40 + 20$ Multiply.

$= 60$ Simplify.

In Method 1, you found that $5(8 + 4) = 60$. In Method 2, you found that $5 \cdot 8 + 5 \cdot 4 = 60$. So, $5(8 + 4) = 5 \cdot 8 + 5 \cdot 4$.

CHOOSE Your Method

Draw a model showing that each equation is true.

a. $2(4 + 6) = (2 \cdot 4) + (2 \cdot 6)$

b. $4(3 + 2) = (4 \cdot 3) + (4 \cdot 2)$

c. $7(10 + 8) = (7 \cdot 10) + (7 \cdot 8)$

d. $6(20 + 3) = (6 \cdot 20) + (6 \cdot 3)$

Analyze the Results

1. Refer to Check Your Progress c above. How could you use the Distributive Property to evaluate 7(18) mentally?
2. Use the Distributive Property to evaluate 9(33) mentally.

Activity 1 modeled the Distributive Property with numbers. You can also use algebra tiles to model the Distributive Property with variables. Refer to the set of algebra tiles below.

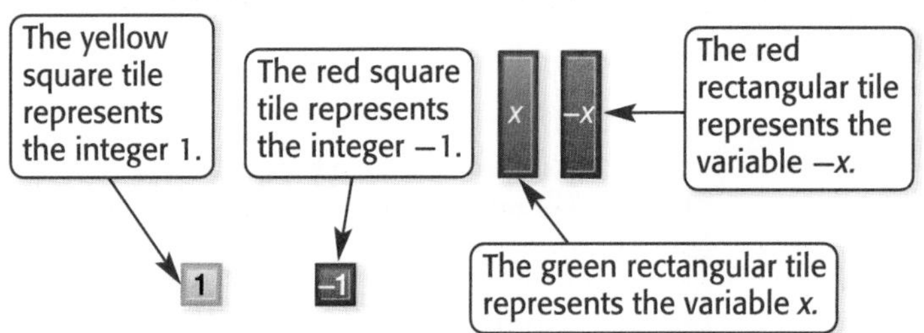
The yellow square tile represents the integer 1.
The red square tile represents the integer −1.
The red rectangular tile represents the variable −x.
The green rectangular tile represents the variable x.
1
−1
x
−x

ACTIVITY

2 **Use algebra tiles to tell whether the equation $2(2x - 1) = 4x - 2$ is *true* or *false*.**

Study Tip

To model the expression $2x - 1$, rewrite as $2x + (-1)$. Then use one red square tile to represent the integer −1.

Model the left side of the equation, $2(2x - 1)$.

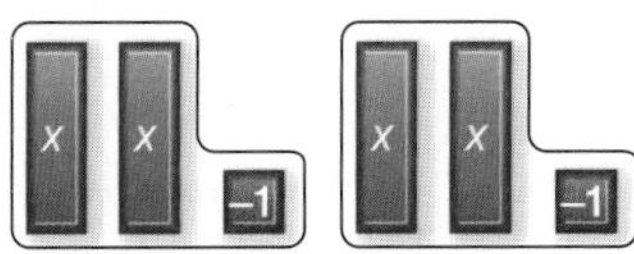
x
x
−1
x
x
−1

There are two groups with $2x - 1$ in each group.

Rearrange the tiles.

x
x
x
x
−1
−1

No tiles were added or taken away from the original expression, $2(2x - 1)$. The equation $2(2x - 1) = 4x - 2$ is true.

Analyze the Results

Tell whether each statement is *true* or *false*. Justify your answer with tiles or a drawing.

1. $3(x + 1) = 3x + 3$
2. $4(x + 1) = 4x + 1$
3. $3(2x - 1) = 6x - 2$
4. $2(3x - 2) = 6x - 4$

5. **MAKE A CONJECTURE** Use what you learned in this lab to make a conjecture about the expressions $5(2x + 3)$ and $10x + 15$.

6. **REASONING** Use what you learned in this lab to rewrite the expressions below without parentheses.

   $2(x + 1)$ $6(x - 4)$ $3(5x + 6)$

7. **WRITING IN MATH** A friend decides that $4(x + 3) = 4x + 3$. How would you explain to your friend that $4(x + 3) = 4x + 12$? Include drawings in your explanation.

12-1 The Distributive Property

MAIN IDEA

Use the Distributive Property to compute multiplication problems mentally and to rewrite algebraic expressions.

New Vocabulary

Distributive Property

Math Online

glencoe.com
- Extra Examples
- Personal Tutor
- Self-Check Quiz

GET READY for the Lesson

BASEBALL Three friends went to a baseball game. Each ticket cost \$20 and each friend also bought a baseball hat for \$15 each.

1. What does the expression $3(20 + 15)$ represent?
2. Evaluate the expression in Exercise 1.
3. Evaluate the expression $3 \cdot 20 + 3 \cdot 15$. What do you notice?

The expressions $3(20 + 15)$ and $3 \cdot 20 + 3 \cdot 15$ show how the **Distributive Property** combines addition and multiplication.

Distributive Property — Key Concept

Words To multiply a sum by a number, multiply each addend by the number outside the parentheses.

Example

Numbers	Algebra
$2(7 + 4) = 2 \times 7 + 2 \times 4$	$a(b + c) = ab + ac$
$(5 + 6)3 = 5 \times 3 + 6 \times 3$	$(b + c)a = ba + ca$

You can use the Distributive Property to compute some multiplication problems mentally.

EXAMPLE Use the Distributive Property

1 Find 7×62 mentally using the Distributive Property.

$7 \times 62 = 7(60 + 2)$	Write 62 as 60 + 2.
$= 7(60) + 7(2)$	Distributive Property
$= 420 + 14$	Multiply 7 and 60 mentally.
$= 434$	Add.

CHECK Your Progress

Find each product mentally. Show the steps you used.

a. 5×84 b. 12×32 c. 2×3.6

EXAMPLE Apply the Distributive Property

Real-World Career
How Does a Jewelry Designer Use Math? Jewelry designers use math to create detailed drawings, models, or computer simulations of the jewelry design.

Math Online
For more information, go to glencoe.com.

2 **JEWELRY** **Francesca is making earrings and bracelets for four friends. Each pair of earrings uses 4.5 centimeters of wire and each bracelet uses 13 centimeters. How much total wire is needed?**

METHOD 1 **Multiply. Then add.**

$$\underbrace{4(4.5)} + \underbrace{4(13)} = 18 + 52 \text{ or } 70 \text{ centimeters}$$

amount of wire for 4 pairs of earrings; amount of wire for 4 bracelets

METHOD 2 **Add. Then multiply.**

$$\underbrace{4(4.5 + 13)} = 4(17.5) \text{ or } 70 \text{ centimeters}$$

amount of wire for one pair of earrings and one bracelet

Using either method, Francesca needs 70 centimeters of wire.

CHOOSE Your Method

d. **EXERCISE** Each day, Martin lifts weights for 10 minutes and runs on the treadmill for 25 minutes. Use the Distributive Property to find the total minutes that Martin exercises for 7 days.

The Distributive Property can also be used to rewrite algebraic expressions containing variables.

EXAMPLES Rewrite Algebraic Expressions

Use the Distributive Property to rewrite each algebraic expression.

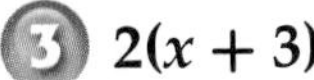

3 $\mathbf{2(x + 3)}$

$2(x + 3) = 2(x) + 2(3)$ Distributive Property

$= 2x + 6$ Multiply.

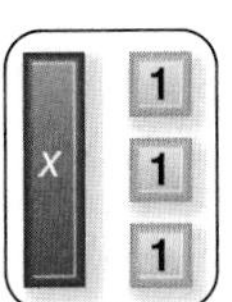

4 $\mathbf{4(x - 1)}$

$4(x - 1) = 4[x + (-1)]$ Rewrite $x - 1$ as $x + (-1)$.

$= 4(x) + 4(-1)$ Distributive Property

$= 4x + (-4)$ Multiply.

$= 4x - 4$ Rewrite $4x + (-4)$ as $4x - 4$.

CHECK Your Progress

e. $8(x + 3)$ f. $5(x + 9)$ g. $2(x - 3)$

CHECK Your Understanding

Example 1 (p. 632)

Find each product mentally. Show the steps you used.

1. 4×38
2. 9×82
3. 11×27
4. 15×34
5. 3×1.6
6. 8×5.4

Example 2 (p. 633)

7. **STATE FAIR** Six friends will go to the state fair to ride the Ferris wheel. The cost of one admission ticket to the fair is $9.50, and the cost for one person to ride the Ferris wheel is $1.50. Use the Distributive Property to find the total cost.

Examples 3, 4 (p. 633)

Use the Distributive Property to rewrite each algebraic expression.

8. $3(x + 1)$
9. $5(x + 8)$
10. $4(x + 6)$
11. $7(x + 4)$
12. $2(x - 1)$
13. $9(x - 2)$

Practice and Problem Solving

HOMEWORK HELP

For Exercises	See Examples
14–25	1
26, 27	2
28–39	3, 4

Find each product mentally. Show the steps you used.

14. 5×56
15. 3×47
16. 6×28
17. 9×44
18. 11×72
19. 14×25
20. 13×22
21. 12×43
22. 4×5.6
23. 3×3.9
24. 6×2.3
25. 7×3.8

26. **DIGITAL CAMERAS** A camera shop charges $0.08 cents to process a digital photo and $0.17 cents to print a digital photo. Use the Distributive Property to find the total cost to process and print 12 digital photos.

27. **ANIMALS** A coyote can run up to 43 miles per hour while a rabbit can run up to 35 miles per hour. Use the Distributive Property to find how many more miles a coyote will run in six hours than a rabbit at these rates.

Use the Distributive Property to rewrite each algebraic expression.

28. $8(x + 7)$
29. $6(x + 11)$
30. $5(x + 1)$
31. $7(x + 3)$
32. $8(x + 1)$
33. $4(x + 2)$
34. $7(x - 2)$
35. $6(x - 3)$
36. $2(x - 9)$
37. $5(x - 11)$
38. $3(x - 4)$
39. $7(x - 5)$

EXTRA PRACTICE
See pages 703, 717.

ALGEBRA For Exercises 40 and 41, find the value of x that makes each equation true.

40. $3(7 + 4) = x + 12$
41. $6(x - 8) = 36 - 48$

H.O.T. Problems

42. **CHALLENGE** Evaluate the expression 0.1(3.7) mentally. Justify your response using the Distributive Property.

43. **OPEN ENDED** Write an equation involving decimals that illustrates the Distributive Property.

44. **WRITING IN MATH** A friend rewrote the expression $5(x - 2)$ as $5x - 2$. Write a few sentences to your friend explaining the error. Then, rewrite the expression $5(x - 2)$ correctly.

TEST PRACTICE

45. Which of the following equations represents the Distributive Property?

A $7x + 1 = 7(x + 1)$

B $7(x + 1) = 7x + 7$

C $7x + 7 = 7(x + 7)$

D $7(x + 1) = x + 7$

46. Four friends ate lunch together at a restaurant. Each friend ordered a deli sandwich for $2.75 each and a beverage for $1.25 each. Which expression represents the total cost of the four meals?

F 4($4)

G 4($3)

H 4($2.50)

J 4($1.50)

Spiral Review

For Exercises 47 and 48, refer to the graph at the right.

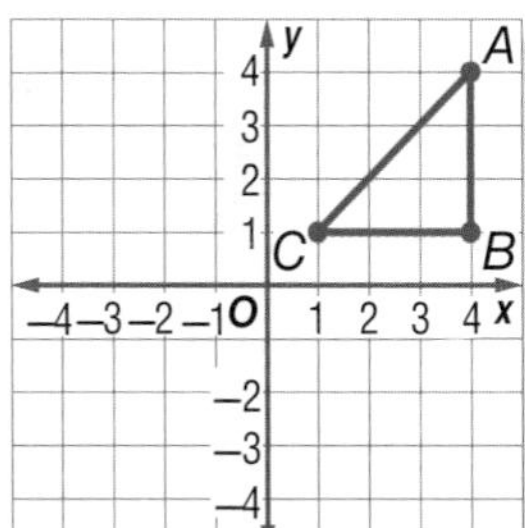

47. Reflect $\triangle ABC$ across the y-axis. What are the coordinates of $\triangle A'B'C'$? (Lesson 11-9)

48. Rotate $\triangle ABC$ 90° counter clockwise around the origin. What are the coordinates of $\triangle A'B'C'$? (Lesson 11-10)

49. **TEMPERATURE** At 8 A.M. the temperature was 5°F. By noon, the temperature had dropped 9 degrees. What was the temperature at noon? (Lesson 11-2)

50. Order the set of integers −8, −12, 0, 11, 9, −7 from least to greatest. (Lesson 11-12)

GET READY for the Next Lesson

PREREQUISITE SKILL Find the value of each expression. (Lesson 1-5)

51. $15 + (3 + 24)$ 52. $(37 + 18) + 2$ 53. $4 \times (6 \times 8)$ 54. $(12 \times 3) \times 5$

12-2 Simplifying Algebraic Expressions

MAIN IDEA

Use the Commutative and Associative properties to simplify expressions.

New Vocabulary

equivalent expressions
Commutative Property
Associative Property
like terms

Math Online

glencoe.com

- Extra Examples
- Personal Tutor
- Self-Check Quiz

GET READY for the Lesson

CAT FOOD Andrew bought one can of Cat Cuisine and one can of Feline Feast. Then he decided to also buy one can of Meow Meals. The expression $(40 + 30) + 50$ represents the total cost, in cents.

Canned Cat Food	
Brand	**Cost per Can**
Cat Cuisine	40¢
Feline Feast	30¢
Meow Meals	50¢

1. Evaluate the expression $(40 + 30) + 50$.
2. Evaluate the expression $40 + (30 + 50)$.
3. Evaluate the expression $40 + (50 + 30)$.
4. What do you notice about your answers to Exercises 1–3?
5. What can you conclude about the order in which you add any three numbers?

The three expressions you evaluated in Exercises 1–3 above are **equivalent expressions** because they have the same value, 120. These equivalent expressions illustrate the Commutative and Associative properties.

Commutative and Associative Properties — Key Concept

Commutative Properties

Words The order in which numbers are added or multiplied does not change the sum or product.

Example	**Numbers**	**Algebra**
	$2 + 4 = 4 + 2$	$a + b = b + a$
	$2 \cdot 4 = 4 \cdot 2$	$a \cdot b = b \cdot a$

Associative Properties

Words The way in which numbers are grouped when they are added or multiplied does not change the sum or product.

Example	**Numbers**	**Algebra**
	$(2 + 4) + 6 = 2 + (4 + 6)$	$(a + b) + c = a + (b + c)$
	$(2 \cdot 4) \cdot 6 = 2 \cdot (4 \cdot 6)$	$(a \cdot b) \cdot c = a \cdot (b \cdot c)$

You can use the Commutative and Associative properties to simplify algebraic expressions.

Sometimes you will use zero pairs to solve equations. Remember, you can add or subtract zero pairs from the mat because adding or subtracting zero pairs does not change the value of an expression.

Study Tip

Look Back To review zero pairs, see Explore 11-2.

ACTIVITY

2 **Solve $x + 3 = -6$ using models.**

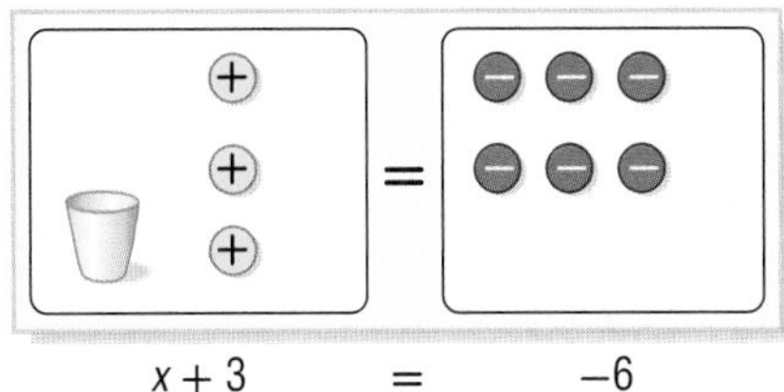

Model the equation.

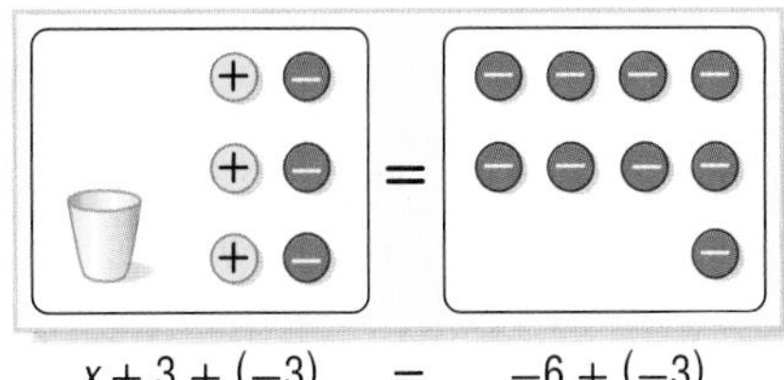

You cannot remove 3 positive counters from each side. Add 3 negative counters to each side of the mat. The left side now has three zero pairs.

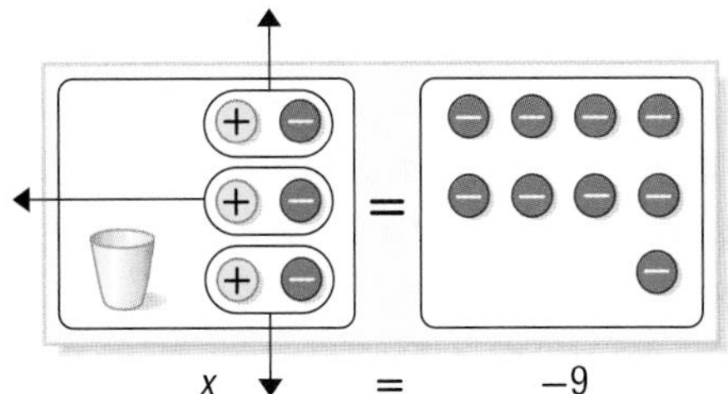

Remove the zero pairs from the left side of the mat. There are 9 negative counters on the right side, so $x = -9$.

The solution is -9.

Check $x + 3 = -6$ Write the original equation.

$-9 + 3 \stackrel{?}{=} -6$ Replace x with -9.

$-6 = -6$ ✓ This sentence is true.

CHECK Your Progress **Solve each equation using models.**

d. $x + 3 = -7$ e. $2 + x = -5$ f. $-3 = x + 3$

ANALYZE THE RESULTS

1. Explain how you decide how many counters to add or subtract from each side.
2. Write an equation in which you need to remove zero pairs in order to solve it.
3. Model the equation *some number plus 5 is equal to* -2. Then solve the equation.
4. **MAKE A CONJECTURE** Write a rule that you can use to solve an equation like $x + 5 = -7$ without using models.

12-3 Solving Addition Equations

MAIN IDEA

Solve addition equations.

New Vocabulary

inverse operations
Subtraction Property of Equality

Math Online

glencoe.com

- Extra Examples
- Personal Tutor
- Self-Check Quiz
- Reading in the Content Area

GET READY for the Lesson

FOOTBALL After gaining 4 yards on a play, the football team was still 5 yards short of the goal line. The number line represents this situation.

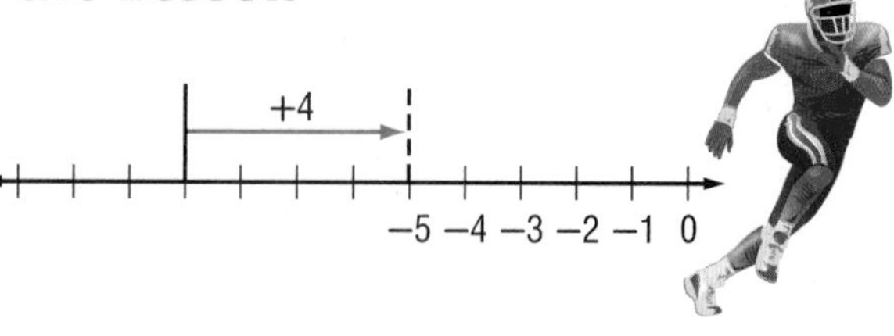

1. Write an expression to represent the gain of 4 yards.
2. Write an addition equation you could use to find the yards needed before gaining 4 yards.
3. You could solve the addition equation by counting back on the number line. What operation does counting back suggest?

In Lesson 1-8, you mentally solved equations. Another way is to use **inverse operations**, which *undo* each other. For example, to solve an addition equation, use subtraction.

EXAMPLE Solve an Equation By Subtracting

1 **Solve $8 = x + 3$.**

METHOD 1 Use models.

Model the equation.

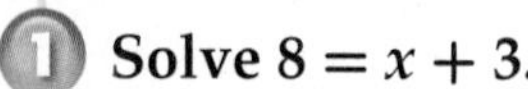

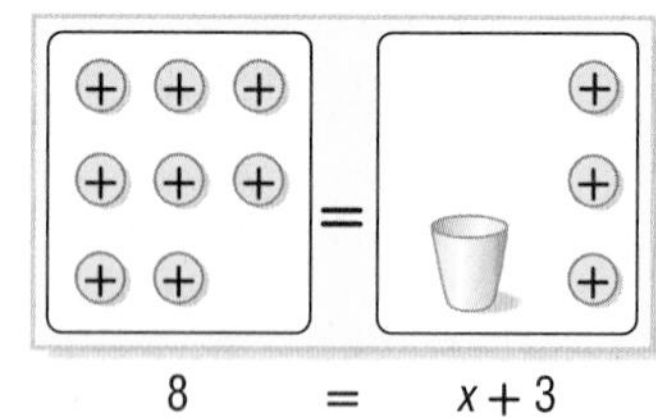

$8 = x + 3$

Remove 3 counters from each side.

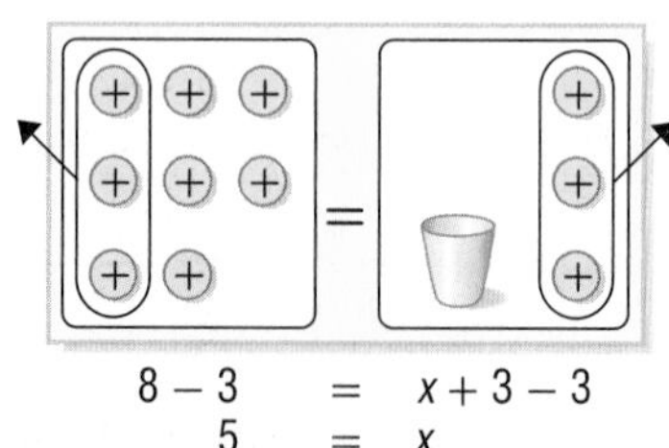

$8 - 3 = x + 3 - 3$
$5 = x$

METHOD 2 Use symbols.

$$\begin{aligned} 8 &= x + 3 \\ -3 &= \quad -3 \\ \hline 5 &= x \end{aligned}$$

Write the equation.

Subtract 3 from each side to "undo" the addition of 3 on the right.

Check $8 = 5 + 3$ ✓

The solution is 5.

CHOOSE Your Method

Solve each equation. Use models if necessary.

a. $c + 2 = 5$ b. $6 = x + 5$ c. $3 + y = 12$

EXAMPLE Solve an Equation Using Zero Pairs

2 **Solve $b + 5 = 2$. Check your solution.**

METHOD 1 Use models.

Model the equation.

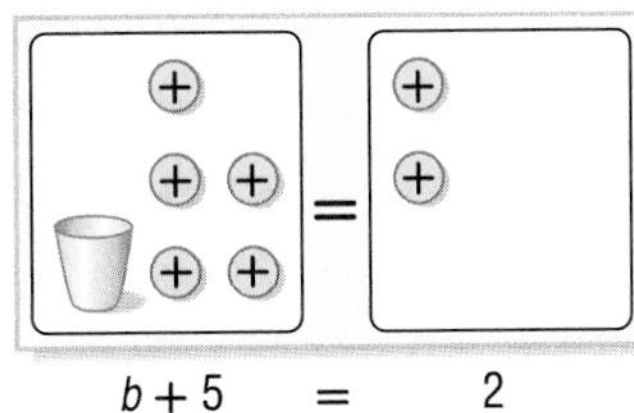

Add 3 zero pairs to the right side so there are 5 positive counters.

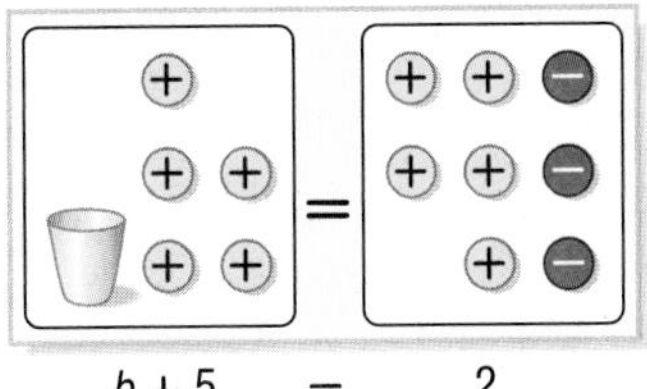

Remove 5 positive counters from each side.

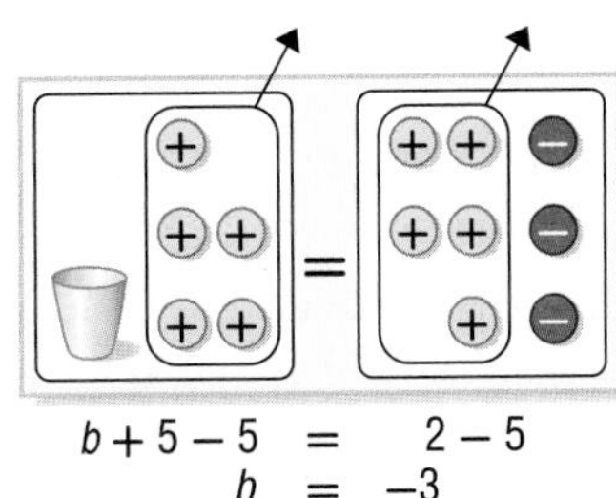

METHOD 2 Use symbols.

$$\begin{aligned} b + 5 &= 2 \\ -5 &= -5 \\ \hline b &= -3 \end{aligned}$$

Write the equation. Subtract 5 from each side.

Check

$b + 5 = 2$ Write the equation.

$-3 + 5 \stackrel{?}{=} 2$ Replace b with -3.

$2 = 2$ ✓ This sentence is true.

The solution is -3. **Check** $-3 + 5 = 2$ ✓

Study Tip

Checking Solutions You should always check your solution. You will know immediately whether your solution is correct or not.

CHOOSE Your Method

Solve each equation. Use models if necessary.

d. $2 = x + 6$ e. $c + 4 = 3$ f. $x + 3 = -2$

When you solve an equation by subtracting the same number from each side of the equation, you are using the **Subtraction Property of Equality**.

Subtraction Property of Equality

Key Concept

Words If you subtract the same number from each side of an equation, the two sides remain equal.

Examples

Numbers	Algebra
$5 = 5$	$x + 2 = 3$
$-3 = -3$	$-2 = -2$
$2 = 2$	$x = 1$

Real-World Link Gorillas typically travel no more than 20 feet on two legs. The two-legged upright stance is generally used for chest-beating or to observe or reach something.

Source: Sea World

Real-World EXAMPLE

3 **ANIMALS** **A male gorilla weighs 379 pounds on average. This is 181 pounds more than the weight of the average female gorilla. Write and solve an addition equation to find the weight of an average female gorilla.**

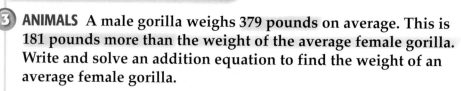

Words	181 pounds plus	the weight of an average female gorilla	is	379 pounds.
Variable	Let w represent the weight of an average female gorilla.			
Equation	$181 +$	w		$= 379$

$$181 + w = 379 \quad \text{Write the equation.}$$
$$-181 \quad = -181 \quad \text{Subtract 181 from each side.}$$
$$w = 198 \quad 379 - 181 = 198$$

So, an average female gorilla weighs 198 pounds.

CHECK Your Progress

g. **INTERNET** In a recent year, there were 171 million home Internet users in the United States and Japan. Of that total, 36 million users were in Japan. Write and solve an addition equation to find the number of Internet users in the United States.

CHECK Your Understanding

Examples 1, 2 (pp. 644–645)

Solve each equation. Use models if necessary. Check your solution.

1. $x + 3 = 5$
2. $2 + m = 7$
3. $c + 6 = -3$
4. $-4 = 6 + e$

Example 3 (p. 646)

5. **TEMPERATURE** The highest recorded temperature in North Carolina is 110 degrees Fahrenheit. This is 144 degrees more than the lowest recorded temperature. Write and solve an addition equation to find the lowest recorded temperature.

EXAMPLES Use Properties to Simplify Expressions

1 **Simplify the expression $3 + (5 + x)$.**

$3 + (5 + x) = (3 + 5) + x$ Associative Property

$= 8 + x$ Add 3 and 5.

Study Tip

Equivalent Expressions The answer to Example 2 could also be written as $36 + x$. The Commutative Property states that $x + 36$ and $36 + x$ are equivalent expressions.

2 **Simplify the expression $(14 + x) + 22$.**

$(14 + x) + 22 = (x + 14) + 22$ Commutative Property

$= x + (14 + 22)$ Associative Property

$= x + 36$ Add 14 and 22.

3 **Simplify the expression $4(6x)$.**

$4(6x) = 4 \cdot (6 \cdot x)$ Parentheses indicate multiplication.

$= (4 \cdot 6) \cdot x$ Associative Property

$= 24x$ Multiply 4 and 6.

CHECK Your Progress

Simplify each expression. Justify each step.

a. $7 + (8 + x)$ b. $(3 \cdot x) \cdot 11$ c. $9(3x)$

You can also use models, such as algebra tiles, to simplify algebraic expressions. **Like terms** contain the same variables, such as x, $2x$, $3x$, and $4x$. Using algebra tiles, like terms are represented by like models that have the same shape.

EXAMPLE Use Models to Simplify Expressions

4 **Simplify the expression $2x + 4 + 3x$.**

Use two x-tiles to model $2x$, four 1-tiles to model 4, and three x-tiles to model $3x$.

x x + 1 1 1 1 + x x x

The like terms are $2x$ and $3x$ because the x-tiles have the same shape. There are five x-tiles and four 1-tiles. So, $2x + 4 + 3x = 5x + 4$.

Study Tip

Simplifying Expressions The expression $5x + 4$ is not equivalent to $9x$. The expression $9x$ would be modeled by nine x-tiles while the expression $5x + 4$ is modeled by five x-tiles and four 1-tiles.

CHECK Your Progress

Simplify each expression using models or a drawing.

d. $5x + x$ e. $4x + 7 + 2x$ f. $x + 1 + 3x$

Real-World Link
King Tutankhamun, known as King Tut, was mummified over 3,300 years ago. The mask that was placed over his face shows what the boy-king may have looked like.
Source: *National Geographic*

Real-World EXAMPLE Write Algebraic Expressions

5 MUSEUMS Three friends will go to the history museum to see the new Egyptian mummy exhibit. The cost of admission is $\$x$ each plus an additional \$1 each to view the mummy exhibit. A fourth friend will go to the museum but will not view the mummy exhibit. Write and simplify an expression that represents the total cost for the four friends.

The cost of admission to the museum and the mummy exhibit can be represented by the expression $x + 1$.

The expression $\underbrace{3(x + 1)} + x$ represents the total cost.

cost of admission and exhibit for three friends — $3(x + 1)$

cost of admission for the fourth friend — x

$3(x + 1) + x = 3x + 3 + x$	Distributive Property
$= 3x + x + 3$	Commutative Property
$= 4x + 3$	The model shows that $3x$ and x are like terms.

So, the total cost for the four friends is $\$4x + \$3$.

CHECK Your Progress

g. **MUSEUMS** Write and simplify an expression for the total cost of six friends to go to the museum if only four friends will view the mummy exhibit.

CHECK Your Understanding

Examples 1–3 (p. 637)

Simplify each expression below. Justify each step.

1. $1 + (6 + x)$
2. $7 \cdot (9 \cdot x)$
3. $(18 + x) + 11$
4. $(12 \cdot x) \cdot 4$
5. $5(6x)$
6. $11(3x)$

Example 4 (p. 637)

Simplify each expression below using models or a drawing.

7. $6x + 2x$
8. $3x + 6 + 2x$
9. $x + 5 + 4x$

Example 5 (p. 638)

Use the Distributive Property to rewrite each algebraic expression.

10. **FASHION** Mikayla bought five skirts at $\$x$ each. Three of the five skirts came with a matching top for an additional \$9 each. Write and simplify an expression that represents the total cost of her purchase.

Practice and Problem Solving

HOMEWORK HELP	
For Exercises	See Examples
6–9	1
10–17	2
18, 19	3

Solve each equation. Use models if necessary. Check your solution.

6. $y + 7 = 10$
7. $x + 5 = 11$
8. $9 = 2 + x$
9. $7 = 4 + y$
10. $9 + a = 7$
11. $6 + g = 5$
12. $d + 3 = -5$
13. $x + 4 = -2$
14. $-5 = 3 + f$
15. $-1 = g + 7$
16. $b + 4 = -3$
17. $h + 6 = 2$
18. **SNAKES** The average length of a King Cobra is 118 inches, which is 22 inches longer than a Black Mamba. Write and solve an addition equation to find the average length of a Black Mamba.
19. **MONEY** Gary and Paz together have $756. If Gary has $489, how much does Paz have? Write and solve an addition equation to find how much money belongs to Paz.
20. Find the value of x if $x + 3 = 7$.
21. If $c + 6 = 2$, what is the value of c?

Solve each equation. Check your solution.

22. $t + 1.9 = 3.8$
23. $1.8 + n = -0.3$
24. $a + 6.1 = -2.3$
25. $7.8 = x + 1.5$
26. $m + \frac{1}{3} = \frac{2}{3}$
27. $t + \frac{1}{4} = -\frac{1}{2}$
28. **GAMES** Suppose your friend had a score of −7 in the second round of a certain game. This made her total score after two rounds equal to −3. What was her score in the first round?
29. **TRUCKS** The table shows the heights of the five biggest monster trucks. Bigfoot 5 is 4.9 feet taller than Bigfoot 2. Write and solve an addition equation to find the height of Bigfoot 2.

World's Biggest Monster Trucks

Truck	Height (feet)
Bigfoot 5	15.4
Swamp Thing	12.2
Godzilla	12.0
Bigfoot 2	■
Black Stallion	11.0

Source: *Scholastic Book of World Records*

30. **SCHOOL** Mr. Vonada gives four quizzes each quarter for a total of 100 points. Cleveland earned scores of 22 points, 19 points, and 24 points on the first three quizzes. Write and solve an equation to find the number of points Cleveland earned on the fourth quiz if his total quiz points for the quarter were 90 points.

EXTRA PRACTICE
See pages 704, 717.

H.O.T. Problems

31. **CHALLENGE** In the equation $x + y = -3$, the value of x is an integer greater than 1, but less than 6. Determine the possible solutions for y in this equation.
32. **OPEN ENDED** Write two different addition equations that have 12 as the solution.
33. **WRITING IN MATH** Without solving, decide whether the solution to $a + 14 = -2$ will be positive or negative. Explain your reasoning.

TEST PRACTICE

34. The model represents the equation $x + 4 = 7$.

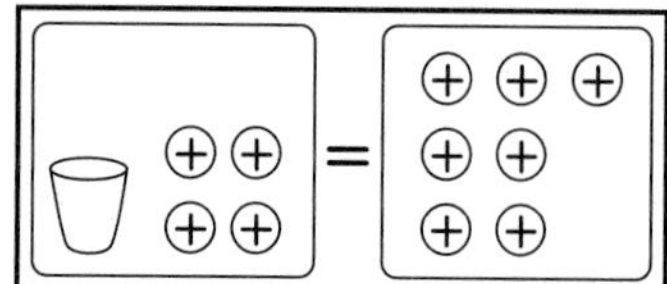

What is the first step in finding the value of x?

A Add 4 positive tiles to each side of the model.

B Subtract 7 positive tiles from each side of the model.

C Add 7 positive tiles to each side of the model.

D Subtract 4 positive tiles from each side of the model.

35. Niko wants to buy a skateboard that costs \$85. He has already saved \$15. Which equation represents the amount of money Niko needs to buy the skateboard?

F $t - 15 = 85$

G $t + 15 = 85$

H $15 - t = 85$

J $t = 15 + 85$

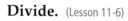

Spiral Review

36. BASKETBALL A family of six bought tickets to a basketball game. Each ticket cost \$$x$. Three family members also bought a beverage for \$2 each. Write and simplify an expression that represents the total cost. (Lesson 12-2)

37. HOBBIES Each day, Andrea plays the piano for 35 minutes and reads for 25 minutes. Use the Distributive Property to find how much time Andrea spends on these two activities in five days. (Lesson 12-1)

Refer to the coordinate plane to identify the point for each ordered pair. (Lesson 11-7)

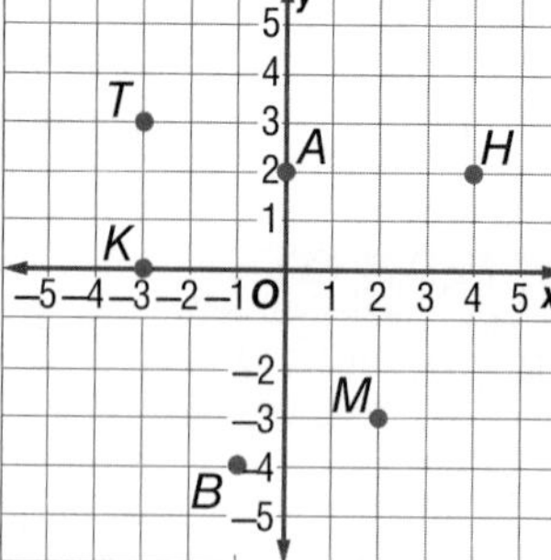

38. $(4, 2)$ **39.** $(-3, 0)$ **40.** $(-1, -4)$

Refer to the coordinate plane to write the ordered pair that names each point. Then identify the quadrant where each point is located. (Lesson 11-7)

41. T **42.** M **43.** A

Divide. (Lesson 11-6)

44. $10 \div (-2)$ **45.** $-24 \div 6$ **46.** $-36 \div (-6)$ **47.** $-81 \div (-9)$

GET READY for the Next Lesson

PREREQUISITE SKILL **Add.** (Lesson 11-2)

48. $-2 + 6$ **49.** $-9 + 3$ **50.** $-8 + 5$ **51.** $-7 + 9$

CHAPTER 12

Mid-Chapter Quiz

Lessons 12-1 through 12-3

Find each product mentally. Show the steps you used. (Lesson 12-1)

1. 8×24
2. 7×62
3. 12×38
4. 9×1.7

5. **MOVIES** Five friends went to the movies and they each bought a small popcorn and medium drink. Use the Distributive Property to find the total cost. (Lesson 12-1)

Drinks		Popcorn	
Small	\$2.50	Small	\$2.75
Medium	\$3.50	Medium	\$3.50
Large	\$4.75	Large	\$4.25

Use the Distributive Property to rewrite each algebraic expression. (Lesson 12-1)

6. $4(x + 5)$
7. $2(x - 3)$
8. $6(x + 7)$
9. $8(x - 5)$

10. **MULTIPLE CHOICE** Each day this week Natalie spent 8 minutes practicing her serve and 12 minutes practicing her forehand at tennis. Which expression represents the total time Natalie spent practicing tennis this week? (Lesson 12-1)

    A 7(4)

    B 7(20)

    C 8(20)

    D 8(19)

Simplify each expression below. Justify each step. (Lesson 12-2)

11. $2 + (13 + x)$
12. $(x \cdot 7) \cdot 5$
13. $(19 + x) + 16$
14. $12(4x)$

15. **TICKETS** A family of six bought airplane tickets. The price of each ticket was \$$x$. Two of the family members paid an extra \$50 for first class tickets. Write and simplify an expression that represents the total cost of the tickets. (Lesson 12-2)

Simplify each expression below using a model or a drawing. (Lesson 12-2)

16. $2x + 2x$
17. $x + 2 + 4x$
18. $3x + 5 + 3x$
19. $x + 6x + 3$

20. **KITES** A kite is 70 feet in the air. A few minutes later it ascends to 120 feet. Write and solve an addition equation to find the change in altitude of the kite. (Lesson 12-3)

Solve each equation. Use models if necessary. Check your solution. (Lesson 12-3)

21. $y + 4 = 9$
22. $15 = x + 12$
23. $7 + h = -3$
24. $a + 6 = -1$

25. **MULTIPLE CHOICE** David spent a total of 90 minutes this week completing his chores. Which of the following equations represents the number of minutes David spent washing the dishes? (Lesson 12-3)

Chore	Time (min)
Vacuuming	42
Dishes	■

    F $m = 42 + 90$

    G $42 - m = 90$

    H $m + 42 = 90$

    J $90 = m - 42$

Explore 12-4

Algebra Lab
Solving Subtraction Equations Using Models

MAIN IDEA

Solve subtraction equations using models.

Recall that subtracting an integer is the same as adding its opposite. For example, $4 - 7 = 4 + (-7)$ or $x - 3 = x + (-3)$.

ACTIVITY

Solve $x - 5 = -4$ using models.

$x - 5 = -4 \longrightarrow x + (-5) = -4$ Rewrite the equation.

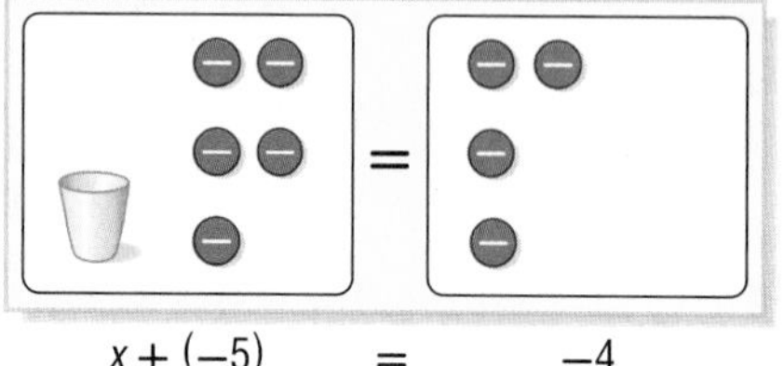

Model the addition equation.

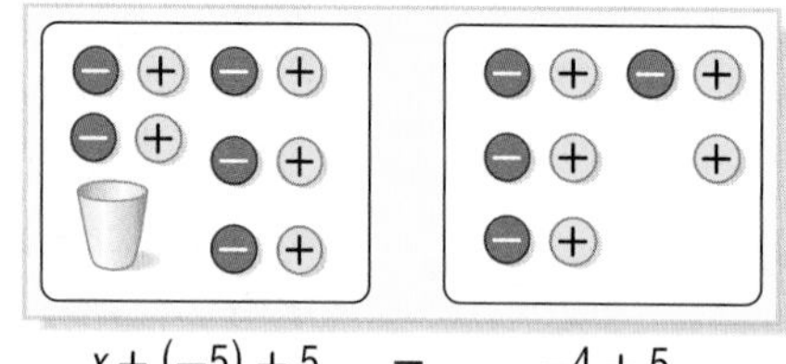

Add 5 positive counters to each side of the mat to make 5 zero pairs on the left side.

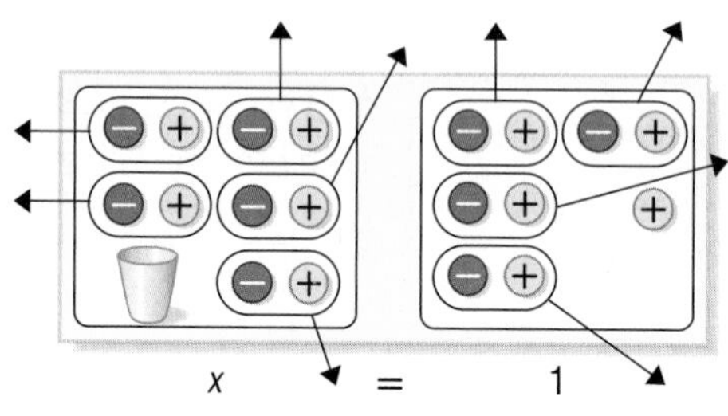

Remove 5 zero pairs from the left side and 4 zero pairs from the right side. There is one positive counter on the right side. So, $x = 1$.

The solution is 1. **Check** $1 - 5 = 1 + (-5)$ or -4 ✓

Solve each equation using models.

a. $x - 4 = 2$ b. $-3 = x - 1$ c. $x - 5 = -1$

Analyze the Results

1. Explain why it is helpful to rewrite a subtraction problem as an addition problem when solving equations using models.
2. **MAKE A CONJECTURE** Write a rule for solving equations like $x - 7 = -5$ without using models.

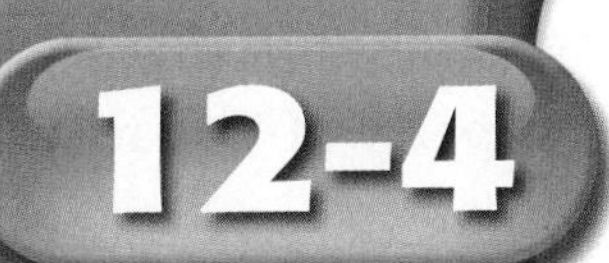

Solving Subtraction Equations

MAIN IDEA

Solve subtraction equations.

New Vocabulary

Addition Property of Equality

Math Online

glencoe.com

- Concepts in Motion
- Extra Examples
- Personal Tutor
- Self-Check Quiz

GET READY for the Lesson

BOWLING Meghan's bowling score was 36 points less than Charmaine's. Meghan's score was 109.

1. Let s represent Charmaine's score. Write an equation for *36 points less than Charmaine's score is equal to 109.*
2. Find Charmaine's score by counting forward. What operation does counting forward suggest?

Because addition and subtraction are inverse operations, you can solve a subtraction equation by adding.

EXAMPLE Solve an Equation by Adding

1 **Solve $x - 3 = 2$.**

METHOD 1 **Use models.**

Model the equation.

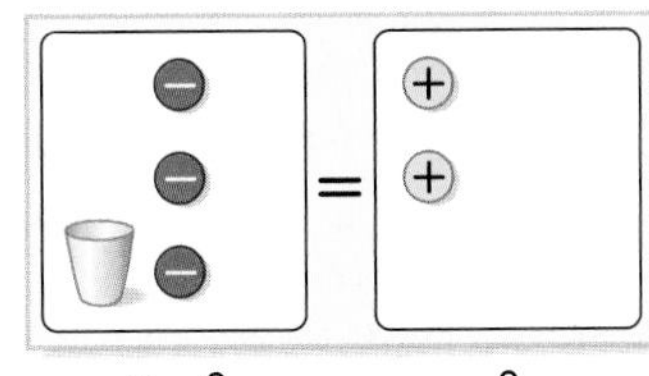

$x - 3 \quad = \quad 2$

Add 3 positive counters to each side of the mat. Remove the zero pairs.

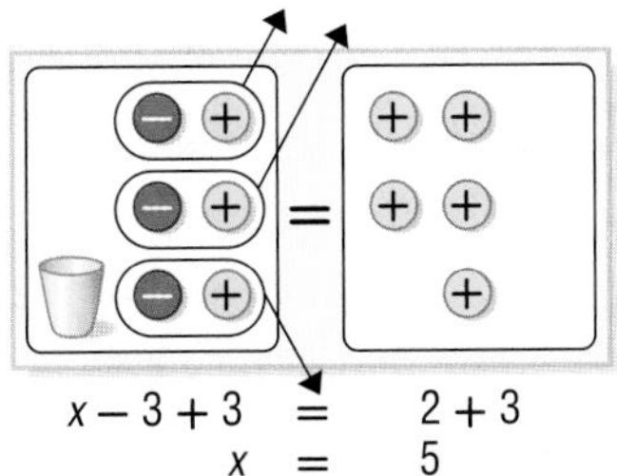

$x - 3 + 3 = 2 + 3$

$x = 5$

METHOD 2 **Use symbols.**

$x - 3 = 2$	Write the equation.
$+3 = +3$	Add 3 to each side.
$x = 5$	Simplify.

Check $5 - 3 = 2$ ✓

The solution is 5.

CHOOSE Your Method **Solve. Use models if necessary.**

a. $x - 7 = 4$

b. $y - 6 = -2$

c. $9 = a - 5$

When you solve an equation by adding the same number to each side of the equation, you are using the **Addition Property of Equality**.

Addition Property of Equality

Key Concept

Words If you add the same number to each side of an equation, the two sides remain equal.

Examples

Numbers	Algebra
$5 = 5$	$x - 2 = 3$
$+3 = +3$	$+2 = +2$
$8 = 8$	$x = 5$

EXAMPLE Solve a Subtraction Equation

2 Solve $-10 = y - 4$. Check your solution.

$-10 = y - 4$ Write the equation.
$+4 = +4$ Add 4 to each side.
$-6 = y$ Simplify.

The solution is -6.

Check $-10 = y - 4$
$-10 \stackrel{?}{=} -6 - 4$
$-10 = -10$ ✓

CHECK Your Progress **Solve each equation. Check your solution.**

d. $b - 4 = -2$ e. $-5 = t - 5$ f. $c - 2 = -6$

Real-World EXAMPLE

3 WEATHER The difference between the record high and low temperatures in South Carolina is 130°F. The record low temperature is −19°F. What is the record high temperature?

Words	Record high temperature	minus	record low temperature	is	130°F.
Variable	Let h represent the record high temperature.				
Equation	h	$-$	(-19)	$=$	130

$h - (-19) = 130$ Write the equation.
$h + 19 = 130$ Definition of subtraction
$-19 = -19$ Subtract 19 from each side.
$h = 111$ Simplify.

The record high temperature is 111°F.

Study Tip

Writing Algebraic Expressions Choose a variable to represent the unknown. The term "difference" represents subtraction. The word "is" is usually represented by the equals sign, =.

CHECK Your Progress

g. **GROWTH** Georgia's height is 4 inches less than Sienna's height. Georgia is 58 inches tall. Write and solve a subtraction equation to find Sienna's height.

CHECK Your Understanding

Examples 1, 2 (pp. 651–652)

Solve each equation. Use models if necessary. Check your solution.

1. $a - 5 = 9$
2. $b - 3 = 7$
3. $4 = y - 8$
4. $x - 4 = -1$
5. $x - 2 = -7$
6. $-3 = n - 2$

Example 3 (p. 652)

7. **AGES** Devon is 13 years old. This is 4 years younger than his older brother, Todd. Write and solve a subtraction equation to find Todd's age.

Practice and Problem Solving

HOMEWORK HELP

For Exercises	See Examples
8–11	1
12–19	2
20, 21	3

Solve each equation. Use models if necessary. Check your solution.

8. $c - 1 = 8$
9. $f - 1 = 5$
10. $2 = e - 1$
11. $1 = g - 3$
12. $r - 3 = -1$
13. $t - 2 = -2$
14. $t - 4 = -1$
15. $h - 2 = -9$
16. $-3 = u - 8$
17. $-5 = v - 6$
18. $x - 3 = -5$
19. $y - 4 = -7$
20. **DIVING** A diver is swimming below sea level. A few minutes later the diver descends 35 feet until she reaches a depth of 75 feet below sea level. Write and solve a subtraction equation to find the diver's original position.
21. **SOCIAL STUDIES** The difference between the number of electoral votes for Florida and North Carolina is 12 votes. Write and solve a subtraction equation to find the number of electoral votes for the state of Florida.

Electoral Votes

State	Number of Votes
Florida	■
North Carolina	15

22. **ALGEBRA** Find the value of t if $t - 7 = -12$.
23. **ALGEBRA** If $b - 10 = 5$, what is the value of $b + 6$?

Solve each equation. Check your solution.

24. $-6 + a = -8$
25. $a - 1.1 = 2.3$
26. $-4.6 = e - 3.2$
27. $-4.3 = f - 7.8$
28. $m - \frac{1}{3} = \frac{2}{3}$
29. $n - \frac{1}{4} = -\frac{1}{2}$
30. **FOOTBALL** A football team gained 10 yards after running two plays. The first play resulted in a loss of 8 yards. Write and solve a subtraction equation to find the total number of yards gained in the second play.
31. **SHOPPING** Alejandra spent her birthday money on a video game that cost \$24, a new controller for \$13, and a memory card for \$16. The tax on the purchase was \$3. Write and solve a subtraction equation to find how much money Alejandra gave the cashier if she received \$4 back in change.

EXTRA PRACTICE
See pages 704, 717.

H.O.T. Problems

32. **NUMBER SENSE** Tell what you know about the value of x in the equation $x - 8 = -3$ without solving it.
33. **OPEN ENDED** Write a problem that can be represented by the equation $x - 12 = 2$. Explain the meaning of the equation.

34. **FIND THE ERROR** Ramón and Bron are explaining how to solve the equation $d - 6 = 4$. Who is correct? Explain your reasoning.

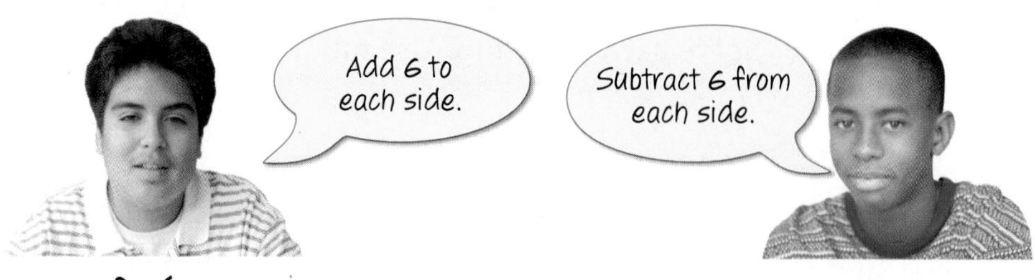

35. **WRITING IN MATH** Determine whether the solution of the equation $x - 8 = 1$ is positive or negative without solving it. Explain your reasoning.

TEST PRACTICE

36. The model represents $x - 5 = 3$.

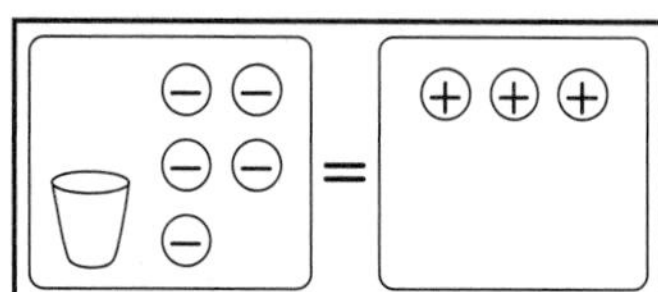

What is the value of x?

A $x = -8$

B $x = -2$

C $x = 2$

D $x = 8$

37. Indiana became a state in 1816. Arizona became a state 96 years later. Which equation can be used to find the year y Arizona became a state?

F $y = 1816 - 96$

G $y + 96 = 1816$

H $y - 1816 = 96$

J $1816 - y = 96$

Spiral Review

38. **BASEBALL** Refer to the graph. Write and solve an addition equation to find how many fewer people can be seated at Dodger Stadium than at Yankee Stadium. (Lesson 12-3)

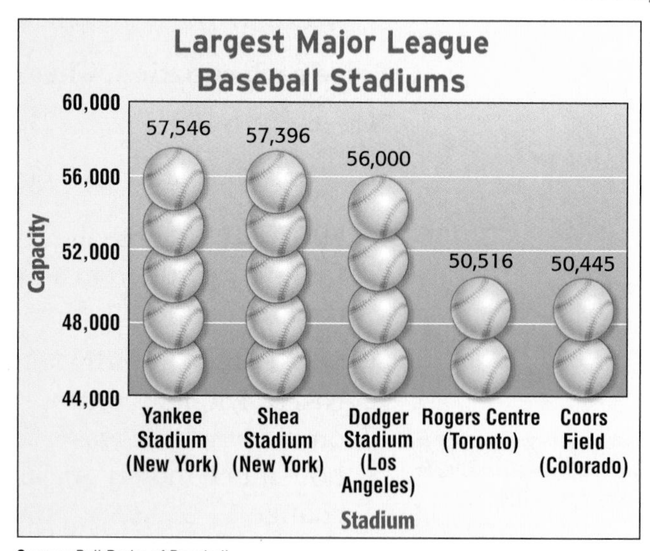

Source: Ball Parks of Baseball

39. **ALGEBRA** Simplify the expression $6x + 3 + 5x$. (Lesson 12-2)

40. Graph and label $X(2, -3)$ and $Y(-3, 2)$ on a coordinate plane. (Lesson 11-7)

GET READY for the Next Lesson

PREREQUISITE SKILL **Divide.** (Lesson 11-6)

41. $-8 \div 2$

42. $42 \div 3$

43. $-36 \div (-3)$

44. $-24 \div 8$

Extend 12-4

Algebra Lab
Solving Inequalities Using Models

MAIN IDEA

Use models to solve simple addition and subtraction inequalities.

Recall from Lesson 3-2, that an inequality states that two expressions are *not* equal. An inequality like $x < 7$ or $x > 5$ can be written to express how a variable compares to a number.

A balance scale can be used to demonstrate an inequality. On the balance scale below, a cup containing an unknown number of counters is placed on the left scale. Three positive counters are placed on the right scale.

Note that the left side of the scale weighs more than the right scale. So, $x > 3$. The number of counters in the cup is greater than 3.

Check possible values of x.

$x > 3$ Write the inequality.

$2 \overset{?}{>} 3$ No

$3 \overset{?}{>} 3$ No

$4 \overset{?}{>} 3$ Yes

$5 \overset{?}{>} 3$ Yes

$6 \overset{?}{>} 3$ Yes

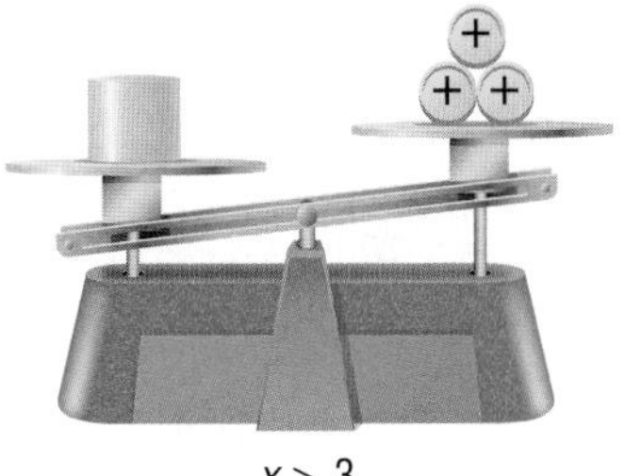

$x > 3$

So, the solution of the inequality is any number greater than 3.

If two positive counters are added to each side of the balance scale, note that the left scale still weighs more than the right scale. So, $x + 2 > 5$.

Check possible values of x.

$x + 2 > 5$ Write the inequality.

$2 + 2 \overset{?}{>} 5$ No

$3 + 2 \overset{?}{>} 5$ No

$4 + 2 \overset{?}{>} 5$ Yes

$5 + 2 \overset{?}{>} 5$ Yes

$6 + 2 \overset{?}{>} 5$ Yes

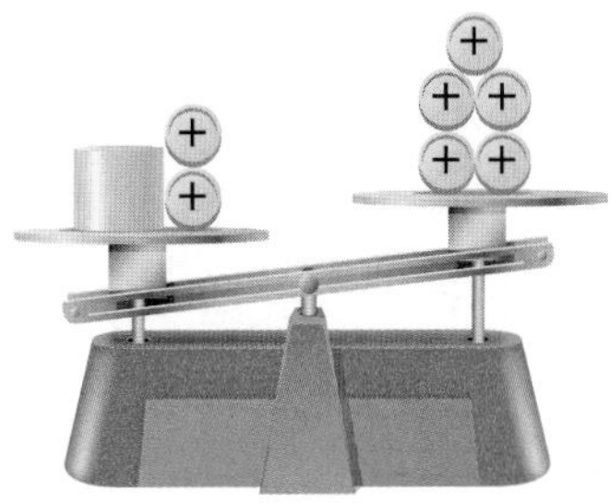

$x + 2 > 5$

Possible values of x include 4, 5, and 6 but do not include 2 and 3. So, the solution to the inequality $x + 2 > 5$ is $x > 3$.

Note that the possible values of x did not change when the same number of counters are added to each side.

In Lessons 12-3 and 12-4, you learned that you can add or subtract the same quantity to each side of an equation when solving it. This is also true for inequalities.

Study Tip

Balance Scales Even though the left scale is *higher* than the right scale, $x + 1$ is *not greater* than 6. Since a balance scale represents weight, $x + 1$ weighs less than 6. So, $x + 1 < 6$.

ACTIVITY Solve an Addition Inequality

1 **Solve the inequality $x + 1 < 6$ using a model.**

The model shows one cup and one positive counter on the left scale and six positive counters on the right scale.

To get the cup by itself, remove one positive counter from each side of the scale. There are five positive counters remaining on the right scale. So, $x < 5$.

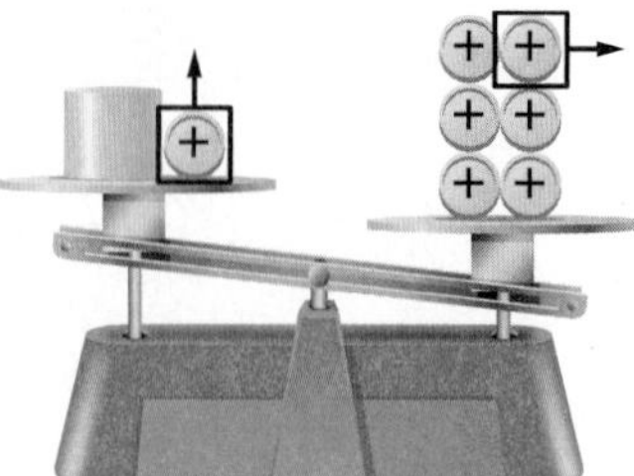

$x + 1 < 6$

You can also solve an inequality algebraically. Since addition and subtraction are inverse operations, you can use addition to solve a subtraction inequality.

ACTIVITY Solve a Subtraction Inequality

2 **Solve the inequality $x - 2 \geq 4$ algebraically.**

$x - 2 \geq 4$	Write the inequality.
$+2 \quad +2$	Add 2 to each side.
$x \geq 6$	Simplify.

So, $x \geq 6$.

Study Tip

Inequalities The inequality $x \geq 6$ means that solutions for x can be greater than 6 or equal to 6.

CHECK Your Progress

Solve each of the following inequalities either algebraically or by using a model.

a. $x + 4 < 7$ b. $x - 2 > 3$ c. $x - 5 \leq 1$

ANALYZE THE RESULTS

1. Write two different inequalities, one involving addition and one involving subtraction, both of whose solutions are $x > 4$.
2. **WRITING IN MATH** Explain how you could solve the inequality $x + 7 \leq 12$ algebraically.

Solving Multiplication Equations

MAIN IDEA

Solve multiplication equations.

New Vocabulary

coefficient

Math Online

glencoe.com

- Extra Examples
- Personal Tutor
- Self-Check Quiz

GET READY for the Lesson

CELL PHONES Max is downloading ringtones on his cell phone. The cost to download each ringtone is \$2. When Max is finished, he has spent a total of \$10.

1. Let x represent the number of ringtones. Explain how the equation $2x = 10$ represents the situation.

The equation $2x = 10$ is a multiplication equation. In $2x$, 2 is the **coefficient** of x because it is the number by which x is multiplied. Multiplication and division are inverse operations. So, to solve a multiplication equation, use division.

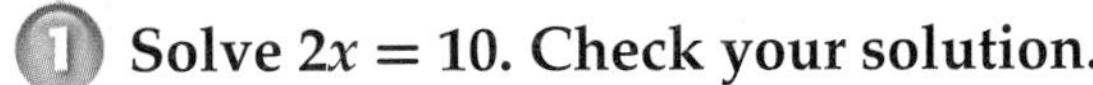

EXAMPLE Solve a Multiplication Equation

1 **Solve $2x = 10$. Check your solution.**

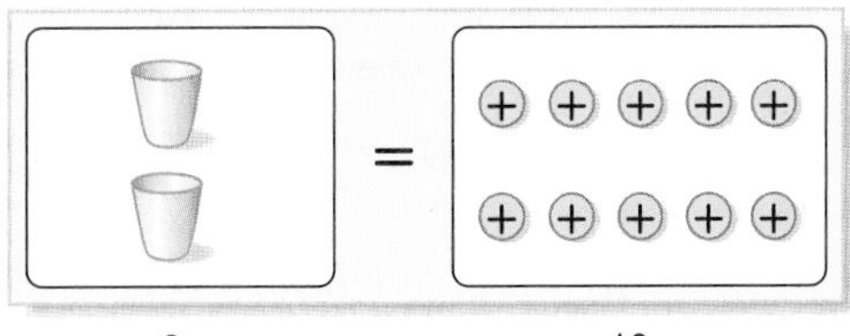

Model the equation.

$$2x = 10$$

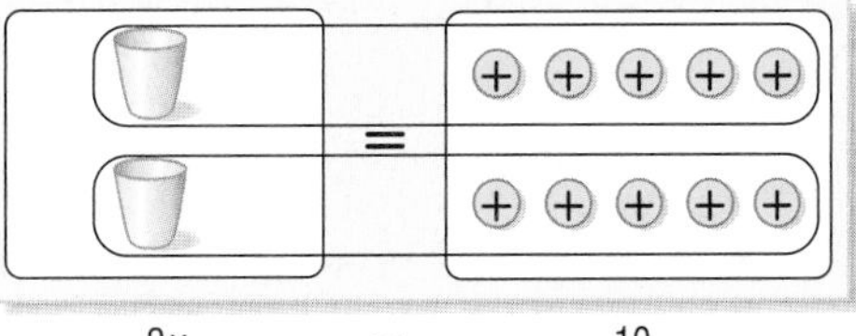

Divide the 10 counters equally into 2 groups. There are 5 in each group.

$$\frac{2x}{2} = \frac{10}{2}$$

$$x = 5$$

Check

$2x = 10$	Write the original equation.
$2(5) \stackrel{?}{=} 10$	Replace x with 5.
$10 = 10$	This sentence is true. ✓

The solution is 5.

CHECK Your Progress

Solve each equation. Use models if necessary.

a. $3x = 15$

b. $8 = 4x$

c. $2x = -14$

EXAMPLE Solve a Multiplication Equation

2 Solve $-3x = 18$.

$-3x = 18$ Write the equation.

$\frac{-3x}{-3} = \frac{18}{-3}$ Divide each side by -3.

$1x = -6$ $-3 \div (-3) = 1$, $18 \div (-3) = -6$

$x = -6$ $1x = x$

The solution is -6. Check this solution.

Study Tip

Division Property of Equality If you divide each side of an equation by the same number, the two sides remain equal.

CHECK Your Progress

Solve each equation. Check your solution.

d. $-2x = 12$

e. $-4t = -16$

f. $24 = -3c$

The equation $d = r \cdot t$ or $d = rt$ shows the relationship between the variables d (distance), r (rate or speed), and t (time).

Real-World EXAMPLE

3 EXERCISE Tyrese jogged 3 miles on a treadmill at a rate of 5 miles per hour. How long did he jog on the treadmill?

Use the formula distance = rate × time.

$d = rt$ Write the equation.

$3 = 5t$ Replace d with 3 and r with 5.

$\frac{3}{5} = \frac{5t}{5}$ Divide each side by 5.

$0.6 = t$ Simplify.

Tyrese jogged on the treadmill for 0.6 hour.

CHECK Your Progress

g. **MUSIC** Ariel has 234 songs downloaded on her MP3 player. She can listen to about 18 songs per hour. Write and solve an equation to find about how many hours of music Ariel has downloaded.

CHECK Your Understanding

Examples 1, 2 (pp. 657–658)

Solve each equation. Use models if necessary.

1. $2a = 6$
2. $20 = 4c$
3. $16 = 8b$
4. $-4d = 12$
5. $-6c = 24$
6. $-3g = -21$

Example 3 (p. 658)

7. **MEASUREMENT** The length of an object in feet is equal to 3 times its length in yards. The length of a waterslide is 48 feet. Write and solve a multiplication equation to find the length of the waterslide in yards.

Practice and Problem Solving

HOMEWORK HELP	
For Exercises	See Examples
8–13	1
14–19	2
20, 21	3

Solve each equation. Use models if necessary.

8. $5d = 30$
9. $4c = 16$
10. $36 = 6e$
11. $21 = 3g$
12. $3f = -12$
13. $4g = -24$
14. $-5a = 15$
15. $-6x = 12$
16. $-5t = -25$
17. $-6n = -36$
18. $-32 = -4s$
19. $-7 = -14x$

20. **JEWELRY** A jewelry store is selling a set of 4 pairs of gemstone earrings for \$58, including tax. Neva and three of her friends bought the gift set so each could have one pair of earrings. Write and solve a multiplication equation to find how much each person should pay.

21. **EXPLORATION** In the winter of 2004, Pen Hadow and Simon Murray walked 680 miles to the South Pole. The trip took 58 days. Write and solve a multiplication equation to find about how many miles they traveled each day.

Real-World Link On their trip to the South Pole, both Pen Hadow and Simon Murray started out pulling a sled with 400 pounds of food and gear.

Source: *Time Almanac*

Solve each equation. Check your solution.

22. $1.5x = 3$
23. $2.5y = 5$
24. $8.1 = 0.9a$
25. $39 = 1\frac{3}{10}b$
26. $\frac{1}{2}e = \frac{1}{4}$
27. $\frac{2}{5}g = -\frac{3}{5}$
28. $-1.2k = -4.8$
29. $-0.3w = -0.12$
30. $-2.6 = -1.3m$

31. **MUSIC** The total time to burn a CD is 18 minutes. Last weekend, Josiah spent 90 minutes burning CDs. Write and solve a multiplication equation to find the number of CDs Josiah burned last weekend. Explain how you can check your solution.

32. **TRAVEL** The Raimonde family drove 1,764 miles across the United States on their vacation. If it took a total of 28 hours, what was their average speed, in miles per hour?

FOOTBALL For Exercises 33 and 34, use the table at the right.

33. George Blanda played in the NFL for 26 years. Write and solve an equation to find how many points he averaged each year.

34. Norm Johnson played in the NFL for 16 years. Write and solve an equation to find how many points he averaged each year.

Top NFL Kickers

Player	Career Points
Gary Anderson	2,434
Morten Andersen	2,437
George Blanda	2,002
John Carney	1,749
Norm Johnson	1,736

Source: *Scholastic Book of World Records*

35. **HEARTBEATS** An average person's heart beats about 103,680 times a day. Write and solve an equation to find about how many times the average person's heart beats in one minute.

36. **BLINKING** On average, a person blinks his or her eyes about 20,000 times a day. About how many times is this per minute?

EXTRA PRACTICE
See pages 705, 717.

H.O.T. Problems

37. **Which One Doesn't Belong?** Identify the equation that does not belong with the other three. Explain your reasoning.

$5x = 20$ | $4b = 7$ | $8w = 32$ | $12y = 48$

38. **CHALLENGE** Explain how you know that the equations $\frac{1}{4} = 2x$ and $\frac{1}{4} \div x = 2$ have the same solution. Then, find the solution.

39. **WRITING IN MATH** Write a problem about a real-world situation that can be represented by the equation $4x = 24$.

TEST PRACTICE

40. **SHORT RESPONSE** Use the formula $A = \ell w$ to find the length in feet of the rectangle shown below.

13 ft | Area = 364 ft^2

41. If Mr. Solomon bikes at a constant speed of 12 miles per hour, which method can be used to find the number of hours it will take him to bike 54 miles?

A Add 12 to 54.

B Subtract 12 from 54.

C Multiply 54 by 12.

D Divide 54 by 12.

Spiral Review

Solve each equation. (Lessons 12-3 and 12-4)

42. $b - 5 = -2$
43. $g - 6 = -7$
44. $p + 3 = -2$
45. $7 + q = -1$

46. **MONEY** Eight people borrowed a total of $56. If each borrowed the same amount, how much did each person borrow? (Lesson 11-6)

GET READY for the Next Lesson

47. **PREREQUISITE SKILL** Micah used 120 minutes of his monthly cell phone minutes the first week. The second week he used 95 and the third week he used 212. If he had 73 minutes left to use, how many minutes did he have for the month? Use the *work backward* strategy. (Lesson 11-5)

Problem Solving in Art

School Days It's time to complete your project. Use your understanding of integers, transformations, properties, and equations to design a school logo or mascot. Prepare a class demonstration and include drawings and coordinate graphs with your project.

Math Online **Unit Project at** glencoe.com

Problem-Solving Investigation

MAIN IDEA: Solve problems by choosing the best method of computation.

P.S.I. TEAM +

e-Mail: CHOOSE THE BEST METHOD OF COMPUTATION

HAO: I collected five insects over the weekend. Their lengths are recorded in the table. About how long was the average length?

Insect	Length (mm)
Beetle	27
Butterfly	41
Cricket	38
Dragonfly	42
Grasshopper	52

YOUR MISSION: Choose the best method of computation to solve the problem.

Understand	Since you don't need an exact answer, estimate.
Plan	Estimate the length of each insect. Add the total estimates and divide by the number of insects, five.
Solve	Beetle 27 → 30 mm Butterfly 41 → 40 mm Cricket 38 → 40 mm Dragonfly 42 → 40 mm Grasshopper 52 → + 50 mm 200 mm Since $200 \div 5 = 40$, the average insect length is 40 millimeters.
Check	Look at the insect lengths. The insect lengths are relatively close to 40 millimeters. So, an average length of 40 millimeters is reasonable.

Analyze The Strategy

1. Explain when you would use estimation as the method of computation.
2. Describe how to mentally find the product of 40 and 3.
3. **WRITING IN MATH** Write a problem in which you would use a calculator as the method of computation. Explain your reasoning.

Mixed Problem Solving

See pages 705, 717.

Choose the best method of computation to solve Exercises 4–6. Explain your reasoning.

4. **GEOGRAPHY** The area of Alaska is 591,004 square miles. The area of Texas is 261,797 square miles. About how many times larger is Alaska than Texas?

5. **MEASUREMENT** The width of Mario's bedroom is $4\frac{7}{12}$ feet shorter than its length. If the length of his bedroom is $13\frac{3}{4}$ feet, what is the width?

6. **SHOPPING** Lily bought 8 gel pens and a notebook folder. If each gel pen cost $0.79 and the notebook folder cost $1.29, about how much did she spend in all?

Use any strategy to solve Exercises 7–15. Some strategies are shown below.

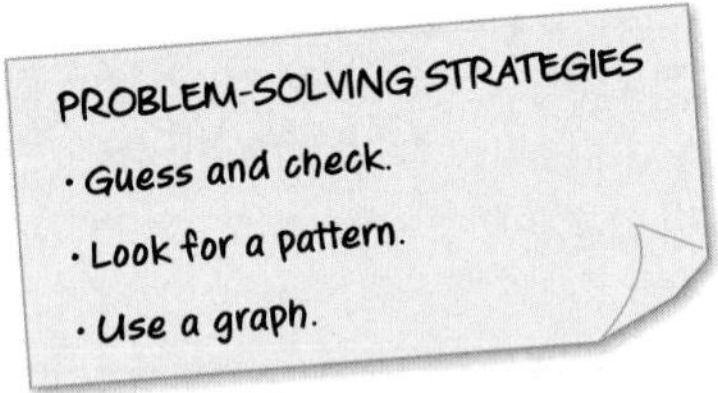

7. **PATTERNS** What are the next two figures in the pattern?

8. **MONEY** Mrs. Perez wants to save an average of $150 per week over six weeks. Use the graph to find how much she must save during the sixth week to meet her goal.

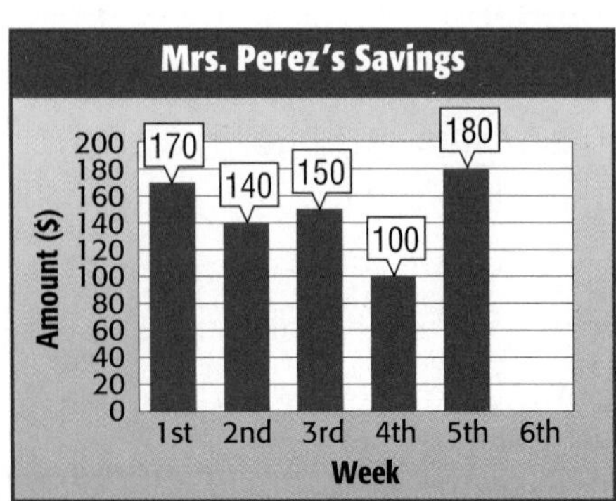

9. **PATTERNS** Refer to the table. If the pattern continues, what will be the output when the input is 4?

Input	Output
0	0
1	1.5
2	3
3	4.5

10. **DVDs** Erica bought three DVDs. The price of each DVD was $12.99. Not including tax, what was the total cost?

11. **SCHOOL** The first class at Harrison Junior High School begins every weekday at 8:07 A.M. The last class ends at 2:56 P.M. About how much total time is spent in class each day?

12. **EGGS** The table below gives the average price of a dozen eggs in a recent year for two states.

State	Price ($)
Nebraska	0.34
Alabama	1.62

Which state was closer to the national average of $0.73?

13. **PUZZLES** In a magic square, each row, column, and diagonal have the same sum. Copy and complete the magic square.

−2	■	■
−3	−1	1
■	■	■

14. **SCHOOL** A multiple choice test has 10 questions. The scoring is +3 for correct answers, −1 for incorrect answers, and 0 for questions not answered. Meg scored 23 points on the test. She did not answer one of the questions. How many did she answer correctly? How many did she answer incorrectly?

15. **GEOMETRY** Find the difference in the areas of the square and rectangle.

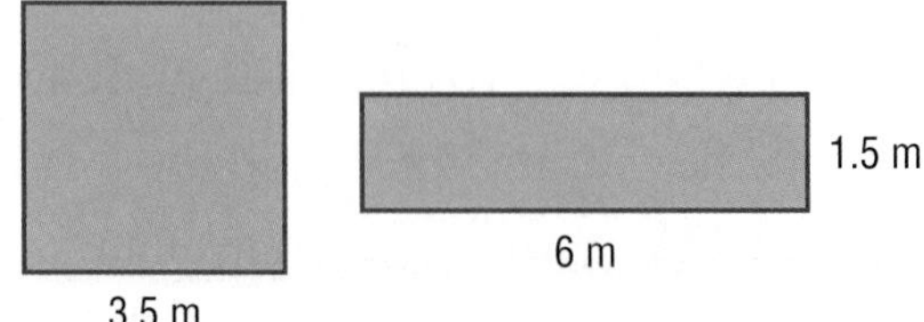

CHAPTER 12

Study Guide and Review

Math Online glencoe.com
• STUDY TO GO
• Vocabulary Review

GET READY to Study

Be sure the following Big Ideas are noted in your Foldable.

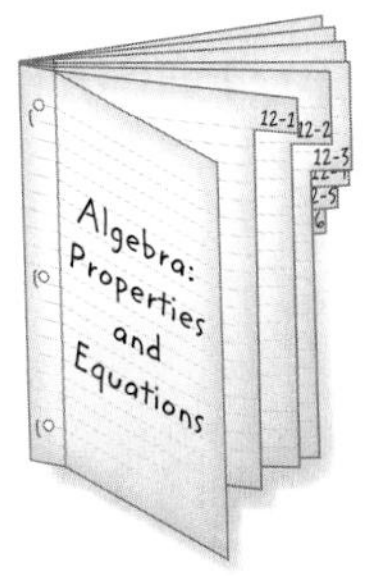

BIG Ideas

Distributive Property (Lessons 12-1 and 12-2)

- To multiply a sum by a number, multiply each addend by the number outside the parentheses.

$3(4 + 5) = 3(4) + 3(5)$

$2(6x + 1) = 2(6x) + 2(1)$

Simplifying Expressions (Lesson 12-2)

- **Commutative Properties**
  The order in which numbers are added or multiplied does not change the sum or product.
- **Associative Properties**
  The way in which numbers are grouped does not change the sum or product.
- To simplify an expression, use models or like terms.

Equations (Lessons 12-3 through 12-5)

- **Subtraction Property of Equality**
  If you subtract the same number from each side of an equation, the two sides remain equal.
- **Addition Property of Equality**
  If you add the same number to each side of an equation, the two sides remain equal.
- **Division Property of Equality**
  If you divide each side of an equation by the same number, the two sides remain equal.

Key Vocabulary

Addition Property of Equality (p. 652)
Associative Properties (p. 636)
coefficient (p. 657)
Commutative Properties (p. 636)
Distributive Property (p. 632)
equivalent expressions (p. 636)
inverse operations (p. 644)
like terms (p. 637)
Subtraction Property of Equality (p. 645)

Vocabulary Check

State whether each sentence is *true* or *false*. If *false*, replace the underlined word or number to make a true sentence.

1. The numerical factor of a term that contains a variable is called a <u>quadrant</u>.
2. The <u>Distributive Property</u> can be used to rewrite the expression $3(2x + 5)$ as $6x + 15$.
3. <u>Inverse operations</u> undo each other.
4. The expression $6x + 3 + 2x$, when simplified, becomes <u>$11x$</u>.
5. The like terms of the expression $5x + 8 + 4x$ are $5x$ and <u>8</u>.
6. Addition and <u>multiplication</u> are inverse operations.
7. The expressions $2(x + 7)$ and $2x + 14$ are <u>equivalent</u>.
8. The solution to the equation $x + 8 = 11$ is $x =$ <u>19</u>.

Lesson-by-Lesson Review

12-1 The Distributive Property (pp. 632–635)

Find each product mentally. Show the steps you used.

9. 7×39
10. 9×77
11. 6×5.4
12. 8×4.6
13. **CAR SHOW** Admission to a local car show costs \$9.50. Lunch at the car show's Snack Shop costs \$5.50. Find the total cost for four admissions to the car show and four lunches at the Snack Shop.

Use the Distributive Property to rewrite each expression.

14. $4(x + 3)$
15. $8(x + 1)$
16. $3(3x - 6)$
17. $7(x - 5)$

Example 1 **Find 5 × 6.4 mentally.**

$5 \times 6.4 = 5(6 + 0.4)$ Write 6.4 as 6 + 0.4.
$= 5(6) + 5(0.4)$ Distributive Property
$= 30 + 2$ Multiply.
$= 32$ Add.

Example 2 **Use the Distributive Property to rewrite the expression $6(x + 7)$.**

$6(x + 7) = 6(x) + 6(7)$ Distributive Property
$= 6x + 42$ Multiply.

12-2 Simplifying Algebraic Expressions (pp. 636–641)

Simplify each expression below. Justify each step.

18. $2 + (x + 3)$
19. $(17 + x) + 12$
20. $7 \cdot (x \cdot 4)$
21. $(5 \cdot x) \cdot 8$
22. $4(16x)$
23. $15(3x)$

Simplify each expression below using models or a drawing.

24. $3x + 3x$
25. $x + 2x$
26. $2x + 2 + 3x$
27. $x + 6 + 4x$
28. **BUFFET** Five friends ate at a buffet restaurant. The cost of each buffet was \$$x$. Three of the friends also bought a beverage for \$2 each. Write and simplify an expression for the total cost.

Example 3 **Simplify the expression $(9 + x) + 13$.**

$(9 + x) + 13 = (x + 9) + 13$ Distributive Property
$= x + (9 + 13)$ Associative Property
$= x + 22$ Add 9 and 13.

Example 4 **Simplify the expression $2x + x$.**

Model $2x$ with two x-tiles and x with one x-tile.

There are a total of three x-tiles. So, $2x + x = 3x$.

Mixed Problem Solving
For mixed problem-solving practice, see page 717.

12-3 Solving Addition Equations (pp. 644–648)

Solve each equation. Use models if necessary.

29. $c + 8 = 11$
30. $x + 15 = 14$
31. $54 = m + 9$
32. $-5 = 2 + w$
33. $g + 13 = -25$
34. $17 + d = -2$
35. $23 = h + 11$
36. $19 + r = 11$

37. **HEIGHT** When Marco stands on a box, he is 10 feet tall. If the box is 4 feet tall, write and solve an addition equation to find Marco's height.

38. **SCUBA DIVING** A scuba diver was 13 meters below the ocean's surface. She ascended m meters to a depth 8 meters below the ocean's surface. Write and solve an addition equation to find m.

Example 5 **Solve $x + 8 = 10$.**

$x + 8 = 10$
$-8 = -8$ Subtract 8 from each side.
$x = 2$ Simplify.

Example 6 **On the last hole, Tim's golf score was 3 points higher than his score on the first hole. Write and solve an equation to find Tim's score on the first hole if his score on the last hole was 5.**

Let y represent Tim's score on the first hole.

$y + 3 = 5$ Write the equation.
$-3 = -3$ Subtract 3 from each side.
$y = 2$ Simplify.

So, Tim's score on the first hole was 2.

12-4 Solving Subtraction Equations (pp. 651–654)

Solve each equation. Use models if necessary.

39. $z - 7 = 11$
40. $s - 9 = -12$
41. $-4 = y - 9$
42. $-6 = g - 4$
43. $14 = m - 5$
44. $h - 2 = -9$
45. $p - 22 = -7$
46. $d - 3 = -14$

47. **ALGEBRA** Find the value of k if $k - 5 = -4$.

48. **MONEY** Julia has $39. She has $8 less than her brother. Write and solve a subtraction equation to find how much money her brother has.

Example 7 **Solve $a - 5 = -3$.**

$a - 5 = -3$
$+5 = +5$ Add 5 to each side.
$a = 2$ Simplify.

Example 8 **Solve $4 = m - 8$.**

$4 = m - 8$
$+8 = +8$ Add 8 to each side.
$12 = m$ Simplify.

Study Guide and Review

12-5 Solving Multiplication Equations (pp. 657–660)

Solve each equation. Use models if necessary.

49. $4b = 32$
50. $-3m = 21$
51. $60 = 5y$
52. $28 = -2d$
53. $-18 = -6c$
54. $-6p = -24$
55. $-4x = 12$
56. $7a = -35$

57. **CDs** A store is selling blank CDs in packages of 25 for $5. Write and solve a multiplication equation to find the cost of one blank CD.

Example 9 **Solve $-6y = 24$.**

$-6y = 24$ Write the equation.

$\frac{-6y}{-6} = \frac{24}{-6}$ Divide each side by -6.

$y = -4$ Simplify.

12-6 PSI: Choose the Best Method of Computation (pp. 661–662)

Solve. Choose the best method of computation.

58. **BIRTHDAYS** Alita, Alisa, and Alano are sharing the cost of their mother's birthday gift, which is $50. About how much will each girl need to contribute?

59. **TREES** The largest known living tree, the General Sherman sequoia in California, weighs 6,167 tons. One ton equals 2,000 pounds. What is the weight of the tree in pounds?

60. **ANIMALS** Gina's rabbit, Lucky, weighs 9.6 pounds. Troy's hamster, Ben, weighs about a fifth of Lucky's weight. About how much does Ben weigh?

Example 10 **Kent ran 4.2 miles on Saturday. If he ran each mile in 8 minutes, how long did it take him to run 4.2 miles?**

Use paper and pencil because an exact answer is needed. The calculations have fairly small numbers.

$$\begin{array}{r} {}^{1}\\ 4.2 \\ \times\ 8 \\ \hline 33.6 \end{array}$$

4.2 ← one decimal point

33.6 ← one decimal point

So, it took Kent 33.6 minutes to run 4.2 miles.

CHAPTER 12 Practice Test

Math Online glencoe.com
• Chapter Test

Find each product mentally. Show the steps you used.

1. 5×37
2. 9×4.2
3. 6×1.6
4. 8×43

5. **FIELD TRIPS** The 30 students in Megan's social studies class are planning to visit Washington, D.C. The hotel costs \$115 per student and the transportation costs \$45 per student. Use the Distributive Property to find the total hotel and transportation costs for the 30 students.

Use the Distributive Property to rewrite each algebraic expression.

6. $3(x + 4)$
7. $12(x - 1)$
8. $8(x + 6)$
9. $5(x - 2)$
10. $4(x - 7)$
11. $2(x + 3)$

Simplify each expression. Justify each step.

12. $(x + 6) + 3$
13. $2 + (x + 8)$
14. $7 \cdot (x \cdot 5)$
15. $9(4x)$

16. **MULTIPLE CHOICE** The price of each ticket to a football game is \$21. The expression $21x$ represents the total cost, in dollars, of x tickets. Which of the following expressions is *not* equivalent to $21x$?

    A $7(x + 3)$

    B $7x + 14x$

    C $15x + 3x + 3x$

    D $(3 \cdot 7) \cdot x$

Simplify each expression using models or a drawing.

17. $x + 2x$
18. $4x + 2x$
19. $3x + 1 + x$
20. $2x + 3 + 4x$

Solve each equation. Use models if necessary. Check your solution.

21. $w + 8 = 5$
22. $-8 = d - 9$
23. $15 = 3n$
24. $-3 + x = 11$
25. $z - 3 = -6$
26. $-5m = -30$
27. $y + 7 = -1$
28. $48 = -12g$
29. $-1 = h + 2$
30. $6b = -18$

31. **SCIENCE** For a class project, Reynaldo and Beth are raising tadpoles. They have 12 tadpoles and some frogs. They have a total of 20 tadpoles and frogs. Write and solve an equation to find how many frogs they have.

32. **MULTIPLE CHOICE** Reshma drove 385 miles to visit her grandmother. If she drove at a speed of 55 miles per hour, which equation represents the time t that it took Reshma to drive 385 miles?

    F $t + 55 = 385$

    G $t - 55 = 385$

    H $385 = 55t$

    J $385t = 55$

33. **MEASUREMENT** Cody's fish tank measures 19.75 inches long, 10.25 inches wide, and 12.8 inches high. What is the approximate volume of the fish tank? Choose the best method of computation.

Chapter 12 Test Practice

Cumulative, Chapters 1–12

Math Online glencoe.com
• Test Practice

PART 1 Multiple Choice

Read each question. Then fill in the correct answer on the answer document provided by your teacher or on a sheet of paper.

1. A garden in the shape of a regular polygon is made of congruent equilateral triangles. The table shows the relationship between the area of the triangle and the area of the garden it is part of. Which expression can be used to find the area of a similar garden made of triangles with an area of n square units each?

Area of Triangle (square units)	Area of Garden (square units)
7	49
8	56
9	63
10	70
n	■

A $1n$
C $7n$
B $n + 7$
D $n + 42$

2. The drawing shows two circles that share a common center point. Which expression can be used to find the circumference of the outer circle in inches?

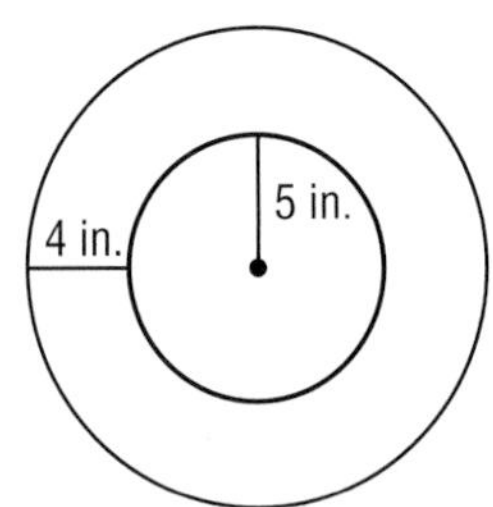

F $\pi(5 + 4)$
H $2\pi(5 + 4)$
G $2(5 + 4)$
J $\frac{1}{2}(5 + 4\pi)$

TEST-TAKING TIP

Question 2 If you are not permitted to write in your test booklet, copy the figure onto paper.

3. Each figure below is divided into sections of equal size. Which figure has 87.5% of its total area shaded?

A
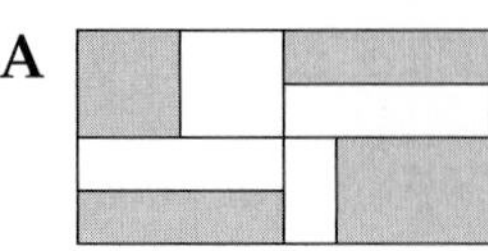

B
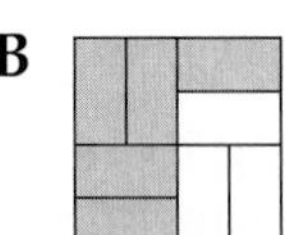

C
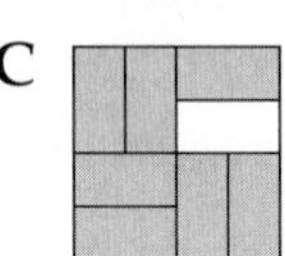

D
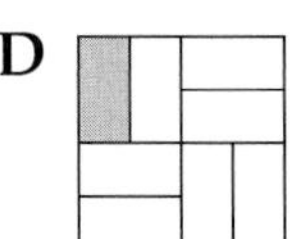

4. Which of the following equations illustrate the Distributive Property?

F $3(2x + 4) = 5x + 4$
G $3(2x + 4) = 5x + 7$
H $3(2x + 4) = 6x + 4$
J $3(2x + 4) = 6x + 12$

5. Find the area of the rectangle.

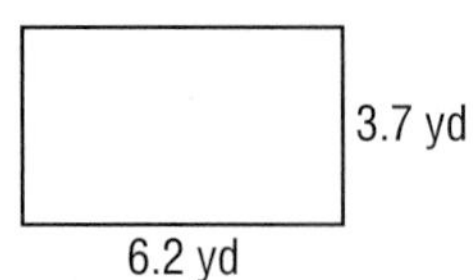

A 6.5 yd^2
C 19.8 yd^2
B 9.4 yd^2
D 22.94 yd^2

6. Which of the following expressions is equivalent to $6x + 18$?

F $6(x + 18)$
H $(3 \cdot x) \cdot 2$
G $4x + 18 + 2x$
J $6x + 18x$

Preparing for Standardized Tests
For test-taking strategies and practice, see pages 718–735.

7. Triangle XYZ is reflected across the y-axis. What are the coordinates of the image after the reflection?

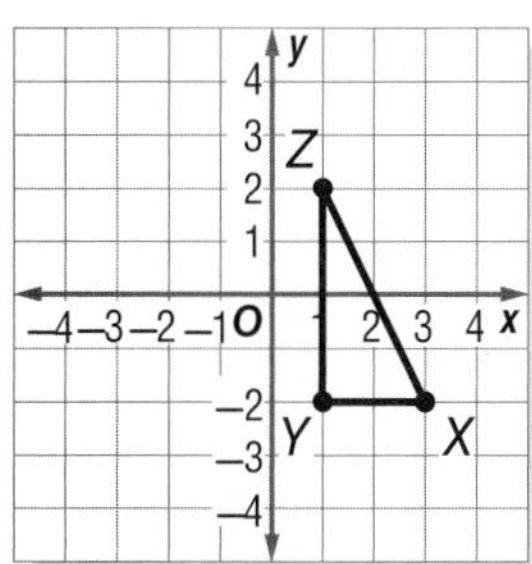

A $X'(3, -2), Y'(1, -2), Z'(1, 2)$

B $X'(-3, 2), Y'(-1, 2), Z'(-1, -2)$

C $X'(-3, -2), Y'(-1, -2), Z'(-1, 2)$

D $X'(3, 2), Y'(1, 2), Z'(1, -2)$

8. A student in Rob's math class chose a letter at random from the letter cards shown below. What is the probability that the letter chosen was a vowel?

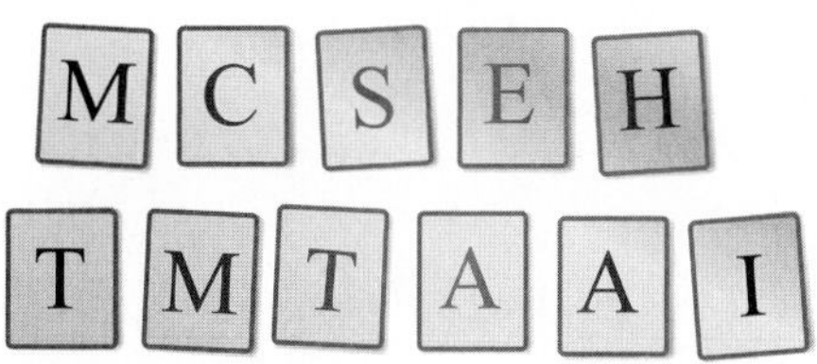

F $\frac{1}{3}$ H $\frac{2}{9}$

G $\frac{4}{11}$ J $\frac{1}{2}$

9. Find the greatest common factor of 12, 32, and 36.

A 2 C 6

B 4 D 8

PART 2 Short Response/Grid In

Record your answers on the answers on the answer sheet provided by your teacher or on a sheet of paper.

10. Marta drew a triangle on the sidewalk for a game. Find the perimeter of the triangle in inches.

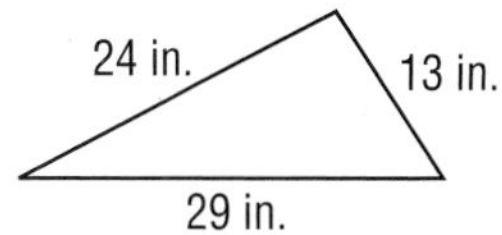

11. A movie theater has 26 rows of seats with an equal number of seats in each row. The theater can seat a total of 390 people. Solve the equation $26x = 390$ to find the number of seats, x, that are in each row.

PART 3 Extended Response

Record your answers on the answer sheet provided by your teacher or on a sheet of paper. Show your work.

12. A submarine, 22 meters below the surface of the ocean, ascended m meters. The new depth of the submarine was -18 meters in relation to the surface of the ocean.
   a. Write an equation that can be used to solve for m.
   b. Solve this equation for m.
   c. A second submarine is at a depth of p meters in relation to the surface of the ocean. It descended 9 meters to a depth of -15 meters. Write an equation that can be used to solve for p.
   d. Solve this equation for p.

NEED EXTRA HELP?												
If You Missed Question...	1	2	3	4	5	6	7	8	9	10	11	12
Go to Lesson...	6-6	10-2	7-1	12-1	11-2	12-2	11-9	7-4	4-1	10-1	12-5	12-3

Looking Ahead

to Next Year

Let's Look Ahead

Negative Rational Numbers

MAIN IDEA

Model, compare, and order negative fractions and decimals, and determine the absolute value of rational numbers.

New Vocabulary

rational number
absolute value

Math Online

glencoe.com

- Extra Examples
- Personal Tutor
- Self-Check Quiz

GET READY for the Lesson

You know that a number line can be used to compare integers. You can also use a number line to compare positive and negative fractions. The number line below shows that $-\frac{5}{8} < \frac{1}{8}$.

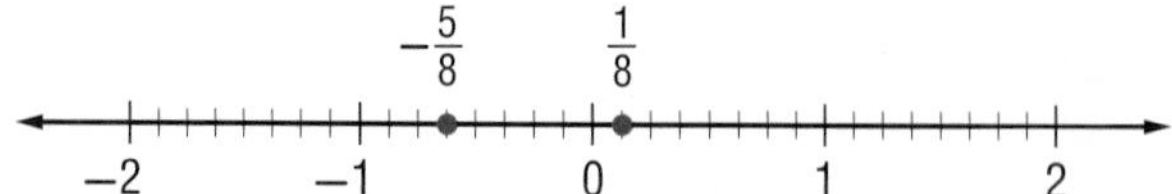

Graph each pair of fractions on a number line. Then determine which fraction is less.

1. $-\frac{7}{8}, -\frac{3}{8}$
2. $-\frac{1}{2}, -\frac{3}{4}$
3. $1\frac{1}{4}, -1\frac{1}{4}$
4. $-\frac{5}{8}, -\frac{3}{8}$

A **rational number** is a number that can be expressed as a fraction. Fractions, terminating and repeating decimals, percents, and integers are all rational numbers.

You can use models to graph negative fractions on a number line.

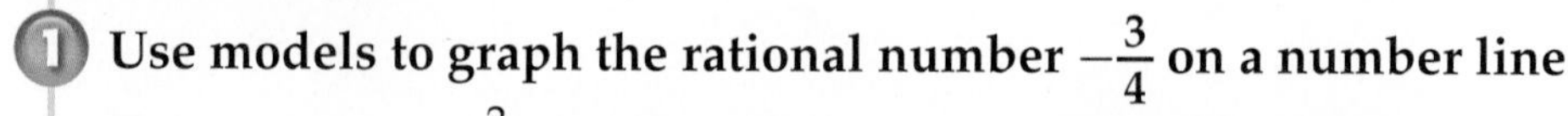

EXAMPLE Model Negative Rational Numbers

1 Use models to graph the rational number $-\frac{3}{4}$ on a number line.

Step 1 Model $-\frac{3}{4}$ using a fraction strip. Draw a number line from −1 to 0. Place the fraction strip above the number line.

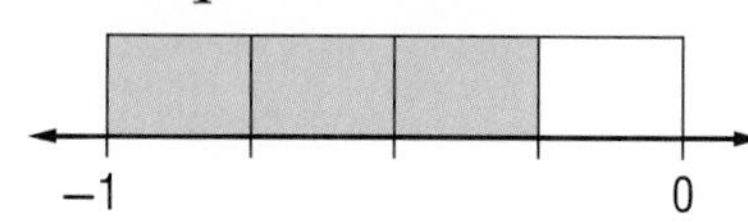

Step 2 Each section in the fraction strip represents one fourth. Label the number line with $-\frac{1}{4}$, $-\frac{2}{4}$, and $-\frac{3}{4}$.

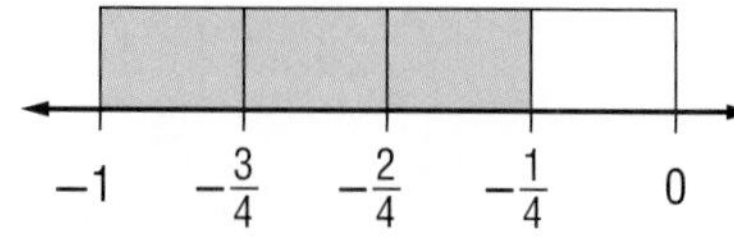

Step 3 Remove the fraction strip. Draw a dot at $-\frac{3}{4}$.

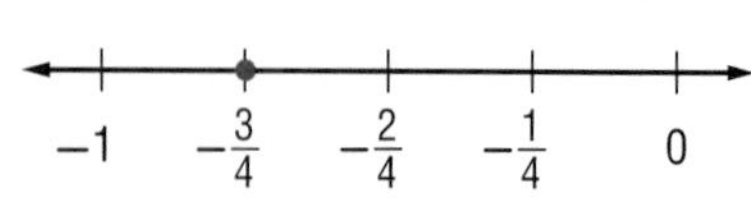

CHECK Your Progress

a. Graph −0.4 on a number line. Use models if necessary.

You can use number lines to compare two negative rational numbers.

EXAMPLE Compare Negative Rational Numbers

2 Compare −0.25 and −0.7 using <, >, or =.

METHOD 1

Graph each rational number on a number line.

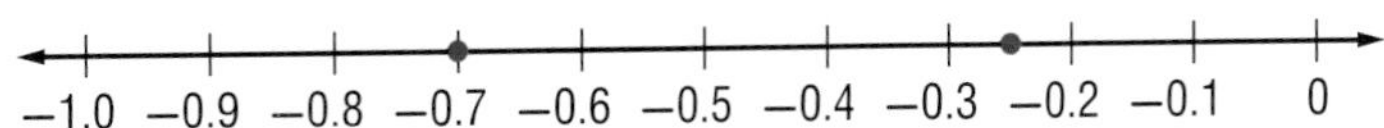

Since −0.7 is to the left of −0.25, −0.7 < −0.25.

METHOD 2

You can also use place value to compare the digits in the tenths place, since the digit 0 in the ones place is the same. Since −7 < −2, −0.7 < −0.25.

Study Tip

Comparing Negative Numbers A common error is to say that because 7 > 2, −0.25 > −0.7. Both numbers are negative. Include the sign of each number when comparing by place value.

CHOOSE Your Method

Compare each pair of rational numbers.

b. $-\frac{4}{7}, -\frac{2}{7}$

c. −1.4, −1.8

You can also use number lines to order negative rational numbers.

Real-World EXAMPLE Order Negative Rational Numbers

3 OCEAN DEPTH Four different sea creatures were observed at the following depths: −4.8 meters, −4.2 meters, −4.7 meters, and −4.5 meters. Order these depths from least to greatest.

To order the depths, you need to first compare the depths. Use place value or graph each decimal on a number line.

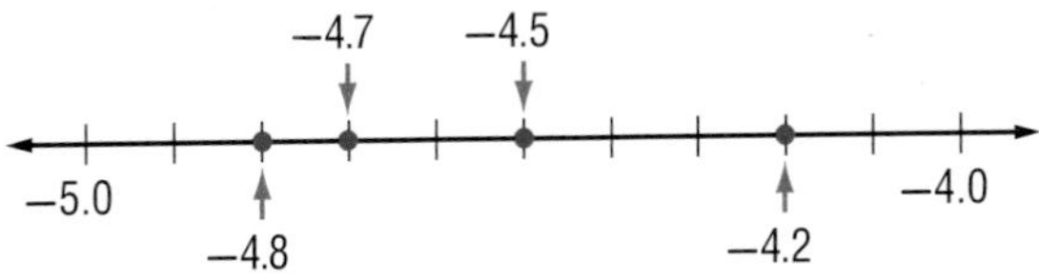

Since −4.8 < −4.7 < −4.5 < −4.2, the order from least to greatest is −4.8, −4.7, −4.5, −4.2.

CHECK Your Progress

d. **TEMPERATURE** Juanita recorded the outside temperature over a four-hour period. The results were −17.8°F, −22.5°F, −12.9°F, and −16.4°F. Order these temperatures from least to greatest.

Every negative rational number is a certain distance away from zero on the number line.

The number line below shows that −2.5 and 2.5 are each 2.5 units from 0, even though they are on opposite sides of 0.

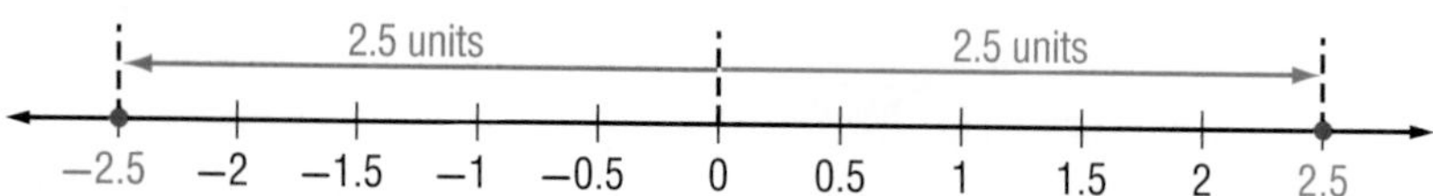

The **absolute value** of a number is the distance between that number and zero on a number line. So, −2.5 and 2.5 have the same absolute value, 2.5.

> **Study Tip**
> **Absolute Value** It is not always true that the absolute value of a number is the opposite of the number. Remember that absolute value is always positive or zero.

The symbol for the absolute value of a number includes two vertical bars on each side of the number.

$|\mathbf{2.5}| = \mathbf{2.5}$ The absolute value of 2.5 is 2.5.

$|\mathbf{-2.5}| = \mathbf{2.5}$ The absolute value of −2.5 is 2.5.

Absolute Value — Key Concept

Words The absolute value of a number is the distance the number is from zero on the number line. The absolute value of a number is always greater than or equal to zero.

Examples $|5.25| = 5.25$ $\left|-\frac{2}{3}\right| = \frac{2}{3}$ $|x| = x$

You can use number lines to evaluate expressions involving absolute value.

EXAMPLE Evaluate Expressions with Absolute Value

4 **Evaluate $\left|-\frac{1}{2}\right|$.**

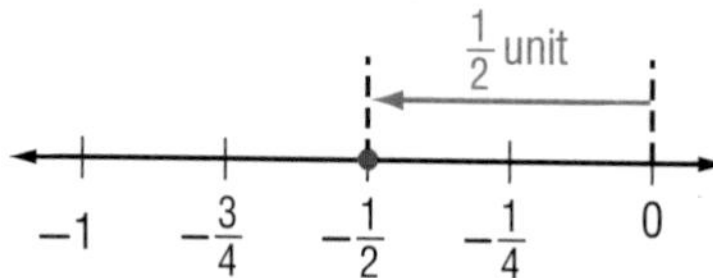

Draw $-\frac{1}{2}$ on the number line.

So, $\left|-\frac{1}{2}\right| = \frac{1}{2}$. The number $-\frac{1}{2}$ is $\frac{1}{2}$ unit from 0.

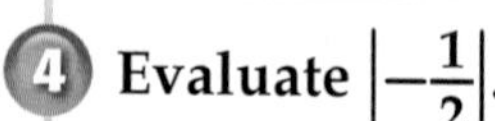

CHECK Your Progress

Evaluate each expression below.

e. $|-15.7|$

f. $\left|\frac{5}{13}\right|$

g. $|4.65|$

h. $\left|-\frac{2}{5}\right|$

CHECK Your Understanding

Example 1 (p. LA2)

Graph each rational number on a number line. Use models if necessary.

1. $-\frac{3}{8}$
2. -5.75
3. $-\frac{1}{6}$

Example 2 (p. LA2)

Replace each ● with <, >, or = to make a true sentence.

4. $-\frac{1}{2}$ ● $-\frac{3}{4}$
5. -0.25 ● $-\frac{1}{4}$
6. -8.04 ● -8.40

Example 3 (p. LA3)

7. **SNACKS** A package of granola bars is advertised to contain 3.5 grams of protein per serving. Four granola bars were tested. The changes from the advertised number of grams of protein per serving to the number of grams of protein per serving were −0.2, −0.15, −0.25, and −0.05. Order these numbers from least to greatest.

Example 4 (p. LA4)

Evaluate each expression.

8. $\left|-\frac{4}{9}\right|$
9. $|-12.4|$
10. $\left|\frac{3}{10}\right|$

Practice and Problem Solving

For Exercises	See Examples
11–16	1
17–22	2
23, 24	3
25–30	4

Graph each rational number on a number line. Use models if necessary.

11. $-\frac{4}{5}$
12. $-\frac{1}{4}$
13. -0.1
14. -4.5
15. $-\frac{2}{3}$
16. -8.75

Replace each ● with <, >, or = to make a true sentence.

17. -3.5 ● $-3\frac{1}{2}$
18. -0.16 ● 0.16
19. $-\frac{7}{12}$ ● $-\frac{5}{12}$
20. $\frac{1}{6}$ ● $-\frac{1}{6}$
21. -18.28 ● -18.82
22. $-2\frac{3}{8}$ ● $-\frac{3}{8}$

23. **STOCKS** The *net change* for a certain stock is the dollar value change in the stock's closing price from the previous day's closing price. The net changes for four different stocks on Tuesday were −1.90, −0.09, −1.09, and −1.91. Order these numbers from least to greatest.

24. **DIVING** The depths of each scuba diver in a scuba diving class are $-12\frac{1}{4}$ feet, $-12\frac{1}{8}$ feet, $-12\frac{3}{4}$ feet, and $-12\frac{5}{8}$ feet. Order the depths from highest depth to lowest depth.

Evaluate each expression.

25. $|-12.11|$
26. $\left|-7\frac{1}{6}\right|$
27. $|0.8|$
28. $|6.95|$
29. $\left|-5\frac{4}{9}\right|$
30. $\left|\frac{11}{12}\right|$

31. **SCIENCE** *Absolute zero* is defined as the temperature at which all motion stops because it is so cold. It corresponds to $-459.67°F$. Find $|-459.67|$ and explain what it means in the context of this problem.

ALGEBRA Evaluate each expression if $a = 3$, $b = -5$, and $c = -2$.

32. $|a| + |b|$
33. $|b| + |c|$
34. $|a + b|$
35. $|b + c|$
36. $|a| - |c|$
37. $|a - c|$

POPULATION For Exercises 38 and 39, use the table that shows the percent of change, from 2000 to 2004, in the populations of four large U.S. counties.

Population Change (2000–2004)	
County	**Percent of Change**
Cook County, IL	−0.9
Wayne County, MI	−2.2
Philadelphia County, PA	−3.1
Middlesex County, MA	−0.1

Source: *The World Almanac*

38. Order the numbers from least to greatest.

39. Order the counties, from least to greatest, according to the amount of decrease in their populations. Explain why this order is different from the order in Exercise 38.

Replace each ● with <, >, or = to make a true sentence.

40. $|-14.8|$ ● $|14|$
41. $\left|-\frac{2}{9}\right|$ ● $\left|-\frac{4}{9}\right|$
42. $|0.7|$ ● $\left|-\frac{7}{10}\right|$

SPACE For Exercises 43 and 44, use the following information.

The average temperature of Pluto is $-369°F$, and the average temperature of Neptune is $-330°F$.

43. Suppose the high temperatures for Pluto on four consecutive days are $-369.28°F$, $-368.79°F$, $-369.17°F$, and $-367.99°F$. Order these temperatures from highest to lowest.

44. Use Neptune's average temperature to write four daily temperatures to the hundredths place. Then write two different inequalities comparing these temperatures.

H.O.T. Problems

45. **CHALLENGE** Determine whether $|x| = |-x|$ is *always*, *sometimes*, or *never* true. Explain your reasoning.

46. **REASONING** Explain how you know whether a number is a rational number.

47. **NUMBER SENSE** Find all possible values of m that make the equation $|m| = 7.2$ true.

48. **WRITING IN MATH** Write a word problem about a real-world situation in which you would need to compare rational numbers. Then solve the problem.

Two-Step Equations

MAIN IDEA

Translate two-step verbal sentences into algebraic equations and solve using inverse operations.

New Vocabulary

two-step equation

Math Online

glencoe.com

- Extra Examples
- Personal Tutor
- Self-Check Quiz

GET READY for the Lesson

BOOKS Suppose you order two identical paperback books online for a total price of \$11, including a total shipping charge of \$3.

1. Let x = the cost of one book. How does the equation $2x + 3 = 11$ represent the situation?
2. How many arithmetic operations are on the left side of the equation? State them.

Equations like $2x + 3 = 11$ that have two different operations are called **two-step equations**. To solve a two-step equation, you can use models or work backward using the order of operations.

EXAMPLE Use Models to Solve a Two-Step Equation

1 **Solve $2x + 3 = 11$.**

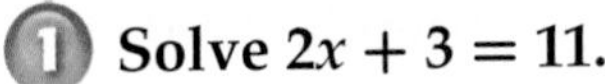

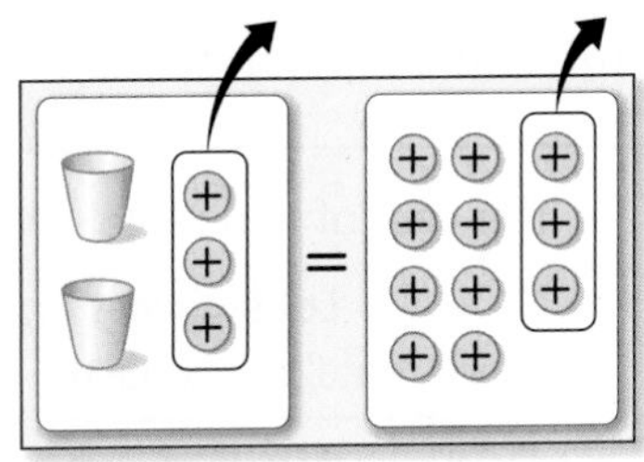

$2x + 3 - 3 = 11 - 3$

Model the equation. Then remove 3 counters from each side to get the variable by itself.

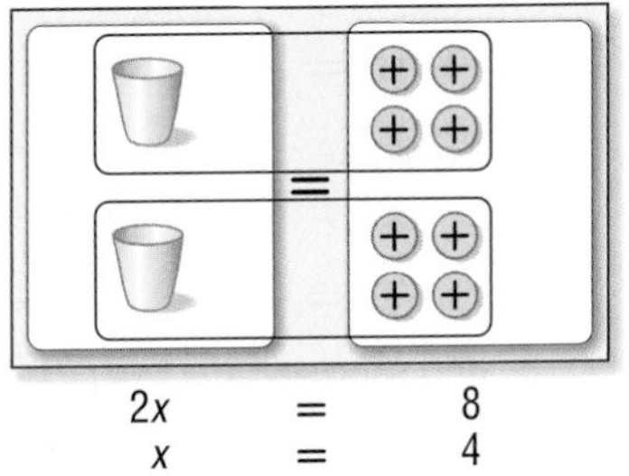

$2x = 8$
$x = 4$

Divide the 8 counters equally into 2 groups. There are 4 in each group.

The solution is 4.

CHECK Your Progress

Solve each equation. Use models if necessary.

a. $3a + 2 = 14$

b. $10 = 4c - 2$

EXAMPLE Solve a Two-Step Equation Algebraically

2 **Solve $3x - 2 = 7$. Check your solution.**

$$\begin{aligned} 3x - 2 &= 7 && \text{Write the equation.} \\ +2 &= +2 && \text{Add 2 to each side.} \\ 3x &= 9 && \text{Simplify.} \\ \frac{3x}{3} &= \frac{9}{3} && \text{Divide each side by 3.} \\ x &= 3 && \text{Simplify.} \end{aligned}$$

Check

$$\begin{aligned} 3x - 2 &= 7 && \text{Write the original equation.} \\ 3(3) - 2 &= 7 && \text{Replace } x \text{ with 3.} \\ 9 - 2 &= 7 && \text{Multiply.} \\ 7 &= 7 && \text{Simplify. The solution is correct.} \end{aligned}$$

The solution is 3.

CHECK Your Progress

Solve each equation algebraically. Check your solution.

c. $4c - 3 = 5$

d. $1 = 3a + 4$

Real-World EXAMPLE Write and Solve an Equation

3 **SNOWBOARDING** **Three friends went snowboarding. The admission price was $5 each. Two of the friends rented boards. If they spent a total of $19, what was the cost of one board rental?**

Words	The cost of two rentals and three admissions is $19.		
Variable	Let s = cost for snowboard rental.		
	Two rentals at \$$s$ each	plus admission	equals $19.
Equation	$2s$	$+ 3(5)$	$= 19$

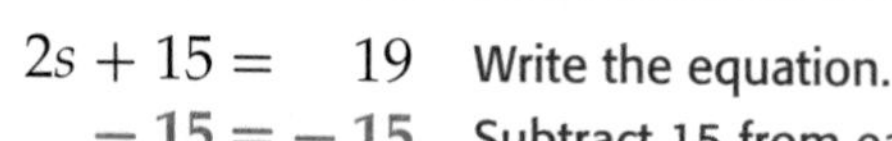

$$\begin{aligned} 2s + 15 &= 19 && \text{Write the equation.} \\ -15 &= -15 && \text{Subtract 15 from each side.} \\ 2s &= 4 && \text{Simplify.} \\ \frac{2s}{2} &= \frac{4}{2} && \text{Divide each side by 2.} \\ s &= 2 && \text{Simplify.} \end{aligned}$$

So, snowboard rental is $2. Check your solution.

Real-World Link In a recent year, snowboarding was the fastest growing sport in the United States, with over 7.2 million participants.

CHECK Your Progress

Write an equation to represent each situation. Then solve.

e. Ten is two less than four times a number. What is the number?

f. Three times a number n plus 8 is 44. What is the value of n?

Check Your Understanding

Example 1 (p. LA7)

Solve each equation. Use models if necessary.

1. $2a + 5 = 13$
2. $2b - 6 = 12$
3. $4t + 4 = 8$

Example 2 (p. LA8)

Solve each equation. Check your solution.

4. $4r - 1 = 11$
5. $5c - 3 = 17$
6. $3 = 4h - 5$

Example 3 (p. LA8)

ZOOS For Exercises 7 and 8, use the information below.

Jenna visited the zoo with both of her parents. They paid a total of $24 and Jenna's admission cost $6.

7. Let p = the cost of admission for each parent. Write an equation to represent the total cost of admission for Jenna and her parents.
8. What was the cost of admission for each of Jenna's parents?

Practice and Problem Solving

HOMEWORK HELP	
For Exercises	See Examples
9–17	1, 2
18–21	3

Solve each equation. Use models if necessary. Check your solution.

9. $3a + 4 = 7$
10. $10 = 2r - 8$
11. $6s + 3 = 15$
12. $12 = 4 + 2z$
13. $4h - 10 = 6$
14. $5 + 2r = 23$
15. $13 = 6 + 7c$
16. $8 = 2n - 4$
17. $3e - 8 = 7$

18. **GEOMETRY** The perimeter of a rectangle is 48 inches. Find its length if its width is 5 inches.

19. **FUNDRAISING** If expenses for a fundraiser are $210 and tickets cost $5 each, how many tickets must be sold to raise $500 more than the expenses?

For Exercises 20 and 21, write an equation. Then solve the equation.

20. **TENNIS** While on vacation, Daniella played tennis. Racket rental was $7, and court time cost $27 per hour. If the total cost was $88, how many hours did Daniella play?

21. **GROCERIES** Carter bought 6.75 pounds of fruit and four potatoes that each weighed the same amount. If he bought a total of 7.75 pounds, how much did each potato weigh?

H.O.T. Problems

22. **CHALLENGE** Use what you know about the Distributive Property and solving two-step equations to solve the equation $2(n - 9) = -4$.

23. **OPEN ENDED** Write a two-step equation using multiplication and addition. Solve your equation.

24. **WRITING IN MATH** Tell which operation to undo first in the equation $19 = 4 + 5x$. Explain.

Angle and Line Relationships

MAIN IDEA

Identify the relationships of angles formed by two parallel lines cut by a transversal, and identify the relationships of vertical, adjacent, complementary, and supplementary angles.

New Vocabulary

parallel lines
transversal
interior angles
exterior angles
alternate interior angles
alternate exterior angles
corresponding angles
adjacent angles

Math Online

glencoe.com

- Extra Examples
- Personal Tutor
- Self-Check Quiz

MINI Lab

Darken two of the horizontal lines on a sheet of notebook paper. Then draw another line that intersects both lines. Label the angles as shown.

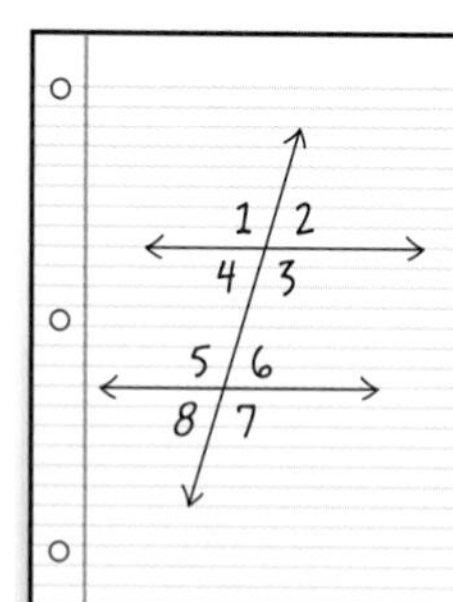

1. Find the measure of each angle.
2. What do you notice about the measures of the angles?
3. Which angles have the same measure?
4. What do you notice about the measures of the angles that share a side?

In geometry, two lines in a plane that never intersect and are the same distance apart are **parallel lines**.

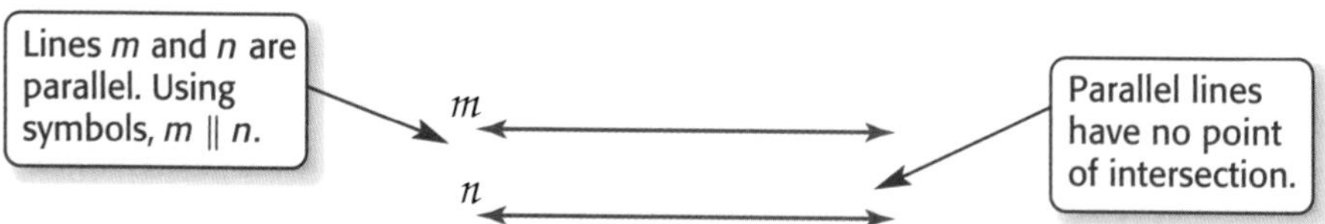

When two parallel lines are intersected by a third line called a **transversal**, eight angles are formed.

Names of Special Angles — Key Concept

- **Interior angles** lie inside the parallel lines.
  $\angle 3$, $\angle 4$, $\angle 5$, $\angle 6$
- **Exterior angles** lie outside the parallel lines.
  $\angle 1$, $\angle 2$, $\angle 7$, $\angle 8$
- **Alternate interior angles** are on opposite sides of the transversal and inside the parallel lines.
  $\angle 3$ and $\angle 5$, $\angle 4$ and $\angle 6$

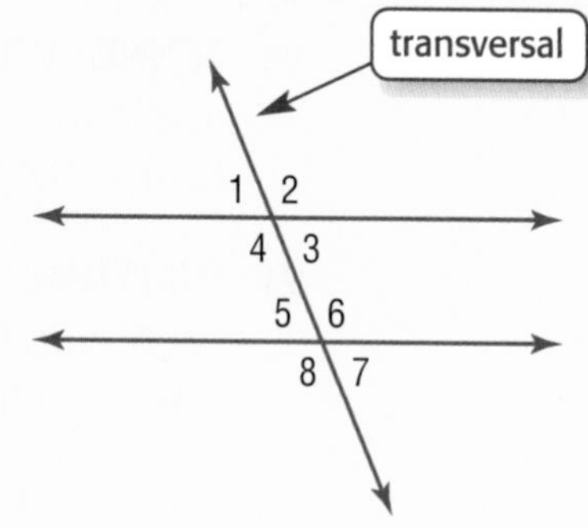

- **Alternate exterior angles** are on opposite sides of the transversal and outside the parallel lines.
  $\angle 1$ and $\angle 7$, $\angle 2$ and $\angle 8$
- **Corresponding angles** are in the same position on the parallel lines in relation to the transversal.
  $\angle 1$ and $\angle 5$, $\angle 2$ and $\angle 6$, $\angle 3$ and $\angle 7$, $\angle 4$ and $\angle 8$

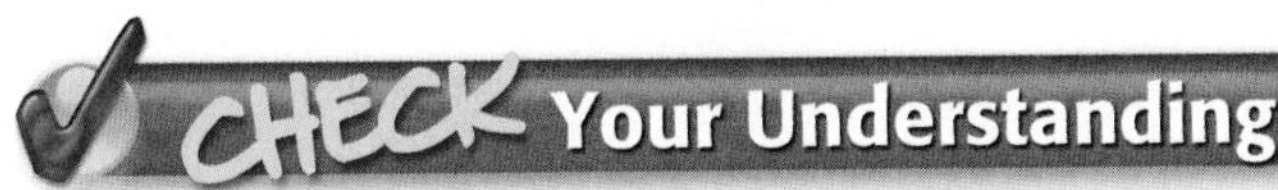

CHECK Your Understanding

Example 1 (p. LA11) **In the figure at the right, $m \parallel n$ and k is a transversal. If $m\angle 1 = 56°$, find the measure of each angle.**

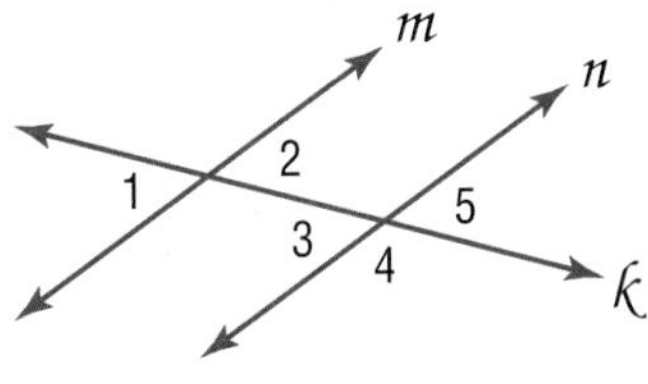

1. $\angle 2$

2. $\angle 3$
3. $\angle 4$
4. $\angle 5$

Example 2 (p. LA12) **Find the value of x in each figure.**

5. 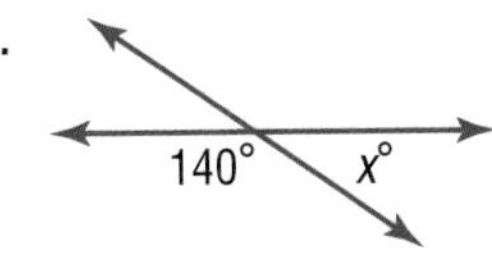

6. 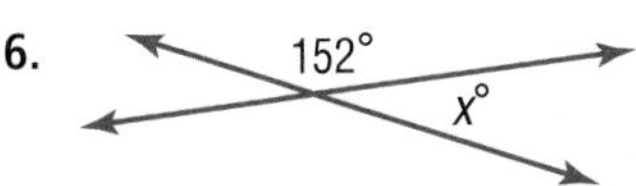

Example 3 (p. LA12) **CONSTRUCTION For Exercises 7 and 8, use the diagram shown and the following information.**

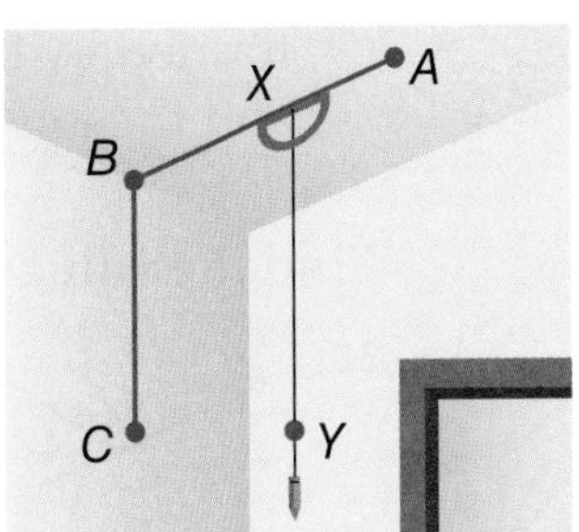

To measure the angle between a sloped ceiling and a wall, a carpenter uses a plumb line (a string with a weight attached). The plumb line is parallel to the wall.

7. If $m\angle AXY = 68°$, what is $m\angle XBC$?
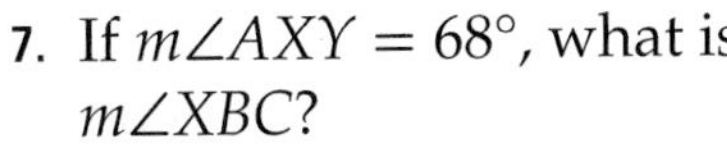
8. What type of angles are $\angle AXY$ and $\angle XBC$?
9. If $m\angle BXY = 84°$, what is $m\angle AXY$?

Practice and Problem Solving

HOMEWORK HELP	
For Exercises	See Examples
10–15	1
16–19	2
20, 21	3

In the figure at the right, $g \parallel h$ and t is a transversal. If $m\angle 4 = 53°$, find the measure of each angle.

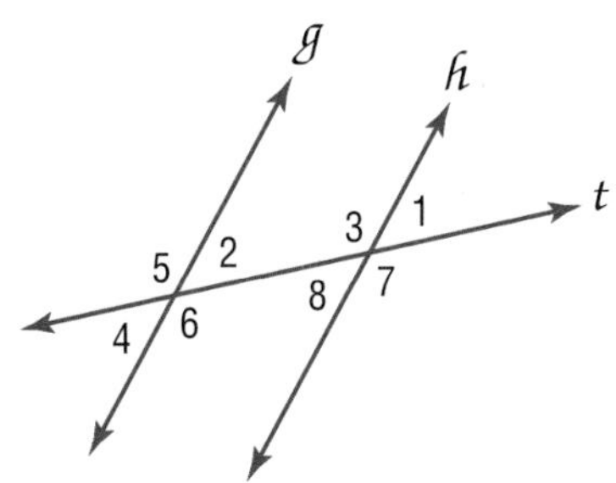

10. $\angle 1$
11. $\angle 5$
12. $\angle 7$
13. $\angle 8$
14. $\angle 2$
15. $\angle 3$

Find the value of x in each figure.

16. 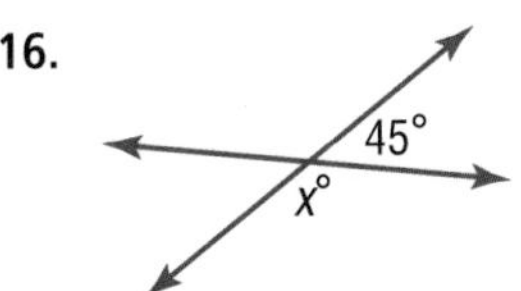

17. 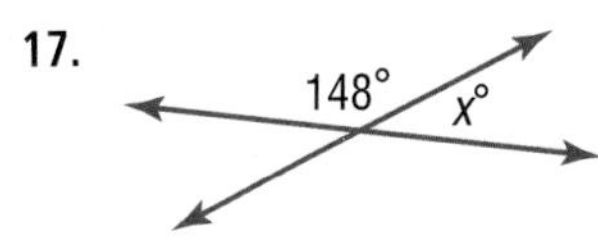

18.

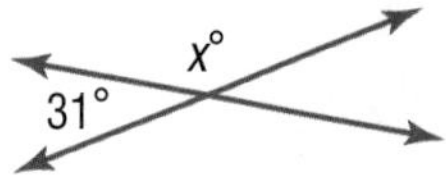

19.

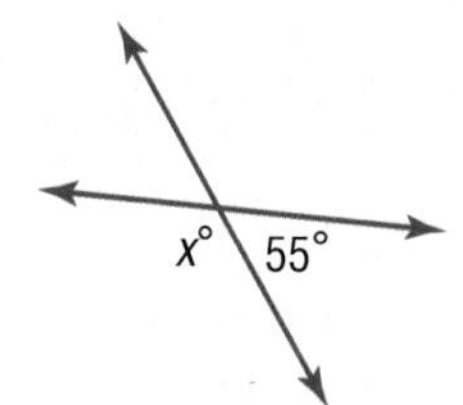

STAIRS **For Exercises 20 and 21, refer to the diagram at the right.**

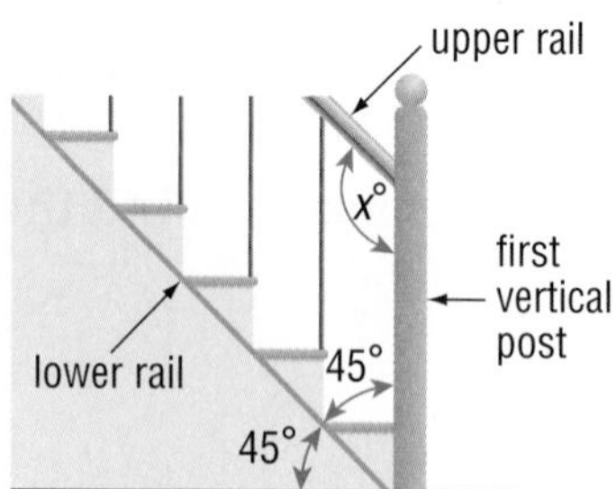

20. Suppose the upper rail is parallel to the lower rail. What is the measure of the angle formed by the upper rail and the first vertical post?

21. What is the measure of the angle formed by the second vertical post and the lower rail?

Find the value of x in each figure.

22.

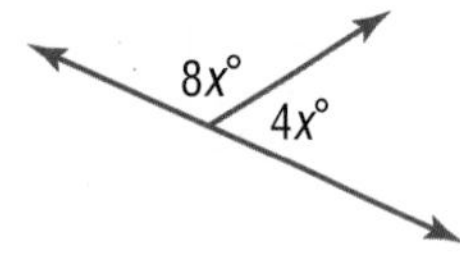

23.

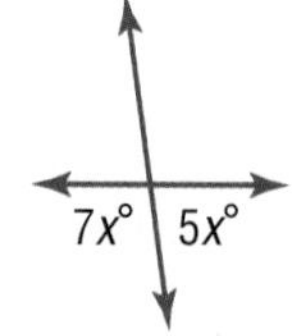

24. **ALGEBRA** Find $m\angle E$ if $\angle E$ and $\angle F$ are adjacent angles formed by two intersecting lines, $m\angle E = 2x + 15$, and $m\angle F = 5x - 38$.

25. **ALGEBRA** Find $m\angle G$ if $\angle H$ and $\angle G$ are alternate interior angles formed by two parallel lines cut by a transversal, $m\angle G = 5x - 8$, and $m\angle H = 3x + 12$.

H.O.T. Problems

26. **CHALLENGE** Suppose two parallel lines are cut by a transversal. How are the interior angles on the *same side* of the transversal related? Explain.

27. **REASONING** Explain the difference between adjacent and vertical angles.

28. **OPEN ENDED** Draw and label a diagram to show that $\angle MNP$ and $\angle PNQ$ are adjacent and supplementary angles.

29. **WRITING IN MATH** Explain how parallel lines and angles are related. Include the following in your answer:

- a drawing of parallel lines intersected by a transversal, and
- a list of the congruent and supplementary angles.

Area of Circles

MAIN IDEA

Apply strategies and formulas to find the area of a circle, and calculate the area of a sector of a circle, given the measure of a central angle and the radius of the circle.

New Vocabulary

central angle
sector

Math Online

glencoe.com

- Extra Examples
- Personal Tutor
- Self-Check Quiz

MINI Lab

Step 1 Fold a paper plate into eighths.

Step 2 Unfold the plate and cut along the creases.

Step 3 Arrange the pieces to form the figure shown.

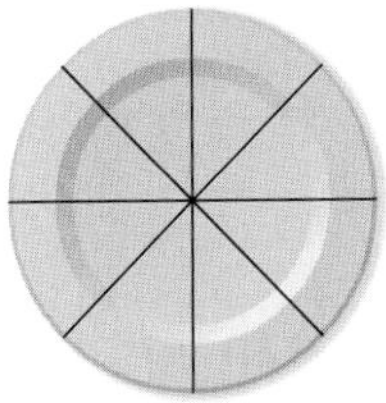

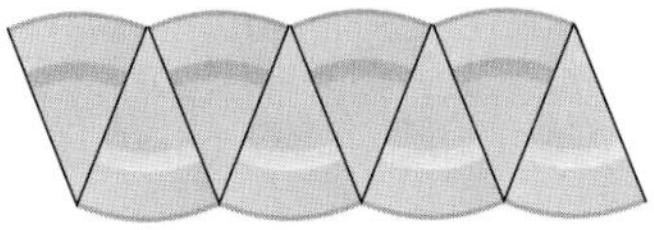

1. What shape does the figure look like?
2. What part of the circle represents the figure's height?
3. Relate the circle's circumference to the base of the figure.
4. How would you find the area of the figure?
5. What formula would you use in Exercise 4?

A circle can be separated into parts as shown. The parts can then be arranged to form a figure that resembles a parallelogram.

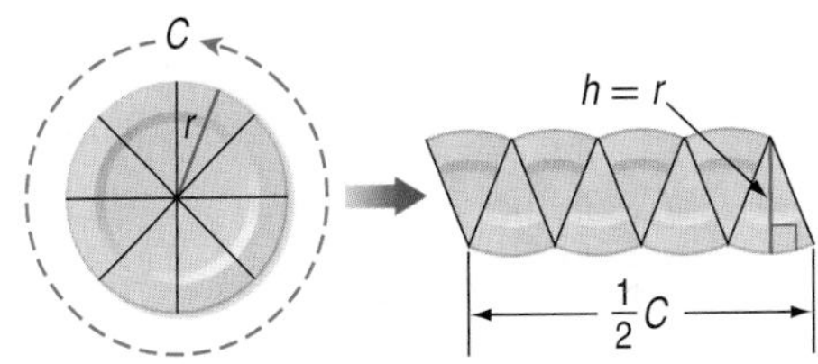

You can use the formula for the area of a parallelogram to find the formula for the area of a circle.

$A = bh$	Area of a parallelogram
$A = \left(\frac{1}{2}C\right)r$	The base is one half the circumference. The height is the radius.
$A = \frac{1}{2}(2\pi r)r$	Replace C with $2\pi r$, the formula for circumference.
$A = \pi \cdot r \cdot r$	Simplify. $\frac{1}{2} \cdot 2 = 1$
$A = \pi r^2$	Simplify. $r \cdot r = r^2$

Area of a Circle

Key Concept

Words The area A of a circle is the product of π and the square of the radius r.

Example $A = \pi r^2$

Model (circle with radius r)

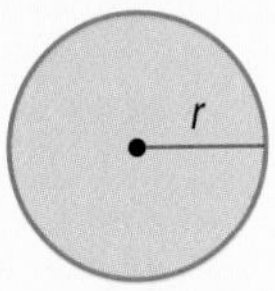

EXAMPLES Find Areas of Circles

Study Tip

Pi The numbers 3.14 and $\frac{22}{7}$ are often used as approximations for π.

Find the area of each circle. Use 3.14 for π. Round to the nearest tenth.

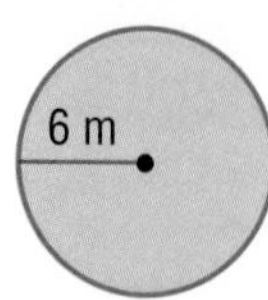

Estimate $3.14 \times (6)^2 \approx 3 \times 36 \approx 3 \times 40$, or 120

$A = \pi r^2$ Area of a circle

$A \approx 3.14 \times (6)^2$ Replace π with 3.14 and r with 6.

$A \approx 3.14 \times 36$ Evaluate $(6)^2$.

$A \approx 113.04$ Multiply.

Round to the nearest tenth. The area is about 113.0 square meters.

Check $113.0 \approx 120$ ✓

2

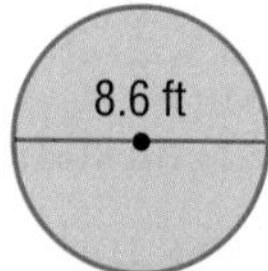

Study Tip

Area of Circles Check to see whether you are given the circle's radius or diameter. The formula for the area of a circle uses its radius. So, if you are given the circle's diameter, you must first find its radius.

The diameter is 8.6 feet. So, the radius is half of the diameter, or 4.3 feet.

Estimate $3.14 \times (4.3)^2 \approx 3 \times 16$, or 48

$A = \pi r^2$ Area of a circle

$A \approx 3.14 \times (4.3)^2$ Replace π with 3.14 and r with 4.3.

$A \approx 3.14 \times 18.49$ Evaluate $(4.3)^2$.

$A \approx 58.0586$ Multiply.

Round to the nearest tenth. The area is about 58.1 square feet.

Check $58.1 \approx 48$ ✓

CHECK Your Progress

Find the area of each circle. Use 3.14 for π. Round to the nearest tenth.

a. 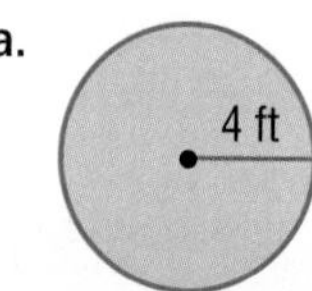

b. 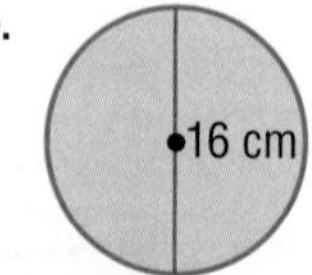

c. 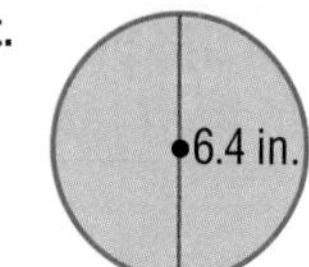

d. 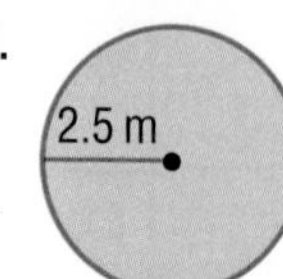

Real-World EXAMPLE

3 **CRATERS The Meteor Crater is located in Arizona. This crater is circular and has a radius of about 0.4 mile. About how much land does this crater cover?**

$A = \pi r^2$ Area of a circle

$A \approx 3.14 \times (0.4)^2$ Replace π with 3.14 and r with 0.4.

$A \approx 3.14 \times 0.16$ Evaluate $(0.16)^2$.

$A \approx 0.5024$ Use a calculator.

About 0.5024 square mile of land is covered by the crater.

Real-World Link Scientists believe that the crater in Arizona was caused by a meteorite that collided with Earth at about 40,000 miles per hour.

Source: Meteor Crater Enterprises

CHECK Your Progress

e. **SCIENCE** An earthquake's epicenter is the point from which the shock waves radiate. What is the area of the region affected by an earthquake whose shock waves radiated 29 miles from its epicenter?

A **central angle** is an angle of a circle that has the center of the circle as its vertex, and its sides contain two radii of the circle. The region of the circle bounded by a central angle is called a **sector** of the circle.

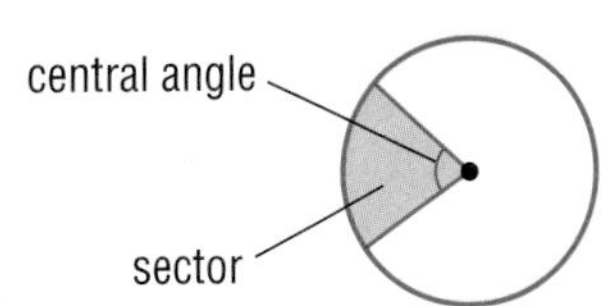

Proportional reasoning can be used to find the formula for the area of a sector. You know that the number of degrees in a circle is 360°.

$$\frac{\text{area of a sector}}{\text{area of a circle}} = \frac{N°}{360°}$$ ← degrees in a sector, ← degrees in a circle

$$\text{area of a sector} \cdot 360° = \text{area of a circle} \cdot N°$$ Find the cross products.

$$\frac{\text{area of a sector} \cdot 360°}{360°} = \frac{\text{area of a circle} \cdot N°}{360°}$$ Divide each side of the equation by 360°.

$$\text{area of a sector} = \frac{\pi r^2 \cdot N°}{360°} \text{ or } \frac{N°}{360°}\pi r^2$$ area of circle $= \pi r^2$

Area of a Sector

Key Concept

Words If a sector of a circle has an area of A square units, a central angle measuring $N°$, and a radius of r units, then $A = \frac{N°}{360°}\pi r^2$.

Model (circle with center O, radius r, central angle $N°$)

EXAMPLE Area of a Sector

4 **Find the area of the sector in the circle shown. Use 3.14 for π. Round to the nearest tenth.**

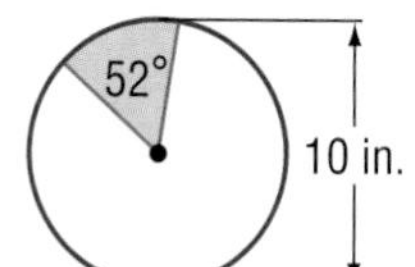
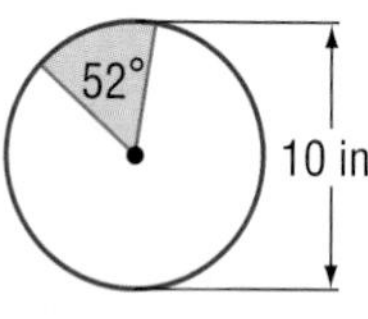

$A = \frac{N°}{360°}\pi r^2$ Area of a sector

$= \frac{52}{360}\pi(5^2)$ $N = 52, r = 5$

≈ 11.3 Simplify.

The area is about 11.3 square inches.

CHECK Your Progress

Find the area of each sector. Use 3.14 for π. Round to the nearest tenth.

f. 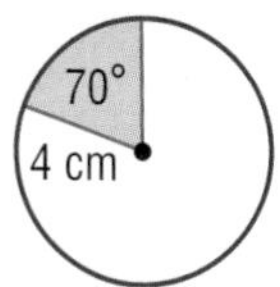

g. 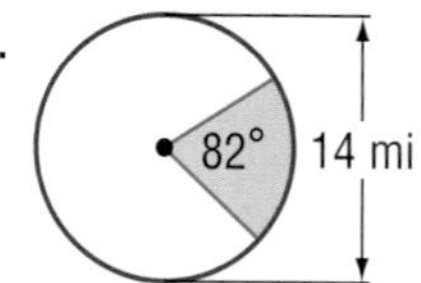

h. 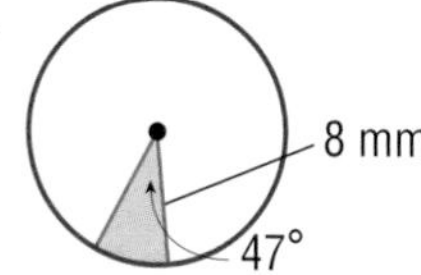

CHECK Your Understanding

Examples 1, 2 (p. LA16) **Find the area of each circle to the nearest tenth. Use 3.14 for π. Round to the nearest tenth.**

1. 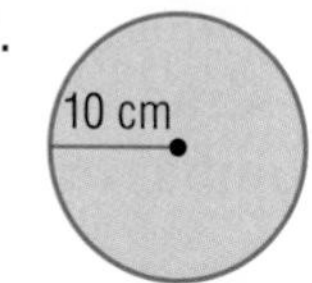

2. 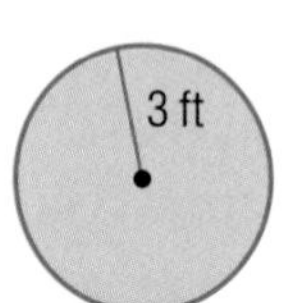

3. 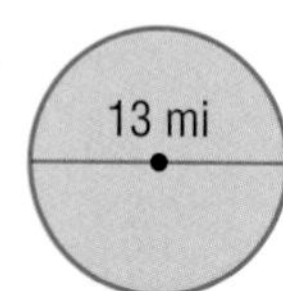

4. 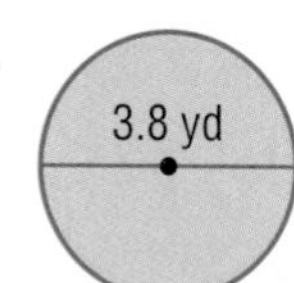

5. 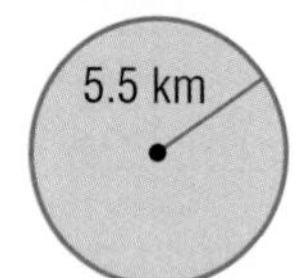

6. 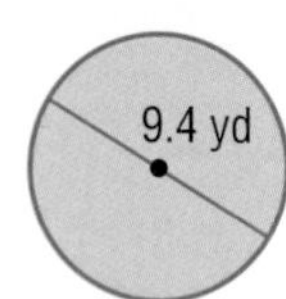

Example 3 (p. LA17)

7. **WRESTLING** A wrestling mat is a square measuring 12 meters by 12 meters. Within the square, there is a circular ring with a radius of 4.5 meters. Find the area within the circle to the nearest tenth.

8. **PIZZA** Find the area of a circular pizza with a diameter of 16 inches.

Example 4 (p. LA18) **Find the area of each sector. Use 3.14 for π. Round to the nearest tenth.**

9. 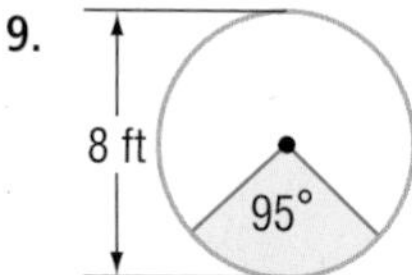

10. 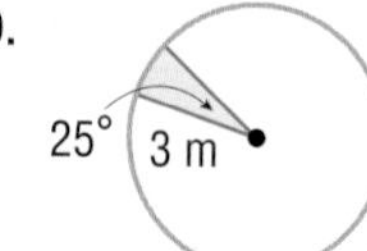

11. 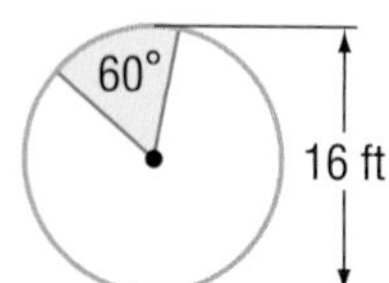

Practice and Problem Solving

HOMEWORK HELP	
For Exercises	See Examples
12–17	1, 2
18, 19	3
20–22	4

Find the area of each circle to the nearest tenth. Use 3.14 for π. Round to the nearest tenth.

12. 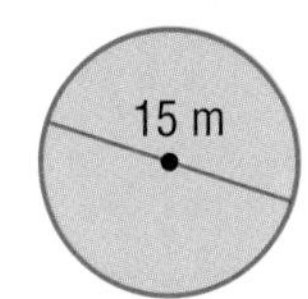

13. 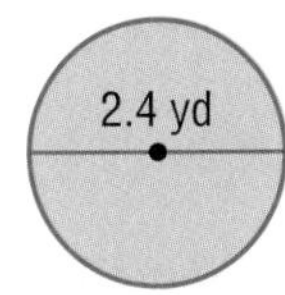

14. 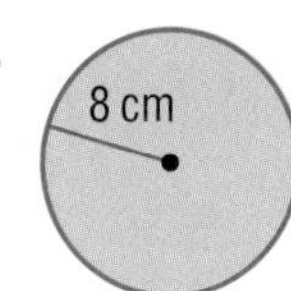

15. 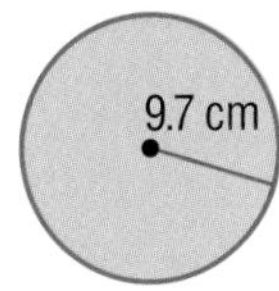

16. 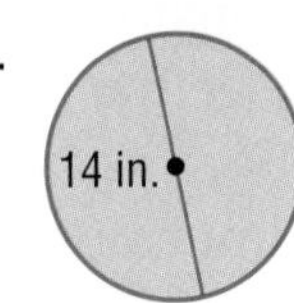

17. 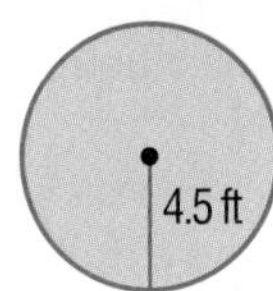

18. **TOOLS** A sprinkler that sprays water in a circular area can be adjusted to spray a radius up to 30 feet. What is the maximum area of lawn that can be watered by the sprinkler?

19. **HOT TUBS** Find the area of a circular hot tub cover with a diameter that measures 6.5 feet.

Find the area of each sector. Use 3.14 for π. Round to the nearest tenth.

20. 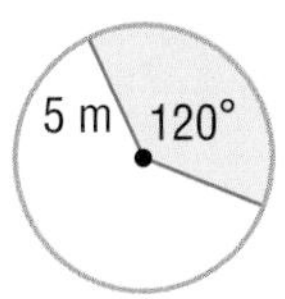

21. 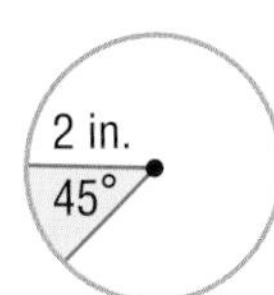

22. 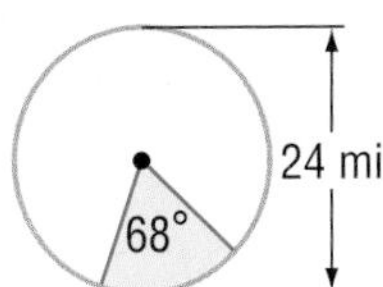

23. **PIE** An apple pie with a diameter of 9 inches is cut into 6 equal-size pieces. The central angle formed by each piece of pie measures 60°. Two pieces of pie have been eaten. Approximately what area of the pie is left?

ESTIMATION Estimate to find the approximate area of each circle.

24. 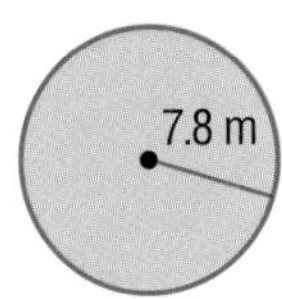

25. 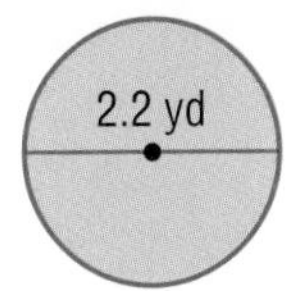

26. 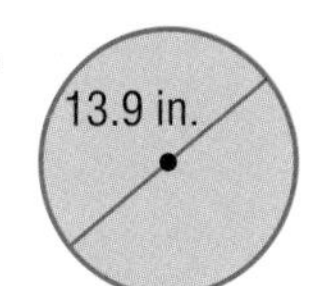

H.O.T. Problems

27. **CHALLENGE** Find the area of the shaded region of the figure shown. Use 3.14 for π.

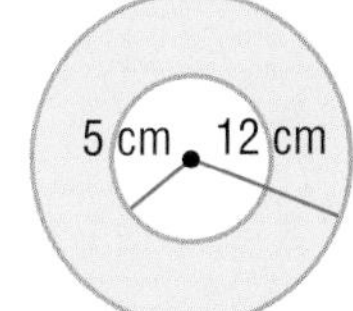

28. **REASONING** Suppose you double the radius of a circle. How is the area affected?

29. **WRITING IN MATH** Explain how to find the area of a sector of a circle.

Surface Area and Volume of Pyramids and Cylinders

MAIN IDEA

Use models and formulas to find the surface area and volume of pyramids and cylinders.

New Vocabulary

net
lateral face
slant height
lateral area
cylinder

Math Online

glencoe.com

- Extra Examples
- Personal Tutor
- Self-Check Quiz

MINI Lab

Nets are two-dimensional patterns of three-dimensional figures. Nets can help you see the faces that make up a figure. So, you can use a net to build a three-dimensional figure.

Step 1 Copy the net on a piece of paper. Use scissors to cut out the net.

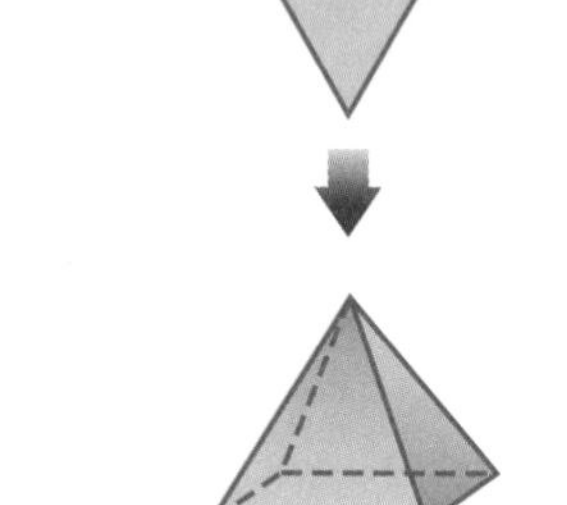

Step 2 Fold on the dashed lines and tape the sides together.

Step 3 Use your model to sketch the figure and to draw its top, side, and front views.

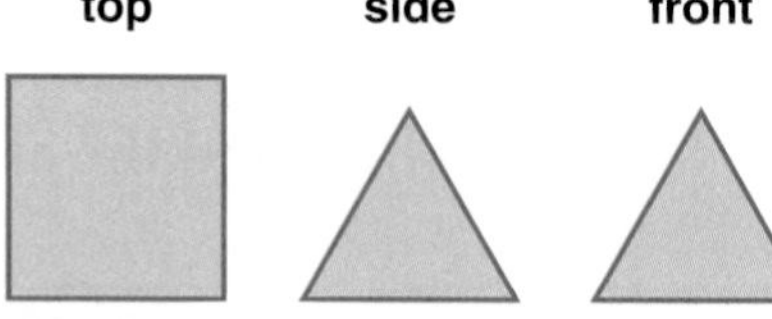

The triangular sides of a pyramid are called **lateral faces**. The triangles intersect at the vertex. The altitude or height of each lateral face is called the **slant height**.

Model of Square Pyramid

vertex
lateral face
slant height
base

Net of Square Pyramid

lateral face
base
slant height

The sum of the areas of the lateral faces is the **lateral area**. The surface area of a pyramid is the lateral area plus the area of the base.

Surface Area of a Pyramid — Key Concept

Words The total surface area S of a regular pyramid is the lateral area L plus the area of the base B.

Symbols $S = L + B$, where L is the lateral area and B is the area of the base.

Study Tip

Slant Height Be careful not to confuse the height of a pyramid with its slant height. A slant height lies along a pyramid's lateral surface.

EXAMPLE Surface Area of a Pyramid

1 **Find the surface area of the pyramid shown.**

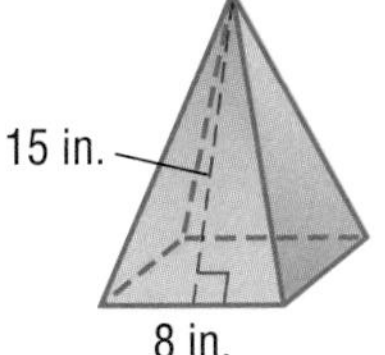

First, find the lateral area and the area of the base.

Area of each lateral face:

$A = \frac{1}{2}bh$ Area of a triangle

$A = \frac{1}{2}(8)(15)$ or 60 Replace b with 8 and h with 15.

There are 4 faces, so the lateral area is 4(60), or 240 square inches.

Area of base:

$A = s^2$ Area of a square

$A = 8^2$ or 64 Replace s with 8.

The surface area of the pyramid is the sum of the lateral area and the area of the base, 240 + 64 or 304 square inches.

CHECK Your Progress

Find the surface area of each pyramid. Round to the nearest whole number.

a. 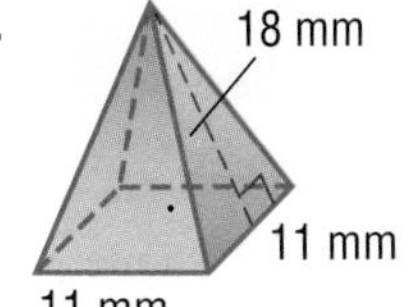

b. 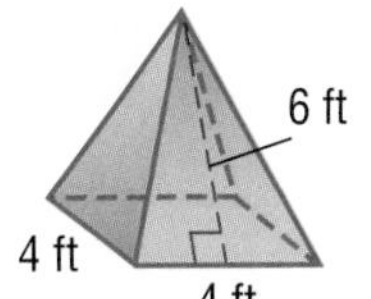

A **cylinder** is a solid with bases that are congruent, parallel circles, connected to a curved side. The net of a cylinder is two circles and a rectangle. The area of the rectangle is the product of the circle's circumference and the cylinder's height, $2\pi rh$. You can find the surface area of a cylinder by finding the area of its two circular bases and adding the area of its curved side.

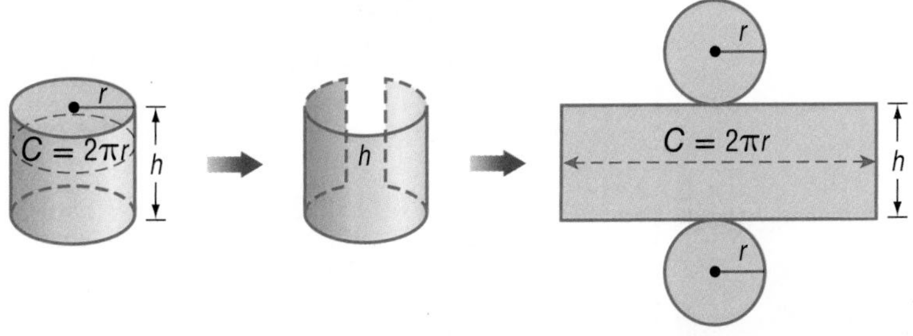

Surface Area of a Cylinder

Key Concept

Words The surface area S of a cylinder with height h and radius r is the area of the two bases plus the area of the curved surface.

Symbols $S = 2\pi r^2 + 2\pi rh$

EXAMPLE Surface Area of a Cylinder

2 **Find the surface area of the cylinder. Use 3.14 for π. Round to the nearest tenth.**

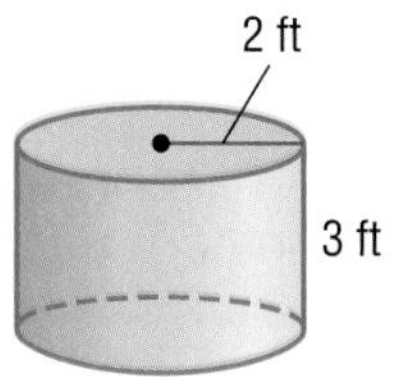

$S = 2\pi r^2 + 2\pi rh$	Surface area of a cylinder
$S = 2\pi(2)^2 + 2\pi(2)(3)$	Replace r with 2 and h with 3.
$S \approx 62.8$	Simplify.

The surface area is about 62.8 square feet.

CHECK Your Progress

Find the surface area of each cylinder. Use 3.14 for π. Round to the nearest tenth.

c. 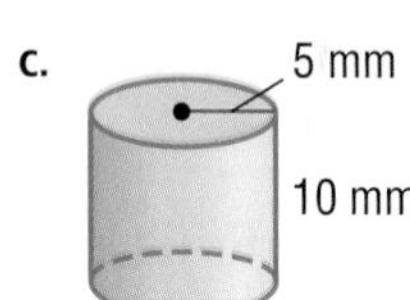

d. 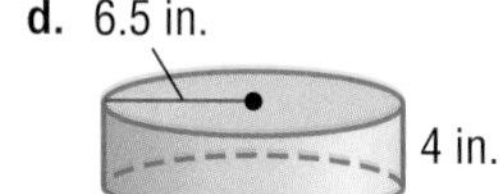

The volume of a pyramid is one third the volume of a prism with the same base area and height.

Volume of a Pyramid — Key Concept

Words The volume V of a pyramid is one third the area of the base B times the height h.

Symbols $V = \frac{1}{3}Bh$

Model

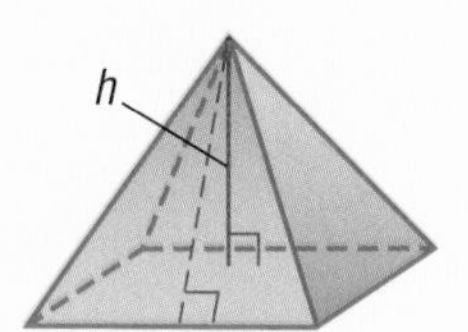

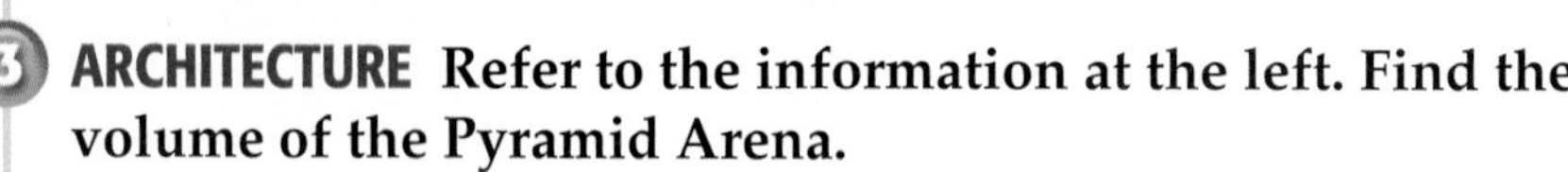

Real-World EXAMPLE Volume of a Pyramid

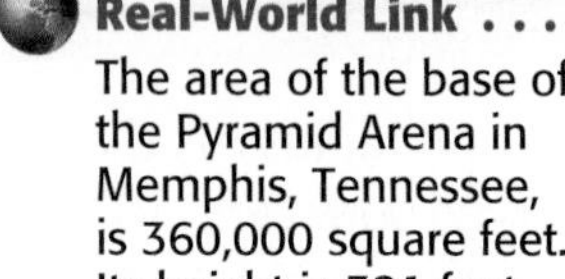

Real-World Link The area of the base of the Pyramid Arena in Memphis, Tennessee, is 360,000 square feet. Its height is 321 feet.

Source: Pyramid Arena

3 **ARCHITECTURE Refer to the information at the left. Find the volume of the Pyramid Arena.**

$V = \frac{1}{3}Bh$	Volume of a pyramid
$V = \frac{1}{3}(360{,}000)321$	Replace B with 360,000 and h with 321.
$V = 38{,}520{,}000$	Simplify.

The volume of the Pyramid Arena is 38,520,000 cubic feet.

CHECK Your Progress

e. **ARCHITECTURE** The Louvre pyramid in Paris has a square base with sides 116 feet long and a height of 71 feet. Find the volume of the Louvre pyramid.

You used the formula $V = Bh$ to find the volume of a prism in Lesson 10-6. Similarly, you can use the formula $V = Bh$ to find the volume of a cylinder, where the base is a circle.

Volume of a Cylinder

Key Concept

Words The volume V of a cylinder with radius r is the area of the base B times the height h.

Symbols $V = Bh$ or $V = \pi r^2 h$, where $B = \pi r^2$

EXAMPLE Volume of a Cylinder

Test-Taking Tip

Volume of Cylinders Even though a cylinder is not a prism, you can use the same volume formula as for a prism. This is because a cylinder has two parallel and congruent bases, and its curved surface unrolls as a rectangle.

4 Find the volume of the cylinder. Use 3.14 for π. Round to the nearest tenth.

6 ft

20 ft

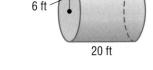

$V = \pi r^2 h$	Volume of a cylinder
$V = \pi \times 6^2 \times 20$	Replace r with 6 and h with 20.
$V \approx 2{,}261.9$	Simplify.

The volume is 2,261.9 cubic feet.

f.

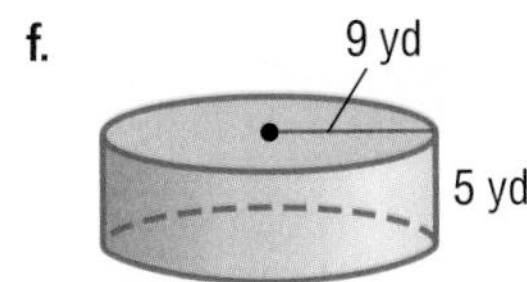

g.

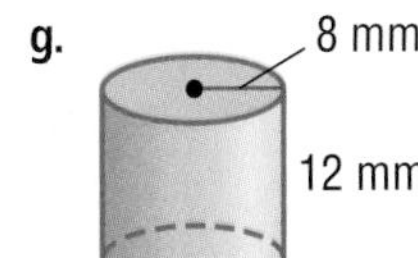

CHECK Your Understanding

Example 1 (p. LA21)

Find the surface area of each pyramid. Round to the nearest tenth.

1.

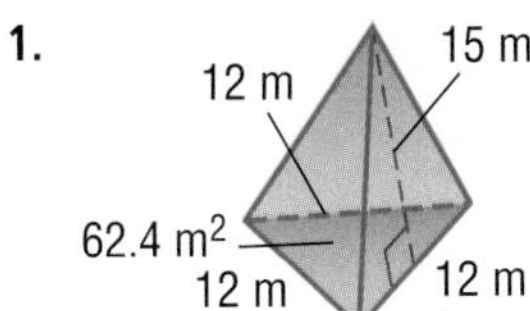

2.

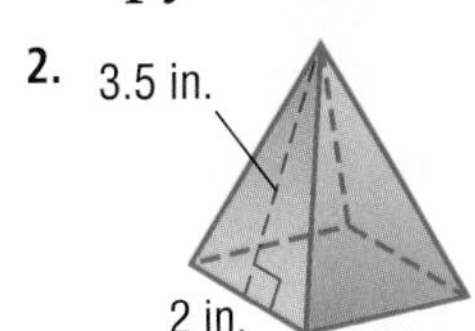

3.

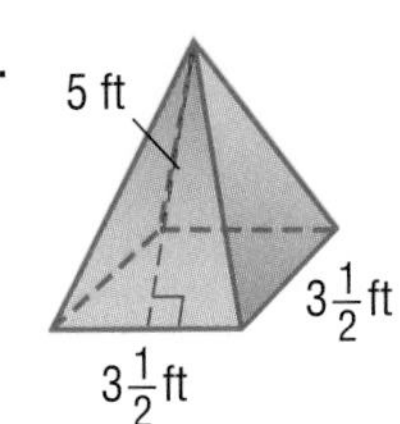

Example 2 (p. LA22)

Find the surface area of each cylinder. Use 3.14 for π. Round to the nearest tenth.

4.

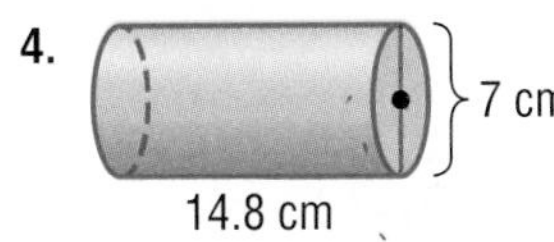

5.

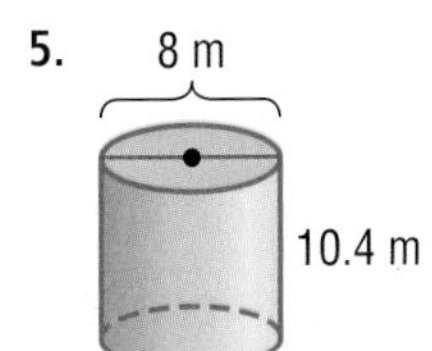

6.

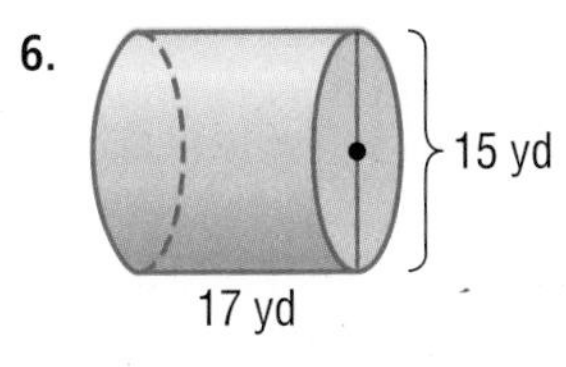

Example 3 (p. LA22)

7. **SCHOOL** Mario is building a model pyramid. The rectangular base of the pyramid is 8 inches by 14 inches, and the height is 11 inches. What is the volume of Mario's pyramid to the nearest tenth?

Example 4 (p. LA23) **Find the volume of each cylinder. Round to the nearest tenth.**

8. 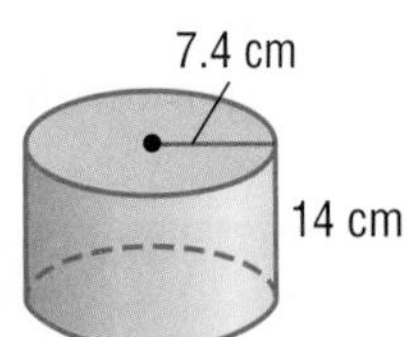

9. 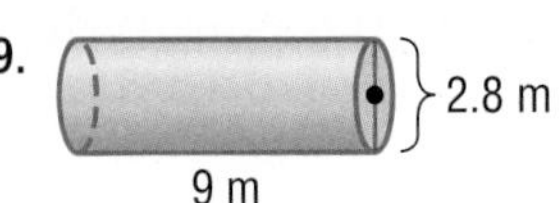

10. 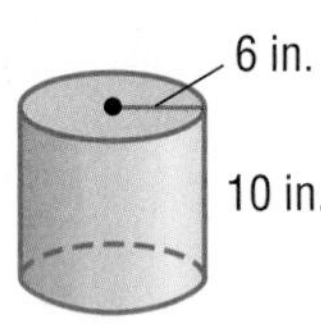

Practice and Problem Solving

HOMEWORK HELP	
For Exercises	See Examples
11–13	1
14–16	2
17, 18	3
19–21	4

Find the surface area of each pyramid. Round to the nearest tenth.

11. 6 m, 8.3 m, 6 m, 6 m, $A = 15.6\text{ m}^2$

12. 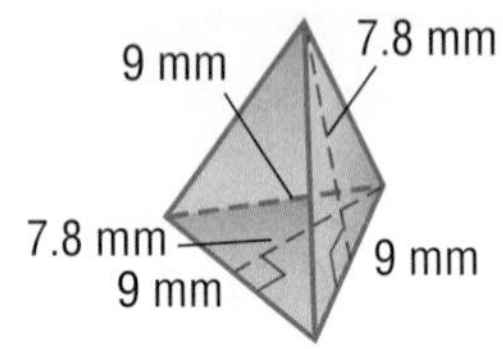

13. 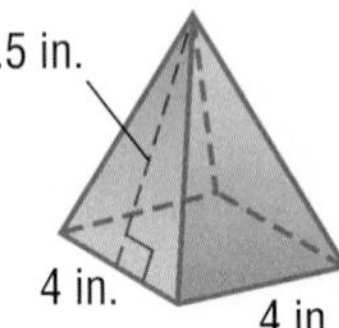

Find the surface area of each cylinder. Round to the nearest tenth.

14. 12 m, 18.4 m 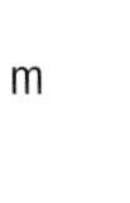

15. 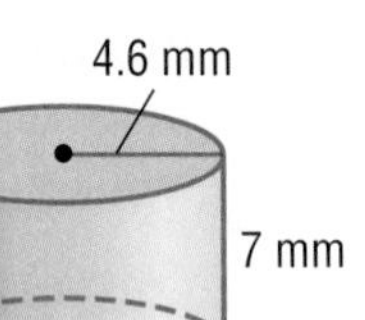

16. 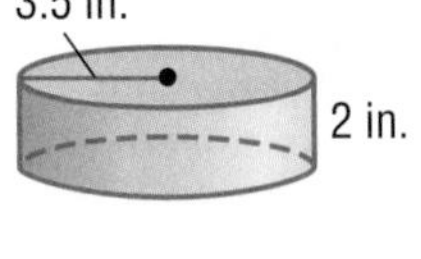

17. **ARCHITECTURE** The Aquarium Pyramid in Galveston, Texas, is 128 feet tall and the area of its rectangular base is 70,840 square feet. What is the volume of the Aquarium Pyramid?

18. **CANDLES** Latisha bought a pyramid-shaped candle at a craft show. The area of its base was 144 square centimeters and it was 18 centimeters tall. Find the volume of the candle.

Find the volume of each cylinder. Round to the nearest tenth.

19. 4.2 m, 8 m

20. 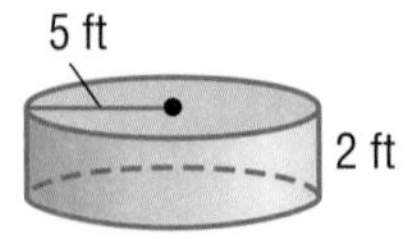

21. 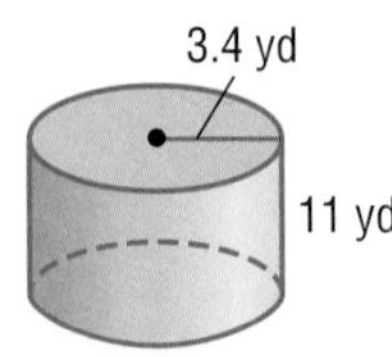

H.O.T. Problems

22. **CHALLENGE** Will the surface area of a cylinder increase more if you double the height or double the radius? Explain your reasoning.

23. **REASONING** Explain the difference between surface area and volume.

24. **WRITING IN MATH** Write about a real-life situation that can be solved by finding the volume of a pyramid or cylinder. Then solve the problem.

Histograms

MAIN IDEA

Collect, organize, analyze, and display data to solve problems with histograms, and use these data displays to show frequency distribution.

New Vocabulary

histogram
frequency distribution

Math Online

glencoe.com

- Extra Examples
- Personal Tutor
- Self-Check Quiz

GET READY for the Lesson

CONCERTS Alicia researched the average price of concert tickets. The frequency table shows the results.

1. What does each tally mark represent?
2. How is the frequency for each price range determined?

Average Ticket Prices of Top 20 Money Earning Concerts

Price	Tally	Frequency
$25.00–$49.99	𝍸 ||||	9
$50.00–$74.99	𝍸 ||	7
$75.00–$99.99	|	1
$100.00–$124.99	||	2
$125.00–$149.99		0
$150.00–$174.99	|	1

Data from a frequency table can be displayed as a histogram. A **histogram** is a type of bar graph used to display numerical data that have been organized into equal intervals. These intervals allow you to see the **frequency distribution** of the data, or how many pieces of data are in each interval.

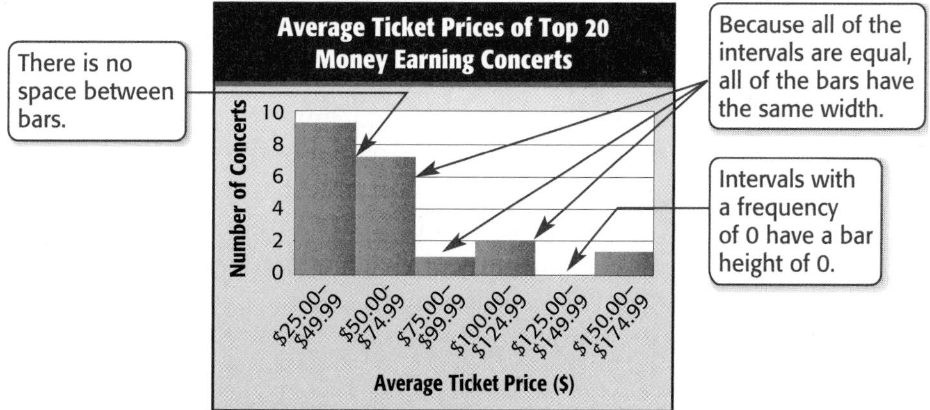

EXAMPLE Interpret Data

1 **CONCERTS Refer to the histogram above. How many concerts had ticket prices that were at least $100?**

Two concerts had prices between $100.00–$124.99 and one concert had a price between $150.00–$174.99. So, 2 + 1, or 3 concerts had prices that were at least $100.

CHECK Your Progress

a. Refer to the histogram above. How many concerts had ticket prices less than $75?

You can use data from a frequency table to construct a histogram.

EXAMPLE Construct a Histogram

2 **PARKS The frequency table at the right shows the number of daily visitors to selected state parks. Draw a histogram to represent the data.**

Daily Visitors to Selected State Parks		
Visitors	Tally	Frequency
100–149	||	2
150–199	~~||||~~ ||	7
200–249	~~||||~~ |||	8
250–299	|	1
300–349	|	1
350–399	|	1

Step 1 Draw and label a horizontal and vertical axis. Include a title.

Step 2 Show the intervals from the frequency table on the horizontal axis. Label the vertical axis to show the frequencies.

Step 3 For each interval, draw a bar whose height is given by the frequencies.

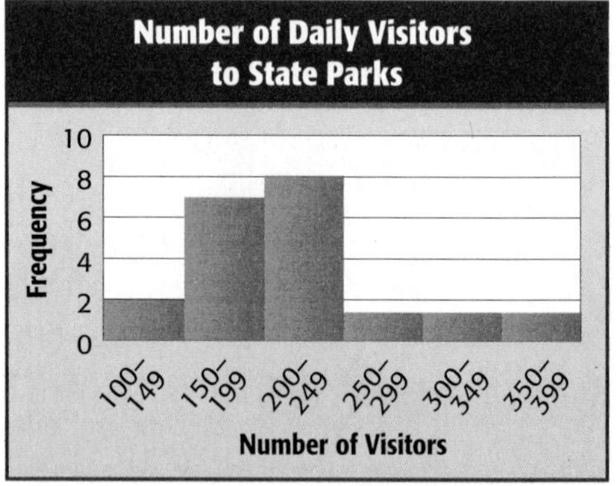

CHECK Your Progress

b. **SCHOOL** The list at the right shows a set of test scores. Choose intervals, make a frequency table, and construct a histogram to represent the data.

Test Scores						
72	97	80	86	92	98	88
76	79	82	91	83	90	76
81	94	96	92	72	83	85
65	91	92	68	86	89	97

CHECK Your Understanding

Example 1 (p. LA25) **SHOES For Exercises 1 and 2, use the histogram at the right.**

1. How many prices were at least $70?

2. How many prices were in the interval $80.00–$89.99?

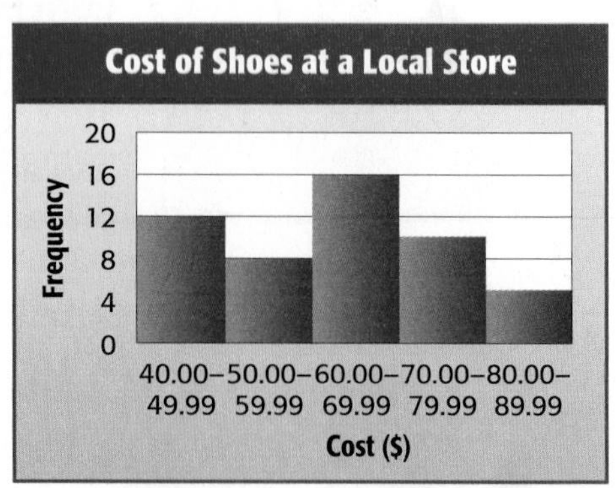

Example 2 (p. LA26)

3. **READING** The frequency table below shows the number of books read on vacation by the students in Mrs. Angello's class.

Number of Books Read		
Books	**Tally**	**Frequency**
0–2	卌 I	6
3–5	卌 卌	10
6–8	卌 II	7
9–11	III	3
12–14	IIII	4

Draw a histogram to represent the data.

Practice and Problem Solving

HOMEWORK HELP	
For Exercises	See Examples
4–7	1
8, 9	2

OLYMPICS For Exercises 4–7, use the histogram at the right.

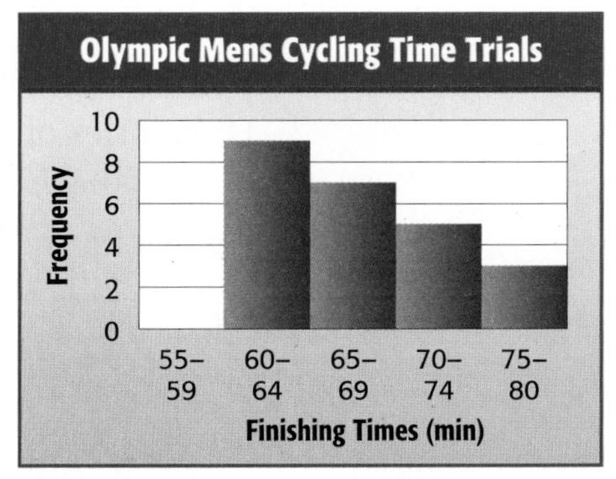

4. Which interval represents the most number of cyclists?
5. Which interval has 7 cyclists?
6. How many cyclists had a time less than 70 minutes?
7. How many cyclists had times in the interval 70–74 minutes?

Draw a histogram to represent each set of data.

8.

Number of States Visited by Students in Marty's Class		
Number of States	**Tally**	**Frequency**
0–4	卌 IIII	9
5–9	III	3
10–14	卌	5
15–19	III	3
20–24	卌 I	6
25–29	I	1

9.

Number of Homeruns in a Season		
Homeruns	**Tally**	**Frequency**
0–9	卌 卌 II	12
10–19	卌 卌	10
20–29	卌 IIII	9
30–39	卌 IIII	9
40–49	卌 I	6
50+	卌	5

FUNDRAISER For Exercises 10–13, refer to the histograms below.

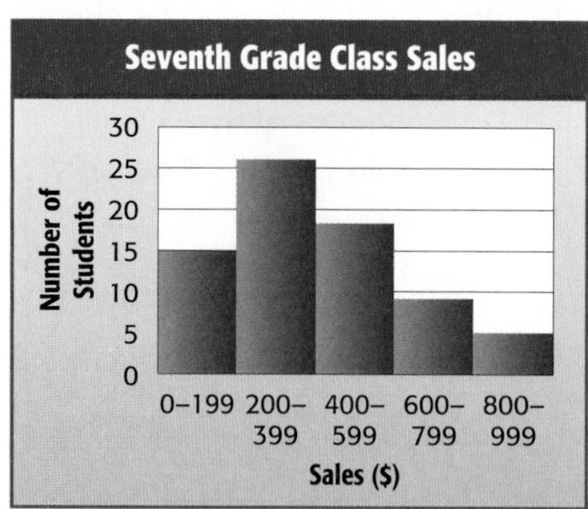

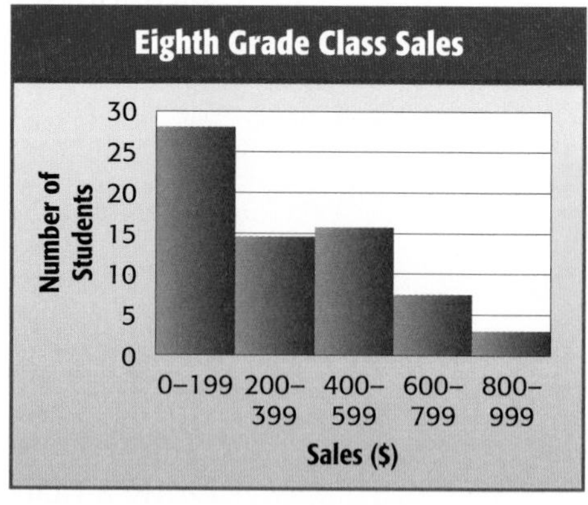

10. Which grade had more students earn between $400 and $599?
11. Which grade had fewer students earn less than $200?
12. About how many students from both grades earned $600 or more?
13. Which grade had the most sales overall?

Draw a histogram to represent each set of data.

14. calories of various types of frozen bars: 25, 35, 200, 280, 80, 80, 90, 40, 45, 50, 50, 60, 90, 100, 120, 40, 45, 60, 70, 350
15. maximum height in feet of various species of trees in the United States: 278, 272, 366, 302, 163, 161, 147, 223, 219, 216

H.O.T. Problems

16. **OPEN ENDED** Give a set of data that could be represented by the histogram at the right.

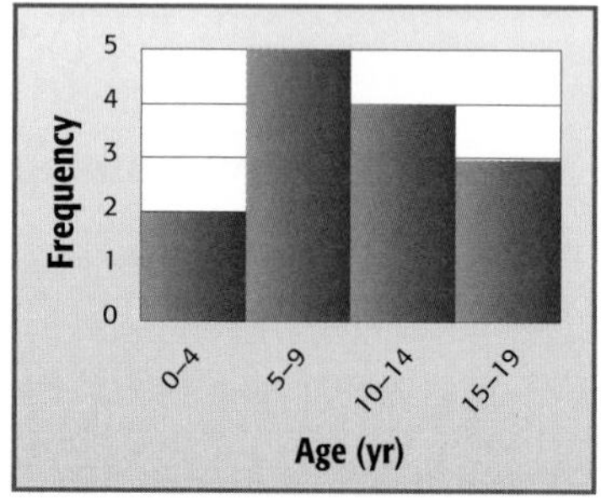

17. **RESEARCH** Use the Internet or other resource to find the populations of each county, census division, or parish in your state. Make a histogram using your data. How does your county, census division, or parish compare with others in your state?
18. **REASONING** Identify the interval that is not equal to the other three. Explain your reasoning.

15–19	30–34	40–45	45–49

19. **WRITING IN MATH** Compare and contrast histograms with tables of individual data. When is a histogram more useful than a table with individual data? When is a table with individual data more useful than a histogram?

Student Handbook

Built-In Workbooks

Reference

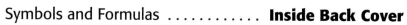

How to Use the Student Handbook

The Student Handbook is the additional skill and reference material found at the end of books. The Student Handbook can help answer these questions.

What If I Need More Practice?

You, or your teacher, may decide that working through some additional problems would be helpful. The **Extra Practice** section provides these problems for each lesson so you have ample opportunity to practice new skills.

What If I Have Trouble with Word Problems?

The **Mixed Problem Solving** portion of the book provides additional word problems that use the skills presented in each chapter. These problems give you real-world situations where the math can be applied.

What If I Need to Prepare for a Standardized Test?

The **Preparing for Standardized Tests** section provides worked-out examples and practice problems for multiple-choice, gridded-response, short-response, and extended-response questions.

What If I Forget What I Learned Last Year?

Use the **Concepts and Skills** section to refresh your memory about topics you have learned in other math classes or to prepare for next year.

What If I Forget a Vocabulary Word?

The **English-Spanish Glossary** provides a list of important, or difficult, words used throughout the textbook. It provides a definition in English and Spanish as well as the page number(s) where the word can be found.

What If I Need to Check a Homework Answer?

The answers to the odd-numbered problems are included in **Selected Answers**. Check your answers to make sure you understand how to solve all of the assigned problems.

What If I Need to Find Something Quickly?

The **Index** alphabetically lists the subjects covered throughout the entire textbook and the pages on which each subject can be found.

What If I Forget a Formula?

Inside the back cover of your math book is a list of **Formulas and Symbols** that are used in the book.

Extra Practice

Extra Practice

Lesson 1-1

Pages 24–27

Use the four-step plan to solve each problem.

1. **MONEY** Sylvia has $102. If she made three purchases of $13, $37, and $29, how much money does she have left?
2. **PATTERNS** Complete the pattern: 1, 8, 15, 22, ■, ■, ■
3. **SPORTS** Sarah is conditioning for track. On the first day, she ran 2 laps. The second day she ran 3 laps. The third day she ran 5 laps, and on the fourth day she ran 8 laps. If this pattern continues, how many laps will she run on the seventh day?
4. **SCHOOL** Tickets to a school dance cost $4 each. If $352 was collected, how many tickets were sold?

Lesson 1-2

Pages 28–31

Tell whether each number is *prime, composite,* or *neither.*

1. 65	2. 37	3. 26	4. 54
5. 155	6. 201	7. 0	8. 93
9. 121	10. 29	11. 53	12. 57

Find the prime factorization of each number.

13. 72	14. 88	15. 32	16. 86
17. 120	18. 576	19. 68	20. 240
21. 24	22. 70	23. 102	24. 121
25. 164	26. 225	27. 54	28. 460

Lesson 1-3

Pages 32–36

Write each product using an exponent.

1. $4 \times 4 \times 4 \times 4 \times 4$	2. $10 \times 10 \times 10$	3. 14×14
4. $3 \times 3 \times 3 \times 3$	5. $2 \times 2 \times 2$	6. $6 \times 6 \times 6 \times 6 \times 6$
7. $8 \times 8 \times 8$	8. $7 \times 7 \times 7 \times 7 \times 7 \times 7$	9. $9 \times 9 \times 9$

Write each power as a product of the same factor. Then find the value.

10. 9^4	11. 2^3	12. 3^5	13. 4^3
14. 6^5	15. 5^4	16. 8^3	17. 12^2

18. nine cubed
19. eight to the fourth power
20. six to the fifth power
21. eleven squared

Write the prime factorization of each number using exponents.

22. 9	23. 20	24. 18	25. 63
26. 44	27. 45	28. 243	29. 175

Lesson 1-4

Pages 37–40

Find the value of each expression.

1. $14 - 5 + 7$
2. $12 + 10 - 5 - 6$
3. $50 - 6 + 12 + 4$
4. $12 - 2 \times 3$
5. $16 + 4 \times 5$
6. $5 + 3 \times 4 - 7$
7. $2 \times 3 + 9 \times 2$
8. $6 \times 8 + 4 \div 2$
9. $7 \times 6 - 14$
10. $8 + 12 \times 4 \div 8$
11. $13 - 6 \times 2 + 1$
12. $80 \div 10 \times 8$
13. $14 - 2 \times 7 + 0$
14. $156 - 6 \times 0$
15. $30 - 14 \times 2 + 8$
16. $54 \div (8 - 5)$
17. $4^2 + 3^3$
18. $(11 - 7) \times 3 - 5$
19. $25 - 9 + 4$
20. $100 \div 10 \times 2$
21. 3×4^3
22. $11 + 4 \times (12 - 7)$
23. $6^2 - 7 \times 4$
24. $12 + 5^2 - 9$
25. Find two to the fifth power divided by 8 times 2.

Lesson 1-5

Pages 42–46

Evaluate each expression if $m = 2$ and $n = 4$.

1. $m + m$
2. $n - m$
3. mn
4. $3m + 5$
5. $2n + 2m$
6. $m \cdot 0$
7. $64 \div n$
8. $12 - m$
9. $5n \div m$
10. $6mn$
11. $4n - 3$
12. $n \div m + 8$

Evaluate each expression if $a = 3$, $b = 4$, and $c = 12$.

13. $a + b$
14. $c - a$
15. $a + b + c$
16. $b - a$
17. $c - a \cdot b$
18. $a + 2 \cdot b$
19. $b + c \div 2$
20. ab
21. $25 + c \div b$
22. $c \div a + 10$
23. $2b - a$
24. $2ab$

Lesson 1-6

Pages 49–53

Copy and complete each function table.

1.

Input (n)	Output ($n - 4$)
5	■
7	■
9	■

2.

Input (n)	Output ($3n$)
1	■
0	■
2	■

3.

Input (n)	Output ($n + 7$)
4	■
1	■
3	■

4.

Input (n)	Output ($n \div 4$)
12	■
16	■
24	■

Find the rule for each function table.

5.

n	■
1	6
0	5
3	8

6.

n	■
6	3
0	0
8	4

7.

n	■
4	20
2	10
6	30

Lesson 1-7

Pages 54–55

Use the *guess and check* strategy to solve Exercises 1–5.

1. David has 5 coins that total \$1.10. What are the coins?
2. Emma is thinking of 3 numbers from 1 through 9 with a product of 30. Find the numbers.
3. Kyle bought pens and markers at an office supply store. Each pen cost \$1.50, and each marker cost \$2.25. If Kyle spent a total of \$9 on 5 writing devices, how many of each did he buy?
4. Tamika has saved \$20 in cash to spend at the mall. If she has 8 bills, how many of each kind of bill does she have?
5. Use the symbols $+$, $-$, $\times$, or $\div$ to make the following sentence true. Use each symbol only once.

$$2 \blacksquare 5 \blacksquare 3 \blacksquare 7 = 14$$

Lesson 1-8

Pages 57–60

Identify the solution of each equation from the list given.

1. $7 + a = 10$; 3, 13, 17
2. $14 + m = 24$; 7, 10, 34
3. $20 = 24 - n$; 2, 3, 4
4. $x + 4 = 19$; 14, 15, 16
5. $23 - p = 7$; 16, 17, 18
6. $11 = w + 6$; 3, 4, 5
7. $73 + m = 100$; 26, 27, 28
8. $44 + s = 63$; 17, 18, 19

Solve each equation mentally.

9. $b + 7 = 12$
10. $s + 10 = 23$
11. $b - 3 = 12$
12. $d + 7 = 19$
13. $23 - q = 9$
14. $21 + p = 45$
15. $17 = 23 - t$
16. $g - 13 = 5$
17. $14 - m = 6$
18. $x - 3 = 11$
19. $16 = h + 9$
20. $50 + z = 90$

Lesson 1-9

Pages 63–67

Find the area of each rectangle.

1. 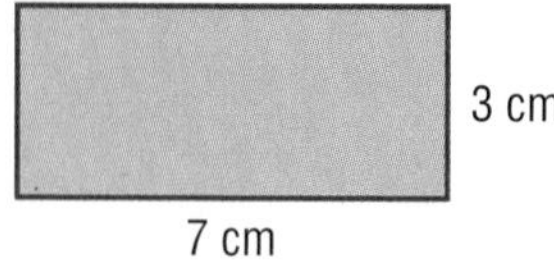

2. 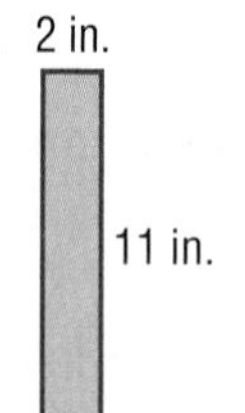

3. 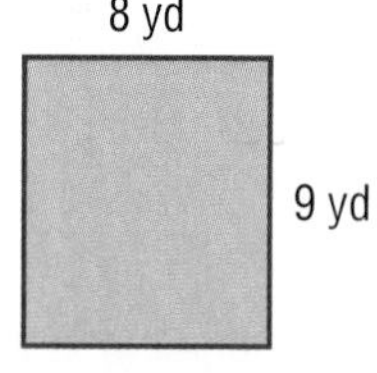

Find the area of each square.

4. 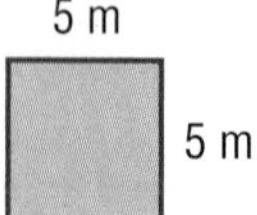

5. 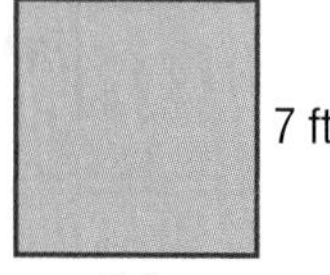

6. 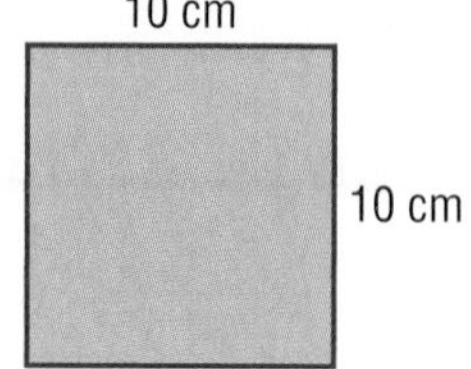

7. Find the area of a rectangle with a length of 18 inches and a width of 21 inches.

Lesson 2-1

Pages 78–79

Use the *make a table* strategy to solve Exercises 1 and 2.

1. **DAYS** Make a frequency table of the data below. How many more students chose Saturday as their favorite day than Sunday?

Favorite Day of the Week				
Sat.	Sat.	Fri.	Sun.	Fri.
Wed.	Sun.	Fri.	Thurs.	Sat.
Sat.	Sat.	Fri.	Sun.	Thurs.

2. **FAMILY** Make a frequency table of the data below. What is the most common number of siblings per student? the least?

Number of Siblings Per Student						
1	0	2	2	3	0	1
2	2	1	1	4	3	0
1	2	0	3	3	2	1
1	1	1	0	0	3	2
2	2	0	2	1	1	1

Lesson 2-2

Pages 81–85

Make a bar graph for each set of data.

1.

Favorite Subject	
Subject	**Frequency**
Math	4
Science	6
History	2
English	8
Phys. Ed.	12

2.

Final Grades	
Subject	**Score**
Math	88
Science	82
History	92
English	94

Use the graph to solve each problem.

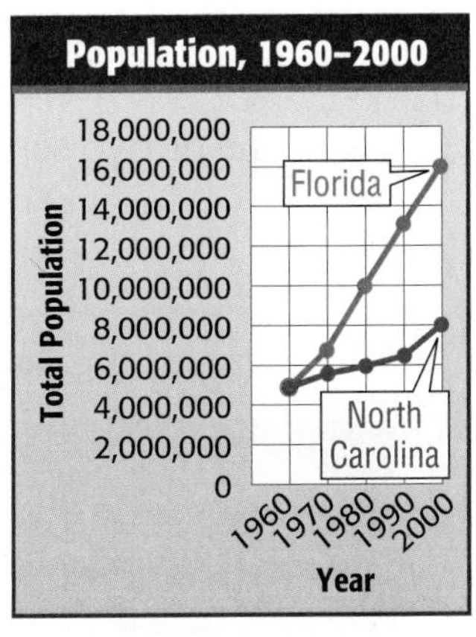

3. Give the approximate population in 1960 for:
   a. Florida
   b. North Carolina
4. How much greater was the population of Florida than North Carolina in 1980?
5. Which state will have the greater population in 2010?
6. How much greater was the difference in population in 2000 than in 1990?
7. Make a prediction for the population of Florida in 2010.

Lesson 2-3

Pages 88–91

Make a line graph for each set of data.

1.

Test Scores	
Test	**Score**
1	62
2	75
3	81
4	83
5	78
6	92

2.

Homeroom Absences	
Day	**Absences**
Mon.	3
Tue.	6
Wed.	2
Thur.	1
Fri.	8

Lesson 2-4

Pages 92–95

Make a stem-and-leaf plot for each set of data.

1. 23, 15, 39, 68, 57, 42, 51, 52, 41, 18, 29
2. 5, 14, 39, 28, 14, 6, 7, 18, 13, 28, 9, 14
3. 189, 182, 196, 184, 197, 183, 196, 194, 184
4. 71, 82, 84, 95, 76, 92, 83, 74, 81, 75, 96
5. 65, 72, 81, 68, 77, 70, 59, 63, 75, 68
6. 18, 23, 35, 21, 28, 32, 17, 25, 30, 37, 25, 24

Lesson 2-5

Pages 96–100

Make a line plot of each set of data.

1.

Number of Points Scored						
8	10	5	9	12	14	5
6	3	7	0	1	14	16
20	2	5	10	18	0	6

2.

Number of Books Read				
30	25	26	29	18
15	30	25	16	15
18	20	20	22	24
30	18	17	15	11

TEMPERATURE For Exercises 3–6, use the line plot shown.

Top 20 Record High Temperatures (°F) by State

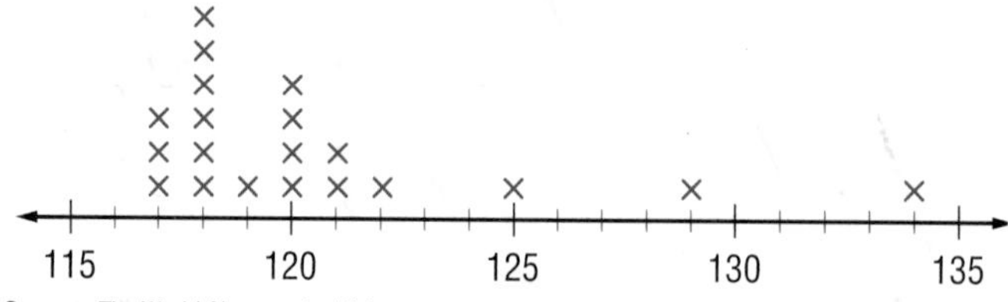

Source: *The World Almanac for Kids*

3. How many states have a record high temperature of higher than 120°F?
4. Which temperature is the most common record high temperature?
5. What is the difference between the lowest and highest temperature represented in the line plot?
6. Write one or two sentences that analyze the data.

Lesson 2-6

Pages 102–106

Find the mean of the data represented in each model.

1.

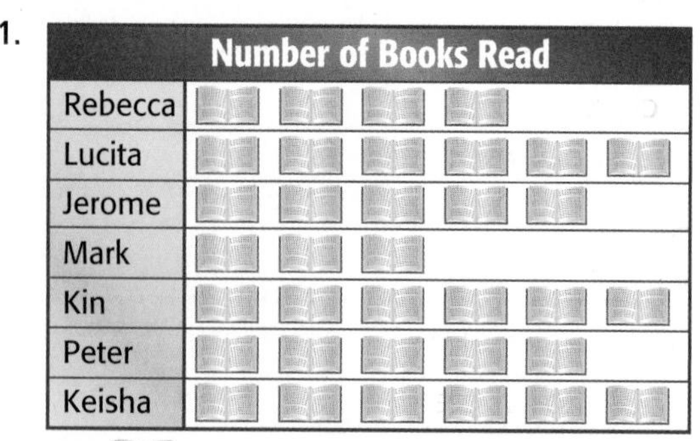

2.

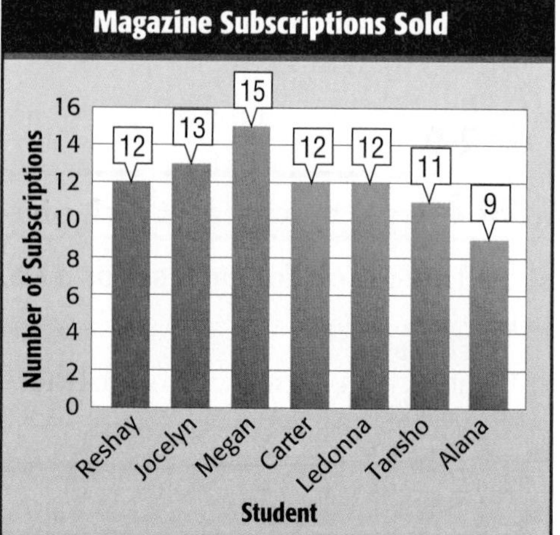

Find the mean for each set of data.

3. Number of birds identified by each student: 1, 5, 9, 1, 2, 4, 8, 2
4. Money raised by each class: \$957, \$562, \$462, \$848, \$721
5. Fliers handed out by each club member: 46, 54, 66, 54, 50, 66

Lesson 2-7

Pages 108–113

Find the median, mode, and range for each set of data.

1. 16, 12, 20, 15, 12
2. 42, 38, 56, 48, 43, 43
3. 8, 3, 12, 5, 2, 9, 3
4. 85, 75, 93, 82, 73, 78
5. 25, 32, 38, 27, 35, 25, 28
6. 112, 103, 121, 104
7. 57, 63, 53, 67, 71, 67
8. 21, 25, 20, 28, 26
9. 57, 42, 86, 76, 42, 57
10. 215, 176, 194, 223, 202
11.

Students Per Class				
18	21	22	19	22
26	24	18	24	26

12.

Stem	Leaf
1	0 5
2	1 6 9
3	1 2 8
4	1 2 2 4
5	0 5 *4|1 = 41*

Lesson 2-8

Pages 114–118

1. **CAPS** Which display makes it easier to determine which type of animal has the most number of endangered species?

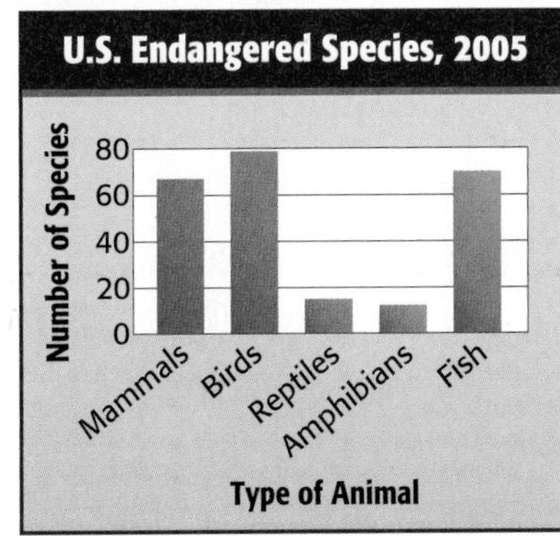

Number of U.S. Endangered Species

Stem	Leaf
1	2 4
2	
3	
4	
5	
6	8
7	1 7 *6|8 = 68 species*

Select an appropriate type of display for data gathered about each situation.

2. gas mileage of different vehicles
3. the number of days of winter with one or more inches of snow
4. the weight of a baby from birth to 18 months of age
5. exact test scores of all students in a science class

Lesson 2-9

Pages 121–125

Write an integer to represent each piece of data.

1. Mara's account had a loss of 15 dollars.
2. It was 9 degrees below zero.
3. The city is 456 feet above sea level.
4. The toy increased in value by $30.

Graph each integer on a number line.

5. −3
6. 3
7. −1
8. −8
9. 9
10. 10
11. Make a line plot of the data represented in the table below.

Quiz Scores						
5	7	−2	−1	10	5	8
−3	8	−4	7	8	−2	−3
−2	−2	9	−1	7	10	6

Lesson 3-1

Pages 138–141

Write each decimal in word form.

1. 0.8
2. 5.9
3. 1.34
4. 0.02
5. 9.21
6. 4.3
7. 13.42
8. 0.006
9. 65.083
10. 0.0072

Write each decimal in standard form and expanded form.

11. two hundredths
12. sixteen hundredths
13. four tenths
14. two and twenty-seven hundredths
15. nine and twelve hundredths
16. fifty-six and nine tenths
17. twenty-seven thousandths
18. one hundred ten-thousandths

Lesson 3-2

Pages 142–145

Use >, <, or = to compare each pair of decimals.

1. 6.0 ● 0.06
2. 1.19 ● 11.9
3. 9.4 ● 9.40
4. 1.04 ● 1.40
5. 1.2 ● 1.02
6. 8.0 ● 8.01
7. 12.13 ● 12.31
8. 0.03 ● 0.003
9. 6.15 ● 6.151
10. 0.112 ● 0.121
11. 0.556 ● 0.519
12. 8.007 ● 8.070
13. 6.005 ● 6.0050
14. 18.104 ● 18.140
15. 0.0345 ● 0.0435

Order each set of decimals from least to greatest.

16. 15.65, 51.65, 51.56, 15.56
17. 2.56, 20.56, 0.09, 25.6
18. 1.235, 1.25, 1.233, 1.23
19. 50.12, 5.012, 5.901, 50.02

Order each set of decimals from greatest to least.

20. 13.66, 13.44, 1.366, 1.633
21. 26.69, 26.09, 26.666, 26.9
22. 1.065, 1.1, 1.165, 1.056
23. 2.014, 2.010, 22.00, 22.14

Lesson 3-3

Pages 146–149

Round each decimal to the indicated place-value position.

1. 5.64; tenths
2. 0.05362; thousandths
3. 6.17; ones
4. 15.298; tenths
5. 0.0026325; ten-thousandths
6. 758.999; hundredths
7. 32.6583; thousandths
8. 0.025; ones
9. 1.0049; thousandths
10. 9.25; tenths
11. 67.492; hundredths
12. 26.96; tenths
13. 4.00098; hundredths
14. 34.065; hundredths
15. 18.999; tenths
16. 74.00065; ten-thousandths

Lesson 3-4

Pages 150–154

Estimate using rounding.

1. 0.245 + 0.256
2. 2.45698 – 1.26589
3. 0.5962 + 1.2598
4. 0.256 + 0.6589
5. 1.2568 – 0.1569
6. 12.999 + 5.048

Estimate using clustering.

7. 4.5 + 4.95 + 5.2 + 5.49
8. 2.25 + 1.69 + 2.1 + 2.369
9. 12.15 + 11.63 + 12 + 11.89
10. 0.569 + 1.005 + 1.265 + 0.765
11. 9.85 + 10.32 + 10.18 + 9.64
12. 18.32 + 17.92 + 18.04
13. 16.3 + 15.82 + 16.01 + 15.9
14. 0.775 + 0.9 + 1.004 + 1.32

Estimate using front-end estimation.

15. 73.41 + 24.08
16. 88.42 – 63.59
17. 106.08 + 90.83
18. 143.24 – 43.65
19. 64.98 + 52.43
20. 96.51 – 41.32

Lesson 3-5

Pages 156–160

Find each sum or difference.

1. 0.46 + 0.72
2. 13.7 + 2.6
3. 17.9 + 7.41
4. 0.51 + 0.621
5. 12.56 – 10.21
6. 2.3 – 1.02
7. 1.025 – 0.58
8. 2.35 + 5
9. 20 – 5.98
10. 15.256 + 0.236
11. 3.7 + 1.5 + 0.2
12. 0.23 + 1.2 + 0.36
13. 0.89 – 0.256
14. 25.6 – 2.3
15. 13.5 – 2.84
16. 1.265 + 1.654
17. 24.56 – 24.32
18. 0.256 – 0.255

Lesson 3-6

Pages 163–166

Multiply.

1. 0.2 × 6
2. 0.73 × 5
3. 0.65 × 3
4. 9.6 × 4
5. 12.15 × 6
6. 0.91 × 8
7. 0.265 × 7
8. 2.612 × 4
9. 0.013 × 5
10. 0.67 × 2
11. 9 × 0.111
12. 1.65 × 7
13. 9.6 × 3
14. 4 × 1.201
15. 6 × 7.5
16. 0.001 × 6
17. 5 × 0.0135
18. 9.2 × 7
19. 14.1235 × 4

Write each number in standard form.

20. 3×10^4
21. 2.6×10^6
22. 3.81×10^2
23. 8.4×10^3
24. 6.05×10^5
25. 7.32×10^4
26. 4.2×10^3
27. 5.06×10^7
28. 9.25×10^5

Lesson 3-7

Pages 169–172

Multiply.

1. 9.6×10.5
2. 3.2×0.1
3. 1.5×9.6
4. 5.42×0.21
5. 7.42×0.2
6. 0.001×0.02
7. 0.6×542
8. 6.7×5.8
9. 3.24×6.7
10. 9.8×4.62
11. 7.32×9.7
12. 0.008×0.007
13. 0.001×56
14. 4.5×0.2
15. 9.6×2.3
16. 8.3×4.9
17. 12.06×5.9
18. 0.04×8.25
19. 5.63×8.1
20. 10.35×9.1
21. 28.2×3.9

Evaluate each expression if $x = 2.6$, $y = 0.38$, and $z = 1.02$.

22. $xy + z$
23. $y \times 9.4$
24. xyz
25. $z \times 12.34 + y$
26. $yz \times 0.8$
27. $xz - yz$

Lesson 3-8

Pages 173–176

Divide. Round to the nearest tenth if necessary.

1. $1.2 \div 6$
2. $23.2 \div 8$
3. $89.4 \div 6$
4. $55.5 \div 15$
5. $128.7 \div 13$
6. $2.583 \div 9$
7. $9.4 \div 47$
8. $33.8 \div 26$
9. $37.8 \div 14$
10. $5.88 \div 4$
11. $3.7 \div 5$
12. $41.4 \div 18$
13. $9.87 \div 3$
14. $8.45 \div 25$
15. $26.5 \div 4$
16. $46.25 \div 8$
17. $19.38 \div 9$
18. $8.5 \div 2$
19. $90.88 \div 14$
20. $23.1 \div 4$
21. $19.5 \div 27$
22. $26.5 \div 19$
23. $46.25 \div 25$
24. $46.25 \div 25$

Lesson 3-9

Pages 179–183

Divide. Round to the nearest hundredth if necessary.

1. $18.45 \div 0.5$
2. $5.2 \div 0.08$
3. $0.65 \div 2.6$
4. $12.831 \div 1.3$
5. $5.133 \div 0.87$
6. $24.13 \div 2.54$
7. $35.89 \div 3.7$
8. $32.5 \div 26$
9. $5.88 \div 0.4$
10. $3.7 \div 0.5$
11. $6.72 \div 2.4$
12. $9.87 \div 0.3$
13. $8.45 \div 2.5$
14. $90.88 \div 14.2$
15. $33.6 \div 8.4$
16. $0.1185 \div 7.9$
17. $0.384 \div 9.6$
18. $5.364 \div 2.3$
19. $12.68 \div 3.2$
20. $43.1 \div 8.65$
21. $26.08 \div 3.3$

Evaluate each expression if $x = 1.8$, $y = 0.6$, and $z = 0.3$.

22. $x \div y$
23. $y \div z$
24. $x \div z$
25. $xy \div z$
26. $xz \div y$
27. $yz \div x$

Lesson 3-10

Pages 184–185

Determine reasonable answers for Exercises 1–4.

1. **MONEY** Brandy wants to buy 2 science fiction books for $3.95 each, 3 magazines for $2.95 each, and 1 bookmark for $0.39 at the book fair. Does she need to bring $20 or will $15 be enough? Explain your reasoning.
2. **GRADUATION** The high school gym will hold 2,800 guests and the 721 seniors who are graduating. Is it reasonable to offer each graduate three, four, or five tickets for family and friends? Explain your reasoning.
3. **BANQUET** One hundred fifty students are being honored at an awards banquet. Each student is permitted to bring a maximum of 3 guests. Is it reasonable to set up 40 tables that seat 12 people each? Explain your reasoning.
4. **POPCORN** The data show the number of bags of popcorn sold by each student in Ms. Taylor's class. Is 1,300 bags a reasonable goal for the class to set for next year's popcorn fundraiser? Explain your reasoning.

Bags of Popcorn Sold								
10	250	35	68	102	50	28	18	98
12	75	60	185	11	27	33	87	100

Lesson 4-1

Pages 197–201

Identify the common factors of each set of numbers.

1. 20, 40
2. 8, 28
3. 32, 80
4. 9, 54
5. 27, 81
6. 18, 34, 54

Find the GCF of each set of numbers.

7. 64 and 32
8. 14 and 22
9. 12 and 27
10. 17 and 51
11. 48 and 60
12. 54 and 72
13. 60 and 75
14. 54 and 27
15. 14, 28, 42
16. 27, 45, 63
17. 12, 18, 30
18. 16, 40, 72

Lesson 4-2

Pages 204–208

Replace each ■ with a number so the fractions are equivalent.

1. $\frac{12}{16} = \frac{■}{4}$
2. $\frac{7}{8} = \frac{■}{32}$
3. $\frac{3}{4} = \frac{75}{■}$
4. $\frac{8}{16} = \frac{■}{2}$
5. $\frac{6}{18} = \frac{1}{■}$
6. $\frac{27}{36} = \frac{3}{■}$
7. $\frac{1}{4} = \frac{16}{■}$
8. $\frac{9}{18} = \frac{■}{2}$
9. $\frac{9}{45} = \frac{■}{15}$
10. $\frac{2}{3} = \frac{18}{■}$
11. $\frac{24}{32} = \frac{■}{4}$
12. $\frac{48}{72} = \frac{6}{■}$

Write each fraction in simplest form. If the fraction is already in simplest form, write *simplest form*.

13. $\frac{50}{100}$
14. $\frac{24}{40}$
15. $\frac{2}{5}$
16. $\frac{8}{24}$
17. $\frac{20}{27}$
18. $\frac{4}{10}$
19. $\frac{3}{5}$
20. $\frac{14}{19}$
21. $\frac{9}{12}$
22. $\frac{6}{8}$

Lesson 4-3

Pages 209–212

Write each mixed number as an improper fraction.

1. $3\frac{1}{16}$
2. $2\frac{1}{4}$
3. $1\frac{3}{8}$
4. $1\frac{5}{12}$
5. $7\frac{3}{5}$
6. $6\frac{5}{8}$
7. $3\frac{1}{3}$
8. $1\frac{7}{9}$
9. $2\frac{3}{16}$
10. $1\frac{2}{3}$
11. $2\frac{3}{4}$
12. $1\frac{5}{9}$
13. $3\frac{3}{8}$
14. $4\frac{1}{5}$
15. $6\frac{2}{3}$

Write each improper fraction as a mixed number or a whole number.

16. $\frac{33}{10}$
17. $\frac{103}{25}$
18. $\frac{22}{5}$
19. $\frac{13}{2}$
20. $\frac{29}{6}$
21. $\frac{101}{100}$
22. $\frac{21}{8}$
23. $\frac{19}{6}$
24. $\frac{23}{5}$
25. $\frac{99}{50}$
26. $\frac{39}{8}$
27. $\frac{13}{4}$
28. $\frac{26}{9}$
29. $\frac{18}{5}$
30. $\frac{11}{11}$

Lesson 4-4

Pages 214–215

Use the *make an organized list* strategy to solve Exercises 1–4.

1. **UNIFORMS** The volleyball team is ordering new uniforms. They can choose from 3 styles of shirts and 3 styles of shorts. How many combinations are possible for the complete uniform?
2. **SCRAPBOOKS** Maria is making a scrapbook for her friend's birthday. She has 6 pictures to put on one page with 3 rows of 2 pictures in each row. How many different ways can she arrange the pictures?
3. **FOOD** A local restaurant has a 99¢ menu which includes 2 sandwiches, 3 side dishes, and 1 size of soft drink. How many different meals are possible if a meal includes 1 sandwich, 1 side dish, and a drink?
4. **CARS** Jeremy is arranging five model cars on a shelf in his room. How many ways can he choose to arrange them in a single row on the shelf?

Lesson 4-5

Pages 216–219

Identify the first three common multiples of each set of numbers.

1. 4, 12
2. 3, 9
3. 2, 5
4. 10, 15
5. 4, 12, 16
6. 2, 6, 10
7. 3, 8, 12
8. 9, 15, 30
9. 5, 8, 10

Find the LCM for each set of numbers.

10. 9 and 6
11. 4 and 14
12. 14 and 49
13. 10 and 45
14. 18 and 12
15. 12 and 30
16. 5 and 7
17. 6 and 15
18. 20 and 8
19. 18 and 24
20. 11 and 33
21. 2, 3, and 6
22. 6, 2, and 22
23. 15, 12, and 8
24. 15, 24, and 30

Lesson 4-6

Pages 220–224

Replace each ● with <, >, or = to make a true sentence.

1. $\frac{1}{2}$ ● $\frac{1}{3}$
2. $\frac{2}{3}$ ● $\frac{3}{4}$
3. $\frac{5}{9}$ ● $\frac{4}{5}$
4. $\frac{3}{5}$ ● $\frac{6}{12}$
5. $\frac{12}{23}$ ● $\frac{15}{19}$
6. $\frac{9}{27}$ ● $\frac{13}{39}$
7. $\frac{7}{8}$ ● $\frac{9}{13}$
8. $\frac{5}{9}$ ● $\frac{7}{8}$
9. $\frac{25}{100}$ ● $\frac{3}{8}$
10. $\frac{6}{7}$ ● $\frac{8}{15}$
11. $\frac{5}{9}$ ● $\frac{19}{23}$
12. $\frac{17}{25}$ ● $\frac{68}{100}$
13. $\frac{5}{7}$ ● $\frac{2}{3}$
14. $\frac{9}{36}$ ● $\frac{7}{28}$
15. $\frac{2}{5}$ ● $\frac{2}{6}$
16. $\frac{5}{9}$ ● $\frac{12}{13}$
17. $\frac{3}{5}$ ● $\frac{5}{8}$
18. $\frac{8}{9}$ ● $\frac{3}{4}$
19. $\frac{11}{15}$ ● $\frac{20}{30}$
20. $\frac{15}{24}$ ● $\frac{6}{8}$

Order the fractions from least to greatest.

21. $\frac{1}{3}, \frac{1}{2}, \frac{2}{7}, \frac{2}{5}$
22. $\frac{3}{8}, \frac{3}{4}, \frac{1}{2}, \frac{5}{6}$
23. $\frac{5}{12}, \frac{3}{5}, \frac{3}{4}, \frac{1}{2}$
24. $\frac{7}{8}, \frac{5}{6}, \frac{8}{9}, \frac{3}{4}$
25. $\frac{7}{10}, \frac{5}{8}, \frac{3}{4}, \frac{4}{5}$
26. $\frac{5}{9}, \frac{7}{12}, \frac{3}{4}, \frac{13}{18}$

Lesson 4-7

Pages 225–228

Write each decimal as a fraction or mixed number in simplest form.

1. 0.5
2. 0.8
3. 0.32
4. 0.875
5. 0.54
6. 0.38
7. 0.744
8. 0.101
9. 0.303
10. 0.486
11. 0.626
12. 0.448
13. 0.074
14. 0.008
15. 9.36
16. 10.18
17. 0.06
18. 0.75
19. 0.48
20. 0.9
21. 0.005
22. 0.4
23. 1.875
24. 5.08
25. 0.46
26. 0.128
27. 0.08
28. 6.96
29. 3.625
30. 0.006
31. 12.05
32. 24.125

Lesson 4-8

Pages 229–232

Write each fraction or mixed number as a decimal.

1. $\frac{3}{16}$
2. $\frac{1}{8}$
3. $\frac{7}{16}$
4. $\frac{14}{25}$
5. $\frac{7}{10}$
6. $\frac{5}{8}$
7. $\frac{11}{20}$
8. $\frac{8}{25}$
9. $\frac{15}{16}$
10. $\frac{1}{10}$
11. $\frac{7}{20}$
12. $\frac{5}{16}$
13. $\frac{9}{10}$
14. $\frac{11}{25}$
15. $\frac{9}{20}$
16. $4\frac{4}{25}$
17. $1\frac{3}{8}$
18. $\frac{9}{16}$
19. $2\frac{3}{10}$
20. $6\frac{3}{5}$
21. $\frac{3}{4}$
22. $3\frac{4}{5}$
23. $2\frac{1}{10}$
24. $\frac{17}{25}$
25. $5\frac{13}{20}$
26. $\frac{37}{50}$
27. $4\frac{9}{25}$
28. $\frac{7}{4}$

Lesson 4-9

Pages 233–237

Use the coordinate plane at the right to name the ordered pair for each point.

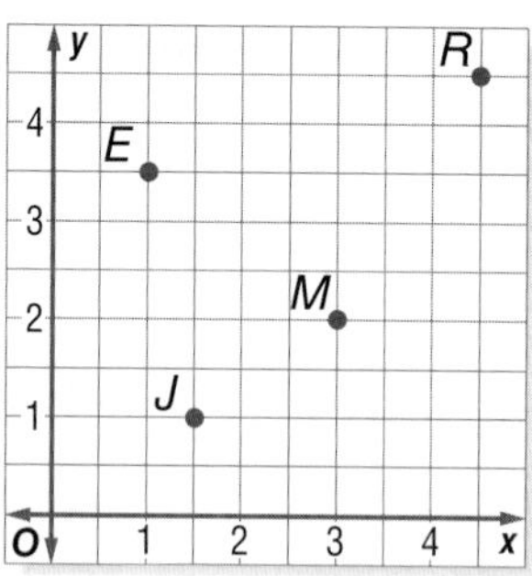

1. E
2. M
3. R
4. J

Graph and label each point on a coordinate plane.

5. $H(0, 2)$
6. $K(3, 1)$
7. $S(4, 2.5)$
8. $Z(1.5, 0)$

MONEY For Exercises 9 and 10, use the following information.
Tanya is saving $1.50 each day to buy a new sweater. The table at the right shows this relationship.

Days	Total Saved ($)
1	1.5
2	3
3	4.5
4	6

9. List this information as ordered pairs (number of days, total amount saved).
10. Graph the ordered pairs. Then describe the graph.

Lesson 5-1

Pages 249–253

Round each number to the nearest half.

1. $\frac{11}{12}$
2. $\frac{5}{8}$
3. $\frac{2}{5}$
4. $\frac{1}{10}$
5. $\frac{1}{6}$
6. $\frac{2}{3}$
7. $\frac{9}{10}$
8. $\frac{1}{8}$
9. $\frac{4}{9}$
10. $1\frac{1}{8}$
11. $\frac{7}{9}$
12. $2\frac{4}{5}$
13. $\frac{7}{9}$
14. $7\frac{1}{10}$
15. $10\frac{2}{3}$
16. $\frac{1}{3}$
17. $\frac{7}{15}$
18. $\frac{5}{7}$
19. $1\frac{2}{7}$
20. $9\frac{4}{5}$
21. $3\frac{9}{11}$
22. $2\frac{3}{7}$
23. $7\frac{3}{8}$
24. $4\frac{17}{20}$

Lesson 5-2

Pages 254–255

Use the *act it out* strategy to solve Exercises 1–4.

1. **SUNDAES** The student council is setting up an ice cream sundae bar for a fundraiser. They plan to offer 3 flavors of ice cream and 4 toppings. How many different sundaes can be made with one flavor of ice cream and one topping?
2. **COLORS** Paco is coloring the design shown for a poster. How many ways can he color it if he only uses two colors?

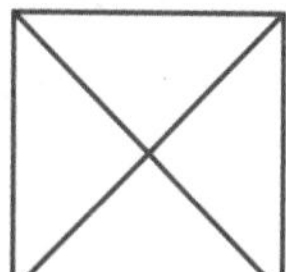

3. **DESKS** Mr. Johnson is rearranging the desks in his classroom. If the desks have to be set up in rows all facing the same direction, how many arrangements are possible for 24 desks? Describe all possible arrangements.
4. **BRACELETS** Rachel uses $7\frac{1}{2}$ inches of string to make a bracelet. She had 4 feet of string, but already used 16 inches for a necklace. Does she have enough string left to make one bracelet for each of her 3 friends? Explain your reasoning.

Lesson 5-3

Pages 256–260

Add or subtract. Write in simplest form.

1. $\frac{2}{5}+\frac{2}{5}$
2. $\frac{5}{8}+\frac{3}{8}$
3. $\frac{9}{11}-\frac{3}{11}$
4. $\frac{3}{14}+\frac{5}{14}$
5. $\frac{7}{8}-\frac{3}{8}$
6. $\frac{3}{4}-\frac{1}{4}$
7. $\frac{15}{27}-\frac{7}{27}$
8. $\frac{1}{36}+\frac{5}{36}$
9. $\frac{2}{9}-\frac{1}{9}$
10. $\frac{7}{8}-\frac{5}{8}$
11. $\frac{9}{16}-\frac{5}{16}$
12. $\frac{6}{8}+\frac{4}{8}$
13. $\frac{1}{2}+\frac{1}{2}$
14. $\frac{1}{3}-\frac{1}{3}$
15. $\frac{8}{9}+\frac{7}{9}$
16. $\frac{5}{6}-\frac{3}{6}$
17. $\frac{3}{9}+\frac{8}{9}$
18. $\frac{8}{40}+\frac{12}{40}$
19. $\frac{56}{90}-\frac{26}{90}$
20. $\frac{2}{9}+\frac{8}{9}$
21. $\frac{7}{15}-\frac{4}{15}$
22. $\frac{3}{8}+\frac{7}{8}$
23. $\frac{11}{20}-\frac{5}{20}$
24. $\frac{7}{9}-\frac{2}{9}$

Lesson 5-4

Pages 263–268

Add or subtract. Write in simplest form.

1. $\frac{1}{3}+\frac{1}{2}$
2. $\frac{2}{9}+\frac{1}{3}$
3. $\frac{1}{2}+\frac{3}{4}$
4. $\frac{1}{4}+\frac{3}{12}$
5. $\frac{5}{9}-\frac{1}{3}$
6. $\frac{5}{8}-\frac{2}{5}$
7. $\frac{3}{4}-\frac{1}{2}$
8. $\frac{7}{8}-\frac{3}{16}$
9. $\frac{9}{16}+\frac{13}{24}$
10. $\frac{8}{15}+\frac{2}{3}$
11. $\frac{5}{14}+\frac{11}{28}$
12. $\frac{11}{12}+\frac{7}{8}$
13. $\frac{2}{3}-\frac{1}{6}$
14. $\frac{9}{16}-\frac{1}{2}$
15. $\frac{5}{8}-\frac{11}{20}$
16. $\frac{14}{15}-\frac{2}{9}$
17. $\frac{9}{20}+\frac{2}{15}$
18. $\frac{5}{6}+\frac{4}{5}$
19. $\frac{23}{25}-\frac{27}{50}$
20. $\frac{19}{25}-\frac{1}{2}$
21. $\frac{1}{5}+\frac{2}{3}$
22. $\frac{4}{9}-\frac{1}{6}$
23. $\frac{1}{3}+\frac{4}{5}$
24. $\frac{7}{8}-\frac{5}{12}$
25. $\frac{1}{12}+\frac{3}{10}$
26. $\frac{3}{15}+\frac{1}{25}$
27. $\frac{5}{12}-\frac{1}{6}$
28. $\frac{9}{24}-\frac{3}{8}$

Lesson 5-5

Pages 270–274

Add or subtract. Write in simplest form.

1. $5\frac{1}{2}+3\frac{1}{4}$
2. $2\frac{2}{3}+4\frac{1}{9}$
3. $7\frac{4}{5}+9\frac{3}{10}$
4. $9\frac{4}{7}-3\frac{5}{14}$
5. $13\frac{1}{5}-10$
6. $3\frac{3}{4}+5\frac{5}{8}$
7. $3\frac{2}{5}+7\frac{6}{15}$
8. $10\frac{2}{3}+5\frac{6}{7}$
9. $15\frac{6}{9}-13\frac{5}{12}$
10. $13\frac{7}{12}-9\frac{1}{4}$
11. $5\frac{2}{3}-3\frac{1}{2}$
12. $17\frac{2}{9}+12\frac{1}{3}$
13. $6\frac{5}{12}+12\frac{5}{8}$
14. $8\frac{3}{5}-2\frac{1}{5}$
15. $23\frac{2}{3}-4\frac{1}{2}$
16. $7\frac{1}{8}+2\frac{5}{8}$
17. $18\frac{1}{5}-6\frac{1}{4}$
18. $4-1\frac{2}{3}$
19. $26-4\frac{1}{9}$
20. $3\frac{1}{2}-1\frac{3}{4}$
21. $4\frac{3}{8}-2\frac{5}{6}$
22. $18\frac{1}{6}-10\frac{3}{4}$
23. $12\frac{4}{9}-7\frac{5}{6}$
24. $4\frac{1}{15}-2\frac{3}{5}$
25. $7\frac{1}{14}-4\frac{3}{7}$
26. $12\frac{3}{20}-7\frac{7}{15}$
27. $8\frac{5}{12}-5\frac{9}{10}$
28. $8\frac{1}{5}-4\frac{2}{3}$

Lesson 5-6

Pages 276–279

Estimate each product.

1. $\frac{5}{6} \times 8$
2. $\frac{1}{3} \times 46$
3. $\frac{4}{5}$ of 21
4. $\frac{1}{9} \times 35$
5. $\frac{5}{9}$ of 20
6. $\frac{1}{8} \times 30$
7. $\frac{2}{3} \times \frac{4}{5}$
8. $\frac{1}{6} \times \frac{2}{5}$
9. $\frac{4}{9} \times \frac{3}{7}$
10. $\frac{5}{12} \times \frac{6}{11}$
11. $\frac{3}{8} \times \frac{8}{9}$
12. $\frac{3}{5} \times \frac{5}{12}$
13. $\frac{2}{5} \times \frac{5}{8}$
14. $\frac{4}{5} \times \frac{11}{12}$
15. $\frac{5}{7} \times \frac{7}{8}$
16. $\frac{1}{20} \times \frac{8}{9}$
17. $\frac{9}{11} \times \frac{14}{15}$
18. $\frac{2}{5} \times \frac{18}{19}$
19. $5\frac{3}{7} \times \frac{4}{5}$
20. $2\frac{5}{9} \times \frac{1}{8}$
21. $3\frac{9}{10} \times \frac{15}{16}$
22. $\frac{11}{12} \times 2\frac{1}{3}$
23. $3\frac{14}{15} \times \frac{3}{8}$
24. $\frac{1}{10} \times 3\frac{1}{2}$
25. $9\frac{13}{15} \times \frac{1}{2}$
26. $6\frac{7}{8} \times 2\frac{1}{5}$
27. $7\frac{1}{4} \times 4\frac{3}{4}$
28. $6\frac{2}{3} \times 5\frac{4}{5}$

Lesson 5-7

Pages 282–286

Multiply. Write in simplest form.

1. $\frac{1}{8} \times \frac{1}{9}$
2. $\frac{4}{7} \times 6$
3. $\frac{7}{10} \times 5$
4. $\frac{3}{8} \times 6$
5. $4 \times \frac{5}{9}$
6. $\frac{9}{10} \times \frac{3}{4}$
7. $\frac{8}{9} \times \frac{2}{3}$
8. $\frac{6}{7} \times \frac{4}{5}$
9. $\frac{7}{11} \times \frac{12}{15}$
10. $\frac{8}{13} \times \frac{2}{11}$
11. $\frac{4}{7} \times \frac{2}{9}$
12. $\frac{3}{7} \times \frac{5}{8}$
13. $\frac{5}{6} \times \frac{15}{16}$
14. $\frac{6}{14} \times \frac{12}{18}$
15. $\frac{2}{3} \times \frac{3}{13}$
16. $\frac{4}{9} \times \frac{1}{6}$
17. $\frac{3}{4} \times \frac{5}{6}$
18. $\frac{8}{11} \times \frac{11}{12}$
19. $\frac{5}{6} \times \frac{3}{5}$
20. $\frac{6}{7} \times \frac{7}{21}$
21. $\frac{8}{9} \times \frac{9}{10}$
22. $\frac{7}{9} \times \frac{5}{7}$
23. $\frac{4}{9} \times \frac{24}{25}$
24. $\frac{1}{9} \times \frac{6}{13}$
25. $\frac{5}{9} \times \frac{3}{10}$
26. $\frac{2}{3} \times \frac{7}{8}$
27. $\frac{7}{12} \times \frac{4}{9}$
28. $\frac{11}{15} \times \frac{3}{10}$

Lesson 5-8

Pages 287–290

Multiply. Write in simplest form.

1. $\frac{4}{5} \times 2\frac{3}{4}$
2. $2\frac{3}{10} \times \frac{3}{5}$
3. $8\frac{5}{6} \times \frac{2}{5}$
4. $\frac{3}{4} \times 9\frac{5}{7}$
5. $6\frac{2}{3} \times 7\frac{3}{5}$
6. $7\frac{1}{5} \times 2\frac{4}{7}$
7. $8\frac{3}{4} \times 2\frac{2}{5}$
8. $4\frac{1}{3} \times 2\frac{1}{7}$
9. $4\frac{3}{5} \times 2\frac{1}{2}$
10. $5\frac{5}{6} \times 4\frac{2}{7}$
11. $6\frac{8}{9} \times 3\frac{5}{6}$
12. $2\frac{1}{9} \times 1\frac{1}{2}$
13. $4\frac{7}{15} \times 3\frac{3}{4}$
14. $5\frac{7}{9} \times 6\frac{3}{8}$
15. $1\frac{1}{4} \times 3\frac{2}{3}$
16. $2\frac{3}{5} \times 1\frac{4}{7}$
17. $4\frac{1}{5} \times 12\frac{2}{9}$
18. $3\frac{5}{8} \times 4\frac{1}{2}$
19. $6\frac{1}{2} \times 2\frac{1}{3}$
20. $3\frac{4}{5} \times 2\frac{3}{8}$
21. $5\frac{1}{4} \times 10\frac{3}{7}$
22. $1\frac{5}{9} \times 6\frac{1}{4}$
23. $2\frac{1}{6} \times 1\frac{3}{4}$
24. $3\frac{5}{7} \times 1\frac{1}{2}$
25. $2\frac{3}{5} \times 4\frac{5}{8}$
26. $6\frac{1}{8} \times 5\frac{1}{7}$
27. $2\frac{2}{3} \times 2\frac{1}{4}$
28. $2\frac{1}{2} \times 3\frac{1}{3}$

Lesson 5-9

Pages 293–297

Find the reciprocal of each number.

1. $\frac{12}{13}$ 2. $\frac{7}{11}$ 3. 5 4. $\frac{1}{4}$ 5. $\frac{7}{9}$ 6. $\frac{9}{2}$ 7. $\frac{1}{5}$

Divide. Write in simplest form.

8. $\frac{2}{3} \div \frac{1}{2}$ 9. $\frac{3}{5} \div \frac{2}{5}$ 10. $\frac{7}{10} \div \frac{3}{8}$ 11. $\frac{5}{9} \div \frac{2}{3}$

12. $4 \div \frac{2}{3}$ 13. $8 \div \frac{4}{5}$ 14. $9 \div \frac{5}{9}$ 15. $\frac{2}{7} \div 7$

16. $\frac{1}{14} \div 7$ 17. $\frac{2}{13} \div \frac{5}{26}$ 18. $\frac{4}{7} \div \frac{6}{7}$ 19. $\frac{7}{8} \div \frac{1}{3}$

20. $15 \div \frac{3}{5}$ 21. $\frac{9}{14} \div \frac{3}{4}$ 22. $\frac{8}{9} \div \frac{5}{6}$ 23. $\frac{4}{9} \div 36$

24. $\frac{15}{16} \div \frac{5}{8}$ 25. $\frac{3}{5} \div \frac{7}{10}$ 26. $\frac{5}{9} \div \frac{3}{8}$ 27. $\frac{5}{6} \div \frac{3}{8}$

Lesson 5-10

Pages 298–301

Divide. Write in simplest form.

1. $\frac{3}{5} \div 1\frac{2}{3}$ 2. $2\frac{1}{2} \div 1\frac{1}{4}$ 3. $7 \div 4\frac{9}{10}$ 4. $1\frac{3}{7} \div 10$

5. $3\frac{3}{5} \div \frac{4}{5}$ 6. $8\frac{2}{5} \div 4\frac{1}{2}$ 7. $6\frac{1}{3} \div 2\frac{1}{2}$ 8. $5\frac{1}{4} \div 2\frac{1}{3}$

9. $4\frac{1}{8} \div 3\frac{2}{3}$ 10. $2\frac{5}{8} \div \frac{1}{2}$ 11. $1\frac{5}{6} \div 3\frac{2}{3}$ 12. $21 \div 5\frac{1}{4}$

13. $12 \div 3\frac{3}{5}$ 14. $18 \div 2\frac{1}{4}$ 15. $1\frac{7}{9} \div 2\frac{2}{3}$ 16. $2\frac{1}{15} \div 3\frac{1}{3}$

17. $1\frac{1}{8} \div 2\frac{2}{3}$ 18. $5\frac{1}{3} \div 2\frac{1}{2}$ 19. $1\frac{1}{4} \div 1\frac{7}{8}$ 20. $2\frac{3}{5} \div 1\frac{7}{10}$

21. $6\frac{3}{4} \div 3\frac{1}{2}$ 22. $4\frac{1}{2} \div \frac{3}{8}$ 23. $3\frac{1}{2} \div 1\frac{7}{9}$ 24. $12\frac{1}{2} \div 5\frac{5}{6}$

25. $8\frac{1}{4} \div 2\frac{3}{4}$ 26. $2\frac{3}{8} \div 5\frac{3}{7}$ 27. $4\frac{5}{9} \div 5\frac{1}{3}$ 28. $3\frac{1}{2} \div 5\frac{1}{4}$

Lesson 6-1

Pages 314–319

Write each ratio as a fraction in simplest form.

1. 10 girls in a class of 25 students
2. 7 striped ties out of 21 ties
3. 12 golden retrievers out of 20 dogs
4. 8 red marbles in a jar of 32 marbles
5. 6 roses in a bouquet of 21 flowers
6. 21 convertibles out of 75 cars

Write each rate as a unit rate.

7. \$2 for 5 cans of tomato soup
8. \$200 for 40 hours of work
9. 540 parts produced in 18 hours
10. \$2.16 for one dozen cookies
11. 228 words typed in 6 minutes
12. 558 miles in 9 hours

Lesson 6-2

Pages 322–327

For Exercises 1–3, use the ratio tables given to solve each problem.

1. **DANCE** A principal needs 2 chaperones for every 25 students at the school dance. How many chaperones does he need if 225 students are expected to be at the dance?

Chaperones	2	■
Students	25	225

2. **PIZZA** Nyoko is having a pizza party. If two large pizzas serve 9 people, how many pizzas should she order to serve 27 guests at the party?

Pizzas	2	■
Guests	9	27

3. **SALE** A store is having a sale on CDs. Olivia buys 5 CDs and spends $40. If she decides to go back to the sale and buy 2 more CDs, how much will she pay?

Lesson 6-3

Pages 329–333

Determine if the quantities in each pair of ratios or rates are proportional. Explain your reasoning and express each proportional relationship as a proportion.

1. 8 out of 10 boys play a sport; 25 out of 30 boys play a sport
2. $16 for 2 tickets; $40 for 5 tickets
3. 3 teachers for 63 students; 7 teachers for 147 students
4. 315 minutes to read 9 pages; 50 minutes to read 30 pages
5. **BABYSITTING** Lusita earned $85 for 8 hours of babysitting. Abigail earned $125 for 11 hours of babysitting. Are their earnings proportional? Explain your reasoning.
6. **ATTENDANCE** Miss Ferguson recorded 17 absences in her class during the first 30 days of school. During the last 60 days of school, she recorded 42 absences. Are these attendance rates proportional? Explain your reasoning.

Lesson 6-4

Pages 334–339

Solve each proportion.

1. $\frac{15}{21} = \frac{5}{b}$
2. $\frac{22}{25} = \frac{n}{100}$
3. $\frac{24}{48} = \frac{h}{96}$
4. $\frac{9}{27} = \frac{y}{81}$
5. $\frac{4}{7} = \frac{16}{x}$
6. $\frac{4}{6} = \frac{a}{24}$
7. $\frac{6}{14} = \frac{18}{m}$
8. $\frac{3}{7} = \frac{21}{d}$
9. $\frac{4}{10} = \frac{8}{e}$
10. $\frac{9}{10} = \frac{27}{f}$
11. $\frac{a}{3} = \frac{16}{24}$
12. $\frac{5}{9} = \frac{35}{w}$
13. **SURVEY** A class survey showed that 8 out of 25 students had plans to travel over spring break. How many students out of the 400 students in the entire school would you expect to be traveling over spring break?
14. **INSECTS** The world's fastest insect is the dragonfly, which can fly up to 36 miles per hour. At this rate, how long would it take a dragonfly to travel 54 miles?

Lesson 6-5

Pages 341–342

Use the *look for a pattern* strategy to solve Exercises 1–3.

1. **GEOMETRY** Draw the next two figures in the pattern below.

2. **RUNNING** Mandi is training to run a marathon. She begins by running 3 miles each day for the first week. If she increases her distance by 3 miles each week, how many weeks will it take her to reach 20 miles a day? Explain.

3. **NUMBER SENSE** Describe the pattern below. Then find the next three numbers in the pattern.

3, 5, 8, 12, 17, 23, ■, ■, ■, . . .

Lesson 6-6

Pages 343–348

Use words and symbols to describe the value of each term as a function of its position. Then find the value of the eleventh term in the sequence.

1.

Position	5	6	7	8	n
Value of Term	15	18	21	24	■

2.

Position	3	4	5	6	n
Value of Term	10	11	12	13	■

3. **MEASUREMENT** There are 16 cups in 1 gallon. Make a table and write an algebraic expression relating the number of cups to the number of gallons. Then find the number of cups of milk Adam has if he bought 4 gallons.

Lesson 6-7

Pages 349–353

Write an equation to represent the function displayed in each table.

1.

Input, x	0	1	2	3	4
Output, y	0	9	18	27	36

2.

Input, x	1	2	3	4	5
Output, y	2	4	6	8	10

3.

Input, x	4	8	12	16	20
Output, y	1	2	3	4	5

4.

Input, x	6	12	18	24	30
Output, y	1	2	3	4	5

WATER Use the following information for Exercise 5–7.

The average person uses about 12 gallons of water each day for showering.

5. Make a table to show the relationship between the number of gallons g a person uses for showering in days d.

6. Write an equation to find g, the number of gallons of water a person uses for showering in d days.

7. How many gallons of water does a person use for showering each week?

Extra Practice

Lesson 7-1

Pages 365–369

Write each percent as a fraction or mixed number in simplest form.

1. 13% 2. 25% 3. 8% 4. 105%
5. 60% 6. 70% 7. 80% 8. 45%
9. 14% 10. 5% 11. 2% 12. 450%

Write each fraction or mixed number as a percent.

13. $\frac{77}{100}$ 14. $\frac{3}{4}$ 15. $\frac{17}{20}$ 16. $\frac{3}{25}$
17. $3\frac{3}{10}$ 18. $\frac{27}{50}$ 19. $\frac{2}{5}$ 20. $\frac{3}{50}$
21. $\frac{8}{5}$ 22. $2\frac{19}{20}$ 23. $\frac{11}{25}$ 24. $\frac{5}{4}$

Lesson 7-2

Pages 370–375

1. **HOBBIES** Of Andrea's classmates, 32% play a musical instrument, 40% play a sport, 18% play an instrument and a sport, and 10% have some other hobby. Sketch a circle graph to display the data.
2. **TRANSPORTATION** The table shows the activities students choose most often as the first thing they do when they get home from school. Sketch a circle graph to display the data.

After-School Activities	
Watch TV	42%
Play video games	20%
Play a sport	8%
Talk on the phone	5%
Have a snack	25%

POPULATION For Exercises 3–5, use the graph at the right.

3. In which age group are the most Americans?
4. Which age group makes up the smallest portion of the U.S. population?
5. Which two age groups make up exactly half of the total U.S. population?

U.S. Population by Age, 2000

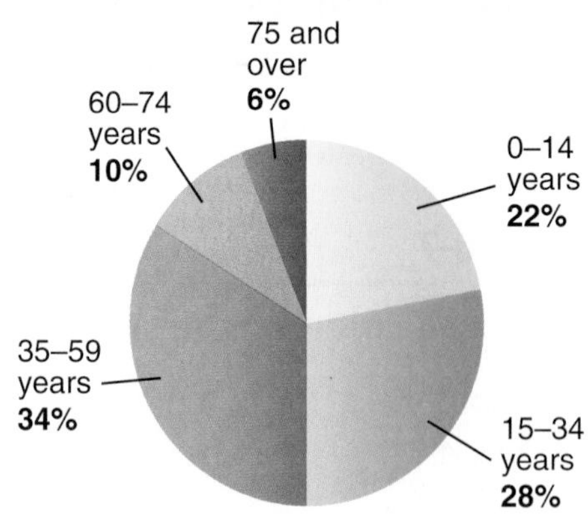

Source: *The World Almanac*

Lesson 7-3

Pages 377–380

Write each percent as a decimal.

1. 5% 2. 22% 3. 50% 4. 420%
5. 75% 6. 1% 7. 100% 8. 37%
9. 19% 10. 9% 11. 90% 12. 900%

Write each decimal as a percent.

13. 0.02 14. 0.2 15. 0.03 16. 1.02
17. 0.66 18. 0.11 19. 0.35 20. 0.31
21. 0.09 22. 5.2 23. 2.22 24. 0.08

Lesson 7-4

Pages 381–386

A set of 30 tickets are placed in a bag. There are 6 baseball tickets, 4 hockey tickets, 4 basketball tickets, 2 football tickets, 3 symphony tickets, 2 opera tickets, 4 ballet tickets, and 5 theater tickets. One ticket is selected without looking. Find each probability. Write each answer as a fraction.

1. *P*(basketball)
2. *P*(sports event)
3. *P*(opera or ballet)
4. *P*(soccer)
5. *P*(*not* symphony)
6. *P*(theater)
7. *P*(basketball or hockey)
8. *P*(*not* a sports event)
9. *P*(*not* opera)

Lesson 7-5

Pages 389–393

For Exercises 1–3, draw a tree diagram to show the sample space for each situation. Tell how many outcomes are possible.

1. tossing a quarter and rolling a number cube
2. a choice of a red, blue, or green sweater with a white, black, or tan skirt
3. a choice of a chicken, ham, turkey, or bologna sandwich with coffee, milk, juice, or soda
4. How many ways can a person choose two books from a shelf of four books?

For Exercises 5–7, use the Fundamental Counting Principle to determine the number of possible outcomes for each situation.

5. spinning a spinner with six different sections and tossing a coin
6. rolling a number cube and selecting a letter from the word *tiger*
7. selecting one sweater from three different sweaters and one pair of pants from two different pairs of pants

Lesson 7-6

Pages 394–398

BASKETBALL For Exercises 1 and 2, use the following information. In basketball, Daniel made 12 of his last 18 shots.

1. Find the probability of Daniel making a shot on his next attempt.
2. Suppose Daniel takes 30 shots during his next game. About how many of the shots will he make?

SCHOOL APPAREL For Exercises 3–6, use the table to predict the number of students out of 1,500 that would be most likely to purchase each type of school apparel.

3. T-Shirt
4. hat
5. jacket
6. denim shirt

School Apparel Most Likely to Purchase	
Apparel	**Students**
Hat	20
T-Shirt	37
Sweatshirt	25
Jacket	12
Denim Shirt	6

Lesson 7-7

Pages 399–400

Use the *solve a simpler problem* strategy to solve Exercises 1–4.

1. **SHOPPING** Tyler has $70 to spend on clothes. He has chosen 4 items that cost $16.50, $12.99, $24.99, and $19.99. Does he have enough money to buy all four items?
2. **HAIR** Grace wants to tip her hair stylist 20% on a $31 cut and style. About how much money should she tip?
3. **BALLOONS** Two students can blow up 16 balloons in 5 minutes. How many balloons can 6 students blow up at the same rate in 20 minutes?
4. **HEART RATE** The heart rate of a hummingbird is about 1,258 beats per minute. About how many times does a hummingbird's heart beat per second?

Lesson 7-8

Pages 401–405

Estimate each percent.

1. 38% of 150
2. 20% of 75
3. 25% of 78
4. 10% of 90
5. 16% of 30
6. 39% of 40
7. 6% of 86
8. 9% of 29
9. 3% of 46
10. 89% of 47
11. 25% of 48
12. 5% of 420
13. 55% of 134
14. 28% of 4
15. 12% of 40
16. 11% of 14
17. 90% of 140
18. 40% of 45

Lesson 8-1

Pages 418–423

Draw a line segment of each length.

1. $1\frac{1}{4}$ inches
2. $\frac{5}{8}$ inch
3. $1\frac{3}{4}$ inches
4. $1\frac{1}{2}$ inches
5. $3\frac{1}{8}$ inches
6. $2\frac{3}{8}$ inches
7. $1\frac{5}{8}$ inches
8. $2\frac{1}{2}$ inches
9. $3\frac{3}{4}$ inches

Measure the length of each line segment to the nearest half, fourth, or eighth inch.

10. •————•
11. •————•
12. •————•
13. •————•

Complete.

14. 3 yd = ▪ in.
15. 12 ft = ▪ yd
16. 9 ft = ▪ in.
17. 48 in. = ▪ ft
18. 2 yd = ▪ in.
19. 2 mi = ▪ ft
20. 5 ft = ▪ in.
21. 9 yd = ▪ ft
22. 21,120 ft = ▪ mi
23. 99 in. = ▪ ft
24. 8 yd = ▪ in.
25. 72 in. = ▪ ft

Lesson 8-2

Pages 424–429

Complete.

1. 3 gal = ■ pt
2. 24 pt = ■ gal
3. 20 lb = ■ oz
4. 2 gal = ■ fl oz
5. 20 pt = ■ qt
6. 18 qt = ■ pt
7. 2,000 lb = ■ T
8. 3 T = ■ lb
9. 9 lb = ■ oz
10. 15 qt = ■ gal
11. 4 pt = ■ c
12. 4 qt = ■ fl oz
13. 12 pt = ■ c
14. 24 fl oz = ■ c
15. 5 pt = ■ c
16. 4 lb = ■ oz
17. 5 T = ■ lb
18. 8 pt = ■ qt
19. 48 fl oz = ■ c
20. 6 gal = ■ qt
21. 9 qt = ■ c
22. 2 gal = ■ c
23. 16 c = ■ qt
24. 2 qt = ■ fl oz

Lesson 8-3

Pages 432–436

Write the metric unit of length that you would use to measure each of the following.

1. length of a paper clip
2. width of a classroom
3. distance from school to home
4. length of a hockey skate
5. length of a school bus
6. distance from Cleveland to Columbus
7. thickness of a calculator
8. length of a blade of grass

Estimate the metric length of each of the following. Then measure to find the actual length.

9. •————————•
10. •——————————————•
11. •—————•
12. •—————————•

Lesson 8-4

Pages 437–441

Write the metric unit of mass or capacity that you would use to measure each of the following. Then estimate the mass or capacity.

1. bag of sugar
2. pitcher of fruit punch
3. mass of a dime
4. amount of water in a drinking glass
5. a vitamin
6. pencil
7. mass of a puppy
8. bottle of perfume
9. grain of sand
10. mass of a car
11. baseball
12. paperback book

Lesson 8-5

Pages 442–443

Use a benchmark to solve Exercises 1–3.

1. **BULLETIN BOARD** Mr. Diaz wants to cover a bulletin board with black paper. He needs to know the approximate dimensions of the bulletin board in feet. He has a sheet of notebook paper that he knows is $8\frac{1}{2}$ inches wide and 11 inches long. Describe a way Mr. Diaz could estimate the dimensions of the bulletin board in feet.
2. **PUNCH** Mrs. King is preparing for a dinner party. She needs to know if her favorite dish will hold enough dessert for each guest to have a serving size of 1 cup each. She has a pitcher that she knows holds 2 quarts. Describe a way she can determine if the dish will hold enough dessert for 8 guests.
3. **CARPET** Julia is ordering new carpet for her bedroom. She needs to know the length and width of the bedroom in feet. She has a jump rope and she knows that the door to her bedroom is 2.5 feet wide. Describe a way she can estimate the dimensions of her bedroom.

Lesson 8-6

Pages 445–449

Complete.

1. 400 mm = ■ cm	2. 4 kg = ■ g	3. 660 cm = ■ m
4. 3 L = ■ mL	5. 30 mm = ■ cm	6. 84 g = ■ kg
7. ■ m = 54 cm	8. ■ L = 563 mL	9. ■ mg = 21 g
10. 4 L = ■ mL	11. 61 mg = ■ g	12. 4,497 mL = ■ L
13. ■ mm = 45 cm	14. 632 mL = ■ L	15. 61 g = ■ mg
16. ■ mg = 5 kg	17. 6 L = ■ mL	18. 18 km = ■ cm
19. ■ m = 36 cm	20. 5 kg = ■ g	21. 3,250 mL = ■ L
22. 7 km = ■ m	23. 453 g = ■ kg	24. 9 L = ■ mL

Lesson 8-7

Pages 450–454

Add or subtract.

1. 6 h 14 min − 2 h 8 min	2. 5 h 35 min 25 s + 45 min 35 s	3. 5 h 4 min 45 s − 2 h 40 min 5 s
4. 15 h 16 min − 8 h 35 min 16 s	5. 9 h 20 min 10 s + 1 h 39 min 55 s	6. 2 h 40 min 20 s + 3 h 5 min 50 s
7. 3 h 24 min 10 s − 2 h 30 min 5 s	8. 9 h 12 min + 2 h 51 min 15 s	9. 4 h 9 min 15 s − 4 h 3 min 20 s

Find the elapsed time.

10. 1:10 P.M. to 4:45 P.M.
11. 9:40 A.M. to 11:18 A.M.
12. 10:30 A.M. to 6:00 P.M.
13. 8:45 P.M. to 1:30 A.M.
14. 8:05 A.M. to 3:25 P.M.
15. 10:30 P.M. to 1:45 A.M.
16. 11:20 P.M. to 12:15 A.M.
17. 12:40 A.M. to 10:25 A.M.

Lesson 8-8

Pages 455–458

Choose the more reasonable temperature for each.

1. soup: 35°C or 65°C
2. ice pop: −3°C or 3°C
3. family room: 58°F or 72°F
4. hot tea: 140°F or 215°F
5. pitcher of lemonade: 8°C or 25°C
6. school cafeteria: 10°C or 23°C
7. snowman: 28°F or 36°F
8. lasagna baking in oven: 180°F or 350°F

Give a reasonable estimate of the temperature in degrees Fahrenheit and degrees Celcius for each activity.

9. riding a motorcycle
10. visiting the zoo
11. sledding
12. washing a car by hand
13. water skiing
14. riding a snowmobile
15. raking leaves
16. shopping at the mall

Lesson 9-1

Pages 470–473

Use a protractor to find the measure of each angle. Then classify the angle as *acute, obtuse, right,* or *straight.*

1. 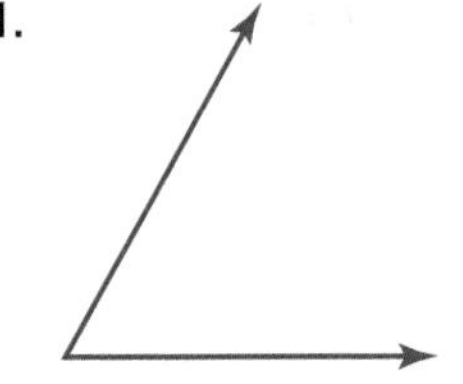

2. 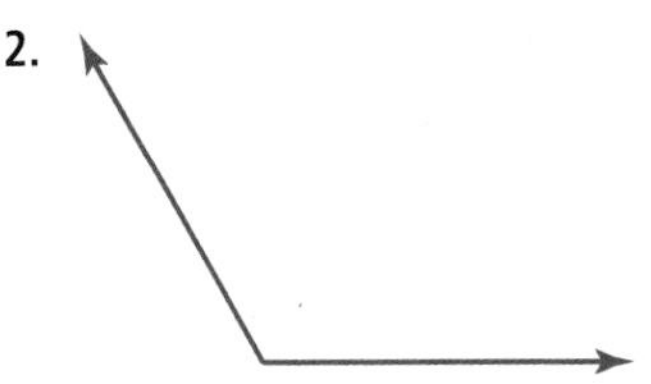

3. 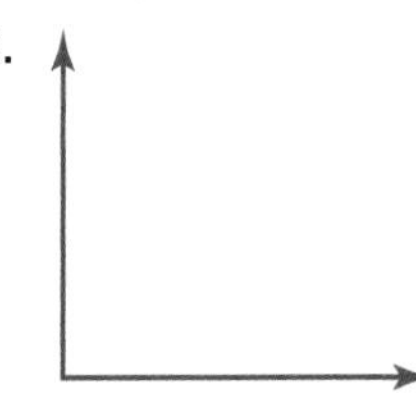

4. 

5. 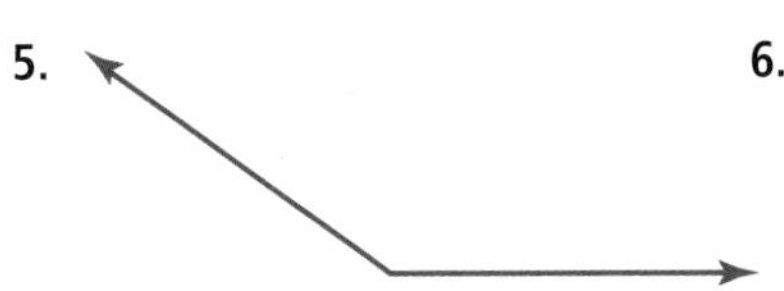

6. 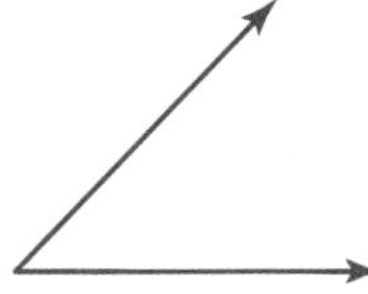

Lesson 9-2

Pages 474–478

Estimate the measure of each angle.

1. 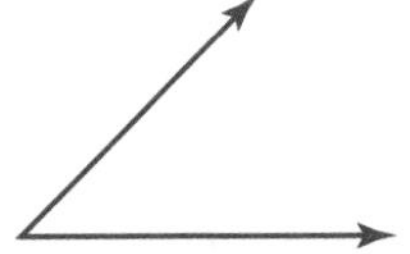

2. 

3. 

4. 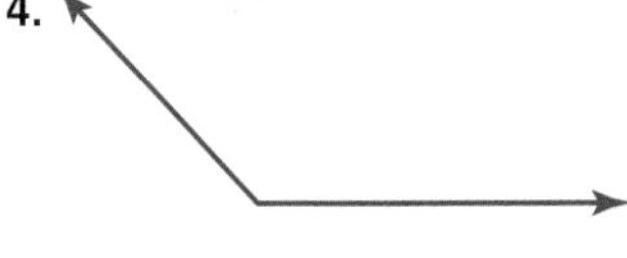

Use a protractor and a straightedge to draw angles having the following measurements.

5. 165°
6. 20°
7. 90°
8. 41°
9. 75°
10. 180°
11. 30°
12. 120°
13. 15°
14. 55°
15. 100°
16. 145°

Lesson 9-3

Pages 479–484

Classify each pair of angles as *complementary, supplementary,* or *neither.*

1. 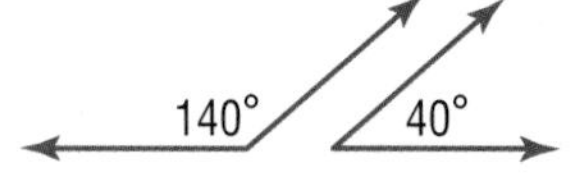

2. 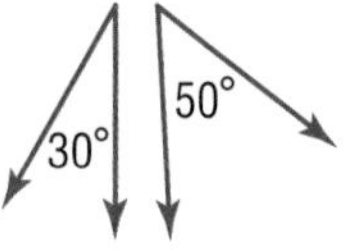

3. 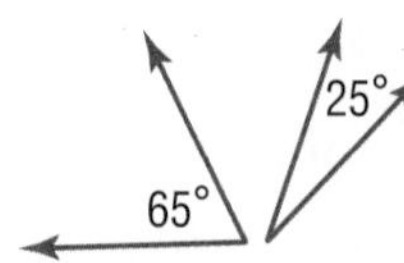

Find the value of x in each figure.

4. 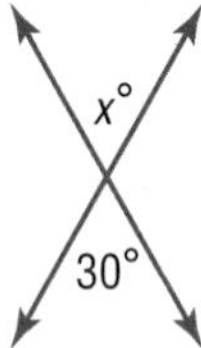

5. 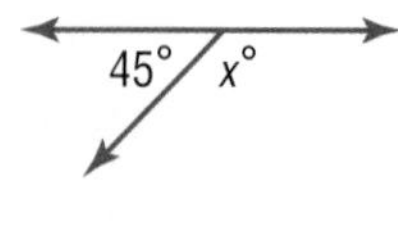

6. 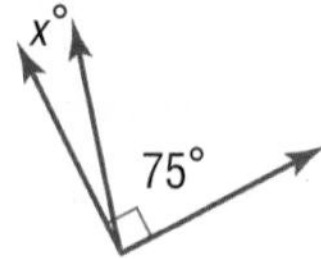

Lesson 9-4

Pages 486–491

Classify each triangle drawn or having the given angle measures as *acute, right,* or *obtuse.*

1. 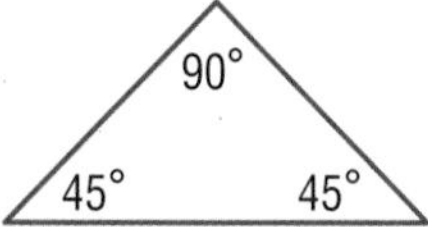

2. 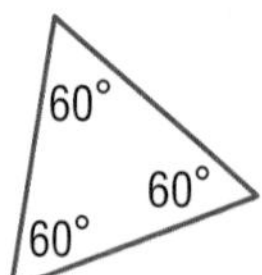

3. 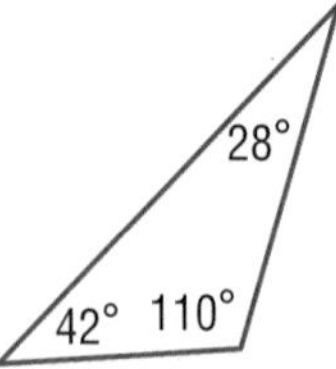

Find the value of x in each triangle having the given angle measures.

4. 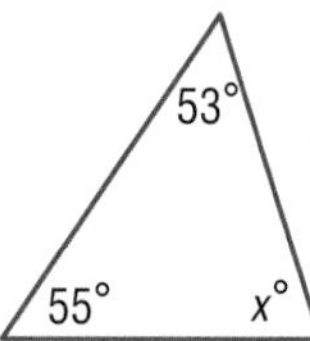

5. 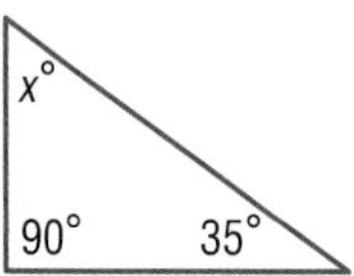

6. 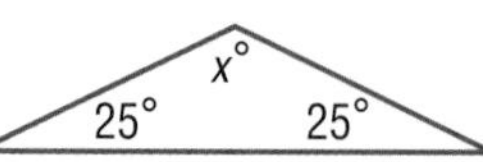

Lesson 9-5

Pages 494–499

Find the value of x in each quadrilateral.

1. 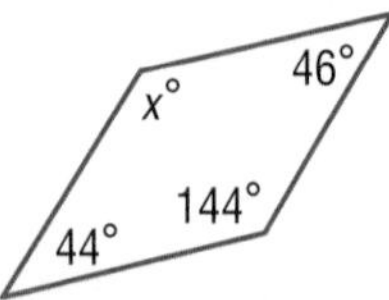

2. 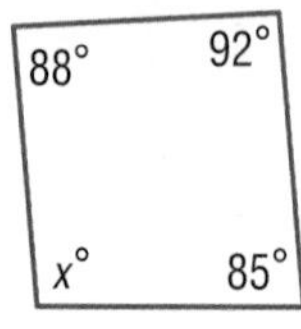

3. 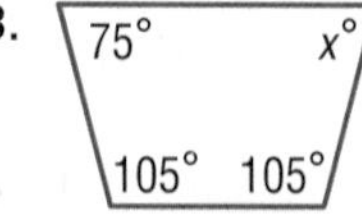

Classify each quadrilateral.

4. 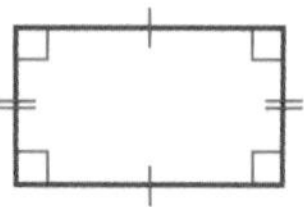

5. 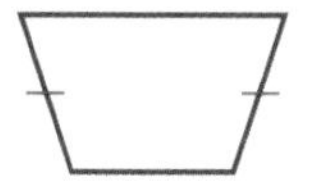

6. 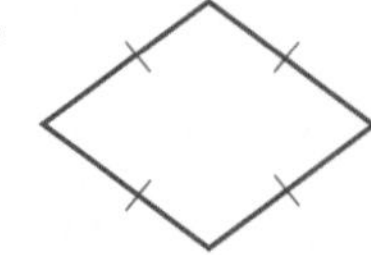

Lesson 9-6

Pages 500–501

Solve Exercises 1–4. Use the *draw a diagram* strategy.

1. **CHEERLEADING** Twenty-four cheerleaders are working on a new dance routine. The routine requires groups arranged in triangular formations with each triangle having 3 people in the last row. If all 24 cheerleaders must participate, how many triangles are there?
2. **PIZZA** Lucy's mom is making personal size pizzas for a party. Each pizza is 5 inches square. How many pizzas will fit on a baking sheet that is 14 inches by 20 inches if the pizzas are placed 1 inch apart?
3. **FLAGS** Jared is designing a flag that is 2 feet wide by 4 feet long. He wants the entire flag to be a checkerboard pattern of squares. If he uses 6 rows and 12 columns to create the checkerboard, what will be the side length of each square?
4. **CAKE** Isabel needs to cut a lasagna that is 9 inches by 13 inches into 12 equal-size pieces. What will be the dimensions of each serving?

Lesson 9-7

Pages 502–507

Tell whether each pair of polygons is *congruent, similar,* or *neither.*

1. 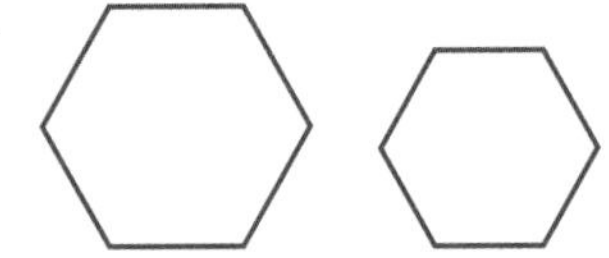
2. 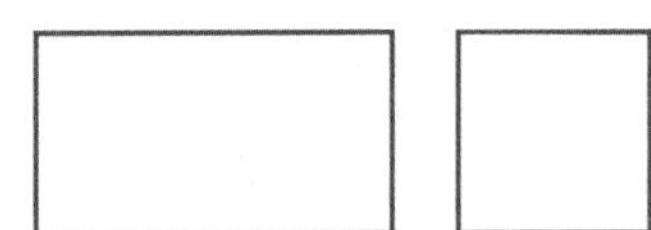
3. 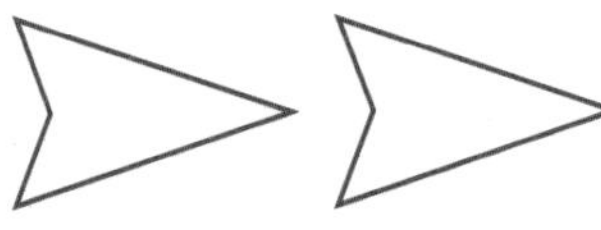
4. 
5. 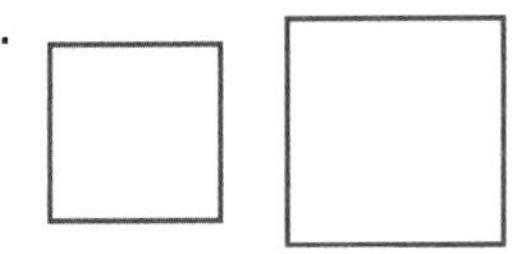
6. 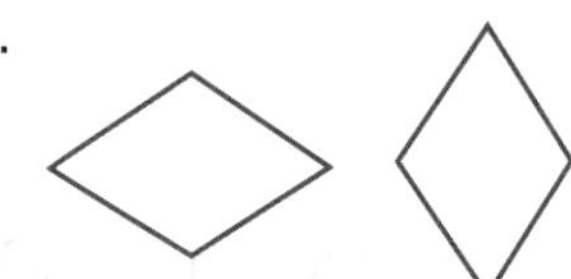
7. 
8. 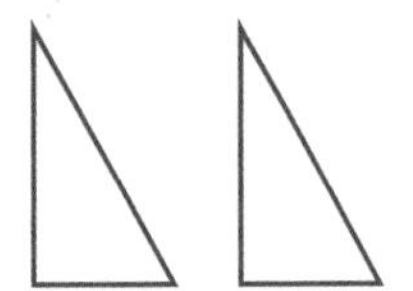
9. 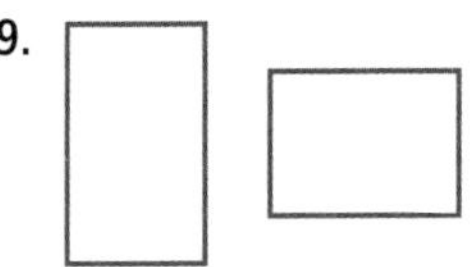

The triangles shown are similar.

10. What side of triangle ABC corresponds to $\overline{RT}$?
11. What side of triangle RST corresponds to $\overline{BA}$?

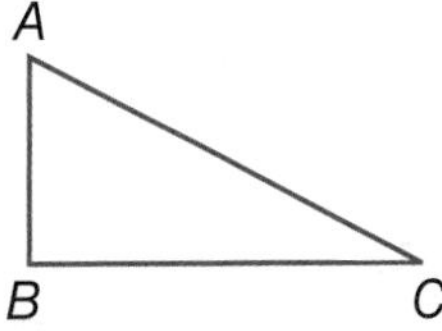

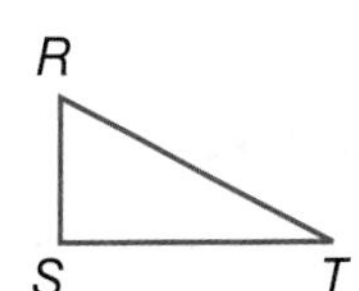

12. **Which rectangle below is similar to rectangle *ABCD*?**

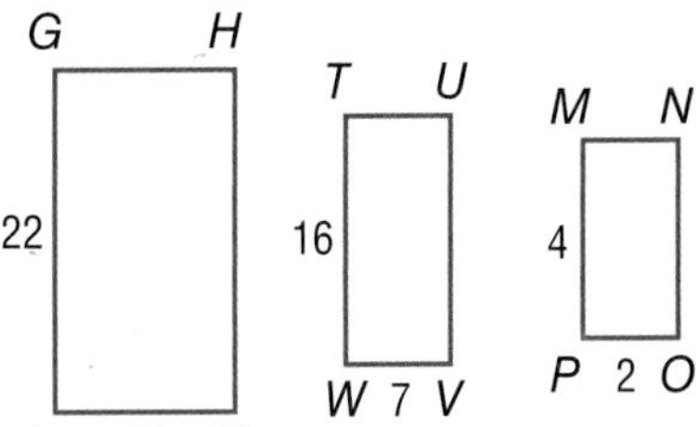

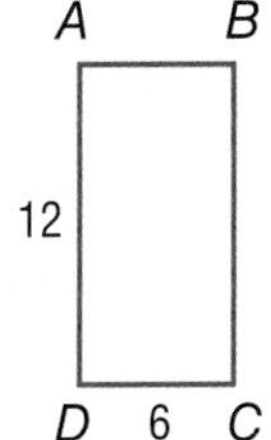

Lesson 10-1

Pages 522–526

Find the perimeter of each figure.

1. 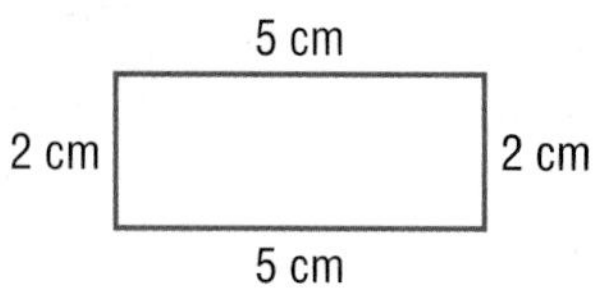

2. 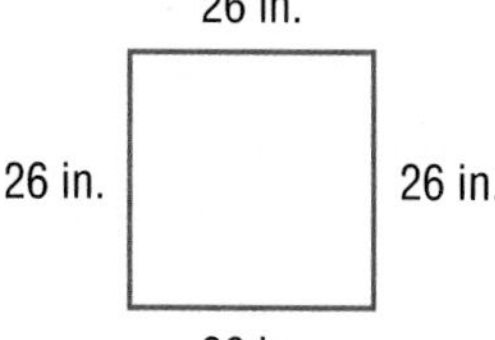

3. 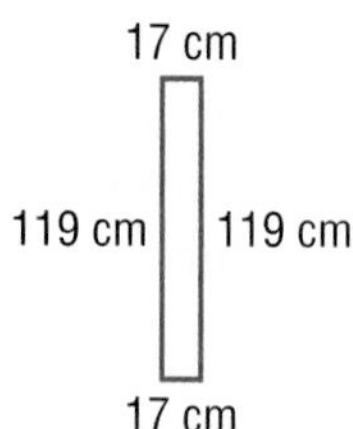

4. 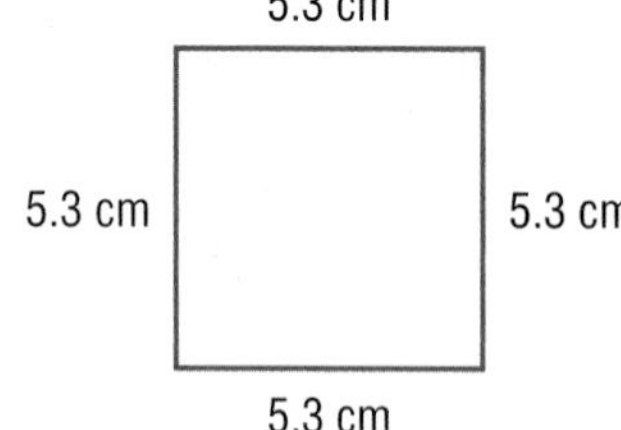

5. 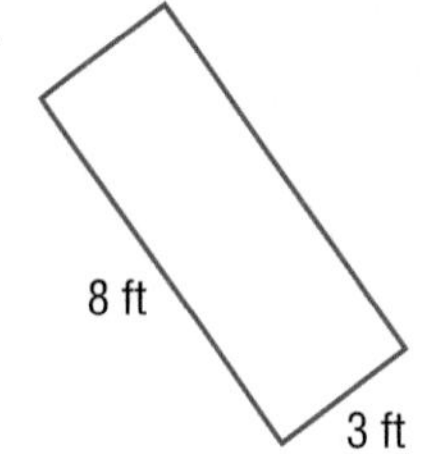

6. 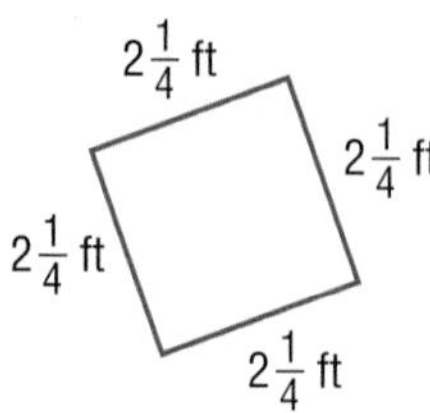

Lesson 10-2

Pages 528–533

Find the radius or diameter of each circle with the given dimension.

1. $d = 20$ cm
2. $r = 4$ in.
3. $d = 11$ yd

Estimate the circumference of each circle.

4. 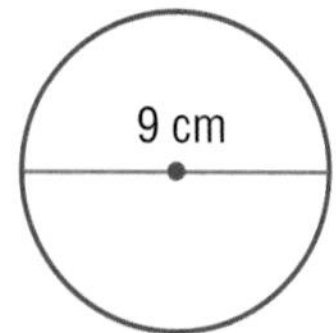

5. 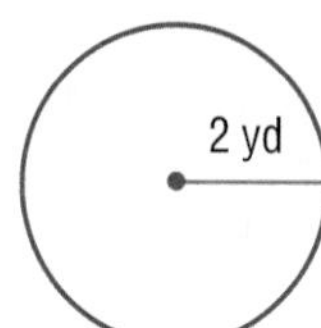

6. 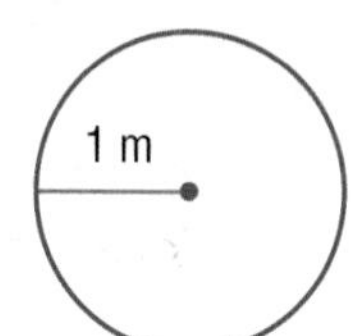

7. 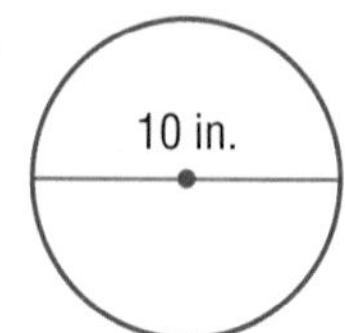

Find the circumference of each circle. Round to the nearest tenth.

8. 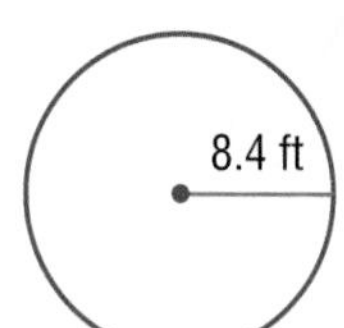

9. 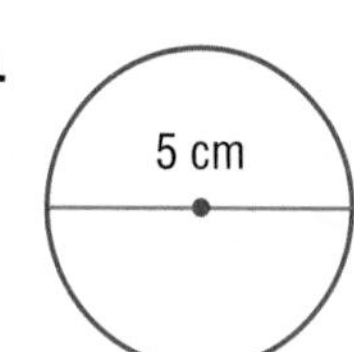

10. 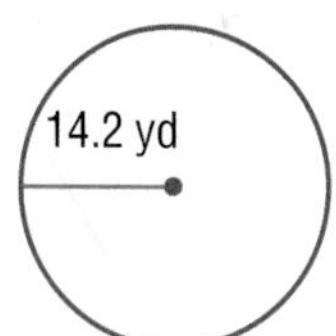

11. 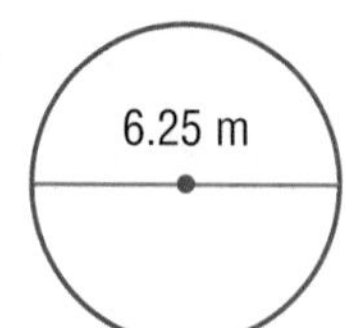

Lesson 10-3

Pages 534–538

Find the area of each parallelogram.

1. 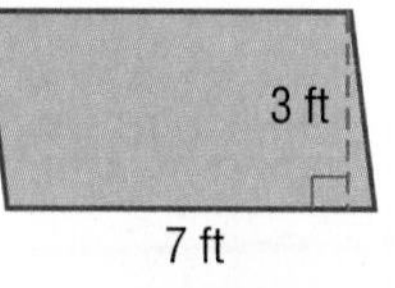

2. 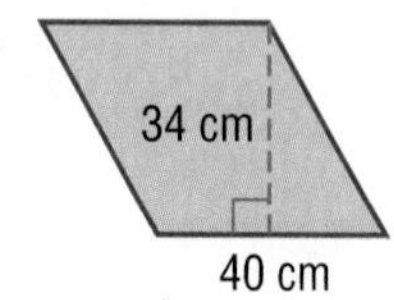

3. 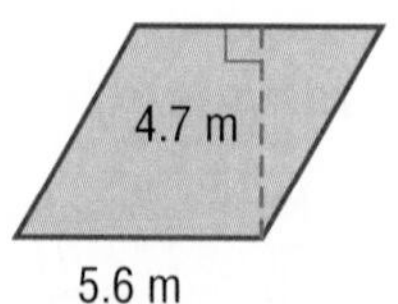

4. 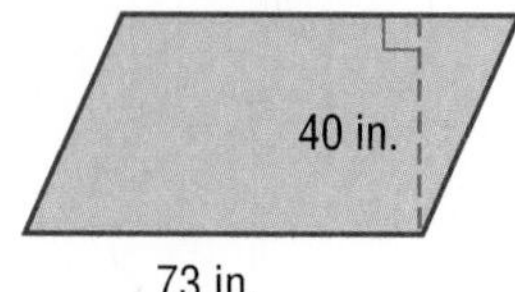

5. 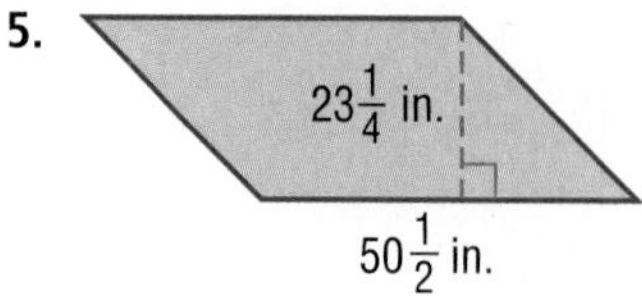

6. 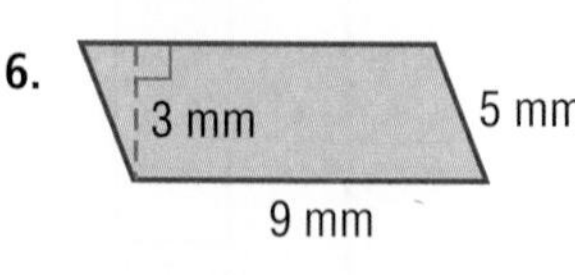

Lesson 10-4

Pages 540–544

Find the area of each triangle.

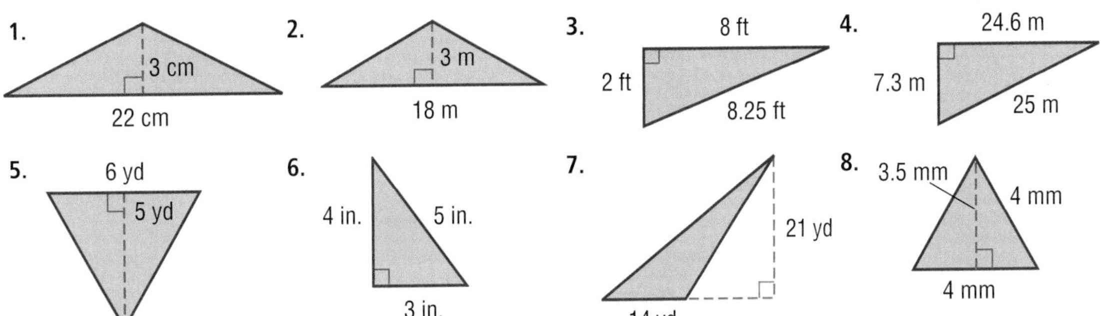

9. base, 6 ft
   height, 3 ft

10. base, 4.25 in.
    height, 7.5 in.

11. base, 9 m
    height, 7 m

12. base, 13 cm
    height, 16 cm

Lesson 10-5

Pages 546–547

Solve Exercises 1–3. Use the *make a model* strategy.

1. **ART** A teacher is displaying 24 paintings on the wall to form a collage. Each painting is the same size and shape. If she arranges them in a rectangular shape with the least perimeter possible, how many paintings will be in each row?
2. **BLOCKS** Latanya has 28 blocks. She wants to use them all to build a pyramid by stacking the blocks in single rows on top of each other. How many rows will there be in the pyramid?
3. **BOOKS** A librarian is organizing a seasonal display of books using 4 shelves and 36 books. If each shelf has 2 more books than the previous one, how many books are on each shelf?

Lesson 10-6

Pages 548–553

Find the volume of each prism.

1. 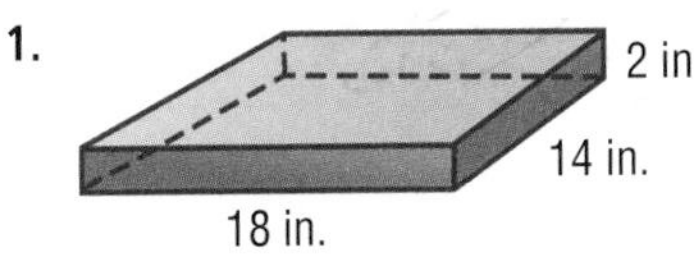

2. 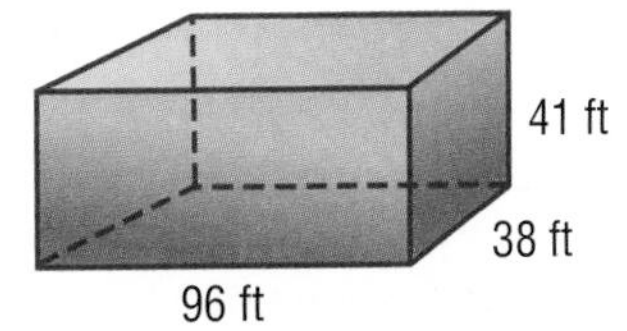

3. 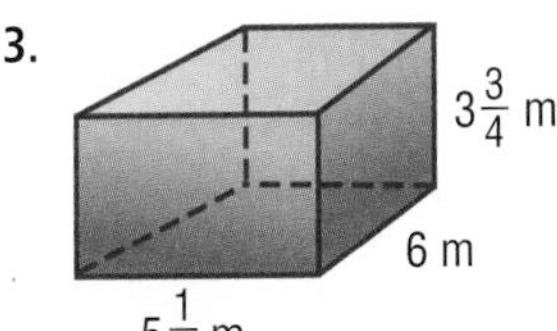

4. 9 mm
   9 mm
   9 mm

5. 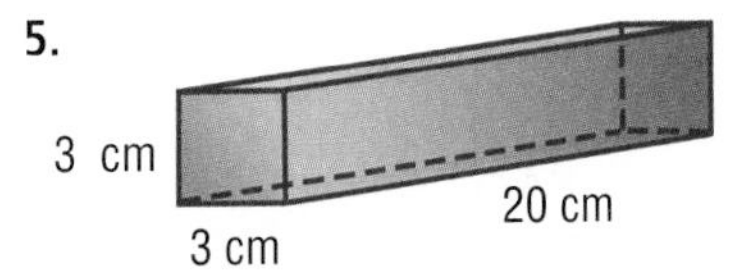

6. 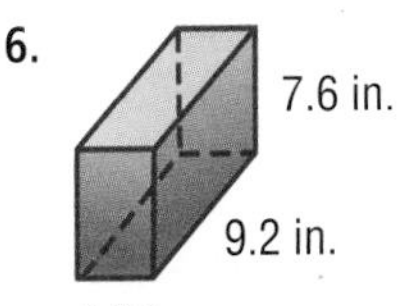

7. 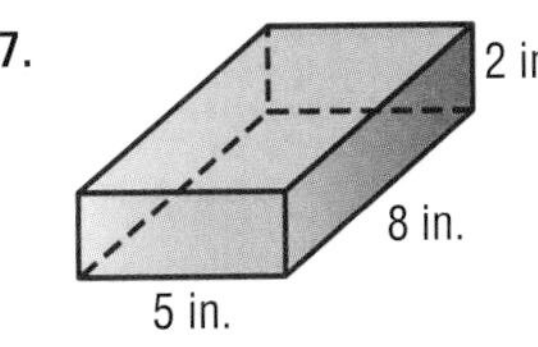

8. 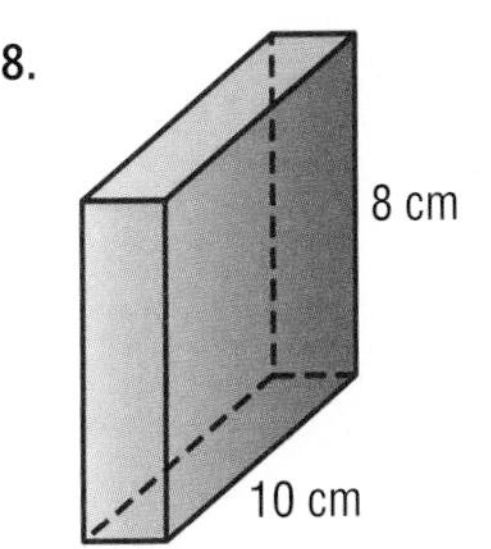

9. 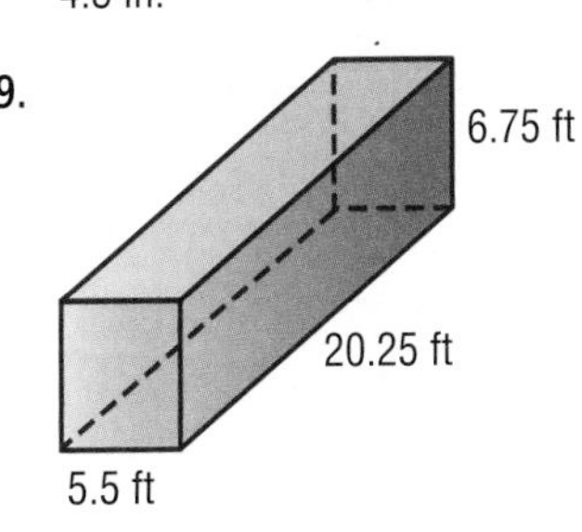

Lesson 10-7

Pages 555–559

Find the surface area of each rectangular prism.

1. 2 in., 12 in., 14 in.

2. 41 ft, 60 ft, 30 ft

3. 3.2 m, 5.4 m, 3.2 m

4. 7 mm, 9 mm, 9 mm

5. 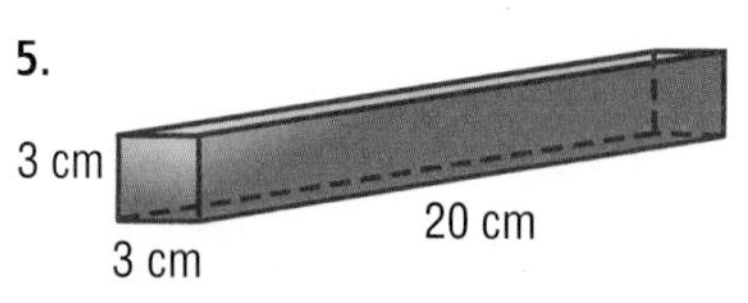

6. 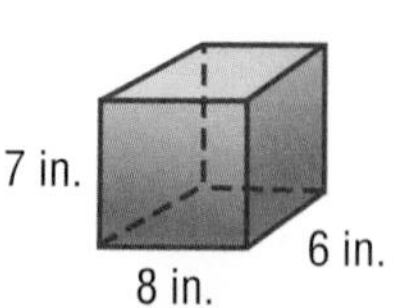

7. 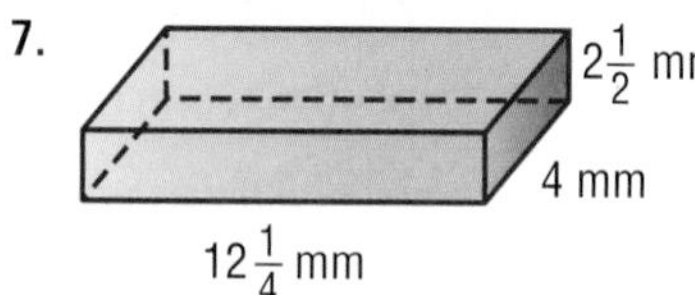

8. 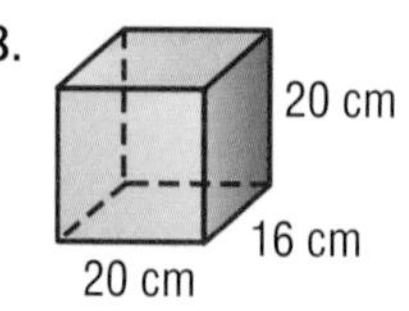

9. 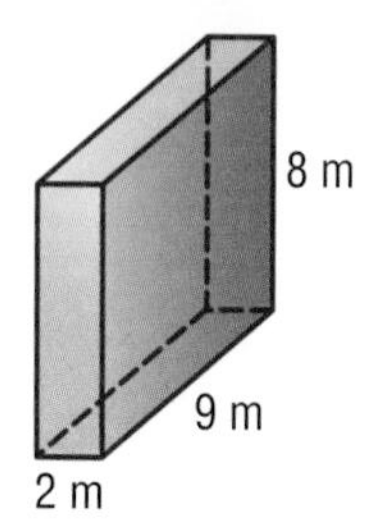

Lesson 11-1

Pages 572–575

Replace each ● with <, >, or = to make a true sentence.

1. −5 ● −55
2. 4 ● −66
3. −777 ● −77
4. −75 ● −75
5. −898 ● −99
6. 0 ● 44
7. 56 ● −1
8. −82 ● −9
9. −6 ● −7
10. 90 ● 101
11. 4 ● −2,000
12. −3 ● 0
13. 8 ● 6
14. −5 ● −7
15. −2 ● 0
16. 3 ● −2

Order each set of integers from least to greatest.

17. 0, 3, −21, 9, −89, 8, −65, −56
18. 70, −9, 67, −78, 0, 45, −36, −19
19. 12, 8, −9, −12, 10, 16
20. 65, 34, −50, 28, −64, −45
21. −4, 39, −14, 22, −30, 33, −70
22. −3, 77, 0, 41, −48, 6, −19

Lesson 11-2

Pages 577–581

Add. Use counters or a number line if necessary.

1. $-4 + (-7)$
2. $-1 + 0$
3. $7 + (-13)$
4. $-20 + (+2)$
5. $4 + (-6)$
6. $-12 + 9$
7. $-12 + (-10)$
8. $5 + (-15)$
9. $+17 + (+9)$
10. $18 + (-18)$
11. $-4 + (-4)$
12. $0 + (-9)$
13. $-12 + (-9)$
14. $-8 + (+7)$
15. $3 + (-6)$
16. $-9 + 16$
17. $-5 + (-3)$
18. $-5 + 5$
19. $-3 + (-3)$
20. $-11 + 6$
21. $-10 + (+6)$
22. $-5 + (-9)$
23. $+18 + (-20)$
24. $-4 + (-8)$
25. $2 + (-4)$
26. $-3 + (-11)$
27. $-17 + 9$
28. Evaluate $a + b$ if $a = 14$ and $b = -5$.
29. Find the value of $m + n$ if $m = -5$ and $n = -6$.

Lesson 11-3

Pages 582–586

Subtract. Use counters if necessary.

1. $7 - (-4)$
2. $-4 - (-9)$
3. $13 - (-3)$
4. $2 - (-5)$
5. $-9 - 5$
6. $-11 - (-18)$
7. $-4 - (-7)$
8. $-6 - (-6)$
9. $-6 - 6$
10. $17 - 9$
11. $-12 - (-9)$
12. $0 - (-4)$
13. $-7 - 0$
14. $-12 - (-10)$
15. $-2 - (-1)$
16. $3 - (-5)$
17. $5 - (-1)$
18. $-5 - (-6)$
19. $9 - (-1)$
20. $1 - 9$
21. $-5 - 1$
22. $-1 - 4$
23. $0 - (-7)$
24. $8 - 13$
25. $-4 - (-6)$
26. $9 - 9$
27. $-7 - (-7)$
28. $7 - 5$
29. $8 - (-5)$
30. $5 - 8$
31. $1 - 6$
32. $-8 - (-8)$

Lesson 11-4

Pages 587–590

Multiply.

1. $3 \times (-5)$
2. -5×1
3. $-8 \times (-4)$
4. $6 \times (-3)$
5. -3×2
6. $-1 \times (-4)$
7. $8 \times (-2)$
8. $-5 \times (-7)$
9. $3 \times (-9)$
10. -9×4
11. $-4 \times (-5)$
12. $5 \times (-2)$
13. $-8(3)$
14. $-9(-1)$
15. $7(-3)$
16. $2(3)$
17. $-6(0)$
18. $-5(-1)$
19. $5(-5)$
20. $-2(-3)$
21. $8(-4)$
22. $-2(4)$
23. $-4(-4)$
24. $2(9)$
25. $-2(-12)$
26. $7 \times (-4)$
27. $-5 \times (-9)$
28. -2×11
29. $4(-2)$
30. $4(-4)$
31. $-3(-11)$
32. $-3(3)$
33. Find the product of 2 and -3.
34. Evaluate qr if $q = -3$ and $r = -3$.

Lesson 11-5

Pages 592–593

For Exercises 1–5, solve using the *work backward* strategy.

1. **MONEY** Mario has \$2.50 left after spending \$4.25 at the arcade and \$5.50 on lunch. How much money did Mario have originally?
2. **NUMBER SENSE** A number is multiplied by 3, and then 4 is subtracted from the product. Then, 2 is added to the difference. If the result is 16, what is the number?
3. **CANDY** Julia sold 23 chocolate bars, 15 peanut butter bars, and 8 caramel bars. If she had 9 candy bars left, how many candy bars did she have to start?
4. **NUMBER SENSE** A number is divided by 4. Next, the quotient is multiplied by 2. Then, 4 is added to the product. If the result is 14, what is the number?
5. **HOMEWORK** Alex finished his homework at 8:15 P.M. He spent 30 minutes working on science, 45 minutes working on math, and 25 minutes working on history. If he worked without taking any breaks, at what time did he begin working on his homework?

Lesson 11-6

Pages 594–598

Divide.

1. $12 \div (-6)$
2. $-7 \div (-1)$
3. $-4 \div 4$
4. $6 \div (-6)$
5. $0 \div (-4)$
6. $45 \div (-9)$
7. $15 \div (-5)$
8. $-6 \div 2$
9. $-28 \div (-7)$
10. $20 \div (-2)$
11. $-40 \div (-8)$
12. $12 \div (-4)$
13. $-18 \div 6$
14. $9 \div (-1)$
15. $-30 \div 6$
16. $-54 \div (-9)$
17. $28 \div (-7)$
18. $-24 \div 8$
19. $24 \div (-4)$
20. $-14 \div 7$
21. $9 \div 3$
22. $-18 \div (-6)$
23. $-9 \div (-1)$
24. $18 \div (-9)$

Lesson 11-7

Pages 599–603

Write the ordered pair that names each point. Then identify the quadrant where each point is located.

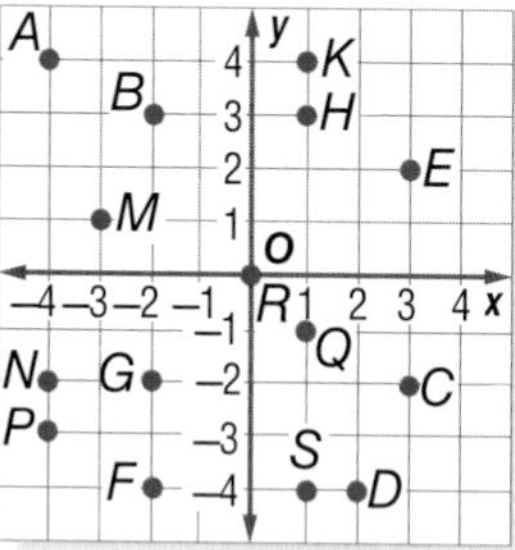

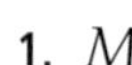

1. *M*
2. *A*
3. *D*
4. *E*
5. *P*
6. *Q*
7. *B*
8. *C*
9. *F*
10. *G*
11. *N*
12. *R*
13. *K*
14. *H*
15. *S*

Graph and label each point on a coordinate plane.

16. $S(4, -1)$
17. $T(-3, -2)$
18. $W(2, 1)$
19. $Y(-5, 3)$
20. $Z(-1, -3)$
21. $U(3, -3)$
22. $V(1, 2)$
23. $X(-1, 4)$

Lesson 11-8

Pages 604–609

1. Translate $\triangle ABC$ 3 units right and 4 units up.

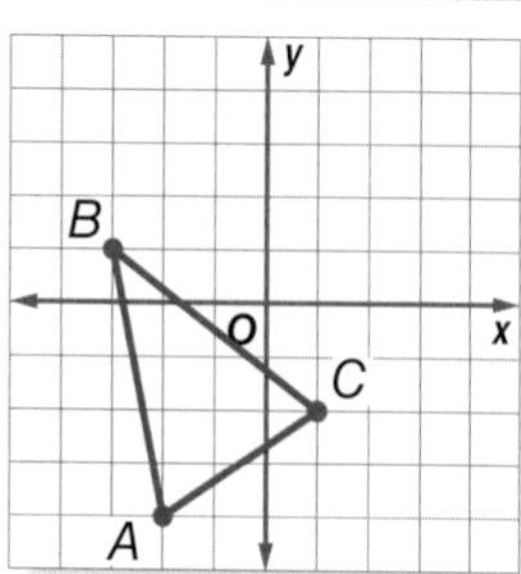

2. Translate quadrilateral *WXYZ* 2 units left and 3 units down.

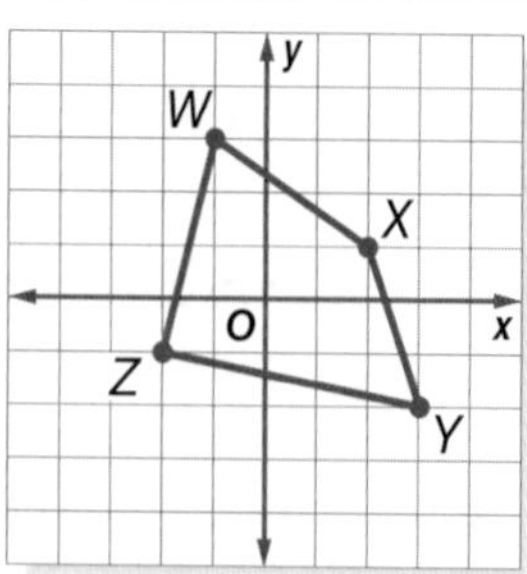

Triangle *LMN* has vertices *L*(1, 4), *M*(4, 1), and *N*(2, −1). Find the vertices of *L′M′N′* after each translation. Then graph the figure and its translated image.

3. 2 units left, 1 unit up
4. 5 units right, 3 units down
5. 3 units left, 2 units down
6. 4 units right, 2 units up

Lesson 11-9

Pages 610–614

Graph each figure and its reflection over the x-axis. Then find the coordinates of the reflected image.

1. Quadrilateral $QRST$ with vertices $Q(-3, 1)$, $R(-1, 4)$, $S(2, 4)$, and $T(2, 1)$.
2. Triangle ABC with vertices $A(-3, -1)$, $B(3, -2)$, and $C(-1, -4)$.
3. Parallelogram $JKLM$ with vertices $J(-2, -1)$, $K(2, -2)$, $L(3, -4)$, and $M(-1, -3)$.

Graph each figure and its reflection over the y-axis. Then find the coordinates of the reflected image.

4. Triangle FGH with vertices $F(-4, -4)$, $G(-2, 3)$, and $H(-1, -1)$.
5. Quadrilateral $MNOP$ with vertices $M(1, 0)$, $N(2, 4)$, $O(4, 1)$, and $P(3, -1)$.
6. Triangle STU with vertices $S(0, 3)$, $T(-2, -2)$, and $U(-3, 0)$.

Lesson 11-10

Pages 615–619

Graph triangle ABC and its image after each rotation. Then give the coordinates of the vertices for triangle $A'B'C'$.

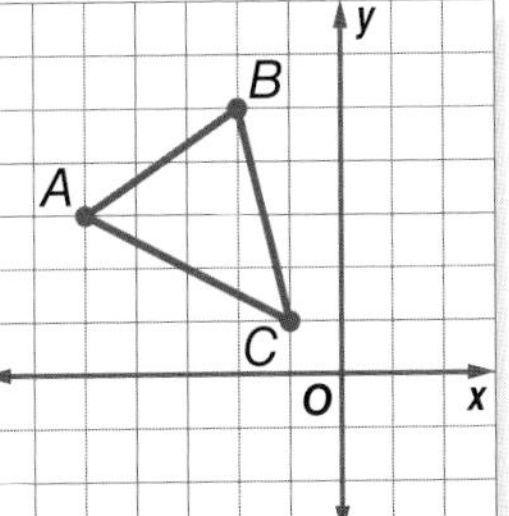

1. 90° counterclockwise
2. 180° clockwise
3. 270° counterclockwise
4. 270° clockwise

Determine whether each sign has rotational symmetry. Write *yes* or *no*. If *yes*, name its angle(s) of rotation.

5. 

6. 

Lesson 12-1

Pages 632–635

Find each expression mentally.

1. 3×24
2. 8×67
3. 5×39
4. 9×48
5. 4×52
6. 6×75
7. 2×3.7
8. 7×4.2
9. 3×5.4

Use the Distributive Property to rewrite each algebraic expression.

10. $4(x + 3)$
11. $7(x + 4)$
12. $2(x + 8)$
13. $9(x + 6)$
14. $3(x + 11)$
15. $5(x + 7)$
16. $6(x - 4)$
17. $8(x - 9)$
18. $4(x - 5)$
19. $2(x - 8)$
20. $7(x - 3)$
21. $6(x - 12)$

Lesson 12-2

Pages 636–641

Simplify each expression below. Justify each step.

1. $(2 + x) + 6$
2. $8 + (3 + x)$
3. $4 + (x + 5)$
4. $(5 \cdot x) \cdot 9$
5. $4 \cdot (x \cdot 7)$
6. $(6 \cdot x) \cdot 8$
7. $4(6x)$
8. $12(3x)$
9. $7(2x)$

Simplify each expression below.

10. $5x + 3x$
11. $2x + x$
12. $6x + 8x$
13. $9x + 3x$
14. $7x + 4x$
15. $10x + x$
16. $3x + 2 + 4x$
17. $15x + 3 + x$
18. $2x + 5 + 6x$
19. $7x + 2x + 4$
20. $18x + 3x + 9$
21. $6x + x + 8$

Lesson 12-3

Pages 644–648

Solve each equation. Use models if necessary. Check your solution.

1. $x + 4 = 14$
2. $b + (-10) = 0$
3. $-2 + w = -5$
4. $k + (-3) = -5$
5. $6 = -4 + h$
6. $-7 + d = -3$
7. $9 = m + 11$
8. $f + (-9) = -19$
9. $p + 66 = 22$
10. $-34 + t = 41$
11. $-24 = e + 56$
12. $-29 + a = -54$
13. $17 + m = -33$
14. $b + (-44) = -34$
15. $w + (-39) = 55$
16. $6 + a = 13$
17. $-5 = m + 3$
18. $w + (-9) = 12$
19. $8 = p + 7$
20. $-4 + c = -9$
21. $y + 11 = 8$
22. $16 = t + 5$
23. $-3 + x = 1$
24. $14 + c = 6$
25. $-9 = -12 + w$
26. $q + 6 = 4$
27. $5 + z = 13$
28. $9 = h + -5$
29. Find the value of w if $-8 + w = -11$.
30. Find the value of g if $g + (-7) = 19$.

Lesson 12-4

Pages 651–654

Solve each equation. Use models if necessary. Check your solution.

1. $y - 7 = 2$
2. $a - 10 = -22$
3. $g - 1 = 9$
4. $c - 8 = 5$
5. $z - 2 = 7$
6. $n - 1 = -87$
7. $j - 15 = -22$
8. $x - 12 = 45$
9. $y - 65 = -79$
10. $q - 16 = -31$
11. $q - 6 = 12$
12. $j - 18 = -34$
13. $k - 2 = -8$
14. $r - 76 = 41$
15. $n - 63 = -81$
16. $b - 7 = 4$
17. $-5 = g - 3$
18. $y - 2 = -6$
19. $8 = m - 3$
20. $x - 5 = 2$
21. $-6 = p - 8$
22. $h - 9 = -6$
23. $12 = w - 8$
24. $a - 6 = -1$
25. $-11 = t - 5$
26. $c - 4 = 8$
27. $18 = q - 7$
28. $r - 2 = -5$
29. $1 = z - 9$
30. $g - 10 = -4$
31. $-6 = d - 4$
32. $s - 4 = 10$
33. Find the value of c if $c - 5 = -2$.
34. Find the value of t if $4 = t - 7$.

Lesson 12-5

Pages 657–660

Solve each equation. Use models if necessary.

1. $5x = 30$
2. $2w = 18$
3. $2a = 7$
4. $2d = -28$
5. $-3c = 6$
6. $11n = 77$
7. $3z = 15$
8. $9y = -63$
9. $6m = -54$
10. $5f = -75$
11. $20p = 5$
12. $4x = 16$
13. $4t = -24$
14. $7b = 21$
15. $19h = 0$
16. $22d = -66$
17. $3m = -78$
18. $8x = -2$
19. $9c = -72$
20. $5p = 35$
21. $-5k = 20$
22. $33y = 99$
23. $6z = -9$
24. $6m = -42$
25. $18 = 9x$
26. $-5p = 4$
27. $-32 = 4r$
28. $3w = 27$
29. $-12 = 16a$
30. $-4t = 6$
31. $16 = -5b$
32. $-2c = -13$
33. Solve the equation $3d = 21$.
34. What is the solution of the equation $-4x = 65$?

Lesson 12-6

Pages 661–662

Choose the best method of computation and solve Exercises 1–5. Explain your reasoning.

1. **MONEY** Allie needs to save $20 to buy the DVD she wants. If she already has $14.65, how much more does she need to save?
2. **POPULATION** The populations of four large cities in California are 1,263,756; 744,230; 904,522; and 3,845,541. What is the total number of people who live in these four cities?
3. **FOOD** If 50 students each contribute 3 cans of food for a food drive, how many cans of food will be collected in all?
4. **MONEY** Tasha spent $17.00, $9.85, $5.10, and $12.99 at four stores in the mall. About how much did she spend in all?
5. **PRECIPITATION** The average monthly precipitation for Columbus, Ohio, is shown in the table. What is the average total amount of precipitation for the year?

Average Monthly Precipitation for Columbus, OH (in.)											
Jan.	Feb.	Mar.	Apr.	May	June	July	Aug.	Sept.	Oct.	Nov.	Dec.
2.5	2.2	2.9	3.3	3.9	4.1	4.6	3.7	2.9	2.3	3.2	2.9

Mixed Problem Solving

Chapter 1 Algebra: Number Patterns and Functions

Pages 22–75

1. **BOOKS** The table below shows the number of pages of a book Elias read during the past 5 days. Use the four-step plan to find how many pages Elias will read on Saturday if he continues at this rate. (Lesson 1-1)

Pages Elias Read	
Day	**Pages Read**
Monday	5
Tuesday	11
Wednesday	19
Thursday	29
Friday	41
Saturday	■

2. **GEOMETRY** Draw the next figure in the pattern shown below. (Lesson 1-1)

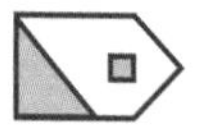 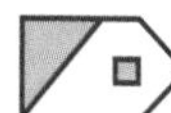 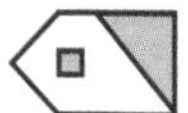  

3. **GEOGRAPHY** The United States of America has 50 states. Write 50 as a product of primes. (Lesson 1-2)

4. **ATTENDANCE** The number of students attending Blue Hills Middle School this year can be written as 5^4. How many students are enrolled at Blue Hills Middle School? (Lesson 1-3)

RETAIL For Exercises 5 and 6, use the table below. It shows the cost of several items sold at The Clothes Shack. (Lesson 1-4)

The Clothes Shack	
Item	**Price ($)**
Jeans	30
T-Shirt	15
Sweatshirt	20
Shorts	10

5. Write an expression that can be used to find the total cost of 2 pairs of jeans, 3 T-shirts, and 4 pairs of shorts.

6. Find the total cost for the purchase.

7. **ARCHITECTURE** The perimeter of a rectangle can be found using the expression $2\ell + 2w$, where ℓ represents length and w represents width. Find the perimeter of the front of a new building whose design is shown below. (Lesson 1-5)

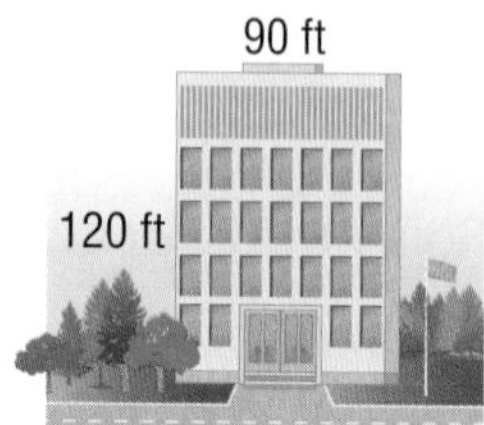

FUNDRAISING For Exercises 8 and 9, use the following information.

The school chorale is selling T-shirts and sweatshirts to raise money for a local charity. The cost of each item is shown. (Lesson 1-6)

8. Write a function rule that represents the total selling price of t T-shirts and s sweatshirts.

9. What would be the total amount collected if 9 T-shirts and 6 sweatshirts are sold?

10. **NUMBERS** Jen is thinking of three even numbers less than 10 with a sum of 18. Find the numbers. (Lesson 1-7)

11. **AGE** The equation $13 + a = 51$ describes the sum of the ages of Elizabeth and her mother. If a is Elizabeth's mother's age, what is the age of Elizabeth's mother? (Lesson 1-8)

12. **GARDENING** Rondell has a rectangular garden that measures 18 feet long and 12 feet wide. Suppose one bag of topsoil covers 36 square feet. How many bags of topsoil does Rondell need for his garden? (Lesson 1-9)

Chapter 2 Statistics and Graphs

1. **CAR WASH** Make a frequency table of the data below. How many more vans were washed than trucks? (Lesson 2-1)

Vehicles Washed				
S	V	C	V	V
T	C	S	S	C
V	C	T	S	V

C = car T = truck V = van S = SUV

2. **SWIMMING** Use the graph below to predict how many laps Emily will swim during week 6 of training. (Lesson 2-2)

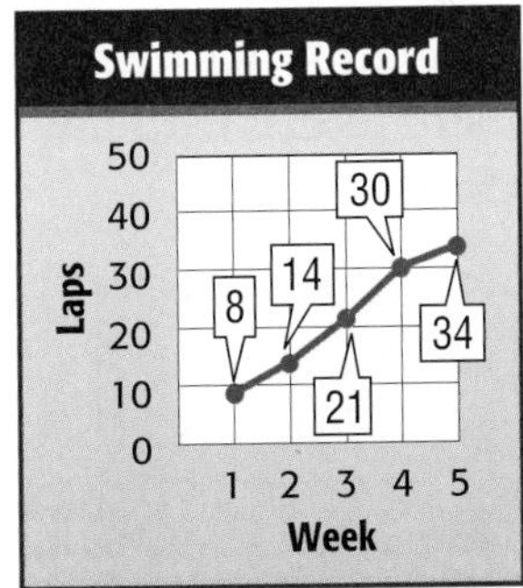

SALES For Exercises 3–5, use the table below. (Lesson 2-3)

Bookworm Book Shop	
Month	**Total Sales ($)**
January	9,750
February	8,200
March	7,875
April	12,300
May	10,450
June	9,900

3. Make a line graph for the data.

4. Which month had the greatest change in sales from the previous month?

5. Which month had the greatest decrease in sales from the previous month?

6. **JOBS** Sarah earned $64, $88, $96, $56, $48, $62, $75, $55, $50, $68, and $42. Make a stem-and-leaf plot of the data. (Lesson 2-4)

7. **LIBRARY** The table shows the ages of children attending story hour at the library. Make a line plot of the data. Which age group was the most common? (Lesson 2-5)

Age of Children at Story Hour (years)			
2	7	4	5
6	3	2	4
8	6	7	5
3	2	4	4

SCHOOL For Exercises 8–10, use the following information.

David's scores for five math tests are 83, 92, 88, 54, and 93. (Lesson 2-6)

8. Identify the outlier.

9. Find the mean of the data with and without the outlier.

10. Describe how the outlier affects the mean.

WEATHER For Exercises 11 and 12, use the following information.

The daily high temperatures during one week were 67°, 68°, 64°, 64°, 69°, 92°, and 66°. (Lesson 2-7)

11. Find the mean, median, mode, and range.

12. Which measure best describes the average temperature?

13. **SKIING** Display the data in the bar graph using another type of display. Which display is more appropriate? (Lesson 2-8)

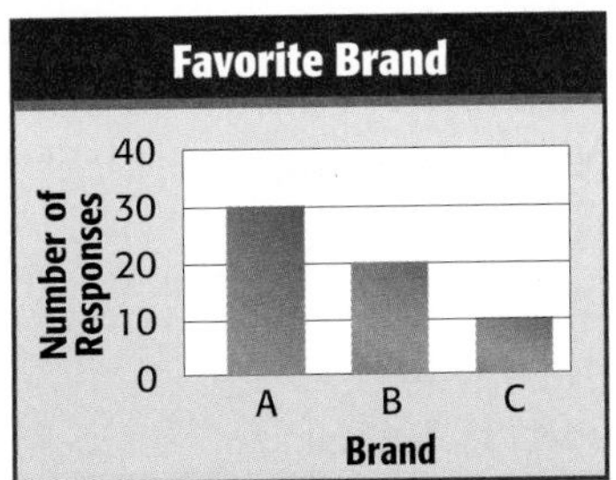

14. **HIKING** Delmar hikes on a trail that has its highest point at an altitude of 5 miles above sea level. Write this number as an integer. (Lesson 2-9)

15. **BANKING** Alicia has overdrawn her checking account by $12. Write her account balance as an integer. (Lesson 2-9)

Chapter 3 Operations with Decimals

1. **CHEMISTRY** The mass of a particular compound is 1.0039 grams. Express this mass in words. (Lesson 3-1)

2. **PACKAGING** The length of a rectangular box of cereal measures 12.58 inches. Express this measurement in expanded form. (Lesson 3-1)

3. **TYPING** Julia finished her typing assignment in 18.109 minutes. Teresa took 18.11 minutes to complete the same assignment. Who completed the assignment faster? (Lesson 3-2)

4. **TRACK AND FIELD** The table below lists the finishing times for runners of a particular event. Order these finish times from least to greatest. (Lesson 3-2)

Finishing Times	
Runner	**Time (s)**
Brent	43.303
Trey	43.033
Cory	43.230
Jonas	43.3003

5. **SCHOOL** The average enrollment at a middle school in a certain county is 429.86 students. Round this figure to the nearest whole number. (Lesson 3-3)

6. **SCIENCE** It takes 87.97 days for the planet Mercury to revolve around the Sun. Round this number to the nearest tenth. (Lesson 3-3)

7. **MONEY** Marta plans to buy the items listed at the right. Estimate the cost of these items before tax is added. (Lesson 3-4)

8. **GEOMETRY** The perimeter of a figure is the distance around it. Estimate the perimeter of the triangle. (Lesson 3-4)

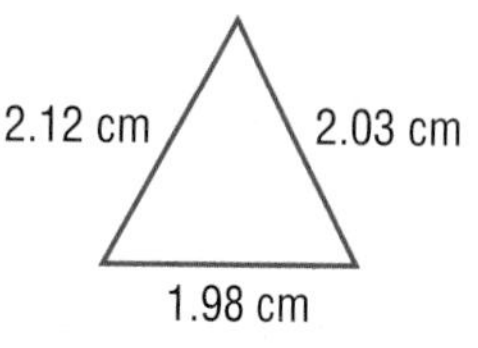

9. **SHOPPING** Kyle spent \$41.38 on a new pair of athletic shoes and \$29.86 on a new pair of jeans. Find the total of Kyle's purchases. (Lesson 3-5)

10. **MONEY** The current balance of Tami's checking account is \$237.80. Find the new balance after Tami writes a check for \$29.95. (Lesson 3-5)

11. **MONEY** Nick is buying a sports magazine that costs \$4.95. What will it cost him for a year if he buys one magazine every month? (Lesson 3-6)

12. **DECORATING** Selena is considering buying new carpeting for her living room. How many square feet of carpet will Selena need? (Lesson 3-7)

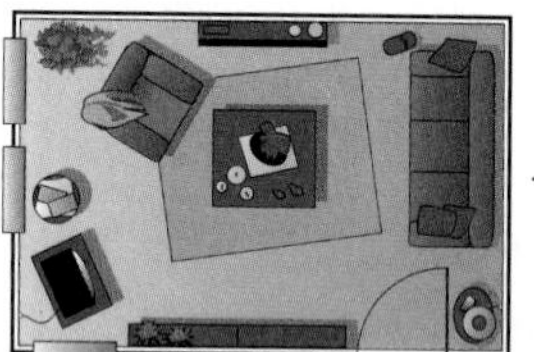

13. **SHIPPING** The costs for shipping the same package using three different shipping services are given in the table below. What is the average shipping cost? (Lesson 3-8)

Shipping Services Costs	
Shipping Service	**Cost**
A	\$6.42
B	\$8.57
C	\$7.48

14. **SHOPPING** Logan paid \$13.30 for 2.8 pounds of roast beef. Find the price per pound of the roast beef. (Lesson 3-9)

15. **GOLF** Juanita is going to play miniature golf with her friends. She estimates it will cost \$5.50 to play golf, \$8.00 for dinner, and \$3 for ice cream afterwards. Which is a more reasonable amount for Juanita to take with her if she also wants to play arcade games: \$15 or \$20? Explain your reasoning. (Lesson 3-10)

1. **TICKETS** Mrs. Cardona collected money from her students for tickets to the school play. She recorded the amounts of money collected in the table below. What is the most a ticket could cost per student? (Lesson 4-1)

Ticket Money Collection	
Day	**Amount**
Monday	$15
Tuesday	$12
Wednesday	$18
Thursday	$21

2. **TEACHERS** Twenty-eight of the forty-two teachers on the staff at Central Middle School are female. Write this fraction in simplest form. (Lesson 4-2)

3. **WEATHER** In September, 4 out of 30 days were rainy. Express this fraction in simplest form. (Lesson 4-2)

4. **CARPENTRY** The measurement of a board being used in a carpentry project is $\frac{53}{8}$ feet. Write this improper fraction as a mixed number. (Lesson 4-3)

5. **BAKING** A bread recipe calls for $4\frac{2}{3}$ cups of flour. Write this mixed number as an improper fraction. (Lesson 4-3)

6. **PROJECT** Tamika is working on a group project with 3 other students. They must decide who will be the researcher, the artist, the writer, and the presenter. How many ways can they choose to assign the jobs? (Lesson 4-4)

7. **BICYCLES** The front gear of a bicycle has 54 teeth, and the back gear has 18 teeth. How many complete rotations must the smaller gear make for both gears to be aligned in the original starting position? (Lesson 4-5)

8. **PATTERNS** Which three common multiples for 5 and 7 are missing from the list below? (Lesson 4-5)

   35, 70, 105, ■, ■, ■, 245, 280, ...

9. **CANDLES** Three candles have widths of $\frac{3}{8}$ inch, $\frac{5}{6}$ inch, and $\frac{2}{3}$ inch. What is the measure of the candle with the greatest width? (Lesson 4-6)

10. **HOME REPAIR** Dan has a wrench that can open to accommodate three different sizes of washers: $\frac{5}{16}$ inch, $\frac{3}{8}$ inch, and $\frac{2}{5}$ inch. Which washer is the smallest? (Lesson 4-6)

11. **TRAVEL** The road sign shows the distances from the highway exit to certain businesses. What fraction of a mile is each business from the exit? (Lesson 4-7)

12. **SPIDERS** The diagram shows the range in the length of a tarantula. What is one length that falls between these amounts? Write the amount as a mixed number in simplest form. (Lesson 4-7)

13. **TRACK AND FIELD** Monifa runs a race covering $4\frac{5}{8}$ miles. Express this distance as a decimal. (Lesson 4-8)

14. **SEWING** Dana buys $3\frac{3}{4}$ yards of fabric to make a new dress. What decimal does this represent? (Lesson 4-8)

15. **MAPS** Josh is drawing a map of his walk to school. His house is located at (3, 5). To get to school he walks one block south and three blocks west. At what coordinates on the map should Josh draw the school? (Lesson 4-9)

1. **PICTURE FRAMES** A picture frame is to be wrapped in the box shown below. To the nearest half-inch, how large can the frame be? (Lesson 5-1)

2. **MONEY** For every \$5 Marta earns mowing lawns, she puts \$2 in her savings account. How much money will she have to earn in order to deposit \$30 into her savings account? (Lesson 5-2)

3. **COFFEE** A coffee pot was $\frac{7}{8}$ full at the beginning of the day. By the end of the day, the pot was only $\frac{1}{8}$ full. How much of the coffee was used during the day? (Lesson 5-3)

4. **SCHOOL** Of the students in a class, $\frac{5}{16}$ handed in their report on Monday, and $\frac{7}{16}$ handed their report in on Tuesday. What fraction of the class had completed the work after two days? (Lesson 5-3)

5. **MEASUREMENT** How much more is $\frac{2}{3}$ pound than $\frac{1}{2}$ pound? (Lesson 5-4)

6. **CHEMISTRY** A chemist has $\frac{3}{4}$ quart of a solution. The chemist uses $\frac{1}{5}$ quart of the solution. How much of the solution remains? (Lesson 5-4)

7. **REPAIRS** A plumber repaired two sinks in an apartment. The first job took $1\frac{1}{3}$ hours, and the second job took $2\frac{3}{4}$ hours. What was the total time it took the plumber to repair both sinks? (Lesson 5-5)

8. **BANKING** The table below shows the interest rates advertised by three different banks for a 30-year fixed rate mortgage. Find the difference between the highest and the lowest rate. (Lesson 5-5)

30-Year Fixed Mortgage Rates	
Bank	**Interest Rate**
A	$7\frac{1}{4}$%
B	$6\frac{3}{5}$%
C	$6\frac{5}{6}$%

9. **CARPET** Seth is buying carpeting for his basement. The room's dimensions are shown below. About how much carpet will he need to buy? (Lesson 5-6)

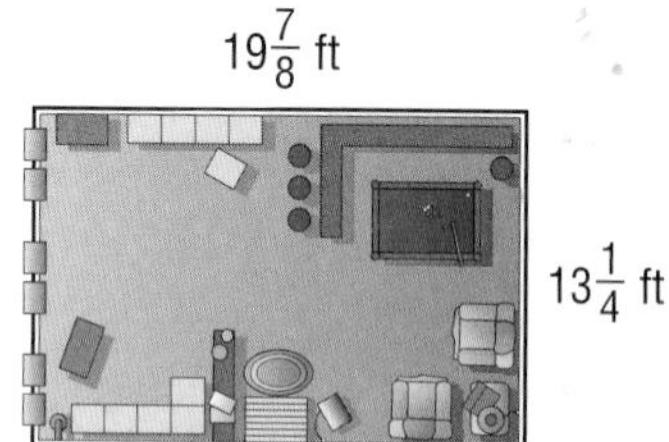

10. **CARS** According to a survey, $\frac{2}{5}$ of people prefer tan cars. Of those people who prefer tan cars, $\frac{1}{3}$ prefer a two-door car. What fraction of the people surveyed would prefer a tan two-door car? (Lesson 5-7)

11. **COOKING** A cookie recipe calls for $3\frac{3}{8}$ cups of sugar. Tami wants to make $1\frac{1}{2}$ times the recipe. How much sugar will Tami need? (Lesson 5-8)

12. **SNACKS** Mika buys $\frac{3}{4}$ pound of peanuts and divides it evenly among 5 friends. Find the amount of peanuts that each friend will get. (Lesson 5-9)

13. **BOATING** On a recent fishing trip, a boat traveled $53\frac{5}{8}$ miles in $2\frac{3}{4}$ hours. How many miles per hour did the boat average? (Lesson 5-10)

1. **SNACKS** A 16-ounce bag of potato chips costs $2.08, and a 40-ounce bag costs $4.40. Which bag is less expensive per ounce? (Lesson 6-1)

2. **NATURE** Write the ratio that compares the number of leaves to the number of acorns. (Lesson 6-1)

3. **TRAVELING** The speed limit on a highway is 65 miles per hour. Suppose you travel this highway at a constant speed of 65 miles per hour without any stops. Use a ratio table to determine how long it will take you to travel 260 miles. (Lesson 6-2)

4. **BABYSITTING** Janelle babysits 2 babies who eat 6 jars of baby food each day. Use a ratio table to determine how many jars of baby food Janelle will need to babysit them for 10 days. (Lesson 6-2)

5. **BLANKETS** Kyla made two blankets, one 20 inches by 34 inches and the other 48 inches by 60 inches. Are the dimensions of the two blankets proportional? Explain your reasoning. (Lesson 6-3)

6. **SURVEY** A survey showed that 12 out of 27 students in one class and 16 out of 36 students in another class had visited Disney World for a vacation. Are the results from the two classes proportional? Explain your reasoning. (Lesson 6-3)

7. **GARBAGE** If a family of 4 produces 3 bags of trash each week, how many of the same sized bags of trash would you expect a family of 6 to produce each week? (Lesson 6-4)

8. **CALORIES** A 100-pound person burns 7 Calories per minute playing basketball. How many Calories per minute would a 150-pound person burn playing basketball? (Lesson 6-4)

9. **PATTERNS** Draw the next two figures in the pattern below. (Lesson 6-5)

10. **NUMBER SENSE** Describe the pattern below. Then find the missing number. (Lesson 6-5)

2, 6, ■, 54, 162

11. **TIME** There are 24 hours in 1 day. Make a table and write an algebraic expression relating the number of days to the number of hours. Then find the number of hours Parker must wait for his birthday presents if his birthday is 13 days away. (Lesson 6-6)

12. **MEASUREMENT** There are 100 centimeters in 1 meter. Make a table and write an algebraic expression relating the number of meters to the number of centimeters. Then find the length in meters of a snake that is 225 centimeters long. (Lesson 6-6)

13. **TICKETS** A movie theater is offering a special rate for its grand opening. Ticket prices are shown in the table. Write a sentence and an equation to describe the data. How much will it cost for a group of 5 friends to see a movie? (Lesson 6-7)

Number of Tickets, t	Total Cost, c
1	$6
2	$12
3	$18

14. **BUSINESS** A company offers a $750 bonus to its employees for each person hired that is referred to the company by a current employee. Write an equation to find b, the amount of bonus pay awarded to an employee who refers p people. How much would an employee receive who refers 3 friends who get hired? (Lesson 6-7)

SCHOOL For Exercises 1 and 2, use the following information.

Ben has been on time for school 17 times and tardy 3 times during the past month. (Lesson 7-1)

1. What percent of the time has Ben been tardy?

2. What is the fraction of days that Ben has been on time?

3. **SPORTS** The table shows the results of a survey of middle school students about their favorite sport. Sketch a circle graph to display the data. (Lesson 7-2)

Favorite Sports	
Football	30%
Basketball	24%
Baseball	20%
Volleyball	15%
Track	11%

4. **BANKING** A local bank advertises a home equity loan with an interest rate of 0.1275. Express this interest rate as a percent. (Lesson 7-3)

GAMES For Exercises 5–7, use the following information.

To win a prize at a carnival, a player must choose a green beanbag from a box without looking. (Lesson 7-4)

5. What is the probability of selecting a yellow beanbag?

6. Find the probability of selecting a blue beanbag.

7. What is the probability of selecting a green beanbag?

SUNDAES For Exercises 8 and 9, use the menu shown below. (Lesson 7-5)

8. How many different ways can you choose an ice cream, a topping, and nuts?

9. Find the probability that a person will randomly choose vanilla ice cream with hot fudge and pecans.

SEASONS For Exercises 10–12, use the table that shows the results of a survey. (Lesson 7-6)

Favorite Season Survey	
Season	**Students**
Fall	4
Winter	7
Summer	9
Spring	5

10. Find the probability that a student prefers summer.

11. What is the probability that a student prefers winter?

12. Suppose there are 500 students in the school. How many students can be expected to prefer spring?

13. **TAX** Emma's mom buys a television set for $350. About how much will she pay in tax if the tax rate is 6.75%? (Lesson 7-7)

14. **FOOD** About what percent of the eggs have been used? (Lesson 7-8)

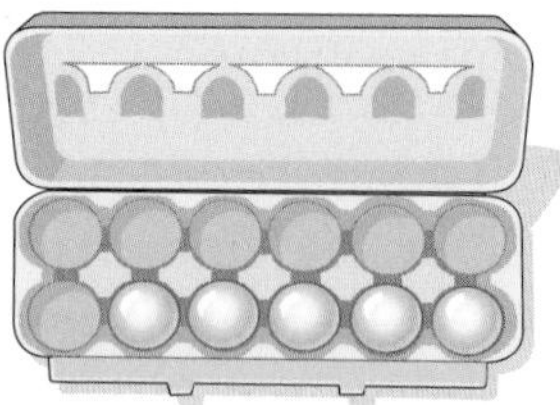

15. **ALLOWANCES** According to a survey, 58% of all teenagers receive a weekly allowance for doing household chores. Estimate how many teenagers out of 453 would receive a weekly allowance for doing household chores. (Lesson 7-8)

1. **FURNITURE** The length of a desk is 48 inches. What is the measure of the length of the desk in feet? (Lesson 8-1)

2. **TYPING** Measure the length of the keyboard key shown to the nearest half, fourth, or eighth inch. (Lesson 8-1)

3. **BAKING** A recipe for a sponge cake calls for 1 pint of milk. How many fluid ounces of milk will be needed for the recipe? (Lesson 8-2)

4. **SNACKS** Based on the facts shown below, how many ounces are in each serving of potato chips? (Lesson 8-2)

5. **TRAVEL** Kenji is planning to travel from Cincinnati, Ohio, to Miami, Florida. When estimating the distance, should he be accurate to the nearest centimeter, meter, or kilometer? (Lesson 8-3)

6. **ARCHITECTURE** An architect's plans for a new house shows the thickness of the doors to be used on each room. Which metric unit of length is most appropriate to use to measure the thickness of the doors? (Lesson 8-3)

7. **FOOD** Mustard comes in a 397-gram container or a 0.34-kilogram container. Which container is smaller? (Lesson 8-4)

8. **SHOPPING** A grocery store sells apple juice in a 1.24-liter bottle and in a 685-milliliter bottle. Which bottle is larger? (Lesson 8-4)

9. **BOOKSHELVES** Cameron's uncle is going to build him a new bookshelf for his room. Cameron needs to know the height of the wall from the floor to the ceiling in feet. He has a poster that he knows is 12 inches wide. Describe a way Cameron can estimate the maximum height of the bookshelf for his uncle. (Lesson 8-5)

10. **CLEANING** A bottle of household cleaner contains 651 milliliters. How many liters does the bottle contain? (Lesson 8-6)

11. **SWIMMING** As part of swim team practice, Opal swims 24,000 centimeters of the breaststroke. Write this distance in meters. (Lesson 8-6)

12. **AUTOMOBILES** Diego is considering buying a new sport utility vehicle which is advertised as having a weight of 2,500 kilograms. What is the weight in grams? (Lesson 8-6)

13. **HOMEWORK** Jerome spends 2 hours and 20 minutes on his homework after school. If he starts working at 3:45 P.M., what time will he be done with his homework? (Lesson 8-7)

14. **RACING** A sail boat race starts at 9:35 A.M., and the first boat crosses the finish line at 11:12 A.M. Find the racing time for the winning boat. (Lesson 8-7)

15. **BASEBALL** Use the information at the right. How much longer was the Sharks baseball game than the Cardinals baseball game? (Lesson 8-7)

Length of Baseball Games

Team	Time
Cougars	2 h 30 min
Sharks	3 h 15 min
Cardinals	2 h 58 min

16. **SNOW** Ricardo and Jamal are shoveling snow. Ricardo states that it must be about −5°C outside. Jamal argues that 15°C is probably closer to the actual temperature. Who is correct? Explain. (Lesson 8-8)

Mixed Problem Solving

1. **WINDOWS** The plans for a new home use several windows in the shape shown. Classify the angle in the window as acute, obtuse, right, or straight. (Lesson 9-1)

2. **ALGEBRA** If the angles shown are supplementary angles, what is the measure of $\angle B$? (Lesson 9-1)

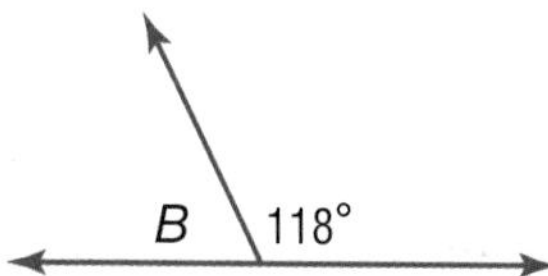

3. **ALGEBRA** Angles M and N are complementary angles. Find $m\angle N$ if $m\angle M = 62°$. (Lesson 9-1)

4. **CLOCKS** Estimate the measure of the angle made by the hour hand and the minute hand of a clock when it is 3:35 P.M. (Lesson 9-2)

5. **QUILTING** A quilt pattern uses pieces of fabric cut into triangles with an angle of 35°. Use a protractor and a straightedge to draw an angle with this measure. (Lesson 9-2)

6. **REASONING** Determine whether the statement below is *sometimes, always,* or *never* true. Explain your reasoning. (Lesson 9-3)

   Supplementary angles form a straight line.

7. **ANGLES** Classify the relationship between $\angle 1$ and $\angle 2$ below. Explain your reasoning. (Lesson 9-3)

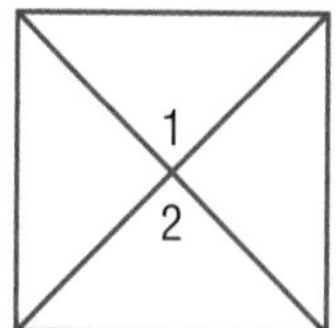

8. **GEOMETRY** What is the measure of the third angle of a triangle if one angle measures 105° and the other angle measures 28°? (Lesson 9-4)

9. **CARPENTRY** Santiago is making a parallelogram shaped sign to hang outside a store. He has marked off three of the signs angles on the wood. They measure 85°, 95°, and 85°. What should be the measure of the fourth angle? (Lesson 9-5)

10. **TREES** The Fraileys are planting trees along each side of their driveway. The driveway is straight and 52 feet long. If there is 4 feet between each tree, and the trees are planted directly across from each other on each side of the driveway, how many trees can the Fraileys plant in all? (Lesson 9-6)

11. **MAPS** On two different-sized maps, the city park is shown using the two figures below. Tell whether the pair of figures is congruent, similar, or neither. (Lesson 9-7)

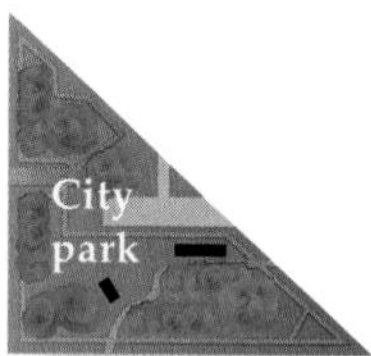

12. **GARDENING** Jessica works in two different garden areas. The first area is rectangular in shape with a length of 10 feet and a width of 6 feet. The second garden area is square in shape with sides measuring 6 feet. Tell whether these two garden plots are congruent, similar, or neither. (Lesson 9-7)

1. **SCHOOL** The perimeter of a desk is 60 inches. If the length of the desk measures 18 inches, what is the measure of the width? (Lesson 10-1)

2. **FENCING** The dimensions of the garden are shown below. To keep out small animals, a fence will surround the garden. If the fencing costs $2.70 per yard, what will be the total cost of the fencing? (Lesson 10-1)

3. **SWIMMING** A new swimming pool is being installed at the Hillside Recreation Center. The pool is circular in shape and has a diameter of 32 feet. Find the circumference of the pool to the nearest tenth. (Lesson 10-2)

4. **RIDES** A carousel has a diameter of 36 feet. How far do passengers travel on each revolution? (Lesson 10-2)

5. **PIZZA** The Pizza Shop claims to make the largest pizza in town. The radius of their largest pizza is 15 inches. Find the circumference of the pizza. (Lesson 10-2)

PAINTING For Exercises 6 and 7, use the following information.

Van is going to paint a wall in his home. The wall is in the shape of a parallelogram with a base of 8 feet and a height of 15 feet. (Lesson 10-3)

6. What is the area of the wall?

7. If one quart of paint will cover 70 square feet, how many quarts of paint does Van need to buy?

8. **WINDOWS** A window is divided into 4 triangular sections. Each section has a base that is 18 inches and a height of 12 inches. What is the area of each triangular section of the window? (Lesson 10-4)

9. **KITES** Alex is making a kite like the one shown. What will be the area of the kite? (Lesson 10-4)

20 in.
36 in.

10. **DOLLS** Molly is arranging her doll collection on 5 shelves. If the top shelf has 2 dolls on it and each shelf has 3 more dolls than the one above it, how many dolls does Molly have in all? (Lesson 10-5)

11. **PHOTOS** Sofie is filling a picture frame that is 8 inches by 10 inches with pictures of her friends. Each picture is 2 inches by 2 inches and she has 21 pictures. Does she have enough pictures to fill the frame with no empty space between pictures? (Lesson 10-5)

12. **TOY BOX** Jamila is given the toy box shown below for her birthday. What is the volume of the toy box? (Lesson 10-6)

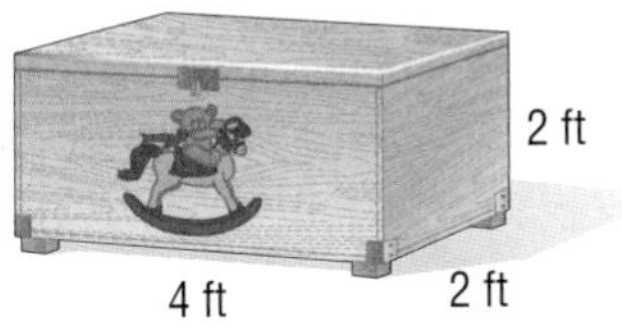

13. **MAIL** A package has a length of 9 inches, a width of 5 inches, and a height of 6 inches. Find the volume of the package if the height is increased by 3 inches. (Lesson 10-6)

14. **GIFT WRAP** Olivia is placing a gift inside a box that measures 15 centimeters by 8 centimeters by 3 centimeters. What is the surface area of the box? (Lesson 10-7)

15. **CONSTRUCTION** Find the surface area of the concrete block shown below. (Lesson 10-7)

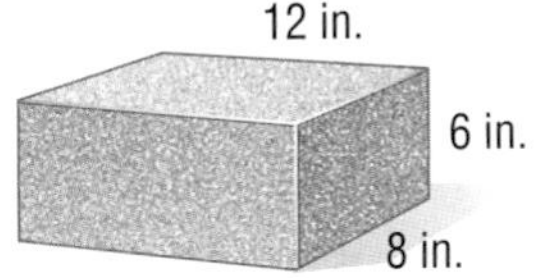

Chapter 11 Integers and Transformations

1. **GEOGRAPHY** The table below lists the record low temperatures for five states. Order these temperatures from least to greatest. (Lesson 11-1)

Record Low Temperatures	
State	**Low Temp. (°F)**
South Carolina	−19
Texas	−23
Alabama	−27
Florida	−2
Michigan	−51

Source: *The World Almanac*

2. **BUSINESS** The table shows the results of a company's operations over a period of three months. Find the company's overall profit or loss over this time period. (Lesson 11-2)

Month	Profit	Loss
April		\$15,000
May	\$19,000	
June		\$12,000

3. **TEMPERATURE** Find the difference in temperature between 41°F in the day in Anchorage, Alaska, and −27°F at night. (Lesson 11-3)

4. **FOOTBALL** The Riverdale football team lost 4 yards on each of 3 consecutive plays. Write the integer that represents the change in yards. (Lesson 11-4)

5. **AGES** Perry is 2 years younger than his sister, who is one third the age of his mother. His mother is 2 years older than his father who is 46. How old is Perry? (Lesson 11-5)

6. **FLYING** An airplane coming in for a landing descended 21,000 feet over a period of 7 minutes. Write the integer that represents the average altitude change per minute. (Lesson 11-6)

MONEY For Exercises 7 and 8, use the following information.

Latisha earns \$3 per hour for babysitting. The expression $3x$ represents the total amount earned where x is the number of hours she babysat. (Lesson 11-7)

7. Copy and complete the table to find the ordered pairs (hours, total amount earned) for 1, 2, 3, and 4 hours.

x (hours)	$3x$	y (amount earned)	(x, y)
1			
2			
3			
4			

8. Graph the ordered pairs. Then describe the graph.

For Exercises 9–11, use triangle *ABC* shown below.

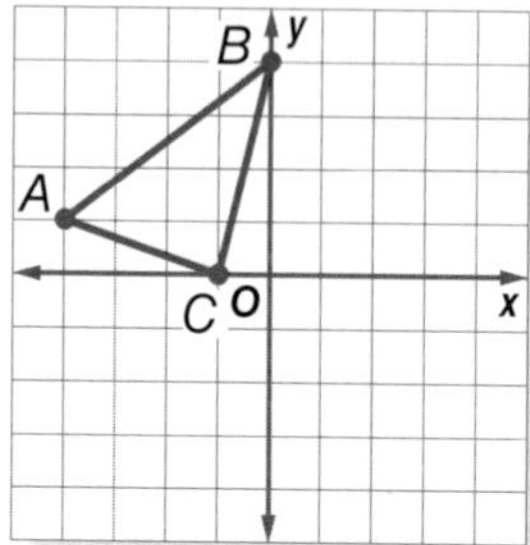

9. Describe the translation that will occur if point A is moved to (1, −1). Then graph triangle $A'B'C'$ using this translation. (Lesson 11-8)

10. Find the coordinates of the vertices of triangle ABC after a reflection over the y-axis. The graph the reflection. (Lesson 11-9)

11. Graph triangle ABC and its image after a clockwise rotation of 180° about the origin. Then give the coordinates of the vertices for triangle $A'B'C'$. (Lesson 11-10)

1. **MOVIES** Five friends are going to the movies. Each ticket costs \$5.50, and they each purchase a drink for \$2.75. Use the Distributive Property to find the total cost of the trip to the movies for all five friends. (Lesson 12-1)

2. **DRIVING** Mr. Reynolds is driving at a constant rate of 60 miles per hour. Ms. Santiago is driving at a constant rate of 65 miles per hour. Use the Distributive Property to find how many more miles Ms. Santiago will drive in 3 hours than Mr. Reynolds. (Lesson 12-1)

3. **SWIMMING** Eight friends went to the pool. The cost to enter the pool is $\$x$. The table shows the prices of snacks and beverages at the snack bar. Three of the friends each bought a popcorn and a bottle of water. The remaining five friends did not buy any snacks or beverages. Write and simplify an expression that represents the total cost of tickets and snacks or beverages. (Lesson 12-2)

Item	Price
popcorn	\$3.50
juice pop	\$3.00
small soda	\$1.50
bottle of water	\$2.00

4. **AGE** Megan is x years old. Her brother, Greg, is four years older than she is. Their father is three times as old as Greg. Their uncle is x years older than their father. Write and simplify an expression that represents the uncle's age in years. (Lesson 12-2)

5. **PICNICS** For a summer picnic, 37 guests were invited. By 2 P.M., 18 guests had arrived. Write and solve an equation to find how many guests had not arrived. (Lesson 12-3)

6. **CARPENTRY** A board that measures 19 meters in length is cut into two pieces. The shorter of the two pieces measures 7 meters. Write and solve an equation to find the length of the longer piece. (Lesson 12-3)

7. **BANKING** After Sean withdrew \$55 from his bank account, the balance was \$123. Write and solve an equation to find the balance of the bank account before the withdrawal. (Lesson 12-4)

8. **WEATHER** As a cold front moved through town, the temperature dropped 21°F. The temperature after the cold front moved in was 36°F. Write and solve an equation to find the temperature before the cold front came through town. (Lesson 12-4)

9. **SCHOOL** The number of questions Juan answered correctly on an exam is 2 times as many as the amount Ryan answered correctly. Juan correctly answered 20 questions. Write a multiplication equation that can be used to find how many questions Ryan answered correctly. (Lesson 12-5)

10. **UNIFORMS** Members of the track team collected donations for new uniforms. They received \$48 the second week of taking donations, \$62 the third week, and \$18 the fourth week. If their grand total was \$256, how much did they collect the first week? (Lesson 12-6)

11. **ROLLER COASTERS** The five fastest roller coasters at an amusement park have top speeds of 78, 72, 70, 68, and 66 miles per hour. Find the average top speed. (Lesson 12-6)

Preparing for Standardized Tests

Throughout the year, you may be required to take several standardized tests, and you may have some questions about them. Here are some answers to help you get ready.

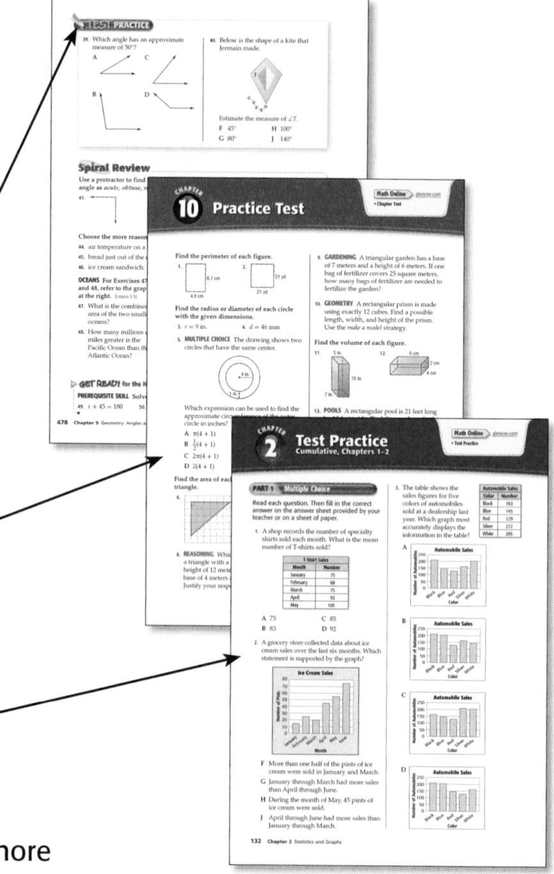

How Should I Study?

The good news is that you've been studying all along—a little bit every day. Here are some of the ways your textbook has been preparing you.

- **Every Day** Each lesson had multiple-choice practice questions.
- **Every Week** The Mid-Chapter Quiz and Practice Test also had several multiple-choice practice questions.
- **Every Month** The Test Practice pages at the end of each chapter had even more questions, including short-response/grid-in and extended-response questions.

Are There Other Ways to Review?

Absolutely! The following pages contain even more practice for standardized tests.

Tips for SUCCESS

Before the Test

- Go to bed early the night before the test. You will think more clearly after a good night's rest.
- Become familiar with common formulas and when they should be used.
- Think positively.

During the Test

- Read each problem carefully. Underline key words and think about different ways to solve the problem.
- Watch for key words like NOT. Also look for order words like *least, greatest, first,* and *last.*
- Answer questions you are sure about first. If you do not know the answer to a question, skip it and go back to that question later.
- Check your answer to make sure it is reasonable.
- Make sure that the number of the question on the answer sheet matches the number of the question on which you are working in your test booklet.

Whatever you do...

- Don't try to do it all in your head. If no figure is provided, draw one.
- Don't rush. Try to work at a steady pace.
- Don't give up. Some problems may seem hard to you, but you may be able to figure out what to do if you read each question carefully or try another strategy.

Multiple-Choice Questions

Multiple-choice questions are the most common type of question on standardized tests. These questions are sometimes called *selected-response questions*. You are asked to choose the best answer from four or five possible answers.

Incomplete shading

Ⓐ Ⓑ Ⓒ Ⓓ

Too light shading

Ⓐ Ⓑ Ⓒ Ⓓ

Correct shading

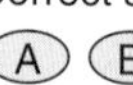  Ⓒ Ⓓ

To record a multiple-choice answer, you may be asked to shade in a bubble that is a circle or an oval or just to write the letter of your choice. Always make sure that your shading is dark enough and completely covers the bubble.

The answer to a multiple-choice question may not stand out from the choices. However, you may be able to eliminate some of the choices. Another answer choice might be that the correct answer is not given.

TEST EXAMPLE

1 **The graph shows the maximum life span in years for the five birds having the greatest life spans. Which ratio compares the life span of a bald eagle to the life span of an eletus parrot?**

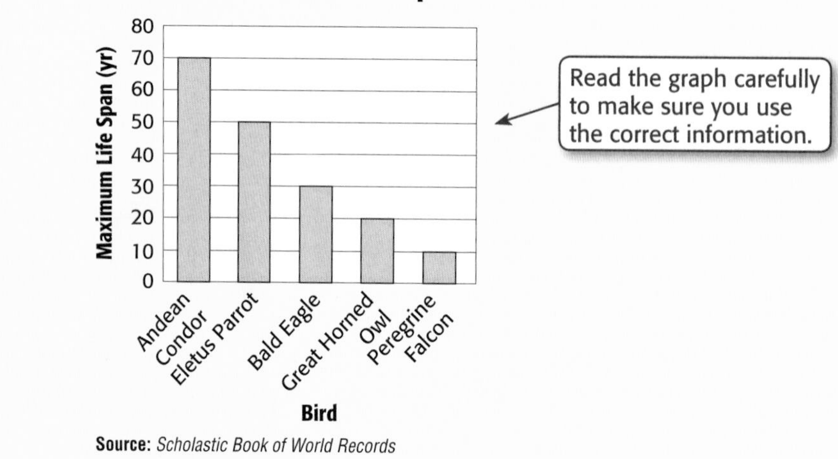

Source: *Scholastic Book of World Records*

A $\frac{5}{2}$ **B** $\frac{5}{3}$ **C** $\frac{3}{5}$ **D** $\frac{2}{5}$

STRATEGY

Elimination
Can you eliminate any of the choices?

Before finding the ratio, look at the answers. The bald eagle has a shorter life span than the eletus parrot, so the ratio cannot be greater than 1. Answers A and B are greater than 1. The answer must be C or D.

Compare the life span for the bald eagle to the life span for the eletus parrot. Write the ratio and simplify.

$$\frac{\text{life span of bald eagle}}{\text{life span of eletus parrot}} = \frac{30}{50}$$ Write the ratio.

$$= \frac{30 \div 10}{50 \div 10}$$ The GCF of 30 and 50 is 10.

$$= \frac{3}{5}$$ Simplify.

The correct choice is C.

Preparing for Standardized Tests

Some problems are easier to solve if you draw a diagram. If you cannot write in the test booklet, draw a diagram on scratch paper.

TEST EXAMPLE

STRATEGY

Diagrams
Draw a diagram for the situation.

2 **The Jung family is building a fence around a rectangular section of their backyard. The section measures 21 feet by 27 feet. They are going to place a post at each corner and then place posts three feet apart along each side. How many posts will they need?**

F 28 **H** 36

G 32 **J** 63

To solve this problem, draw a diagram of the situation. Label the important information from the problem. The width of the rectangle is 21 feet and the length of the rectangle is 27 feet.

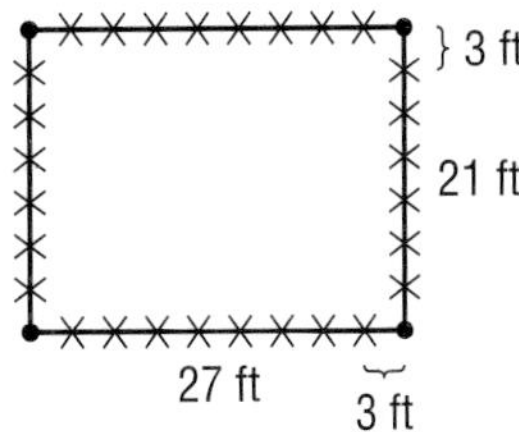

In the diagram, the corner posts are marked with points. The other posts are marked with x's and are 3 feet apart.

Add the number of posts on each side to the four corner posts.

4 points + 6 x's + 8 x's + 6 x's + 8 x's = 32 posts

The correct choice is G.

Often multiple-choice questions require you to convert measurements to solve. Pay careful attention to each unit of measure in the question and the answer choices.

TEST EXAMPLE

STRATEGY

Units
Do you have the correct unit of measure?

3 **The Art Club is selling 12-ounce boxes of chocolate candy for a fundraiser. There are 36 boxes of candy in a case. How many pounds of chocolate are in one case?**

A 864 lb **B** 432 lb **C** 216 lb **D** 27 lb

Each box of candy weighs 12 ounces. So, 36 boxes weigh 36 • 12 or 432 ounces. However, Choice B is not the correct answer. The question asks for *pounds* of chocolate.

432 oz = __?__ lb **THINK** 16 oz = 1 lb

$432 \div 16 = 27$ Divide to change ounces to pounds.

So, there are 27 pounds in one case. The correct choice is D.

Multiple-Choice Practice

Choose the best answer.

Preparing for Standardized Tests

Number and Operations

1. About 125,000,000 people in the world speak Japanese. About 100,000,000 people speak German. How many more people speak Japanese than German?

   A 25,000,000 C 12.5
   B 25,000 D 1.25

2. It is estimated that there are 40,000,000 pet dogs and 500,000 pet parrots in the United States. How many times more pet dogs are there than pet parrots?

   F 80 H 3,500
   G 800 J 39,500,000

3. The graph shows the elements found in Earth's crust. About what fraction of Earth's crust is silicon?

Elements in Earth's Crust

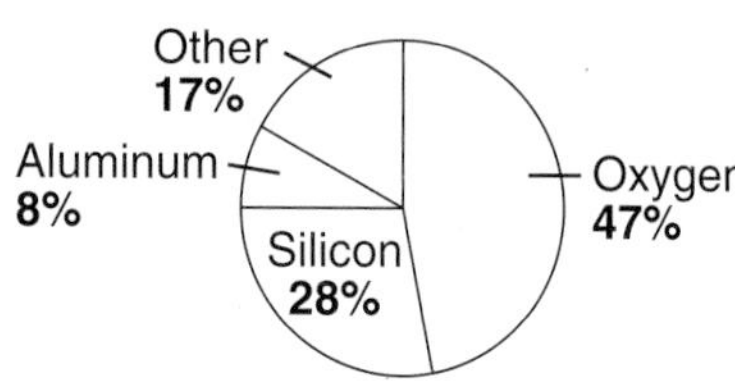

Source: *The World Almanac for Kids*

   A $\frac{1}{28}$ B $\frac{1}{4}$ C $\frac{2}{5}$ D $\frac{1}{2}$

4. In Israel, the average consumption of sugar per person is 6^3 pounds per year. How much sugar is this?

   F 18 lb H 216 lb
   G 180 lb J 729 lb

Algebra

5. The largest ball of twine is in Cawker City, Kansas. It weighs 17,400 pounds. A ball of twine in Jackson, Wyoming, weighs 12,100 pounds less than this. How much does the ball of twine in Jackson weigh?

   A 5,300 lb C 29,500 lb
   B 14,750 lb D 32,150 lb

6. The fastest fish in the world is the sailfish. If a sailfish could maintain its speed as shown in the table, how many miles could the sailfish travel in 6 hours?

Hours Traveled	0	1	2	3	4	5	6
Miles Traveled	0	68	136	204	272	340	?

   F 6 mi H 408 mi
   G 68 mi J 476 mi

7. The table shows the cost to rent a carpet cleaner. Which function rule describes the relationship between total cost y and rental time in hours x?

Rental Time (x)	Total Cost (y)
0	0
1	7
2	14
3	21

   A $7y = x$ C $y = x + 7$
   B $y = 7x$ D $y = x \div 7$

8. What is the area of the triangle?

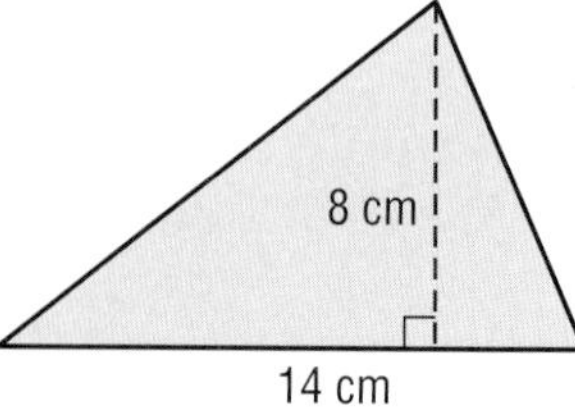

   F 22 cm^2
   G 44 cm^2
   H 56 cm^2
   J 112 cm^2

Geometry

9. Which term does *not* describe the figure in the center of the quilt section?

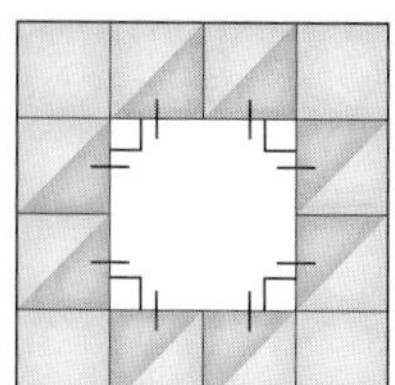

   A square
   B rectangle
   C parallelogram
   D trapezoid

10. The circle graph shows government spending in a recent year. Which is the *best* term to describe the angle that represents Defense?

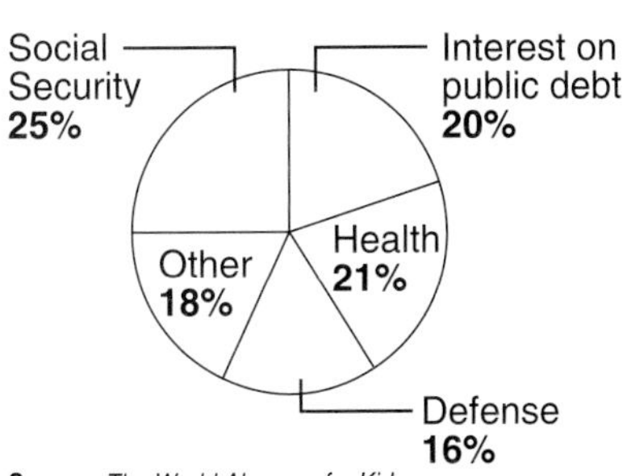

F acute
G obtuse
H right
J straight

11. The diagram shows a map of Wild Horse hiking area. What is the measure of $\angle 1$?

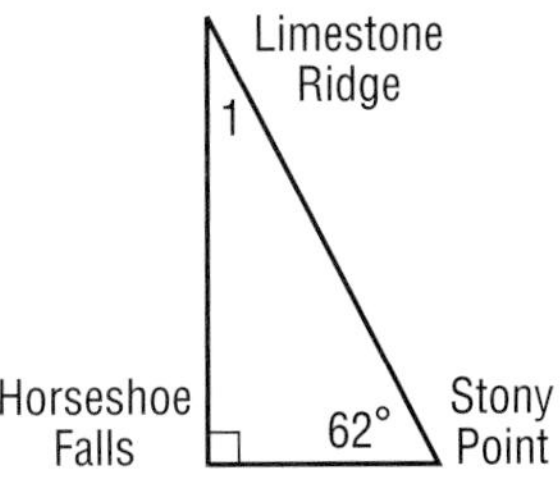

A 10°
B 28°
C 62°
D 90°

12. Jerome is using a regular hexagon in a logo for the Writer's Club. How many lines of symmetry does the hexagon have?

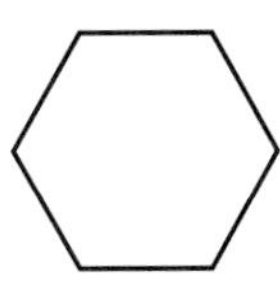

F 0
G 1
H 3
J 6

Measurement

13. Claire and her friends are making friendship bracelets. Each bracelet uses 8 inches of yarn. How many feet of yarn will they need for 12 bracelets?

A 4 ft
B 8 ft
C 32 ft
D 96 ft

14. A spider can travel a maximum of 1.2 miles per hour. If there are 5,280 feet in a mile, how many feet can a spider travel in a minute?

F $35\frac{1}{5}$ ft
G $73\frac{1}{3}$ ft
H 88 ft
J $105\frac{3}{5}$ ft

TEST-TAKING TIP

Questions 13 and 14 The units of measure asked for in the answer may be different than the units given in the question. Check that your answer is in the correct units.

15. Mail Magic sells two different mailing boxes with the dimensions shown in the table. What is the volume of box B?

	Length (in.)	Width (in.)	Height (in.)
Box A	16	10	8
Box B	10	8	8

A 78 in^3
B 448 in^3
C 640 in^3
D 1,280 in^3

Data Analysis and Probability

16. The table shows Justin's test scores in math class for the first quarter. Find the mean of his test scores.

Test	Score
Chapter 1	85
Chapter 2	91
Chapter 3	86
Chapter 4	78

F 13 G 85 H 86 J 91

17. One marble is selected without looking from the set of marbles below. What is the probability that the marble is black?

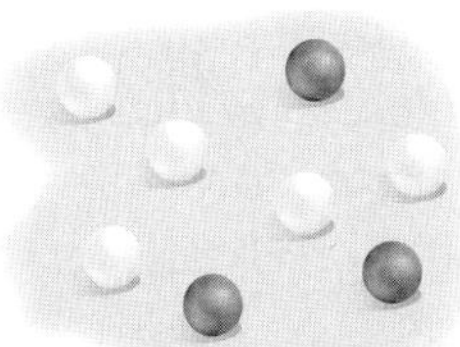

A $\frac{1}{8}$ B $\frac{3}{8}$ C $\frac{5}{8}$ D $\frac{3}{4}$

Preparing for Standardized Tests

Gridded-Response Questions

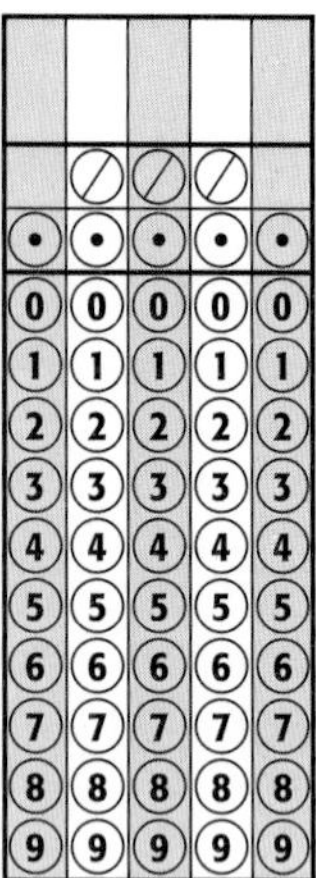

Gridded-response questions are another type of question on standardized tests. These questions are sometimes called *student-produced response* or *grid in.*

For gridded response, you must mark your answer on a grid printed on an answer sheet. The grid contains a row of four or five boxes at the top, two rows of ovals or circles with decimal and fraction symbols, and four or five columns of ovals, numbered 0–9. An example of a grid from an answer sheet is shown.

TEST EXAMPLE

1 **The bar graph shows the average elevation of five states. How much higher in feet is the average elevation of Colorado than the average elevation of Nevada?**

States with Highest Average Elevation

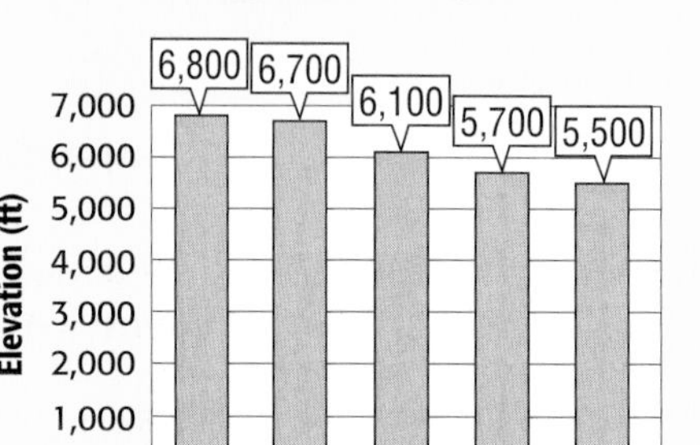

Source: *Scholastic Book of World Records*

What do you need to find?

You need to find the difference between the elevations for Colorado and Nevada.

Read the graph to find the elevations for the two states.

The elevation for Colorado is 6,800 feet, and the elevation for Nevada is 5,500 feet.

Subtract Nevada's elevation from Colorado's elevation.

$$6{,}800 - 5{,}500 = 1{,}300$$

How do you fill in the answer grid?

- Write your answer in the answer boxes.
- Write only one digit or symbol in each answer box.
- Do not write any digits or symbols outside the answer boxes.
- You may write your answer with the first digit in the left answer box, or with the last digit in the right answer box. You may leave blank any boxes you do not need on the right or the left side of your answer.

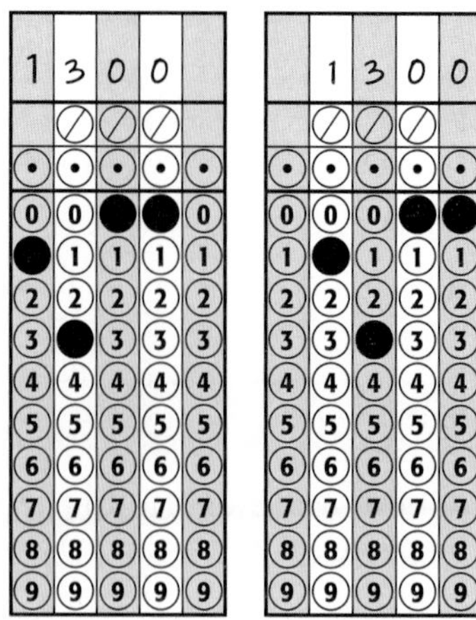

- Fill in only one bubble for every answer box that you have written in. Be sure not to fill in a bubble under a blank answer box.

Preparing for Standardized Tests

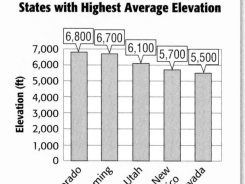

Gridded-response answers may be fractions or decimals.

TEST EXAMPLE

2 **Jamica's recipe for a batch of bread calls for $1\frac{1}{4}$ cups of flour. How much flour in cups does she need for $\frac{1}{2}$ of a batch of bread?**

$1\frac{1}{4} \times \frac{1}{2} = \frac{5}{4} \times \frac{1}{2}$ Write $1\frac{1}{4}$ as $\frac{5}{4}$.

$= \frac{5 \times 1}{4 \times 2} = \frac{5}{8}$ She needs $\frac{5}{8}$ cup of flour.

5	/	8		

0	.	6	2	5

	.	6	2	5

Do not leave a blank answer box in the middle of an answer.

How do you fill in the answer grid?

You can either grid the fraction $\frac{5}{8}$, or rewrite it as 0.625 and grid the decimal. Be sure to write the decimal point or fraction bar in the answer box. Acceptable answer responses that represent $\frac{5}{8}$ and 0.625 are shown.

Preparing for Standardized Tests

Never change the improper fraction to a mixed number. Instead, grid either the improper fraction or the equivalent decimal.

TEST EXAMPLE

3 **The dwarf pygmy goby is the smallest fish in the world. It measures $\frac{1}{2}$ inch in length. A clownfish measures 5 inches in length. How much longer is a clownfish than a dwarf pygmy goby?**

$5 \rightarrow 4\frac{2}{2}$ Rename 5 as $4\frac{2}{2}$.

$-\frac{1}{2} \rightarrow -\frac{1}{2}$ Subtract.

$4\frac{1}{2} = \frac{9}{2}$ Before filling in the grid, change the mixed number to an equivalent improper fraction or decimal.

You can either grid the improper fraction $\frac{9}{2}$, or rewrite it as 4.5 and grid the decimal. Do not enter 41/2, as this will be interpreted as $\frac{41}{2}$.

9	/	2		

		4	.	5

Gridded-Response Practice

Solve each problem. Then copy and complete a grid like the one shown on page 724.

Number and Operations

1. Mario is going to the amusement park with three friends. Each person needs $22 to pay for admission to the park, $6 for lunch, and $3 to rent a locker. Find the total amount of money in dollars needed by Mario and his friends.

2. The elevation of Death Valley, California, is 282 feet below sea level. New Orleans, Louisiana, has an elevation of 8 feet below sea level. What is the difference in feet between the two elevations?

3. Jordan got 18 out of 25 questions correct on his science test. What percent of the questions did he get correct?

4. The Geography Club is selling wrapping paper for a fundraiser. The sponsor ordered 16 cases. If each student in the club takes $\frac{2}{3}$ of a case and there are no cases left over, how many students are in the club?

Algebra

5. Kareem is studying for a social studies exam. The table shows the number of minutes he plans to study each day. If the pattern continues, how many minutes will he study on Friday?

Day	Mon.	Tues.	Wed.	Thurs.	Fri.
Minutes	15	20	30	45	?

6. The table shows the cost of renting an inner tube to use at the Wave-a-Rama Water Park. Suppose an equation of the form $y = ax$ is written for the data in the table. What is the value of a?

Hours (x)	Cost (y)
1	$5.50
2	$11.00
3	$16.50

7. The graph shows the cost of renting a car from Speedy Rental for various miles driven. What is the cost in dollars of renting the car and driving 300 miles?

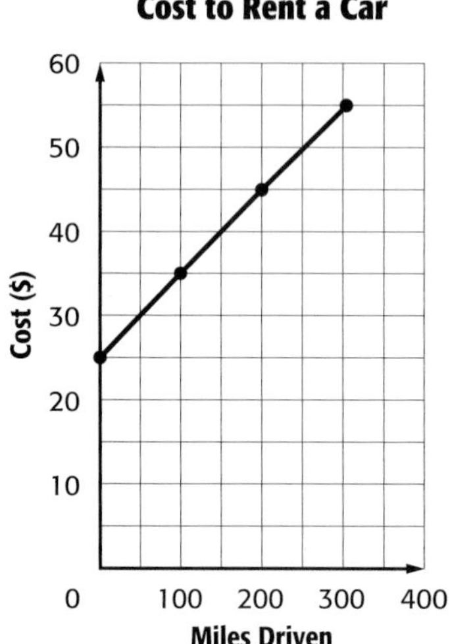

8. During a vacation, Mike decided to try fly-fishing. He rented a rod for $12 and a driftboat for $8 per hour. If he spent a total of $52, for how many hours did he rent the boat?

Geometry

9. Roberto is using a regular pentagon in a design he is making for the art fair. How many lines of symmetry does the figure have?

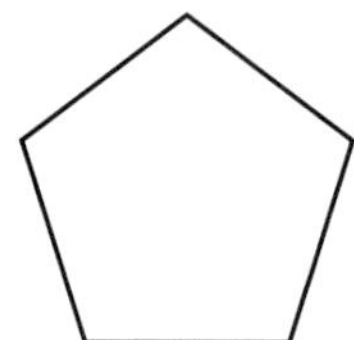

10. If ∠1 and ∠2 are complementary, find the measure of ∠2 in degrees.

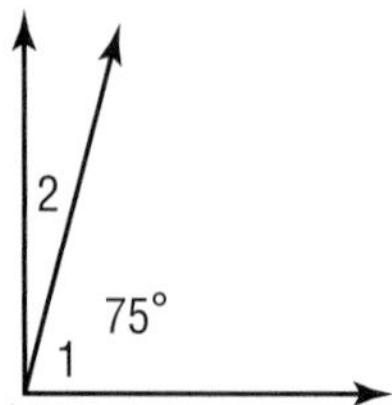

11. The diagram shows the shape of a park in Elm City. What is the measure of ∠3 in degrees?

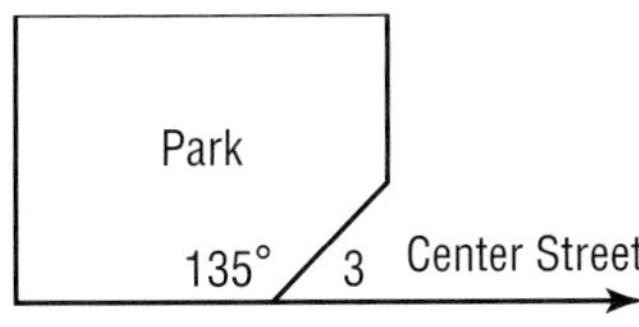

12. What is the area of the parallelogram in square meters?

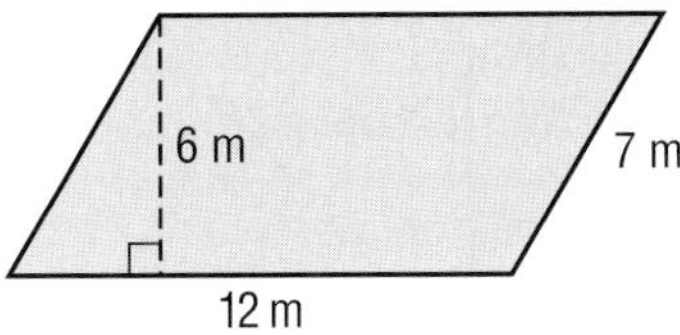

13. A square has a side length of 5 inches. A larger square has a side length of 10 inches. As a fraction, what is the ratio that compares the area of the smaller square to the area of the larger square?

Measurement

14. The B737-400 aircraft uses 784 gallons of fuel per hour. At this rate, how many gallons of fuel does the aircraft use in three hours?

15. Madison can run at a rate of 9 miles per hour. If there are 5,280 feet in a mile, how far in feet can she run in one minute?

16. A sprinkler for a large garden waters a circular region. The diameter of the circular region is 25 feet. In square feet, what is the area of the region covered by the sprinkler? Use 3.14 for π and round to the nearest square foot.

17. At Valley Amusement Park, the three rides shown form a triangle. Find the perimeter of the triangle in yards.

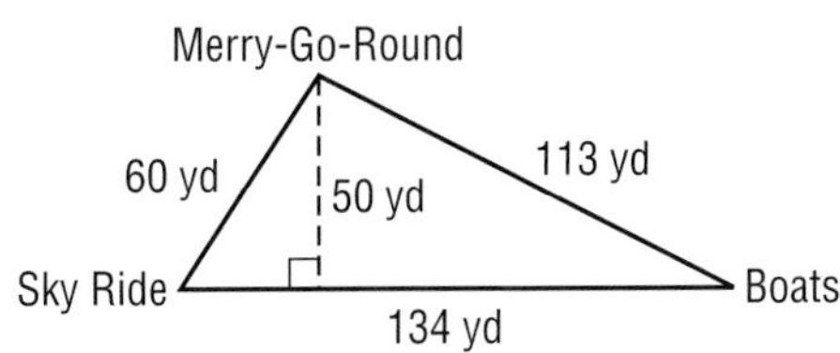

18. Mei Ying ordered a new computer monitor. The monitor arrived in a box shaped as a rectangular prism measuring 2.5 feet by 1 foot by 1.5 feet. What is the surface area in square feet of the box?

Data Analysis and Probability

19. The table shows the number of points scored in a season by five players on a basketball team. Find the mean of the data.

Player	Points
Danielle	114
Mia	93
Tyra	91
Sophia	84
Julia	63

20. Lakendra finds that a particular spiral notebook has the following prices at five different stores. What is the median price of the data in cents?

Store	A	B	C	D	E
Price	57¢	98¢	42¢	85¢	68¢

21. For a certain city, the probability that the temperature in June will reach 100°F is 15%. What is the probability as a percent that the temperature will *not* reach 100°F in June?

22. Find the probability that a randomly thrown dart will land in the shaded region of the dartboard.

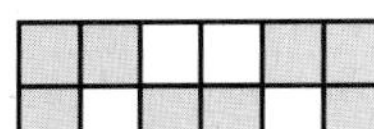

TEST-TAKING TIP

Questions 13 and 22 Fractions do not have to be written in simplest form. Any equivalent fraction that fits the grid is correct.

Preparing for Standardized Tests

Short-Response Questions

Short-response questions require you to provide a solution to the problem as well as any method, explanation, and/or justification you used to arrive at the solution. These are sometimes called *constructed-response, open-response, open-ended, free-response*, or *student-produced questions*.

The following is a sample rubric, or scoring guide, for scoring short-response questions.

Credit	Score	Criteria
Full	2	Full credit: The answer is correct and a full explanation is provided that shows each step in arriving at the final answer.
Partial	1	Partial credit: There are two different ways to receive partial credit. • The answer is correct, but the explanation provided is incomplete or incorrect. • The answer is incorrect, but the explanation and method of solving the problem is correct.
None	0	No credit: Either an answer is not provided or the answer does not make sense.

On some standardized tests, no credit is given for a correct answer if your work is not shown.

Preparing for Standardized Tests

TEST EXAMPLE

Jennie wants to buy five CDs that are on sale at a local store for \$12.50 each. She has also found the same CDs on a Web site for \$10 each. However, on the Web site, she must pay a handling fee of \$4 for any order and an additional \$1.50 in shipping for each CD. Should Jennie buy them at the store or on the Web site? Explain.

Full Credit Solution

First, I will find the cost of five CDs at the store.

$5 \times \$12.50 = \$62.50$

Now I will find the cost of ordering five CDs from the Web site, including the handling and shipping fees.

The steps, calculations, and reasoning are clearly stated.

$5 \times \$10 = \$50$	cost of CDs
$\$50 + \$4 = \$54$	cost of handling on any order
$5(\$1.50) = \$7.50$	cost of shipping 5 CDs
$\$54 + \$7.50 = \$61.50$	total cost

The solution of the problem is clearly stated.

The cost of the CDs from the Web site, including all fees, is \$61.50. Since this is cheaper, Jennie should buy the CDs from the Web site.

Partial Credit Solution

In this sample solution, the answer is correct. However, there is no justification for any of the calculations.

There is no explanation of how \$62.50, or later \$61.50, was obtained.

\$62.50 to buy at the store

$$\begin{array}{r} 50 \\ 4 \\ +\ 7.50 \\ \hline 61.50 \end{array}$$

Jennie should buy them at the Web site.

Partial Credit Solution

In this sample solution, the answer is incorrect. However, the calculations and reasoning are correct, if Jennie was planning to buy 1 CD, not 5.

A CD at the store will cost Jennie \$12.50.

A CD on the Web site will cost Jennie

\$10	cost of CD
4	handling cost
+ 1.50	shipping
15.50	

The student only found the cost of one CD from each place, not 5 CDs.

It will cost Jennie a lot more to buy a CD on the Web site. She should buy it at the store.

No Credit Solution

The student does not understand the problem and merely adds all the available numbers and multiplies by 5. Also, the question asked is not answered.

$$\begin{array}{r} 12.50 \\ 10 \\ 4 \\ +\ 1.50 \\ \hline 28.00 \end{array}$$

$$\begin{array}{r} 28.00 \\ \times\ \ 5 \\ \hline 140.00 \end{array}$$

Jennie needs to pay \$140.00.

Short-Response Practice

Solve each problem. Show all your work.

Number and Operations

1. The table shows the average amount of ice cream eaten per person in one year in three countries. How much more ice cream is eaten by an average person in Australia than by a person in the United States?

Country	Amounts Eaten (pints)
Australia	29.2
Italy	25.0
United States	24.5

Source: *Top 10 of Everything*

2. Rafe and two friends are planning to spend a day at the Science Museum and Aquarium. Admission costs $15 per person for a half day or $25 per person for the full day. Each person also wants to have $18 to spend for food and souvenirs. If the friends have $100 total to spend, which admission can they get, the half or the full day?

3. For a craft activity at a day care, each child will need $1\frac{3}{4}$ yards of ribbon. If there are 25 yards of ribbon available, is there enough for the 15 children who will participate? If not, how much more ribbon is needed?

4. Below are two specials at a store for the same size of cans of fruit. Which special is the better buy?

5. Jacob multiplied 30 by 0.25 and got 7.5. Tanner looked at his answer and said that it was wrong because when you multiply two numbers, the product is always greater than both of the numbers. Explain who is correct.

Algebra

6. The table shows J.T.'s schedule for training for a marathon. If the pattern continues, how many minutes will he run on Day 8?

Day	Running Time (min)
1	20
2	22
3	24
4	26

7. Write a function rule for the values in the table. Be sure to use the variables x and y in your rule.

Input (x)	Output (y)	(x, y)
0	10	(0, 10)
1	11	(1, 11)
2	12	(2, 12)
3	13	(3, 13)

8. The area of a trapezoid can be found by using the formula $A = \frac{b_1 + b_2}{2} \cdot h$. Find the area of the trapezoid.

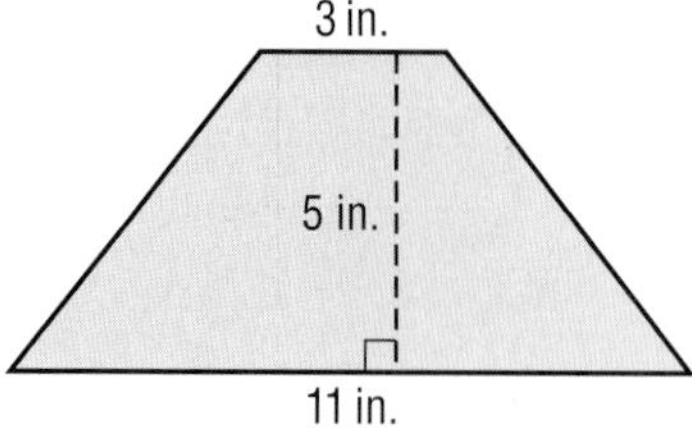

9. Solve $-3 = 2a + 5$.

10. Kaya works for a local park. The park wants to plant 58 trees and has already planted 10 of them. If Kaya plants the rest of the trees at 6 trees each day, how many days will it take her to plant all of the trees?

TEST-TAKING TIP

Question 10 Be sure to completely and carefully read the problem before beginning any calculations. If you read too quickly, you may miss a key piece of information.

Geometry

11. Fernando is using the following design for a tiling pattern. Angle 1 and angle 3 each measure 28°. What is the measure of angle 2?

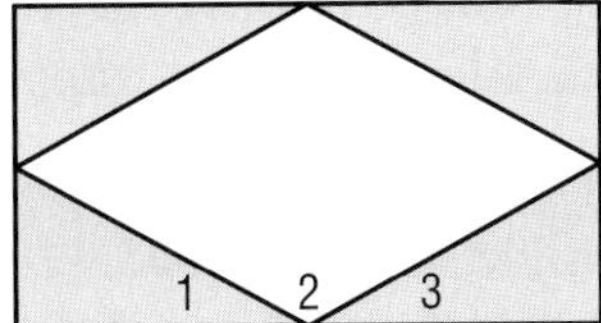

12. Each interior angle in a regular hexagon has the same measure. If the sum of the measures of all the interior angles is 720°, what is the measure of one angle?

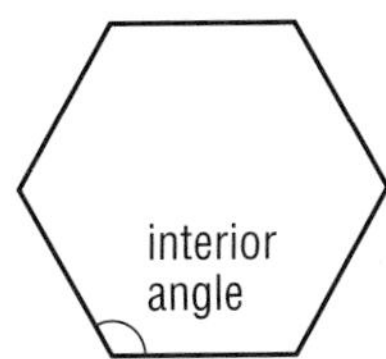

13. Collin is using this star shape in a computer graphics design. How many lines of symmetry does the shape have?

14. The area of the rectangle is 112.5 square meters. Find the perimeter.

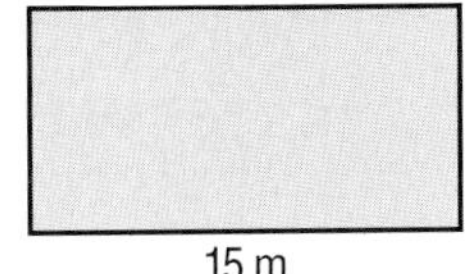

Measurement

15. Ms. Larson paints designs on blouses and sells them at a local shopping center. Usually, she can complete 3 blouses in 8 hours. At this rate, how long will it take her to complete 24 blouses?

16. The three-toed sloth is the world's slowest land mammal. The top speed of a sloth is 0.07 mile per hour. How many feet does a sloth travel in one minute?

17. How many meters long is a 5K, or 5-kilometer, race?

18. A container for laundry soap is a rectangular prism with dimensions of 8 inches by 6 inches by 5 inches. Find the surface area of the container.

19. A company is experimenting with two new boxes for packaging merchandise. Each box is a cube with the side lengths shown. What is the ratio of the volume of the smaller box to the volume of the larger box?

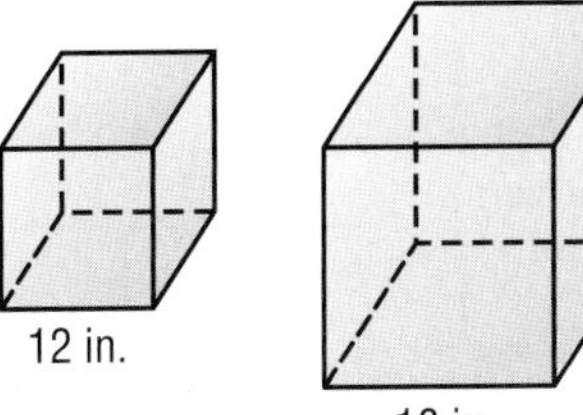

Data Analysis and Probability

20. The table shows the five smallest states and their areas. Find the mean of the areas.

State	Area (mi^2)
Rhode Island	1,545
Delaware	2,489
Connecticut	5,544
New Jersey	8,722
New Hampshire	9,354

Source: *Scholastic Book of World Records*

21. A six-sided number cube is rolled with faces numbered 1 through 6. What is the probability that the number cube shows an even number?

22. The mean of Chance's last five bowling scores is 125. If his next three scores are 140, 130, and 145, what will be the mean for the eight scores?

23. A bag contains 3 red, 4 green, 1 black, and 2 yellow jelly beans. Elise selects a jelly bean without looking. What is the probability that the jelly bean she selects is green?

Extended-Response Questions

Extended-response questions are often called *open-ended* or *constructed-response questions*. Most extended-response questions have multiple parts. You must show all your work and answer all parts to receive full credit.

A rubric is used to determine if you receive full, partial, or no credit. The following is a sample rubric for scoring extended-response questions.

Credit	Score	Criteria
Full	4	Full credit: A correct solution is given that is supported by well-developed, accurate explanations.
Partial	3, 2, 1	Partial credit: A generally correct solution is given that may contain minor flaws in reasoning or computation, or an incomplete solution is given. The more correct the solution, the greater the score.
None	0	No credit: An incorrect solution is given indicating no mathematical understanding of the concept, or no solution is given.

On some standardized tests, no credit is given for a correct answer if your work is not shown.

Make sure that when the problem says to *Show your work*, you show figures, graphs, and any explanations for your calculations.

Preparing for Standardized Tests

TEST EXAMPLE

Bear World is a company that manufactures stuffed animals. The graph shows the total number of bears completed per hour.

a. Make a function table for the graph.

b. Write a function rule for the data.

c. How many bears can Bear World complete in 10 hours?

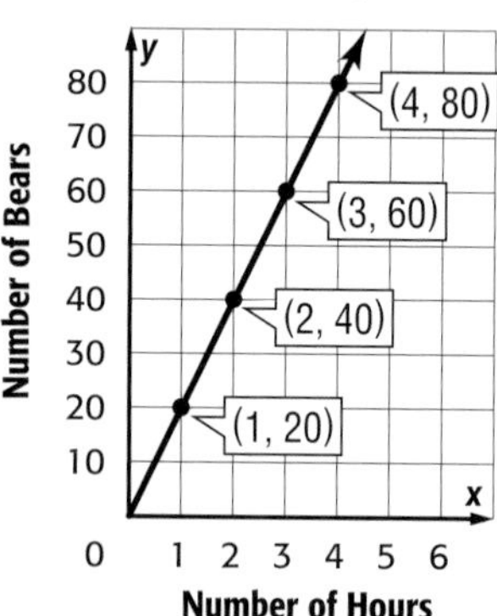

Full Credit Solution

Part a A complete table includes labeled columns for the x- and y-values shown in the graph. There may also be a column for the ordered pairs (x, y).

I used the ordered pairs on the graph to make a table.

Hours (x)	Bears (y)	(x, y)
1	20	(1, 20)
2	40	(2, 40)
3	60	(3, 60)
4	80	(4, 80)

Part b The student explains how they found the rule.

To find the function rule, I looked for a pattern in the values of x and y.

Hours	Bears
1	20
2	40
3	60

I see that the number of hours is multiplied by 20 to get the number of bears.

A rule is $y = 20x$.

Part c The student knows how to use the equation.

To find the number of bears completed in 10 hours, I will substitute 10 for x in the function rule.

$y = 20x$

$= 20(10)$ which is 200

There can be 200 bears completed in 10 hours.

Partial Credit Solution

Part a The student made an incomplete table.

1	2	6
20	40	120

Part b This part receives full credit because the student gives his explanation for writing the function rule.

I noticed that the number of bears increased by 20 after each hour. I tried some formulas and then found that $y = 20x$ works, where x is the number of hours and y is the number of bears. For example, for 3 hours, there are $3(20) = 60$ bears.

Part c Partial credit is given for part c because, although the answer is correct, there is no explanation given.

I think there will be 200 bears after 10 hours.

No Credit Solution

A solution for this problem that will receive no credit may include inaccurate or incomplete tables, incorrect function rules, and an incorrect answer to the question in part c.

Extended-Response Practice

Solve each problem. Show all your work.

Number and Operations

1. Refer to the table below.

Williams Family Budget	
House expenses	43%
Transportation	16%
Clothing and entertainment	12%
Food	11%
Savings account	10%
Other	8%

   a. What fraction of their budget do the Williams spend on transportation? Write the fraction in simplest form.

   b. If the Williams family has a monthly income of $3,000, how much do they spend on clothing and entertainment?

   c. If the Williams spend $385 a month on food, what is their monthly income?

2. Marco purchased 75 ride tickets at the fair. The table shows the number of tickets required for his favorite rides.

Ride	Tickets Needed
Sky Rocket	7
Flying Coaster	6
Dragon Drop	5
Spinning Fear	5
Ferris Wheel	4

   a. Can Marco use all of his tickets riding only the Sky Rocket? Explain.

   b. How can Marco use all of his tickets and ride only two different rides?

   c. Can Marco ride each ride at least twice and use all of his tickets? Explain.

TEST-TAKING TIP

Question 5 After finding the solution to each part, go back and read that part of the problem again. Make sure your solution answers what the problem is asking.

Algebra

3. The table shows the average farm size in acres in the United States.

Year	1950	1960	1970	1980	1990	2000
Size	213	297	374	426	460	434

Source: *Time Almanac*

   a. Make a line graph of the data.

   b. During which two consecutive decades did the number of acres in the average farm change the most? Explain.

   c. Compare the change in the number of acres from 1990 to 2000 to the change between other decades.

4. Mr. Martin and his son are planning a fishing trip to Alaska. Wild Fishing, Inc., will fly them both to the location for $400 and then charge $50 per day. Fisherman's Service will fly them both to the location for $300 and then charge $60 per day.

   a. Which company offers a better deal for a three-day trip? What will be the cost?

   b. Which company charges $700 for 6 days?

   c. Write an equation to represent the cost of the trip for each company.

Geometry

5. Carlos designed the pattern shown to use for a swimming pool.

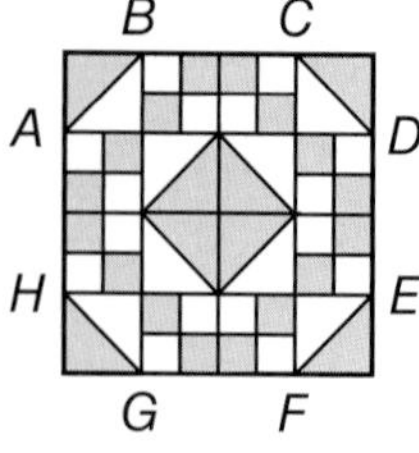

   a. Identify polygon *ABCDEFGH*. Which sides appear to be of equal lengths?

   b. Describe the four figures on the corners of the pattern.

   c. Name a pair of figures that appear to be congruent. Explain.

6. Heta is a landscaper. She is comparing the two square gardens shown in the diagram.

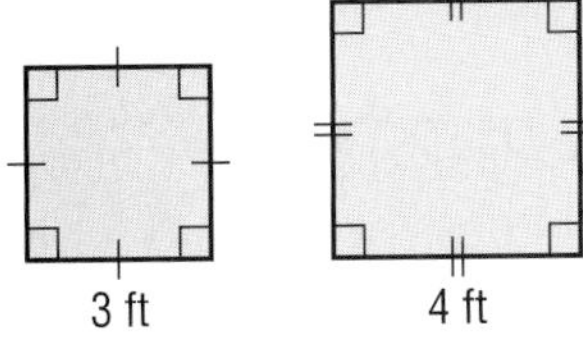

a. What is the ratio of the side length of the smaller garden to the larger garden?

b. What is the ratio of the perimeter of the smaller garden to the larger garden?

c. What is the ratio of the area of the smaller garden to the larger garden?

d. One of Heta's customers would like a square garden that has four times the area of the smaller garden. What should be the side length of this square garden?

Measurement

7. The table shows the maximum speeds of four aircraft.

Aircraft	Speed (mph)
SR-71 Blackbird	2,193
Atlas Cheetah	1,540
CAC J-711	1,435
BAe Hawk	840

Source: *Scholastic Book of World Records*

a. If the Atlas Cheetah could maintain its maximum speed for 3 hours, how many miles would it fly?

b. The circumference of Earth is about 29,000 miles. How long would it take the SR-71 Blackbird to circle Earth at maximum speed?

8. The table shows the radius of some of the moons in our solar system.

Moon (celestial body)	Radius (mi)
Ganymede (Jupiter)	1,635
Titan (Saturn)	1,600
Moon (Earth)	1,080
Oberon (Uranus)	1,020
Charon (Pluto)	364

Source: *Scholastic Book of World Records*

a. Find the circumference of each moon. Round to the nearest mile. Use 3.14 for π.

b. How many times greater is the circumference of Ganymede than the circumference of Charon? Express as a decimal rounded to the nearest hundredth. Use 3.14 for π.

Data Analysis and Probability

9. The table shows the size in square feet and square meters of the world's largest shopping malls.

Mall	Square Feet	Square Meters
West Edmonton Mall, Canada	5,300,000	490,000
Mall of America, USA	4,200,000	390,000
Sawgrass Mills Mall, USA	2,300,000	210,000
Austin Mall, USA	1,100,000	100,000
Suntec City Mall, Singapore	890,000	80,000

Source: *Scholastic Book of World Records*

a. Make a bar graph of the data using either square feet or square meters.

b. How many times greater in area is the largest mall compared to the fifth largest mall?

c. Determine the approximate number of square feet in one square meter.

10. Taj is designing a game for the school carnival. People playing the game will choose a marble without looking from the bag shown.

a. What is the probability that a person will draw a gray marble?

b. For which color would you award the most expensive prize? Explain.

c. How could Taj change the contents of the bag so that the probability of drawing a black marble is 50%?

Preparing for Standardized Tests

Concepts and Skills Bank

1 Divisibility Patterns

A whole number is **divisible** by another number if the remainder is 0 when the first number is divided by the second. The divisibility rules for 2, 3, 4, 5, 6, 9, and 10 are stated below.

Divisibility Rules	
Rule	**Examples**
A whole number is divisible by:	
• 2 if the ones digit is divisible by 2.	2, 4, 6, 8, 10, 12, . . .
• 3 if the sum of the digits is divisible by 3.	3, 6, 9, 12, 15, 18, . . .
• 4 if the number formed by the last two digits is divisible by 4.	4, 8, 12, . . . 104, 112, . . .
• 5 if the ones digit is 0 or 5.	5, 10, 15, 20, 25, . . .
• 6 if the number is divisible by both 2 and 3.	6, 12, 18, 24, 30, . . .
• 9 if the sum of the digits is divisible by 9.	9, 18, 27, 36, 45, . . .
• 10 if the ones digit is 0.	10, 20, 30, 40, 50, . . .

A whole number is **even** if it is divisible by 2. A whole number is **odd** if it is not divisible by 2.

EXAMPLES Use Divisibility Rules

1 Tell whether 2,320 is divisible by 2, 3, 5, or 10. Then classify the number as *even* or *odd*.

2: Yes; the ones digit, 0, is divisible by 2.

3: No; the sum of the digits, 7, is not divisible by 3.

5: Yes; the ones digit is 0.

10: Yes; the ones digit is 0.

The number 2,320 is even.

2 COUNTY FAIRS The number of tickets needed to ride certain rides at a county fair is shown. If Alicia has 51 tickets, can she use all of the tickets by riding only the Ferris wheel?

Canfield County Fair	
Ride	**Tickets**
Roller Coaster	5
Ferris Wheel	3
Bumper Cars	2
Scrambler	4

Use divisibility rules to check whether 51 is divisible by 3.

$51 \longrightarrow 5 + 1 = 6$ and 6 is divisible by 3.

So, 51 is divisible by 3.

Alicia can use all of the tickets by riding only the Ferris wheel.

3 Tell whether 564 is divisible by 4, 6, or 9.

4: Yes; the number formed by the last two digits, 64, is divisible by 4.
6: Yes; the number is divisible by both 2 and 3.
9: No; the sum of the digits, 15, is not divisible by 9.

Exercises

Tell whether each number is divisible by 2, 3, 4, 5, 6, 9, or 10. Then classify each number as *even* or *odd*.

1. 48
2. 64
3. 125
4. 156
5. 216
6. 330
7. 225
8. 524
9. 80
10. 60
11. 110
12. 315
13. 405
14. 243
15. 1,233
16. 2,052
17. 78
18. 45
19. 284
20. 585
21. 1,620
22. 3,213
23. 5,103
24. 6,950
25. 138
26. 489
27. 8,505
28. 1,986
29. 605
30. 900
31. 8,001
32. 9,948
33. 11,292
34. 15,000
35. 3,135
36. 9,270
37. 25,365
38. 27,027
39. 14,980

40. Find a number that is divisible by both 3 and 5.

41. Find a number that is divisible by 2, 9, and 10.

42. **SEATING** Sue is planning a holiday gathering for 117 people. She can use tables that seat 5, 6, or 9 people. If all of the tables must be the same size and all of the tables must be full, what size table should she use?

43. **RULERS** List the number of ways 318 rulers can be packaged for shipping to an office supply store so that each package has the same number of rulers.

44. **FLAGS** Each star on the United States flag represents a state. If another state joins the Union, could the stars be arranged in rows with each row having the same number of stars? Explain.

45. **LEAP YEAR** Any year that is divisible by 4 and does not end in 00 is a leap year. Years ending in 00 are leap years only if they are divisible by 400. Were you born in a leap year? Explain.

TOYS For Exercises 46 and 47, use the information below.

A toy company sells the sleds and snow tubes listed in the table. One of their shipments weighed 189 pounds and contained only one type of sled or show tube.

Toy	Weight (lb)
Super Snow Sled	4
Phantom Sled	6
Tundra Snow Tube	3
Snow Tube Deluxe	5

46. Which sled or show tube does the shipment contain?

47. How many sleds or snow tubes does the shipment contain?

Concepts and Skills Bank

Place Value and Whole Numbers

The number system we use is based on units of 10. A number like 7,825 is a **whole number**. A digit and its **place-value** position name a number. For example, in 7,825, the digit 7 is in the thousands place, and its value is 7,000.

Place-Value Chart

1,000,000,000	100,000,000	10,000,000	1,000,000	100,000	10,000	1,000	100	10	1
one billion	hundred millions	ten millions	one million	hundred thousands	ten thousands	thousands	hundreds	tens	ones
						7,	8	2	5

Each set of three digits is called a period. Periods are separated by a comma.

EXAMPLE Identify Place Value

1 Identify the place-value position of the digit 9 in 597,240,618.

1,000,000,000	100,000,000	10,000,000	1,000,000	100,000	10,000	1,000	100	10	1
one billion	hundred millions	ten millions	one million	hundred thousands	ten thousands	thousands	hundreds	tens	ones
	5	9	7,	2	4	0,	6	1	8

The digit 9 is in the ten millions place.

Numbers written as 7,825 and 597,240,618 are written in **standard form**. Numbers can also be written in **word form**. When writing a number in word form, use place value. At each comma, write the name for the period.

EXAMPLES Write a Whole Number in Word Form

Write each number in word form.

2 4,567,890

Standard Form 4,567,890

Word Form four million five hundred sixty-seven thousand eight hundred ninety

3 804,506

Standard Form 804,506

Word Form eight hundred four thousand five hundred six

Concepts and Skills Bank

Numbers can also be written in **expanded notation**.

EXAMPLE Write a Whole Number in Expanded Notation

4 **Write 28,756 in expanded notation.**

Step 1 Write the product of each digit and its place value.

$20{,}000 = (2 \times 10{,}000)$ The digit 2 is in the ten thousands place.
$8{,}000 = (8 \times 1{,}000)$ The digit 8 is in the thousands place.
$700 = (7 \times 100)$ The digit 7 is in the hundreds place.
$50 = (5 \times 10)$ The digit 5 is in the tens place.
$6 = (6 \times 1)$ The digit 6 is in the ones place.

Step 2 Write the sum of the products.

This is 28,756 written in expanded notation.

$$28{,}756 = (2 \times 10{,}000) + (8 \times 1{,}000) + (7 \times 100) + (5 \times 10) + (6 \times 1)$$

Exercises

Identify each underlined place-value position.

1. $4\underline{3}8$
2. $6{,}8\underline{4}5{,}085$
3. $\underline{4}13{,}467$
4. $3{,}74\underline{5}$
5. $7\underline{2}{,}567{,}432$
6. $\underline{9}04{,}784{,}126$
7. Write a number that has the digit 7 in the billions place, the digit 8 in the hundred thousands place, and the digit 4 in the tens place.

Write each number in word form.

8. 263
9. 2,013
10. 54,006
11. 47,900
12. 567,460
13. 551,002
14. 7,805,261
15. 1,125,678
16. 102,546,165
17. 582,604,072
18. 8,146,806,835
19. 67,826,657,005

Write each number in standard form.

20. forty-two
21. seven hundred fifty-one
22. three thousand four hundred twelve
23. six thousand nine hundred five
24. six hundred thousand
25. sixteen thousand fifty-two
26. seventy-six million
27. two hundred twenty-four million
28. three million four hundred thousand
29. ten billion nine hundred thousand

Write each number in expanded notation.

30. 86
31. 398
32. 620
33. 5,285
34. 4,002
35. 7,500
36. 85,430
37. 524,789
38. 8,043,967
39. **PIANOS** There are over ten million pianos in American homes, businesses, and institutions. Write ten million in expanded notation.
40. **BOOKS** The Library of Congress in Washington, D.C., has 24,616,867 books. Write this number in word form.

3 Comparing and Ordering Whole Numbers

When comparing the values of two whole numbers, the first number is either less than, greater than, or equal to the second number. You can use place value or a number line to compare two whole numbers.

Words	Symbol
less than	$<$
greater than	$>$
equal to	$=$

METHOD 1 **Use place value.**

- Line up the digits at the ones place.
- Starting at the left, compare the digits in each place-value position. In the first position where the digits differ, the number with the greater digit is the greater whole number.

METHOD 2 **Use a number line.**

- Numbers to the right are greater than numbers to the left.
- Numbers to the left are less than numbers to the right.

EXAMPLE Compare Whole Numbers

1 Replace the ● in 25,489 ● 25,589 with $>$, $<$, or $=$ to make a true sentence.

METHOD 1 **Use place value.**

25,489 Line up the digits.
25,589 Compare.

The digits in the hundreds place are not the same. Since $4 < 5$, $25{,}489 < 25{,}589$.

METHOD 2 **Use a number line.**

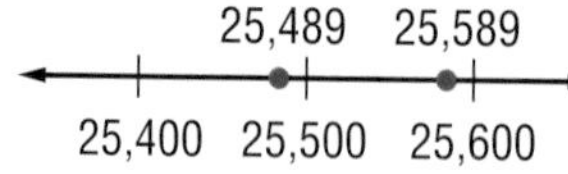

Graph and then compare the numbers.

Since 25,489 is to the left of 25,589, $25{,}489 < 25{,}589$.

EXAMPLE Order Whole Numbers

2 Order 8,989, 8,957, and 8,984 from least to greatest.

8,957 is less than both 8,989 and 8,984 since $5 < 8$ in the tens place.
8,984 is less than 8,989 since $4 < 9$ in the ones place.

So, the order from least to greatest is 8,957, 8,984, and 8,989.

Exercises

Replace each ● with $<$ or $>$ to make a true sentence.

1. 496 ● 489
2. 5,602 ● 5,699
3. 3,455 ● 3,388
4. 13,405 ● 14,003
5. 75,495 ● 75,606
6. 501,222 ● 510,546
7. 245,000 ● 24,500
8. 675,656 ● 6,756,560
9. 1,000,000 ● 989,566

Order each list of whole numbers from least to greatest.

10. 678, 699, 610
11. 4,654, 4,432, 4,678
12. 42,000, 4,200, 41,898

13. **STATES** Alabama has a total area of 52,237 square miles. Arkansas has an area of 53,182 square miles. Which state is larger?

4 Estimating with Whole Numbers

When an exact answer to a math problem is not needed, or when you want to check the reasonableness of an answer, you can use **estimation**. There are several methods of estimation. One common method is **rounding**.

To round a whole number, look at the digit to the right of the place being rounded.

- If the digit is 4 or less, the underlined digit remains the same.
- If the digit is 5 or greater, add 1 to the underlined digit.

EXAMPLES Estimate by Rounding

Estimate by rounding.

1 **30,798 + 4,115 + 1,891**

30,798	→	31,000	In this case, each number is rounded to the same place value.
4,115	→	4,000	
+ 1,891	→	+ 2,000	
		37,000	

So, the sum is about 37,000.

2 **478 × 12**

478	→	500	In this case, each number is rounded to its greatest place value.
× 12	→	× 10	
		5,000	

So, the product is about 5,000.

Clustering can be used to estimate sums if all of the numbers are close to a certain number.

EXAMPLES Estimate by Clustering

Estimate by clustering.

3 **97 + 102 + 99 + 104 + 101 + 98**

All of the numbers are clustered around 100. There are 6 numbers. So, the sum is about 6 × 100 or 600.

4 **748 + 751 + 753 + 747**

All of the numbers are clustered around 750. There are 4 numbers. So, the sum is about 4 × 750 or 3,000.

Compatible numbers are two numbers that are easy to compute mentally.

EXAMPLES Estimate by Using Compatible Numbers

Estimate by using compatible numbers.

5 **102 ÷ 24**

$24\overline{)102}$ → $\overset{4}{25\overline{)100}}$

So, the quotient is about 4.

6 **71 + 19 + 28 + 83**

71 + 19 + 28 + 83 → 70 + 20 + 30 + 80

= (70 + 30) + (20 + 80)

= 100 + 100 or 200

The sum is about 200.

Another strategy that works well for some addition and subtraction problems is **front-end estimation**. In this strategy, you add or subtract the left-most column of digits.

EXAMPLES Use Front-End Estimation

Estimate by using front-end estimation.

7 **739 + 259**

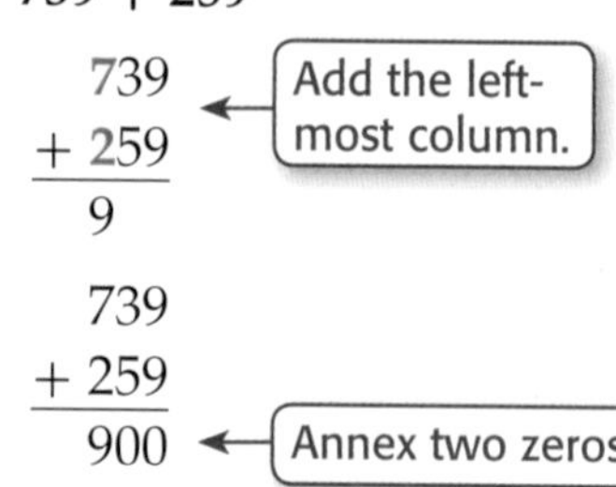

739
+ 259
900 ← Annex two zeros.

So, the sum is about 900.

8 **3,542 − 1,280**

3,542
− 1,280
2 ← Subtract the left-most column.

3,542
− 1,280
2,000 ← Annex three zeros.

So, the difference is about 2,000.

Exercises

Estimate by rounding.

1. 82 × 3
2. 435 + 219 + 121
3. 4286 − 2089
4. 99 × 12
5. 106 ÷ 9
6. 192 × 39

Estimate by clustering.

7. 19 + 18 + 21 + 20 + 19
8. 498 + 499 + 502 + 503
9. 82 + 79 + 77 + 81
10. 4 + 3 + 6 + 5 + 7 + 4 + 6 + 7

Estimate by using compatible numbers.

11. 31 ÷ 4
12. 17 + 82 + 58 + 43 + 75
13. 122 + 79 + 232 + 68
14. 205 ÷ 6
15. 540 + 470 + 325 + 70
16. 813 ÷ 9

Estimate by using front-end estimation.

17. 852 − 312
18. 65 × 4
19. 120 × 3
20. 895 − 246
21. 340 × 2
22. 673 − 344

Use any method to estimate.

23. 680 − 290
24. 647 ÷ 9
25. 68 × 11
26. 99.6 ÷ 18.25
27. 43 + 62 + 22 + 77
28. 69 + 72 + 68 + 70 + 73
29. 390 ÷ 52
30. 102 × 6
31. 498 × 19
32. 493 − 183
33. 199 + 205 + 198
34. 999 ÷ 52
35. 1,209 ÷ 403
36. 2,005 × 4
37. 490 + 514 + 520 + 483
38. 3,486 ÷ 5
39. 7,654 − 3,254
40. 3,045 × 124
41. 188 + 320
42. 12,250 ÷ 4,008
43. 874 − 523

5 Adding and Subtracting Whole Numbers

To add or subtract whole numbers, add or subtract the digits in each place-value position. Start at the ones place.

EXAMPLES Add or Subtract Whole Numbers

Find each sum or difference.

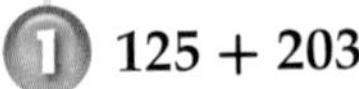

1 125 + 203

```
  125   Line up the digits
+ 203   at the ones place.
  328 ← Add the ones.
        Add the tens.
        Add the hundreds.
```

2 587 − 162

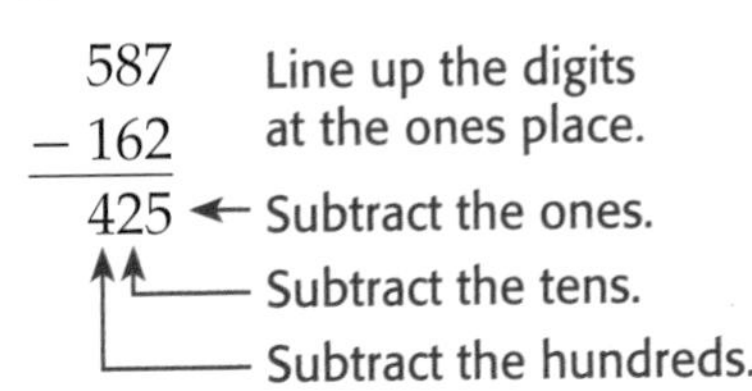

587
− 162
425
Line up the digits at the ones place.
Subtract the ones.
Subtract the tens.
Subtract the hundreds.

You may need to regroup when adding and subtracting whole numbers.

EXAMPLES Add or Subtract with Regrouping

Find each sum or difference.

3 387 + 98

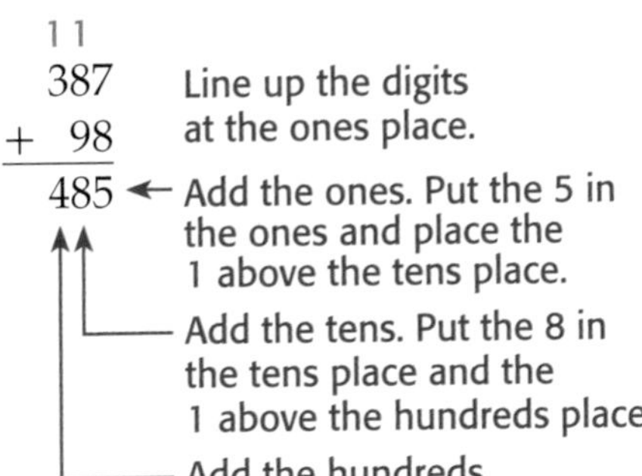

1 1
387
+ 98
485
Line up the digits at the ones place.
Add the ones. Put the 5 in the ones and place the 1 above the tens place.
Add the tens. Put the 8 in the tens place and the 1 above the hundreds place.
Add the hundreds.

4 612 − 59

```
 5 10 12
  612    Line up the digits
−  59    at the ones place.
  553
```

Since 9 is larger than 2, rename 2 as 12. Rename the 1 in the tens place as 10 and the 6 in the hundreds place as 5. Then subtract.

Exercises

Find each sum or difference.

1. 506 + 30
2. 315 + 583
3. 1,342 + 627
4. 5,042 + 2,143
5. 71,235 + 27,563
6. 468 − 21
7. 895 − 472
8. 1,872 − 460
9. 6,056 − 5,052
10. 74,618 − 23,311
11. 2,680 + 945
12. 5,126 + 896
13. 2,973 + 1,689
14. 1,089 + 5,239
15. 16,999 + 25,509
16. 982 − 36
17. 487 − 199
18. 6,052 − 5,456
19. 64,205 − 3,746
20. 215,000 − 12,999
21. **PLANETS** The diameter of Earth is 7,926 miles. The diameter of Mars is 4,231 miles. How much greater is the diameter of Earth than the diameter of Mars?

6 Multiplying and Dividing Whole Numbers

To multiply a whole number by a 1-digit whole number, multiply from right to left, regrouping as necessary. When you multiply by a number with two or more digits, write individual products and then add.

EXAMPLES Multiply Whole Numbers

Find each product.

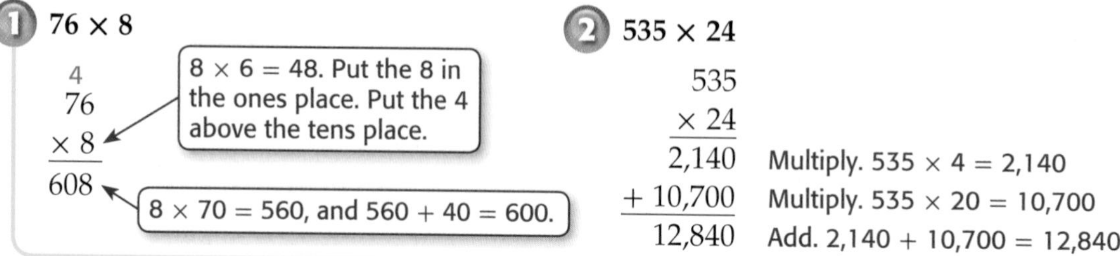

1 76 × 8
4
76
× 8
608
8 × 6 = 48. Put the 8 in the ones place. Put the 4 above the tens place.
8 × 70 = 560, and 560 + 40 = 600.
2 535 × 24
535
× 24
2,140
+ 10,700
12,840
Multiply. 535 × 4 = 2,140
Multiply. 535 × 20 = 10,700
Add. 2,140 + 10,700 = 12,840

When dividing whole numbers, divide in each place-value position from left to right. Recall that in the statement $50 \div 2 = 25$, 50 is the **dividend**, 2 is the **divisor**, and 25 is the **quotient**.

EXAMPLES Divide Whole Numbers

Find each quotient.

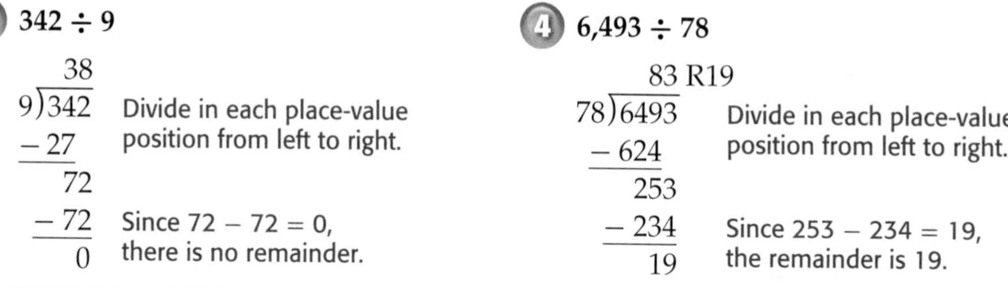

3 342 ÷ 9
38
9)342
− 27
72
− 72
0
Divide in each place-value position from left to right.
Since 72 − 72 = 0, there is no remainder.
4 6,493 ÷ 78
83 R19
78)6493
− 624
253
− 234
19
Divide in each place-value position from left to right.
Since 253 − 234 = 19, the remainder is 19.

Exercises

Find each product or quotient.

1. 40×5
2. 23×3
3. 172×3
4. 361×7
5. $2{,}821 \times 4$
6. $3\overline{)72}$
7. $4\overline{)96}$
8. $6\overline{)918}$
9. $12\overline{)60}$
10. $34\overline{)204}$
11. 28×10
12. 54×27
13. 414×22
14. 321×34
15. 525×150
16. $31\overline{)1{,}953}$
17. $27\overline{)56}$
18. $29\overline{)210}$
19. $50\overline{)12{,}575}$
20. $400\overline{)1{,}632}$
21. **FOOD** The average person in the U.S. eats about 36 pounds of frozen food per year. How many pounds would an average family of four eat?
22. **MEASUREMENT** One acre is the same as 43,560 square feet. It is also equal to 4,840 square yards. How many square feet are there in one square yard?

7 Square Roots

Numbers that can be represented as areas of squares are called **perfect squares**.

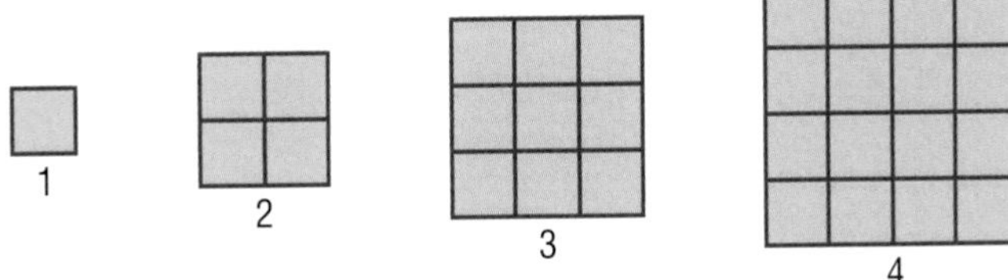

The inverse operation of squaring a number is finding the **square root** of a number.

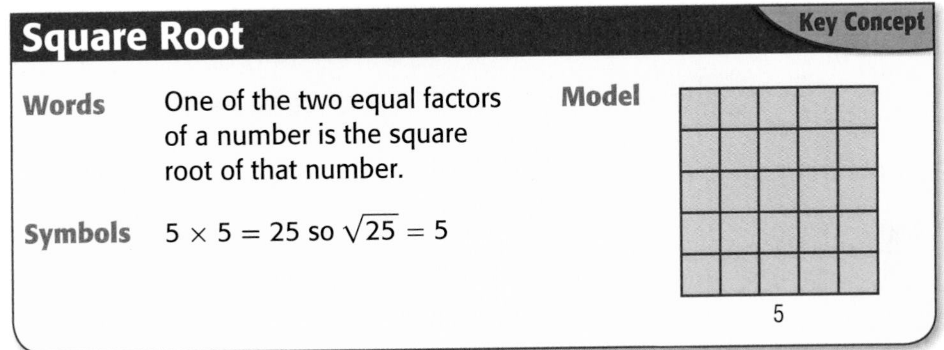

Square Root

Key Concept

Words One of the two equal factors of a number is the square root of that number.

Symbols $5 \times 5 = 25$ so $\sqrt{25} = 5$

Model

EXAMPLE Find the Square Root of a Number

1 Find $\sqrt{49}$.

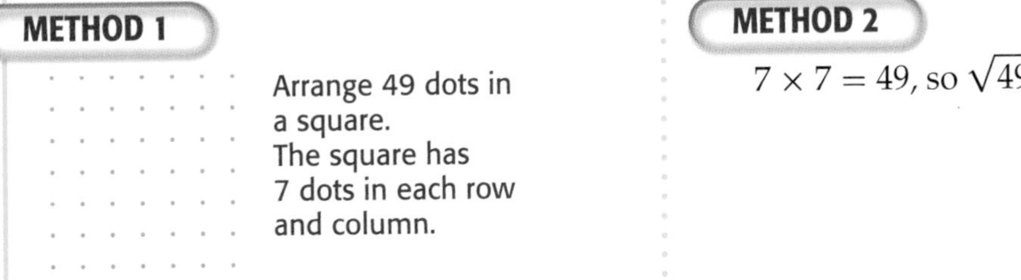

METHOD 1

Arrange 49 dots in a square.
The square has 7 dots in each row and column.

So, $\sqrt{49} = 7$.

METHOD 2

$7 \times 7 = 49$, so $\sqrt{49} = 7$

THINK What number times itself equals 100?

Exercises

Find each square root.

1. $\sqrt{4}$	2. $\sqrt{36}$	3. $\sqrt{64}$	4. $\sqrt{121}$
5. $\sqrt{81}$	6. $\sqrt{9}$	7. $\sqrt{16}$	8. $\sqrt{144}$
9. $\sqrt{1}$	10. $\sqrt{169}$	11. $\sqrt{400}$	12. $\sqrt{225}$
13. $\sqrt{289}$	14. $\sqrt{196}$	15. $\sqrt{256}$	16. $\sqrt{324}$

17. **GEOMETRY** A square has an area of 361 square centimeters. What is the length of one of its sides?

18. **GAMES** A checkerboard is a large square made up of 32 small red squares and 32 small black squares. How many small squares are along one side of the checkerboard?

8 Solving Percent Problems

To find the percent of a number you can use one of the following methods.

Write the percent as a fraction and then multiply, or

Write the percent as a decimal and then multiply.

EXAMPLES Find the Percent

1 Find 12% of 75.

METHOD 1

Write the percent as a fraction.

$12\% = \frac{12}{100}$ or $\frac{3}{25}$

$\frac{3}{25}$ of $75 = \frac{3}{25} \times 75$ or 9

So, 12% of 75 is 9.

METHOD 2

Write the percent as a decimal.

$12\% = \frac{12}{100}$ or 0.12

0.12 of $75 = 0.12 \times 75$ or 9

2 Find 150% of 120.

METHOD 1

Write the percent as a fraction.

$150\% = \frac{150}{100}$ or $1\frac{1}{2}$

$1\frac{1}{2}$ of $120 = 1\frac{1}{2} \times 120$

$= \frac{3}{2} \times 120$

$= \frac{3}{2} \times \frac{120}{1}$ or 180

So, 120% of 75 is 90.

METHOD 2

Write the percent as a decimal.

$150\% = \frac{150}{100}$ or 1.5

1.5 of $120 = 1.2 \times 75$ or 90

Exercises

Find each percent.

1. 10% of 40	2. 25% of 200	3. 36% of 450	4. 15% of 50
5. 220% of 10	6. 175% of 350	7. 200% of 45	8. 250% of 200
9. 42% of 42	10. 12% of 120	11. 58% of 230	12. 77% of 95
13. 234% of 100	14. 14% of 540	15. 90% of 68	16. 85% of 239

17. **SCHOOL** Wiley's latest math test score was 80%. If there were 60 points on the test, how many points did Wiley earn?

18. **SHOPPING** A store is advertising 30% off all its backpacks. Sondra wants a backpack that originally cost $24.95. Find the amount of money she will save by shopping at the sale.

19. **SAVINGS** LaDona wants to save 35% of her babysitting money to buy a new cell phone. If she earns $50 babysitting one weekend, how much money should she save?

9 Scale and Proportion

Scale drawings and *scale models* are used to represent objects that are too large or too small to be drawn or built at actual size. A map is an example of a scale drawing.

The **scale** gives the ratio that compares the measurements on the drawing or model to the measurements of the real object. The measurements on a drawing or model are proportional to the measurements on the actual object.

EXAMPLE Find Actual Measurements

1 On a map of New York, the distance between Albany and Poughkeepsie is $3\frac{1}{2}$ inches. If the scale on the map is 1 inch = 20 miles, what is the actual distance between Albany and Poughkeepsie?

Let d represent the actual distance.

map distance → / actual distance → $\frac{1}{20} = \frac{3.5}{d}$ ← map distance / ← actual distance

$1 \times d = 20 \times 3.5$ Find the cross products.

$d = 70$ Multiply.

The distance between Albany and Poughkeepsie is 70 miles.

Exercises

For Exercises 1–4, use the following information.

On a set of blueprints, $\frac{1}{4}$ inch = 1 foot.

1. Find the actual length of a bedroom if the blueprint measures 3 inches long.
2. Find the length on a blueprint for a window that is 3 feet wide.
3. Find the lengths on a blueprint for a living room that is 15 feet wide by 18 feet long.
4. Find the actual length of a garage if the blueprint measures 4.5 inches wide by 5.5 inches long.
5. **MODELS** A truck has an overall length of 210 inches. If a toy model of the truck has a scale of 1:75, what is the length of the toy?
6. **TRAINS** A model train has a scale of 1:87. If the model caboose measures 11.7 centimeters, find the length of the actual caboose.

For Exercises 7–10, use the following information.

A map of North Carolina has a scale of 1 inch = 40 miles.

7. Find the actual distance between Greensboro and Columbia if the distance on the map is $2\frac{3}{4}$ inches.
8. Find the distance on the map if the actual distance between Raleigh and Wilmington is 130 miles.
9. On the map, the border between North Carolina and Virginia is 6 inches. Find the actual length of the border.

10 Algebraic Properties

In algebra, **properties** are mathematical rules that apply to any number. You have already learned several properties. In addition, several new properties are outlined below.

Properties of Numbers — Key Concept

Property	Words	Examples
Additive Inverse	When a number and its opposite are added, the sum is 0.	$4 + (-4) = 0$ $-4 + 4 = 0$
Additive Identity	When 0 is added to any number, the sum is the number.	$5 + 0 = 5$ $0 + 5 = 5$
Multiplicative Inverse	When any nonzero number is multiplied by its reciprocal, the product is 1.	$5 \times \frac{1}{5} = 1$ $\frac{1}{5} \times 5 = 1$
Multiplicative Identity	When any number is multiplied by 1, the product is the number.	$8 \cdot 1 = 8$ $1 \cdot 8 = 8$
Multiplicative Property of Zero	When any number is multiplied by 0, the product is 0.	$3 \cdot 0 = 0$ $0 \cdot 3 = 0$

EXAMPLES Identify Properties

1 Identify the property shown by the equation $63 + 0 = 63$.

You are adding zero to a number and the sum is the original number. So, this is the Additive Identity Property.

2 Simplify the expression $24 \times 1 + 5 \times 0$. Justify each step.

$24 \times 1 + 5 \times 0 = 24 + 5 \times 0$ Multiplicative Identity Property

$= 24 + 0$ Multiplication Property of Zero

$= 24$ Additive Identity Property

Exercises

Identify the property shown by each equation.

1. $87 + 0 = 87$
2. $0 = 54 \times 0$
3. $123 \times 1 = 123$
4. $36 \times \frac{1}{36} = 1$
5. $44 \times 0 = 0$
6. $-63 + 63 = 0$
7. $0 = 58 + (-58)$
8. $125 = 0 + 125$
9. $1 = \frac{1}{20} \times 20$

Simplify each expression. Justify each step with an algebraic property.

10. $120 \times 0 + 1 \times 63$
11. $12 \times \frac{1}{12} + (-1)$
12. $32 \times (-12 + 12)$
13. $45 + (-45) + 0$
14. $68 \times 0 \times \frac{1}{68}$
15. $0 \times (36 + 0)$

11 Solving Inequalities

You have already solved one- and two-step equations. You can apply what you learned about equations to solve one- and two-step inequalities. **Inequalities** are sentences that compare quantities that are not equal. Some inequality symbols are $<$, $>$, $\leq$, and $\geq$.

Inequality Symbols — Key Concept

Symbols	$<$	$>$	$\leq$	$\geq$
Words	• less than • fewer than	• greater than • more than	• less than or equal to • no more than • at most	• greater than or equal to • no less than • at least

EXAMPLES Write an Inequality

Write an inequality for each sentence.

1 Four times a number is at least 28.

Words	Four times a number is at least 28.
Variable	Let n represent the number.
Inequality	$4n \geq 28$

2 The sum of 5 times a number and 17 is less than 42.

Words	The sum of five times a number and 17 is less than 42.
Variable	Let n represent the number.
Inequality	$5n + 17 < 42$

You can use your knowledge of solving equations to solve an inequality.

EXAMPLE Solve an Inequality

3 Solve $5n + 17 < 42$.

$5n + 17 < 42$	Write the inequality.
$5n + 17 - 17 < 42 - 17$	Subtract 17 from each side.
$5n < 25$	Simplify.
$\frac{5n}{5} < \frac{25}{5}$	Divide each side by 5.
$n < 5$	Simplify.

Unlike the solution of an equation which often represents just one number, the solution $n < 5$ represents a set of numbers. The solution of an inequality can be graphed on a number line.

EXAMPLE Graph an Inequality

4 **Graph $n < 5$**

Draw a number line.

Draw an empty circle at 5. The circle on the graph is empty to show that the numbers up to 5, but not including 5, are included in the solution.

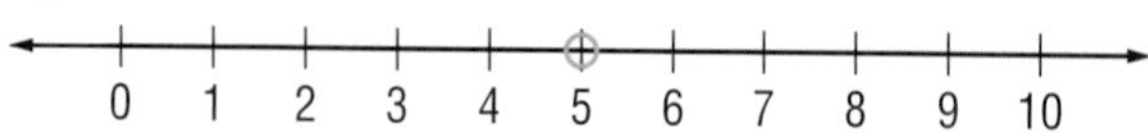

When $\geq$ or $\leq$ is used, the circle on the graph is solid to show that the number is included in the solution.

Test numbers on either side of 5 to determine the direction of the graph.

n	$5n + 17 < 42$	Solution?
8	$5(8) + 17 \overset{?}{<} 42$ $40 + 17 \overset{?}{<} 42$ $57 \overset{?}{<} 42$	No
2	$5(2) + 17 \overset{?}{<} 42$ $10 + 17 \overset{?}{<} 42$ $27 \overset{?}{<} 42$	Yes

Draw an arrow pointing in the direction of 2. Use other numbers in this set to verify the solution.

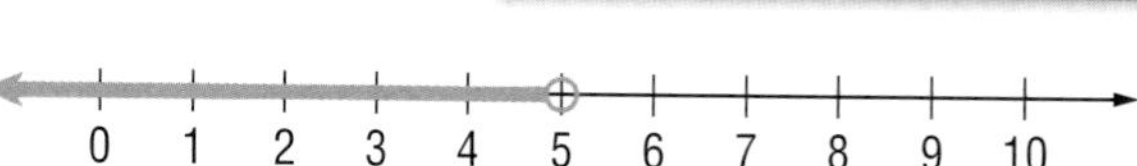

Exercises

Solve each inequality. Graph each solution on a number line.

1. $2x > 32$
2. $3n < 18$
3. $y - 0.54 \leq 2.6$
4. $p + 4.7 \geq 12.9$
5. $2x - 9 \geq 31$
6. $3r + 10 < 34$
7. $4.5t + 0.1 \leq 5.5$
8. $3.8s - 3.8 > 13.3$
9. $\frac{1}{2}x + \frac{1}{3} \geq \frac{2}{3}$
10. $\frac{2}{5}p - \frac{3}{10} \geq \frac{9}{10}$
11. $2\frac{1}{2}n - \frac{1}{2} < 1\frac{1}{2}$
12. $1\frac{3}{4}d + \frac{1}{7} \leq \frac{1}{2}$

13. **SAVINGS** Damon wants to buy a new MP3 player that costs at least $110. He has already saved $50, and he can save an additional $15 a week. Write and solve an inequality to find how many weeks Damon must save in order to buy the MP3 player.

14. **ARCADE** Mariel spends $0.75 every time she plays her favorite video game. She has $6.75 to spend. Write and solve an inequality that shows how many times Mariel can play the video game.

15. **SCHOOL** Heather spends at least $3\frac{1}{4}$ hours each week on homework. Reading takes a total of $\frac{3}{4}$ hour each week. If she has homework 5 nights a week and spends the same amount on homework each night, write and solve an inequality that shows the amount of time per night she spends on other assignments.

12 Geometric Elements

In Chapter 9, you learned about angles and polygons. Angles and polygons are made up of the three basic geometric figures described below.

Points, Lines, and Planes
Key Concept

	Point	Line	Plane
Model	• *P*	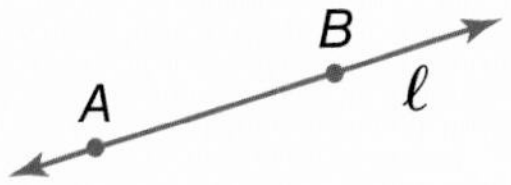	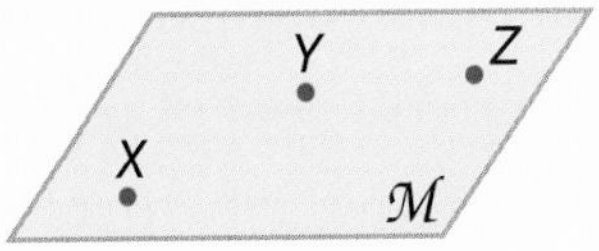
Symbols	point *P*	line ℓ, line *AB*, or $\overleftrightarrow{AB}$, line *BA* or $\overleftrightarrow{BA}$	plane *M*, plane *XYZ*, plane *YZX*, plane *ZXY*, plane *ZYX*
Words	A location that has neither shape nor size.	A collection of points that extends forever in both directions and that has no thickness or width.	A two-dimensional flat surface that extends in all directions.

Other geometry terms are related to points, lines, and planes. A **line segment** is a part of a line that has two endpoints. A **ray** is a part of a line having one endpoint and extending forever in one direction.

Parts of a Line
Key Concept

Words	A line segment is named by its endpoints.	A ray is named first by its endpoint, then by a point in the direction of the ray.
Model		
Symbols	$\overline{XY}$	$\overrightarrow{RS}$

EXAMPLES Name Lines and Planes

Use the figure below to name each of the following.

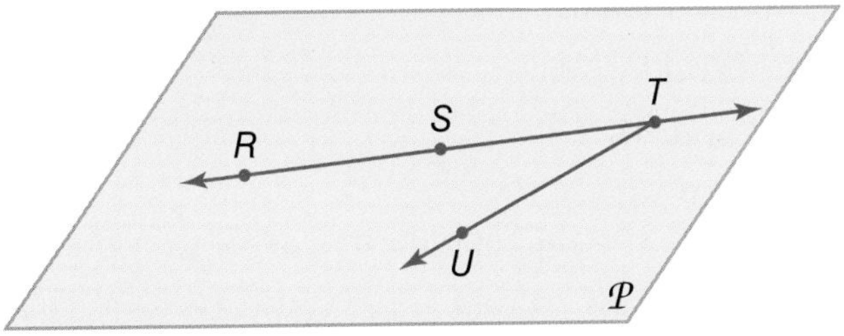

1 a line containing point *S*

There are 3 points on the line. Any two of the points can be used to name the line.
$\overleftrightarrow{RS}$, $\overleftrightarrow{SR}$, $\overleftrightarrow{ST}$, $\overleftrightarrow{TS}$, $\overleftrightarrow{TR}$, $\overleftrightarrow{RT}$

2 a ray containing point *U*

There is only one ray shown containing point *U* and only one way to name it. $\overrightarrow{TU}$

Concepts and Skills Bank

In the figure for Examples 1 and 2, $\overrightarrow{TU}$ intersects $\overleftrightarrow{RT}$ at point T. In any plane, it is possible for lines, line segments and rays to not intersect or to intersect in one point, a collection of points, or an infinite number of points.

EXAMPLES Find the Intersection of Figures in a Plane

Use the diagram at the right to name each of the following.

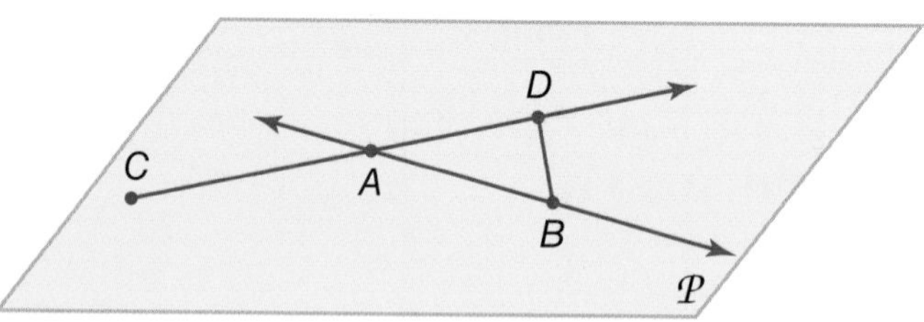

3 the intersection of $\overrightarrow{CD}$ and $\overleftrightarrow{AB}$

Ray CD and line AB intersect at point A.

4 the intersection of point A and $\overline{DB}$

Point A and line segment DB do not intersect.

Exercises

Use the diagram at the right to name each of the following.

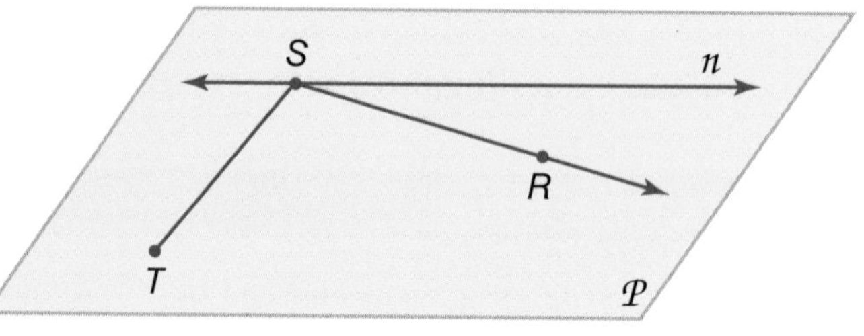

1. A line that contains point S.
2. A ray that contains segment RS.
3. A plane that contains point R.
4. A segment that contains point S.

Use the diagram at the right to name each of the following intersections.

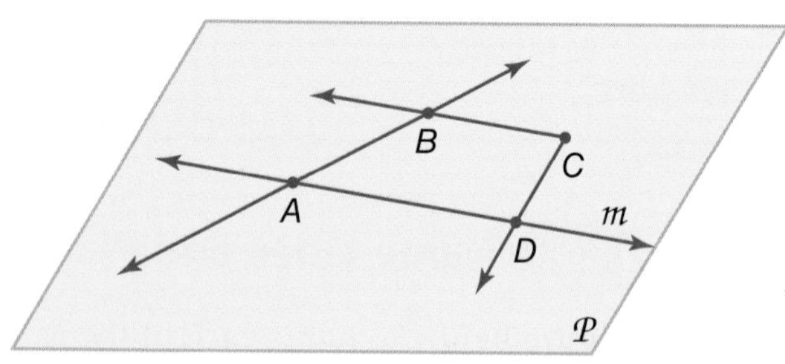

5. $\overleftrightarrow{AB}$ and line m
6. $\overrightarrow{CB}$ and $\overrightarrow{CD}$
7. $\overline{AB}$ and $\overline{CD}$
8. line m and $\overline{AD}$
9. $\overrightarrow{CD}$ and line m
10. $\overrightarrow{CB}$ and $\overrightarrow{AB}$
11. $\angle BCD$ and $\angle BAD$

Draw a diagram to illustrate each of the following.

12. line ℓ and line m that intersect at point P
13. $\overline{XY}$ and $\overline{YZ}$ that intersect at point Y
14. $\angle CDB$ and $\overline{FG}$ that intersect at point A
15. $\angle ABC$ and $\angle DEF$ that do not intersect

13 Composite Figures

A **composite** figure is made of triangles, squares, rectangles, and other two-dimensional figures.

You can find the perimeter of a composite figure by finding the sum of the side lengths around the figure. To find the area of a composite figure, separate it into figures with areas you know how to find. Then, add the areas.

EXAMPLES Find the Perimeter and Area of a Composite Figure

1 Find the perimeter of the figure to the right.

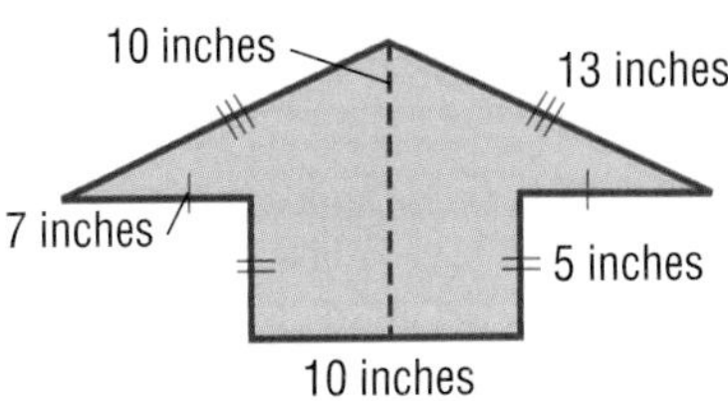

Start at the top vertex and add clockwise to find the perimeter of the figure.

$13 + 7 + 5 + 10 + 5 + 7 + 13 = 60$ inches

2 Find the area of the figure in Example 1.

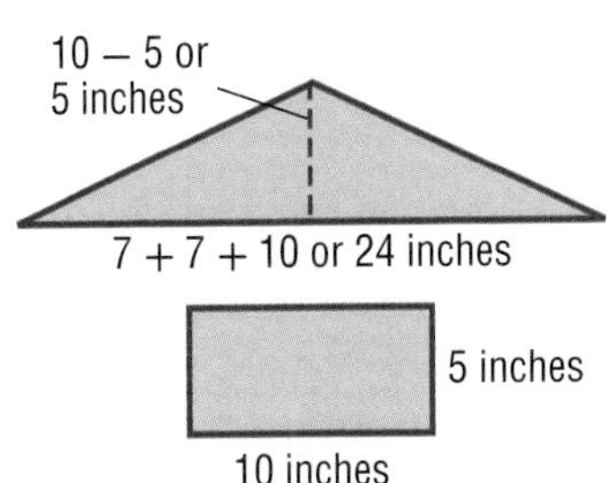

The figure can be separated into a triangle and a rectangle. Find the area of each figure. Then, add the areas.

Area of triangle $= \frac{1}{2}bh$

$= \frac{1}{2} \times 24 \times 5$ or 60 square inches

Area of rectangle $= \ell w$

$= 5 \times 10$ or 50 square inches

Area of composite figure $= 50 + 60$, or 110 square inches

EXAMPLE Find the Perimeter and Area of an Irregular Figure

3 Each square on the grid represents 1 square foot. Estimate the perimeter and area of the shaded figure.

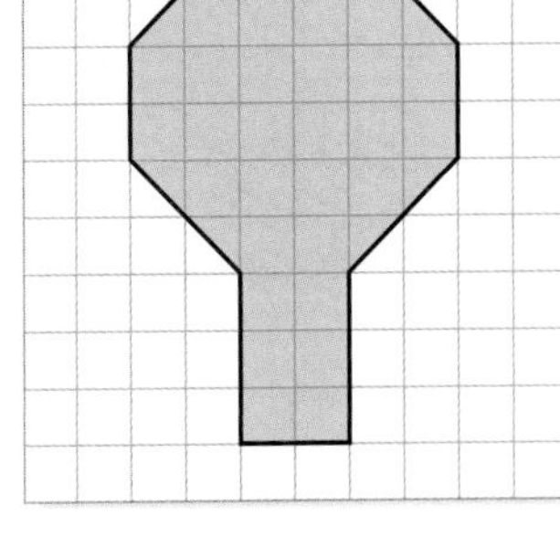

Count the number of vertical or horizontal square lengths. There are 16. Count the number of slanted lengths. There are 6. The slant length of one square is longer than one side length and less than two side lengths. Estimate the length at 1.5 feet. So, $6 \times 1.5 = 9$.

To find the perimeter, add the two quantities. Since $16 + 9 = 25$, the perimeter is about 25 feet.

To find the area, count the number of whole squares covered. There are 28. Count the partial squares. Six partial squares are about 3 square feet.

$28 + 8 = 36$ Add the whole squares and partial squares.

The area is about 36 square feet.

Exercises

Find the perimeter and area of each composite figure. Round to the nearest tenth if necessary.

1. 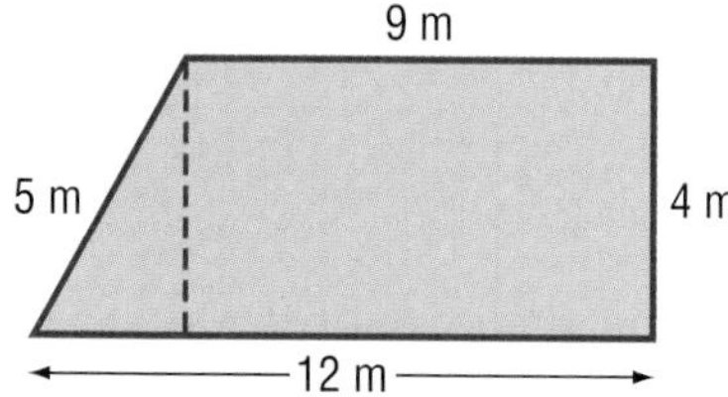

2. 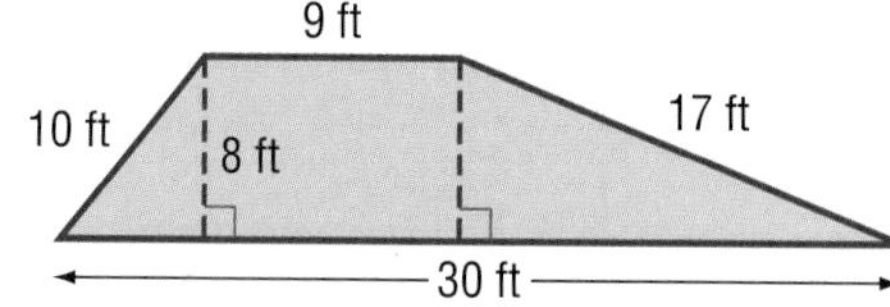

3. 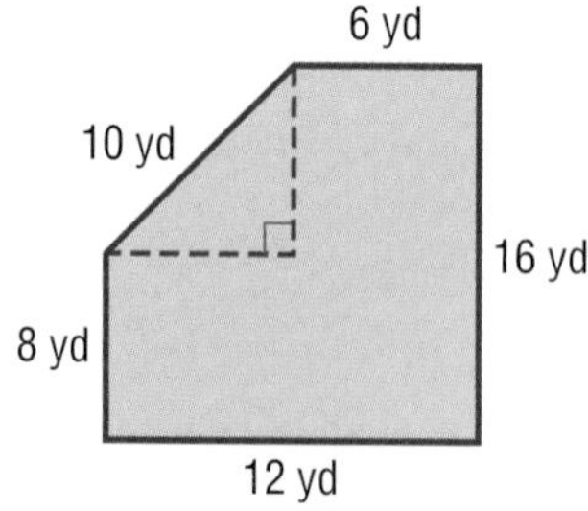

4. 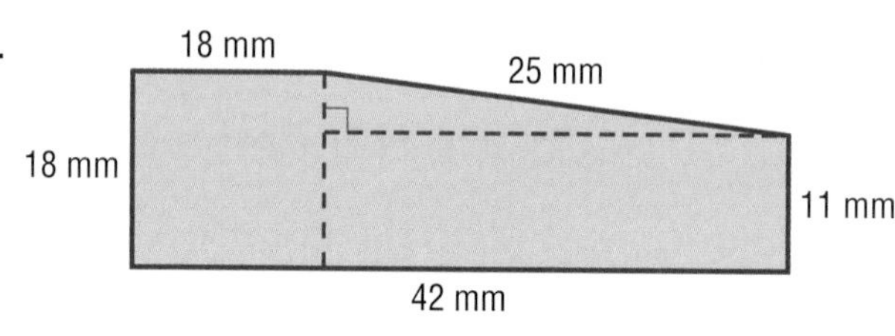

Estimate the perimeter and area of each irregular figure. Unless otherwise noted, each square = 1 ft².

5. 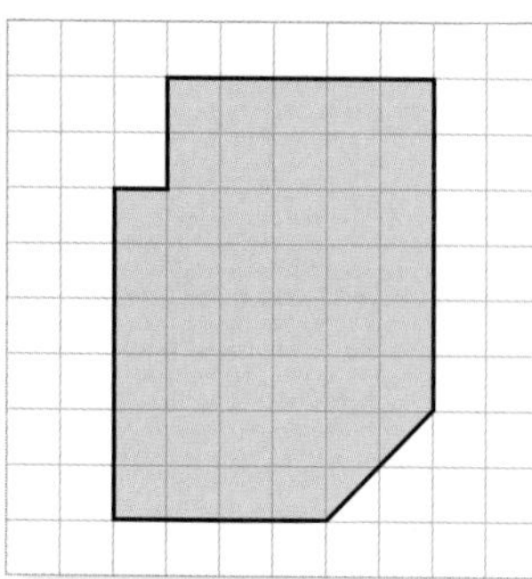

6. 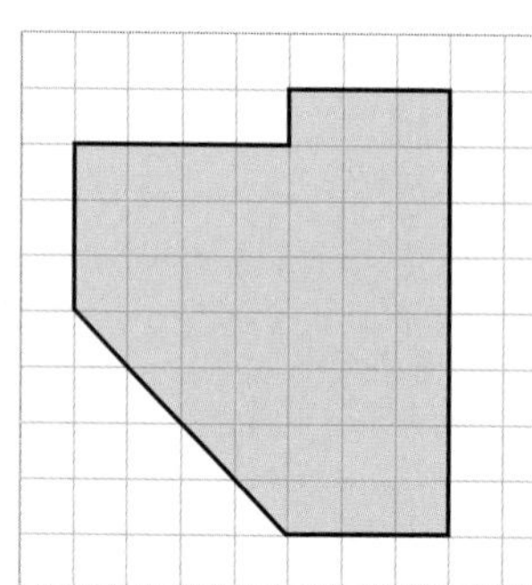

7. 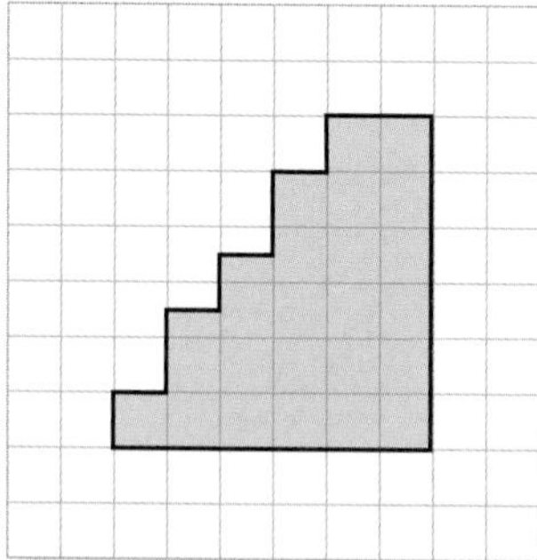

8. 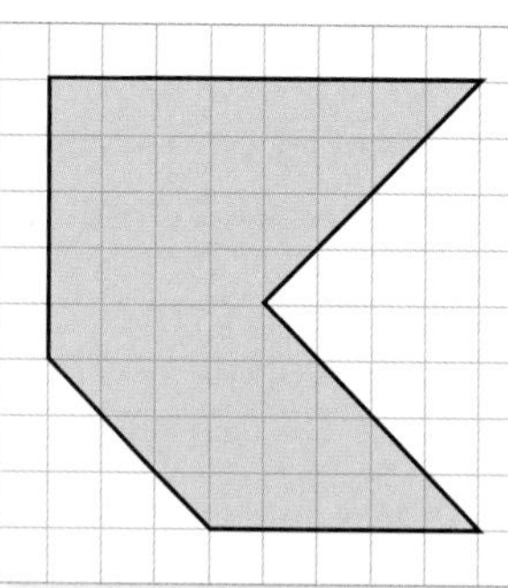

9. 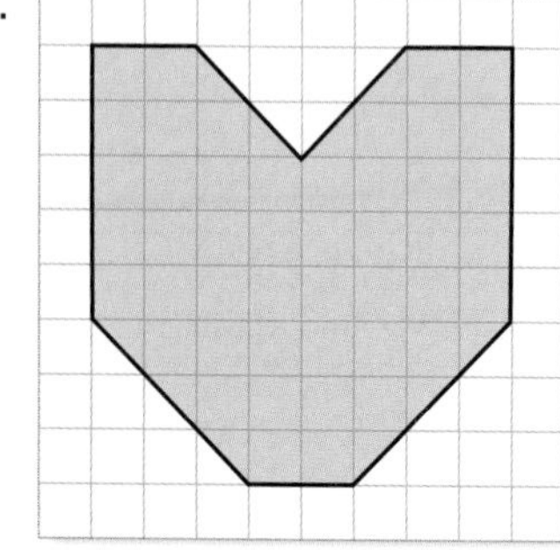

10. 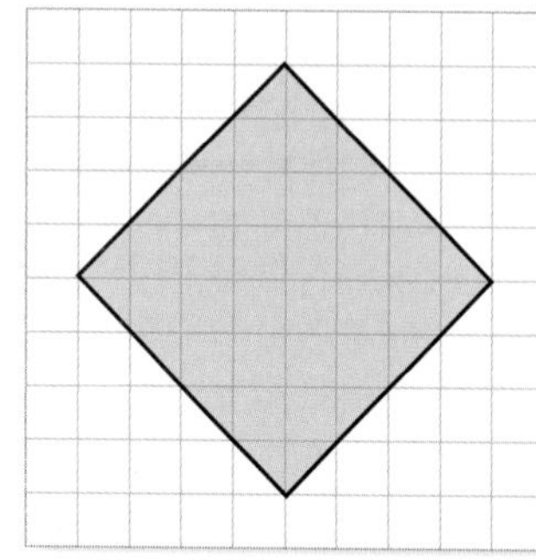

Three-Dimensional Figures

Two-dimensional figures have length and width. **Three-dimensional figures** have length, width, and depth (or height).

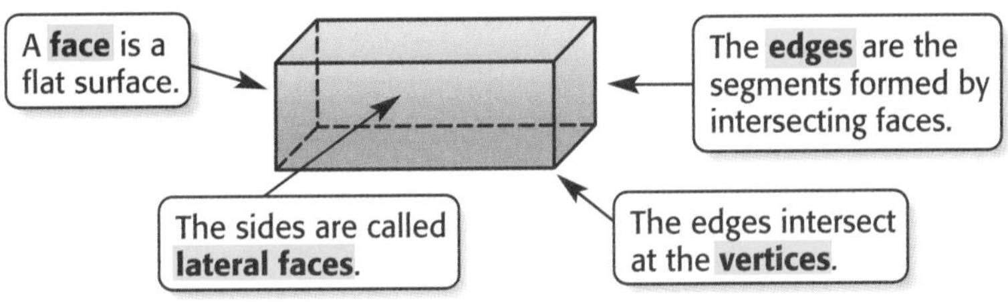

A face is a flat surface.
The edges are the segments formed by intersecting faces.
The sides are called lateral faces.
The edges intersect at the vertices.

Three-Dimensional Figures

Key Concept

Figure	Characteristics	Examples
Prism	• At least three lateral faces are rectangles. • The two faces that are parallel and congruent are the *bases*. • The name of the prism reflects the shape of the base.	rectangular prism triangular prism square prism or cube
Pyramid	• At least three lateral faces are triangles. • Has one base that is any closed figure with three or more sides. • The name of the pyramid reflects the shape of the base.	triangular pyramid square pyramid
Cone	• One base is a circle. • There is one vertex and there are no edges.	diameter of the base
Cylinder	• Two bases are circles. • There are no vertices and no edges.	diameter of the base
Sphere	• All of the points on a sphere are the same distance from the *center*. • There are no faces, bases, edges, or vertices.	diameter of the sphere center of the sphere

Concepts and Skills Bank

EXAMPLES Identify Properties of a Figure

1 **Use the rectangular prism at the right. Name the vertices, faces, edges and bases. Then find the measures of the angles of the base.**

This figure has two parallel congruent bases that are rectangles, *PQRS* and *TWVU* so it is a rectangular prism.

vertices: *P, Q, R, S, T, U, V, W*

faces: *PSUT, QRVW, PQWT, SRVU, PQRS, TWVU*

edges: $\overline{PS}, \overline{SR}, \overline{RQ}, \overline{QP}, \overline{TU}, \overline{UV}, \overline{VW}, \overline{WT}, \overline{PT}, \overline{SU}, \overline{RV}, \overline{QW}$

Since the prism is a rectangular prism, the bases are rectangles. The measure of the angles are each 90°.

2 **Find the perimeters of the faces.**

The faces are rectangles so use the formula $P = 2\ell + 2w$.

Face	$P = 2\ell + 2w$	Perimeter
PQRS	$P = 2 \times 24 + 2 \times 12$	72 mm
TUVW	$P = 2 \times 24 + 2 \times 12$	72 mm
PSUT	$P = 2 \times 12 + 2 \times 15$	54 mm
QRVW	$P = 2 \times 12 + 2 \times 15$	54 mm
PQWT	$P = 2 \times 15 + 2 \times 24$	78 mm
SRVU	$P = 2 \times 15 + 2 \times 24$	78 mm

Exercises

Identify each solid. Name the bases, faces, edges, and vertices.

1. 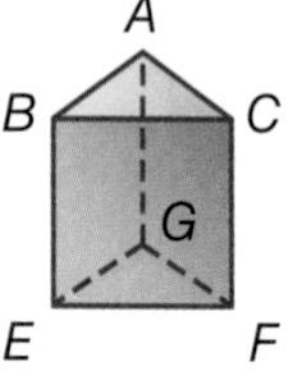

2. 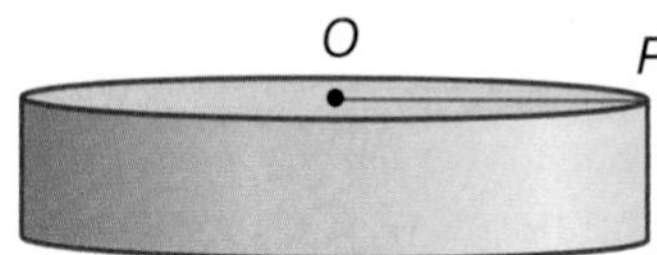

3. 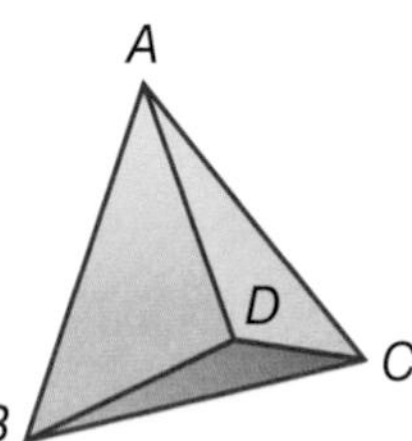

4. 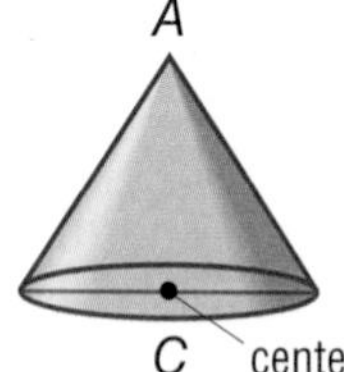

Find the angle measures of the bases for each prism. Then, find the perimeters of the faces and bases.

5. 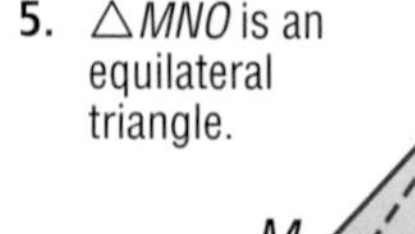
$\triangle MNO$ is an equilateral triangle.

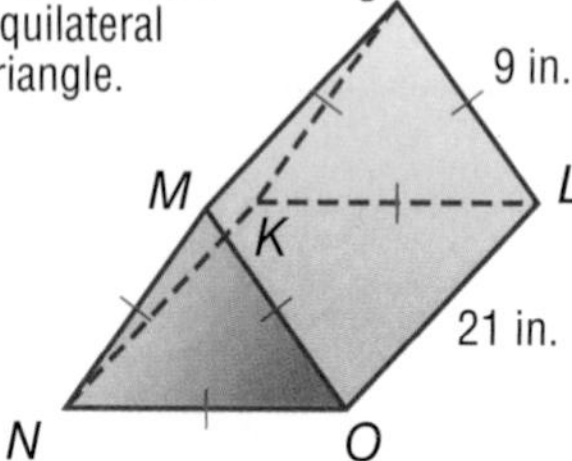

6. 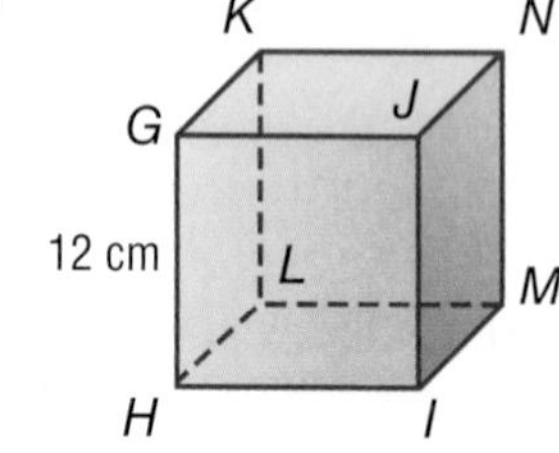

The figure is a cube.

Frequency Tables

Statistics involves collecting, organizing, analyzing, and presenting data. **Data** are pieces of information that are often numerical. The heights of New York's champion trees are listed in the table.

Heights (ft) of New York's State Champion Trees						
61	112	75	96	72	76	88
77	95	95	91	102	70	98
96	114	72	83	82	83	108
68	93	99	104	85	58	100
72	112	114	56	110	90	95
120	63	91	110	105	135	75

Source: NYS Big Tree Register, 2001

This data can be organized in a frequency table. A **frequency table** shows the number of pieces of data that fall within given intervals.

The **scale** includes the least number, 56, and the greatest, 135. Here the scale is 51 to 140.

Heights (ft) of New York's State Champion Trees		
Height	**Tally**	**Frequency**
51–80	HHT HHT III	13
81–110	HHT HHT HHT HHT III	23
111–140	HHT I	6

The scale is separated into equal parts called **intervals**. The interval is 30.

Tally marks are counters used to record items in a group.

EXAMPLE Make a Frequency Table

1 ARCHITECTURE The heights of the tallest buildings in Miami, Florida are shown. Make a frequency table of the data.

Tallest Buildings (ft) in Miami, Florida			
400	450	625	420
794	480	484	510
425	456	405	520
764	400	487	487

Source: *The World Almanac*

Step 1 Choose an appropriate scale and interval for the data.

scale: 400 to 799
interval: 100

The scale includes all of the data, and the interval separates the scale into equal parts.

Step 2 Draw a table with three columns and label the columns *Height (ft), Tally,* and *Frequency*.

Step 3 In the first column, list the intervals. In the second column, tally the data. In the third column, add the tallies.

Tallest Buildings (ft) in Miami, FL		
Height (ft)	**Tally**	**Frequency**
400–499	HHT HHT I	11
500–599	II	2
600–699	I	1
700–799	II	2

EXAMPLE Interpret a Frequency Table

2 ENGLISH The frequency table shows how often the five most common words in English appeared in a magazine article. What do you think is the most common word in English? How did you reach your conclusion?

Most Common Words		
Word	**Tally**	**Frequency**
to	卌 卌 卌 卌 I	21
of	卌 卌 III	13
the	卌 卌 卌 卌 卌 卌 卌 卌 卌 II	47
and	卌 卌 卌 卌 卌 I	26
a	卌 卌 IIII	14

According to the frequency table, the word "the" was used most often. This and other such articles may suggest that "the" is the most common word in English.

Exercises

Make a frequency table for each set of data.

1.

Cost ($) of Various Skateboards				
99	67	139	63	75
56	89	59	70	78
99	55	125	64	110

2.

Movies Viewed Per Month					
2	4	10	6	4	5
8	1	3	5	6	4
4	9	2	10	6	7

3.

Favorite Color					
R	G	B	R	R	K
B	P	P	Y	R	B
P	R	Y	K	B	Y
B	B	P	B	P	R

R = red B = blue Y = yellow
G = green P = purple K = pink

4.

Favorite Type of Music							
A	K	R	A	A	C	C	P
C	C	C	A	C	K	A	C
K	O	P	C	P	O	O	C
C	K	A	R	O	A	A	C

C = country P = pop A = alternative
R = rap K = rock O = other

5. The high temperatures in Indiana cities on March 13 are listed.

52 57 48 53 52 49 48 52 51 47 51 49 57 53 48 52 52 49

Make a frequency table of the data.

6. Which scale is more appropriate for the data set 56, 85, 23, 78, 42, 63: 0 to 50 or 20 to 90? Explain your reasoning.

7. What is the best interval for the data set 132, 865, 465, 672, 318, 940, 573, 689: 10, 100, or 1,000? Explain your reasoning.

For Exercises 8 and 9, use the frequency table at the right.

8. Describe the data shown in the table.

9. Which flavor should the ice cream shop stock the most for this July? Explain your reasoning.

Ice Cream Flavors Sold in July of Previous Year		
Flavor	**Tally**	**Frequency**
vanilla	卌 卌 卌 卌 IIII	24
chocolate	卌 卌 卌 III	18
strawberry	卌 卌 II	12
chocolate chip	卌 卌 卌 I	16
peach	卌 III	8
butter pecan	卌 卌 I	11

Concepts and Skills Bank

16 Probability of Compound Events

In Chapter 7, you learned how to find the probability of a simple event. **Compound events** consist of two or more simple events. When the outcome of one event does not affect the outcome of the other event, the events are independent events.

EXAMPLE Probability of Independent Events

1 Sondra has two shirts, white and green, and 3 pairs of pants, blue, tan and black. If she chooses a shirt and a pair of pants at random, what is the probability that she will choose the white shirt and black pants?

Draw a tree diagram showing the clothing choices and the probabilities.

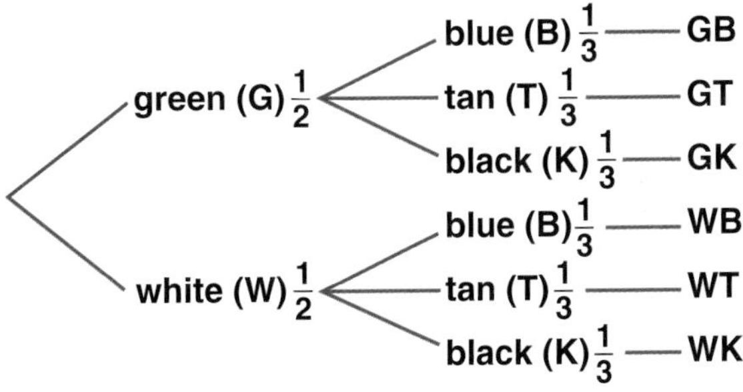

List the sample space. GB GT GK WB WT WK

$P(\text{white, black}) = \dfrac{\text{number of times white and black occur}}{\text{number of possible outcomes}}$

$P(\text{white, black}) = \dfrac{1}{6}$

So, the probability of Sondra choosing a white shirt and black pants is $\frac{1}{6}$ or about 17%. The probability in Example 1 can also be found by multiplying the probabilities of each event. $P(\text{White}) = \frac{1}{2}$ and $P(\text{Black}) = \frac{1}{3}$, so $P(\text{White and Black}) = \frac{1}{2} \cdot \frac{1}{3}$ or $\frac{1}{6}$. This leads to the following.

Probability of Independent Events — Key Concept

Words The probability of two independent events can be found by multiplying the probability of the first event by the probability of the second event.

Symbols $P(A \text{ and } B) = P(A) \cdot P(B)$

If the outcome of one event affects the outcome of a second event, the events are called **dependent events**. Just as in independent events, the probabilities of dependent events can by found by multiplying the probabilities of each event. However, now the probability of the second event depends on the fact that the first event has already occurred.

Probability of Dependent Events

Key Concept

Words If two events, A and B are dependent, then the probability of both events occurring is the product of the probability of A and the probability of B after A occurs.

Symbols $P(A \text{ and } B) = P(A) \cdot P(B \text{ following } A)$

EXAMPLE Probability of Dependent Events

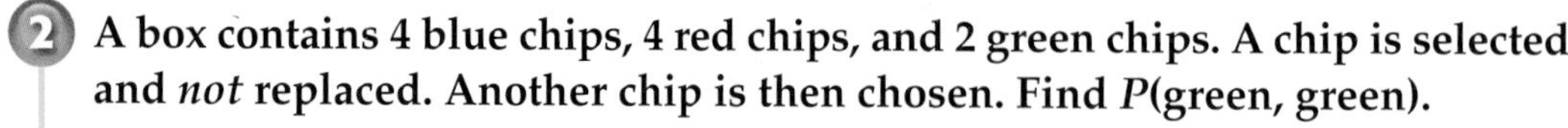

2 **A box contains 4 blue chips, 4 red chips, and 2 green chips. A chip is selected and *not* replaced. Another chip is then chosen. Find P(green, green).**

$P(\text{first chip is green}) = \frac{2}{10}$ ← number of green chips / ← total number of chips

$P(\text{second chip is green}) = \frac{1}{9}$ $\frac{\text{number of green chips after one green chip is removed}}{\text{total number of chips after one green chip is removed}}$

$P(\text{green, green}) = \frac{\overset{1}{\cancel{2}}}{\underset{5}{\cancel{10}}} \times \frac{1}{9} \text{ or } \frac{1}{45}$

So, the probability of choosing a green chip followed by another green chip is $\frac{1}{45}$, or about 4%.

Exercises

A card is drawn from the cards shown and replaced. Then, a second card is drawn. Find each probability.

1. P(two even numbers)
2. P(two odd numbers)
3. P(two numbers less than 4)
4. P(a number less than 5 followed by a number greater than 5)

A drawer contains 4 blue socks, 5 white socks, and 8 black socks. After a sock is chosen, it is *not* replaced. Find each probability.

5. P(blue sock, blue sock)
6. P(white sock, white sock)
7. P(black sock, black sock)
8. P(blue sock, black sock)
9. **PROJECTS** Five boys and 2 girls are volunteering for a class project. Their names are written on slips of paper and placed in a bowl. The teacher will choose two names without replacing the first name chosen. What is the probability that two girls will be chosen?
10. **BOARD GAMES** The letter tiles shown at the right are placed in a bag. One tile is chosen at random and replaced. A second tile is then chosen. What is the probability that two consonants will be chosen?

11. Find the probability that two consonants will be chosen if the first tile is *not* replaced. Compare this to your answer to Exercise 14.

Concepts and Skills Bank

Photo Credits

Cover: Travis VanDenBerg/Almay; **v** Digital Vision/Getty Images; **vi** Aaron Haupt; **vii** (t)Aaron Haupt, (c)courtesy Beatrice Luchin, (bl)courtesy Dinah Zike; **x–xi** CORBIS; **xii–xiii** Martin Ruegner/Masterfile; **xiv–xv** Steve Shott/Dorling Kindersley/Getty Images; **xvi–xvii** Arthur Tilley/Taxi/Getty Images; **xviii–xix** Gail Mooney/Masterfile; **xx–xxi** Mpia-hd, Birkle, Slawik/Photo Researchers, Inc.; **xxii–xxiii** Guillen Photography/Alamy; **xxiv** Digital Vision/Getty Images; **1** John Evans; **3** Yva Momatiuk/John Eastcott/Minden Pictures; **4** Masterfile Royalty Free; **5** Alan Majchrowicz/age fotostock; **6, 7** Swerve/Alamy; **8** Momatiuk-Eastcott/CORBIS; **9** Tim Fitzharris/Minden Pictures; **10** Geoff Daniels/Eye Ubiquitous/Alamy; **11** (t)Mark Karrass/CORBIS, (b)Joseph Sohm/Visions of America, LLC/Alamy; **12** Kim Karpeles/Alamy; **13** Emily Riddell/Alamy; **14** Lloyd Sutton/Masterfile; **15** Masterfile Royalty Free; **16** (t)Carmel Studios/SuperStock, (b)RideZone/Joel Styer; **17** (t)Roy Ooms/Masterfile, (b)Joel W. Rogers/CORBIS; **18** Stephen Alvarez/National Geographic/Getty Images; **19** Paul Souders/CORBIS; **20–21** Liane Cary/age fotostock; **22** Scott Boehm/Getty Images Sport; **25** John Hrusa/epa/CORBIS; **32** Horizons Companies; **33** Tony Freeman/PhotoEdit; **36** (l)Cindy Charles/PhotoEdit, (r)Robert W. Ginn/PhotoEdit; **38** Photodisc/Getty Images; **40** (l)David Young-Wolff/PhotoEdit, (r)Tony Freeman/PhotoEdit; **42** Myrleen Ferguson Cate/PhotoEdit; **44** Leo Dennis Productions/Brand X/CORBIS; **45** Sandy Felsenthal/CORBIS; **49** William Leaman/Alamy; **50** graficart.net/Alamy; **52** (l)LWA/Dann Tardif/Blend Images/Getty Images, (r)Pierre Arsenault/Masterfile; **54** Amos Morgan/Photodisc/Getty Images; **56** Image100/CORBIS; **58** Digital Vision Ltd./SuperStock; **64** Mike Powell/Getty Images; **66** (l)Cindy Charles/PhotoEdit, (r)Myrleen Ferguson Cate/PhotoEdit; **76** Stephen J. Krasemann/Photo Researchers; **78** Jack Hollingsworth/Photodisc/Getty Images; **81** Tim Fitzharris/Minden Pictures; **83** NASA; **88** Bill Frymire/Masterfile; **89** David Young-Wolff/PhotoEdit; **90** Delly Carr/epa/CORBIS; **93** JTB Photo Communications, Inc./Alamy; **94** (l)David Young-Wolff/PhotoEdit, (r)Michael Newman/PhotoEdit; **96** Anup Shah/npl/Minden Pictures; **97** LWA-Stephen Welstead/CORBIS; **108** NOAA; **110** Martin Ruegner/Masterfile; **115** Marcus Mok/Asia Images/Getty Images; **117** Daryl Benson/Masterfile; **122** AP Photo/Rob Carr; **124** Georgette Douwma/Photo Researchers; **134–135** Kauko Helavuo/Stone/Getty Images; **136** David Stoecklein/CORBIS; **140** Stephen Dunn/Getty Images Sport; **142** Art Stein/ZUMA/CORBIS; **145** (l)Michael Newman/PhotoEdit, (r)David Young-Wolff/PhotoEdit; **146** G.K. and Vikki Hart/Brand X Pictures/Jupiter Images; **147** (t)Michael Barley/CORBIS, (b)Louis Fox/Stone/Getty Images; **148** Tim de Waele/CORBIS; **153** Travis Lindquist/Getty Images Sport; **157** Philippe Psaila/Science Photo Library; **159** Gunter Marx Photography/CORBIS; **160** (l)David Young-Wolff/PhotoEdit, (r)Cleve Bryant/PhotoEdit; **163** age fotostock/SuperStock; **164** NASA/Photo Researchers; **165** Microzoa/The Image Bank/Getty Images; **169** F. Lukasseck/Masterfile; **170** Shoula/Photonica/Getty Images; **176** (l)Simon Watson/Stone/Getty Images, (r)Thinkstock/CORBIS; **181** Jeremy Woodhouse/Masterfile; **184** Brad Wilson/Stone/Getty Images; **194** Anthony Johnson/The Image Bank/Getty Images; **200** Mark Ransom Photography; **204** Bruce Coleman Inc./Alamy; **206** Brooklyn Production/CORBIS; **207** D. Hurst/Alamy; **210** SuperStock, Inc./SuperStock; **211** Gary Vestal/Photographer's Choice/Getty Images; **214** Brad Wilson/Stone/Getty Images; **217** Steve Terrill/CORBIS; **218** Steve Vidler/SuperStock; **219** (l)Cleve Bryant/PhotoEdit, (r)Michael Newman/PhotoEdit; **223** George H. H. Huey/CORBIS; **225** (t)C Squared Studios/Photodisc/Getty Images, (b)Stockbyte/Getty Images; **226** Tetra Images/CORBIS; **227** age fotostock/SuperStock; **228** (l)Stockdisc/PunchStock, (r)Pete Leonard/zefa/CORBIS; **230** BananaStock/SuperStock; **231** franzfoto.com/Alamy; **234** Matt Meadows; **246** Tom McHugh/Photo Researchers; **251** (t)Ariel Skelley/Getty Images, (bl)Leroy Simon/Visuals Unlimited, (br)Matt Meadows; **254** Blend Images/Alamy; **258** Photodisc/SuperStock; **263** Thomas Northcut/Photodisc/Getty Images; **265** Susumu Nishinaga/Photo Researchers; **267** (l)Randy Faris/CORBIS, (r)Plush Studios/Blend Images/CORBIS; **270** Doug Martin; **273** James Watt/AnimalsAnimals-Earth Scenes; **274** (l)Stockbyte/Getty Images, (r)Matthew Donaldson/ImageState; **276** Stockbyte/SuperStock; **282** Stephen Dalton/Photo Researchers; **287** Bob Cranston/Animals Animals-Earth Scenes; **288** Sal Maimone/SuperStock; **289** Alfredo Dagli Orti/The Art Archive/CORBIS; **291** Gary Rhijnsburger/Masterfile; **294** Richard Lord/PhotoEdit; **296** (t)Steve Shott/Dorling Kindersley/Getty Images, (bl)Tony Latham/Getty Images, (br)Brad Wilson/Photonica/Getty Images; **299** Gerry Ellis/Globio/Minden Pictures; **310–311** IT Stock Free/age fotostock; **312** Andre Jenny/Alamy; **316** Masterfile Royalty-Free; **318** CJ Gunther/epa/CORBIS; **319** (l)Big Cheese Photo/SuperStock, (r)Photodisc/First Light; **323** Lucas Jackson/Reuters/CORBIS; **324** Raimund Koch/Photonica/Getty Images; **326** (l)Daybreak Imagery/Animals Animals, (r)Rick Gomez/CORBIS; **329** George Doyle & Ciaran Griffin/Stockbyte/Getty Images; **331** George B. Diebold/CORBIS; **332** Daryl Benson/Masterfile; **335** Masterfile Royalty-Free; **337** Bob Langrish/Animals Animals; **338** (l)Reza Estakhrian/Taxi/Getty Images, (r)LWA/Dann Tardif/Blend Images/Getty Images; **341** Blend Images/Alamy; **343** Foodcollection/Getty Images; **347** Lon C. Diehl/PhotoEdit; **350** Daniel J. Cox/CORBIS; **352** Roger de la Harpe/Animals Animals; **362** Joe McBride/CORBIS; **366** ThinkStock/SuperStock; **378** Benn Mitchell/The Image Bank/Getty Images; **385** (l)BananaStock/SuperStock, (r)Jim Arbogast/Digital Vision/Getty Images; **391** Doug Martin; **395** age fotostock/SuperStock; **396** Arthur Tilley/Taxi/Getty Images; **397** (l)RubberBall/Alamy Images, (r)Michael Newman/PhotoEdit; **399** Stockbyte/Getty Images; **402** Thomas Mangelsen/Minden Pictures; **414–415** Gail Mooney/Masterfile; **416** Jose Azel/Getty Images; **422** (t)LWA-Dann Tardif/CORBIS, (b)"Close to Home" ©1995 John McPherson. Reprinted with permission of Universal Press Syndicate. All rights reserved.; **423** (l)Barbara Penoyar/Photodisc/Getty Images, (r)ed-imaging; **426** fStop/SuperStock; **433** Altrendo Nature/Altrendo/Getty Images; **435** Jack Novak/SuperStock; **438** Jeff Gross/Getty Images Sport; **440** Kennan Ward/CORBIS; **442** Brad Wilson/Getty Images; **446** David Young-Wolff/PhotoEdit; **448** (l)ed-imaging, (r)Dann Tardif/LWA/Blend Images/Getty Images; **451** Digital Vision/Getty Images; **456** K. Hackenberg/zefa/CORBIS; **459** Siede Preis/Photodisc/Getty Images; **460** Masterfile Royalty-Free; **468** coasterimage.com; **476** John Gress/Reuters/CORBIS; **477** Photodisc/SuperStock; **482** (l)Masterfile Royalty-Free, (r)John Kelly/The Image Bank/Getty Images; **483** Walter Bibikow/age fotostock; **488** C Squared Studios/Photodisc/Getty Images; **489** Jan Cook/Botonica/PictureQuest; **490** (l)Greg Garza/dallassky.com, (r)National Trail Parks And Recreation District; **492** Randy Lincks/Masterfile; **496** (l)David Pollack/CORBIS, (r)Mark Gibson; **500** Rubberball Productions/Getty Images; **506** Wald Frerck/AP Images; **510** (t)Richard Klune/CORBIS, (b)David Frazier/CORBIS; **512** Wes Thompson/CORBIS; **515** (l)Mark Richards/PhotoEdit, (r)Jane Grushow/Grant Heilman Photography; **518** Owaki-Kulla/CORBIS; **523** Jorge Silva/Reuters/CORBIS; **524** David Pollack/CORBIS; **525** (t)Douglas Fisher/Alamy, (bl)Mike Kemp/Rubberball/Getty Images, (br)Plush Studios/Blend Images/Getty Images; **527** Matt Meadows; **532** (t)Michael T. Sedam/CORBIS, (bl)Jeff Greenberg/PhotoEdit, (br)Ken Fisher/Stone/Getty Images; **540** Joe Sohm/Visions of America, LLC/Alamy; **541** Stockbyte/Getty Images; **543** (l)Big Cheese Photo/SuperStock, (r)Christopher S. O'Meal/age fotostock; **546** David Young-Wolff/PhotoEdit; **550** Ryan McVay/Getty Images; **551** David Muench/CORBIS; **556** E.R. Degginger/Animals Animals-Earth Scenes; **568-569** SW Productions/Brand X/CORBIS; **570** Lester Lefkowitz/CORBIS; **573** Kendra Knight/age fotostock; **574** Mpia-hd, Birkle, Slawik/Photo Researchers; **579** Comstock/SuperStock; **588** Jose Luis Pelaez Inc/Getty Images; **592** Jose Luis Pelaez/Iconica/Getty Images; **596** Mark E. Gibson/CORBIS; **597** (t)Masterfile Royalty-Free, (bl)Dougal Waters/Digital Vision/Getty Images, (br)Getty Images; **602** Cartesia/Photodisc/Getty Images; **606** alfredo barni/Marka/age fotostock; **608** Daryl Benson/Masterfile; **610** Dorling Kindersley/Getty Images; **613** Ben Mangor/SuperStock; **628** Guillen Photography/Alamy; **632** David Young-Wolff/PhotoEdit; **633** Sebastien Dolidon/CORBIS; **636** Russell Glenister/image100/CORBIS; **638** Robert Harding/Robert Harding World Imagery/Getty Images; **639** Stockbyte/Getty Images; **640** (t)Marvy!/CORBIS, (bl)Design Pics Inc./Alamy, (br)Mel Curtis/Digital Vision/Getty Images; **646** Photo 24/Getty Images; **651** Digital Vision/age fotostock; **654** (l)Michael Newman/PhotoEdit, (r)Myrleen Ferguson Cate/PhotoEdit; **657** BananaStock/SuperStock; **659** ImageState/Alamy Images; **661** Sean Justice/Riser/Getty Images; **719** Tim Fuller; **LA0–LA1** Brand X Pictures/Punchstock; **LA7** Image Source/SuperStock; **LA8** Greg Ceo/Getty Images; **LA12** Nora Good/Masterfile; **LA17** Charles & Josette Lenars/CORBIS; **LA22** Ilene MacDonald/Alamy

Glossary/Glosario

Math Online A mathematics multilingual glossary is available at glencoe.com. The glossary includes the following languages.

Arabic	Cantonese	Korean	Tagalog
Bengali	English	Russian	Urdu
Brazilian Portuguese	Haitian Creole	Spanish	Vietnamese
	Hmong		

Cómo usar el glosario en español:

1. Busca el término en inglés que desees encontrar.
2. El término en español, junto con la definición, se encuentran en la columna de la derecha.

English | Español

A

acute angle (p. 471) An angle with a measure greater than 0° and less than 90°.

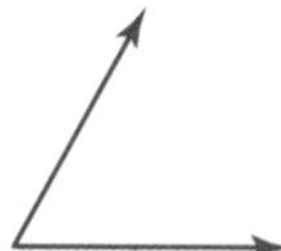

ángulo agudo Ángulo que mide más de 0° y menos de 90°.

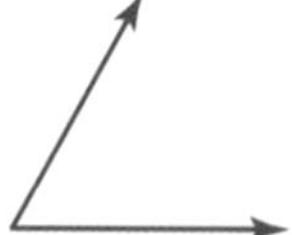

acute triangle (p. 486) A triangle having three acute angles.

triángulo acutángulo Triángulo con tres ángulos agudos.

Addition Property of Equality (p. 652) If you add the same number to each side of an equation, the two sides remain equal.

propiedad de adición de la igualdad Si sumas el mismo número a ambos lados de una ecuación, los dos lados permanecen iguales.

algebra (p. 42) A mathematical language that uses symbols, usually letters, along with numbers. The letters stand for numbers that are unknown.

álgebra Lenguaje matemático que usa símbolos, por lo general, además de números. Las letras representan números desconocidos.

algebraic expression (p. 42) A combination of variables, numbers, and at least one operation.

expresión algebraica Combinación de variables, números y, por lo menos, una operación.

angle (p. 470) Two rays with a common endpoint form an angle.

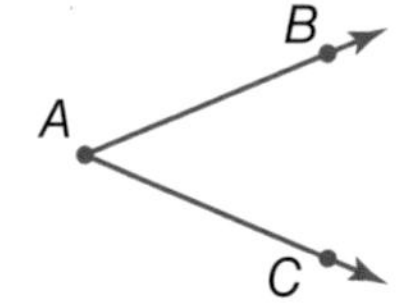

$\angle BAC$, $\angle CAB$, or $\angle A$

ángulo Dos rayos con un extremo común forman un ángulo.

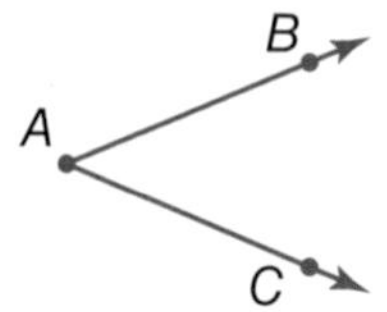

$\angle BAC$, $\angle CAB$ o $\angle A$

Glossary/Glosario

angle of rotation (p. 617) The degree measure of the angle through which a geometric figure is rotated.

ángula de rotación El ángulo a través del cual se rota una figura geométrica.

area (p. 63) The number of square units needed to cover the surface enclosed by a geometric figure.

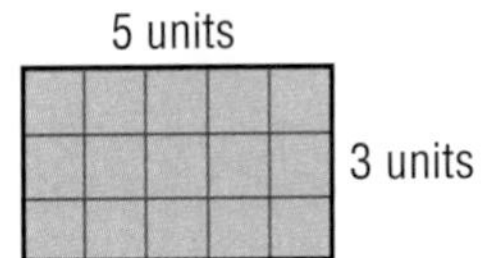

$A = 5 \times 3 = 15$ square units

área El número de unidades cuadradas necesarias para cubrir una superficie cerrada por una figura geométrica.

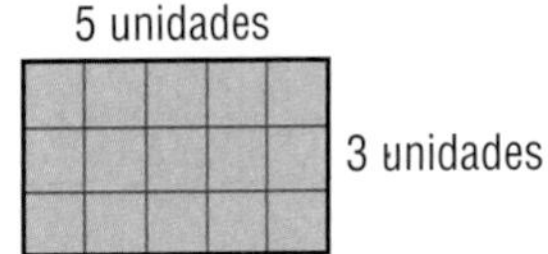

$A = 5 \times 3 = 15$ unidades cuadradas

Associative Properties (p. 636) The way in which numbers are grouped does not change the sum or product.

propiedad asociativa La forma en que se agrupan tres números al sumarlos o multiplicarlos no altera su suma o producto.

arithmetic sequence (p. 343) A sequence in which each term is found by adding the same number to the previous term.

sucesión aritmética Sucesión en que cada término se puede hallar al sumar el mismo número al término previo.

average (p. 102) The sum of two or more quantities divided by the number of quantities; the mean.

promedio La suma de dos o más cantidades dividida entre el número de cantidades; la media.

B

bar graph (p. 81) A graph using bars to compare quantities. The height or length of each bar represents a designated number.

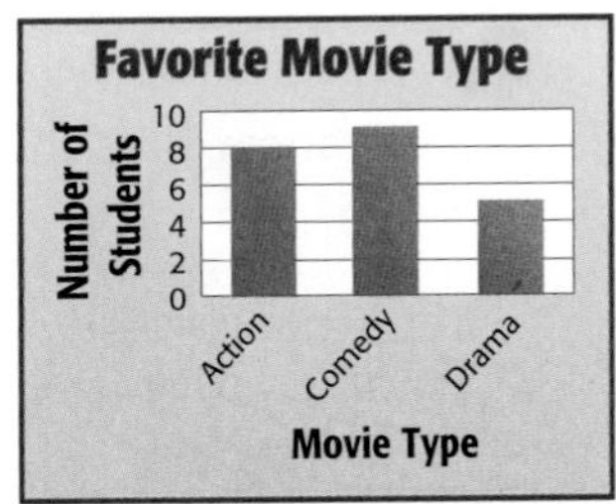

gráfica de barras Gráfica que usa barras para comparar cantidades. La altura o longitud de cada barra representa un número designado.

base (p. 32) In a power, the number used as a factor. In 10^3, the base is 10. That is, $10^3 = 10 \times 10 \times 10$.

base En una potencia, el número usado como factor. En 10^3, la base es 10. Es decir, $10^3 = 10 \times 10 \times 10$.

base (p. 535) Any side of a parallelogram.

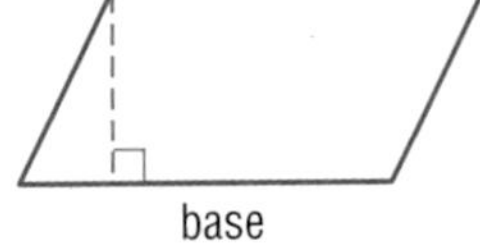

base Cualquier lado de un paralelogramo.

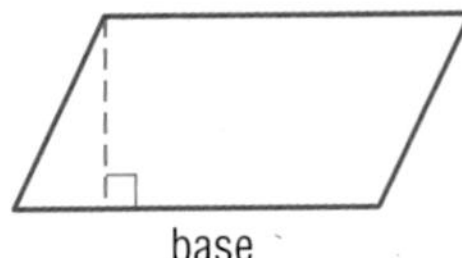

C

capacity (p. 424) The amount that can be held in a container.

capacidad Cantidad que puede contener un recipiente.

Celsius (p. 455) The metric measure of temperature.

center (p. 528) The given point from which all points on a circle are the same distance.

centimeter (p. 432) A metric unit of length. One centimeter equals one-hundredth of a meter.

chord (p. 528) A segment with endpoints that are on a circle.

circle (p. 528) The set of all points in a plane that are the same distance from a given point called the center.

circle graph (p. 370) A graph used to compare parts of a whole. The circle represents the whole and is separated into parts of the whole.

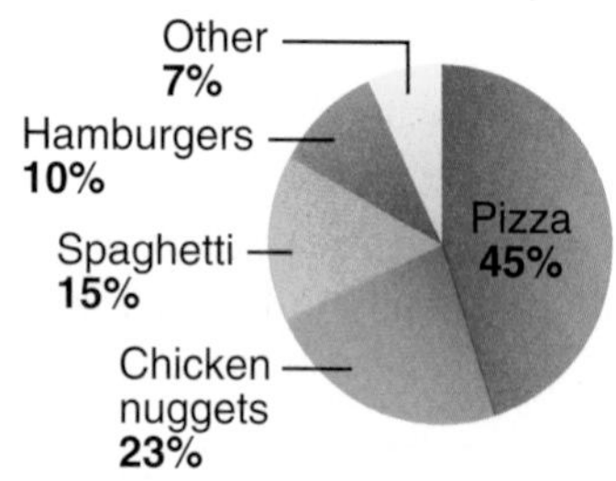

circumference (p. 528) The distance around a circle.

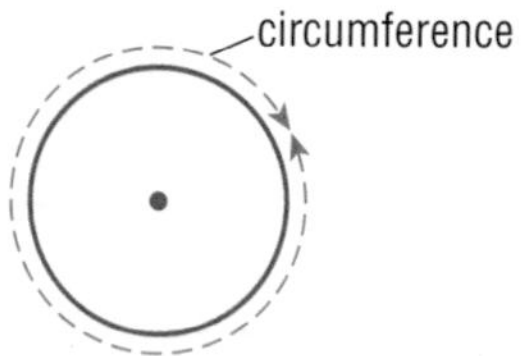

clustering (p. 151) An estimation method in which a group of numbers close in value are rounded to the same number.

coefficient (p. 654) The numerical factor of a term that contains a variable.

common factor (p. 197) Factors that are shared by two or more numbers.

common multiples (p. 216) Multiples that are shared by two or more numbers. For example, some common multiples of 2 and 3 are 6, 12, and 18.

Celsius Medida de temperatura del sistema métrico.

centro Un punto dado del cual equidistan todos los puntos de un círculo o de una esfera.

centímetro Unidad métrica de longitud. Un centímetro es igual a la centésima parte de un metro.

cuerda Segmento cuyos extremos están sobre un círculo.

círculo Conjunto de todos los puntos en un plano que equidistan de un punto dado llamado centro.

gráfica circular Tipo de gráfica estadística que se usa para comparar las partes de un todo. El círculo representa el todo y éste se separa en partes.

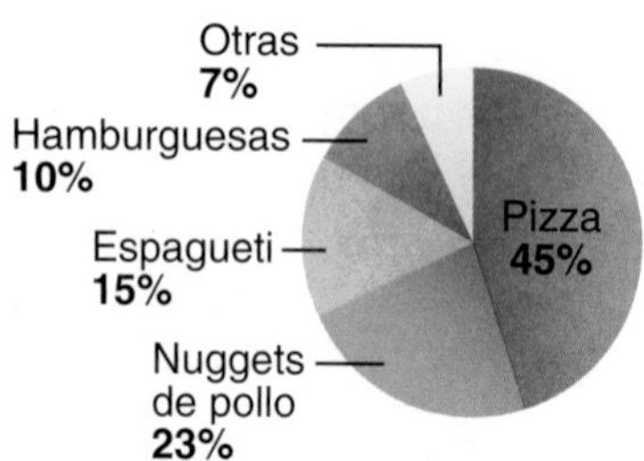

circunferencia La distancia alrededor de un círculo.

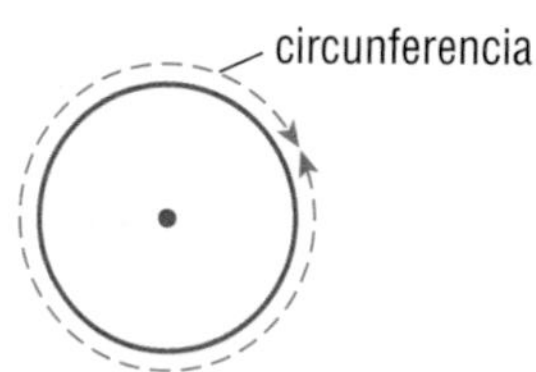

agrupamiento Método de estimación en que un grupo de números cuyo valor está estrechamente relacionado, se redondean al mismo número.

coeficiente El factor numérico de un término que contiene una variable.

factor común Un número entero factor de dos o más números.

múltiplos comunes Múltiplos compartidos por dos o más números. Por ejemplo, algunos múltiplos comunes de 2 y 3 son 6, 12 y 18.

Commutative Properties (p. 636) The order in which numbers are added or multiplied does not change the sum or product.

propiedad commutativa La forma en que se suman o multiplican dos números no altera su suma o producto.

compatible numbers (p. 276) Numbers that are easy to divide mentally.

números compatibles Números que son fáciles de dividir mentalmente.

complementary angles (p. 480) Two angles are complementary if the sum of their measures is 90°.

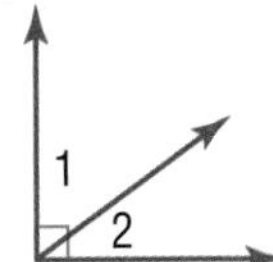

∠1 and ∠2 are complementary angles.

ángulos complementarios Dos ángulos son complementarios si la suma de sus medidas es 90°.

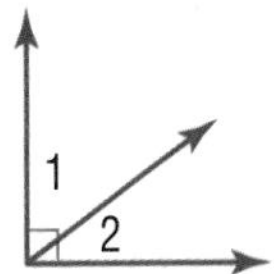

∠1 y ∠2 son complementarios.

complementary events (p. 383) Two events in which either one or the other must take place, but they cannot both happen at the same time. The sum of their probabilities is 1.

eventos complementarios Dos eventos tales, que uno de ellos debe ocurrir, pero ambos no pueden ocurrir simultáneamente. La suma de sus probabilidades es 1.

composite number (p. 28) A number greater than 1 with more than two factors.

número compuesto Un número entero mayor que 1 con más de dos factores.

congruent angles (p. 479) Angles with the same measure.

ángulos congruentes Ángulos con la misma medida.

congruent figures (p. 502) Figures that are the same shape and size.

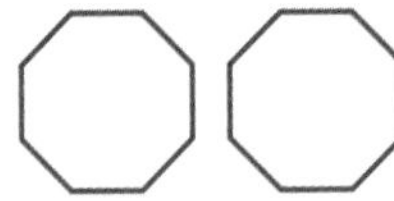

figuras congruentes Figuras que tienen la misma forma y tamaño.

congruent segments (p. 488) Line segments that have the same length.

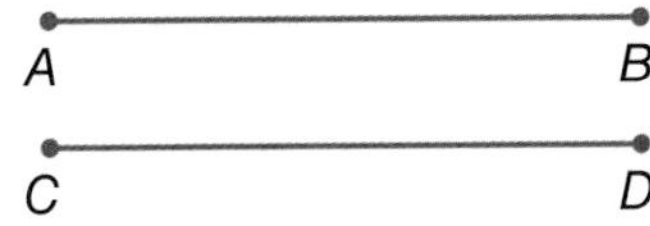

$\overline{AB}$ is congruent to $\overline{CD}$.

segmentos congruentes Segmentos de recta que tienen la misma longitud.

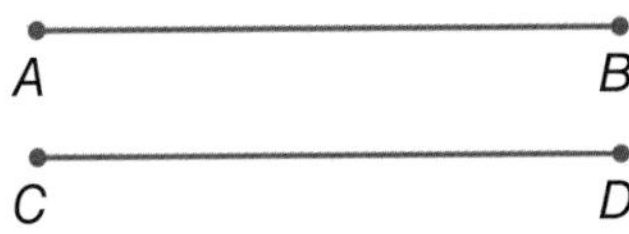

$\overline{AB}$ es congruente con $\overline{CD}$.

coordinate plane (p. 233) A plane in which a horizontal number line and a vertical number line intersect at their zero points.

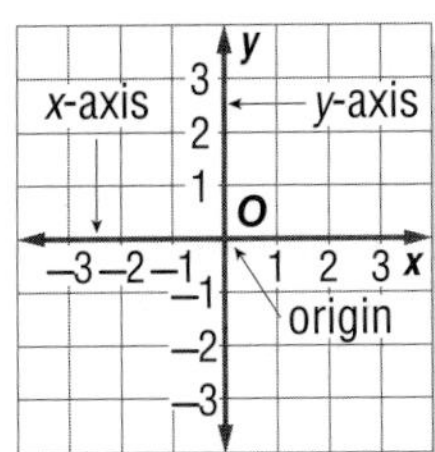

plano de coordenadas Plano en que una recta numérica horizontal y una recta numérica vertical se intersecan en sus puntos cero.

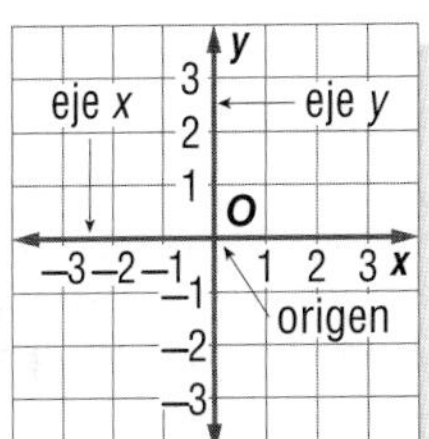

corresponding sides (p. 503) The parts of congruent or similar figures that match.

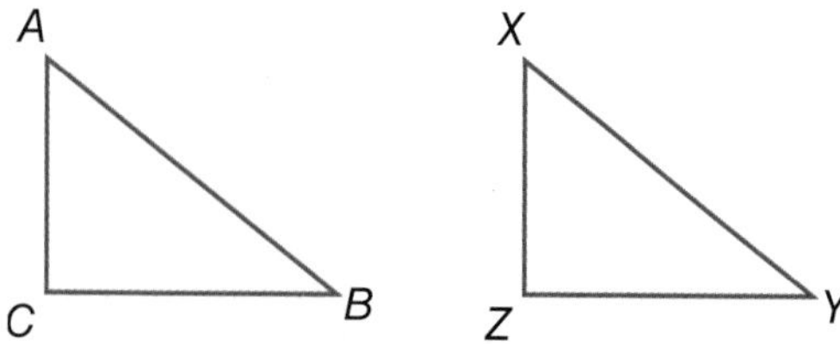

Side $\overline{AB}$ corresponds to side $\overline{XY}$.

lados correspondientes Las partes de figuras congruentes que coinciden.

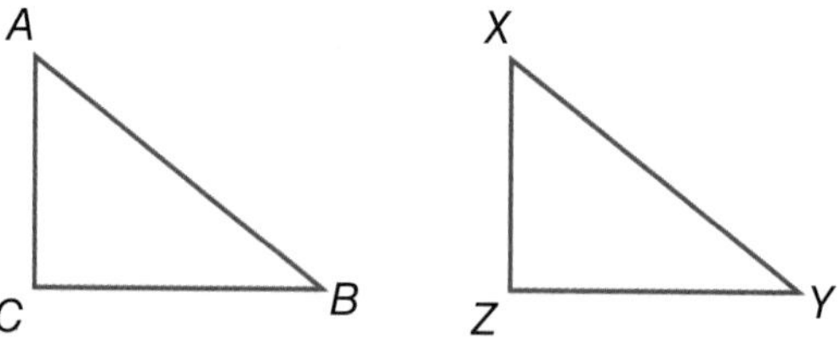

El lado $\overline{AB}$ y el lado $\overline{XY}$ son correspondientes.

counterexample (p. 201) An example that disproves a statement.

contraejemplo Ejemplo que demuestra que un enunciado no es verdadero.

cubed (p. 33) The product in which a number is a factor three times. Two cubed is 8 because $2 \times 2 \times 2 = 8$.

al cubo El producto de un número por sí mismo, tres veces. Dos al cubo es 8 porque $2 \times 2 \times 2 = 8$.

cubic units (p. 548) Used to measure volume. Tells the number of cubes of a given size it will take to fill a three-dimensional figure.

3 cubic units

unidades cúbicas Se usan para medir el volumen. Indican el número de cubos de cierto tamaño que se necesitan para llenar una figura tridimensional.

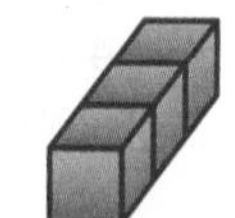

3 unidades cúbicas

cup (p. 424) A customary unit of capacity equal to 8 fluid ounces.

taza Unidad de capacidad del sistema inglés de medidas que equivale a 8 onzas líquidas.

D

data (p. 81) Information, often numerical, which is gathered for statistical purposes.

datos Información, con frecuencia numérica, que se recoge con fines estadísticos.

defining the variable (p. 50) Choosing a variable to represent the input when writing a function rule.

definir la variable Elección de la variable que representa el valor de entrada al escribir la regla de una función.

degree (p. 470) The most common unit of measure for angles.

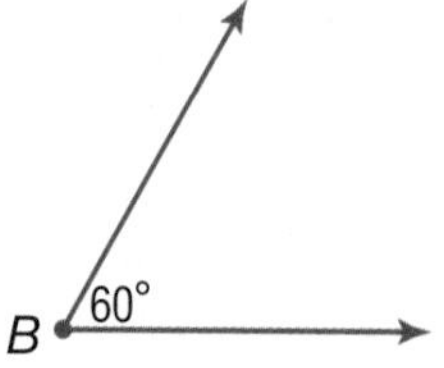

The measure of $\angle B$ is 60°.

grado La unidad más común para medir ángulos.

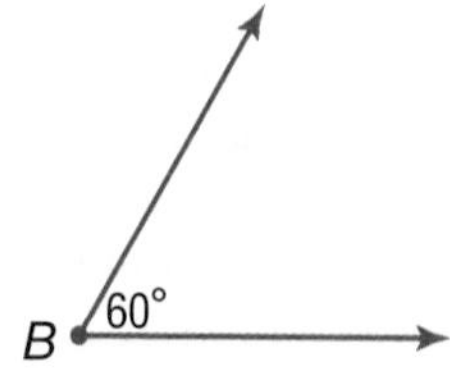

$\angle B$ mide 60°.

degrees (p. 455) The measure of temperature.

grados Medida de la temperatura.

decimal (p. 138) Numbers that have digits in the tenths place and beyond.

decimal Número que tiene un dígito en el lugar de las décimas y más allá.

diameter (p. 528) The distance across a circle through its center.

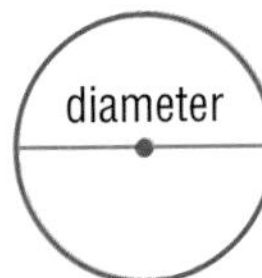

diámetro La distancia a través de un círculo pasando por el centro.

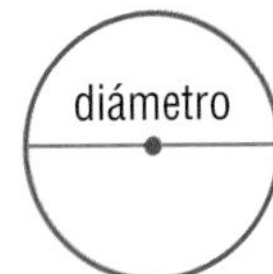

Distributive Property (p. 632) To multiply a sum by a number, multiply each addend by the number outside the parentheses.

propiedad distributiva Para multiplicar una suma por un número, multiplica cada sumando por el número fuera de los paréntesis.

E

elapsed time (p. 452) The amount of time that has passed from beginning to end.

tiempo transcurrido Cantidad de tiempo que ha pasado entre el principio y el fin.

equals sign (p. 57) A symbol of equality, =.

signo de igualdad Símbolo que indica igualdad, =.

equation (p. 57) A mathematical sentence that contains an equals sign, =.

ecuación Un enunciado matemático que contiene el signo de igualdad, =.

equilateral triangle (p. 488) A triangle having all three sides congruent and all three angles congruent.

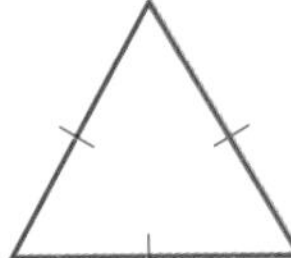

triángulo equilátero Triángulo cuyos tres lados y tres ángulos son congruentes.

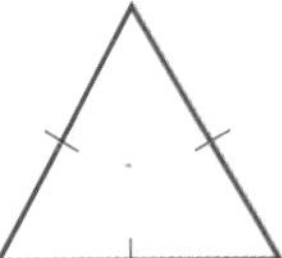

equivalent decimals (p. 143) Decimals that name the same number.

decimales equivalentes Decimales que representan el mismo número.

equivalent expressions (p. 636) Expressions that have the same value.

expresiones equivalentes Expresiones que poseen el mismo valor, sin importer los valores de la(s) variable(s).

equivalent fractions (p. 204) Fractions that name the same number.

fracciones equivalentes Fracciones que representan el mismo número.

equivalent ratios (p. 322) Ratios that have the same value.

razones equivalentes Dos razones que tienen el mismo valor.

evaluate (p. 42) To find the value of an algebraic expression by replacing variables with numerals.

evaluar Calcular el valor de una expresión sustituyendo las variables por número.

expanded form (p. 139) The sum of the products of each digit and its place value of a number.

forma desarrollada La suma de los productos de cada dígito y el valor de posición del número.

experimental probability (p. 387) A probability that is found by doing an experiment.

probabilidad experimental Probabilidad que se calcula realizando un experimento.

exponent (p. 32) In a power, the number of times the base is used as a factor. In 5^3, the exponent is 3. That is, $5^3 = 5 \times 5 \times 5$.

exponente En una potencia, el número de veces que la base se usa como factor. En 5^3, el exponente es 3. Es decir, $5^3 = 5 \times 5 \times 5$.

F

factor (p. 28) A number that divides into a whole number with a remainder of zero.

factor Número que al dividirlo entre un número entero tiene un residuo de cero.

Fahrenheit (p. 455) The customary measure of temperature.

Fahrenheit Medida de temperatura del sistema inglés.

fluid ounce (p. 424) A customary unit of capacity.

onzas líquidas Unidad de capacidad del sistema inglés de medidas.

foot (p. 418) A customary unit of length equal to 12 inches.

pie Unidad de longitud del sistema inglés de medidas que equivale a 12 pulgadas.

formula (p. 63) An equation that shows a relationship among certain quantities.

fórmula Ecuación que muestra una relación entre ciertas cantidades.

frequency (p. 81) The number of times an item occurs.

frecuencia El número de veces que sucede un artículo.

front-end estimation (p. 151) An estimation method in which the front digits are added or subtracted.

estimación frontal Método de estimación en que se suman o restan los dígitos del frente.

function (p. 49) A relation in which each element of the input is paired with exactly one element of the output according to a specified rule.

función Relación en que cada elemento de entrada es apareado con un único elemento de salida, según una regla específica.

function machine (p. 47) Takes a number called the input and performs one or more operations on it to produce a new value called the output.

máquina de funciones Recibe un número conocido como entrada, al que le realiza una o más operaciones para producir un nuevo número llamado salida.

function rule (p. 49) An expression that describes the relationship between each input and output.

regla de funciones Expresión que describe la relación entre cada valor de entrada y de salida.

function table (p. 49) A table organizing the input, rule, and output of a function.

tabla de funciones Tabla que organiza las entradas, la regla y las salidas de una función.

Fundamental Counting Principle (p. 390) Uses multiplication of the number of ways each event in an experiment can occur to find the number of possible outcomes in a sample space.

Principio Fundamental de Contar Este principio usa la multiplicación del número de veces que puede ocurrir cada evento en un experimento para calcular el numero de posibles resultados en un espacio muestral.

G

gallon (p. 424) A customary unit of capacity equal to 4 quarts.

galón Unidad de capacidad del sistema inglés de medidas que equivale a 4 cuartos de galón.

gram (p. 437) A metric unit of mass. One gram equals one-thousandth of a kilogram.

gramo Unidad de masa del sistema métrico. Un gramo equivale a una milésima de kilogramo.

graph (p. 81) A visual way to display data.

gráfica Manera visual de representar datos.

graph (p. 234) Place a point named by an ordered pair on a coordinate grid.

graficar Trazar un punto determinado por un par ordenado, en un plano de coordenadas.

graph (p. 122) To graph an integer on a number line, draw a dot at the location on the number line that corresponds to the integer.

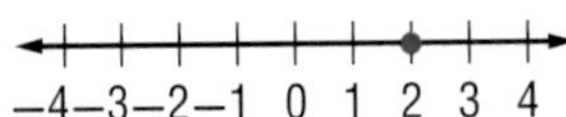

graficar Para graficar un entero sobre una recta numérica, dibuja un punto en la ubicación de la recta numérica correspondiente al entero.

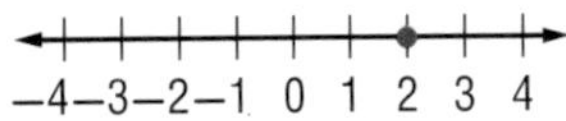

greatest common factor (GCF) (p. 197) The greatest of the common factors of two or more numbers. The GCF of 24 and 30 is 6.

máximo común divisor (MCD) El mayor factor común de dos o más números. El MCD de 24 y 30 es 6.

H

height (p. 535) The shortest distance from the base of a parallelogram to its opposite side.

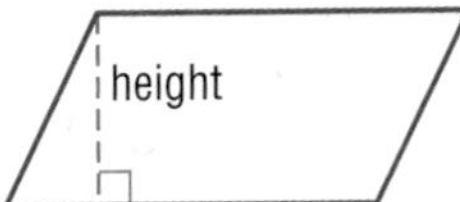

altura La distancia más corta desde la base de un paralelogramo hasta su lado opuesto.

horizontal axis (p. 81) The axis on which the categories or values are shown in a bar and line graph.

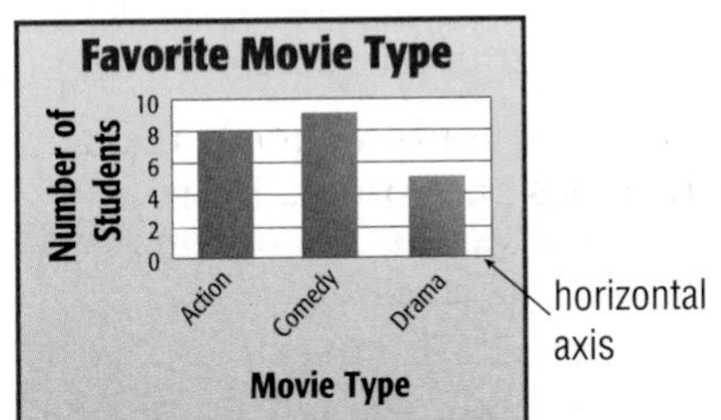

eje horizontal El eje sobre el cual se muestran las categorías o valores en una gráfica de barras o gráfica lineal.

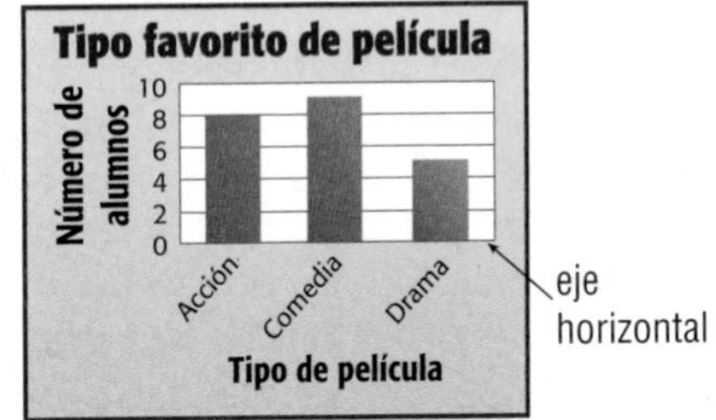

I

image (p. 604) The resulting figure after a transformation.

imagen La posición de una figura después de una transformación.

improper fraction (p. 209) A fraction that has a numerator that is greater than or equal to the denominator. The value of an improper fraction is greater than or equal to 1.

fracción impropia Fracción cuyo numerador es mayor que o igual a su denominador. El valor de una fracción impropia es mayor que o igual a 1.

inch (p. 418) A customary unit of length. Twelve inches equal one foot.

pulgada Unidad de longitud del sistema inglés de medidas. Doce pulgadas equivalen a un pie.

inequality (p. 142) A mathematical sentence indicating that two quantities are not equal.

desigualdad Enunciado matemático que indica que dos cantidades no son iguales.

input (p. 47) The number on which a function machine performs one or more operations to produce an output.

entrada El número que una máquina de funciones transforma mediante una o más operaciones para producir un resultado de salida.

integer (p. 121) The whole numbers and their opposites. ..., −3, −2, −1, 0, 1, 2, 3, ...

entero Los números enteros y sus opuestos. ..., −3, −2, −1, 0, 1, 2, 3, ...

interval (p. 81) The difference between successive values on a scale.

intervalo La diferencia entre valores sucesivos de una escala.

inverse operations (p. 644) Operations which *undo* each other. For example, addition and subtraction are inverse operations.

operaciones inversas Operaciones que *se anulan* mutuamente. La adición y la sustracción son operaciones inversas.

isosceles triangle (p. 488) A triangle in which at least two sides are congruent.

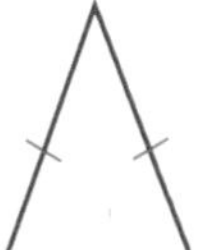

triángulo isósceles Triángulo que tiene por lo menos dos lados congruentes.

K

key (p. 92) A sample data point used to explain the stems and leaves in a stem-and-leaf plot.

clave Punto de muestra de los datos que se usa para explicar los tallos y las hojas en un diagrama de tallo y hojas.

kilogram (p. 437) The base unit of mass in the metric system. One kilogram equals one thousand grams.

kilogramo Unida básica de masa del sistema métrico. Un kilogramo equivale a mil gramos.

kilometer (p. 432) A metric unit of length. One kilometer equals one thousand meters.

kilómetro Unidad métrica de longitud. Un kilómetro equivale a mil metros.

L

least common denominator (LCD) (p. 220) The least common multiple of the denominators of two or more fractions.

mínimo común denominador (mcd) El menor múltiplo común de los denominadores de dos o más fracciones.

least common multiple (LCM) (p. 217) The least of the common multiples of two or more numbers. The LCM of 2 and 3 is 6.

mínimo común múltiplo (mcm) El menor múltiplo común de dos o más números. El mcm de 2 y 3 es 6.

leaves (p. 92) The units digit written to the right of the vertical line in a stem-and-leaf plot.

hojas El dígito de las unidades que se escribe a la derecha de la recta vertical en un diagrama de tallo y hojas.

like fractions (p. 256) Fractions with the same denominator.

fracciones semejantes Fracciones que tienen el mismo denominador.

like terms (p. 637) Terms that contain the same variables.

términus semejantes Terminos que contienen la(s) misma(s) variable(s).

line graph (p. 82) A graph used to show how a set of data changes over a period of time.

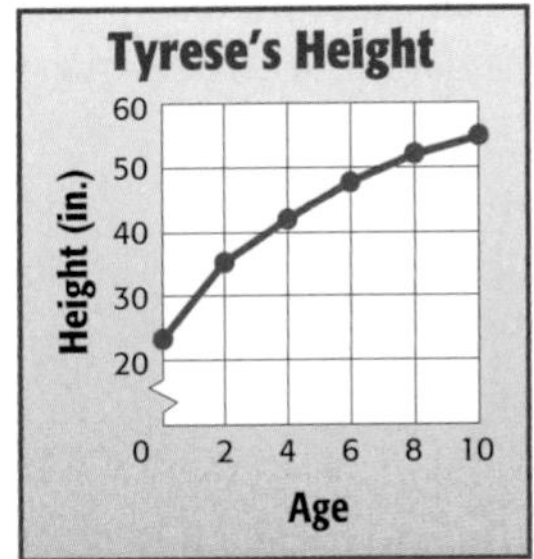

gráfica lineal Gráfica que se usa para mostrar cómo cambian los valores durante un período de tiempo.

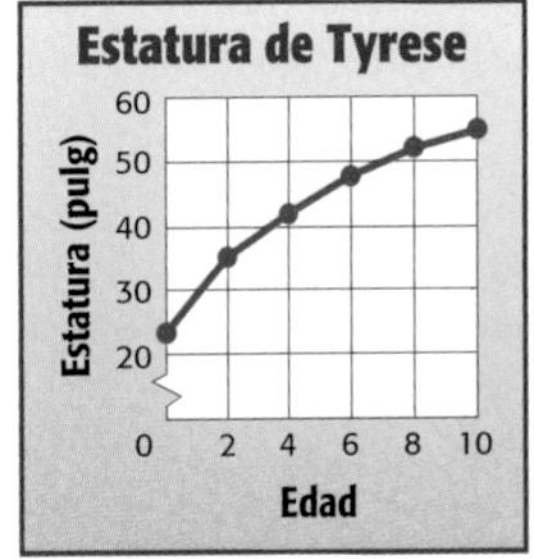

line plot (p. 96) A diagram that shows the frequency of data on a number line.

esquema lineal Diagrama que muestra la frecuencia de los datos sobre una recta numérica.

line segment (p. 488) A straight path between two endpoints.

segmento de recta Línea recta entre dos puntos.

liter (p. 438) The basic unit of capacity in the metric system. A liter is a little more than a quart.

litro Unidad básica de capacidad del sistema métrico. Un litro es un poco más de un cuarto de galón.

mass (p. 437) The amount of material an object contains.

masa Cantidad de materia que contiene un cuerpo.

mean (p. 102) The sum of the numbers in a set of data divided by the number of pieces of data.

media La suma de los números en un conjunto de datos dividida entre el número total de datos.

measures of central tendency (p. 108) Numbers that are used to describe the center of a set of data. These measures include the mean, median, and mode.

medidas de tendencia central Numéros que se usan para describir el centro de un conjunto de datos. Estas medidas incluyen la media, la mediana y la moda.

median (p. 108) The middle number in a set of data when the data are arranged in numerical order. If the data has an even number, the median is the mean of the two middle numbers.

mediana Número central de un conjunto de datos, una vez que los datos han sido ordenados numéricamente. Si hay un número par de datos, la mediana es el promedio de los dos datos centrales.

meter (p. 432) The base unit of length in the metric system. One meter equals one-thousandth of a kilometer.

metro Unidad básica de longitud del sistema métrico. Un metro equivale a la milésima parte de un kilómetro.

metric system (p. 432) A decimal system of weights and measures. The meter is the base unit of length, the kilogram is the base unit of weight, and the liter is the base unit of capacity.

sistema métrico Sistema decimal de pesos y medidas. El metro es la unidad básica de longitud, el kilogramo es la unidad básica de masa y el litro es la unidad básica de capacidad.

mile (p. 418) A customary unit of length equal to 5,280 feet or 1,760 yards.

milla Unidad de longitud del sistema inglés que equivale a 5,280 pies ó 1,760 yardas.

milligram (p. 437) A metric unit of mass. One milligram equals one-thousandth of a gram.

miligramo Unidad métrica de masa. Un miligramo equivale a la milésima parte de un gramo.

milliliter (p. 438) A metric unit of capacity. One milliliter equals one-thousandth of a liter.

mililitro Unidad métrica de capacidad. Un mililitro equivale a la milésima parte de un litro.

millimeter (p. 432) A metric unit of length. One millimeter equals one-thousandth of a meter.

milímetro Unidad métrica de longitud. Un milímetro equivale a la milésima parte de un metro.

mixed number (p. 209) The sum of a whole number and a fraction. $1\frac{1}{2}$, $2\frac{3}{4}$, and $4\frac{5}{8}$ are mixed numbers.

número mixto La suma de un número entero y una fracción. $1\frac{1}{2}$, $2\frac{3}{4}$, y $4\frac{5}{8}$ son números mixtos.

mode (p. 108) The number(s) or item(s) that appear most often in a set of data.

moda Número(s) de un conjunto de datos que aparece(n) más frecuentemente.

multiple (p. 216) The product of the number and any whole number.

múltiplo El producto de un número y cualquier número entero.

N

negative number (p. 121) A number that is less than zero.

número negativo Número menor que cero.

net (p. 554) A two-dimensional figure that can be used to build a three-dimensional figure.

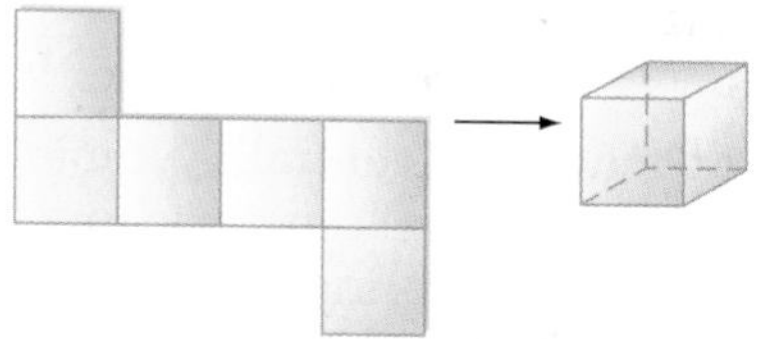

red Figura bidimensional que sirve para hacer una figura tridimensional.

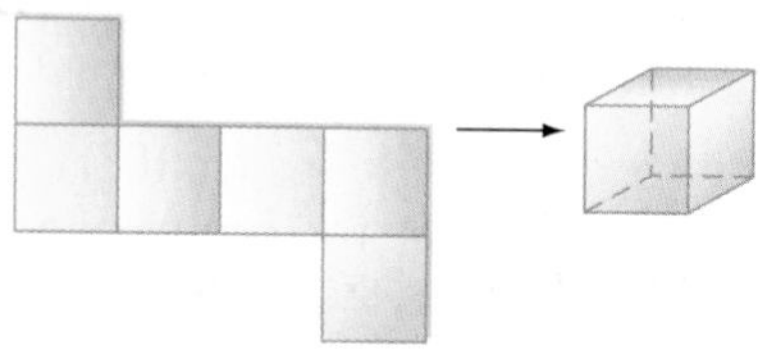

numerical expression (p. 37) A combination of numbers and operations.

expresión numérica Una combinación de números y operaciones.

O

obtuse angle (p. 471) An angle that measures greater than 90° but less than 180°.

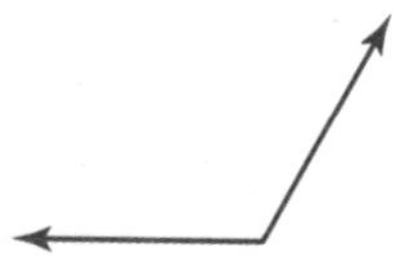

ángulo obtuso Ángulo que mide más de 90° pero menos de 180°.

obtuse triangle (p. 486) A triangle having one obtuse angle.

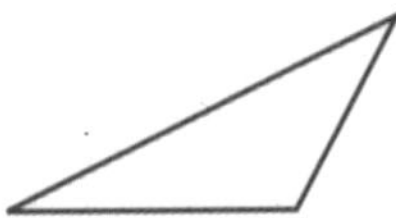

triángulo obtusángulo Triángulo que tiene un ángulo obtuso.

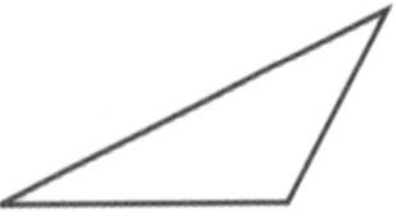

odds (p. 385) A ratio that compares the number of ways an event can occur to the number of ways the event cannot occur.

posibilidades Proporción en la que se compara el número de veces en que un evento puede ocurrir, con el número de veces en que el evento no puede ocurrir.

opposites (p. 121) Numbers that are the same distance from zero in opposite directions.

opuestos Números ubicados a una misma distancia del cero, pero en direcciones opuestas.

ordered pair (p. 233) A pair of numbers used to locate a point in the coordinate system. The ordered pair is written in this form: (x-coordinate, y-coordinate).

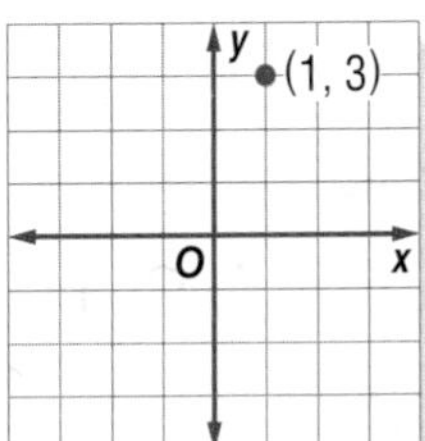

par ordenado Par de números que se utiliza para ubicar un punto en un plano de coordenadas. Se escribe de la siguiente forma: (coordenada x, coordenada y).

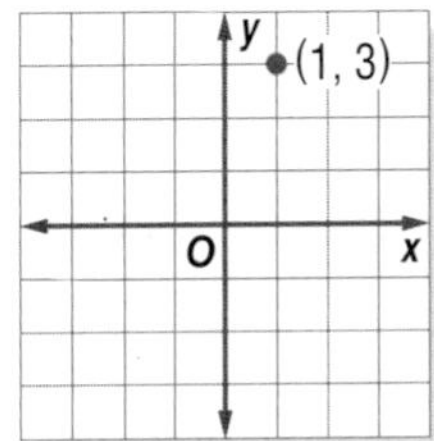

order of operations (p. 37) The rules that tell which operation to perform first when more than one operation is used.

1. Simplify the expressions inside grouping symbols, like parentheses.
2. Find the value of all powers.
3. Multiply and divide in order from left to right.
4. Add and subtract in order from left to right.

orden de las operaciones Reglas que establecen cuál operación debes realizar primero, cuando hay más de una operación involucrada.

1. Primero ejecuta todas las operaciones dentro de los símbolos de agrupamiento.
2. Evalúa todas las potencias.
3. Multiplica y divide en orden de izquierda a derecha.
4. Suma y resta en orden de izquierda a derecha.

origin (p. 233) The point of intersection of the x-axis and y-axis in a coordinate system.

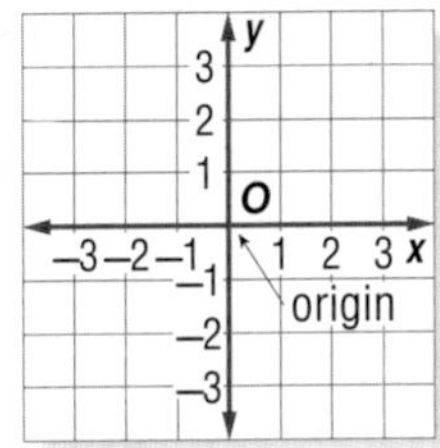

origen Punto en que el eje x y el eje y se intersecan en el sistema de coordenadas.

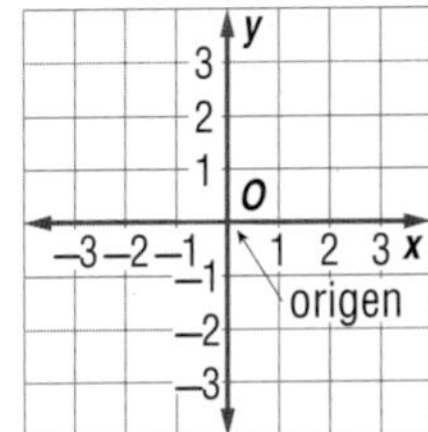

ounce (p. 425) A customary unit of weight. 16 ounces equals one pound.

onza Unidad de peso del sistema inglés de medidas. 16 onzas equivalen a una libra.

outcomes (p. 381) Possible results of a probability event. For example, 4 is an outcome when a number cube is rolled.

resultado Uno de los resultados posibles de un evento probabilístico. Por ejemplo, 4 es un resultado posible cuando se lanza un dado.

outlier (p. 103) A value that is much higher or much lower than the other values in a set of data.

valor atípico Dato que se encuentra muy separado de los otros valores en un conjunto de datos.

output (p. 47) The resulting value when a function machine performs one or more operations on an input value.

salida Valor que se obtiene cuando una máquina realiza una o más operaciones en un valor de entrada.

P

parallelogram (p. 495) A quadrilateral that has both pairs of opposite sides congruent and parallel.

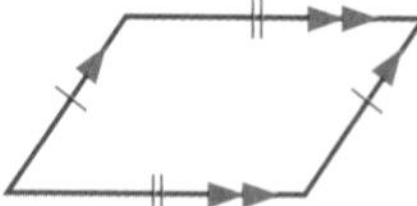

paralelogramo Cuadrilátero con ambos pares de lados opuestos paralelos y congruentes.

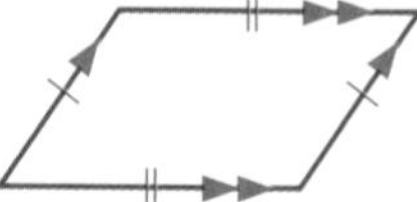

percent (p. 365) A ratio that compares a number to 100.

por ciento Razón en que se compara un número a 100.

perimeter (p. 522) The distance around any closed geometric figure.

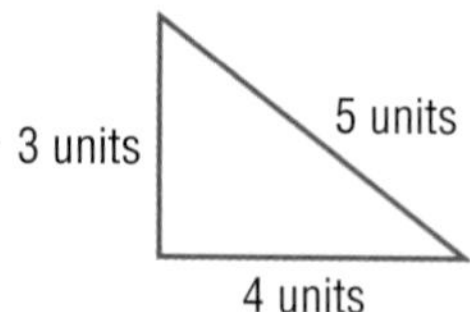

$P = 3 + 4 + 5 = 12$ units

perímetro La distancia alrededor de una figura geométrica cerrada.

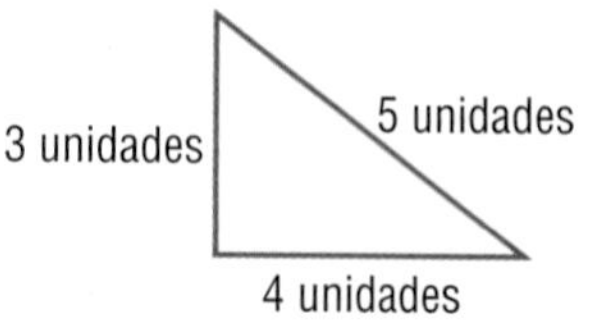

$P = 3 + 4 + 5 = 12$ unidades

pint (p. 424) A customary unit of capacity equal to two cups.

pinta Unidad de capacidad del sistema inglés de medidas que equivale a dos tazas.

polygon (p. 497) A simple closed figure in a plane formed by three or more line segments.

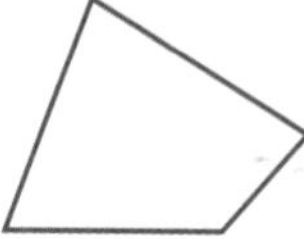

polígono Figura simple cerrada en un plano, formada por tres o más segmentos de recta.

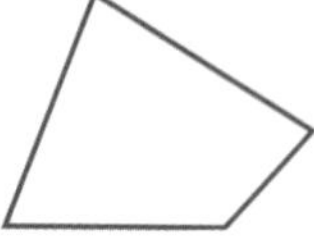

population (p. 394) The entire group of items or individuals from which the samples under consideration are taken.

población El grupo total de individuos o de artículos del cual se toman las muestras bajo estudio.

positive number (p. 121) A number that is greater than zero.

número positivo Número mayor que cero.

pound (p. 425) A customary unit of weight equal to 16 ounces.

libra Unidad de peso del sistema inglés de medidas que equivale a 16 onzas.

power (p. 33) A number that can be expressed using an exponent. The power 3^2 is read *three to the second power*, or *three squared*.

potencias Números que se expresan usando exponentes. La potencia 3^2 se lee *tres a la segunda potencia* o *tres al cuadrado*.

prime factorization (p. 29) A composite number expressed as a product of prime numbers.

factorización prima Un número compuesto expresado como el producto de números primos.

prime number (p. 28) A whole number that has exactly two unique factors, 1 and the number itself.

número primo Número entero que tiene exactamente dos factores, 1 y sí mismo.

probability (p. 381) The chance that some event will occur.

probabilidad Posibilidad de que suceda algún evento.

proper fraction (p. 209) A fraction that has a numerator less than the denominator.

fracción propia Fracción cuyo numerador es menor que el denominador.

proportion (p. 329) An equation stating that two ratios or rates are equivalent. $\frac{a}{b} = \frac{c}{d}$, $b \neq 0, d \neq 0$.

proporción Ecuación que establece la igualdad de dos razones o tasas. $\frac{a}{b} = \frac{c}{d}$, $b \neq 0, d \neq 0$.

proportional (p. 329) Two quantities are proportional if they have a constant ratio or rate.

proporcional Dos cantidades son proporcionales si poseen una tasa o razón constante.

Q

quadrants (p. 659) The four regions in a coordinate plane separated by the x-axis and y-axis.

cuadrantes Las cuatro regiones de un plano de coordenadas separadas por el eje x y el eje y.

quadrilateral (p. 494) A polygon with four sides.

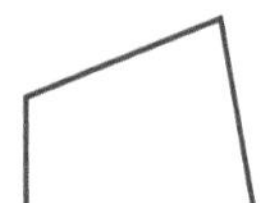

cuadrilátero Un polígono con cuatro lados.

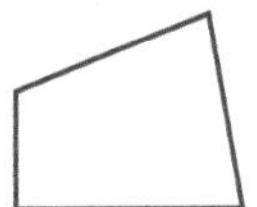

quart (p. 424) A customary unit of capacity equal to two pints.

cuarto de galón Unidad de capacidad del sistema inglés de medidas que equivale a dos pintas.

R

radius (p. 528) The distance from the center of a circle to any point on the circle.

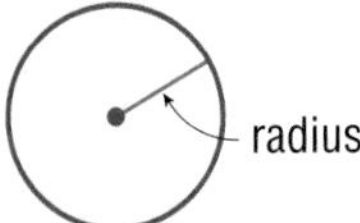

radio Distancia desde el centro de un círculo hasta cualquier punto del mismo.

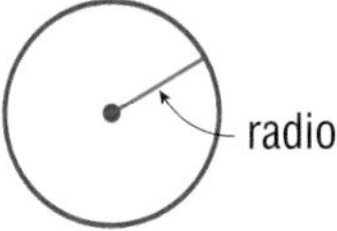

random (p. 382) Outcomes occur at random if each outcome is equally likely to occur.

aleatorio Los resultados ocurren al azar si la posibilidad de ocurrir de cada resultado es equiprobable.

range (p. 109) The difference between the greatest number and the least number in a set of data.

rango La diferencia entre el número mayor y el número menor en un conjunto de datos.

rate (p. 315) A ratio comparing two quantities with different kinds of units.

tasa Razón que compara dos cantidades que tienen diferentes tipos de unidades.

ratio (p. 314) A comparison of two quantities by division. The ratio of 2 to 3 can be stated as 2 out of 3, 2 to 3, 2:3, or $\frac{2}{3}$.

razón Comparación de dos cantidades mediante división. La razón de 2 a 3 puede escribirse como 2 de cada 3, 2 a 3, 2:3 o $\frac{2}{3}$.

rational number (p. 225) A number than can be written as a fraction.

número racional Número que se puede expresar como fracción.

ratio table (p. 322) A table with columns filled with pairs of numbers that have the same ratio.

tabla de razones Tabla cuyas columnas contienen pares de números que tienen una misma razón.

reciprocals (p. 293) Any two numbers that have a product is 1. Since $\frac{5}{6} \times \frac{6}{5} = 1$, $\frac{5}{6}$ and $\frac{6}{5}$ are reciprocals.

recíproco Cualquier par de números cuyo producto es 1. Como $\frac{5}{6} \times \frac{6}{5} = 1$, $\frac{5}{6}$ y $\frac{6}{5}$ son recíprocos.

rectangle (p. 495) A quadrilateral with opposite sides congruent and parallel and all angles are right angles.

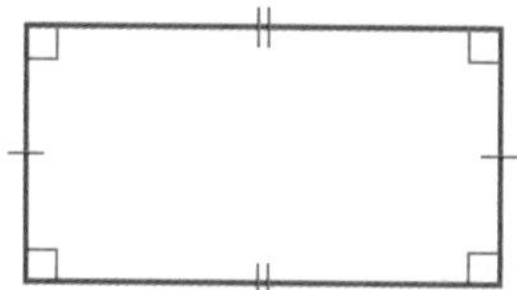

rectángulo Cuadrilátero cuyos lados opuestos son congruentes y paralelos y cuyos ángulos son todos ángulos rectos.

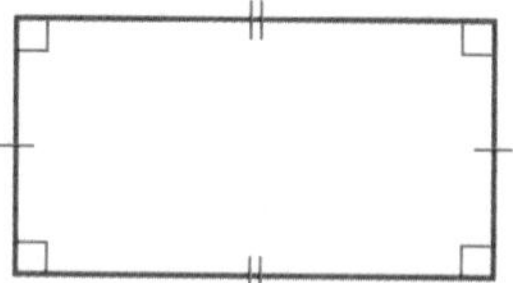

rectangular prism (p. 548) A three-dimensional figure that has two parallel and congruent bases in the shape of polygons and at least three lateral faces shaped like rectangles. The shape of the bases tells the name of the prism.

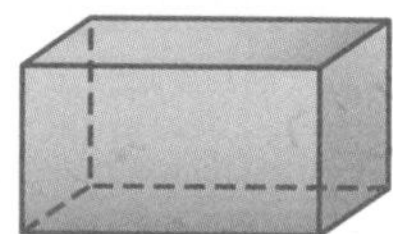

prisma rectangular Figura tridimensional que tiene dos bases paralelas y congruentes en forma de polígonos y posee por lo menos tres caras laterales en forma de rectángulos. La forma de las bases indica el nombre del prisma.

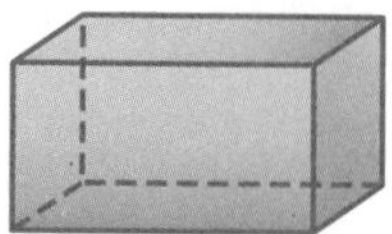

reflection (p. 610) The mirror image that is created when a figure is flipped over a line.

reflexión Figura que se vuelca sobre una linea para crear una imagen especular de la figura.

rhombus (p. 494) A parallelogram with all sides congruent.

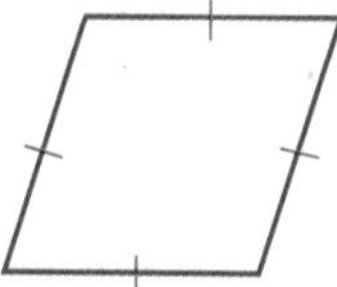

rombo Paralelogramo cuyos lados son todos congruentes.

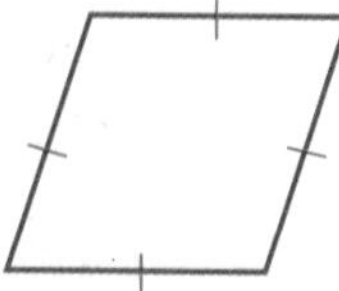

right angle (p. 471) An angle that measures 90°.

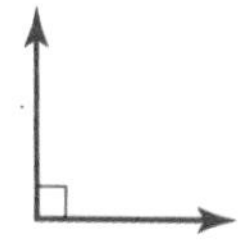

ángulo recto Ángulo que mide 90°.

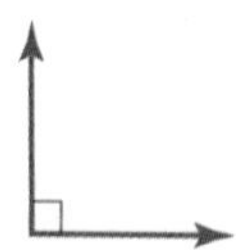

right triangle (p. 486) A triangle having one right angle.

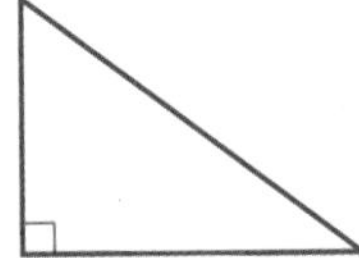

triángulo rectángulo Triángulo que tiene un ángulo recto.

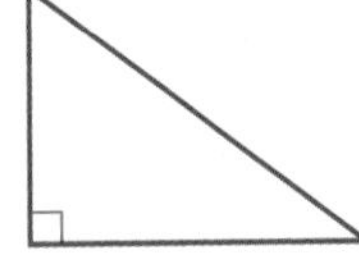

rotation (p. 615) A rotation occurs when a figure is rotated about a point.

rotación Rotar una figura alrededor de un punto.

rotational symmetry (p. 617) A figure has rotational symmetry if it can be rotated a certain number of degrees about its center and still look like the original figure.

simetria rotacional Una figura posee simetría rotacional si se puede girar menos de 360° en torno a su centro sin que esto cambia su apariencia con respecto a la figura original.

S

sample (p. 394) A randomly-selected group that is used to represent a whole population.

muestra Grupo escogido al azar o aleatoriamente y que se usa para representar la población entera.

sample space (p. 389) The set of all possible outcomes.

espacio muestral Conjunto de todos los resultados posibles.

scale (p. 81) The set of all possible values of a given measurement, including the least and greatest numbers in the set, separated by the intervals used.

escala Conjunto de todos los valores posibles de una medida dada, incluyendo el número menor y el mayor del conjunto, separados por los intervalos usados.

scalene triangle (p. 488) A triangle in which no sides are congruent.

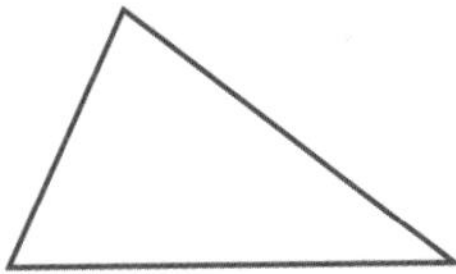

triángulo escaleno Triángulo sin lados congruentes.

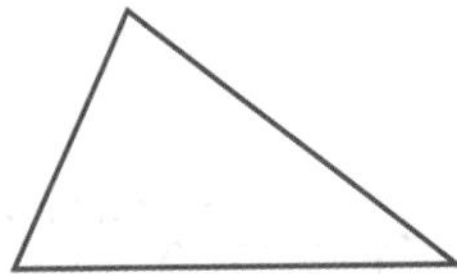

scaling (p. 323) To multiply or divide two related quantities by the same number.

homotecia Multiplicar o dividir dos cantidades relacionadas entre un mismo número.

sequence (p. 343) A list of numbers in a specific order, such as 0, 1, 2, 3, or 2, 4, 6, 8.

sucesión Lista de números en un orden específico como, por ejemplo, 0, 1, 2, 3 ó 2, 4, 6, 8.

side (p. 470) A ray that is part of an angle.

lado Rayo que es parte de un ángulo.

similar figures (p. 502) Figures that have the same shape but different sizes.

figuras semejantes Figuras que tienen la misma forma, pero diferente tamaño.

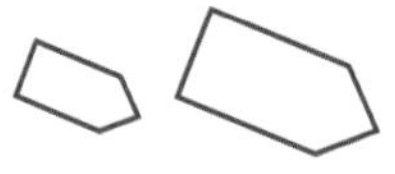

simple event (p. 381) A specific outcome or type of outcome.

evento simple Resultado específico o tipo de resultado.

simplest form (p. 205) The form of a fraction when the GCF of the numerator and the denominator is 1. The fraction $\frac{3}{4}$ is in simplest form because the GCF of 3 and 4 is 1.

forma reducida La forma de una fracción cuando el MCD del numerador y del denominador es 1. La fracción $\frac{3}{4}$ está en forma reducida porque el MCD de 3 y 4 es 1.

solution (p. 57) The value of a variable that makes an equation true. The solution of $12 = x + 7$ is 5.

solución Valor de la variable de una ecuación que hace verdadera la ecuación. La solución de $12 = x + 7$ es 5.

solve (p. 57) To replace a variable with a value that results in a true sentence.

resolver Reemplazar una variable con un valor que resulte en un enunciado verdadero.

square (p. 495) A parallelogram with all sides congruent, all angles right angles, and opposite sides parallel.

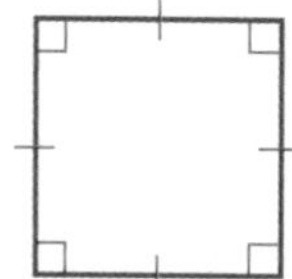

cuadrado Paralelogramo con todos los lados congruentes, todos los ángulos rectos y lados opuestos paralelos.

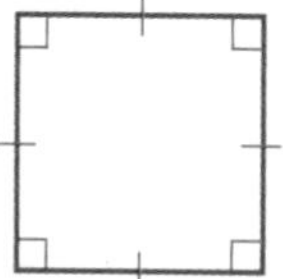

squared (p. 33) A number multiplied by itself; 4×4, or 4^2.

cuadrado El producto de un número multiplicado por sí mismo; 4×4, o 4^2.

standard form (p. 139) Numbers written without exponents.

forma estándar Números escritos sin exponentes.

stem-and-leaf plot (p. 92) A system used to condense a set of data where the greatest place value of the data forms the stem and the next greatest place value forms the leaves.

Stem	Leaf
2	3 8
3	8 9
4	1 4 5 6 7 9
5	0 0 5 8
6	3 $4|1 = 41$

diagrama de tallo y hojas Sistema que se usa para condensar un conjunto de datos y en el cual el mayor valor de posición de los datos forma el tallo y el siguiente valor de posición mayor forma las hojas.

Tallo	Hojas
2	3 8
3	8 9
4	1 4 5 6 7 9
5	0 0 5 8
6	3 $4|1 = 41$

stems (p. 92) The greatest place value common to all the data that is written to the left of the line in a stem-and-leaf plot.

tallo El mayor valor de posición común a todos los datos que se escribe a la izquierda de la línea en un diagrama de tallo y hojas.

straight angle (p. 471) An angle that measures exactly 180°.

ángulo llano Ángulo que mide exactamente 180°.

Subtraction Property of Equality (p. 645) If you subtract the same number from each side of an equation, the two sides remain equal.

propiedad de sustracción de la igualdad Si sustraes el mismo número de ambos lados de una ecuación, los dos lados permanecen iguales.

supplementary angles (p. 480) Two angles are supplementary if the sum of their measures is 180°.

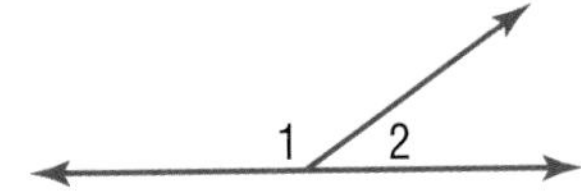

∠1 and ∠2 are supplementary angles.

ángulos suplementarios Dos ángulos son suplementarios si la suma de sus medidas es 180°.

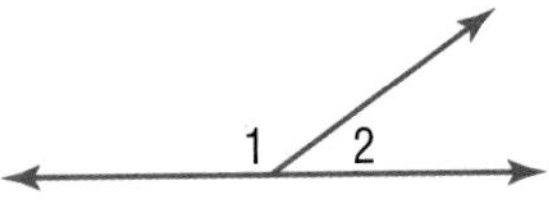

∠1 y ∠2 son suplementarios.

surface area (p. 555) The sum of the areas of all the surfaces (faces) of a three-dimensional figure.

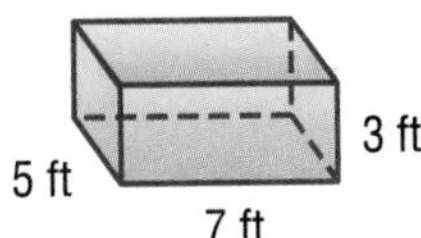

$S = 2(7 \times 5) + 2(7 \times 3) + 2(5 \times 3)$
$\cdots = 142$ square feet

área de superficie La suma de las áreas de todas las superficies (caras) de una figura tridimensional.

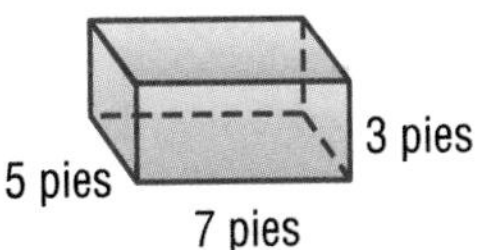

$S = 2(7 \times 5) + 2(7 \times 3) + 2(5 \times 3)$
$\cdots = 142$ pies cuadrados

survey (p. 394) A question or set of questions designed to collect data about a specific group of people.

encuesta Pregunta o conjunto de preguntas diseñadas para recoger datos sobre un grupo específico de personas.

T

temperature (p. 455) The measure of the hotness or coldness of an object or environment.

temperatura Medida del grado de calor o frío de un cuerpo o ambiente.

term (p. 343) Each number in a sequence.

término Cada uno de los números de una sucesión.

tessellation (p. 508) A pattern formed by repeating figures that fit together without gaps or overlaps.

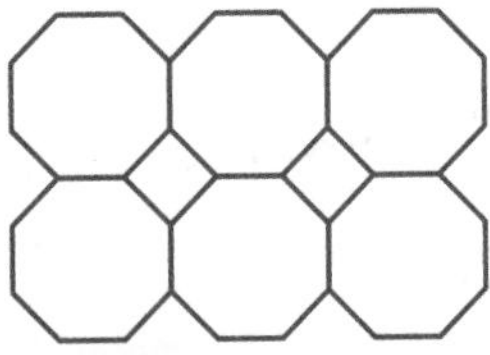

teselado Patrón formado por figuras repetidas que no se traslapan y que no dejan espacios entre sí.

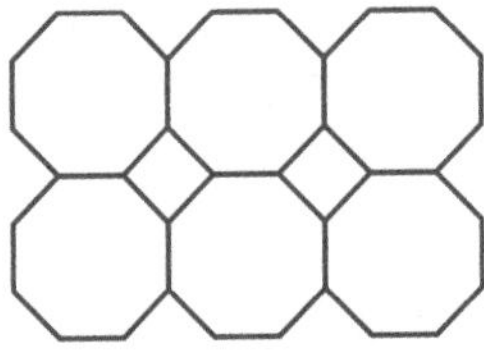

theoretical probability (p. 387) A probability based on what should happen under perfect conditions.

probabilidad teórica Probabilidad basada en los resultados que se deben obtener bajo condiciones perfectas.

ton (p. 425) A customary unit of weight equal to 2,000 pounds.

tonelada Unidad de peso del sistema inglés que equivale a 2,000 libras.

translation (p. 604) Sliding a figure without turning it.

traslación Transformatión en que una figura se desliza horizontal o verticalmente o de ambas maneras.

transformation (p. 604) A movement of a geometric figure.

transformación Movimiento de una figura geométrica.

trapezoid (p. 495) A quadrilateral with one pair of opposite sides parallel.

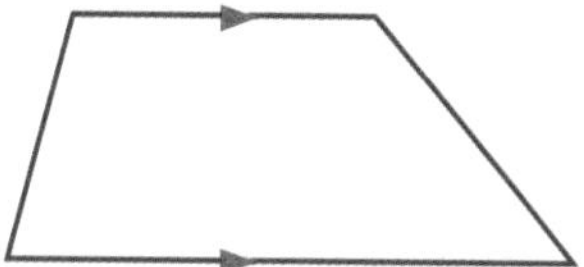

trapecio Cuadrilátero que tiene un solo par de lados opuestos paralelos.

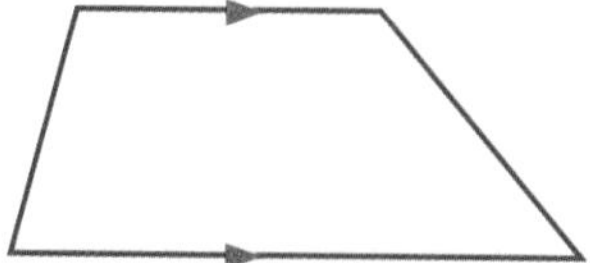

tree diagram (p. 390) A diagram used to show the total number of possible outcomes in a probability experiment.

Spinner	Coin	Outcome
red (R)	heads (H)	RH
	tails (T)	RT
green (G)	heads (H)	GH
	tails (T)	GT

diagrama de árbol Diagrama que se usa para mostrar el número total de resultados posibles en un experimento probabilístico.

Girador	Moneda	Salida
rojo (R)	cara (C)	RC
	escudo (E)	RE
verde (V)	cara (C)	VC
	escudo (E)	VE

unlike fractions (p. 263) Fractions with different denominators.

unit rate (p. 315) A rate that has a denominator of 1.

fracciones con distinto numerador Fracciones que tienen diferentes denominadores.

tasa unitaria Una tasa con un denominador de 1.

variable (p. 42) A symbol, usually a letter, used to represent a number.

Venn diagram (p. 197) A diagram that uses circles to display elements of different sets. Overlapping circles show common elements.

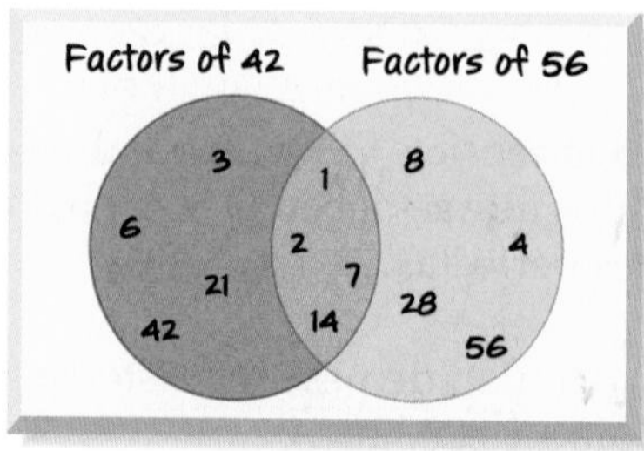

vertex (p. 470) The common endpoint of the two rays that form an angle.

variable Un símbolo, por lo general, una letra, que se usa para representar un número.

diagrama de Venn Diagrama que usa círculos para mostrar elementos de diferentes conjuntos. Círculos sobrepuestos indican elementos comunes.

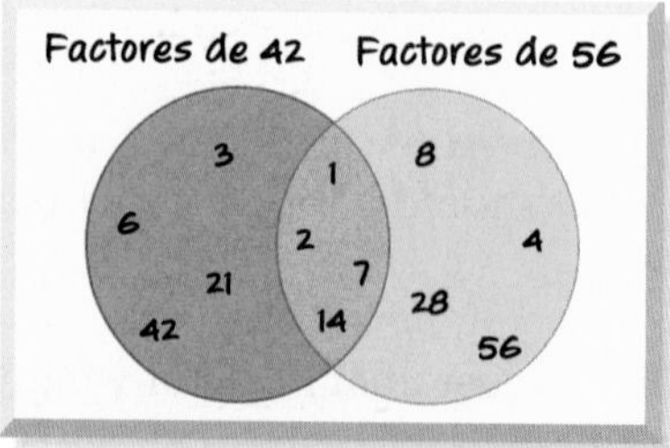

vértice El extremo común de dos rayos que forman un ángulo.

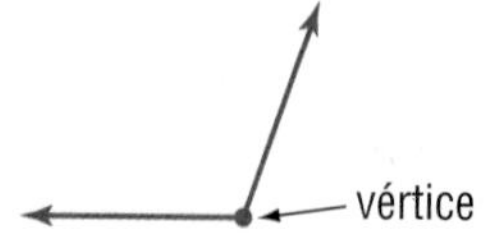

vertical angles (p. 479) Nonadjacent angles formed by a pair of lines that intersect.

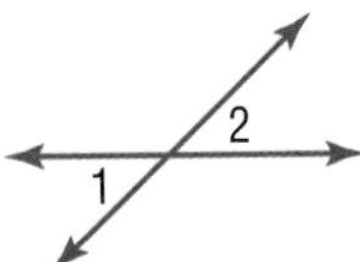

ángulos opuestos por el vértice Ángulos no adyacentes formados por dos rectas que se intersecan.

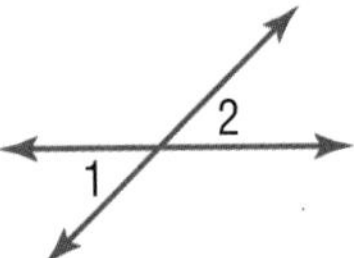

vertical axis (p. 81) The axis on which the scale and interval are shown in a bar or line graph.

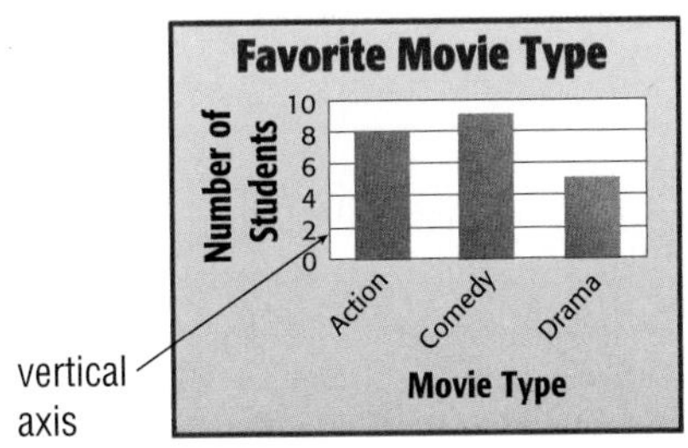

eje vertical Eje sobre el cual se muestran la escala y el intervalo en una gráfica de barras o en una gráfica lineal.

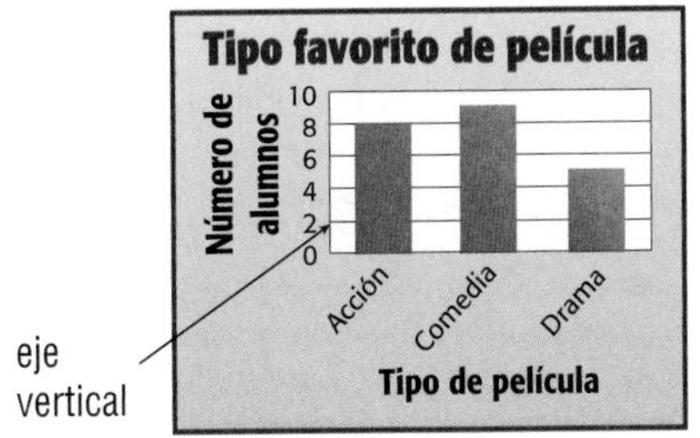

volume (p. 548) The amount of space that a three-dimensional figure contains. Volume is expressed in cubic units.

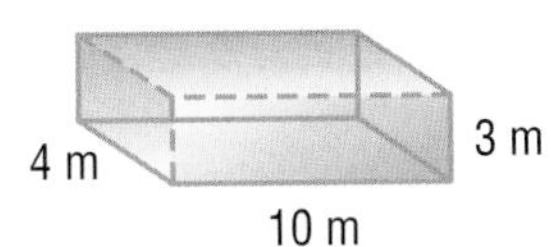

$V = 10 \times 4 \times 3 = 120$ cubic meters

volumen Cantidad de espacio que contiene una figura tridimensional. El volumen se expresa en unidades cúbicas.

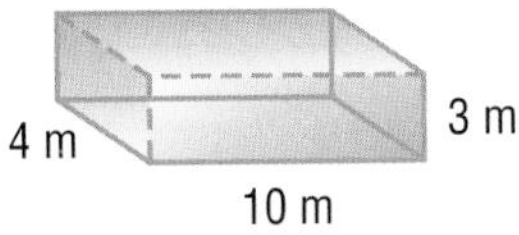

$V = 10 \times 4 \times 3 = 120$ metros cúbicos

X

***x*-axis** (p. 233) The horizontal line of the two perpendicular number lines in a coordinate plane.

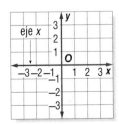

eje *x* La recta horizontal de las dos rectas numéricas perpendiculares en un plano de coordenadas.

***x*-coordinate** (p. 233) The first number of an ordered pair.

coordenada *x* El primer número de un par ordenado.

Y

yard (p. 418) A customary unit of length equal to 3 feet, or 36 inches.

yarda Unidad de longitud del sistema inglés que equivale a 3 pies o 36 pulgadas.

***y*-axis** (p. 233) The vertical line of the two perpendicular number lines in a coordinate plane.

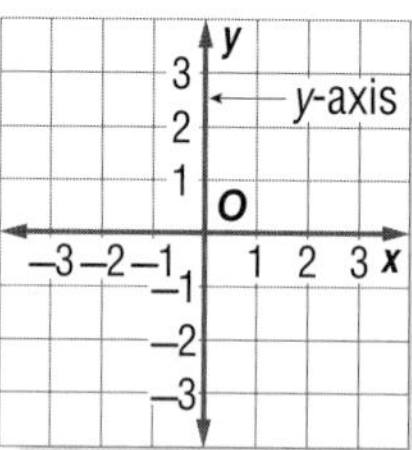

eje *y* La recta vertical de las dos rectas numéricas perpendiculares en un plano de coordenadas.

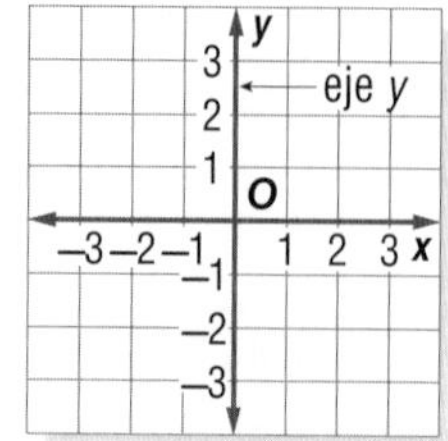

***y*-coordinate** (p. 233) The second number of an ordered pair.

coordenada *y* El segundo número de un par ordenado.

Z

zero pair (p. 576) The result when one positive counter is paired with one negative counter.

par nulo Resultado que se obtiene cuando una ficha positiva se aparea con una ficha negativa.

Selected Answers

Chapter 1 Number Patterns and Functions

Page 23 **Chapter 1** **Getting Ready**

1. 212 **3.** 109 **5.** 175 **7.** 36 **9.** 94 **11.** 26 **13.** \$34 **15.** 540 **17.** 918 **19.** 1,034 **21.** 14 **23.** 73 **25.** 63

Pages 26–27 **Lesson 1-1**

1. 750 lb **3.** 1,592 mi **5.** 29, 35, 41 **7.** \$16,800 **9.** 360 **11.** B **13.** 14 **15.** 7

Pages 30–31 **Lesson 1-2**

1. composite **3.** neither **5.** $2 \times 2 \times 3 \times 3$ **7.** 5×13 **9.** 2×23 **11.** neither **13.** composite **15.** composite **17.** prime **19.** composite **21.** prime **23.** $2 \times 3 \times 3$ **25.** $3 \times 5 \times 5$ **27.** $2 \times 2 \times 2 \times 2 \times 2$ **29.** 5×5 **31.** $2 \times 2 \times 2 \times 13$ **33.** 7×11 **35.** 81 **37.** all **39.** composite **41.** prime **43.** Sample answer: 2 packs with 10 cards each, 4 packs with 5 cards each, 5 packs with 4 cards each **45.** Sample answer: 7, 23, and 29; $7 + 23 + 29 = 59$ **47.** 2; It is a prime number because it has two factors, 1 and itself. **49.** C **51.** D **53.** 6 h **55.** 25 **57.** 1,000

Pages 34–36 **Lesson 1-3**

1. 2^4 **3.** $2 \times 2 \times 2 \times 2 \times 2 \times 2$; 64 **5.** 243 **7.** $2^2 \times 5$ **9.** $2 \times 3^2 \times 5$ **11.** 8^4 **13.** 5^5 **15.** 7^6 **17.** 3×3; 9 **19.** $10 \times 10 \times 10 \times 10 \times 10$; 100,000 **21.** $6 \times 6 \times 6 \times 6 \times 6$; 7,776 **23.** $1 \times 1 \times 1 \times 1 \times 1 \times 1 \times 1$; 1 **25.** 256 lb **27.** $2^3 \times 7$ **29.** $2^2 \times 17$ **31.** 2×7^2 **33.** $2 \times 3^3 \times 7$ **35.** 7×7; 49 **37.** $4 \times 4 \times 4 \times 4 \times 4$; 1,024 **39.** 20^3; 8,000 **41.** 3^5; $3^5 = 243$ and $5^3 = 125$ because the base in 3^5 is used as a factor more times than in 5^3 **43.** The next value is found by dividing the previous power by 3; 1. **45.** The next value is found by dividing the previous power by 10; 10; 1. **47.** A **49.** C **51.** P **53.** 86,400 s **55.** 9 **57.** 20

Pages 39–40 **Lesson 1-4**

1. 7 **3.** 47 **5.** 29 **7.** $4 \times \$24 + 2 \times \$10$; \$116 **9.** 6 **11.** 13 **13.** 61 **15.** 199 **17.** 117 **19.** 35 **21.** 99 **23.** $5 \times \$7 + 5 \times \$3 + 5 \times \$2$; \$60 **25.** 22 **27.** 38 **29.** $7 \times 6 - 2$; 40 **31.** Sample answer: $25 \div 5 + 10 \div 2$ **33.** Sample answer: At the basketball game, Jared made 7 field goals and 3 free throws. If field goals are two points and free throws are 1 point, find the total number of points Jared scored; 17 **35.** 256 **37.** $5 \times 5 \times 3$ **39.** $2 \times 5 \times 13$ **41.** 39 **43.** 60

Pages 44–46 **Lesson 1-5**

1. 7 **3.** 5 **5.** 14 **7.** B **9.** 24 **11.** 6 **13.** 6 **15.** 12 **17.** 18 **19.** 1 **21.** 4 **23.** 21 **25.** 18 **27.** 127 mph **29.** 180 **31.** 12 **33.** 72 **35.** 48 **37.** 34 **39.** 24 **41.** 6 **43.** 112 ft^2 **45.** $9 \times c \div 5 + 32$; 18°F **47.** yes; 30 **49.** 6 + 8; It contains no variable. **51.** B **53.** J **55.** 28 **57.** 1,000,000,000 people **59.** 9 **61.** 21

Pages 51–53 **Lesson 1-6**

1.

Input (x)	Output ($x + 3$)
0	3
2	5
4	7

3. $x - 1$ **5.** Let p represent the number of pounds; \$$3p$.

7.

Input (x)	Output ($x + 3$)
0	0
3	1
9	3

9. $x - 5$ **11.** $5x$ **13.** $x \div 2$ **15.** Let g represent the number of guests; $30 \div g$. **17.** $6x + 1$ **19.** Let h represent the number of hours; $6h$; 12 **21.** Let s represent the number of songs; $1s + 10$; \$56 **23.** $8s + 3a$; \$75 **25.** Nadia; 3 less than a number is represented by the expression $x - 3$. **27.** Paper/pencil; she can use paper and pencil to find the value of $43 \cdot 4$ and then subtract 6 from the product. The output is 166. **29.** A **31.** 3 **33.** 72 **35.** 25

Pages 54–55 **Lesson 1-7**

1. Sample answer: when you are trying to find the solution of an equation. **3.** 2 used packages and 2 new packages **5.** Sample answer: 2, 4, 5, and 7 **7.** 610 ft **9.** Sample answer: 7 and 13 **11.** $3 \times 4 + 6 \div 1 = 18$ **13.** 9

Pages 59–60 **Lesson 1-8**

1. 8 **3.** 2 **5.** 12 **7.** 2 **9.** 8 **11.** 10 **13.** 5 **15.** 9 **17.** 6 **19.** 8 **21.** 3 **23.** 5 **25.** 11 **27.** 9 **29.** 5 games **31.** 8 ft **33.** No; replacing x with 3 results in $4 \cdot 3 + 3$ or 15, not 18. **35.** always **37.** False; this is an equation, so both sides of the equation must equal the same value. Therefore $m + 8$ must equal 12 and m can only have one solution, 4. **39.** A **41.** Let n represent the number of weeks; $4n$; \$32 **43.** 18 **45.** 96 **47.** 72

Pages 65–67 **Lesson 1-9**

1. 15 cm^2 **3.** 16 ft^2 **5.** 117 in^2 **7.** 90 in^2 **9.** 512 m^2 **11.** 816 ft^2 **13.** 1,200 cm^2 **15.** 100 in^2 **17.** 484 ft^2 **19.** 49 in^2 **21.** 52 m^2 **25.** 3 guinea pigs **27.** Sample answer: 3 ft by 8 ft and 2 ft by 12 ft **29.** No, the area would not increase by 10 square units. Sample model:

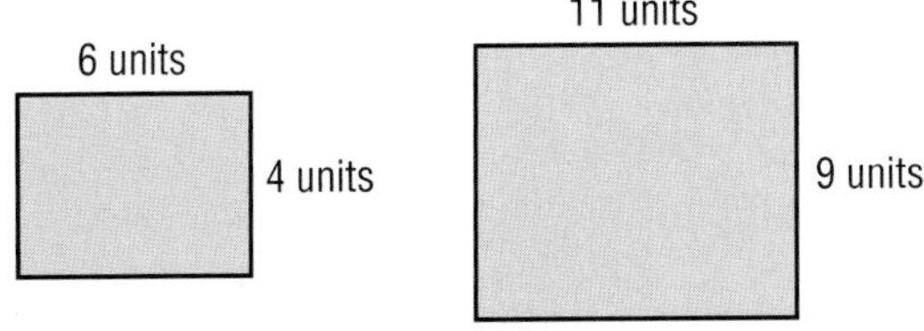

The area of the original rectangle is 24 units^2 and the area of the larger rectangle is 99 units^2. The difference in the areas is 75 units^2. **31.** D **33.** 8 **35.** 28 **37.**

Input (x)	Output ($x \div 2$)
2	1
4	2
8	4

39. 100,000

Pages 68–72 **Chapter 1** **Study Guide and Review**

1. true **3.** false; area **5.** false; function **7.** \$45 **9.** prime **11.** $3 \times 5 \times 5$ **13.** 3, 7, and 13 **15.** 12^3; 1,728 **17.** 30 **19.** 7 **21.** $3 \times 5 + 7$; 22 **23.** 54 **25.** 26 **27.** 456 **29.** 144

31.

Input (x)	Output ($3x$)
0	0
5	15
7	21

33. $x + 6$ **35.** Let h represent the number of hours driven; $60h$ **37.** 27 red cars, 17 black cars **39.** 11 catfish **41.** 5 **43.** 6 **45.** 8 **47.** 26 cm^2

Chapter 2 Statistics and Graphs

Page 77 **Chapter 2** **Getting Ready**

1. 44 **3.** 109 **5.** 81 **7.** \$120 **9.** 12 **11.** 17 **13.** 53 **15.** 13 **17.** 6 **19.** 30

Pages 78–79 **Lesson 2-1**

1. Sample answer: Any time there is a multiple number of data. **3.** Sample answer: You're given a list of people and the songs that they like. The list of songs they like are: D D A B C B D D B A A B A B B D D B B C B B. You are asked what the most popular song is. Make a table that lists the songs, tally marks, and the total for each song. The most popular song was song B.

Favorite Songs		
Song	**Tally**	**Frequency**
A	||||	4
B	~~||||~~ ~~||||~~	10
C	||	2
D	~~||||~~ |	6

5.

Number of Songs	Tally	Frequency
less than 10	~~||||~~ ~~||||~~ ~~||||~~ ||	17
at least 10	~~||||~~ ||	7

7 students

7. 7 cars **9.** Sample answer: 2,000 pieces/day × 6 days/wk × 50 wk/yr × 5 yr or about 3,000,000 pieces

11.

13. 45 students

Pages 83–85 **Lesson 2-2**

1.

Types of Roller Coasters in the United States

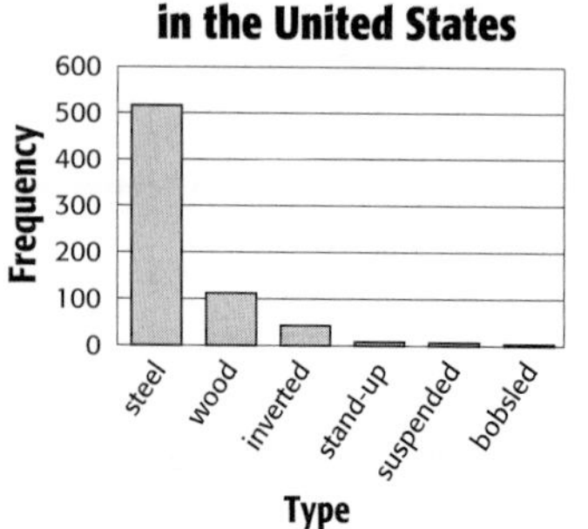

Sample answer: The number of steel coasters is about five times the number of wood coasters.

3.

U.S. Gulf Coast Shoreline

Amount (mi)
8,000
7,000
6,000
5,000
4,000
3,000
2,000
1,000
0
AL FL LA MS TX
State

Florida has about 2,500 miles less shoreline than Louisiana.

5.

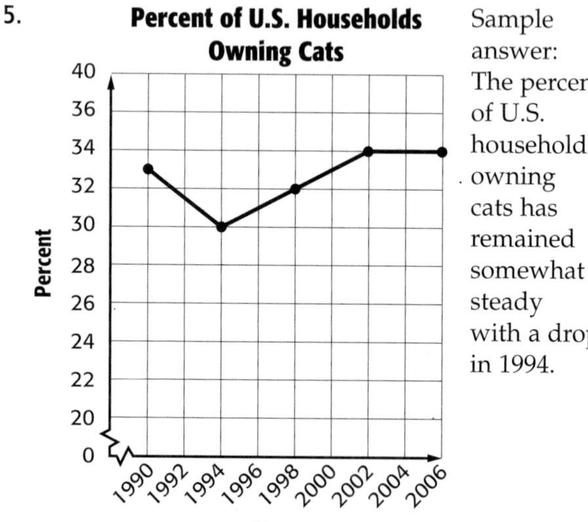

Sample answer: The percent of U.S. households owning cats has remained somewhat steady with a drop in 1994.

7. Sample answer: scale: 50–95; interval: 5

9.

Average Monthly Temperatures Minneapolis, Minnesota

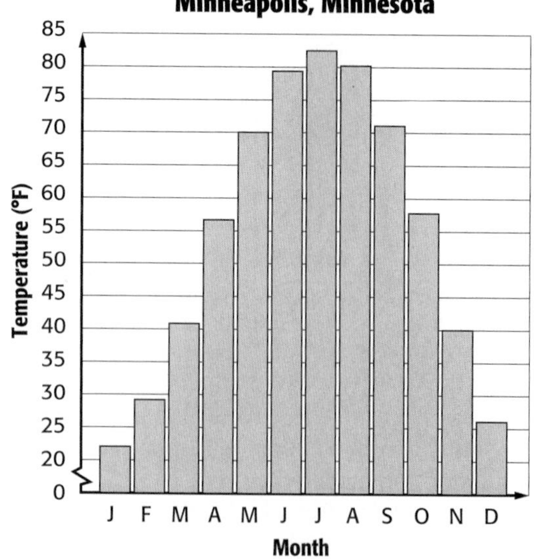

13. Sample answer: Bar graphs and line graphs are ways to display data. Both display the categories on the horizontal axis and the scale on the vertical axis. Whereas a bar graph shows the frequency of each category, a line graph shows how data change over time.

15.

Favorite Color		
Color	**Tally**	**Frequency**
red	𝍸 I	6
green	I	1
blue	𝍸 II	7
purple	𝍸	5
yellow	III	3
pink	II	2

4

17. 84 yd^2 **19.** 88 **21.** 86

Pages 89–91 ***Lesson 2-3***

1. Sample answer: The size of the rainforests is decreasing. **3.** about 50 million acres **5.** Sample answer: The rainforests are decreasing in size. You can expect this pattern to continue. **7.** about 325; sales are increasing by about 25 skateboards per year **9.** Sample answer: Most years, the winning times decreased. From 2003 to 2004, the winning time increased. **11.** 2016 **13.** about 250 mi **15.** about 6 h

17. **Baseball Wins**

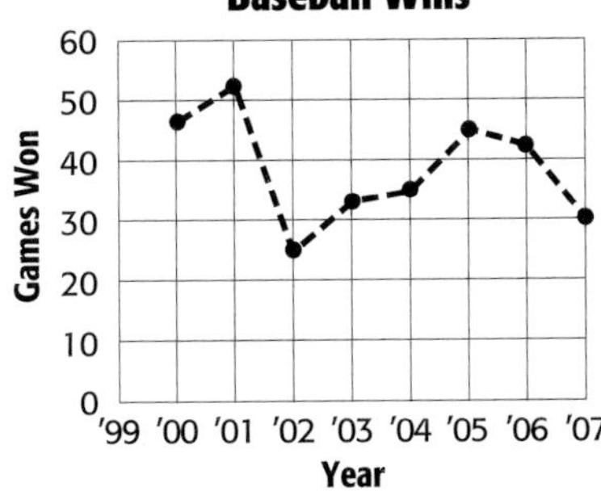

19. There are no obvious trends in the graph. **21.** Sample answer: The point at which the lines cross represent the time when both cars will have traveled the same distance. **23.** B

25. **Depth of Scuba Diver**

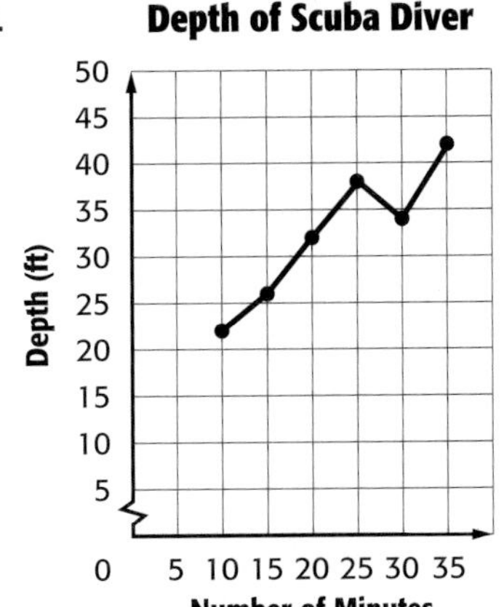

27. 25, 40, 45, 50, 55, 60, 75, 80

Pages 93–95 ***Lesson 2-4***

1. **Minutes Spent on Homework**

Stem	Leaf
2	5 8 9
3	1 1 2 7 9
4	5 6 *3|1 = 31 minutes*

3. 7 **5.** Sample answer: The snack food listed with the least amount of Calories has 240 Calories. Most of the snack foods had fewer than 260 Calories.

7. **Video Game Score**

Stem	Leaf
1	2 5 6
2	2 3
3	0 3 4 5 5 8
4	2
5	3
6	3 4 8

2|3 = 23

9. **Test Scores**

Stem	Leaf
6	3
7	0 1 3 4 5 6 9
8	0 2 2
9	0 1 2 3 3 5

8|2 = 82

11. Sample answer: The tunnels range from 6.3 to 15.2 miles in length. Most of the tunnels were between 6 and 9 miles long. **13.** Sample answer: The least number of shows performed by the top 25 tours was 23, while the most was 99 shows. The majority of the tours performed under 60 shows. **15.** Sample answer: The Tigers had more games in which they scored more than 70 points. The Eagles had more games in which they scored fewer than 70 points. **19.** Sample answer: The data can be easily read. **21.** about 130 min **23.** 1950 to 1960 **25.** 51, 52, 54, 55, 63, 64, 65, 77, 78

Pages 98–100 ***Lesson 2-5***

1. **World's Fastest Wooden Roller Coasters (miles per hour)**

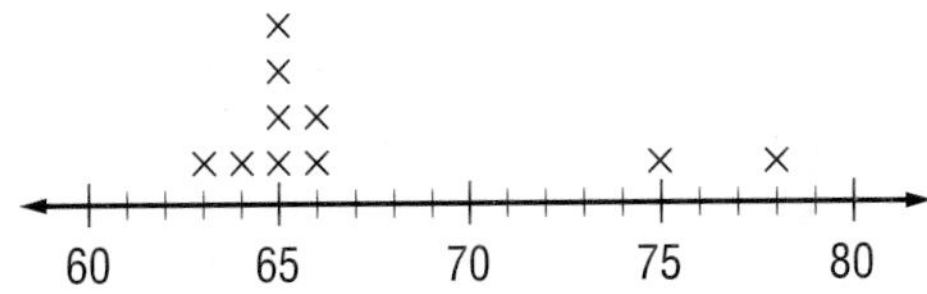

3. 5 **5.** **Test Scores**

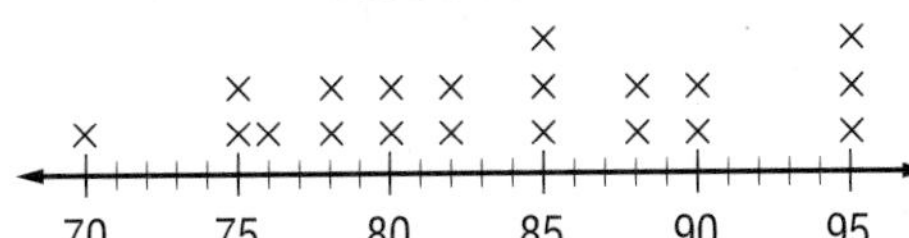

7. 2 **9.** 24 yr **11.** 3 **13.** Sample answer: The majority of meats have 20 to 24 grams of protein. The greatest number of grams of protein on the line plot is 27 grams. **15.** Sample answer: The world's tallest building has 110 stories. Only four buildings on the plot have 100 stories or more. **17.** Sample answer: Line plot; you can identify 29 on the number line and simply count the number of ×'s. **21.** There is a cluster between 11 and 13. **23.** B

25. **Points Scored**

Stem	Leaf
3	8
4	5 5 7 9
5	0 2 5 5 8
6	0 2 4 *4|5 = 45 points*

27. 49.50 s; Based on the trend from 1992 to 2004, the winning time decreased. **29.** 5

Pages 104–106 ***Lesson 2-6***

1. 2 **3.** 11,960 ft **5.** Sample answer: Such an extremely low outlier causes the mean of the data to be considerably less than the average depth of the majority of the oceans. Thus, the mean is less representative of the data. **7.** 7 **9.** 88 **11.** 320 ft **13.** Sample answer: An outlier such as this that is higher than the other values causes the mean of the data to be higher than most of the values in the table. So, the mean is less representative of the data. **15.** \$80 **17.** 120; Explanations will vary. **19.** 15; Explanations will vary. **21.** 19.2°F; The mean high temperature was 60°F and the mean low temperature was 40.8°F, therefore the difference is 19.2°F. **23.** Calculator; Since the data are large, I would use a calculator to find the mean; 296 people. **25.** Sample answer: Pages read: 27, 38, 26, 39, 40 **27.** C
29.

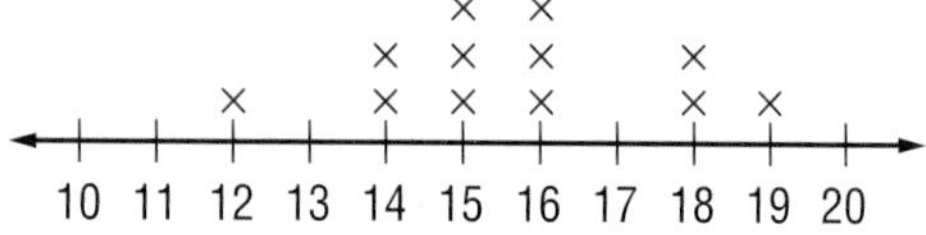

31. 176 ft^2 **33.** 63 **35.** 397

Pages 111–113 ***Lesson 2-7***

1. 17; 17; 10 **3.** mean: \$16; median: \$15.5; mode: \$12; range: \$14 **5.** B **7.** 18; 18; 11 **9.** 25; 24 and 26; 4 **11.** mean: 84; median: 85; mode: 85; range: 35 **13.** mean: 14; median: 12.5; mode: none; range: 18 **15.** Mode; The mode of the temperatures in Louisville is 70°, and the mode for Lexington's temperatures is 76°. Since 76° − 70° = 6°, the mode was used to make this claim. **19.** True; the mode of the data set is 1. **21.** Sample answer: The median or mode best represents the data. The mean, 8, is greater than all but one of the data values. The range will increase. The mode will not be affected. **23.** 4
25.

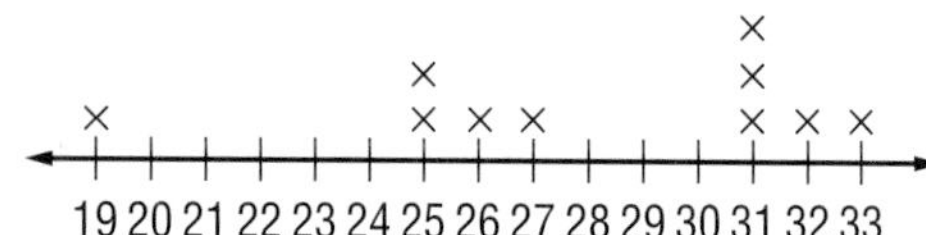

27. 70 **29.** Asia **31.** about 10,000 ft

Pages 116–118 ***Lesson 2-8***

1. stem-and-leaf plot **3.** line graph **5.** bar graph **7.** line graph **9.** bar graph **11.** Sample answer: bar graph

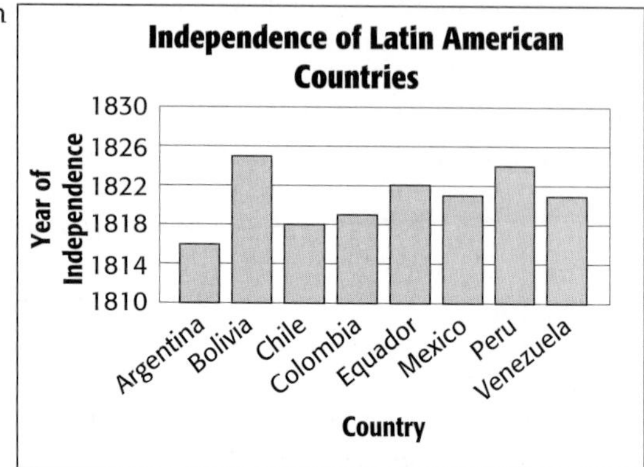

13. Sample answer:

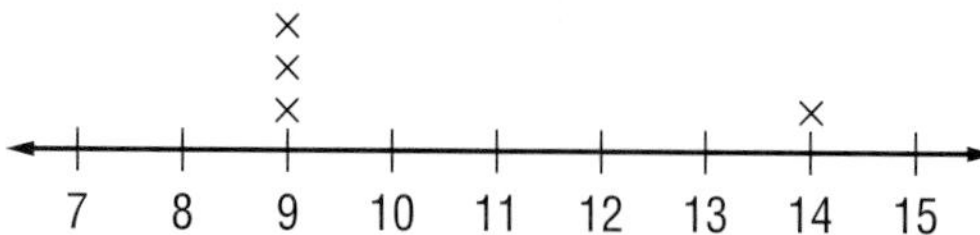

The line plot allows you to easily see how many countries have a given number of neighbors. The bar graph however allows you to see the number of neighbors each given country has. **15.** False; counterexample: to compare the price of five different cell phones, a line graph would not be appropriate because these data do not show change over a period of time **19.** F **21.** 23
23.

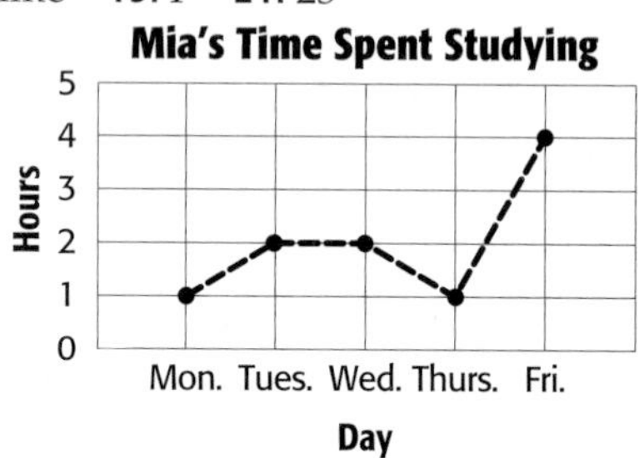

Pages 123–125 ***Lesson 2-9***

1. +3 or 3
3–6.

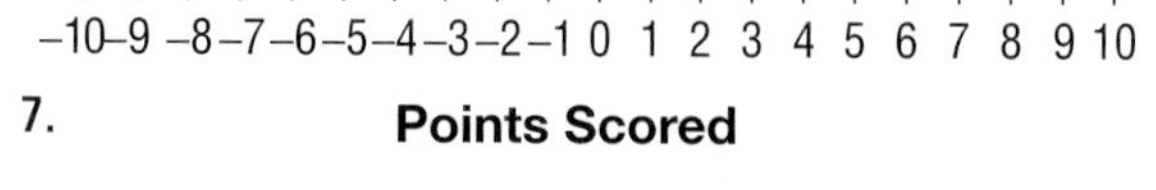

7.

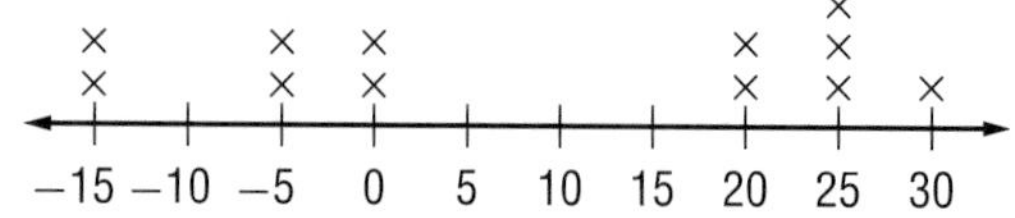

9. +30 or 30 **11.** −6
12–19.

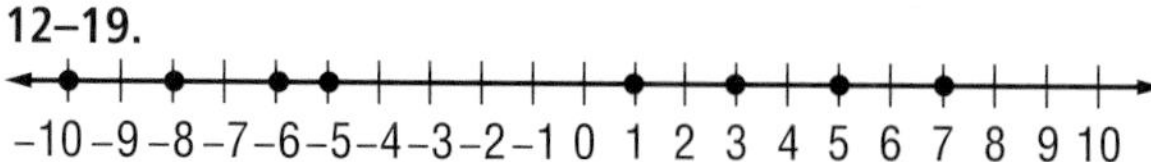

21.

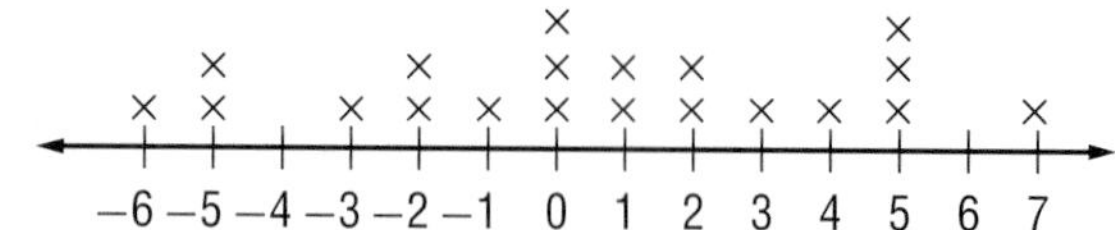

23. +3 **25.** −42 **27.** +212
29.

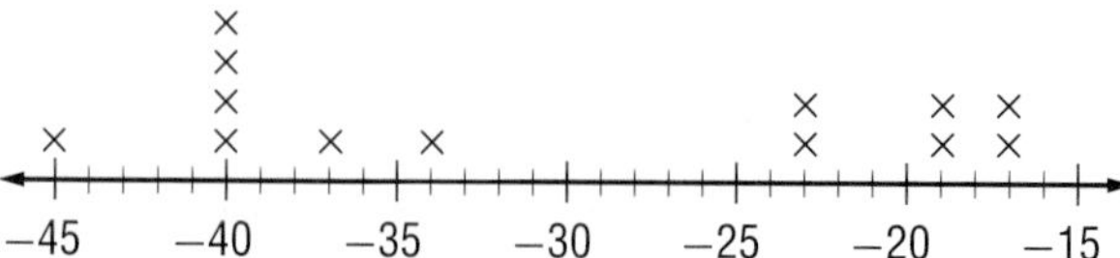

The line plot allows you to see that a temperature of −40°F has the most ×'s and therefore is the most common. **31.** Sample answer: A student may have used integers to keep track of scoring for a card game or a board game. **33.** 10 **35.** Sample answer: Negative integers are to the left of zero on the number line,

positive integers are to the right of zero on the number line and zero is neither positive nor negative.
37. G **39.** 57; 55; 5 **41.** 8 **43.** 3 **45.** 10,000

Pages 126–130 ***Chapter 2*** ***Study Guide and Review***

1. false; mean **3.** true **5.** false; positive numbers **7.** false; line graph

9.

Hours Spent on the Internet		
Number of Hours	Tally	Frequency
0	|||	3
1	||||	4
2	~~||||~~ ||	7
3	|||	3
4	|	1

4 students

11. **Production of Food**

Production (in millions): 0, 10, 20, 30, 40, 50, 60, 70
Food: Bananas, Tomatoes, Apples, Oranges, Watermelon

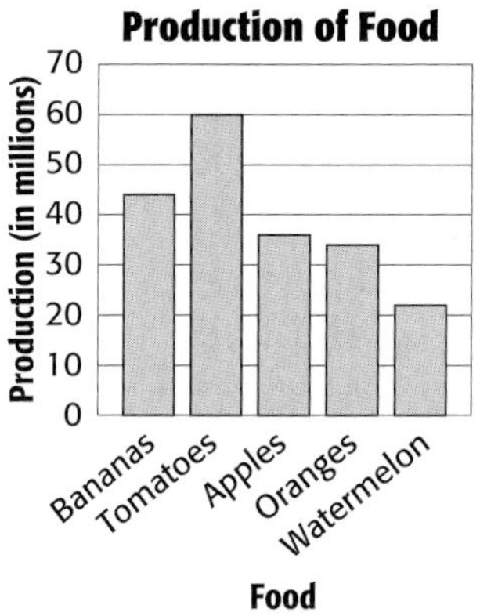

The watermelon was produced about half as much in quantity as the banana.

13. Sample answer: The climber's height is decreasing. **15.** Sample answer: From 7:30 P.M. to 8 P.M., the number of students increased, peaking at 8 P.M., and then decline steadily from 8 P.M. to 9 P.M.

17. **Days Until Vacation**

Stem	Leaf
0	2 8
1	0
2	0 7 7
3	9 9
4	0 3 7

4|0 = 40 days

19. **Ages of Attendees**

Stem	Leaf
1	8 9
2	2 7 8 9
3	1 5
4	1
5	6

2|2 = 22 years old

21. 5 **23.** 27 days **25.** 72 mph **27.** 42; 36; 46 **29.** bar graph **31.** stem-and-leaf plot **33.** −20 ft **35.** −7
37. (number line: −10 −8 −6 −4 −2 0 2 4 6 8 10)

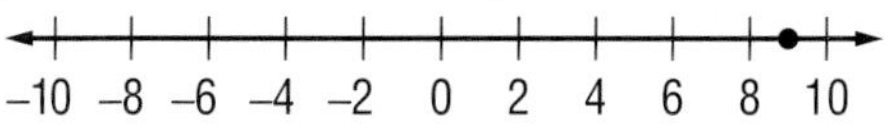

Chapter 3 Adding and Subtracting Decimals

Page 137 ***Chapter 3*** ***Getting Ready***

1. 476 **3.** 1,526 **5.** 4,332 **7.** 2,920 h **9.** 19 **11.** 32 **13.** 288 **15.** > **17.** >

Pages 140–141 ***Lesson 3-1***

1. seven tenths **3.** five and thirty-two hundredths **5.** thirty-four and five hundred forty-two thousandths **7.** 0.9; 9 × 0.1 **9.** 3.22; (3 × 1) + (2 × 0.1) + (2 × 0.01) **11.** eighteen and seventy-five hundredths; (1 × 10) + (8 × 1) + (7 × 0.1) + (5 × 0.01) **13.** nine tenths **15.** one and three hundredths **17.** four and ninety-four hundredths **19.** three hundred eighty-seven thousandths **21.** twenty and fifty-four thousandths **23.** nine and seven hundred sixty-nine ten-thousandths **25.** 11.3; (1 × 10) + (1 × 1) + (3 × 0.1) **27.** 34.16; (3 × 10) + (4 × 1) + (1 × 0.1) + (6 × 0.01) **29.** 0.0102; (0 × 0.1) + (1 × 0.01) + (0 × 0.001) + (2 × 0.0001) **31.** 52.01; (5 × 10) + (2 × 1) + (0 × 0.1) + (1 × 0.01) **33.** thirty-four dollars and sixty-seven cents **35.** three hundred one and nineteen ten-thousandths **37.** 0.0048 **39.** 0.932 **41.** Sample answer: When you read or hear the word form of a decimal you can use cues to determine how to write the decimal in standard form. For example, when you say *three and five tenths*, the word *and* is the cue for where the decimal point goes. Anything before the word *and* is to the left of the decimal point and anything after the word *and* goes to the right of the decimal point. **43.** 284.12 **45.** 0 **47.** *F* **49.** *A* **51.** *E*

Pages 143–145 ***Lesson 3-2***

1. < **3.** > **5.** Botswana **7.** < **9.** < **11.** < **13.** < **15.** > **17.** > **19.** Carly Patterson **21.** 15.99, 16, 16.02, 16.2 **23.** 4.45, 4.9945, 5.545, 5.6 **25.** 2.111, 2.11, 2.1, 2.01 **27.** 32.32, 32.302, 32.032, 3.99 **29.** 321.5, 321.53, 321.539 **31.** Sample answer: Number sense; since 115.6 > 115.4, Syracuse receives more snowfall than Takeetna during one year. Therefore, Syracuse would receive more in 10 years than Takeetna. **33.** Carlos, 0.4 < 0.49 < 0.5 **35.** Sample answer: Who finished the race first, Ariel or Nelia? **37.** H **39.** (1 × 100) + (0 × 10) + (1 × 1) + (5 × 0.1)
40–43. (number line: −10 −9 −8 −7 −6 −5 −4 −3 −2 −1 0 1 2 3 4 5)
45. tenths **47.** ten-thousandths

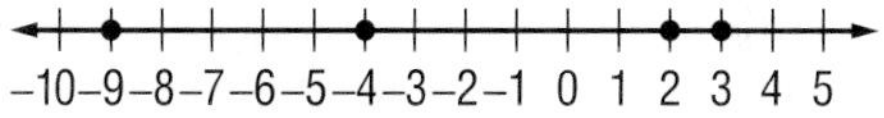

Pages 147–149 ***Lesson 3-3***

1. 0.3 **3.** 45.52 **5.** 7.6760 **7.** 11.9 s, 12.0 s, 11.8 s, 11.9 s **9.** 7.4 **11.** 6 **13.** 2.50 **15.** 5.457 **17.** 150 mph **19.** $4 **21.** 0.249 **23.** 21.251 **25.** 988.081 **27.** 1,567.8930 **29.** Sample answer: 14.998 **31.** Sample answer: 6.0827
33. (number line: 6.7 6.71 6.72 6.73 6.74 6.75 6.76 6.77 6.78 6.79 6.8)
To the nearest tenth, 6.73 rounds to 6.7 because 6.73 is closer to 6.7 than to 6.8. **35.** H **37.** < **39.** 32.05 **41.** 58 **43.** 62

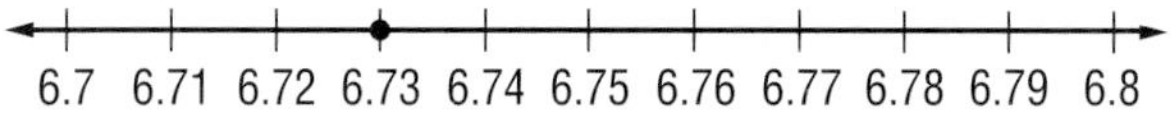

Pages 152–154 ***Lesson 3-4***

1. 0 + 1 = 1 **3.** 4 − 3 = 1 **5.** 3 × 5 = 15 **7.** C **9.** $300 **11.** 30 + 90 = 120 **13.** 60 − 10 = 50 **15.** 9 + 1 + 7 = 17 **17.** $50 **19.** 4 × $3 = $12 **21.** 3 × $55 = $165 **23.** 4 × 100 = 400 **25.** 10 **27.** 300 **29.** $60 **31.** Using rounding, the Flathead rail tunnel is 8 − 6 or 2 miles longer than the Moffat rail tunnel. Using front-end estimation, the Flathead rail tunnel is 7 − 6 or 1 mile longer than the Moffat rail tunnel. Therefore, the estimates would not be the same. When using rounding, the numbers were rounded to the nearest whole number. With front-end estimation, however, the values of the digits in the front place were used. **33.** \$144 **35.** Maximum: \$15.49; Minimum: \$14.50 **37.** C **39.** 45.5 carats **41.** 477 **43.** 465

Pages 158–160 ***Lesson 3-5***

1. 8.7 **3.** 38.34 **5.** 7.32 **7.** 2.22 **9.** 28.7 **11.** 2.5 million **13.** 16.7 **15.** 3.34 **17.** 103.01 **19.** 2.1 **21.** 80.02 **23.** 29.95 **25.** 0.13 second **27.** 117.65 **29.** 106.865 **31.** 12.073 **33.** 4.196 **35.** 10.4 **37.** 8.9 billion **41.** $0.8642 + 0.7531 = 1.6173$ **43.** Yoko is correct. Sample answer: she annexed a zero for the first number. **45.** A **47.** Sample answer: $4 + 4 = 8$ **49.** Sample answer: $9 - 7 = 2$ **51.** 135,099

Pages 165–166 ***Lesson 3-6***

1. 16.2 **3.** 1.56 **5.** 0.45 **7.** 1.17 **9.** 40.6 **11.** 8.4 **13.** 6.3 **15.** 2.6 **17.** 7.2 **19.** 0.06 **21.** 0.0684 **23.** 82.35 **25.** 52 **27.** 1,500 **29.** 25.0 **31.** 3,450 **33.** 12.8 ft^2 **35.** 8,850 m **37.** 9.1 **39.** 64.2 **41.** 119.35 **43.** 32 mm; $20 \times 1.95 = 39$; $4 \times 1.75 = 7$; $39 - 7 = 32$ **45.** Sample answer: First evaluate 1.17×100 to be 117. Then multiply 117 by 5.4 to get the answer of 631.8. Or first evaluate 5.4×100 to be 540. Then multiply 540 by 1.17 to get the answer of 631.8. **47.** C **49.** \$5,398.58 million **51.** \$70 – \$50 = \$20 **53.** 1,075 **55.** 2,970

Pages 170–172 ***Lesson 3-7***

1. 0.3 **3.** 29.87127 **5.** 1.092 **7.** 3.645 **9.** 0.043 **11.** 0.28 **13.** 1.48 **15.** 7.154 **17.** 0.186 **19.** 166.992 **21.** 0.0224 **23.** 23.22 **25.** 36.685 **27.** 75.722 **29.** 84.474 ft **31.** 75.2452 **33.** 20.48512 **35.** 5.8199 **37.** 0.109746 **39.** 9.7 **43.** Sample answer: If $ab = 0.63$, then a could be 0.9 and b could be 0.7 which are both less than 1; if $ab = 0.5$, then a could be 0.5 which is less than 1, but b would be 1.0 which is *not* less than 1. **45.** 0.672 **47.** Sample answer: 0.1×0.6 **49.** Sample answer: Counting Method: Find the sum of the number of decimal places in each factor. The product has the same number of decimal places.; Estimation Method: Estimate the number of places. **51.** J **53.** 348.8 **55.** 47.04 **57.** 24,847.86 mi **59.** 9 **61.** 9

Pages 175–176 ***Lesson 3-8***

1. 0.9 **3.** 1.4 **5.** 0.6 **7.** 0.49 **9.** 18.4 **11.** 13.8 **13.** 1.6 **15.** 0.7 **17.** 1.9 **19.** 30.0 **21.** 1.493 thousand feet **23.** Brand B; The cost of each bottled water for Brand B is about \$0.44. For Brand A the cost is about \$0.58 and for Brand C the cost is about \$0.46. So Brand B has the best cost per bottle. **25.** 24.925 **27.** Sample answer: {5.4, 5.5, 5.6} **29.** Tabitha; the decimal point is placed directly above. **31.** 62.46 **33.** 13.68 **35.** 2.592 **37.** 49.872 **39.** 6×6; 36 s **41.** 5 **43.** 8.2

Pages 181–183 ***Lesson 3-9***

1. 12.3 **3.** 1.5 **5.** 250 **7.** 0.8025 **9.** 3 **11.** 0.2 **13.** 2.3 **15.** 0.0492 **17.** 0.4 **19.** 420 **21.** 0.605 **23.** 20 steps **25.** 4.9 **27.** 29.4 **29.** 15.1 **31.** 5.8 **33.** 6.8 **35.** 2.2 times **37.** 3,800 **39.** Sample answer: If $a < 1$ and $b < 1$, then $a \div b < 1$. If $a = 0.08$ and $b = 0.2$, then $a \div b = 0.4$ which is less than 1; If $a = 0.8$ and $b = 0.02$, then $a \div b = 40$ which is not less than 1. **41.** $1.92 \div 0.51 \approx 2 \div 0.5$ or 4. The number line shows that there are 4 halves in 2. **43.** Sample answer: How many times more people live in India than in Indonesia? Round to the nearest tenth. 4.8 **45.** J **47.** 47.04 **49.** 467.9304 **51.** –4 **53.** –1 **55.** 3

Pages 184–185 ***Lesson 3-10***

1. Sample answer: checking on an answer for the distance to a planet **3.** \$100 is enough **5.** 270 ft **7.** 5 magnets and 7 key chains **9.** 2004 **11.** 4 and 12

Pages 186–190 ***Chapter 3*** ***Study Guide and Review***

1. false; expanded form **3.** false; thousandths **5.** false; clustering **7.** false; $(2 \times 100) + (4 \times 10) + (5 \times 1)$ **9.** false; greater than **11.** 0.13, $(1 \times 0.1) + (3 \times 0.01)$ **13.** 83.005, $(8 \times 10) + (3 \times 1) + (0 \times 0.1) + (0 \times 0.01) + (5 \times 0.001)$ **15.** > **17.** = **19.** 0.0951, 0.9051, 9.501, 90.51 **21.** \$0.99, \$9, \$9.99, \$19.99 **23.** 0.0004 **25.** 50 mi^2 **27.** \$72 – \$30 = \$42 **29.** $130 - 10 = 120$ **31.** \$24 **33.** 30 **35.** 20 **37.** 41.96 **39.** 1.406 **41.** 58.12 s **43.** 8.4 **45.** 1.394 **47.** 0.24 **49.** 106.4 **51.** \$8.95 **53.** 0.78 **55.** 0.204 **57.** 0.1 **59.** 73.08 ft^2 **61.** 6.37 **63.** 2.65 **65.** 3.37 **67.** 1.6 **69.** 0.46 **71.** 36.9 **73.** 7.3 mph **75.** \$16; Sample answer: \$23.80 is almost \$24. \$24 × 0.67 is \$16.08.

Chapter 4 Fractions and Decimals

Page 195 ***Chapter 4*** ***Getting Ready***

1. none **3.** 5 **5.** Yes; Sample answer: 78 is divisible by 6. **7.** $2 \times 2 \times 2 \times 3 \times 11$ **9.** $2 \times 2 \times 7$ **11.** 5.3 **13.** 0.2

Pages 199–201 ***Lesson 4-1***

1. 1, 11 **3.** 8 **5.** 3 **7.** 7 **9.** 1, 3, 5, 15 **11.** 1, 3 **13.** 6 **15.** 12 **17.** 7 **19.** 4 **21.** 1 **23.** 4 photos; The GCF of 8, 12, and 16 is 4. **25.** 9 boxes **27.** Sample answer: 12, 18, 24 **29.** Sample answer: 30, 45, 60 **31.** \$24 **33.** Sample answer: when one of the numbers is a factor of the other number. **35.** True. An odd number does not have a factor of 2. So, the GCF of two odd numbers will not have a factor of 2 and is therefore odd. **37.** Sample answer: 3, 9, 12 **39.** Sample answer: With larger numbers, it is easier to find the prime factorizations to find the GCF. **41.** D **43.** 8,000 **45.** 3.9, 7, 8.3, 9.85 **47.** 3 **49.** none

Pages 206–208 ***Lesson 4-2***

1. 9 **3.** 5 **5.** $\frac{1}{5}$ **7.** $\frac{5}{19}$ **9.** $\frac{6}{25}$ **11.** 9 **13.** 5 **15.** 4 **17.** 4 **19.** $\frac{2}{5}$ **21.** $\frac{1}{2}$ **23.** simplest form **25.** $\frac{3}{20}$ **27.** $\frac{4}{25}$ **29.** Sample answer: $\frac{2}{5}$ and $\frac{8}{20}$ **31.** Sample answer: $\frac{6}{10}$ and $\frac{3}{5}$ **33.** $\frac{6}{25}$ **35.** The fraction $\frac{4}{20}$ does not belong, as it is not equivalent to $\frac{2}{5}$. **37.** Sample answer: Either multiply or divide the numerator and denominator by the same number. **39.** H **41.** 15 **43.** about \$2.50 **45.** 89 **47.** 3 R1 **49.** 7 R4

Pages 211–212 Lesson 4-3

1. $\frac{33}{8}$ **3.** $\frac{17}{3}$ **5.** $5\frac{1}{6}$ **7.** 1 **9.** $\frac{26}{3}$ **11.** $\frac{13}{8}$ **13.** $\frac{23}{4}$ **15.** $\frac{25}{6}$ **17.** $\frac{9}{5}$ million km^2 **19.** $5\frac{2}{5}$ **21.** $2\frac{3}{8}$ **23.** 7 **25.** 1 **27.** $\frac{87}{5}$ **29.** $1\frac{1}{4}$ h **31.** Sample answer: Draw a model to show how many sixths there are in $4\frac{1}{6}$.

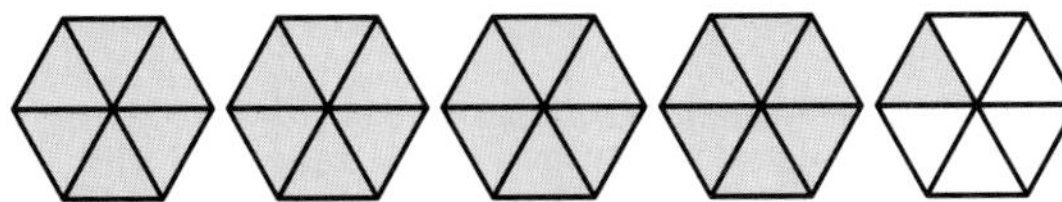

So, $4\frac{1}{6}$ can be written as $\frac{25}{6}$. **33.** If the numerator is less than the denominator, the fraction is less than 1. If the numerator is equal to the denominator, the fraction is equal to 1. If the numerator is greater than the denominator, the fraction is greater than 1. **35.** H **37.** simplest form **39.** 3 **41.** 3 **43.** about 150

Pages 214–215 Lesson 4-4

1. Students should confirm the solution, which accounts for all possibilities. **3.** 18 combinations **5.** on the bottom right of the rectangle **7.** 7 **9.** 6 codes **11.** $14.76 **13.** 10 mi

Pages 218–219 Lesson 4-5

1. 14, 28, 42 **3.** 30 **5.** 15 weeks **7.** 7, 14, 21 **9.** 24, 48, 72 **11.** 18, 36, 54 **13.** 63 **15.** 60 **17.** 180 **19.** 2020 **21.** Sample answer: $y = 2$ and $z = 18$; $y = 2$ and $z = 9$ **23.** Trina is correct. Trina found the LCM. D.J. found the GCF. **25.** Sample answer: Balloons are sold in packages of 20. Balloon weights are sold in packages of 8. What is the least number of balloons and weights Samantha should buy to not have any left over? **27.** H **29.** $1\frac{1}{2}$ dozen **31.** 51 **33.** 5 **35.** C

Pages 222–224 Lesson 4-6

1. > **3.** < **5.** $6\frac{1}{4}, 6\frac{3}{8}, 6\frac{2}{3}, 6\frac{5}{6}$ **7.** < **9.** = **11.** > **13.** < **15.** $\frac{5}{8}$ ft **17.** $\frac{1}{4}, \frac{1}{2}, \frac{2}{3}, \frac{5}{6}$ **19.** $9\frac{1}{6}, 9\frac{2}{5}, 9\frac{3}{7}, 9\frac{3}{5}$ **21.** $\frac{3}{4}$ in. **23.** > **25.** < **27.** $\frac{2}{10}, \frac{2}{5}, \frac{1}{2}, 1\frac{4}{10}, \frac{7}{2}$ **31.** $\frac{3}{9}, \frac{3}{8}$, and $\frac{3}{7}$; Since the numerators are the same, the larger the denominator, the smaller the fraction. **33.** A **35.** C **37.** $\frac{43}{8}$ **39.** 0.7 **41.** 0.89

Pages 227–228 Lesson 4-7

1. $\frac{2}{5}$ **3.** $\frac{16}{25}$ **5.** $\frac{21}{40}$ **7.** $2\frac{3}{4}$ **9.** $23\frac{3}{4}$ mpg **11.** $\frac{7}{10}$ **13.** $\frac{1}{2}$ **15.** $\frac{21}{100}$ **17.** $\frac{41}{50}$ **19.** $\frac{17}{40}$ **21.** $\frac{1}{250}$ **23.** $\frac{17}{20}$ mi **25.** $12\frac{1}{10}$ **27.** $17\frac{3}{100}$ **29.** $42\frac{24}{25}$ **31.** $50\frac{121}{200}$ **33.** $\frac{1}{5}$ lb **35.** Sample answer: $\frac{1}{5}$ in. and $\frac{7}{20}$ in. **37.** Always; a decimal that ends in the thousandths place can have a denominator of 1,000. Since 1,000 is divisible by 2 and 5, the denominator of every terminating decimal is divisible by 2 and 5. **39.** Sample answer: Since 6 is in the hundredths place, write 0.36 as a fraction with a denominator of 100. So, $0.36 = \frac{36}{100}$. Then simplify by using the GCF, 4: $\frac{36}{100} = \frac{9}{25}$. Therefore, $0.36 = \frac{9}{25}$. **41.** H **43.** > **45.** < **47.** 20 combinations **49.** 18 **51.** 21

Pages 231–232 Lesson 4-8

1. 0.9 **3.** 3.5 **5.** 0.36 **7.** 3.7 **9.** 4.225 **11.** 0.05 **13.** 0.385 **15.** 0.625 **17.** 0.5625 **19.** 6.0625 **21.** 12.5375 **23.** 5.8125 in. **25.** < **27.** < **29.** 3.6 s **31.** 0.3333... **33.** 0.4444... **35.** Sample answer: $\frac{7}{11} = 0.636363...$ **37.** Method 1: For fractions whose denominators are factors of 10, 100, or 1,000, you can write equivalent fractions with these denominators. Then use place value to write the fraction as a decimal. Method 2: For fractions whose denominators are not factors of 10, 100, or 1,000, use paper and pencil to divide the numerator by the denominator. **39.** F **41.** $\frac{73}{100}$ **43.** $11\frac{7}{50}$ **45.** $\frac{5}{6}$ **46–49.**

Pages 235–237 Lesson 4-9

1. (1, 0) **3.** (2, 1.25) **5–8.**

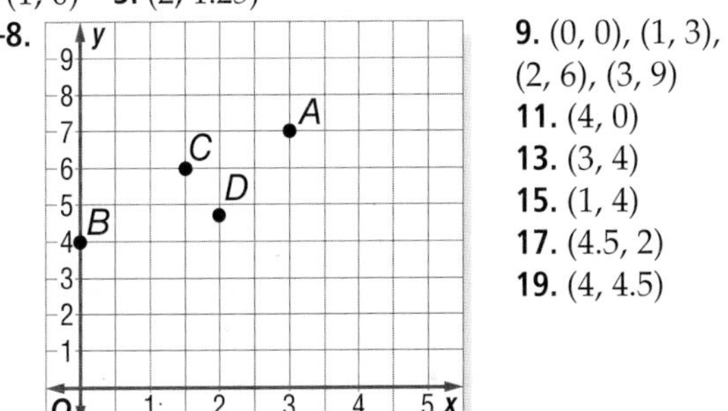

9. (0, 0), (1, 3), (2, 6), (3, 9) **11.** (4, 0) **13.** (3, 4) **15.** (1, 4) **17.** (4.5, 2) **19.** (4, 4.5)

21–26. **27–28.**

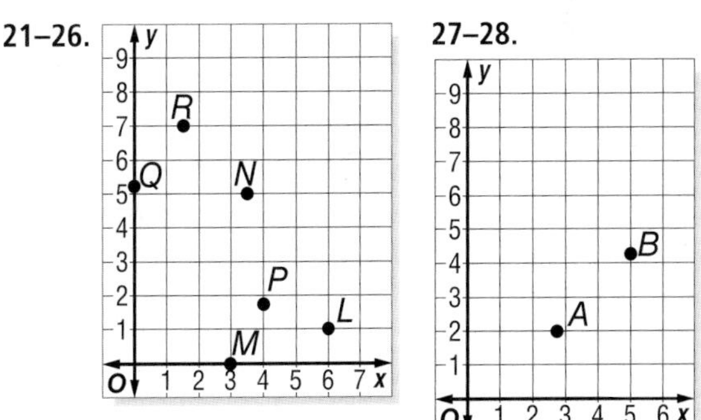

29. (5, 20), $(5\frac{1}{4}, 21)$ $(5\frac{1}{2}, 22)$, $(5\frac{3}{4}, 23)$ **31.** (0, 0), (1, 4), (2, 8), (3, 12) **33.** (6, 2) **35.** Sample answer: (5, 0) **37.** Sample answer: The point (3, 2) is 3 units to the right of the origin on the x-axis and 2 units up the y-axis. The point (2, 3) is 2 units over on the x-axis and 3 units up the y-axis. **39.** J **41.** $1\frac{17}{50}$ **43.** $13\frac{1}{125}$ **45.** line graph

Pages 238–242 Chapter 4 Study Guide and Review

1. true **3.** true **5.** false; greater **7.** false; GCF **9.** false; equivalent fractions **11.** 3 **13.** 14 **15.** 9 **17.** 56 **19.** 9 **21.** $\frac{3}{16}$ **23.** $\frac{13}{4}$ **25.** $5\frac{3}{4}$ **27.** $1\frac{1}{3}$ ft **29.** 24 **31.** 50 **33.** 48 **35.** 5 bags of clothespins, 8 bags of plastic eyes

37. < **39.** < **41.** $\frac{1}{2}, \frac{5}{9}, \frac{2}{3}, \frac{3}{4}$ **43.** $\frac{3}{4}$ dollar **45.** $\frac{7}{20}$
47. $\frac{1}{8}$ **49.** $9\frac{63}{200}$ **51.** $\frac{6}{125}$ **53.** 0.875 **55.** 0.84 **57.** 12.75
59. 0.75 h **61.** (3, 3) **63.** (1, 2.5)

64–66.

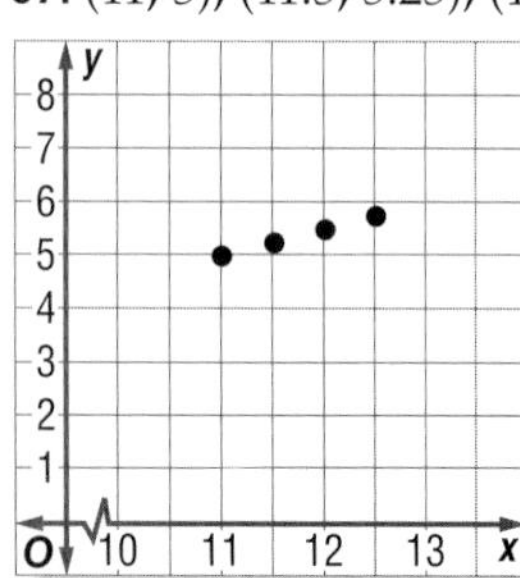

67. (11, 5), (11.5, 5.25), (12, 5.5), (12.5, 5.75)

The points appear to fall on a line.

Chapter 5 Adding and Subtracting Fractions

Page 247 ***Chapter 5*** ***Getting Ready***

1. $1 + 7 = 8$ **3.** $8 - 5 = 3$ **5.** $\$18 + \$4 = \$22$ **7.** $\frac{3}{4}$
9. $\frac{3}{19}$ **11.** $1\frac{1}{10}$ **13.** $1\frac{2}{5}$

Pages 251–253 ***Lesson 5-1***

1. 1 **3.** $\frac{1}{2}$ **5.** 0 **7.** $2\frac{1}{2}$ in. **9.** up; by rounding $4\frac{3}{8}$ gallons up, the gardener will have enough fertilizer. **11.** 3
13. 9 **15.** 3 **17.** $5\frac{1}{2}$ **19.** $3\frac{1}{2}$ **21.** 2 in. **23.** 2 in. **25.** up; by rounding $14\frac{3}{8}$ inches up to $14\frac{1}{2}$ inches, then his gift will fit in any box he selects that is at least $14\frac{1}{2}$ tall
27. $6\frac{1}{2}$ **29.** $4\frac{1}{2}$ **31.** $6\frac{1}{2}$ in. by $4\frac{1}{2}$ in. **33.** $3\frac{3}{14}, 3\frac{5}{9}, 3\frac{6}{7}$
35. The most students in any one category is $\frac{13}{54}$, which does not round up to $\frac{1}{2}$. Therefore, more than half of the students are not represented by any one category.
37. $\frac{75}{100} = \frac{3}{4}$ and $\frac{79}{100} \approx \frac{75}{100}$, so $\frac{79}{100} \approx \frac{3}{4}$. **39.** $4\frac{4}{5}$; the other numbers round to 4 and $4\frac{4}{5}$ rounds to 5.
41. Sample answer: If the numerator and denominator are close in value, round to 1. If the denominator is about twice as much as the numerator, round to $\frac{1}{2}$. If there is a great difference in the value of the numerator and denominator, round to 0. **43.** G **45.** 0.125
47. 0.4 **49.** $1\frac{3}{25}$ mi

Pages 254–255 ***Lesson 5-2***

1. Sample answer: If after you act it out the answer seems reasonable compared to the solution you came up with, then it is probably correct. **3.** 8; chicken and coffee, chicken and tea, chicken and lemonade, chicken and water, fish and coffee, fish and tea, fish and lemonade, fish and water **5.** team 3 **7.** \$41 **9.** 6 **11.** about 5 times **13.** \$12.50

Pages 258–260 ***Lesson 5-3***

1. $\frac{4}{5}$ **3.** $1\frac{1}{2}$ **5.** $\frac{3}{5}$ **7.** $\frac{5}{14}$ **9.** $1\frac{4}{7}$ **11.** $\frac{2}{3}$ **13.** $1\frac{3}{8}$ **15.** $\frac{1}{4}$
17. $\frac{1}{3}$ **19.** $\frac{1}{9}$ **21.** $\frac{1}{2}$ c **23.** $\frac{67}{100}$ **25.** $1\frac{3}{8}$
27. $\frac{3}{7} + \frac{2}{7} = \frac{5}{7}$ **29.** $\frac{3}{10}; \frac{1}{2}$
31. Sample answer: $\frac{3}{11} + \frac{6}{11} = \frac{9}{11}$

33. Sample answer:
$\frac{4}{9} + \frac{7}{9} = \frac{11}{9}$ or $1\frac{2}{9}$

35. $\frac{21}{15} = 1\frac{2}{5}$ **37.** D **39.** 6 **41.** 0 **43.** 275 in^2
45. 6 **47.** 45

Pages 266–268 ***Lesson 5-4***

1. $\frac{8}{9}$ **3.** $\frac{1}{6}$ **5.** $\frac{1}{2}$ **7.** $\frac{5}{8}$ **9.** $\frac{3}{16}$ in. **11.** $\frac{9}{20}$ **13.** $\frac{9}{10}$
15. $\frac{3}{8}$ **17.** $\frac{11}{12}$ **19.** $\frac{7}{20}$ **21.** $1\frac{3}{14}$ **23.** $\frac{1}{8}$ **25.** $1\frac{1}{4}$ **27.** $\frac{7}{22}$
29. $\frac{49}{120}$ **31.** $\frac{3}{10}$ **33.** $2\frac{1}{24}$ **35.** $\frac{1}{3} + \frac{2}{5} = \frac{11}{15}$
37. Sample answer:

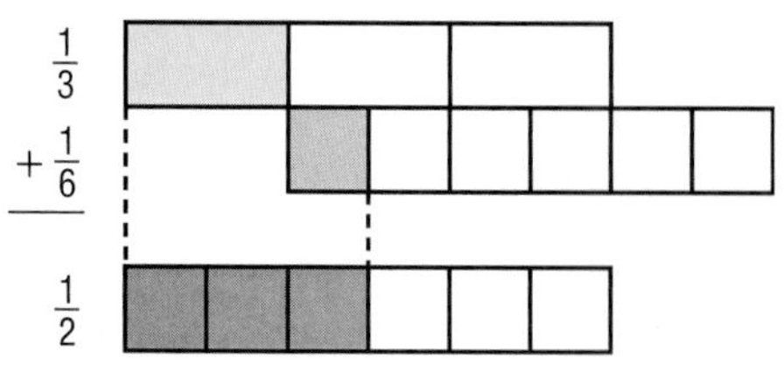

39. Sample answer:

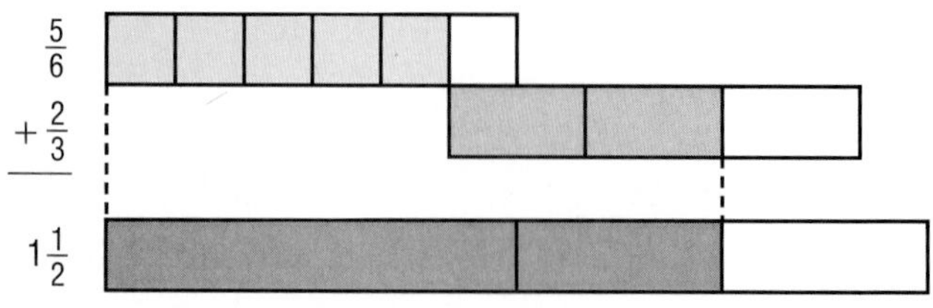

41. $\frac{1}{2}$ **43.** $\frac{5}{24}$ **45.** Sample answer:

$\frac{1}{2} + \frac{2}{5}$ $\frac{1}{2} + \frac{2}{5} = \frac{9}{10}$

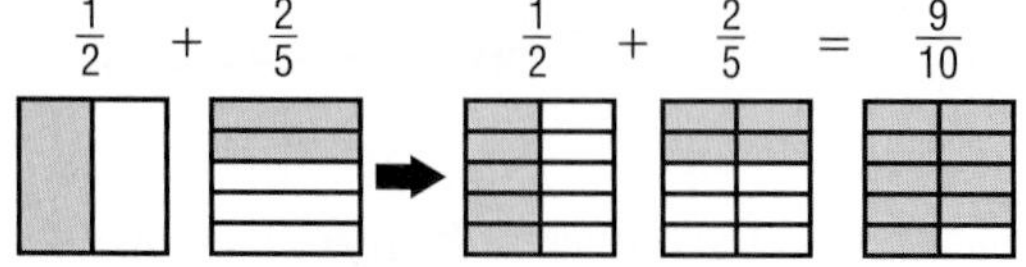

47. Sometimes; Sample answer: For example, $\frac{1}{2}, \frac{1}{4}$, and $\frac{3}{4}$ are each less than 1. The sum of $\frac{1}{2} + \frac{1}{4}$ is $\frac{3}{4}$, which is less than 1; the sum of $\frac{1}{2} + \frac{3}{4}$ is $1\frac{1}{4}$, which is greater than 1; the sum of $\frac{1}{4} + \frac{3}{4}$ is equal to 1. **49.** Sample answer: Jessica ran $\frac{4}{5}$ mile in 10 minutes. Rosa ran $\frac{3}{4}$ mile in the same amount of time. How much farther did Jessica run than Rosa in the 10 minutes? **51.** H **53.** $\frac{1}{4}$ **55.** $\frac{2}{5}$ **57.** 39 points **59.** 9 **61.** 4

Pages 272–274 ***Lesson 5-5***

1. $4\frac{1}{2}$ **3.** $8\frac{3}{10}$ **5.** $\frac{13}{15}$ **7.** B **9.** 11 **11.** $5\frac{2}{5}$ **13.** 10 **15.** $2\frac{1}{2}$ **17.** $5\frac{2}{5}$ **19.** $2\frac{3}{4}$ **21.** $6\frac{1}{2}$ **23.** $6\frac{8}{9}$ **25.** $10\frac{7}{12}$ gal **27.** $6\frac{7}{8}$ blocks **29.** $2\frac{1}{3} - 1\frac{2}{3} = \frac{2}{3}$ **31.** Sample answer: $1\frac{3}{4} + 2\frac{1}{2} = 4\frac{1}{4}$ **33.** A **35.** $\frac{2}{3}$ **37.** $\frac{1}{20}$ **39.** \$2 **41.** $7\frac{1}{2}$ **43.** 3

Pages 277–279 ***Lesson 5-6***

1. Sample answer: $\frac{1}{8} \times 16 = 2$ **3.** Sample answer: $\frac{2}{5} \times 25 = 10$ **5.** Sample answer: $0 \times 1 = 0$ **7.** Sample answer: $7 \times 4 = 28$ **9.** Sample answer: 10 ft × 4 ft or 40 ft^2 **11.** Sample answer: $\frac{1}{4} \times 20 = 5$ **13.** Sample answer: $\frac{1}{3} \times 42 = 14$ **15.** Sample answer: $\frac{1}{7} \times 21 = 3$, $3 \times 5 = 15$ **17.** Sample answer: $\frac{2}{3} \times 9 = 6$ **19.** Sample answer: $12 \times \frac{1}{4} = 3$ pizzas **21.** Sample answer: $\frac{1}{2} \times 0 = 0$ **23.** Sample answer: $1 \times \frac{1}{2} = \frac{1}{2}$ **25.** Sample answer: $4 \times 3 = 12$ **27.** Sample answer: $5 \times 9 = 45$ **29.** Sample answer: $6 \times 8 = 48$ ft^2 **31.** about 3 people; $\frac{1}{20} \times 58 \approx \frac{1}{20} \times 60$, or 3 **33.** Sample answer: $\frac{1}{3} \times 9 = 3$ oz **35.** $\frac{7}{10} \times 90 \approx \frac{7}{10} \times 100$; $\frac{1}{10} \times 100 = 10$; $7 \times 10 = 70$ **37.** Sample answer: Using rounding to the nearest half, $\frac{5}{12} \times 78 \approx \frac{1}{2} \times 80$ which is 60; using compatible numbers, $\frac{5}{12} \times 78 \approx \frac{1}{3} \times 75$ which is 25; the difference between the estimates is 40 – 25, or 15. **39.** D **41.** $\frac{11}{12}$ lb **43.** $\frac{5}{9}$ **45.** $\frac{2}{5}$ **47.** 714 yd^2 **49.** 2 **51.** 8

Pages 284–286 ***Lesson 5-7***

1. $\frac{1}{16}$ **3.** 8 **5.** $\frac{1}{4}$ **7.** $2\frac{2}{5}$ in. **9.** $\frac{2}{15}$ **11.** $\frac{15}{32}$ **13.** $1\frac{1}{2}$ **15.** $12\frac{1}{2}$ **17.** $\frac{1}{6}$ **19.** $\frac{1}{6}$ **21.** $\frac{3}{10}$ **23.** $\frac{1}{5}$ **25.** $22\frac{2}{5}$ yr **27.** 146 days **29.** $\frac{1}{24}$ **31.** $\frac{3}{16}$ **33.** $\frac{28}{75}$ **35.** $1\frac{3}{10}$ **37.** 40,000 mi^2 **39.** 30; Sample answer: $\frac{1}{6}$ of 26 is not a whole number and $\frac{1}{6}$ of 18 is 3. If only 3 students have been to France, it is impossible that 4 students have been to Paris. **41.** The overlapping shaded area is $\frac{2}{6}$ or $\frac{1}{3}$ of the whole.

$\frac{1}{2}$

$\frac{2}{3}$

43. false; $4\frac{9}{10} \times \frac{9}{10} = 4\frac{41}{100}$ **45.** Sample answer: $a = \frac{3}{8}$ and $b = \frac{5}{7}$; $a = \frac{5}{8}$ and $b = \frac{3}{7}$; $a = \frac{5}{14}$ and $b = \frac{3}{4}$ **47.** The fraction $\frac{a}{b} \times \frac{b}{c} \times \frac{c}{d} \times \frac{d}{e}$ can be simplified before multiplying by crossing out the factors that appear in both a numerator and a denominator. The factors that can be crossed out are b, c, and d. Thus, the only factors remaining are a in a numerator and e in a denominator, or $\frac{a}{e}$. **49.** F **51.** Sample answer: $2 \times 5 = 10$ **53.** Sample answer: $\frac{1}{2} \times 1 = \frac{1}{2}$ **55.** 36 wk **57.** \$14.75, \$14.78, \$14.87, \$15.24, \$15.42 **59.** $\frac{17}{3}$ **61.** $\frac{53}{8}$

Pages 288–290 ***Lesson 5-8***

1. $1\frac{3}{16}$ **3.** $4\frac{9}{10}$ **5.** $1\frac{1}{5}$ **7.** $2\frac{1}{8}$ **9.** $1\frac{1}{2}$ **11.** $\frac{17}{20}$ **13.** $10\frac{2}{15}$ **15.** $12\frac{3}{4}$ **17.** $19\frac{1}{2}$ **19.** $\frac{7}{8}$ **21.** $\frac{1}{12}$ **23.** $\frac{3}{8}$ mi **25.** $\frac{3}{5}$ **27.** $1\frac{1}{24}$ **29.** about 69 million mi **31.** about 483 million mi **35.** $25\frac{5}{32}$ **37.** Sample answer: $1\frac{1}{5} \times 1\frac{1}{2} = 1\frac{4}{5}$ **39.** Sample answer: If $a = 3\frac{1}{2}$ and $b = 2\frac{1}{2}$, then $ab = 8\frac{3}{4}$ and $a + b = 6$; since $ab > a + b$, the statement is true. If $a = 1\frac{1}{4}$ and $b = 1\frac{3}{4}$, then $ab = 2\frac{3}{16}$ and $a + b = 3$; since $a + b > ab$, the statement is false. **41.** C **43.** $\frac{15}{28}$ **45.** $\frac{3}{20}$ **47.** $\frac{2}{5} \times 300$ or about 120 million **49.** $\frac{3}{32}$ **51.** $\frac{1}{12}$

Pages 295–297 ***Lesson 5-9***

1. $\frac{3}{2}$ **3.** $\frac{5}{2}$ **5.** $\frac{1}{2}$ **7.** 6 **9.** $\frac{2}{5}$ **11.** 110 horses **13.** 10 **15.** $\frac{9}{7}$ **17.** 1 **19.** $\frac{3}{4}$ **21.** $\frac{5}{6}$ **23.** $3\frac{1}{3}$ **25.** 14 **27.** $\frac{1}{6}$ **29.** $\frac{2}{9}$ **31.** $\frac{8}{27}$ yd **33.** 22 **35.** $12 \div 3 = 4$; $4 \div \frac{3}{8} = 10\frac{2}{3}$; 10 shirts **37.** $\frac{5}{6}$ **39.** $1\frac{1}{3}$ **43.** Tom; To divide by 4, multiply by the reciprocal, which is $\frac{1}{4}$. **45.** $\frac{a}{c}$; Sample answer: If the denominators of two fractions are the same, then the quotient of the first fraction divided by the second fraction will be a fraction whose numerator is the numerator of the first fraction and whose denominator is the numerator of the second fraction. **47.** D **49.** 8 **51.** $8\frac{1}{7}$ **53.** $\frac{3}{10}$ **55.** $\frac{5}{3}, \frac{3}{5}$ **57.** $\frac{9}{2}, \frac{2}{9}$ **59.** $\frac{34}{5}, \frac{5}{34}$

Pages 299–301 ***Lesson 5-10***

1. $1\frac{3}{4}$ **3.** $11\frac{1}{5}$ **5.** 28 slices **7.** $\frac{5}{12}$ **9.** $2\frac{2}{3}$ **11.** 39 **13.** $4\frac{1}{26}$ **15.** $\frac{2}{3}$ **17.** $2\frac{2}{5}$ **19.** $\frac{6}{11}$ **21.** $\frac{4}{5}$ **23.** $1\frac{7}{8}$ **25.** 8 photos **27.** 15 strips **29.** 20 mph **31.** Sample answer: $8\frac{2}{3} \div 3\frac{1}{4}$ **33.** less than; Sample answer: Since $3\frac{5}{8} > 2\frac{2}{5}$, the quotient $5\frac{1}{6} \div 3\frac{5}{8} < 5\frac{1}{6} \div 2\frac{2}{5}$. The expression $5\frac{1}{6} \div 3\frac{3}{8}$ represents $5\frac{1}{6}$ being divided into a greater number of parts than the expression $5\frac{1}{6} \div 2\frac{2}{5}$. If $5\frac{1}{6}$ is divided into a greater number of parts, each part will be smaller. **35.** A **37.** $\frac{1}{2}$ qt **39.** $1\frac{2}{5}$ **41.** 6

Selected Answers

Pages 302–306 ***Chapter 5*** ***Study Guide and Review***

1. true **3.** true **5.** true **7.** false; LCD **9.** false; 40 **11.** $4\frac{1}{2}$ **13.** $\frac{1}{2}$ **15.** $9\frac{1}{2}$ **17.** 24 **19.** 3 **21.** $\frac{2}{3}$ **23.** $\frac{4}{7}$ **25.** $1\frac{2}{9}$ **27.** $\frac{3}{4}$ **29.** $\frac{7}{8}$ **31.** $\frac{13}{20}$ **33.** 1 **35.** 5 **37.** $17\frac{7}{12}$ **39.** $1\frac{1}{6}$ ft **41.** $2\frac{13}{24}$ **43.** $6\frac{7}{8}$ **45.** Sample answer: $\frac{1}{5} \times 20 = 4$ **47.** Sample answer: $1 \times 13 = 13$ **49.** Sample answer: $5 \times 8 = 40$ **51.** Sample answer: $\frac{5}{6} \times 60 = 50$ min **53.** $\frac{1}{6}$ **55.** $\frac{1}{3}$ **57.** $26\frac{1}{2}$ **59.** $12\frac{5}{6}$ ft^2 **61.** $\frac{1}{6}$ **63.** $2\frac{1}{4}$ c **65.** $3\frac{1}{5}$

Chapter 6 Ratio, Proportion, and Functions

Page 313 ***Chapter 6*** ***Getting Ready***

1. $\frac{2}{3}$ **3.** $\frac{3}{5}$ **5.** $\frac{13}{25}$ **7.** 8 **9.** 5 **11.** 6 **13.** 9 **15.** 78, 82, 86 **17.** 75 min

Pages 316–319 ***Lesson 6-1***

1. $\frac{3}{4}$; for every 3 pens, there are 4 pencils. **3.** $\frac{3}{2}$; for every 3 action thrillers showing, there are 2 romantic comedies showing. **5.** $\frac{\$3}{1\text{ case}}$ **7.** 82 times **9.** $\frac{2}{5}$; for every 2 sandwiches, there are 5 milk cartons. **11.** $\frac{2}{9}$; for every 2 motorcycles, there were 9 cars. **13.** $\frac{1}{3}$; for every 1 puppy, there are 3 kittens available for adoption. **15.** $\frac{2}{9}$, 2 to 9, or 2:9; every 2 out of 9 cell phone covers sold last week was black. **17.** $\frac{2}{7}$, 2 to 7, or 2:7; two out of every 7 food items donated was a can of fruit. **19.** $\frac{\$9}{1\text{ ticket}}$ **21.** $\frac{\$0.25}{1\text{ egg}}$ **23.** 17 trees **25.** $\frac{23}{17}$; The Maple Leafs have made 23 Stanley Cup appearances to every 17 Bruins' appearances. **27.** The 6-pack of water costs \$0.50 per bottled water. The 24-pack costs \$0.42 per bottled water. So, the 24-pack is less expensive per bottled water. **29.** 32 tickets **31.** Sample answer: A ratio is a comparison of two quantities by division; for example, $\frac{2\text{ students}}{13\text{ students}}$ and $\frac{5\text{ birds}}{27\text{ pets}}$. A rate also compares two quantities by division; however, the quantities of a rate have different kinds of units. For example, $\frac{60\text{ miles}}{6\text{ hours}}$ and $\frac{\$45}{2\text{ tickets}}$. **33.** J **35.** $\frac{3}{4}$ **37.** $2\frac{1}{4}$ **39.** 6 **41.** $\frac{5}{6}$ **43.** $\frac{5}{7}$

Pages 325–327 ***Lesson 6-2***

1. \$28 **3.** 15 tsp **5.** 8 **7.** \$9 **9.** 3 balls **11.** 285 mi **13.** If 20 lb ≈ 9 kg, then 60 lb is about 27 kg. Since half of 60 is 30, a 30 lb dog weighs half of 27 kg or 27 ÷ 2, which is 13.5 kg. **15.**

People Served	24			
Liters of Soda	4			
Pints of Sherbet	2			
Cups of Ice	6			

17. 3 L soda, 1.5 pt sherbet, 4.5 c ice; Since 18 is half of 36, half the recipe that serves 36 people. 6 L ÷ 2 = 3 L, 3 pt ÷ 2 = 1.5 pt, and 9 c ÷ 2 = 4.5 c. **19.** No; if 5 girls and 5 boys are added, there would be 15 girls and 13 boys in the class. Using the ratio table below, you can see that there should be 12 boys for 15 girls.

Number of Girls	10	5	15
Number of Boys	8	4	12

21. C **23.** G **25.** $1\frac{8}{15}$ **27.** $1\frac{7}{18}$ **29.** $\frac{\$8}{1\text{ hat}}$ **31.** $\frac{29\text{ students}}{1\text{ teacher}}$

Pages 331–333 ***Lesson 6-3***

1. No; Since the unit rates, $\frac{\$8}{1\text{ wk}}$ and $\frac{\$7.43}{1\text{ wk}}$, are not the same, the rates are not equivalent. **3.** Yes; Since $\frac{3\text{ h} \cdot 3}{\$12 \cdot 3} = \frac{9\text{ h}}{\$36}$, the fractions are equivalent; $\frac{3\text{ h}}{\$12} = \frac{9\text{ h}}{\$36}$. **5.** No; Micah's unit rate is $\frac{25\text{ push-ups}}{1\text{ min}}$ and Eduardo's unit rate is $\frac{26\text{ push-ups}}{1\text{ min}}$. **7.** No; Since the unit rates, $\frac{4\text{ points}}{1\text{ game}}$ and $\frac{6\text{ points}}{1\text{ game}}$, are not the same, the rates are not equivalent. **9.** No; Since the unit rates, $\frac{\$0.50}{1\text{ bagel}}$ and $\frac{\$0.38}{1\text{ bagel}}$, are not the same, the rates are not equivalent. **11.** Yes; Since $\frac{15\text{ computers} \cdot 3}{45\text{ computers} \cdot 3} = \frac{45\text{ computers}}{135\text{ students}}$, the fractions are equivalent; $\frac{15\text{ computers}}{45\text{ students}} = \frac{45\text{ computers}}{135\text{ students}}$. **13.** No; Since $\frac{16\text{ students}}{28\text{ students}} \neq \frac{240\text{ students}}{560\text{ students}}$, the ratios are not proportional. **15.** No; sample answer: By looking for equivalent fractions, you notice that 3 × 4 results in 12, the numerator of the second fraction, however 5 × 4, or 20 is not the denominator of the second fraction. **17.** No; the ratios for each player do not form equivalent fractions. The simplified ratio for Hideki Matsui is $\frac{15\text{ at bats}}{4\text{ hits}}$ and the simplified ratio for Mark Teixeira is $\frac{4\text{ at bats}}{1\text{ hit}}$. **19.** No; Both Brad Eldred and Ramon Santiago tied with the best record of $\frac{3.3\text{ at bats}}{1\text{ hit}}$, but Brad Eldred has more hits than Ramon Santiago. **21.** Yes; sample answer: $\frac{\$35}{5\text{ weeks}} = \frac{\$7}{1\text{ week}}$ and $\frac{\$56}{56\text{ days}} = \frac{\$56}{8\text{ weeks}}$ or $\frac{\$7}{1\text{ week}}$. **23.** no; Sample answer: the product of the means is 7 × 5, or 35. The product of the extremes is 2 × 21, or 42. Since the products are not equal, the ratios do not form a proportion. **25.** yes; Sample answer: the product of the means is 9 × 12, or 108. The product of the extremes is 4 × 27, or 108. Since the products are equal, the ratios form a proportion. **27.** B **29.** \$78 **31.** 3 × 5 **33.** 2 × 3 × 17 **35.** 7 wins per year **37.** \$7 per hour

Pages 336–339 ***Lesson 6-4***

1. 15 **3.** 10 **5.** 38 **7.** \$72 **9.** 21 **11.** 7 **13.** 3 **15.** 5 **17.** 840 gal **19.** \$30 **21.** 2 min **23.** 16 **25.** 16,000

27. 48 breaths **29.** $\frac{2}{5} = \frac{x}{340,00}$; 136,000 **31.** Sample answer: Equivalent fractions: $\frac{18}{20} = \frac{9}{n}$; $n = 10$. Unit rates: $\frac{18}{6} = \frac{3}{1} = \frac{9}{n}$; $n = 3$. **33.** Always; In order for the ratios to form a proportion, they must be equivalent fractions, therefore reducing to the same fraction.

35. Write the proportion: $\frac{3 \text{ laps}}{24 \text{ minutes}} = \frac{x \text{ laps}}{50 \text{ minutes}}$. Since $24 \times 2 \approx 50$, multiply 3 by 2 to get an estimate of 6 laps. **37.** 27 **39.** No; Since the unit rates, $\frac{\$9}{1 \text{ baseball hat}}$ and $\frac{\$8}{1 \text{ baseball hat}}$, are not the same, the rates are not equivalent. **41.** $75

43.

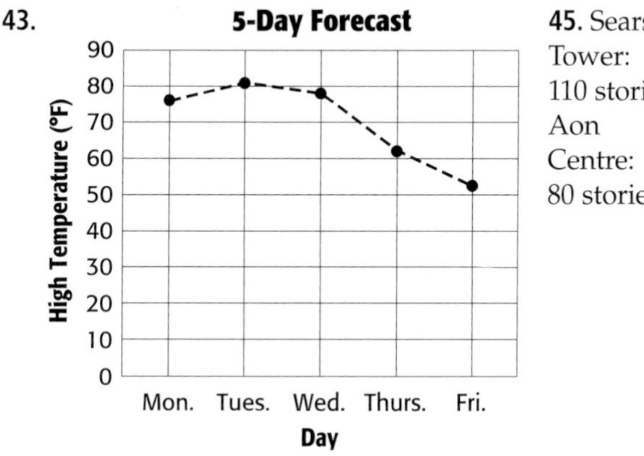

45. Sears Tower: 110 stories; Aon Centre: 80 stories

Pages 341–342 ***Lesson 6-5***

1. Sample answer: Use this strategy when the change between events is the same. **3.** 2013 **5.** Add 3, then increase the number added by 1 for each number; 28, 36, 45 **7.** 2 adults, 2 senior citizens **9.** 1, 5, 10, 10, 5, 1 **11.** 25 **13.** 60 mph

Pages 346–348 ***Lesson 6-6***

1. multiply the position number by 2; $2n$; 30
3. 160 ounces

Ounces	Pounds
16	1
32	2
48	3
64	4
16*n*	*n*

5. add 9 to the position number; $n + 9$; 21 **7.** multiply the position number by 5; $5n$; 60

9. 4 hours

Minutes	Hours
60	1
120	2
180	3
n	$n \div 60$

11. The amount increases by $8. **13.** add 3; 13, 16 **15.** add 0.9; 5.9, 6.8 **17.** add $1\frac{1}{2}$; $7\frac{1}{2}$, 9 **19.** $11\frac{1}{2}$ **21.** $24\frac{1}{2}$

23. 168 hours

Hours	Days
24	1
48	2
72	3
96	4
24*n*	*n*

25. Sample answer: 1, $2\frac{1}{4}$, $3\frac{1}{2}$, $4\frac{3}{4}$, ... **27.** The value of each term is the square of its position; n^2; 10,000. **29.** B **31.** 13 **33.** 7 **35.** 15 **37.** $1\frac{1}{2}$ **39.** $\frac{2}{3}$

Pages 351–353 ***Lesson 6-7***

1. $y = 4x$ **3.**

Number of Lunches, *n*	Multiply by 3	Total Cost, *t*
1	1×3	$3
2	2×3	$6
3	3×3	$9
4	4×3	$12

5. $60 **7.** $y = 6x$ **9.** $y = 15x$

11.

Coins Collected, *c*	Multiply by 15	Points Earned, *p*
1	1×15	15
2	2×15	30
3	3×15	45
4	4×15	60

13. 315 points **15.** $v = 400d$ **17.** The disc jockey charges $35 per hour; $t = 35h$; $175 **19.** Sample answer: A harbor seal eats an average of 14 lb of food per day; $f = 14d$. **21.** $y = x \div 3$ **23.** $t = 1m$; 4 in.; Coronado receives 3 in. of precipitation in 4 months. So, Burbank averages 1 in. more precipitation in 4 months than Coronado. **25.** $y = \frac{x}{2} - 3$ **27.** C **29.** add 8; 35, 43

Pages 355–358 ***Chapter 6*** ***Study Guide and Review***

1. false; division **3.** true **5.** false; multiplying or dividing **7.** false; adding **9.** $\frac{3}{5}$ **11.** $\frac{6}{7}$ **13.** $\frac{15.75 \text{ lb}}{1 \text{ wk}}$ **15.** $\frac{1}{3}$, 1 to 3, or 1:3; one out of 3 DVDs Rick owns is an action DVD. **17.** 48 **19.** Yes; since the unit rates are the same, $\frac{8.8 \text{ ft}}{1 \text{ in.}}$, the rates are equivalent; $\frac{220 \text{ ft}}{25 \text{ in.}} = \frac{88 \text{ ft}}{10 \text{ in.}}$. **21.** Yes; since the unit rates are the same, $\frac{6 \text{ min}}{1 \text{ necklace}}$, rates are equivalent; $\frac{48 \text{ min}}{8 \text{ necklaces}} = \frac{24 \text{ min}}{4 \text{ necklaces}}$. **23.** 100 **25.** 35 **27.** 64 **29.** Subtract 7 from the position number; $n - 7$; 9 **31.** $12n$; $540 **33.** $y = 7x$

Chapter 7 Percent and Probability

Page 363 ***Chapter 7*** ***Getting Ready***

1. $\frac{1}{4}$ **3.** $\frac{3}{10}$ **5.** 3 **7.** 48 **9.** 12 **11.** 6 lb **13.** 10 **15.** 70 **17.** 10

Pages 367–369 ***Lesson 7-1***

1. $\frac{3}{20}$ **3.** $1\frac{4}{5}$ **5.** 25% **7.** 225% **9.** 90% **11.** 125% **13.** $\frac{47}{100}$ **15.** $\frac{1}{5}$ **17.** $2\frac{4}{5}$ **19.** $\frac{27}{50}$ **21.** 35% **23.** 140% **25.** 5% **27.** 92% **29.** 75% **31.** 84% **33.** 210% **35.** $\frac{9}{50}$ **37.** Sample answer: $\frac{11}{20} = \frac{55}{100}$ or 55%, $\frac{3}{5} = \frac{60}{100}$ or 60%, $\frac{7}{10} = \frac{70}{100}$ or 70% **39.** $\frac{8}{45}$; The others are equal to 45%. **41.** C **43.** $t = \$7h$ **45.** add 3; 17, 20 **47.** $\frac{13}{50}$ **49.** $\frac{1}{10}$

Pages 373–375 **Lesson 7-2**

1. **Have You Started Going Out or Dating?**

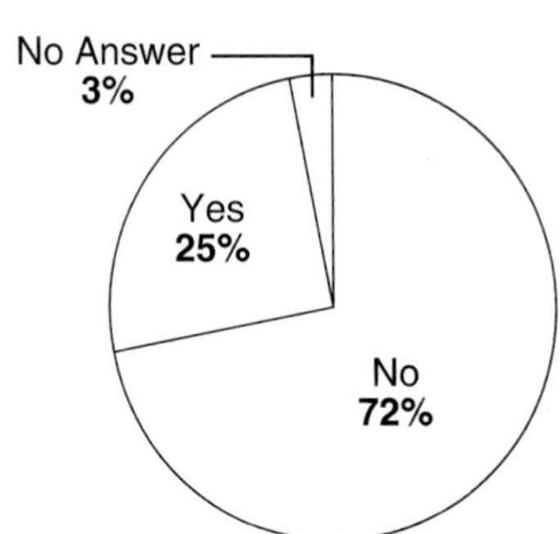

3. 42% 5. **Class President Ballots**

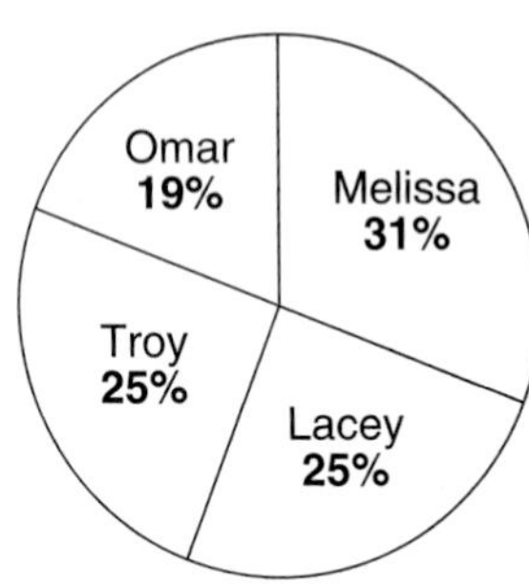

7. oxygen 9. Oxygen and carbon make up about 82% of the entire body. 11. 18% 13. The number of students that walk to school is about half the number of students that ride the school bus. 17. Sample answer: Estimation; Since he is sketching the circle graph he does not need exact percents for the size of each sector.

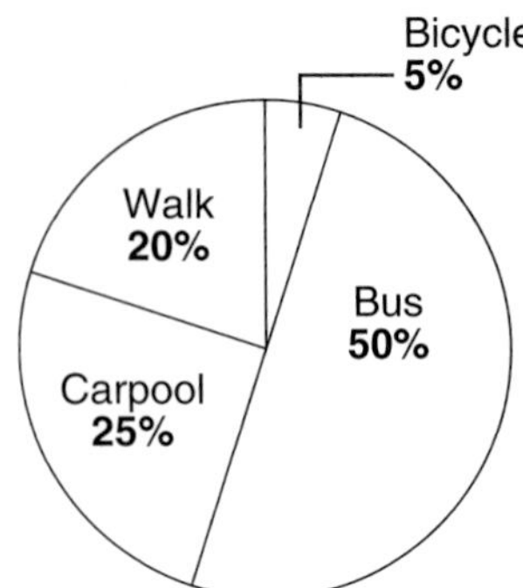

19. A 21. 90% 23. $c = 3d$ 25. 3.8, 3.5, 3.05, 0.39 27. 0.125 29. 0.2

Pages 379–380 **Lesson 7-3**

1. 0.27 3. 0.04 5. 1.15 7. 32% 9. 91% 11. 291% 13. 70% 15. 0.35 17. 0.03 19. 1.04 21. 0.95 23. 0.96 25. 99% 27. 355% 29. 60% 31. 87% 33. 12% 35. < 37. > 39. 0.01537, 10.37%, 15.37%, 1.537 41. Sample answer: Niko scored a 92% on his math test. Express this percent as a decimal. 43. 0.25 45. $\frac{6}{25}$ 47. $1\frac{1}{4}$ 49. 12 years old 51. $\frac{3}{8}$ 53. $\frac{3}{7}$

Pages 384–386 **Lesson 7-4**

1. $\frac{1}{9}$ 3. $\frac{2}{9}$ 5. $\frac{5}{9}$ 7. The complement of selecting a "Go Back 1 Space" card is selecting a card other than that card. The probability of the complement is 75%, $\frac{3}{4}$, or 0.75. 9. 0 or $\frac{0}{1}$ 11. $\frac{3}{4}$ 13. $\frac{5}{8}$ 15. $\frac{1}{2}$ 17. $\frac{7}{10}$ 19. $\frac{1}{2}$ 21. $\frac{3}{5}$ 23. 83%, 0.83, or $\frac{83}{100}$ 25. The chances of picking a purple jelly bean or not choosing a purple jelly bean are equally likely to happen since the probability of picking a purple jelly bean is 50%. 27. Picking a green jelly bean is not likely to happen since the probability of picking a green jelly bean is 10%. 29. Spinner A: $\frac{1}{2}$, 0.5, 50%; Spinner B: $\frac{1}{8}$, 0.125, 12.5%; Spinner C: $\frac{1}{4}$, 0.25, 25%; Sample answer: In Spinner A, the green section is half of the circle. In Spinner B, the green section is one-eighth of the circle. In Spinner C, the green section is one-fourth of the circle. 31. No; Sample answer: The sizes of the angles formed by each section are not equal. 33. 4 : 2 or 2 : 1 35. 3 : 3 or 1 : 1
37. Sample answer:

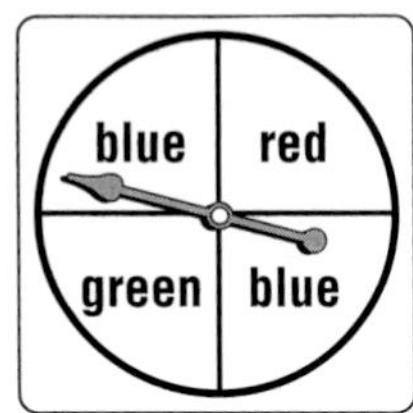

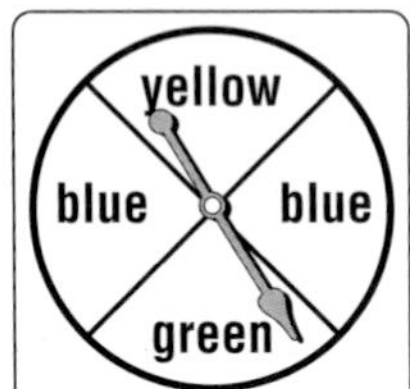

The first spinner is a possibility because $\frac{2}{4}$ or $\frac{1}{2}$ of the sections are blue. The second spinner is a possibility because $\frac{2}{4}$ of the whole spinner is blue. 39. A 41. 0.93 43. $11\frac{1}{2}$ 45. $3\frac{1}{10}$ 47. heads, tails 49. Jan., Feb., Mar., April, May, June, July, Aug., Sept., Oct., Nov., Dec.

Pages 391–393 **Lesson 7-5**

1. Let R = Ramiro, G = Garth, and L = Lakita. The different ways are RGL, RLG, GRL, GLR, LRG, and LGR. So, there are 6 ways the three students can line up. 3. 42 5. Let 1 = CD 1, 2 = CD 2, 3 = CD 3, and 4 = CD 4 4. The different ways are 1234; 1243; 1324; 1342; 1423; 1432; 2134; 2143; 2314; 2341; 2413; 2431; 3124; 3142; 3214; 3241; 3412; 3421; 4123; 4132; 4213; 4231; 4312; and 4321. So, Kame can listen to four CDs 24 ways.

9.

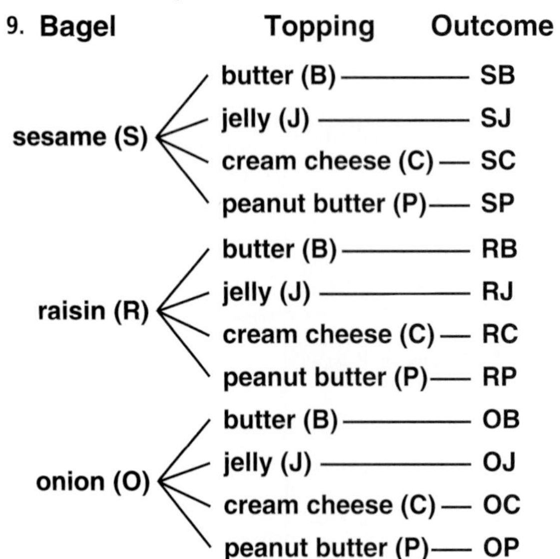

11.

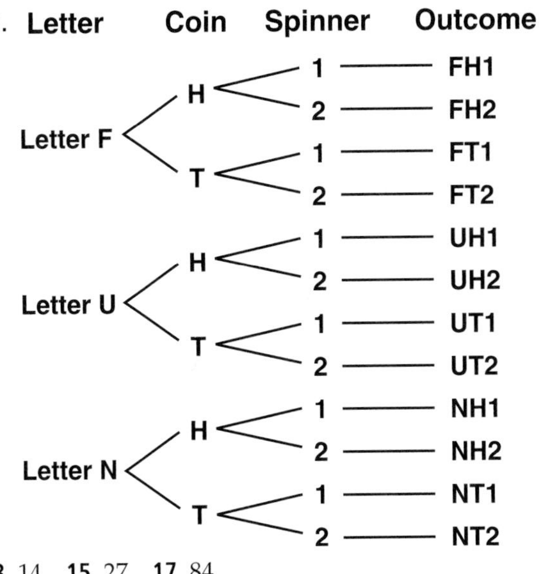

13. 14 **15.** 27 **17.** 84

19. $\frac{1}{12}$

Multiple-Choice	True/False	True/False	Outcome
A	true (T)	true (T)	ATT
		false (F)	ATF
	false (F)	true (T)	AFT
		false (F)	AFF
B	true (T)	true (T)	BTT
		false (F)	BTF
	false (F)	true (T)	BFT
		false (F)	BFF
C	true (T)	true (T)	CTT
		false (F)	CTF
	false (F)	true (T)	CFT
		false (F)	CFF

21. $\frac{2}{9}$ **23a.** 10 groups **23b.** 30 ways; Sample answer: By just choosing a three-student group, the order in which the names are drawn does not matter. When the order matters, such as in part b, the sample space is increased. A three-person group of Kayla, Jeremy, and Chi-Wei is the same as a three-person group of Jeremy, Chi-Wei, and Kayla. However, these two groupings would reflect different orderings of captain, co-captain, and secretary in part b. **25.** Sample answer: the results when a number cube is rolled and a coin is tossed **27.** J **29.** $\frac{4}{7}$ **31.** 0.35 **33.** 2 **35.** 7

Pages 396–398 ***Lesson 7-6***

1. $\frac{1}{8}$, 0.125, or 12.5% **3.** $\frac{3}{10}$, 0.3, or 30% **5.** $\frac{3}{5}$, 0.6, or 60% **7.** 180 **9.** 40 **11.** about 560,000 **13.** about 9,000 **15.** about 6 free-throws **17.** about 60 times **19.** Sample answer: Paper/pencil; Since Nolan is finding the probability, he just needs to write $\frac{4}{14}$ as a reduced fraction, $\frac{2}{7}$. **21.** A **23.** J **25.** $\frac{8}{15}$ **27.** $\frac{8}{15}$ **29.** $\frac{2}{3}$ **31.** $2\frac{4}{9}$ **33.** $8\frac{1}{3}$ **35.** $150

Pages 399–400 ***Lesson 7-7***

1. Sample answer: When there is a way to solve the problem in which you can arrive at the answer by using simpler numbers. **3.** Sample answer: When multiplying 60 by 1,100, first multiply 6 by 11, which is 66. Then add three zeroes. So, $60 \times 1{,}100 = 66{,}000$. **5.** Sample answer: 10 movies **7.** 168 **9.** 32 mi **11.** 35 s **13.** 4.5 times **15.** times 3; 36

Pages 403–405 ***Lesson 7-8***

1. Sample answer: $\frac{1}{5}$ of $50 is $10

3. Sample answer: $\frac{1}{2}$ of $120 is $60

5. Sample answer: $\frac{3}{5}$ of 15 is 9

7. Sample answer: $\frac{1}{5}$ of $25 is $5

9. Sample answer: $\frac{1}{5}$ of 100 is 20

11. Sample answer: $\frac{1}{5}$ of 70 is 14

13. Sample answer: $\frac{1}{4}$ of 120 is 30

15. Sample answer: $\frac{4}{5}$ of 80 is 64

17. Sample answer: $\frac{9}{10}$ of 200 is 180

19. Sample answer: $\frac{3}{4}$ of 8 h is 6 h

21. Sample answer: $\frac{2}{5}$ of 650,000 is 260,000

23. Sample graph:

Communicating With Grandparents

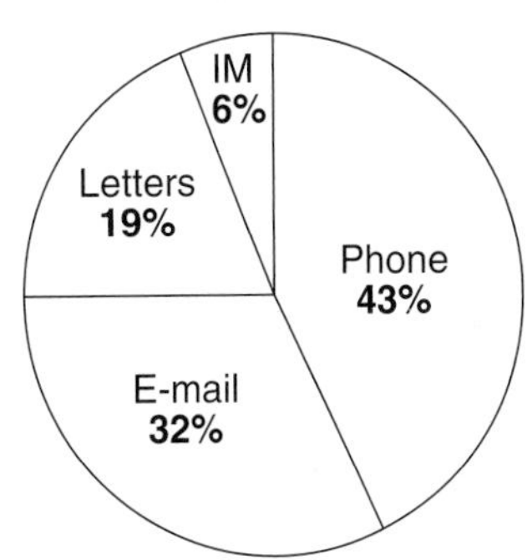

25. about 25% **27.** about 40% **29.** More; Rachel rounded $32 down to $30, so the actual amount she will save will be more than $12. **31.** B **33.** G **35.** 15 T-shirts **37.** 45% **39.** 36.2%

Pages 406–410 ***Chapter 7*** ***Study Guide and Review***

1. false; sample space **3.** true **5.** false; complementary events **7.** $\frac{3}{100}$ **9.** $1\frac{1}{5}$ **11.** 160% **13.** $\frac{13}{100}$ **15.** comedy **17.** The section representing comedy is 4 times the size of the section representing romance. **19.** 0.38 **21.** 0.9 **23.** 130% **25.** 51% **27.** $\frac{1}{2}$ **29.** $\frac{1}{4}$ **31.** $\frac{2}{7}$ **33.** $\frac{4}{7}$ **35.** $\frac{6}{14}$ or $\frac{3}{7}$

37.

Color	Style
black	classic fit
black	stretch
black	bootcut
blue	classic fit
blue	stretch
blue	bootcut

6 outcomes

39.

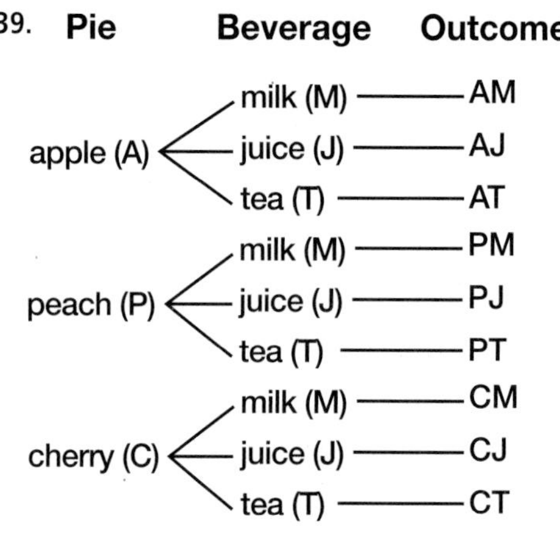

41. 12 **43.** $\frac{1}{4}$, 0.25, or 25% **45.** 92 people
47. Sample answer: $\frac{2}{5}$ of 80 is 32
49. Sample answer: $\frac{1}{4} \times 120 = 30$
51. Sample answer: $\frac{1}{2} \times 50 = 25$
53. Sample answer: $\frac{1}{4}$ of 20,000,000 is 5,000,000 or $\frac{3}{10}$ of 20,000,000 is 6,000,000
55. Sample answer: $\frac{1}{5}$ of 3,000 is 600 students

Chapter 8 Systems of Measurement

Page 417 ***Chapter 8*** ***Getting Ready***

1. 20.69 **3.** 12.96 **5.** 16.52 **7.** 9.5 mi **9.** 43.67 **11.** 5.14 **13.** 18.38 **15.** 3,800 **17.** 6,750 **19.** 71,800 **21.** $56,000

Pages 421–423 ***Lesson 8-1***

1. **3.** $1\frac{1}{2}$ in. **5.** 12 **7.** 2 **9.** B
11.
13.

15. $\frac{5}{8}$ in. **17.** $1\frac{1}{4}$ in. **19.** $1\frac{1}{2}$ in. **21.** 18 **23.** 15,840 **25.** $3\frac{1}{3}$ **27.** $42\frac{1}{2}$ **29.** 152 yd **31.** 54 in.; $4\frac{1}{3}$ ft is the same as 52 in., which is less than 54 in.
33. Not reasonable; 30 inches is 2.5 feet, which would be long for a backpack. **35.** miles; Sample answer: The distance from my home to school is longer than several yards, so I would use miles. **43.** Forty-five yards does not equal 90 feet. Since 1 yard equals 3 feet, multiply by 3. So, 45 yards equals 135 feet. **45.** 192 sixteenths; 72 half inches **47.** Sample answer: No, each foot has 12 inches, so 24 feet has 24 × 12 or 288 inches.
49. H **51.** Sample answer: $\frac{1}{3}$ of 120 is 40 **53.** Sample answer: $\frac{3}{4}$ of 80 is 60 **55.** line graph **57.** 80 **59.** 50

Pages 426–429 ***Lesson 8-2***

1. 14 **3.** 2 **5.** 8 **7.** 14,000 **9.** 20 **11.** 10 **13.** 3 **15.** 12 **17.** 72 **19.** $\frac{3}{4}$ **21.** $5\frac{1}{4}$ **23.** 286,000 lb **25.** 5 batches **27.** 24 **29.** 14 **31.** fluid ounces **33.** tons **35.** 4 pt; 4 pt = 64 fl oz, which is greater than 60 fl oz. **41.** 192 oz
43. Multiply $\frac{1}{3}$ by 16, or divide 16 by 3. **45.** Sample answer: Mental math; since she is at the grocery, she can mentally convert the quantities to determine how much to buy. Two pints is equal to 4 cups. Ten ounces is a little more than one cup, and 24 ounces is 3 cups. So buying a 10-ounce and 24-ounce container, Antonia will have a little more than 4 cups, which will be enough for the recipe. **47.** Same capacity; both quantities equal 1 cup; however, a cup of sand would weigh more than a cup of cotton balls. **49.** 19.5 **51.** 66 in. **53.** $\frac{3}{4}$, 0.75, or 75% **55.** 13.3, 8.5, 6.1, 4.7, 3.4 **57.** Sample answer: 1 yd

Pages 434–436 ***Lesson 8-3***

1. millimeter **3.** meter **5.** Sample answer: 2 cm; 19 mm **7.** millimeter **9.** meter **11.** centimeter **13.** kilometer **15.** Sample answer: 35 mm; 36 mm **17.** Sample answer: 5 cm; 5.3 cm **19.** Sample answer: 10 cm; 8 cm **21.** meter **23.** Sample answer: 8.5 cm **25.** Sample answer: 7 mm **29.** yard **31.** 3 cm; 15 mm is the same as 1.5 cm, which is less than 3 cm. **33.** 2 km; 1 km is about 0.6 mi. So, 2 km is about 0.6 + 0.6 or 1.2 mi, which is greater than 1 mi. **35.** centimeter; Sample answer: To build a proper fence line, one needs to be accurate to the nearest centimeter. **37.** Sample answers: pencil, book, eraser; centimeter **39.** Sample answer: millimeter, height of a comma; centimeter, width of a straw; meter, distance from a person's nose to the end of their outstretched arm; kilometer, distance from a museum to a restaurant
41. H **43.** 48 **45.** 4 **47.** $\frac{3}{5}$ **49.** $\frac{3}{7}$ **51.** *Shrek 2*: $436.5; *Finding Nemo*: $339.7; *The Lion King*: $328.5; *Shrek*: $267.7; *The Incredibles*: $261.4 **53.** Sample answer: can of soda **55.** Sample answer: orange juice

Pages 439–441 ***Lesson 8-4***

1. gram; Sample answer: 10 g **3.** kilogram; Sample answer: 4 kg **5.** gram; Sample answer: 60 g **7.** less **9.** more; Sample answer: 270 g × 4 = 1,080 g; 1 kg = 1,000 g and 1,080 g > 1,000 g **11.** gram; Sample answer: 8 g **13.** kilogram; Sample answer: 400 kg **15.** liter; Sample answer: 80 L **17.** gram; Sample answer: 400 g **19.** milligram; Sample answer: 21 mg **21.** milliliter; Sample answer: 0.1 mL **23.** Cinnamon Teal, Hottentot Teal, and Marbled Teal; Their combined mass is 991 grams which is close to 1,000 grams or 1 kilogram. **25.** 243 milliliters; 1.36 L is more than 1 L or 1,000 mL. Since 1,000 mL > 243 mL, the 243-mL bottle has less mass. **27.** 10 bathtubs **29.** Sample answer: a large water bottle **31.** False; The amount of feathers needed to fill a container and the amount of marbles needed to fill a container of the same size will both have different masses. The marbles will have a greater mass. **33.** A **35.** centimeter **37.** 80 oz **39.** Sample answer: Yes; an 85% chance of rain means that it is likely to rain. So, you should carry an umbrella.

Pages 442–443 ***Lesson 8-5***

1. Sample answer: 3 feet is equal to 1 yard, so 3 heel-to-toe steps would be about 1 yard. **3.** Sample answer: Cut the string so that the length is the same as the width of the door. Determine how many string lengths are needed from the floor to the bottom of the cabinets. **5.** mean **7.** Sample answer: Divide the height of the doorway into thirds. Have each student stand next to the doorway to determine whether he or she is taller than the $\frac{2}{3}$ mark. **9.**
11. 20 games

Pages 447–449 ***Lesson 8-6***

1. 95,000 **3.** 380 **5.** 0.205 **7.** 44.3 km **9.** 1.9 **11.** 3.54 **13.** 238,000 **15.** 18,000 **17.** 0.007 **19.** 0.45 **21.** 5 laps **23.** 0.00025 **25.** 300,000 **27.** 55 cm, 560 mm, 5.6 km **29.** 8,500 mm, 80 m, 8.2 km **31.** 3,750 m **33.** 15.4 **35.** 22,400 cm; 201 m; 198,000 mm; 19,600 cm; 192 m **37.** 202.2 m, 198 m, none; median **39.** $(x \div 1{,}000{,}000)$ kg **41.** Shayla; to change from a smaller unit to a larger unit, you need to divide. **43.** 3,000 **45.** H **47.** 360 mL **49.** $\frac{15}{8}$ **51.** $\frac{48}{7}$ **53.** 4 feet **55.** 8.12 **57.** 106.4

Pages 452–454 ***Lesson 8-7***

1. 11 h 15 min **3.** 15 h 30 min 29 s **5.** 8 h 10 min **7.** 1 h 5 s **9.** 1 h 33 min 37 s **11.** 8 h 9 min 8 s **13.** 2 h 50 min **15.** 4 h 26 min **17.** 6 h 9 min **19.** 9 h 10 min **21.** 8 h 1 min 15 s **23.** aerobics **25.** 1:57 P.M. **27.** Never; Sample answer: movies usually run for at least $1\frac{1}{2}$ hours. Stopwatches are better equipped to time a short distance race where seconds or even tenths of seconds are important. **29.** 3 h 40 min 65 s; 3 h 40 min 65 s = 3 h 41 min 5 s, but the others equal 3 h 42 min 24 s. **31.** C **33.** F **35.** 6,500 **37.** liter **39.** 304 **41.** 125

Pages 457–458 ***Lesson 8-8***

1. 350°F **3.** 101°F **5.** Sample answer: 40°F and 5°C **7.** Sample answer: 70°F and 25°C **9.** Sample answer: 0°F **11.** 5°C **13.** 200°F **15.** −10°C **17.** The friend; in the Fahrenheit scale, water freezes at 32°F, so 40°F would be too warm. **19.** Sample answer: 130°F or 55°C **21.** Sample answer: 20°F or −10°C **23.** Sample answer: 90°F or 30°C **25.** Sample answer: 127°F **27.** 25°C **29.** 90°C **31.** A **33.** 40 min 57 s **35.** 8.4

Pages 461–464 ***Chapter 8*** ***Study Guide and Review***

1. one hundredth **3.** gram **5.** liter **7.** longer **9.** 60 **11.** 2 **13.**
15. 10 ft **17.** 11 **19.** 16 **21.** 12 **23.** meter **25.** millimeter **27.** centimeter **29.** liter; 2 L **31.** kilogram; 1,500 kg **33.** 550 mL **35.** Sample answer: Divide the quart into 4 equal parts. Then fill the quart with orange juice until it is three-fourths filled. **37.** 0.001 **39.** 5,000 **41.** 0.023 **43.** 3,500 **45.** 7 h 36 min **47.** 59 min 52 s **49.** 10 h 8 min 10 s **51.** 45 min **53.** 212°F **55.** 58°C

Chapter 9 Geometry: Angles and Polygons

Page 469 ***Chapter 9*** ***Getting Ready***

1. 4 **3.** 58 **5.** 19 **7.** 58 **9.** 181 **11.** Yes

Pages 472–473 ***Lesson 9-1***

1. 125°; obtuse **3.** 90°; right **5.** 90°; right **7.** 30°; acute **9.** 130°; obtuse **11.** acute **13.** 90° **15.** 110° **17.** acute **19.** Lorena is correct. Nathan read the wrong scale on the protractor. **21.** Sample answer: First determine if it is a 90° angle, or the size of the angle on a corner of a piece of paper. If so, it is a right angle. If the angle is smaller than this, it is acute. If it is larger, but less than a straight line, it is obtuse. **23.** G **25.** 41 h 35 min

27. **29.**

Pages 475–478 ***Lesson 9-2***

1. about 120° **3.** about 160° **5.** about 75° **7.** **9.** about 45° **11.** about 10° **13.** about 70°

15. **17.** **19.** **21.**

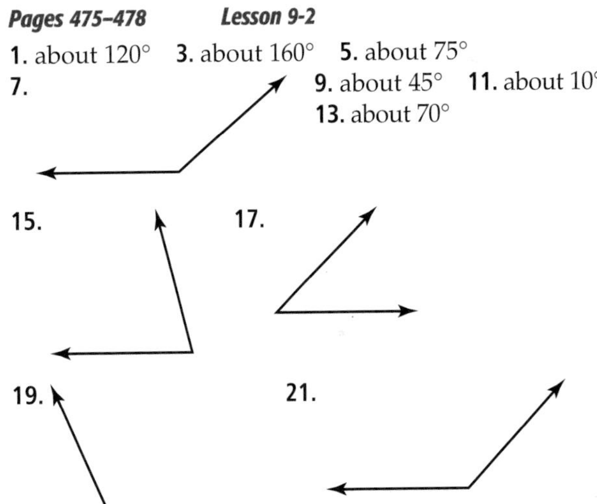

23. about 150° **25.** about 210° **27.** about 185° **29.** Sample Answer; Ladder A is at about 60°, Ladder B is at about 15°, and Ladder C is at about 45°. Only Ladder B is safe. **31.** about 40°; prune

33.

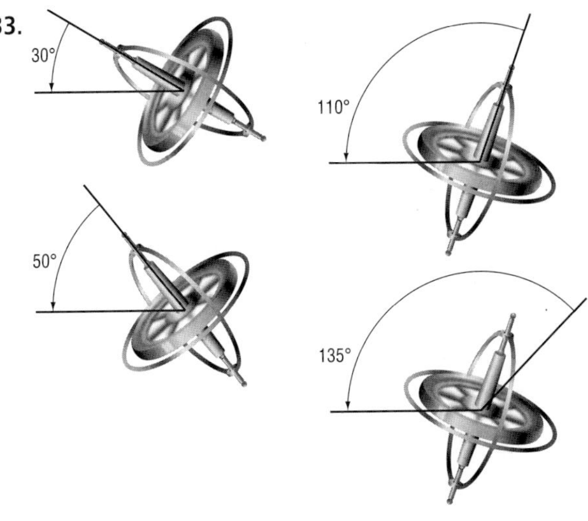

35. Sample answer: $m\angle A \approx 110°$, $m\angle B \approx 70°$, $m\angle C \approx 110°$, $m\angle D \approx 70°$. Opposite angles of the figure have the same angle measures. **37.** Sample answer: K 1 $\angle 1$ measures about 80–90°. **39.** C

41. 90°; right **43.** 150°; obtuse **45.** 140°C **47.** 18.4 million mi^2 **49.** 135 **51.** 50

Pages 481–484 ***Lesson 9-3***

1. supplementary **3.** complementary **5.** 150 **7.** 70 **9.** supplementary **11.** complementary **13.** supplementary **15.** 88 **17.** 155 **19.** 35 **21.** Sample answer: If the measure of $\angle 1$ is 50°, what is the measure of $\angle 2$? **23.** 115° **25.** vertical angles **27.** vertical angles **29.** Always; Sample answer: Vertical angles are congruent. **31.** Never; Sample answer: Since an obtuse angle is greater than 90°, the sum of two obtuse angles must be greater than, not equal to, 180°. **33.** 25 **35.** Two; Sample answer: The 65° angle and the angle labeled $x°$ are complementary and the 30°angle and the angle labeled $y°$ are complementary. **37.** Sample answer: a. an obtuse angle, b. a right angle, c. no, two acute angles will never add up to 180° because each acute angle, by definition has a measure that is less than 90°. Two angles that are both less than 90° will never add up to 180°. **39.** The measures of the two angles must be equal. If two angles are supplementary to the same angle, say $x°$, then each angle has a measure of $(180 - x)°$ and thus have the same measure themselves. **41.** 95

43. **45.**

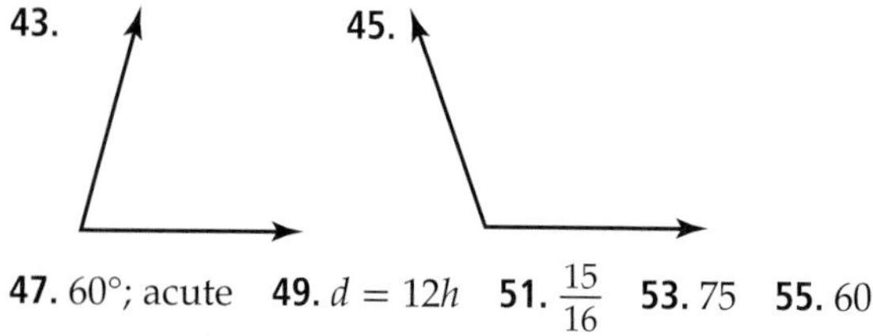

47. 60°; acute **49.** $d = 12h$ **51.** $\frac{15}{16}$ **53.** 75 **55.** 60

Pages 489–491 ***Lesson 9-4***

1. obtuse **3.** 45 **5.** 90 **7.** scalene **9.** acute **11.** obtuse **13.** obtuse **15.** 100 **17.** 50 **19.** 120 **21.** 65 **23.** equilateral; also isosceles **25.** scalene **27.** 105° **29.** They are complementary. **31.** 90°, 60°, 30°; right scalene **33.** 70°, 60°, 50°; acute scalene **35.** $x = 25$; $y = 50$ **37.** C **39.** 55°

41. **43.** 18 yd

45. **47.**

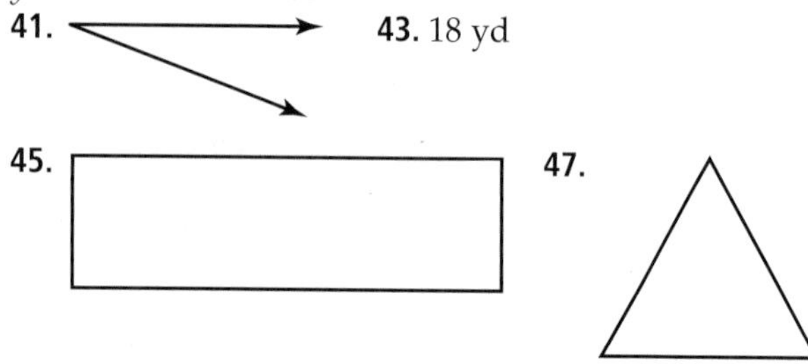

Pages 496–499 ***Lesson 9-5***

1. 75 **3.** The first figure is a rectangle. The second figure is a square. **5.** 95 **7.** 67 **9.** 135 **11.** rectangle **13.** parallelogram **15.** quadrilateral **17.** The first figure is a rectangle. The second figure is a parallelogram. **19.** Polygon 3 is a square. Polygon 5 is a parallelogram. Polygon 3 is a regular polygon. **21.** 57.2 **23.** Sample answer: The shapes have all sides congruent. **25.** Sample answer: chalkboard: quadrilateral or rectangle; floor tile: square. **27.** Sometimes; When a rhombus has all angles congruent, it is a square. **29.** Sometimes; When the rectangle has all sides congruent, it is a square. **31.** See students' work for drawings. They should have drawn an equilateral triangle and a square. Side lengths may vary by drawing, but the side lengths on each drawing should be equivalent. The angle measures for the equilateral triangle should be 60°, 60°, and 60°. The angle measures for the square should be 90°, 90°, 90°, and 90°. **33.** B **35.** 105 **37.** 74 **39.** supplementary **41.** $6\frac{3}{5}$ mi **43.** yes **45.** yes

Pages 500–501 ***Lesson 9-6***

1. Sample answer: The students probably drew a diagram to solve the problem because a diagram would help them to understand and picture the information. **3.** 5 ways **5.** 12 brownies

7.

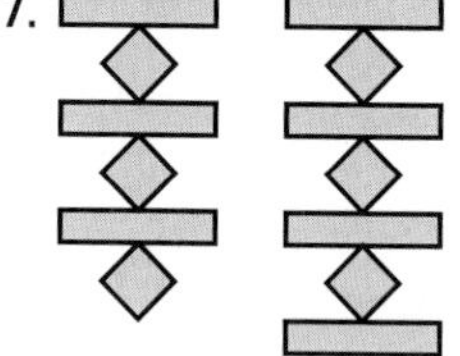

9. 12 years old **11.** about 1.7 times **13.** 10 streamers

Pages 505–507 ***Lesson 9-7***

1. congruent **3.** neither **5.** $\overline{ST}$ **7.** similar **9.** similar **11.** neither **13.** similar **15.** $\overline{DE}$ **17.** $\overline{LN}$ **19.** similar **21.** not similar **23.** not similar **25.** 132 ft; 112 ft **27.** 14

29. Sample answer:

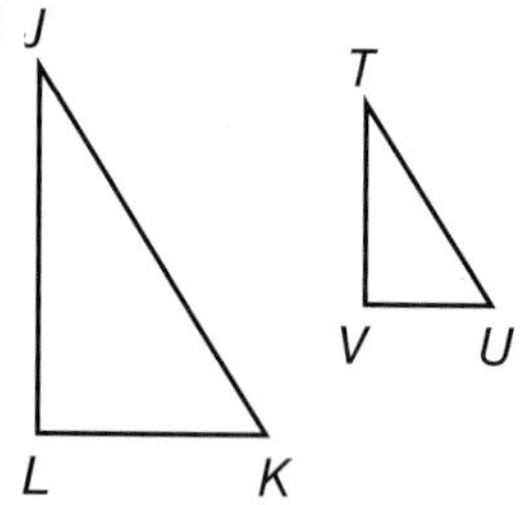

Triangle *JKL* is similar to triangle *TUV*.

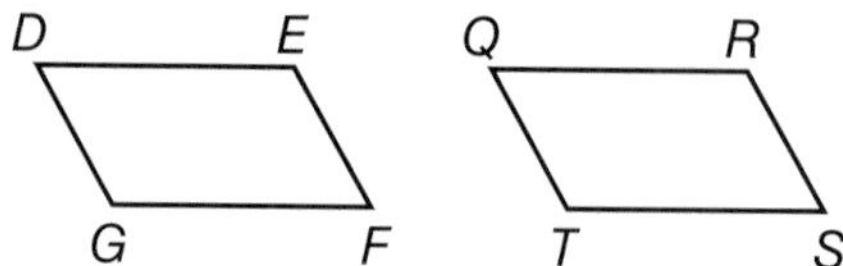

Parallelogram *DEFG* is congruent to parallelogram *QRST*.

31. blue triangle $1\frac{1}{2}$ in.; red triangle: 3 in.; Sample answer: The ratio of the perimeters, 3: $1\frac{1}{2}$ or 2:1, is equal to the ratio of the side lengths, 1: $\frac{1}{2}$, or 2:1. **33.** C **35.** There were six signals sent. **37.** 80 **39.** The probability that it will not snow is 40%.

Pages 509–514 **Chapter 9** **Study Guide and Review**

1. false, vertical angles **3.** false, acute **5.** true **7.** 145°; obtuse **9.** [figure] **11.** [figure]

13. about 120° **15.** Sample answer: about 17° **17.** neither **19.** 44 **21.** 78 **23.** acute **25.** 49 **27.** scalene **29.** isosceles **31.** 64 **33.** square **35.** 15 lines **37.** similar **39.** rectangle *WXYZ*

Chapter 10 Measurement: Perimeter, Area, and Volume

Page 519 **Chapter 10** **Getting Ready**

1. 36 **3.** 26 **5.** \$70 **7.** 37.7 **9.** 81.7 **11.** 115 **13.** 77 **15.** 1,155 **17.** 208

Pages 524–526 **Lesson 10-1**

1. 64 in. **3.** 42 m **5.** 3,000 mm **7.** 52 cm **9.** 61.4 in. **11.** 400 mm **13.** 8 in. **15.** 30 ft **17.** 150 cm **19.** $14\frac{5}{8}$ in. **21.** 4 **23.** No; a 4 by 2 rectangle and a 5 by 1 rectangle both have a perimeter of 12 units **25.** Each rectangle has the same perimeter: 14 ft Sample answer: 1 ft by 9 ft, 2 ft by 8 ft, and 3 ft by 7 ft. **27.** B **29.** similar **31.** congruent **33.** $\frac{3}{10}$ of 160 is 48 **35.** $\frac{1}{3}$ of 90 is 30 **37.** 43.4 **39.** 37.6

Pages 531–533 **Lesson 10-2**

1. 1.5 m **3.** 10 in. **5.** 63 ft **7.** 40.8 cm **9.** 69.1 in. **11.** 2.5 mm **13.** 34 cm **15.** 24 ft **17.** 54 mi **19.** 39 ft **21.** 100.5 in. **23.** 37.7 cm **25.** 131.9 mm **27.** 37.7 cm **29.** 19 people **31.** Greater than; Sample answer: Since the radius is 4 feet, the diameter is 8 feet. Since π is a little more than 3, the circumference will be a little more than 3 times 8, or 24 feet. **33.** no; Sample answer: The circumference of each candle is about 12.6 inches. So, 12.6 × 8 or 100.8 inches of ribbon are needed to make all of the candles. Since 2 yards is equal to 72 inches, and 72 < 100.8, she does not have enough ribbon for all of the candles. **35.** Bena entered the correct keystrokes. Orlando did not multiply the radius by 2. **37.** Using compatible numbers, divide the circumference by π; 15 ÷ 3 = 5 m **39.** H **41.** 64 ft **43.** 80 mi **45.** not similar **47.** $\frac{7}{26}$ **49.** 102 **51.** 180

Pages 536–538 **Lesson 10-3**

1. 18 units^2 **3.** 77 m^2 **5.** 13.26 cm^2 **7.** 24 units^2 **9.** 48 m^2 **11.** 814 ft^2 **13.** 35 m^2 **15.** 3,000 mi^2 **17.** 84 cm^2 **19.** Patio 1: $9\frac{1}{3}$ ft; Patio 2: $12\frac{1}{2}$ ft; Patio 3: $14\frac{3}{4}$ ft

21.

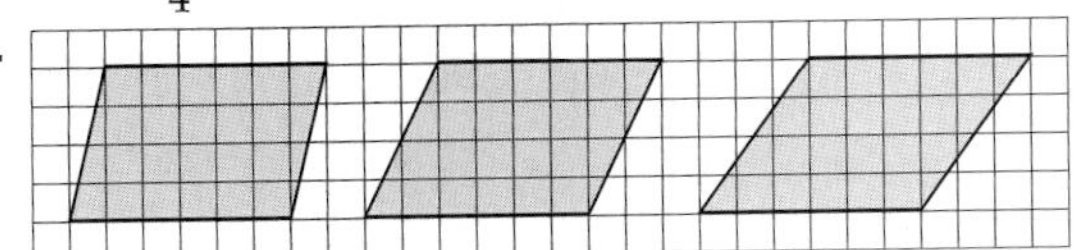

Sample answer: Each parallelogram has the same base, height and area, but each parallelogram has a different slant. **23.** The formula for the area of a parallelogram $A = bh$ corresponds to the formula for the area of a rectangle $A = \ell w$ in that the base b corresponds to the length ℓ and the height h corresponds to the width w. **25.** G **27.** 114 m **29.** 612 ft **31.** 30 **33.** 84

Pages 542–544 **Lesson 10-4**

1. 6 units^2 **3.** 87.75 m^2 **5.** 24 units^2 **7.** 45 in^2 **9.** 87.5 m^2 **11.** 245 in^2 **13.** 14 yd^2 **15.** 27 ft^2; 3 bags **17.** 64 in^2 **19.** 16 in.; 14 in^2 **21.** Dolores; The height is 17 meters, not 18. **23.** 4 in^2; 64 in^2 **25.** 144 in^2; Sample answer: The total area of the triangles and the smaller squares is 64 in^2 + 80 in^2 or 144 in^2. The area of the large square is 12 in. × 12 in. or 144 in^2. The areas are equal. **27.** Sample answer:

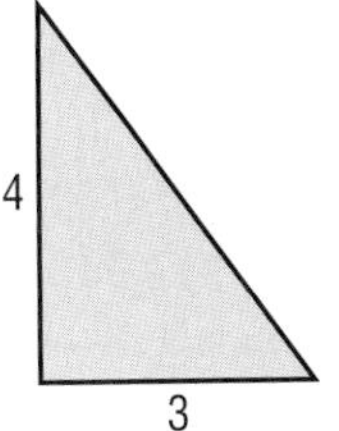

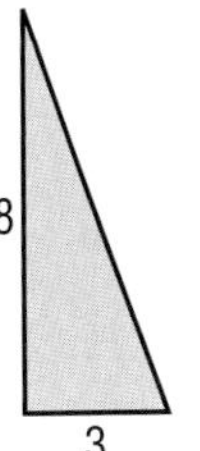

Area = 6 units^2 Area = 12 units^2; 1:2

29. G **31.** 31.4 m **33.** 24 ways

Pages 546–547 **Lesson 10-5**

1. Sample answer: By making a model, it allowed D.J. to first see if he had enough chairs before setting them all up. **3.** 3 ft × 11 ft; 6 ft × 8 ft **5.** 64 pieces **7.** 70 parents **9.** 4 holes **11.** \$170 **13.** 5 boys

Pages 550–553 **Lesson 10-6**

1. 15 ft^3 **3.** 126.54 yd^3 **5.** 4,986.875 in^3 **7.** 120 m^3 **9.** 600 yd^3 **11.** 1,430 ft^3 **13.** 2,702.5 in^3 **15.** 17 m **17.** < **19.** = **21.** 16.875 in^3 **23.** They both have the same volume. Volume of the first prism: 5 × 4 × 10 or 200 in^3. Volume of the second prism: 10 × 5 × 4 or 200 in^3. **25.** 3-bedroom moving truck **27.** No; rounding down will give an underestimate. The volume is greater than 5 × 3 × 12 or 180 cm^3. **29.** Prism C; All the other prisms have a volume of 96 units^3, while Prism C has a volume of 72 units^3. **31.** Calculator; he can calculate the volume first in cubic inches and then convert this to gallons. $V = 36 \times 13 \times 16 = 7{,}488$ in^3; 7,488 in^3 ÷ 231 in^3 ≈ 32 gallons. So, Basilio will need about 32 gallons of water to fill his fish tank. **33.** Each of the three dimensions being multiplied is expressed in a unit of measure. And just like the product 5 × 5 × 5 can be expressed using exponents as 5^3, feet × feet × feet can be expressed with exponents as feet^3. **35.** F **37.** 988 ft^2 **39.** 125 **41.** 3.5 **43.** 54 cm^2 **45.** 322 in^2

Pages 557–559 **Lesson 10-7**

1. 292 m^2 **3.** 298 cm^2 **5.** 256 in^2 **7.** $384\frac{5}{8}$ cm^2 **9.** 3,668.94 m^2 **11.** 390 in^2 **13.** Yes; The approximate surface area of the rectangular prism is (2 × 13 × 6) + (2 × 13 × 8) + (2 × 6 × 8) or 460 ft^2. **15.** Area; Sample answer: The length and width of the house will determine the land area needed to build the house.; m^2 **17.** Area; Sample answer: the length and width of the floor will determine the number of tiles; ft^2 **19.** Volume; Sample answer: capacity is the amount of cereal inside the box; ounces **21.** 316.5 in^2 **23.** 207.75 in^2

25. 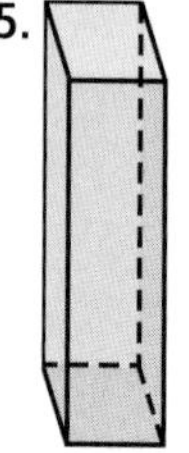 **27.** 48 in^2; 144 in^2 **29.** Sample answer: Angelina is wrapping a gift box that measures 8 inches by 2 inches by 12 inches. What is the minimum amount of wrapping paper she will need? **31.** H **33.** perimeter: 52 ft; area: 88 ft^2 **35.** 6; Let F represent the first friend, S represent the second friend, and T represent the third friend; FST, FTS, STF, SFT, TFS, and TSF. **37.** 535% **39.** 210%

Pages 561–564 Chapter 10 Study Guide and Review

1. volume **3.** perimeter **5.** volume **7.** diameter **9.** 90 ft **11.** $r = 29$ cm **13.** $d = 18$ ft **15.** 54 yd **17.** 53.4 cm **19.** 125.6 ft **21.** 1,395 ft^2 **23.** 40 in^2 **25.** 7.5 units^2 **27.** 6 m^2 **29.** $9\frac{9}{16}$ ft^2 **31.** 16 **33.** 108 ft^3 **35.** 266 in^2 **37.** 94 in^2

Chapter 11 Integers and Transformations

Page 571 Chapter 11 Getting Ready

1. 27 **3.** 12 **5.** 36 **7.** 5 **9.** 5 **11.** 42 **13.** 45 **15.** 16 **17.** 15 **19.** 9 **21.** 2 **23.** 15

Pages 573–575 Lesson 11-1

1. $<$ **3.** $>$ **5.** −13, −8, −5, 1, 9 **7.** 9, 7, −8, −14, −33, −54 **9.** Kate; −2 **11.** $<$ **13.** $<$ **15.** $<$ **17.** 1, 5, 6, 14, 14, 23 **19.** −221, −89, −71, −10, 54, 63 **21.** Gary, Beth, Sindhu **23.** Sun **25.** −9 **27.** D; Since $-4 < -3 < 2 < 6$, player D had the most strokes over par. **29.** Sample answer: Since −7 is to the right of −11 on a number line, $-11 < -7$. **31.** Sample answer: Graph them on a number line and then write the integers as they appear from right to left. **33.** F **35.** 2,450 ft^3 **37.** $2\frac{1}{3}$ **39.** $5\frac{1}{8}$ **41.** 2 **43.** 7

Pages 579–581 Lesson 11-2

1. +4 or 4 **3.** −8 **5.** −3 **7.** $90 **9.** +6 or 6 **11.** +7 or 7 **13.** +13 or 13 **15.** −9 **17.** −6 **19.** −16 **21.** 0 **23.** −5 **25.** −10 **27.** 27 + (−14); 13 messages **29.** −5 **31.** 1 **33.** 3 **35.** −$22 **37.** New York City: 8 A.M.; Jakarta: 3 A.M.; Paris: 4 A.M. **39.** 9 ft **41.** Ramón is correct. He used 4 positive counters and 6 negative counters to correctly model 4 + (−6). After forming zero pairs, he is left with the correct answer, −2. **43.** zero **45.** negative **47.** negative **49.** Sample answer: $x = 24$, $y = 8$ **51.** D **53.** −10, 0, 3, 5 **55.** 72 cm **57.** 39 in. **59.** 130 ft **61.** 2 **63.** 7

Pages 585–586 Lesson 11-3

1. 2 **3.** −5 **5.** −6 **7.** $7 **9.** 1 **11.** 7 **13.** 7 **15.** −4 **17.** −3 **19.** 0 **21.** −8 **23.** −9 **25.** −20 **27.** 17°F **29.** −7 **31.** Sample answer: $a = -8$ and $b = -10$ **33.** 2; Sample answer: $x + (-y) = x - y$ and $x - y = 2$. **35.** 8; Sample answer: $x - (-y) = x + y$ and $x + y = 8$. **37.** no; Sample answer: A positive integer minus a negative integer is positive. **39.** J **41.** 0 **43.**

−10 −9 −8 −7 −6 −5 −4 −3 −2 −1 0 1 2 3 4 5 6 7 8 9 10

45. 150% **47.** 140% **49.** 12 **51.** 56 **53.** 72

Pages 589–590 Lesson 11-4

1. −28 **3.** 16 **5.** 7 **7.** −72 m **9.** −9 **11.** −20 **13.** −18 **15.** −54 **17.** 36 **19.** 63 **21.** 45 **23.** 9 **25.** −8,800 ft **27.** 60 **29.** −162; −486; Each term is multiplied by 3. **31.** 24 **33.** −9°F **35.** Sample answer: −6 and 3, 9 and −2, and −18 and 1 **37.** false; Sample answer: The product of any two negative integers is always positive. **39.** true; Sample answer: The product of a positive integer and a negative integer is always negative. **41.** Sample answer: Mr. Morris wrote three checks for $25. Write and integer to represent the amount of money he has in his checking account after writing the three checks. **43.** J **45.** 2 **47.** −13 **49.** 1 **51.** −1 **53.** 0.82 m **55.** 5

Pages 592–593 Lesson 11-5

1. Sample answer: Start with the answer that you found and then work the problem backward to see if you arrive at the number you were given in the problem. **3.** 6:55 A.M. **5.** 3 **7.** 127 pictures **9.** 61,197,430 **11.** 2 qt **13.** 8,192,000 bits

Pages 596–598 Lesson 11-6

1. −3 **3.** 5 **5.** −8 **7.** C **9.** −3 **11.** −8 **13.** 5 **15.** −9 **17.** −9 **19.** 9 **21.** −65 m **23.** −3 **25.** 4 **29.** −25 ft/min **31.** Sample answer: −45 ÷ 9; An elevator descends 45 floors in 9 seconds. Which integer represents the number of floors the elevator descends each second? **33.** Sample answer: Algebra tiles; manipulatives would allow Melissa's brother to see how to perform this division problem.

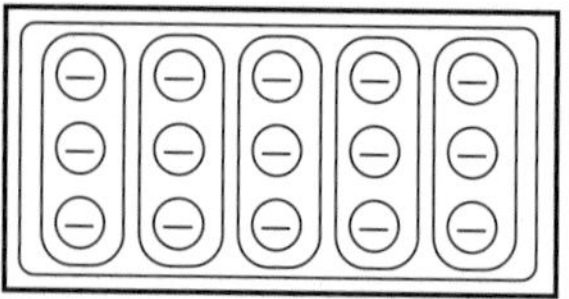

35. negative; Sample answer: The quotient of x and y integers is negative only when x and y have different signs. Thus, if x and y have different signs, the product of x and y will also be negative. **37.** A **39.** $0.55 **41.** −1°F **43.** 8 min 21 s **45.** 12 min 39 s **47.** $\frac{7}{25}$ **49.** $\frac{17}{20}$ **50–53.**

−10 −9 −8 −7 −6 −5 −4 −3 −2 −1 0 1 2 3 4 5 6 7 8 9 10

Pages 601–603 *Lesson 11-7*

1. *H* **3.** *B* **5.** (0, –2); none **7.** gym
9–11.

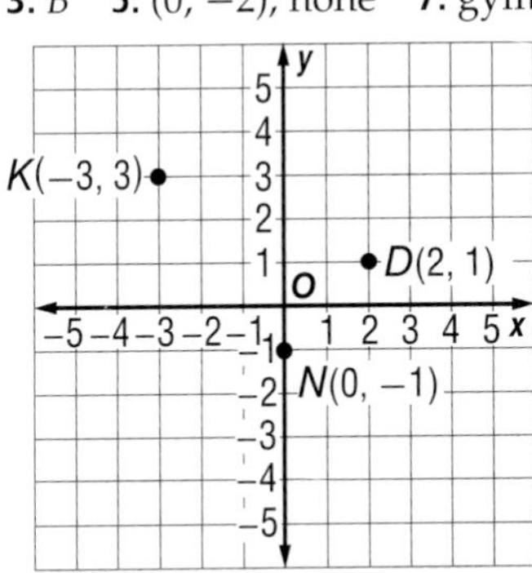

13. *U* **15.** *S* **17.** *J* **19.** (5, 0); none **21.** (3, –5); IV
23. (5, 4); I
24–29.

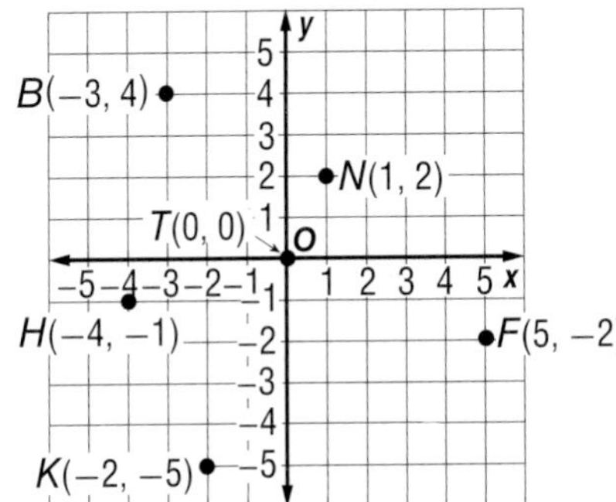

31. Pirate Ship **33.** The Clock
34–36.

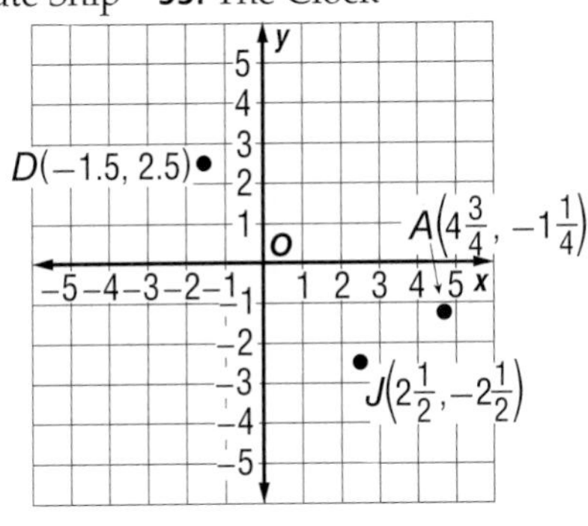

37. Brazil **39.** *x*-axis **41.** Quadrants I and III; Sample answer: For the product of two numbers to be positive, the numbers must either be both positive or both negative. In Quadrant I, both coordinates are positive and in Quadrant III, both coordinates are negative.
43. Sample answer: (–1, 2), (2, 5), (5, 8) **45.** B
47. 5 **49.** 4 **51.** –36 **53.** 45 **55.** 52 **57.** –3
59. –11

Pages 606–609 *Lesson 11-8*

1.

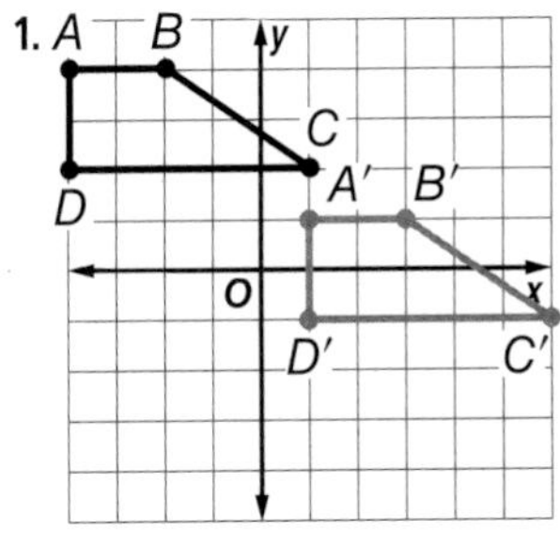

3.

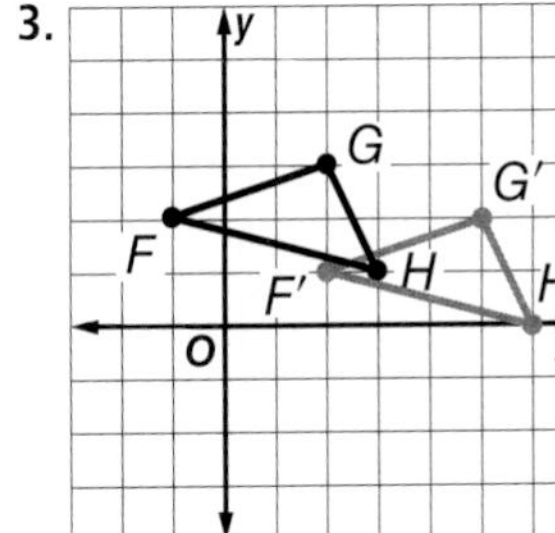

5.

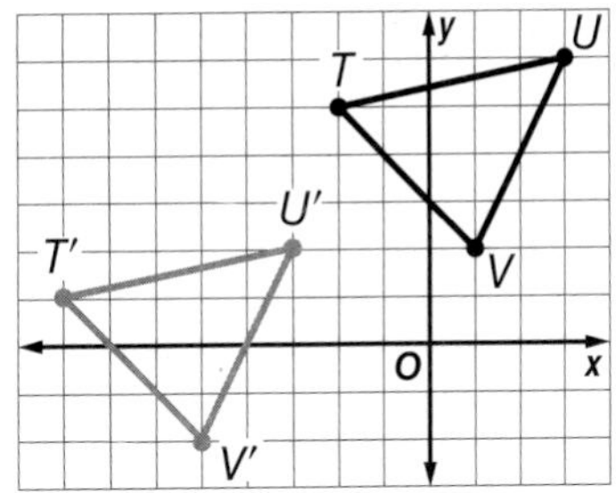

7.

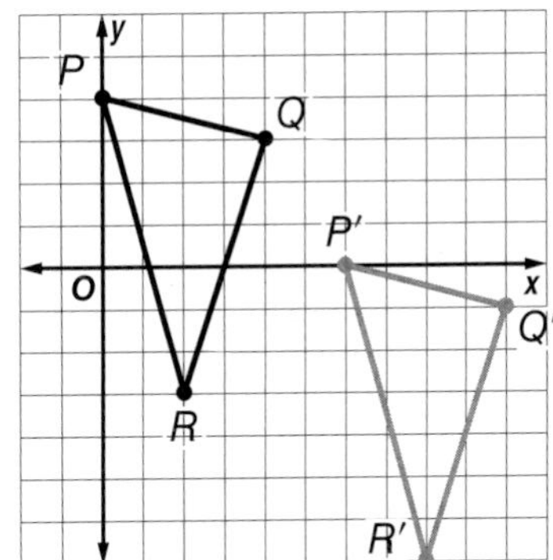

9.

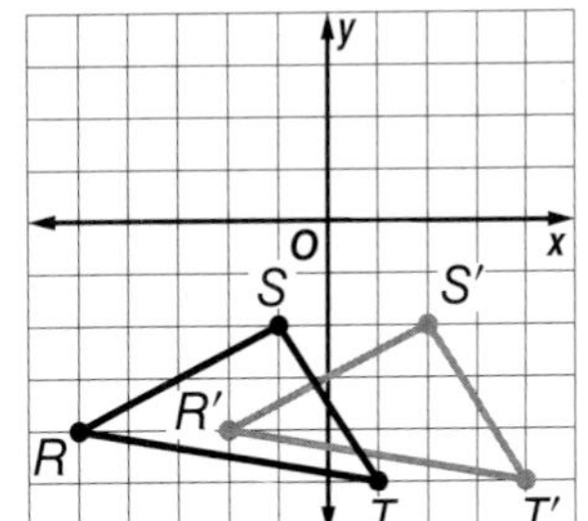

11.

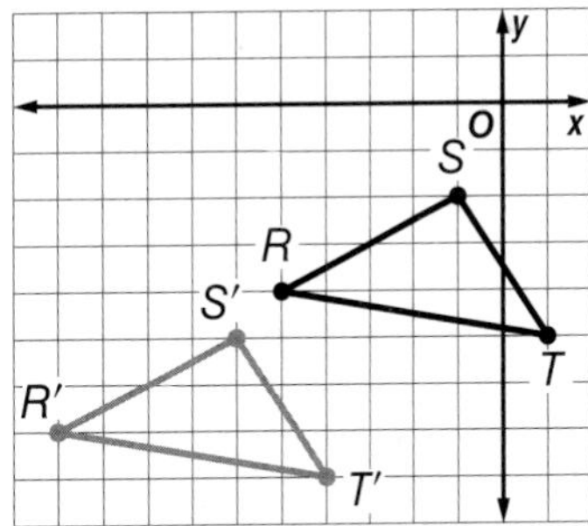

13.

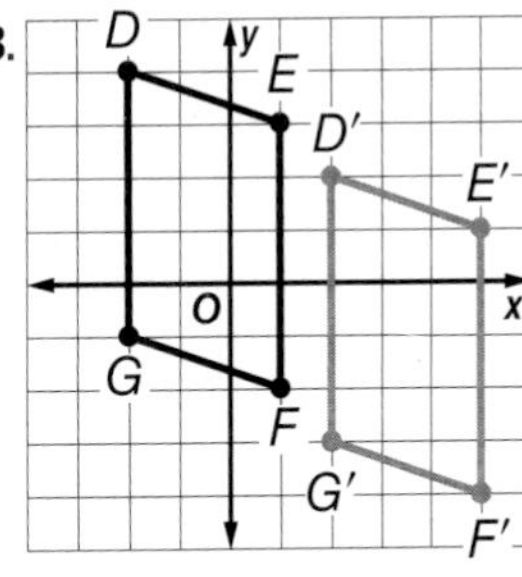

15.

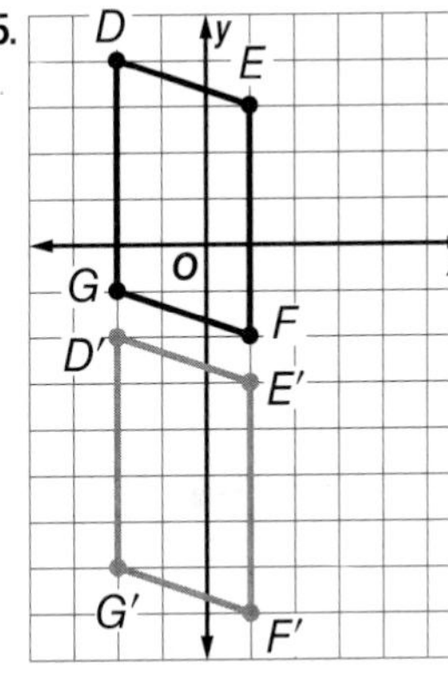

17. (3, 7), (5 , 7), (5, 10), and (3, 10)
19. *A*(–7, 6), *B*(–2, 6), *C*(–2, 0), *D*(–7, 0)
21. *I*(1, –3), *J*(6, –3), *K*(6, –5), *L*(1, –5)

Selected Answers

23.

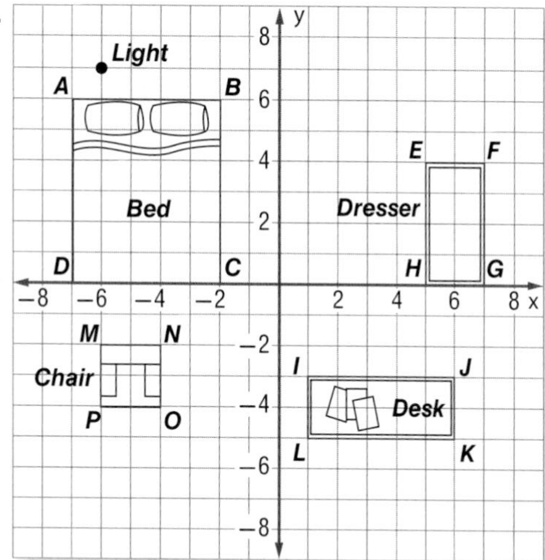

25.

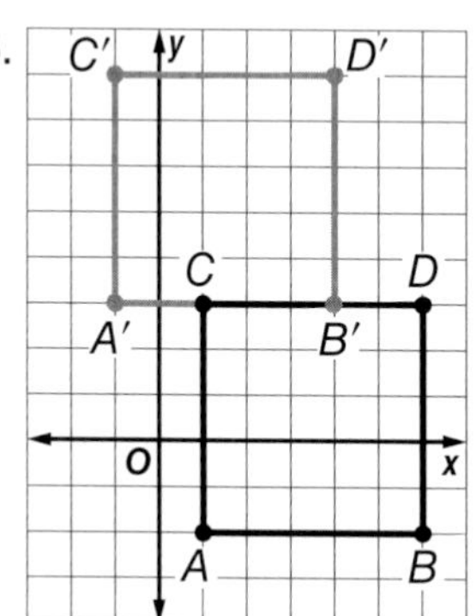

27.

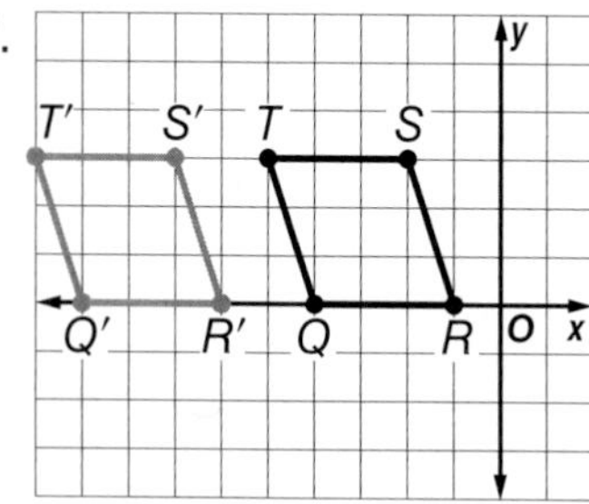

29. $A'(2, 7)$, $B'(6, 1)$, $C'(1, 0)$ **31.** $A'(-1, 1)$, $B'(3, -5)$, $C'(-2, -6)$ **33.** Sample answer: First find the coordinates of rectangle *QRST*. Then find the new coordinates of rectangle *Q′R′S′T′* by adding 7 to each x-coordinate and subtracting 4 from each y-coordinate. Finally graph rectangle *QRST* and rectangle *Q′R′S′T′* on the same coordinate plane. **35.** J **37.** *G* **39.** *C* **41.** yes **43.** yes

Pages 612–614 ***Lesson 11-9***

1.

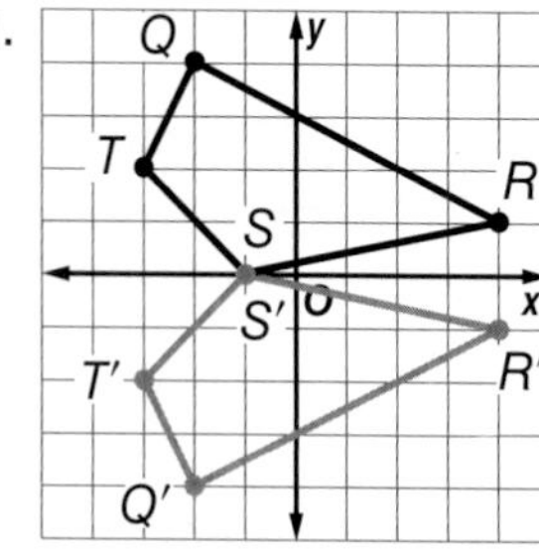

$Q'(-2, -4)$, $R'(4, -1)$, $S'(0, -1)$, and $T'(-3, -2)$

3.

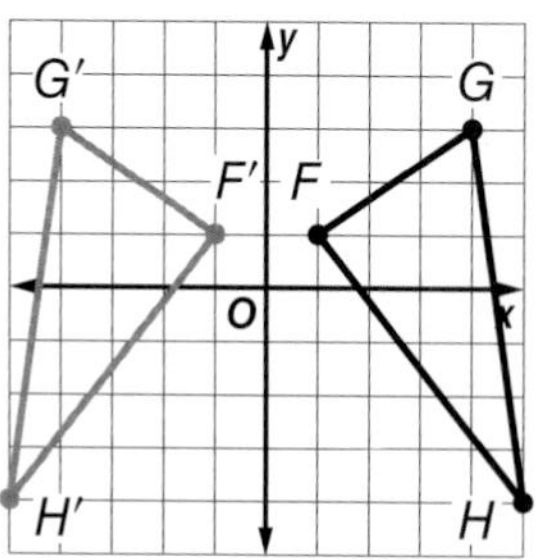

$F'(-1, 1)$, $G'(-4, 3)$, and $H'(-5, -4)$

5.

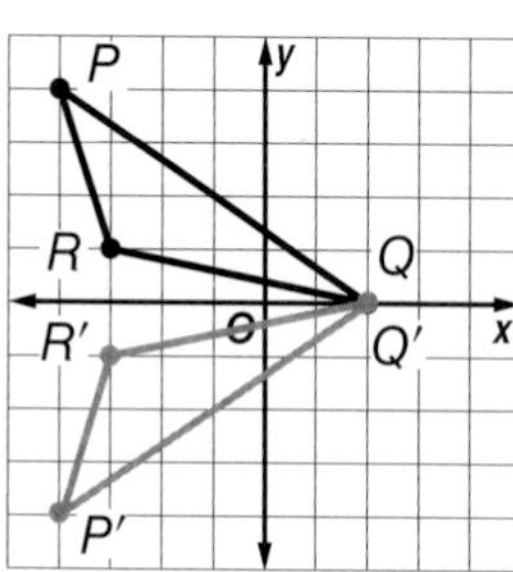

$P'(-4, -4)$, $Q'(2, 0)$, and $R'(-3, -1)$

7.

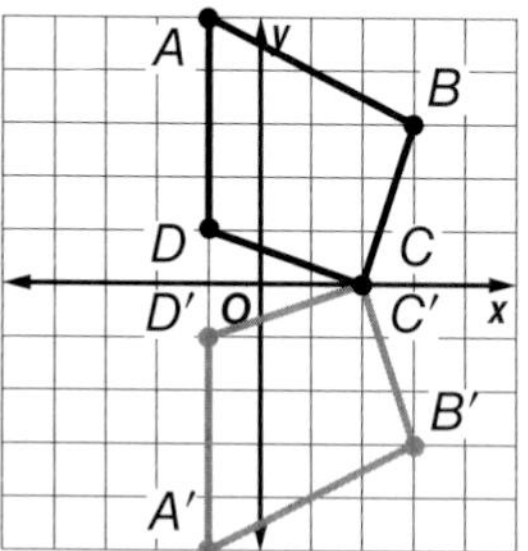

9.

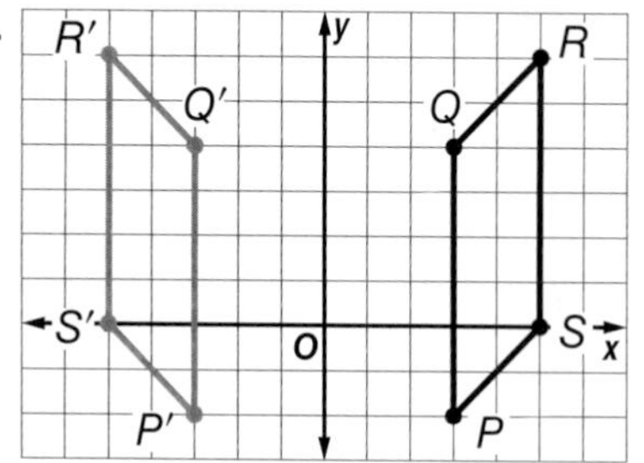

$P'(-3, -2)$, $Q'(-3, 4)$, $R'(-5, 6)$, and $S'(-5, 0)$

11.

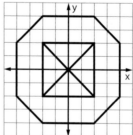

13.

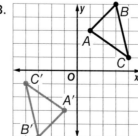

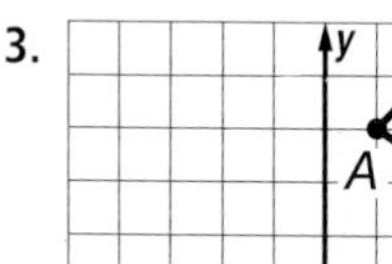

$A''(-1, -3)$, $B''(-3, -5)$, and $C''(-4, -1)$

15. $A'(3, 0)$, $B'(1, -2)$, and $C'(0, 2)$ **17.** A and B, C and D **19.** figure C **21.** Sample answer:

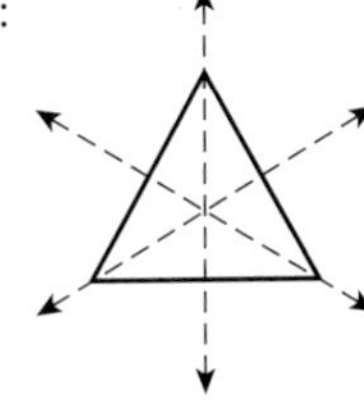

25.

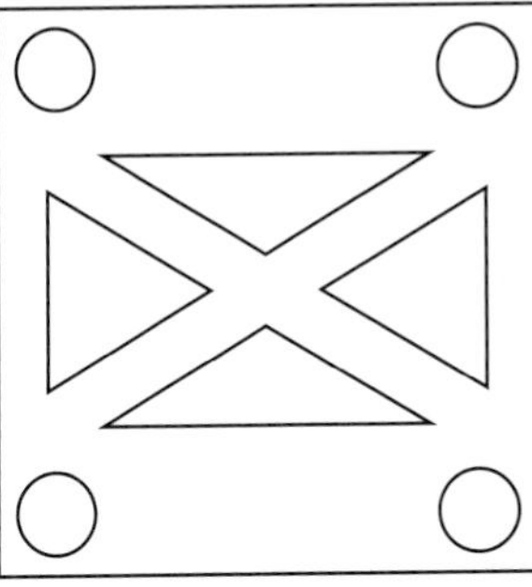

27. *B*

29.

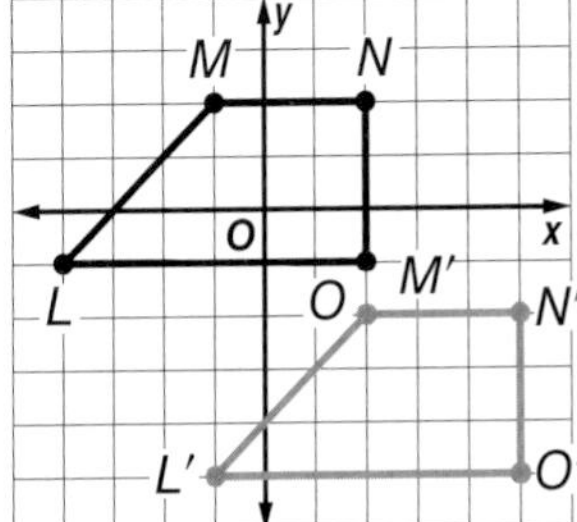

21–34.

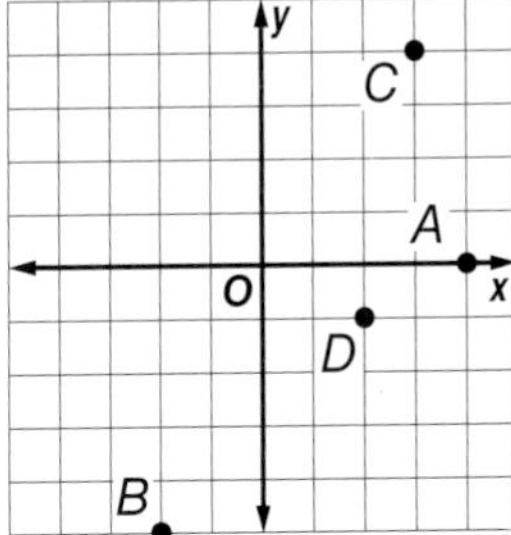

35. −6 **37.** −3

Pages 617–619 ***Lesson 11-10***

1.

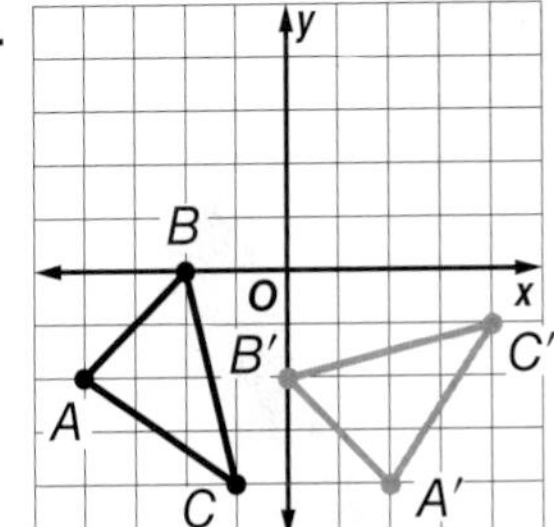

$A'(2, -4)$, $B'(0, -2)$, and $C'(4, -1)$

3.

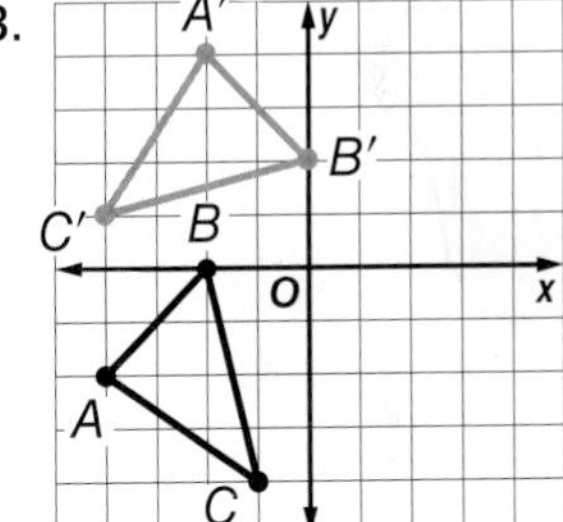

$A'(-2, 4)$, $B'(0, 2)$, and $C'(-4, 1)$

5. yes; 45°, 90°, 135° 180°, 225°, 270°, 315°, and 360°

7.

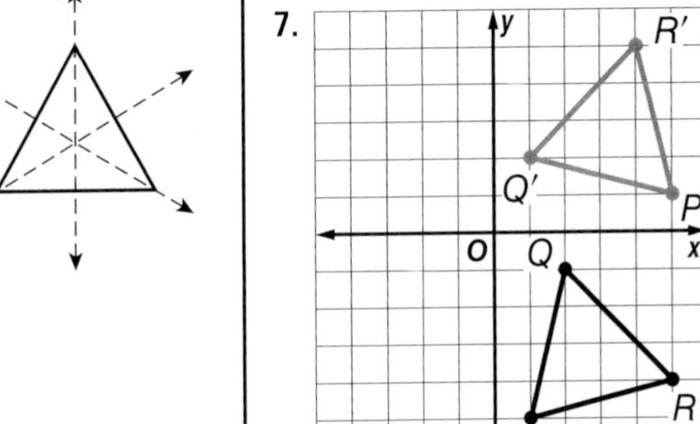

$P'(5, 1)$, $Q'(1, 2)$, and $R'(4, 5)$

9.

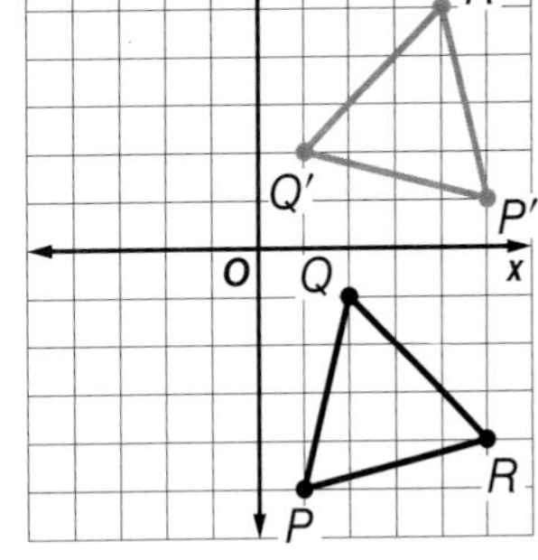

$P'(5, 1)$, $Q'(1, 2)$, and $R'(4, 5)$

11.

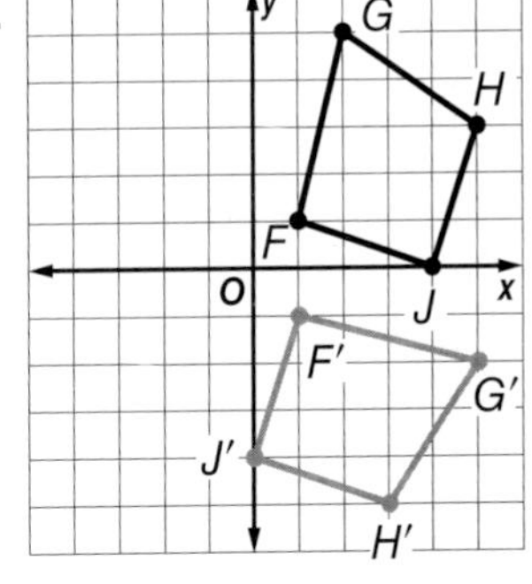

$F'(1, -1)$, $G'(5, -2)$, $H'(3, -5)$, and $J'(0, -4)$

13.

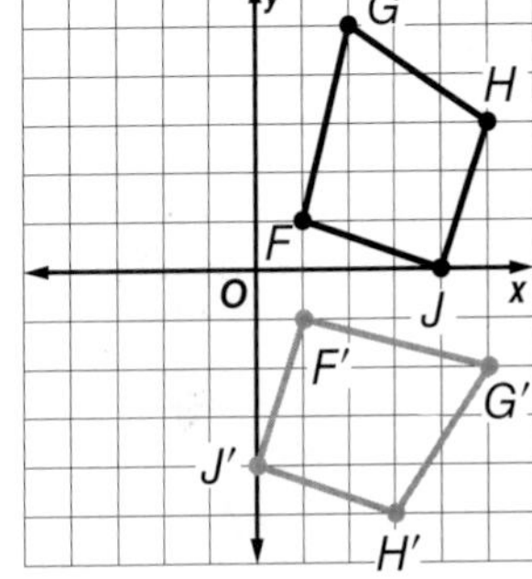

$F'(1, -1)$, $G'(5, -2)$, $H'(3, -5)$, and $J'(0, -4)$

15. yes; 45°, 90°, 135°, 180°, 225°, 270°, 315°, and 360°

17. yes: 120°, 240°, and 360°

19. Sample answer:

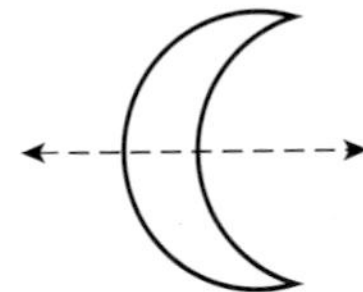

21. no; Sample answer: if a figure has rotational symmetry, then it has at least one line of symmetry.

8 lines of symmetry

6 lines of symmetry

3 lines of symmetry

23. Samuel; Sample answer: Elena rotated the image 90° clockwise.

25.

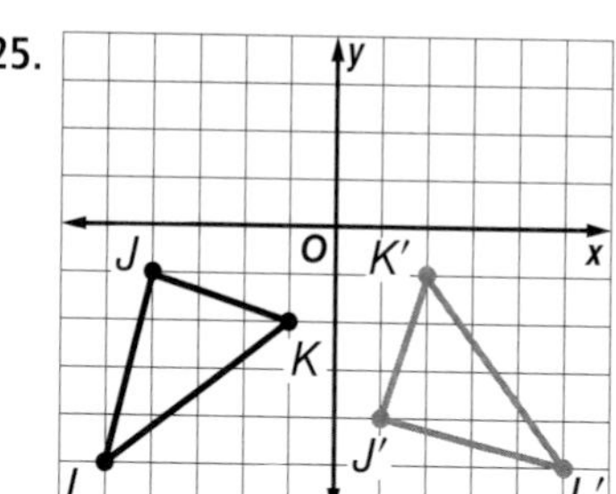

$J'(1, -4)$, $K'(2, -1)$, and $L'(5, -5)$

27.

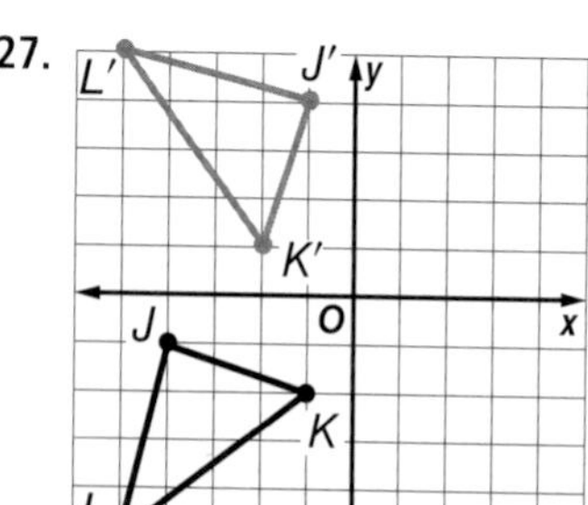

$J'(-1, 4)$, $K'(-2, 1)$, and $L'(-5, 5)$

29. A **31.**

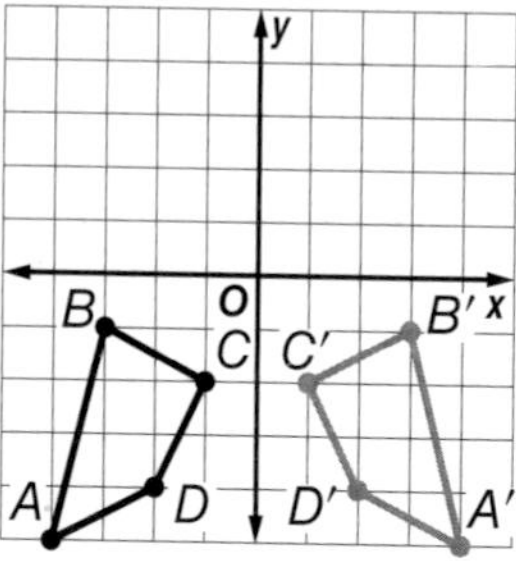

33.

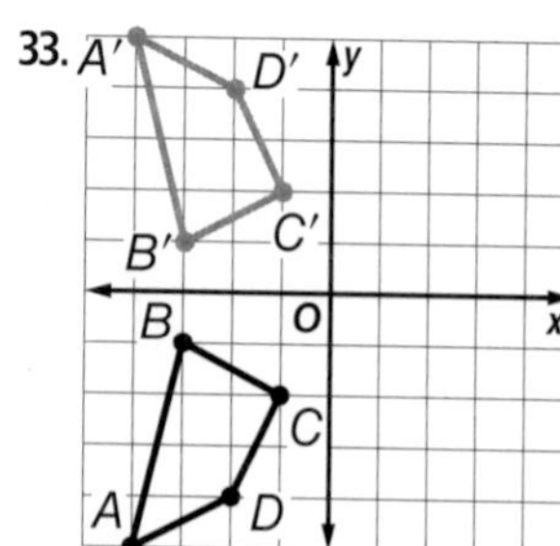

35.

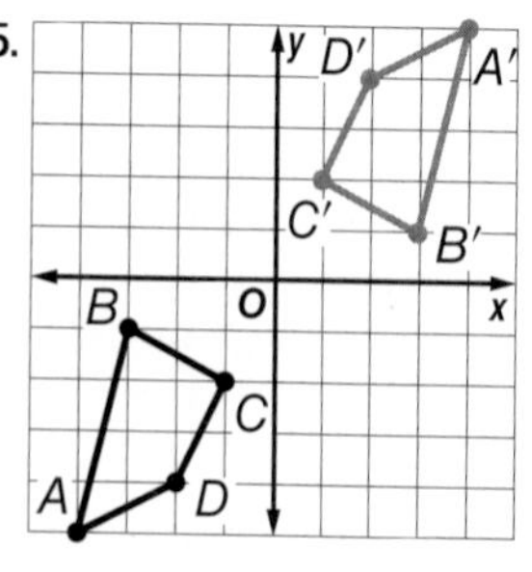

Pages 620–624 *Chapter 11* *Study Guide and Review*

1. true **3.** false; greater **5.** false; positive **7.** false; quadrants **9.** false; negative **11.** < **13.** < **15.** 8, −5, −7, −12 **17.** −6 **19.** −3 **21.** −4 **23.** 5 **25.** −5 **27.** 9 **29.** 2 **29.** −12 stairs **31.** −24

33. 35 **35.** −54 **37.** 4:45 P.M. **39.** −4 **41.** 9 **43.** −3 **45.** 7 **47.** −4 **49.** (−4, 3)

50–53.

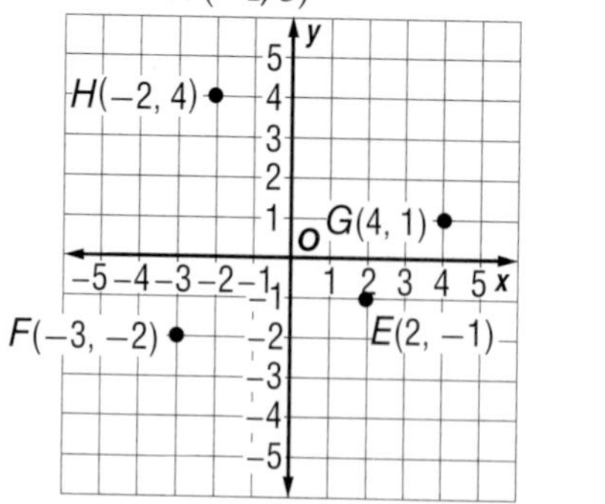

55.

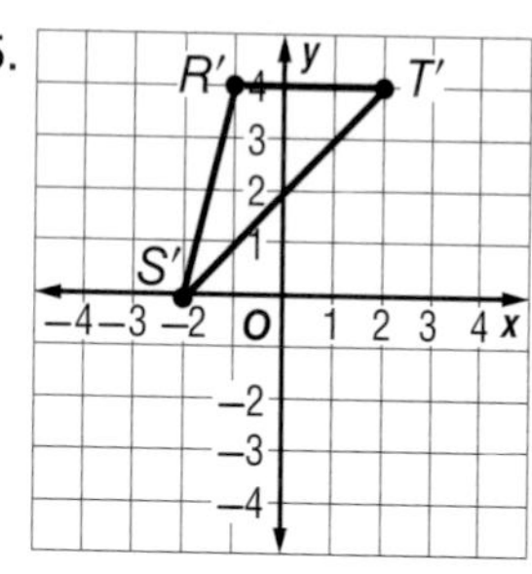

$R'(-1, 4)$, $S'(-2, 0)$, and $T'(2, 4)$

57. (1, −3)

59.

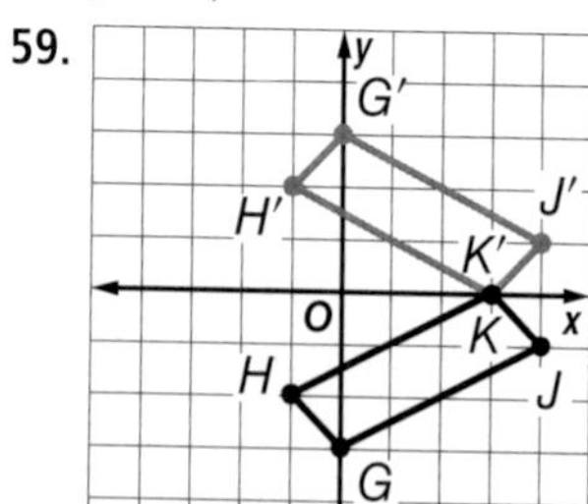

$G'(0, 3)$, $H'(-1, 2)$, $K'(3, 0)$, and $J'(4, 1)$

61.

63.

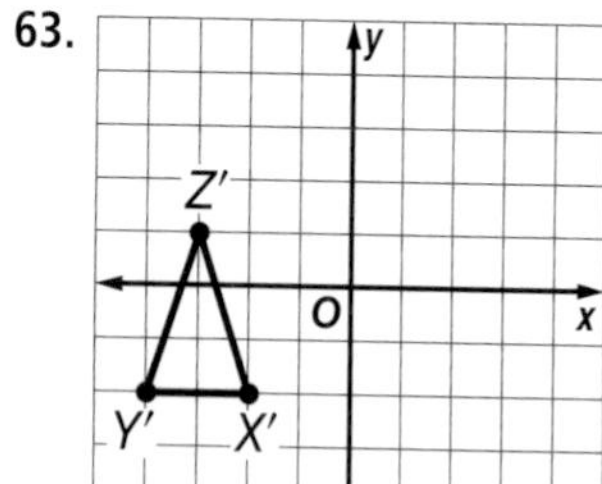

$X'(-2, -2)$, $Y'(-4, -2)$, and $Z'(-3, 1)$

65. yes; 45°, 90°, 135°, 180°, 225°, 270°, 315°, 360°

Chapter 12 Algebra: Properties and Equations

Page 629 *Chapter 12* *Getting Ready*

1. 26 **3.** −16 **5.** $29 **7.** 15 **9.** −1 **11.** −45 **13.** −42 **15.** −$35

Pages 634–635 **Lesson 12-1**

1. $4(30) + 4(8) = 152$ **3.** $11(20) + 7(20) = 297$ **5.** $3(1) + 3(0.6) = 4.8$ **7.** $6(\$9.50) + 6(\$1.50) = 6(\$11) = \$66$ **9.** $5x + 40$ **11.** $7x + 28$ **13.** $9x - 18$ **15.** $3(40) + 3(7) = 141$ **17.** $9(40) + 9(4) = 396$ **19.** $14(20) + 14(5) = 350$ **21.** $12(40) + 12(3) = 516$ **23.** $3(3) + 3(0.9) = 11.7$ **25.** $7(3) + 7(0.8) = 26.6$ **27.** $6(43) - 6(35) = 6(8) = 48$ mi **29.** $6x + 66$ **31.** $7x + 21$ **33.** $4x + 8$ **35.** $6x - 18$ **37.** $5x - 55$ **39.** $7x - 35$ **41.** 6 **43.** Sample answer: $3(4.8) = 3(4) + 3(0.8)$ **45.** B

47.

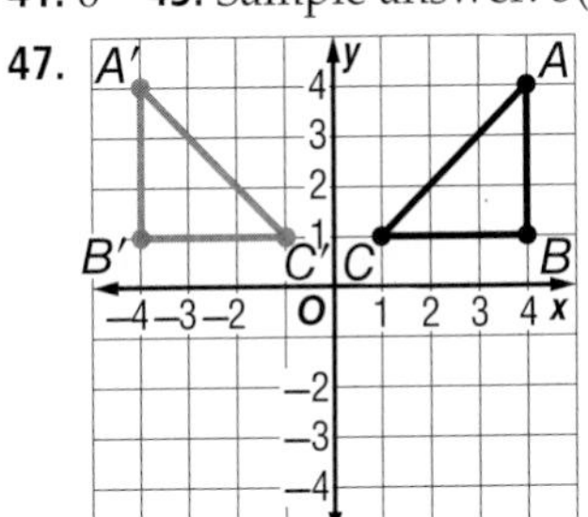

; $A'(-4, 4)$, $B'(-4, 1)$, $C'(-1, 1)$ **49.** −4°F **51.** 42 **53.** 192

Pages 638–641 **Lesson 12-2**

1. $1 + (6 + x) = (1 + 6) + x$ Associative Property
$= 7 + x$ Add 1 and 6.

3. $(18 + x) + 11 = (x + 18) + 11$ Commutative Property
$= x + (18 + 11)$ Associative Property
$= x + 29$ Add 18 and 11.

5. $5(6x) = 5 \cdot (6 \cdot x)$ Parentheses indicate multiplication.
$= (5 \cdot 6) \cdot x$ Associative Property
$= 30x$ Multiply 5 and 6.

7. $8x$

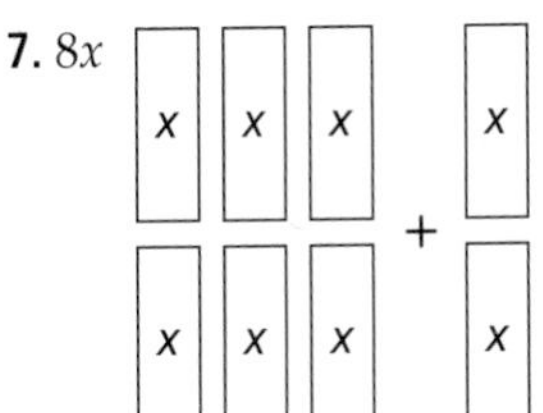

9. $5x + 5$

11. $12 + (8 + x) = (12 + 8) + x$ Associative Property
$= 20 + x$ Add 12 and 8.

13. $11 \cdot (6 \cdot x) = (11 \cdot 6) \cdot x$ Associative Property
$= 66x$ Multiply 11 and 6.

15. $(6 + x) + 9 = (x + 6) + 9$ Commutative Property
$= x + (6 + 9)$ Associative Property
$= x + 15$ Add 6 and 9.

17. $(8 \cdot x) \cdot 4 = (x \cdot 8) \cdot 4$ Commutative Property
$= x \cdot (8 \cdot 4)$ Associative Property
$= x \cdot 32$, or $32x$ Multiply 8 and 4.

19. $9(5x) = 9 \cdot (5 \cdot x)$ Parentheses indicate multiplication.
$= (9 \cdot 5) \cdot x$ Associative Property
$= 45x$ Multiply 9 and 5.

21. $12(6x) = 12 \cdot (6 \cdot x)$ Parentheses indicate multiplication.
$= (12 \cdot 6) \cdot x$ Associative Property
$= 72 \cdot x$, or $72x$ Multiply 12 and 6.

23. $3x$ **25.** $4x$

27. $6x + 1$

29. $7x + 4$

31. $7x + 4$

33. $2(x + 6) + 2x$; $4x + 12$ **35.** $3(x + 4) + 4x$; $\$7x + \$12$ **37.** $3x + 2 + 4x$; $7x + 2$ **39.** 5 **41.** 12 **43.** $6(x + 3.75) + 2x$; $\$8x + \$22.50$ **45.** Sample answer: $8x + 7 + 7x$ **47.** $6x + (-21)$ or $6x - 21$ **49.** $3x$ and x; Sample answer: $8x$ and x are like terms because they contain the same variable, x. The drawing shows that $3x$ and x are like terms because their models have the same shape.

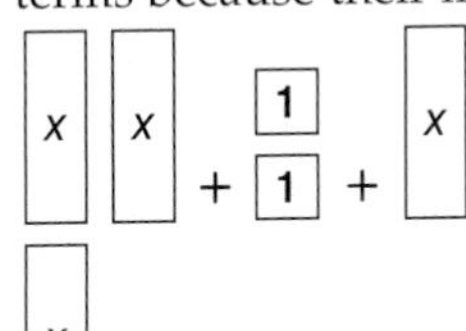

51. J **53.** $4x + 8$ **55.** $4x - 12$ **57.** 7 **59.** {−22, −17, 18, 36} **61.** {−28, −14, 24, 32} **63.** 5 **65.** −15

Selected Answers

Pages 646–648 Lesson 12-3

1. 2 **3.** −9 **5.** $t + 144 = 110$; −34°F **7.** 6 **9.** 3 **11.** −1 **13.** −6 **15.** −8 **17.** −4 **19.** $489 + p = 756$; \$267 **21.** −4 **23.** −2.1 **25.** 6.3 **27.** $-\frac{3}{4}$ **29.** $15.4 = x + 4.9$; 10.5 ft **31.** −5, −6, −7, or −8 **33.** Negative; when you add 14 to the number *a*, the result is less than 0. **35.** G **37.** 5(35) + 5(25) − 5(60); 300 min **39.** *K* **41.** (−3, 3); II **43.** (0, 2); none **45.** −4 **47.** 9 **49.** −6 **51.** 2

Pages 653–654 Lesson 12-4

1. 14 **3.** 12 **5.** −5 **7.** $13 = a - 4$; 17 **9.** 6 **11.** 4 **13.** 0 **15.** −7 **17.** 1 **19.** −3 **21.** $15 = v - 12$; 27 **23.** 21 **25.** 3.4 **27.** 3.5 **29.** $-\frac{1}{4}$ **31.** $x - 56 = 4$; \$60 **33.** Sample answer: The difference in weight between Marc's two dogs is 2 pounds. If one dog weighs 12 pounds, what is the weight of the other dog?; The equation means "some number take away 12 equals 2." **35.** positive; Sample answer: When you take away 8 from the number *x*, the result is positive. So, *x* must be positive. **37.** H **39.** $11x + 3$ **41.** −4 **43.** 12

Pages 658–660 Lesson 12-5

1. 3 **3.** 2 **5.** −4 **7.** $3x = 48$; 16 yd **9.** 4 **11.** 7 **13.** −6 **15.** −2 **17.** 6 **19.** 0.5 **21.** $58m = 680$; about 12 mi **23.** 2 **25.** 30 **27.** $-1\frac{1}{2}$ **29.** 0.4 **31.** $18x = 90$; 5; To check, multiply 18 by 5. The result should be 90. **33.** $26p = 2{,}002$; 77 points **35.** $1{,}440x = 103{,}680$; 72 beats **37.** $4b = 7$; The solution for the other equations is 4. **39.** Sample answer: Jacob bought four t-shirts for \$24 at a souvenir shop. How much did each t-shirt cost? **41.** D **43.** −1 **45.** −8 **47.** 500 min.

Pages 661–662 Lesson 12-6

1. Use estimation when you do *not* need an exact answer. **3.** Sample answer: the population of a certain city was 128,507 in 2007 and 238,091 in 2008. What was the increase in population? A calculator would be used here since the numbers are too large to compute with mentally and an exact answer is needed. **5.** Since an exact answer is required and the numbers are small, use paper and pencil; $9\frac{1}{6}$ feet

7. **9.** 6 **11.** estimation; 7 h

13.

−2	3	−4
−3	−1	1
2	−5	0

15. 3.25 m^2

Pages 663–666 Chapter 12 Study Guide and Review

1. false; coefficient **3.** true **5.** false; $4x$ **7.** true **9.** 7(30) + 7(9) = 273 **11.** 6(5) + 6(0.4) = 32.4 **13.** 4(\$9.50) + 4(\$5.50) = 4(\$15) = \$60 **15.** $8x + 8$ **17.** $7x - 35$

19. $(17 + x) + 12 = (x + 17) + 12$ Commutative Property
$= x + (17 + 12)$ Associative Property
$= x + 29$ Add 17 and 12.

21. $(5 \cdot x) \cdot 8 = (x \cdot 5) \cdot 8$ Commutative Property
$= x \cdot (5 \cdot 8)$ Associative Property
$= x \cdot 40$, or $40x$ Multiply 5 and 8.

23. $15(3x) = 15 \cdot (3 \cdot x)$ Parentheses indicate multiplication.
$= (15 \cdot 3) \cdot x$ Associative Property
$= 45x$ Multiply 15 and 3.

25. $3x$

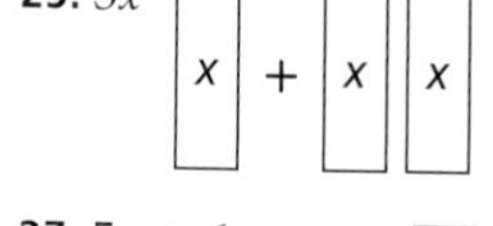

27. $5x + 6$

29. 3 **31.** 45 **33.** −38 **35.** 12 **37.** $4 + x = 10$; 6 ft **39.** 18 **41.** 5 **43.** 19 **45.** 15 **47.** 1 **49.** 8 **51.** 12 **53.** 3 **55.** −3 **57.** $25t = 5$; \$0.20 **59.** calculator; 12,334,000 pounds

Index

Index

Index

Index

Index

Index

Index

Symbols

Number and Operations

$+$	plus or positive
$-$	minus or negative
$a \cdot b$, $a \times b$, ab or $a(b)$	a times b
$\div$	divided by
$=$	is equal to
$\neq$	is not equal to
$>$	is greater than
$<$	is less than
$\geq$	is greater than or equal to
$\leq$	is less than or equal to
$\approx$	is approximately equal to
%	percent
$a:b$	the ratio of a to b, or $\frac{a}{b}$
$0.7\overline{5}$	repeating decimal 0.75555...

Algebra and Functions

$-a$	opposite or additive inverse of a
a^n	a to the nth power
$\lvert x \rvert$	absolute value of x
$\sqrt{x}$	principal (positive) square root of x

Geometry and Measurement

$\cong$	is congruent to
$\sim$	is similar to
$^\circ$	degree(s)
$\overleftrightarrow{AB}$	line AB
$\overrightarrow{AB}$	ray AB
$\overline{AB}$	line segment AB
AB	length of $\overline{AB}$
⊾	right angle
$\perp$	is perpendicular to
$\parallel$	is parallel to
$\angle A$	angle A
$m\angle A$	measure of angle A
$\triangle ABC$	triangle ABC
(a, b)	ordered pair with x-coordinate a and y-coordinate b
O	origin
π	pi $\left(\text{approximately } 3.14 \text{ or } \frac{22}{7}\right)$

Probability and Statistics

$P(A)$	probability of event A

Formulas

Perimeter	square	$P = 4s$
	rectangle	$P = 2\ell + 2w$ or $P = 2(\ell + w)$
Circumference	circle	$C = 2\pi r$ or $C = \pi d$
Area	square	$A = s^2$
	rectangle	$A = \ell w$
	parallelogram	$A = bh$
	triangle	$A = \frac{1}{2}bh$
	circle	$A = \pi r^2$
Surface Area	cube	$S = 6s^2$
	rectangular prism	$S = 2\ell w + 2\ell h + 2wh$
	cylinder	$S = 2\pi rh + 2\pi r^2$
Volume	cube	$V = s^3$
	rectangular prism	$V = \ell wh$ or Bh
	cylinder	$V = \pi r^2 h$ or Bh
	pyramid	$V = \frac{1}{3}Bh$

Measurement Conversions

Length	1 kilometer (km) = 1,000 meters (m) 1 meter = 100 centimeters (cm) 1 centimeter = 10 millimeters (mm)	1 foot (ft) = 12 inches (in.) 1 yard (yd) = 3 feet or 36 inches 1 mile (mi) = 1,760 yards or 5,280 feet
Volume and Capacity	1 liter (L) = 1,000 milliliters (mL) 1 kiloliter (kL) = 1,000 liters	1 cup (c) = 8 fluid ounces (fl oz) 1 pint (pt) = 2 cups 1 quart (qt) = 2 pints 1 gallon (gal) = 4 quarts
Weight and Mass	1 kilogram (kg) = 1,000 grams (g) 1 gram = 1,000 milligrams (mg) 1 metric ton = 1,000 kilograms	1 pound (lb) = 16 ounces (oz) 1 ton (T) = 2,000 pounds
Time	1 minute (min) = 60 seconds (s) 1 hour (h) = 60 minutes 1 day (d) = 24 hours	1 week (wk) = 7 days 1 year (yr) = 12 months (mo) or 52 weeks or 365 days 1 leap year = 366 days
Metric to Customary	1 meter ≈ 39.37 inches 1 kilometer ≈ 0.62 mile 1 centimeter ≈ 0.39 inch	1 kilogram ≈ 2.2 pounds 1 gram ≈ 0.035 ounce 1 liter ≈ 1.057 quarts